ENCAPSULATIONS

ENCAPSULATIONS

Nanotechnology in the Agri-Food Industry, Volume 2

Edited by

ALEXANDRU MIHAI GRUMEZESCU

Department of Science and Engineering of Oxide Materials and Nanomaterials, Faculty of Applied Chemistry and Materials Science, University Politehnica of Bucharest, Bucharest, Romania

ELSEVIER

AMSTERDAM • BOSTON • HEIDELBERG • LONDON • NEW YORK • OXFORD
PARIS • SAN DIEGO • SAN FRANCISCO • SINGAPORE • SYDNEY • TOKYO

Academic Press is an imprint of Elsevier

Academic Press is an imprint of Elsevier
125 London Wall, London EC2Y 5AS, United Kingdom
525 B Street, Suite 1800, San Diego, CA 92101-4495, United States
50 Hampshire Street, 5th Floor, Cambridge, MA 02139, United States
The Boulevard, Langford Lane, Kidlington, Oxford OX5 1GB, UK

British Library Cataloguing-in-Publication Data
A catalogue record for this book is available from the British Library

Library of Congress Cataloging-in-Publication Data
A catalog record for this book is available from the Library of Congress

ISBN: 978-0-12-804307-3

For information on all Academic Press publications
visit our website at https://www.elsevier.com/

Publisher: Nikki Levy
Acquisition Editor: Patricia Osborn
Editorial Project Manager: Karen Miller
Production Project Manager: Caroline Johnson
Designer: Mark Rogers

Typeset by Thomson Digital

CONTENTS

LIST OF CONTRIBUTORS

M. Angela A. Meireles
UNICAMP (University of Campinas), LASEFI/DEA/FEA (School of Food Engineering), Campinas, São Paulo, Brazil

Mohamed H. Abd El-Salam
Dairy Department, National Research Centre, Cairo, Egypt

Sulaiman O. Aljaloud
North Carolina Agricultural and Technical State University, Food Microbiology and Biotechnology Laboratory, Greensboro, NC, United States; King Saud University, Department of Exercise Physiology, College of Sport Sciences and Physical Activity, Riyadh, Saudi Arabia

Maria S. Anokhina
N. M. Emanuel Institute of Biochemical Physics of Russian Academy of Sciences, Chemistry of Food Substances and Biotechnology Department, Moscow, Russian Federation

Anna S. Antipova
N. M. Emanuel Institute of Biochemical Physics of Russian Academy of Sciences, Chemistry of Food Substances and Biotechnology Department, Moscow, Russian Federation

Mohd. Aqil
Jamia Hamdard, Department of Pharmaceutics, Faculty of Pharmacy, New Delhi, India

Bojana Balanč
University of Belgrade, Department of Chemical Engineering, Faculty of Technology and Metallurgy, Belgrade, Serbia

Santanu Basu
Panjab University, Dr. S. S. Bhatnagar University Institute of Chemical Engineering and Technology, Chandigarh, India

Larisa E. Belyakova
N. M. Emanuel Institute of Biochemical Physics of Russian Academy of Sciences, Chemistry of Food Substances and Biotechnology Department, Moscow, Russian Federation

Bharat Bhanvase
Chemical Engineering Department, Laxminarayan Institute of Technology, Rashtrasant Tukadoji Maharaj Nagpur University, Nagpur, Maharashtra, India

Bhaswati Bhattacharya
National Institute of Food Technology Entrepreneurship and Management, Kundli, Haryana, India

Vladimir I. Binyukov
N. M. Emanuel Institute of Biochemical Physics of Russian Academy of Sciences, Chemistry of Food Substances and Biotechnology Department, Moscow, Russian Federation

Natalia G. Bogdanova
N. M. Emanuel Institute of Biochemical Physics of Russian Academy of Sciences, Chemistry of Food Substances and Biotechnology Department, Moscow, Russian Federation

Tarik Bor
North Carolina Agricultural and Technical State University, Food Microbiology and Biotechnology Laboratory, Greensboro, NC, United States

Branko Bugarski
University of Belgrade, Department of Chemical Engineering, Faculty of Technology and Metallurgy, Belgrade, Serbia

Gladis Yakeline Cortes-Mazatán
Department of Polymerization Processes, Center for Research in Applied Chemistry, Saltillo, Coahuila, Mexico

Verica Đorđević
University of Belgrade, Department of Chemical Engineering, Faculty of Technology and Metallurgy, Belgrade, Serbia

Safinaz El-Shibiny
Dairy Department, National Research Centre, Cairo, Egypt

Alba Nidia Estrada-Ramírez
Doctoral Program in Polymer Technology, Center for Research in Applied Chemistry, Saltillo, Coahuila, Mexico

Bożena Fenert
Koszalin University of Technology, Section of Toxicology and Bioanalytics, Koszalin, Poland

Eva Fenyvesi
CycloLab Cyclodextrin Research & Development Laboratory Ltd., Budapest, Hungary

Sergio Enrique Flores-Villaseñor
Doctoral Program in Polymer Technology, Center for Research in Applied Chemistry, Saltillo, Coahuila, Mexico

Sophie Fourmentin
University of the Littoral Opal Coast (ULCO), Unit of Environmental Chemistry and Interaction on the Living (UCEIV), Dunkirk, France

Gulnara Gaynanova
A.E. Arbuzov Institute of Organic and Physical Chemistry, Kazan Scientific Center, Russian Academy of Sciences; Kazan National Research Technological University, Kazan, Russian Federation

Shailesh Ghodke
North Maharashtra University, University Institute of Chemical Technology, Jalgaon, Maharashtra, India

Bronisław K. Głód
Siedlce University of Natural Sciences and Humanities, Department of Analytical Chemistry, Institute of Chemistry, Faculty of Science, Siedlce, Poland

Alexandru Mihai Grumezescu
University Politehnica of Bucharest, Department of Science and Engineering of Oxide Materials and Nanomaterials, Faculty of Applied Chemistry and Materials Science, Bucharest, Romania

Ajai Prakash Gupta
Quality Control & Quality Assurance, Indian Institute of Integrative Medicine, Jammu, Jammu and Kashmir, India

Sumit Gupta
Food Technology Division, Bhabha Atomic Research Centre, Mumbai, Maharashtra, India

Suphla Gupta
Plant Biotechnology Department, Indian Institute of Integrative Medicine, Jammu, Jammu and Kashmir, India

Rabin Gyawali
North Carolina Agricultural and Technical State University, Food Microbiology and Biotechnology Laboratory, Greensboro, NC, United States

Daniel Ioan Hădărugă
Polytechnic University of Timişoara, Department of Applied Chemistry, Organic and Natural Compounds Engineering, Timişoara, Romania

Nicoleta Gabriela Hădărugă
Banat's University of Agricultural Sciences and Veterinary Medicine, Department of Food Science, Timişoara, Romania

Salam A. Ibrahim
North Carolina Agricultural and Technical State University, Food Microbiology and Biotechnology Laboratory, Greensboro, NC, United States

Anatol Jaworek
Institute of Fluid Flow Machinery, Polish Academy of Sciences, Department of Electrohydrodynamics, Gdańsk, Poland

Leslie Violeta Vidal Jiménez
University of Concepcion, Faculty Agricultural Engineering, Chillan, Chile

Kalpana Joshi
Department of Biotechnology, Sinhgad College of Engineering, Pune, Maharashtra, India

Ana Kalušević
University of Belgrade, Department of Food Technology and Biochemistry, Faculty of Agriculture, Belgrade-Zemun, Serbia

Ruslan Kashapov
A.E. Arbuzov Institute of Organic and Physical Chemistry, Kazan Scientific Center, Russian Academy of Sciences; Kazan National Research Technological University, Kazan, Russian Federation

Valerii V. Kasparov
N. M. Emanuel Institute of Biochemical Physics of Russian Academy of Sciences, Chemistry of Food Substances and Biotechnology Department, Moscow, Russian Federation

Miriana Kfoury
University of the Littoral Opal Coast (ULCO), Unit of Environmental Chemistry and Interaction on the Living (UCEIV), Dunkirk, France

Saima Khan
Plant Biotechnology Department, Indian Institute of Integrative Medicine, Jammu, Jammu and Kashmir, India

Nisha Kumari
Jamia Hamdard, Department of Pharmaceutics, Faculty of Pharmacy, New Delhi, India

Manoj Kushwaha
Quality Control & Quality Assurance, Indian Institute of Integrative Medicine, Jammu, Jammu and Kashmir, India

Steva Lević
University of Belgrade, Department of Food Technology and Biochemistry, Faculty of Agriculture, Belgrade-Zemun, Serbia

Daibing Luo
Key Laboratory of Photochemical Conversion and Optoelectronic Materials, Technical Institute of Physics and Chemistry, Chinese Academy of Sciences, Beijing; Sichuan University, Analytical & Testing Center, Chengdu, P.R. China

Antonio Martínez-Férez
Research and Development Centre for Functional Food (CIDAF), Health Science Technological Park; University of Granada, Chemical Engineering Department, Granada, Spain

Elena I. Martirosova
N. M. Emanuel Institute of Biochemical Physics of Russian Academy of Sciences, Chemistry of Food Substances and Biotechnology Department, Moscow, Russian Federation

Alla Mirgorodskaya
A.E. Arbuzov Institute of Organic and Physical Chemistry, Kazan Scientific Center, Russian Academy of Sciences; Kazan National Research Technological University, Kazan, Russian Federation

Satyendra Mishra
North Maharashtra University, University Institute of Chemical Technology, Jalgaon, Maharashtra, India

Malik Muzafar
Plant Biotechnology Department, Indian Institute of Integrative Medicine, Jammu, Jammu and Kashmir, India

Viktor Nedović
University of Belgrade, Department of Food Technology and Biochemistry, Faculty of Agriculture, Belgrade-Zemun, Serbia

Juan Felipe Osorio-Tobón
UNICAMP (University of Campinas), LASEFI/DEA/FEA (School of Food Engineering), Campinas, São Paulo, Brazil

Nadezhda P. Palmina
N. M. Emanuel Institute of Biochemical Physics of Russian Academy of Sciences, Chemistry of Food Substances and Biotechnology Department, Moscow, Russian Federation

Shweta Pandey
Jamia Hamdard, Department of Pharmaceutics, Faculty of Pharmacy, New Delhi, India

Jayamanti Pandit
Jamia Hamdard, Department of Pharmaceutics, Faculty of Pharmacy, New Delhi, India

Tatiana Pashirova
A.E. Arbuzov Institute of Organic and Physical Chemistry, Kazan Scientific Center, Russian Academy of Sciences; Kazan National Research Technological University, Kazan, Russian Federation

René Darío Peralta-Rodríguez
Department of Polymerization Processes, Center for Research in Applied Chemistry, Saltillo, Coahuila, Mexico

Sandra Pimentel-Moral
University of Granada, Department of Analytical Chemistry; Research and Development Centre for Functional Food (CIDAF), Health Science Technological Park, Granada, Spain

Yurii N. Polikarpov
N. M. Emanuel Institute of Biochemical Physics of Russian Academy of Sciences, Chemistry of Food Substances and Biotechnology Department, Moscow, Russian Federation

Nishant Rai
University of Delhi, Department of Pharmacology, Vallabhbhai Patel Chest Institute, Delhi, India

Jorge Carlos Ramirez-Contreras
Department of Polymerization Processes, Center for Research in Applied Chemistry, Saltillo, Coahuila, Mexico

Purnima Rawat
Jamia Hamdard, Department of Pharmaceutics, Faculty of Pharmacy, New Delhi, India

Paz Robert
University of Chile, Department of Food Science and Chemical Technology, Santiago, Región Metropolitana, Chile

Antonio Segura-Carretero
University of Granada, Department of Analytical Chemistry; Research and Development Centre for Functional Food (CIDAF), Health Science Technological Park, Granada, Spain

Maria G. Semenova
N. M. Emanuel Institute of Biochemical Physics of Russian Academy of Sciences, Chemistry of Food Substances and Biotechnology Department, Moscow, Russian Federation

Harshita Sharma
Jamia Hamdard, Department of Pharmaceutics, Faculty of Pharmacy, New Delhi, India

Eric Keven Silva
UNICAMP (University of Campinas), LASEFI/DEA/FEA (School of Food Engineering), Campinas, São Paulo, Brazil

Siddhartha Singha
National Institute of Food Technology Entrepreneurship and Management, Kundli, Haryana, India

Shirish Sonawane
National Institute of Technology, Warangal, Telangana, India

Yasmin Sultana
Jamia Hamdard, Department of Pharmaceutics, Faculty of Pharmacy, New Delhi, India

Oleg Synyashin
A.E. Arbuzov Institute of Organic and Physical Chemistry, Kazan Scientific Center, Russian Academy of Sciences; Kazan National Research Technological University, Kazan, Russian Federation

Lajos Szente
CycloLab Cyclodextrin Research & Development Laboratory Ltd., Budapest, Hungary

Sushama Talegaonkar
Jamia Hamdard, Department of Pharmaceutics, Faculty of Pharmacy, New Delhi, India

Kata Trifković
University of Belgrade, Department of Chemical Engineering, Faculty of Technology and Metallurgy, Belgrade, Serbia

Prasad S. Variyar
Food Technology Division, Bhabha Atomic Research Centre, Mumbai, Maharashtra, India

Elmira Vasilieva
A.E. Arbuzov Institute of Organic and Physical Chemistry, Kazan Scientific Center, Russian Academy of Sciences; Kazan National Research Technological University, Kazan, Russian Federation

Vito Verardo
University of Almería, Agrifood Campus of International Excellence, Department of Chemistry and Physics (Analytical Chemistry Area), Research Centre for Agricultural and Food Biotechnology (BITAL), Almería, Spain

Liangzhuan Wu
Key Laboratory of Photochemical Conversion and Optoelectronic Materials, Technical Institute of Physics and Chemistry, Chinese Academy of Sciences, Beijing, P.R. China

Arvind Kumar Yadav
Quality Control & Quality Assurance, Indian Institute of Integrative Medicine, Jammu, Jammu and Kashmir, India

Lucia Zakharova
A.E. Arbuzov Institute of Organic and Physical Chemistry, Kazan Scientific Center, Russian Academy of Sciences; Kazan National Research Technological University, Kazan, Russian Federation

Paweł K. Zarzycki
Koszalin University of Technology, Section of Toxicology and Bioanalytics, Koszalin, Poland

Darya V. Zelikina
N. M. Emanuel Institute of Biochemical Physics of Russian Academy of Sciences, Chemistry of Food Substances and Biotechnology Department, Moscow, Russian Federation

Jinfang Zhi
Key Laboratory of Photochemical Conversion and Optoelectronic Materials, Technical Institute of Physics and Chemistry, Chinese Academy of Sciences, Beijing, P.R. China

Yuriy Zuev
Kazan Institute of Biochemistry and Biophysics, Kazan Scientific Center, Russian Academy of Sciences, Kazan, Russian Federation

SERIES FOREWORD

The emergence of nanotechnology has reached impressive heights in recent years and the development of special nanodevices and nanomaterials has found intriguing applications in agriculture and food sector. Most of the investigated nanotechnological approaches initially aimed to solve evolving problems in the agri-food industry in order to impact on the economic potential. Soon after the implementation of new technologies and approaches that were using nanostructured materials, the worldwide concern was rapidly extended to numerous applications that could be developed by using the science of nanosized materials. Smart materials, biosensors, packaging materials, nutraceuticals, and nanodevices have been designed to address numerous agri-food related issues with direct impact in health, economy, ecology, and industry. As the engineering of nanostructures has constantly progressed and extended its applications, there is virtually unlimited potential in this sector. However, the widely differing opinions on the applicability and usefulness of nanotechnology between both specialists and the general public has hampered progress. The main concern manifested by people is related to the potential risk for health and the environmental impact of the recently developed nanoengineered materials and devices. Therefore, current approaches are strictly considering these concerns when designing nanotechnological solutions for agriculture and food sectors.

This multivolume series was developed by the constant need to discover current inquiries and approaches on the field of agri-food science and also to learn about the most recent progress, approaches, and applications that have emerged through nanotechnology.

As agriculture is the backbone of most developing countries, nanotechnology has the potential to revolutionize the agriculture and food sector by promoting productivity through genetic improvement of plant and animal foods. It can also ensure the delivery of drugs, genes, and pesticides to specific sites at cellular levels in targeted plants and animals, by limiting side effects. Nanotechnology can be used to evaluate gene expression under different stress condition for both plant and animal foods through the development of nanoarray-based gene-technologies. Additionally, this technology can detect fertilizers, pesticides with high precision by smart nanosenors for an adequate management of the natural resources. Moreover, numerous industrial-related applications with direct impact on economy have emerged. For example, nano- and micro-structured arrays can detect the early presence of pathogens, contaminants, and food spoilage factors. Other applications for this technology are smart integration systems for food processing and packaging, as well as nanoemulsion-based decontaminants for food equipment and storage compartments, and nanoparticles that facilitate the bioavailability and delivery of nutrients directly to cells.

The potential benefits of nanotechnology for agriculture, food, fisheries, and aquaculture were identified and supported by many countries, which invested a significant amount of money in the development of applications. Also, numerous campaigns are currently trying to increase awareness on the developing process and recent technologies in order to influence the acceptance of customers. Although nanoagri-food industrialized concept could help to find a sustainable solution for the current global food crisis, the offered advantages should balance the concerns regarding soil, water, environment, and health related issues that such approach could bring.

The series entitled *Nanotechnology in the Agri-Food Industry* brings comprehensive and recent knowledge regarding the impact of the science of nanometer-sized materials on the field of agriculture and food industry, but also discuss the current inquiries regarding risks of these applications in all relevant fields such as environment and health, aiming to increase awareness to a wider amount of readers.

Dr Alina Maria Holban
Department of Microbiology and Immunology, Faculty of Biology, University of Bucharest, Romania

Dr Monica Cartelle Gestal
Harvill Lab, University Of Georgia, Athens, GA, United States

SERIES PREFACE

About the Series (Volumes I–X)

In a permanently changing society, health and well being remain the key drivers for the food industry. Despite the technological progress made in the agri-food industry, a true food crisis emerges in several areas of the globe. This can be explained by insufficient food but mostly by inadequate food for a very distinct range of consumers. In this context, innovative technologies represent the core throughout the whole food chain from raw materials/ingredient sourcing, food processing, quality control of finished products, and packaging. Nanotechnology, coupled with novel interdisciplinary approaches and processing methods, has enabled some important advances recently flourishing in many of these areas. The science of nanosized materials can improve and even resolve the huge challenges faced by the food and bioprocessing industries for developing and implementing systems that can produce qualitative and quantitative foods that are safe, sustainable, environment friendly, and efficient. This emerging tool finds its applications in various fields and represents an endless approach for the development of innovative strategies in food development, processing, and packaging.

This multivolume set aims to bring together the most recent and innovative applications of nanotechnology in the agri-food industry, but also to present the future perspectives in the design of new or alternative foods.

The series contains 200 chapters organized in 10 volumes, prepared by outstanding research groups that made significant impacts on the field of nanotechnology and food-related research sectors. This comprehensive set represents an updated and highly structured material for undergraduate and postgraduate students in food science, biotechnological, engineering fields, but also a valuable resource of recent scientific progress, along with most known applications of nanomaterials on the food industry to be used by researchers, engineers, and academia. Moreover, novel opportunities and ideas for developing or improving technologies in the agri-food industry by innovative companies, biotechnological industries, and other economical structures are highlighted and their potential is widely dissected. This series may be also valuable for the wide audience interested in recent nanotechnological progress in the agri-food field worldwide.

These 10 volumes cover almost all aspects related to the applications of *Nanotechnology in the Agri-Food Industry* and are named as:

Volume I Novel Approaches
Volume II Encapsulations
Volume III Emulsions
Volume IV Nutraceuticals
Volume V Nutrient Delivery
Volume VI Food Preservation
Volume VII Food Packaging
Volume VIII NanoBioSensors
Volume IX Water Purification
Volume X New Pesticides and Soil Sensors

Each volume contains 20 chapters, which were carefully composed and illustrated to highlight the most innovative and intensively investigated applications of nanotechnology on particular wide interest domains of the agri-food industry field.

Alexandru Mihai Grumezescu
University Politehnica of Bucharest,
Department of Science and Engineering
of Oxide Materials and Nanomaterials,
Faculty of Applied Chemistry and
Materials Science, Bucharest, Romania
http://grumezescu.com/

VOLUME PREFACE

Food flavor and aroma represent key elements in establishing food quality. Therefore, it is imperative to choose a method for protecting them from reduction in flavor and aroma thresholds. Nanoencapsulation is a new technique to protect food ingredients, such as flavor and aroma via isolation and protection from the environment with a matrix or a nanometer scale shell. The major benefits of nanoencapsulation for food ingredients include improvement in bioavailability of flavor and aroma ingredients, improvement in solubility of poor water soluble ingredients, higher ingredient retention during production process, higher activity levels of encapsulated ingredients, improved shelf life, and controlled release of flavor and aroma. However, nanoencapsulation is not widely utilized in the food industry due to costs and complexity of this approach. This volume discusses the main nanoencapsulation processes such as: spray drying, melt injection, extrusion, coacervation, and emulsification. The materials used in nanoencapsulation including: lipids, proteins, carbohydrates, cellulose, gums, food grade polymers; and applications that benefit from this technology such as controlled release, protections, and taste masking are explained in detail. Volume II contains 20 chapters, prepared by outstanding international researchers from Brazil, Chile, China, Egypt, France, Hungary, India, Mexico, Poland, Romania, Russia, Serbia, Spain, and USA.

In Chapter 1, *Electrohydrodynamic Microencapsulation Technology*, Anatol Jaworek present an up-to-date overview regarding electrospray drying, electrospray extrusion, electrospray coextrusion, electrospray mixing, electrospray cooling, submerged electrospray, and electrospray encapsulation in reactive gas. Various applications of this technology in the food industry are discussed.

Sushama Talegaonkar et al., in Chapter 2, *Exploring Nanoencapsulation of Aroma and Flavors as New Frontier in Food Technology*, review the most investigated nanocapsules, versatile matrices, methods, and analytical techniques of nanoencapsulation of flavors and aromas.

Chapter 3, prepared by Juan Felipe Osorio-Tobón et al., *Nanoencapsulation of Flavors and Aromas by Emerging Technologies*, offers an up-to-date overview related to the application of emerging technologies based on supercritical fluids and ultrasonication to form nanoparticles/nanoemulsions of essential oils with applications in the design of flavor and aroma agents in various food products, besides to add value to these products and to promote innovation in food industry through the obtaining of flavorings considered safe obtained by applying clean technologies.

In Chapter 4, *Cyclodextrins as Encapsulation Material for Flavors and Aroma,* Miriana Kfoury et al., provides a brief history of aroma and cyclodextrins (CD) and discuss encapsulation techniques as well as characterization methods for CD/aroma or flavor inclusion complexes. This chapter also discusses parameters controlling the stability binding of aroma and flavors to CDs and reports the beneficial effects of encapsulation on the properties of aroma.

Maria G. Semenova et al., in Chapter 5, *Structural and Thermodynamic Insight Into the Potentiality of Food Biopolymers to Behave as Smart Nanovehicles for Essential Polyunsaturated Lipids,* review the key structural and thermodynamic properties underlying the novel functionality of the complex nanosized particles formed by biopolymers (individual caseins; sodium caseinate and covalent conjugates of sodium caseinate with maltodextrins) and polyunsaturated lipids. This new functionality includes both the protection against oxidation for the lipids, and their controllable release under the action of digestive enzymes *in vitro.*

Chapter 6, *Encapsulation: Entrapping Essential Oil/Flavors/Aromas in Food,* prepared by Suphla Gupta et al., highlights new approaches related to the encapsulation of essential oil/flavor or aroma compounds in food and the importance of these approaches in food industry.

Tarik Bor et al., in Chapter 7, *Antimicrobials From Herbs, Spices, and Plants,* present the current knowledge about the role of several natural products and nanoencapsulation strategies to enhance the efficacy of these products and several methods of encapsulation processes and their activity in growth media and food systems.

Lucia Zakharova et al., in Chapter 8, *Supramolecular Strategy of the Encapsulation of Low-Molecular-Weight Food Ingredients,* describe the recent progress in the field of engineering of delivery and storage systems for food ingredients. Several lines of investigations have been reviewed, including the amphiphile and polymer-based nanocontainers, supramolecular guest–host strategies involving macrocycle platforms, as well as binding/release behavior of loads.

Kata Trifković et al., in Chapter 9, *Novel Approaches in Nanoencapsulation of Aromas and Flavors,* discuss the current state of knowledge on nanoencapsulation methods of aromas and flavors, overviewing the processes and techniques utilized for coacervation, nanoprecipitation, molecular inclusion, and production of nanoparticulate formulations such as nanoemulsions, liposomes, solid–lipid nanoparticles (SLNs), and nanostructure lipid carriers (NLCs). Furthermore, the chapter gives insights into physicochemical and morphological characteristics of aroma nanoencapsulates, summarizing advantages, and limitations of aroma nanoscale formulations versus microparticle formulations produced by conventional microencapsulation technologies. Finally, a critical prospect of potential application of aroma nanoencapsulates in real food products is given, supported by examples available in the recent literature.

Daibing Luo et al., in Chapter 10, *Nanocomposite for Food Encapsulation Packaging*, present new concepts of nanotechnology in the overall food industry and its benefits in providing rich nutritional value, quality packaging, smart sensing and relevant research for safer techniques for incorporation of nanotechnology in the food industry.

Leslie Violeta Vidal Jiménez, in Chapter 11, *Microencapsulated Bioactive Components as a Source of Health*, reveals the current knowledge about microcapsules with antioxidant capacity from a maqui (*Aristotelia chilensis*) leaf extract by emulsification and subsequent retention after microencapsulation.

Sergio Enrique Flores-Villaseñor et al., in Chapter 12, *Biocompatible Microemulsions for the Nanoencapsulation of Essential Oils and Nutraceuticals*, report an up-to-date overview about food-grade microemulsions, examined and compared with the examples based on physical characteristics: pH, conductivity, viscosity, particle diameter, shape, zeta potential, and surface tension.

Sandra Pimentel-Moral et al., in Chapter 13, *Nanoencapsulation Strategies Applied to Maximize Target Delivery of Intact Polyphenols*, present the state of the art in encapsulation technologies for delivery of bioactive compounds, focused on polyphenols. For each type of delivery system the authors describe its properties, advantages, and limitations.

Jayamanti Pandit et al., in Chapter 14, *Nanoencapsulation Technology to Control Release and Enhance Bioactivity of Essential Oils*, report an up-to-date overview about various nanoencapsulation systems used for essential oils with their methodology, properties, advantages, and limitations. This review also gives a brief account of various products available on market.

Sumit Gupta and Prasad S. Variyar, in Chapter 15, *Nanoencapsulation of Essential Oils for Sustained Release: Application as Therapeutics and Antimicrobials*, highlight the various categories of nanocarriers, that is, polymeric nanoparticulate formulations, lipid-based nanoparticles such as nano- and microemulsions, liposomes, and solid lipid nanoparticles. Apart from polymeric and lipid nanocarriers, cyclodextrin-based molecular inclusion complexes are also discussed. Various uses of nanoencapsulated flavors for therapeutics, antimicrobials, and sustained release for fragrances are also detailed.

Shailesh Ghodke et al., in Chapter 16, *Nanoencapsulation- and Nanocontainer-Based Delivery Systems for Drugs, Flavors, and Aromas*, outline the different types of nanocontainers, the mechanism of response, and the different active molecules such as drugs, flavors, and aromas for delivery purposes. The potential further developments of sustainable biocompatible nanocontainers are also broadly covered.

Paweł K. Zarzycki et al., in Chapter 17, *Cyclodextrins-Based Nanocomplexes for Encapsulation of Bioactive Compounds in Food, Cosmetics, and Pharmaceutical Products: Principles of Supramolecular Complexes Formation, Their Influence on the Antioxidative Properties of Target Chemicals, and Recent Advances in Selected Industrial Applications*, summarize the literature search concerning general information about cyclodextrins and related macrocycles, their supramolecular nanocomplexes with low-molecular mass compounds, physicochemical properties of host-guest complexes, encapsulation of bioactive target components in food and cosmetics products as well as pharmaceutical formulations, based on the research communications, mainly published over a period of the past 3 years (2012–15). In addition, reviews concerning the use of macrocyclic nanoadditives as the ingredient of food, pharmaceutical formulations, cosmetics, and related products that are considered theoretically important as well as present on the market are reported.

Eva Fenyvesi and Lajos Szente, in Chapter 18, *Nanoencapsulation of Flavors and Aromas by Cyclodextrins*, present an up-to-date overview regarding the technological advantages, such as stable, standardized compositions, improved wettability, simple dosing, and handling of dry powders, reduced packaging and storage costs, more economical manufacturing processes and reduced labor costs, which justify cyclodextrins application in the food industry. The possible methods for formulation of flavors and aromas are discussed and the analytical methods for their characterization are described. Several examples of the application in foods and beverages are given.

Mohamed H. Abd El-Salam and Safinaz El-Shibiny, in Chapter 19, *Natural Biopolymers as Nanocarriers for Bioactive Ingredients Used in Food Industries*, present updated knowledge about natural biopolymers certified as GRAS (generally recognized safe) either in use or of potential use as nanocarriers for bioactive ingredients, methods of their fabrication, and applications in encapsulation of different groups of bioactive food ingredients.

Siddhartha Singha et al., in Chapter 20, *Process Technology of Nanoemulsions in Food Processing*, present an up-to-date understanding of emulsions involving nanosized dispersed phase. This chapter also address some specific issues to implement them into various applications in food systems. Thermodynamics of both kinetically stable emulsions and so-called spontaneous nanosized emulsions have been included in the scope of the text. However, main attention has been given to the production technology of kinetically stable emulsions both via high-energy and low-energy routes. Critical examination of established methods like high-pressure homogenization, microfluidization, and ultrasonic homogenization as well as some emerging techniques has been done. A brief survey of the nanosized emulsions (both commercial and promising one) applicable in food processing concludes the chapter.

Alexandru Mihai Grumezescu
University Politehnica of Bucharest,
Department of Science and Engineering
of Oxide Materials and Nanomaterials,
Faculty of Applied Chemistry and
Materials Science, Bucharest, Romania
http://grumezescu.com/

1

ELECTROHYDRODYNAMIC MICROENCAPSULATION TECHNOLOGY

Anatol Jaworek
Institute of Fluid Flow Machinery, Polish Academy of Sciences, Department of Electrohydrodynamics, Gdańsk, Poland

1 Introduction

Electrohydrodynamic microencapsulation is a method of fabrication of microcapsules by employing the same physical processes as in electrohydrodynamic atomization (EHDA), and can be regarded as a version of EHDA. Electrohydrodynamic atomization (also known as electrospraying) is a process in which liquid flowing from a capillary nozzle is subjected to electrical forces due to a strong electric field generated nearby the nozzle, usually by imposing a high electric potential. Driven by an electric field, the electric charge carriers move within the liquid and distribute under the liquid surface of the meniscus. First, the electric forces cause deformation of the natural spherical meniscus to a conical shape, known as the Taylor cone. This cone results as an effect of equilibrium of surface tension, and electrical and gravitational forces. When the electric charge accumulated at the apex of this cone generates an electric field sufficiently high to produce electrodynamic pressure that overcomes the liquid surface tension, a thin liquid jet emerges from the apex, which removes the excess charge. This electrically charged jet is accelerated by an electric field and disrupts into droplets at its free end due to electrical repulsion of charges placed at its surface. The process of droplets production can be a pulsating or continuous one, depending on liquid physical properties, liquid flow rate, and the magnitude and polarity of high potential imposed to the nozzle. In order to maintain the liquid flow from this meniscus, a continuous liquid supply with flow rate equal to that of expelled liquid is necessary. No additional mechanical energy, other than that from the electric field, is applied for electrohydrodynamic

Encapsulations. http://dx.doi.org/10.1016/B978-0-12-804307-3.00001-6

atomization. Droplets produced by electrohydrodynamic atomization bear an electric charge of magnitude usually about half of the Rayleigh limit. The Rayleigh limit is the magnitude of electric charge on a drop that produces repulsive force equal to the surface tension force.

Devices used for electrospraying comprise of a liquid feeder, usually a syringe pump, a capillary nozzle, usually a blunt hypodermic needle, and high voltage supply. Typically, the voltage applied to capillary for electrospraying is within 5 and 30 kV of positive polarity, and the liquid flow rate varies between 0.1 and 10 mL/h depending on liquid physical properties.

Electrospraying has many advantages over conventional mechanical spraying systems with droplets charged by induction:

1. Droplets size distribution is usually narrow, with low standard deviation, and in some spraying modes (dripping and microdripping, or for cone-jet mode with jet breakup due to varicose instabilities), the droplets can be of equal size.
2. Droplets' sizes can be smaller than those available from conventional mechanical atomizers, and can be smaller than 1 μm.
3. Droplets are electrically charged to a higher magnitude that causes their self-dispersion in the space and lack of droplet agglomeration and coagulation.
4. The motion of charged droplets can be easily controlled (including deflection or focusing) by electric field that allows increasing deposition efficiency on an object.

Employing the same electrohydrodynamic processes, electrohydrodynamic atomization has been employed as technology from microcapsules production.

2 Electrohydrodynamic Atomization

The meniscus and jet at the outlet of capillary nozzle can behave in various ways depending on voltage magnitude at the capillary, the flow rate at which the liquid is fed to the nozzle, nozzle inner and outer diameters, and liquid properties: surface tension, viscosity, electrical conductivity, electric permittivity, and density. Those behaviors have been classified as electrospraying modes (Zeleny, 1915; Hayati et al., 1987a,b; Cloupeau and Prunet-Foch, 1990, 1994; Grace and Marijnissen, 1994; Shiryaeva and Grigorev, 1995; Jaworek and Krupa, 1999a,b). There is still a lack of general theory predicting the modes of electrospraying and properties of electrosprayed droplets, and phenomenological classifications, based on visual observations of geometrical forms of meniscus, jet, and droplet, are the most frequently used. The modes of spraying can be grouped into two main categories:

- Dripping modes, producing fragments of liquid ejected directly from the capillary nozzle due to electrohydrodynamic disturbance of the meniscus. These fragments can be in the form of regular large drops (dripping mode), micron-size droplets (microdripping mode), or elongated spindles (spindle or multi-spindle modes). When irregular fragments of liquid are ejected in random directions from elongated irregular meniscus, this mode is classified as ramified meniscus. At some distance from the nozzle outlet, however, these fragments of liquids contract into spherical droplets under the action of surface tension force.
- Jet modes, which comprise those modes in which a fine liquid jet or a few jets emerge from the meniscus. The jet can be smooth and stable at a certain distance, much larger than its diameter, stretching from the meniscus apex along the nozzle axis (cone-jet mode), can oscillate in one plane lying on the capillary axis (oscillating mode), or rotate around the capillary axis (precession mode). In certain circumstances, a few jets are simultaneously ejected at the circumference of the capillary, and this mode is known as *multijet mode*. In each case, the jet at its end disintegrates into fine droplets, which are one or two orders of magnitude smaller than in the case of dripping modes.

In practice, the most useful are those modes that generate droplets of equal size in regular manner, that is, dripping, microdripping, and cone-jet modes. These modes have been employed for electrohydrodynamic microencapsulation processes. In the cone-jet mode, the liquid meniscus assumes the form of regular, axisymmetric cone, known as the Taylor cone (Taylor, 1964; Taylor and van Dyke, 1969), with fine jet (<100 μm in diameter) formed at its apex. The free end of the jet undergoes electrohydrodynamic instabilities of one of two types: varicose or kink (Cloupeau and Prunet-Foch, 1990, 1994). The produced droplets are of nearly monodisperse size.

In the case of dripping and microdripping modes the droplets are monodisperse. However, in some cases, a droplet detaching the meniscus can be connected to it with a thin thread, which breaks into one or a few droplets due to electric charge accumulated on it. These droplets are called satellite droplets, and are much smaller than the main drop. At higher electric fields, a similar thread can be produced at the front side of the droplet, which assumes a conic form, generating thus additional plume of sibling droplets, resembling a fine mist. Measurements have shown that in the presence of satellite or sibling droplets, the size distribution of droplets is bimodal or multimodal. In order to remove these satellite droplets, a screening plate electrode with circular opening coaxial with capillary nozzle, placed some distance from the nozzle, is used (Hong et al., 2008).

The net charge on the droplet detaching the jet is that on the jet length from which the droplet was formed, which equals to the wavelength λ of waves generated on the jet (Schneider et al., 1967; Brandenberger et al., 1999):

$$q = \frac{2\pi\varepsilon_0 U\lambda}{\ln\left(\frac{R_{cyl}}{R_{jet}}\right)} \quad (1.1)$$

where R_{cyl} is the radius of charging cylindrical electrode, U is the potential of this electrode and R_{jet} is the mean radius of the jet within the charging electrode.

For inviscid liquid, the wavelength is:

$$\lambda = \pi\sqrt{2}d_{jet} \quad (1.2)$$

where d_{jet} is the diameter of the undisturbed jet.

The size of droplets in electrospray is of fundamental importance for micro- and nanoparticle or microcapsules production. The mean size of the droplets produced by the cone-jet mode was determined theoretically and was confirmed by experimental results. Fernandez de la Mora and Loscertales (1994) determined the following equation for the droplet diameter produced in cone-jet mode of electrospraying:

$$d = \alpha\left(\frac{Q_l\varepsilon_0\varepsilon_r}{\kappa_l}\right)^{1/3} \quad (1.3)$$

where α is a constant depending on spray conditions and liquid permittivity, κ_l is the liquid conductivity, ε_0 is the permittivity of the free space, ε_r is the dielectric constant of liquid, and Q_l is the liquid flow rate.

Based on the scaling laws, Gañan-Calvo (1997, 1999) have obtained another equation for the size of droplet in cone-jet mode:

$$d = \alpha\left(\frac{Q_l^3\varepsilon_0\rho_l}{\pi^4\sigma_l\kappa_l}\right)^{1/6} \quad (1.4)$$

where ρ_l is the liquid density, σ_l is the surface tension of the liquid, and α was assumed to be equal 2.9 in this case.

Chen and Pui (1997) developed a model, which predicts the droplet size in the cone-jet mode, depending on permittivity of the liquid:

$$d = \alpha(\varepsilon_r)\left(\frac{Q_l \varepsilon_0 \varepsilon_r}{\kappa_l}\right)^{1/3} \tag{1.5}$$

The variable $\alpha(\varepsilon_r)$ depends on the relative permittivity of liquid (Chen and Pui, 1997):

$$\alpha(\varepsilon_r) = -10.9\varepsilon_r^{-6/5} + 4.08\varepsilon_r^{-1/3} \tag{1.6}$$

Hartman et al. (2000), for the cone-jet mode obtained:

$$d = \alpha\left(\frac{Q_l^3 \varepsilon_0 \rho_l}{\sigma_l \kappa_l}\right)^{1/6} \tag{1.7}$$

Regardless of the model developed, it can be concluded that the size of droplets is proportional to the square or cube root of liquid flow rate Q_l, and can be decreased by increasing liquid surface tension or conductivity.

These scaling laws of droplets size distribution were confirmed by many experiments for a single and two coaxial jets (Gañan-Calvo, 1999; Lopez-Herrera et al., 2003; Barrero et al., 2004; Bocanegra et al., 2005; Mei and Chen, 2007), and can be applied for all techniques of electrospray encapsulation.

The minimum flow rate at which the cone-jet mode can operate at steady-state is (Barrero and Loscertales, 2007):

$$Q_{min} \approx \frac{\sigma_l \varepsilon_0 \varepsilon_r}{\rho_l \kappa_l} \tag{1.8}$$

In this case, the size of droplets can be on the order of 1 μm, when the liquid's electrical conductivity is 10^{-3} S/m (water). The droplet's size decreases to about 10 nm when the conductivity is of 1 S/m (liquid metal).

Scheideler and Chen (2014) found the scaling law for the minimum flow rate of highly viscous liquids:

$$Q_{min} \sim \frac{\sigma_l D^2}{\mu_l} \tag{1.9}$$

Poncelet et al. (1999a,b) assumed that the effect of voltage can be considered as decreasing the effective surface tension of liquid:

$$\sigma_l(U) = \sigma_{l0}\left[1 - \left(\frac{U}{U_{cr}}\right)^2\right] \tag{1.10}$$

With this assumption, the effect of voltage on droplet diameter in microdripping mode was determined by Poncelet et al. (1999a,b):

$$d(U) = d_0 \left[1 - \left(\frac{U}{U_{cr}} \right)^2 \right]^{1/3} \tag{1.11}$$

where d_0 is the droplet diameter for $U = 0$, and the critical voltage U_{cr} of transition from dripping mode to microdripping mode is (Poncelet et al., 1999a,b):

$$U_{cr} = \left(\frac{d_c \sigma_1}{k \varepsilon_0} \right)^{1/2} \tag{1.12}$$

where d_c is the inner diameter of the capillary, k is the parameter whose value depends on characteristic times of the formation of liquid drop.

The effect of voltage polarity on the critical voltage for microdrops formation and on the size of droplets has been explained by Poncelet et al. (1999a,b) for the case of sodium alginate solution. For negative polarity of the nozzle, the negative alginate polyelectrolyte ions migrate from the solution to the surface due to repulsive electric field. The mobility of these ions at negative polarity of nozzle is lower than Na^+ ions at positive polarity, and their migration to the meniscus surface is much longer than Na^+ ions. For the same magnitude of voltage and flow rate, the surface charge density for the negatively charged nozzle is lower, and, consequently, the effective surface tension of liquid will be higher than for positive polarity that results in larger droplets and higher critical voltage of droplets formation. For positive polarity, the Na^+ ions build faster the surface charge on the meniscus surface, which reduces the effective surface tension that causes smaller droplets generation at lower critical voltage.

Physical restrictions regarding cone-jet or microdripping modes generation result from liquid properties. However, because of complexity of these phenomena and the fact that parameters of liquid important to electrospraying (electric conductivity, viscosity, density, surface tension, electric permittivity) cannot be changed independently, the regimes of electrospraying modes operation have not been studied systematically for all parameters, and have not been determined unambiguously. For example, the effect of liquid conductivity on cone-jet mode generation determined by different authors varies as follows: $10^{-5} - 10^{-11}$ S/m by Mutoh et al. (1979), $10^{-1} - 10^{-11}$ S/m by Smith (1986), or $10^{-1} - 10^{-9}$ S/m by

Cloupeau and Prunet-Foch (1988). Nowadays, it is assumed that semiconducting liquids, which can be sprayed in cone-jet mode by electrohydrodynamic method, are those of conductivity ranging from 10^{-4} to 10^{-8} S/m. Smith (1986) also found that liquids with surface tension higher than 50 mN/m cannot be atomized by electrostatic forces.

3 Micro- and Nanoencapsulation Techniques

Encapsulation is a method of capturing of solid particles, liquid droplets, or gas bubbles, in a solid or liquid envelope, made of another immiscible material. Two types of capsules are produced: core/shell and particle/matrix (Fig. 1.1). In the first case, the core material in the form of continuous phase is surrounded by a membrane-like envelope, called shell, while in the second case the micro/nano particles or micro/nano droplets, forming a dispersed phase, are uniformly distributed within a homogeneous solid-phase matrix. Material forming the shell in microcapsule is called also envelope, coating, membrane, or external phase. Micro- and nanoencapsulation within a solid phase is frequently considered as a "conversion" of liquid phase to solid phase (or a liquid to powder) with physical and chemical properties of the shell or matrix. Such particles can be fluidized in order to their pneumatic transportation or further processing. Under certain circumstances, the core material can be released by diffusion through the shell or matrix, or via dissolution, volatilization, or removing the shell material by another way. Microcapsules find application mainly in food, pharmaceutical, or cosmetic industries, and in medicine therapy.

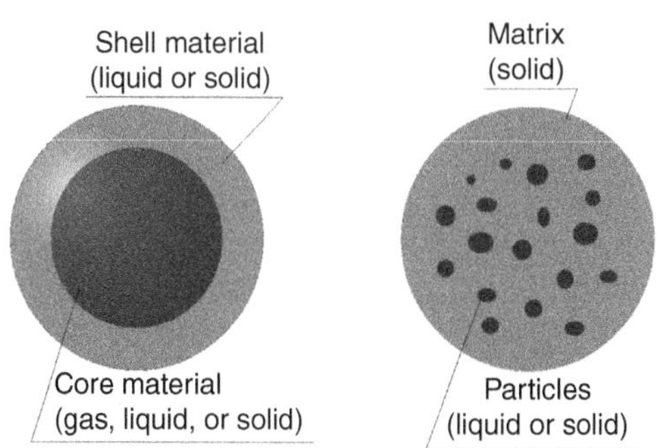

Figure 1.1. Core/shell (a) and particles/matrix (b) types of microcapsules.

An important problem with microcapsule formation is proper selection of shell material for a specific application of the capsules. With respect to intra- and intermolecular interactions of micro- and nanocapsules, the responsive chemical composition of monomer unit, its molecular weight and its physical and chemical characteristics, such as hydrogen bonding, electrostatic interactions, and hydrophobic interactions, have to be taken into account (de Vos et al., 2009). Usually, various types of polymers, natural or synthetic, are used for microcapsules production.

In food industry, by microencapsulation an environment-sensitive core material is protected by shell that allows slowing down of its degradation and prolongs the shelf life of the product (Desai and Park, 2005). Microencapsulation is usually used for the protection of cultures, vitamins, flavors, dyes, enzymes, salts, sweeteners, acidulates, nutrients, or preservatives against the environment. Depending on shell material physical properties, the transfer rate of core material to the outside environment can be controlled by temperature, mechanical stress, shell dissolution, changing pH of liquid environment, or the activity of enzymes (Desai and Park, 2005).

In case of drug delivery by micro- or nanoencapsulation, the core material containing the medicine can be transported by blood-vessel system to a target organ where it is released in a controlled manner (Jono et al., 2000; Mishra et al., 2010). Of particular interest for drug delivery applications are porous capsules. Active substances housed within shell or matrix diffuse through the pores of microcapsule and can modify physical and chemical properties of such particle, which can be used as a novel medicine material tailored to a specific therapy. Due to their designed porous microstructure and large specific surface area, microcapsules can be applied as drug carriers or scaffolds in tissue engineering (Wu and Clark, 2007, 2008). The goal of targeted drug delivery is to minimize the risk-to-benefit ratio and eliminate side effects (Zamani et al., 2013).

In medicine therapy, via bioencapsulation, the tissue, DNA, or other biologically active substances are captured within semipermeable membrane to protect the enclosed biological structures from potential hazardous processes in environment (de Vos et al., 2009; Orive et al., 2003a,b, 2004, 2014). Microencapsulation of cells is a method of immobilization of bioactive cells within a microparticle formed by a polymer membrane. Such a membrane permits free transport of nutrients and oxygen to the cells with sufficient rate, and release of therapeutic protein or insulin and waste products outside. Encapsulation allows protection of entrapped cells from mechanical stress and, in the case of allergenic tissue, from host's immune response. Microencapsulation of cells is used in regenerative medicine in order to overcome the difficulties with organ graft rejection and minimize the side effects associated with the use of immunomodulatory protocols or immunosuppressive drugs, for acceleration the process of organs regeneration, or for drug delivery for therapeutic purposes (Orive et al., 2014).

In agrochemistry, microcapsules are applied in order to limit environment pollution by controlling the release rate of insecticides or pesticides. In this way, the quantity of active substance necessary for a given application can be decreased (Hirech et al., 2003). In textile industry, fragrances added to textile and skin softeners enclosed in microcapsules can be released stepwise, elongating the time of their activity (Nelson, 2002). Phase-change materials used for latent heat storage/recovery were also produced in the form of microcapsules (Jin et al., 2008; Li and Li, 2007).

Conventional techniques of micro- and nanoencapsulation encompass (Gouin, 2004; Desai and Park, 2005); Ye et al., 2010; Jyothi et al., 2010; Mishra et al., 2010; Nedović et al., 2001; Nedović et al., 2011):

1. Spray drying (Edris and Bergnstahl, 2001; Kaushik and Dureja, 2015). In this process, the encapsulated material is homogenized with the material forming the matrix, and next this mixture is atomized by any type of atomizer. The solid particle/matrix structure is formed after solvent evaporation in hot air.
2. Extrusion (sol–gel encapsulation). By this process a colloidal suspension (sol) is gelatinized after spraying into a gelling bath, to form a network in a continuous phase (matrix) bearing the dispersed phase (Gouin, 2004; Herrero et al., 2006).
3. Coextrusion, which is a process of simultaneous spraying of two liquids from two coaxial nozzles. Core liquid is pumped through the inner nozzle, while the shell material is pumped through the outer one (Sze et al., 2002). A version of coextrusion is the *flow-focusing* method developed by Gañan-Calvo et al. (2015) and Lopez-Herrera et al. (2003).
4. Spray cooling (spray chilling). This process proceeds by hardening the matrix material at normal temperature after dispersion of the core and matrix mixture at a higher temperature (Desai and Park, 2005). Some kinds of paraffin, wax, diacylglycerols, fat, or stearin are used as matrix material in this process.
5. Fluidized-bed coating. By this process, larger particles (core material of the size of about 100 μm) are fluidized in gaseous phase, and in this state are mixed with smaller particles (usually about 1 μm) or fine droplets (De et al., 2002; Szafran

et al., 2012). These small particles form a self-assembled monolayer on the surface of a core particle due to electrostatic forces, when a contact potential is formed between them because of the different forbidden energy band or different work functions of those materials. The particle layer can next be stabilized by spraying a coating formulation. When fine droplets are used, they also form a thin layer on the particle surface and the shell is obtained after solvent evaporation.

6. Emulsification. Core liquid in emulsified in a continuous phase, which will be the shell material, or will form a suspension in the matrix. The shell material can next be hardened via gelling or solvent evaporation from the atomized suspension (Tiarks et al., 2001; Jafari et al., 2006). Two combinations of emulsions, water/oil emulsion and oil/water emulsions, are the most commonly used as liquid/liquid microcapsules.
7. Coacervation. In this process, two immiscible soles are separated from the initial solution after addition of an electrolyte. In the next step, the separated hydrocolloids are deposited around an active ingredient suspended in the same media, forming a shell over it. The film thickness depends on pH, temperature, ionic strength, and humidity (Fery et al., 2004; Gouin, 2004; Mishra et al., 2010).
8. Solvent extraction/evaporation. The process consists of four major steps: (1) dissolution or dispersion of core material (bioactive compound) in an organic solvent containing matrix material, (2) emulsification of the dissolved compound in another liquid immiscible with the first one, (3) removal of the solvent from the dispersed phase, and (4) filtration or centrifugation and drying of the microcapsules (Freitas et al., 2005).
9. Rapid expansion of supercritical solution of core and shell materials in a supercritical fluid (usually CO_2) (Fages et al., 2004; Dos Santos et al., 2002; Wang et al., 2004; Lee et al., 2008). In this process, a rapid pressure drop of the solution flowing from a nozzle causes desolvation of shell material, and formation of a coating layer around the core.

Electrohydrodynamic atomization has been adapted to some of these techniques of microencapsulation in order to increase the process efficiency and decrease the size of microcapsules (cf. Electrohydrodynamic Micro- and Nanoencapsulation).

The following parameters are used for the characterization of the encapsulation process, independently of the method used:

Encapsulation efficiency is defined as the ratio of mass m_{ec} of encapsulated core material to the mass m_{dc} of the material used for the encapsulation (Leo et al., 2006; Xie et al., 2006b; Xu and Hanna, 2006, 2007).

$$EE = \frac{m_{\mathrm{ec}}}{m_{\mathrm{dc}}} 100\% \tag{1.13}$$

Encapsulation yield is the ratio of mass m_{p} of produced capsules to the total mass of core m_{dc} and shell m_{ds} materials used for the encapsulation process (Xu and Hanna, 2006, 2007).

$$EY = \frac{m_{\mathrm{p}}}{m_{\mathrm{dc}} + m_{\mathrm{ds}}} 100\% \tag{1.14}$$

Loading capacity is the ratio of mass m_{ec} of encapsulated core material to the mass m_{p} of produced capsules (Leo et al., 2006; Xu and Hanna, 2006, 2007).

$$LC = \frac{m_{\mathrm{ec}}}{m_{\mathrm{p}}} 100\% \tag{1.15}$$

In the case of living cells, the *material viability* is defined as the ratio of live cells n_{l} to the total number of cells in the capsules specimen (live n_{l} and dead n_{d}) (Townsend-Nicholson and Jayasinghe, 2006; Stankus et al., 2006):

$$V = \frac{n_{\mathrm{l}}}{n_{\mathrm{l}} + n_{\mathrm{d}}} 100\% \tag{1.16}$$

In the case of enzyme particle/matrix system, the *efficiency of immobilization* is defined as (Knezevic et al., 2002):

$$EI = \frac{E_0 V_0 - E_{\mathrm{f}} V_{\mathrm{f}}}{E_0 V_0} 100\% \tag{1.17}$$

where E_0 and E_{f} are the initial and filtered immobilized enzyme activity in the International Unit of enzyme activity, respectively, and V_0 and V_{f} are the initial and filtered volume of the enzyme, respectively. The international unit of enzyme activity (1 IU) is defined as the amount of enzyme required to produce 1 μmol of free fatty acid per minute.

The release kinetics of core material is characterized by Avrami's equation (Desai and Park, 2005):

$$R = \exp\left(-(kt)^n\right) \tag{1.18}$$

where R is the retention of core material during release, t is the time, k is the release rate constant, and n is a parameter specific to the release mechanism.

4 Electrohydrodynamic Micro- and Nanoencapsulation

The following electrospray techniques of micro- and nano-encapsulation have been developed. Those processes are called electro-microencapsulation (Jaworek, 2008).

1. *Electrospray drying* (Fig. 1.2) (Ding et al., 2005; Xie et al., 2006a; Ho and Lee 2011). A colloidal suspension of core material in a solution containing an envelope formulation (polymer) is electrosprayed and next the shell is solidified by solvent evaporation. Usually a surfactant is added to colloidal suspension prepared for electrospraying to prevent particle coagulation, aggregation, or flocculation. By this method particle/matrix microcapsules are formed. Electrospray drying via spraying of colloidal suspension requires low concentration of particles in order to generate core/shell microcapsules from droplets with only one particle inside, and the loading capacity is not very high. Depending on the evaporation rate of shell-material liquid, capsules of various morphologies can be produced. When the solvent evaporates too fast, irregular and porous capsules are obtained. For slow drying, the capsules are spherical, uniform in size, and with a smooth shell surface (Xie et al., 2006a).
2. *Electrospray extrusion* (Fig. 1.3) (Lewińska et al., 2004, 2006; Sato et al., 1996a). A colloidal suspension or solution of core material is electrosprayed into a gelling bath with gelatinizing or polymerizing agent. The agent forms a hard envelope on the core material. Ultraviolet light has also been tested for surface

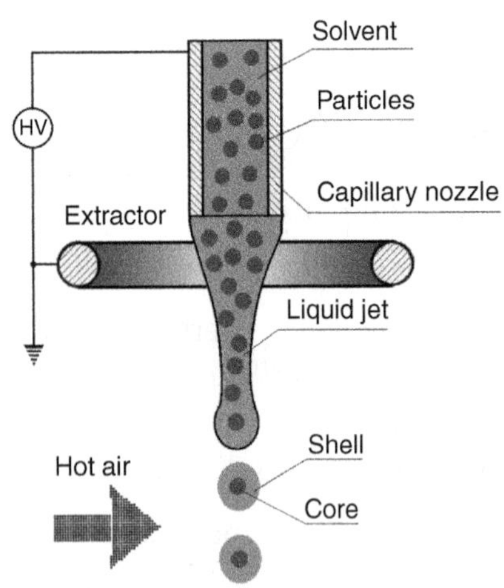

Figure 1.2. Electrohydrodynamic microencapsulation by electrospray drying.

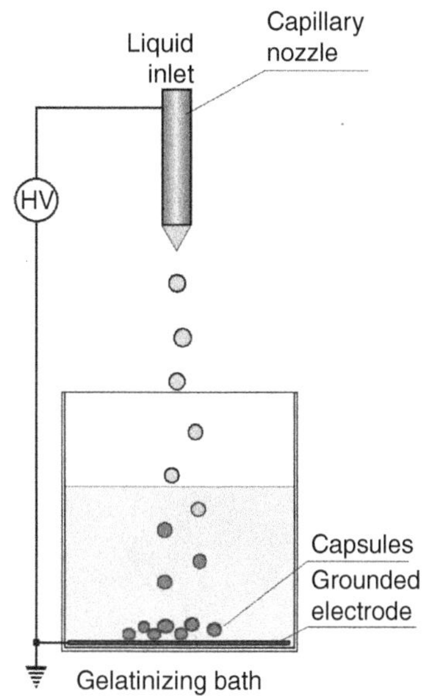

Figure 1.3. Electrohydrodynamic microencapsulation by electrospray extrusion and gelling.

polymerization in order to form a hardened envelope. In order to obtain capsules with only one particle within a shell or a single droplet, sufficiently low concentration of core particles in colloidal suspension is required. But in that case, some of the shell droplets are not loaded with particles, and they have to be removed. That results in low encapsulation yield of this method. In order to prevent particle coagulation, aggregation or flocculation, a surfactant, which spontaneously forms micelles around the particle and separates it from the other ones, is added to colloidal suspension (Perez-Masia et al., 2014). For example, for pure carbohydrate solutions, unstable jetting and dripping modes occurred, but with the addition of surface active molecules (Tween-20, Span-20, or Lecithin), micelles were formed on droplet's surface that, above certain critical micelle concentration, facilitated regular microcapsules production. With surfactant concentration increasing, smaller and more homogeneous microcapsules have been formed that was attributed to lower surface tension and higher conductivity of the solutions (Perez-Masia et al., 2014).

3. *Electrospray coextrusion* (Fig. 1.4) (Bocanegra et al., 2005; Lopez-Herrera et al., 2003; Loscertales et al., 2001, 2002; Chen et al. 2005; Xie et al., 2008). In this process, two different

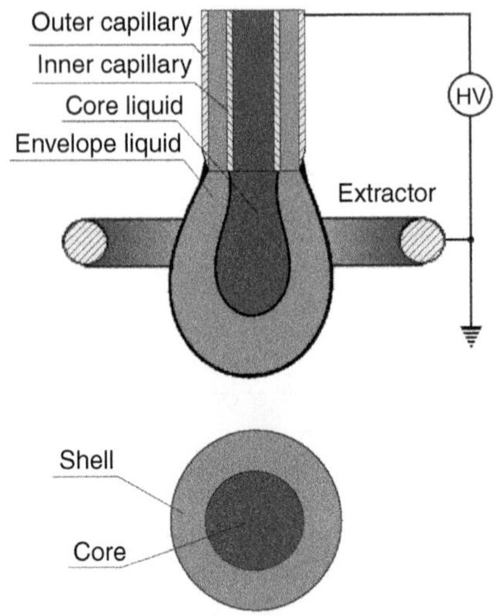

Figure 1.4. Electrohydrodynamic microencapsulation by electrospray coextrusion.

liquids are simultaneously electrosprayed from two coaxial capillary nozzles. In this case, the capillaries are at the same potential. The core liquid flows from the central capillary, while the envelope liquid flows through the annular nozzle between the capillaries. Electrospray mode in the coextrusion depends mainly on the properties of outer liquid, its viscosity and electric conductivity (Chen et al., 2005). Because of that, electrospray coextrusion technique allows spraying of core liquid of high resistivity, if the envelope has sufficiently high conductivity (Bocanegra et al., 2005). The properties and morphology of such capsules depend on the liquid of higher conductivity, called in the literature the *driving liquid* (Lopez-Herrera et al., 2003). The opposite arrangement, that is, a conducting core liquid and dielectric envelope, is also feasible (Bocanegra et al., 2005). By this method core/shell microcapsules are formed with high encapsulation efficiency and loading capacity. For effective microencapsulation by this technique, the flow rate of liquid flowing through the outer nozzle (shell liquid) should be larger than the flow rate of inner-nozzle liquid (core liquid) (Townsend-Nicholson and Jayasinghe, 2006).

4. *Electrospray cooling* (Fig. 1.5) (Milanovic et al., 2010). The core material is dispersed in a molten matrix and electrosprayed into a cooling bath where it is solidified. This technique allows particle/matrix microcapsules production.

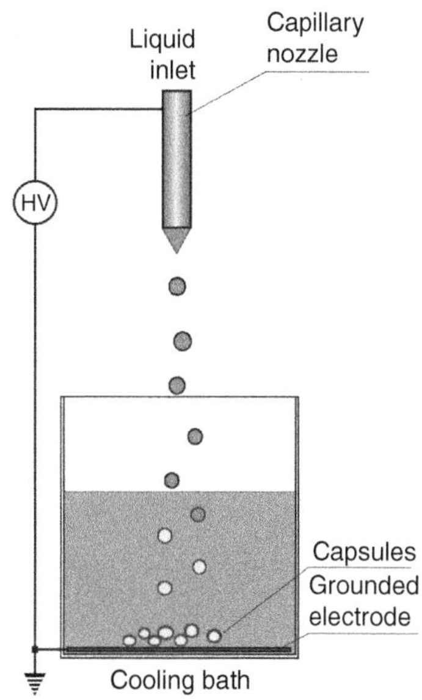

Figure 1.5. Electrohydrodynamic microencapsulation by electrospray cooling.

5. *Electrospray mixing* (Fig. 1.6). By this method, two streams of oppositely charged droplets of two different materials, one the core and the second the shell, are electrosprayed from two neighbor capillary nozzles at opposite polarities. Droplets of opposite charge collide due to their Coulomb attraction, and after collision merge forming core/shell microcapsules via submerging the droplet of higher surface tension within that of smaller surface tension, when Langer and Yamate's (1969) conditions are satisfied. A hard envelope can be obtained by chemical hardening or after solvent evaporation from the outer liquid. By this method core/shell microcapsules are formed. The method is characterized by low encapsulation efficiency and for this reason is not frequently used (Langer and Yamate, 1969; Borra et al., 1997, 1999). By mixing two oppositely charged droplets of different reacting compounds, this method can be applied as an aerosol-phase chemical reactor (Borra et al., 1997).
6. *Submerged electrospray* (Fig. 1.7) (Sakai et al., 1991; Barrero et al., 2004; Jayasinghe, 2007; Young et al., 2012). In this technique, the liquid to be encapsulated is electrosprayed directly into host liquid. This method is a variant of electrospray gelling with the difference that core material is directly electrosprayed into a gelatinizing bath with the omission of the gaseous phase. The liquid can be electrosprayed by a nozzle made as a bore in a plate placed at the vessel bottom, with grounded extractor electrode

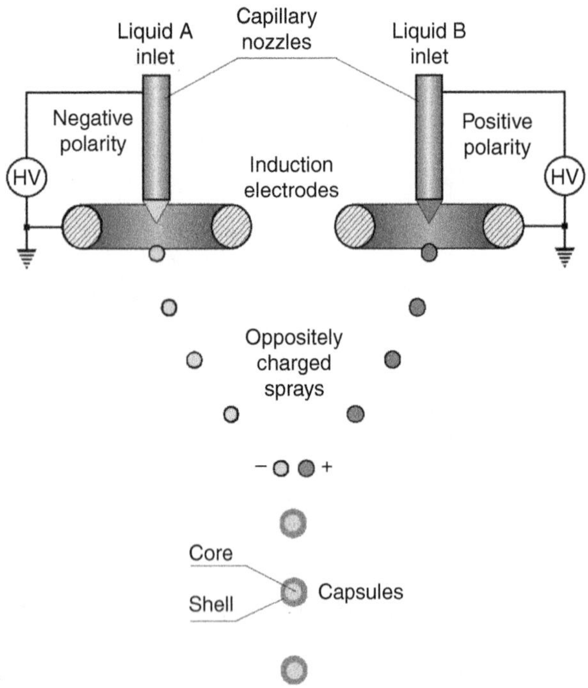

Figure 1.6. Electrohydrodynamic microencapsulation by electrospray mixing.

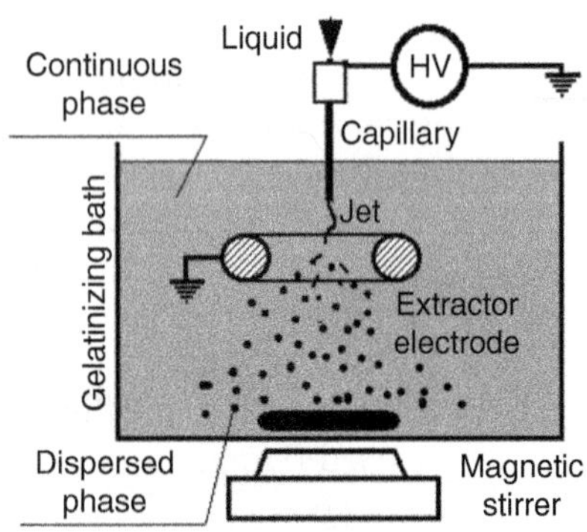

Figure 1.7. Electrohydrodynamic microencapsulation via submerged electrospraying.

positioned above it (Sakai et al., 1991), or from a capillary nozzle, similar to electrospraying in gaseous phase (Barrero et al., 2004; Jayasinghe, 2007; Young et al., 2012). A modified nozzle for submerged electrospraying, which operated more effectively for conducting liquids, has been designed by Sakai et al. (1991). In that device, an insulating plate with a small opening within it was placed between the nozzle and extractor electrode. Water jet flowing through this opening was smoother than that directly flowing out of the capillary nozzle. The system for microcapsules generation can also be supplied with an AC superimposed on DC voltage for synchronous droplet generation (Sakai et al., 1991).

7. *Electrospray reaction* (Fig. 1.8). Another technique employing electrospray is the *gas-aerosol reactive deposition* (GARED) technique developed by Salata (2005). Solution of polymer containing metal ions (polyvinyl alcohol + metal nitrate solution in a water/methanol mixture) was electrosprayed into gas containing reactive compounds (nitrogen with H_2X, where X = S, Se, Te, or As). MeX nanoparticles within polymer shell were produced after reaction of metal ions with H_2X. The polymer capsules containing some of II–VI and III–V group materials, as well as metal nanoparticles, were produced in this way. Via changing the concentration of metal ions and the size of droplets, it was able to control the number of reactive ions and the size of solid particles. The process is schematically illustrated in Fig. 1.8.

One of the versions of electrospray coextrusion technique is that using three coaxial capillaries. By this way uniform double-walled microcapsules with controllable size and shell thickness can be produced with larger flexibility than in the case of two coaxial nozzles. Double-layer shell allows separation of core material from the outer shell material. Such a double-layered structure can provide encapsulation of nanoparticles, liquids and/or gaseous bubbles in core/double-shell or particle/matrix/shell forms,

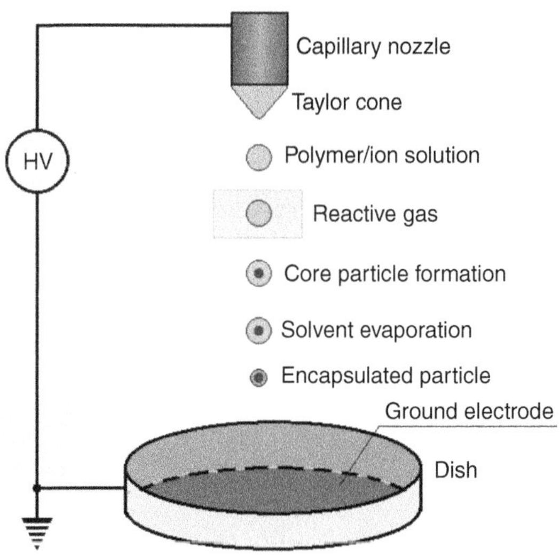

Figure 1.8. Electrohydrodynamic microencapsulation in reactive gas.

with multistage controlled release of the core material (Berkland et al., 2004, 2007; Chen et al., 2005; Ahmad et al., 2008, 2009; Kalra et al., 2009; Lee et al., 2011; Labbaf et al., 2013; Cao et al., 2014).

Various physical conditions have to be met in order to generate microcapsules by electrohydrodynamic method. The first requirements regarding microcapsules produced from two droplets have been formulated by Langer and Yamate (1969):

1. The droplets materials have to be immiscible, but must wet each other,
2. The surface tension of core material has to be higher,
3. The two droplets have to be of nearly the same size.

In the case of inner dielectric liquid (oil) encapsulated in liquid of higher electric conductivity (water, ethanol, glycerol or their mixtures), the formation of outer liquid jet is faster due to electric shear stresses than the inner liquid jet, which develops due only to viscous stress between outer and inner liquids (Chen et al., 2005). The generation of inner jet can be characterized with two dimensionless parameters, Weber and Ohnesorge numbers:

$$We_{\mathrm{in}} = \frac{\alpha d_j v_0^2 \rho_{\mathrm{out}}}{\gamma_{\mathrm{inter}}} \tag{1.19}$$

which is the ratio of hydrodynamic force to capillary force, and

$$Oh_{\mathrm{in}} = \frac{\mu_{\mathrm{in}}}{\left(\rho_{\mathrm{in}} \alpha d_j \gamma_{\mathrm{inter}}\right)^{1/2}} \tag{1.20}$$

which is the ratio of viscose characteristic time to breaking characteristic time constants, where d_j is the jet diameter, v_{in} and v_{out} are the inner and outer liquid flow velocities, respectively, α is the ratio of the inner jet diameter to the outer jet diameter, ρ_{out} is the outer liquid density, γ_{inter} is the interface tension between the jets, μ_{in} is the viscosity of inner liquid, and ρ_{in} is the inner liquid density (Chen et al., 2005).

In electrospray coextrusion systems, the capsules are formed only when the ratio of charge relaxation lengths (r^*) of inner liquid to outer liquid are less than 500, as defined by Mei and Chen (2007):

$$r^* = \frac{r^*_{\mathrm{out}}}{r^*_{\mathrm{in}}} < 500 \tag{1.21}$$

and the ratio of inertial break-up lengths (R^*) of the inner to outer jets is less than 0.015:

$$R^* = \frac{R^*_{\mathrm{out}}}{R^*_{\mathrm{in}}} < 0.015 \tag{1.22}$$

The charge relaxation length (r^*) and inertial breakup length (R^*) are defined by the following equations (Mei and Chen, 2007):

$$r^*_{\text{out}} = \left(\frac{Q_{l(\text{out})}\varepsilon_0\varepsilon_{\text{r(out)}}}{\kappa_{\text{l(out)}}} \right)^{1/3} \quad r^*_{\text{in}} = \left(\frac{Q_{\text{l(in)}}\varepsilon_0\varepsilon_{\text{r(in)}}}{\kappa_{\text{l(in)}}} \right)^{1/3} \tag{1.23}$$

$$R^*_{\text{out}} = \left(\frac{\rho_{\text{l(out)}}Q^2_{\text{l(out)}}}{\sigma_{\text{l}}} \right)^{1/3} \quad R^*_{\text{in}} = \left(\frac{\rho_{\text{l(in)}}Q^2_{\text{l(in)}}}{\sigma_{\text{l}}} \right)^{1/3} \tag{1.24}$$

where: Q_{l} is the liquid flow rate, ε_0 is the permittivity of the free space, ε_{r} is the dielectric constant of the liquid, ρ_{l} is the liquid density, κ_{l} is the liquid conductivity, and σ_{l} is the surface tension of outer liquid. Following the results of Mei and Chen, ethylene glycol or water in mineral oil could not be electrosprayed in stable cone-jet mode. Two unwanted situations can be distinguished when the ratios of $r^*_{\text{inner}}/r^*_{\text{outer}}$ and $R^*_{\text{inner}}/R^*_{\text{outer}}$ are beyond the limits (1.21) and (1.22) (Mei and Chen, 2007):

1. When the outer-liquid jet breaks up faster than the inner one, the outer liquid can form a droplet free of core material.
2. When the inner-liquid jet moves faster than the outer one, the inner liquid forms a separate droplet before the outer liquid can surround it, and thus the inner liquid escapes encapsulation.

In these two cases, separate droplets from inner and outer liquids are formed instead of microcapsules.

In the case of microbubbles (capsules of gaseous core), the bubble diameter is (Farook et al., 2007):

$$d_{\text{bubl}} = 1.1 D_{\text{inn}} \left(\frac{Q_{\text{g}}}{Q_{\text{l}}} \right)^{0.4} \tag{1.25}$$

where Q_{g} and Q_{l} are the gas and liquid flow rates, respectively, and D_{inn} is the inner diameter of inner capillary. The equation is valid for the Reynolds number from 40 to 1000.

The bubble diameter scales with gas pressure p and liquid flow rate and viscosity μ_{l} as follows (Garstecki et al., 2004, 2005):

$$d_{\text{bubl}} = k \frac{p}{Q_{\text{l}}\mu_{\text{l}}} \tag{1.26}$$

where k is a constant.

Microcapsules can also be generated synchronously by pulse or AC voltage excitation (Sakai et al., 1991; Sato et al., 1996a; Yeo et al., 2005). This technique has been originally developed for microparticles production (Vonnegut and Neubauer, 1952; Sample and Bollini, 1972; Bollini et al., 1975; Sato, 1984; Sato et al., 1996a,b, 1998, 1999; Balachandran et al., 1994). The size of droplets can be controlled via tuning pulse frequency, flow rate of liquid, and DC-bias and pulse or AC-voltage amplitude. The droplets generation rate can be controlled by liquid flow rate and high voltage frequency of excitation. The frequency of excitation voltage should be close to the frequency of natural mechanical vibration of the jet (Sakai et al., 1991). The frequency of synchronous droplets generation increases with an increase of the liquid flow rate (Sato et al., 1996b; Jaworek et al., 2000) and the range of frequencies is wider for liquids of low viscosity (Jaworek et al., 2000, 2003; Paine et al., 2007). Capsules produced by pulse excitation can be smaller than 1 μm and uniform in size. Application of alternating current allows the production of capsules of zero net charge with a frequency of 20 kHz (Yeo et al., 2005). The capsules produced by this method are smaller than 10 μm in diameter. In contrast to DC electroencapsulation, by which only polar liquids with free electric charge carriers can be dispersed, the AC encapsulation technique allows also the use of dielectric liquids based on organic solvents.

Possible degradation of core or shell materials containing very sensitive biological or any other fragile samples during electrospraying is of particular concern. The material degradation can be caused by electrochemical processes initiated by electric current flowing through the liquid (van Berkel and Kertesz, 2007), by UV radiation emitted by corona discharge accompanying the electrospraying, or OH radicals produced in this discharge (Jaworek et al., 2014). Teer and Dole (1975) electrosprayed polystyrene beads suspended in a liquid and noticed that only about 10% of them were degraded when negative voltage higher than –24 kV was applied to the capillary. This percent was decreased for positive polarity of the nozzle, when the voltage was lower than +20 kV, or in SF_6 atmosphere. The degradation was probably caused by the bombardment of droplets by electrons produced in glow discharge in nitrogen, but the SF_6 gas was able to quench the discharge, reducing thus the degradation effect. Viability of living cells after electrospraying was investigated in many laboratories (Ramsden et al., 1992; Bugarski et al., 1993; Al-Hajry et al., 1999; Goosen, 1999; Nedović et al., 2001; Orive et al., 2003b; Zhou et al., 2005; Lewińska et al., 2006; Salalha et al., 2006; Patel et al., 2008; Workman et al., 2014). From those investigations it can be concluded that typically at least 90% of living cells are viable after electrospray, and these cells are able to survive in electrosprayed capsules for a long time, provided oxygen, nutrient media, and metabolites could diffuse throughout the capsule walls. Laboratory tests also confirmed that degradation of electrosprayed material such as drugs, proteins, or other biological or pharmaceutical specimens is not significant (Uematsu et al., 2004; Gomez et al., 1998; Xie and Wang, 2007a,b). Physical experiments using optical emission spectroscopy have shown that electrical discharge is inherent to the process of electrospraying of conducting or semiconducting liquids. However, the production of OH radicals, which could be dangerous for biological samples, is negligible at lower voltages, when only glow and onset streamers discharges are generated, but can occur for discharges of higher currents such as streamers or arc (Jaworek et al., 2014).

5 Electrohydrodynamic Microencapsulation for Food Processing

In food industry, microcapsules are used for enveloping food ingredients such as flavors, dyes, enzymes, salts, sweeteners, acidulates, vitamins, etc. as core material, in calcium alginate, starch, gums, fats, waxes, oils, dextrans, glucoses, polysaccharides, casein, gelatin, chitosan, or proteins as shell (Gouin, 2004; Given, 2009; Nedović et al., 2011). Oil-in-water emulsions are most frequently used as carrier for core ingredients (Abu-Ali and Barringer, 2005; Bocanegra et al., 2005). Calcium alginate is the most frequently used for encapsulation in food or drug processing because it is nontoxic, biodegradable, and biocompatible. Calcium alginate is obtained from sodium alginate by its gelling in a bath containing 0.05–1.5 M calcium chloride solution. Proteins are used as encapsulation matrices because they are natural biopolymers of amphiphilic properties that facilitate food formulations stabilization.

Microencapsulation in food industry is applied for the following goals (Desai and Park, 2005; Nedović et al., 2011; Xie et al., 2015):

1. Protection of core material from degradation by preventing its contact with a reactive outside environment (eg, moisture, air, light);
2. Reduction of transfer rate of core material to outside environment via evaporation or diffusion;
3. Modification of physical properties of the core material for handling purposes or for further processing;
4. Control of the release rate of core material to expected specific environmental conditions by properly chosen shell;
5. Masking of flavor or aroma of core material, which can cause unpleasant feelings during eating;

6. Stabilization of an ingredient in core material during its handling, processing and storage;
7. Separation of the core material from other reactive ingredients in a final food product mixture;
8. Immobilization of cells or enzymes for food processing applications, such as fermentation or metabolite production processes.

Encapsulation of powder or liquid ingredients can improve the quality of food such as its aroma, taste, and appearance, as well as shelf life. Electrohydrodynamic technology was used for various food ingredients encapsulation, such as antioxidants, colorants, enzymes, flavors, lipids, living cells of bacteria or yeast, proteins, or vitamins. The details of the encapsulation process parameters for various ingredients are presented in Tables 1.1–1.8.

Antioxidants (Table 1.1), as molecules that inhibit the oxidation of other molecules, were encapsulated in calcium alginates (Belščak-Cvitanović et al., 2011; Stojanovic et al., 2012), gelatin (Gomez-Mascaraque et al., 2015), or zein (Torres-Giner et al., 2010). Belščak-Cvitanović et al. (2011) have encapsulated polyphenolic extracts of six different herbs: raspberry leaf, hawthorn, ground ivy, yarrow, nettle, and olive leaf, as sources of antioxidants, within alginate–chitosan copolymer shell by electrospray coextrusion method. Thyme aqueous extract was encapsulated within alginate beads by Stojanovic et al. (2012), in order to improve thyme functionality and stability in food products. Encapsulation efficiency varied in the range from 50 to 80% depending on the encapsulation method. The electrospray encapsulation process did not degrade the antioxidants, and the total antioxidant content remained unchanged. Gomez-Mascaraque et al. (2015) encapsulated (–)–Epigallocatechin gallate (EGCG), the most abundant and biologically active antioxidant in green tea, in gelatin matrix. The experimental results shown that EGCG microencapsulated in gelatin was stabilized against its degradation in aqueous solution (pH = 7.4), that means, the antioxidant activity was better preserved in the microencapsulated state than in its free form. This stabilization was attributed to both the delay of its dissolution in aqueous media and to the intermolecular interactions between the active molecule and its encapsulating matrix. Torres-Giner et al. (2010) have encapsulated docosahexaenoic acid (DHA—a long chain polyunsaturated fatty acid of the Omega-3 series) in zein using electrospray drying method. The FTIR analysis showed that the degradation kinetics of DHA is much slower when encapsulated in zein matrix than unencapsulated one. For example, the complete degradation of unprotected DHA was after the first 20 h of exposure to environmental conditions,

Table 1.1 Electrohydrodynamic Technology for antioxidants Microencapsulation

Core Material	Shell/Matrix Material	Method/Spray System (Capsule Type)	Nozzle	Voltage (Electrode Distance, Electrospray Mode)	Flow Rate (Spray Current)	Capsule Size (Encapsulation Efficiency/Yield/ Viability)	Authors, Paper
Polyphenolic extracts of six different herbs	Sodium alginate-chitosan copolymer	Electrospray extrusion (gelling bath 2 w/v% $CaCl_2$ + 0.5 w/v% chitosan in 2 w/v% ascorbic acid, pH = 2.65)	Stainless steel capillary 23 gauge	7.3 kV (30 mm to bath)	25.2 mL/h	780–1785 μm (particles); (efficiency = 80–89%, loading = 59%)	Belščak-Cvitanović et al. (2011)
Thyme (Thymus serpyllum L.) (10 g dried thyme in 200 mL water)	Sodium alginate (16 mg/mL in thyme solution + 0.2 g/mL sucrose) or sodium alginate (16 mg/mL in thyme solution + 0.05 g/mL inulin)	Electrospray extrusion (particle/matrix), (gelling bath 2% w/v $CaCl_2$)	Stainless steel capillary 22 gauge	6.5 kV	25.2 mL/h	730 μm (capsules); (efficiency = 50–80%)	Stojanovic et al. (2012)
(−)-Epigallocatechin gallate (EGCG) (10 wt% in matrix liquid)	Gelatin (5, 8, 10, 20 v/v% in 20 v/v% acetic acid)	Electrospray drying (particle/matrix)	Stainless steel capillary 0.9 mm i.d.	15–28 kV (100 mm to metal collector)	0.15 or 0.5 mL/h		Gomez-Mascaraque et al. (2015)
Docosahexaenoic acid (Omega-3 series) (20 wt% in 85% ethanol) + zein (1:1, 1:2 or 1:3)	Zein	Electrospray drying (horizontal configuration)	Stainless steel capillary 0.9 mm i.d.	12 kV (150 mm to Al collector)	0.2 mL/h	0.49 ±0.2 μm (capsules)	Torres-Giner et al. (2010)

Table 1.2 Electrohydrodynamic Technology for Colorants Microencapsulation

Core Material	Shell/ Matrix Material	Method–Spray System (Capsule Type)	Nozzle	Voltage (Electrode Distance, Electrospray Mode)	Flow Rate (Spray Current)	Capsule Size (Encapsulation Efficiency Yield–Viability)	Authors, Paper
Curcumin (1:500 to 1:10 solution in zein solution)	Zein 25–30 kDa (2.5 w/w% in 80 w/w% ethanol)	Electrospray drying (particle/ matrix)	Stainless steel capillary	14 kV (70 mm to metal collector), (cone-jet mode)	0.15 mL/h	0.45–0.65 μm (capsules); (efficiency = 85–90%)	Gomez-Estaca et al. (2012)
Curcumin (250 mg in gelatin solution)	Gelatin (2.5 g in 25 mL 96% ethanol + 12.5 mL water + 12.5 mL acetic acid)	Electrospray drying (particle/ matrix)	Stainless steel capillary	14 kV (100 mm to metal collector), (cone-jet mode)	0.15 or 0.5 mL/h	0.5 μm (capsules); (efficency = 100%)	Gomez-Estaca et al. (2015)
β-carotene (5 mg/mL in glycerol)	Whey protein (40 wt% in water + 10–20% of glycerol)	Electrospray extrusion (particle/ matrix)	Stainless steel capillary 0.9 mm	14 kV (70 mm to collector)	0.3 mL/h		Lopez-Rubio and Lagaron (2012)
Lycopene (1.25 w/v% in soybean oil) 16 w/v% in 10 w/v% Tween-20 water solution	Whey protein (30 w/v% in water + 5 w/v% Tween-20)	Electrospray coextrusion (core/shell)	Stainless steel capillaries	12–18 kV (90–200 mm to collector)	0.15 mL/h	(efficiency = 75%)	Perez-Masia et al. (2015b)
Lycopene (1.25 w/v% in soybean oil) 16 w/v% in 10 w/v% Tween-20 water solution	Dextran (Mw = 70k) (20 w/v% in water + 5 w/v% Tween-20 in water)	Electrospray coextrusion (core/shell)	Stainless steel capillaries	12–18 kV (90–200 mm to collector)	0.15 mL/h	(efficiency = 57%)	Perez-Masia et al. (2015b)

Table 1.2 Electrohydrodynamic Technology for Colorants Microencapsulation (*cont.*)

Core Material	Shell/ Matrix Material	Method–Spray System (Capsule Type)	Nozzle	Voltage (Electrode Distance, Electrospray Mode)	Flow Rate (Spray Current)	Capsule Size (Encapsulation Efficiency Yield–Viability)	Authors, Paper
Lycopene	Chitosan (Mw = 50–190k) (2 w/v% + 1.5% w/w lycopene in 90% v/v acetic acid + 5 w/v% Tween-20 in water 10 w/v%)	Electrospray drying (particle/ matrix)	Stainless steel capillary	12–18 kV (90–200 mm to collector)	0.15 mL/h		Perez-Masia et al. (2015b)

Table 1.3 Electrohydrodynamic Technology for Enzymes Microencapsulation

Core Material	Shell/ Matrix Material	Method–Spray System (Capsule Type)	Nozzle	Voltage (Electrode Distance, Electrospray Mode)	Flow Rate (Spray Current)	Capsule Size (Encapsulation Efficiency–Yield–Viability)	Authors, Paper
Lipase from Candida rugosa	Sodium alginate solution (2 or 6 w/v%)	Electrospray extrusion (particle/ matrix), (gelling bath 2 w/v% $CaCl_2$)	Stainless steel capillary 21 gauge	4.9 kV (15 mm to bath)	25.2 mL/h	650 μm (capsules); (efficiency = 98–99%)	Knezevic et al. (2002)

Table 1.4 Electrohydrodynamic Technology for Flavors Microencapsulation

Core Material	Shell/Matrix Material	Method–Spray System (Capsule Type)	Nozzle	Voltage (Electrode Distance, Electro-spray Mode)	Flow Rate (Spray Current)	Capsule Size (Encapsulation Efficiency–Yield–Viability)	Authors, Paper
Sugar (49.5%) in water (50%) + colorant (0.5%) or oil (66%) in water (33%) + emulsifier	Cocoa butter	Electrospray coextrusion (core/shell)	Stainless steel capillaries, 0.075 mm i.d., 0.36 mm o.d., 0.75 mm i.d., 1 mm o.d.	2 kV, (2 mm to flat plate extractor electrode)	50–250 μL/h (inner liquid), 200–1000 μL/h (outer liquid); (91–147 nA for aqueous solution, 126–210 nA for o/w emulsion)	12.4–22.5 μm (aqueous solution); 9–21 μm (o/w emulsion)	Bocanegra et al. (2005)
Maltol flavor (1.2–2.5 mg/mL in matrix liquid)	Stearic acid and ethylcellulose (5 w/v%, 10–50 mg/mL in ethanol)	Electrospray drying (particle/matrix)	Stainless steel capillary, 0.45 mm i.d., 0.87 mm o.d.	13–15 kV (100–150 mm to collector) (cone-jet mode)	0.6–0.9 mL/h	10–100 nm (capsules); (efficiency = 70%, yield = 69%)	Eltayeb et al. (2013a)
Peppermint oil	Sodium alginate + pectin mixture (1.5 g in 100 mL water + 0.2% Tween-80 + 0.2% xanthan gum + 5% glycerol)	Electrospray coextrusion (core/shell), (gelling bath 2% $CaCl_2$)	Stainless steel capillary 0.2 mm i.d., 0.51 mm o.d.	5 kV (100 mm to bath)	3 mL/h (core liquid), 7.8 mL/h (shell liquid)	1.58–3.24 μm (capsules); (efficiency = 85%)	Koo et al. (2014)
Vanillin, ethylmaltol, maltol (solution in matrix liquid)	Ethylocellulose and stearic acid in ethanol	Electrospray drying (core/shell)	Stainless steel capillary, 0.45 mm i.d., 0.87 mm o.d.	12–19 kV (200 mm to collection bath) (cone-jet mode)	0.6 mL/h	65 nm (capsules)	Eltayeb et al. (2013b)

Table 1.4 Electrohydrodynamic Technology for Flavors Microencapsulation (*cont.*)

Core Material	Shell/Matrix Material	Method–Spray System (Capsule Type)	Nozzle	Voltage (Electrode Distance, Electrospray Mode)	Flow Rate (Spray Current)	Capsule Size (Encapsulation Efficiency–Yield–Viability)	Authors, Paper
Ethyl vanilin (0.1–0.2 g/mL in matrix liquid)	Sodium alginate (0.02 g/mL in water) + poly (vinyl alcohol) (0.1 g/mL in water) 1:4 wt/wt	Electrospray gelling (particle/matrix), (gelling bath: 15 mg/mL $CaCl_2$ in water)	Stainless steel capillary 1.6. mm i.d.	4.5 kV (25 mm to bath)	70 mL/h	<1500 μm (capsules)	Levic et al. (2013)
Ethyl vanillin (0.1 g/mL in matrix liquid)	Sodium alginate (0.02 g/mL in water)	Electrospray gelling (particle/matrix), (gelling bath: 15 mg/mL $CaCl_2$ in water)	Stainless steel capillary 22 gauge	4.5 kV (25 mm to bath)	25.2 mL/h	450 μm (capsules)	Manojlovic et al. (2008)
Ethyl vanillin (10 w/w% in matrix)	Carnauba wax (melted at 95°C) 8 w/w% in water + 1% (Tween-20 + Span-40, 0.53:047) (surfactants)	Electrospray cooling (particle/matrix), (in cooling water bath 2–5°C)	Stainless steel capillary 22 gauge			210–360 μm (capsules)	Milanovic et al. (2010)

while for DHA microencapsulated in zein, the same degradation level required about 72 h.

Colorants (Table 1.2), food additives used to improve optical properties of product for a longer time than its natural hue does, were encapsulated in zein (Gomez-Estaca et al., 2012), gelatin (Gomez-Estaca et al., 2012, 2015), whey protein (Lopez-Rubio and Lagaron, 2012; Perez-Masia et al., 2015b), dextran (Perez-Masia et al., 2015b), or chitosan (Perez-Masia et al., 2015b). Gomez-Estaca et al. (2012) encapsulated curcumin (bright-yellow food

Table 1.5 Electrohydrodynamic Technology for Lipids Microencapsulation

Core Material	Shell/Matrix Material	Method–Spray System (Capsule Type)	Nozzle	Voltage (Electrode Distance, Electrospray Mode)	Flow Rate (Spray Current)	Capsule Size (Encapsulation Efficiency–Yield–Viability)	Authors, Paper
Cooking oil	Ethanol + glycerol + Tween-80 (3:1:0.04, or 2:1:0.03) mixture	Electrospray coextrusion (core/shell)	Stainless steel capillaries 0.2 mm i.d., 0.55 mm o.d., 0.7 mm i.d., 1.2 mm o.d.	4.92 kV (microdripping mode), 6 kV (cone-jet mode), 6.15 kV (pulsed mode), (20 mm to ground grid electrode 75 mm dia.)	0.44 mL/h (inner liquid), 0.72 mL/h (outer liquid) - dripping mode, 0.74 mL/h (inner liquid), 0.5 mL/h (outer liquid)–cone-jet mode, 0.35 mL/h (inner liquid), 1 mL/h (outer liquid)- pulsed mode	6.4 μm (core), 9.6 μm (capsule)–cone-jet mode	Chen et al. (2005)
Fish oil (30 w/w% in matrix liquid)	Zein (10 and 20 w/w% in 70 w/w% ethanol or 70 w/w% isopropanol)	Electrospray extrusion (particle/matrix)	Single-needle spinneret (vertical inverted configuration)	20 kV (200 mm to collector)	1 mL/h	0.4–0.5 μm (10% in ethanol), 0.6–0.8 μm (10% in isopropanol)	Moomand and Lim (2015)
Lipid (10 w/w% trimyristin + 6 w/w% tyloxapol in water + 0.005 w/w% thiomersal as preservative + 2.25% anhydrous glycerol)	Sodium alginate (1% in water)	Electrospray extrusion (particle/matrix), (gelling bath 5% w/w $CaCl_2$, and subsequent freeze-drying in liquid nitrogen)	Stainless steel capillary 0.16, 0.23, 0.26, 0.33, 0.40 or 0.52 mm i.d.	5 kV (25 mm to bath)		330–1350 μm (depending on nozzle diameter)	Strasdat and Bunjes (2013)

Table 1.6 Electrohydrodynamic Technology for Living Cells Microencapsulation

Core Material	Shell/Matrix Material	Method–Spray System (Capsule Type)	Nozzle	Voltage (Electrode Distance, Electro-spray Mode)	Flow Rate (Spray Current)	Capsule Size (Encapsulation Efficiency–Yield–Viability)	Authors, Paper
Brewer yeast (*S. uvarum*) (2×10^7 cells/mL in shell liquid)	Sodium alginate (2 wt% in water) + poly (vinyl alcohol) (10 wt% in water)	Electrospray gelling (particle/matrix), (gelling bath: 2% $CaCl_2$ in water)	Stainless steel capillary 22 gauge	7 kV (25 mm to bath)	25.2 mL/h	806 μm (beads)	Bezbradica et al. (2004)
Brewer yeast (*S. uvarum*)	Sodium alginate solution (2% w/v)	Electrospray extrusion (particle/matrix), (gelling bath 2% w/v $CaCl_2$)	Stainless steel capillary 20, 22, 27 gauge	6.5 kV (20, 40, or 80 mm to bath)	25.2 mL/h	250–2000 μm (>10^9 cells/mL loading) 500–600 μm (27 gauge)	Nedović et al. (2001)
Brewing yeast cells (*S. uvarum*), (2×10^7 cells/mL in shell liquid)	Sodium alginate (2 wt% in water) + poly (vinyl alcohol) (10 wt% in water)	Electrospray gelling (particle/matrix), (gelling bath: 2% $CaCl_2$ in water)	Stainless steel capillary 27 gauge	8 kV (25 mm to bath)		500–600 μm (capsules)	Leskošek-Ćukalović and Nedović (2005)
L. acidophilus in 1.4 w/v% sodium alginate in water +8 w/v% glycerol	Zein modified with citric acid (0.1–0.15%)	Electrospray gelling (core/shell), (gelling bath: 1.5 w/v% $CaCl_2$ in 60 v/v% ethanol + 0.1 w/v% citric acid)	Stainless steel capillary 0.69 mm i.d.	4–10 kV (60 mm to bath)	10 mL/h	259 μm (at 10 kV)	Laelorspoen et al. (2014)
Living *Bifidobacterium animalis* (10^{10} cells/mL in 130 mM NaCl + 10 mM Na_3PO_4 or skimmed milk)	Whey protein (30–40 wt% in 130 mM NaCl + 10 mM Na_3PO_4) or pullulan polysaccharide (Mw = 100k) (in 15–20 wt% in 130 mM NaCl + 10 mM Na_3PO_4)	Electrospray extrusion (particle/matrix)	Stainless steel capillary 0.9 mm	12–14 kV (70 mm to collector)	0.3 mL/h	1.1 – 4.7 μm (capsules)	Lopez-Rubio et al. (2012)

(*Continued*)

Table 1.6 Electrohydrodynamic Technology for Living Cells Microencapsulation (*cont.*)

Core Material	Shell/Matrix Material	Method–Spray System (Capsule Type)	Nozzle	Voltage (Electrode Distance, Electro-spray Mode)	Flow Rate (Spray Current)	Capsule Size (Encapsulation Efficiency–Yield–Viability)	Authors, Paper
S. cerevisiae var. *ellipsoideus* yeast cells in matrix liquid (25 mL in 100 mL sodium alginate)	Sodium alginate (2 w/w% in water)	Electrospray extrusion (particle/matrix), (gelling bath 2% w/v $CaCl_2$)	Stainless steel capillary 1 mm o.d.		15 mL/h	800 μm (capsules) (10^7 cells/capsule)	Rakin et al. (2009)
Bifidobacterium longum BIOMA 5920 (5 mL in liquid matrix 4:1)	Sodium alginate–humanlike collagen [20 mL of alginate (1.5%, w/v) + collagen (0, 1, 2 and 3% w/v) in water]	Electrospray extrusion (particle/matrix), (gelling bath 2% w/v $CaCl_2$)	Stainless steel capillary 0.4 mm	8–12 kV (50 mm to bath)		300–600 μm (capsules); (efficiency = 87–97%)	Su et al. (2011)

colorant) in zein matrix for its prevention from environmental conditions. Curcumin, which has low water solubility, after microencapsulation, presented good dispersion in aqueous food (semiskimmed milk). Curcumin, which normally degrades under neutral or alkaline pH conditions, or when it is exposed to light, remained in the amorphous state in the zein matrix, and the size and morphology of the nanocapsules were practically unchanged after three months of storage at 23°C and 43% relative humidity in the dark. Curcumin was also encapsulated in gelatin matrix (Gomez-Estaca et al., 2015) with encapsulation efficiency close to 100%. Lopez-Rubio and Lagaron (2012) encapsulated β-carotene (strongly colored red-orange pigment) in whey protein matrix. Perez-Masia et al. (2015b) encapsulated lycopene (bright red carotene and carotenoid pigment) in whey protein or dextran using electrospray coextrusion, or in chitosan dissolved in acetic acid using electrospray drying technique for the formation particle/matrix microcapsules. The electrosprayed dextran-based solution of lycopene provided homogeneous microcapsules, while

Table 1.7 Electrohydrodynamic Technology for Proteins Microencapsulation

Core Material	Shell/Matrix Material	Method–Spray System (Capsule Type)	Nozzle	Voltage (Electrode Distance, Electro-spray Mode)	Flow Rate (Spray Current)	Capsule Size (Encapsulation Efficiency–Yield–Viability)	Authors, Paper
Bovine serum albumin (65 kDa)	Poly(lactide) (175 kDa), (1–4% w/v in 1,2-dichloroethane or dichloromethane + acetone, or dichloromethane + N,N-Dimethylformamide)	Electrospray drying	Stainless steel capillary 0.84 mm i.d. 1.24 mm o.d.	10–15 kV (100 mm to bath and ring extractor of 110 mm i.d.)	0.5–3 mL/h	4.77 μm (capsules), 1.64 μm (with acetone); (efficiency = 23–80%, yield = 64–80%, loading = 74–91%)	Xu et al. (2006)
Bovine serum albumin (65 kDa)	Poly(lactide) (175 kDa), (1–4% w/v in 1,2–dichloroethane)	Electrospray extrusion (water + phosphate buffer saline 0.067 M bath)	Stainless steel capillary 0.84 mm i.d. 1.24 mm o.d.	10–15 kV (100 mm to bath and ring extractor of 110 mm i.d.)	0.5–3 mL/h	4.77 μm (capsules), 1.64 μm (with acetone); (efficiency = 23–80%, yield = 64–80%, loading = 74–91%)	Xu and Hanna (2006)
Bovine serum albumin (1 or 2% in water, 65 kDa)	Tripolyphosphate cross-linked to chitosan	Electrospray extrusion (water + tripolyphosphate 5 or 10% w/v bath)	Stainless steel capillary 0.84 mm i.d. 1.24 mm o.d.	7.5–11.5 kV (for 1% concentration), 18–21 kV (for 2% concentration), (100 mm to bath)	4, 6, 8 mL/h	5–10 μm (capsules) (efficiency = 55–80%, yield = 51–83%, loading = 32–60%)	Xu and Hanna (2007)

coaxial electrospraying produced more aggregated microcapsules. This effect was attributed to noncomplete encapsulation of core material (soybean oil/lycopene solution) during the co-extrusion process and leakage of core material from the dextran shell that can result from improper flow rates of inner and outer liquids. When whey protein was used as shell material, more heterogeneous structures were obtained through electrospraying, while smoother and more regularly sized capsules were produced through conventional spray drying technology. This result was

Table 1.8 Electrohydrodynamic Technology for Vitamins Microencapsulation

Core Material	Shell/ Matrix Material	Method–Spray System (Capsule Type)	Nozzle	Voltage (Electrode Distance, Electro-spray Mode)	Flow Rate (Spray Current)	Capsule Size (Encapsulation Efficiency–Yield–Viability)	Authors, Paper
Folic acid	Sodium alginate	Electrospray gelling, particle/ matrix, (gelling bath 0.45M $CaCl_2$, for 1800 s)	Stainless steel capillary, 0.45 mm i.d., 0.84 mm o.d.	9–12 kV (120–150 mm to bath) (cone-jet mode)	0.6 mL/h	4.2 ± 1.2 μm (droplets), 50–200 nm (capsules)	Bakhshi et al. (2013)
Folic acid	Whey protein or starch (20 w/v% in water + 0.5 wt% guar gum + Span-20)	Electrospray drying (particle/ matrix)	Stainless steel capillary	10 kV (90–110 mm to collector)	0.15 mL/h		Perez-Masia et al. (2015a)

explained by higher electrical conductivity of the protein solutions that could destabilize the electrospray jet, producing microcapsules of broader size distribution. Microcapsules produced from chitosan matrix by conventional spray drying were smaller than those obtained by electrospray coextrusion due to high conductivity of the solution containing acetic acid. From FTIR analysis came results showing that conventional spray-drying technique provides microcapsules of higher encapsulation yield compared to electrospray techniques. However, the encapsulation efficiency was higher for the electrospray techniques than for conventional spray drying; for example, for electrospray coextrusion it was about 57% for lycopene in dextran, and 75% for lycopene in whey protein, while for conventional spray drying these figures were 16 and 27%, respectively. An advantage of electrospray techniques compared to spray drying is that capsules are not subjected to the high temperatures required for drying process that could degrade core material.

Enzymes (Table 1.3) have been encapsulated by Knezevic et al. (2002), which used electrohydrodynamic droplet generation method for immobilization of lipase from Candida Rugosa in calcium alginate. Immobilization efficiency was in the range of 98.2% to 99.2% and depended only slightly on the alginate concentration. The maximal obtained lipase activity by that process was obtained for 2 w/v% of sodium alginate, and was about 752 IU/g, that was 75% of that of free lipase.

Flavors (Table 1.4), additives used to enhance aroma and/or taste impression of food product, have been encapsulated in cocoa butter (Bocanegra et al., 2005), ethylcellulose (Eltayeb et al., 2013a,b), calcium alginate (Koo et al., 2014; Levic et al., 2013; Manojlovic et al., 2008), or Carnauba wax (Milanovic et al., 2010). Aroma is one of the most important characteristics of food products. Substances used as aromas are usually composed of many volatile and odorous organic species. They are usually thermally sensitive chemical compounds, which require special treatment during food processing. Consequently, the degradation of aroma compounds during production, storage and/or transport could be critical in terms of their stability and food quality. Encapsulation might be one of the methods that could be used in order to improve these parameters. Bocanegra et al. (2005) encapsulated aqueous solution of sugar, or oil-in-water emulsion in cocoa butter as shell material, using electrospray coextrusion technique. The authors noticed that submicron droplets of the dispersed phase forming an oil-in-water emulsion as a core liquid are needed in order to produce stable capsules. For too-large droplets of the dispersed phase, the core liquid is not sufficiently homogeneous to form a stable Taylor cone (Taylor, 1964; Taylor and van Dyke, 1969), and electrospray instabilities can occur. In order to neutralize the electric charge of droplets, a high-voltage needle electrode at negative polarity was used for negative ions emission (Bocanegra et al., 2005). Eltayeb et al. (2013a,b) have encapsulated maltol, ethylmaltol, and vanillin flavors in stearic acid and ethylcellulose matrix using electrospray drying technique, in order to limit flavor degradation or its loss during food processing and storage, or to protect them against chemical reactions such as oxidation. The size of microcapsules was below 100 nm after solvent evaporation. Koo et al. (2014) produced microcapsules of peppermint oil in alginate–pectin matrix using electrospray coextrusion system. For the alginate/pectin ratio 80:20 the capsules size was 1.58 μm, and for 100% pectin was 3.24 μm. For alginate/pectin ratio 80:20, the encapsulation efficiency was about 85%. Levic et al. (2013) have encapsulated ethyl vanillin in calcium alginate and calcium alginate/poly(vinyl alcohol) microcapsules to protect flavor from its volatilization. TGA measurements of the stored samples confirmed that those formulations were stable for a period of one month and no chemical interactions between ethyl vanillin and calcium alginate that could alter the functional groups in the flavor compound has been noticed. Manojlovic et al. (2008) encapsulated ethyl vanillin in calcium alginate in order to its application in thermally processed foods. Although the vanillin was encapsulated successfully in the matrix, the TG and DTG analysis showed that all the encapsulated vanilla had been completely released from the capsules during the heating process at a temperature in the range from 220 to 325°C, that would be an undesired effect during the baking process. This effect was attributed to the small size of the capsules (450 μm) and high moisture content (~88% w/w), and/or absence of oil solvent, which could increase the viscosity of the surrounding media and prevents volatilization. Milanovic et al. (2010) encapsulated also ethyl vanillin in Carnauba wax in order to improve its functionality and stability in food products. Carnauba wax (8 w/w%) was melted in hot water (at 95°C) with the addition of 1% emulsifiers (Tween-20 + Span 40, 0.53:047) and electrosprayed. The atomized droplets were solidified in a cooling water bath at a temperature of 2–5°C. TG analysis indicated that decomposition process proceeds in several steps: vanilla evaporation occurred at around 200°C, while matrix degradation started at 250°C, also indicating further maxima at about 360, 440, and 520°C.

Lipids (Table 1.5), group of naturally occurring molecules that include fats, waxes, sterols, and fat-soluble vitamins, have been encapsulated in ethanol + glycerol mixture (Chen et al., 2005), zein (Moomand and Lim, 2015), or calcium alginate (Strasdat and Bunjes, 2013). Chen et al. (2005) used ethanol/glycerol/Tween-80 mixture as shell liquid for the encapsulation of cooking oil using electrospraying technique. Moomand and Lim (2015) encapsulated fish oil in zein nanofiber and microparticles. For 10% (w/w) concentration of zein in ethanol, the sizes of microcapsules ranged from 0.4 to 0.5 μm, whereas for zein in isopropanol the sizes were between 0.6 and 0.8 μm. For 20% w/w concentration of oil in zein, smooth fibers were formed in ethanol, due to high viscosity of the matrix, whereas beads of 1 μm in diameter homogenously mixed with fibers were generated in isopropanol. Strasdat and Bunjes (2013) compared the properties of lipid nanoparticles encapsulated into calcium chloride shell by electrospray droplet generation technique and conventional spraying method using a two-fluid spray nozzle with subsequent spray drying. DSC analysis indicated that lipid microencapsulated by electrospray extrusion have different melting patterns than those prepared by spray drying method that could be an effect of negative influence of the electrospraying process on lipid nanoparticles.

Living cells (Table 1.6), such as yeast or bacteria, have been encapsulated in calcium alginate (Bezbradica et al., 2004; Nedović et al., 2001; Leskošek-Ćukalović and Nedović, 2005; Rakin et al., 2009; Su et al., 2011), zein (Laelorspoen et al., 2014), or whey protein (Lopez-Rubio et al., 2012). Bezbradica et al. (2004), Leskošek-Ćukalović and Nedović (2005) have immobilized brewing yeast cells (*Saccharomyces uvarum*) in calcium alginate/poly(vinyl alcohol) microcapsules. The electrospray gelling procedure has no significant effect on viability and fermentation activity of the immobilized yeast cells. Nedović et al. (2001) immobilized brewer yeast in alginate microcapsules of the size in the range from 250 μm to 2.0 mm. The electrohydrodynamic droplet generation resulted in highly uniform microcapsules size distribution with standard deviation less than 10%. There were no noticeable differences in viability of yeast cells after electrospraying. Rakin et al. (2009) carried out experiments with immobilization of *Sacharomyces cerevisiae* var. *ellipsoideus* yeast cells in polyvinyl alcohol and calcium alginate for bioethanol production from corn meal hydrolyzates using electrohydrodynamic microencapsulation method. Laelorspoen et al. (2014) enclosed *Lactobacillus acidophilus* in calcium alginate with citric acid-modified zein core/shell microcapsules, which were dried at room temperature. The citric acid did not affect the microcapsules size, but survival of living cell significantly decreased due to acidity of the zein solution. The results showed that the number of bacteria that survived decreased slightly from 8.85 to 8.31 log CFU/mL with the voltage increasing from 4 to 10 kV. Bifidobacteria are incorporated into a range of functional foods, like yogurt, cheese, ice cream, or milk, but they are anaerobes and sensitive to low pH. To prevent their exposure to oxygen and for enhancing the survival of bifidobacteria in gastrointestinal juice, Su et al. (2011) encapsulated those bacteria colonies in alginate–human-like collagen microcapsules. Lopez-Rubio et al. (2012) have encapsulated *Bifidobacterium* strains in whey protein or pullulan matrix for increasing their viability and functional food applications due to its gelling and emulsification properties. *Bifidobacterium animalis* was dissolved in phosphate-buffered saline (130 mM sodium chloride, 10 mM sodium phosphate, of pH = 7.2) or in skimmed milk before encapsulation. The electrospraying process did not affect the viability of the living cells.

Proteins (Table 1.7) have been encapsulated by Xu et al. (2006) and Xu and Hanna (2006, 2007) in various shells. Xu et al. (2006), and Xu and Hanna (2006, 2007) have encapsulated bovine serum albumin (BSA) in poly(lactide) (PLA) or tripolyphosphate (TPP) cross-linked chitosan using electrospray drying of protein suspension. Poly(lactide) was used due to its high biocompatibility and biodegradable properties. The size of microcapsules increased with an increase of the PLA concentration from 1 to 3%. When the concentration of PLA was 4%, the electrosprayed liquid viscosity increased to the value at which fibers and beads were generated simultaneously. That occurred when organic to aqueous phase ratio and/or BSA/PLA mass ratio decreased. For too-low PLA concentration (<2%), microcapsules shrank after drying. Encapsulation efficiency increased with an organic/aqueous phase ratio increasing, and decreased with increasing BSA/PLA mass ratio. The microcapsules size was insensitive to the changes in BSA/PLA mass ratio.

Vitamins (Table 1.8) have been encapsulated by Bakhshi et al. (2013) and Perez-Masia et al. (2015a). Bakhshi et al. (2013) encapsulated folic acid within calcium alginate matrix in order to improve its stability and retard its degradation by light and oxygen. By electrohydrodynamic method it was possible to encapsulate 40 mg/mL (calcium alginate), however, electrospray operation in stable cone-jet mode for any voltage and flow rate was possible only for 10 mg/mL of folic acid concentration. Encapsulation yield was 70% and loading capacity 96%. Perez-Masia et al. (2015a) have demonstrated encapsulation of folic acid in whey protein or starch. Guar gum was added as a thickening agent to the matrix solution that stabilized electrospray cone-jet mode, facilitated more regular microcapsules formation and prevented a premature folic acid precipitation in syringe. The whey folic-acid/protein capsules kept their bioactive stability at a level of almost 100% in dark conditions after 15 days, compared to 40% degradation of nonencapsulated folic acid. Perez-Masia et al. (2015a) compared size distribution of the microcapsules produced by electrospraying of whey protein matrix, with those attained through conventional spray drying. Although the morphology of both products was similar, and both types of capsules were spherical, the capsules generated by spray drying method were larger than those electrosprayed, and their size distribution was broader. Smaller capsules obtained by electrospraying technique can be easier incorporated and dispersed within food products without affecting their textural characteristics. When resistant starch with guar gum was used as a matrix, irregular large microcapsules were obtained. This was attributed to guar gum present in the solution that retained water causing more unstable electrospraying process.

6 Conclusions

Electrohydrodynamic microencapsulation method is an effective technique of capturing various food ingredients within solid or liquid envelope (shell or matrix). The following techniques of electrohydrodynamic microencapsulation have been applied for microencapsulation of various food ingredients: electrospray drying, electrospray extrusion, electrospray coextrusion, electrospray cooling, electrospray mixing, submerged electrospray, electrospray reaction. Microencapsulation is used for the protection of environment-sensitive core materials from adverse environmental conditions, which increases the life of encapsulated products. Ingredients requiring encapsulation are flavors, dyes, enzymes, salts, sweeteners, lipids, acidulates, vitamins, or living cells. They are enclosed within a liquid or solid-phase shell or matrix, made of material depending on core ingredient, required properties of microcapsules and expected environment. Most popular shell materials are calcium alginate, starch, gum, fat, wax, oil, dextran, glucose, polysaccharide, or proteins. In each case it is assumed that core material will have to be released from the shell or matrix by a specific environment conditions due to diffusion through the shell walls or after shell dissolution.

References

Abu-Ali, J., Barringer, S.A., 2005. Method for electrostatic atomization of emulsions in an EHD system. J. Electros. 63, 361–369.

Ahmad, Z., Thian, E.S., Huang, J., Edirisinghe, M.J., Best, S.M., Jayasinghe, S.N., Bonfield, W., Brooks, R.A., Rushton, N., 2008. Deposition of nanohydroxyapatite particles utilising direct and transitional electrohydrodynamic processes. J. Mater. Sci. 19, 3093–3104.

Ahmad, Z., Nangrejo, M., Edirisinghe, M., Stride, E., Colombo, P., Zhang, H.B., 2009. Engineering a material for biomedical applications with electric field assisted processing. Appl. Phys. A 97, 31–37.

Al-Hajry, H.A., Al-Maskry, S.A., Al-Kharousi, L.M., El-Mardi, O., Shayya, W.H., Goosen, M.F.A., 1999. Electrostatic encapsulation and growth of plant cell cultures in alginate. Biotechnol. Prog. 15, 768–774.

Bakhshi, P.K., Nangrejo, M.R., Stride, E., Edirisinghe, M., 2013. Application of electrohydrodynamic technology for folic acid encapsulation. Food Bioprocess Technol. 6, 1837–1846.

Balachandran, W., Machowski, W., Ahmad, C.N., 1994. Electrostatic atomization of conducting liquids using AC superimposed on DC fields. IEEE Trans. Ind. Appl 30 (4), 850–855.

Barrero, A., Loscertales, I.G., 2007. Micro- and nanoparticles via capillary flows. Annu. Rev. Fluid Mech. 39, 89–106.

Barrero, A., Lopez-Herrera, J.M., Boucard, A., Loscertales, I.G., Marquez, M., 2004. Steady cone-jet electrosprays in liquid insulator baths. J. Colloid Interf. Sci. 272, 104–108.

Belščak-Cvitanović, A., Stojanović, R., Manojlović, V., Komes, D., Juranović Cindrić, I., Nedović, V., Bugarski, B., 2011. Encapsulation of polyphenolic antioxidants from medicinal plant extracts in alginate–chitosan system enhanced with ascorbic acid by electrostatic extrusion. Food Res. Int. 44, 1094–1101.

Berkland, C., Pollauf, E., Pack, D.W., Kim, K.(Kevin), 2004. Uniform double-walled polymer microspheres of controllable shell thickness. J. Controlled Release 96, 101–111.

Berkland, C., Pollauf, E., Varde, N., Pack, D.W., Kim, K.(Kevin), 2007. Monodisperse liquid-filled biodegradable microcapsules. Pharm. Res. 24 (5), 1007–1013.

Bezbradica, D., Matic, G., Obradovic, B., Nedović, V., Leskosek-Cukalovic, I., Bugarski, B., 2004. Immobilization of brewing yeast in PVA/alginate micro beads using lectrostatic droplet generation. Hemijska Industrija 58, 118–120.

Bocanegra, R., Gaonkar, A.G., Barrero, A., Loscertales, I.G., Pechack, D., Marquez, M., 2005. Production of cocoa butter microcapsules using an electrospray process. J. Food Sci. 70 (8), E492–E497.

Bollini, R., Sample, S.B., Seigal, S.D., Boarman, J.W., 1975. Production of monodisperse charged metal particles by harmonic electrical spraying. J. Colloid Interf. Sci. 51 (2), 272–277.

Borra, J.P., Camelot, D., Marijnissen, J.C.M., Scarlet, B., 1997. A new production process of powders with defined properties by electrohydrodynamic atomization of liquids and post-production electrical mixing. J. Electrostatics 40 (41), 633–638.

Borra, J.P., Camelot, D., Chou, K.L., Kooyman, P.J., Marijnissen, J.C.M., Scarlett, B., 1999. Bipolar coagulation for powder production: micro-mixing inside droplets. J. Aerosol. Sci. 30 (7), 945–958.

Brandenberger, H., Nüssli, D., Piëch, V., Widmer, F., 1999. Monodisperse particle production: a method to prevent drop coalescence using electrostatic forces. J. Electrost. 45 (3), 227–238.

Bugarski, B., Smith, J., Wu, J., Goosen, M.F.A., 1993. Methods for animal cell immobilization using electrostatic droplet generation. Biotechnol. Tech. 7 (9), 677–682.

Cao, L., Luo, J., Tu, K., Wang, L.Q., Jiang, H., 2014. Generation of nanosized core–shell particles using a coaxial tri-capillary electrospray-template removal method. Colloids and Surf. B 115, 212–218.

Chen, D.-R., Pui, D.Y.H., 1997. Experimental investigation of scaling laws for electrospraying: dielectric constant effect. Aerosol Sci. Techn. 27 (3), 367–380.

Chen, X., Jia, L., Yin, X., Cheng, J., Lu, J., 2005. Spraying modes in coaxial jet electrospray with outer driving liquid. Phys. Fluids 17 (Paper no. 032101).

Cloupeau, M., Prunet-Foch, B., 1988. Research on electrohydrodynamic spraying. 4th International Conference of Liquid Atomization and Spray Systems. August 22–24, 1988, Sendai, Japan.

Cloupeau, M., Prunet-Foch, B., 1990. Electrostatic spraying of liquids: main functioning modes. J. Electrostat. 25, 165–184.

Cloupeau, M., Prunet-Foch, B., 1994. Electrohydrodynamic spraying functioning modes: a critical review. J. Aerosol Sci. 25 (6), 1121–1136.

De, S., Pritchett, M., Mazumder, M. K., Yurteri, C. U., Egorov, O., 2002. Electrostatic microencapsulation technique for producing composite particles. Part. Sci. Techn. 20, 169–185.

de Vos, P., Bučko, M., Gemeiner, P., Navratil, M., Švitel, J., Faas, M., Strand, B.L., Skjak-Braek, G., Morch, Y.A., Vikartovska, A., Lacik, I., Kollarikova, G., Orive, G., Poncelet, D., Pedraz, J.L., Ansorge-Schumacher, M.B., 2009. Multiscale requirements for bioencapsulation in medicine and biotechnology. Biomaterials 30, 2559–2570.

Desai, K.G.H., Park, H.J., 2005. Recent developments in microencapsulation of food ingredients. Drying Tech. 23, 1361–1394.

Ding, L., Lee, T., Wang, Ch.H., 2005. Fabrication of monodispersed Taxol-loaded particles using electrohydrodynamic atomization. J. Contr. Release 102, 395–413.

Dos Santos, I.R., Richard, J., Pech, B., Thies, C., Benoit, J.P., 2002. Microencapsulation of protein particles within lipids using a novel supercritical fluid process. Int. J. Pharm. 242, 69–78.

Edris, A., Bergnstahl, B., 2001. Encapsulation of orange oil in a spray dried double emulsion. Nahrung/Food 45 (2), 133–137.

Eltayeb, M., Bakhshi, P.K., Stride, E., Edirisinghe, M., 2013a. Preparation of solid lipid nanoparticles containing active compound by electrohydrodynamic spraying. Food Res. Int. 53, 88–95.

Eltayeb, M., Stride, E., Edirisinghe, M., 2013b. Electrosprayed core–shell polymer–lipid nanoparticles for active component delivery. Nanotechnology 24 (Paper no. 465604), 9 pp.

Fages, J., Lochard, H., Letourneau, J.-J., Sauceau, M., Rodier, E., 2004. Particle generation for pharmaceutical applications using supercritical fluid technology. Powder Technol. 141, 219–226.

Farook, U., Stride, E., Edirisinghe, M.J., Moaleji, R., 2007. Microbubbling by coaxial electrohydrodynamic atomization. Med. Bio. Eng. Comput. 45, 781–789.

Fernandez de la Mora, J., Loscertales, I.G., 1994. The current emitted by highly conducting Taylor cones. J. Fluid Mech. 260, 155–184.

Fery, A., Dubreuil, F., Mohwald, H., 2004. Mechanics of artificial microcapsules. New J. Phys. 6 (18), 1–13.

Freitas, S., Merkle, H.P., Gander, B., 2005. Microencapsulation by solvent extraction/evaporation: reviewing the state of the art of microsphere preparation process technology. J. Control. Release 102, 313–332.

Gañan-Calvo, A.M., 1997. On the theory of electrohydrodynamically driven capillary jets. J. Fluid Mech. 335, 165–188.

Gañan-Calvo, A.M., 1999. The surface charge in electrospraying: its nature and its universal scaling laws. J. Aerosol Sci. 30 (7), 863–872.

Gañan-Calvo, A.M., Castro-Hernandez, E., Flores-Mosquera, M., Martin-Banderas, L., 2015. Massive, generic, and controlled microencapsulation by flow focusing: some physicochemical aspects and new applications. J. Flow Chem. 5 (1), 48–54.

Garstecki, P., Gitlin, I., Diluzio, W., Whitesides, G.M., Kumachewa, E., Stone, H.A., 2004. Formation of monodisperse bubbles in a microfluidic flow-focusing device. Appl. Phys. Lett. 85, 2649–2651.

Garstecki, P., Ganan-Calvo, A.M., Whitesides, G.M., 2005. Formation of bubbles and droplets in microfluidic systems. Bull. Pol. Acad. Sci. 53 (4), 361–372.

Given, Jr., P.S., 2009. Encapsulation of flavors in emulsions for beverages. Curr. Opin. Colloid In. 14, 43–47.

Gomez, A., Bingham, D., Juan de, L., Tang, K., 1998. Production of protein nanoparticles by electrospray drying. J. Aerosol Sci. 29 (5/6), 561–574.

Gomez-Estaca, J., Balaguer, M.P., Gavara, R., Hernandez-Munoz, P., 2012. Formation of zein nanoparticles by electrohydrodynamic atomization: effect of the main processing variables and suitability for encapsulating the food coloring and active ingredient curcumin. Food Hydrocolloids 28, 82–91.

Gomez-Estaca, J., Gavara, R., Hernandez-Munoz, P., 2015. Encapsulation of curcumin in electrosprayed gelatin microspheres enhances its bioaccessibility and widens its uses in food applications. Innov. Food Sci. Emerg. 29, 302–307.

Gomez-Mascaraque, L.G., Lagaron, J.M., Lopez-Rubio, A., 2015. Electrosprayed gelatin submicroparticles as edible carriers for the encapsulation of polyphenols of interest in functional foods. Food Hydrocolloids 49, 42–52.
Goosen, M.F.A., 1999. Physico-chemical and mass transfer considerations in microencapsulation. Annals of the New York Academy of Sciences. 875, Bioartificial Organs II: Technology, Medicine, and Materials, pp. 84–104.
Gouin, S., 2004. Microencapsulation: industrial appraisal of existing technologies and trends. Trends Food Sci. Technol. 15, 330–347.
Grace, J.M., Marijnissen, J.C.M., 1994. A review of liquid atomization by electrical means. J. Aerosol Sci. 25 (6), 1005–1019.
Hartman, R.P.A., Brunner, D.J., Camelot, D.M.A., Marijnissen, J.C.M., Scarlett, B., 2000. Jet break-up in electrohydrodynamic atomization in the cone-jet mode. J. Aerosol Sci. 31 (1), 65–95.
Hayati, I., Bailey, A.I., Tadros, Th.F., 1987a. Investigations into the mechanisms of electrohydrodynamic spraying of liquids. Pt. I. Effect of electric field and the environment on pendant drop and factors affecting the formation of stable jets and atomization. J. Colloid Interf. Sci. 117 (1), 205–221.
Hayati, I., Bailey, A.I., Tadros, Th.F., 1987b. Investigations into the mechanisms of electrohydrodynamic spraying of liquids. Pt. II. Mechanism of stable jet formation and electrical forces acting on a liquid cone. J. Colloid Interf. Sci. 117 (1), 222–230.
Herrero, E.P., Del Valle, E.M.M., Galan, M.A., 2006. Development of a new technology for the production of microcapsules based in atomization processes. Chem. Eng. J. 117, 137–142.
Hirech, K., Payan, S., Carnelle, G., Brujes, L., Legrand, J., 2003. Microencapsulation of an insecticide by interfacial polymerisation. Powder Technol. 130, 324–330.
Ho, H., Lee, J., 2011. PEG/PLA core/shell particles from coaxial electrohydrodynamic spray drying. Macromol. Res. 19 (8), 815–821.
Hong, Y., Li, Y., Yin, Y., Li, D., Zou, G., 2008. Electrohydrodynamic atomization of quasi-monodisperse drug-loaded spherical/wrinkled microparticles. J. Aerosol Sci. 39, 525–536.
Jafari, S.M., He, Y., Bhandari, B., 2006. Nanoemulsion production by sonication and microfluidization—a comparison. Int. J. Food Properties 9, 475–485.
Jaworek, A., 2008. Electrostatic micro- and nanoencapsulation and electroemulsification: a brief review. J. Microencapsul. 25 (7), 443–468.
Jaworek, A., Krupa, A., 1999a. Jet and drop formation in electrohydrodynamic spraying of liquids: a systemic approach. Exp. Fluids 27 (1), 43–52.
Jaworek, A., Krupa, A., 1999b. Classification of the modes of EHD spraying. J. Aerosol Sci. 30 (7), 873–893.
Jaworek, A., Machowski, W., Krupa, A., Balachandran, W., 2000. Viscosity effect on EHD spraying using AC superimposed on DC Electric Field. IEEE Industry Applications Society Annual Meeting, Rome. vol. 8–12, pp. 770–776.
Jaworek, A., Balachandran, W., Krupa, A., Kulon, J., Machowski, W., 2003. Electrohydrodynamic atomization of viscous liquids. Inst. Phys. Conf. Ser (178), 181–186.
Jaworek, A., Sobczyk, A.T., Czech, T., Krupa, A., 2014. Corona discharge in electrospraying. J. Electrostat. 72, 166–178.
Jayasinghe, S.N., 2007. Submerged electrosprays: a versatile approach for microencapsulation. J. Microencapsul. 24 (5), 430–444.
Jin, Zh., Wang, Y., Liu, J., Yang, Zh., 2008. Synthesis and properties of paraffin capsules as phase change materials. Polymer 49, 2903–2910.
Jono, K., Ichikawa, H., Miyamoto, M., Fukumori, Y., 2000. A review of particulate design for pharmaceutical powders and their production by spouted bed coating. Powder Technol. 113, 269–277.

Jyothi, N.V.N., Prasanna, P.M., Sakarkar, S.N., Prabha, K.S., Ramaiah, P.S., Srawan, G.Y., 2010. Microencapsulation techniques, factors influencing encapsulation efficiency. J. Microencapsul. 27 (3), 187–197.

Kalra, V., Lee, J.H., Park, J.H., Marquez, M., Joo, Y.L., 2009. Confined assembly of asymmetric block-copolymer nanofibers via multiaxial jet electrospinning. Small 5, 2323–2332.

Kaushik, D., Dureja, H., 2015. Taste masking of bitter pharmaceuticals by spray drying technique. J. Chem. Pharmac. Res. 7 (4), 950–956.

Knezevic, Z., Bobic, S., Milutinovic, A., Obradovic, B., Mojovic, L., Bugarski, B., 2002. Alginate-immobilized lipase by electrostatic extrusion for the purpose of palm oil hydrolysis in lecithin/isooctane system. Process Biochem. 38, 313–318.

Koo, S.Y., Cha, K.H., Song, D.G., Chung, D., Pan, C.H., 2014. Microencapsulation of peppermint oil in an alginate–pectin matrix using a coaxial electrospray system. Int. J. Food Sci. Tech. 49, 733–739.

Labbaf, S., Deb, S., Cama, G., Stride, E., Edirisinghe, M., 2013. Preparation of multicompartment sub-micron particles using a triple-needle electrohydrodynamic device. J. Colloid Interf. Sci. 409, 245–254.

Laelorspoen, N., Wongsasulak, S., Yoovidhya, T., Devahastin, S., 2014. Microencapsulation of Lactobacillus acidophilus in zein–alginate core–shell microcapsules via electrospraying. J. Funct. Foods 7, 342–349.

Langer, G., Yamate, G., 1969. Encapsulation of liquid and solid aerosol particles to form dry powders. J. Colloid Interf. Sci. 29 (3), 450–455.

Lee, L.Y., Wang, C.H., Smith, K.A., 2008. Supercritical antisolvent production of biodegradable micro- and nanoparticles for controlled delivery of paclitaxel. J. Control. Release 125, 96–106.

Lee, Y.H., Bai, M.Y., Chen, D.R., 2011. Multidrug encapsulation by coaxial tri-capillary electrospray. Colloid. Surf. B 82, 104–110.

Leo, E., Ruozi, B., Tosi, G., Vandelli, M.A., 2006. PLA-microparticles formulated by means a thermoreversible gel able to modify protein encapsulation and release without being co-encapsulated. Int. J. Pharm. 323, 131–138.

Leskošek-Čukalović, I.J., Nedović, V.A., 2005. Immobilized cell technology in beer brewing—current experience and results. Proc. Nat. Sci, Matica Srpska Novi Sad. 109, 129–141.

Levic, S., Djordjevic, V., Rajic, N., Milivojevic, M., Bugarski, B., Nedović, V., 2013. Entrapment of ethyl vanillin in calcium alginate and calcium alginate/poly(vinyl alcohol) beads. Chem. Pap. 67 (2), 221–228.

Lewińska, D., Rosiński, S., Weryński, A., 2004. Influence of process conditions during impulsed electrostatic droplet formation on size distribution of hydrogel beads. Artiff. Cells Blood Sub. Biotechnol. 32 (1), 41–53.

Lewińska, D., Bukowski, J., Kinasiewicz, A., Weryński, A., 2006. Electrostatic microencapsulation of hepatocytes using an impulsed voltage droplet generator. ESAO Conf. Paper.

Li, F., Li, Y., 2007. A computational analysis for effects of fibre hygroscopicity on heat and moisture transfer in textiles with PCM microcapsules. Model. Simul. Mater. Sci. Eng. 15, 223–235.

Lopez-Herrera, J.M., Barrero, A., Lopez, A., Loscertales, I.G., Marquez, M., 2003. Coaxial jets generated from electrified Taylor cones: scaling laws. J. Aerosol Sci. 34, 535–552.

Lopez-Rubio, A., Lagaron, J.M., 2012. Whey protein capsules obtained through electrospraying for the encapsulation of bioactives. Innov. Food Sci. Emerg. 13, 200–206.

Lopez-Rubio, A., Sanchez, E., Wilkanowicz, S., Sanz, Y., Lagaron, J.M., 2012. Electrospinning as a useful technique for the encapsulation of living bifidobacteria in food hydrocolloids. Food Hydrocolloids 28, 159–167.

Loscertales, I.G., Cortijo-Bon, R., Barrero, A., Guerrero, I., Gañan-Calvo, A.M., 2001. A novel technique to produce multicomponent micro/nano capillary jets and micro/nano capsules by electrohydrodynamic forces. J. Aerosol Sci. 32, S611–S612.

Loscertales, I.G., Barrero, A., Guerrero, I., Cortijo, R., Marquez, M., Gañan-Calvo, A.M., 2002. Micro/nano encapsulation via electrified coaxial liquid jets. Science 295, 1695–1698.

Manojlovic, V., Rajic, N., Djonlagic, J., Obradovic, B., Nedović, V., Bugarski, B., 2008. Application of electrostatic extrusion—flavour encapsulation and controlled release. Sensors 8, 1488–1496.

Mei, F., Chen, D.R., 2007. Investigation of compound jet electrospray: particle encapsulation. Phys. Fluids 19 (Paper No. 103303).

Milanovic, J., Manojlovic, V., Levic, S., Rajic, N., Nedović, V., Bugarski, B., 2010. Microencapsulation of flavors in Carnauba wax. Sensors 10, 901–912.

Mishra, B., Patel, B.B., Tiwari, S., 2010. Colloidal nanocarriers: a review on formulation technology, types and applications toward targeted drug delivery. Nanomed.-Nanotechnol. 6, 9–24.

Moomand, K., Lim, L.T., 2015. Properties of encapsulated fish oil in electrospun zein fibres under simulated in vitro conditions. Food Bioprocess. Technol. 8, 431–444.

Mutoh, M., Kaieda, S., Kamimura, K., 1979. Convergence and disintegration of liquid jets induced by an electrostatic field. J. Appl. Phys. 50 (5), 3174–3179.

Nedović, V.A., Obradović, B., Leskošek-Čukalević, I., Trifunović, O., Pešić, R., Bugarski, B., 2001. Electrostatic generation of alginate microbeads loaded with brewing yeast. Process Biochem. 37, 17–22.

Nedović, V.A., Kalusevic, A., Manojlovic, V., Levic, S., Bugarski, B., 2011. An overview of encapsulation technologies for food applications. Procedia Food Sci. 1, 1806–1815.

Nelson, G., 2002. Application of microencapsulation in textiles. Int. J. Pharmac. 242, 55–62.

Orive, G., Gascon, A.R., Hernandez, R.M., Igartua, M., Pedraz, J.L., 2003a. Cell microencapsulation technology for biomedical purposes: novel insights and challenges. Trends Pharma. Sci. 24 (5), 207–210.

Orive, G., Hernandez, R.M., Gascon, A.R., Igartua, M., Pedraz, J.L., 2003b. Survival of different cell lines in alginate-agarose microcapsules. Euro. J. Pharm. Sci. 18, 23–30.

Orive, G., Hernandez, R.M., Gascon, A.R., Calafiore, R., Chang, Th.M.S., de Vos, P., Hortelano, G., Hunkeler, D., Lacik, I., Pedraz, J.L., 2004. History, challenges and perspectives of cell microencapsulation. Trends Biotechnol. 22 (2), 87–92.

Orive, G., Santos, E., Pedraz, J.L., Hernandez, R.M., 2014. Application of cell encapsulation for controlled delivery of biological therapeutics. Adv. Drug Deliver. Rev. 67–68, 3–14.

Paine, M.D., Alexander, M.S., Stark, J.P.W., 2007. Nozzle and liquid effects on the spray modes in nanoelectrospray. J. Colloid Interf. Sci. 305, 111–123.

Patel, P., Irvine, S., McEwan, J.R., Jayasinghe, S.N., 2008. Bio-protocols for directly forming active encapsulations containing living primary cells. Soft Matter 4, 1219–1229.

Perez-Masia, R., Lagaron, J.M., Lopez-Rubio, A., 2014. Surfactant-aided electrospraying of low molecular weight carbohydrate polymers from aqueous solutions. Carbohyd. Polym. 101, 249–255.

Perez-Masia, R., Lagaron, J.M., Lopez-Rubio, A., 2015a. Morphology and stability of edible lycopene-containing microand nanocapsules produced through electrospraying and spray drying. Food Bioprocess. Technol. 8, 459–470.
Perez-Masia, R., Lopez-Nikolas, R., Periago, M.J., Ros.F G., Lagaron, J.M., Lopez-Rubio, A., 2015b. Encapsulation of folic acid in food hydrocolloids through nanospray drying and electrospraying for nutraceutical applications. Food Chem. 168, 124–133.
Poncelet, D., Babak, V.G., Neufeld, R.J., Goosen, M.F.A., Bugarski, B., 1999a. Theory of electrostatic dispersion of polymer solutions in the production of microgel beads containing biocatalyst. Adv. Coll. Interface Sci. 79, 213–228.
Poncelet, D., Neufeld, R.J., Goosen, M.F.A., Bugarski, B., Babak, V., 1999b. Formation of microgel beads by electric dispersion of polymer solutions. AICHE J. 45 (9), 2018–2023.
Rakin, M., Mojovic, L., Nikolic, S., Vukasinovic, M., Nedović, V., 2009. Bioethanol production by immobilized *Sacharomyces cerevisiae* var. *ellipsoideus* cells. Afr. J. Biotechnol. 8 (3), 464–471.
Ramsden, K.-T., Phillips, G., Mussenden, P., Bucke, P.C., 1992. Application of electric field for production of immobilized biocatalyst. Biotechnol. Tech. 6 (5), 445–450.
Sakai, T., Sadakata, M., Sato, M., Kimura, K., 1991. Production of uniformly sized dual concentric droplets from coaxial smooth jet under applied AC electric field. Atomiz. Sprays 1 (2), 171–185.
Salalha, W., Kuhn, J., Dror, Y., Zussman, E., 2006. Encapsulation of bacteria and viruses in electrospun nanofibres. Nanotechnology 17, 4675–4681.
Salata, O.V., 2005. Tools of nanotechnology: electrospray. Curr. Nanosci. 1, 25–33.
Sample, S.B., Bollini, R., 1972. Production of liquid aerosols by harmonic electrical spraying. J. Colloid Interf. Sci. 41 (2), 185–193.
Sato, M., 1984. The production of essentially uniform-sized liquid droplets in gaseous or immiscible liquid media under applied AC potential. J. Electrostat. 15, 237–247.
Sato, M., Kato, S., Saito, M., 1996a. Production of oil/water type uniformly sized droplets using a convergent AC electric field. IEEE Trans. Ind. Appl. 32 (1), 138–145.
Sato, M., Kudo, N., Saito, M., 1996b. Surface tension reduction of liquid by applied electric field using vibrating jet method. IEEE Industry Applications Society Annual Meeting, San Diego.
Sato, M., Kudo, N., Saito, M., 1998. Surface tension reduction of liquid by applied electric field using vibrating jet method. IEEE Trans. Ind. Appl. 34 (2), 294–300.
Sato, M., Takahashi, H., Awazu, M., Ohshima, T., 1999. Production of ultra-uniformly-sized silica particles by applying AC superimposed on DC voltage. J. Electrostat. 46 (2–3), 171–176.
Scheideler, W.J., Chen, C.H., 2014. The minimum flow rate scaling of Taylor cone-jets issued from a nozzle. Appl. Phys. Lett. 104 (Paper no. 024103).
Schneider, J.M., Lindblad, N.R., Hendricks, Jr., C.D., Crowley, J.M., 1967. Stability of an electrified liquid jet. J. Appl. Phys. 38 (6), 2599–2605.
Shiryaeva, S.O., Grigorev, A.I., 1995. The semiphenomenological classification of the modes of electrostatic dispersion of liquids. J. Electrostat. 34 (1), 51–59.
Smith, D.P.H., 1986. The electrohydrodynamic atomization of liquids. IEEE Trans. Ind. Appl. 22 (3), 527–535.
Stankus, J.J., Guan, J., Fujimoto, K., Wagner, W.R., 2006. Microintegrating smooth muscle cells into a biodegradable, elastomeric fiber matrix. Biomater. 27, 735–744.

Stojanovic, R., Belscak-Cvitanovic, A., Manojlovic, V., Komes, D., Nedović, V., Bugarski, B., 2012. Encapsulation of thyme (*Thymus serpyllum* L) aqueous extract in calcium alginate beads. J. Sci. Food Agric. 92, 685–696.

Strasdat, B., Bunjes, H., 2013. Incorporation of lipid nanoparticles into calcium alginate beads and characterization of the encapsulated particles by differential scanning calorimetry. Food Hydrocolloids 30, 567–575.

Su, R., Zhu, X.L., Fan, D.D., Mi, Y., Yang, C.Y., Jia, X., 2011. Encapsulation of probiotic *Bifidobacterium longum* BIOMA 5920 with alginate–human-like collagen and evaluation of survival in simulated gastrointestinal conditions. Int. J. Bio. Macromol. 49, 979–984.

Szafran, R.G., Ludwig, W., Kmiec, A., 2012. New spout-fluid bed apparatus for electrostatic coating of fine particles and encapsulation. Powder Technol. 225, 52–57.

Sze, Tu L., Dehghani, F., Foster, N.R., 2002. Micronisation and microencapsulation of pharmaceuticals using a carbon dioxide antisolvent. Powder Technol. 126, 134–149.

Taylor, G.F.R.S., 1964. Disintegration of water drops in an electric field. Proc. Royal Soc. A280, 383–397.

Taylor, G.F.R.S., Dyke van, M.D., 1969. Electrically driven jets. Proc. Royal Soc. A313, 453–475.

Teer, D., Dole, M., 1975. Electrospray mass spectroscopy of macromolecule degradation in the electrospray. J. Polym. Sci. 13 (5), 985–995.

Tiarks, F., Landfester, K., Antonietti, M., 2001. Preparation of polymeric nanocapsules by miniemulsion polymerization. Langmuir 17, 908–918.

Torres-Giner, S., Martinez-Abad, A., Ocio, M.J., Lagaron, J.M., 2010. Stabilization of a nutraceutical Omega-3 fatty acid by encapsulation in ultrathin electrosprayed zein prolamine. J. Food Sci. 75 (6), N69–N79.

Townsend-Nicholson, A., Jayasinghe, S.N., 2006. Cell electrospinning: a unique biotechnique for encapsulating living organisms for generating active biological microthreads/scaffolds. Biomacromol. 7, 3364–3369.

Uematsu, I., Matsumoto, H., Morota, K., Minagawa, M., Tanioka, A., Yamagata, Y., Inoue, K., 2004. Surface morphology and biological activity of protein thin films produced by electrospray deposition. J. Colloid Interf. Sci. 269, 336–340.

van Berkel, G.J., Kertesz, V., 2007. Using the electrochemistry of the electrospray ion source. Anal. Chem. 79, 5510–5520.

Vonnegut, B., Neubauer, R.L., 1952. Production of monodisperse liquid particles by electrical atomization. J. Colloid Interf. Sci. 7, 616–622.

Wang, Y., Dave, R.N., Pfeffer, R., 2004. Polymer coating/encapsulation of nanoparticles using a supercritical antisolvent process. J. Supercrit. Fluid. 28, 85–99.

Workman, V.L., Tezera, L.B., Elkington, P.T., Jayasinghe, S.N., 2014. Controlled generation of microspheres incorporating extracellular matrix fibrils for three-dimensional cell culture. Adv. Funct. Mater. 24, 2648–2657.

Wu, Y., Clark, R.L., 2007. Controllable porous polymer particles generated by electrospraying. J. Colloid Interf. Sci. 310, 529–535.

Wu, Y., Clark, R.L., 2008. Electrohydrodynamic atomization: a versatile process for preparing materials for biomedical applications. J. Biomater. Sci. Polymer Edn. 19 (5), 573–601.

Xie, J., Wang, Ch.-H., 2007a. Encapsulation of proteins in biodegradable polymeric microparticles using electrospray in the Taylor cone-jet mode. Biotechnol. Bioeng. 97 (5), 1278–1290.

Xie, J., Wang, Ch.-H., 2007b. Electrospray in the dripping mode for cell microencapsulation. J. Colloid Interf. Sci. 312, 247–255.

Xie, J., Lim, L.K., Phua, Y., Hua, J., Wang, Ch.-W., 2006a. Electrohydrodynamic atomization for biodegradable polymeric particle production. J. Colloid Interf. Sci. 302, 103–112.

Xie, J., Marijnissen, J.C.M., Wang, Ch.-H., 2006b. Microparticles developed by electrohydrodynamic atomization for the local delivery of anticancer drug to treat C6 glioma in vitro. Biomaterials 27, 3321–3332.

Xie, J., Ng, W.J., Lee, L.Y., Wang, Ch.-H., 2008. Encapsulation of protein drugs in biodegradable microparticles by coaxial electrospray. J. Colloid Interf. Sci. 317, 469–476.

Xie, J., Jiang, J., Davoodi, P., Srinivasan, M.P., Wang, C.-H., 2015. Electrohydrodynamic atomization: a two-decade effort to produce and process micro-/nanoparticulate materials. Chem. Eng. Sci. 125, 32–57.

Xu, Y., Hanna, M., 2006. Electrospray encapsulation of water-soluble protein with polylactide: effects of formulations on morphology, encapsulation efficiency and release profile of particles. Int. J. Pharm. 320, 30–36.

Xu, Y.X., Hanna, M., 2007. Electrosprayed bovine serum albumin-loaded tripolyphosphate cross-linked chitosan capsules: synthesis and characterization. J. Microencapsul. 24 (2), 143–151.

Xu, Y., Skotak, M., Hanna, M.A., 2006. Electrospray encapsulation of water-soluble protein with polylactide I. Effects of formulations and process on morphology and particle size. J. Microencapsul. 23 (1), 69–78.

Ye, M., Kim, S., Park, K., 2010. Issues in long-term protein delivery using biodegradable microparticles. J. Control. Release 146, 241–260.

Yeo, L.Y., Gagnon, Z., Chang, H.-Ch., 2005. AC electrospray biomaterials synthesis. Biomaterials 26, 6122–6128.

Young, C.J., Poole-Varren, L.A., Martens, P.J., 2012. Combining submerged electrospray and UV photopolymerization for production of synthetic hydrogel microspheres for cell encapsulation. Biotechnol. and Bioeng. 109 (6), 1561–1570.

Zamani, M., Prabhakaran, M.P., Ramakrishna, S., 2013. Advances in drug delivery via electrospun and electrosprayed nanomaterials. Int. J. Nanomed. 8, 2997–3017.

Zeleny, J., 1915. On the conditions of instability of electrified drops with applications to the electrical discharge from liquid points. Proc. Cambridge Phil. Soc. 18, 71–83.

Zhou, Y., Sun, T., Chan, M., Zhang, J., Han, Z., Wang, X., Toh, Y., Chen, J.P., Yu, H., 2005. Scalable encapsulation of hepatocytes by electrostatic spraying. J. Biotechnol. 117, 99–109.

2

EXPLORING NANOENCAPSULATION OF AROMA AND FLAVORS AS NEW FRONTIER IN FOOD TECHNOLOGY

Sushama Talegaonkar*, Shweta Pandey*, Nishant Rai, Purnima Rawat*, Harshita Sharma*, Nisha Kumari***

**Jamia Hamdard, Department of Pharmaceutics, Faculty of Pharmacy, New Delhi, India; **University of Delhi, Department of Pharmacology, Vallabhbhai Patel Chest Institute, Delhi, India*

1 Introduction

Nanotechnology involves creating and manipulating organic and inorganic matter at the nanoscale. A wide range of applications are foreseen in the agrifood sector, including nanosensors, tracking devices, targeted delivery of specific components, food safety applications, new product development, precision processing and smart packaging (Chellarama et al., 2014). In recent years the potential applications of nanotechnology ranges from modifications of the natural (organic) protein, carbohydrate, and fat molecules that form part of the normal diet, to achieve added or altered functionality, to the use of inorganic nanoparticulate materials in food packaging and food ingredients including food additives. The predominant food-related use of nanoscience in the short term is in food contact materials including packaging, while in the longer term nanoscale food research appears to be focused on controlled release of nanoscale-encapsulated food ingredients or nutrients (Qureshi et al., 2012). The use of nano-based technologies for alteration of food stability and texture is also of particular interest. Engineered organic nanomaterials (primarily nanoparticulate forms of natural food components) are likely to be the predominant area of interest in the agrifood sector (Martirosyan and

Encapsulations. http://dx.doi.org/10.1016/B978-0-12-804307-3.00002-8

Schneider, 2014). Many applications of nanotechnology currently under research involve modification of the normal structure and assembly of natural biopolymeric food components, for example, modification of proteins or selected carbohydrates to make them suitable as nanostabilizers of bioactive components. Protein-carbohydrate engineering coupled with enzymatic functionalization is being used to construct nanoscale structures which impart new functionality in food. Some of the benefits of nanoscience have already been seen in the agrifood sector (Ali et al., 2014; Kah and Hofmann, 2014), others are still at the research and concept stage. Potential applications of nanoencapsulation in food processing are as follows:

- Methods to enable foods such as soft drinks, ice cream, chocolate, or chips to be marketed as healthy foods by reducing fat, carbohydrate or calorie content or by increasing protein, fiber or vitamin content (Sekhon, 2010).
- Production of stronger flavorings, colorings, and nutritional additives, and processing aids to increase the pace of manufacturing and to lower costs of ingredients and processing (Nedovic et al., 2011).
- Development of foods capable of changing their color, flavor, or nutritional properties according to a person's dietary needs, allergies, or taste preferences (high on the research agenda of food giants including Kraft and Nestle).
- These are based on nanoencapsulation of the substances in the form of liposomes, micelles, or protein-based carriers. These nanocarrier systems are used to mask the undesirable taste of certain additives and supplements, or to protect them from degradation during processing (Wang, 2011; Jeroen and Soest, 2007). The nanoencapsulated nutrients and supplements are also claimed for enhanced bioavailability, antimicrobial activity, and other health benefits (Desai et al., 1996; Cedervall et al., 2007).

Applications of nanotechnology in food industry are summarized in Table 2.1.

Nanotechnology in the recent era has developed into an inclusive, multibillion-dollar global industry. It is also clear from a number of reviews, reports, patent applications, and company products that applications of nanotechnology have also started to make an impact on different aspects of the food and associated industries (Chen et al., 2006). The nanofood sector is currently led by USA, followed by Japan, Australia, and the United Kingdom. However, Asian countries (led by China) are expected to be the biggest market for nanofood. The report estimated that by 2012 the overall market value would reach US$5.8 billion. Generally, the market profile information and analysis

Table 2.1 Application of Nanotechnology in the Food Industry (Luykx et al., 2008; Chellarama et al., 2014; Martirosyan and Schneider, 2014; Chinnamuthu and Boopathi, 2009)

Farming	Reformation of Food Products	Packaging of Food Products	Nutraceuticals
Nanotechnology used as diagnostic tool for assessment of enzyme/substrate interactions by detecting a single desired molecule	Nanostructures for enhancing bioavailability of nutritional elements in standard components such as cooking oils	Food-borne pathogens and chemicals can be detected in packed foods by antibody attached surface modified fluorescent nanoparticles	Nanosized powders for improving absorption of nutraceuticals
Sustained delivery of pesticides, fertilizers, and other growth and nutrition induced agrochemicals by nano-sized structures	Encapsulation of flavor enhancers (eg, nanocapsules, nanospheres, nanoparticles)	Biodegradable nanosensors for temperature, moisture, and time monitoring	Cellulose nanocrystal composites as drug carriers
Controlled delivery of growth hormones	Increase the viscosity and gelation in certain products using nanotubes and nanoparticles	Nanoclays and nanofilms as barrier materials to prevent spoilage and oxygen absorption	Nanoencapsulation of nutraceuticals for better absorption, better stability, and targeted delivery
Soil monitoring by identification, preservation, and tracking of soil conditions (eg, nanosensors and nanochips)	Replacement of meat cholesterol by plant-based steroids using nanocapsule infusions	Electrochemical nanosensors for determining presence of ethylene in food products	Nanocochelates for delivering nutrients more efficiently to cells without affecting texture and taste of food
Nanochips for crop growth monitoring	Nanoparticles and nanodevices that specifically bind and detect to remove chemicals or pathogens from food products (eg, chelating agents, adsorbents, sensors)	Nanoparticle containing antimicrobial and antifungal surface coatings are used as packaging material	Vitamin sprays that disperse nanodroplets with better absorption

from different sources revealed that the functional food market is expanding and has a strong growth trend all over the world. The market value for nanofoods is estimated to increase to more than $20 billion by 2010 and it is predicted that more than 40% of food products would be nanotechnology based by the year 2015. It has been suggested that the number of companies currently applying nanotechnologies to food could be as high as 400. A number of major food and beverage companies are reported to have an interest in nanotechnology. These include Nestlé, Altria, Heinz, Kraft, and Unilever, as well as small nanotech start-up companies. It is also widely anticipated that the number of companies applying nanotechnologies to food will increase dramatically in the coming era. Considering such quick developments in this field, and the global setup of international food companies, it is not unreasonable to expect that more nanofood products will appear in EU markets within the next few years. According to the estimation of nanotechnology analysts, between 150 and 600 nanofoods are already in the market and many more nanofood products are in the process of development. Many of the world's largest food companies, including Nestlé, Heinz, Kraft, and Unilever, are exploring nanotechnology for food processing and packaging (Bernardes et al., 2014). Nanoencapsulation is one of the potential applications of this nanotechnology in which active substances and additives are coated by extremely small capsules. It is a new technology that has been used widely in food industries for encapsulating aroma and flavors (Yurdugul and Mozafari, 2004). The success of this technology is due to the correct choice of the wall material, encapsulation method, and form of the core release. Encapsulated particles can be classified according to their size, shape, and construction. On the basis of size, capsules can be classified macrocapsules (>5000 μm), microcapsules (0.2–5000 μm), and nanocapsules (<0.2 μm), and in terms of their shape and construction, capsules can be divided into two groups: nanocapsules and nanospheres. Nanocapsules are core–shell type structures where the core acts as reservoir for several molecules or drugs, and the shell as a protective polymeric membrane. On the basis of division in core part these are further divided into mononuclear and polynuclear nanocapsules (Ranjan et al., 2014). In contrast, nanospheres are matrix systems with polymeric network in which the core and drug remain dispersed or dissolved uniformly. Nanospheres may be homogeneous or heterogeneous depending on whether the core is in the dissolved molecular state or in the form of suspended particles, respectively (Marcuzzo et al., 2010) (Table 2.2).

2 Nanoencapsulation of Aroma and Flavors

Nanoencapsulation is a potential application of this nanotechnology for delivering bioactive components in food sciences (Nedovic et al., 2011). Among these bioactive components flavor

Table 2.2 Nanoencapsulated Food Products in the Market

Type of Product	Product Name	Nanoformulation	Purpose	References
Beverage	Nano Tea	Nanoparticles (160 nm)	Immune booster, antiaging, antioxidant activity	Patent No.: 01100033.3—The three-step preparation method and its application for nanotea
Beverage	"Daily Vitamin Boost" fortified fruit juice	300 nm iron (SunActive Fe)	For providing nutrition of 22 essential vitamins and minerals	Ijabadeniyi (2012)
Nutritional drink	Oat Chocolate Nutritional Drink Mix, Toddler Health	300 nm particles of iron (SunActive Fe)	Nanosized iron particles have increased reactivity and bioavailability	Ijabadeniyi (2012)
Nutraceuticals	Enzymes, vit. C	W/o microemulsions sol–gel synthesis	Immobility of enzymes, improved stability, and bioavailability	Zuidam and Nedovic (2010)
Food additive	Aquasol preservative, Aqua Nova	Nanoscale micelle of lipophilic or water insoluble substances	Surrounding active ingredients within soluble nanocapsules increases absorption within the body (including individual cells)	Miller and Senjen (2008)
Beverage	Nano Slim™	Nanoparticles	Fat burner	Miller and Senjen (2008)
Generic food additive	Solu™ E 200, BASF	Vitamin E nanosolution using Nova Sol	Solubilisates of fat-soluble vitamins	Miller and Senjen (2008)
Food additive	Bioral™ Omega-3 nanocochleates	Nanocochleates as small as 50 nm	Effective carrier for the incorporation of highly bioavailable Omega-3 fatty acids to muffins, cakes, cookies, pasta, soups, cereals, chips, and confectionary	Miller and Senjen (2008)
Generic food additive	Nutralease	Nanomicelles for encapsulation of nutraceuticals	Improved bioavailability means nutraceuticals are released into membrane between the digestive system and the blood	Chen et al. (2006)

plays an important role in consumer satisfaction and acceptance for further consumption of foods (Poncelet et al., 2011). Flavor is sensitized primarily by the chemical senses of taste and smell and modified with natural or artificial flavorants to affect these senses positively to enhance consumer acceptability (Lin et al., 2005). Manufacturing and storage procedures, packaging materials, and food components and ingredients often cause modifications in overall flavor by reducing aroma compound intensity or producing off-flavor components (Bryksa and Yada, 2012a). Most available aroma compounds are produced via chemical synthesis or extraction. Flavor stability in different foods is an important parameter because of its direct relationship with the quality and acceptability of foods, therefore it has been of increasing interest to maintain and control the stability. Flavors form very complex systems and there are many variables which affect stability. Some are more stable in water-soluble carbohydrates (Fathi et al., 2014) and others show stability in lipid-based wall coatings (Fathi et al., 2012). There are various factors such as physicochemical properties, interactions among volatile aroma molecules and food components which influence the overall quality of the food product; therefore it is beneficial to encapsulate volatile ingredients before incorporating them in foods or beverages to limit aroma degradation or loss during processing and storage (Vos et al., 2010). Nanoencapsulation strategies involve polymer-based nanocarriers, lipid-based nanocarriers, and molecular complexes. Various techniques are employed to form these carrier systems, including emulsification, coacervation, spray drying, freeze drying, and extrusion technologies. For assessing the palatability of food products there have been recent and growing interests in the development of rapid and sensitive instrumental techniques with potential application in correlation to sensory results (Nicolescu, 2007). Therefore, since flavor perception involves the senses of both smell and taste, it is important to evaluate systems that provide information about the overall smell and taste of foods. Electronic instruments are utilized for this purpose (Galmarini et al., 2008; Otles and Yalcin, 2012). The Functional Food market is being driven by a growing consumer understanding of diet and disease links, aging populations, increasing health care costs, and advances in food technology and nutrition (Adley, 2014).

During the manufacturing and at the time of food intake many instant food and beverage products, such as instant coffee and tea, instant soups, instant desserts, and the like are subjected to heating, which results in the loss or deterioration of volatile compounds responsible for contributing desirable aroma and/or flavor of the product. To compensate for such loss during processing and to provide the product with the desired aroma and/or flavor an attempt is made to add natural and synthetic aromas and flavors, which consist of various combinations of alcohols, aldehydes, ketones, esters, and the like, to the finished product. However, such natural and synthetic flavors and aromas are also highly volatile and sensitive to oxidation by atmospheric oxygen, moisture, and heat. As a result, even after incorporating many of these compounds in the food or beverage, there can be loss of much of the original characteristic odors and tastes and the additives can fail to provide the consumer product with the desired characteristics (Garwood et al., 1995).

2.1 Classification of Aroma and Flavors Used in Food Products

Flavors consist of concentrated preparation, with or without solvents and carriers, used to impart specific taste to food (Fahlbusch et al., 2003). Flavor ingredients are the largest single group of direct food additives utilized by the food industry. There are three principal types of flavors used in food, under definitions agreed upon in the European Union and Australia (Table 2.3).

Most artificial flavors are specific and often complex mixtures of singular naturally occurring flavor compounds combined

Table 2.3 Types of Flavoring Substances (Hui, 2006)

Type	Description
Natural flavoring substances	The flavoring substances, which are naturally present in plant and animal source materials, must be isolated via extraction or distillation, processes where specific substances are separated from a natural mixture. They can be either used in their natural state or processed for human consumption, but cannot contain any nature-identical or artificial flavoring substances.
Nature-identical flavoring substances	Natural identical flavoring are substances that are chemically identical to natural substances present in products intended for human consumption, but which are obtained by chemical processes or by chemical modification of other natural substances. They cannot contain any artificial flavoring substances.
Artificial flavoring substances	Artificial flavors are simply chemical mixtures that mimic a natural flavor in some way intended for human consumption. These are typically produced by fractional distillation and additional chemical manipulation of naturally sourced chemicals, crude oil or coal tar. Although they are chemically different from natural flavoring substance sensory characteristics are the same as natural ones.

Table 2.4 Artificial Flavor Enhancers

Chemical	Odor	Chemical	Odor
Diacetyl	Buttery	Limonene	Orange
Isoamyl	Banana	Ethyl decadienoate	Pear
Benzaldehyde	Bitter almond	Allyl hexanoate	Pineapple
Cinnamaldehyde	Cinnamon	Ethyl maltol	Sugar, cotton candy
Ethyl propionate	Fruity	Ethyl vanillin	Vanilla
Methyl anthranilate	Grape	Methyl salicylate	Wintergreen

together to either imitate or enhance a natural flavor (Smith et al., 2003) (Table 2.4).

Food stuffs containing synthetic flavor are often avoided because the consumers suspect that these compounds are toxic or harmful to their health but the substances used to produce artificial flavors are almost similar to those that occur naturally (Gane et al., 2013). It has been recommended that artificial flavors may be safer to use than natural flavors because of the standards of purity and mixture consistency that are obligatory either by the company or by law (Fleming, 2015). In comparison to artificial flavors, natural flavors may contain impurities because they are directly obtained from their sources and used as such in natural form while artificial flavors undergo more testing before being sold for consumption and therefore are typically more pure (Smith et al., 2005). Flavors from food products are generally the result of a combination of natural flavors, which set up the basic smell profile of a food product while artificial flavors modify the smell to accent it (Yuri et al., 2010). Natural flavor substances are obtained by physical separation, enzymatic processes, or microbial processes from vegetable or animal sources, either in the raw state or after processing (drying, torrefaction, and fermentation).

The flavor industry is not a single homogenous entity, but a composite of closely interrelated and somewhat overlapping sectors including essential oils and natural extracts, aroma chemicals, and compounded flavors. Essential oils and natural extracts represent complex aroma mixtures containing hundreds of chemical constituents. They may be used for imparting scent or aroma to consumer products or may be used as raw materials for compounding flavor and fragrance compositions, or they may be the source of isolated aroma chemicals, also used in compounding (Bolger et al., 2011).

Essential oil can be classified into three chemical groups: straight hydrocarbons, oxygenated compounds, and benzene derivatives.

Aroma chemicals comprise organic compounds with a defined chemical structure that are isolated from microbial fermentation, plant or animal sources, or produced by organic synthesis. Aroma chemical used to compound flavors are of two types:

1. *Isolates* which have been physically removed from natural sources that contain them and which may be further chemically modified, and
2. *Synthetic aroma chemicals* that duplicate the structure of aroma characteristics of their counterparts found in nature.

Aroma substances are classified as:

1. *Benzenoids (including naphthalenoids)*: chemicals containing benzene or naphthalene ring including alcohols, acids, esters, aldehydes, ketones, phenols, phenol esters, and lactones
2. *Terpene and terpenoids*: chemicals with (or closely related to) characteristic terpene structure, both acyclic and cyclic, having two or more isoprene moieties and oxygenated derivative of terpene hydrocarbons including alcohols, aldehydes, ketones, and esters
3. *Other aroma chemicals*: includes aliphatic, alicyclic, and heterocyclic compounds and esters of lower fatty acids (Hui, 2006)

The world of aroma is very attractive, especially because it concerns the taste of what we eat, It can be found in food, wine, perfumes, fragrance oils, spices, and essential oils. For example, many are produced biochemically during ripening of fruits and other crops. In wines, flavors are fermentation byproducts (Contis et al., 1997). Many of the aroma compounds are used in the production of flavorants which play a major role in the food industry service to improve and generally increase the appeal of their products to consumers (Matua, 2014) (Table 2.5).

2.2 Materials for Encapsulation

Correct choice and selection of wall material is an important starting step of encapsulation technology because it affects the efficiency of encapsulation and stability of the nanocapsule. The ideal wall material should have nonreactivity with the material to be encapsulated, ability to hold and seal the core inside the capsule, the ability to impart maximum protection to the encapsulated flavor and aromas against environmental conditions (such

Table 2.5 Classification of Aroma Substances (https://en.wikipedia.org/wiki/Aroma compound; Smith et al., 2005; Fleming, 2015; Gane et al., 2013; Glindemann et al., 2005; Hui, 2006)

Compound Name	Fragrance	Natural Occurrence	Chemical Structure
(a) Esters			
Geranyl acetate	Fruity, rose	Rose, floral	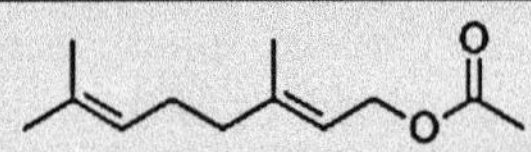
Methyl butyrate, methyl butanoate	Fruity	Pineapple	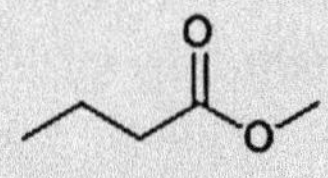
Ethyl acetate	Sweet	Wine	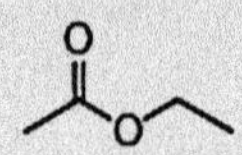
Ethyl butyrate, ethyl butanoate	Fruity	Orange, pineapple	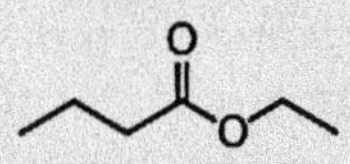
Isoamyl acetate	Fruity, banana, pear	Banana plant	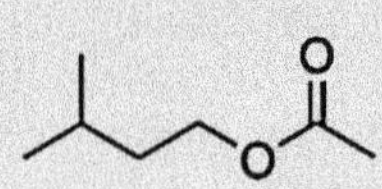
Pentyl butyrate, pentylbutanoate	Fruit	Pear, apricot	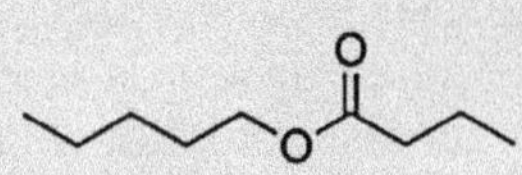
Pentylpentanoate	Fruity	Apple	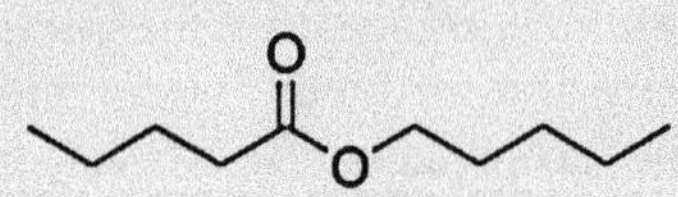
Octyl acetate	Fruity	Orange	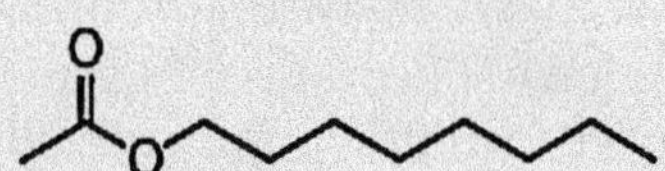
Benzyl acetate	Fruity, strawberry	Strawberries	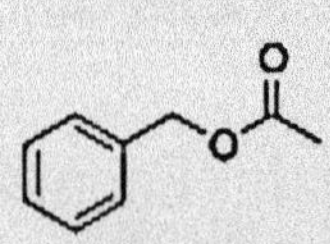
Methyl anthranilate	Fruity	Grape	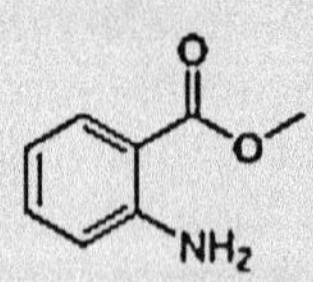

Table 2.5 Classification of Aroma Substances (*cont.*)

Compound Name	Fragrance	Natural Occurrence	Chemical Structure
(b) Linear terpenes			
Myrcene	Woody, complex	Verbena, bay leaf	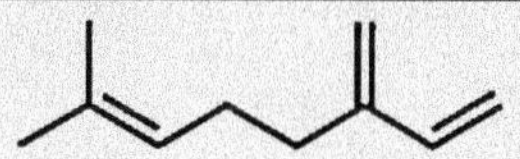
Geraniol	Rose, flowery	Geranium, lemon	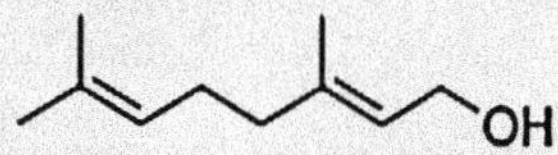
Nerol	Sweet rose, flowery	Neroli, lemongrass	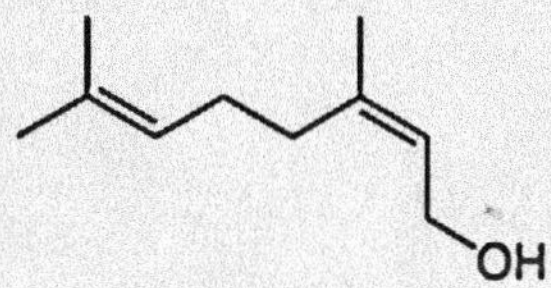
Citral, lemonal, geranial, neral	Lemon	Lemon myrtle, lemongrass	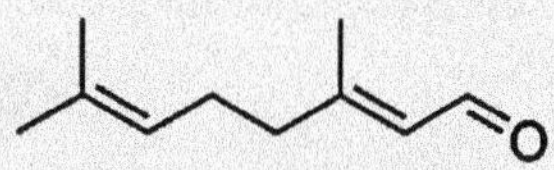
Citronellal	Lemon	Lemongrass	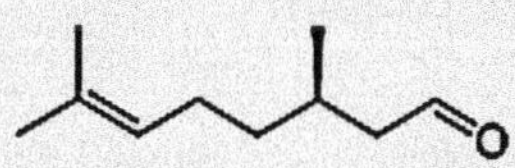
Citronellol	Lemon	Lemongrass, rose, pelargonium	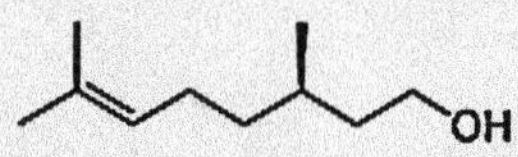
Linalool	Floral, sweet, woody, lavender	Coriander, sweet basil, lavender	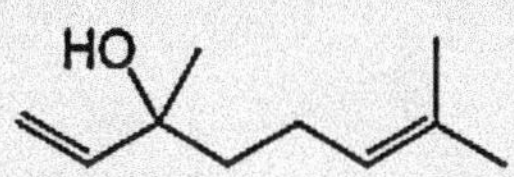
Nerolidol	Woody, fresh bark	Neroli, ginger, jasmine	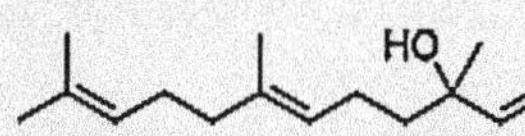
(c) Cyclic terpenes			
Limonene	Orange	Orange, lemon	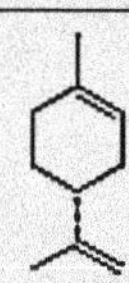
Camphor	Camphor	Camphor laurel	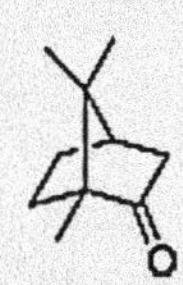

(*Continued*)

Table 2.5 Classification of Aroma Substances (*cont.*)

Compound Name	Fragrance	Natural Occurrence	Chemical Structure
Menthol	Menthol	Mentha	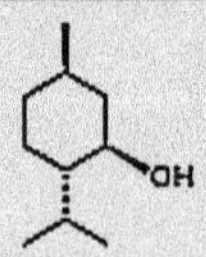
Carvone	Caraway or spearmint	Caraway, dill, spearmint	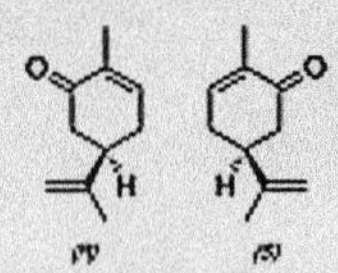
Terpineol	Lilac	Lilac, cajuput	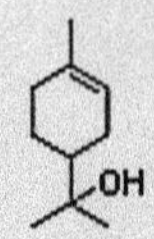
alpha-Ionone	Violet, woody	Violet	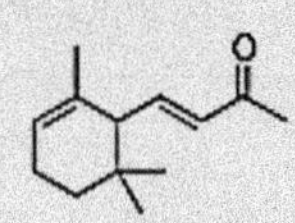 α
Thujone	Minty	Wormwood, lilac, juniper	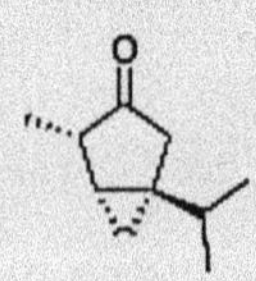
(d) Aromatic			
Benzaldehyde	Almond	Bitter almond	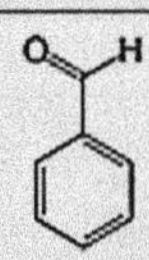
Eugenol	Clove	Clove	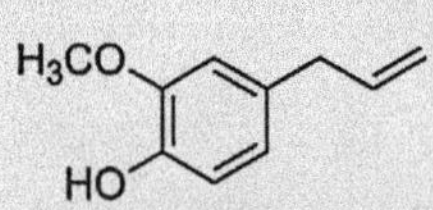
Cinnamaldehyde	Cinnamon	Cassia, cinnamon	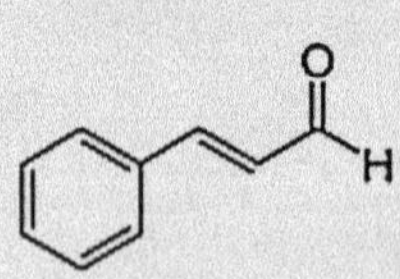

Table 2.5 Classification of Aroma Substances (*cont.*)

Compound Name	Fragrance	Natural Occurrence	Chemical Structure
Vanillin	Vanilla	Vanilla	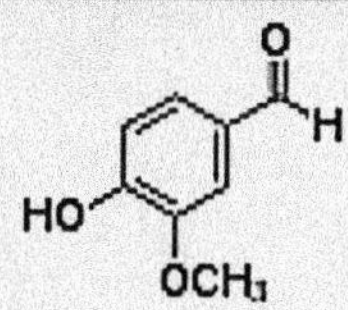
Anisole	Anise	Anise	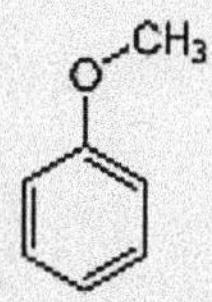
Anethole	Anise	Anise, sweet basil	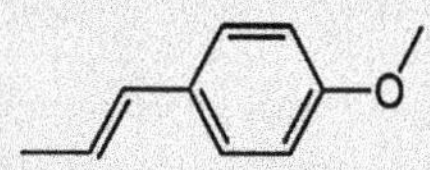
Estragole	Tarragon	Tarragon	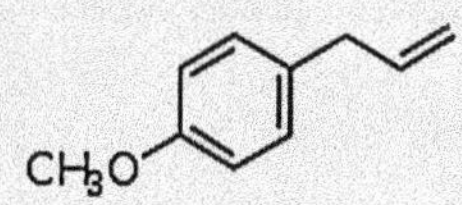
Thymol	Thyme	Thyme	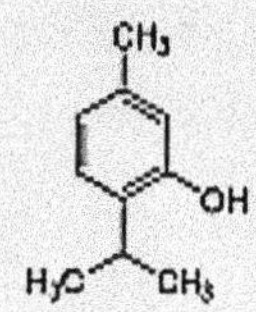
(e) Amines			
Putrescine, diaminobutane	Rotting flesh	Rotting flesh	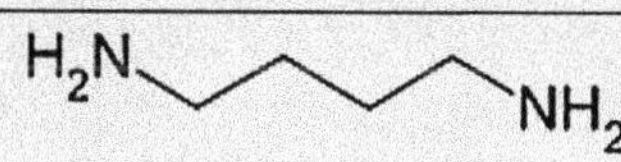
Cadaverine	Rotting flesh	Rotting flesh	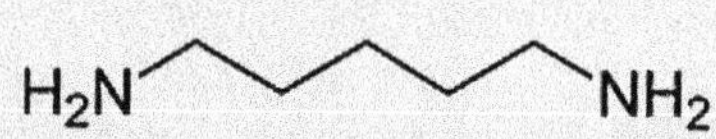
Pyridine	Fishy	Belladonna	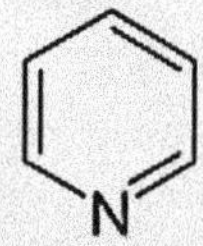

(*Continued*)

Table 2.5 Classification of Aroma Substances (*cont.*)

(f) Other aroma compounds

(i) Alcohols

- Furaneol: strawberry
- 1-Hexanol: herbaceous, woody
- *cis*-3-Hexen-1-ol: fresh cut grass
- Menthol: peppermint

(ii) Aldehydes

High concentrations of aldehydes tend to be very pungent and overwhelming, but low concentrations can evoke a wide range of aromas.

- Acetaldehyde: ethereal
- Hexanal: green, grassy
- *cis*-3-Hexenal: green tomatoes
- Furfural: burnt oats
- Hexyl cinnamaldehyde
- Isovaleraldehyde: nutty, fruity, cocoa-like
- Anisic aldehyde: floral, sweet, hawthorn. It is a crucial component of chocolate, vanilla, strawberry, raspberry, apricot, and other aromas.
- Cuminaldehyde (4-propan-2-ylbenzaldehyde): spicy, cumin-like, green

(iii) Esters

- Fructone: fruity, apple-like
- Hexyl acetate: apple, floral, fruity
- Ethyl methylphenylglycidate: strawberry

(iv) Ketones

- Cyclopentadecanone: musk ketone
- Dihydrojasmone: fruity woody floral
- Oct-1-*en*-3-one: blood, metallic, mushroom-like
- 2-Acetyl-1-pyrroline: fresh bread, jasmine rice
- 6-Acetyl-2,3,4,5-tetrahydropyridine: fresh bread, tortillas, popcorn

(v) Lactones

- gamma-Decalactone: intense peach flavor
- gamma-Nonalactone: coconut odor, popular in suntan lotions
- delta-Octalactone: creamy note
- Jasmine lactone: powerful fatty fruity peach and apricot
- Massoia lactone: powerful creamy coconut
- Wine lactone: sweet coconut odor
- Sotolon: maple syrup, curry, fenugreek

Table 2.5 Classification of Aroma Substances (*cont.*)

(vi) Thiols

- Allyl thiol (2-propenethiol; allyl mercaptan; $CH_2{=}CHCH_2SH$): garlic volatiles and garlic breath
- Grapefruit mercaptan: grapefruit
- Methanethiol, commonly called methyl mercaptan: after eating asparagus
- Furan-2-ylmethanethiol, also called furfurylmercaptan: roasted coffee
- Benzyl mercaptan: leek or garlic-like

(vii) Miscellaneous compounds

- Methylphosphine and dimethylphosphine: garlic-metallic, two of the most potent odorants known
- Diacetyl: butter flavor
- Acetoin: butter flavor
- Nerolin: orange flowers
- Tetrahydrothiophene: added to natural gas
- 2,4,6-Trichloroanisole: cork taint
- Substituted pyrazines

as light, heat, oxygen, and humidity); also, it should not have an unpleasant taste in terms of food applicability and economic viability. There are different organic materials of plant, marine, and microbial or animal origin (Nollet and Toldra, 2010) which are used in food industry for encapsulation and developed many years ago (eg, cyclodextrins), but their potential for flavor encapsulation was realized only recently (eg, cyclodextrins) and they provide exceptional stability to oxidation and evaporative losses (Nur Ain et al., 2011). Due to cost reduction and imminent FDA approval, cyclodextrins will likely find widespread use in foods. The majority of materials used for nanoencapsulation in the food sector are biomolecules. In addition to carbohydrate polymers/polysaccharides, which are the most abundant of the four major classes of biomolecules, proteins, and lipids are also biomolecules suitable for nanoencapsulation in the food sector (Cabuk et al., 2014). Nanocarriers can be structured by a great variety of organic and inorganic materials characterized by high biodegradability and biocompatibility. Organic materials can be classified in polymers, proteins and lipid based. Table 2.6 lists groups of biomolecules, arranged according to their origin, which are found to be most suitable either when used alone or when used in combination with others for nanoencapsulation in the food industry.

Table 2.6 Classification of Materials Suited for Nanodelivery of Aroma and Flavors in the Food Industry (Tiwari and Tiwari, 2014; Joshi and Patel, 2012; Vroman and Tighzert, 2009; Milanovic et al., 2010; Bryksa and Yada, 2012b; Wandrey et al., 2010)

	Natural Polymers			Synthetic/ Semisynthetic Polymers	Inorganic Materials
Origin	Carbohydrate polymer	Protein	Lipid	Polylactic acid (PLA)	Calcium phosphates
Plant	Starch derivatives	Gluten (corn)	Fatty acids/ alcohols	Polyglycolic acid (PGA)	Silica
	Cellulose derivatives	Isolates (pea, soy)	Glycerides	Polyhydroxybutyrate (PHB)	Titanium
	Plant exudates		Waxes	Poly(lactide-*co*-glycolide) (PLGA)	Aluminum
	(Gum arabic, Gum karaya and Mesquite gum)		Phospholipids	Poly(ε-caprolactone) (PCL)	Zirconium
				Polydioxanone (PDS)	Cerium
				Polyanhydrides	
	Plant extracts			Polyamides	
	(Galactomannan, Soluble soybean)				
	Polysaccharide				
Marine	Carrageenan				
	Alginate				
Microbial/ animal	Xanthan	Caseins	Fatty acids/ alcohols		
	Gellan	Whey proteins	Glycerides		
	Dextran	Gelatin	Waxes		
	Chitosan		Phospholipids		
			(Shellac)		

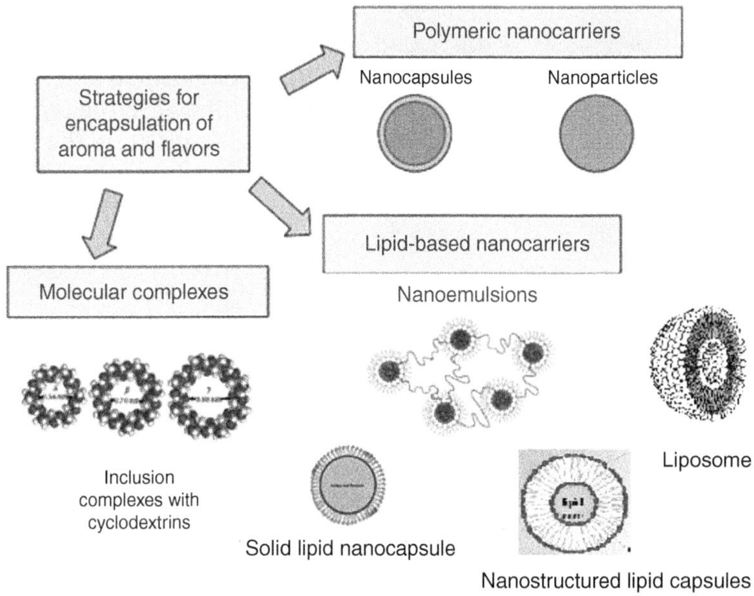

Figure 2.1. Strategies for nanoencapsulation of aroma and flavors. Refs.: https://www.google.co.in/search?rlz=1C1CHZL_en-GBIN655IN655&biw=1517&bih=741&noj=1&tbs=sur%3Afc&tbm=isch&sa=1&q=cyclodextrin&oq=cyclodextrin&gs_l=img.12..0l10.35903.35903.0.38071.1.1.0.0.0.0.167.167.0j1.1.0....0...1c.1.64.img..0.1.167.DZ-PB_UaXgY#imgrc=6JaoCUYKU1FUxM%3A; Bilia, A.R., Guccione, C., Isacchi, B., Righeschi, C., Firenzuoli, F., Bergonzi, M.C., 2014. Essential Oils Loaded in Nanosystems: A Developing Strategy for a Successful Therapeutic Approach. Hindawi Publishing Corporation. 1–14; https://www.google.co.in/search?q=nanoemulsions&biw=1517&bih=741&source=lnms&tbm=isch&sa=X&sqi=2&ved=0ahUKEwjBj8zq887JAhWCuo4KHYJIAjgQ_AUIBigB&dpr=0.9#q=liposome&tbs=sur:fc,itp:lineart&tbm=isch; https://en.wikipedia.org/wiki/Protocell

2.3 Strategies for Nanoencapsulation of Aroma and Flavors

There are different strategies of developing nanodelivery systems such as polymeric-based nanocarriers, lipid-based nanocarriers, and molecular exclusion (Fig. 2.1) (Bilia et al., 2014; Matalanis et al., 2011). Nanodelivery systems can be engineered to possess a number of desirable features for food products, including (1) sustained and controlled release of aroma and flavors, (2) protection from unfavorable environmental conditions, and (3) stability (Naknean and Meenune, 2010). Nanocarriers provide generally two controlled release mechanisms which can be observed during delivery of an aroma and flavors:

1. *Delayed release* is a mechanism by which the release of aroma and flavor substances is delayed from a bounded lag time up to a point where its release is desirable and is no longer hindered.

This mechanism can be used for flavor release in ready-meals (Monedero et al., 2010).

2. *Sustained release* is a mechanism which is engineered to maintain constant concentration of aroma and flavor to the target site. This mechanism can be employed for extending the release of the encapsulated material, including flavors in chewing gum (Fathi et al., 2012; Somasundaran et al., 2006).

2.4 Methods of Nanoencapsulation of Aroma and Flavors

Although nanoencapsulation has found applications in the food industry, the technology remains far from being fully exploited. Exciting new techniques such as cocrystallization, electrospray, and liposome formation will improve the number and quality of encapsulated ingredients.

2.4.1 Coacervation Phase Separation

Coacervation is often considered as the original method of encapsulation. In this method the coacervates are separated from the mixture of two colloidal liquids after agglomeration. Depending on the number and types of polymers used, coacervation can be divided into simple or complex: when a single type of polymer is combined with highly hydrophilic agents in the colloidal solution, then it is called simple coacervation, while in complex coacervation two or more types of polymers are used (Feng et al., 2009). The general common driving force for this method is electrostatic attraction between oppositely charged molecules. In case of simple coacervation this force is induced between charged food components and an oppositely charged capsule wall material; consequently, the food flavors and aroma are trapped within a particle formed by electrostatic complexation of positively charged (eg, chitosan) and a negatively charged (eg, pectin and alginate) biopolymers (complex coacervation). The functional performance of the obtained nanocapsules depends upon:

- the surface characteristics and chemical nature of the biopolymeric shell;
- surface charge, which is a pH-dependent parameter, and higher surface charge, which is responsible for the better performance of nanoencapsulation;
- rate of agitation: aggregation might occur at a very low or very high stirring rate;
- rate of addition of polymer solution: the lower the value or drop wise addition the better will be the performance; and
- solubilities of flavors and aromas and biopolymer in solution: the higher the solubilities, the better will be the performance.

In this process, usually the core material used must be compatible with the recipient polymer and be insoluble or barely soluble in the coacervation medium. Although it is considered as original method of encapsulation but nowadays this technology has not been used in the food industry because it is complex and expensive. Other complications related to this method are optimization of wall material concentration in the emulsification and coacervation process because the concentration essential to obtain a fine emulsion may be different to that required to increase the yield of nanocapsules, dissolution of active compound into the processing solvent and oxidation of product, evaporation of volatiles, because residual core materials sometimes adhere to the exterior of capsule. The complex coacervates are very unstable and toxic chemical agents, such as glutaraldehyde, are essential to stabilize them (Yadav et al., 2015). Steps involved in coacervation phase separation method are represented in Fig. 2.2.

2.4.2 Spray Drying

For the large-scale production of the encapsulated flavors and volatiles, spray drying is one of the most frequently applied technologies because it is fast, relatively cheap, and reproducible method (Christelle et al., 2013). This method is based on

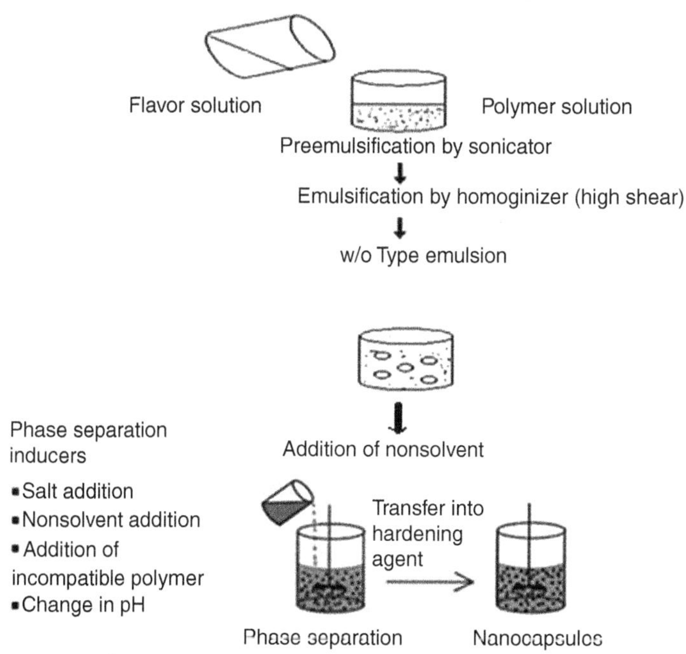

Figure 2.2. Processes involved in coacervation phase separation method of nanoencapsulation.

the principle of dissolving or dispersing the active ingredients in the solution of biopolymer. The dispersion is then atomized in a heated air chamber which quickly removes the solvent and generates a dried particle consisting of the active ingredient embedded in a porous wall material. This method has some drawbacks for volatile or thermo-sensitive bioactives (Mahdi et al., 2008). While in food systems water-based dispersions are typically used, there is a problem with less water-soluble compounds (eg, chitosan and cellulose). There are various factors that may affect the size of produced particles and the efficiency of encapsulation, which are as follows:

1. Core and wall material properties
2. Solution viscosity
3. Particle size (for emulsified active ingredients)
4. Spray dryer features
5. Flow rate
6. Inlet/outlet temperatures
7. Humidity
    - The lower the feed flow rate, the smaller the size and the higher the efficiency
    - The higher the air flow rate, the larger the size, and a moderate air flow rate leads to high efficiency; the lower the humidity, the smaller the size and the lower the efficiency
    - The higher the inlet temperature, the larger the size and the higher the efficiency, due to formation of a hard crust

The uses of carbohydrates like starch may be prohibited in this method due to chances of gelation of solution increases at high temperature. The thermostable carbohydrate biopolymers like cyclodextrins and modified materials like hydroxypropyl cellulose are appropriate in case of spray drying at high temperatures. The spray droplets are formed from common spray dryers by using rotary atomizers or pressure nozzles or two-fluid nozzles. The formation of particles less than 2 mm in traditional spray dryers is small as they cannot be strongly captured in cyclone collectors; therefore particles formed in these dryers are not in the range of submicron particles. Currently, a new type of spray-dryers (Nano Spray Dryer B-90) has been introduced for the production of nanoscale particles. Fabrication of submicron particles is accomplished by effective fluid breakdown combined with extremely efficient particle collectors. A vibrating mesh is used in this method for the formation of fine droplets. Generally, the formation of micron-sized droplets is based on a piezoelectric-driven actuator which causes vibration in a thin, perforated, stainless-steel membrane carried in a small spray cap. The actuator is operated at an ultrasonic frequency, which results in the upward and downward vibration of membrane, and millions of specific size droplets are discharged per second with very fine droplet size distribution. An electrostatic particle collector is used for the collection of dried particles instead of cyclones, which are size dependent. The efficiency of this method is high for generating nanosize particles less than 300 nm). The benefits of the process have ensured its dominance over others; these include accessibility of equipment, a wide variety of carrier solid, good stability of the finished product, continuous large-scale production, excellent retention of volatiles, and low process cost (Xiao and Zhong, 2011). This technique has the advantage of high retention of aroma components during food drying process. This technique is suitable for most of the heat-labile substances because of the use of lower temperature throughout the process. (Anandharamakrishnan, 2014a). Fig. 2.3 is representation of spray dryers with their different units.

2.4.3 Freeze Drying

The freeze-drying technique, or lyophilization, is one of the most useful processes for drying of those materials which are unstable in aqueous solutions and are highly sensitive to temperature. This process involves solidification of the surface of the

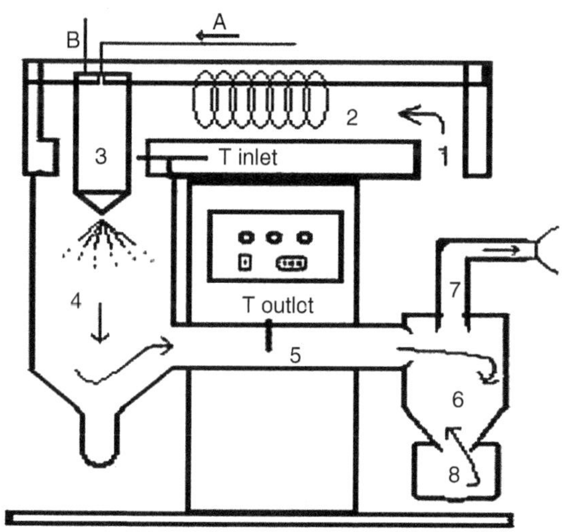

Figure 2.3. Schematic representation of laboratory scale spray dryer unit. A, solution or suspension to be dried in; B, atomization gas in; 1, drying gas in; 2, heating of drying gas; 3, spraying of solution or suspension; 4, drying chamber; 5, part between drying chamber and cyclone; 6, cyclone; 7, drying gas is taken away; 8, collection vessel of product; and arrows mean that this is cocurrent lab-spray dryer. Ref.: https://upload.wikimedia.org/wikipedia/commons/1/17/Labspraydryer.jpg

solution into amorphous solid in which selective diffusion of flavors is possible (Ishwaryaa et al., 2015). Freeze drying is the process that gives the most desirable properties competitively to spray-dried powder when compared with other techniques like tray drying and drum drying. The dried encapsulated powder product produced after freezing and subsequent freeze drying offered an opportunity to attain a product with good resistance to oxidation and maintaining the desired shape of the nanocapsules because of fixation by freezing. Still this drying technique is less attractive than others due to the higher cost which is approximately 50 times more as compared to spray drying, also the storage and transport of particles produced is extremely expensive and it involve long processing time therefore the commercial value and applicability is highly restricted (Barbosa et al., 2015).

2.4.4 Spray Chilling/Spray Cooling

These are the low-cost encapsulation processes and are consistently used for encapsulation of aroma compounds to provide heat stability, delayed release in wet environments, and to transform liquid flavors into free-flowing powders. These technologies are similar to spray drying where the core material is dispersed in a liquefied coating or wall material for emulsification of the flavor compounds followed by atomization to disperse droplets from the feedstock. (Okuro et al., 2013). After that the droplets are immediately mixed with a cooling medium and subsequently solidify into powder form. There is generally no water to be evaporated. In this technique for atomization of molten coating material, a pneumatic nozzle is used, and for chilling process the material is placed in a vessel generally containing a carbon dioxide ice bath (5°C). This results in the adherence of droplets on material to be encapsulated and solidification to form a coat film. The process is suitable for protecting many thermosensitive water soluble materials, and materials that may be volatilized or damaged during thermal processing. Fractionated or hydrogenated vegetable oil with a melting point in the range of 30–42°C is a suitable example of encapsulating material for this process. This is fat-based technology and lipid carriers, such as wax and oil (eg, beeswax, cocoa butter, palm oil, and kernel oil) can be used (Mozafari et al., 2006). This encapsulation technique can conceivably change the functionality, decrease the hygroscopicity, mask taste or odor, change solubility, and provide physical protection and subsequently allowing the controlled release of these ingredients. This low-cost technology is comparatively simple to apply and scale up, and it does not necessitate the use of organic solvents and the application of high temperatures in the process. Spray-chilled products have applications in bakery products, dry soup mixes, and foods containing a high level of fat (Shimoni, 2000).

2.4.5 Extrusion

Encapsulation of flavors via extrusion has been used for volatile and unstable flavors in glassy carbohydrate matrices. These matrices possess very good barrier properties in their glassy state and extrusion is a suitable method for encapsulation of flavors in such matrices (Shimoni, 2000). However, structural defects such as pores, cracks, thin wall formed during or after processing, and other process parameters may enhance the diffusion of flavor from extruded carbohydrates (Madene et al., 2006). A volatile compound is dispersed in a matrix polymer at high temperature (100–120°C) and this mixture is then forced through a die after which filaments are obtained that are plunged into a desiccant liquid that traps the active substances after hardening of the extruded mass. Extrusion of polymer solutions through nozzles to produce either beads or capsules is mainly used on a laboratory scale. The most common liquid used for the dehydration and hardening process is isopropyl alcohol. The strands or filaments of hardened material are broken into small pieces, separated, and dried. Several factors improve the quality of capsules, and these are emulsifier and flavor oil content and emulsification pressure (Raoa and Geckeler, 2011). The major advantage of the extrusion method is to maintain the stability of flavors against oxidation.

2.4.6 Electrospray

Electrospray is a new nanoencapsulation method of liquid atomization that utilizes electrical forces. The droplets obtained by electro spraying are charged, and for certain modes can be of nanometers size. And during the flight of the droplets toward the ground electrode solvent are evaporated. The main reason for the attraction of this electrospray technique is its high encapsulation efficiency and possibility of the production in one step. While a high voltage is exerted, formation of nanodroplet depends on polymeric molecular mass, polymer chains entanglement, and solvent evaporation rate (Anandharamakrishnan, 2014b).

2.4.7 Supercritical Fluid

In recent years, supercritical fluid (SCF) methods have attracted increasing attention for encapsulation of food bioactives. Many compounds can be brought into a supercritical state, such as water, propane, nitrogen, and carbon dioxide. Out of these compounds carbon dioxide is one of the most widely used fluids because carbon dioxide operates relatively at moderate temperatures and pressures (*Tc* ¼ 304.2 K, *Pc* ¼ 7.38 MPa), which might be beneficial for encapsulating temperature-sensitive food materials. Supercritical CO_2 (SCeCO_2) is also suitable for entrapment of easily oxidizable substances (ie, unsaturated fatty acids) because it provides an inert medium to that substance (Zuidam and Heinrich, 2010). On the other hand, it lowers the use of organic solvents, facilitates the separation of the SCF from the product by depressurization, allows the obtaining of solvent-free products, and produces particles with preferred morphology and fine size distribution because of the high solubility of most organic solvents in supercritical fluids. Various SCF encapsulation techniques are available depending upon the properties of the materials to be formulated. Two of the most extensively used methods include the rapid expansion of supercritical solutions (RESS) and supercritical antisolvent (SAS). In the RESS process, the biopolymers and encapsulants are saturated in SCF using high pressure, and then solution is precipitated out. Its expansion through a nozzle or capillary tube leads to high-pressure drop, high degree of supersaturation, and accordingly leads to formation of nanosized particles. On the other hand in SAS process, first SCeCO_2 is pumped to the precipitation cylinder and when suitable pressure and temperature are obtained the organic solvent (containing encapsulant, biopolymer, and organic solvent) is injected through a nozzle and extracted by SCeCO_2, and the biopolymers precipitate on the surface of the encapsulant. But the main setback of this method is that the use of organic solvents is necessary. A number of parameters such as temperature, pressure, encapsulant to biopolymer ratio, and rate of solution flow may control and effect particle size and efficiency of encapsulation. Lower temperature and the solution flow rate combined with higher pressure and higher ratio of encapsulant/biopolymer lead to smaller size of nanocarriers and higher efficiency. In this process carbohydrate biopolymers should have high solubility in organic solvent. Chemical modification could be achieved to improve carbohydrate solubility in organic solvent (eg, increase of degree of substitution of hydroxyl groups in dextran with hydrophobic esters) (Martín et al., 2010). Some examples of application of carbohydrate-based nanocarriers developed using SCF include the encapsulation of lycopene and lutein using cyclodextrins and HPMCP.

The supercritical fluid extraction of emulsions (SFEE) is the other technique which can be used to encapsulate food materials based on SAS. This technique involves preparing an o/w type of emulsion, in which the active compound is dissolved in the

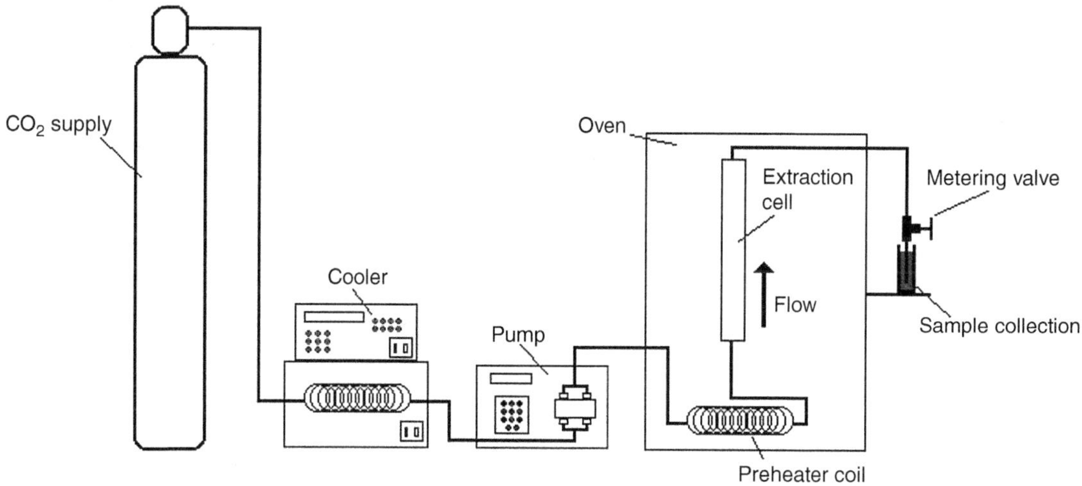

Figure 2.4. Schematic representation of supercritical fluid extraction unit. Refs.: https://www.google.co.in/search?q=Supercritical+Fluid+Extraction+unit&rlz=1C1CHZL_en-GBIN655IN655&source=lnms&tbm=isch&sa=X&ved=0ahUKEwiMloPipcDJAhWGto4KHRchBUwQ_AUIBygB&biw=1517&bih=741&dpr=0.9#q=Supercritical+Fluid+Extraction+unit&tbm=isch&tbs=sur:fc&imgrc=ltSognyKQ3AArM%3A; https://upload.wikimedia.org/wikipedia/commons/c/c7/SFEschematic.jpg

dispersed organic phase, while the aqueous phase holds carrier and stabilizing materials. Then this emulsion is allowed to come in contact with the supercritical fluid to facilitate the organic solvent extraction from the organic phase, resulting in the precipitation of the active material in the form of nanocapsules. The obtained aqueous suspension is then dried to produce a dry powder. This technique enables circumventing high temperatures and interference of the product by formation of gas bubbles throughout organic solvent elimination process, as compared to the alternative methods such as direct evaporation. The SFEE technique has been used to encapsulate β-carotene and lycopene in OSA-modified, and 100 nm particle size was obtained without degradation of the active components (Nagavarma et al., 2012). Fig. 2.4 is representation of supercritical fluid extraction units.

2.4.8 Emulsion Diffusion Method (EDM)

This is an effective method to encapsulate both hydrophilic and lipophilic food components with high encapsulation efficiency, greater reproducibility, lesser physical stress, narrow size distribution, and simple and convenient scaling-up procedure. It is a two-step process, conventional emulsification step followed by a diffusion step, that is, removal of part of the organic phase. In the first emulsification step, the aqueous phase and organic phase (comprising polymer and oil in organic solvent) are prepared separately and then emulsion is obtained using a mechanical shearing method. Presence of encapsulant in either aqueous and/or organic phase depends upon its polarity (Yadav et al., 2015). In the second (diffusion) step, addition of water leads to very fast removal of the organic solvent contained in the oil phase and leads to separation of the polymer from the oil, particle size reduction, polymer precipitation, and finally nanocapsule formation. The remaining organic solvent further can be evaporated under reduced pressure. The organic solvent must have greater solubility in both polymer and oil and partial solubility in water to make diffusion possible after dilution. The nanocapsule size is interrelated to the shear rate and temperature, the chemical compositions and amount of organic phase and emulsifier, the concentration of polymer and the oil-to-polymer ratio, used in the emulsification process. EDM method has been developed for production of carbohydrate-based delivery systems such as eugenol and fish-oil-loaded cyclodextrin, β-carotene, and capsicum oleoresin-loaded poly-ε-caprolactone. This technique provides an easy way for nanoencapsulation of diverse food bioactives, however, there remain a possibility of presence of residual organic solvents in the final product, it is necessary to be aware of their potential toxic effects (Sailaja et al., 2011).

2.4.9 Cocrystallization

Cocrystallization is a relatively simple procedure as compared to various other flavor encapsulation processes, and offers an economical and flexible alternative. From the different studies it can be concluded that there are numerous products that can be encapsulated by cocrystallization process including fruit juices, flavors, essential oils, brown sugar, and so on (Nedovic, 2010). Spontaneous crystallization of supersaturated sucrose syrup is carried out at high temperature (above 120°C) and low moisture and aroma compounds can be added at the time of spontaneous crystallization. The crystal structure of sucrose can be reformed in aggregates of very small crystals that incorporate the flavors either by entrapment or inclusion within the crystals. This aids to enhance flavor stability. The granular product thus obtained has low hygroscopicity and good flow property and dispersibility. A disadvantage of this process is during transformation of liquid flavor into dry granules some heat sensitive compounds may be degraded. On the other hand cocrystallization products retained as much volatile oil as did spray-dried and extruded products, which is advantageous. In addition a strong antioxidant is necessary to retard development of oxidized flavors during storage (Munin and Edwards-Levy, 2011).

3 Advantages of Nanoencapsulation of Flavor and Aroma Compounds

The effectiveness of nutraceutical ingredients depends on preserving and enhancing their bioavailability. Nanosizing or nanoencapsulating food ingredients delivers greater bioavailability, improves solubility, and increases potency compared to those substances in larger or microencapsulated form. How nanoencapsulation is effective in increasing the quality of food products is explained under the following subheadings.

3.1 Stability Enhancement of Flavor and Aroma Compounds in Food and Beverages

Some flavors and aromas are highly prone to oxidation so that when they come in contact with environmental conditions such as oxygen and heat they lose their characteristics. Encapsulation provides stability to such compounds. A well-known example is coffee aroma, which is highly unstable in such environmental conditions. Encapsulation of coffee aroma by gelatin-based nanoemulsions provides protection from loss by evaporation and from the deleterious effects of oxygen (Balassa et al., 1970). Encapsulation gives fully aromatized soluble coffee products which maintain their freshly brewed coffee flavor in the cup, even after the products have been exposed to atmospheric oxygen for an extended period of time. Table 2.7 shows an example of such flavor and aroma compounds showing greater stability after encapsulation.

3.2 Taste and Nutrition Enhancement of Food Products

Encapsulation of flavor and aroma is also beneficial in enhancing taste and nutritional supplements in food products. Being nanosized encapsulates they show better dispersibility of water in soluble additives (eg, colors, flavors, supplements, preservatives) in food products without the use of additional fat or surfactants (Chaudhry et al., 2008). The reason behind the enhancement of taste and flavor is the enlarged surface areas of the nanosized additives, and enhanced absorption and bioavailability in the body compared with conventional forms of bulk products. Currently available examples include vitamins, antioxidants, flavors, colors, and preservatives (Huyghebaert et al., 2010).

The nanoencapsulated nutrients and supplements are also claimed for enhanced bioavailability, antimicrobial activity, and

Table 2.7 Application of Encapsulation in Stability Enhancement of Aroma and Flavors

Encapsulated Compounds	Materials	Strategies/ Methods	References
Coffee aroma	Gelatin	Nanoemulsion	Garwood et al. (1995)
Caraway essential oil	Whey protein, maltodextrin	Spray drying	Bylaite et al. (2001); (Soottitantawat et al., 2015)
Caraway extract, coriander oil	β-Cyclodextrin and modified starches	Molecular inclusion	Partanen et al. (2002a); Dima et al. (2014)
Extracted sea buckthorn kernel oil	Maltodextrin and starch derivative	Nanoemulsion	Partanen et al. (2002b)
Swiss cheese flavor	Gum arabic and maltodextrin	Nanoemulsion	Teixeira et al. (2004)
Menthol	Gum arabic and modified starch	Nanoemulsion	Soottitantawat et al. (2005)
Rice flavor	Gum arabic and maltodextrin	Spray drying	Apintanapong and Noomhorm (2003)
Natural flavorings of oregano, citronella, and marjoram	Skimmed milk powder and whey protein	Nanoemulsion concentrate	Baranauskienė et al. (2007)
Shiitake flavors powder	Cyclodextrin and maltodextrin	Spray drying	Shiga et al. (2004)
Limonene	Silica (SiO_2) sphere	Sol–gel approach	Ciriminna and Pagliaro (2013); Ashraf et al. (2015)
Ethyl vanillin	Carnauba wax	Melt dispersion technique	Milanovic et al. (2010)

other health benefits. Nanotextured foodstuffs can also enable a reduction in the use of fat (Clegg et al., 2009). A typical product of this technology would be a nanotextured ice cream, mayonnaise, or spread which would offer taste and texture attributes similar to the full-fat equivalent, but with a substantial reduction in the fat intake of the consumer. Such products would offer "healthy" but still tasteful food products to the consumer (Chaudhry et al., 2010). Nestlé and Unilever are reported to be developing a nanoemulsion-based ice cream with a lower fat content that retains a fatty texture and flavor (Huyghebaert et al., 2010).

Certain inorganic nanosized additives also have applications that are beneficial to health in the food area. Examples of these include transition metals (eg, silver, iron), alkaline earth metals (eg, calcium, magnesium), and nonmetals (eg, selenium, silica) (Bradley et al., 2011). The use of inorganic nanoadditives is claimed for enhanced tastes and flavors due to enlarged surface areas (Food Safety Authority of Ireland, 2008). An example is nanosalt, the use of which would give more salt particles on a product (eg, chips/crisps) and allow the consumer to taste more salt even when salt were added at a lower level (Fabra et al., 2012). Food supplements in this category are also claimed for enhanced absorption and improved bioavailability compared with conventional equivalents (Chaudhry et al., 2010; Cientifica, 2006).

More immediately, nanonutritional additives are already being used to boost the vitamin and mineral content of some processed foods and to speed up the manufacturing of processed meats. Examples include ongoing R&D in Taiwan and Japan on development of nanosized cellulose, starch, wheat, and rice flour, and a wide range of spices and herbs for herbal medicine and food applications (Miller and Senjen, 2008). Another example is Zn/Fe loaded into alginate nanoparticles. Alginate nanoparticles could decrease the loss of Fe/Zn in comparison control; also, loading efficacy of Zn/Fe was 70–85% and release profile of nanoparticles showed a steady state. Hence, this nanoparticle can be suggestive for the enrichment of ice cream and probably other foods (Sharifi et al., 2013).

3.3 Masking of Undesirable Flavor or Aroma Compounds

For food product acceptance by the consumers, masking of undesirable flavor and aroma is an important step to be considered during food manufacturing and processing. Including all the strategies of nanoencapsulation discussed earlier in this chapter, self-assembly of food proteins (Huang et al., 2009) in nanoscale can be useful for binding components such as enzymes or vitamins, providing protection to encapsulated nutraceuticals and masking undesirable flavor or aroma compounds (des Rieux et al., 2006; Luecha et al., 2010). This application of encapsulation of taste masking had been applied in the huge Italian market for pasta products. Italian consumers are very conservative about pasta products; as a consequence more than 95% of the market is made up of traditional pasta products. The concept to use this national staple food to deliver healthy components is appealing for all pasta producers. On the other hand, there are two main hurdles that have not yet been solved:

- Consumers cannot admit any decrease in the sensorial quality of pasta, and
- The standard pasta cooking procedure (ie, boiling in abundant water) causes the loss of all added hydrophilic bioactive compounds.

Encapsulation technologies solved both these problems.

- It can mask the negative sensorial impact of the various bioactive ingredients (such as the bitterness of silymarin and the off flavor of PUFA).
- It blocks the bioactive ingredients into the starch and gluten matrix avoiding the loss in the boiling water (Kokini, 2010).

4 Quality Assessment by Instrumental Methods to Predict Flavor and Aroma in Food Products

Flavor and aroma compounds perform a crucial role in shaping the organoleptic quality of many food products (Baser and Buchbauer, 2010). An organoleptic quality is equally important as the other qualities for consumers and often considered in the purchase. The analysis of aroma, that is, the presence, composition, and content of volatile substances, can constitute a valuable source of information on the quality of food. A classical approach to the evaluation of organoleptic quality is based on the utilization of sensory analysis, carried out by a group of trained assessors. This analysis is a perfect tool in carrying out marketing tests of consumers, but because of great human participation it has many limitations (Duncan, 2011). Because of these deficiencies a good supplement of the evaluation of organoleptic food properties is instrumental analysis. Appropriate instrumental methods allow a detailed and complex qualitative and quantitative analysis of volatile components, which influence on the flavor composition of food products (Sankarankutty, 2014). The methods employed most often, allowing the analysis and recognition of aromas are chromatographic techniques, in particular gas chromatography. In recent years, intensive studies have been carried out regarding sensory activity of the individual volatile components of various food products and the dependence between the odor and the chemical composition of the volatile fraction of these products, using conventional chromatographic techniques and newly invented biosensor. The consumption of foods and beverages is totally dependent on odor and taste along with its appearance and texture, which is related to the stimulation of the human senses for that particular chemical, odor, and flavor. Gas chromatography, infrared spectroscopy, e-nose, and e-tongue are such commonly used techniques for analyzing flavor and aroma and also are beneficial in identification and comparison of volatile compounds present in food products and attempting to find the relation between flavor compounds content and quality of these products from the time of production to the time of food intake. These techniques are described here briefly with their principal and application.

4.1 Gas Chromatography–Olfactometry

The human nose perception of flavor and aroma compounds, released from foods and fragrances, depends on the extension of the release from the matrix and the odor properties of the compounds. It is acknowledged that only a small portion of the large number of volatiles occurring in a fragrant matrix contributes to its overall distinguished odor (Wardencki et al., 2009). Further, these molecules do not contribute equally to the overall flavor profile of a sample, thus a large peak in GC, generated by a chemical detector, does not necessarily correspond to high odor intensities, due to differences in intensity/concentration relationships (Lehotay and Hajslova, 2002). Accordingly, the general interest of researchers was directed to the determination of the contribution of single constituents to the overall flavor of a product. Moreover, the unpredictable extent of interaction of flavor molecules with each other, and with other food components (carbohydrates lipids, protein, etc.) must also to be considered (Azarnia et al., 2012). GC is the most appropriate analytical solution to such issues, as it enables the assessment of odor-active components in complex mixtures, through the precise correlation with the chromatographic peaks of interest; this is possible because the eluted substances are perceived simultaneously by two detectors, one of them being the human olfactory system. As a result, GC–O provides not only an instrumental, but also a sensorial analysis. The GC system was equipped with a nondestructive thermal conductivity detection (TCD) system with the outlet connected to a sniffing port (also called olfactometry port or transfer line). The sniffing port remains positioned inside a telephone booth; the purpose behind it is the isolation of the evaluator from the potential influences of odorants present in the environs. In 1971, a sophisticated GC–O system was reported in which humid air was added to the GC effluent, for the purpose of avoiding nasal mucosa dry-out. Further developments incorporated the use of a Venturitube, to maintain capillary column resolution and to deliver, ergonomically, the effluent to the evaluator. Over the following years, the sniffing ports began to incorporate design features and, nowadays, well-planned options are available on the market (Zellner et al., 2008). The introduction of GC–O proved to be pivotal for the development in the research field of odor-active compounds, providing valuable information on the chromatogram locations on which to focus attention and resources. GC–O is a unique analytical technique which associates the resolution power of capillary GC with the selectivity and sensitivity of the human nose.

4.2 Infrared Spectroscopy

In IR spectroscopy vibration of chemical bonds present in the organic matrix of food products are recorded and resolved at specific frequencies, and determined by the mass of the constituent atoms, the molecular shape, stiffness of the bonds, and the periods of the associated vibrational coupling (Smyth and Cozzolino, 2012). The infrared (IR) spectral region shows the absorption of specific vibrational bond where diatomic molecules have only a single bond that may stretch (eg, the distance between two atoms may increase or decrease). More complex molecules have many bonds; vibrations can also be conjugated, leading to two possible modes of vibration: stretching and bending. Despite these potential problems, absorption frequencies may be used to identify specific chemical groups, and this capability has traditionally been the main role of Fourier transform (FT) mid-infrared (MIR) (FT-MIR) spectroscopy, 40–43. The MIR region of the electromagnetic spectrum lies between 4000 and 400 cm^{-1} and can be segmented into four broad regions: the X–H stretching region (4000–2500 cm^{-1}), the triple-bond region (2500–2000 cm^{-1}), the double-bond region (2000–1500 cm^{-1}), and the fingerprint region (1500–400 cm^{-1}).Characteristic absorption bands are associated with major components of food. Water is a significant absorber in the MIR spectral region and can interfere with determination of other components present in beverages. Absorptions in the fingerprint region are mainly caused by bending and skeletal vibrations, which are predominantly sensitive to large wavenumber shifts, thereby minimizing against unambiguous identification of specific functional groups. Even in this region, however, the spectrum may be used as a fingerprint of a sample such as a food product or food components. This fingerprint determination of such forms is the basis of many applications of MIR spectroscopy in food analysis. Broader fields of application include constituent quantification and qualification issues for food and food ingredients and substance identification.

4.3 E-Nose

An electronic nose with an array of polymer–carbon black composite sensors (32 in all) has been developed and tested (Sujatha et al., 2012). Polymers with a variety of chemical functionalities are used to make sensors and it is possible to create an array which is able to identify and quantify a broad range of target compounds, such as alcohols and aromatics, and distinguish isomers and enantiomers. A model of polymer–carbon black composite sensors is under development. An interaction between target molecules and sensors enables identification of compounds and aids in selecting the array members. The objective of development of electronic nose is to mimic human olfaction that works on nonseparative mechanism: that is, an aroma or flavor is considered as a global fingerprint. Electronic noses include three major parts: a sample delivery device, a detection system and a computing device (Baldwin et al., 2011). The sample delivery device is responsible to generate volatile compounds of a sample (headspace), which is analyzed by fraction. It is important to maintain constant operating conditions in sample delivery system. This headspace is then injected into the detection system of the electronic nose. The detection system is an essential and reactive part of the instrument, which comprises a sensor set. It works by experiencing a change in electrical properties when the sensors react by coming in contact with volatile compounds. Each of the 32 sensors has specific sensitivity to all volatile molecules in their own way. The adsorption of volatile compounds on the sensor surface causes a physical change in the sensor (Kalit et al., 2014). An electronic device transforms the signal into a digital value. Recorded data are then computed by the help of statistical models. The most commonly used sensors include metal oxide semiconductors (MOS), conducting polymers (CP), quartz crystal microbalance, field effect transistors (MOSFET) and surface acoustic waves (SAW). In the current scenario, advanced types of electronic noses have been developed that incorporate mass spectrometry or gas chromatography as a detection system. The computing system is programmed to combine the responses of all of the sensors, which denotes the input for the data treatment.

4.4 E-Tongue

The term "electronic tongue" (e-tongue) refers to an array of sensors that are immersed in liquids, in order to identify their different physical–chemical characteristics, for example, "tastes." The e-tongue can be used in many sectors and it has widespread application in food and beverage industries and any other trades to monitor the quality of products. In beverage industries, for instance, analysis for taste evaluation is carried out by human tasters; thus they can be assisted by the e-tongue allowing continuous and precise measurements (Ramamoorthy et al., 2014). The advantage of the e-tongue is that there is no decrease of sensitivity during a long period of exposition, which does not occur with human being. As pointed out, one can notice the importance of a system like the e-tongue since it was developed to be portable and compact, which allows the performance of measurements at various places. The e-tongue prevents the exposure of human beings to toxic substances or to awkward tastes. The e-tongue system uses an array of sensors made of ultrathin films of polymers such as Langmuir–Blodgett (LB) films of 16-mer polyaniline, polypyrrole (PPy), stearic acid (SA), and composite films of several polymers. Such films were deposited on top of a glass substrate that holds interdigitated microelectrodes. Sensors prepared from different materials produce different electric responses and the variation is desired since it allows a "fingerprint" of the samples. The taste-sensing system SA402B and the ASTREE e-tongue are the two electronic sensory systems which are commercially available (Gouma and Sberveglieri, 2007). Both investigate liquid samples by measuring changes in electronic potential but the underlying sensor technologies are different. ASTREE uses chemical field effect transistor technology whereas the taste-sensing system SA402B is equipped with lipid membrane sensors. Moreover other taste sensing systems are also under development, as for example a voltammetric e-tongue. Up to date, several studies have been conducted using the e-tongue for sensory analysis of aroma and flavors in food products (Sliwinska et al., 2014).

5 Safety and Risk Assessment of Nanotechnology and Nanofoods

There are so many applications and tremendous benefits of nanotechnology in nanofood development that it can never be overemphasized (Ijabadeniyi, 2012). However, it is important that safety becomes the key word when trying to implement it in the food industry (Pratt et al., 2008). Regulatory bodies should provide a necessary framework to the manufacturers of nanofood products for proper risk assessment (Food Standards Australia New Zealand, 2013). Studies relevant to oral exposure risk assessment are required for particles to be used in food. Governments should set down regulations and appropriate labeling that will help to increase consumer acceptability. Furthermore, it has been recommended that nanomaterials should be used in the food industry only after they have been proven following vigorous testing. There is also the need for more research into the toxicological impact and possible hazard of food nanoparticles to human health and environment (Martirosyan and Schneider, 2014). According to current federal law and regulations, any substance that is "generally recognized as safe" (GRAS) for a particular use may be used in food for that purpose without premarket approval from FDA. General recognition of safety (ie, a GRAS determination) must reflect the views of experts qualified by scientific training and experience to evaluate the safety of substances directly or indirectly added to food. Those expert views must be based on "scientific procedures," supplemented in the case of substances used in food prior to 1958 by experience based on common use in food. According to the Code of Federal Regulations (CFR), the term "safe" means that there is reasonable certainty in the minds of competent scientists that a substance is not harmful under intended conditions of use. The challenge is determining what types of testing and data are necessary for determining this. The FDA has yet to issue formal guidance for nanotechnology in food, and Tarantino encouraged sponsors who are considering developing nanomaterials-based products to engage in early and frequent consultation with the agency. Not only would early consultation benefit manufacturers, by providing them with an indication of what types of testing and data would be required for approval, it would also provide the FDA with information that could be helpful as it develops the necessary guidance.

6 Conclusions

The challenges associated with functional nonfood product development is the efficient encapsulation of high-added-value ingredients, such as polyunsaturated fatty acids, flavors, vitamins, and other ingredients because their volatile permeability is used to improve functionality. Due to the favorable properties of these ingredients, numerous developments have been made in the field of encapsulated food flavors. The choice of a suitable technique of encapsulation depends upon the properties of the ingredients to be encapsulated, which may affect both their stability within the food, as well as altering the properties of the biomolecules in the food. The important properties to be considered after development of food products are stability required during storage and processing, the properties of the food components, the specific release properties required, the maximum obtainable flavor load in the powder, and the production cost. Nanoparticles in food will be exposed to a range of storage and use conditions, and this is an aspect that may need consideration in relation to assessment of the impact of nanotechnology in the food sector. For example, nanoparticles may be used in products that should be stored at low temperatures and then heated for consumption. Nanoscale food components can be encapsulated and mixed with other foods in novel combinations. Foods can be enriched with fruits and vegetables through nanotechnology to deliver higher nutrient density in such foods. Similarly one can make milk taste like cola beverage so that youngsters will have less inhibition in consuming nutritious milk. This technology also seems to be useful in dissolving additives like vitamins, minerals, antioxidants, phytochemicals, and nutritious oils which are not normally soluble. In order for nanotechnology to be used to its full potential, it must be accepted by consumers. Nowadays, the potential risks of nanomaterials to human health and the environment are poorly understood. Thus, further studies on the applications of nanotechnology in food processing and packaging, toxicity, and analysis of risk and benefit are needed to address the lack of knowledge, sustaining the growth of nanotechnology in the food industry and packaging and avoiding any danger to consumer health. The purpose and promise of food nanoscience and technology will be central to providing and improving food safety and security, and more generally, to delivering better overall health and wellbeing.

References

Adley, C.C., 2014. Past, present and future of sensors in food production. Foods 3, 491–510.

Ali, M.A., Rehman, I., Iqbal, A., Din, S., Rao, A.Q., Latif, A., Samiullah, T.R., Azam, S., Husnain, T., 2014. Nanotechnology, a new frontier in agriculture. Int. J. Adv. Life Sci. 1, 129–138.

Anandharamakrishnan, C., 2014a. Techniques for Nanoencapsulation of Food Ingredients. In: Anandharamakrishnan, C. (Ed.), Liquid-based nanoencapsulation techniques. Springer briefs in food, health, and nutrition. Springer, New York, NY, pp. 29–41.

Anandharamakrishnan, C., 2014b. Techniques for Nanoencapsulation of Food Ingredients. In: Anandharamakrishnan, C. (Ed.), Electrospraying and electrospinning techniques for nanoencapsulation. Springer briefs in food, health, and nutrition. Springer, New York, NY, pp. 43–49.

Apintanapong, M., Noomhorm, A., 2003. The use of spray drying to microencapsulate 2-acetyl-1-pyrroline, a major flavor component of aromatic rice. Int. J. Food Sci. Technol. 38, 95–102.

Ashraf, M.A., Khan, A.M., Ahmad, M., Sarfraz, M., 2015. Effectiveness of silica based sol–gel microencapsulation method for odorants and flavors leading to sustainable environment. Front. Chem. 3, 1–15.

Azarnia, S., Boye, J.I., Warkentin, T., Malcolmson, L., 2012. Application of gas chromatography in the analysis of flavour compounds in field peas. Gas Chromatography in Plant Science, Wine Technology, Toxicology and Some Specific Applications, InTech, Croatia, Europe, pp. 1–17.

Balassa, L.L., Tomahawk, L.T., Grove, B., 1970. United States Patent 3,495,988. Encapsulation of aromas and flavors, pp. 1–16.

Baldwin, E.A., Bai, J., Plotto, A., Dea, S., 2011. Electronic noses and tongues: applications for the food and pharmaceutical industries. Sensors 11, 4744–4766.

Baranauskienė, R., Bylaitė, E., Zukauskaitė, J., Venskutonis, R.P., 2007. Flavor retention of peppermint (*Mentha piperita* L.) essential oil spray-dried in modified starches during encapsulation and storage. J. Agric. Food Chem. 55, 3027–3036.

Barbosa, J., Borges, S., Amorim, M., Pereira, M.J., Oliveira, A., Pintado, M.E., Teixeira, P., 2015. Comparison of spray drying, freeze drying, and convective hot-air drying for the production of a probiotic orange powder. J. Funct. Foods 17, 340–351.
Baser, K.H.C., Buchbauer, G., 2010. Sources of essential oils. In: Franz, C., Novak, J. (Eds.), Handbook of Essential Oils: Science, Technology, and Applications. CRC Press, New York, NY, p. 39.
Bernardes, P.C., Andrade, N.J., Soares, N.F.F., 2014. Nanotechnology in the food industry. Biosci. J. 30, 1919–1932.
Bilia, A.R., Guccione, C., Isacchi, B., Righeschi, C., Firenzuoli, F., Bergonzi, M.C., 2014. Essential oils loaded in nanosystems: a developing strategy for a successful therapeutic approach. Evid. Based Complement. Alternat. Med. 2014, 1–14.
Bolger, M., Toledo de Figueiredo, M.C., DiNovi, M., Kawamura,Y., et al., 2011. Evaluation of certain food additives and contaminants. Seventy-Fourth Report of the Joint FAO/WHO Expert Committee on Food Additives, pp. 7–54.
Bradley, E.L., Castle, L., Chaudhry, Q., 2011. Applications of nanomaterials in food packaging with a consideration of opportunities for developing countries. Trends Food Sci. Tech. 22, 604–610.
Bryksa, B.C., Yada, R.Y., 2012a. Nanotechnology: the word is new but the concept is old. An overview of the science and technology in food and food products at the nanoscale level. Int. J. Food Stud. 1, 188–210.
Bryksa, B.C., Yada, R.Y., 2012b. Challenges in food nanoscale science and technology. J. Food Drug Anal. 20, 418–421.
Bylaite, E., Venskutonis, P., Maþdþieriene, R., 2001. Properties of caraway (*Carum carvi* L.) essential oil encapsulated into milk protein-based matrices. Eur. Food Res. Technol. 212, 661–670.
Cabuk, U., Okuklu, B., Stanciuc, N., Harsa, S.T., 2014. Nanoencapsulation of biologically active peptides from whey proteins. J. Nutr. Health Food Sci. 2, 1–4.
Cedervall, T., Lynch, I., Foy, M., Berggard, T., Donnelly, S.C., Cagney, G., Linse, S., Dawson, K.A., 2007. Detailed identification of plasma proteins adsorbed on copolymer nanoparticles. Angew. Chem. Int. Ed. Engl. 46, 5754–5756.
Chaudhry, Q., Castle, L., Watkins, R. (Eds.), 2010. Nanotechnologies in Food. Royal. Society of Chemistry Publishers, Cambridge, UK.
Chaudhry, Q., Scotter, M., Blackburn, J., Ross, B., Boxall, A., Castle, L., Aitken, R., Watkins, R., 2008. Applications and implications of nanotechnologies for the food sector. Food Addit. Contam. 25 (3), 241–258.
Chellarama, C., Murugaboopathib, G., Johna, A.A., Sivakumarc, R., Ganesand, S., Krithikae, S., Priyae, G., 2014. Significance of nanotechnology in the food industry. APCBEE Procedia 8, 109–113.
Chen, L. Y., Remondetto, G. E., Subirade, M., 2006. Food protein based materials as nutraceutical delivery systems. Trends Food Sci. Technol. 17, 272–283.
Chinnamuthu, C.R., Boopathi, P.M., 2009. Nanotechnology and agroecosystem. Madras Agric. J. 96 (1–6), 17–31.
Christelle, T., Elisabeth, C.M., Pierre, G., Elisabeth, D., 2013. Aroma Encapsulation in Powder by Spray Drying, and Fluid Bed Agglomeration and Coating. Food Engineering and SeriesSpringer, New York, NY, pp. 1–6.
Cientifica, 2006. Nanotechnologies in the Food Industry, published August 2006. http://www.cientifica.com/www/details.php?id=47
Ciriminna, R., Pagliaro, M., 2013. Sol–gel microencapsulation of odorants and flavors: opening the route to sustainable fragrances and aromas. Chem. Soc. Rev. 42, 9243–9250.
Clegg, S.M., Knight, A.I., Beeren, C.J.M., Wilde, P.J., 2009. Fat reduction whilst maintaining the sensory characteristics of fat using multiple emulsions. Fifth International Symposium on Food Rheology and Structure, pp. 238–241.

Contis, E.T., Ho, C.T., Mussinan, C.J., Parliament, T.H., Shahidi, F., Spanier, A.M., 1997. Food Flavors: Formation, Analysis and Packaging Influences. Developments in Food Science, vol. 40. Elsevier, Amsterdam, pp. 27–780.
des Rieux, A., Fievez, V., Garinot, M., Schneider, Y.J., Preat, V., 2006. Nanoparticles as potential oral delivery systems of proteins and vaccines: a mechanistic approach. J. Control. Release 116, 1–27.
Desai, M.P., Labhasetwar, V., Amidon, G.L., Levy, R.J., 1996. Gastrointestinal uptake of biodegradeablemicroparticles: effect of particle size. Pharm. Res. 13, 1838–1845.
Dima, C., Cotarlet, M., Tiberius, B., Bahrim, G., Alexe, P., Dima, S., 2014. Encapsulation of coriander essential oil in beta-cyclodextrin: antioxidant and antimicrobial properties evaluation. Rom. Biotechnol. Lett. 19, 9128–9140.
Duncan, T.V., 2011. Applications of nanotechnology in food packaging and food safety: barrier materials, antimicrobials and sensors. J. Colloid Interface Sci. 363, 1–24.
Fabra, M.J., Chambin, O., et al.,2012. Influence of temperature and NaCl on the release in aqueous liquid media of aroma compounds encapsulated in edible films. J. Food Eng. 108 (1), 30–36.
Fahlbusch, K., Hammerschmidt, F., Panten, J., Pickenhagen, W., Schatkowski, D., 2003. Flavors and fragrances. Ullmann's Encyclopedia of Industrial ChemistryWiley-VCH, Weinheim.
Fathi, M., Mozafari, M.R., Mohebbi, M., 2012. Nanoencapsulation of food ingredients using lipid based delivery systems. Trends Food Sci. Technol. 23, 13–27.
Fathi, M., Martın, A., McClements, D.J., 2014. Nanoencapsulation of food ingredients using carbohydrate-based delivery systems. Trends Food Sci. Technol. 39, 18–39.
Feng, T., Xiao, Z., Tian, H., 2009. Recent patents in flavor microencapsulation. Recent Pat. Food Nutr. Agric. 1, 193–202.
Fleming, A., 2015. Fake flavours: why artificial aromas can't compete with real food smells. Food & Drink. The Guardian.
Food Safety Authority of Ireland, 2008. The Relevance for Food Safety of Applications of Nanotechnology in the Food and Feed Industries, pp. 9–20.
Food Standards Australia New Zealand, 2013. Guidance on establishing food-health relationships for general level health claims, version 1.1.
Galmarini, M.V., Zamora, M.C., Baby, R., Chirife, J., Mesina, V., 2008. Aromatic profiles of spray-dried encapsulated orange flavours: influence of matrix composition on the aroma retention evaluated by sensory analysis and electronic nose techniques. Int. J. Food Sci. Technol. 43, 1569–1576.
Gane, S., Georganakis, D., Maniati, K., Vamvakias, M., Ragoussis, N., Skoulakis, E.M.C., Turin, L., 2013. Molecular vibration-sensing component in human olfaction. PLoS One 8, 1–7.
Garwood, R.E., Mandralis, Z.I., Scott, A.W., 1995. United States Patent US005,399,368A. Encapsulation of volatile aroma compounds.
Glindemann, D., Dietrich, A., Staerk, H., Kuschk, P., 2005. The two odors of iron when touched or pickled: (skin) carbonyl compounds and organophosphines. Angew. Chem. Int. Ed. 45, 7006–7009.
Gouma, P., Sberveglieri, G., 2007. Novel materials and applications of electronic noses and tongues. MRS Bull. 29, 697–702.
Huang, Y.P., Yu, H.L., Guo, L., Huang, Q.R., 2009. Preparation, characterization, and applications of octanoyl-chitosan-polyethylene glycol monomethyl ether amphiphile. The 238th ACS National Meeting, August 16–20, 2009, Washington, DC, USA. AGFD220.
Hui, Y.H., 2006. Handbook of Food Science, Technology, and Engineering, vol. 2, CRC Press, Bocá Raton, FL, 83–36.

Huyghebaert, A., Huffel, X.V., Houins, G., 2010. Nanotechnology in the Food Chain Opportunities & Risks. International Symposium, pp. 1–139.
Ijabadeniyi, O.A., 2012. Safety of nanofood: a review. Afr. J. Biotechnol. 11 (87), 15258–15263.
Ishwaryaa, S.P., Anandharamakrishnana, C., Stapley, A.G.F., 2015. Spray-freeze-drying: a novel process for the drying of foods and bioproducts. Trends Food Sci. Technol. 41, 161–181.
Jeroen, J.G., Soest, V., 2007. Encapsulation of fragrances and flavours: a way to control odor and aroma in consumer products. Flavour Frag. J., 439–455.
Joshi, J.R., Patel, R.P., 2012. Role of biodegradable polymers in drug delivery. Int. J. Current Pharm. Res. 4, 74–81.
Kah, M., Hofmann, T., 2014. Nanopesticide research: current trends and future priorities. Environ. Int. 63, 224–235.
Kalit, M.T., Marković, K., Kalit, S., Vahčić, N., Havranek, J., 2014. Application of electronic nose and electronic tongue in the dairy industry. Mljekarstvo 64, 228–244.
Kokini, J.L., 2010. Advances in Nanotechnology as Applied to Food Systems. University of Illinois, Urbana-Champaign, IL.
Lehotay, S.J., Hajslova, J., 2002. Application of gas chromatography in food analysis. Trends Anal. Chem. 21, 686–697.
Lin, D.Y., Zhang, S.Z., Block, E., Katz, L.C., 2005. Encoding social signals in the mouse main olfactory bulb. Nature 434, 470–477.
Luecha, J., Sozer, N., Kokini, J.L., 2010. Synthesis and properties of corn zein/montmorillonite nanocomposite films. J. Mater. Sci. 45 (13), 3529–3537.
Luykx, D.M., Peters, R.J., Van Ruth, S.M., Bouwmeester, H., 2008. A review of analytical methods for the identification and characterization of nano delivery systems in food chemistry. J. Agric. Food Chem. 56, 8231–8247.
Madene, A., Jacquot, M., Scher, J., Desobry, S., 2006. Flavour encapsulation and controlled release—a review. Int. J. Food Sci. Technol. 41, 1–21.
Mahdi, J.S., Elham, A., et al., 2008. Encapsulation efficiency of food flavors and oils during spray drying. Drying Technol. 26, 816–835.
Marcuzzo, E., Sensidoni, A., et al., 2010. Encapsulation of aroma compounds in biopolymeric emulsion based edible films to control flavor release. Carbohydr. Polym. 80 (3), 984–988.
Martín, A., Varona, S., Navarrete, A., Cocero, M.J., 2010. Encapsulation and co-precipitation processes with supercritical fluids: applications with essential oils. Open Chem. Eng. J. 4, 31–41.
Martirosyan, A., Schneider, Y.J., 2014. Engineered nanomaterials in food: implications for food safety and consumer health. Int. J. Environ. Res. Public Health 11, 5720–5750.
Matalanis, A., Jones, O.G., et al., 2011. Structured biopolymer-based delivery systems for encapsulation, protection, and release of lipophilic compounds. Food Hydrocoll. 25 (8), 1865–1880.
Matua, M.A., 2014. Identifying Food Additives. Ministry for Primary Industries, pp. 1–10.
Milanovic, J., Manojlovic, V., Levic, S., Rajic, N., Nedovic, V., Bugarski, B., 2010. Microencapsulation of flavors in carnauba wax. Sensors 10, 901–912.
Miller, G., Senjen, R., 2008. Out of the laboratory and onto our plates. Nanotechnol. Food Agric., 2–58.
Monedero, F.M., Hambleton, A., et al., 2010. Study of the retention and release of *n*-hexanal incorporated into soy protein isolate–lipid composite films. J. Food Eng. 100 (1), 133–138.
Mozafari, M.R., John Flanagan, J., Merino, L.M., Awati, A., Omri, A., Suntres, Z.E., Singh, H., 2006. Recent trends in the lipid-based nanoencapsulation of antioxidants and their role in foods. J. Sci. Food Agric. 86, 1–8.

Munin, A., Edwards-Levy, F., 2011. Encapsulation of natural polyphenolic compounds: A review. Pharmaceutics 3, 793–829.

Nagavarma, B.V.N., Yadav, H.K.S., Ayaz, A., Vasudha, L.S., Shivakumar, H.G., 2012. Different techniques for preparation of polymeric nanoparticles—a review. Asian J. Pharm. Clin. Res. 5, 16–23.

Naknean, P., Meenune, M., 2010. Review article: factors affecting retention and release of flavor compounds in food carbohydrates. Int. Food Res. J. 17, 23–34.

Nedovic, V., 2010. Encapsulation technologies for active food ingredients and food processing. Food Sci. Nutr., 3–160.

Nedovic, V., Kalusevica, A., Manojlovicb, V., Levica, S., Bugarski, B., 2011. An overview of encapsulation technologies for food applications. Procedia Food Sci. 1, 1806–1815.

Nicolescu, C.L., 2007. Application of electronic noses in food analysis. Fascicle VIII, pp. 21–26.

Nollet, L.M.L., Toldra, F., 2010. Sensory Analysis of Foods of Animal Origin. CRC Press, Boca Raton, FL, 389–406.

Nur Ain, A.H., Farah Diyana, M.H., Zaibunnisa, A.H., 2011. Encapsulation of lemongrass (Cymbopogon citratus) oleoresin with β-cyclodextrin: phase solubility study and its characterization. Second International Conference on Biotechnology and Food Science, IPCBEE vol.7, pp. 44–48.

Okuro, P.K., Matos Jr., F.E., Favaro-Trindade, C.S., 2013. Technological challenges for spray chilling encapsulation offunctional food ingredients. Food Technol. Biotech. 51, 171–182.

Otles, S., Yalcin, B., 2012. Review on the application of nanobiosensors. Acta Sci. Pol. Technol. Aliment. 11 (1), 7–18.

Partanen, R., Ahro, M., Hakala, M., Kallio, H., Forssell, P., 2002a. Microencapsulation of caraway extract in β-cyclodextrin and modified starches. Eur. Food Res. Technol. 214, 242–247.

Partanen, R., Yoshii, H., Kallio, H., Yang, B., Forssell, P., 2002b. Encapsulation of sea buckthorn kernel oil in modified starches. J. Am. Oil Chem. Soc. 79, 219–223.

Poncelet, D., Picot, A., Mafadi, S.E., 2011. Encapsulation: an essential technology for functional food applications. Innov. Food Technol., 32–33.

Pratt, D., Adley, C., Anderson, W., Chambers, G., Davoren, M., Dawson, K., Evans, R., Fenelon, M., Harty, T., Hyde, L., Lawrie, S., Lynch, I., Mahony, P., Radmoski, M., Reilly, A., Ryan, M., 2008. The Relevance for Food Safety of Applications of Nanotechnology in the Food and Feed Industries. Food Safety Authority of Ireland, pp. 1–88.

Qureshi, A.M., Karthikeyan, S., Karthikeyan, P., Khan, P.A., Uprit, S., Mishra, U.K., 2012. Application of nanotechnology in food and dairy processing: an overview. Pak. J. Food Sci. 22, 23–31.

Ramamoorthy, H.V., Mohamed, S.N., Devi, D.S., 2014. E-nose and e-tongue: applications and advances in sensor technology. J. Nanosci. Nanotechnol. 2, 2279–2381.

Ranjan, S., Dasgupta, N., Chakraborty, A.R., Samuel, S.M., Ramalingam, C., Shanker, R., Kumar, A., 2014. Nanoscience and nanotechnologies in food industries: opportunities and research trends. J. Nanopart. Res. 16, 24–64.

Raoa, J.P., Geckeler, K.E., 2011. Polymer nanoparticles: preparation techniques and size-control parameters. Prog. Polym. Sci. 36, 887–913.

Sailaja, K.A., Amareshwar, P., Chakravarty, P., 2011. Different techniques used for the preparation of nanoparticles using natural polymers and their application. Int. J. Pharm. Pharm. Sci. 3, 45–50.

Sankarankutty, K.M., 2014. Biosensors and their applications for ensuring food safety. Global J. Pathol. Microbiol. 2, 15–21.

Sekhon, B.S., 2010. Food nanotechnology—an overview. Nanotechnol. Sci. Appl. 3, 1–15.

Sharifi, A., Golestan, L., Baei, M.S., 2013. Studying the enrichment of ice cream with alginate nanoparticles including Fe and Zn salts. J. Nanopart. Res. 2013, 1–5.
Shiga, H., Yoshii, H., Ohe, H., Yasuda, M., Furuta, T., Kuwahara, H., Ohkawara, M., Linko, P., 2004. Encapsulation of shiitake (Lenthinus edodes) flavors by spray drying. Biosci. Biotechnol. Biochem. 68, 66–71.
Shimoni, E., 2000. Nanoencapsulation of food ingredients: from macromolecular nanostructuring to smart delivery systems. Laboratory of Functional Foods, Nutraceuticals, and Food Nanosciences, vol. 43, pp. 317–326.
Sliwinska, M., Wisniewska, P., Dymerski, T., Namiesnik, J., Wardenck, W., 2014. Food analysis using artificial senses. J. Agric. Food Chem. 62, 1423–1448.
Smith, R.L., Cohen, S.M., Doull, J., Feron, V.J., Goodman, J.I., Marnett, L.J., Portoghese, P.S., Waddell, W.J., Wagner, B.M., Adam, T.B., 2003. GRAS flavoring substances 22. Food Technol. 59, 24–62.
Smith, R.L., Cohenb, S.M., Doullc, J., Ferond, V.J., Goodmane, J.I., 2005. A procedure for the safety evaluation of natural flavor complexes used as ingredients in food: essential oils. Food Chem. Toxicol. 43, 345–363.
Smyth, H., Cozzolino, D., 2012. Instrumental methods (spectroscopy, electronic nose, and tongue) as tools to predict taste and aroma in beverages: advantages and limitations. Chem. Rev. 113, 1429–1440.
Somasundaran, P., Chakraborty, S., Deo, N., Somasundaran, T., 2006. Nanoencapsulation for extraction and release of fragrance. Cosmetics and Toiletries Magazine, vol. 121, pp. 47–54.
Soottitantawat, A., Takayama, K., Okamura, K., Muranaka, D., Yoshii, H., Furuta, T., Ohkawara, M., Linko, P., 2005. Microencapsulation of l-menthol by spray drying and its release characteristics. Innov. Food Sci. Emerg. Technol. 6, 163–170.
Soottitantawat, A., Partanen, R., Neoh, T.L., Yoshii, H., 2015. Encapsulation of hydrophilic and hydrophobic flavors by spray drying. Japan J. Food Eng. 16, 37–52.
Sujatha, G., Dhivya, N., Ayyadurai, K., Thyagarajan, D., 2012. Advances in electronic-nose technologies. Int. J. Eng. Res. Appl. 2, 1541–1546.
Teixeira, M.I., Andrade, L.R., Farina, M., Rocha-Leao, M.H.M., 2004. Characterization of short chain fatty acid microcapsules produced by spray drying. Mater. Sci. Eng. C. 24, 653–658.
Tiwari, A., Tiwari, A., 2014. Bioceramic nanomaterials in medical application. In: Kanwar, J.R., Mahidhara, G., Roy, K., Kanwar, R.K. (Eds.), Bioengineered Nanomaterials. CRC Press, Boca Raton, FL, pp. 163–176.
Vos, P.D., Faas, M.M., Spasojevic, M., Sikkema, J., 2010. Encapsulation for preservation of functionality and targeted delivery of bioactive food components. Int. Dairy J. 20, 292–302.
Vroman, I., Tighzert, L., 2009. Biodegradable polymers. Materials 2, 307–344.
Wandrey, C., Bartkowiak, A., Harding, S.E., 2010. Materials for encapsulation. In: Zuidam, N.J., Nedovic, V.A. (Eds.), Encapsulation Technologies for Active Food Ingredients and Food Processing. Springer Science Business Media, New York, NY, pp. 1–100.
Wang, Y., 2011. Improving health-promoting effects and stability of food ingredients using nanoencapsulation. Food Hydrocoll. 25, 1327–1336.
Wardencki, W., Chmiel, T., Dymerski, T., BIernacka, P., Plutowska, B., 2009. Application of gas chromatography, mass spectrometry and olfactometry for quality assessment of selected food products. Ecol. Chem. Eng. 16, 288–298.
Xiao, D., Zhong, Q., 2011. In vitro release kinetics of nisin as affected by Tween 20 and glycerol co-encapsulated in spray-dried zein capsules. J. Food Eng. 106 (1), 65–73.
Yadav, V., Sharma, A., Singh, S.K., 2015. Microencapsulation techniques applicable to food flavours research and development: a comprehensive review. Int. J. Food Nutr. Sci. 4, 119–124.

Yurdugul, S., Mozafari, M.R., 2004. Recent advance in micro- and nanoencapsulation of food ingredients. Cell. Mol. Biol. Lett. 9, 64–65.

Yuri, M., Hironori, S., Lena, A., Ikuo, H., 2010. Expiration: the moment we experience retronasal olfaction in flavor. Neurosci. Lett. 473, 92–96.

Zellner, B.A., Dugo, P., Dugo, G., Mondello, L., 2008. Gas chromatography–olfactometry in food flavor analysis. J. Chromatogr. A 1186, 123–143.

Zuidam, N.J., Heinrich, E., 2010. Encapsulation of aroma. Encapsulation Technologies for Active Food Ingredients and Food ProcessingSpringer Science Business Media, New York, NY, 127-160.

Zuidam, N.J., Nedovic, V.A., 2010. Overview of microencapsulates for use in food products or processes and methods to make them. In: Zuidam, N.J., Shimoni, E. (Eds.), Encapsulation Technologies for Active Food Ingredients and Food Processing. Springer Science Business Media, New York, NY, pp. 3–28.

Internet Resources

https://en.wikipedia.org/wiki/Aroma_compound

https://www.google.co.in/search?rlz(1C1CHZL_en-GBIN655IN655&biw=1517&bih=741&noj=1&tbs=sur%3Afc&tbm=isch&sa=1&q=cyclodextrin&oq=cyclodextrin&gs_l=img.12.0l10.35903.35903.0.38071.1.1.0.0.0.0.167.167.0j1.1.0..0..1c.1.64.img.0.1.167.DZ-PB_UaXgY#imgrc=6JaoCUYKU1FUxM%3A

https://www.google.co.in/search?q=nanoemulsion&noj=1&tbs=sur:fc&tbm=isch&tbas=0&source=lnt&sa=X&ved=0ahUKEwiyxa77-87JAhWFt44KHbizBW8QpwUIFA&dpr=0.9&biw=1517&bih=741#imgrc=6Q3k17Plyp11oM%3A

https://www.google.co.in/search?q=nanoemulsions&biw=1517&bih=741&source=lnms&tbm=isch&sa=X&sqi=2&ved=0ahUKEwjBj8zq887JAhWCuo4KHYJlAjgQ_AUIBigB&dpr=0.9#q=liposome&tbs=sur:fc,itp:lineart&tbm=isch

https://www.google.co.in/search?q=lipid+capsules&biw=1517&bih=741&source=lnms&tbm=isch&sa=X&ved=0ahUKEwiH2J25_M7JAhWHCo4KHZhaCP8Q_AUIBigB&dpr=0.9#q=lipid+capsules&tbm=isch&tbs=sur:fc

https://www.google.co.in/search?q=spray+dryer&rlz=1C1CHZL_en-GBIN655IN655&espv=2&biw=1517&bih=741&source=lnms&tbm=isch&sa=X&ved=0ahUKEwiGocGUosDJAhVScI4KHZXbC4gQ_AUIBigB&dpr=0.9#q=spray+dryer&tbm=isch&tbs=sur:fc&imgrc=0rcGlKWwnKbsNM%3A

https://www.google.co.in/search?q=Supercritical+Fluid+Extraction+unit&rlz=1C1CHZL_en-GBIN655IN655&source=lnms&tbm=isch&sa=X&ved=0ahUKEwiMloPipcDJAhWGto4KHRchBUwQ_AUIBygB&biw=1517&bih=741&dpr=0.9#q=Supercritical+Fluid+Extraction+unit&tbm=isch&tbs=sur:fc&imgrc=ltSognyKQ3AArM%3A

3

NANOENCAPSULATION OF FLAVORS AND AROMAS BY EMERGING TECHNOLOGIES

Juan Felipe Osorio-Tobón, Eric Keven Silva, M. Angela A. Meireles
UNICAMP (University of Campinas), LASEFI/DEA/FEA (School of Food Engineering), Campinas, São Paulo, Brazil

1 Introduction

Modern consumers are increasingly aware of the relationship between their diet and maintaining a healthy way of life. Consumers are also aware about environmental issues related to the processes by which foods are produced. Therefore, these new consumers valorize products obtained from natural sources and if they have been produced using clean technologies.

In this context, the use of flavoring compounds such as essential oils (EOs) extracted from leaves, fruits, and seeds and preserved through emerging technologies of nanoencapsulation is an attractive investment that can bring major innovations in the food industry. Nanoencapsulation involves a set of techniques that allow the formation of particles/emulsions with functional properties, consisting of an encapsulating matrix (carbohydrate, protein, lipid, and others) and an active material (essential oil) distributed within these systems. Nanoencapsulation provides protection to the compounds that make up EOs (triglycerides, hydrocarbons, phenols, ether, and others) against adverse conditions that can promote their volatilization and oxidation while allowing the release of these compounds under controlled conditions of pH, temperature, and desired ambient.

The use of supercritical technologies for nanoencapsulation of EOs has aroused great interest due to the fact that these techniques are suitable for the processing of heat sensitive compounds. The minimum process temperature of these methods is directly linked to the critical conditions (temperature and pressure) of the supercritical fluid used in the process. One of the

Encapsulations. http://dx.doi.org/10.1016/B978-0-12-804307-3.00003-X

supercritical fluids most used in the nanoencapsulation of several types of compounds is carbon dioxide. Supercritical carbon dioxide is recognized by its moderate critical properties (T_c = 304.2 K and P_c = 7.38 MPa), which are relatively easily achieved and enable nanoencapsulation in conditions that do not contribute to the degradation of the bioactive compounds (Silva and Meireles, 2014).

The formation of nanoemulsions assisted by ultrasound is another emerging technology with great potential for EOs nanoencapsulation because it is an effective method in the reduction of the droplet size of the dispersed phase, and thus leads to greater protection of the active compounds present in EOs.

In this context, this chapter discusses the application of emerging technologies based on supercritical fluids and ultrasonication to form nanoparticles/nanoemulsions of EOs with applications as flavor and aroma agents in food products, besides to add value to these products and to promote innovation in food industry through the provision of flavorings considered safe when obtained using clean technologies.

2 Issues Relating to Addition of Flavors and Aromas in Foods

Pursuit of pleasure and well-being is a primitive instinct of human beings. In this context, foods need to be sensorially attractive to awaken the interest of the consumers. For example, flavor and aroma properties are parameters that determine the quality of foodstuffs, and in many respects these characteristics are prioritized over nutritional properties. On the other hand, the growing interest in replacing artificial additives by natural compounds has created the possibility to use compounds that, besides having pleasant characteristics of flavor and aroma, provide benefits such as the ability to prevent disease or to assist in food preservation through the addition of compounds that contribute other functional properties. Particularly, flavoring and aroma agents obtained from flowers, herbs, seeds, and spices have great potential for applications in the food industry and have acceptance by the market. The compounds responsible for flavor and aroma in these raw materials are contained in the EOs. Since ancient times, compounds of EOs extracted from citrus fruits, such as orange, lime, and lemon, have been used as flavoring agents.

As natural products, EOs have attractive functional properties, therefore interest in the research and use of EOs in various areas such as food and pharmaceutical industries has gained an important place in recent years. EOs are used in a variety of processes and products. According to Baser and Buchbauer (2009) products such as perfumes, cosmetics, toiletries, detergents, household chemicals, and related products have been perfumed with EOs. As a flavoring agent, the development of the soft drinks industry has been of great importance because it is a major consumer of EOs, especially those of citrus origin. Other food products such as ice creams, confectionery, bakery, chewing gum, and a variety of fast foods also commonly use EOs in their formulations. Nowadays innovative antimicrobial packaging is being developed in order to extend the shelf-life of food through the antimicrobial activity of the EOs. EOs from citrus, cinnamon, clove, ginger, anise, pepper, pimento, laurel, cardamom, ginger, basil, oregano, dill, and fennel are used in the production of the products mentioned above.

As a result of their functional properties and the wide quantity of products where EOs are used, a great panorama about potential EOs applications in food industry is opened; therefore this chapter will be limited to explore the nanoencapsulation of EOs.

2.1 Classification and Properties

Flavor can be defined as a set of sensory sensations derived from the contact with sensory receptors in the nose and other structures from tactile and taste receptors in the mouth. These sensory receptors are capable of transmitting to the brain the combination of numerous biochemical reactions that occur at the moment of consuming a food. Additionally, flavor perception is influenced by numerous other factors such as color, temperature, and texture, together with other psychological factors such as expectations and appetite. However, flavor is the result of the interaction of two main sensory properties, taste, and flavor. Taste is the result of various biochemical reactions that occur in the tongue, which are received and transmitted by receptors responsible for the taste perception. Five types of taste have been identified: salt, sweet, bitter, sour, and umami. Aroma is the result of perception by nose receptors of volatile chemical compounds that are not related to the nutritional value of food. Sensory receptors in the nose are capable to identify a larger number of compounds than those located in the tongue, especially those with low molecular weight (Cheetham, 2010).

A considerable amount of compounds have been identified as flavors and aromas. A practical way to differentiate between flavor and aroma compounds is by differences in volatility. For example, less volatile compounds contribute more to flavor than aroma.

Nevertheless, chemical structures of these molecules show high diversity and even flavor compounds with highly similar chemical structures may have an entirely different sensorial effect (Cserháti and Forgács, 2003).

EOs are complex mixtures of organic compounds, often composed of more than 100 different terpenic compounds. Fig. 3.1 presents the chemical structures of the main compounds that

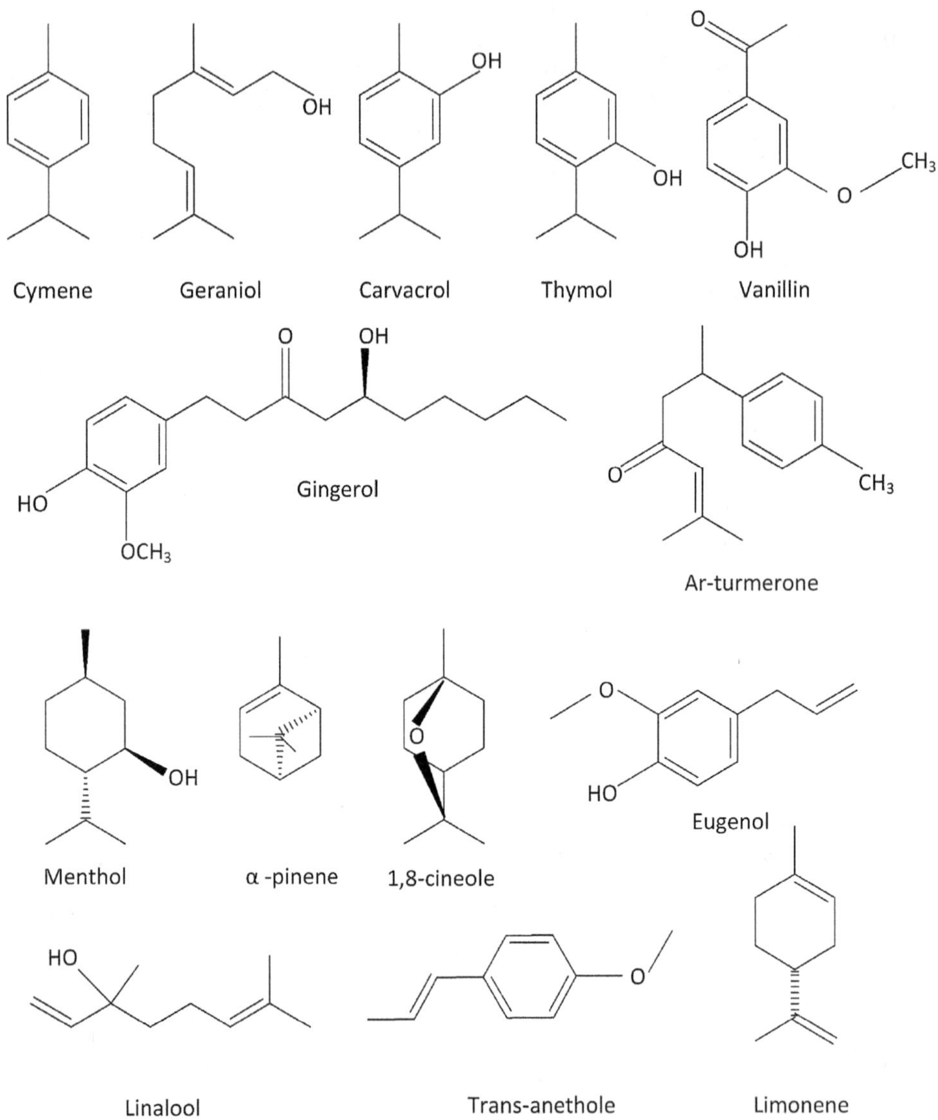

Figure 3.1. Major chemical constituents in EOs.

compose the EOs. In this class of compounds, a considerable diversity of chemical structures has been found; among them, hydrocarbons, alcohols, aldehydes, ketones, esters, acids, phenolic compounds, and heterocyclic compounds are part of the identified compounds. Although classification of these compounds according to their chemical structure is possible, this is irrelevant within the scope of this chapter.

EOs due to its origin are classified as a natural flavor. According to United States code of federal regulation (CRF, 2013), the terms *natural flavor* or *natural flavoring* mean the essential oil, oleoresin, essence, or extractive, protein hydrolysate, distillate, or any product of roasting, heating, or enzymolysis, which contains the flavoring constituents derived from a spice, fruit or fruit juice, vegetable or vegetable juice, edible yeast, herb, bark, bud, root, leaf, or similar plant material, meat, seafood, poultry, eggs, dairy products, or fermentation products thereof, whose significant function in food is flavoring rather than nutrition. On the other hand, the terms *artificial flavor* or *artificial flavoring* means any substance, the function of which is to impart flavor, that is not derived from any of the raw materials listed above. As previously stated, EOs are highlighted among natural flavors and aromas.

EOs usually exist in liquid form at room temperature and usually they are separated from the aqueous phase by a physical method that does not lead to significant change in its chemical composition. Due to their hydrophobic nature and the fact that their density is often lower than that of water, EOs are generally lipophilic, soluble in organic solvents, and immiscible in aqueous medium (Asbahani et al., 2015). Besides providing flavor and aroma, they have several biological properties that can be explored in the development of applications to food and pharmaceutical industries. EOs are recognized by their anticancer, antimicrobial, anti-inflammatory, antioxidant, antiviral, and antinociceptive activities (Baser and Buchbauer, 2009). One of the most important properties of the EOs is antimicrobial activity; this property can be explored in the food packaging industry in order to develop novel antimicrobial packaging. This type of packaging can maintain the nutritional and sensory quality of food while the shelf-life is extended. All plants have the ability to produce volatile compounds, however, particularly herbs and spices have been commonly used as raw materials for obtaining EOs. Table 3.1 presents some examples of herbs and spices most commonly used as a source of EOs and major flavoring compounds.

Due to its chemical characteristics, EOs are unstable and fragile volatile compounds. Consequently, its stability, solubility, and interactions with other food components are related with the

Table 3.1 Major Herbs and Spices: Sources of EOs

Herbs and Spices	Part of Plant Used	Major Flavoring Compounds	References
Cinnamon (*Cinnanamonzeylanicum*)	Leaves	Trans-cinnamaldehyde, 3-methoxy-1,2-propanediol, o-methoxy-cinnamaldehyde, coumarin, and benzeneethanol	(Wang et al., 2009)
Vanilla (*V. fragans*)	Beans of the orchid	Vanillin and *p*-hydroxibenzaldehyde	(Longares-Patrón e; Cañizares-Macías, 2006)
Ginger (*Zingiber officinale* Roscoe)	Rhizomes	Gingerol, zingerone, and shogaol	(Mesomo et al., 2013)
Turmeric (*Curcuma Longa*)	Rhizomes	Ar-turmerone and turmerone	(Ravindran et al., 2007)
Oregano (*Origanum vulgare*)	Flowers and leaves	Carvacrol, thymol, limonene, pinene, ocimene, and caryophyllene	(Almeida et al., 2013)
Mints (*Mentha spicata Huds*)	Leaves	Menthone, menthofuran, and menthol	(Costa et al., 2014)
Rosmary (*Rosmarinus officinalis*)	Leaves	Alpha-pinene, 1,8-cineole, camphor, borneol, trans-caryophyllene, carnosic, and rosmarinic acids, carnosol and rosmanol	(Taylor and Linforth, 2009)
Cardamom (*Elettaria cardamomum*)	Seeds	1,8-cineole, α-pinene, β-pinene, sabinene, myrcene, and α-phellandrene	(Parthasarathy et al., 2008)
Clove (*Syzygium aromaticum* L.)	Dried unopened flower buds	Eugenol, eugenyl acetate, and b-caryophyllene	
Coriander (*Coriandrum sativum*)	Seeds	Linalool, limonene, camphor, and geraniol	(Pavlić et al., 2015)
Pepper black and white	Seeds	α-pinene, β-pinene, 1-α-phellandrene, dllimonene, piperonal, and dihydrocarveol	(Ferreira et al., 1999)
Tea (*Camellia sinensis*)	Leaves	Linalool, genariol, α-damascone, linalool oxide, cis-jasmone, maltol, anethole, α-terpineol, nerolidol	(Pripdeevech and Wongpornchai, 2013)
Aniseed (*Pimpinellaanisum* L.)	Seeds	Trans-anethole, γ-himachalene, cis-isoeugenol and linalool	(Samojlik et al., 2012)

intensity with which they are perceived by senses and defined the conditions under their degradation are triggered. In the next sections some important topics are discussed regarding the behavior of EOs and how the perception of EOs could be changed or how their compounds could be easily degraded if they are not protected from external factors.

2.1.1 Stability

The stability of EOs is determined by physical and chemical factors. Physical factors such as evaporation, separation of the phases in emulsions, and it adsorption in complex matrices leads to loss of their characteristics. Temperature is the most important parameter that could affect the stability of these compounds. Heat causes the loss of the volatile compounds and leads to chemical changes, decreasing the intensity with which they are perceived by the senses (Taylor and Linforth, 2009). On the other hand, in contact with water and air, oxidation and hydrolysis reactions are triggered, causing the degradation of the compounds. Additionally, these compounds can also react with other compounds or simply be rearranged at a molecular level, losing their characteristics (Rowe, 2005).

2.1.2 Solubility

EOs are slightly soluble in water; therefore it is necessary to provide suitable conditions for their incorporation in food, ensuring the quality of the products. The chemical characteristics of the compounds and the system temperature govern the solubility of the compounds of the EOs in water. The solubility of EOs in aqueous solutions is influenced by their chemical structure, which could affect the thermodynamic equilibrium of the compounds in the mixture. The distribution of the compounds that compose the EOs between the different phases of foodstuff governs flavor release from food and therefore the intensity of their perception (Covarrubias-Cervantes et al., 2005). Encapsulation of this type of compounds contributes with the solubility increase of the EOs in food systems.

2.1.3 Interactions with Other Food Components

Natural flavors such as EOs contain various chemical compounds with different polarities; therefore when they are added to food, they may react with other components, causing their degradation and the loss of bioactivity. For example, one of the most important parameters is the fat content. Intensity of flavor and aroma is much greater in fat than in water, therefore it is important to avoid the dispersion of the flavor and aroma compounds in the aqueous phase (Taylor and Linforth, 2009). On the other hand, proteins and to a lesser extent carbohydrates can form chemical bonds with other composites such as reversible weak interactions or strong irreversible covalent interactions. These interactions lead to a dramatic reduction in the intensity of flavor and aroma; as a result, the perception of the compounds by consumers is affected (Wang and Arntfield, 2015).

2.2 EOs Extraction Methods

Among the most widely used methods for obtaining EOs are hydrodistillation, entrainment by water steam, organic solvent extraction and cold pressing (Asbahani et al., 2015). Although the main advantage of hydrodistillation is that, as EOs are immiscible in water and thus after condensation, EOs could be easily separated, hydrodistillation is recognized as a tedious and expensive operation. Entrainment by water steam is a method based on hydrodistillation but without direct contact between plant and water, in a process with lower extraction time. In organic solvent extraction, the EOs are extracted using an organic solvent; then, the extract is concentrated by removing the solvent under reduced pressure. Cold pressing is the traditional method to extract EOs from citrus fruits. During extraction, oil sacs break and release EOs. This oil is removed mechanically by cold pressing, yielding a watery emulsion. However, these techniques cause EOs alterations due to the use of high temperatures. Other drawbacks such as the use of organic solvents or the limitation to being applied only with a particular type of raw material require the development of new extraction techniques. As Nanoencapsulation of EOs shows, a relatively new extraction technique to obtain EOs as supercritical fluid extraction (SFE) has been successfully applied. This technique allows for reducing extraction times and using solvents considered safe (GRAS).

3 Nanoencapsulation of EOs

As previously stated, EOs are susceptible to various types of degradation under action of oxygen, light, and temperature. In addition to that, similarly with almost all bioactive compounds, the EOs are not soluble in water, which limits their application in food. Therefore, through nanoencapsulation technique it is possible increase the solubility of the EOs, decreasing the necessity of use surfactants and enhancing the possible use of EOs as food additives. However, in some applications (eg, food packaging), the strong flavor of the EOs would change or alter the original taste of food. To overcome this limitation, it is possible to entrap the EO compound into a capsule in order to mask their undesirable flavor. For example, a cinnamon EO nanofilm developed by Wen et al. (2016) contributed to the increase the shelf-life of strawberries without affecting significantly the flavor of the product. Thus, the use of encapsulation techniques allows us to take advantage of the bioactive properties of the EOs without affecting the sensorial properties during the product storage.

Additionally, in certain applications the controlled release of the compounds in specific conditions is necessary. In that sense, it is the duty of the food industry to overcome these limitations and provide products with longer shelf-life, maintaining palatability during this period. A feasible alternative is the stabilization of these compounds using encapsulation techniques. Besides protecting the compounds against adverse environmental conditions, the purpose of encapsulation is to potentiate the action of the compounds, controlling their release rate and/or transforming liquid products in solid materials such as particles (Nedovic et al., 2011).

Nanoencapsulation is an important field of nanotechnology, and it can be defined as the isolation process of compounds inside carrier materials with nanoscale dimension. Among the nanoencapsulation systems, nanoemulsions, and lipid nanoparticles particularly appear suitable for food applications (Spigno et al., 2013). According to Cushen et al. (2012), a nanoparticle is defined as a discrete entity that has three dimensions in the range of 100 nm or less. On the other hand, a nanoemulsion can be considered to be a conventional emulsion that contains very small particles. They are characterized as a thermodynamically unstable colloidal dispersion consisting of two immiscible liquids, with one of the liquids being dispersed as small spherical droplets (r < 100 nm) in the other liquid (McClements, 2012).

Besides guaranteeing excellent protection of EOs against degradation or evaporation, due to the subcellular size, nanoencapsulation techniques may increase the passive cellular absorption mechanisms, thus reducing resistance to mass transfer and increasing antimicrobial activity (Donsì et al., 2011). On the other hand, flavor nanoencapsulation also can reduce fat absorption, allowing the delivery of flavor and aroma without an increase in the caloric value of the food, in addition to other undesired effects (Coles and Frewer, 2013).

3.1 Encapsulation Materials

There is an innumerable quantity of materials that have potential to be used as a carrier material in nanoencapsulation of many compounds. Encapsulation materials, besides protecting and releasing appropriately the EOs, should be water soluble, biodegradable, form suspensions of low viscosity, not be reactive, and have a low cost. However, only a limited number of encapsulation materials can be recognized as GRAS materials, and therefore applied in food processing. It is important to highlight this fact because it limits the development of new products by the food industry when compared to the pharmaceutical industry. The development of new pharmaceutical products is less restrictive and could make possible the application of a large diversity of materials in drug encapsulation (Wandrey et al., 2009).

The majority of materials used for encapsulation in the food sector are biomolecules. These materials have to provide maximal protection of EOs against environmental conditions, to maintain the bioactivity of the compounds during processing or storage under various conditions. Regarding nanoencapsulation of EOs, among the materials most commonly used in the encapsulation of EOs, there are several biopolymers such as modified starches, β-cyclodextrins, maltodextrinas, and various gums (Martín et al., 2010).

3.1.1 Carbohydrates

Among all materials, the most widely used for encapsulation in food applications are polysaccharides. In the food industry their consumption exceeds the production of synthetic polysaccharides; therefore these polymeric carbohydrate molecules composed of long chains of monosaccharides became materials of great economic importance. Starches and their derivates such as amylose, amylopectin, dextrins, maltodextrins, polydextrose, syrups, and cellulose and their derivatives are commonly used in encapsulation of all type of compounds (Nedovic et al., 2011).

Compared with lipid carriers, for instance, systems that used carbohydrates as encapsulating material can interact with a wide range of bioactive compounds via their functional groups, which makes them versatile materials to encapsulate hydrophilic and hydrophobic compounds such as EOs or even other food ingredients with bioactive compounds. Additionally, they are considered as a suitable encapsulation material to processes that uses high temperature due to their temperature stability in comparison to lipids or proteins which might be melted or denatured (Fathi et al., 2014). For example, cyclodextrins are carbohydrates recognized by its ability to entrap hydrophobic molecules such as EOs. Several EOs components such as limonene, eucalyptol, linalool and α-pinene, among others, were encapsulated by

Kfoury et al. (2015). After the encapsulation process, the radical scavenging ability of the EOs was enhanced and the volatility of the EOs was significantly reduced. These characteristics would allow reducing the EOs' losses during storage or processing and developing bioactive packaging as well as enhancing the aroma power of the EOs. β-cyclodextrin was used by Wen et al. (2016) to develop a cinnamon EO antimicrobial film with a diameter of 240 nm. To build the film, a highly biocompatible polymer known as polyvinyl alcohol was used to prepare the antimicrobial material. The film showed antimicrobial activity against *Staphylococcus aureus* and *Escherichia coli* and contributed to enhancing the thermal stability of the EO.

Other encapsulation material considered safe for food application is *n*-octenyl succinic anhydride (OSA)-modified starch. OSA has superficial activity, being capable of reducing the superficial tension between water and oil, therefore it can be used as an efficient emulsifier and consequently as a suitable encapsulation material. OSA has been used in the nanoencapsulation of several compounds with application in food such as carotenoids (Santos et al., 2012) and curcuminoids (Abbas et al., 2015). OSA also has been used in EO nanoencapsulation of lavandin, improving performance of the bioactive compounds (Varona et al., 2010, 2011, 2013).

Polysaccharides of microbial or animal origin such as chitosan and dextran, and various gums have also been used successfully in nanoencapsulation. Among them chitosan, a derivative of chitin, gives singular chemical and biological characteristics, such as: biocompatibility, antibacterial properties, and hydrophobicity. Recently, chitosan has attracted great attention in the encapsulation of bioactive compounds because of its general recognition as GRAS and other features such as abundance, low toxicity, biodegradability, and biocompatibility. For example, in order to enhance antifungal activity and stability of the *Zataria multiflora Boiss* EOs against the causal agent of gray mold disease, the EOs were nanoencapsulated using chitosan as encapsulation material (Mohammadi et al., 2015). Woranuch and Yoksan (2013) confirmed the improved thermal stability of encapsulated eugenol compared with naked eugenol. These results suggest that eugenol-loaded chitosan nanoparticles could possibly be used as antioxidants for various thermal processing applications, including bioactive plastics for food packaging. Turmeric and lemongrass EOs were encapsulated in chitosan-alginate nanocapsules with size below 300 nm (Natrajan et al., 2015). The nanocapsules showed good stability with encapsulation efficiency between 71% and 86.5%. In this case, these capsules have a potential use for pharmaceutical applications, due to the antiproliferative activity of the EOs and the controlled release that allows the nanoencapsulation technique.

Nanogels are another important application where polysaccharides are used. Nanogels have several applications, particularly in pharmaceutical industry, mainly because of the fact that they can trap bioactive substances such as EOs in their nanometric net, improving its efficiency at low concentration, stability, and release. In another research, the *Cuminum cyminum* EO, an herbaceous plant used commercially as a flavoring agent, was encapsulated by Zhaveh et al. (2015) in a chitosan–caffeic acid nanogel. After encapsulation, the antimicrobial activity against *Aspergillus flavus* was improved, as well as its stability. The results showed that the chitosan–caffeic acid nanogel is a feasible material to encapsulate the EO because besides showing a slow-release, the process had an encapsulation efficiency of 85%, producing capsules with size below 100 nm. Although the use of nanogels has been focused mainly in the pharmaceutical industry, nowadays some interesting applications are being developed to apply nanogels in vegetables and fruits in order to preserve their quality. Specifically, it is possible to enhance the antifungal effects of the EOs through the controlled release of the EOs from nanogels and extend the storage period. For example, thyme EO nanoencapsulated in a chitosan–benzoic acid nanogel was able to inhibit the growth of *A. flavus* in tomatoes after 1 month of storage under sealed conditions (Khalili et al., 2015). However, due to the high volatility of the EOs, under nonsealed conditions its antifungal activity in nanogels was affected. In tomato under nonsealed conditions, the thyme EO was not able to totally inhibit the growth of the fungi. Therefore, to truly enhance the biological activity of the EOs, the contact with air and other environmental conditions should be avoided or controlled.

Cashew gum is a heteropolysaccharide extracted from the exudate of the Brazilian tree *Anacardium occidentale*, whose structure resembles gum arabic. Nowadays cashew gum has great interest as a suitable substitute for gum arabic, which is more expensive. Among their properties, cashew gum is recognized by its ability to interact with water and thus act as stabilizer, emulsifier, and adhesive. Herculano et al. (2015) obtained nanocapsules of Eucalyptus EOs using cashew gum as encapsulation material. The nanoparticles obtained in this study have potential for use as a natural food preservative due to how they showed a high antimicrobial activity and storage stability over 365 days. In another approach, Abreu et al. (2012) developed nanogel nanoparticles of *Lippiasidoides* EOs using cashew gum and chitosan as encapsulation materials with sizes ranged from 335 nm to 558 nm.

Inulin is another carbohydrate with potential use as carrier material in nanoencapsulation of EOs. Inulin is a natural ingredient commonly used in food industry as a prebiotic compound. It also can be used as excipient and stabilizer, but it has great potential as carrier material for encapsulation and controlled release of EOs. For example, oregano EO has been encapsulated using inulin as carrier material by Beirão-Da-Costa et al. (2013). The particles had a particle size ranging between 3 and 4.5 μm after spray drying using temperatures from 393 to 463 K.The temperature had a significant effect in the structure and the morphology of the capsules and due to these changes in the structure, different profiles of release were observed. Thus, further efforts are necessary in order to obtain feasible formulations using inulin as carrier material and establish the best processing conditions for production of encapsulates suitable for use in food industry.

3.1.2 Proteins

Proteins can be used in their natural state, or they can be chemically, physically, or enzymatically modified to modulate their functional attributes. Therefore, it is possible according the specific application to improve the functional performance of proteins. In the same sense, as proteins are easily digested by the human body, it is possible to take advantage of the bioactive properties of the EOs during an eventual release of EOs after ingestion. However, protein particles are often highly sensitive to alterations in pH, ionic strength, and/or temperature because these trigger changes in their surface charge and hydrophobicity (Joye and Mcclements, 2014).

Zein is a corn protein that has the ability to form films, and is also recognized for being biodegradable and biocompatible. For this reasons, zein has been successfully applied in several researches as well as in food and pharmaceutical industries. Thymol and carvacrol, two predominant compounds in oregano and thyme EOs, were encapsulated in nanoparticles of zein using the liquid-liquid dispersion method developed by Wu et al. (2012). This research allowed the dispersion of both EOs in water, enhancing their potential use in food preservation and control of human pathogenic bacteria. For example, solubility of oregano EOs increased up to fourteen-fold without damaging their ability to scavenge free radicals or to control *E. coli* growth.

In another approach, due to its abundance, nutritional value, and acceptance by consumers, milk proteins have been widely applied in the encapsulation of several compounds due to its versatility and excellent functional properties (Tavares et al., 2014).

Particularly, whey protein isolates (WPI) and whey protein concentrates (WPC) have demonstrated improved heat stability of encapsulated compounds. WPI was used in the obtaining of nano dispersions of thymol (Shah et al., 2012). In this research, upon hydration of the spray-dried powder, transparent and heat stable nano dispersions were formed at thymol concentrations well above its solubility limit. Therefore, this technology is applicable to disperse various lipophilic compounds such as EOs in transparent beverages like clear fruit juices. In other studies, although the nanometric size was not reached, WPI and WPC have been successfully used in the EOs encapsulation. Hundre et al. (2015) used WPI to encapsulate vanillin (3-methoxy-4-hydroxy-benzaldehyde) extracted from the pods of *Vanilla planifolia*. This research indicated that there was no interaction between the encapsulation material (WPI) and core (vanillin) materials. Moreover, microencapsulated vanillin +WPI by spray–freeze drying technique yielded better thermal stability than spray-dried and freeze-dried samples. On the other hand, the mixture of WPC and other proteins or polysaccharides could enhance the characteristics of the emulsions and thus the encapsulation process. Chia (*Salvia hispanica* L.) EO was encapsulated using a mixture of WPC, mesquite gum or gum arabic (Rodea-González et al., 2012). The use of a binary mixture provides better stability to droplet coalescence due to how the interaction among the proteins contributes to produce more stable emulsions. After spray-drying, capsules with encapsulation efficiency higher than 70% were obtained from emulsions made with an EO to carrier material of 1:3.

3.1.3 Lipids

Nanoencapsulation using lipids as encapsulation material is among the more developed nanotechnology fields related to application in food systems. Nanoencapsulation using lipid-based systems has several advantages, for example, hydrophobic and highly unstable compounds as EOs are difficult to incorporate in aqueous systems; however, through using lipids as encapsulation materials it is possible to trap materials that have different solubility, using natural ingredients at industrial scale (Zuidam and Nedovic, 2009). According to Martín et al. (2010), phospholipids are a class of amphiphilic lipids formed by a hydrophilic head (a phosphate group, a diglyceride, and a simple organic molecule) and a hydrophobic tail (long fatty acid) are less toxic than other encapsulation materials and thus an excellent option for EO encapsulation. These compounds may be used in the formation of liposomes, by encapsulating hydrophobic materials such as EOs. Liposomes are closed spherical vesicles arranged in one or more concentric bilayers of phospholipids with an internal aqueous phase. Particularly due to their easy biodegradation and their similarity with biomembranes, the use of liposomes constitutes a suitable system for encapsulation of volatile unstable EOs constituents and it may allow enhance the functional properties of EOs. Donsì et al. (2011) encapsulated a terpenes mixture and D-limonene into nanoemulsions based on food-grade ingredients such as sunflower oil, palm oil, and soy lecithin. According to the results, the addition of low concentrations of the nanoencapsulated terpenes was able to delay the microbial growth or completely inactivate the microorganisms while minimally altering the organoleptic properties of the fruit juices. In another approach, Sebaaly et al. (2015) developed suitable formulations of natural soybean phospholipid vesicles to improve the stability of clove EOs and its main component, eugenol. It was found that liposomes exhibited nanometric spherical shaped vesicles and protected eugenol from degradation induced by UV exposure; maintaining their stability after stored for 2 months at 277 K.

4 Emerging Technologies

4.1 Supercritical Fluids (SCFs)

A pure compound is considered a supercritical fluid (SCF) when its temperature and pressure are above than critical values, T_c and P_c, respectively. The critical temperature is defined as the higher temperature in which a gas can be transformed in a liquid, due to a pressure increase. The critical pressure consists in the higher pressure in which a liquid can be converted to a gas, due a temperature increase. These properties characterize the critical point (CP) (Fig. 3.2). In the supercritical region, there is no equilibrium between liquid and gas phases; however, the physical-chemical properties of the SCFs are intermediaries between gas and liquid. Additionally, around CP, little pressure changes cause huge changes in density, viscosity, solvation power, and diffusivity, allowing a selective precipitation of solute, and thus enabling the solvent recycle (Brunner, 2005).

The SCF more often used is carbon dioxide (CO_2). CO_2 is considered inert, nonflammable, nontoxic, and is available in large quantities, with low cost and high purity, having a relatively low CP, 304.2 K and 3.38 MPa. Carbon dioxide is the most suited solvent for SFE of thermolabile compounds because of its favorable properties (including nontoxic and nonflammable character, high availability at low cost, and high purity) and to its ability to produce isolates with optimal physicochemical, biological, and therapeu-

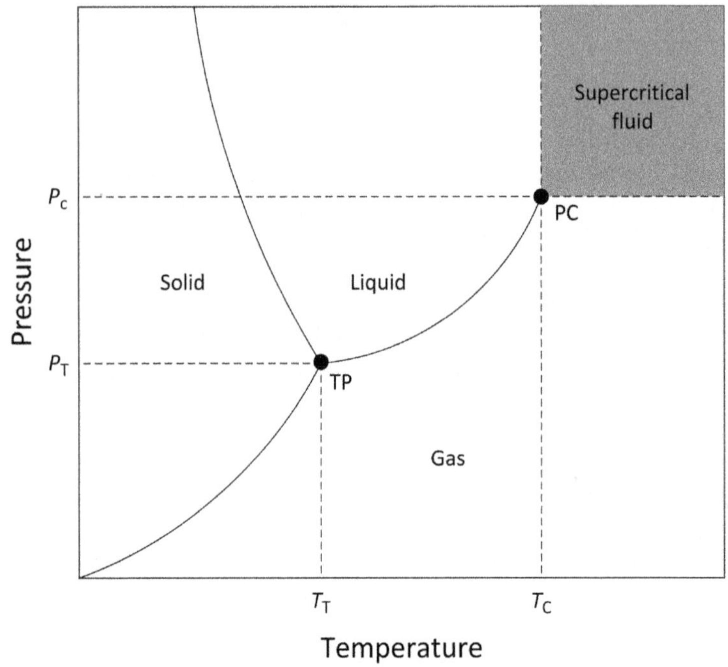

Figure 3.2. Pressure-temperature phase diagram for pure substances.

tic properties. Additionally, due to its moderate critical temperature, it allows for operation near to room temperature with a slight critical pressure (Meireles, 2008). In general, supercritical CO_2 ($scCO_2$) behaves like a lipophilic solvent, but compared with liquid solvents, its main advantage is that its selectivity and solvation power are adjustable and can be controlled through temperature and pressure changes (Reverchon, 1997). For example, because of the high solubility of the EOs in SCFs, these compound have been isolated successfully using SFE. EOs from raw materials such as rosemary (Zabot et al., 2014), hops (Van Opstaele et al., 2012), eucalyptus (Zhao and Zhang, 2014), coriander (Pavlić et al., 2015), ginger (Mesomo et al., 2013), peppermint (Gañán et al., 2015), and turmeric (Carvalho et al., 2014) have been successfully obtained using SFE. Process conditions that use pressures between 8 and 15 MPa and temperatures between 293 and 333 K are feasible for its use as extraction solvent in the extraction of volatile and thermolabile compounds such as EOs, preventing their thermal degradation. On the other hand, as SFE process is performed at the absence of light and oxygen, oxidation reactions, a problem of major importance in antioxidant extraction can be avoided using this technology. Therefore, SCFs display unique characteristics that enable them to be used as a solvent to extract bioactive compounds as well as a feasible antisolvent to precipitate compounds and encapsulate EOs, as shown in the next section, SCFs in Encapsulation of EOs.

4.1.1 SCFs in Encapsulation of EOs

Several techniques have been studied and used to form capsules include spray-drying, spray chilling, jet milling, fluidized bed coating, liposome entrainment, coacervation, thermal and ionic gelation, and so forth. However, some disadvantages such as the difficulty in producing capsules with a narrow particle size distribution as well as use of high operating temperatures, generate the necessity to develop novel techniques to overcome these drawbacks. The use of SCFs as an alternative technique for EOs encapsulation would improve the results obtained with other techniques, overcoming the drawbacks and even creating novel formulations and products. Particular features of the $scCO_2$, such as the possibility of adjusting its solvation power, low viscosity, moderate high diffusion coefficient, and no product contamination; have increased the use of SCFs as a feasible alternative for the encapsulation of EOs.

In the encapsulation processes using an SCF, a solution that contains the solute and the carrier material is dissolved in a SCF. Due to the particular behavior of the solution in the supercritical phase, the solution reaches a higher supersaturation, the solubility of the compounds is reduced drastically, making the carrier material precipitate, and the entrapment of the compounds of interest is caused (Vinjamur et al., 2013).

However, owing to the high sensibility to the changes in the process conditions on the phase equilibrium for this type of systems, a further knowledge of the phase equilibrium of the EOs+ carrier material + CO_2 is a requirement for the development of processes where SCF are used for encapsulation. According to Reverchon et al. (2003) the formation of a single supercritical phase is the key step for the successful production of nanoparticles. In complex systems such as EOs + carrier material + CO_2, the phase diagram of the system could change due to the presence of other soluble solids different from EOs or carrier materials in the solution or emulsion or due to some changes in the proportions of the mixture components. For example, depending on the system temperature and pressure and due to the large differences in size, shape, and polarity among the EOs molecules and the carriers materials, three or more phases could be produced, causing no capsule formation. In the same context, depending on the process conditions a cosolvent effect also could be observed. In that case, the dissolution of some compounds in the mixture could be increased and therefore a higher fraction could be dragged out of the system by the $scCO_2$. A similar behavior was observed by Lévai et al. (2015) in a quercetin encapsulation process using SCF. In the performed process some low molecular weight antioxidant compounds of lecithin were dragged out by the $scCO_2$. In consequence, although the antioxidant activity of the capsules obtained by SFEE was higher than pure quercetin, the antioxidant activity of the SFEE capsules had lower antioxidant activity than the mixture of quercetin and lecithin from the oil in water emulsion.

Consequently, one of the more important aspects of using SCFs in encapsulation processes is that the process must be initially based in the knowledge of phase equilibrium behavior and the solubility of the substrate and the polymer matrix in the SFE. This feature is one of the main limiting factors of supercritical techniques described in SCFs Encapsulation Techniques.

4.1.2 SCFs Encapsulation Techniques

Various techniques using SCFs have been proposed to precipitate and to encapsulate several compounds. According to Silva and Meireles (2014), these techniques can be classified in accordance with the function of the supercritical fluid in the process: solvent (Rapid Expansion of a Supercritical Solution, RESS); antisolvent (Gas Antisolvent, GAS; Supercritical Antisolvent, SAS); cosolvent or solute (Particles from Gas-Saturated Solutions, PGSS); nebulization compound (Carbon Dioxide Assisted Nebulization with a Bubble Dryer, CAN-BD) and extractor and antisolvent techniques (Supercritical Fluid Extraction of Emulsions, SFEE). Many of these processes were originally developed to produce solid compounds, however, with some modifications it is possible to obtain solid-liquid compounds, for example, liposomal nanocapsules of EOs.

4.1.2.1 Rapid Expansion of Supercritical Solution (RESS)

RESS process consists in the saturation of an SCF with a solid substrate in a saturation vessel; afterwards, the saturated solution is pressurized through an expansion heated nozzle into a low pressure chamber or ambient pressure vessel (Fig. 3.3). The pressure change causes a faster nucleation of the substrate in very small particles, which are collected in the gas current. The fast injection of the substrate into gas phase should guarantee the production of very small particles. This process is attractive mainly due to no use of organic solvents, the use of low temperatures, and the narrow particle size distribution (Reverchon and Adami, 2006).

Some disadvantages of the RESS process such as: difficulties for the scale-up and nozzle design, as well as the energy cost associated

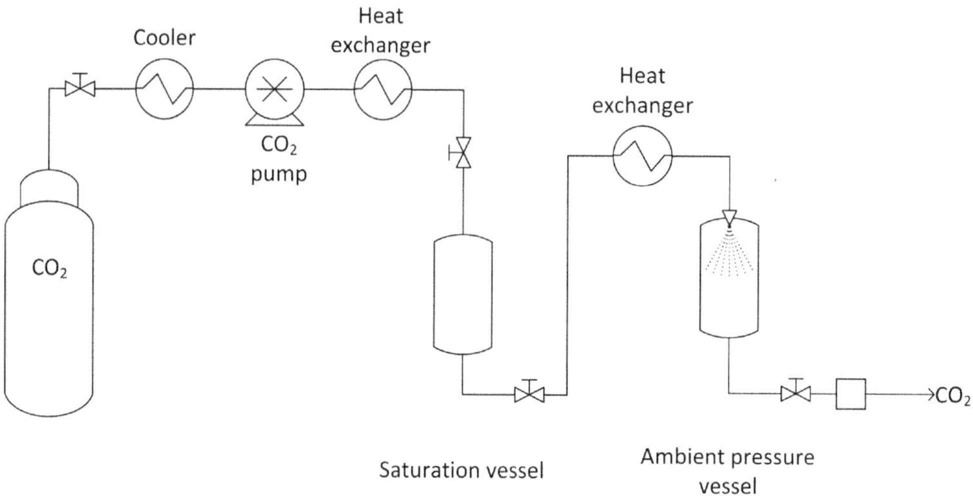

Figure 3.3. Schematic flowsheet of the Rapid Expansion of Supercritical Solution (RESS) process.

with the recompression of CO_2 for its recirculation into the process, could limit its utilization (Rodríguez-Rojo et al., 2013), however, several compounds have been micronized using the RESS process. Most of these micronized compounds are pharmaceutical substances or high economic value materials. The application of the RESS process in the food industry and related products is limited because substances such as fat-soluble vitamins (vitamin E, for instance) are moderately soluble in CO_2 (Weidner, 2009). Additionally, high molecular weight compounds as polymers and phospholipids, which are essential in the encapsulation process of EOs are slightly soluble in CO_2, restricting the use of the RESS process in the encapsulation of EOs. For example, phospholipids are not easily dissolved in pure $scCO_2$. These compounds are used in the formation of liposomes in aqueous medium; therefore the conventional RESS process is not applicable in the encapsulation of EOs using liposomes.

Moreover, to overcome the limitation that the RESS process only is suitable for compounds with low polarity or moderate solubility in $scCO_2$, changes such as the addition of cosolvents such as ethanol or acetone in a process named Rapid Expansion of a Supercritical Solution with a Nonsolvent (RESS-N) has been proposed. The RESS-N process was used by Wen et al. (2010) to form liposomes containing EOs from rhizomes of *Atractylodes macrocephala Koidz*, a Chinese plant commonly used as a supplement in food and folk medicine. In this modified method, the EOs solution and the materials of the liposomes are dissolved in the $scCO_2$-ethanol mixture and are pulverized into a phosphate buffer solution through a coaxial nozzle to incorporate the essential oil into the liposomes. The entrapment efficiency, EOs loading, and average particle size of liposomes were found to be 82.18%, 5.18%, and 173 nm, respectively, under the optimum conditions of at a pressure of 30 MPa, a temperature of 338 K, and a cosolvent mole fraction in $scCO_2$ of 15%.

4.1.2.2 Supercritical Solvent Impregnation (SSI)

This technique is based on the fact that a polymer can be impregnated with another compound. In the beginning, the compound of interest is dissolved into a SCF in a saturation vessel and, immediately, the compound + SFC mixture are put in contact with the polymers particles to be impregnated in an impregnation vessel (Fig. 3.4). The two main items of the setup are a column, in which $scCO_2$ is saturated with the compound of interest, and the impregnation column, in which compound + SFC mixture is brought into contact with the polymer. In the same way as the RESS process, in the SSI process it is also possible to use cosolvents to enhance the solubility or to improve the dispersion of the compound or carrier material in the polymer (Cocero et al., 2009).

Particularly, this technique could benefit from the high solubility of EOs into $scCO_2$, and the high capacity of diffusion of the SCF through the carrier material, which generally is a powdered polymer with a preformed morphology. In fact, this process is used to produce micro and nano compounds, through the introduction of an active compound into preformed particles using diverse carrier materials. Due to this characteristic, the SSI process cannot be

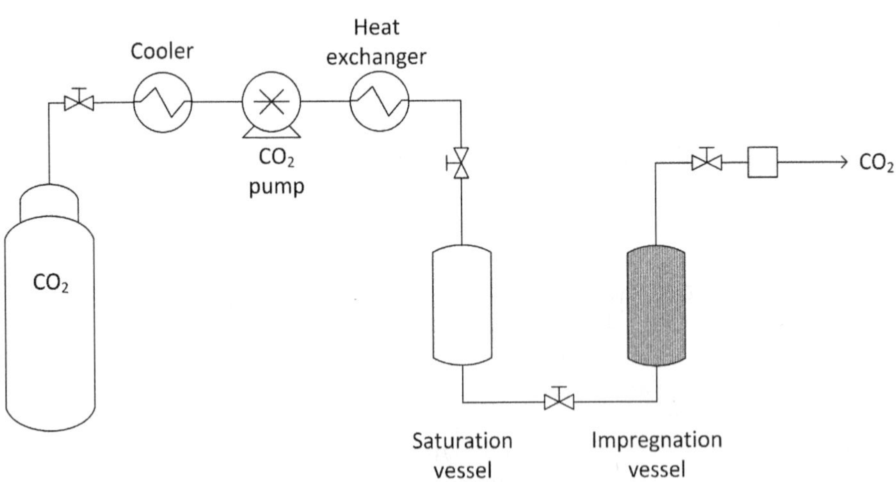

Figure 3.4. Schematic flowsheet of the Supercritical Solvent Impregnation (SSI) process.

considered strictly an encapsulation technique (Rodríguez-Rojo et al., 2013). Nevertheless, it can be particularly advantageous in relation to the conventional encapsulation processes. For example, the EOs encapsulation using spray drying from an oil-water emulsion (o/w) and some carrier material, can trigger the degradation of the EOs due to the high process temperature (above 353 K) and to oxidation of the compounds due to the presence of oxygen in the compressed air using during the process. Particularly, oregano and the compounds of its EOs have been presented as special interest to this technique. Almeida et al. (2013) demonstrated that SSI process is an attractive technique to impregnation of natural matrices with EOs. In this work, oregano EOs was impregnated in microspheres of different types of starch (sorghum and rice) using the SSI process. Mild operating conditions (10 MPa and 313 K) avoid EOs degradation and the high diffusivity of CO_2 in the solid matrix (starch) ensures a deep impregnation of the essential oil. The product was characterized as having a high antioxidant activity even during storage. Besides, the scale-up process is considered simple and therefore SSI process has potential for the production of ingredients for the food industry. In this type of process, depending on the quantity of EOs impregnated, it is possible to change the polymer morphology impregnated due to the formation of new chemical bonds that can change the solid structure of the polymer. Milovanovic et al. (2015) impregnated oregano EOs, containing mainly thymol, in cellulose acetate using the SSI process. Although higher impregnation yield was obtained (72% at 308 K and 20 MPa), the morphology of the samples changed significantly with the high impregnation yields, impregnated near the surface of the bead to induce the change of polymer solid structure via participating in the intermolecular implying the impact of thymol on the cellulose acetate. Enough thymol impregnated near the surface of the bead to induce the change of polymer solid structure via participating in the intermolecular hydrogen bonds, and thus completely changing the morphology of this part of the bead. Lavandin (*Lavandulahybrida*) EOs was impregnated in OSA (Varona et al., 2011). After testing pressures between 10 and 12 MPa and temperatures between 313 and 323 K, it was stated that the distribution coefficient of EOs between the starch and the supercritical phase as well as the EOs load depended on the density of CO_2. The quantity of EOs impregnated increased when temperature was increased and pressure was decreased.

4.1.2.3 Supercritical Antisolvent (SAS)

In SAS process, $scCO_2$ is used as an antisolvent to reduce the solute solubility dissolved in a solvent. In the SAS process a liquid solution that contains the compound to be encapsulated and the carrier material is injected into a SCF. In this process, $scCO_2$ is pumped into a precipitation vessel using specific process parameters such as temperature, pressure, solution flow rate, and CO_2 flow rate. Afterwards, the solution that contains the interest compound, the carrier material, and the organic solvent is pulverized through an expansion nozzle into the precipitation vessel. The solvent diffuses quickly from the drops of solution to supercritical phase. This leads to supersaturation of the solute, which is compensated by nucleation and the compound of interest is precipitated within the carrier material. The formed particles are collected using a filter fixed at the bottom of the precipitation vessel whereas the residual organic solvent and the CO_2 are removed from the system (Fig. 3.5). SAS process is characterized by operating in a semicontinuous mode, where the solution and the $scCO_2$ are continuously injected into the precipitation vessel. For the SAS process to be successful, the solute must be soluble in the organic solvent at the process temperature and must be insoluble in the SCF. Another important aspect is that the solvent must be completely miscible with the SCF, because if the solute is just partially soluble into the SCF, depending on the process conditions, two or more fluid phases could be formed and the solute may remain dissolved or partly dissolved.

A process variation is the operation in batch mode, which is known as gas antisolvent (GAS) process. In GAS process, the precipitation vessel is loaded with a given quantity of the liquid solution and, then, the supercritical antisolvent is added until the final pressure is obtained (Reverchon and Adami, 2006).

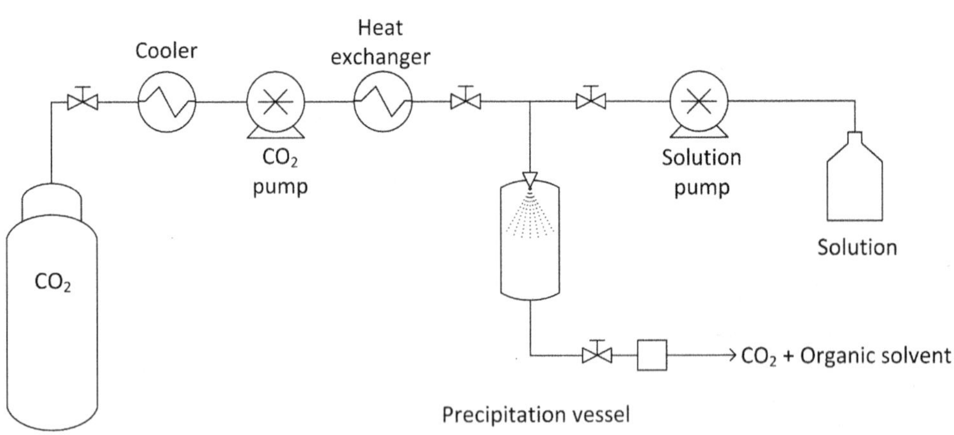

Figure 3.5. Schematic flowsheet of the Supercritical Antisolvent (SAS) process.

In SAS process is possible to obtain nanoparticles due to the fast supersaturation and nucleation caused by the high mass transference rates of the SCF. On the other hand, narrowed size distributions can be obtained by controlling process parameters. For example, using this technique, is possible the formation of dry lipid particles, which is more advantageous than forming liposomes in suspension because of the improved stability upon storage and transport (Beh et al., 2012). According to Kalani and Yunus (2011) the major disadvantage of this technique is the long washing period prior to the agglomeration and aggregation of particles. However, this problem can be minimized by intensively mixing the supercritical antisolvent and the solution, which increases the mass transfer and thus produces smaller particle size. Despite this disadvantage, several compounds such as carotenoids (Mezzomo et al., 2012; Mattea et al., 2009b; Martin et al., 2007; Santos and Meireles, 2013) and various antioxidant compounds (Visentin et al., 2012; Zu et al., 2012; Zhao et al., 2011; Sosa et al., 2011) have been successfully precipitated and encapsulated using the SAS process.

Specifically, SAS process has not been applied in the production of dry particle of EOs. Despite this, the use of the SAS process appears to be an efficient and environmentally friendly process to produce liposomes. Lesoin et al. (2011) developed a process to produce liposomes of lecithin using the SAS process. In this work, the process was performed using a precipitation temperature of 308 K, range of precipitation pressure from 9 to 13 MPa, and CO_2/solvent (ethyl alcohol) molar ratio ranges from 50 to 100. The liposome size distribution was included in the range of 0.1–100 μm and encapsulation efficiency was about 20%. Moreover, when SAS process is compared with conventional process (Bangham method), SAS process is carried out under mild temperature conditions, unlike the Bangham method (308 K for the SAS process and 323 K for the Bangham process), and SAS liposomes were more stable than those obtained by Bangham method. According to these results, SAS process could be a feasible method with potential application in the production of EOs liposomes.

4.1.2.4 Particles from Gas-Saturated Solutions (PGSS)

PGSS is a process that allows for a saturated supercritical fluid solution to obtain nanoparticles. In the PGSS process the solute is initially saturated with $scCO_2$ in a high-pressure vessel denominated as static mixer (Fig. 3.6). Afterward, the saturated solution is expanded at moderate pressure through an expansor nozzle into a spray tower causing the formation of solid particles or liquid particles. The rapid expansion of the saturated solution cause

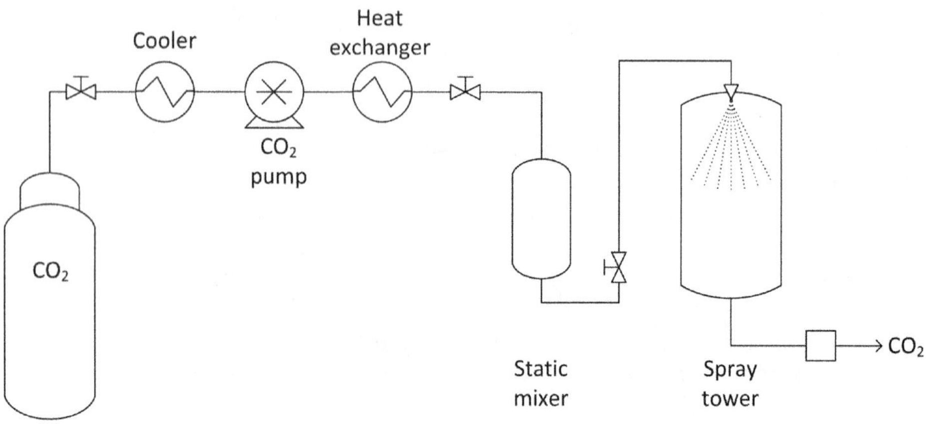

Figure 3.6. Schematic flowsheet of the Particles from Gas-Saturated Solutions (PGSS) process.

an intense cooling effect caused by Joule–Thomson effect (Cocero et al., 2009). For example, in an encapsulation process of β-carotene with poly-(ε-caprolactone) by PGSS process, De Paz et al. (2012) established that when a higher CO_2 content is used, the Joule–Thomson cooling effect, which is the driving force for particle formation by PGSS process, becomes more intense, promoting the formation of smaller particles. Therefore, the knowledge and control of this phenomenon is a key factor to the success of the PGSS process.

The advantages of PGSS process are related with low SFC consumption and low to medium pressure process. Because of simplicity of this process, the processing cost is very low compared to other processes. It can be used with suspensions or emulsion of active ingredients in polymers or other carrier substances leading to composite particles. Among the disadvantages of PGSS process are included the difficulty of producing submicron-sized particles and the control of particle size (Fahim et al., 2014). However, PGSS process has a great potential in the production of particles and capsules of many molten fats, lipids, or polymers due to the high solubility in $scCO_2$ of these compounds at moderate pressures (Martín et al., 2010). For example, a saturated mixture of $scCO_2$ and an emulsion of EOs + carrier material can be formed. Afterward, this saturated solution can be expanded, precipitating the carrier material entrapping the EOs in capsules.

PGSS process was successfully applied in the lavandin EOs encapsulation using as carrier materials polyethylene glycol (PEG) and OSA modified starches (Varona et al., 2010). In this process was evaluated a PGSS process modification named PGSS-drying. The main difference between the two processes is that in PGSS-drying an oil-in-water emulsion is intensively mixed with $scCO_2$ using a static mixer, encapsulating the EOs in starch by removing the water from the emulsion. The EOs were effectively encapsulated through the PGSS-drying process using OSA modified starches as carrier material. Between the two carries materials, PEG had a better performance reaching higher encapsulation efficiencies than OSA (14–66% of initial EOs encapsulated). Besides, spherical particles with narrow particle size distribution were obtained, which is a key factor in order to control the release of the lavandin oil for further applications. In other work about lavandin EOs encapsulation, demonstrated that the antibacterial activity of lavandin EOs against three pathogenic bacteria (*E. coli*, *S. aureus*, and *Bacillus cereus*) could be enhanced by encapsulation, due to the protection and control release of the EOs (Varona et al., 2013). In another approach, menthol was encapsulated using beeswax as the wall material (Zhu et al., 2010). The experiments were performed at 333 K with pressure in the range from 6 to 20 MPa, mass fraction of menthol in the menthol/beeswax mixture from 10 to 40% and flow rate of solution from 0.21 to 0.81 cm^3/min. Results indicated that in the range of studied conditions, increase of the pressure, decrease of the gas-saturated solution flow rate, and decrease of the menthol mass fraction can decrease the particle size and narrow particle size distribution of the produced menthol/bees wax microparticles. Although in this work were not obtained nanoparticles (2–50 μm), the microparticles produced have an obvious protection against menthol volatilization.

Choi et al. (2010) obtained PEG microparticles containing coriander EOs using the PGSS process. In this work, temperatures in the range from 310 to 333 K and pressure from 10 to 25 MPa were evaluated. At these process conditions, the stability of the coriander EOs was improved and microparticles with size between 0.1 and 10 μm were produced. On the other hand, they observed a positive influence on the formation of spherical microparticles and highest entrapment efficiency with increasing temperature and decreasing pressure. Gitin et al. (2011) encapsulated garlic EOs using PEG as wall material. In this study, temperatures between 324 and 335 K and pressures between 15.7 and 20.3 MPa were used. Particularly, although the encapsulation efficiency of the process had good performance (26.10–48.93%), it was observed that when the quantity of the EOs was increased, some fraction of oil that was not encapsulated produced particle agglomeration. Therefore, it was impossible to reach nanometric scale and particle sizes ranging from 71.124 μm to 205.64 μm were obtained.

In summary, PGSS process is one of the emerging methods most used to encapsulate. Particularly, although the encapsulation efficiency of the process had good performance, however, further research in order to achieve nanometric scale and avoid particle agglomeration is necessary. EOs protection and the subsequent controlled release of the bioactive compounds are two characteristics of the PGSS process that could be explore in the encapsulation of EOs and other flavor and aroma compounds in order to develop potential applications to food industry.

4.1.2.5 Supercritical Fluid Extraction of Emulsions (SFEE)

SFEE process is a combination of the conventional emulsion precipitation process with the SAS process, where $scCO_2$ is used to eliminate the organic solvent from the emulsion droplets (Fig. 3.7). Emulsion techniques usually involve large quantities of organic solvents, and the removal of them involves additional separation techniques and the use of high temperatures. On the other hand, the particles obtained using SFC sometimes present agglomeration problems generating large particles or cannot reach nanometric size. However, the application of SCF in the particle technology with the particle formation process from emulsions can overcome the main problems of each separated technology (Cocero et al., 2009).

SFEE can benefit from the excellent transport properties of the SFC and cause the precipitation of the particles inside the emulsion droplets. Therefore, the growth of the particles is limited by the size of the emulsion droplets and agglomeration is reduced thanks to the surfactants forming the emulsion. One of the main drawbacks of the SFEE process is that instead of obtaining dry particles as SAS process, the final product usually consists of a suspension of the desired compound in water (Mattea et al., 2009a). Santos et al. (2012) demonstrated that submicrometer (344–366 nm) particles of carotenoids

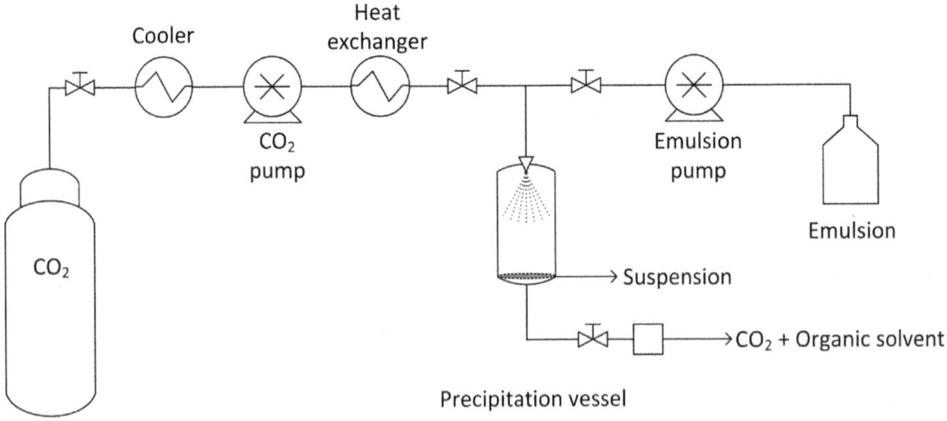

Figure 3.7. Schematic flowsheet of the Supercritical Fluid Extraction of Emulsions (SFEE) process.

(β-carotene and lycopene) with high stability and solubility in aqueous media can successfully be produced by SFEE process. These results indicate that lycopene used presents a higher solubility in CO_2 and stability than β-carotene under the same operating conditions (323 K and pressures between 7 and 13 MPa). Quercetin is another compound with low water solubility that have been successfully encapsulated in the SFEE process (Lévai et al., 2015). After encapsulation, particles with mean size around 100 nm and encapsulation efficiency around 70% were obtained.

In the same way as the SAS process, the SFEE process is in principle suitable to obtain micelles loaded with EOs. For example, the solubility of many EOs compounds in CO_2 at moderate pressures is much lower than that of organic solvents; therefore in any case the fraction of EOs compound extracted during micelle formation can be easily recovered and recycled by depressurization of the gas effluent (Martín et al., 2010). Although at this time there are no specific works about EOs encapsulation, SFEE process could explore the properties of EOs and its behavior in supercritical mixtures to obtain micelles loaded with EOs for food application.

4.2 Ultrasonication

Ultrasonication technology has been evaluated in diverse applications such as extraction of bioactive compounds (Santos et al., 2015), particle formation (Jordens et al., 2015), dehydration of waste oil (Xie et al., 2015), biodiesel production (Sarve et al., 2015), and sulfur removal from bauxite water slurry (Ge et al., 2015), among others.

Depending on the type of process, cavitation is produced due to the formation of high frequency waves in several expansion and contraction cycles, which could cause various effects in the vegetal matrix or liquid medium when ultrasonication is being applied. In most cases these effects are related with enhancing the mass transfer. For example, in the extraction of bioactive compounds, the cell walls of the vegetal matrix are disrupted, the extraction solvent penetration and the mass transfer are favored, and thus the overall yield and the extraction rates are increased (Toma et al., 2001). On the other hand, in precipitation processes, ultrasonication enhances the nucleation rate which results in a larger amount of fine particles and breakage of the already formed particles into smaller particles by the large shock-waves and micro jets (Horst et al., 2007). However, the use ultrasonication for the formation of nanoemulsions with EOs is a technology almost unexplored. Research in this area could lead to significant advances in the supply of new systems for food flavoring agents.

Emulsions by definition are colloidal systems thermodynamically unstable due to interfacial tension (enthalpy) observed between the phases of different chemical nature. The use of emulsifiers and surfactants allows stabilization of these systems by various mechanisms such as adsorption of the molecules in water-oil interface with a reduction in the interfacial tension; electrostatic interaction of the molecules that act as a barrier to coalescence of the droplets in dispersed phase; combination of the two mechanisms mentioned above; viscosity increase of the continuous phase providing a physical barrier to coalescence. However, the emulsification method used is directly related with the reach of the kinetic stability because a reduction in droplet size of the dispersed phase decreases the velocity of system phase separation (McClements, 2012). This effect is related with Stokes law applied to colloidal systems, which indicates that the velocity of the droplet of the dispersed phase is proportional to the square of its radius (Desrumaux and Marcand, 2002).

Emulsification using high-intensity ultrasound with a frequency range of 16–100 kHz and a power of 10–1000 W cm^{-2} has the ability to produce fine emulsions with size distribution of highly uniform drops (Chandrapala et al., 2012; Silva et al., 2015). Fig. 3.8 shows the schematic flowsheet and the type of energy involved in each piece of an ultrasound equipment. Emulsification via ultrasound

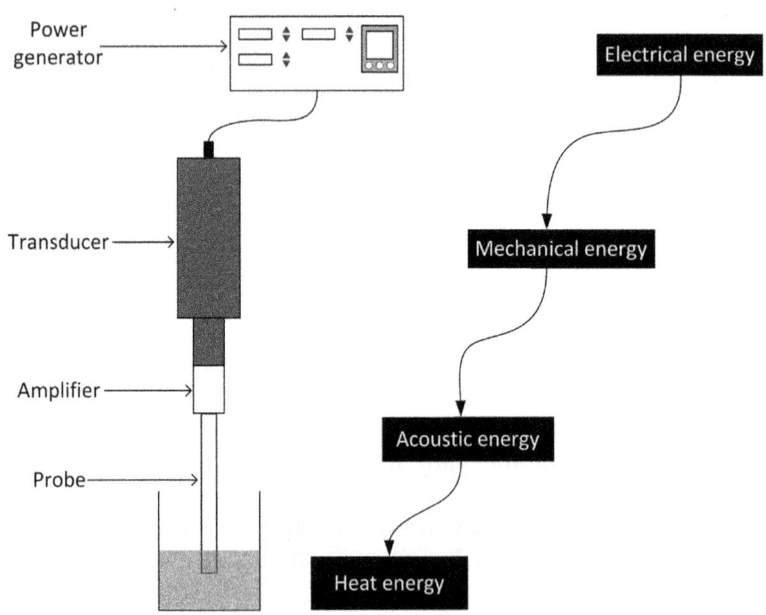

Figure 3.8. Schematic flowsheet and the type of energy involved in ultrasonication.

has distinct effects on the breakdown of the oil droplets in an emulsion due to the mechanisms involved in this technique. Homogenization occurs through two main mechanisms associated with the generation of an acoustic field and the application of low-intensity ultrasound frequency. The acoustic field is responsible for generating unstable interfacial waves that promote the mixture of oil with water (continuous phase) leading to the formation of droplets in the system. The application of low frequencies of ultrasound generates an acoustic cavitation phenomenon, which is primarily responsible for the reduction in the oil droplets size. The rapid formation and collapse of microbubbles are promoted by the intense shear rates associated with cavitation. The collapse causes extreme levels of turbulence to be highly localized and this acts as an effective method to the breakdown of the droplets of the dispersed phase into droplets with sizes that can be lower than the submicron range or even reaching the nanoscale (Li and Fogler, 1978a; Li and Fogler, 1978b). Therefore, the superiority of ultrasound as an emulsification method can be associated with microshear resulting of cavitation. The intense shear involved in this technique along with other effects of cavitation, such as heating, may cause the rupture of chemical bonds and physical changes in the polymers used as emulsifiers (Arzeni et al., 2012; Gülseren et al., 2007).

Therefore, when ultrasound is applied as an emulsification method using polymers with biological activity associated, such as whey proteins, soy, and egg proteins, an evaluation must be made about the maintaining of the activity of these macromolecules after ultrasound processing. This can be considered the only disadvantage associated with use of this technology (Arzeni et al., 2012).

Nanoemulsions obtained can be considered as final process products and are characterized by having diverse possibilities of application in food products. In addition, since most systems are directed to form emulsions type oil-in-water, drying the continuous phase of the EOs nanoemulsions by different techniques (spray-drying and freeze-drying, among others) allows the particle formation with high encapsulation efficiencies of the bioactive compound of interest (Soottitantawat et al., 2003, 2005).

4.2.1 Effects of the Application of Ultrasound in Oils

Besides the modifications in the molecular structure of the emulsifiers (gums, starches, proteins, and others) previously mentioned, the use of ultrasound may also cause changes in oily compounds such as EOs. Physicochemical effects such as appearance of off-flavors, metallic taste, production of free radicals, breakdown of compounds (structure modification), and some changes in physical parameters are associated with the use of ultrasound in food processing. These changes may be related with process conditions (temperature and pressure) and microshear mechanisms produced for the acoustic cavitation phenomenon (Pingret et al., 2012, 2013). For example, after pasteurization of pineapple, grape, and cranberry juices using thermo-sonication, Bermúdez-Aguirre and Barbosa-Cánovas (2012) observed changes in color and pH of the samples. Thus, ultrasound may promote the formation of chemical products that affect the pH or cause the destruction of pigments and generate nonenzymatic browning. In other research, Anese et al. (2013) observed after processing of tomato pulp with ultrasound an increase in the viscosity due to the formation of a stronger fiber network; however this network entraps the lycopene and the lycopene bioaccessibility is decreased. Cravotto et al. (2011) studied the extraction of kiwi seed oil using Soxhlet and other four different nonconventional techniques, including ultrasound-assisted extraction. The authors observed formation of off-flavors and presence of oxidation compounds in oil extracted with ultrasound due to the partial degradation of the kiwi seed oil. On the other hand, Metherel et al. (2009) showed that the cavitation phenomenon produced by the ultrasound device led to the generation of free radicals in flaxseed oil, and although peroxide levels were increased, the fatty acid composition almost was not affected by ultrasound. Therefore, although the application of ultrasound can promote the nanoencapsulation of EOs, the formation of off-flavors, induction of oxidation in lipid chains, and formation of free radicals are key factors that have to be considered in order to avoid some negative effect on the EOs.

4.2.2 Applications of Ultrasonication in Obtaining Nanoemulsions of EOs

Hashtjin and Abbasi (2015) studied the formation of natural orange peel EOs (OPEO) nanoemulsion assisted by ultrasound. OPEO is among the most common important EOs used in the food industry and is recognized for being limonene-rich. To perform the nanoemulsions, Tween 80 and native gums such as Persian gum (PG) and gum tragacanth (GT) were used as emulsifiers. Nanoemulsions were prepared using OPEO (1% w/w), as the oil phase, and mixture of Tween 80 (2% w/w), combined with soluble fractions of PG and GT (0.25% w/w) and deionized water (96.75% w/w), as the aqueous phase. The effects of sonication amplitude (70–100%) and different process times (90–150 s) were evaluated.

OPEO nanoemulsions with size of 13 nm were obtained using sonication amplitude of 94% and process time of 138 s at 310 K. In this work, the authors established that sonication amplitude and process time, as well as their interaction had a significant effect over emulsion droplet size. OPEO nanoemulsions showed Newtonian behavior and were physically stable at 278 K and 298 K over three months of storage. The authors concluded that the obtained results strengthen the potential of ultrasonic technique for application in food and pharmaceutical products. Through ultrasonic technique it is possible to produce nanoscale emulsions of EOs with long-term kinetic stability.

Salvia-Trujillo et al. (2014) evaluated the antimicrobial activity of lemongrass EOs nanoemulsions against *E. coli*. Nanoemulsions were prepared using sodium alginate (1% w/v), Tween 80 (1% v/v), and lemongrass EOs (1% v/v). When the sonication was performed at 400 W and 180 s, nanoemulsions with droplet size of 4.3 ± 0.2 nm were obtained. However, the lowest droplet size of emulsion diminished the antimicrobial potential of lemongrass essential oil nanoemulsions against *E. coli* when compared with nanoemulsions with droplet size above 34.9 ± 8.5 nm obtained at 120 W and 30 s of sonication.

5 Conclusions and Future Perspectives

Flavor and aroma compounds have a fundamental place in relation with food quality and its acceptability by consumers. The change in perception, manner of buying, and how food are produced have prompted consumers to look for food produced using natural additives obtained through environmentally friendly processes that use nontoxic solvents. Besides, herbs and spices have been used since ancestral times as flavoring and aromatic agents in making foods. Due to the characteristics of their EOs, nowadays they have become a valuable source of bioactive compounds and the industrial use of EOs is a very promising area. However, because of their volatility and sensibility to temperature, EOs require greater care for preservation. Nanoencapsulation allows an increase in EOs' solubility and offers protection against factors that trigger their degradation, enhancing EOS' performance as flavoring agents and taking advantage of their pharmacological properties. The developing of novel formulations that enhance its flavoring and aroma characteristics as well as the use of its bioactive properties for increasing shelf-life or even prevent diseases through controlled release allows us to think of a promising future in the field of EO nanocncapsulation. Although currently the application of emerging techniques in the nanoencapsulation of EOs is relatively limited, the production of nanoparticles and nanoemulsions using supercritical fluids as well as nanoemulsion formation assisted by ultrasound both present the potential to develop new food products with enhanced properties. According to the discussion in this chapter, these two technologies have great potential for application in nanoencapsulation of EOs with applications in the food industry. The possibility of using encapsulation materials and solvents recognized as safe (GRAS) in processes performed at operating conditions that guarantee the integrity of the compounds is a valuable opportunity to continue the development of a promising area such as nanoencapsulation of natural flavors and aromas for food industry.

Acknowledgments

The authors are grateful to CNPq (470916/2012-5) and FAPESP (2012/10685-8 2013/04304-4) for financial support. J. Felipe Osorio-Tobón thanks CAPES/DEA/PROEX for a PhD assistantship. Eric Keven Silva thanks CNPq (140275/2014-2) for the PhD assistantship. M.A.A. Meireles thanks CNPq for the productivity grant (301301/2010-7).

References

Abbas, S., Karangwa, E., Bashari, M., Hayat, K., Hong, X., Sharif, H.R., Zhang, X., 2015. Fabrication of polymeric nanocapsules from curcumin-loaded nanoemulsion templates by self-assembly. Ultrason. Sonochem. 23, 81–92.

Abreu, F.O.M.S., Oliveira, E.F., Paula, H.C.B., De Paula, R.C.M., 2012. Chitosan/cashew gum nanogels for essential oil encapsulation. Carbohydr. Polym. 89, 1277–1282.

Almeida, A.P., Rodríguez-Rojo, S., Serra, A.T., Vila-Real, H., Simplicio, A.L., Delgadilho, I., Beirão Da Costa, S., Beirão Da Costa, L., Nogueira, I.D., Duarte, C.M.M., 2013. Microencapsulation of oregano essential oil in starch-based materials using supercritical fluid technology. Innov. Food Sci. Emerg. Technol. 20, 140–145.

Anese, M., Mirolo, G., Beraldo, P., Lippe, G., 2013. Effect of ultrasound treatments of tomato pulp on microstructure and lycopene in vitro bioaccessibility. Food Chem. 136, 458–463.

Arzeni, C., Martínez, K., Zema, P., Arias, A., Pérez, O.E., Pilosof, A.M.R., 2012. Comparative study of high-intensity ultrasound effects on food proteins functionality. J. Food Eng. 108, 463–472.

Asbahani, A.E., Miladi, K., Badri, W., Sala, M., Addi, E.H.A., Casabianca, H., Mousadik, A.E., Hartmann, D., Jilale, A., Renaud, F.N.R., Elaissari, A., 2015. Essential oils: from extraction to encapsulation. Int. J. Pharm. 483, 220–243.

Baser, K.H.C., Buchbauer, G., 2009. Handbook of Essential Oils: Science, Technology, and Applications. CRC Press, Boca Raton.

Beh, C.C., Mammucari, R., Foster, N.R., 2012. Lipids-based drug carrier systems by dense gas technology: a review. Chem. Eng. J. 188, 1–14.

Beirão-Da-Costa, S., Duarte, C., Bourbon, A.I., Pinheiro, A.C., Januário, M.I.N., Vicente, A.A., Beirão-Da-Costa, M.L., Delgadillo, I., 2013. Inulin potential for encapsulation and controlled delivery of oregano essential oil. Food Hydrocolloid. 33, 199–206.

Bermúdez-Aguirre, D., Barbosa-Cánovas, G.V., 2012. Inactivation of Saccharomyces cerevisiae in pineapple, grape, and cranberry juices under pulsed and continuous thermo-sonication treatments. J. Food Eng.V 108, 383–392.

Brunner, G., 2005. Supercritical fluids: technology and application to food processing. J. Food Eng. 67, 21–33.

Carvalho, P.I.N., Osorio-Tobón, J.F., Rostagno, M.A., Petenate, A.J. & Meireles, M.A. A. 2014. Optimization of the ar-turmerone extraction from turmeric (*Curcuma longa* L.) using supercritical carbon dioxide. Fourteenth European Meeting on Supercritical Fluids. Marseilles, France.

Cocero, M.J., Martin, A., Mattea, F., Varona, S., 2009. Encapsulation and coprecipitation processes with supercritical fluids: fundamentals and applications. J. Supercrit. Fluid. 47, 546–555.

Coles, D., Frewer, L.J., 2013. Nanotechnology applied to European food production—a review of ethical and regulatory issues. Trends Food Sci. Tech. 34, 32–43.

Costa, S.S., Gariepy, Y., Rocha, S.C.S., Raghavan, V., 2014. Microwave extraction of mint essential oil—temperature calibration for the oven. J. Food Eng. 126, 1–6.

Covarrubias-Cervantes, M., Bongard, S., Champion, D., Voilley, A., 2005. Temperature effect on solubility of aroma compounds in various aqueous solutions. LWT – Food Sci. Tech. 38, 371–378.

Cravotto, G., Bicchi, C., Mantegna, S., Binello, A., Tomao, V., Chemat, F., 2011. Extraction of kiwi seed oil: Soxhlet versus four different nonconventional techniques. Nat. Prod. Res. 25, 974–981.

CRF, 2013. Code of Federal Regulations. Food and Drugs, Parts 100 to 169 [Online]. Available from: http://www.gpo.gov/.

Cserháti, T., Forgács, E., 2003. Flavor (flavour) compounds: structures and characteristics. In: Caballero, B. (Ed.), Encyclopedia of Food Sciences and Nutrition. second ed. Academic Press, Oxford.

Cushen, M., Kerry, J., Morris, M., Cruz-Romero, M., Cummins, E., 2012. Nanotechnologies in the food industry—recent developments, risks, and regulation. Trends Food Sci. Tech. 24, 30–46.

Chandrapala, J., Oliver, C., Kentish, S., Ashokkumar, M., 2012. Ultrasonics in food processing. Ultrason. Sonochem. 19, 975–983.

Cheetham, P.S.J., 2010. Natural Sources of Flavours: Food Flavour Technology. Wiley-Blackwell, New Jersey.

Choi, J.-A., Lim, G.-B., Ryu, J.-H., 2010. Preparation of PEG microparticles containing coriander essential oil using supercritical PGSS process. KSBB J. 25, 379–386.

De Paz, E., Martín, Á., Duarte, C.M.M., Cocero, M.J., 2012. Formulation of β-carotene with poly-(ε-caprolactones) by PGSS process. Powder Technol. 217, 77–83.

Desrumaux, A., Marcand, J., 2002. Formation of sunflower oil emulsions stabilized by whey proteins with high-pressure homogenization (up to 350 MPa): effect of pressure on emulsion characteristics. Int. J. Food Sci. Tech. 37, 263–269.

Donsì, F., Annunziata, M., Sessa, M., Ferrari, G., 2011. Nanoencapsulation of essential oils to enhance their antimicrobial activity in foods. LWT – Food Sci. Tech. 44, 1908–1914.

Fahim, T.K., Zaidul, I.S.M., Abu Bakar, M.R., Salim, U.M., Awang, M.B., Sahena, F., Jalal, K.C.A., Sharif, K.M., Sohrab, M.H., 2014. Particle formation and micronization using nonconventional techniques—review. Chem. Eng. Process.: Process Intensification 86, 47–52.

Fathi, M., Martín, Á., Mcclements, D.J., 2014. Nanoencapsulation of food ingredients using carbohydrate-based delivery systems. Trends Food Sci. Technol. 39, 18–39.

Ferreira, S.R.S., Nikolov, Z.L., Doraiswamy, L.K., Meireles, M.A.A., Petenate, A.J., 1999. Supercritical fluid extraction of black pepper (*Piper nigrun* L.) essential oil. J. Supercrit. Fluid. 14, 235–245.

Gañán, N.A., Dambolena, J.S., Martini, R.E., Bottini, S.B., 2015. Supercritical carbon dioxide fractionation of peppermint oil with low menthol content—experimental study and simulation analysis for the recovery of piperitenone. J. Supercrit. Fluid. 98, 1–11.

Ge, L., Gong, X., Wang, Z., Zhao, L., Wang, Y., Wang, M., 2015. Sulfur removal from bauxite water slurry (BWS) electrolysis intensified by ultrasonic. Ultrason. Sonochem. 26, 142–148.

Gitin, L., Varona, S., Cocero Alonso, M.J., 2011. Encapsulation of garlic essential oil by batch PGSS process. Innov. Romanian Food Biotech. 9, 45–51.

Gülseren, I., Güzey, D., Bruce, B.D., Weiss, J., 2007. Structural and functional changes in ultrasonicated bovine serum albumin solutions. Ultrason. Sonochem. 14, 173–183.

Hashtjin, A.M., Abbasi, S., 2015. Nano-emulsification of orange peel essential oil using sonication and native gums. Food Hydrocolloid. 44, 40–48.

Herculano, E.D., De Paula, H.C.B., De Figueiredo, E.A.T., Dias, F.G.B., Pereira, V.D.A., 2015. Physicochemical and antimicrobial properties of nanoencapsulated *Eucalyptus staigeriana* essential oil. LWT – Food Sci. Tech. 61, 484–491.

Horst, C., Gogate, P.R., Pandit, A.B., 2007. Ultrasound Reactors: Modeling of Process Intensification. Wiley-VCH Verlag GmbH & Co. KGaA, Weinheim, Germany.

Hundre, S.Y., Karthik, P., Anandharamakrishnan, C., 2015. Effect of whey protein isolate and β-cyclodextrin wall systems on stability of microencapsulated vanillin by spray–freeze drying method. Food Chem. 174, 16–24.

Jordens, J., De Coker, N., Gielen, B., Van Gerven, T., Braeken, L., 2015. Ultrasound precipitation of manganese carbonate: the effect of power and frequency on particle properties. Ultrason. Sonochem. 26, 64–72.

Joye, I.J., Mcclements, D.J., 2014. Biopolymer-based nanoparticles and microparticles: fabrication, characterization, and application. Curr. Opin. Colloid Interface Sci. 19, 417–427.

Kalani, M., Yunus, R., 2011. Application of supercritical antisolvent method in drug encapsulation: a review. Int. J. Nanomed. 6, 1429–1442.

Kfoury, M., Auezova, L., Greige-Gerges, H., Fourmentin, S., 2015. Promising applications of cyclodextrins in food: improvement of essential oils retention, controlled release and antiradical activity. Carbohydr. Polym. 131, 264–272.

Khalili, S.T., Mohsenifar, A., Beyki, M., Zhaveh, S., Rahmani-Cherati, T., Abdollahi, A., Bayat, M., Tabatabaei, M., 2015. Encapsulation of Thyme essential oils in chitosan-benzoic acid nanogel with enhanced antimicrobial activity against *Aspergillus flavus*. LWT – Food Sci. Tech. 60, 502–508.

Lesoin, L., Crampon, C., Boutin, O., Badens, E., 2011. Preparation of liposomes using the supercritical anti-solvent (SAS) process and comparison with a conventional method. J. Supercrit. Fluid. 57, 162–174.

Lévai, G., Martín, Á., De Paz, E., Rodríguez-Rojo, S., Cocero, M.J., 2015. Production of stabilized quercetin aqueous suspensions by supercritical fluid extraction of emulsions. J. Supercrit. Fluid. 100, 34–45.

Li, M.K., Fogler, H.S., 1978a. Acoustic emulsification. Part 1. The instability of the oil-water interface to form the initial droplets. J. Fluid Mech. 88, 499–511.

Li, M.K., Fogler, H.S., 1978b. Acoustic emulsification. Part 2. Breakup of the large primary oil droplets in a water medium. J. Fluid Mech. 88, 513–528.

Longares-Patrón, A., Cañizares-Macías, M.P., 2006. Focused microwaves-assisted extraction and simultaneous spectrophotometric determination of vanillin and p-hydroxybenzaldehyde from vanilla fragans. Talanta 69, 882–887.
Martin, A., Mattea, F., Gutierrez, L., Miguel, F., Cocero, M.J., 2007. Coprecipitation of carotenoids and bio-polymers with the supercritical antisolvent process. J. Supercrit. Fluid. 41, 138–147.
Martín, Á., Varona, S., Navarrete, A., Cocero, M.J., 2010. Encapsulation and ooprecipitation processes with supercritical fluids: applications with essential oils. Open Chem. Eng. J. 4, 31–41.
Mattea, F., Martin, A., Jose Cocero, M., 2009a. Carotenoid processing with supercritical fluids. J. Food Eng. 93, 255–265.
Mattea, F., Martin, A., Matias-Gago, A., Jose Cocero, M., 2009b. Supercritical antisolvent precipitation from an emulsion: beta-carotene nanoparticle formation. J. Supercrit. Fluid. 51, 238–247.
McClements, D.J., 2012. Nanoemulsions versus microemulsions: terminology, differences, and similarities. Soft Matter 8, 1719–1729.
Meireles, M.A.A., 2008. Extracting Bioactive Compounds for Food Products: Theory and Applications. CRC Press, USA.
Mesomo, M.C., Corazza, M.L., Ndiaye, P.M., Dalla Santa, O.R., Cardozo, L., Scheer, A.D.P., 2013. Supercritical CO_2 extracts and essential oil of ginger (*Zingiber officinale* R.): Chemical composition and antibacterial activity. J. Supercrit. Fluid. 80, 44–49.
Metherel, A.H., Taha, A.Y., Izadi, H., Stark, K.D., 2009. The application of ultrasound energy to increase lipid extraction throughput of solid matrix samples (flaxseed). Prostag. Leukotr. Ess. 81, 417–423.
Mezzomo, N., De Paz, E., Maraschin, M., Martin, A., Jose Cocero, M., Ferreira, S.R.S., 2012. Supercritical antisolvent precipitation of carotenoid fraction from pink shrimp residue: Effect of operational conditions on encapsulation efficiency. J. Supercrit. Fluid. 66, 342–349.
Milovanovic, S., Stamenic, M., Markovic, D., Ivanovic, J., Zizovic, I., 2015. Supercritical impregnation of cellulose acetate with thymol. J. Supercrit. Fluid. 97, 107–115.
Mohammadi, A., Hashemi, M., Hosseini, S.M., 2015. Nanoencapsulation of *Zataria multiflora* essential oil preparation and characterization with enhanced antifungal activity for controlling *Botrytis cinerea*, the causal agent of gray mould disease. Innov. Food Sci. Emerg. Technol. 28, 73–80.
Natrajan, D., Srinivasan, S., Sundar, K., Ravindran, A., 2015. Formulation of essential oil-loaded chitosan–alginate nanocapsules. J. Food Drug Anal. 23, 560–568.
Nedovic, V., Kalusevic, A., Manojlovic, V., Levic, S., Bugarski, B., 2011. An overview of encapsulation technologies for food applications. Procedia Food Sci. 1, 1806–1815.
Parthasarathy, V.A., Chempakam, B., Zachariah, T.J., 2008. Chemistry of Spices. CABI Publishers, Oxford, UK.
Pavlic, B., Vidovic, S., Vladic, J., Radosavljevic, R., Zekovic, Z., 2015. Isolation of coriander (*Coriandrum sativum* L.) essential oil by green extractions versus traditional techniques. J. Supercrit. Fluid. 99, 23–28.
Pingret, D., Durand, G., Fabiano-Tixier, A.-S., Rockenbauer, A., Ginies, C., Chemat, F., 2012. Degradation of edible oil during food processing by ultrasound: electron paramagnetic resonance, physicochemical, and sensory appreciation. J. Agr. Food Chem. 60, 7761–7768.
Pingret, D., Fabiano-Tixier, A.-S., Chemat, F., 2013. Degradation during application of ultrasound in food processing: a review. Food Control 31, 593–606.

Pripdeevech, P., Wongpornchai, S., 2013. Odor and flavor volatiles of different types of tea. In: Preedy, V.R. (Ed.), Tea in Health and Disease Prevention. Academic Press, San Diego, USA.

Ravindran, P.N., Babu, K.N., Sivaraman, K., 2007. Turmeric: The Genus *Curcuma*. Taylor & Francis, Boca Raton.

Reverchon, Ernesto, Caputo, Giuseppe, Marco, D., Iolanda, 2003. Role of phase behavior and atomization in the supercritical antisolvent precipitation. ETATS-UNISAmerican Chemical Society, Washington, DC.

Reverchon, E., 1997. Supercritical fluid extraction and fractionation of essential oils and related products. J. Supercrit. Fluid. 10, 1–37.

Reverchon, E., Adami, R., 2006. Nanomaterials and supercritical fluids. J. Supercrit. Fluid. 37, 1–22.

Rodea-González, D.A., Cruz-Olivares, J., Román-Guerrero, A., Rodríguez-Huezo, M.E., Vernon-Carter, E.J., Pérez-Alonso, C., 2012. Spray-dried encapsulation of chia essential oil (*Salvia hispanica* L.) in whey protein concentrate-polysaccharide matrices. J. Food Eng. 111, 102–109.

Rodríguez-Rojo, S., Martín, Á., Cocero, M.J., 2013. Encapsulation Methods with Supercritical Carbon Dioxide: Basis and Applications. Encapsulation Nanotechnologies. John Wiley & Sons, Salem, Massachusetts.

Rowe, D.J., 2005. Chemistry and Technology of Flavors and Fragrances. Blackwell, Boca Raton, USA.

Salvia-Trujillo, L., Rojas-Graü, M.A., Soliva-Fortuny, R., Martín-Belloso, O., 2014. Impact of microfluidization or ultrasound processing on the antimicrobial activity against *Escherichia coli* of lemongrass oil-loaded nanoemulsions. Food Control 37, 292–297.

Samojlik, I., Mijatovic, V., Petkovic, S., ŠKrbic, B., BoŽIn, B., 2012. The influence of essential oil of aniseed (*Pimpinella anisum* L.) on drug effects on the central nervous system. Fitoterapia 83, 1466–1473.

Santos, D.T., Martin, A., Meireles, M.A.A., Jose Cocero, M., 2012. Production of stabilized submicrometric particles of carotenoids using supercritical fluid extraction of emulsions. J. Supercrit. Fluid. 61, 167–174.

Santos, D.T., Meireles, M.A.A., 2013. Micronization and encapsulation of functional pigments using supercritical carbon dioxide. J. Food Process Eng. 36, 36–49.

Santos, P., Aguiar, A.C., Barbero, G.F., Rezende, C.A., Martínez, J., 2015. Supercritical carbon dioxide extraction of capsaicinoids from malagueta pepper (*Capsicum frutescens* L.) assisted by ultrasound. Ultrason. Sonochem. 22, 78–88.

Sarve, A., Sonawane, S.S., Varma, M.N., 2015. Ultrasound assisted biodiesel production from sesame (*Sesamum indicum* L.) oil using barium hydroxide as a heterogeneous catalyst: Comparative assessment of prediction abilities between response surface methodology (RSM) and artificial neural network (ANN). Ultrason. Sonochem. 26, 218–228.

Sebaaly, C., Jraij, A., Fessi, H., Charcosset, C., Greige-Gerges, H., 2015. Preparation and characterization of clove essential oil-loaded liposomes. Food Chem. 178, 52–62.

Shah, B., Ikeda, S., Michael Davidson, P., Zhong, Q., 2012. Nanodispersing thymol in whey protein isolate-maltodextrin conjugate capsules produced using the emulsion–evaporation technique. J. Food Eng. 113, 79–86.

Silva, E.K., Gomes, M.T.M.S., Hubinger, M.D., Cunha, R.L., Meireles, M.A.A., 2015. Ultrasound-assisted formation of annatto seed oil emulsions stabilized by biopolymers. Food Hydrocolloids. 47, 1–13.

Silva, E.K., Meireles, M.A.A., 2014. Encapsulation of food compounds using supercritical technologies: applications of supercritical carbon dioxide as an antisolvent. Food and Public Health 4, 247–258.

Soottitantawat, A., Bigeard, F., Yoshii, H., Furuta, T., Ohkawara, M., Linko, P., 2005. Influence of emulsion and powder size on the stability of encapsulated d-limonene by spray drying. Innov. Food Sci. Emerg. Technol. 6, 107–114.

Soottitantawat, A., Yoshii, H., Furuta, T., Ohkawara, M., Linko, P., 2003. Microencapsulation by spray drying: influence of emulsion size on the retention of volatile compounds. J. Food Sci. 68, 2256–2262.

Sosa, M.V., Rodriguez-Rojo, S., Mattea, F., Cismondi, M., Cocero, M.J., 2011. Green tea encapsulation by means of high pressure antisolvent coprecipitation. J. Supercrit. Fluid. 56, 304–311.

Spigno, G., Donsì, F., Amendola, D., Sessa, M., Ferrari, G., De Faveri, D.M., 2013. Nanoencapsulation systems to improve solubility and antioxidant efficiency of a grape marc extract into hazelnut paste. J. Food Eng. 114, 207–214.

Tavares, G.M., Croguennec, T., Carvalho, A.F., Bouhallab, S., 2014. Milk proteins as encapsulation devices and delivery vehicles: Applications and trends. Trends Food Sci. Technol. 37, 5–20.

Taylor, A.J., Linforth, R., 2009. Food Flavour Technology. Wiley, Singapore.

Toma, M., Vinatoru, M., Paniwnyk, L., Mason, T.J., 2001. Investigation of the effects of ultrasound on vegetal tissues during solvent extraction. Ultrason. Sonochem. 8, 137–142.

Van Opstaele, F., Goiris, K., De Rouck, G., Aerts, G., De Cooman, L., 2012. Production of novel varietal hop aromas by supercritical fluid extraction of hop pellets—part 1: preparation of single variety total hop essential oils and polar hop essences. J. Supercrit. Fluid. 69, 45–56.

Varona, S., Kareth, S., Martín, Á., Cocero, M.J., 2010. Formulation of lavandin essential oil with biopolymers by PGSS for application as biocide in ecological agriculture. J. Supercrit. Fluid. 54, 369–377.

Varona, S., Rodríguez-Rojo, S., Martín, Á., Cocero, M.J., Duarte, C.M.M., 2011. Supercritical impregnation of lavandin (*Lavandula hybrida*) essential oil in modified starch. J. Supercrit. Fluid. 58, 313–319.

Varona, S., Rodríguez Rojo, S., Martín, Á., Cocero, M.J., Serra, A.T., Crespo, T., Duarte, C.M.M., 2013. Antimicrobial activity of lavandin essential oil formulations against three pathogenic food-borne bacteria. Ind. Crop Prod. 42, 243–250.

Vinjamur, M., Javed, M., Mukhopadhyay, M., 2013. Encapsulation of nanoparticles using CO2-expanded liquids. J. Supercrit. Fluid. 79, 216–226.

Visentin, A., Rodriguez-Rojo, S., Navarrete, A., Maestri, D., Cocero, M.J., 2012. Precipitation and encapsulation of rosemary antioxidants by supercritical antisolvent process. J. Food Eng. 109, 9–15.

Wandrey, C., Bartkowiak, A., Harding, S.E., 2009. Materials for Encapsulation. In: Zuidam, N.J., Nedovic, V. (Eds.), Encapsulation Technologies for Active Food Ingredients and Food Processing. Springer, New York.

Wang, K., Arntfield, S.D., 2015. Binding of selected volatile flavour mixture to salt-extracted canola and pea proteins and effect of heat treatment on flavour binding. Food Hydrocolloid. 43, 410–417.

Wang, R., Wang, R., Yang, B., 2009. Extraction of essential oils from five cinnamon leaves and identification of their volatile compound compositions. Innov. Food Sci. Emerg. Technol. 10, 289–292.

Weidner, E., 2009. High-pressure micronization for food applications. J. Supercrit. Fluid. 47, 556–565.

Wen, P., Zhu, D.-H., Wu, H., Zong, M.-H., Jing, Y.-R., Han, S.-Y., 2016. Encapsulation of cinnamon essential oil in electrospun nanofibrous film for active food packaging. Food Control 59, 366–376.

Wen, Z., Liu, B., Zheng, Z., You, X., Pu, Y., Li, Q., 2010. Preparation of liposomes entrapping essential oil from *Atractylodes macrocephala Koidz* by modified RESS technique. Chem. Eng. Res. Des. 88, 1102–1107.

Woranuch, S., Yoksan, R., 2013. Eugenol-loaded chitosan nanoparticles: I. Thermal stability improvement of eugenol through encapsulation. Carbohydr. Polym. 96, 578–585.

Wu, Y., Luo, Y., Wang, Q., 2012. Antioxidant and antimicrobial properties of essential oils encapsulated in zein nanoparticles prepared by liquid–liquid dispersion method. LWT – Food Sci. Tech. 48, 283–290.

Xie, W., Li, R., Lu, X., 2015. Pulsed ultrasound assisted dehydration of waste oil. Ultrason. Sonochem. 26, 136–141.

Zabot, G.L., Moraes, M.N., Meireles, M.A.A., 2014. Influence of the bed geometry on the kinetics of rosemary compounds extraction with supercritical CO_2. J. Supercrit. Fluid. 94, 234–244.

Zhao, C., Wang, L., Zu, Y., Li, C., Liu, S., Yang, L., Zhao, X., Zu, B., 2011. Micronization of *Ginkgo biloba* extract using supercritical antisolvent process. Powder Technol. 209, 73–80.

Zhao, S., Zhang, D., 2014. Supercritical CO_2 extraction of Eucalyptus leaves oil and comparison with Soxhlet extraction and hydro-distillation methods. Sep. Purif. Technol. 133, 443–451.

Zhaveh, S., Mohsenifar, A., Beiki, M., Khalili, S.T., Abdollahi, A., Rahmani-Cherati, T., Tabatabaei, M., 2015. Encapsulation of *Cuminum cyminum* essential oils in chitosan–caffeic acid nanogel with enhanced antimicrobial activity against *Aspergillus* flavus. Ind. Crop Prod. 69, 251–256.

Zhu, L., Lan, H., He, B., Hong, W., Li, J., 2010. Encapsulation of menthol in beeswax by a supercritical fluid technique. Int. J. Chem. Eng. 2010, 7.

Zu, S., Yang, L., Huang, J., Ma, C., Wang, W., Zhao, C., Zu, Y., 2012. Micronization of taxifolin by supercritical antisolvent process and evaluation of radical scavenging activity. Int. J. Mol. Sci. 13, 8869–8881.

Zuidam, N.J., Nedovic, V., 2009. Encapsulation Technologies for Active Food Ingredients and Food Processing. Springer, New York.

4

CYCLODEXTRINS AS ENCAPSULATION MATERIAL FOR FLAVORS AND AROMA

Miriana Kfoury*, Nicoleta Gabriela Hădărugă, Daniel Ioan Hădărugă†, Sophie Fourmentin***

**University of the Littoral Opal Coast (ULCO), Unit of Environmental Chemistry and Interaction on the Living (UCEIV), Dunkirk, France*
***Banat's University of Agricultural Sciences and Veterinary Medicine, Department of Food Science, Timişoara, Romania; †Polytechnic University of Timişoara, Department of Applied Chemistry, Organic and Natural Compounds Engineering, Timişoara, Romania*

1 Introduction

Aroma and flavors are one of the most important ingredients of food products. Nevertheless, they can present problems of instability, losses during heating, or reactions with other food matrix ingredients. In this context, the food industry is aware of the potential of encapsulation technology to overcome these problems and provide as well controlled or targeted delivery or release systems. Encapsulation in cyclodextrins (CDs) is among the most interesting tools in this perspective. CDs are nontoxic naturally occurring oligosaccharides derived from starch. They are available in industrial scale, low prices, food quality, and have been included in the GRAS list of the FDA. CDs possess a hydrophilic external surface and a hydrophobic interior cavity. This unique conformation allows them to encapsulate hydrophobic guests in their internal cavity forming noncovalent inclusion complexes.

The possible benefits of CD encapsulation are various. Interestingly, encapsulation in CDs could be performed both in solution and in solid state using various techniques from laboratory to industrial scale. CDs can control the release of aroma and flavors during food product preparation, storage, and consumption and increase the solubility and retention of these poorly soluble substances. Encapsulation in CDs can also improve the stability of aroma and flavors, extend product shelf life, protect it from oxidation

Encapsulations. http://dx.doi.org/10.1016/B978-0-12-804307-3.00004-1

and isomerization during storage, and create a controlled release system of active substances. Moreover, encapsulation also allows aroma and flavors to remain active as antimicrobial agents despite the harmful environmental conditions and long storage period.

One of the crucial steps when formulating CD encapsulated aroma is to select the appropriate CD and determine optimal conditions for the preparation of solid dosage forms. Proper choice of encapsulation conditions may thus control encapsulation yield, aroma release, and aroma stability during preparation, storage, and consumption of the food product.

The current chapter focuses on the encapsulation of aroma in CDs. This chapter will not stop at the preparation of inclusion complexes in solution and in solid state but it will cover the description of analytical methods used to characterize and bring evidence of the inclusion complexes. Correlations between hydrophobicity and geometry of aroma compounds, the space filling of CDs cavity, and the formation/stability constants (K_f) of inclusion complexes will also be discussed. Moreover, relationships between the solubility, intrinsic photostability, and volatility of aroma with the solubilizing, photoprotector, and retention abilities of CDs will be examined. Additionally, this chapter reviews the effects of encapsulation on the physicochemical and biological properties of aroma.

2 Aroma and Flavors

2.1 Definition/Description

The smell or olfaction is one of the most important senses of vertebrates that identifies and evaluates foodstuffs or their own peers. It furnishes pleasant sensations (eg, the flower perfumes) or warns about various dangers such as degraded food or chemical risks. Smell is a way for communication between humans or animals and nature. The sense of smell is almost all the time related to *perfume, aroma,* or *flavor*. The term *perfume* belongs to the Latin name *per fumum* that refers to the use of the olibanum for religious purposes (Ohloff, 1994). *Aroma* or *aroma substances* (odorants) are volatile compounds that are perceived by the odor receptor sites from the olfactory tissue (by orthonasal and/or retronasal detection). On the other hand, *flavor* is a more complex sensation that especially comprises compounds responsible for odor and taste; furthermore, other human sensations such as visual and mouth feeling are involved in flavor perception (Schudel, 1990; Theimer and Davies, 1967).

The olfaction mechanism is one of the most complex systems. Shortly, it presumes a molecular interaction of the volatile compound at the binding site of the olfactive receptors, followed by a cascade of enzymatic reactions that open the sodium and calcium ion-channels and generates the electric signal at the olfactory cells level. An important role in the olfaction mechanism is related to odorant-binding proteins (OBPs) and odorant-degrading enzymes (ODEs) that regulate the aroma concentration at the olfactory receptors in the mucus (Glusman et al., 2000). There are a lot of classifications of aroma related to the main type of smell descriptors; the latest comprise aromatic, phenolic, floral, fruity, citric, aldehydic, green, herbaceous, coniferous, balsamic, woody, animal, empyreumatic, spicy, and putrid (Ashurst, 1991; Pigott and Paterson, 1994).

In addition to the contribution of aroma, taste compounds also have an important role for the overall flavor of food products. There are five basic tastes named sweetness, saltiness, sourness, bitterness, and *umami* (the meaty taste) (Shallenberger, 1993).

2.2 Extraction and Synthesis

The main sources of aroma compounds are plant materials. Consequently, they are "natural," but other aroma and flavor derivatives (even "nature-identical" or "artificial") exist. They are obtained by fermentation (Gabelman, 1994; McNeil et al., 2013) or by semisynthesis or total synthetic procedures (Hădărugă and Hădărugă, 2003). Thermally generated aromas (the Maillard reaction) are also very important in the food field (Parliment, 1989; Zhang and Ho, 1991). According to the Regulation (EC) 1334/2008 of the European Parliament and of the Council on flavorings and certain food ingredients with flavoring properties for use in and on foods (Anon, 2008), natural flavoring substances are those that have been identified in nature, the source are materials of vegetable, animal or microbiological origin, and the manufacturing process is natural (Baines and Seal, 2012; Parker et al., 2015). Various raw materials as well as plant parts can be used (blossoms, buds, fruits, peels, seeds, leaves, bark, or roots) (Bauer et al., 2001). Hydrodistillation or steam-distillation for essential oils (by direct heating, steam flow in high or low-pressure conditions), extraction by organic solvents or animal fat (concrete oil, oleoresins, resinoids, or balsams, absolute oil, tinctures, and pomades) (Pybus and Sell, 1999), supercritical fluid extraction, cold mechanical pressing, vacuum distillation, cryoconcentration, electromagnetical field concentration, and pervaporation are often used for separation/extraction of aroma and flavors (Bauer et al., 2001;

Guenther, 1972; Martin and Laffort, 1990; Rivera Calo et al., 2015). Natural aroma and flavors are also obtained by means of biotechnology using plant cells, microorganisms such as bacteria, yeasts, fungi, or enzymes (Gabelman, 1994; Van der Schaft, 2015). Numerous ways for semisynthesis and total synthesis of even "nature-identical" or "artificial" aroma compounds exist. Generally, odorant compounds that occur in high quantities are used as raw materials for obtaining other expensive flavors (eg, α- and β-pinene) (Fig. 4.1).

2.3 Physiochemical Properties

Aroma compounds must have some molecular properties in order to generate the smell sensation. They must have moderate water solubility, sufficient vapor pressure for an appropriate concentration in the inspired air, low polarity, and high lipophilicity. Generally, the molar mass of aroma is lower than 400 Da (Kfoury et al., 2015d). One of the main characteristics of odorants is the limit of perception (Weber–Fechner's law). There are two different limits of perception, namely the limit of detection or detection threshold (the lowest concentration at which an odorant can be detected) and the limit of recognition or recognition threshold (the lowest concentration at which an odorant can be recognized) (Bauer et al., 2001). The difference between these two limits of perception is approximately one order of magnitude. The overall flavor of foodstuffs is drastically influenced by so-called "key odorants" (compounds that provide the characteristic aroma of the food, such as citral) and "off-flavoring" compounds (which degrade the food flavor) (Bauer et al., 2001).

The quality and acceptability of these fragrance, aroma, and flavoring products are regulated by various national and international entities such as FDA (Food and Drugs Administration), FEMA (Flavor Extract Manufacturers' Association that approves the list of food additives), GRAS (generally recognized as safe), or EOA (Essential Oil Association). There are two important steps for aroma and flavor analysis: sampling and effective analysis. Classical and modern sampling techniques for aroma and flavoring compound analysis exist: steam-distillation, solvent extraction, or combined methods (hydrodistillation–extraction), static and dynamic headspace analysis (SH and DH), purge-and-trap (PT), pervaporation and microextraction techniques [solid- and liquid-phase microextraction (SPME and LPME) and stir bar sorptive extraction (SBSE)] (Jelen et al., 2012). Various methods are also used for aroma and flavor analysis (Chin and Marriott, 2015; Hădărugă and Hădărugă, 2003; Lawless and Heymann, 1998; Marsili, 1997).

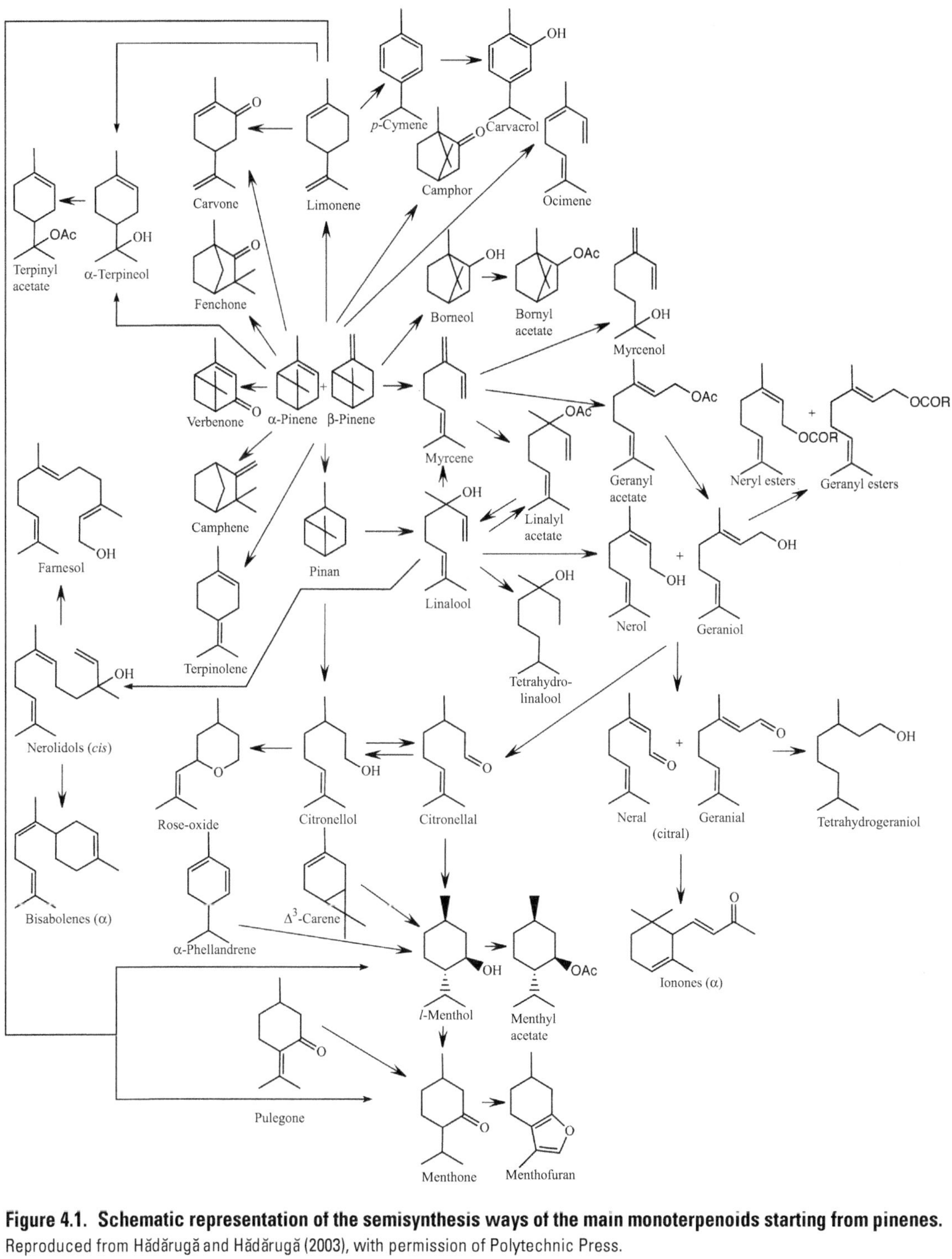

Figure 4.1. Schematic representation of the semisynthesis ways of the main monoterpenoids starting from pinenes. Reproduced from Hădărugă and Hădărugă (2003), with permission of Polytechnic Press.

They consist of classical methods (such as density) as well as advanced physical methods that include gas and liquid chromatography (GC-FID, GC-MS, MDGC, GC-FTIR, HPLC, etc.). Sensorial analysis (even classical or coupled with GC techniques: GC-sniffing and GC-O) allows the identification of "key odorants" or "off-flavoring" compounds even at very low concentrations (lower than the GC limit) by using aroma extract dilution analysis (AEDA), combined hedonic aroma response measurements (CHARM), or OSME specific techniques (Lawless and Heymann, 1998). These are used for determination of flavor dilution factors (FDs) and odor activity values (OAVs). One of the very useful tools for evaluation of authenticity and identification of adulteration of food products by means of flavor analysis is the "electronic nose" (Gardner and Barlett, 1993; Mielle, 1996).

Aroma or odorant compounds belong to hydrocarbons and their derivatives containing oxygen, sulfur, or nitrogen. Classification of odorant compounds is generally performed according to the affiliation of these compounds to terpenoids, the main class of odorant compounds that occur in essential oils. The term *terpene* was used for all compounds having a chemical and architectural relation with C_5H_8, the isoprene unit (eg, monoterpenes, $C_{10}H_{16,}$ and sesquiterpenes, $C_{15}H_{24}$) (Furia and Bellanca, 1975). All terpene derivatives belong to the larger class called terpenoids. Thus, a method of classifying odorant compounds is as terpenoids and nonterpenoid compounds (Fig. 4.2) (Bauer et al., 2001; Parker et al., 2015). On the other hand, the main taste sensations are generated by mono- and disaccharides or other artificial sweeteners for the sweet taste, sodium chloride or other salts for salty taste, and acids (citric, oxalic, or ascorbic) for sour taste. There are many classes of compounds offering a bitter taste, including simple salts ($MgSO_4$), as well as alkaloids (quinine), amino acids and peptides, flavones, and triterpenoids or sulfur-containing compounds (sinigrin from mustard) (Fig. 4.2). Artificial compounds such as monosodium glutamate, disodium 5′-guanilate, or 5′-inosinate and maltol generate the *umami* taste (Shallenberger, 1993).

Due to their hydrophobic character, aromas are not soluble or poorly soluble in water. Moreover, they are generally thermally labile and sensitive to light. Therefore, it is necessary to keep them away from light and moisture. Molecular encapsulation technologies have been developed to convert aroma to different physical aspects: liquids, pastes, powders, granules, spheres, capsules, or microcapsules. These different forms can retain, protect, stabilize, reduce the volatility, and produce controlled release systems of aroma. Various compounds can be used for their encapsulation; among them we can cite starch, gelatin, or cyclodextrins (CDs) (Madene et al., 2006).

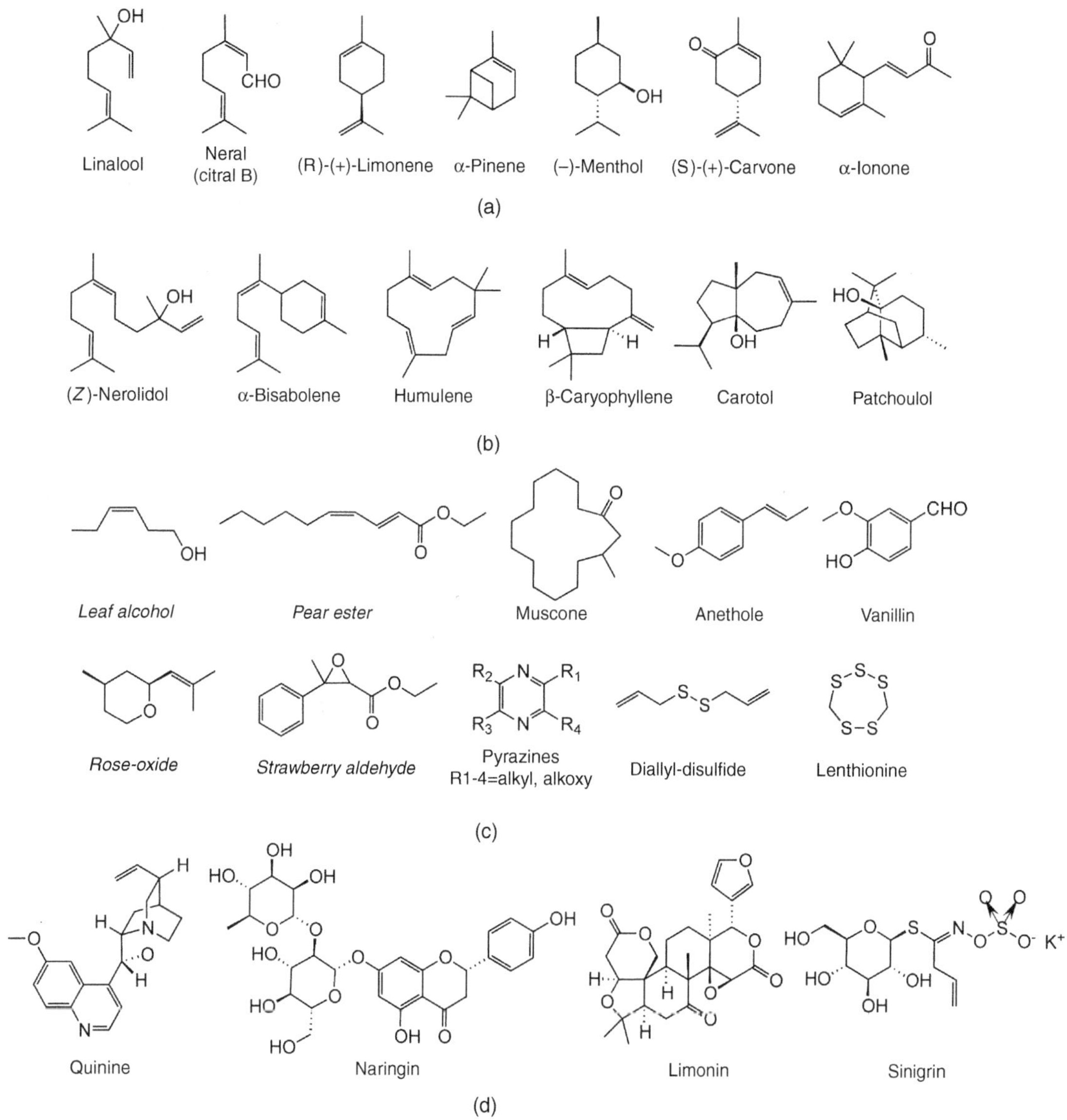

Figure 4.2. Some examples of aroma and flavors from (a) monoterpenoid, (b) sesquiterpenoid, (c) nonterpenoid classes, and (d) flavoring compounds having bitter taste.

3 Cyclodextrins

3.1 History

The development of CDs has been pursued for more than 120 years. It could be mainly divided into three periods: discovery (1891–1935), maturity (1935–70), and application (1970–present)

(Crini, 2014; Szejtli, 1998, 2004). CDs were discovered in 1891 by Villiers, who noted the transformation of potato starch to a novel crystalline carbohydrate under the action of unpure butyric ferment (Villiers, 1891). However, it is generally accepted that Schardinger is the "Founding Father" of CD chemistry (Schardinger, 1903, 1911). He mainly isolated *Bacillus macerans*, the strain of bacteria that produces cyclodextrin glucanotransferase (CGTase), the enzyme responsible for CD formation, and discerned the presence of two forms, α- and β-dextrins, named Schardinger dextrins in his honor, as well as their cyclic structure. γ-Dextrin was discovered later (Freudenberg and Jacobi, 1935). The application of CDs was launched in 1953 when Freudenberg, Cramer, and Plieninger deposed the first patent covering their ability to complex and increase the solubility and stability of physiologically active organic compounds (Freudenberg et al., 1953). Nevertheless, Cramer's work clearly revealed the ability of CDs to act as refuge/host molecules through the encapsulation phenomena and introduced the notion of "inclusion complex" (Cramer, 1954; Cramer and Henglein, 1957). From that time effort has been devoted to produce CDs on a large scale trying to find various applications (Brewster and Loftsson, 2007; Duchêne et al., 2005; Szejtli, 2003; Szejtli and Szente, 2005; Szente and Szejtli, 2004; Wenz, 2009). An exhaustive historical examination of CDs can be found elsewhere (Crini, 2014; Loftsson and Duchêne, 2007; Szejtli, 1998).

3.2 Physicochemical Properties

Production of pharmaceutical and food-grade CDs from starch digestion began effectively at industrial scale since 1979 when genetic engineering progress led to the production of CGTases with increased activity, selectivity, and specificity (Crini, 2014; Li et al., 2009; Liu et al., 2012; Xie et al., 2013). CDs occur as crystalline, homogenous, nonhygroscopic cyclic oligosaccharides (Szejtli, 2004). The most common native CDs contain 6, 7, and 8 D-(+) glucopyranose units bound together by α(1–4) linkages and are referred as α-, β- and γ-CDs (Fig. 4.3).

CDs of over eight glucose units also exist but are hard to produce and purify (Endo and Ueda, 2004; Freudenberg and Cramer, 1948; Larsen, 2002). The chair conformation of the glucose units resulted in a toroidal (truncated, V-shaped, doughnut-shaped, skirt-shaped) shape of CDs with the secondary and primary hydroxyl groups projected on the wide and narrow rim of the torus, respectively. Thus, the external surface of CDs is hydrophilic contrary to the hydrophobic cavity of the torus consisted of C—H groups and

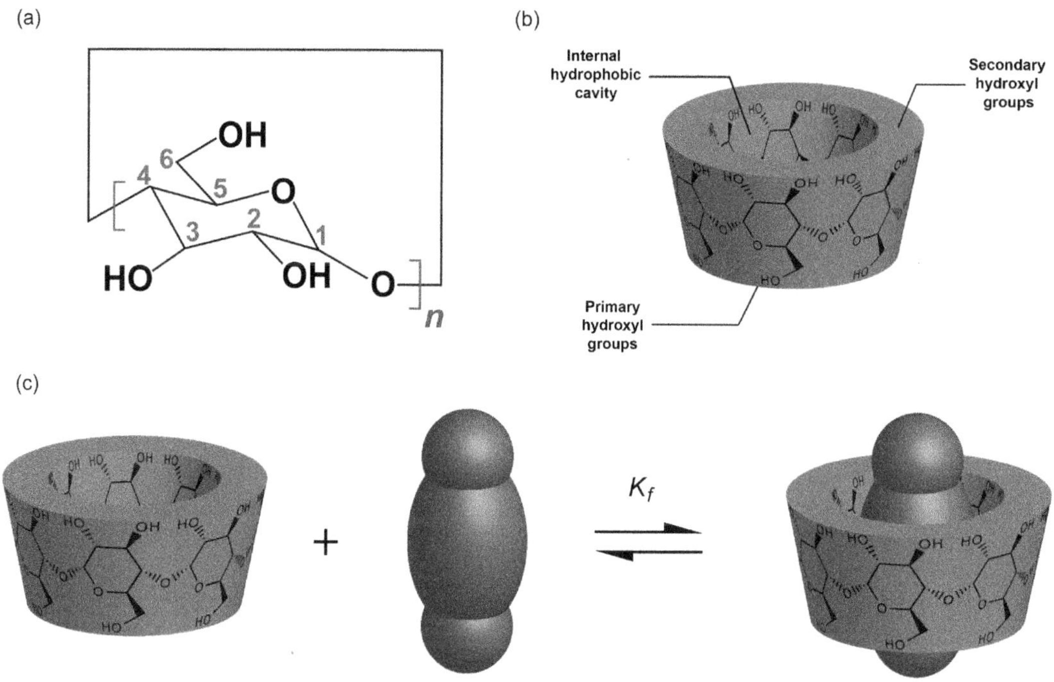

Figure 4.3. Schematic representation of (a) the chemical structure and (b) tridimensional structure of CD and (c) the inclusion phenomena.

glucosidic oxygens (Fig. 4.3). While the height of the cavity is the same for native CDs, the number of glucose units determines the cavity diameter. Physicochemical properties of native α-, β- and γ-CDs are summarized in Table 4.1.

The structure of CDs is stabilized by the formation of hydrogen bonds between C-2 and C-3 hydroxyl groups of adjacent glucose units (Szejtli, 1998). This phenomenon widely affected, in addition to molecular dimensions, the water solubility of CDs. The formation of a complete ring of intramolecular hydrogen bonds in β-CD counteracts its hydration and reduces its solubility as compared to other native CDs (Table 4.1) (Bekiroglu et al., 2003; Szejtli, 1998). Disruption of intermolecular hydrogen bonds (substitution of highly reactive hydroxyl groups with either polar or apolar moieties) generally produces CD derivatives with anomalous increased solubility (Del Valle, 2004; Szente and Szejtli, 1999). Synthesis of derivatives aims also to construct CDs with enhanced complexation ability, selectivity, and CD polymers (Schmidt et al., 2014). More than 15,000 CD derivatives had been obtained by alkylation, hydroxyalkylation, sulfate addition, acetylation,

Table 4.1 Physicochemical Characteristics of α-, β-, and γ-CDs

Physicochemical Properties	α-CD	β-CD	γ-CD
Glucose units	6	7	8
Chemical formula	$C_{36}H_{60}O_{30}$	$C_{42}H_{70}O_{35}$	$C_{48}H_{80}O_{40}$
Chiral carbons	30	35	40
Molecular weight	972	1135	1297
Cavity diameter (Å)	5.7	7.8	9.5
Cycle diameter (Å)	14.6–15	15.4–15.8	17.5–17.9
Cavity volume (Å^3)	173	262	427
Aqueous solubility at 25°C (g 100 mL^{-1})	14.5	1.85	23.2
Torus height (Å)	7.8	7.8	7.8
Internal water molecules	6–8	11–12	13–17
Melting point (°C)	275	280	275
pK_a at 25°C	12.3	12.2	12.1
Log *P* at 25°C	−13	−14	−17
Rotatory power at 25°C	+150.5	+162.5	+177.4
Diffusion coefficient at 40°C	3.4	3.2	3
H donor	18	21	24
H acceptor	30	35	40
Cycle opening, $t_{1/2}$ (h)	33	28	15
Mean, K_f (1:1) (M^{-1})	130 ± 8	490 ± 8	350 ± 9
Hydrolysis by intestinal amylases	Negligible	Slow	Rapid

Brewster and Loftsson, 2007; Kurkov and Loftsson, 2013; Loftsson and Brewster, 2010, 2011; Stella and He, 2008.

amination, esterification, etherification, and so forth (Khan et al., 1998; Szejtli, 2004), but only some particular one such as hydroypropylated derivatives (HP-α-CD, HP-β-CD, HP-γ-CD) (Gould and Scott, 2005), methylated (RAMEα, RAMEB, RAMEγ, CRYSMEB) (Kiss et al., 2007), and sulfobutylated (SBE-β-CD) (Tongiani et al., 2009) are marketed and widely used (Szejtli, 1997). In-depth description on CDs and their derivatives could be found in various literatures (Duchêne, 1991; Hashimoto, 2002; Khan et al., 1998; Kurkov and Loftsson, 2013; Szejtli, 1998; Szente and Szejtli, 1999).

3.3 Inclusion Complex Formation

The growing interest in CDs is due to their amphiphilic structure, which enables them to act as host molecules and encapsulate hydrophobic guests into their apolar cavity through the formation of CD/guest inclusion complexes (Bilensoy and Hincal, 2009; Laza-Knoerr et al., 2010; Pinho et al., 2014; Valente and Söderman, 2014) (Fig. 4.3). This process involves a wide variety of intermolecular interactions (hydrophobic interactions, hydrogen bonds, Van der Waals interactions, steric interactions, etc.). Cramer was the first to clarify the mechanism of inclusion complex formation that he described in five steps: (1) the guest comes near to CD leading to the elimination of water molecules from the CD cavity. Released water molecules get to a gaseous state energy level with enhanced degrees of freedom of translation and rotation and decreased Van der Waals and hydrogen bonds interactions. (2) The guest also acquires a perfect gas state energy level when it liberates water molecules that envelop it. (3) The guest goes into the cavity where Van der Waals interactions and/or hydrogen bonds lead to the stabilization of the subsequent inclusion complex. (4) Water molecules released from guest and CD cavity are rearranged and form hydrogen bonds between each other. (5) The part of the guest that remains outside the cavity is again hydrated by water molecules and integrated into the hydration shell around the CD (Crini, 2014).

The inclusion phenomenon is generally accompanied by an enthalpy–entropy compensation effect. This phenomenon was noticed for the first time by Bender (Bender and Komiyama, 1978) and could be explained by the liberation of water molecules from CD cavity and the formation of hydrogen and Van der Waals interactions between CD and guest (Bender and Komiyama, 1978; Liu and Guo, 2002). Hereto Szejtli (1982) added the importance of steric interactions in the stabilization of inclusion complexes. Nonetheless, the formation of inclusion complexes is reversible and leads to an equilibrium between encapsulated and free guest (Hedges et al., 1995).

Even though the major class of inclusion complexes represent a 1:1 CD:guest stoichiometry, higher stoichiometries might also occur. (1) One CD could encapsulate two guests. (2) One guest could be entrapped by two CDs. (3) Two CDs could encapsulate two guests simultaneously (Landy et al., 2007). This leads to the formation of 1:2, 2:1, and 2:2 inclusion complexes, respectively. These complexes occur generally as a mixture with 1:1 inclusion complex (Fig. 4.4).

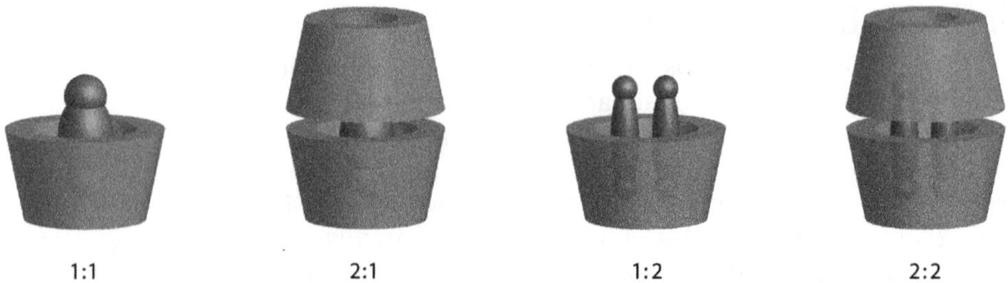

Figure 4.4. Schematic representation of the different stoichiometries of CD/guest inclusion complexes.

3.4 Fields of Application

The negligible toxic effects (Antlsperger and Schmid, 1996; Irie and Uekama, 1997; Stella and He, 2008) and inexpensive costs of CDs (eg, the price of the food grade β-CD is around US$5–6 per kg) resulted in their wide applications in different domains. They are mostly used in food and aroma (López-Nicolás et al., 2014; Marques, 2010; Szejtli, 2004; Szejtli and Szente, 2005; Szente and Szejtli, 2004), cosmetics (Buschmann and Schollmeyer, 2002; Hougeir and Kircik, 2012; Tarimci, 2011), pharmacy or medicine (Liao et al., 2015; Loftsson and Brewster, 2010; Loftsson and Duchêne, 2007), and remediation and decontamination technologies (Blach et al., 2008; Fakayode et al., 2007; Landy et al., 2012; Morin-Crini and Crini, 2013). Nevertheless, several other applications have been developed such as textile finishing (Ammayappan and Jeyakodi Moses, 2009; Lo Nostro et al., 2002), functional textiles (Wang and Chen, 2005), analytical chemistry (Mosinger et al., 2001), chromatography (Juvancz and Szejtli, 2002; Xiao et al., 2012), and catalysis (Hapiot et al., 2006; Komiyama and Monflier, 2006), nanotechnology (Tejashri et al., 2013; Yaméogo et al., 2014), and others.

The use of CD in food formulations has been extensively discussed in the literature (Alonso et al., 2009; Astray et al., 2009; Cravotto et al., 2006; Del Valle, 2004; Jiang et al., 2011; Marques, 2010; Martina et al., 2013; Szejtli, 2004; Szejtli and Szente, 2005; Szente and Szejtli, 2004). CDs are composed of glucose but they are only slightly sweet compounds. A 2.5% β-CD solution is as sweet as a 1.71% solution of sucrose (Toda et al., 1985). Thus, it has been well proved that CDs accomplish the requirements for neutrality in terms of odor and taste and assert consequently a very broad field of use. CDs are generally exploited for their ability to form inclusion complexes, but also as prebiotics.

Readers interested in a detailed description, characteristics, and use of CD encapsulation in various fields are invited to refer to former reviews (Bilensoy, 2011; Dodziuk, 2006; Duchěne et al., 1986; Hashimoto, 2006; Szente and Szejtli, 2004; Szejtli and Szente, 2005).

3.5 Regulatory Status

Regulations for CDs (Table 4.2) differ between countries with Japan being the pioneer producer and consumer of CDs (Mosinger et al., 2001).

The use of α- and β-CDs in food was first authorized in Japan, then in Hungary and Germany in 1976, 1983, and 2000, respectively. In France as in the Netherlands and Belgium, the use of CDs in food was approved in 1986 and generally used for flavors encapsulation. Spanish authorities approved the use of β-CD in foods in 1987. The Joint Expert Committee on Food Additives (JECFA) of the United Nations Food and Agriculture Organization/World Health Organization (UN FAO/WHO) accepted the use of α-, β-, and γ-CDs in 1995, 2000, and 2002, respectively. Owing to their favorable toxicological profile, no acceptable daily intake (ADI) was defined for α- and γ-CDs. Nevertheless, the ADI of β-CD was fixed to 5 mg kg^{-1} body weight in food products [this was allocated based on the no effect level (NOEL) of 470 mg kg^{-1} body weight per day in the diet of dogs during a whole year and a safety factor of 100]. Excess intake of β-CD could result in fewer of the side effects like gas production and soft stools. The assigned E-numbers for α-, β-, and γ-CDs are E457, E459, and E458, respectively (JECFA, 2013; Commission Directive 2003/95/EC, 2003). In addition, α-, β- and γ-CDs were included in the GRAS list of the US Food and Drug Administration (FDA) as food additives in 2000, 2001, and 2004, respectively. In Australia and New Zealand, α- and γ-CDs are considered as Novel Food. It's noteworthy to mention that human beings have been constantly consuming CDs. A study showed that enzyme- and heat-processed starch containing foods include small amounts of native and branched-type (glucosylated and maltosylated) CDs (Szente et al., 2006). Only a few studies attempted to examine the toxicity of CD derivatives (Boulmedarat et al., 2005; Gould and Scott, 2005; Irie and Uekama, 1997; Kiss et al., 2007; Ulloth et al., 2007) and no regulatory status approved till this date their use in food.

Table 4.2 Approval Situation of CDs

CDs	Europe	USA	Australia, New Zealand	Japan
α-CD	Novel Food (2004)	GRAS (2004)	Novel Food (2004)	Natural product
β-CD	Carrier for food additives (<1 g kg^{-1})	GRAS[a] (2001)	—	Natural product
γ-CD	Novel Food (2001)	GRAS (2000)	Novel Food (2003)	Natural product

[a]GRAS as a flavor protector.

3.6 Fate of Cyclodextrins After Ingestion

CDs are generally resistant to the action of amylases (Kurkov and Loftsson, 2013). Unlike linear dextrins and starch, they possess no reducing-end crucial for the action of β-amylases and are very slowly hydrolyzed by α-amylases that digest starch from inside the carbohydrate chain. However, selectivity depends on the amylases and CD type. α- and β-CDs are stable to salivary α-amylases whereas γ-CD is susceptible to the action of both salivary and pancreatic α-amylases (Munro et al., 2004; Szejtli, 1987). Thus, α- and β-CDs are mainly fermented by the colon microflora while the degradation of γ-CD starts in the mouth by salivary α-amylases (Harangi et al., 2012) and proceeds in the gastrointestinal tract by pancreatic α-amylases and bowel bacteria (Antlsperger and Schmid, 1996; De Bie et al., 1998; Irie and Uekama, 1997; Stella and He, 2008; Van Ommen et al., 2004; Zhou et al., 1998). Digestion rate of CDs increases when increasing the cavity size and decreases when CD cavity is occupied by the guest (Buedenbender and Schulz, 2009). Fig. 4.5 illustrates a schematic comparison of the fate of native CDs after oral intake (Kurkov and Loftsson, 2013). Additionally, CDs induce

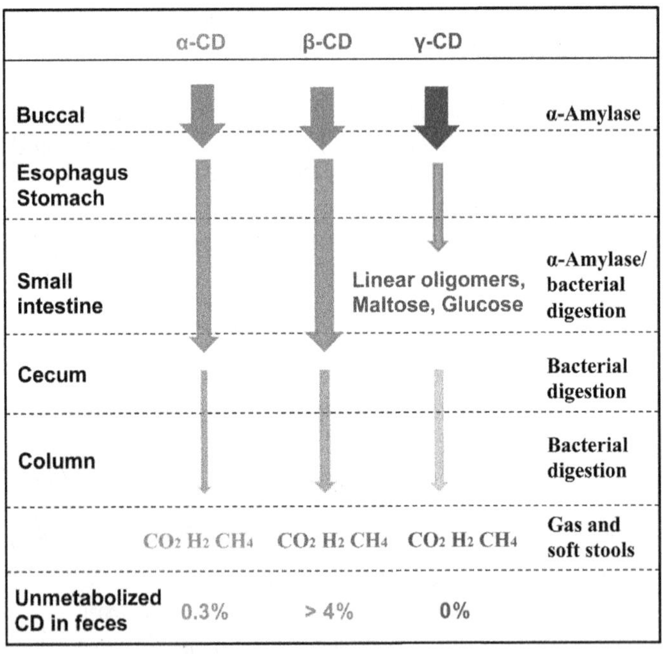

Figure 4.5. Schematic comparison of the digestion of native CDs after ingestion (Kurkov and Loftsson, 2013).

favorable changes in the gastrointestinal microflora supporting the growth and activity of bifidobacteria and lactobacilli and decreasing pathogens such as *Clostridium perfringens* (Pranckute et al., 2014; Spears et al., 2005).

4 Cyclodextrins/Aroma Inclusion Complexes

4.1 Investigation and Characterization of CD/Aroma Inclusion Complexes in Solution

The task to find the appropriate CD and the optimal formulation conditions is challenging for aroma's encapsulation and depends on the demands and the desired application of the final product. An inclusion complex is a dynamic equilibrium and its application relies mainly on the stability of interactions between its components (Dodziuk, 2006). Analytical characterization is fundamental for a correct and efficient use of CD/aroma inclusion complex. In solution, the main parameter that describes the strength of interaction between the inclusion complex components is the formation constant (K_f), also called binding, stability, or association constant (Houk et al., 2003). Although a large number of analytical methods are available for the characterization of inclusion complexes in solution (Marques, 2010; Mura, 2014) few have been applied to CD/aroma complexes. This is mainly attributable to the very low aqueous solubility of aroma. Nonetheless, some techniques could be used such as static headspace coupled to the gas chromatography (SH-GC) (Ciobanu et al., 2012, 2013a,b; Decock et al., 2008; Kfoury et al., 2014a,c, 2015a; Tanemura et al., 1998), UV-Visible (UV-Vis) spectroscopy (Astray et al., 2010; Decock et al., 2006; Zeng et al., 2012), fluorescence spectroscopy (Chen et al., 2010; Jiang et al., 2010), high-performance liquid chromatography (HPLC) (Moeder et al., 1996), and isothermal titration calorimetry (ITC) (Liu et al., 2001, 2004, 2007). Moreover, in order to build more reliable and precise information and surpass the drawbacks of the measurement methods, authors would rather combine and compare results obtained from different techniques (Kfoury et al., 2014c; Yañez and Günther, 2014). The published data regarding the determination of K_f values of CD/aroma inclusion complexes are summarized in Table 4.3. Additionally, to fully characterize inclusion complex in solution, phase solubility studies (Higuchi and Connors, 1965) and nuclear magnetic resonance spectroscopy (NMR) (Schneider et al., 1998) are generally carried out.

Table 4.3 Published Formation Constants (K_f) Values of CD/Aroma Inclusion Complexes

Aroma/Flavor	Log *P*[a]	α-CD	β-CD	γ-CD	CRYSMEB	RAMEB	HPBCD
Trans-anethole	3.096	1,163[b] 927[c] 710[f]	630[b] 1,040[d] 542[c] 497[f]	96[b]	740[b] 1,039[c] 877[f]	1,553[b] 1,815[c] 1,110[f]	1,042[b] 712[e] 845[c] 981[f]
Benzyl alcohol	1.275	52[g]	64[g]	—	56[g]	53[g]	63[g]
Camphene	3.329	598[b]	4,825[b]	360[b]	6,625[b]	6,057[b]	3,033[b]
Camphor	2.160	184[b]	2,058[b]	1,048[b]	1,901[b]	1,194[b]	1,280[b]
Carvacrol	3.810		2,620				
β-Caryophyllene	5.170	—	28,674[h]	4,004[h]	11,488[h]	14,274[h]	4,960[h]
Cinnamaldehyde	2.484	236[g]	450[g] 400[i]	—	595[g]	1,696[g]	969[g] 928[j]
β-Citronellol	3.152	223[g]	3,141[g]	—	3,290[g]	4,048[g]	2,578[g]
p-Cymene	3.898	140[b]	2,505[b]	88[b]	2,549[b]	3,543[b]	2,213[b]
Eucalyptol	2.716	13[b]	615[b]	742[b]	688[b]	673[b]	334[b]
Eugenol	2.100	350[c] 94[g]	462[c] 264[g] 140[d]	—	454[g]	568[g]	436[c] 462[g]
Estragole	2.818	335[c] 478[k]	987[c] 939[k]	108[i]	1,584[c] 1,661[k]	1,916[c] 1,761[k]	1,508[c] 1,581[k]
Geraniol	3.202	908	5,288	—	9,778	11,008	7,128
p-Hydroxybenzaldehyde	1.248	—	138[o]	—	—	—	—
Isoeugenol	2.379	178[c] 85[g]	364[c] 255[g]	—	263[g]	514[g]	418[c] 441[g]
Isomethylionone	4.160	71[g]	9,869[g]	—	15,632[g]	13,176[g]	9,789[g]
Lilial	4.389	4,387[g]	56,567[g]	—	14,7617[g]	166,338[g]	112,205[g]
Limonene	3.615	1,289[b]	3,162[b] 2,230[d]	116[b]	3,668[b]	4,386[b]	2,787[b] 4,730[l]
Linalool	3.213	32[b]	366[b]	138[b]	816[b]	833[b]	596[b] 940[l] 958[e] 720[m]
Menthol	3.335	82[b] 10[n]	1,731[b] 2,240[d]	105[b]	2,396[b]	1,928[b]	1,079[b]

Table 4.3 Published Formation Constants (K_f) Values of CD/Aroma Inclusion Complexes (*cont.*)

Aroma/Flavor	Log P[a]	α-CD	β-CD	γ-CD	CRYSMEB	RAMEB	HPBCD
Menthone	3.149	35[b]	656[b] 546[d]	83[b]	989[b]	748[b]	664[b]
Myrcene	3.994	212[b]	1,431[b]	138[b]	959[b]	1,286[b]	575[b] 1,240[l]
Methyl heptine carbonate	3.220	2,905[g]	226[g]	—	539[g]	485[g]	325[g]
Cis-ocimene	3.970	42[h]	432[h]	20[h]	622[h]	593[h]	538[h]
Trans-ocimene	3.970	46[h]	538[h]	26[h]	789[h]	640[h]	627[h]
α-Pinene	3.542	1,778[b]	2,588[b]	214[b]	2,999[b]	2,395[b]	1,637[b] 5,780[l] 1,842[o]
β-Pinene	3.329	1,018[b]	4,587[b]	633[b]	5,141[b]	4,450[b]	3,151[b] 7,360[l] 1,671[o]
Pulegone	2.516	30[b]	331[b]	82[b]	1,025[b]	796[b]	676[b] 798[o]
Sabinene hydrate	2.320	108[h]	2,108[h]	708[h]	1,308[h]	1,882[h]	772[h]
γ-Terpinene	3.360	37[h]	1,309[h]	40[h]	1,950[h]	2,066[h]	1,488[h]
α-Terpineol	2.600	126[h]	1,143[h]	89[h]	1,223[h]	1,287[h]	761[h]
Thymol	3.342	—	1,467	—	—	—	806[o]
Vanillin	1.067	—	90[p] 100[q]	—	—	—	—

[a]http://www.molinspiration.com/cgi-bin/properties
[b]Ciobanu et al. (2013a).
[c]Kfoury et al. (2014c).
[d]Donze and Coleman (1993).
[e]Demian (2000).
[f]Kfoury et al. (2014b).
[g]Decock et al. (2008).
[h]Kfoury et al. (2015b).
[i]Jiang et al. (2010).
[j]Chen et al. (2010).
[k]Kfoury et al. (2015c).
[l]Tanemura et al. (1998).
[m]Numanoğlu et al. (2007).
[n]Astray et al. (2010).
[o]Kfoury et al. (2014a).
[p]Ferrazza et al. (2014).
[q]Zeng et al. (2012).

4.1.1 Static Headspace-Gas Chromatography

Static headspace coupled to gas chromatography (SH-GC) is employed in numerous fields to detect volatile compounds (Kolb, 1999; Snow and Slack, 2002). It allows direct analysis of a volatile present in a gaseous phase in equilibrium with a coexisting aqueous or solid phase in a closed vessel without any interference of the nonvolatile matrix (Kolb and Ettre, 2006). The SH-GC is sensitive and allows the detection of the encapsulation of volatile aroma and flavors in CD at very low concentrations, where they are, as well as their inclusion complexes, soluble in aqueous solutions (Bicchi et al., 2012). In this case the nonvolatile matrix is an aqueous solution of CD. This technique has been widely applied to determine K_f values for CD/aroma inclusion complexes (Ciobanu et al., 2013a; Decock et al., 2008; Kfoury et al., 2015a; Matsuda et al., 1991; Saito et al., 1999). The determination of a K_f value relies on the comparison of the amount of aroma present in the gaseous phase when evenly introduced in water or CD solutions at different concentrations. Fig. 4.6 illustrates the experimental procedure used in SH-GC.

The effect of increased concentrations of β-CD on the chromatographic peak of γ-terpinene is shown as an example in Fig. 4.7. The magnitude of the variation reflects the strength of binding between the two components and leads to the determination of the K_f value.

Recent studies have developed and validated a new SH-GC method to determine the binding potential of CDs toward volatiles directly present in complex mixtures like essential oils (Fourmentin et al., 2013; Kfoury et al., 2015a). This method is based on the investigation of a panel of equilibriums that take place in a heterogeneous mixture and provides the binding potential of CD to each

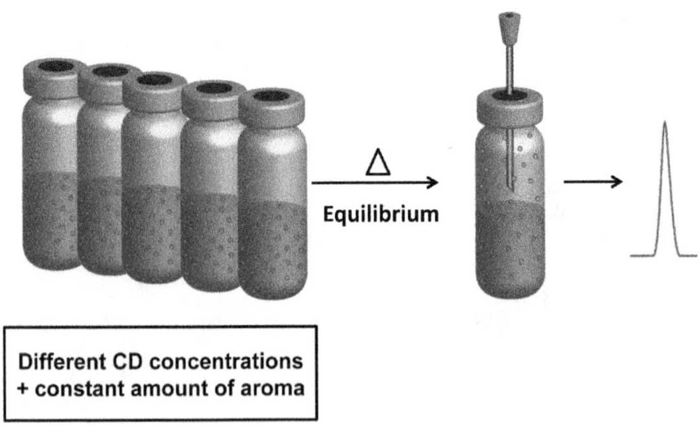

Figure 4.6. Illustration of the static headspace-gas chromatography experiment.

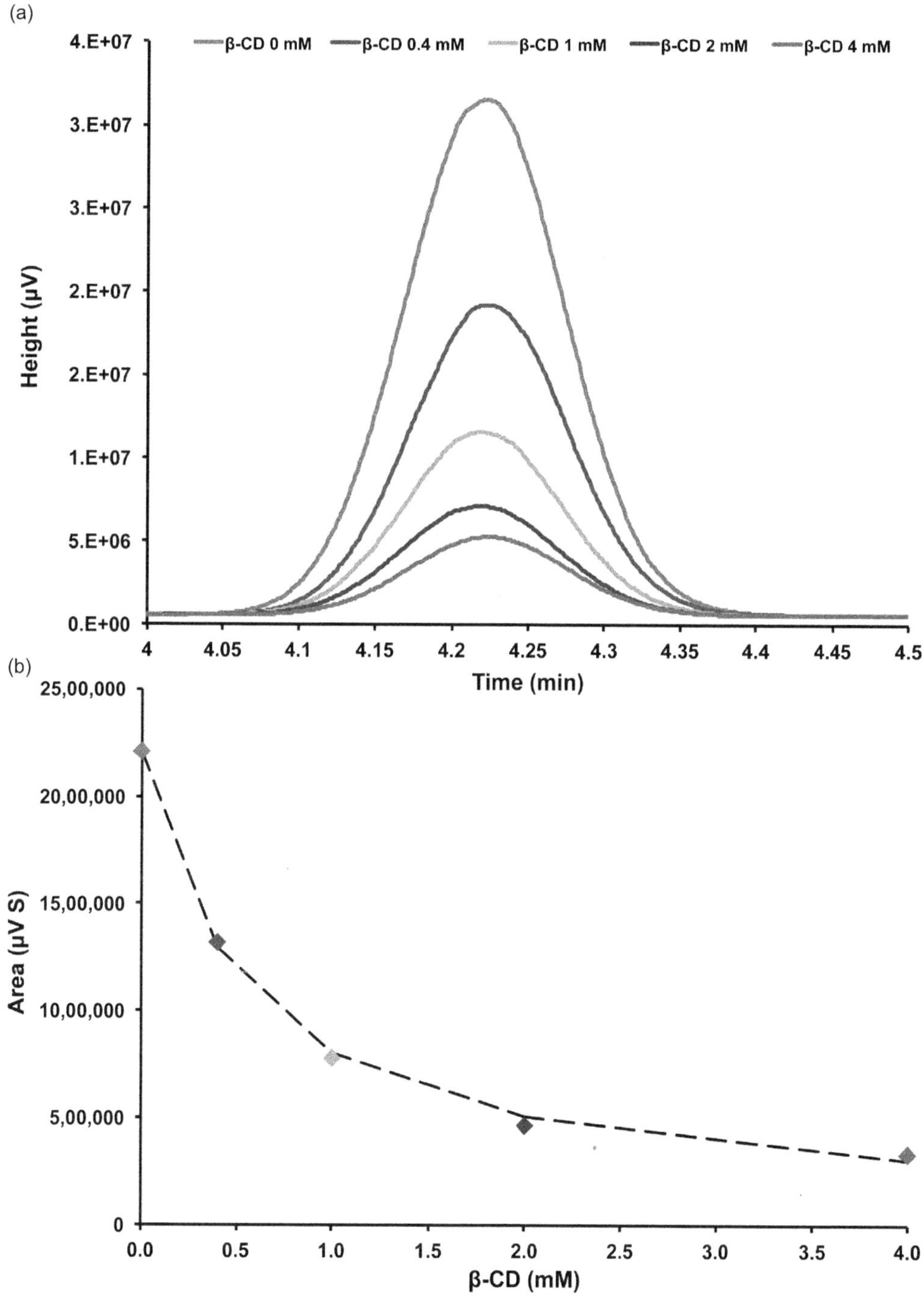

Figure 4.7. Representation of (a) the variation of γ-terpinene chromatogram with various concentrations of CD and (b) the experimental peak areas fit (*filled diamonds*) with theoretical titration curve (*dashed lines*) for a 1:1 inclusion complex.

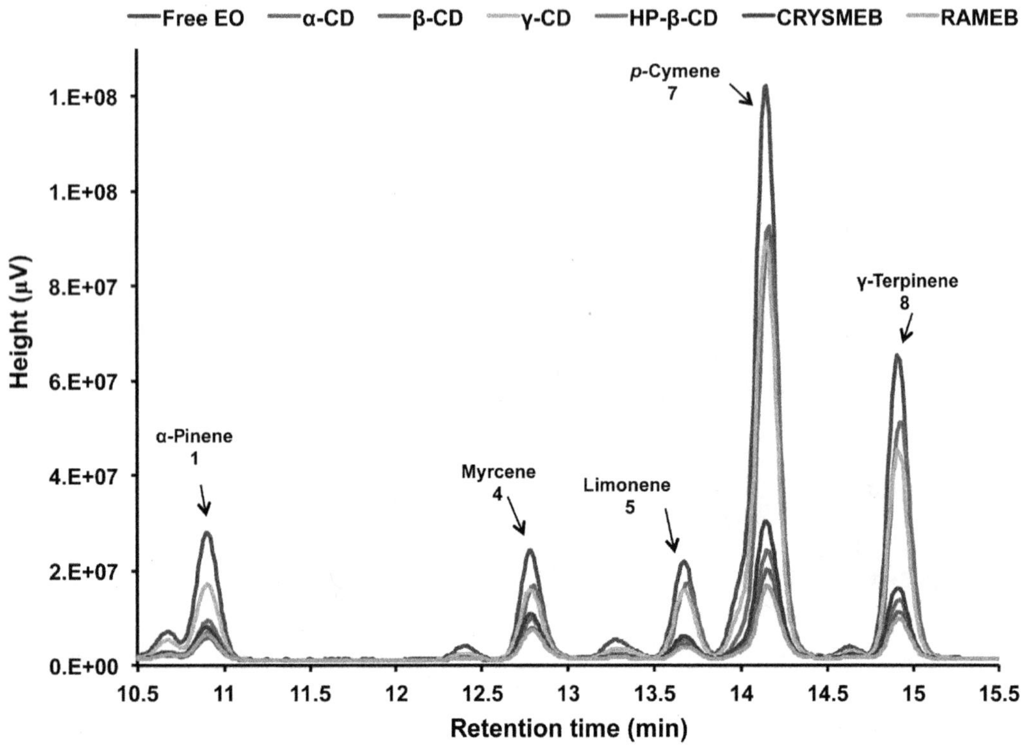

Figure 4.8. Representation of the variation of the chromatogram of *Satureja montana* EO in the presence of various CDs. Reprinted from Kfoury, M., et al., 2015a. Promising applications of cyclodextrins in food: improvement of essential oils retention, controlled release and antiradical activity. Carbohydr. Polym. 131, 264–272, with permission from Elsevier.

individual constituent of the essential oil. Fig. 4.8 illustrates the effect of different CDs (0.4, 1, 2, and 4 mM) on the main volatile components of *Satureja montana* essential oil (Kfoury et al., 2015a).

4.1.2 UV-Visible and Fluorescence Spectroscopies

These two methods are generally based on the measurement of a variation in the absorbance or fluorescence intensity of a fixed aroma's concentration in the presence of different CD concentrations. K_f value can be calculated from spectral changes using different approaches such as Benesi–Hildebrand, Scatchard, or Scott (Benesi and Hildebrand, 1949; Connors, 1987). Several examples for CD/aroma inclusion complexes are found in the literature such as eucalyptol, α-pinene, β-pinene, limonene, thymol, linalool, pulegone, and geraniol (Kfoury et al., 2014a), vanillin and *p*-hydroxybenzaldehyde (Zeng et al., 2012), and so forth. However, several molecules lack a chromophore or a fluorophore structure. This led researchers to improve current classic methods and develop new experimental procedures such as the spectral displacement method (Landy et al., 2000, 2007). It allows the study of compounds that possess weak or no UV chromophore or which are too poorly soluble in aqueous solution to give observable signal (Decock et al., 2006; Kfoury et al., 2014b; Tutaj et al., 2003). Nonetheless this approach can also be applied with fluorescence, ^{1}H NMR, circular dichroism, or other techniques (Landy et al., 2007).

4.1.3 Isothermal Titration Calorimetry

A very useful method used for the investigation of inclusion complexes in solution is the isothermal titration calorimetry (ITC) (Bouchemal and Mazzaferro, 2012; Wszelaka-Rylik and Gierycz, 2013). It allows in a single experiment the determination of the K_f value as well as of the thermodynamic parameters of the inclusion process. Thus, it evaluates the enthalpy and entropy changes and provides a complete thermodynamic profile of the interactions between CD and aroma (Giordano et al., 2001; Liu et al., 2007). Fig. 4.9 illustrates an example of an ITC titration of β-CD water solution against (−)-borneol water solution. Hereto,

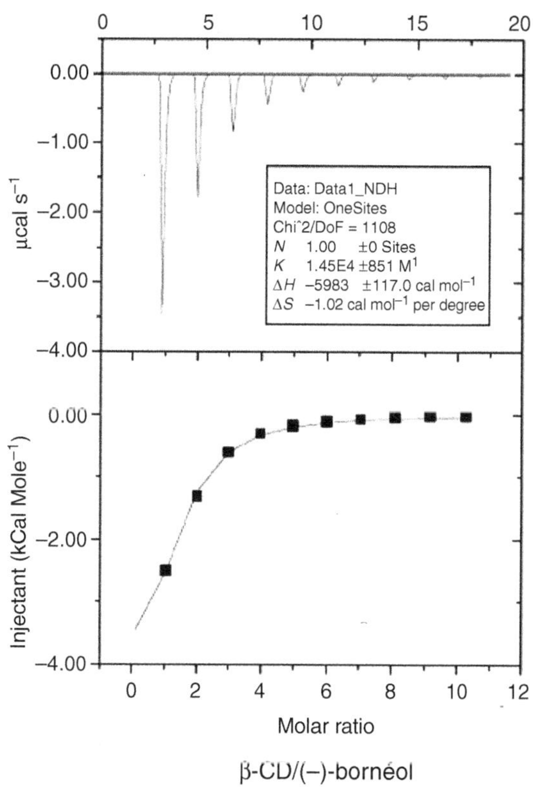

Figure 4.9. ITC results for a titration of β-CD water solution against (−)-borneol water solution.

this technique is still not widely applied to aroma owing to their very poor aqueous solubility. Alternative nonconventional approaches suitable for such compounds are now being explored (Bertaut and Landy, 2014).

4.1.4 Phase Solubility Studies

To describe how the CD concentration influences aroma's solubility, phase solubility studies are generally carried out (Higuchi and Connors, 1965). These studies allow, besides the determination of a K_f value for the inclusion complex, the evaluation of the increase in aroma's solubility, the complexation efficiency, the optimal aroma:CD ratio, and the increase in formulation bulk, crucial factors for the preparation of solid inclusion complexes (Loftsson et al., 2005). Fig. 4.10 illustrates as an example the phase solubility profiles obtained for *trans*-anethole with the three native CDs (Kfoury et al., 2014c).

As we can see, different profiles could be obtained: α-CD/*trans*-anethole showed an A_L-type diagram indicating a linear increase in *trans*-anethole aqueous solubility with CD's concentration; B-type profile was observed for β-CD/*trans*-anethole representative for the formation of inclusion complexes with limited solubility in the aqueous medium; γ-CD didn't show any considerable

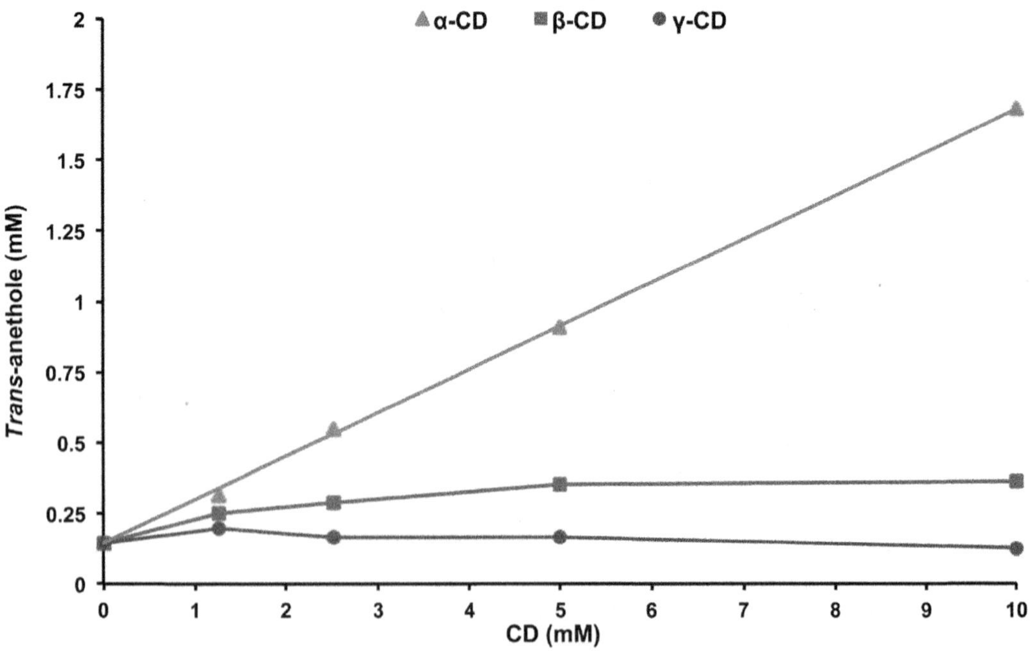

Figure 4.10. Phase solubility profiles of CD/*trans*-anethole inclusion complexes.

increase in *trans*-anethole solubility, probably due to its poor ability to encapsulate this aroma. The assessment of the phase solubility profiles is based on the classification made by Higuchi and Connors (1965). This example confirms that selecting the appropriate CD is of great importance to explore successful encapsulation of the aroma and application of the inclusion complex. Phase solubility studies have been widely performed to investigate the solubilizing potential of CDs toward different aromas such as eugenol, isoeugenol, *trans*-anethole, and estragole (Kfoury et al., 2014c), linalool and 2-pentanoylfuran (Liang et al., 2012), carvacrol (Santos et al., 2015), thymol (Tao et al., 2014), α-terpineol (Mazzobre et al., 2011), α-bisabolol (Waleczek et al., 2003), curcumin (Ansari et al., 2014), cinnamaldehyde (Carlotti et al., 2007; Hill et al., 2013), vanillin (Karathanos et al., 2007; Zeng et al., 2012), *p*-hydroxybenzaldehyde (Zeng et al., 2012), geraniol (Mourtzinos et al., 2008), and so forth. Additionally, authors attempted to develop method aiming to determine the increase in solubility of each single compound present in an initial mixture such as major garlic oil components, for example (Bai et al., 2010).

4.1.5 NMR Spectroscopy

To distinguish between inclusion and noninclusion complexes, NMR spectroscopy plays a crucial role and permits the structural elucidation of inclusion complexes (Fielding, 2000; Schneider et al., 1998). ^{1}H NMR or ^{13}C NMR experiments provide a wealth of information on the inclusion complex and are based respectively on the observation of the difference in the proton or carbon chemical shifts and relaxation rates between free and complexed species (CD and aroma). Analysis of chemical shift changes of the aroma's nuclei as a function of increasing CD concentration allows the determination of the K_f value and provides insights on the mechanism of inclusion, the degree of penetration of the aroma in CD's cavity, and the geometry of the complex in solution (Djedaini et al., 1990; Rekharsky et al., 1995). Additionally, 2D NMR spectroscopy offers a rapid and clear evidence for the spatial proximity of atoms following the observation of intermolecular dipolar cross-correlations (Meyer and Peters, 2003). Two protons closely located in space can produce a nuclear Overhauser effect (NOE) cross-correlation in NOE spectroscopy (NOESY), or rotating-frame NOE spectroscopy (ROESY). These NOE cross-peaks between protons of CD and guest molecules point to spatial contacts within 0.4 nm. The observed NOE effect is particularly useful to clarify the supramolecular structure of the inclusion complex and establish the binding mode and penetration of the aroma into the CD's cavity. Fig. 4.11 shows a section of the contour plot of

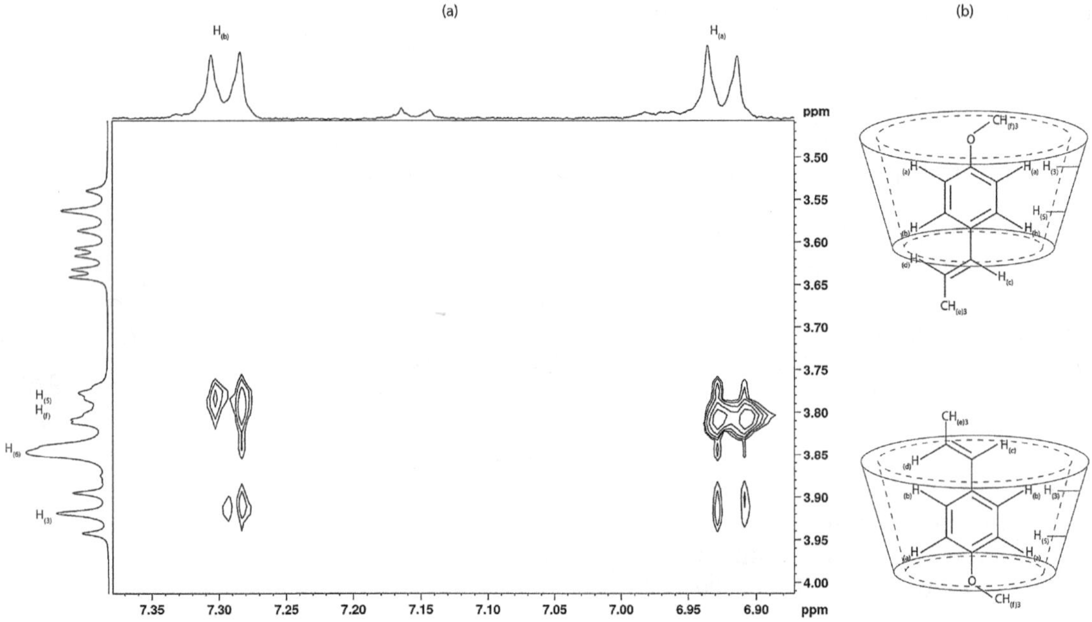

Figure 4.11. (a) ROESY spectrum of β-CD/*trans*-anethole complex in D_2O and (b) possible inclusion mode. Reprinted from Kfoury, M., et al., 2014b. Cyclodextrin, an efficient tool for *trans*-anethole encapsulation: chromatographic, spectroscopic, thermal, and structural studies. Food Chem. 164, 454–461, with permission from Elsevier.

the ROESY spectrum for β-CD/*trans*-anethole inclusion complex (Kfoury et al., 2014b). The ROESY spectrum showed correlation between the protons of the aromatic cycle of *trans*-anethole and those of the cavity of β-CD (H-3 and H-5) indicating that the phenyl ring of *trans*-anethole is embedded inside the cavity.

NMR spectroscopy has also been applied to study CD inclusion complexes with eugenol (Garg et al., 2010), thymol (Del Toro-Sánchez et al., 2010), curcumin (Jahed et al., 2014), vanillin (Ferrazza et al., 2014), *p*-hydroxybenzaldehyde (Zeng et al., 2012), cinnamaldehyde (Wu et al., 2011), linalool and benzyl acetate (Numanoğlu et al., 2007), fenchone (Nowakowski and Ejchart, 2014), or carvacrol (Locci et al., 2004).

4.2 Investigation and Characterization of CD/Aroma Inclusion Complexes in Solid State

CDs are some of the most appropriate matrices for food formulations that need controlled release of bioactive compounds when they are consumed (Crini, 2014). This is the case for aroma and flavors, vitamins, or antioxidants (Daruházi et al., 2008, 2013; Folch-Cano et al., 2010; Hădărugă et al., 2012a; Kfoury et al., 2015a,d; Kurkov and Loftsson, 2013; Li et al., 2014; Pinho et al., 2014). Generally, the inclusion complex is obtained in solid state in order to be used in food (especially in solid-like foodstuffs) (Astray et al., 2009; Del Valle, 2004; Kfoury et al., 2015a; Li et al., 2014; López-de-Dicastillo et al., 2011; Szejtli and Szente, 2005). The main advantages of such CD encapsulation are related to the protection against degradation of labile compounds under temperature, light, water, and/or oxygen action (Hădărugă et al., 2006, 2008a, 2010b, 2014) as well to the controlled release of bioactive compounds (long time action in the human body such as antioxidant flavors) (Del Toro-Sánchez et al., 2010; Numanoğlu et al., 2007; Reineccius et al., 2002), enhancing of water solubility and bioavailability (especially volatile aroma and flavors that are hydrophobic molecules) (Del Valle, 2004; Mazzobre et al., 2011; Mohan et al., 2012; Nxumalo et al., 2013). For example, limonene is easily oxidized to diepoxy-limonene, which was proved to be a carcinogenic compound; its degradation is significantly reduced by nanoencapsulation in CD (Fig. 4.12). Solid CD complexes can also be easily handled and quantified at industrial level that implies reducing costs of production, packaging, and storage (Duchêne, 2011; Szejtli, 1988).

Due to the high lipophilicity (low polarity) and high vapor pressure, as well as the geometrical compatibility of the majority of

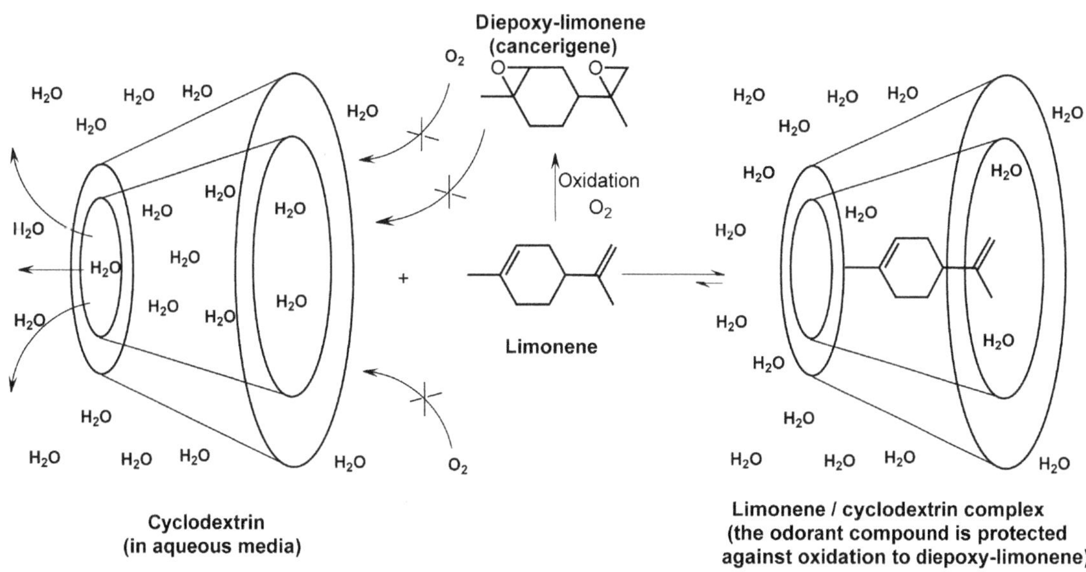

Figure 4.12. Degradation of limonene (the main compound from citrus essential oils) and protection of this labile aroma compound by CD complexation.

aroma compounds, the interaction with the inner cavity of CDs and the stability of the final complex are enhanced (Brewster and Loftsson, 2007; Duchêne, 2011; Loftsson and Duchêne, 2007; Szejtli, 1988). The main aroma and flavors that well interact with the CD cavity belong to acyclic or cyclic terpenes and their oxygenated derivatives, nonterpenoidic aliphatic compounds, nonphenolic or phenolic aromatic compounds, and N-, O-, S-containing heterocyclic compound classes (Marques, 2010; Hădărugă et al., 2012b; Kfoury et al., 2015b). If aroma compound mixtures are used for CD complexation (as is the case for all essential oils), the relative concentrations of the complexed compounds are more or less modified, depending on the CD nanoencapsulation competitivity (Costescu et al., 2008; Hădărugă et al., 2012b, 2014).

4.2.1 Synthesis of Solid Inclusion Complexes

The techniques for solid inclusion complex preparation can be classified as follows: (1) *cocrystallization* or *coprecipitation*, especially as laboratory technique, is one of the oldest methods that assumes the mixing of CD and aroma solutions (or even suspensions) at appropriate temperatures for a due time, followed by crystallization by cooling at low temperatures, filtering and drying (Hădărugă et al., 2012b; Hegheş et al., 2015; Moreira da Silva et al., 1999; Ponce Cevallos et al., 2010; Riviş et al., 2008; Zhang et al., 2007). Various solvents and complexation temperatures can be used according to the aroma properties. The main disadvantage of this technique is that related to the yield of encapsulation (Hădărugă et al., 2012a,b). An important part of the compounds remains in the filtrate and this technique is more useful for low water soluble CDs such as β-CD. (2) *Cogrinding* comprises slurry, dump, paste, and kneading complexation, according to the synthesis conditions (De Souza Siqueira Quintans et al., 2013; Guimarães et al., 2015; Songkro et al., 2012; Tao et al., 2014). This method is used on both laboratory and industrial scales. It presumes the simultaneous mixing of the main components of the complex in the presence of various water/solvent mixture quantities and ratios. The encapsulation conditions that include the temperature, encapsulation time, the nature of CD, as well as other additives will give the appropriate cogrinding complexation particularities. The final product is obtained by drying at normal pressure or vacuum and specific drying atmosphere and temperature that depend on the thermal and oxidative stability of the encapsulated compounds, as well as the quality of encapsulation and complex stability. The resulted solid complexes have no well-formed crystals in comparison with the cocrystallization method, which furnish more reproducible complex crystals. On the contrary, the recovering yields are higher in comparison with cocrystallization method. (3) *Coevaporation* is a technique that needs the mixing of aroma with CD in the presence of an appropriate solvent for a determined time. The solvent is progressively removed at various temperatures (normal, high, or low values) according to the compounds or their complex stability. (4) *Spray-drying* and *freeze-drying* are techniques that belong to coevaporation method (Arana-Sánchez et al., 2010; Hill et al., 2013; Kfoury et al., 2015a; Nuchuchua et al., 2009; Tao et al., 2014; Yuan et al., 2014; Zhu et al., 2015). They presume the strong mixing of appropriate compounds for complexation in the presence of a solvent and the solution or suspension is sprayed in specific conditions in order to obtain micro- or nanoparticles of complexes that are more or less amorphous (Duchêne, 2011). (5) *Sealed-heating* is a technique that works in a well-sealed reactor at high pressure and temperature during few hours. The product is separated by known procedure, but the complex crystals are not well formed (Duchêne, 2011). (6) *Complexation by using supercritical carbon dioxide method* is less used in the case of aroma or other volatile compound. The equipment is composed by cells containing the compound and the CD. The cells are maintained in supercritical conditions in the presence of carbon dioxide (73.8 bar and 31.1°C). After depressurization, the obtained complex is grounded (De Marco and Reverchon, 2008; Duchêne, 2011). (7) *Microwave method* allows rapid attaining of the desired temperature for complex formation in all CD/aroma mixtures in the presence of the solvent and other additives (Duchêne, 2011). The complexation time is drastically reduced and the complex has uniform properties in all bulk volume. Thus, the complexation time is generally 1.5 min and the temperature 60°C (at a maximum power of 150 W). These microwave complexation parameters can be modified according to the stability and properties of guests.

A synopsis of the main CD complexes containing aroma and flavors (or bioactive systems such as essential oils) that were obtained even at laboratory or industrial level, is presented in Table 4.4.

4.2.2 Characterization of Solid Inclusion Complexes

CDs and their complexes can be analyzed by various methods that allow determining both the content of included compounds and the bonding of these compounds in solid CD complexes (Astray et al., 2009; Buschmann and Schollmeyer, 2002; Marques, 2010; Del Valle, 2004; Mura, 2015). Indirect methods for the evaluation of inclusion quality are related to the determination of water (or other solvents) content of complexes (Hădărugă

Table 4.4 A Selection of Aroma Compounds and Essential Oils Obtained as CD Complexes in Solid State

Aroma/Flavoring Compound/ Essential Oil	Type of Cyclodextrin, Methods of Complexation, and Analysis of Solid Complexes	References
Aroma compounds		
2-Acetyl-1-pyrroline (rice flavor)	α-Cyclodextrin and 2-hydroxypropyl-β-cyclodextrin solid complexes, obtained by spray-drying method, and analyzed by SEM and GC-FID	Kawakami et al. (2009)
Allyl isothiocyanate	β-Cyclodextrin solid complex, obtained by coprecipitation method, and analyzed by UV-Vis	Zhang et al. (2007)
Trans-anethole	β-Cyclodextrin solid complex, obtained by coprecipitation and freeze-drying methods, and analyzed by UV-Vis, FT-IR, TGA, SEM, and XRD	Zhang et al. (2015)
Benzaldehyde	β-Cyclodextrin solid complex, obtained by coprecipitation method, and analyzed by GC-MS and TG	Riviş et al. (2008)
Benzyl acetate	2-Hydroxypropyl-β-cyclodextrin solid complex, obtained by slurry and freeze-drying methods, and analyzed by ^{1}H-NMR and circular dichroism spectroscopy (CD)	Numanoğlu et al. (2007)
Camphene/fenchene	β-Cyclodextrin solid complex, obtained by coprecipitation method, and analyzed by GC-FID, ^{1}H-NMR, and XRD. Application for the separation of camphene from fenchene	Ceborska et al. (2013a)
Carvacrol	β-Cyclodextrin solid complex, obtained by slurry method, and analyzed by TG, DSC, and SEM. Evaluated for its anticancerigene activity	Guimarães et al. (2015)
(R)-(−)- and (S)-(+)-carvone	β-Cyclodextrin solid complex, obtained by coprecipitation method, and analyzed by ^{1}H-NMR	Moreira da Silva et al. (1999)
β-Caryophyllene	β-Cyclodextrin solid complex, obtained by coprecipitation method, and analyzed by GC-MS and TG	Riviş et al. (2008)
Cinnamaldehyde	β-Cyclodextrin solid complex, obtained by freeze-drying method, and analyzed by oxidative DSC, TEM, and UV-Vis. Evaluated for its antimicrobial activity	Hill et al. (2013)
Cinnamaldehyde	β-Cyclodextrin solid complex, obtained by coprecipitation method, and analyzed by DSC	Ponce Cevallos et al. (2010)
Citral (geranial and neral)	Monochlorotriazinyl-β-cyclodextrin solid complex, obtained by freeze-drying method, and analyzed by SEM, TEM, FT-IR, TG, and MM	Zhu et al. (2015)

Table 4.4 A Selection of Aroma Compounds and Essential Oils Obtained as CD Complexes in Solid State (*cont.*)

Aroma/Flavoring Compound/ Essential Oil	Type of Cyclodextrin, Methods of Complexation, and Analysis of Solid Complexes	References
Citronellal	β-Cyclodextrin solid complex, obtained by kneading method, and analyzed by GC-MS, SEM, FT-IR, and DSC. Evaluated for its mosquito repellent activity	Songkro et al. (2012)
Citronellol	β-Cyclodextrin solid complex, obtained by kneading method, and analyzed by GC-MS, SEM, FT-IR, and DSC. Evaluated for its mosquito repellent activity	Songkro et al. (2012)
p-Cymene	β-Cyclodextrin solid complex, obtained by kneading method. Evaluated for its antinociceptive and antiinflammatory activities	De Souza Siqueira Quintans et al. (2013)
γ-Decalactone	β-Cyclodextrin solid complex, obtained by coprecipitation method, and analyzed by GC-MS and TG	Riviş et al. (2008)
Estragole (methyl chavicol)	α-, β-, γ-, 2-Hydroxypropyl-β-, randomly methylated-β-, and low methylated-β-cyclodextrin solid complexes, obtained by freeze-drying method, and analyzed by UV-Vis, DSC, FT-IR, ^{1}H-NMR, and SHGC. Evaluated for their antioxidant activity	Kfoury et al. (2015b)
Ethyl benzoate	2-Hydroxypropyl-β-cyclodextrin solid complex, obtained by freeze-drying method, and analyzed by UV and FT-IR	Yuan et al. (2014)
Eugenol	β-Cyclodextrin solid complex, obtained by freeze-drying method, and analyzed by oxidative DSC, TEM, and UV. Evaluated for its antimicrobial activity	Hill et al. (2013)
Eugenol	α-, β-, γ-, and 2-hydroxypropyl-β-cyclodextrin solid complexes, obtained by freeze-drying method, and analyzed by FT-IR, DSC, and TG	Nuchuchua et al. (2009)
Eugenol	β-Cyclodextrin solid complex, obtained by freeze-drying method, and analyzed by FT-IR, XRD, and UV-Vis. Evaluated for its antibacterial activity	Wang et al. (2011b)
Geraniol	β-Cyclodextrin solid complex, obtained by kneading and slurry methods, and analyzed by DSC, KFT, FTIR, and SEM	Menezes et al. (2012)
Hydroxycitronellal	β-Cyclodextrin solid complex, obtained by coprecipitation method, and analyzed by GC-MS and TG	Riviş et al. (2008)
α-Ionone	β-Cyclodextrin solid complex, obtained by coprecipitation method, and analyzed by GC-MS and TG	Riviş et al. (2008)

(*Continued*)

Table 4.4 A Selection of Aroma Compounds and Essential Oils Obtained as CD Complexes in Solid State (*cont.*)

Aroma/Flavoring Compound/ Essential Oil	Type of Cyclodextrin, Methods of Complexation, and Analysis of Solid Complexes	References
(+)- and (−)-Isopulegole	β-Cyclodextrin solid complex, obtained by coprecipitation, and analyzed by HPLC and XRD	Ceborska et al. (2013b)
(−)-Limonene	β-Cyclodextrin solid complex, obtained by coprecipitation method, and analyzed by GC-MS and TG	Riviş̧ et al. (2008)
(+)-Limonene	β-Cyclodextrin solid complex, obtained by coprecipitation method, and analyzed by GC-MS and TG	Riviş et al. (2008)
D-Limonene	β-Cyclodextrin solid complex, obtained by kneading method, and analyzed by vacuum oven drying for the water content	Fang et al. (2013)
(−)-Linalool	β-Cyclodextrin solid complex, obtained by kneading and slurry methods, and analyzed by DSC, KFT, FTIR, SEM, XRD, and GC-MS/FID	Menezes et al. (2013)
(−)-Linalool	α-, β-Cyclodextrin solid complexes, obtained by coprecipitation method, and analyzed by GC-MS and TG	Riviş et al. (2008)
(+)-Linalool	α-, β-Cyclodextrin solid complexes, obtained by coprecipitation method, and analyzed by GC-MS and TG	Riviş et al. (2008)
Linalool	2-Hydroxypropyl-β-cyclodextrin solid complex, obtained by slurry and freeze-drying methods, and analyzed by ^{1}H-NMR and circular dichroism spectroscopy (CD)	Numanoğlu et al. (2007)
(+/−)-Linalyl acetate	β-Cyclodextrin solid complex, obtained by coprecipitation method, and analyzed by GC-MS and TG	Riviş et al. (2008)
(−)-Menthol	β-Cyclodextrin solid complex, obtained by coprecipitation method, and analyzed by GC-MS and TG	Riviş et al. (2008)
α-Terpineol	β-Cyclodextrin solid complex, obtained by coprecipitation method, and 2-hydroxypropyl-β-cyclodextrin, obtained by freeze-drying; they were analyzed by DSC and SEM	Dos Santos et al. (2011, 2012)
α-Terpineol	β-Cyclodextrin solid complex, obtained by coprecipitation method, and analyzed by DSC and water content	Mazzobre et al. (2011)
Thymol	β-Cyclodextrin solid complex, obtained by coprecipitation method, and analyzed by DSC	Ponce Cevallos et al. (2010)
Thymol	β-Cyclodextrin solid complex, obtained by freeze-drying and kneading methods, and analyzed by UV-Vis, TEM, and DSC. Evaluated by its antimicrobial activity	Tao et al. (2014), Riviş et al. (2008)

Table 4.4 A Selection of Aroma Compounds and Essential Oils Obtained as CD Complexes in Solid State (*cont.*)

Aroma/Flavoring Compound/ Essential Oil	Type of Cyclodextrin, Methods of Complexation, and Analysis of Solid Complexes	References
Vanillin	β-Cyclodextrin solid complex, obtained by coprecipitation method, and analyzed by GC-MS and TG	Rivis et al. (2008)
Essential oils		
Abies alba (fir, Pinaceae)	β-Cyclodextrin solid complex, obtained by coprecipitation method, and analyzed by GC-MS and TG	Hădărugă et al. (2005)
Allium sativum L. (garlic, Liliaceae)	β-Cyclodextrin solid complex, obtained by coprecipitation method, and analyzed by GC-MS, TG, DSC, SEM, and KFT	Hădărugă et al. (2007b, 2012b)
A. sativum L. (garlic, Liliaceae)	β-Cyclodextrin solid complex, obtained by coprecipitation method, and analyzed by UV, FT-IR, DSC, and XRD	Wang et al. (2011a,b)
A. sativum L. (garlic, Liliaceae)	β-Cyclodextrin solid complex, obtained by coprecipitation method, and analyzed by GC-MS, FT-IR, and moisture content. Evaluated for its antifungal activity	Ayala-Zavala et al. (2008)
Anethum graveolens L. (dill, Apiaceae)	β-Cyclodextrin solid complex, obtained by coprecipitation method, and analyzed by GC-MS, TG, DSC, SEM, and KFT	Hădărugă et al. (2005, 2012b)
Artemisia dracunculus (tarragon, Asteraceae)	α-, β-, γ-, 2-Hydroxypropyl-β-, randomly methylated-β-, and low methylated-β-cyclodextrin solid complexes, obtained by freeze-drying method, and analyzed by UV-Vis, DSC, FT-IR, ^{1}H-NMR, and SH-GC. Evaluated for their antioxidant activity	Kfoury et al. (2015b)
Carum carvi L. (caraway, Apiaceae)	β-Cyclodextrin solid complex, obtained by coprecipitation method, and analyzed by GC-MS, TG, DSC, SEM, and KFT	Hădărugă et al. (2005, 2012b)
Capsicum annuum (pepper, Solanaceae)	β-Cyclodextrin solid complex, obtained by coprecipitation method, and analyzed by HPLC, TG, DSC, and KFT	Hegheş et al. (2015)
Cinnamomum casia L. (cinnamon, Lauraceae)	β-Cyclodextrin solid complex, obtained by freeze-drying method, and analyzed by oxidative DSC, TEM, and UV-Vis. Evaluated for its antimicrobial activity	Hill et al. (2013)
Cinnamomum zeylanicum (cinnamon, Lauraceae)	β-Cyclodextrin solid complex, obtained by coprecipitation method, and analyzed by GC-MS, FT-IR, and moisture content. Evaluated for its antifungal activity	Ayala-Zavala et al. (2008)
Citrus sinensis L. Osbeck (sweet orange, Rutaceae)	β-Cyclodextrin solid complex, obtained by coprecipitation and kneading methods, and analyzed by TG, KFT, FT-IR, SEM, and GC-FID. Evaluated for its larvicidal activity	Galvão et al. (2015)

(*Continued*)

Table 4.4 A Selection of Aroma Compounds and Essential Oils Obtained as CD Complexes in Solid State (*cont.*)

Aroma/Flavoring Compound/ Essential Oil	Type of Cyclodextrin, Methods of Complexation, and Analysis of Solid Complexes	References
Cymbopogon winterianus (citronella, Poaceae)	β-Cyclodextrin solid complex, obtained by kneading method, and analyzed by GC-MS, SEM, FT-IR, and DSC. Evaluated for its mosquito repellent activity	Songkro et al. (2012)
Citrus aurantium sp. *sinensis* L. (orange, Rutaceae)	β-Cyclodextrin solid complex, obtained by coprecipitation method, and analyzed by GC-MS and TG	Costescu et al. (2008)
C. aurantium ssp. bergamia Risso (bergamot, Rutaceae)	β-Cyclodextrin solid complex, obtained by coprecipitation method, and analyzed by GC-MS and TG	Costescu et al. (2008)
Citrus lemon L. (lemon, Rutaceae)	β-Cyclodextrin solid complex, obtained by coprecipitation method, and analyzed by GC-MS and TG	Costescu et al. (2008)
Coriandrum sativum L. (coriander, Apiaceae)	β-Cyclodextrin solid complex, obtained by coprecipitation method, and analyzed by GC-MS, TG, DSC, SEM, and KFT	Hădărugă et al. (2005, 2012b)
Eucalyptus globulus (eucalyptus, Myrtaceae)	β-Cyclodextrin solid complex, obtained by coprecipitation method, and analyzed by GC-MS and TG	Costescu et al. (2008)
Eugenia caryophyllata (clove, Myrtaceae)	β-Cyclodextrin solid complex, obtained by coprecipitation method, and analyzed by GC-MS and TG	Costescu et al. (2008)
E. caryophyllata (clove, Myrtaceae)	β-Cyclodextrin solid complex, obtained by freeze-drying method, and analyzed by oxidative DSC, TEM, and UV-Vis. Evaluated for its antimicrobial activity	Hill et al. (2013)
Foeniculum vulgare (fennel, Apiaceae)	β-Cyclodextrin solid complex, obtained by coprecipitation method, and analyzed by GC-MS, TG, DSC, SEM, and KFT	Hădărugă et al. (2005, 2012b)
Juniperus communis L. (juniper, Pinaceae)	β-Cyclodextrin solid complex, obtained by coprecipitation method, and analyzed by GC-MS, TG, DSC, SEM, and KFT	Costescu et al. (2008), Hădărugă et al. (2012b)
Lavandula augustifolia (lavender, Lamiaceae)	β-Cyclodextrin solid complex, obtained by coprecipitation method, and analyzed by GC-MS, TG, DSC, and SEM	Hădărugă et al. (2007b)

Table 4.4 A Selection of Aroma Compounds and Essential Oils Obtained as CD Complexes in Solid State (*cont.*)

Aroma/Flavoring Compound/ Essential Oil	Type of Cyclodextrin, Methods of Complexation, and Analysis of Solid Complexes	References
Lippia gracilis H.B.K. (alecrim-da-chapada, Brasil, Verbenaceae)	2-Hydroxypropyl-β-cyclodextrin solid complex, obtained by slurry and kneading methods, and analyzed by GC-MS, GC-FID, KFT, TG, and XRD	Marreto et al. (2008)
Lippia graveolens H. B. K. (Mexican oregano, Verbenaceae)	β-Cyclodextrin solid complex, obtained by spray-drying method, and analyzed by SEM and GC-FID. Evaluated for its antioxidant and antimicrobial activities	Arana-Sánchez et al. (2010)
Litsea cubeba Pers. (mountain pepper (May Chang-China), Lauraceae)	β-Cyclodextrin solid complex, obtained by coprecipitation method, and analyzed by UV-Vis, FT-IR, and GC-MS	Wang et al. (2009)
Majorana hortensis (marjoram, Lamiaceae)	β-Cyclodextrin solid complex, obtained by coprecipitation method, and analyzed by GC-MS, TG, DSC, and SEM	Hădărugă et al. (2007b)
Mentha x villosa Hudson (small-leaved mint, Lamiaceae)	β-Cyclodextrin solid complex, obtained by coprecipitation and kneading methods, and analyzed by GC-MS, total oil and surface oil contents, TA, XRD, and EGD	Martins et al. (2007)
Ocimum basilicum L. (basil, Lamiaceae)	β-Cyclodextrin solid complex, obtained by coprecipitation method, and analyzed by GC-MS, TG, DSC, and SEM	Hădărugă et al. (2007b)
O. basilicum var. *basilicum* (basil, Lamiaceae)	α-, β-, γ-, 2-Hydroxypropyl-β-, randomly methylated-β-, and low methylated-β-cyclodextrin solid complexes, obtained by freeze-drying method, and analyzed by UV-Vis, DSC, FT-IR, ^{1}H-NMR, and SH-GC. Evaluated for their antioxidant activity	Kfoury et al. (2015b)
Picea excelsa L. (spruce, Pinaceae)	β-Cyclodextrin solid complex, obtained by coprecipitation method, and analyzed by GC-MS and TG	Hădărugă et al. (2005)
Pistacia terebinthus (turpentine, Pinaceae)	β-Cyclodextrin solid complex, obtained by coprecipitation method, and analyzed by GC-MS and TG	Hădărugă et al. (2005)
Salvia sclarea L. (sage, Lamiaceae)	β-Cyclodextrin solid complex, obtained by coprecipitation method, and analyzed by GC-MS, TG, and DSC	Hădărugă et al. (2007a)
S. sclarea L. (sage, Lamiaceae)	β-Cyclodextrin solid complex, obtained by coprecipitation method, and analyzed by UV	Tian et al. (2008)

(*Continued*)

Table 4.4 A Selection of Aroma Compounds and Essential Oils Obtained as CD Complexes in Solid State (*cont.*)

Aroma/Flavoring Compound/ Essential Oil	Type of Cyclodextrin, Methods of Complexation, and Analysis of Solid Complexes	References
Satureja montana L. (Winter savory, Lamiaceae)	β-Cyclodextrin solid complex, obtained by coprecipitation method, and analyzed by GC-MS, total oil and surface oil extraction, and photochemiluminiscence (PCL). Evaluated for its antifungal activity	Haloci et al. (2014)
Thymus vulgaris L. (thyme, Lamiaceae)	β-Cyclodextrin solid complex, obtained by coprecipitation method, and analyzed by GC-FID, GC-MS, FTIR, and ^{1}H-NMR. Evaluated for its antifungal activity	Del Toro-Sánchez et al. (2010)
T. vulgaris L. (thyme, Lamiaceae)	β-Cyclodextrin solid complex, obtained by coprecipitation method, and analyzed by GC-MS, TG, DSC, and SEM	Hădărugă et al. (2007b)
T. vulgaris L. (thyme, Lamiaceae)	β-Cyclodextrin solid complex, obtained by freeze-drying and kneading methods, and analyzed by UV-Vis, TEM, and DSC. Evaluated by its antimicrobial activity	Tao et al. (2014)

et al., 2012a,b, 2013; Hegheş et al., 2015). Other analytical methods allow evaluating the crystallinity or amorphous state, dimensions, and morphology of the micro- or nanocrystals of CD/aroma complexes (He et al., 2008; Lu et al., 2012). Analytical methods can be classified as follows: (1) chromatographic, (2) spectroscopic, (3) microscopic, (4) thermoanalytical methods as well as (5) specific methods (eg, for volatile compounds).

4.2.2.1 Chromatographic Methods

The chromatographic methods are used for both "pure" CD analysis and for determination of the concentration and composition of CD/aroma complexes (such as CD/essential oil complexes). Paper and thin layer chromatography were the first chromatographic methods used for separation and analysis of CDs. Today, they were replaced by modern techniques such as high-pressure liquid chromatography (HPLC) (Ceborska et al., 2013b; Hădărugă et al., 2010a; Hegheş et al., 2015; López-Nicolás and García-Carmona, 2008) or gas chromatography (GC) (Marques, 2010). A derivatization of CDs to more volatile compounds for GC analysis is needed. Gas chromatography coupled with mass spectrometry (GC-MS) is often used to identify and quantify the CD-encapsulated volatiles such as aroma compounds (Arana-Sánchez et al., 2010; Hădărugă et al., 2012b; Marreto et al., 2008; Menezes et al., 2013). High accuracy for the quantification of aroma and flavors is obtained for gas chromatography-flame ionization detector system (GC-FID) as well as for evaluation of the competitivity of aroma compounds from mixtures (eg, essential oils) to the molecular encapsulation in CDs (Ayala-Zavala et al., 2008). The relative composition and CD (α-, β-, γ- and 2-hydroxypropyl-β-CD) molecular encapsulation competitivity of essential oils from Apiaceae, Liliaceae, Lauraceae, Lamiaceae, Myrtaceae, Rutaceae, and Pinaceae botanical families have been determined by GC-MS (Hădărugă et al., 2005, 2007a,b,c, 2012a,b, 2014). For example, limonene from caraway essential oil was better encapsulated in β-CD in comparison with (S)–(+)-carvone, which was encapsulated in lower relative concentration than in the raw essential oil (Hădărugă et al., 2012b). The same behavior was observed for linalool from coriander, anethole from fennel, methyl-chavicol from basil, γ-terpinene from marjoram, *p*-cymene from eucalyptus, pinenes and camphene from fir, spruce, juniperus, and turpentine, as well as acyclic sulfides from garlic essential oils that were encapsulated in β-CD and analyzed by GC-MS (Hădărugă et al., 2005, 2007c, 2012b). On the other hand, linalyl acetate from lavender and sage (Hădărugă et al., 2007a), as well as cinnamaldehyde from cinnamon essential oils were encapsulated in β-CD in similar relative concentrations such as in the raw products (Hădărugă et al., 2012b; Hill et al., 2013; Ponce Cevallos et al., 2010). It was not the case of carvacrol from thyme, menthol from mint, eugenol from clove, or eucalyptol from eucalyptus essential oils (Costescu et al., 2008). They were encapsulated in β-CD in lower relative concentrations, according to GC-MS analysis of the raw and recovered essential oils.

4.2.2.2 Thermoanalytical Methods

The thermoanalytical methods are useful for the evaluation of the quality of encapsulation of aroma and flavors in CDs (Hegheş et al., 2015; Marreto et al., 2008; Mura, 2015; Riviş et al., 2008). It is possible to determine the mass loss of CD complexes by using thermogravimetry (TG) (Menezes et al., 2012; Mura, 2015; Wang et al., 2011b), which indicates the loss of water, other solvents used for complexation and volatile compounds from aroma. Unfortunately, it is difficult to discriminate between the volatile compounds, even if the water is generally released in the first part of heating. An example is presented in Fig. 4.13a.

The mass loss of β-CD is 13.8% up to 100°C and less than 0.5% up to 260°C, while the β-CD/linalyl acetate (the main compound

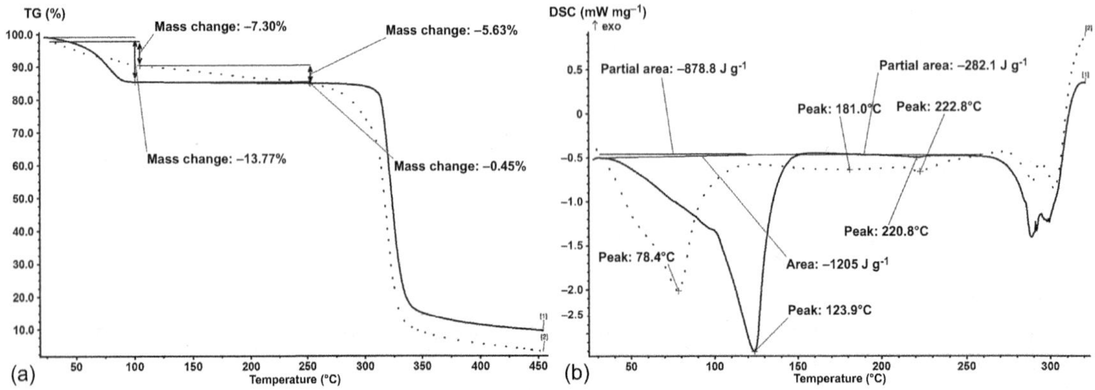

Figure 4.13. (a) Thermogavimetric analysis of β-CD (*continuous line*) and β- CD/linalyl acetate complex (*dotted line*); (b) Differential scanning calorimetry analysis of β-CD (*continuous line*) and β-CD/*Matricaria chamomilla* L. (chamomile) essential oil complex (*dotted line*).

from lavender essential oil) complex had a mass loss of only 7.3% for the first range corresponding to the water (and other volatiles) release and 5.6% for the range of 100–260°C, which corresponds especially to linalyl acetate dissociation. The last interval corresponds to the decomposition of β-CD.

Another thermal method that is often used for the analysis of CD complexes is differential scanning calorimetry (DSC) (Hădărugă et al., 2007a,b, 2012b; Menezes et al., 2013; Mura, 2015). Generally, physical mixture, only the single compounds, as well as the complex are analyzed and compared for evaluating the formation of the inclusion compound. These differences are related to the disappearance of the peak corresponding to the boiling of aroma compound and the modification of the peak corresponding to the water of hydration. Other peaks can appear corresponding to the dissociation of the host–guest complex. For example, the DSC analysis of β-CD and its complex with *Matricaria chamomilla* L. (chamomile) essential oil indicates an important peak corresponding to the dissociation of the hydration water from β-CD (with an endothermal effect of 1205 J g^{-1}) that is shifted to lower temperature (from 123.9 to 78.4°C) and a lower calorimetric effect for the complex (878.8 J g^{-1}). Moreover, a large peak that corresponds to dissociation of chamomile essential oil components appears in the range of ~120–260°C (282.1 J g^{-1} and peak temperature of 181°C) (Fig. 4.13b).

Other thermoanalytical methods such as evolved gas analysis (EGA), drying/sublimation, TAS-chromatography or thermofractography are also used for the analysis of CD/aroma and flavors inclusion complexes (Marques, 2010).

4.2.2.3 Spectroscopic Methods

The CD complex structures in solid state can be determined by X-ray diffraction (XRD) (Mura, 2015). The reduction or modification of the diffractogram shapes after CD complexation indicates the formation of amorphous compounds and thus the formation of the inclusion complex. The crystallinity or amorphous character of the complexes significantly depends on the synthesis method (eg, kneading and lyophilization/freeze-drying furnish more amorphous complexes in comparison with cocrystallization method, which allows to obtain CD complexes with high crystallinity) (Ceborska et al., 2013a; Marreto et al., 2008; Menezes et al., 2013; Wang et al., 2011a).

Other spectroscopic methods such as Fourier transform–infrared spectroscopy (FT–IR) (Duchêne, 2011; Menezes et al., 2012, 2013; Mura, 2015; Nuchuchua et al., 2009) and Raman spectroscopy (Mohan et al., 2012; Mura, 2015) are also used to investigate CD/aroma inclusion complexes.

4.2.2.4 Microscopic Methods

Microscopic methods such as scanning electron microscopy (SEM) (Arana-Sánchez et al., 2010; De Souza Siqueira Quintans et al., 2013; Hădărugă et al., 2007a,b,c, 2012a,b; Menezes et al., 2013; Mura, 2015) and transmission electron microscopy (TEM) (Hill et al., 2013; Tao et al., 2014) are often used to evaluate the morphology of micro- and nanoparticles of complexes as well as the dimensions and uniformity of such materials (He et al., 2008; Zhang et al., 2015; Zhou et al., 2013). An example of SEM analysis for β-CD and its complex with *Silybum marianum* L. (milk thistle) oil is presented in Fig. 4.14. It can be observed that β-CD have hexagonal-like crystal shape, while the corresponding thistle oil complex reveals acicular- or prismatic-like morphology.

There are few other methods used to analyze the CD complexes that are more or less specific for encapsulated volatile compounds. It is the case of Karl Fischer water titration (KFT) that is a selective method for the analysis of water and very useful for CD/aroma complexes (where the bioactive compounds have higher volatility and cannot allow the correct evaluation of water content in such complexes by classical drying or thermal methods) (Hădărugă et al., 2012b; Hegheş et al., 2015). Furthermore, KFT analysis allows discrimination between "surface" water and "strongly retained" water molecules in CD/aroma complexes (Hădărugă et al., 2012a, 2013). This aspect furnishes indirect information about the formation of the molecular inclusion complex (the inner water molecules—strongly retained—are partially replaced by

(a) (b)

Figure 4.14. Scanning electron microscopy analysis of β-CD (a) and β-CD/*Silybum marianum* L. (milk thistle) extract complex (b).

hydrophobic aroma or flavors during CD complexation). Capillary and gelelectrophoresis (Mura, 2014), cyclic voltammetry, circular dichroism spectroscopy (Mura, 2014; Stancanelli et al., 2007), Warburg oxidation (for unsaturated guest compounds), solubility, dissolution, and humidity tests (Numanoğlu et al., 2007; Szejtli, 1988) are also used to study CD/aroma and flavors inclusion complexes. Furthermore, molecular modeling and docking studies allow researchers to evaluate the geometric compatibility and the efficiency of complexation (by means of the calculated interaction energy) (Hădărugă, 2011; Hădărugă et al., 2008a,b, 2009; Pérez-Garrido et al., 2009; Pînzaru et al., 2011).

4.3 Factors Controlling Encapsulation of Aroma and Flavors in CDs

4.3.1 Hydrophobic Effect

Encapsulation of aroma in CDs is in reality a complex process. Although predominantly hydrophobic interactions take place during inclusion complex formation, other requirements are to be considered such as the CD and aroma molecular properties. To demonstrate this aspect, multiple studies looked for factors controlling the encapsulation of flavors and aroma in CDs. This mainly allows a more precise analysis of the influence of aroma's properties (size, geometry, hydrophobicity, solubility) on its encapsulation in CDs aiming to optimize inclusion complex preparation and to select the appropriate application. Many studies showed a satisfactory correlation between hydrophobicity (log *P*) and the K_f values with correlation coefficients equal to 0.986 for β-CD (Decock et al., 2008), 0.8548 and 0.8824, for α-CD and β-CD, respectively (Astray et al., 2010), and 0.9232 for HP-β-CD (Kfoury et al., 2014a). This indicates that the most important criterion in the inclusion complex formation is the hydrophobic character of aroma or flavor. However, considering the stability constants of the overall aromatic compounds present in Table 4.3 with β-CD, although we could observe the same tendency, the correlation becomes weaker (Fig. 4.15).

The deviation from the ideality indicates that the linear correlation between the stability of inclusion complex and the hydrophobic character of aroma might be right for analogous family and cannot be universally applied. This proved that the hydrophobic interactions are not the sole driving forces for inclusion complex formation. Thus, other factors must be taken into consideration to predict a theoretical approach for inclusion complex stability.

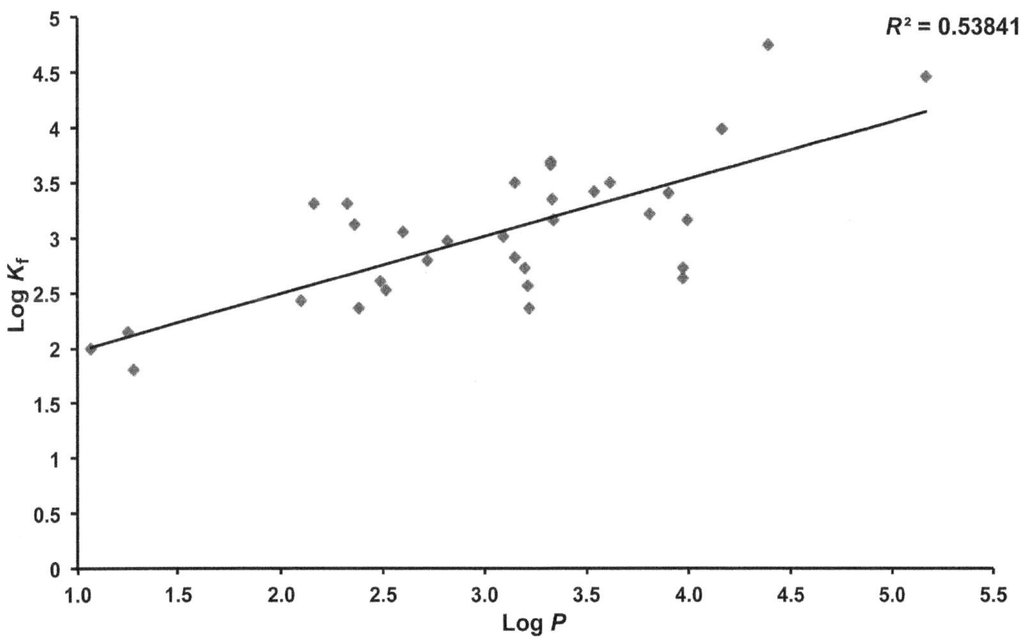

Figure 4.15. Relationship between formation constants (log K_f) and hydrophobic character (log *P*) of aroma from Table 4.3.

4.3.2 Solubility Effect

The aroma or flavor's intrinsic solubility is an important parameter to be considered particularly when formulation with CDs aims for solubility enhancement. Data collected from literature shows a strong correlation between the solubility enhancement of aroma in the presence of HP-β-CD and their intrinsic aqueous solubility (Fig. 4.16) (Kfoury et al., 2014a,c; Liang et al., 2012; Yuan et al., 2014; Zhang et al., 2009).

This proves that the formation of inclusion complexes also occurs through desolvation/dissolution of the aroma and widely depends on their solubility. Here, we should note that this correlation could not be universally applicable for estimating inclusion complex stability because solubilization by CDs not only includes inclusion complex formation but also surfactant-like effects and molecular aggregation (Jansook and Loftsson, 2009; Loftsson et al., 2005).

4.3.3 Steric Effects

The space filling of CD cavity also plays a major role in interpreting the inclusion stability. This is defined by the molecular

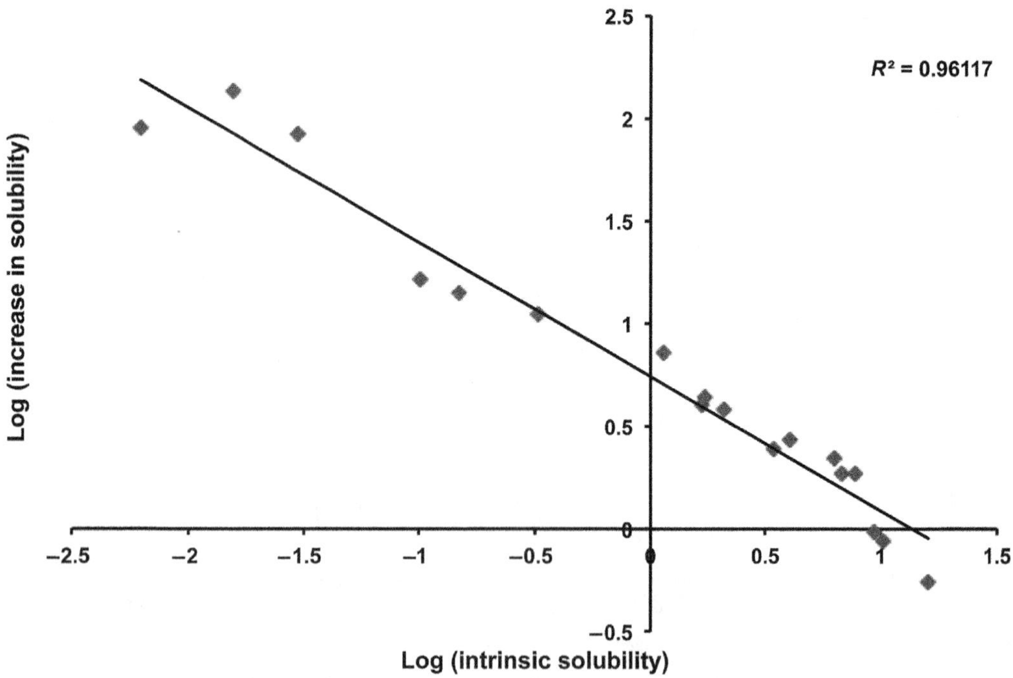

Figure 4.16. Relationships between solubility enhancement upon encapsulation and intrinsic solubility of aroma.

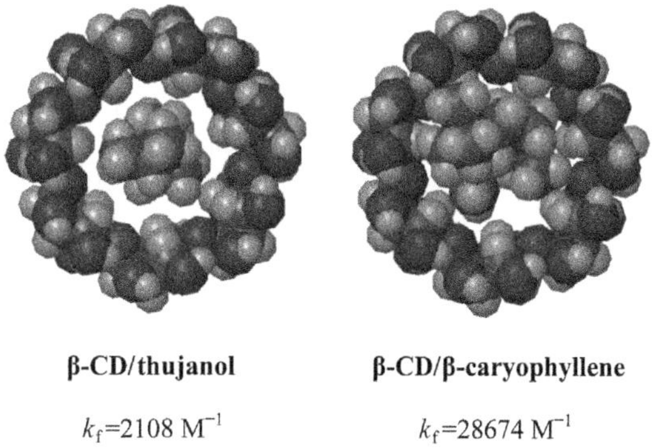

Figure 4.17. Illustration of the space filling of β-CD cavity by thujanol and β-caryophyllene.

volume, shape, and geometry of the aroma that describe its fit into the cavity without any steric hindrance. The ability of molecular modeling programs to visualize the space filling of CD cavity and the correlation with experimental K_f values confirm that the stability of encapsulation depends on the geometric accommodation of the aroma in the CD and increases dramatically with the amount of space filling of the cavity (Decock et al., 2008; Fourmentin et al., 2013). An example is illustrated in Fig. 4.17.

The weak empty space observed between β-CD cavity and β-caryophyllene as compared to β-CD cavity and thujanol suggests that the former provides sufficient contact with the wall of the cavity thus is more tightly encapsulated (Eftink et al., 1989). Results are confirmed by higher K_f value for β-CD/β-caryophyllene inclusion complex (Kfoury et al., 2015c). Another important parameter that has to be taken into consideration is the encapsulation of isomers. For example, E- and Z-isomers, are not likewise recognized by different CDs (Decock et al., 2008; Miron et al., 2012; Ruktanonchai et al., 2011) as well as α- and β-isomers (Ciobanu et al., 2013a; Kfoury et al., 2014a), (+) and (−) enantiomers (Nowakowski and Ejchart, 2014), constitutional isomers (Bethanis et al., 2013; Ciobanu et al., 2013a; Decock et al., 2006, 2008; Kfoury et al., 2014c, 2015b), and aromatic cycle with different substitution pattern (vanillin and *p*-hydroxybenzaldehyde) (Zeng et al., 2012).

Taking into account the different correlations, this indicated that a multiparametric relationship takes hold and influences the inclusion complex formation. The strength of binding between CD and the aroma involves balance between hydrophobic interactions, solvation processes, and hydrogen bonds as well as Van der Waals and steric interactions. Thus, no unique correlation could be implemented probing that several parameters should be taken into account to presume the inclusion complex stability. Finally, the binding magnitude could be classified as very weak, weak, moderate, strong, and very strong when K_f values fell within the respective ranges 0–500 M^{-1}, 500–1,000 M^{-1}, 1,000–5,000 M^{-1}, 5,000–20,000 M^{-1}, and greater than 20,000 M^{-1} (Carrier et al., 2007). This generally facilitates the selection of the proper CD for each specific aroma depending on the desired characteristics of the inclusion complex.

5 Effects of Encapsulation

Encapsulation in CDs aims mainly to improve aroma functionalities, control sensory quality and organoleptic behavior of food, and design new functional systems allowing aroma differentiation and burst. The possible benefits of encapsulated aroma in CDs are various. Table 4.5 is a recap chart of major published works regarding the improved properties of aroma and flavors upon encapsulation in CDs.

5.1 Solubility Enhancement

Aqueous solubility is a crucial requirement to achieve desired concentration of a substance in a food product (Bagade et al., 2014). The lipophilic nature of aroma and flavors is an obstacle in their direct incorporation in food because they are hardly soluble in aqueous media (Vohs, 2013). To overcome this problem, CDs arise as a potential tool in food formulation (Astray et al., 2009; Cravotto et al., 2006; Szente and Szejtli, 2004). Encapsulation in CDs has the ability to increase aroma's solubility, reduce dosage form volume, and standardize the composition (Munin and Edwards-Lévy, 2011). Moreover, inclusion complexes lead to disperse active ingredients into aqueous phases. Another important income of encapsulation is that, after ingestion, encapsulated aroma could be more bioavailable because CDs are able to improve their dissolution in gastrointestinal fluids (Carrier et al., 2007; Salústio et al., 2011; Saravana Kumar et al., 2013; Uekaji et al., 2013). This certainly is of major interest for the encapsulation of nutraceuticals and bioactive aroma. From the literature, it is clear that the use of CDs greatly enhances the solubility of encapsulated agents such as naringenin and natural flavonoid responsible for the bitter taste of grapefruit (Shulman et al., 2011), and increase the bioavailability of flavors such as rhubarb extract (Hsu et al., 2013), tea extract (Haidong et al., 2011), thymol (Nieddu et al., 2014), and curcumin (Patro et al., 2014).

Table 4.5 Examples of Improved Properties of Flavors and Aroma After Encapsulation With Different CDs

Aroma	Cyclodextrins	Improved Properties	References
Allyl isothiocyanate	β-CD	Stability, antibacterial	Aytac et al. (2014)
Trans-anethole	α-CD, β-CD, HP-β-CD, RAMEB, CRYSMEB	Solubility, stability	Kfoury et al. (2014b)
	α-CD, β-CD, HP-β-CD, RAMEB, CRYSMEB	Solubility, antioxidant	Kfoury et al. (2014c)
Borneol	β-CD	Solubility, stability	Su et al. (2012)
Camphor	α-CD, β-CD, γ-CD	Solubility	Yu et al. (2003)
Carvacrol	α-CD, β-CD, HP-β-CD	Solubility, antibacterial, antifungal	Liang et al. (2012)
	β-CD	Solubility, antibacterial	Santos et al. (2015)
	β-CD	Pain management	Guimarães et al. (2015)
Carvone	β-CD	Stability	Partanen et al. (2002)
	β-CD	Solubility, stability	Ajisaka et al. (2000)
β-Caryophyllene	β-CD	Dissolution, bioavailability	Liu et al. (2013)
Cinnamaldehyde	β-CD	Solubility, antibacterial	Hill et al. (2013)
	α-CD, β-CD	Antibacterial	Chun et al. (2015)
Citronellal	β-CD	Solubility, stability	Ajisaka et al. (2000)
Curcumin	β-CD	Solubility	Jahed et al. (2014)
	β-CD	Solubility, stability	Mangolim et al. (2014)
	β-CD	Solubility, release	Ansari and Ahmed (2015)
p-Cymene	β-CD	Analgesic, antiinflammatory	De Souza Siqueira Quintans et al. (2013)
Eugenol	β-CD	Stability	Seo et al. (2010)
	β-CD, HP-β-CD	Stability	Choi et al. (2009)
	HP-β-CD	Solubility	Garg et al. (2010)
	β-CD	Stability, antibacterial	Wang et al. (2011b)
	α-CD, β-CD, HP-β-CD	Solubility, antibacterial, antifungal	Liang et al. (2012)
	β-CD	Solubility, antibacterial	Hill et al. (2013)
	α-CD, β-CD, HP-β-CD, RAMEB, CRYSMEB	Solubility, antioxidant	Kfoury et al. (2014c)
Estragole	α-CD, β-CD, γ-CD, HP-β-CD, RAMEB, CRYSMEB	Stability, antioxidant	Kfoury et al. (2015b)
	α-CD, β-CD, HP-β-CD, RAMEB, CRYSMEB	Solubility, antioxidant	Kfoury et al. (2014c)

(Continued)

Table 4.5 Examples of Improved Properties of Flavors and Aroma After Encapsulation With Different CDs (*cont.*)

Aroma	Cyclodextrins	Improved Properties	References
Ethyl benzoate	HP-β-CD	Solubility, stability	Yuan et al. (2014)
Geraniol	β-CD	Solubility, stability	Mourtzinos et al. (2008)
	β-CD	Solubility, stability	Ajisaka et al. (2000)
Isoeugenol	α-CD, β-CD, HP-β-CD, RAMEB, CRYSMEB	Solubility, antioxidant	Kfoury et al. (2014c)
Limonene	β-CD	Stability	Partanen et al. (2002)
	β-CD	Solubility, stability	Ajisaka et al. (2000)
Linalool	β-CD, HP-β-CD	Solubility, stability	Numanoğlu et al. (2007)
	β-CD	Analgesic, antiinflammatory	Quintans-Júnior et al. (2013)
	α-CD, β-CD, HP-β-CD	Solubility, antibacterial, antifungal	Liang et al. (2012)
	β-CD	Pain management	Nascimento et al. (2014)
Menthol	β-CD	Solubility, stability	Ajisaka et al. (2000)
Menthone	β-CD	Solubility, stability	Ajisaka et al. (2000)
Myrcene	β-CD	Solubility, stability	Ajisaka et al. (2000)
Nerol	β-CD	Solubility, stability	Ajisaka et al. (2000)
Pulegone	β-CD, γ-CD	Stability	Moon et al. (2008)
Terpineol	β-CD, HP-β-CD	Stability	Dos Santos et al. (2012)
	β-CD	Solubility, stability	Mazzobre et al. (2011)
	β-CD	Solubility, stability	Ajisaka et al. (2000)
		Stability, antibacterial	Yang et al. (2015)
Thymol	β-CD	Release	Del Toro-Sánchez et al. (2010)
	β-CD	Solubility, stability	Mourtzinos et al. (2008)
	β-CD	Antifungal	Del Toro-Sánchez et al. (2010)
	β-CD	Pharmacokinetic properties	Nieddu et al. (2014)
Vanillin	α-CD, β-CD, γ-CD	Stability	Kayaci and Uyar (2011)
	β-CD	Stability	Hundre et al. (2015)

5.2 Protection of Aroma and Flavors

Flavors are generally expensive and therefore it is an ultimate concern to prevent their loss or degradation (Govindasamy et al., 2013). Thanks to CDs aroma and flavors could be covered with a protective barrier that reduces their volatility, prevents their degradation, evaporation, and reaction with other food matrix ingredients during storage and strong thermal or mechanical treatments improving consequently the food shelf life (Marques, 2010; Szente and Szejtli, 2004). CDs are well proved to be useful in stabilizing aroma that cannot be effectively stabilized using other techniques (Reineccius et al., 2002). For example, encapsulation in β-CD improved the retention of citral (26-fold enhancement) and menthol (86-fold enhancement) in fruit leathers and hard candies, respectively, (Reineccius et al., 2004) as well as the caraway essential oil components (carvone and limonene) during heat treatment (Partanen et al., 2002). Encapsulation of flavors in CDs could also inhibit off note development, maintain true aromatic profile, minimize interactions with packaging materials, and enhance photostability without adding antioxidants. Increase in the global sensory quality and reduction in the browning of pear juice (Andreu-Sevilla et al., 2011; López-Nicolás et al., 2009) and lemon-flavored juice beverages (Strassburger et al., 2010) were observed upon treatment with CDs. Similar promising light, heat, or oxidative stability results were obtained with other CD encapsulated flavors such as citral (Szente and Szejtli, 2004), *trans*-anethole (Kfoury et al., 2014b), estragole (Kfoury et al., 2015b), geraniol and thymol (Mourtzinos et al., 2008), or curcumin (Paramera et al., 2011) (Table 4.5). Encapsulation in CDs might be also one of the tools to preserve the integrity of aroma, maintain or improve their antioxidant, antibacterial, and antifungal activities (Costa et al., 2015; De Vos et al., 2010; Hill et al., 2013; Liang et al., 2012; Pinho et al., 2014) and prevent the deterioration of food quality by reducing interactions between phenols and proteins (Budryn et al., 2015).

5.3 Ensuring Controlled Release

Aroma release must be controlled because it strongly affects aroma perception and food intake (Ramaekers et al., 2014; Ruijschop and Kok, 2009). A slow release leads to a weak perception whereas a rapid one promotes only a brief burst of flavor leading to an unbalanced flavor profile of food (Charles et al., 2015; Guichard et al., 2013). The ultimate goal of CDs is to allow a controlled release of encapsulated aroma at desired rate during processing, storage, and food consumption (Marques, 2010). CDs play the role of a flavor reservoir and the dissociation of the inclusion complex requires generally the presence of moisture normally found in food or heating (Ho et al., 2011). As seen in other studies, adding β-CD to fat-free yogurt resulted in an extended period of lemon aroma release making it closer to regular fat yogurt (Kant et al., 2004) and encapsulating menthol in chewing gums ensured higher aromatic content and allowed it to release steadily when chewed (Yoshii et al., 2007). Other studies also described a controlled release of flavors such as linalool and benzyl acetate (Numanoğlu et al., 2007), terpineol (Mazzobre et al., 2011; Yang et al., 2015), ethyl benzoate (Yuan et al., 2014), estragole (Kfoury et al., 2015b), or thymol (Del Toro-Sánchez et al., 2010) upon encapsulation (Table 4.5).

5.4 Improving Organoleptic Behavior and Masking Off Flavors

The importance of adding aroma and flavors to food also lies in their powerful bioactivities and beneficial health effects (Amorati et al., 2013; Lucera et al., 2012; Tongnuanchan and Benjakul, 2014). However, application must be done at important concentrations to achieve good antioxidant or antimicrobial efficacy against pathogens and toxin-producing microorganisms and to ensure this activity for an extended storage period (Calo et al., 2015; Radulovic et al., 2013). This complicates their use and negatively affects sensory acceptability of food due to the high pronounced odor or flavor (Sánchez-González et al., 2011). The use of encapsulated aroma can effectively alleviate these deficiencies because there is always a dynamic equilibrium between the free and the encapsulated aroma (Qi and Hedges, 1995). Thus, a part of the active aroma will be encapsulated, masking odor and flavor, and will be progressively released. This reduces the strong flavors caused by the direct incorporation of aroma and allows the addition of higher amount than conventional dosing without affecting the sensory acceptability of food and maintains their activity for a long period of time (Appendini and Hotchkiss, 2002; Hanušová et al., 2009; López-Nicolás et al., 2014). Encapsulation in CDs could also reduce off-flavors, first by avoiding contact between aroma and oxygen or ions and preventing direct exposure to light, and second by encapsulating generated off-flavors (Szejtli and Szente, 1979, 2005). Studies showed that incorporation of CDs reduced up to 50% the bitterness intensity of ginseng in a model energy drink (Tamamoto et al., 2010) and soybean antioxidant hydrolysates (Hou et al., 2013). The addition of HP-β-CD improved also the nutritional and some physicochemical parameters of carrot-orange juice (Karangwa et al., 2012).

5.5 Active Packaging

Another technology that drives researchers to invest their knowledge in CD encapsulation is the design of biobased active food-packaging materials. Incorporating aroma for flavoring purposes or as bioactive agents as a form of inclusion complexes in food packaging materiel allows their gradual release into food (Martina et al., 2013). This leads to an extended self-life due to the preservation of organoleptic properties and the prevention of microbial growth and food spoilage (Arfat et al., 2015; Lee et al., 2015; Morelli et al., 2015). When inclusion complex is exposed to water molecules (high humidity normally presents in food), interactions between CD and aroma are weakened, and the latter is passively released into the food (Mascheroni et al., 2013). β-CD ensured, for example, a gradual release of carvacrol (Lavoine et al., 2014), eugenol, and *trans*-cinnamaldehyde (Sun et al., 2014) from the packaging material into the food. A multilayered antimicrobial packaging using a β-CD/*trans*-cinnamaldehyde inclusion complex improved the shelf life of fresh-cut melon (Moreira et al., 2014) and papaya (Brasil et al., 2012) and CD-encapsulated allyl isothiocyanate showed better antimicrobial activity for fresh-cut onions treatment as compared to free allyl isothiocyanate (Piercey et al., 2012). This provides a postharvest handling procedure that maintains optimal flavor and nutritional quality of fruits and vegetables until consumption (Kader, 2008).

5.6 Improving Handling and Dosage

Another goal of employing encapsulation is to provide a superior ease in handling aroma during storage or incorporation into food products (JadhavAmbadas et al., 2013; Nokhodchi et al., 2011). Encapsulation in CDs could be used to convert an oily aroma into a powder, which offers a standardized, dosable, and nearly odorless form (Marques, 2010; Riviş et al., 2008). Additionally, encapsulation can avoid negative effects on the texture, flavor, and solubility during the rehydration of a dehydrated food product. For example, CDs are efficient in the preparation of freeze-dried pear juice and offer a superior retention of aroma (Tobitsuka et al., 2005). To facilitate storage and dosage standardized solid inclusion complexes are commercialized at industrial scale. CAVAMAX® W7/CITRAL and CAVAMAX® W7/L-MENTHOL are β-CD inclusion complexes of citral and menthol produced by Wacker Chemie AG for direct application.

5.7 Aroma Differentiation/Burst

Finally, encapsulation in CDs may improve aroma differentiation and aroma burst (Taylor et al., 2009). Aroma differentiation is generally experienced when aroma diffuses slowly out of CDs into the food during cooking or heating process leading to perception change (Romani et al., 2012). The study of the release of menthol, ethyl butyrate, ethyl hexanoate, benzaldehyde, citral, and methyl anthranilate from CDs coupled to sensory evaluation suggested that CDs may alter the food sensory profile in a temperature-dependent manner (Reineccius et al., 2003). Also, when inclusion complexes are placed in excess water or in the mouth, a burst release of encapsulated aroma takes place and released aroma is perceived. Encapsulated aroma in pear juice that is not perceived during direct smelling is released once the juice is in the mouth (López-Nicolás et al., 2009). This phenomenon could also be experienced with tea where the release of encapsulated aroma into hot water is a burst release upon the addition of tea bags to hot water.

6 Conclusions

The need for tasty and healthy foodstuffs with an extended shelf life requires the development of new aroma delivery systems. The improvement of encapsulation technologies may overcome possible limitations of aroma and flavors use and maintain the maximum of their functionalities in a food matrix. CDs are food-grade encapsulant materials available in large scale, low prices, food quality, and included in the GRAS list of the FDA. Encapsulation in CDs is of particular importance and is an economical and simple technological process. CDs improve aqueous solubility of poorly soluble substances, offer a protective effect for sensitive ingredients (eg, aroma and flavors), ensure controlled release, reduce loss and volatility and provide superior handling of aroma and flavors. This leads to maintaining food organoleptic properties and extending shelf life. The challenge consists in adopting the appropriate CD for the selected aroma to achieve maximum encapsulation yield/efficiency and optimal characteristics based on the desired performance properties of the inclusion complex. This chapter provided a brief history of aroma and CDs and discussed encapsulation techniques as well as characterization methods for CD/aroma or flavors inclusion complexes. It also investigated parameters controlling the stability binding of aroma and flavors to CDs and reported the beneficial effects of encapsulation on aroma's properties.

References

Ajisaka, N., Hara, K., Mikuni, K., Hara, K., Hashimoto, H., 2000. Effects of branched cyclodextrins on the solubility and stability of terpenes. Biosci. Biotechnol. Biochem. 64 (4), 731–734.

Alonso, L., Cuesta, P., Fontecha, J., Juarez, M., Gilliland, S.E., 2009. Use of β-cyclodextrin to decrease the level of cholesterol in milk fat. J. Dairy Sci. 92 (3), 863–869.

Ammayappan, L., Jeyakodi Moses, J., 2009. An overview on application of cyclodextrins in textile product enhancement. J. Textile Assoc. 70 (1), 9–18.

Amorati, R., Foti, M.C., Valgimigli, L., 2013. Antioxidant activity of essential oils. J. Agric. Food Chem. 61 (46), 10835–10847.

Andreu-Sevilla, A.J., López-Nicolás, J.M., Carbonell-Barrachina, Á.A., García-Carmona, F., 2011. Comparative effect of thea of α-, β-, or γ-cyclodextrin on main sensory and physico-chemical parameters. J. Food Sci. 76 (5), 347–353.

Anon, 2008. Regulation (EC) No. 1334/2008 of the European Parliament and of the Council on flavourings and certain food ingredients with flavouring properties for use in and on foods. L354. Official Journal of the European Union, Strasbourg.

Ansari, M.J., Ahmed, M.M., 2015. Physicochemical characterizations, dissolution behavior and release kinetics of curcumin and β-cyclodextrin molecular inclusion complexes. Int. J. Pharm. Bio. Sci. 6 (1), 785–795.

Ansari, M.J., Kohli, K., Ali, J., Anwer, M.K., Jamil, S., Ahmed, M.M., 2014. Physicochemical characterizations and dissolution behavior of curcumin and α-cyclodextrin molecular inclusion complexes. Der Pharm. Lett. 6 (6), 1–7.

Antlsperger, G., Schmid, G., 1996. Toxicological comparison of cyclodextrins. In: Szejtli, J., Szente, L. (Eds.), Proceedings of the Eighth International Symposium on Cyclodextrins. Springer, The Netherlands, pp. 149–155.

Appendini, P., Hotchkiss, J.H., 2002. Review of antimicrobial food packaging. Innov. Food Sci. Emerg. 3 (2), 113–126.

Arana-Sánchez, A., Estarrón-Espinosa, M., Obledo-Vázquez, E.N., Padilla-Camberos, E., Silva-Vázquez, R., Lugo-Cervantes, E., 2010. Antimicrobial and antioxidant activities of Mexican oregano essential oils (*Lippia graveolens* H. B. K.) with different composition when microencapsulated in β-cyclodextrin. Lett. Appl. Microbiol. 50, 585–590.

Arfat, Y.A., Benjakul, S., Vongkamjan, K., Sumpavapol, P., Yarnpakdee, S., 2015. Shelf-life extension of refrigerated sea bass slices wrapped with fish protein isolate/fish skin gelatin-ZnO nanocomposite film incorporated with basil leaf essential oil. J. Food Sci. Technol. 52, 6182–6193.

Ashurst, P.R., 1991. Food Flavourings. Blackie, Glasgow.

Astray, G., Gonzalez-Barreiro, C., Mejuto, J.C., Rial-Otero, R., Simal-Gándara, J., 2009. A review on the use of cyclodextrins in foods. Food Hydrocoll. 23 (7), 1631–1640.

Astray, G., Mejuto, J.C., Morales, J., Rial-Otero, R., Simal-Gándara, J., 2010. Factors controlling flavors binding constants to cyclodextrins and their applications in foods. Food Res. Int. 43 (4), 1212–1218.

Ayala-Zavala, J.F., Soto-Valdez, H., González-León, A., Álvarez-Parrilla, E., Martín-Belloso, O., González-Aguilar, G.A., 2008. Microencapsulation of cinnamon leaf (*Cinnamomum zeylanicum*) and garlic (*Allium sativum*) oils in β-cyclodextrin. J. Incl. Phenom. Macrocycl. Chem. 60, 359–368.

Aytac, Z., Dogan, S.Y., Tekinay, T., Uyar, T., 2014. Release and antibacterial activity of allyl isothiocyanate/β-cyclodextrin complex encapsulated in electrospun nanofibers. Colloids Surf. B. 120, 125–131.

Bagade, O.M., Kad, D.R., Bhargude, D.N., Bhosale, D.R., Kahane, S.K., 2014. Consequences and impose of solubility enhancement of poorly water soluble drugs. Res. J. Pharm. Technol. 7 (5), 598–607.

Bai, Y., Yu, B., Xu, X., Jin, Z., Tian, Y., Lu, L., 2010. Comparison of encapsulation properties of major garlic oil components by hydroxypropyl β-cyclodextrin. Eur. Food Res. Technol. 231 (4), 519–524.

Baines, D., Seal, R. (Eds.), 2012. Natural Food Additives, Ingredients and Flavourings. Woodhead Publishing Limited, Oxford.

Bauer, K., Garbe, D., Surburg, H., 2001. Common Fragrance and Flavor Materials: Preparation, Properties and Uses. Wiley-VCH, Weinheim.

Bekiroglu, S., Kenne, L., Sandström, C., 2003. ^{1}H NMR studies of maltose, maltoheptaose, α-, β-, and γ-cyclodextrins, and complexes in aqueous solutions with hydroxy protons as structural probes. J. Org. Chem. 68 (5), 1671–1678.

Bender, M.L., Komiyama, M., 1978. Cyclodextrin Chemistry. Springer-Verlag, Berlin.

Benesi, H.A., Hildebrand, J.H., 1949. A spectrophotometric investigation of the interaction of iodine with aromatic hydrocarbons. J. Am. Chem. Soc. 71, 2703–2707.

Bertaut, E., Landy, D., 2014. Improving ITC studies of cyclodextrin inclusion compounds by global analysis of conventional and non-conventional experiments. Beilstein J. Org. Chem. 10, 2630–2641.

Bethanis, K., Tzamalis, P., Tsorteki, F., Kokkinou, A., Christoforides, E., Mentzafos, D., 2013. Structural study of the inclusion compounds of thymol, carvacrol and rugenol in β-cyclodextrin by X-ray crystallography. J. Incl. Phenom. Macro. 77 (1–4), 163–173.

Bicchi, C., Cordero, C., Liberto, E., Sgorbini, B., Rubiolo, P., 2012. Headspace sampling in flavor and fragrance field. Comprehensive Sampling and Sample Preparationvol. 4Elsevier, Amsterdam, pp. 1–25.

Bilensoy, E., 2011. Cyclodextrins in Pharmaceutics, Cosmetics, and Biomedicine: Current and Future Industrial Applications. John Wiley & Sons, Hoboken, NJ.

Bilensoy, E., Hincal, A.A., 2009. Recent advances and future directions in amphiphilic cyclodextrin nanoparticles. Expert Opin. Drug Deliv. 6 (11), 1161–1173.

Blach, P., Fourmentin, S., Landy, D., Cazier, F., Surpateanu, G., 2008. Cyclodextrins: a new efficient absorbent to treat waste gas streams. Chemosphere 70 (3), 374–380.

Bouchemal, K., Mazzaferro, S., 2012. How to conduct and interpret ITC experiments accurately for cyclodextrin-guest interactions. Drug Discov. Today 17 (11–12), 623–629.

Boulmedarat, L., Bochot, A., Lesieur, S., Fattal, E., 2005. Evaluation of buccal methyl-β-cyclodextrin toxicity on human oral epithelial cell culture model. J. Pharm. Sci. 94 (6), 1300–1309.

Brasil, I.M., Gomes, C., Puerta-Gomez, A., Castell-Perez, M.E., Moreira, R.G., 2012. Polysaccharide-based multilayered antimicrobial edible coating enhances quality of fresh-cut papaya. LWT Food Sci. Technol. 47 (1), 39–45.

Brewster, M.E., Loftsson, T., 2007. Cyclodextrins as pharmaceutical solubilizers. Adv. Drug Deliv. Rev. 59, 645–666.

Budryn, G., Zaczyńska, D., Rachwał-Rosiak, D., Oracz, J., 2015. Changes in properties of food proteins after interaction with free and β encapsulated hydroxycinnamic acids. Eur. Food Res. Technol. 240, 1157–1166.

Buedenbender, S., Schulz, G.E., 2009. Structural base for enzymatic cyclodextrin hydrolysis. J. Mol. Biol. 385 (2), 606–617.

Buschmann, H.J., Schollmeyer, E., 2002. Applications of cyclodextrins in cosmetic products: a review. J. Cosmet. Sci. 53 (3), 185–191.

Calo, J.R., Crandall, P.G., O'Bryan, C.A., Ricke, S.C., 2015. Essential oils as antimicrobials in food systems—a review. Food Control 54, 111–119.
Carlotti, M.E., Sapino, S., Cavalli, R., Trotta, M., Trotta, F., Martina, K., 2007. Inclusion of cinnamaldehyde in modified γ-cyclodextrins. J. Incl. Phenom. Macro. 57 (1–4), 445–450.
Carrier, R.L., Miller, L.A., Ahmed, I., 2007. The utility of cyclodextrins for enhancing oral bioavailability. J. Control. Release 123 (2), 78–99.
Ceborska, M., Asztemborska, M., Luboradzki, R., Lipkowski, J., 2013a. Interactions with β-cyclodextrin as a way for encapsulation and separation of camphene and fenchene. Carbohydr. Polym. 91, 110–114.
Ceborska, M., Szwed, K., Suwinska, K., 2013b. β-Cyclodextrin as the suitable molecular container for isopulegol enantiomers. Carbohydr. Polym. 97, 546–550.
Charles, M., Romano, A., Yener, S., Barnabà, M., Navarini, L., Märk, T.D., Biasoli, F., Gasperi, F., 2015. Understanding flavour perception of espresso coffee by the combination of a dynamic sensory method and in-vivo nosespace analysis. Food Res. Int. 69, 9–20.
Chen, H., Ji, H., Zhou, X., Wang, L., 2010. Green synthesis of natural benzaldehyde from cinnamon oil catalyzed by hydroxypropyl-β-cyclodextrin. Tetrahedron 66 (52), 9888–9893.
Chin, S.-T., Marriott, P.J., 2015. Review of the role and methodology of high resolution approaches in aroma analysis. Anal. Chim. Acta 854, 1–12.
Choi, M.J., Soottitantawat, A., Nuchuchua, O., Min, S.G., Ruktanonchai, U., 2009. Physical and light oxidative properties of eugenol encapsulated by molecular inclusion and emulsion-diffusion method. Food Res. Int. 42 (1), 148–156.
Chun, J.Y., Jo, Y.J., Bjrapha, P., Choi, M.J., Min, S.G., 2015. Antimicrobial effect of α- or β-cyclodextrin complexes with trans-cinnamaldehyde against *Staphylococcus aureus* and *Escherichia coli*. Dry. Technol. 33 (3), 377–383.
Ciobanu, A., Mallard, I., Landy, D., Brabie, G., Nistor, D., Fourmentin, S., 2012. Inclusion interactions of cyclodextrins and crosslinked cyclodextrin polymers with linalool and camphor in *Lavandula angustifolia* essential oil. Carbohydr. Polym. 87 (3), 1963–1970.
Ciobanu, A., Landy, D., Fourmentin, S., 2013a. Complexation efficiency of cyclodextrins for volatile flavor compounds. Food Res. Int. 53 (1), 110–114.
Ciobanu, A., Mallard, I., Landy, D., Brabie, G., Nistor, D., Fourmentin, S., 2013b. Retention of aroma compounds from *Mentha piperita* essential oil by cyclodextrins and crosslinked cyclodextrin polymers. Food Chem. 138 (1), 291–297.
Commission Directive 2003/95/EC, 2003. Commission Directive 2003/95/EC of October 27, 2003 amending Directive 96/77/EC laying down specific purity criteria on food additives other than colours and sweeteners (Text with EEA relevance).
Connors, K.A., 1987. Binding Constants: The Measurements of Molecular Complex Stability. Wiley, New York, NY.
Costa, P., Medronho, B., Gonçalves, S., Romano, A., 2015. Cyclodextrins enhance the antioxidant activity of essential oils from three Lamiaceae species. Ind. Crop. Prod. 70, 341–346.
Costescu, C.I., Hădărugă, N.G., Hădărugă, D.I., Riviş, A., Ardelean, A., Lupea, A.X., 2008. Bionanomaterials: synthesis, physico-chemical and multivariate analyses of the dicotyledonatae and pinatae essential oil/β-cyclodextrin nanoparticles. Rev. Chim. 59, 739–744.
Cramer, F., 1954. Einschlussverbindungen. Springer-Verlag, Berlin.
Cramer, F., Henglein, F.M., 1957. Gesetzmäßigkeiten bei der bildung von adukten der cyclodextrine. Chem. Ber. Recl. 90, 2561–2571.
Cravotto, G., Binello, A., Baranelli, E., Carraro, P., Trotta, F., 2006. Cyclodextrins as food additives and in food processing. Curr. Nutr. Food Sci. 2 (4), 343–350.

Crini, G., 2014. Review: a history of cyclodextrins. Chem. Rev. 114 (21), 10940–10975.

Daruházi, Á.E., Szente, L., Balogh, B., Mátyus, P., Béni, S., Takács, M., Gergely, A., Horváth, P., Szõke, É., Lemberkovics, É., 2008. Utility of cyclodextrins in the formulation of genistein Part 1. Preparation and physicochemical properties of genistein complexes with native cyclodextrins. J. Pharm. Biomed. Anal. 48, 636–640.

Daruházi, Á.E., Kiss, T., Vecsernyés, M., Szente, L., Szõke, É., Lemberkovics, É., 2013. Investigation of transport of genistein, daidzein and their inclusion complexes prepared with different cyclodextrins on Caco-2 cell line. J. Pharm. Biomed. Anal. 84, 112–116.

De Bie, A.T.H.J., Van Ommen, B., Bär, A., 1998. Disposition of [14C]γ-cyclodextrin in germ-free and conventional rats. Regul. Toxicol. Pharm. 27 (2), 150–158.

De Marco, I., Reverchon, E., 2008. Supercritical antisolvent micronization of cyclodextrins. Powder Technol. 183, 239–246.

De Souza Siqueira Quintans, J., Menezes, P.P., Santos, M.R.V., Bonjardim, L.R., Almeida, J.R.G.S., Gelain, D.P., Araújo, A.A.D.S., Quintans, Jr., L.J., 2013. Improvement of *p*-cymene antinociceptive and anti-inflammatory effects by inclusion in β-cyclodextrin. Phytomedicine 20 (5), 436–440.

De Vos, P., Faas, M.M., Spasojevic, M., Sikkema, J., 2010. Encapsulation for preservation of functionality and targeted delivery of bioactive food components. Int. Dairy J. 20 (4), 292–302.

Decock, G., Fourmentin, S., Surpateanu, G.G., Landy, D., Decock, P., Surpateanu, G., 2006. Experimental and theoretical study on the inclusion compounds of aroma components with β-cyclodextrins. Supramol. Chem. 18 (6), 477–482.

Decock, G., Landy, D., Surpateanu, G., Fourmentin, S., 2008. Study of the retention of aroma components by cyclodextrins by static headspace gas chromatography. J. Incl. Phenom. Macro. 62 (3–4), 297–302.

Del Toro-Sánchez, C.L., Ayala-Zavala, J.F., Machi, L., Santacruz, H., Villegas-Ochoa, M.A., Alvarez-Parrilla, E., González-Aguilar, G.A., 2010. Controlled release of antifungal volatiles of thyme essential oil from β-cyclodextrin capsules. J. Incl. Phenom. Macro. 67 (3), 431–441.

Del Valle, E.M.M., 2004. Cyclodextrins and their uses: a review. Process Biochem. 39 (9), 1033–1046.

Demian, B.A., 2000. Correlation of the solubility of several aromatics and terpenes in aqueous hydroxypropyl-β-cyclodextrin with steric and hydrophobicity parameters. Carbohydr. Res. 328, 635–639.

Djedaini, F., Lin, S.Z., Perly, B., Wouessidjewe, D., 1990. High-field nuclear magnetic resonance techniques for the investigation of a β-cyclodextrin:indomethacin inclusion complex. J. Pharm. Sci. 79 (7), 643–646.

Dodziuk, H., 2006. Cyclodextrins and Their Complexes: Chemistry, Analytical Methods, Applications. John Wiley & Sons, Weinheim.

Donze, C., Coleman, A.W., 1993. β-CD inclusion complexes: relative selectivity of terpene and aromatic guest molecules studied by competitive inclusion experiments. J. Incl. Phenom. Mol. 16, 1–15.

Dos Santos, C., del Pilar Buera, M., Mazzobre, M.F., 2011. Phase solubility studies of terpineol with β-cyclodextrins and stability of the freeze-dried inclusion complex. Procedia Food Sci. 1, 355–362.

Dos Santos, C., Buera, M.P., Mazzobre, M.F., 2012. Influence of ligand structure and water interactions on the physical properties of β-cyclodextrins complexes. Food Chem. 132 (4), 2030–2036.

Duchêne, D., 1991. New trends in cyclodextrins and derivatives. Éditions de Sante, Paris.

Duchêne, D., 2011. Cyclodextrins and their inclusion complexes. In: Bilensoy, E. (Ed.), Cyclodextrins in Pharmaceutics, Cosmetics, and Biomedicine: Current and Future Industrial Applications. John Wiley & Sons, Hoboken, NJ, pp. 3–18.

Duchêne, D., Ponchel, G., Bochot, A., 2005. New uses of cyclodextrins. Eur. J. Pharm. Sci. 25, S1–S2.

Duchěne, D., Vaution, C., Glomot, F., 1986. Cyclodextrins, their value in pharmaceutical technology. Drug Dev. Ind. Pharm. 12 (11–13), 2193–2215.

Eftink, M.R., Andy, M.L., Bystrom, K., Perlmutter, H.D., Kristol, D.S., 1989. Cyclodextrin inclusion complexes: studies of the variation in the size of slicyclic guests. J. Am. Chem. Soc. 111 (17), 6765–6772.

Endo, T., Ueda, H., 2004. Large ring cyclodextrins—recent progress. Fabad J. Pharm. Sci. 29 (1), 27–38.

Fakayode, S.O., Lowry, M., Fletcher, K.A., Huang, X., Powe, A.M., Warner, I.M., 2007. Cyclodextrins host-guest chemistry in analytical and environmental chemistry. Curr. Anal. Chem. 3 (3), 171–181.

Fang, Z., Comino, P.R., Bhandari, B., 2013. Effect of encapsulation of D-limonene on the moisture adsorption property of β-cyclodextrin. LWT Food Sci. Technol. 51, 164–169.

Ferrazza, R., Rossi, B., Guella, G., 2014. DOSY-NMR and Raman investigations on the self-aggregation and cyclodextrin complexation of vanillin. J. Phys. Chem. B 118 (25), 7147–7155.

Fielding, L., 2000. Determination of association constants (Ka) from solution NMR data. Tetrahedron 56 (34), 6151–6170.

Folch-Cano, C., Jullian, C., Speisky, H., Olea-Azar, C., 2010. Antioxidant activity of inclusion complexes of tea catechins with β-cyclodextrins by ORAC assays. Food Res. Int. 43, 2039–2044.

Fourmentin, S., Ciobanu, A., Landy, D., Wenz, G., 2013. Space filling of β-cyclodextrin and β-cyclodextrin derivatives by volatile hydrophobic guests. Beilstein J. Org. Chem. 9, 1185–1191.

Freudenberg, K., Cramer, F., 1948. Die Konstitution der Schardinger-dextrine α, β und γ. Z. Naturforsch. 3b, 464.

Freudenberg, K., Jacobi, R., 1935. Uber Schardinger dextrine aus starke. Liebigs Ann. Chem. 518, 102–108.

Freudenberg, K., Cramer, F., Plieninger, H., 1953. Verfahren zur herstellung von Einschlusverbindungen physiologisch wirksamer organischer verbindungen, Knoll A.-G. Chemische Fabriken, Germany, Patent No. 895,769, 5.

Furia, T.E., Bellanca, N. (Eds.), 1975. Fenaroli's Handbook of Flavour Ingredients. CRC Press, Cleveland, OH.

Gabelman, A. (Ed.), 1994. Bioprocess Production of Flavor, Fragrance, and Color Ingredients. John Wiley & Sons, New York, NY.

Galvão, J.G., Silva, V.F., Ferreira, S.G., França, F.R.M., Santos, D.A., Freitas, L.S., Alves, P.B., Araújo, A.A.S., Cavalcanti, S.C.H., Nunes, R.S., 2015. β-Cyclodextrin inclusion complexes containing *Citrus sinensis* (L.) Osbeck essential oil: an alternative to control Aedes aegypti larvae. Thermochim. Acta 608, 14–19.

Gardner, J.W., Barlett, P.N., 1993. A brief history of electronic noses. Sens. Actuat. B 18, 211–220.

Garg, A., Gupta, B., Prakash, R., Singh, S., 2010. Preparation and characterization of hydroxypropyl-β-cyclodextrin inclusion complex of eugenol: differential pulse voltammetry and ^{1}H-NMR. Chem. Pharm. Bull. 58 (10), 1313–1319.

Giordano, F., Novak, C., Moyano, J.R., 2001. Thermal analysis of cyclodextrins and their inclusion compounds. Thermochim. Acta. 380 (2), 123–151.

Glusman, G., Bahar, A., Sharon, D., Pilpel, Y., White, J., Lancet, D., 2000. The olfactory receptor gene superfamily: data mining, classification, and nomenclature. Mamm. Genome 11 (11), 1016–1023.

Gould, S., Scott, R.C., 2005. 2-Hydroxypropyl-beta-cyclodextrin (HP-Beta-CD): a toxicology review. Food Chem. Toxicol. 43 (10), 1451–1459.
Govindasamy, R., Arumugam, S., Simon, J.E., 2013. An assessment of the essential oil and aromatic plant industry with a focus on Africa. ACS Symp. Ser. 1127, 289–321.
Guenther, E., 1972. The Essential Oils. Krieger Publishing Co., Malabar, FL.
Guichard, E., Boisard, L., Salles, C., 2013. Flavour release and sensory perception in cheeses. Handbook on Cheese: Production, Chemistry, and Sensory PropertiesNova Science Publishers, France, pp. 313–340.
Guimarães, A.G., Oliveira, M.A., Alves, R.D.S., Menezes, P.D.P., Serafini, M.R., De Souza Araújo, A.A., Bezerra, D.P., Quintans, L.J., 2015. Encapsulation of carvacrol, a monoterpene present in the essential oil of oregano, with β-cyclodextrin, improves the pharmacological response on cancer pain experimental protocols. Chem. Biol. Interact. 227, 69–76.
Hădărugă, D.I., 2011. Anti-inflammatory oxicam/cyclodextrin supramolecular systems: molecular modeling and docking experiments. J. Agroalim. Proc. Technol. 17, 456–465.
Hădărugă, D.I., Hădărugă, N.G., 2003. Odorant and Flavouring Compounds. Polytechnic Press, Timişoara.
Hădărugă, N.G., Hădărugă, D.I., Lupea, A.X., Păunescu, V., Tatu, C., 2005. Bioactive nanoparticles: 7. Essential oil from Apiaceae and Pinaceae family plants/β-cyclodextrin supramolecular systems. Rev. Chim. 56, 876–882.
Hădărugă, N.G., Hădărugă, D.I., Păunescu, V., Tatu, C., Ordodi, V.L., Bandur, G., Lupea, A.X., 2006. Thermal stability of the linoleic acid/α- and β-cyclodextrin complexes. Food Chem. 99, 500–508.
Hădărugă, D.I., Hădărugă, N.G., Resiga, D., Pode, V., Dumbravă, D., Lupea, A.X., 2007a. Obtaining and characterization of sage (*Salvia sclarea* L.) essential oil/β-cyclodextrin supramolecular systems. Rev. Chim. 58, 566–573.
Hădărugă, N.G., Hădărugă, D.I., Riviş, A., Gruia, A., Pînzaru, I.A., 2007b. Thermal and oxidative stability of the *Allium sativum* L. bioactive compounds/α- and β-cyclodextrin nanoparticles. Rev. Chim. 58, 1009–1015.
Hădărugă, N.G., Hădărugă, D.I., Riviş, A., Păunescu, V., Costescu, C., Lupea, A.X., 2007c. Bioactive nanoparticles. Essential oil from Lamiaceae family plants/β-cyclodextrin supramolecular systems. Rev. Chim. 58, 909–914.
Hădărugă, D.I., Hădărugă, N.G., Hermenean, A.O., Riviş, A., Pâslaru, V., Codina, G., 2008a. Bionanomaterials: thermal stability of the oleic acid/α- and β-cyclodextrin complexes. Rev. Chim. 59, 994–998.
Hădărugă, D.I., Hădărugă, N.G., Mureşan, S., Bandur, G.N., Lupea, A.X., Păunescu, V., Riviş, A., Tatu, C., 2008b. Fatty acid/β-cyclodextrin nanoparticles: thermal analyses and molecular modeling studies. J. Agroalim. Proc. Technol. 14, 43–49.
Hădărugă, D.I., Hădărugă, N.G., Riviş, A., Pârvu, D., 2009. Molecular modeling and docking studies on Compositae biocompounds—cyclodextrin interactions. J. Agroalim. Proc. Technol. 15, 273–282.
Hădărugă, D.I., Hădărugă, N.G., Bandur, G.N., Riviş, A., Costescu, C., Ordodi, V.L., Ardelean, A., 2010a. *Berberis vulgaris* extract/β-cyclodextrin nanoparticles synthesis and characterization. Rev. Chim. 61, 669–675.
Hădărugă, D.I., Hădărugă, N.G., Butnaru, G., Tatu, C., Gruia, A., 2010b. Bioactive microparticles (10): Thermal and oxidative stability of nicotine and its complex with β-cyclodextrin. J. Incl. Phenom. Macrocycl. Chem. 68, 155–164.
Hădărugă, D.I., Hădărugă, N.G., Bandur, G.N., Isengard, H.-D., 2012a. Water content of flavonoid/cyclodextrin nanoparticles: relationship with the structural descriptors of biologically active compounds. Food Chem. 132, 1651–1659.

Hădărugă, N.G., Hădărugă, D.I., Isengard, H.-D., 2012b. Water content of natural cyclodextrins and their essential oil complexes: a comparative study between Karl Fischer titration and thermal methods. Food Chem. 132, 1741–1748.

Hădărugă, N.G., Hădărugă, D.I., Isengard, H.-D., 2013. "Surface water" and "strong-bonded water" in cyclodextrins: a Karl Fischer titration approach. J. Incl. Phenom. Macrocycl. Chem. 75, 297–302.

Hădărugă, D.I., Hădărugă, N.G., Costescu, C.I., David, I., Gruia, A.T., 2014. Thermal and oxidative stability of the *Ocimum basilicum* L. essential oil/β-cyclodextrin supramolecular system. Beilstein J. Org. Chem. 10, 2809–2820.

Haidong, L., Fang, Y., Zhihong, T., Changle, R., 2011. Study on preparation of β-cyclodextrin encapsulation tea extract. Int. J. Biol. Macromol. 49 (4), 561–566.

Haloci, E., Toska, V., Shkreli, R., Goci, E., Vertuani, S., Manfredini, S., 2014. Encapsulation of *Satureja montana* essential oil in β-cyclodextrin. J. Incl. Phenom. Macrocycl. Chem. 80, 147–153.

Hanušová, K., Dobiáš, J., Klaudisová, K., 2009. Effect of packaging films releasing antimicrobial agents on stability of food products. Czech J. Food Sci. 27, 347–349.

Hapiot, F., Tilloy, S., Monflier, E., 2006. Cyclodextrins as supramolecular hosts for organometallic complexes. Chem. Rev. 106 (3), 767–781.

Harangi, J., Béke, G., Harangi, M., Mótyán, J.A., 2012. The digestable parent cyclodextrin. J. Incl. Phenom. Macro. 73 (1–4), 335–339.

Hashimoto, H., 2002. Present status of industrial application of cyclodextrins in Japan. J. Incl. Phenom. 44 (1–4), 57–62.

Hashimoto, H., 2006. Applications other than in the pharmaceutical industry: CyD applications in food, cosmetic, toiletry, textile and wrapping material fields. Cyclodextrins and Their Complexes: Chemistry, Analytical Methods, ApplicationsJohn Wiley & Sons, Hoboken, NJ, pp. 459–468.

He, Y., Fu, P., Shen, X., Gao, H., 2008. Cyclodextrin-based aggregates and characterization by microscopy. Micron 39, 495–516.

Hedges, A.R., Shieh, W.J., Sikorski, C.T., 1995. Use of cyclodextrins for encapsulation in the use and treatment of food products. In: Risch, S.J., Reineccius, G.A. (Eds.), Encapsulation and Controlled Release of Food Ingredients, ACS Symp. Ser. 590. American Chemical Society, Washington, DC, pp. 60–71.

Hegheş, A., Hădărugă, N.G., Fuliaş, A.-V., Bandur, G.N., Hădărugă, D.I., Dehelean, C.-A., 2015. *Capsicum annuum* extracts/β-cyclodextrin complexes. Thermal analyses—Karl Fischer water titration correlations and antioxidant activity. J. Therm. Anal. Calorim. 120, 603–615.

Higuchi, T., Connors, A.K., 1965. Phase solubility techniques. Adv. Anal. Chem. Instrum. 4, 117–212.

Hill, L.E., Gomes, C., Taylor, T.M., 2013. Characterization of beta-cyclodextrin inclusion complexes containing essential oils (*trans*-cinnamaldehyde, eugenol, cinnamon bark, and clove bud extracts) for antimicrobial delivery applications. LWT Food Sci. Technol. 51 (1), 86–93.

Ho, B.T., Joyce, D.C., Bhandari, B.R., 2011. Release kinetics of ethylene gas from ethylene-α-cyclodextrin inclusion complexes. Food Chem. 129 (2), 259–266.

Hou, L., Wang, J., Zhang, D., 2013. Optimization of debittering of soybean antioxidant hydrolysates with β-cyclodextrins using response surface methodology. J. Food Sci. Technol. 50 (3), 521–527.

Hougeir, F.G., Kircik, L., 2012. A review of delivery systems in cosmetics. Dermatol. Ther. 25 (3), 234–237.

Houk, K.N., Leach, A.G., Kim, S.P., Zhang, X., 2003. Binding affinities of host-guest, protein-ligand, and protein-transition-state complexes. Angew. Chem. Int. Ed. 42 (40), 4872–4897.

Hsu, C.M., Yu, S.C., Tsai, F.J., Tsai, Y., 2013. Enhancement of rhubarb extract solubility and bioactivity by 2-hydroxypropyl-β-cyclodextrin. Carbohydr. Polym. 98 (2), 1422–1429.

Hundre, S.Y., Karthik, P., Anandharamakrishnan, C., 2015. Effect of whey protein isolate and β-cyclodextrin wall systems on stability of microencapsulated vanillin by spray-freeze drying method. Food Chem. 174, 16–24.

Irie, T., Uekama, K., 1997. Pharmaceutical applications of cyclodextrins. III. Toxicological issues and safety evaluation. J. Pharm. Sci. 86 (2), 147–162.

JECFA (Joint FAO/WHO Expert Committee on Food Additives). 78th JECFA November 2013.

JadhavAmbadas, S., Kankudte Ashish, D., Wadikar Jagdish, C., Chintale Ashwini, G., Karde Vaibhav, K., Puri Sachin, G., 2013. Liquisolid technique a novel approach to enhance solubility and dissolution rate: a review. Res. J. Pharm. Technol. 6 (8), 819–824.

Jahed, V., Zarrabi, A., Bordbar, A.K., Hafezi, M.S., 2014. NMR (^{1}H, ROESY) spectroscopic and molecular modelling investigations of supramolecular complex of β-cyclodextrin and curcumin. Food Chem. 165, 241–246.

Jansook, P., Loftsson, T., 2009. CDs as solubilizers: effects of excipients and competing drugs. Int. J. Pharm. 379 (1–2), 32–40.

Jelen, H.H., Majcher, M., Dziadas, M., 2012. Microextraction techniques in the analysis of food flavor compounds: a review. Anal. Chim. Acta 738, 13–26.

Jiang, S., Li, J.N., Jiang, Z.T., 2010. Inclusion reactions of β-cyclodextrin and its derivatives with cinnamaldehyde in *Cinnamomum loureirii* essential oil. Eur. Food Res. Technol. 230 (4), 543–550.

Jiang, H., Zhang, S., Shi, Q., 2011. Removal of cholesterol by β-cyclodextrin. Asian J. Chem. 23 (9), 3783–3786.

Juvancz, Z., Szejtli, J., 2002. The role of cyclodextrins in chiral selective chromatography. Trends Anal. Chem. 21 (5), 379–388.

Kader, A.A., 2008. Flavor quality of fruits and vegetables. J. Sci. Food Agric. 88 (11), 1863–1868.

Kant, A., Linforth, R.S.T., Hort, J., Taylor, A.J., 2004. Effect of β-cyclodextrin on aroma release and flavor perception. J. Agric. Food Chem. 52 (7), 2028–2035.

Karangwa, E., Hayat, K., Rao, L., Nshimiyimana, D.S., Foh, M.B.K., Li, L., Ntwali, J., Raymond, L.V., Xia, S., Zhang, X., 2012. Improving blended carrot-orange juice quality by the addition of cyclodextrins during enzymatic clarification. Food Bioprocess Technol. 5 (6), 2612–2617.

Karathanos, V.T., Mourtzinos, I., Yannakopoulou, K., Andrikopoulos, N.K., 2007. Study of the solubility, antioxidant activity and structure of inclusion complex of vanillin with β-cyclodextrin. Food Chem. 101 (2), 652–658.

Kawakami, K., Fujita, A., Mikami, T., Yoshii, H., Paramita, V., Neoh, T.L., Furuta, T., 2009. Formation of rice flavor powder with α-cyclodextrin by spray drying. Eur. Food Res. Technol. 229, 239–245.

Kayaci, F., Uyar, T., 2011. Solid inclusion complexes of vanillin with cyclodextrins: their formation, characterization, and high-temperature stability. J. Agric. Food Chem. 59 (21), 11772–11778.

Kfoury, M., Auezova, L., Fourmentin, S., Greige-Gerges, H., 2014a. Investigation of monoterpenes complexation with hydroxypropyl-β-cyclodextrin. J. Incl. Phenom. Macro. 80 (1–2), 51–60.

Kfoury, M., Auezova, L., Greige-Gerges, H., Ruellan, S., Fourmentin, S., 2014b. Cyclodextrin, an efficient tool for *trans*-anethole encapsulation: chromatographic, spectroscopic, thermal and structural studies. Food Chem. 164, 454–461.

Kfoury, M., Landy, D., Auezova, L., Greige-Gerges, H., Fourmentin, S., 2014c. Effect of cyclodextrin complexation on phenylpropanoids' solubility and antioxidant activity. Beilstein J. Org. Chem. 10, 2322–2331.

Kfoury, M., Auezova, L., Greige-Gerges, H., Fourmentin, S., 2015a. Promising applications of cyclodextrins in food: improvement of essential oils retention, controlled release and antiradical activity. Carbohydr. Polym. 131, 264–272.

Kfoury, M., Auezova, L., Ruellan, S., Greige-Gerges, H., Fourmentin, S., 2015b. Complexation of estragole as pure compound and as main component of basil and tarragon essential oils with cyclodextrins. Carbohydr. Polym. 118, 156–164.

Kfoury, M., Balan, R., Landy, D., Nistor, D., Fourmentin, S., 2015c. Investigation of the complexation of essential oil components with cyclodextrins. Supramol. Chem. 27, 1–10.

Kfoury, M., Ciobanu, A., Hădărugă, N.G., Hădărugă, D.I., Fourmentin, S., 2015d. Cyclodextrines comme agents d'encapsulation des arômes. In: Morin-Crini, N., Fourmentin, S., Crini, G. (Eds.), Cyclodextrines. Histoire, Propriétés, Chimie and Applications. Presses Universitaires de Franche-Comté, Besançon, pp. 259–278, (Chapitre 12).

Khan, A.R., Forgo, P., Stine, K.J., D'Souza, V.T., 1998. Methods for selective modifications of cyclodextrins. Chem. Rev. 98 (5), 1977–1996.

Kiss, T., Fenyvesi, F., Pasztor, N., Feher, P., Varadi, J., Kocsan, R., Szente, L., Fenyvesi, E., Szabo, G., Vecsernyes, M., Bacskay, I., 2007. Cytotoxicity of different types of methylated β-cyclodextrins and ionic derivatives. Pharmazie 62 (7), 557–558.

Kolb, B., 1999. Headspace sampling with capillary columns. J. Chromatogr. A. 842 (1–2), 163–205.

Kolb, B., Ettre, L.S., 2006. Static Headspace-Gas Chromatography: Theory and Practice, second ed. John Wiley & Sons, Hoboken, NJ.

Komiyama, M., Monflier, E., 2006. Cyclodextrin catalysis. Cyclodextrins, Their Complexes: Chemistry, Analytical Methods, Applications. John Wiley & Sons, Hoboken, NJ, pp. 93–105.

Kurkov, S.V., Loftsson, T., 2013. Cyclodextrins. Int. J. Pharm. 453 (1), 167–180.

Landy, D., Fourmentin, S., Salome, M., Surpateanu, G., 2000. Analytical improvement in measuring formation constants of inclusion complexes between β-cyclodextrin and phenolic compounds. J. Incl. Phenom. 38 (1–4), 187–198.

Landy, D., Tetart, F., Truant, E., Blach, P., Fourmentin, S., Surpateanu, G., 2007. Development of a competitive continuous variation plot for the determination of inclusion compounds stoichiometry. J. Incl. Phenom. Macro. 57 (1–4), 409–413.

Landy, D., Mallard, I., Ponchel, A., Monflier, E., Fourmentin, S., 2012. Remediation technologies using cyclodextrins: an overview. Environ. Chem. Lett. 10 (3), 225–237.

Larsen, K.L., 2002. Large cyclodextrins. J. Incl. Phenom. 43 (1–2), 1–13.

Lavoine, N., Givord, C., Tabary, N., Desloges, I., Martel, B., Bras, J., 2014. Elaboration of a new antibacterial bio-nano-material for food-packaging by synergistic action of cyclodextrin and microfibrillated cellulose. Innov. Food Sci. Emerg. 26, 330–340.

Lawless, H.T., Heymann, H., 1998. Sensory Evaluation of Food: Principles and Practices. Chapman & Hall, New York, NY.

Laza-Knoerr, A.L., Gref, R., Couvreur, P., 2010. Cyclodextrins for drug delivery. J. Drug Target. 18 (9), 645–656.

Lee, J.H., Lee, J., Song, K.B., 2015. Development of a chicken feet protein film containing essential oils. Food Hydrocoll. 46, 208–215.

Li, Z., Zhang, J., Wang, M., Gu, Z., Du, G., Li, J., Wu, J., Chen, J., 2009. Mutations at subsite -3 in cyclodextrin glycosyltransferase from *Paenibacillus macerans* enhancing α-cyclodextrin specificity. Appl. Microbiol. Biotechnol. 83 (3), 483–490.

Li, Z., Chen, S., Gu, Z., Chen, J., Wu, J., 2014. Alpha-cyclodextrin: enzymatic production and food applications. Trends Food Sci. Technol. 35, 151–160.

Liang, H., Yuan, Q., Vriesekoop, F., Lv, F., 2012. Effects of cyclodextrins on the antimicrobial activity of plant-derived essential oil compounds. Food Chem. 135 (3), 1020–1027.

Liao, R., Liu, M., Liao, X., Yang, B., 2015. Cyclodextrin-based smart stimuli-responsive drug carriers. Prog. Chem. 27 (1), 79–90.

Liu, L., Guo, Q.X., 2002. The driving forces in the inclusion complexation of cyclodextrins. J. Incl. Phenom. 42 (1–2), 1–14.

Liu, Y., Li, B., Wada, T., Inoue, Y., 2001. Studies on molecular recognition in supramolecular systems. Part 31: Circular dichroism spectral studies of molecular and chiral recognition of aliphatic alcohols by 6-modified β-cyclodextrins. Tetrahedron 57, 7153–7161.

Liu, Y., Li, L., Zhang, H.Y., Yang, Y.W., Ding, F., 2004. Correlation between thermodynamic behavior and structure in the complexation of modified β-cyclodextrins and bile salts. Supramol. Chem. 16 (5), 371–379.

Liu, Y., Zhang, Q., Chen, Y., 2007. Spectrophotometric and calorimetric titration studies on molecular recognition of camphor and borneol by nucleobase-modified β-cyclodextrins. J. Phys. Chem. B 111 (42), 12211–12218.

Liu, H., Li, J., Du, G., Zhou, J., Chen, J., 2012. Enhanced production of α-cyclodextrin glycosyltransferase in *Escherichia coli* by systematic codon usage optimization. J. Ind. Microbiol. Biotechnol. 39 (12), 1841–1849.

Liu, H., Yang, G., Tang, Y., Cao, D., Qi, T., Qi, Y., Fan, G., 2013. Physicochemical characterization and pharmacokinetics evaluation of β-caryophyllene/β-cyclodextrin inclusion complex. Int. J. Pharm. 450 (1–2), 304–310.

Lo Nostro, P., Fratoni, L., Baglioni, P., 2002. Modification of a cellulosic fabric with β-cyclodextrin for textile finishing applications. J. Incl. Phenom. 44 (1–4), 423–427.

Locci, E., Lai, S., Piras, A., Marongiu, B., Lai, A., 2004. 13C-CPMAS and 1H-NMR study of the inclusion complexes of β-cyclodextrin with carvacrol, thymol, and eugenol prepared in supercritical carbon dioxide. Chem. Biodivers. 1 (9), 1354–1366.

Loftsson, T., Brewster, M.E., 2010. Pharmaceutical applications of cyclodextrins: basic science and product development. J. Pharm. Pharmacol. 62 (11), 1607–1621.

Loftsson, T., Brewster, M.E., 2011. Pharmaceutical applications of cyclodextrins: effects on drug permeation through biological membranes. J. Pharm. Pharmacol. 63 (9), 1119–1135.

Loftsson, T., Duchêne, D., 2007. Cyclodextrins and their pharmaceutical applications. Int. J. Pharm. 329 (1–2), 1–11.

Loftsson, T., Hreinsdóttir, D., Másson, M., 2005. Evaluation of cyclodextrin solubilization of drugs. Int. J. Pharm. 302 (1–2), 18–28.

López-de-Dicastillo, C., Catalá, R., Gavara, R., Hernández-Muñoz, P., 2011. Food applications of active packaging EVOH films containing cyclodextrins for the preferential scavenging of undesirable compounds. J. Food Eng. 104, 380–386.

López-Nicolás, J.M., García-Carmona, F., 2008. Rapid, simple and sensitive determination of the apparent formation constants of *trans*-resveratrol complexes with natural cyclodextrins in aqueous medium using HPLC. Food Chem. 109, 868–875.

López-Nicolás, J.M., Andreu-Sevilla, A.J., Carbonell-Barrachina, Á.A., García-Carmona, F., 2009. Effects of addition of α-cyclodextrin on the sensory quality,

volatile compounds, and color parameters of fresh pear juice. J. Agric. Food Chem. 57 (20), 9668–9675.
López-Nicolás, J.M., Rodríguez-Bonilla, P., García-Carmona, F., 2014. Cyclodextrins and antioxidants. Crit. Rev. Food Sci. 54 (2), 251–276.
Lu, Z., Chen, R., Fu, R., Xiong, J., Hu, Y., 2012. Cytotoxicity and inhibition of lipid peroxidation activity of resveratrol/cyclodextrin inclusion complexes. J. Incl. Phenom. Macrocycl. Chem. 73, 313–320.
Lucera, A., Costa, C., Conte, A., Del Nobile, M.A., 2012. Food applications of natural antimicrobial compounds. Front. Microbiol. 3, 287.
Madene, A., Jacquot, M., Scher, J., Desobry, S., 2006. Flavour encapsulation and controlled release—a review. Int. J. Food Sci. Technol. 41 (1), 1–21.
Mangolim, C.S., Moriwaki, C., Nogueira, A.C., Sato, F., Baesso, M.L., Neto, A.M., Matioli, G., 2014. Curcumin-β-cyclodextrin inclusion complex: stability, solubility, characterisation by FT-IR, FT-Raman, X-ray diffraction and photoacoustic spectroscopy, and food application. Food Chem. 153, 361–370.
Marques, H.M.C., 2010. A review on cyclodextrin encapsulation of essential oils and volatiles. Flavour Frag. J. 25 (5), 313–326.
Marreto, R.N., Almeida, E.E.C.V., Alves, P.B., Niculau, E.S., Nunes, R.S., Matos, C.R.S., Araújo, A.A.S., 2008. Thermal analysis and gas chromatography coupled mass spectrometry analyses of hydroxypropyl-β-cyclodextrin inclusion complex containing *Lippia gracilis* essential oil. Thermochim. Acta 475, 53–58.
Marsili, R. (Ed.), 1997. Techniques for Analyzing Food Aroma. Marcel Dekker, New York, NY.
Martin, G., Laffort, P., 1990. Odeurs and Désodorisation dans L'environnement. IRC-Obac, Paris.
Martina, K., Binello, A., Lawson, D., Jicsinszky, L., Cravotto, G., 2013. Recent applications of cyclodextrins as food additives and in food processing. Curr. Nutr. Food Sci. 9 (3), 167–179.
Martins, A.P., Craveiro, A.A., Machado, M.I.L., Raffin, F.N., Moura, T.F., Novák, C., Éhen, Z., 2007. Preparation and characterization of *Mentha x villosa* Hudson oil–β-cyclodextrin complex. J. Therm. Anal. Calorim. 88, 363–371.
Mascheroni, E., Fuenmayor, C.A., Cosio, M.S., Silvestro, G.D., Piergiovanni, L., Mannino, S., Schiraldi, A., 2013. Encapsulation of volatiles in nanofibrous polysaccharide membranes for humidity-triggered release. Carbohydr. Polym. 98 (1), 17–25.
Matsuda, H., Ito, K., Fujiwara, Y., Tanaka, M., Taki, A., Uejima, O., Sumiyoshi, H., 1991. Complexation of various fragrance materials with 2-hydroxypropyl-β-cyclodextrin. Chem. Pharm. Bull. 39 (4), 827–830.
Mazzobre, M.F., dos Santos, C.I., Buera, M., 2011. Solubility and stability of β-cyclodextrin-terpineol inclusion complex as affected by water. Food Biophys. 6 (2), 274–280.
McNeil, B., Archer, D., Giavasis, I., Harvey, L. (Eds.), 2013. Microbial Production of Food Ingredients, Enzymes and Nutraceuticals. Woodhead Publishing Ltd, Oxford.
Menezes, P.P., Serafini, M.R., Santana, B.V., Nunes, R.S., Quintans, Jr., L.J., Silva, G.F., Medeiros, I.A., Marchioro, M., Fraga, B.P., Santos, M.R.V., Araújo, A.A.S., 2012. Solid-state β-cyclodextrin complexes containing geraniol. Thermochim. Acta 548, 45–50.
Menezes, P.P., Serafini, M.R., Quintans-Júnior, L.J., Silva, G.F., Oliveira, J.F., Carvalho, F.M.S., Souza, J.C.C., Matos, J.R., Alves, P.B., Matos, I.L., Hădăr ugă, D.I., Araújo, A.A.S., 2013. Inclusion complex of (−)-linalool and β-cyclodextrin. J. Therm. Anal. Calorim. 115, 2429–2437.

Meyer, B., Peters, T., 2003. NMR spectroscopy techniques for screening and identifying ligand binding to protein receptors. Angew. Chem. Int. Ed. 42 (8), 864–890.

Mielle, P., 1996. Electronic noses: towards the objective instrumental characterization of food aroma. Trends Food Sci. Technol. 7, 432–438.

Miron, D., Battisti, F., Caten, C.S.T., Mayorga, P., Schapoval, E.E.S., 2012. Spectrophotometric simultaneous determination of citral isomers in cyclodextrin complexes with partial least squares supported approach. Curr. Pharm. Anal. 8 (4), 401–408.

Moeder, C., O'Brien, T., Thompson, R., Bicker, G., 1996. Determination of stoichiometric coefficients and apparent formation constants for α- and β-CD complexes of terpenes using reversed-phase liquid chromatography. J. Chromatogr. A. 736 (1–2), 1–9.

Mohan, P.R.K., Sreelakshmi, G., Muraleedharan, C.V., Joseph, R., 2012. Water-soluble complexes of curcumin with cyclodextrins: characterization by FT-Raman spectroscopy. Vib. Spectrosc. 62, 77–84.

Moon, T.W., Ji, W.L., Jhee, K.H., Khang, K., Jeong, H.S., Yang, S.A., Kim, H.J., 2008. Supramolecular encapsulation of pulegone from oriental herb, *Schizonepeta tenuifolia* briquet by β- and γ-cyclodextrins. B. Kor. Chem. Soc. 29 (8), 1579–1582.

Moreira da Silva, A., Empis, J., Teixeira Dias, J.J.C., 1999. Inclusion of enantiomeric carvones in β-cyclodextrin: a variable temperature ^{1}H NMR study in aqueous solution. J. Incl. Phenom. Macrocycl. Chem. 33, 81–97.

Moreira, S.P., de Carvalho, W.M., Alexandrino, A.C., de Paula, H.C.B., Rodrigues, M.C.P., de Figueiredo, R.W., Maia, G.A., de Figueiredo, E.M.A.T., Brasil, I.M., 2014. Freshness retention of minimally processed melon using different packages and multilayered edible coating containing microencapsulated essential oil. Int. J. Food Sci. Technol. 49 (10), 2192–2203.

Morelli, C.L., Mahrous, M., Belgacem, M.N., Branciforti, M.C., Bretas, R.E.S., Bras, J., 2015. Natural copaiba oil as antibacterial agent for bio-based active packaging. Ind. Crop. Prod. 70, 134–141.

Morin-Crini, N., Crini, G., 2013. Environmental applications of water-insoluble β-cyclodextrin–epichlorohydrin polymers. Prog. Polym. Sci. 38 (2), 344–368.

Mosinger, J., Tománková, V., Němcová, I., Zýka, J., 2001. Cyclodextrins in analytical chemistry. Anal. Lett. 34 (12), 1979–2004.

Mourtzinos, I., Kalogeropoulos, N., Papadakis, S.E., Konstantinou, K., Karathanos, V.T., 2008. Encapsulation of nutraceutical monoterpenes in β-cyclodextrin and modified starch. J. Food Sci. 73 (1), 89–94.

Munin, A., Edwards-Lévy, F., 2011. Encapsulation of natural polyphenolic compounds: a review. Pharmaceutics 3 (4), 793–829.

Munro, I.C., Newberne, P.M., Young, V.R., Bär, A., 2004. Safety assessment of γ-cyclodextrin. Regul. Toxicol. Pharm. 39, S3–S13.

Mura, P., 2014. Analytical techniques for characterization of cyclodextrin complexes in aqueous solution: a review. J. Pharm. Biomed. Anal. 101, 238–250.

Mura, P., 2015. Analytical techniques for characterization of cyclodextrin complexesin the solid state: a review. J. Pharm. Biomed. Anal. 113, 226–238.

Nascimento, S.S., Camargo, E.A., DeSantana, J.M., Araújo, A.A.S., Menezes, P.P., Lucca-Júnior, W., Albuquerque-Júnior, R.L.C., Bonjardim, L.R., Quintans-Júnior, L.J., 2014. Linalool and linalool complexed in β-cyclodextrin produce anti-hyperalgesic activity and increase fos protein expression in animal model for fibromyalgia. Naunyn Schmiedebergs Arch. Pharmacol. 387 (10), 935–942.

Nieddu, M., Rassu, G., Boatto, G., Bosi, P., Trevisi, P., Giunchedi, P., Carta, A., Gavini, E., 2014. Improvement of thymol properties by complexation with cyclodextrins: in vitro and in vivo studies. Carbohydr. Polym. 102 (1), 393–399.

Nokhodchi, A., Hentzschel, C.M., Leopold, C.S., 2011. Drug release from liquisolid systems: speed it up, slow it down. Expert Opin. Drug Deliv. 8 (2), 191–205.
Nowakowski, M., Ejchart, A., 2014. Complex formation of fenchone with α-cyclodextrin: NMR titrations. J. Incl. Phenom. Macro. 79 (3–4), 337–342.
Nuchuchua, O., Saesoo, S., Sramala, I., Puttipipatkhachorn, S., Soottitantawat, A., Ruktanonchai, U., 2009. Physicochemical investigation and molecular modeling of cyclodextrin complexation mechanism with eugenol. Food Res. Int. 42, 1178–1185.
Numanoğlu, U., Şen, T., Tarimci, N., Kartal, M., Koo, O.M.Y., Önyüksel, H., 2007. Use of cyclodextrins as a cosmetic delivery system for fragrance materials: linalool and benzyl acetate. AAPS PharmSciTech 8 (4), 85.
Nxumalo, E.N., Msomi, P.F., Mhlanga, S.D., Mamba, B.B., 2013. Production of N-doped carbon nanotubes using α- and β-cyclodextrins: the effect of solubility. Mater. Lett. 100, 66–69.
Ohloff, G., 1994. Scent and Fragrances: The Fascination of Odors and Their Chemical Perspectives. Springer-Verlag, Berlin.
Paramera, E.I., Konteles, S.J., Karathanos, V.T., 2011. Stability and release properties of curcumin encapsulated in *Saccharomyces cerevisiae*, β-cyclodextrin and modified starch. Food Chem. 125 (3), 913–922.
Parker, J.K., Elmore, J.S., Methven, L. (Eds.), 2015. Flavour Development, Analysis and Perception in Food and Beverages. Elsevier, Amsterdam.
Parliment, T.H., 1989. Thermal generation of aromas. In: Parliment, T.H., McGorrin, R.J., Ho, C.-T. (Eds.), Thermal Generation of Aromas. American Chemical Society, Washington, DC, pp. 2–11.
Partanen, R., Ahro, M., Hakala, M., Kallio, H., Forssell, P., 2002. Microencapsulation of caraway extract in β-cyclodextrin and modified starches. Eur. Food Res. Technol. 214 (3), 242–247.
Patro, N.M., Sultana, A., Terao, K., Nakata, D., Jo, A., Urano, A., Ishida, Y., Gorantl, R.N., Pandit, V., Devi, K., Rohit, S., Grewal, B.K., Sophia, E.M., Suresh, A., Ekbote, V.K., 2014. Comparison and correlation of in vitro, in vivo and in silico evaluations of alpha, beta and gamma cyclodextrin complexes of curcumin. J. Incl. Phenom. Macro. 78 (1–4), 471–483.
Pérez-Garrido, A., Helguera, A.M., Guillén, A.A., Cordeiro, M.N.D.S., Escudero, A.G., 2009. Convenient QSAR model for predicting the complexation of structurally diverse compounds with β-cyclodextrins. Bioorg. Med. Chem. 17, 896–904.
Piercey, M.J., Mazzanti, G., Budge, S.M., Delaquis, P.J., Paulson, A.T., Truelstrup Hansen, L., 2012. Antimicrobial activity of cyclodextrin entrapped allyl isothiocyanate in a model system and packaged fresh-cut onions. Food Microbiol. 30 (1), 213–218.
Pigott, J.R., Paterson, A., 1994. Understanding Natural Flavors. Blackie Academic & Professional, London.
Pinho, E., Grootveld, M., Soares, G., Henriques, M., 2014. Cyclodextrins as encapsulation agents for plant bioactive compounds. Carbohydr. Polym. 101, 121–135.
Pînzaru, I.A., Hădărugă, D.I., Hădărugă, N.G., Peter, F., 2011. Rutin-saturated fatty acid bioconjugate/cyclodextrin supramolecular systems: molecular modeling and docking studies. J. Agroalim. Proc. Technol. 17, 123–129.
Ponce Cevallos, P.A., Buera, M.P., Elizalde, B.E., 2010. Encapsulation of cinnamon and thyme essential oils components (cinnamaldehyde and thymol) in β-cyclodextrin: effect of interactions with water on complex stability. J. Food Eng. 99, 70–75.
Pranckute, R., Kaunietis, A., Kuisiene, N., Čitavičius, D., 2014. Development of synbiotics with inulin, palatinose, α-cyclodextrin and probiotic bacteria. Pol. J. Microbiol. 63 (1), 33–41.

Pybus, D., Sell, C., 1999. The Chemistry of Fragrances. The Royal Society of Chemistry, Cambridge.

Qi, Z.H., Hedges, A.R., 1995. Use of cyclodextrins for flavours. In: Ho, C.T., Tan., C.H. (Eds.), Glavur Technology: Physical Chemistry, Modification, Process, ACS Symp. Ser. 610. American Chemical Society, Washington, DC, pp. 231–243.

Quintans-Júnior, L.J., Barreto, R.S.S., Menezes, P.P., Almeida, J.R.G.S., Viana, A.F.S.C., Oliveira, R.C.M., Oliveira, A.P., Gelain, D.P., de Lucca Júnior, W., Araújo, A.A.S., 2013. β-Cyclodextrin-complexed (−)-linalool produces antinociceptive effect superior to that of (−)-linalool in experimental pain protocols. Basic Clin. Pharmacol. Toxicol. 113 (3), 167–172.

Radulovic, N.S., Blagojevic, P.D., Stojanovic-Radic, Z.Z., Stojanovic, N.M., 2013. Antimicrobial plant metabolites: structural diversity and mechanism of action. Curr. Med. Chem. 20 (7), 932–952.

Ramaekers, M.G., Luning, P.A., Ruijschop, R.M.A.J., Lakemond, C.M.M., Bult, J.H.F., Gort, G., Van Boekel, M.A.J.S., 2014. Aroma exposure time and aroma concentration in relation to satiation. Br. J. Nutr. 111 (3), 554–562.

Reineccius, T.A., Reineccius, G.A., Peppard, T.L., 2002. Encapsulation of flavors using cyclodextrins: comparison of flavor retention in alpha, beta, and gamma types. J. Food Sci. 67 (9), 3271–3279.

Reineccius, T. A., Reineccius, G. A., Peppard, T. L., 2003. Flavor release from cyclodextrin complexes: Comparison of alpha, beta, and gamma types. J. Food Sci. 68(4), 1234–1239.

Reineccius, T.A., Reineccius, G.A., Peppard, T.L., 2004. Utilization of β-cyclodextrin for improved flavor retention in thermally processed foods. J. Food Sci. 69 (1), 58–62.

Rekharsky, M.V., Goldberg, R.N., Schwarz, F.P., Tewari, Y.B., Ross, P.D., Yamashoji, Y., Inoue, Y., 1995. Thermodynamic and nuclear magnetic resonance study of the interactions of .alpha.- and .beta.-cyclodextrin with model substances: phenethylamine, ephedrines, and related substances. J. Am. Chem. Soc. 117 (34), 8830–8840.

Rivera Calo, J., Crandall, P.G., O'Bryan, C.A., Ricke, S.C., 2015. Essential oils as antimicrobials in food systems—a review. Food Control 54, 111–119.

Riviş, A., Hădărugă, N.G., Hadăruga, D.I., Trasca, T., Druga, M., Pinzaru, I., 2008. Bioactive nanoparticles: the complexation of odorant compounds with α-and β-cyclodextrin. Rev. Chim. 59 (2), 149–153.

Romani, S., Balestra, F., Angioloni, A., Rocculi, P., Rosa, M.D., 2012. Physico-chemical and electronic nose measurements on the study of biscuit baking kinetics. Ital. J. Food Sci. 24 (1), 32–40.

Ruijschop, R., Kok, P.D., 2009. Induction of satiation via aroma in foods. Food Sci. Technol. 23 (1), 25–27.

Ruktanonchai, U.R., Srinuanchai, W., Saesoo, S., Sramala, I., Puttipipatkhachorn, S., Soottitantawat, A., 2011. Encapsulation of citral isomers in extracted lemongrass oil with cyclodextrins: molecular modeling and physicochemical characterizations. Biosci. Biotechnol. Biochem. 75 (12), 2340–2345.

Saito, Y., Tanemura, I., Sato, T., Ueda, H., 1999. Interaction of fragrance materials with 2-hydroxypropyl-β-cyclodextrin by static and dynamic head-space methods. Int. J. Cosmet. Sci. 21 (3), 189–198.

Salústio, P.J., Pontes, P., Conduto, C., Sanches, I., Carvalho, C., Arrais, J., Marques, H.M.C., 2011. Advanced technologies for oral controlled release: cyclodextrins for oral controlled release. AAPS PharmSciTech 12 (4), 1276–1292.

Sánchez-González, L., Vargas, M., González-Martínez, C., Chiralt, A., Cháfer, M., 2011. Use of essential oils in bioactive edible coatings: a review. Food Eng. Rev. 3 (1), 1–16.

Santos, E.H., Kamimura, J.A., Hill, L.E., Gomes, C.L., 2015. Characterization of carvacrol beta-cyclodextrin inclusion complexes as delivery systems for

antibacterial and antioxidant applications. LWT Food Sci. Technol. 60 (1), 583–592.
Saravana Kumar, K., Sushma, M., Prasanna Raju, R., 2013. Dissolution enhancement of poorly soluble drugs by using complexation technique—a review. J. Pharm. Sci. Res. 5 (5), 120–124.
Schardinger, F., 1903. Uber thermophile bakterien aus verschiedenen speisen und milch, sowie uber einige umsetzungsprodukte derselben in kohlenhydrathaltigen Nahrlosungen, darunter krystallisierte polysaccharide (dextrine) aus starke. Z. Untersuch. Nahr. u. Genussm. 6, 865–880.
Schardinger, F., 1911. Bildung kristallisierter polysaccharide (dextrine) aus starkekleister durch microben. Zentralbl. Bakteriol. Parasitenk. Abt. II 29, 188–197.
Schmidt, B.V.K.J., Hetzer, M., Ritter, H., Barner-Kowollik, C., 2014. Complex macromolecular architecture design via cyclodextrin host/guest complexes. Prog. Polym. Sci. 39 (1), 235–249.
Schneider, H.J., Hacket, F., Rüdiger, V., Ikeda, H., 1998. NMR Studies of cyclodextrins and cyclodextrin complexes. Chem. Rev. 98 (5), 1755–1785.
Schudel, P., 1990. Odour, a fascinating phenomenon. Pure Appl. Chem. 62 (7), 1381–1383.
Seo, E.J., Min, S.G., Choi, M.J., 2010. Release characteristics of freeze-dried eugenol encapsulated with β-cyclodextrin by molecular inclusion method. J. Microencapsul. 27 (6), 496–505.
Shallenberger, R.S., 1993. Taste Chemistry. Blackie Academic & Professional, London.
Shulman, M., Cohen, M., Soto-Gutierrez, A., Yagi, H., Wang, H., Goldwasser, J., Lee-Parsons, C.W., Benny-Ratsaby, O., Yarmush, M.L., Nahmias, Y., 2011. Enhancement of naringenin bioavailability by complexation with hydroxypropoyl-β-cyclodextrin. PLoS One 6 (4), e18033.
Snow, N.H., Slack, G.C., 2002. Head-space analysis in modern gas chromatography. Trends Anal. Chem. 21 (9–10), 608–617.
Songkro, S., Hayook, N., Jaisawang, J., Maneenuan, D., Chuchome, T., Kaewnopparat, N., 2012. Investigation of inclusion complexes of citronella oil, citronellal and citronellol with β-cyclodextrin for mosquito repellent. J. Incl. Phenom. Macrocycl. Chem. 72, 339–355.
Spears, J.K., Karr-Lilienthal, L.K., Fahey, Jr., G.C., 2005. Influence of supplemental high molecular weight pullulan or γ-cyclodextrin on ileal and total tract nutrient digestibility, fecal characteristics, and microbial populations in the dog. Arch. Anim. Nutr. 59 (4), 257–270.
Stancanelli, R., Mazzaglia, A., Tommasini, S., Calabro, M.L., Villari, V., Guardo, M., Ficarra, P., Ficarra, R., 2007. The enhancement of isoflavones water solubility by complexation with modified cyclodextrins: a spectroscopic investigation with implications in the pharmaceutical analysis. J. Pharm. Biomed. Anal. 44, 980–984.
Stella, V.J., He, Q., 2008. Cyclodextrins. Toxicol. Pathol. 36 (1), 30–42.
Strassburger, K., Startup, W., Levey, V., Mattingly, T., Briggs, J., Harrison, J., Wilson, T., 2010. Enhanced stability of citral in juice beverages by applying cyclodextrin micro emulsion technology. ACS Symp. Ser. 1036, 143–158.
Su, J.Y., Chen, J.P., Chen, L., Li, L., Zhu, L., Xu, Z.B., 2012. Effect of hydroxypropyl-β-cyclodextrin inclusion on characteristics of D-borneol. J. South China Univ. 40 (8), 122–127.
Sun, X., Sui, S., Ference, C., Zhang, Y., Sun, S., Zhou, N., Zhu, W., Zhou, K., 2014. Antimicrobial and mechanical properties of β-cyclodextrin inclusion with essential oils in chitosan films. J. Agric. Food Chem. 62 (35), 8914–8918.
Szejtli, J., 1982. Cyclodextrins and Their Inclusion Complexes. Akadémiai Kiadó, Budapest.

Szejtli, J., 1987. The metabolism, toxicity and biological effects of cyclodextrins. In: Duchêne, D. (Ed.), Cyclodextrins and Their Industrial Uses. Editions de Santé, Paris, pp. 173–210, Chapter 5.
Szejtli, J., 1988. Cyclodextrin Technology. Kluwer Academic Publisher, Dordrecht.
Szejtli, J., 1997. Utilization of cyclodextrins in industrial products and processes. J. Mater. Chem. 7 (4), 575–587.
Szejtli, J., 1998. Introduction and general overview of cyclodextrin chemistry. Chem. Rev. 98 (5), 1743–1753.
Szejtli, J., 2003. Cyclodextrins in the textile industry. Starch 55 (5), 191–196.
Szejtli, J., 2004. Past, present, and future of cyclodextrin research. Pure Appl. Chem. 76 (10), 1825–1845.
Szejtli, J., Szente, L., 1979. Stabilization of volatile, oxidizable flavour substances by β cyclodextrin. Planta Med. 36 (3), 292–293.
Szejtli, J., Szente, L., 2005. Elimination of bitter, disgusting tastes of drugs and foods by cyclodextrins. Eur. J. Pharm. Biopharm. 61 (3), 115–125.
Szente, L., Szejtli, J., 1999. Highly soluble cyclodextrin derivatives: chemistry, properties, and trends in development. Adv. Drug Deliv. Rev. 36 (1), 17–28.
Szente, L., Szejtli, J., 2004. Cyclodextrins as food ingredients. Trends Food Sci. Technol. 15 (3–4), 137–142.
Szente, L., Harangi, J., Greiner, M., Mandel, F., 2006. Cyclodextrins found in enzyme- and heat-processed starch-containing foods. Chem. Biodivers. 3 (9), 1004–1014.
Tamamoto, L.C., Schmidt, S.J., Lee, S.Y., 2010. Sensory properties of ginseng solutions modified by masking agents. J. Food Sci. 75 (7), 341–347.
Tanemura, I., Saito, Y., Ueda, H., Sato, T., 1998. Solubility method using static head-space gas chromatography for determination of the stability constants of fragrance materials with 2- hydroxypropyl-β-cyclodextrin. Chem. Pharm. Bull. 46 (3), 540–541.
Tao, F., Hill, L.E., Peng, Y., Gomes, C.L., 2014. Synthesis and characterization of β-cyclodextrin inclusion complexes of thymol and thyme oil for antimicrobial delivery applications. LWT Food Sci. Technol. 59 (1), 247–255.
Tarimci, N., 2011. Cyclodextrins in the cosmetic field. In: Bilensoy, E. (Ed.), Cyclodextrins in Pharmaceutics, Cosmetics, Biomedicine. John Wiley & Sons, London, p. 131, Chapter 7.
Taylor, A.J., Pearson, K., Hollowood, T.A., Linforth, R.S.T., 2009. Aroma release at the nano- and microscale: molecules to droplets. ACS Symp. Ser. 1007, 246–258.
Tejashri, G., Amrita, B., Darshana, J., 2013. Cyclodextrin-based nanosponges for pharmaceutical use: a review. Acta Pharm. 63 (3), 335–358.
Theimer, E.T., Davies, J.T., 1967. Olfaction, musk odor, and molecular properties. J. Agric. Food Chem. 15 (1), 6–14.
Tian, X.-N., Jiang, Z.-T., Li, R., 2008. Inclusion interactions and molecular microcapsule of *Salvia sclarea* L. essential oil with β-cyclodextrin derivatives. Eur. Food Res. Technol. 227, 1001–1007.
Tobitsuka, K., Miura, M., Kobayashi, S., 2005. Interaction of cyclodextrins with aliphatic acetate esters and aroma components of la France pear. J. Agric. Food Chem. 53 (13), 5402–5406.
Toda, J., Misaki, M., Konno, A., Wada, T., Yasumatsu, K., 1985. Interaction of cyclodextrins with taste substances. In: Inglett, G. (Ed.), Proceedings of the Second International Flavour Conference, vol. 1. Academic Press, New York, pp. 19–35.
Tongiani, S., Ozeki, T., Stella, V.J., 2009. Sulfobutyl ether-alkyl ether mixed cyclodextrin derivatives with enhanced inclusion ability. J. Pharm. Sci. 98 (12), 4769–4780.

Tongnuanchan, P., Benjakul, S., 2014. Essential oils: extraction, bioactivities, and their uses for food preservation. J. Food Sci. 79 (7), 1231–1249.
Tutaj, B., Kasprzyk, A., Czapkiewicz, J., 2003. The spectral displacement technique for determining the binding constants of β-cyclodextrin–alkyltrimethylammonium inclusion complexes. J. Incl. Phenom. 47 (3–4), 133–136.
Uekaji, Y., Jo, A., Urano, A., Terao, K., 2013. Application of α-cyclodextrin in nanomedicinal foods and cosmetics. Bio-Nanotechnology: A Revolution in Food, Biomedical and Health Sciences. CycloChem Co. Ltd, Kobe, Japan, pp. 179–211.
Ulloth, J.E., Almaguel, F.G., Padilla, A., Bu, L., Liu, J.W., De Leon, M., 2007. Characterization of methyl-β-cyclodextrin toxicity in NGF-differentiated PC12 cell death. Neurotoxicology 28 (3), 613–621.
Valente, A.J.M., Söderman, O., 2014. The formation of host-guest complexes between surfactants and cyclodextrins. Adv. Colloid Interface Sci. 205, 156–176.
Van der Schaft, P., 2015. Approaches to production of natural flavours. In: Parker, J.K., Elmore, J.S., Methven, L. (Eds.), Flavour Development, Analysis and Perception in Food and Beverages. Elsevier, Amsterdam, pp. 235–248.
Van Ommen, B., De Bie, A.T.H.J., Bär, A., 2004. Disposition of 14C-α-cyclodextrin in germ-free and conventional rats. Regul. Toxicol. Pharm. 39, 57–66.
Villiers, A., 1891. Sur la fermentation de la fécule par l'action du ferment butyrique. Compt. Rend. Acad. Sci. 112, 536–538.
Vohs, J.K., 2013. Chemistry of cooking: a course for non-science majors. ACS Symp. Ser. 1130, 23–35.
Waleczek, K.J., Cabral Marques, H.M., Hempel, B., Schmidt, P.C., 2003. Phase solubility studies of pure (−)-α-bisabolol and camomile essential oil with β-cyclodextrin. Eur. J. Pharm. Biopharm. 55 (2), 247–251.
Wang, C.X., Chen, Sh.L., 2005. Fragrance-release property of β-cyclodextrin inclusion compounds and their application in aromatherapy. J. Ind. Text. 34 (3), 157–166.
Wang, Y., Jiang, Z.-T., Li, R., 2009. Complexation and molecular microcapsules of *Litsea cubeba* essential oil with β-cyclodextrin and its derivatives. Eur. Food Res. Technol. 228, 865–873.
Wang, J., Cao, Y., Sun, B., Wang, C., 2011a. Physicochemical and release characterisation of garlic oil-β-cyclodextrin inclusion complexes. Food Chem. 127, 1680–1685.
Wang, T., Li, B., Si, H., Lin, L., Chen, L., 2011b. Release characteristics and antibacterial activity of solid state eugenol/β-cyclodextrin inclusion complex. J. Incl. Phenom. Macrocycl. Chem. 71, 207–213.
Wenz, G., 2009. Recognition of monomers and polymers by cyclodextrins. Adv. Polym. Sci. 222 (1), 1–54.
Wszelaka-Rylik, M., Gierycz, P., 2013. Isothermal titration calorimetry (ITC) study of natural cyclodextrins inclusion complexes with drugs. J. Therm. Anal. Calorim. 111 (3), 2029–2035.
Wu, H., Fang, Y., Ji, H., 2011. Effect of co-solvent interaction on cinnamaldehyde-β-cyclodextrin inclusion complex. CIESC J. 62, 168–173.
Xiao, Y., Ng, S.C., Tan, T.T.Y., Wang, Y., 2012. Recent development of cyclodextrin chiral stationary phases and their applications in chromatography. J. Chromatogr. A 1269, 52–68.
Xie, T., Yue, Y., Song, B., Chao, Y., Qian, S., 2013. Increasing of product specificity of γ-cyclodextrin by mutating the active domain of α-cyclodextrin glucanotransferase from Paenibacillus macerans sp. 602-1. Chin. J. Biotechnol. 29 (9), 1234–1244.

Yaméogo, J.B.G., Gèze, A., Choisnard, L., Putaux, J.L., Semdé, R., Wouessidjewe, D., 2014. Progress in developing amphiphilic cyclodextrin-based nanodevices for drug delivery. Curr. Top. Med. Chem. 14 (4), 526–541.

Yañez, C., Günther, G., 2014. Is the determination of the association constant of cyclodextrin inclusion complexes dependent on the technique. J. Chil. Chem. Soc. 59 (2), 2523–2525.

Yang, Z., Xiao, Z., Ji, H., 2015. Solid inclusion complex of terpinen-4-Ol/β-cyclodextrin: kinetic release, mechanism and its antibacterial activity. Flavour Frag. J. 30 (2), 179–187.

Yoshii, H., Sakane, A., Kawamura, D., Neoh, T.L., Kajiwara, H., Furuta, T., 2007. Release kinetics of (−)-menthol from chewing gum. J. Incl. Phenom. Macro. 57 (1–4), 591–596.

Yu, S.C., Bochot, A., Le Bas, G., Chéron, M., Mahuteau, J., Grossiord, J.L., Seiller, M., Duchêne, D., 2003. Effect of camphor/cyclodextrin complexation on the stability of O/W/O multiple emulsions. Int. J. Pharm. 261 (1–2), 1–8.

Yuan, C., Lu, Z., Jin, Z., 2014. Characterization of an inclusion complex of ethyl benzoate with hydroxypropyl-β-cyclodextrin. Food Chem. 152, 140–145.

Zeng, Z., Fang, Y., Ji, H., 2012. Side chain influencing the interaction between β-cyclodextrin and vanillin. Flavour Frag. J. 27 (5), 378–385.

Zhang, Y., Ho, C.-T., 1991. Comparison of the volatile compounds formed from the thermal reaction of glucose with cysteine and glutathione. J. Agric. Food Chem. 39, 760–763.

Zhang, Q.-F., Jiang, Z.-T., Li, R., 2007. Complexation of allyl isothiocyanate with β-cyclodextrin and its derivatives and molecular microcapsule of allyl isothiocyanate in β-cyclodextrin. Eur. Food Res. Technol. 225, 407–413.

Zhang, M., Li, J., Zhang, L., Chao, J., 2009. Preparation and spectral investigation of inclusion complex of caffeic acid with hydroxypropyl-β-cyclodextrin. Spectrochim. Acta A Mol. 71 (5), 1891–1895.

Zhang, W., Li, X., Yu, T., Yuan, L., Rao, G., Li, D., Mu, C., 2015. Preparation, physicochemical characterization and release behavior of the inclusion complex of *trans*-anethole and β-cyclodextrin. Food Res. Int. 74, 55–62.

Zhou, H., Goldman, M., Wu, J., Woestenborghs, R., Hassell, A.E., Lee, P., Baruch, A., Pesco-Koplowitz, L., Borum, J., Wheat, L.J., 1998. A pharmacokinetic study of intravenous itraconazole followed by oral administration of itraconazole capsules in patients with advanced human immunodeficiency virus infection. J. Clin. Pharmacol. 38 (7), 593–602.

Zhou, Q., Wei, X., Dou, W., Chou, G., Wang, Z., 2013. Preparation and characterization of inclusion complexes formed between baicalein and cyclodextrins. Carbohydr. Polym. 95, 733–739.

Zhu, G., Feng, N., Xiao, Z., Zhou, R., Niu, Y., 2015. Production and pyrolysis characteristics of citral–monochlorotriazinyl-β-cyclodextrin inclusion complex. J. Therm. Anal. Calorim. 120, 1811–1817.

5

STRUCTURAL AND THERMODYNAMIC INSIGHT INTO THE POTENTIALITY OF FOOD BIOPOLYMERS TO BEHAVE AS SMART NANOVEHICLES FOR ESSENTIAL POLYUNSATURATED LIPIDS

Maria G. Semenova, Anna S. Antipova, Maria S. Anokhina, Larisa E. Belyakova, Yurii N. Polikarpov, Darya V. Zelikina, Elena I. Martirosova, Nadezhda P. Palmina, Vladimir I. Binyukov, Valerii V. Kasparov, Natalia G. Bogdanova

N. M. Emanuel Institute of Biochemical Physics of Russian Academy of Sciences, Chemistry of Food Substances and Biotechnology Department, Moscow, Russian Federation

1 Introduction

Research activities in the area of nanoparticle delivery systems for the essential micronutrients and nutraceuticals, which can be incorporated into food products, have increased almost exponentially during the past decade. Moreover, current market projections for these technologies suggest a multifold increase in their commercial potential over the next decade, as functional foods become increasingly popular among consumers as a result of increased awareness of bioactive food ingredients and their impact on the prophylactics of human health and physiological functions (Ransley et al., 2001; McClements et al., 2008; McClements et al., 2009; McClements, 2014; Faulks and Southon, 2008; Augustin and Hemar, 2009; Semenova and Dickinson, 2010, Chapter 2).

Actually, it is commonly recognized nowadays that reducing the size of the delivery systems for prophylactic and therapeutic

Encapsulations. http://dx.doi.org/10.1016/B978-0-12-804307-3.00005-3

biologically active compounds to the nanoscale, especially below 500 nm, do exhibit enhanced delivery over the more conventional dosage forms because of enhancements in the following factors: (1) the apparent solubility of the active ingredients; (2) the rate of mass transfer through biological barriers, leading to an increase in cellular uptake; (3) the gastrointestinal retention time in the mucus covering the intestinal epithelium; (4) the rate of release (due to the large surface area); and (5) the direct uptake of particles by the intestinal epithelium (Jani et al., 1990; Horn and Rieger, 2001; Hussain et al., 2001; Chen et al., 2006; Medina et al., 2007; Lai et al., 2007; Gibbs et al., 1999; Ratnam et al., 2006; Acosta, 2009; McClements, 2014; Semenova and Dickinson, 2010, Chapter 2; Moghimi et al., 2001; Kreuter, 2001; Nishioka and Yoshino, 2001; Merisko-Liversidge et al., 2003; McClements et al., 2008).

The aim of the nanoencapsulation of the bioactive compounds will be, on the one hand, their protection against degradation/oxidation along with undesirable interactions with other food components during food processing and storage and, on the other hand, the enhancement of their bioavailability in vivo (Gouin, 2004; Acosta, 2009; Dickinson, 2014; Lalush et al., 2005; Chen et al., 2006; Sanguansri and Augustin, 2006; Bouwmeester et al., 2007; Letchford and Burt, 2007; Semenova et al., 2008; Semenova and Dickinson, 2010, Chapter 1). Such nanoencapsulation utilizes a combination of the "top-down" and "bottom-up" nanotechnological approaches to produce more sophisticated nanoparticle systems (Horn and Rieger, 2001). As this takes place, the more promising "bottom-up" way is based on the self- and coassembly of molecules or particles. These spontaneous processes are under strong thermodynamic control, that is, rather than requiring large amounts of energy they involve a series of optimizations that utilize the tendency of a system to minimize its overall free energy, thereby minimizing any required activation energies, in order to perform an efficient nanoencapsulation (Föster and Konrad, 2003; Sanguansri and Augustin, 2006; Acosta, 2009; Semenova and Dickinson, 2010, Chapter 1). In the "bottom-up" nanotechnological approach, a large variety of ordered nanostructures may be obtained by altering either the environmental conditions (temperature, pH, ionic strength, etc.) or the concentration of the interacting molecules/particles that is followed by changing of the balance of all the attractive and repulsive forces between the structure-forming molecules or particles in the system (Min et al., 2008; Semenova and Dickinson, 2010, Chapter 1).

Biopolymers are especially important in respect to both the nanoencapsulation of the health-promoting biologically active molecules and formation of smart, that is, stimuli-sensitive (switchable) self-assembled delivery nanovehicles for the latter owing to a number of the following reasons (Veerman et al., 2003; Murray and Ettelaie, 2004; Dickinson, 2004, 2006a,b; Graveland-Bikker and de Kruif, 2006; van der Linden, 2006; Morris, 2005, 2006; Weiss et al., 2006; Bolder et al., 2006; Zhang et al., 2006; Manski et al., 2007; Huppertz and de Kruif, 2008; Semenova and Dickinson, 2010; Gibis et al., 2013):

1. a biological origin;
2. an abundant availability from renewable natural sources;
3. biocompatibility;
4. biodegradability;
5. nontoxicity in any combinations;
6. the nanoscale dimensions;
7. the amphiphilic nature of their macromolecules, having a variety of both polar and nonpolar functional groups, which facilitates either their self-assembly or coassembly with distinct in nature molecules in a bulk aqueous media or at interfaces as a result of the changes in pH, temperature, ionic strength, or addition of specific ions, enzymes, etc. As this takes place, various kinds of inter- and intramolecular physical interactions can be involved into these self and coassembly processes. They are hydrogen bonding; electrostatic attraction between opposite charges; van der Waals dipole–dipole attractions; and hydrophobic attraction between nonpolar groups in an aqueous medium (McClements, 2006; Morris, 2006; Min et al., 2008; Semenova and Dickinson, 2010, Chapter 4). It is worthy to note here that the individual intermolecular interaction energies could be rather small in magnitude, but their multiple combination in regular assembly evidently can make the nanoscale structures stable;
8. high solubility in an aqueous medium;
9. responsiveness to the environmental changes in pH, ionic strength, temperature, specific ions, enzymes, and so forth;
10. a wide diversity in both molecular conformations and architecture of their associate/aggregates (globular, random coil, helical, worm-like, tubular), which can be favorable for biopolymer encapsulating abilities such as the coating, incorporation, entrapment, or absorption of different bioactive molecules;
11. involvement into the nanoscale building blocks of particles, aggregates, fibers, complexes, and networks commonly found in food gels and dispersions.

Therefore, the elaboration of the smart stimuli-sensitive (switchable) nanoscale biopolymer vehicles for biologically active molecules must rely first on both an in-depth thermodynamic understanding of the molecular mechanisms of their formation and on the structural and thermodynamic characteristics of the ultimate coassembled nanostructures. Once this information is available, it can be used to design, stabilize, or manipulate such nanostructures rationally, so as to enhance functionality under the conditions encountered in food manufacture and in product usage by consumers (Weiss et al., 2006; Semenova and Dickinson, 2010).

The plant [α-linolenic acid (ALA), ϖ-3, linoleic acid (LA) ω-6] and fish [ϖ-3: eicosapentaenoic acid (EPA), and docosahexaenoic acid (DHA)] polyunsaturated fatty acids (PUFAs) are one of the essential bioactive compounds, which are not synthesized from other food components in a human body, and which are acknowledged to be important for the human health (Hibbeln et al., 2006; Cleland et al., 1988; Kremer et al., 1985; Geusens et al., 1994; Semenova and Dickinson, 2010, Chapter 2; Lee and Ying, 2008; McClements, 2014; Phang and Garg, 2014).

Such natural molecules as triacylglycerols and phospholipids are generally natural delivery systems for these PUFAs (McClements, 2014; Phang and Garg, 2014). In particular, the polyunsaturated soy phosphatidylcholine liposomes PC (Lipoid S100) along with supplying of the essential PUFAs (~57% LA and 3÷7% ALA) are able, for example, to present both the antiaging and liver protection functions (Kidd, 1996, 2000).

However, the high susceptibility of the PUFAs to oxidative degradation during product preparation, transport, and storage under extreme environmental conditions (light exposure, high temperature, pH, ionic strength, etc.), which is followed by the perception of off-flavors and unpleasant odor, as well as their low solubility in water pose a real challenge for their use in a food manufacture, in particular under the elaboration of the low-saturated fat foods (McClements, 2014; Semenova and Dickinson, 2010).

One of the general ways of improving the chemical stability of these bioactive lipids is their encapsulation within colloidal micro- and nanosized delivery systems that inhibit their oxidative degradation (Gibbs et al., 1999; Mayer et al., 2013; McClements, 2014; McClements and Decker, 2000; Salminen et al., 2014; Waraho et al., 2011).

Another way to protect polyunsaturated lipids against oxidation is their encapsulation within the nanosized either biopolymer macromolecules or biopolymer associated/aggregated particles. The advantage of this way lies in the maximal retention of functionality of both biologically active and biopolymer molecules in the complex particles, which could further be used as health-promoting structure-forming ingredients.

In particular, by now, the encapsulation ability of casein molecules/associates to behave as delivery nanovehicles for hydrophobic nutraceuticals has been clearly demonstrated for: (1) polyphenols (Esmaili et al., 2011; Hasni et al., 2011; Rahimi Yazdi and Corredig, 2012; Sahu et al., 2008); (2) a vitamin (D2) (Semo et al., 2007); (3) a polyunsaturated soy phosphatidylcholine (Semenova 2007; Semenova et al., 2008, 2012, 2014a,b,c, 2016); and (4) for the chemotherapeutic drugs (Sahu et al., 2008; Shapira et al., 2010), under the elaboration of nonfat or low-fat functional food products (Livney, 2010; Semenova and Dickinson, 2010). Furthermore, it was found that incorporation of the vitamin D_2, for example, had relatively little effect on the morphology of the reassembled casein micelles (Semo et al., 2007).

In addition, the promising abilities of nanosized [sodium caseinate (SC) + maltodextrin (MD)] covalent conjugates as delivery vehicles for hydrophobic nutraceuticals were also clearly demonstrated (Grigorovich et al., 2012; Markman and Livney, 2012; Semenova et al., 2014c, 2016). Both the high protection against oxidation for the nutraceuticals studied [soy phosphatidylcholine (PC), soy lysophosphatidylcholine (LPC)] (Semenova et al., 2014c, 2016); vitamin D, and epigallocatechin gallate (Markman and Livney, 2012) and their high solubility in the wide range of pH (including the protein isoelectric point, pI) in the complexes with the conjugates in an aqueous medium were revealed. Moreover, it was shown that the covalent conjugation of the maltodextrins to the protein can be used in order to control the release of PC under the action of the gastric and intestinal enzymes in the simulated conditions of the gastrointestinal tract (GIT) in vitro (Semenova et al., 2014c). It is significant to note here that in our recent (Semenova et al., 2016) and current research work, we have mainly studied the covalent conjugates of SC with maltodextrins MD, having different dextrose equivalent (DE) = 2 (SA2) and 10 (MD10), at the same R_{weight} = MD:SC = 2. For these samples of the conjugates it was established (on the loss of the free amino groups of lysine in SC, as a result of the covalent binding with the maltodextrins) that maltodextrin MD10, having the shorter molecular chain (DE = 10, 10 glucose units) as compared with maltodextrin SA2 (DE = 2, 50 glucose units) (Harkema, 1998), is covalently bound in a greater amount (6 molecules MD10 as compared with 4 molecules SA2) to the individual mass-averaged casein molecules (24 kDa), comprising the original sodium caseinate particles (Semenova et al., 2016). This result could be attributed to the less steric hindrance for the further MD attachment, caused by the already bound shorter maltodextrin MD10 molecules. Moreover, as this took place, many molecules of the maltodextrin MD10 (21 from the 27 moles), having the lower molar weight as compared with maltodextrin SA2 (1 from the 5 moles), remained free at the equal weight ratio of MD to SC (R_{weight} = 2) studied for both maltodextrins (Semenova et al., 2016).

This chapter reviews our recent findings on the key generic structural and thermodynamic parameters that determine the potentiality of such food biopolymers as individual caseins, sodium caseinate (SC), and covalent conjugates of sodium caseinate with maltodextrins produced by the Maillard reaction, to behave as smart nanovehicles for the polyunsaturated lipids containing the essential PUFAs (liposomes of the soy PC (Lipoid S100: having about ~59% of ω-6 LA, and about ~3% of ω-3 ALA) and micelles of soy lysophosphatidylcholine (LPC) (Lipoid LPC: having about ~44 ÷ 48% of ω-6 LA, and about ~1% of ω-3 ALA) alone or in their combination with mutually complementary in the ω-3 ALA content triglycerides of a cold-pressed flaxseed oil (FO), (having about ~55% of ω-3 ALA, and about 7% of ω-6 LA) under different experimental conditions (pH, ionic strength, temperature, specific ions, enzymatic action), namely:

1. to provide high traditional biopolymer functionality (solubility in an aqueous medium) to the final coassembled particles produced by the nanoencapsulation of the polyunsaturated phospholipids by the food biopolymers;
2. to protect the polyunsaturated lipids against oxidation under different environmental conditions (pH, ionic strength, temperature, specific ions);
3. to control the release of the polyunsaturated lipids under the simulated conditions of the GIT in vitro.

It is worthy of noting here, that the unique advantage of caseins as one of the basic biopolymers for the elaboration of the smart [ie, stimuli-sensitive (switchable)] nanovehicles is directly governed by the difference of their individual properties in various respects, including net molecular charge ($\alpha_{s1} > \alpha_{s2} > \beta > \kappa$), sensitivity to precipitation by calcium ions ($\alpha_{s2} > \alpha_{s1} > \beta >> \kappa$), and distribution of hydrophobic and hydrophilic amino acids within their primary structures (Dickinson, 2006a,b; Dickinson and Golding, 1998; Swaisgood, 2003; Semenova and Dickinson, 2010, Chapter 6; Dickinson et al., 2001). Moreover, it has been found that α_{s1}-casein, β-casein and κ-casein each shows a strong tendency towards self-association in aqueous medium (Schmidt, 1982; Thurn et al., 1987a,b; Burchard, 1994; Leclerc and Calmettes, 1997; Dickinson et al., 1998; de Kruif et al., 2002). As with the case of whole casein(ate), this self-association is controlled by a delicate balance of specific electrostatic and hydrophobic interactions (de Kruif et al., 2002, 2012; Dickinson, 2006b). Thus, the self- and coassembly of the individual caseins may be readily manipulated by environmental changing that is governed mainly by their highly disordered conformations due to the unusually high proline content, which is fairly uniformly distributed along the polypeptide chain (Swaisgood, 2003). That is the caseins have been described as "rheomorphi" proteins (Holt and Sawyer, 1993), indicating that they adopt molecular structures in solution that are dictated by the local environment; that is, their structures are flexible enough to "go with the flow" (Dickinson, 2006b).

By this means, in our experimental work we have used the following different ways for triggering self- and coassembly of individual caseins, sodium caseinate, and covalent conjugates of sodium caseinate with maltodextrins:

- both temperature (in the range from 10 to 40°C) and pH (5.5, 7.0) in the case of more temperature sensitive β-casein (Semenova et al., 2012), for which it is well known that in an aqueous solution below 4°C, the protein is monomeric, but above this temperature it exhibits endothermic reversible self-assembly (Belitz and Grosch, 1982; Payens and Vreeman, 1982)
- an addition of calcium ions (Semenova et al., 2014a) in the case of the most calcium-sensitive α-casein molecules (Belitz and Grosch, 1982; Nagy et al., 2010; Dickinson, 2006b);
- both pH (5.5, 6.0, 7.0) and (ionic strength = 0.001M, 0.01M, 0.1M) for the whole sodium caseinate (Semenova et al, 2014b);
- pH (4.8, 7.0) in the case of the covalent conjugates of sodium caseinate with maltodextrins (Semenova et al., 2016).

2 Intermolecular Forces Underlying the Encapsulation of the Polyunsaturated Lipids by the Biopolymers

First and foremost, the high extent (>92% within the 10% experimental error) of the encapsulation of both pure forms of the studied lipids (either PC liposomes or LPC micelles) and their mixed variants [either (PC + FO) or (LPC + FO)] by all studied biopolymers was observed in an aqueous medium in the region of pH from neutrality to the close vicinity to the casein's isoelectric point (pI) (pH = 4.8, as in the case of the covalent conjugates of the SC with maltodextrins) at the weight ratio (10:1) of the biopolymers (≤1 wt/v%) to the lipids (≤0.1 wt/v%) used. It is interesting to note, that it was found in spite of the experimental facts that in this region of pH zwitterionic phospholipids micelles, individual caseins, sodium caseinate, and the covalent conjugates of sodium caseinate with maltodextrins had similar negative charges, lowering with approaching the casein's pI (Dickinson et al., 1998; Semenova et al., 2014b, 2016) (Table 5.1).

Table 5.1 The ζ-Potential of Both Pure and Mixed Phospholipid Micelles as well as Biopolymer Particles and Their Complexes Measured in an Aqueous Medium at Different pH and Ionic Strength = 0.001M

	ζ-Potential (mV)		
Samples	**pH = 4.8**	**pH = 5.5**	**pH = 7.0**
PC	−2.3	−7.9	−21.7
LPC	−9.2	−19.1	−23.6
PC + FO	—	—	−19.1
LPC + FO	—	—	−27.2
SC	—	−24.1	−30.5
SC + PC	—	−31.6	−31.1
Conjugate (SC+ MD-SA2)	−6.0	−14.7	−21.5
Conjugate (SC+ MD-SA2) + PC	−4.8	−15.4	−22.8
Conjugate (SC+ MD-10)	−8.0	−14.6	−26.9
Conjugate (SC+ MD-10) + PC	−1.4	−17.1	−25.6
Conjugate (SC+ MD-SA2) + LPC	−4.5	−14.8	−25.4
Conjugate (SC+ MD-10) + LPC	−10.9	−17.6	−23.4
Conjugate (SC+ MD-SA2) + (PC + FO)	—	—	−41.7
Conjugate (SC+ MD-SA2) + (LPC +FO)	—	—	−51.5

R_{weight} = MD:SC = 2:1 for the conjugates. Electrophoretic mobilities of the tested samples were determined with a Zetasizer Nano ZS Malvern (UK) calibrated against a standard latex dispersion.

The detail description of both the extraction/separation and chemical/spectrophotometric determination of the free (ie, non-encapsulated) PC and LPC are given in our earlier published works (Istarova et al., 2005; Semenova et al., 2012, 2014a,b,c, 2016). The free triglycerides of FO were determined spectrophotometrically by the enzymatic assay using the Triglycerides-Novo kit (Best Vector, Russia).

In addition, the complex formation between sodium caseinate (SC) and PC liposomes in an aqueous medium was proved earlier by the determined binding isotherms in the pH range from 7.0 to

5.0 at the same ionic strength (I = 0.05 M) of the buffers (Istarova et al., 2005). Besides, the shape of the Scatchard plots indicated the cooperative character of their binding generally (Istarova et al., 2005).

Relying on both literature and our experimental data, it may be safely suggested that different kinds of the interactions (electrostatic attraction between opposite charges, hydrogen bonding and hydrophobic attraction) contribute into the encapsulation (complex formation) of the studied pure (PC and LPC) and mixed [(PC + FO) or (LPC + FO)] micelles of the phospholipids by the casein-based biopolymer particles in an aqueous medium. As this takes place, the prevailing role of one or another type of the interactions depends on pH, ionic strength, and the concentration of the lipids (Istarova et al., 2005; Bai et al., 2010; Strömstedt et al., 2010; Antunes et al., 2009; Semenova et al., 2012, 2014b; Schulz et al., 2012). These findings agree well with the amphiphilic nature of both the phospholipids and the casein-based biopolymer particles. In addition, an increase in temperature up to 313–333 K leads generally to the predominant contribution of the hydrophobic attraction into the biopolymer–lipids coassemblies (Istarova et al., 2005; Semenova et al., 2012, 2014b). Besides, the contribution of the calcium ionic bridges into the encapsulation of the PC liposomes by the α_S-casein molecules is also suggested on the basis of both the well-known high responsiveness of these molecules to the presence of Ca^{2+} (Dickinson, 2006b; Shoemaker and Vanderlick, 2003) and the experimentally determined (>88%) percentage of the Ca^{2+} binding by both the individual PC liposomes as well as α_S-casein molecules and their complex particles (Semenova et al., 2014a).

By way of illustration, the data of the isothermal mixing microcalorimetry, which are presented in Table 5.2, show the positive values of the molar enthalpy, $\Delta H_{protein\text{–}PC}$, of the pair interactions β-casein + PC and SC + PC at the temperature (313K ≡ 40°C) used generally for the complex formation between the biopolymers and the lipids in our work (Semenova et al., 2012, 2014a,b,c, 2016). We have chosen this temperature for the encapsulation of the studied pure and mixed phospholipid micelles by all casein-based biopolymers in order to strengthen the hydrophobic attraction between their unipolar parts and, in doing so, to facilitate ultimately the location of the polyunsaturated hydrocarbon chains of the lipids in the hydrophobic interior of the complex coassembled (biopolymer + lipids) particles, forming in an aqueous medium.

The revealed positive values of the $\Delta H_{protein\text{–}PC}$ point to the predominantly endothermic in character protein–lipid interactions that is characteristic of hydrophobic attraction between nonpolar parts of PC and casein molecules accompanied by the simultaneous release of counterions and water molecules structured primarily

Table 5.2 The Molar Enthalpy, $\Delta H_{protein-PC}$, of the Pair Biopolymer–PC Interactions Measured in an Aqueous Medium (10% Ethanol) at Different pH (Ionic Strength = 0.001M, 313 K)

	$\Delta H_{protein-PC}$ (kJ mol^{-1})[a]		
System	pH = 5.5	pH = 6.0	pH = 7.0
SC (1 wt/v%) + PC (0.1% wt/v)	10.4	18.9	12.5
β-Casein (0.5 wt/v%) + PC (0.05 wt/v %)	38.0	—	22.6

Calorimetric measurements were made using a LKB 2277 flow calorimeter set. The sensitivity of the calorimetric measurement was no less than 3×10^{-6} J s^{-1}.
[a]The molar mass of the individual casein molecule (23–24 kDa) (Swaisgood, 2003) has been taken into account under the calculation of the molar enthalpy $\Delta H_{protein-PC}$.

near of these nonpolar parts. Thus, such kind of the interactions was evidently driven by the gain in the translational entropy of the system. This hydrophobic attraction was found to be most pronounced in the case of the β-casein associates formed in an aqueous medium. This result could be attributed to the expected the surfactant-like micelle structure of the β-casein associates in an aqueous medium under the experimental conditions (Belitz and Grosch, 1982; Payens and Vreeman, 1982; Schmidt and Payens, 1972; Semenova and Dickinson, 2010, Chapter 6; Schmidt, 1982; Leclerc and Calmettes, 1997; Burchard, 1994). Such structure has evidently the more hydrophobic core in comparison with the whole SC particles, comprising four caseins (38% α_{s1}–, 10% α_{s2}–, 36% β–, and 13% κ-casein) differing in various respects, including net molecular charge ($\alpha_{s1} > \alpha_{s2} > \beta > \kappa$), and distribution of hydrophobic and hydrophilic amino acids within their primary structures (Swaisgood, 2003; Semenova and Dickinson, 2010, Chapter 6).

3 Protection Ability of the (Biopolymer + Lipids) Complexes Against Oxidation of the Encapsulated Lipids

The degree of the oxidation of the polyunsaturated lipids in the tested samples was generally determined by the quantitative measurements of the formation of the secondary product [malonic dialdehyde (MDA) (Gutteridge, 1977)] of the peroxidation of the PUFAs. The quantitative estimation of MDA was performed using the TBARS method (by the reaction of MDA with 2-thiobarbituric acid (TBA) in the presence of trichloroacetic acid) (Fernández et al., 1997; Kwon et al., 1965; Heath and Packer, 1968; Fu and Huang, 2001). The optical density of the colored TBA–MDA compounds was measured spectrophotometrically at two different wave lengths in order to prevent the effect of any slight turbidity of the tested samples, namely, at $\lambda = 532$ (the maximum of the absorbance by the TBARS) (Gutteridge, 1977; Fernández et al., 1997; Kwon et al., 1965; Gutteridge and Tickner, 1978) and at $\lambda = 580$ (the minimum of the absorbance by the TBARS) (Gutteridge and Tickner, 1978). The experimental error of the TBARS method, which was estimated on the basis of not less than three experimental repetitions, was equal to 15%.

We have revealed that all studied biopolymers provide the great protection to the encapsulated polyunsaturated lipids against oxidation (Semenova et al., 2012, 2014a,b,c, 2016) in the concentration region and at the weight ratio of the biopolymers to the lipids = 10:1 used in our experimental work.

For example, the Table 5.3 shows the protective abilities provided by both individual caseins and the whole sodium caseinate to the PC liposomes under the different specific environmental triggering of their coassembly.

It is worthy to note here that in our experimental work with both individual caseins and sodium caseinate we have used the simple injection of the required amount of PC dissolved preliminary in the pure ethanol (EtOH) into the protein solutions followed by their shaking at a designated temperature (generally at 40°C) for an hour. The ultimate concentration of ethanol in the mixed solution was equal to 10 v/v%.

In addition, it is significant to note here that for these systems we have generally used the addition of the low concentration of Cu^{2+} ions (10^{-5} M $CuSO_4$) into the tested sample solutions in order to both accelerate the PC oxidation and to make the experimental work less time consuming. It was shown preliminarily that this concentration of Cu^{2+} ions did not influence on the properties of the tested samples, but accelerated the PC oxidation essentially. In addition, we have heated the SC tested samples at 70°C (1 ÷ 3 h) (Semenova et al., 2014b).

It is also worthy of note that for the triggering of the PC coassembly with α_S-casein we have used two different ways in the addition of calcium ions under the formation of the ternary complex particles (protein + phospholipid + calcium ions). In the first way, the ternary complex particles were prepared by the addition of

Table 5.3 Effect of the Coassembly at the Specific Environmental Triggering of the Soy PC Liposomes With Both Individual Caseins and the Whole Sodium Caseinate in an Aqueous Medium (Ionic Strength 0.001M, 10 v/v% EtOH) on the Protection of the PC Liposomes Against Oxidation, Accelerated by the Presence of 10^{-5} M $CuSO_4$ Under Storage

	Malonic Dialdehyde (nmol mL^{-1})		
Tested Samples	PC Liposomes	Complex: Protein + PC Liposomes	References
pH 7.0, ionic strength = 0.001M			
β-Casein, 10°C, 6 days of the storage	3	0.0	Unpublished data
β-Casein, 20°C, 6 days of the storage	8	0.0	
β-Casein, 40°C, 1 day of the storage	18	0.3	
SC, 22°C, 2 days of the storage	11	0.2	Semenova et al. (2014b)
α_S-Casein 22°C, 2 days of the storage	11	0.0	Semenova et al. (2014a)
pH 7.0, the total ionic strength with calcium ions = 0.001M			
α_S-Casein 22°C, 2 days of the storage, I way of Ca^{2+} addition	11	0.5	
α_S-Casein 22°C, 2 days of the storage, II way of Ca^{2+} addition	11	0.0	Semenova et al. (2014a)
pH 5.5, ionic strength = 0.001M			
β-Casein, 10°C, 6 days of the storage	18	0.0	Semenova et al. (2012)
β-Casein, 20°C, 6 days of the storage	6	0.7	
β-Casein, 40°C, 1 day of the storage	11	0.0	
SC, 22°C, 2 days of the storage	12	0.0	

calcium ions to the preliminary formed binary complex particles (α_S-casein + PC) using the equilibrium dialysis. By the second way, the ternary complex particles were prepared by the addition of PC to the preliminary formed binary particles (α_S-casein + Ca^{2+} ions), for which Ca^{2+} ions were added by the equilibrium dialysis (Semenova et al., 2014a).

To take the next illustration, Table 5.4 presents the great protection abilities of the covalent conjugates produced by the Maillard reaction relative to both the pure phospholipids (PC liposomes as well as LPC micelles) and their mixed variants with triglycerides of FO, having the equimass ALA to LA weight ratio (Semenova et al., 2016).

It is valuable to note here that both the pure PC and mixed (PC + FO) vesicles were prepared in the three stages: the first stage was the mechanical homogenization (Heidolph) at 22,000 r/min (2 min); the second stage was the ultrasound sonication (VCX-130; Sonics & Materials, USA) of the samples placed into the ice bath (three times for 5 min (30 s —operation/30 s—rest); on the third stage the prepared dispersion was passed 19 times through the membrane filter (a pore size is 100 nm), using the AVANTI Polar Lipid (USA). Mixtures of the prepared vesicles were mixed with the conjugates in a shaker at 40°C for 1 h that was followed by additional shaking at 60°C (pasteurization) for 1 h.

Fig. 5.1 shows an example of the size distributions of the pure PC liposomes (Fig. 5.1a) and mixed (PC + FO) liposomes (Fig. 5.1b), obtained by such method in an aqueous medium at pH 4.8. In addition, the size distributions of the pure conjugate (SC + MD-SA2) and its complexes with the liposomes are shown for a comparison.

The data presented in both Table 5.4 and Fig. 5.1 clearly show that the covalent conjugates have the major advantage over both the pure SC and the simple mixture of SC with MD–SA2 in their rather high solubility at the pH of the protein isoelectric point (pH 4.8) in an aqueous medium (Grigorovich et al., 2012) that underlies the efficient encapsulation of the lipids by the conjugates under these environmental conditions.

A correlation of the data on the protection of the polyunsaturated lipids against oxidation with the structural and thermodynamic features of their supramolecular complexes with the biopolymers allows distinguishing the key generic structural and thermodynamic properties of the complex particles that provide such protection.

Table 5.4 Effect of the Coassembly of Both the Pure Phospholipids (PC Liposomes and LPC Micelles) and Their Mixed Variants With Triglycerides of FO With the Covalent Conjugates of Sodium Caseinate With Maltodextrins in an Aqueous Medium (Ionic Strength 0.001M) on the Long-Term Protection of the Lipids Against Autooxidation Under Their Storage at Room Temperature Under the Light Exposure

Tested Samples	Malonic Dialdehyde (nmol mL^{-1}): Polyunsaturated Lipids	Malonic Dialdehyde (nmol mL^{-1}): Complexes of the Polyunsaturated Lipids With the Biopolymers
PC		
Conjugate (SC + MD SA-2), pH 7.0, 3 days of the storage	7.3	0.06
Conjugate (SC + MD SA-2), pH 4.8, 7 days of the storage	15.1	6.7
PC + FO		
Conjugate (SC + MD SA-2), pH 7.0, 3 days of the storage	34.3	3.70
LPC		
Conjugate (SC + MD SA-2), pH 7.0	1.1	0.00
Conjugate (SC + MD 10), pH 7.0		0.00
SC pH 4.8, 7 days of the storage	4.1	4.04
Conjugate (SC + MD SA-2), pH 4.8, 7 days of the storage		0.33
Conjugate (SC + MD 10), pH 4.8, 7 days of the storage		0.00
LPC + FO		
Conjugate (SC + MD SA-2), pH 7.0, 20 days of storage	58.9	5.21

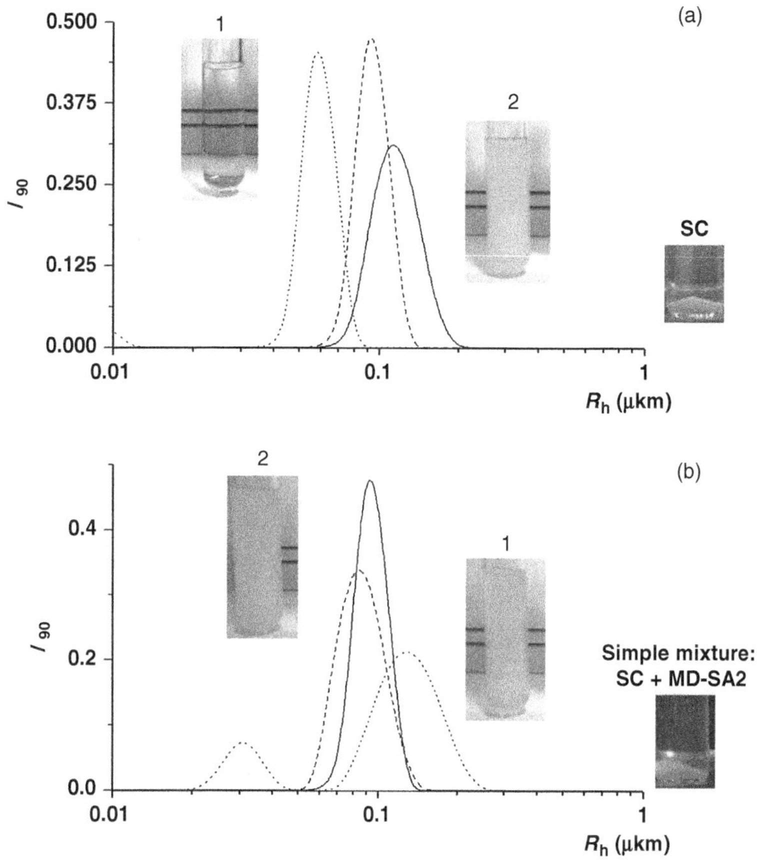

Figure 5.1. The size distributions of the systems involving the pure PC liposomes (dot) (a) and mixed (PC + FO) (dot) liposomes (b), the pure conjugate (SC + MD-SA2, $R_{weight} = 2$) (dash) and their complexes (solid) in the buffered aqueous medium (pH = 4.8, ionic strength = 0.001M). The insertion are the photos of the solutions of the pure PC liposomes (1a), mixed (PC + FO) liposomes (1b) and their complexes with the conjugate (2a) and (2b), respectively. The photos of the solutions of the pure SC and the simple mixtures of SC with MD-SA2 ($R_{weight} = 2$) relate to pH = 4.6, ionic strength = 0.001M.

4 Structural and Thermodynamic Parameters of the (Biopolymer + Lipids) Complex Particles Underlying Their Protective Abilities Against Oxidation of the Encapsulated Lipids

4.1 The Density of the Complex Particles

In the first place, it seems to be the rather high density of the complex particles formed (Semenova et al., 2012, 2014a,b,c, 2016).

The density of the particles has been calculated on the basis of the light-scattering data using the following equation (Tanford, 1961):

$$d = M_w / (N_A V), \tag{5.1}$$

where M_w is the average molar mass of the particles, N_A is the Avogadro's number, V is the volume of the particles, which in our calculations were generally approximated by the spheres relying on the intrinsic for spheres values of the structure-sensitive parameter $\rho = R_G/R_h$ ($1 < \rho \leq 2$) measured using multiangle laser light scattering (LS-01 apparatus; Scientific Instruments, St. Petersburg, Russia) (Semenova et al., 2014b).

Really, for the all cases of the revealed rather high protection of the lipids against oxidation we have generally found either a marked increase in the density of the complex particles as compared with the pure biopolymers, or the retention of the high values of the density of the complex particles that are characteristic of original biopolymer particles, how it was observed, for example, in the cases of either β-casein particles or covalent conjugates at pH 4.8 (Tables 5.5 and 5.6).

Table 5.5 A Comparison Between Densities of the Pure Biopolymer Particles and Their Complexes With PC Liposomes in an Aqueous Medium in the Presence of 10 v/v% EtOH

	d (mg mL^{-1})		
Tested Samples	**Protein**	**Complex: Protein + PC Liposomes**	**References**
pH 7.0, ionic strength = 0.001M			
β-Casein at 10°C	0.65	3.2	
β-Casein at 20°C	0.18	5.8	Unpublished data
β-Casein at 40°C	0.06	38.0	
SC at 22°C	0.72	1.1	Semenova et al. (2014b)
α_S-Casein at 22°C	0.4	1.2	Semenova et al. (2014a)
pH 7.0, the total ionic strength with calcium ions = 0.001M			
α_S-Casein at 22°C I way of Ca^{2+} addition	0.4	75.5	Semenova et al. (2014a)
α_S-Casein at 22°C II way of Ca^{2+} addition		45.5	
pH 5.5, ionic strength = 0.001M			
β-Casein at 10°C	5.6	41	Semenova et al. (2012)
β-Casein at 20°C	6.3	37	
β-Casein, 40°C, 1 day of the storage	49.3	20	
SC, 22°C, 2 days of the storage	0.89	2.56	

Table 5.6 A Comparison Between Densities of the Pure Conjugate Particles and Their Complexes With Both Pure PC Liposomes/LPC Micelles and Their Mixed With Triglycerides of FO Variants in an Aqueous Medium (Ionic Strength = 0.001M)

	d (mg mL^{-1})	
Tested Samples	Conjugate	Complexes of the Polyunsaturated Lipids With the Covalent Conjugates
PC		
Conjugate (SC + MD SA-2), pH 7.0	1.2	1.9
Conjugate (SC + MD SA-2), pH 4.8	30.7	10.8
PC + FO		
Conjugate (SC + MD SA-2), pH 7.0	1.2	5.7
LPC		
Conjugate (SC + MD SA-2), pH 7.0	1.2	2.9
Conjugate (SC + MD 10), pH 7.0	0.7	1.5
Conjugate (SC + MD SA-2), pH 4.8	30.7	21.7
Conjugate (SC + MD 10), pH 4.8	13.0	11.3
LPC + FO		
Conjugate (SC + MD SA-2), pH 7.0	1.2	7.5

Most likely, that the relatively high values of the density of the complex particles can keep out the diffusion of small molecules such as oxygen towards the polyunsaturated hydrocarbon chains of the lipids placed in the interior of them.

On the ground of the static and dynamic laser-light-scattering measurements the found marked increase in the density of the complex particles can be attributed to the shrinkage of the original biopolymer particles as a result of the encapsulation of both the pure and mixed phospholipids micelles that was accompanied generally with the biopolymer simultaneous association. That is manifested itself by the lower extent of the increase in the radii of gyration (R_G) of the complex particles as compared with the increase in their weight-averaged molar masses (M_w) relative

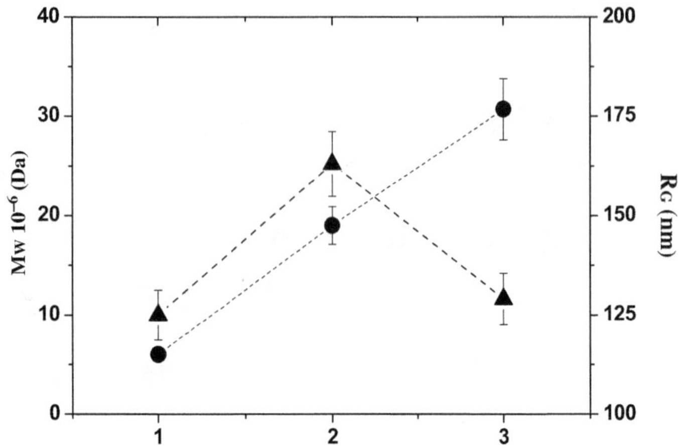

Figure 5.2. Weight average molar weight, M_w, (•) and radius of gyration, R_G, (▲) (pH 7.0, ionic strength 0.001M) for the covalent conjugate (SC + MD-SA2, R_w = 2) (1), complex particles of the conjugate (SC + MD-SA2, R_w = 2) + PC liposomes (2), complex particles of the conjugate (SC + MD-SA2, R_w = 2) + mixed (PC liposomes + triglycerides of FO) (3).

to the pure biopolymers (Semenova et al., 2012, 2014a,b,c, 2016). Fig. 5.2 shows an example of such alterations.

By way of visualization, Tables 5.7 and 5.8 show the 3D-atomic force microscopy (AFM) images and the averaged measured parameters of the complex particles of the conjugate (SC + MD-SA-2) with both pure PC liposomes (Table 5.7) and the mixed (PC + FO) liposomes (Table 5.8) in comparison with the parameters of both the conjugate and the liposomes alone.

On the strength of the data observed (Tables 5.5–5.8 and Fig. 5.2) it may be inferred that the zwitterionic PC liposomes and LPC micelles in their both pure form and in the mixture with triglycerides of FO could form many new structurally important physical intra- and intercross bonds both within and between original caseins-based biopolymer molecules studied.

In addition, an example of the dependencies of the structural parameters of the complex particles on PC concentration (10^{-6}M ÷10^{-3}M), shown in Fig. 5.3 for the complex (SC + PC) particles, gives the supporting evidence of the increasing inter- and intra-cross-bonding of the protein particles by PC with increasing PC concentration. That is manifested itself in the successive increase in the values of the density of the (SC + PC) complex particles with increasing PC concentration. As this takes place, the values of the structure-sensitive parameter $1 < \rho < 2$ indicate the retention of the spherical form of the SC particles in their complexes with PC liposomes.

Table 5.7 The Area of the Cross-Section (S), the Volume (V), and the Height (Z) of the Pure PC Liposomes, the Conjugate (SC + MD–SA2) $R_w = 2$ Particles and Their Complexes (pH 7.0) Measured by Atomic-Force Microscopy (an AF Microscope SOLVER P47 (SMENA, NT-MDT, Russia), Where *N* is a Number of the Particles Under the Consideration and Averaging

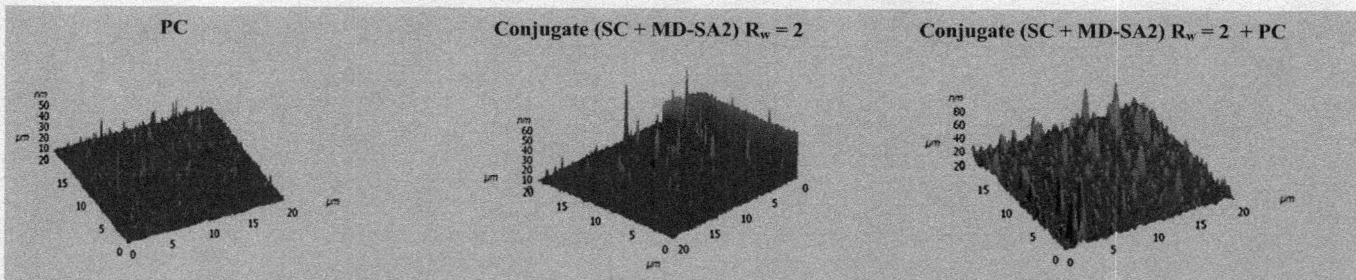

Samples	N	S $(\mu m)^2$	$V \times 10^3$ $(\mu m)^3$	Z (nm)
PC	2459	0.00353 ± 6E-5	0.02950 ± 6E-5	7.2073 ± 0.0688
Conjugate	881	0.01791 ± 0.00266	0.008666 ± 0.00521	6.54225 ± 0.09108
Conjugate + PC	760	0.04184 ± 0.00413	0.60239 ± 0.06658	13.17596 ± 0.19213

Table 5.8 The Area of the Cross-Section (S), the Volume (V), and the Height (Z) of the Pure PC Liposomes and Mixed (PC + FO) Liposomes as well as the Complex Particles of the Latter With the Conjugate (SC + MD-SA2) $R_w = 2$ (pH 7.0) Measured by Atomic-Force Microscopy (an AF Microscope SOLVER P47 (SMENA, NT-MDT, Russia), Where *N* is a Number of the Particles Under the Consideration and Averaging

PC

PC + FO

Conjugate (SC + MD-SA2) $R_w = 2$ + (PC + FO)

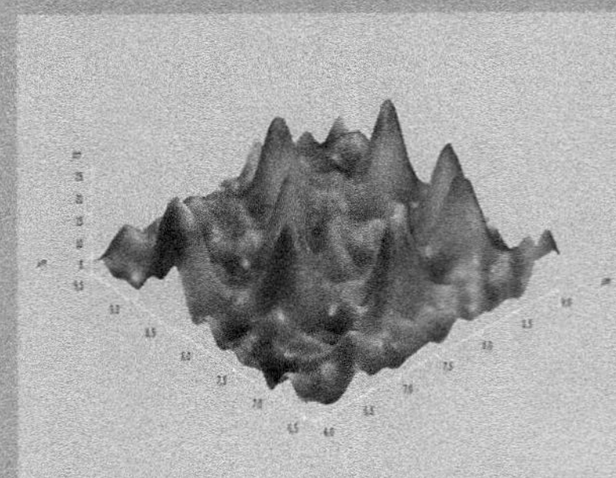

Samples	N	S $(\mu m)^2$	V × 10^3 $(\mu m)^3$	Z (nm)
PC	2459	0.00353 ± 6E-5	0.02950 ± 6E-5	7.2073 ± 0.0688
PC + FO	3569	0.00662 ± 1.7E-4	0.05521 ± 0.0016	7.34354 ± 0.0515
Conjugate + (PC + FO)	3000	0.0265 ± 0.0005	0.5214 ± 0.0111	17.5393 ± 0.1291

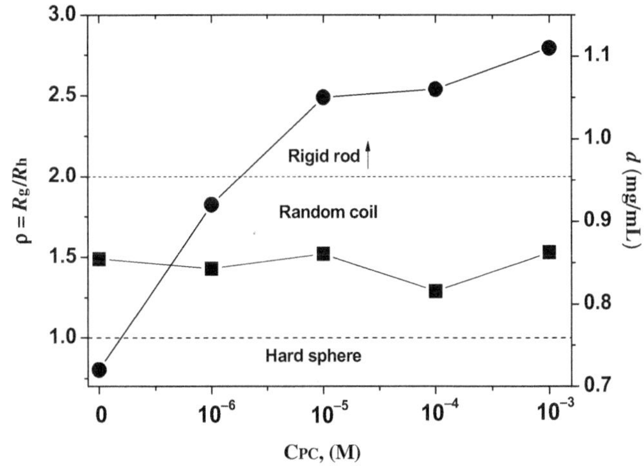

Figure 5.3. Effect of PC concentration on both the structure sensitive parameter ρ (■), and the density (•) of the (SC + PC) nanoparticles at pH 7.0 and ionic strength = 0.001 M. The characteristic values of ρ, which are inherent for the different architecture of the polymer associates, are shown by the dotted lines. The given value of the ionic strength is indicative of the ionic strength of the buffer.

4.2 Thermodynamic Stability of the PC Liposome Bilayers

In addition, in the case of the PC liposomes, the found general increase in the thermodynamic stability of the PC bilayers involved into the complex particles (reflected by the increase in the enthalpy of the phase transition, ΔH_{tr}, from the solid-like gel state of the bilayer to its fluid liquid-crystalline state) (Fig. 5.4) seem facilitated to some extent by the revealed protective ability of the complex particles against PC oxidation (Semenova et al., 2012, 2014a,b, 2016).

In order to determine ΔH_{tr} we have used the differential scanning calorimetry (DSC) measurements on the complex particles involving a model saturated phosphatidylcholine, namely, dipalmitoylphosphatidylcholine (DPPC). The choice of DPPC was dictated by the difference in the phase behavior of the bilayers of the saturated and unsaturated phosphatidylcholine liposomes in an aqueous medium. By this means DPPC bilayer shows generally the phase transition from the solid-like gel state to the fluid liquid-crystalline state in the vicinity to 40°C (Semenova et al., 2012, 2014a,b, 2016), whereas the polyunsaturated PC bilayer does not, because it has already been in the fluid liquid-crystalline state at any temperature exceeding 0°C (Menger et al., 2005). It is worthy of note here that under the temperature used for the

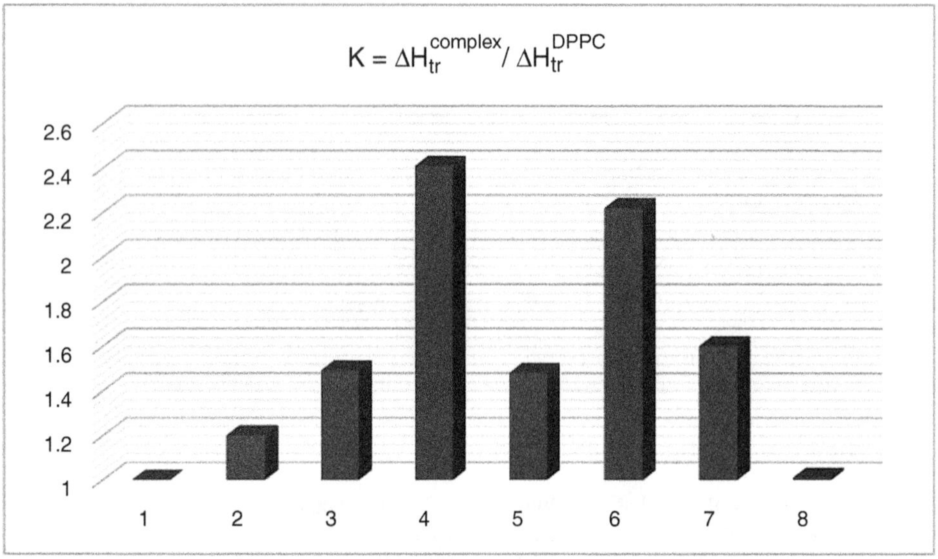

Figure 5.4. The coefficient, k, of the increase in the enthalpy of the transition of the phospholipid bilayer from the solid-like gel state to the fluid liquid-crystalline state as a result of the complex formation between the model dipalmitoylphosphatidylcholine (DPPC) and the following biopolymers in an aqueous medium at different pH and the same ionic strength of the buffers = 0.001M. pH = 7.0: (1) SC; (2) β-casein; (3) Conjugate (SC + MD-SA2, R_{weight} = MD:SC =2); (4) α_S-casein; (5) α_S-casein + Ca^{2+} ions (the I way of the calcium ions addition); (6) α_S-casein + Ca^{2+} ions (the II way of the calcium ions addition). pH = 5.5: (7) SC. pH = 4.8: (8) Conjugate (SC + MD-SA2, R_{weight} = MD:SC =2). DSC measurements were performed by the high sensitivity differential scanning calorimetry using a DASM-4M model, Differential Scanning Calorimeter (Pushino, Russia).

complex formation between PC/DPPC liposomes and the biopolymers, namely 40°C, the model DPPC bilayer apparently approaches maximally to the fluid liquid-crystalline state of the real polyunsaturated PC bilayer that is evidence in favor of the usefulness of DPPC as the model substance.

The identified positive impact of the complex formation between DPPC and the casein-based biopolymers on the thermodynamic stability of the DPPC bilayer could be attributable, on the one hand, to the neutralization of the uncompensated charge of the DPPC polar heads as a result of the attractive interactions with the oppositely charged functional groups of the caseins (Shoemaker and Vanderlick, 2003), and, on the other hand, to strengthening of the hydrophobic attraction between hydrocarbon chains of the DPPC molecules induced some optimization of their packaging caused, in turn, by the incorporation of the hydrophobic parts of casein molecules into the bilayers of the DPPC liposomes (Gennis, 1989).

4.3 Microviscosity of the Bilayers of the Lipid Particles

Moreover, the investigation of the state of both the pure and, mixed with FO triglycerides, bilayers of the PC liposomes as well as LPC micelles, using electron paramagnetic resonance spectroscopy (EPRS), indicates the increase in their microviscosity as a result of the interactions with the biopolymers. It is significant to note that this increase has been seen even in spite of the decreasing effect of the added FO triglycerides molecules (Table 5.9) that was most pronounced for the case of the PC liposomes.

First and foremost, the found results agree well with the literature data, which indicate that the increase in the extent of the total unsaturation of the lipid bilayers causes generally the decrease in their microviscosity, whereas the addition of proteins leads generally to the increase in it (Wassall et al., 2004). Moreover, the found results correlate well with the increase in the thermodynamic stability of the phospholipid bilayers involved into the complex particles (Fig. 5.4) detected by the DSC measurements. In particular, the similar larger effect of the conjugates on both the microviscosity of the bilayers of the lipids (Table 5.9) and the ΔH_{tr} (Fig. 5.4) was detected as compared with that of SC.

Most likely, the disclosed increase in the microviscosity of the bilayers of the encapsulated PC liposomes could be attributed to the more dense packing of the hydrocarbon chains of PC as a result of the location of the protein hydrophobic groups in the phospholipid bilayers (Wassall et al., 2004).

The less pronounced effect of the complex formation on the microviscosity of LPC micelles could be attributed to their probable transformation into the new clusters within the interior of the complex nanoparticles with biopolymers that facilitates somehow the attraction between LPC hydrocarbon chains.

Thus, it is likely that this biopolymer effect on the microviscos ity of the encapsulated phospholipid-based particles could also contribute to their protection against oxidation in their supramolecular complexes with the biopolymers studied.

5 Thermodynamic Parameters Controlling Solubility of the (Biopolymer + Lipids) Complex Particles in an Aqueous Medium

The behavior of the biopolymer particles in a bulk and at the interfaces of the biopolymer solutions is driven thermodynamically by the values of their chemical potentials, which can be

Table 5.9 Effect of Both the Addition of FO Triglycerides and the Encapsulation by the Biopolymers (SC, Conjugates [SC + MD] at R_{weight} = MD:SC = 2) (1 wt/v%) on the Microviscosity of Both the Bilayers of the PC Liposomes and LPC Micelles, Measured by a Spectrometer Bruker EMX (Germany) Using 16-Doxylstearic Acid Spin Probe C_{16} With the Depth of Localization of 20 Å (pH 7.0 of an Aqueous Medium)

Sample	$\tau \times 10^{10}$ (s)	±	Effect (%)	±
PC	9.85	0.17	0	0
PC + FO	8.78	0.10	−10.9	1
PC + SC	10.13	0.10	+2.8	0.3
PC + conjugate (SC + MD-SA2)	13.68	0.30	+38.9	0.4
PC + conjugate (SC + MD-10)	12.88	0.10	+30.8	0.3
PC + FO + conjugate (SC + MD-SA2)	14. 10	0.20	+43.1	0.4
LPC	13.41	0.14	0	0
LPC + FO	10.88	0.47	−18.9	3.4
LPC + conjugate (SC + MD-SA2)	15.99	0.14	+19.2	0.9
LPC + conjugate (SC + MD-10)	15.17	0.48	+13.1	3.5
LPC + FO + conjugate (SC + MD-SA2)	11.57	0.11	−13.8	0.1

presented as a virial series in biopolymer concentration, as developed by Ogston (1962) at the level of an approximation of just pairwise molecular interactions:

$$\mu_1 = \mu_1^0 - (RT/m_1) \times (m_2 + \tfrac{1}{2} A_2^* m_2^2) \tag{5.2}$$

$$\mu_2 = \mu_2^0 + RT[\ln(m_2/m^0) + A_2^* m_2] \tag{5.3}$$

Here, μ_i^0 and m_i are the standard chemical potential and concentration (molal scale) of the i-component ($i = 1$ for solvent, $i = 2$ for biopolymer); A_2^* is the second virial coefficient (in molal scale units of cm^3/mol, that is, taking the polymer molar mass into account); and m^0 is the standard-state molality for the polymer.

Thus, the sign and magnitude of the second virial coefficient provides information on how the behavior of the biopolymer solution deviates from that of the thermodynamically ideal state, thereby reflecting the nature and intensity of the intermolecular pair interactions (both biopolymer–biopolymer and biopolymer–solvent) (Prigogine and Defay, 1954; Tanford, 1961; Ogston 1962; Nagasawa and Takahashi, 1972; Semenova and Dickinson, 2010).

In order to make the notion of the second virial coefficient more clear it is worthy of note here that the sign of A_2 provides a simple indicator of the type of interactions present in a biopolymer solution. Hence, a positive value of A_2 indicates thermodynamically unfavorable biopolymer–biopolymer interactions in the solution (an increase in the magnitude of the chemical potential μ_i of the biopolymer ($i = 2$) in solution)—in other words, a mutual biopolymer repulsion. A positive A_2 also indicates thermodynamically favorable biopolymer–solvent interactions (a decrease in the magnitude of μ_i of the solvent ($i = 1$) in the presence of the biopolymer in solution)—in other words, a mutual attraction. The exact opposite is the case for a negative value of the second virial coefficient (Semenova and Dickinson, 2010).

The advantage of a static light-scattering, used in our work, over some other methods, determining the second virial coefficient, is the capability to measure both thermodynamic (A_2) and structural parameters (M_w, R_G) in a single experiment. This advantage has provided us with the possibility to estimate the contributions from the excluded volume effects, A_2^{exc}, to the pair interactions between biopolymer particles in an aqueous solution. The repulsive "steric/excluded volume" interactions arose from the highly thermodynamically unfavorable overlap of full electron clouds leading to the restriction in the occupation of the same volume in solution by two different particles (Semenova and Dickinson, 2010). Thus, both the size and the shape of the biopolymer particles, as determined by both their macromolecular conformation/flexibility and their ultimate architecture, are of prime importance to the excluded volume repulsive interactions.

In turn, the difference between measured A_2^* and the excluded volume term A_2^{exc} allows an estimation of the total contributions from the other kinds of the interactions [electrostatic (A_2^{el}), hydrogen bonding ($A_2^{h/b}$) and hydrophobic (A_2^{h})] as follows:

$$A_2^* - A_2^{exc} = A_2^{el} + A_2^{h/b} + A_2^{h} \tag{5.4}$$

Under the assumption of the spherical architecture of the particles found in our experiments generally ($1 < \rho \leq 2$) we have used the simplest case of the interacting solid spheres and the following equation for the calculation of the A_2^{exc} (Tanford, 1961):

$$A_2^{exc} = 10^{-3}\, 4\, N_A / 3(2R)^3 \tag{5.5}$$

The parameter R in the Eq. (5.5) is the radius of the equivalent hard sphere representing the biopolymer particle. The equivalent hard sphere corresponds to the space occupied in the aqueous medium by a single biopolymer particle, which is completely inaccessible to both other biopolymer particles and solvent molecules. In our calculations, we have generally used the R_h, measured by the dynamic light scattering (using multi-angle laser light scattering (LS-01 apparatus, Scientific Instruments, St. Petersburg, Russia), as the radius of the equivalent hard sphere.

First and foremost, it is important to note that the general noticeable trend in the change of the second virial coefficient of biopolymers as a result of their complex formation with the phospholipid particles studied is both the less negative and more positive values. It is interesting that such alteration occurs in spite of the simultaneous, as a rule, decrease in the positive values of the excluded volume terms (A_2^{exc}) (Semenova et al., 2014a,b, 2016). This result evidently indicates the intensification of the electrostatic repulsions (a positive contribution from the A_2^{el} to the A_2^{*}) between the particles that cause either a decrease in the absolute negative values of the summary term ($A_2^{el} + A_2^{h/b} + A_2^{h}$) or even give rise to its positive values (Semenova et al., 2014a,b, 2016). Actually, by way of illustration, Fig. 5.5 a and b shows the dominant contribution of the electrostatic repulsions in the pair interactions of the complex particles that, in addition, agrees well with their higher zeta-potential as compared with that of the pure conjugate particles (Fig. 5.5a). This gives the impression that more hydrophobic patches of both biopolymer and polyunsaturated lipids were hidden in the interior of their supramolecular complexes, while their charged groups were exposed simultaneously at the surface of them (Semenova et al., 2014b). One might expect that because of such distribution, the hydrophilic–hydrophobic balance of the surface properties of the coassembled complex particles shifted evidently toward more hydrophilicity, that is, the greater number of the polar and charged groups, which were hydrophilic in their nature, became exposed into the aqueous medium. Besides, relying on this result, it may be safely suggested that the hydrophobic hydrocarbon tails of the lipid molecules are hidden apparently in the interior of the complex biopolymer-based particles (Semenova et al., 2014a,b, 2016).

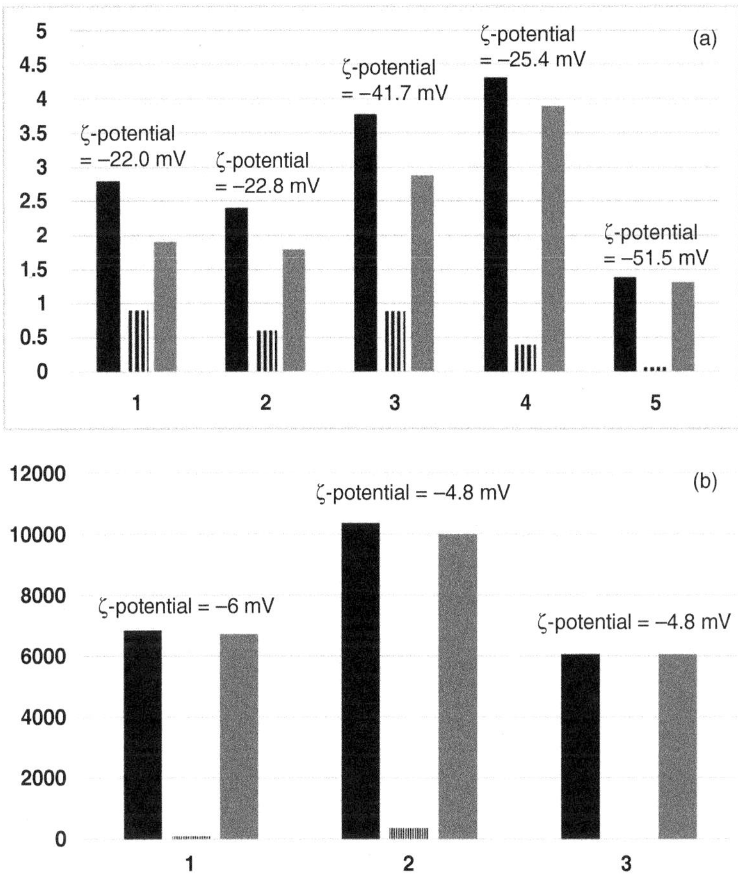

Figure 5.5. Contributions of the different kinds of the pair interactions into the value of the molal second virial coefficient A_2^* (m³ mol⁻¹). The black columns refer to the A_2^* (m³ mol⁻¹); the striped columns refer to the contributions from the excluded volume effects, A_2^{exc}, the gray columns refer the summary term $A_2^{el} + A_2^{h/b} + A_2^{h}$ combining the contributions from the electrostatic interactions (A_2^{el}), hydrogen bonding ($A_2^{h/b}$) and hydrophobic attractions (A_2^{h}). A: (pH 7.0, ionic strength = 0.001M) Conjugate (SC + MD SA-2, R_{weight} = MD:SC = 2) (1) alone; (2) with PC liposomes; (3) with mixed (PC + FO) liposomes – for a better comparison the presented values reduced by tenfold; (4) with LPC; (5) with (LPC + FO) – for a better comparison the presented values reduced by 10^4. B: (pH 4.8, ionic strength = 0.001M) Conjugate (SC + MD SA-2, R_{weight} = MD:SC = 2) (1) alone; (2) with PC liposomes; (3) with mixed (PC + FO) liposomes – for a better comparison the presented values increased by 10-fold.

Thus, the positive value of the second virial coefficient underlies, generally, rather high solubility of the complex particles formed in an aqueous medium even at pH = 4.8 (Fig. 5.5b), that is, at the protein isoelectric point that, in turn, can be favorable to the bioavailability of the complex particles in vivo (Ratnam et al., 2006; Acosta, 2009; Amidon et al., 1995; Horter and Dressman, 2001; Hecq et al., 2005; Semenova and Dickinson, 2010, Chapter 2; McClements et al., 2008; Velikov and Pelan, 2008; Singh et al., 2009).

6 Structural Parameters of the Biopolymer Nanovehicles Controlling Release of the Polyunsaturated Lipids in Vitro

6.1 The Density and Architecture of the Complex Particles

The generality of the importance of such structural parameters as the density of the complex particles and their architecture for their susceptibility to the action of the specific for both proteins and maltodextrins digestive enzymes are supported by the data in vitro obtained for the complex particles formed between PC liposomes and the following biopolymers: β-casein (Semenova et al., 2012), α_S-casein (Semenova et al., 2014a), SC (Semenova et al., 2014b), and both covalent (SC + MD-SA2) and electrostatic (SC + dextran sulfate) complexes (Semenova et al., 2014c).

Based on these data, Fig. 5.6 shows, as an example, the relationship between the initial rate of the hydrolysis in vitro, V_0, and both the density and architecture of the tested particles. It was observed that the higher is the original density of the particles at their similar original architecture (the close values of ρ) and the lower is the initial rate of their hydrolysis by the digestive enzyme, and vice versa.

In addition, it was established that both the chemical and physical modification of food proteins through the covalent or electrostatic interactions with polysaccharides, respectively, could regulate release of PC from the complex particles essentially decreasing, for example, the release of PC up to 80% in the simulated gastro intestinal conditions that could be mainly attributed to the steric effects from the polysaccharides attached to the protein (Semenova et al., 2014c). However, further detailed studies, using as elaborated recently the standard protocol for the simulated conditions of the digestion in the GIT (Minekus et al., 2014), are required in order to get a more penetrating insight into the impact of the protein-polysaccharide electrostatic and covalent interactions on the elaboration of smart biopolymer nanovehicles for a wide variety of nutraceuticals and in particular for the essential polyunsaturated lipids.

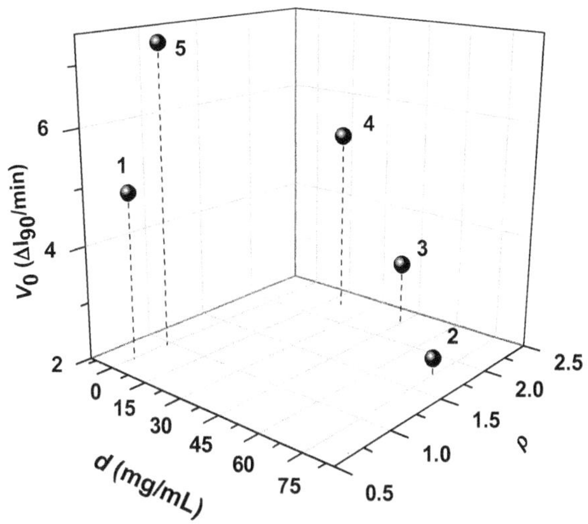

Figure 5.6. Impact of the structural parameters (*ρ* and *d*) of the complex (biopolymer + PC liposomes) particles, formed in an aqueous medium at ionic strength = 0.001M: 1: pH = 7.0, α_S-casein; 2: pH = 7.0, α_S-casein + PC + Ca^{2+} (I way); 3: pH = 7.0, α_S-casein + PC + Ca^{2+} (II way); 4: pH = 5.5, 40°C, β-casein + PC; 5: pH = 7.0, covalent conjugate (SC + MD-SA2, R_{weight} = MD : SC = 2) + PC, on the initial rate (V_0) of their hydrolysis by pepsin in vitro (pH 2.0, I = 0.01 M of the buffer, 37°C; Enzyme:Substrate = 1:1000). The V_0 was determined from the laser light scattering measurements (LS-01, Scientific Instruments, St. Petersburg, Russia) of the changes of the intensity of the light scattering, ΔI_{90}, from the tested solutions with time of the hydrolysis at the angle of scattering Θ = 90.

7 Conclusions

On the strength of the data observed, we could suggest that the following physicochemical properties underlay the potentiality of such food biopolymers like individual caseins, whole sodium caseinate and its covalent conjugates with maltodextrins to behave as smart nanovehicles for the essential polyunsaturated lipids:

1. the amphiphilic nature of the proteins that provided, on the one hand, the high level (>92%) of the encapsulation of the lipids via the different kinds of the interactions (electrostatic and hydrophobic attractions, hydrogen bonding) and, on the other hand, the high level of solubility in an aqueous medium;
2. the ability to form nanosized complex particles that could facilitate both the bioavailability and bioaccessibility of the essential lipids in vivo;
3. enough flexibility to the shrinkage as a result of the many new intra- and interphysical bonds formation with the lipids that led to the marked increase in the density of the complex

particles that, in turn, provided the protection to the polyunsaturated lipids against oxidation;

4. the ability to increase the thermodynamic stability of the bilayers of PC liposomes and microviscosity of both PC liposomes and LPC micelles in the complex particles that could also contribute into the protection of the lipids against oxidation;
5. the possibility to form a wide diversity of the complex particles having different density and architecture that could be used in order to control the release of the essential lipids under the enzymatic action of the digestive enzymes in GIT.

Acknowledgments

The part of the research on the physicochemical properties of the complex particles, which were formed by the covalent conjugates with both pure and mixed with triglycerides of FO PC liposomes and LPC micelles, was financially supported by the Russian Science Foundation (Grant no. 14-16-00102). Another part of the research was supported by the Russian national budget, including experimental work of Maria S. Anokhina, Larisa E. Belyakova, Yurii N. Polikarpov, Vladimir I. Binukov, and Valerii V. Kasparov. In particular, Vladimir I. Binukov and Valerii V. Kasparov were supported financially by the Federal Agency of the Scientific Organizations in their kind assistance in conducting measurements using AFM and EPRS in the IBPC Centers of the collective use of scientific equipment, respectively.

References

Acosta, E., 2009. Bioavailability of nanoparticles in nutrient and nutraceutical delivery. Curr. Opin. Colloid In. 14, 3–5.

Amidon, G.L., Lennernas, H., Shah, V.P., Crison, J.R., 1995. A theoretical basis for a biopharmaceutic drug classification: the correlation of in vitro drug product dissolution and in vivo bioavailability. Pharmaceut. Res. 12, 413–420.

Antunes, F.E., Marques, E.F., Miguel, M.G., Lindman, B., 2009. Polymer–vesicle association. Adv. Colloid Interfac. 147–148, 18–35.

Augustin, M.A., Hemar, Y., 2009. Nano- and micro-structured assemblies for encapsulation of food ingredients. Chem. Soc. Rev. 38, 902–912.

Bai, G., Nichifor, M., Bastos, M., 2010. Association and phase behavior of cholic acid-modified dextran and phosphatidylcholine liposomes. J. Phys. Chem. Lett. 1 (6), 932–936.

Belitz, H.D., Grosch, W., 1982. Lehrbuch der Lebensmittelchemie. Springer-Verlag, Heidelberg.

Bolder, S.G., Hendrickx, H., Sagis, L.M.C., van der Linden, E., 2006. Ca^{2+}-induced cold-set gelation of whey protein isolate fibrils. Appl. Rheol. 16, 258–264.

Bouwmeester, H., Dekkers, S., Noordam, M., Hagens, W., Bulder, A., de Heer, C., ten Voorde, S., Wijnhoven, S., Sips, A., 2007. Health impact of nanotechnologies in food production. Report 2007, a coproduction of RIKILT and RIVM (Internet).

Burchard, W., 1994. Light scattering. In: Ross-Murphy, S.B. (Ed.), Physical Techniques for the Study of Food Biopolymers. Blackie, Glasgow, pp. 151–214.

Chen, H., Weiss, J., Shahidi, F., 2006. Nanotechnology in nutraceuticals and functional foods. Food Technol. 60 (3), 30–36.

Chen, L., Remondetto, G.E., Subirade, M., 2006a. Food protein-based materials as nutraceutical delivery systems. Trends Food Sci. Tech. 17, 272–283.
Cleland, L.G., French, J.K., Betts, W.H., Murphy, G.A., Elliott, M.J., 1988. Clinical and biochemical effects of dietary fish oil supplements in rheumatoid-arthritis. J. Rheumatol. 15, 1471–1475.
de Kruif, C.G., Tuinier, R., Holt, C., Timmins, P.A., Rollema, H.S., 2002. Physicochemical study of k- and b-casein dispersions and the effect of cross-linking by transglutam-inase. Langmuir 18, 4885–4891.
de Kruif, C.G., Huppertz, T., Urban, V.S., Petukhov, A.V., 2012. Casein micelles and their internal structure. Adv. Colloid Interfac. 171-172, 36–52.
Dickinson, E., 2004. Food colloids: the practical application of protein nanoscience in extreme environments. Editorial overview. Curr. Opin. Colloid In. 9, 295–297.
Dickinson, E., 2006a. Colloid science of mixed ingredients. Soft Matter 2, 642–652.
Dickinson, E., 2006b. Structure formation in casein-based gels, foams, and emulsions. Colloids Surf. A 288, 3–11.
Dickinson, E., 2014. Understanding food structures: the colloid science approach. In: Boland, M., Golding, M., Singh, H. (Eds.), Food Structures, Digestion and Health. Elsevier Academic Press, London, pp. 3–49.
Dickinson, E., Golding, M., 1998. Influence of calcium ions on creaming and rheology of emulsions containing sodium caseinate. Colloids Surf. A 144, 167–177.
Dickinson, E., Semenova, M.G., Antipova, A.S., 1998. Salt stability of casein emulsions. Food Hydrocolloids 12, 227–235.
Dickinson, E., Semenova, M.G., Belyakova, L.E., Antipova, A.S., Il'in, M.M., Tsapkina, E.N., Ritzoulis, C., 2001. Analysis of light scattering data on the calcium ion sensitivity of caseinate solution thermodynamics: relationship to emulsion flocculation. J. Colloid Interf. Sci. 239, 87–97.
Esmaili, M., Ghaffari, M.S., Moosavi-Movahedi, Z., Atri, M.S., Sharifizadeh, A., Farhadi, M., Yousefi, R., Chobert, J.-M., Haertlé, T., Moosavi-Movahedi, A.A., 2011. Beta casein-micelle as a nano vehicle for solubility enhancement of curcumin: food industry application. LWT – Food Sci. Technol. 44, 2166–2172.
Faulks, R.M., Southon, S., 2008. Assessing the bioavailability of nutraceuticals. In: Garti, N. (Ed.), Delivery and Controlled Release of Bioactives in Foods and Nutraceuticals. Woodhead, Cambridge, UK, pp. 3–25.
Fernández, J., Pérez-Álvarez, J.A., Fernández-López, J.A., 1997. Thiobarbituric acid test for monitoring lipid oxidation in meat. Food Chem. 59 (3), 345–353.
Föster, S., Konrad, M., 2003. From self-organising polymers to nano- and biomaterials. J. Mater. Chem. 13, 2671–2688.
Fu, J., Huang, B., 2001. Involvement of antioxidants and lipid peroxidation in the adaptation of two cool-season grasses to localized drought stress. Environ. Exp. Bot. 45, 105–114.
Gennis, R.B., 1989. Biomembranes: molecular structure and function. In: Cantor, R. Ch. (Ed.), Advanced Text in Chemistry. Springer, Berlin.
Geusens, P., Wouters, C., Nijs, J., Jiang, Y.B., Dequeker, J., 1994. Long-term effect of omega-3-fatty-acid supplementation Zn active rheumatoid-arthritis—a 12-month, double-blind, controlled study. Arthritis Rheum. 37, 824–829.
Gibbs, B.F., Kermasha, S., Alli, I., Mulligan, C.N., 1999. Encapsulation in the food industry. Int. J. Food Sci. Nutr. 50, 213–224.
Gibis, M., Rahn, N., Weiss, J., 2013. Physical and oxidative stability of uncoated and chitosan-coated liposomes containing grape-seed extract. Pharmaceutics 5, 421–433.
Gouin, S., 2004. Microencapsulation: industrial appraisal of existing technologies and trends. Trends Food Sci. Tech. 15, 330–347.

Graveland-Bikker, J.F., de Kruif, C.G., 2006. Unique milk protein-based nanotubes: food and nanotechnology meet. Trends Food Sci. Tech. 17, 196–203.

Grigorovich, N.V., Moiseenko, D.V., Antipova, A.S., Anokhina, M.S., Belyakova, L.E., Polikarpov, Yu. N., Korica, N., Semenova, M.G., Baranov, B.A., 2012. Structural and thermodynamic features of covalent conjugates of sodium caseinate with maltodextrins underlying their functionality. Food Funct. 3, 283–289.

Gutteridge, J.M., 1977. The measurement of malondialdehyde in peroxidised ox-brain phospholipid liposomes. Anal. Biochem. 82, 76–82.

Gutteridge, J.M., Tickner, T.R., 1978. The thiobarbituric acid-reactivity of bile pigments. Biochem. Med. 19, 127–132.

Harkema, Ir.J., 1998. Paselli SA2 and Paselli Excel. Fat substitutes. In: Dalzell, J.M. (Ed.), Ingredients Handbook. Leatherhead Food RA, Surrey, pp. 103–133.

Hasni, I., Bourassa, P., Hamdani, S., Samson, G., Carpentier, R., Tajmir-Riahi, H.-A., 2011. Interaction of milk α- and β-caseins with tea polyphenols. Food Chem. 126, 630–639.

Heath, R.L., Packer, L., 1968. Photoperoxidation in isolated chloroplasts: I. Kinetics and stoichiometry of fatty acid peroxidation. Arch. Biochem. Biophys. 125, 189–198.

Hecq, J., Deleers, M., Fanara, D., Vranckx, H., Amighi, K., 2005. Preparation and characterization of nanocrystals for solubility and dissolution rate enhancement of ni-fedipine. Int. J. Pharmaceut. 299, 167–177.

Hibbeln, J.R., Nieminen, L.R.G., Blasbalg, T.L., Riggs, J.A., Lands, W.E.M., 2006. Healthy intakes of n-3 and n-6 fatty acids: estimations considering worldwide diversity. Am. J. Clin. Nutr. 83, 1483S–1493S.

Holt, C., Sawyer, L., 1993. Caseins as rheomorphic proteins: interpretation of the primary and secondary structures of the α_{s1}–, β- and κ-caseins. J. Chem. Soc., Faraday Trans. 89, 2683–2692.

Horn, D., Rieger, J., 2001. Organic nanoparticles in the aqueous phase—theory, experiment, and use. Angewandte Chemie Int. ed. 40, 4330–4361.

Horter, D., Dressman, J.B., 2001. Influence of physicochemical properties on dissolution of drugs in the gastrointestinal tract. Adv. Drug Deliver. Rev. 46, 75–87.

Huppertz, T., de Kruif, C.G., 2008. Structure and stability of nanogel particles prepared by internal cross-linking of casein micelles. Int. Dairy J. 18, 556–565.

Hussain, N., Jaitley, V., Florence, A.T., 2001. Recent advances in the understanding of uptake of microparticulates across the gastrointestinal lymphatics. Adv. Drug Deliver. Rev. 50, 107–142.

Istarova, T.A., Semenova, M.G., Sorokoumova, G.M., Selishcheva, A.A., Belyakova, L.E., Polikarpov, Yu. N., Anokhina, M.S., 2005. Effect of pH on the interactions of sodium caseinate with soy phospholipids in relation to the foaming ability of their mixtures. Food Hydrocolloids 19, 429–440.

Jani, P., Halbert, G.W., Langridge, J., Florence, A.T., 1990. Nanoparticle uptake by the rat gastrointestinal mucosa: quantitation and particle size dependency. J. Pharm. Pharmacol. 42, 821–826.

Kidd, P.M., 1996. Phosphatidylcholine: a superior protectant against liver damage. Altern. Med. Rev. 1 (4), 258–274.

Kidd, P.M., 2000. Dietary phospholipids as anti-aging nutraceuticals. In: Klatz, R.A., Goldman, R. (Eds.), Anti-Aging Medical Therapeutics. Health Quest Publications, Chicago, pp. 283–301.

Kremer, J.M., Michalek, A.V., Lininger, L., Huyck, C., Bigauoette, J., Timchalk, M.A., Rynes, R.I., Zieminski, J., Bartholomew, L.E., 1985. Effects of manipulation of dietary fatty acids on clinical manifestations of rheumatoid-arthritis. Lancet 325, 184–187.

Kreuter, J., 2001. Nanoparticulate systems for brain delivery of drugs. Adv. Drug Deliver. Rev. 47, 65–81.
Kwon, T.W., Menzel, D.B., Olcott, H.S., 1965. Reactivity of malonaldehide with food constituents. J. Food Sci. 30, 808–813.
Lai, S.K., O'Hanlon, D.E., Harrold, S., Man, S.T., Wang, Y.-Y., Cone, R., Hanes, J., 2007. Rapid transport of large polymeric nanoparticles in fresh undiluted human mucus. Proc. Natl. Acad. Sci. USA 104, 1482–1487.
Lalush, I., Bar, H., Zakaria, I., Eichler, S., Shimoni, E., 2005. Utilization of amylose–lipid complexes as molecular nanocapsules for conjugated linoleic acid. Biomacromolecules 6, 121–130.
Leclerc, E., Calmettes, P., 1997. Interactions in micellar solutions of b-casein. Phys. Rev. Lett. 78, 150–153.
Lee, S.J., Ying, D.Y., 2008. Encapsulation of fish oils. In: Garti, N. (Ed.), Delivery and Controlled Release of Bioactives in Foods and Nutraceuticals. CRC Press, Cambridge, pp. 370–403.
Letchford, K., Burt, H., 2007. A review of the formation and classification of amphiphilic block copolymer nanoparticulate structures: micelles, nanospheres, nanocapsules and polymersomes. Eur. J. Pharm. Biopharm. 65, 259–269.
Livney, Y.D., 2010. Milk proteins as vehicles for bioactives. Curr. Opin. Colloid In. 15, 73–83.
Manski, J.M., van der Goot, A.J., Boom, R.M., 2007. Advances in structure formation of anisotropic protein-rich foods through novel processing concepts. Trends Food Sci. Tech. 18, 546–557.
Markman, G., Livney, Y.D., 2012. Maillard-conjugate based core-shell coassemblies for nanoencapsulation of hydrophobic nutraceuticals in clear beverages. Food Funct. 3, 262–270.
Mayer, S., Weiss, J., McClements, D.J., 2013. Behavior of vitamin E acetate delivery systems under simulated gastrointestinal conditions: Lipid digestion and bioaccessibility of low-energy nanoemulsions. J. Colloid Interf. Sci. 404, 215–222.
McClements, D.J., 2006. Noncovalent interactions between proteins and polysaccharides. Biotechnol. Adv. 24, 621–625.
McClements, D.J., 2014. Nanoparticle- and microparticle-based delivery systems: encapsulation. Protection and Release of Active CompoundsCRC Press Taylor and Francis Group, New York.
McClements, D.J., Decker, E.A., 2000. Lipid oxidation in oil-in-water emulsions: impact of molecular environment on chemical reactions in heterogeneous food systems. J. Food Sci. 65 (8), 1270–1282.
McClements, D.J., Decker, E.A., Park, Y., Weiss, J., 2008. Designing food structure to control stability, digestion, release and absorption of lipophilic food components. Food Biophys. 3, 219–228.
McClements, D.J., Decker, E.A., Park, Y., Weiss, J., 2009. Structural design principles for delivery of bioactive components in nutraceuticals and functional foods. Crit. Rev. Food Sci. 49, 577–606.
Medina, C., Santos-Martinez, M.J., Radomski, A., Corrigan, O.I., Radomski, M.W., 2007. Nanoparticles: pharmacological and toxicological significance. Brit. J. Pharmacol. 150, 552–558.
Menger, F.M., Chlebowski, M.E., Galloway, A.L., Lu, H., Seredyuk, V.A., Sorrells, J.L., Zhang, H., 2005. A tribute to the phospholipid. Langmuir 21, 10336–10341.
Merisko-Liversidge, E., Liversidge, G.G., Cooper, E.R., 2003. Nanosizing: a formulation approach for poorly water-soluble compounds. Eur. J. Pharm. Sci. 18, 113–120.

Min, Y., Akbulut, M., Kristiansen, K., Golan, Y., Israelachvili, J., 2008. The role of interparticle and external forces in nanoparticle assembly. Nat. Mater. 7, 527–538.

Minekus, M., Avito, P., Balance, S., Bohn, T., Bourlieu, C., Carriére, F., Boutrou, R., Corredig, M., Dupont, D., Dufour, C., Egger, L., Golding, M., Karakaya, S., Kirkhus, B., Feunteun, S.Le., Lesmes, U., Macierzanka, A., Mackie, A., Marze, S., McClements, D.J., Ménard, O., Recio, J., Santos, C.N., Singh, R.P., Vagarud, G.E., Wickham, M.S.G., Weitschies, W., Brodkorb, A., 2014. A standardised static in vitro digestion method suitable for food—an international consensuns. Food Funct. 5, 1113–1124.

Moghimi, S.M., Hunter, A.C., Murray, J.C., 2001. Long-circulating and target-specific nanoparticles: theory to practice. Pharma. Rev. 53, 283–318.

Morris, V.J., 2005. Is nanotechnology going to change the future of food technology? Int. Rev. Food Sci. Technol. 3, 16–18.

Morris, V.J., 2006. Nanotechnology and its future in new product development. J. Inst. Food Sci. Technol. 20, 15–17.

Murray, B.S., Ettelaie, R., 2004. Foam stability: proteins and nanoparticles. Curr. Opin. Colloid In. 9, 314–320.

Nagasawa, M., Takahashi, A., 1972. Light scattering from polyelectrolyte solutions. In: Huglin, M.B. (Ed.), Light Scattering from Polymer Solutions. Academic Press, London, pp. 671–723.

Nagy, K., Pilbat, A.-M., Groma, G., Szalontai, B., Cuisinier, F.J.G., 2010. Casein aggregates built step-by-step on charged polyelectrolyte film surfaces are calcium phosphate-cemented. J. Biol. Chem. 2 (50), 38811–38817.

Nishioka, Y., Yoshino, H., 2001. Lymphatic targeting with a nanoparticulate system. Adv. Drug Deliver. Rev. 47, 55–64.

Ogston, A.G., 1962. Some thermodynamic relationships in ternary systems, with special reference to the properties of systems containing hyaluronic acid and protein. Arch. Biochem. Biophys. Suppl 1, 39–51.

Payens, T.A.J., Vreeman, H., 1982. Casein micelles and micelles of β- and κ-casein. Mittal, K.L., Fendler, E.J. (Eds.), Solution Behavior of Surfactants, vol. 1, Plenum, New York, pp. 543–571.

Phang, M., Garg, M., 2014. Combined phytosterol and fish oil therapy for lipid lowering and cardiovascular health. In: Boland, M., Golding, M., Singh, H. (Eds.), Food Structures, Digestion and Health. Elsevier Academic Press, London, pp. 437–463.

Prigogine, I., Defay, R., 1954. Chemical Thermodynamics. Longmans Green, London, UK.

Rahimi Yazdi, S., Corredig, M., 2012. Heating of milk alters the binding of curcumin to casein micelles. A fluorescence spectroscopy study. Food Chem. 132, 1143–1149.

Ransley, J.K., Donnelly, J.K., Read, N.W., 2001. Food and Nutritional Supplements: Their Role in Health and Disease. Springer-Verlag, Berlin, Heidelberg.

Ratnam, D.V., Ankola, D.D., Bhardwaj, V., Sahana, D.K., Kumar, M.N.V.R., 2006. Role of antioxidants in prophylaxis and therapy: a pharmaceutical perspective. J. Control. Release 113, 189–207.

Sahu, A., Kasoju, N., Bora, U., 2008. Fluorescence study of the curcumin-casein micelle complexation and its application as a drug nanocarrier to cancer cells. Biomacromolecules 9, 2905–2912.

Salminen, H., Aulbach, S., Leuenberger, B.H., Tedeschi, C., Weiss, J., 2014. Influence of surfactant composition on physical and oxidative stability of Quillaja saponin-stabilized lipid particles with encapsulated omega-3fish oil. Colloids Surf. B 122, 46–55.

Sanguansri, P., Augustin, M.A., 2006. Nanoscale materials development—a food industry perspective. Trends Food Sci. Tech. 17, 547–556.
Schmidt, D.G., 1982. Association of caseins and casein micelle structure. In: Fox, P.F. (Ed.), Developments in Dairy Chemistry—1. Applied Science Publications, London, pp. 61–86.
Schmidt, D.G., Payens, T.A., 1972. The evaluation of positive and negative contributions to the second virial coefficient of some milk proteins. J. Colloid Interf. Sci. 39, 655–662.
Schulz, M., Olubummo, A., Binder, W.H., 2012. Beyond the lipid-bilayer: interaction of polymers and nanoparticles with membranes. Soft Matter 8, 4849–4864.
Semenova, M.G., 2007. Thermodynamic analysis of the impact of molecular interacions on the functionality of food biopolymers in solution and in colloidal systems. Food Hydrocolloids 21, 23–45.
Semenova, M.G., Dickinson, E., 2010. Biopolymers in Food Colloids: Thermodynamics and Molecular Interactions. Brill, Leiden.
Semenova, M.G., Belyakova, L.E., Polikarpov, Yu. N., Antipova, A.S., Anokhina, M.S., 2008. Utilization of sodium caseinate nanoparticles as molecular nanocontainers for delivery of bioactive lipids to food systems: relationship to the retention and controlled release of phospholipids in the simulated digestion conditions. In: Williams, P.A., Phillips, G.O. (Eds.), Gums and Stabilisers for the Food Industry 14. Royal Society of Chemistry, Cambridge, UK, pp. 326–333.
Semenova, M.G., Antipova, A.S., Anokhina, M.S., Belyakova, L.E., Polikarpov, Yu. N., Grogorovich, N.V., Tsapkina, E.N., 2012. Thermodynamic and structural insight into the underlying mechanisms of the phosphatidylcholine liposomes-Casein associates coassembly and functionality. Food Funct. 3 (3), 271–282.
Semenova, M.G., Anokhina, M.S., Antipova, A.S., Belyakova, L.E., Polikarpov, Yu. N., 2014a. Effect of calcium ions on both the coassembly of α_s-casein with soy phosphatidylcholine and the novel functionality of their complex particles. Food Hydrocolloids 34, 22–33.
Semenova, M.G., Antipova, A.S., Belyakova, L.E., Polikarpov, Yu. N., Anokhina, M.S., Grigorovich, N.V., Moiseenko, D.V., 2014b. Structural and thermodynamic properties underlying the novel functionality of sodium caseinate as delivery nanovehicle for biologically active lipids. Food Hydrocolloids 42, 149–161.
Semenova, M.G., Moiseenko, D.V., Grigorovich, N.V., Anokhina, M.S., Antipova, A.S., Belyakova, L.E., Polikarpov, Yu. N., Tsapkina, E.N., 2014c. Protein – polysaccharide interactions and digestion of the complex particles. In: Boland, M., Golding, M., Singh, H. (Eds.), Food Structures, Digestion and Health. Elsevier Academic Press, London, pp. 169–192.
Semenova, M.G., Zelikina, D.V., Antipova, A.S., Martirosova, E.I., Grigorovich, N.V., Obushaeva, R.A., Shumilina, E.A., Ozerova, N.S., Palmina, N.P., Maltseva, E.L., Kasparov, V.V., Bogdanova, N.G., Krivandin, A.V., 2016. Impact of the structure of polyunsaturated soy phospholipids on the structural parameters and functionality of their complexes with covalent conjugates combining sodium caseinate with maltodextrins. Food Hydrocolloids 52, 144–160.
Semo, E., Kesselman, E., Danino, D., Livney, Y.D., 2007. Casein micelle as a natural nanocapsular vehicle for nutraceuticals. Food Hydrocolloids 21, 936–942.
Shapira, A., Assaraf, Y.G., Epstein, D., Livney, Y.D., 2010. Beta-casein nanoparticles as an oral delivery system for chemotherapeutic drugs: impact of drug structure and properties on coassembly. Pharmaceut. Res. 27 (10), 2175–2186.

Shoemaker, S.D., Vanderlick, T.K., 2003. Calcium modulates the mechanical properties of anionic phospholipid membranes. J. Colloid Interf. Sci. 266, 314–321.

Singh, H., Ye, A., Horne, D.S., 2009. Structuring food emulsions in the gastrointestinal tract to modify lipid digestion. Prog. Lipid Res. 48, 92–100.

Strömstedt, A.A., Ringstad, L., Schmidtchen, A., Malmsten, M., 2010. Interaction between amphiphilic peptides and phospholipid membranes. Curr. Opin. Colloid In. 15, 467–478.

Swaisgood, H.E., 2003. Chemistry of the caseins. In: Fox, P.F., McSweeney, P.L.H. (Eds.), Advanced Dairy Chemistry, vol. 1, Proteins. Third ed. Kluwer Academic/Plenum. Part A, New York, pp. 139–202.

Tanford, C., 1961. Physical Chemistry of Macromolecules. Wiley, New York.

Thurn, A., Burchard, W., Niki, R., 1987a. Structure of casein micelles. I. Small angle neutron scattering and light scattering from b- and k-casein. Colloid Polym. Sci. 265 (8), 653–666.

Thurn, A., Burchard, W., Niki, R., 1987b. Structure of casein micelles. II. a_{s1}-casein. Colloid Polym. Sci. 265 (8), 897–902.

van der Linden, E., 2006. Innovations with protein nanofibres. World. Poultry Sci. J. 62, 439–442.

Veerman, C., Baptist, H., Sagis, L.M.C., van der Linden, E., 2003. A new multistep Ca2 + -induced cold gelation process for β-lactoglobulin. J. Agr. Food Chem. 51, 3880–3885.

Velikov, K.P., Pelan, E., 2008. Colloidal delivery systems for micronutrients and nutraceuticals. Soft Matter 4 (10), 1964–1980.

Waraho, T., McClements, D.J., Decker, E.A., 2011. Mechanisms of lipid oxidation in food dispersions. Trends Food Sci. Tech. 22 (1), 3–13.

Wassall, S.R., Caffrey, M., Cherezov, V., Bizustowics, M.R., Shaikh, C.R., Stillwell, W., 2004. The role of polyunsaturated lipids in membrane raft formation. Chem. Phys. Lipids 132, 79–88.

Weiss, J., Takhistov, P., McClements, D.J., 2006. Functional materials in food nanotechnology. J. Food Sci. 71, 107–115.

Zhang, C., Ding, Y., Ping, Q.E., Yu, L.L., 2006. Novel chitosan-derived nanomaterials and their micelle-forming properties. J. Agr. Food Chem. 54, 8409–8416.

6

ENCAPSULATION: ENTRAPPING ESSENTIAL OIL/FLAVORS/AROMAS IN FOOD

Suphla Gupta*, Saima Khan*, Malik Muzafar*, Manoj Kushwaha, Arvind Kumar Yadav**, Ajai Prakash Gupta****

**Plant Biotechnology Department, Indian Institute of Integrative Medicine, Jammu, Jammu and Kashmir, India; **Quality Control & Quality Assurance, Indian Institute of Integrative Medicine, Jammu, Jammu and Kashmir, India*

1 Introduction

Flavor substances of natural and synthetic origin have been widely used in human nutrition. Flavor plays an important role in food quality and influences consumers' satisfaction and food consumption. Direct use of the raw materials for this purpose poses numerous practical difficulties like inconsistency in the composition of raw materials, flavor/aroma content, size and growing site of the product, and so forth. Microbiological contaminations and parasitic infestation, and undesirable alteration in the ratio of certain components during storage are also a few parameters which affect the final result. Besides limited shelf life, the natural sources of certain aromas present serious storage problems. To avoid these disadvantages, concentrates, spice oleoresins (solvent extracts of spices), essential oils, and flavors (extracted by steam distillation) have been used in the food industry for over a century. Extracted or distilled ingredients could be easily converted to solid powder-like products, simply by spreading the oleoresins or oils onto salt or starch, dextrins, or microencapsulating them with gum Arabic, cellulose ethers, or starch derivatives. The powdered flavors offer the food manufacturer a clean, standardized form of what was once a variable and unreliable ingredient, difficult to measure for accurate dosage. Gradually, technology was applied in food processing to improve use, storage, and effectiveness. However, the performance of fragrances tends to fade by evaporation, interactions with other components, oxidation, and chemical degradation. Encapsulation is an elegant way of improving

Encapsulations. http://dx.doi.org/10.1016/B978-0-12-804307-3.00006-5

the performance, such as substantivity, tenacity, and endurance (Porzio, 2007, 2009, 2012). Technological advantages that include stability, standardizable compositions, simple dosing, and so forth has taken the flavor business to immense profitability. Significant attention has been paid to improving flavor retention due to the instability of volatile flavors in the presence of air, light, moisture, or high temperature (Manojlovic et al., 2008; Bhandari et al., 1998; Given, 2009).

Aroma is due to volatile and odorous organic molecules, mostly in gaseous or liquid forms, but some solid materials may also contain a distinct smell (vanillin and menthol). These molecules are lipophilic and low molecular weight (between 100 and 250), belonging to hydrocarbons, alcohols, aldehydes, ketones, esters, acids, and sulfides group of compounds. Aroma molecules can be either added during the processing of the food product or are generated during cooking of the food product. For reducing or eliminating any losses through evaporation, during the process the volatile compounds are stabilized by complex formation. The reduction of volatility can be demonstrated by a rise in the boiling point of solutions, or in the sublimation for solids (Duchene et al., 1989). Complexation has been used to avoid the obliteration of certain flavors, colors or vitamins related with certain ingredients by processing or in storage. For example, from the flavored complex, the flavor may be released in the warm moisture of the mouth.

The aroma characteristic depends on the composition of the molecule and the type of the matrix. The composition of the molecule determines the macroscopic variables in thermodynamic and kinetic parameters of the molecule, which decide the volatility and resistance to mass transfer between different phases, especially from the product to air of the compound (Druaux and Voilley, 1997; van Ruth and Roozen, 2002; van den Oever et al., 2009). Proper choice of food composition and microstructures may maneuver the aroma release during its preparation and consumption.

The flavor and fragrance industry had a turnover of about US$22.09 billion in 2012 and increased to US$24.89 at the rate of 8.69% in 2014. According to an assumption (Ubbink and Schoonman, 2003) about half of this turnover is from flavor industry while about 20–25% of all flavors are estimated to be sold in an encapsulated form. Major part (80–90%) of the encapsulates are prepared using spray-dried method, followed by spray chilling (5–10%), melt extrusion (2–3%), and preparation by melt injection (the remaining 2%) (Ubbink and Schoonman, 2003; Porzio, 2007, 2009, 2012, 2013) (Fig. 6.1).

The physical and chemical properties of the encapsulates depend on the aroma properties as well as the carrier material. As

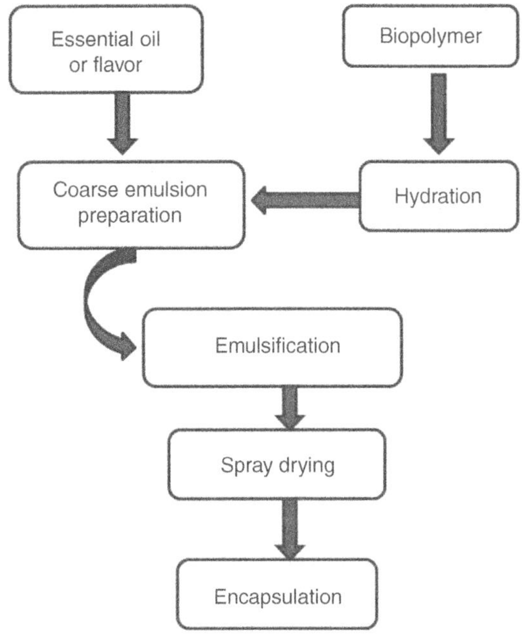

Figure 6.1. Latest trend in the flavor and fragrance industry.

majority of aroma molecules are lipophilic, and due to their inability to mix well with hydrophilic carrier material, they form oil droplets. When more hydrophobic carrier material is used, aroma mixes with the carrier material forming single-matrix morphology, similar to when entrapping water-soluble aroma molecules (like vanillin) in hydrophilic carrier materials. Once encapsulated, the desirable quality is required that the aroma should diffuse out in the right amount at the appropriate time. Research has shown that the molecular size of the carrier material decides the diffusion rate. In general, the diffusion through carrier material decreases with increasing molecular size (stearic hindrance, related to molecular weight), decreasing vapor pressure (volatility), and increasing log P of the aromas (Goubet et al., 1998). Among the carbohydrate carrier materials the order of retention is alcohols > ketones = esters > acids, with many exceptions (Goubet et al., 1998). These observations indicate that the presence of chemical groups not only influences the retention, but also other characteristics like polarity, having an inverse relation in which the higher the polarity, the lower the retention.

Other factors that influence the encapsulation are the physical and chemical properties of coating material. Permeability of the aroma molecule is influenced by the molecular weight of the coating material, conformation, chemical composition, and its physical state. Higher molecular weight and the highest degree of polymerization of the carrier material limit aroma diffusion, as was found for maltodextrins (Goubet et al., 1998). Very high polymerization probably lowers the degrees of interactions between aroma and carrier material, thereby lowering its diffusion rate. Physical state of the carrier molecule also influences the aroma release. When an amorphous coating material is in a glassy state, molecules have restricted relative mobility, similar to those of a crystalline phase (Ubbink and Schoonman, 2003; Goubet et al., 1998; Soottitantawat et al., 2004; Carolina et al., 2007). The temperature plays a crucial role in transition from glassy to crystalline state. Below the glass transition temperature, the diffusion through carrier material is limited due to crystalline phase, while above the transition temperature, the carrier material is in rubbery state and diffusion is relatively faster. Also the high molecular weight of the carrier molecules like carbohydrate (Ubbink and Schoonman, 2003) have increased transition temperature but low storage stability due to poor porosity (Ubbink and Schoonman, 2003). Usually, a compromise has to be made between a high glass transition temperature of the carrier material (which is favored by high molecular weight material) and low porosity (which increases with the use of low molecular weight materials). Also, the glass transition temperature decreases with increasing percentages of plasticizers and exposure to relatively high humidity. High humidity leads to greater mobility of the carrier material, which could be sufficient to associate and crystallize, thereby forcing out the aroma and thus causing loss of aroma (Goubet et al., 1998). The choice of carrier molecule becomes more complex and restricted when lipophilic aromas and hydrophilic carrier material is to be used, due to compatibility between the two. In that case, hydrophilic carriers as encapsulation carriers are mostly selected. To understand their association and retention characteristics, Hildebrand's solubility parameters for aromas and polymers could be useful to predict their compatibility.

Keeping these parameters under consideration, various techniques have been studied and proposed for encapsulation of aroma, broadening the application range of encapsulation technique. Since long it has been used in pharmaceutical industry for controlled release and delivery of drugs (Kosaraju, 2005). Now, its use in foods and consumer products (Gibbs et al., 1999; Jackson and Lee, 1991b; Gouin, 2004; Reineccius, 1996, 1998; Zeller et al., 1999; Pothakamury and Barbosa-Canovas, 1995) is widely accepted with many advantages being recognized, especially in encapsulation of aroma chemicals, flavors, and fragrances. The demand for fragranced products has been growing with diverse applications and promise of further refinement in the future. The applicability of the technique in consumer products includes air fragrances, bath additives, candles, cosmetics, household, oral hygiene, laundry products, and so forth.

2 New Technologies: Future Course

Modern advancement is based on traditional platforms. The sol–gel process, used commonly now for encapsulation, originated in the traditional ceramic industry (Gouin, 2004). During sol–gel encapsulation, an inorganic gel network is formed by gelation of a sol (a colloidal suspension) like metal alkoxides, which can react and undergo the sol–gel transition in an aqueous environment similar to inorganic interfacial polymerization encapsulation.

Development of newer encapsulation methods is time- and effort-consuming, requiring a multidisciplinary approach, particularly with foods materials. However, for fragrance encapsulation, no extensive legislation is required for approval, thereby making the use of new materials as matrix materials easier to apply for commercial uses. However, cost considerations in the food industry are much more stringent than in the pharmaceutical or cosmetic industries.

3 Nanotechnology

Under the umbrella of nanotechnology properties, methodology involving scaling down of structural features of materials to the nanometer range is utilized. Two approaches are used to make nanostructures materials: (1) top-down (break-down of larger structures) or (2) bottom-up (building from individual atoms or molecules capable of self-assemblage). Biomineralization, a naturally occurring process, has inspired researchers for creating artificial nanocomposites based on nanoclays. Naturally occurring nanoclays are hydrophilic and chemical modifications have presented them as a desirable candidate for nanoencapsulation, with good retention and dispersion properties even with hydrophobic particles. Manipulation of materials at the nanometer level opens up the door to improved functionality of aroma chemicals. A very recent development is encapsulation of actives in colloidosomes (Gouin, 2004; Dinsmore et al., 2002) similar to liposome entrapment. Here, colloidal particles are formed due to the surface tension on the surface of an inner phase or active ingredient in a water-in-oil or oil-in-water emulsion, resulting in selectively permeable capsules. Another technique with great potential for controlled granule disintegration includes gelatinization of native starches under high-pressure treatment (Gouin, 2004; Douzals et al., 1996). Several overviews (patent) of the art of the encapsulation of various materials, such as flavors and fragrances, can be found in the literature (Risch, 1995; Popplewell and Porzio, 1998). New synthetic-based matrices are being developed for enhanced deposition and longevity which are also temperature and environment sensitive, especially in the case of fabric cleaner and deodorant (Liu and Hu, 2005). Mixtures of various carriers can also be used to tailor desired properties such as release, deposition, and substantivity (van Soest, 2007). Also matrices containing inorganic materials have been developed that are suitable for transforming fragrances into free-flowing powder with improved deposition properties for laundry application (van Soest, 2007). A particular example of a new material used for fragrance encapsulation is the use of polysaccharide esters such as starch acetate (Vedantam and Yong, 2004).

Encapsulation of food ingredients into coating materials can be achieved by several methods (Table 6.1). The physical and chemical properties of the core matrix and the coating materials determine the process for microencapsulation of food ingredients. Coating materials, which are basically film-forming materials, can be selected from a diversity of natural or synthetic polymers, depending on the material to be coated and the quality needed in the final microcapsules. The composition of the coating material is the main determinant of the functional properties of the microcapsule and of how it may be used to improve the performance of a particular ingredient. In general, three precautions need to be considered for developing microcapsules: (1) wall formation around the material, (2) be sure that leakage does not occur, and (3) be sure undesired materials must be kept out. In optimizing the process, at least four criteria may be considered: (1) properties of the wall materials, (2) features of the core materials, (3) condition of the infeed emulsion, and (4) circumstances of the spray drying.

An ideal coating material should exhibit the following characteristics (Desai and Park, 2005):

1. Good rheological properties at high concentration and easy to handle and use during encapsulation.
2. Ability to disperse or emulsify the active material and stabilize the emulsion produced.
3. Nonreactivity with the material to be encapsulated both during processing and in prolonged storage.
4. Ability to seal and hold the active material within its structure during processing or storage.

Table 6.1 Various Microencapsulation Techniques and the Processes Involved in Each Technique

S. No.	Microencapsulation Technique	Major Steps in Encapsulation
1.	Centrifugal suspension separation	a. Mixing of core in a coating material b. Pouring the mixture over a rotating disc to obtain encapsulated tiny particles c. Drying
2.	Coacervation	a. Formation of a three-immiscible chemical phases b. Deposition of the coatings; solidification of the coating
3.	Cocrystallization	a. Preparation of supersaturated sucrose solution b. Adding of core into supersaturated solution c. Emission of substantial heat after solution reaches the sucrose crystallization temperature
4.	Centrifugal extrusion	a. Preparation of core solution b. Preparation of coating material solution c. Coextrusion of core and coat solution through nozzles
5.	Extrusion	a. Preparation of molten coating solution b. Dispersion of core into molten polymer c. Cooling or passing of core–coat mixture through dehydrating liquid
6.	Fluidized-bed coating	a. Preparation of coating solution b. Fluidization of core particles c. Coating of core particles
7.	Inclusion complexation	Preparation of complexes by mixing or grinding or spray drying
8.	Liposomal entrapment	a. Microfluidization b. Ultrasonication c. Reverse-phase evaporation
9.	Lyophilization	a. Mixing of core in a coating solution b. Freeze-drying of the mixture
10.	Spray chilling	a. Preparation of the dispersion b. Homogenization of the dispersion c. Atomization of the infeed dispersion
11.	Spray cooling	a. Preparation of the dispersion b. Homogenization of the dispersion c. Atomization of the infeed dispersion
12.	Spray drying	a. Preparation of the dispersion b. Homogenization of the dispersion c. Atomization of the infeed dispersion d. Dehydration of the atomized particles

5. Ability to completely release the solvent or other materials used during the process of encapsulation under drying or other desolventization conditions.
6. Ability to provide maximum protection to the active material against environmental conditions (eg, oxygen, heat, light, humidity).
7. Solubility in solvents acceptable in the food industry (eg, water, ethanol).
8. Chemical nonreactivity with the active core materials.
9. Inexpensive, food-grade status.
10. Nontoxic, nonallergen, biodegradable, environmentally friendly matrix.

Complex and advanced shell materials and technologies have been developed having a wide variety of utilities through microencapsulation, for example, use of a trigger to prompt the release of the encapsulated ingredient. The triggering agent could be pH change (enteric and antienteric coating), mechanical stress, temperature, enzymatic activity, time, osmotic force, and so forth. Since none of the single coating materials can meet all of the desired criteria mentioned previously for microencapsulation, to fulfill maximum desired qualities, the coating materials are either employed in combination or with modifiers (oxygen scavengers, antioxidants, chelating agents, and surfactants). Also, chemical modifications/derivatives of coating material are prepared to improve its quality for better/desirable encapsulating properties. Some commonly used biocompatible and food-grade coating materials are listed in Table 6.2(Desai and Park, 2005).

4 Matrix or Coating Materials

The coating or matrix material is referred to as wall, membrane/carrier/shell or capsule. Based on the size of the particles, matrix coating can be of three types:

1. Macrocoated powders with sizes larger than 0.1 mm
2. Matrix microparticles or microcapsules with sizes in the range 0.1–100 μm
3. Nanoparticles or nanocapsules with sizes smaller than 0.1 μm

Macrocoating is used to stabilize fragrances or to transform them from liquid to free-flowing solid powder, while in microencapsulation and nanoencapsulation, the particle is entrapped inside a miniature capsule—the microcapsule or the nanocapsule, respectively. The substance inside the capsule in encapsulation can be a gas, liquid, or solid and the wall of the capsule (single or multiple layers) can be of various materials, such a wax, plastic, or biopolymers like proteins or polysaccharides. The encapsulates

Table 6.2 Commonly Used Biocompatible and Food-Grade Coating Materials

S. No.	Category	Coating Materials	Widely Used Methods	References
1.	Carbohydrate	Starch, maltodextrins, chitosan, corn syrup solids, dextrin, modified starch, cyclodextrins	Spray- and freeze-drying, extrusion, coacervation, inclusion complexation	[10,17,18]
2.	Cellulose	Carboxymethyl cellulose, methyl cellulose, ethyl cellulose, cellulose acetate phthalate, cellulose acetate butylate-phthalate	Coacervation, spray drying, and edible films	[19]
3.	Gum	Gum acacia, agar, sodium alginate, carrageenan	Spray-drying syringe method (gel beads)	[20]
4.	Lipids	Wax, paraffin, beeswax, diacylglycerols, oils, fats	Emulsion, liposomes, film formation	[21]
5.	Protein	Gluten, casein, gelatin, albumin, peptides	Emulsion, spray drying	[22]

can have a variety of shapes, such as spherical, oblong, or irregular, and can be monolithic or aggregates.

The most commonly used matrix materials are:

1. Lipids (waxes, paraffin, oils, fats, etc.)
2. Inorganics (silicates, clays, calcium sulfate, etc.)
3. Polysaccharides and sugars (gums, starches, celluloses, cyclodextrin, dextrose, etc.)
4. Proteins (gelatin, casein, soy protein, etc.)
5. Synthetics [acrylic polymers, poly(vinylpyrrolidone), etc.]

Literature details differences between matrix encapsulation and true encapsulation of a substance. In matrix encapsulation the resulting particles are found as aggregates of actives (molecule) in a matrix material, having their significant portion on the surface of the particles. True encapsulation is used for processes leading to core-shell-type products, where actives/particles are enclosed in a shell. Environmentally friendly biodegradable polymers have gained more attention as carriers matrix because of their biocompatibility and biodegradability. They may be of synthetic origin such as polyesters, poly(*ortho*-esters), polyanhydrides, and polyphosphazenes, or natural like polysaccharides such as chitosan, hyaluronic acid, and alginates.

Cyclodextrin is one of the most common and simplest encapsulant systems that can form inclusion complexes with different flavors depending on their hydrophobicity molecular size and geometry (Piel et al., 2001; Faucci et al., 2002; Cabral Marques, 2010). The advantages of cyclodextrin inclusion complexes include high encapsulation yield and long retention time (Reineccius et al., 2002, 2003b; Madene et al., 2006). Over the years, work has progressed on characterizing the CyD family of compounds, and Szejtli et al. (1979) were the first to find applications for them in the flavor industry (Reineccius, 2009). In the USA and Europe, three principal manufacturers of α-, β-, and γ-CyD are present. β- and γ-CyDs have both received self-affirmed GRAS status (generally recognized as safe) in the USA (Reineccius et al., 2002). β-CyD is generally approved for use in Europe (as food additive E459). CyD is permitted for use in Japan, though the regulations do not specify the precise nature of the chemical entity. The United Nations Food and Agriculture Organization/World Health Organization (FAO/WHO), Joint Expert Committee on Food Additives (JECFA) have established an acceptable daily intake (ADI) of 0–5 mg/kg body weight in the case of β-CyD while an ADI "not specified" has been allocated in the case of both α- and γ-CyD (UN FAO/WHO, 1995, 2000, 2001) (Reineccius et al., 2002). Besides, a number of enzymatically or chemically modified CyDs have also been described, such as branched β-CyDs (Ajisaka et al., 2000) and hydroxypropyl-β-CyD (Qi and Hedges, 1995), some of which are claimed to have superior properties over the parent CyD. At present, however, none are approved for use as food additives. The current higher cost of CyDs ($\gamma > \alpha > \beta$) relative to that of conventional encapsulation carrier materials remains an area of concern.

Unfortunately, as mentioned earlier under the legislation in Europe, the use of β-cyclodextrin for encapsulation of flavor substances is only permitted to a certain extent, and for other cyclodextrins, hardly at all, so other excipients for encapsulation have been widely investigated and reported (Paramita et al., 2012; Mourtzinos et al., 2008). Some of these excipients, in particular starch, maltodextrins, glucose, and sucrose, are used by the body for energy as well. The problems associated with high consumption of energy in food has associated symptoms like obesity, diabetes, cardiocirculatory disorders, and muscular-skeletal and locomotion system complaints, therefore introduction and promotion and high consumption of these substances should be minimized. Cellulose, the most abundant and renewable polysaccharide, is a preferred alternative for high-energy foods. Cellulose is biocompatible, of hydrophilic nature, widely accepted, safe, and biodegradable in nature (Rusli et al., 2013; Reza Fareghi et al., 2013; Bagheri and Shateri, 2012), besides having a neutral inherent taste and a regulatory effect on digestion. However, solubility of cellulose remains an issue, as it is hard to dissolve cellulose in aqueous solutions due to the presence of strong inter- and intramolecular hydrogen bonds and considerable van der Waals forces, limiting its application in flavor industry. Therefore, the hunt for substitutes or cellulose derivatives is on to support the application of cellulose in flavor encapsulation (Luo et al., 2013; Porzio and Popplewell, 1994). Wen Lou and Popplewell (2003) employed a matrix containing hydroxyethyl cellulose (HEC) to encapsulate flavor or fragrance materials as food products and laundry applications. Roberts et al. (1996) studied the effect of viscosity and thickener (sucrose, guar gum, and carboxymethyl cellulose) on dynamic flavor release and found that the increase of viscosity and thickener of CMC resulted in lower release in flavor. Sansukcharearnpon et al. (2010) prepared a polymer-blend of ethylcellulose (EC), hydroxypropyl methylcellulose (HPMC), and poly (vinyl alcohol) to encapsulate flavors, and found that menthol shows the slowest release. Recently, cellulose-based flavoring substance was produced using a novel approach of regenerated porous cellulose particles (RPC) for flavor encapsulation (Siegel, 2010).

In this method, the encapsulation of flavor was done by absorption of flavor in the voids located between cellulose chains. In employing this process, though a high-yield encapsulation could be obtained, the retention of flavor in the flavoring substance was poor due to less adsorptive capability of cellulose. To improve the retention of flavors, CMC was used as the coating layer of RPC. This approach promises to have potential applications in flavor entrapment. Commercial encapsulation is generally carried out in practice using a number of processes, including spray drying, spray cooling/chilling, freeze-drying, fluidized bed coating, extrusion, coacervation, cocrystallization, and molecular inclusion (Risch and Reineccius, 1995; Gibbs et al., 1999; Shahidi and Han, 1993). All these processes, except for molecular inclusions process, are macroprocesses having particle size diameters in the range 3–800 μm. An encapsulated material particle may be in the form of droplets of core material dispersed in a continuous matrix of carrier material, or the continuous core, surrounded by a shell of carrier. However, in molecular inclusion, the process occurs at the molecular level, whereby individual molecules of food or flavor ingredient are trapped or included within cavities present in individual molecules of carrier as seen in entrapment using the cyclodextrins (CyDs).

5 Encapsulation Processes

5.1 Spray Drying

Spray drying is one of the oldest and most commonly used processes to encapsulate aroma. It is so common in foods that one often tends to forget it is a form of encapsulation. The technique for producing encapsulated flavoring was discovered by A. Boake Roberts in 1937 (Gaonkar et al., 2014), when acetone was accidentally added to tomato puree which helped him to maintain the color and flavor of tomato powder during spray drying. Subsequently, spray drying has become the most important commercial process for making dry flavorings. About 80–90% of the encapsulated aroma is prepared by this method (Porzio, 2007, 2009, 2012, 2013; Ubbink and Schoonman, 2003). In spray drying, dispersion of the aroma is performed in an aqueous solution of carrier material, followed by atomization and spraying of the emulsion into a hot chamber. In the hot chamber, the cocurrent hot air stream, at predetermined temperature (160–220°C), heats the droplets almost instantaneously to 100°C (Porzio, 2007, 2009) (Figs. 6.2 and 6.3). During this step a film is formed at the droplet surface and the concentration of ingredients in the drying droplet keeps increasing, thereby retarding the larger aroma molecules while the smaller water molecules continue to diffuse to the exterior at a substantial rate (Porzio, 2007, 2009; Ubbink and Schoonman, 2003; Re, 1998; Jafari et al., 2008a,b). The final product comes out in the form of powder of dry particles containing

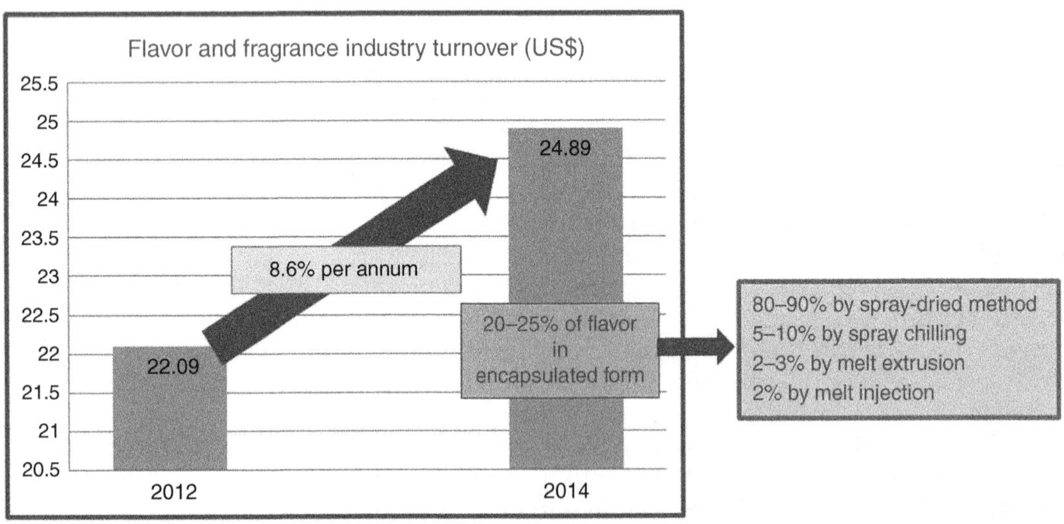

Figure 6.2. Flow diagram of spray-drying encapsulation of essential oil or flavor.

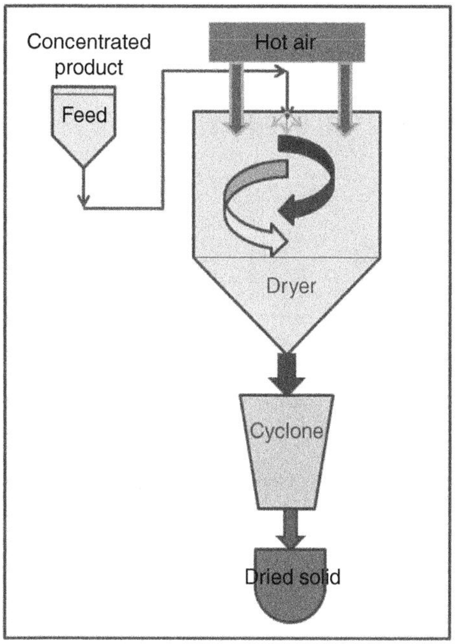

Figure 6.3. Schematic diagram of spray drying.

a dispersion of fine aroma droplets and having a particle size of 10–150 μm. The aroma encapsulates and their load, particle size, morphology, and release mechanism varies considerably. Lists of some technologies in this area are given in Table 6.3.

Spray-drying technique has many advantages including

1. Continuous production: the dried particles can be collected continuously at the bottom of the spray dryer
2. Easy operation: constant quality is possible when drying conditions are held constant and automated
3. Inexpensive
4. Time tested
5. Wide choice of carrier material and equipment

A disadvantage of spray drying might be that very volatile aromas (eg, fresh top-notes like ethyl acetate) can be (partially) lost during the process (Reineccius et al., 2003a; Reineccius and Mei, 2004), thereby changing the balance of some flavor. Also, some aromas might be oxidized during the spray-drying process. In practical use, due to their small size, spray-dried aroma powders might be so small that their utility is hampered. For example, tea bags are too porous to hold aroma powder and are therefore not preferred for this application (Porzio, 2007, 2009). In general, spray-dried aromas are water-soluble, which might either be

Table 6.3 Various Parameters of the Technologies Employed in Food Industry

S. No.	Technology	Particle Size (μm)	Morphology	Release Mechanism Based Upon	Load (%)
1.	Cocrystallization/ coprecipitation	Various	Crystals	Dissolution	Various
2.	Coextrusion	800–8000	Core with wall	Diffusion, heat	70–95
3.	Complex coacervates	10–800	Core with wall	Diffusion	20–90
4.	Cyclodextrin	0.002	Molecular inclusion	Dissolution	8–10
5.	Fluid bed coating	200–5000	Droplets in matrix	Dissolution	10–50
6.	(Fluidized) spray drying	10–400	Droplets in matrix	Dissolution	10–50
7.	Granulation/agglomeration/ compaction	200–3000	Droplets in matrix	Dissolution	5–40
8.	Melt injection	200–2000	Droplets in matrix	Dissolution	5–25
9.	Melt extrusion	300–5000	Droplets in matrix	Swelling + dissolution, heat	5–40
10.	Microspheres	10–800	Matrix of droplets	Diffusion, dissolution	20–40
11.	Silica particles	10–1000	Matrix	Diffusion, heat	5–50
12.	Spray chilling/cooling	20–200	Matrix	Diffusion, heat	10–20
13.	Yeast cells	25–30	In cell membrane	Diffusion, heat	1–70

desirable or not depending on the state of affairs. For example, in chewing gum the aroma is spray-dried to allow faster release upon hydration in the mouth, than when it is in the gum matrix (Szente and Szejtli, 2004). However, spray-dried aroma powders dissolves quickly in water-containing food products and can then not provide a controlled release benefit over nonencapsulated aroma during cooking or eating. A delayed release of spray-dried aroma in water might only be achieved when the carrier material has been irreversibly changed during the preparation process. An example is protein denaturation (Porzio, 2007, 2009). Spray drying is considered to be among the most ancient processes in the flavor industry which still has immense applicability. The process is applied in close to 90% of encapsulated flavorings present in the market. Microencapsulation by spray drying offers advantages over conventional microencapsulation techniques by producing microcapsules via a relatively simple, continuous process. The spray-drying equipment used is the same as is used for the production of dry milk. During drying, the flavoring particle never exceeds the exit air temperature. Vitamins, minerals, colorants, fat and oil flavor, aroma compounds, oleoresins, and enzymes have been encapsulated using this technique. It is an economical, as well as an effective method for protecting materials and is most widely employed, particularly for flavors for which specialized equipment is not required. For a good quality spray-dried encapsulation, the encapsulating matrix should possess the following qualities:

- The matrix must be water soluble.
- The infeed emulsion must be of low viscosity and high solids concentrations (50–70%). This is required for proper pumping and controlled atomization.
- Encapsulation matrix must yield temporary emulsion (if not stable emulsion), which should at least stay from the time of homogenization until it is atomized in the spray dryer.
- The encapsulation matrix must have good drying capacity in a normal spray dryer even at high temperatures.
- It should not be hygroscopic after drying.

The crux of the matter is that the matrix must ultimately produce a good-quality flavoring, protect it from degradation or evaporation during storage, and finally provide release in the finished product in the desired amount and at the required time. The requirements discussed earlier limit matrix materials to very few matrices, like maltodextrins, corn syrup solids, modified starches, and gum acacia, the traditional encapsulation matrix (Porzio, 2007; Re, 1998; Reineccius et al., 2003a).

Encapsulation of food or flavor ingredients within a solid matrix of materials such as food-approved starches or derivatives, gums, proteins, lipids, and so forth has the following advantages (Shahidi and Han, 1993; Risch and Reineccius, 1995; Gibbs et al., 1999):

- Conversion of liquid product into a more conveniently handled solid form.
- Encapsulated material has higher shelf life.
- Protection against loss by evaporation, aerial oxidation, exposure to light, moisture, or pH, other component reactivity.

5.2 Spray Chilling

Modification in spray-drying process in atomization of air, employing chilled air instead of hot air, resulted in spray-chilling process. Normally, materials used in this process are vegetable oils or hydrogenated oils. Frozen liquids, heat-sensitive materials, and those which are not soluble in the usual solvents can be encapsulated by spray chilling/spray cooling. It is the least expensive encapsulation technology and is routinely used for the encapsulation of a number of organic and inorganic salts like ferrous sulfate, vitamin, minerals, or acidulants as well as textural ingredients, enzymes, flavors, and other functional ingredients to improve heat stability, delay release in wet environments, and/or convert liquid hydrophilic ingredient into free flowing powders.

5.3 Spray Cooling

Spray cooling is encapsulation of the particle in the matrix. In this technique the particles are found buried in the fat matrix. This process is different from true encapsulation, which involves a core/shell type of microencapsules. In matrix encapsulation process, a considerable fraction of the active ingredient lies on the surface of the microcapsules/protruding out of the fat matrix, thus having direct access to the environment. The disadvantage of the process seems to be strong binding of the ingredient to the fat matrix, that can prevent the release of the ingredient if the fat matrix is melted and/or damaged during the processing (Gouin, 2004). Particles produced by a matrix encapsulation process usually do not retain their flavoring qualities for a longer time and release their entire content within a few minutes after being incorporated in the food. However, in a core/shell type of microcapsule, the bulk of the ingredient is encapsulated and much slower release kinetics is obtained. Even though the spray cooling/chilling process does not escort to a perfect encapsulate, but the properties obtained could achieve the desired delayed release of the ingredient in the actual application.

5.4 Extrusion Process

The traditional extrusion process involves mixing of a low-moisture carbohydrate melt (c. 15% moisture) and flavor (10–20% on a dry weight basis) to form an emulsion, and then extruding the melt through a dye (1/64 in. holes) under low pressure. This extruded flavoring is dropped into a cold isopropanol bath to form an amorphous glass structure. The glass structure is then broken into small pieces by a mixer and dried in hot air to yield the finished encapsulated product. Noteworthy developments have occurred in this process, like use of scraped surface heat exchangers in combination (or a twin screw extruder) to permit continuous processing, use of low moisture contents, and alternatives to cooling in isopropanol solutions. While there have been numerous patents outlining the progress, the following provide the key references in this technology (Benczedi and Bouquerand, 2001; Clark McIver et al., 2004; Subramaniam et al., 2003). The process is utilized for the incorporation of a flavor or fragrance ingredient or composition into a carbohydrate matrix (Popplewell and Porzio, 1998; Porzio and Popplewell, 1994; Porzio and Zasypkin, 2002). The advancement in this process has made it very competitive (cost-wise) with spray drying. Considering these changes, scraped surface heat exchanger (twin screw extruder) was a major innovation. This permitted use of a continuous process that offered inherent advantages in cost, process control, and quality. Quality improvements came partly in being able to add flavoring late in the process, thereby exposing it to less heat. Quality improvements also came in being able to use much lower amounts of water in the system; also, the scraped surface heat exchangers could work at higher viscosities, as could the extrusion process. The continuous improvement in the process has yielded, in more recent forms, no water to be removed from the system, eliminating a processing step (drying) and improving flavor quality through better retention of flavorings (improved retention from 75–85% to >90%) (Subramaniam et al., 2003). The latest innovation in the process was eliminating the washing of the formed particles in cold isopropanol. This step traditionally involved washing the surface to remove any surface flavoring, which rapidly solidified the molten extrudate (rapidly moving it in the glassy state), and reduced the moisture content (to some extent). The step was problematic as it was slow and costly, and involved working with an organic solvent, thereby raising a possibility of legal issues from residue and hazards in handling. Solidifying extrudate in liquid nitrogen or a nitrogen-cooled environment offered significant improvement. The encapsulating material used in this process has similar requirements to those used in spray drying. The encapsulating matrix need not be atomized but it must be soluble at 85% or higher solids at 110°C or higher temperatures. Emulsification is also less of an issue since synthetic emulsifiers are added to the matrix (c. 2% on a total weight basis), and there is little opportunity for phase separation during manufacturing. This is due to the short time between emulsion formation and solidification, and the extremely high viscosity of the carbohydrate melt. The critical criteria for the success of the process were however, that matrix must form a good amorphous structure and be nonhygroscopic. It must also result in good retention of flavor compounds during manufacturing, offer protection during storage, and release flavor when placed in water. The extrusion microencapsulation technology has been used almost exclusively for the encapsulation of volatile and unstable flavors in glassy carbohydrate matrices. The main advantage of this process is impermeability of oxygen due to hydrophilic glassy matrix and very long shelf life (up to 5 years). Carbohydrate matrices in the glassy states have very good barrier properties and the process of extrusion enables the encapsulation of flavors in such matrices (Zasypkin and Porzio, 2004).

Centrifugal extrusion encapsulation technique also used for a number of food-approved coating for example, seasonings, vitamins, and flavorings. This is a liquid coextrusion process having rotating encapsulating cylinder with nozzles on the outer circumference of the head of the cylinder. The head consists of a concentric feed tube through which coating and core materials are pumped separately to the many nozzles mounted on the outer surface of the device. While the core material passes through the center tube, coating material flows through the outer tube. The entire device is attached to a rotating shaft such that the head rotates around its vertical axis. As the head rotates, the core and coating materials are coextruded through the concentric orifices of the nozzles as a fluid rod of the core sheathed in coating material. The centrifugal force impels the rod outward, causing it to break into tiny particles. By the action of surface tension, the coating material envelops the core material, thus accomplishing encapsulation. The microcapsules are collected on a moving bed of fine-grained starch, which cushions their impact and absorbs unwanted coating moisture. Particles produced by this method have a diameter ranging from 150 to 2000 mm (Risch and Reineccius, 1995; Desai and Park, 2005). The wall materials used in this technique include gelatin, sodium alginate, carrageenan, starches, cellulose derivatives, gum acacia, fats, fatty acids, waxes, and polyethylene glycol.

5.5 Fluidized Bed Coating

In fluidized bed technology, uniform layer of shell material coated onto solid particles. This technology is capable of coating particles with all kind of shell material like polysaccharides, proteins, powder coatings, enteric coating, yeast cell extract, emulsifiers, complex formulations, fats, and so forth which help in dispelling the aroma in a controlled release pattern. A variety of substances like gums and proteins, aqueous solutions of hydrocolloids and melted fats/waxes and so forth. Modifications in the technique for better understanding and interesting concepts having potential applicability are being continuously investigated.

In fluidized bed coating technique solid particles and coating material are suspended in humidity and temperature controlled chamber with high velocity air and atomized (Kim and Morr, 1995;

Risch, 1995; Risch and Reineccius, 1995). Optimal coating and size depends upon particle size distribution and the duration of time the particles are in the chamber. This technique is applicable for hot-melt coatings such as emulsifiers, fatty acids, hydrogenated vegetable oil, stearines and waxes or solvent-based coatings such as gums, maltodextrin, and starches (Tsutsumi et al., 1998; Matsuda et al., 2001; Gouin, 2004).

5.6 Liposomal Entrapment

A liposome (lipid vesicle) is composed of lipid bilayers, enclosing an aqueous or liquid compartment. The lipid bilayer is arranged in such a way that the hydrophilic portion of the lipids is oriented towards the aqueous phase and the hydrophobic groups associate with the hydrophobic ones of other lipid molecules. Folding of the lipid sheet into a spherical shape forms a very stable capsule due to absence of any interaction of the lipids with water. Aqueous or lipid-soluble materials, but not both, are entrapped in these membranes. While applying this technique, the first consideration is that the layers of lipids should be nontoxic and acceptable food-grade quality. The physical properties of the liposomes like, permeability, stability, surface activity, and affinity can be varied through size and lipid composition variations depending on the requirements. The liposome bilayer does not allow sugars and large polar molecules to permeate through but small lipophilic molecules can do so (Kim and Baianu, 1991). Usually, phospholipids forms the outer layer, and size of the particle can vary from 25 nm to several micrometers in diameter. Liposomes are used for delivery of vaccines, hormones, enzymes, and vitamins into the body. Food applications of liposomes in cheese-making is a well-known process in the food industry (Kirby and Needs, 1996).

5.7 Lyophilization

Another simple technique particularly suitable for the encapsulation of aromatic materials is lyophilization or freeze-drying. In this process dehydration of material is done under cold conditions (between −40 and 80°C). The retention of volatile compounds during the lyophilization is dependent upon the chemical nature of the system (Kopelman et al., 1977). This method is suitable for the dehydration of almost all heat sensitive materials and aromas. It has been used to encapsulate water-soluble essences and natural aromas as well as drugs. The advantage of the method include convenient transport and preservation however, long dehydration period required (commonly 20 h) for freeze-drying is a little detrimental for the technique.

With the advancement in the technology in the food engineering and processing sector, newer tools are being investigated, improvized, and automated for faster and better encapsulation.

5.8 Centrifugal Suspension Separation

Centrifugal suspension is a more recent microencapsulation process. The process involves premixing the core and wall materials and adding them to a rotating disk. The time required, mixing speed and content of the coating material needs optimization for optimal production of the final particle. The coated materials along with residual liquid leave the disk. The microcapsules, so formed, are then dried or chilled after removal from the disk. The whole process can be completed between a few seconds to minutes. Encapsulation of solids, liquids, or suspensions of the size, 30–2 mm can be performed in this manner using this technique. The thickness of coatings can vary between 1 and 200 mm and include fats, polyethylene glycol (PEG), diglycerides, and other meltable substances. Since this is a continuous, high-speed method that can coat particles, it is highly suitable for foods. One application of the technique is to protect foods that are sensitive to or readily absorb moisture, such as aspartame, vitamins, or methionine (Soper and Thomas, 1998).

5.9 Cocrystallization

Cocrystallization is an encapsulation process utilizing sucrose for entrapping the core materials. In this method, supersaturated sucrose syrup, maintained at a temperature high enough to prevent crystallization, is added to predetermine the amount of core material with vigorous mechanical agitation, thus providing nucleation for the sucrose/ingredient mixture to crystallize. As the syrup reaches the temperature at which transformation and crystallization begin, a substantial amount of heat is emitted. Agitation is critical and is required continuously during the process to facilitate and extend transformation/crystallization until the agglomerates are discharged from the vessel. The encapsulated products are then dried to the desired moisture, if necessary, and screened to a uniform size. Studies have highlighted the importance of controlling the rates of nucleation and crystallization as well as the thermal balance during various phases (Rizzuto et al., 1984).

5.10 Inclusion Complexation

Molecular inclusion is another technique that operates at a molecular level generally utilizing α-cyclodextrin as the encapsulating medium. Inclusion complexes can be formed between flavor compounds and some food components like the interactions with starch and CyDs. α-Cyclodextrin is a cyclic derivative of starch made up of seven glucopyranose units. It is prepared from partially hydrolyzed starch (maltodextrin) by an enzymatic process. Studies have demonstrated that amylose rich starch forms helical structures that can complex small molecules. Some early work by Wyler and Solms (1981) demonstrated complexation model of volatiles (limonene, menthone, and decanal) with gelatinized potato starch (Reineccius, 2009). The process of complexation is time dependent ranging from 1 min (decanal) to several days (limonene), depending upon the compound. The amount complexed also depends upon temperature; less gets complexed at higher temperatures. A subsequent study by Wyler and Solms (1982) demonstrated that the same volatiles on encapsulation gets stable to evaporative losses and oxidation when included in the starch complex. Saldarini and Doerig (1981) were awarded a patent on flavor encapsulation in starch hydrolysates (10–13 DE). They found that they could produce a powder containing 4.6% acetaldehyde in this manner which retained 4.1% after 1 week storage at 40°C. There has been continued interest in starch–flavor complexation and several reports on interactions of starch between several classes of compounds, effect of temperatures, starch treatment, starch source, and so forth (Delarue and Giampaoli, 2000; Rutschmann and Solms, 1990; Arvisenet et al., 2002; Conde-Petit et al., 2006; Tapanapunnitikul et al., 2008). Wulff et al. (2005) optimized these interactions and proposed that segments of 8–16 anhydroglucose units were involved in binding. Also, the association constant of a given volatile compound was found to be strongly dependent on origin and chain length of amylose.

CyDs are well known for their ability to form inclusion complexes. The cyclodextrins are a series of cyclic oligosaccharides that are produced by enzymatic reaction of starch employing cyclodextrin transglycosylase. The advantages of the molecules are that they are nontoxic ingredients and are not absorbed in the upper gastrointestinal tract. They are safe and are completely metabolized by the colon microflora. In aqueous solution, each CyD forms a thick-walled bucket with a hydrophobic cavity and the hydrophilic exterior forming an inclusion complex, entrapping the whole, or part, of a guest molecule inside its cavity. The guest molecules, which are apolar, can be entrapped into the apolar internal cavity through a hydrophobic interaction (Pagington, 1986a,b; Desai and Park, 2005) generally on an equimolar molar basis having one mole of CyD (host) will include 1 mol of a guest molecule. Entrapment of the molecule inside the cavity is through weak chemical forces, like van der Waals forces, dipole-dipole interactions, and hydrogen bonding. The cavity size varies accommodating a range of sizes of guest molecule (Hedges and McBride, 1999) that form an equilibrium between free and complexed guest molecules. The equilibrium constant is crucial factor and it depends on the nature of the CyD, guest molecule, food composition, as well as on physical factors like temperature and moisture level. Reports citing dry CyD inclusion complexes being stable for periods of up to 10 years at room temperature (Szente et al., 1988; Qi and Hedges, 1995) are present. The presence of water or high temperature is required to liberate guest molecules once the complex has been formed. Generally the dry microcrystalline cyclodextrin complexes appear wettable, nearly odorless, not hygroscopic powders. The crystalline nature, flowing properties and other mechanical properties of flavor inclusion complexes depend on the condition of the complexation procedure (in particular on the drying processes of the wet complexes).

The flavor/β-cyclodextrin complexes, prepared by cocrystallization, kneading, and by suspension technology show remarkable resistance toward moisture sorption and clumping upon high humidity storage. The aroma and flavor load of these complexes varies in most cases between 6 and 15% w/w, more often is in the range of 8–10% (Szejtli, 1982; Szejtli et al., 1979; Szente et al., 1988; Szente and Szejtli, 1988, 2004). Encapsulation of flavor molecule in the CyDs could result in flavor modification (such as by masking off-notes), flavor stabilization and solubilization. Goubet et al. (2001) addressed a very interesting topic of competitive binding by CyDs. CyDs and some of their derivatives have also been employed as processing aids, for example, for removal of cholesterol from eggs or excess bitterness (naringenin, limonene) from citrus juices (Hedges and McBride, 1999; Qi and Hedges, 1995). The application of CD-assisted molecular encapsulation in foods offers numerous advantages (Szente et al., 1988). A few of them are listed in Table 6.4. It ensures protection of active ingredients against oxidation, light-induced reactions, heat-promoted decomposition, and supports the elimination (or reduction) of undesired tastes/odors. It prevents loss by volatility, sublimation microbiological contaminations, fibers/other undesired components, hygroscopicity, and so forth. They became accepted during the 1980s as a common ingredient for food manufacturers; their production and consumption have been steady over the past decade.

According to Mintel's Global New Products Database (GNPD), since 2001 more than 265 food products having CyDs have been listed, with a more than 30% share of beverages (Cabral Marques, 2010). Molecular encapsulation on CyDs derivatives

Table 6.4 CD-Assisted Molecular Encapsulation in Foods and Approved in Various Countries

Country or Organization	α-CyD	β-CyD	γ-CyD
Canada	Filed for novel food status, Jul. 2006		
EU	Novel food, approved 2008	Carrier for food additives (<1 g/kg)	Novel food, filed Jan. 2010
FSANZ	Novel food, Jan. 2004		Novel food, 2003
Japan	Natural product	Natural product	Natural product
Korea	Approved for dietary supplement	Approved for dietary supplement	Approved for dietary supplement
Mercosur states	Food approved		
Mexico	Follow FDA approvals with an import license	Follow FDA approvals with an import license	Follow FDA approvals with an import license
Philippines	Food approved	—	Food approved
Taiwan	Approved for dietary supplement	—	—
Thailand		—	Approved for dietary supplement
USA	GRAS[a], Jan. 2004	GRAS[b], Oct. 2001	GRAS[a], Sep. 2000
WHO/FAO	ADI = not specified Jun. 2001 and Jun. 2004	ADI = 5 mg/kg per day, Jan. 1995	ADI = not specified 1999 and 2000

ADI, acceptable daily intake; EU, European Union; FDA, US Food and Drug Administration; FSANZ, Food Standards Australia New Zealand; GRAS, generally regarded as safe; WHO/FAO, World Health Organization/Food and Agriculture Organization of the United Nations; The Mercosur states are Argentina, Brazil, Paraguay, Uruguay, and Venezuela.
[a] GRAS in a wide range of intended use in food.
[b] GRAS as a flavor protectant.

(alkylated and CyD polymers) are preferable to design better carriers. To further improve the desirable qualities of CyD in food encapsulation, it has been variously chemically modified. Methylated, ethylated, acetylated, and their respective hydroxyl and sulfo forms are being investigated. Polymeric β-CyDs have been employed in the rational design of water-soluble or insoluble carrier systems.

Pagington (1986a,b) listed several methods of preparing flavor/CyD inclusion complexes, including stirring or shaking a solution of CyD with the guest and filtering off the precipitated complex, blending solid CyD with the guest in a mixer and drying, and passing the vapor of a guest flavor through a CyD solution. Qi and Hedges (1995) provided experimental details of a coprecipitation method deemed most suitable for laboratory evaluation. However, large-scale production is generally accomplished by the paste method (which may be called a slurry method) since less water must subsequently be removed during drying.

5.11 Coacervation

Coacervation is commonly called phase separation. In this technique the core material is completely entrapped by the matrix. This technique involves the precipitation or separation of a colloidal phase from an aqueous phase (Dziezak, 1988). Coacervation technique involves the separation of a liquid phase of coating material from a polymeric solution, coating of that phase as a uniform layer around suspended core particles, followed by coating solidification. Commonly, the batch-type coacervation processes consist of three steps with continuous agitation. (Pagington, 1986b; Kirby and Needs, 1996; Kirby et al., 1991). First step includes formation of a three-immiscible chemical phase, followed by deposition of the coating on the particle and finally the solidification of the coating material to form the final product. Since the coacervation microencapsulation methodology involves chemical interactions between different molecules, a lot of research has been done to evaluate various coating materials suitable for various purposes. Coating systems—such as gliadin, heparin/gelatin, carrageenan, chitosan, soy protein, polyvinyl alcohol, gelatin/carboxymethylcellulose, β-lactoglobulin/gum acacia, and guar gum/dextran—have been widely studied for this purpose (Gouin, 2004). The most studied and well-understood coating system among them is the gelatin/gum acacia system. However, in recent years, modified coacervation processes have also been developed that can overcome some of the problems encountered during a typical gelatin/gum acacia complex coacervation process, especially when dealing with encapsulation of heat-sensitive food ingredients such as volatile flavor oils (Shahidi and Han, 1993; Ijichi et al., 1997; Soper and Thomas, 1998). Another type of method in coacervation encapsulation, the complex coacervation, is used in the food industry. Here, two hydrocolloids (one polycationic and the other polyanionic) interact through ionic bonding, to form a complex on the surface of hydrophobic droplets (flavoring), thereby serving as the capsule wall. This process has changed very little from the time it was patented by Green and Schleicher (Reineccius, 2009). The process starts by dissolving one of the hydrocolloids in water and adding flavoring (hydrophobic liquid) onto it with continuous mixing, to form an emulsion. The second hydrocolloid is then added as a solution, and pH is adjusted to a point where the two hydrocolloids bear a net opposite charge for interaction. The system is diluted with water so that when it is cooled to gel, the entire mass does not form a gel. At this time, the two hydrocolloids interact through ionic bonding to form the final wall. Finally, the capsules are harvested by filtering and freeze-dried or mixed with a drying agent such as silicates (Reineccius, 2009). Since, in the earlier stages, the particle is the template for formation of the capsule wall, its size is the primary determinant of the final capsule size. Literature cites their size to be extremely small (in the nanometer range); theoretically, it is difficult to make food-grade coacervates of less than 100 μm without rigorous agglomeration. Further, to make the capsule more durable in handling and insoluble in application, it can be chemically cross-linked to harden the wall structure. Cross-linking may be done with glutaraldehyde, transglutaminase (enzyme), or other cross-linking chemicals. All other major encapsulation systems are soluble (or slowly soluble) in water and, therefore their flavoring is released quickly on hydration. The slow-release property of coacervates by diffusion through the insoluble capsule wall on hydration, makes it an attractive process where slow release is required, for example, in thermally processed foods (Reineccius, 2009). Due to these qualities coacervation has found application in food industry.

6 Edible Films Formed in Encapsulation

6.1 Materials

Just as the encapsulation process is chosen on the basis of several factors, so are the materials used in this process. Materials are chosen on the basis of compatibility with a process, cost, flavor type, stability during storage, legal and religious constraints, and functionality in the final application. Depending on the market demand, the flavoring used in an all-natural food should be of a natural wall material such as gum acacia or pectin or a nonemulsifying wall material such as maltodextrin for water soluble flavoring. However, if the flavoring is insoluble, emulsifying food polymer such as modified starch, gum acacia, or protein is employed. Generally, various considerations are to be taken into account in selecting the raw material for encapsulation. The most common raw material used in flavor encapsulation is water-soluble carbohydrates that provide different functions in different encapsulation processes. The most commonly used carbohydrates are listed in Table 6.5.

Table 6.5 Food Polymers Used for Encapsulation

S. No.	Encapsulation Material	Compound Class	Primary Function	Encapsulation Process
1.	Chemically modified starch	Polysaccharide	Emulsification (film former)	Spray drying
2.	Corn syrup solids	Starch hydrolysate	Bulking agent (oxygen barrier)	Spray drying
3.	Carrageenan, alginates, pectins, and so forth	Polysaccharide	Polycation polymer	Coacervation
4.	Chitosan	Polysaccharide	Polyanionic polymer	Coacervation and spray drying
5.	Cyclodextrins	Oligosaccharide	Host molecule	Inclusion complex formation
6.	Gelatin	Hydrolyzed protein	Polycation polymer	Coacervation
7.	Glucose, sucrose, sugar alcohols	Mono- and disaccharides and alcohols	Filler (oxygen barrier)	Spray drying and extrusion
8.	Gum acacia	Polysaccharide	Emulsification (film former) polyanion polymer	Spray drying and coacervation
9.	Maltodextrin	Starch hydrolysate	Bulking agent	Spray drying and extrusion
10.	Whey proteins, caseinates	Protein	Polycation polymer and emulsification	Coacervation and spray drying

6.1.1 Gum Acacia

Gum acacia is the material traditionally used for the encapsulation of food flavors. It is derived from the gum of the acacia tree, which is grown in the semidesert region of North Central Africa. While there are over 1350 species of the acacia tree, only a few species are used for gum manufacture (primarily *Acacia senegal* and *Acacia seyal*). Gum acacia is a polymer made up primarily of D-glucuronic acid, L-rhamnose, D-galactose, and L-arabinose having a small amount of protein (from traces to >2%, depending on species). The protein is believed to be primarily responsible for its emulsification properties. The tree is unique among the plant gums since it exhibits low viscosity at relatively high solids (c. 300 cps at 30–35% solids), and is a very good emulsifier (Reineccius et al., 2003a). *A. senegal* is produced by the tree in response to artificial injury which are gathered from under the trees, dried on the tree before harvesting by hand, sorting, grading, and distributing to processors. Processors remove foreign materials by grinding the gum, solubilizing it in water, filtering, or centrifugation, pasteurization followed by spray drying. While gum may be sold in the crude form, conditions of harvesting and storage result in contamination by microbes and foreign materials, so it must be cleaned prior to use as a food ingredient. Production of gum from the tree is a costly process due to the amount of work put in to making the gum more functional. Researchers have searched for the alternatives by looking into of this raw material or improving the performance of lesser quality gums so they may be used in place of the expensive *A. senegal.* A series of articles from the Phillips Hydrocolloids Research Centre (United Kingdom) has described a method to age gums to increase their emulsification performance (Reineccius et al., 2003a). According to them aged gums or gums from older tress offer better performance, and they propose in these articles and related patent that this can be done artificially to all gums. However, they did not evaluate their products for an encapsulation application (Reineccius et al., 2003a). Further, chemical modifications/additions like addition of an octenylsuccinate anhydride to the gum have been proposed to offer better emulsification and good encapsulation performance compared to the original gum (Reineccius et al., 2003a).

6.1.2 Modified Food Starch

Starches are not natural emulsifier however, it make a contribution to emulsion stability, by affecting its viscosity. The emulsification properties in encapsulation matrices are required for food application, as in beverage flavor emulsions or cloud emulsions. This encouraged chemical modifications in the starch as in chemical addition of octenylsuccinate to partially hydrolyzed starch. Octenyl succinate-derivatized starches are excellent emulsifiers, and display low viscosity at high solids (40–50% solids) and spray dry exceptionally well. Octenyl succinate at a 0.02 degree of substitution on the starch polymer may be used to attain the desirable properties in starch and still be approved for food use. The area of concern however, is that the octenyl succinate is not considered to be a natural product, and this may affect claims for food labels (eg, one cannot put a "100% natural" label on the product). The maltodextrins and corn syrup solids, produced by the treatment of starch using acid, enzyme or acid/enzyme, do not have emulsification properties. If product has a dextrose equivalent (DE) of less than 20, it is labeled as maltodextrin otherwise it is corn syrup solids. Emulsions are essential during encapsulation of lipophilic flavorings via spray drying, but are not an issue in encapsulation of hydrophilic flavorings. In fact, they are often the wall material of choice when spray drying water soluble flavorings. However, for water insoluble flavorings, they may be used at sufficiently high solids levels to generate high viscosities for preparing temporary emulsions for spray drying, resulting instability to the finished product. Also, their use in encapsulation of water insoluble flavorings can be dangerous as it might end up feeding pure flavoring into a spray dryer, an explosion may occur in the dryer. For reducing the final cost of the product, generally, a secondary emulsifier is added for the sole purpose of emulsification, while nonemulsifying cheap sources like maltodextrins, corn syrup can make up a substantial proportion of wall material and perform the needed function as in encapsulation via extrusion, emulsification, basically. Literature has many references to cite blends of either gum acacia or modified starch with maltodextrins or corn syrup solids to take advantage of emulsification properties of gum acacia or modified starch, and then accomplishing a cost reduction through use of the maltodextrins/corn syrup solids. Also, a point to be taken into consideration is that higher DE products do not dry well (spray drying) and are prone to caking on storage. Thus, the selection of DE product, as well as substitution level, must be chosen carefully (Reineccius et al., 2003a) while selecting the matrix.

6.1.3 Mono- and Disaccharides

Simple sugars (glucose, sucrose, lactose and maltose, and their hydrogenated products) also find use in encapsulation of flavorings (extrusion and spray-drying processes). They are used to some extent as fillers (starch hydrolysates) and, more important, are used to confer increased oxidative stability to encapsulated flavorings. They might be used as 10–40% of wall solids; higher levels result in poor drying properties and caking during storage. Other polysaccharides like simple sugars, pectin, carrageenan, alginate, and chitosan are used in flavor encapsulation via coacervation. Pectin, carrageenan, and alginate find uses as anionic hydrocolloids (in place of gum acacia), while chitosan serves as the cationic hydrocolloid (in place of protein), however, their use is limited due to their high viscosity which make them difficult to atomize in a spray-drying operation at solids level required for economics in operation.

The flavor/aromatizing substances generally are rich in several components. Therefore, it is imperative that all the components must be integrated into the complex without shifting its composition. For example, the most hydrophobic guest molecule will be complexed first when several potential guest molecules are present in solution. Many essential oils, natural and synthetic coffee

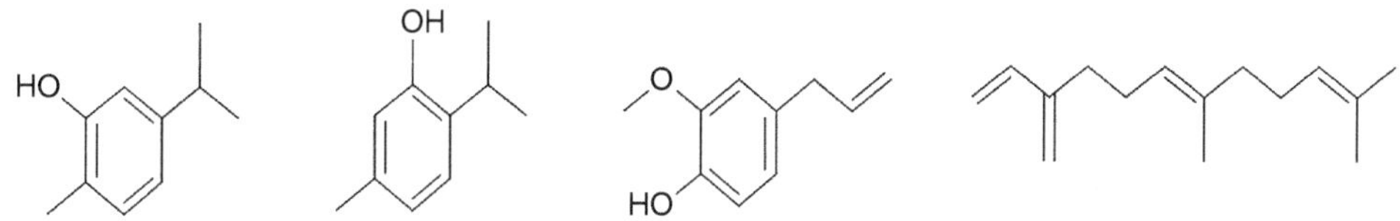

Figure 6.4. Structures of some components of essential oils of vegetable origin.

flavors are stabilized with CyDs to avoid its loss during storage or exposure to light or oxygen. These encapsules, in proximity with water, immediately released the complex-bound flavor substances (Szente and Szejtli, 1988, 2004; Jackson and Lee, 1991a; Szente et al., 1988; Mourtzinos et al., 2008; Pothakamury and Barbosa-Canovas, 1995). These inclusion complexes last for a longer period and are highly stable at the high temperatures encountered during industrial food processing. Encapsulation can also overcome sensitivity due to radiation, as seen in citral that cyclizes under UV irradiation, like other cyclic mono-terpenes, resulting in major taste modifications. This can be prevented by complexation with β-CyD. Astray et al. (2009) have shown that encapsulation protective effect is more marked on solid-state materials than in aqueous solutions, where a portion of the guest species could leak from the complex. The suitability of starches for retaining volatile meat flavor encapsulation was proved effectively by Jeon et al. (2003). β-Farnesene, the most volatile component of chamomile flowers, could be partially protected from volatilization during freeze-drying (Fig. 6.4). It has been experienced that during storage, the β-CyD inclusion complex shields volatile substances from evaporation more efficiently, whereas microcapsules made up of modified starches, as wall material, were more heat tolerant (Tsotsas and Mujumdar, 2011). Inclusion of carvone, a highly volatile compound used as sprouting inhibitor, into β-CyD was found to reduce volatility and improves stability. This was applied for its enhanced and efficient use during potato storage (Cabral Marques, 2010). The vegetable-based essential oils are highly sensitive to oxidation, decomposition, and evaporation, when exposed to the air, light, or heat. The necessary functions of aroma materials are to provide a pleasing odor, to camouflage the original unpleasant smell of the product, and to give the product an identity. However, since fragrance materials are hydrophobic or insoluble compounds and usually exist in a liquid state, the perfuming process is not straightforward (Cabral Marques, 2010). However, they can be stabilized by β-CyD inclusion complexes that significantly reduce the volatility, oxidation, and heat-decomposition (Locci et al., 2004) Due to reduced oxidation property, CyDs have been effectively used as antibrowning agents in different fruit juices as browning inhibitors. The color preservation of fruit juice during processing and storage was proved by Lopez-Nicolas et al. (2007). CyD complexation has been widely used in the food industry for components stability, preservation of certain flavors, colors, or vitamins associated with certain ingredients during processing or on storage. The controlled release of 1-methylcyclopropene (1-MCP), gaseous inhibitor of ethylene perception, is being used widely for apples and ornamental products, and can further be extended to postharvest packaging and storage of perishable horticultural products.

CyD-encapsulated fragrance materials have higher solubility and reduced evaporation. The interaction of the guest molecule with CyDs creates a higher energy barrier to overcome volatilization, thus producing long-lasting fragrances (Cabral Marques, 2010). Numanoğlu et al. experimented extensively to demonstrate possibility of increased stability, water solubility, and decreased volatility, and of fragrance materials in linalool and benzyl acetate. He was able to convert these substances from liquid to powder form by preparing their inclusion complexes with β-CyDs and 2-HPβCyD and showed their controlled release. Also, the stability of these compounds in gel formulations can be increased by complex formation.

6.2 Film Performance

As stated earlier, the wall matrix (film) used for flavor encapsulation must meet several criteria, some of which are: it must form and stabilize an emulsion, retain flavors during encapsulation, protect flavor during storage from evaporation and reaction, and then release flavor to the final food product on consumption. Based on these criteria, various edible films used in the encapsulation of flavorings are outlined later.

6.3 Emulsification Properties

The importance of imparting emulsifying properties depends upon the type of flavoring encapsulated, the encapsulation process, and the final application of the encapsulated flavor. As noted earlier, hydrophilic flavoring do not require emulsions. Though a flavoring labeled as water soluble is generally a mix of alcohol or propylene glycol at 0.1% or slightly higher usage levels. If flavors, or any part thereof, are insoluble in the system being used for encapsulation, then an emulsifying matrix is required. Emulsification is required to minimize flavor losses during the encapsulation process (spray drying and extrusion processes). There is ample data in the literature showing that retention of water insoluble flavorings are substantially improved if a good-quality emulsion is prepared and used during the encapsulation process (Trubiano and Lacourse, 1988; Reineccius, 1998). The work of Soottitantawat et al. (2004) illustrated the benefit of a good-quality emulsion for the retention of water insoluble volatiles (limonene) during spray drying. One should bear this in mind that small mean particle sizes for more water-soluble volatiles does not necessarily improves the retention. Products such as dry savory mixes, baked goods, or confectionery products do not require the edible film to provide significant emulsification properties. However, a substantial portion of dry flavorings is used in dry beverage mixes. Since, maltodextrins and corn syrup solids impart no emulsion stability, other than that due to their high viscosity, so an encapsulated flavor is based on maltodextrin or corn syrup solids cannot be used for a dry beverage flavor application, unless a secondary emulsifier is incorporated. Although an emulsifying agent such as a modified starch or gum acacia may be used in combination with the maltodextrin or corn syrup solid flavor carrier, secondary emulsifiers are otherwise, not commonly used with spray-dried flavors. They are however, essential for manufacture and stability of extruded flavorings. Both gum acacia and the modified food starches are excellent emulsifiers. An advantage of using modified food starches for flavor encapsulation is their ability to form a stable flavor emulsion.

6.4 Flavor Retention

For the retention of the flavor, ability of an edible film to trap or hold onto flavor compounds, during the drying process is critical as loss of flavor, strength, and potentially will lead to imbalanced in character. During the encapsulation process, lighter and more volatile constituents gets preferentially lost resulting in dry flavor, lacking in the very volatile light fresh notes. Another area of concern during encapsulation is lost volatiles upon exit, either in spray-drying process through drier, or into the environment, which must be removed by a costly scrubbing process to protect the environment. Therefore, encapsulation materials that offer poor retention result in increased processing costs and decreased product quality. As mentioned earlier, modified food starches are excellent emulsifiers, and emulsion quality has a strong influence on flavor retention during spray drying (Risch, 1995; Baranauskiene et al., 2007). Thus, although both gum acacia and modified food starches yield good emulsions, their best flavor retention property is partially attributed to higher infeed solids levels, which also improves flavor retention (Reineccius et al., 2003a). Maltodextrins, corn syrup solids, or simple sugars (and their alcohols) typically can be used at high infeed solids levels, but their poor emulsification properties result in poor flavor retention during drying. Since solids content of the infeed slurry is *the* major determinant of flavor retention during drying, one would get a totally different ranking of carriers and retention if the experimental design involved use of carriers at constant viscosity. Dronen (2004) evaluated the retention of a volatile model flavor system (water soluble components) during spray drying prepared on a constant viscosity basis.

6.5 Flavor Release

Not only is flavor retention important, but flavor release from the encapsulated material is also essential. Preferably, all encapsulated flavorings should offer controlled release; however, sometimes a slow or delayed release is desirable. For example, requirement could be no release of an encapsulated flavoring during early stages of thermal processing, so a delayed release may result in less flavor loss, since flavor would be protected from heat until late in processing. In the case of an encapsulated flavoring for a dry beverage mix, one desires a rapid release on reconstitution. Thus, the desired flavor release will be dependent upon the application. Currently, these controlled release properties can be attained directly through the use of coacervation, extrusion, and inclusion complex formation. Since spray-dried particles are water soluble, controlled release properties may be imparted to them through application of secondary coatings, for example, coating with a fat or shellac. Secondary coatings (eg, fats, oils, and shellacs) are costly and problematic to apply, and therefore it is desirable to accomplish controlled release by choosing the appropriate encapsulation technique.

As mentioned earlier, coacervation, extrusion, and CyDs all may be used to impart controlled release properties to a flavoring. Coacervation does this through use of capsule materials that may be insoluble in the final application. Further, solubility and insolubility is also governed by cross linking. Cross-linked capsules will be insoluble irrespective of the food system, and noncross-linked capsules may be “soluble” or insoluble depending upon the sample environment. It was observed that glutaraldehyde cross-linked batches (small and large capsules) gave the lowest release intensity. If capsules are insoluble and retain their integrity in a food application, release is governed by diffusion through the capsule wall. Yeo et al. (2005) had detailed on the controlled release of flavor oil prepared by complex coacervation. They reported on how formulation, freeze–thaw cycles, and ionic strength can affect release of encapsulated oil. Generally, in essence, dissolution of the capsule results in rapid release, while systems in which the wall matrix remained intact gave slower, diffusion-controlled release. Weinbreck et al. (2004) on their work on the encapsulation of sunflower oil, lemon oil, and orange oil in a gum acacia–whey protein coacervate, observed that larger capsules gave strongest flavor release (while eating).

Mourtzinos et al. (2008) demonstrated a similar disparity in release of geraniol and thymol from β-CyD: geraniol was rapidly released (nearly 100%), while thymol was only slowly and incompletely released (30% into aqueous solution).

7 Retention of Volatiles

It has been found that during spray-drying encapsulation, loss of some volatiles including flavors is unavoidable. During the processing, many physical parameters including molecular weight, size of the core material and the vapor pressure of flavor compounds may also affect the retention process (Reineccius, 1996, 1998, 1999; Re, 1998; Goubet et al., 1998). Molecular weight can be correlated to molecular size, which influences the process of diffusion (Goubet et al., 1998; Leahy et al., 1983). Large molecular size generally results in slower diffusion rate, thereby molecules take longer time to reach the atomized droplet surface during drying, resulting in increased retention. Secondly, entrapping of large molecule inside makes it impervious, promoting the retention. The same trend was noticed by Voilley (1995) in a mixture of 16 aroma compounds encapsulated in glucose, maltose, or corn syrup solids. He found that the retention rate of isoamyl butyrate (MW = 158) was higher than that of ethyl butyrate (MW = 116) or ethyl propionate (MW = 102) in all tested wall materials, except in maltose and corn syrup solid with dextrose equivalent 28.5.

7.1 Relative Volatility

The ability of a pure compound to reach the gaseous phase is defined as volatility which can be measured by the vapor pressure (Goubet et al., 1998; Reineccius and Mei, 2004). Relative volatility of a compound is calculated with respect to water (Bhandari, 2005). Higher the relative volatility, the lower the retention. The flavor retention of an encapsulated compound will be highly influenced by its polarity and permeability rate from the core matrix. Bangs and Reineccius (1988) have shown that retention of octenol, octanon, and octanal were related to their relative volatility when they were encapsulated with maltodextrin and spray dried. Also, more polar compounds show less retention due to greater solubility of polar compounds in water (Re, 1998; Rosenberg et al., 1990; Voilley, 1995). As the water solubility of the volatile increases, the volatile losses increase, due to the permeability of water from the selective membrane, even at late stages of the drying process. Goubet et al. (1998) revealed that the retention of aroma compounds with various functional groups is in the order of acids $<$ aldehydes $<$ esters $\leq$ ketones $\leq$ alcohols with acids having the minimum retention. Therefore, it can be concluded that retention of volatiles depends on their molecular weight, relative volatility, polarity, and type of compound. These different parameters act on the capacity of the volatile to diffuse through the droplet surface, and on its ability to form small pools. The final result is that small, very volatile, and water-soluble flavors are lost to a greater extent than the larger, less volatile, and water-insoluble flavorings (Reineccius, 1998, 1999, 2000, 2007, 2009). Besides the aforesaid factors in spray-drying microencapsulation, interactions between the volatiles and the wall material could be a possibility (Re, 1998). This interaction could be physical or physicochemical, resulting in the formation of insoluble complexes and molecular complexation of the wall material with the volatile through hydrogen bonds. These interactions may affect the retention indirectly by stabilizes the emulsion through the formation of the interfacial film.

8 Storage Stability

The most common problem encountered during storage of flavors is worsening due to oxidation. Flavor encapsulation films diverge greatly in their capacity to protect a flavoring from oxygen. For example, dextrose equivalent in maltodextrins coating provides varying protection. The higher the dextrose equivalent, the better the protection against oxidation (Arvisenet et al., 2002). However, higher dextrose equivalent materials are poor in drying, yield poor flavor retention, and are very hygroscopic. These observations led to blending of emulsifying wall materials with higher dextrose equivalent maltodextrins, corn syrup solids, or simple sugars. Thus, sufficient emulsifying wall materials can be added to impart necessary emulsifying capacity, and corn syrup solids or simple sugars may be mixed to impart oxidative stability; fortunately, this also lowers product costs. While modified food starches produce an encapsulated flavoring that has excellent flavor retention and emulsion stability, they have traditionally provided very poor protection against oxidation, the starch hydrolysates being inexpensive may be added for desired quality (Reineccius, 2000, 2009; Reineccius et al., 2003a).

9 Conclusions

Currently, there is a growing awareness by consumers regarding what they eat and what benefits certain ingredients have in maintaining good health (Poshadri and Kuna, 2010). With time, ingredients in food systems start oxidizing and degrading slowly, losing their activity and becoming unsafe. Long-time storage also results in interingredients reactions, which may limit their bioavailability, or may change the physical properties (color and taste) of the food item. Most of these limitations can be overcome by microencapsulation. Encapsulation offers the possibility of using volatiles without encountering the aforesaid problems. An easy route to developing microencapsulated material is to modify existing methods developed for pharmacy, foods, agriculture, or cosmetics, keeping cost-competitiveness and the market in mind. Currently, the focus on the microencapsulation of food, oils, and flavors is on improving encapsulation efficiency and extending the shelf life of the products with an aim to produce high-quality encapsulated powders. For this, the factors affecting the efficiency of encapsulation, like the properties of the wall and core materials, the emulsion characteristics and drying parameters are being optimized for better results. The key advantage of smaller emulsion sizes is a better retention of volatiles in the spray-dried powder. As less flavor is lost during drying, therefore less raw material is required for the finished product to achieve the same flavor level. A second advantage is that smaller emulsion sizes also yield dried powders which have less extractable surface oil. Finer emulsions have the advantage of being more stable.

References

Ajisaka, N., Hara, K., Mikuni, K., Hara, K., Hashimoto, H., 2000. Effects of branched cyclodextrins on the solubility and stability of terpenes. Biosci. Biotechnol. Biochem. 64, 731–734.

Arvisenet, G., Voilley, A., Cayot, N., 2002. Retention of aroma compounds in starch matrices: competitions between aroma compounds toward amylose and amylopectin. J. Agric. Food Chem. 50, 7345–7349.

Astray, G., Gonzalez-Barreiro, C., Mejuto, J.C., Rial-Otero, R., Simal-Gandara, J., 2009. A review on the use of cyclodextrins in foods. Food Hydrocolloid. 23, 1631–1640.

Bagheri, M., Shateri, S., 2012. Thermosensitive nanosized micelles from cholesteryl-modified hydroxypropyl cellulose as a novel carrier of hydrophobic drugs. Iran. Polym. J. 21, 365–373.

Bangs, W.E., Reineccius, G.A., 1988. Corn starch derivatives: possible wall materials for spray-dried flavor manufacture. ACS Symp. Ser. 370, 12–28.

Baranauskiene, R., Bylaite, E., Zukauskaite, J., Venskutonis, R.P., 2007. Flavor retention of peppermint (*Mentha piperita* L.) essential oil spray-dried in modified starches during encapsulation and storage. J. Agric. Food Chem. 55, 3027–3036.

Benczedi, D., Bouquerand, P.-E., 2001. Process for the preparation of granules for the controlled release of volatile compounds. WO2001017372A1, 2001.
Bhandari, B., 2005. Spray Drying: An Encapsulation Technique for Food Flavors. Science Publishers, Enfield, CT.
Bhandari, B.R., D'Arcy, B.R., Bich, L.L.T., 1998. Lemon oil to β-cyclodextrin ratio effect on the inclusion efficiency of β-cyclodextrin and the retention of oil volatiles in the complex. J. Agric. Food Chem. 46, 1494–1499.
Cabral Marques, H.M., 2010. A review on cyclodextrin encapsulation of essential oils and volatiles. Flavour Frag. J. 25, 313–326.
Carolina, B.C., Carolina, S., Zamora, M.C., Jorge, C., 2007. Glass transition temperatures and some physical and sensory changes in stored spray-dried encapsulated flavors. LWT Food Sci. Technol. 40, 1792–1797.
Clark McIver, R., Leresche, J.-P., Neffah, B., 2004. Continuous process for the incorporation of a flavor or fragrance ingredient or composition into a carbohydrate matrix. US20040185159A1, 2004.
Conde-Petit, B., Escher, F., Nuessli, J., 2006. Structural features of starch-flavor complexation in food model systems. Trends Food Sci. Technol. 17, 227–235.
Delarue, J., Giampaoli, P., 2000. Study of interaction phenomena between aroma compounds and carbohydrate matrixes by inverse gas chromatography. J. Agric. Food Chem. 48, 2372–2375.
Desai, K.G.H., Park, H.J., 2005. Recent developments in microencapsulation of food ingredients. Dry. Technol. 23, 1361–1394.
Dinsmore, A.D., Hsu, M.F., Nikolaides, M.G., Marquez, M., Bausch, A.R., Weitz, D.A., 2002. Colloidosomes: selectively permeable capsules composed of colloidal particles. Science 298, 1006–1009.
Douzals, J.P., Marechal, P.A., Coquille, J.C., Gervais, P., 1996. Comparative study of thermal and high-pressure treatments upon wheat starch suspensions. Prog. Biotechnol. 13, 433–438.
Dronen, D., 2004. Characterization of Volatile Loss From Dry Food Polymer Materials. University of Minnesota, Minneapolis.
Druaux, C., Voilley, A., 1997. Effect of food composition and microstructure on volatile flavor release. Trends Food Sci. Technol. 8, 364–368.
Duchene, D., Vaution, C., Glomot, F., 1989. Cyclodextrins: Their Value in Pharmaceutical Technology. Horwood. *Drug Dev. Ind. Pharm.* **1986**, 12(11–13), 2193–215.
Dziezak, J.D., 1988. Microencapsulation and encapsulated ingredients: use of microencapsulation can improve ingredient functionality. Food Technol. 42, 136–140, 140–143, 146–148, 151.
Faucci, M.T., Melani, F., Mura, P., 2002. Computer-aided molecular modeling techniques for predicting the stability of drug: cyclodextrin inclusion complexes in aqueous solutions. Chem. Phys. Lett. 358, 383–390.
Gaonkar, A., Vasisht, N., Khare, A., Sorel, R., 2014. Microencapsulation in the Food Industry. Academic Press, Elsevier, San Diego, CA.
Gibbs, B.F., Kermasha, S., Alli, I., Mulligan, C.N., 1999. Encapsulation in the food industry: a review. Int. J. Food Sci. Nutr. 50, 213–224.
Given, P.S., 2009. Encapsulation of flavors in emulsions for beverages. Curr. Opin. Colloid Interface Sci. 14, 43–47.
Goubet, I., Le Quere, J.L., Voilley, A.J., 1998. Retention of aroma compounds by carbohydrates: influence of their physicochemical characteristics and of their physical state: a review. J. Agric. Food Chem. 46, 1981–1990.
Gouin, S., 2004. Microencapsulation industrial appraisal of existing technologies and trends. Trends Food Sci. Technol. 15, 330–347.
Hedges, A., Mcbride, C., 1999. Utilization of β-cyclodextrin in food. Cereal Foods World 44, 700–702, 704.

Ijichi, K., Yoshizawa, H., Uemura, Y., Hatate, Y., Kawano, Y., 1997. Multilayered gelatin/acacia microcapsules by complex coacervation method. J. Chem. Eng. Japan 30, 793–798.
Jackson, L.S., Lee, K., 1991a. Microencapsulated iron for food fortification. J. Food Sci. 56, 1047–1050.
Jackson, L.S., Lee, K., 1991b. Microencapsulation and the food industry. LWT Food Sci. Technol. 24, 289–297.
Jafari, S.M., Assadpoor, E., Bhandari, B., He, Y., 2008a. Nano-particle encapsulation of fish oil by spray drying. Food Res. Int. 41, 172–183.
Jafari, S.M., Assadpoor, E., He, Y., Bhandari, B., 2008b. Re-coalescence of emulsion droplets during high-energy emulsification. Food Hydrocoll. 22, 1191–1202.
Jeon, Y.-J., Vasanthan, T., Temelli, F., Song, B.-K., 2003. The suitability of barley and corn starches in their native and chemically modified forms for volatile meat flavor encapsulation. Food Res. Int. 36, 349–355.
Kim, H.H.Y., Baianu, I.C., 1991. Novel liposome microencapsulation techniques for food applications. Trends Food Sci. Technol. 2, 55–61.
Kim, Y.D., Morr, C.V., 1995. Evaluation of Gum Arabic and Several Food Proteins as Orange Oil Encapsulants by Dynamic Headspace Analysis, Pt. 1 ed. American Chemical Society: 1995; pp. AGFD–075.
Kirby, C.J., Needs, E.C., 1996. Microemulsions containing functional substances. GB2297759A, 1996.
Kirby, C.J., Whittle, C.J., Rigby, N., Coxon, D.T., Law, B.A., 1991. Stabilization of ascorbic acid by microencapsulation in liposomes. Int. J. Food Sci. Technol. 26, 437–449.
Kopelman, I.J., Meydav, S., Weinberg, S., 1977. Storage studies of freeze dried lemon crystals. J. Food Technol. 12, 403–410.
Kosaraju, S.L., 2005. Colon targeted delivery systems: review of polysaccharides for encapsulation and delivery. Crit. Rev. Food Sci. Nutr. 45, 251–258.
Leahy, M.M., Anandaraman, S., Bangs, W.E., Reineccius, G.A., 1983. Spray drying of food flavors. II. A comparison of encapsulating agents for the drying of artificial flavors. Perfum. Flavor 8, 49–52, 55–56.
Liu, B., Hu, J., 2005. The application of temperature-sensitive hydrogels to textiles: a review of Chinese and Japanese investigations. Fibres Textiles East. Eur. 13, 45–49.
Locci, E., Lai, S., Piras, A., Marongiu, B., Lai, A., 2004. 13C-CPMAS and 1H-NMR study of the inclusion complexes of β-cyclodextrin with carvacrol, thymol, and eugenol prepared in supercritical carbon dioxide. Chem. Biodivers. 1, 1354–1366.
Lopez-Nicolas, J.M., Perez-Lopez, A.J., Carbonell-Barrachina, A., Garcia-Carmona, F., 2007. Use of natural and modified cyclodextrins as inhibiting agents of peach juice enzymatic browning. J. Agric. Food Chem. 55, 5312–5319.
Luo, R., Venkatraman, S.S., Neu, B., 2013. Layer-by-layer polyelectrolyte-polyester hybrid microcapsules for encapsulation and delivery of hydrophobic drugs. Biomacromolecules 14, 2262–2271.
Madene, A., Jacquot, M., Scher, J., Desobry, S., 2006. Flavor encapsulation and controlled release: a review. Int. J. Food Sci. Technol. 41, 1–21.
Manojlovic, V., Rajic, N., Djonlagic, J., Obradovic, B., Nedovic, V., Bugarski, B., 2008. Application of electrostatic extrusion—flavour encapsulation and controlled release. Sensors 8, 1488–1496.
Matsuda, S., Hatano, H., Kuramoto, K., Tsutsumi, A., 2001. Fluidization of ultrafine particles with high G. J. Chem. Eng. Japan 34, 121–125.
Mourtzinos, I., Kalogeropoulos, N., Papadakis, S.E., Konstantinou, K., Karathanos, V.T., 2008. Encapsulation of nutraceutical monoterpenes in beta-cyclodextrin and modified starch. J. Food Sci. 73, S89–S94.
Pagington, J.S., 1986a. Beta-cyclodextrin. Perfum. Flavor 11, 49–52, 55–56, 58.

Pagington, J.S., 1986b. β-Cyclodextrin and its uses in the flavor industry. Developments in Food FlavorsElsevier Applied Science Publishers, London.
Paramita, V., Furuta, T., Yoshii, H., 2012. High-oil-load encapsulation of medium-chain triglycerides and D-limonene mixture in modified starch by spray drying. J. Food Sci. 77, E38–E44.
Piel, G., Dive, G., Evrard, B., Van, H.T., De, H.S.H., Delattre, L., 2001. Molecular modeling study of beta- and gamma-cyclodextrin complexes with miconazole. Eur. J. Pharm. Sci. 13, 271–279.
Popplewell, L.M., Porzio, M.A., 1998. Fat-coated encapsulation compositions and method for preparing the same. WO9818338A1, 1998.
Porzio, M., 2007. Spray drying. Perfum. Flavor 32, 34–39.
Porzio, M., 2009. Functionalized flavors: formulation and challenges. Perfum. Flavor 34, 28–30.
Porzio, M., 2012. Advances in flavor encapsulation. Food Technol. 66, 52–54, 57–58, 61, 63–64.
Porzio, M., 2013. Two challenging flavor systems: citrus oils and vanilla extracts. Perfum. Flavor 38, 24, 26–27.
Porzio, M.A., Popplewell, L.M., 1994. Encapsulation of flavoring agents. WO9423593A1, 1994
Porzio, M.A., Zasypkin, D., 2002. Encapsulation compositions comprising an encapsulate in a glassy matrix. US20020189493A1, 2002.
Poshadri, A., Kuna, A., 2010. Microencapsulation technology: a review. J. Res. Angrau 38, 86–102.
Pothakamury, U.R., Barbosa-Canovas, G.V., 1995. Fundamental aspects of controlled release in foods. Trends Food Sci. Technol. 6, 397–406.
Qi, Z.H., Hedges, A.R., 1995. Use of cyclodextrins for flavors. ACS Symp. Ser. 610, 231–243.
Re, M.I., 1998. Microencapsulation by spray drying. Dry. Technol. 16, 1195–1236.
Reineccius, G.A., 1996. Instrumental means of monitoring the flavor quality of foods. ACS Symp. Ser. 631, 241–252.
Reineccius, G.A., 1998. Kinetics of flavor formation during Maillard browning. Flavor ChemistrySpringer, New York, NY. pp. AGFD–094.
Reineccius, G.A., 1999. Kinetics of flavor formation during Maillard browning. Flavor ChemistryKluwer Academic/Plenum Publishers, New York, NY.
Reineccius, G.A., 2000. Recent developments in academic flavor research. ACS Symp. Ser. 756, 13–21.
Reineccius, G.A., 2007. Flavour-Isolation Techniques. Springer GmbH, Heidelberg, Germany.
Reineccius, G.A., 2009. Edible films and coatings for flavor encapsulation. In: Embuscado, M.E., Huber, K.C. (Eds.), Edible Films and Coatings for Food Applications. Springer Science + Business Media, Saint Paul, MN.
Reineccius, G.A., Liardon, R., Luo, Z., 2003a. The retention of aroma compounds in spray dried matrices during encapsulation and storage. Flavour Research at the Dawn of the Twenty First CenturyLavoisier, Tec & Doc, Cachan, France.
Reineccius, G.A., Mei, J.B., 2004. The Influence of Food Texture on Aroma Release from Dairy Foods. American Chemical Society. pp. AGFD-166.
Reineccius, T.A., Reineccius, G.A., Peppard, T.L., 2002. Encapsulation of flavors using cyclodextrins: comparison of flavor retention in alpha, beta, and gamma types. J. Food Sci. 67, 3271–3279.
Reineccius, T.A., Reineccius, G.A., Peppard, T.L., 2003b. Utilization of β-cyclodextrin for improved flavor retention in thermally processed foods. J. Food Sci. 69, FCT58–FCT62.
Reza Fareghi, A., Najafi Moghaddam, P., Akbar Entezami, A., Ensafi Avval, M., 2013. Modification of hydrophilic cellulose fibers by monolayer growth of polystyrene chains using ATRP. Iran. Polym. J. 22, 361–367.

Risch, S.J., 1995. Review of patents for encapsulation and controlled release of food ingredients. ACS Symp. Ser. 590, 196–203.
Risch, S.J., Reineccius, G.A. (Eds.), 1995. Encapsulation and Controlled Release of Food Ingredients. In: ACS Symposium Series, vol. 590. ACS, Washington, DC. (Developed from a Symposium sponsored by the Division of Agricultural and Food Chemistry at the 206th National Meeting of the American Chemical Society, Chicago, Illinois, August 22–27, 1994.)
Rizzuto, A.B., Chen, A.C., Veiga, M.F., 1984. Modification of the sucrose crystal structure to enhance pharmaceutical properties of excipient and drug substances. Pharm. Technol. 8, 32, 34, 36, 38–39.
Roberts, D.D., Elmore, J.S., Langley, K.R., Bakker, J., 1996. Effects of sucrose, guar gum, and carboxymethylcellulose on the release of volatile flavor compounds under dynamic conditions. J. Agric. Food Chem. 44, 1321–1326.
Rosenberg, M., Kopelman, I.J., Talmon, Y., 1990. Factors affecting retention in spray-drying microencapsulation of volatile materials. J. Agric. Food Chem. 38, 1288–1294.
Rusli, H., Gandasasmita, S., Amran, M.B., 2013. Cellulose acetate-silica fume membrane: characterization and application for separation of starch and maltose. Iran. Polym. J. 22, 335–340.
Rutschmann, M.A., Solms, J., 1990. Formation of inclusion complexes of starch with different organic compounds. IV. Ligand binding and variability in helical conformations of V amylose complexes. LWT Food Sci. Technol. 23, 84–87.
Saldarini, A.V., Doerig, R., 1981. Fixing a volatile flavoring agent in starch hydrolysate. US4285983A, 1981.
Sansukcharearnpon, A., Wanichwecharungruang, S., Leepipatpaiboon, N., Kerdcharoen, T., Arayachukeat, S., 2010. High loading fragrance encapsulation based on a polymer-blend: preparation and release behavior. Int. J. Pharm. 391, 267–273.
Shahidi, F., Han, X.Q., 1993. Encapsulation of food ingredients. Crit. Rev. Food Sci. Nutr. 33, 501–547.
Siegel, S., 2010. Flavoring substance-included cellulose. US20100129516A1, 2010.
Soottitantawat, A., Yoshii, H., Furuta, T., Ohgawara, M., Forssell, P., Partanen, R., Poutanen, K., Linko, P., 2004. Effect of water activity on the release characteristics and oxidative stability of D-limonene encapsulated by spray drying. J. Agric. Food Chem. 52, 1269–1276.
Soper, J.C., Thomas, M.T., 1998. Preparation of protein-encapsulated oil particles using enzyme-catalyzed crosslinking. EP856355A2, 1998.
Subramaniam, A., Mciver, R.C., Vlad, F.J., Benczedi, D., 2003. Spray-dried compositions prepared with active ingredients within wall-forming carbohydrates. WO2003032749A1, 2003.
Szejtli, J. (Ed.), 1982. Proceedings of the First International Symposium on Cyclodextrins, Budapest, Hungary, September 30–October 2, 1981; p. 544.
Szejtli, J., Szente, L., Banky-Elod, E., 1979. Molecular encapsulation of volatile, easily oxidizable labile flavor substances by cyclodextrins. Acta Chim. Acad. Sci. Hung. 101, 27–46.
Szente, L., Harangi, J., Szejtli, J., 1988. Long-Term Storage Stability Studies on Flavor-β-Cyclodextrin Complexes. Kluwer, Dordrecht, The Netherlands.
Szente, L., Szejtli, J., 1988. Stabilization of flavors by cyclodextrins. ACS Symp. Ser. 370, 148–157.
Szente, L., Szejtli, J., 2004. Cyclodextrins as food ingredients. Trends Food Sci. Technol. 15, 137–142.
Tapanapunnitikul, O., Chaiseri, S., Peterson, D.G., Thompson, D.B., 2008. Water solubility of flavor compounds influences formation of flavor inclusion complexes from dispersed high-amylose maize starch. J. Agric. Food Chem. 56, 220–226.

Trubiano, P.E., Lacourse, N.L., 1988. In: Risch, S.J., Reineccius, G.A. (Eds.), Flavor Encapsulation. American Chemical Society, Washington, DC.

Tsotsas, E., Mujumdar, A.S. (Eds.), 2011. Modern Drying Technology: Product Quality and Formulation, vol. 3, Wiley-VCH Verlag GmbH & Co. KGaA, Weinheim, Germany.

Tsutsumi, A., Hasegawa, H., Mineo, T., Yoshida, K., 1998. Coating Granulation by Rapid Expansion of Supercritical Fluid Solutions. Institution of Chemical Engineers, Rugby, UK.

Ubbink, J., Schoonman, A., 2003. Flavor delivery systems. In: Kirk-Othmer Encyclopedia of Chemical Technology. Wiley Online Library.

United Nations Food and Agriculture Organization/World Health Organization (FAO/WHO), Joint Expert Committee on Food Additives (JECFA), 1995. Evaluation of certain food additives and contaminants: 44th report of JECFA. WHO Technical Report Series nr 859. Geneva: World Health Organization, 54 p.

United Nations Food and Agriculture Organization/World Health Organization (FAO/WHO), Joint Expert Committee on Food Additives (JECFA), 2000. Evaluation of certain food additives and contaminants: 53rd report of JECFA. WHO Technical Report Series nr 896. Geneva: World Health Organization, 128 p.

United Nations Food and Agriculture Organization/World Health Organization (FAO/WHO), 2001. Joint Expert Committee on Food Additives, 57th Meeting; Rome; June 5–14, p. 2.

Van Den Oever, G.J., Busch, J., Van Der Linden, E., Smit, G., Zuidam, N.J., 2009. Design of foods for the optimal delivery of basic tastes. Designing Functional Foods, Woodhead Publishing Series in Food Science, Technology and Nutrition, vol. 177, Woodhead Publishing, Cambridge, UK, pp. 453–480.

Van Ruth, S.M., Roozen, J.P., 2002. Delivery of Flavors from Food Matrices. Sheffield Academic Press, Sheffield, UK.

Van Soest, J.J.G., 2007. Encapsulation of Fragrances and Flavours: A Way to Control Odour and Aroma in Consumer Products. Springer GmbH, Heidelberg, Germany.

Vedantam, V.K., Yong, T.T., 2004. Fragrance delivery method during laundry. WO2004083356A1, 2004.

Voilley, A.J., 1995. Flavor encapsulation. Influence of encapsulation media on aroma retention during drying. ACS Symp. Ser. 590, 169–179.

Weinbreck, F., Minor, M., De Kruif, C.G., 2004. Microencapsulation of oils using whey protein/gum arabic coacervates. J. Microencapsul. 21, 667–679.

Wen Lou and Popplewell, L., 2003. Hydroxypropyl cellulose encapsulation material IN USPTO.(Ed.). US 20030077378 A1, 2003.

Wulff, G., Avgenaki, G., Guzmann, M.S.P., 2005. Molecular encapsulation of flavors as helical inclusion complexes of amylose. J. Cereal Sci. 41, 239–249.

Wyler, R., Solms, J., 1982. Starch-flavor complexes. III. Stability of dried starch-flavor complexes and other dried flavor preparations. LWT Food Sci. Technol. 15, 93–96.

Yeo, Y., Bellas, E., Firestone, W., Langer, R., Kohane, D.S., 2005. Complex coacervates for thermally sensitive controlled release of flavor compounds. J. Agric. Food Chem. 53, 7518–7525.

Zasypkin, D., Porzio, M., 2004. Glass encapsulation of flavours with chemically modified starch blends. J. Microencapsul. 21, 385–397.

Zeller, B.L., Saleeb, F.Z., Ludescher, R.D., 1999. Trends in development of porous carbohydrate food ingredients for use in flavor encapsulation. Trends Food Sci. Technol. 9, 389–394.

7

ANTIMICROBIALS FROM HERBS, SPICES, AND PLANTS

Tarik Bor*, Sulaiman O. Aljaloud*,, Rabin Gyawali*, Salam A. Ibrahim***

**North Carolina Agricultural and Technical State University, Food Microbiology and Biotechnology Laboratory, Greensboro, NC, United States; **King Saud University, Department of Exercise Physiology, College of Sport Sciences and Physical Activity, Riyadh, Saudi Arabia*

1 Nanoencapsulation of Natural Antimicrobial Products

1.1 Challenges in Applying Lipophilic Antimicrobials in Foods

Natural antimicrobial compounds including plant-derived EOs are generally recognized as safe (GRAS), which allows their use in food preservation (Gyawali et al., 2011, 2014a,b; Gyawali and Ibrahim, 2012). Nevertheless, EOs are soluble in ethanol but only partially soluble in water up to 0.01% or 1.0 g/L (Burt, 2004). This raises an issue when these compounds are used in complex food models as natural antimicrobials. Fat and protein content of the foods may possibly bind with EOs such as thymol and eugenol and antimicrobial efficiency will be lost or reduced (Gyawali and Ibrahim, 2014).

Although EOs perform well in antibacterial assays in vitro, more EO is needed to see the same antibacterial effect in food samples. The fat, protein, water, antioxidant, preservative, salt content, and pH of the food can influence bacterial sensitivity of EOs with extrinsic factors such as temperature, packaging in vacuum/gas/air, and characteristics of microorganisms. The antimicrobial effect of EOs seems to be increased with a decrease in pH and storage temperature of the food and the amount of oxygen used for packaging. High levels of fat and/or protein content of foodstuffs protect the bacteria from the action of EOs (Burt, 2004). When the EO concentrations are increased up to the higher levels

Encapsulations. http://dx.doi.org/10.1016/B978-0-12-804307-3.00007-7

required to inhibit or inactivate microorganisms, they may exceed regulatory levels and change the sensory threshold levels with intense color, flavor, and appearance impacts on foods. EO compounds are highly volatile and aromatic. This could have negative impact on the sensory properties when added into food products. For example, lettuce samples treated with 1.0 g/L thymol and lemon balm were not acceptable by panelist in sensory evaluation. Eugenol may be required to add >1% in cheese and meat systems (Tassou et al., 1995). Therefore, if these EO compounds are encapsulated that could help solve these issues associated with the direct addition of the free compounds (Gutierrez et al., 2008; Shah, 2011).

2 Encapsulation Process

Encapsulation is used for delivery of food ingredients within capsules when the direct addition of food ingredient causes the risk for the quality of food product. These capsules are classified as nanocapsules if the sizes of these capsules are less than 100 nm while sizes of microcapsules are measured in micrometers. Nano- and microstructured clusters comprising emulsion systems produced with food-grade ingredients including food biopolymers (proteins, carbohydrates), fats, low molecular weight surfactants, and copolymers (protein–carbohydrate conjugates) have been used to deliver functional ingredients into food products (Augustin and Hemar, 2009).

2.1 Design of Encapsulated Food Ingredients

Encapsulated functional ingredients have to be designed by thinking its end application because functionality is dependent to its end application. Encapsulation is specifically used to protect and maintain quality and stability of functional food ingredients in a food product and it is required to overcome physical or chemical instability of the functional ingredients, undesired interactions of the ingredient with other components of the food matrix, or early release of flavor or bioactive ingredients. It is important to know the purpose of encapsulation and the mechanism for releasing ingredients. Because this information helps to design encapsulated ingredients from the appropriate encapsulant matrix and formulation to choose the appropriate encapsulation process to produce a structure that protects the core and respond to external stimulus for release of the functional or bioactive ingredient (Augustin and Hemar, 2009)

2.2 Active Core

The flavoring agents, food acids and bases, lipids, food additives, minerals, colors, and vitamins are among the traditional encapsulated food ingredients. Recently, the encapsulation of bioactive ingredients, specifically omega-3 oils, phytonutrients, and probiotics gained attention due to their health benefits. The stability of food ingredient core in microcapsule is dependent upon the chemical structure, molecular weight, polarity, and volatility of food ingredient (Augustin and Hemar, 2009).

2.3 Matrix Materials for Encapsulation

The materials used as encapsulant can be selected from natural biomaterials or allowed food additives with generally recognized as safe (GRAS) status. Commonly used ingredients in encapsulation are food biopolymers (proteins, carbohydrates), fats, low molecular weight surfactants, and copolymers. Their emulsifying properties and capacity to form gel networks and viscosity make them useful matrix materials (Augustin and Hemar, 2009).

2.3.1 Food Proteins

Soy proteins, milk proteins-caseins, and whey proteins, egg proteins, zein or hydrolysates of these proteins are used as encapsulant matrices. Due to their amphiphilic nature, proteins tend to self-assemble. Their assembly into gelled structures can be triggered by acidification which neutralizes the charge of the protein as the pH reaches the isoelectric point (pI) of the protein, or by heating which causes unfolding and exposure of hydrophobic groups. Proteins can assemble at interfaces and they may serve as an effective transporter of bioactive molecules because of their ligand binding features. Another milk protein, β-lactoglobulin, has high affinity for hydrophobic molecules such as fatty acids and retinol. Heating stimulates the production of protein–carbohydrate conjugates by the Maillard reaction. The Maillard reaction affects the solubility, gelling, and emulsifying properties of proteins. The protein in emulsions is anchored at the interface and the carbohydrate protrudes into solution and this enhances the colloidal stability of the protein-based emulsion (Augustin and Hemar, 2009).

2.3.2 Food Carbohydrates

Sugars (eg, glucose, sucrose, oligosaccharide, glucose syrup) and polysaccharides (eg, starch and starch products—low and high-amylose starch, dextrins; nonstarch polysaccharides—alginate, pectin, carrageenan, gum arabic, chitosan, cellulose derivatives, cyclodextrin) are generally used as components of encapsulant matrix. Food carbohydrates can produce glassy solids where they can give structural support to delivery system. Based on their structures, they are able to bind specific molecules (Augustin and Hemar, 2009).

Starch and maltodextrins have applications as encapsulant matrices. Maltodextrins are useful to produce dehydrated encapsulated systems. Maltodextrins provide structural integrity to the final product and their incorporation in place of simple sugars in formulations reduces stickiness during drying. There are two main polysaccharides in starch: amylose and amylopectin. The majority of food carbohydrates are not surface active except for gum arabic and they have to be used by combining with other ingredients with high emulsifying capacity. Gum arabic is surface active in addition to providing structural stability to both and wet dry encapsulation systems (Augustin and Hemar, 2009).

2.3.3 Lipids

Lipids (eg, natural fats and oils, monoglycerides, diglycerides, phospholipids, glycolipids, waxes—beeswax, and carnauba wax) may be used for encapsulation. The majority of food fats (eg, milk fat, soybean oil, cocoa butter) have triglycerides as a major component (~98%). These nonpolar lipids carry lipophilic bioactive molecules in emulsion systems. Active cores can be embedded in solid fat matrix and released with increasing temperature. Fats are protective moisture barriers as encapsulant matrices in systems. Polar lipids (eg, monoglycerides, phospholipids, glycolipids) are surface active due to their amphiphilic character and they can be used for stabilizing emulsions including active food ingredients (Augustin and Hemar, 2009).

3 Encapsulation Techniques

Several different processes can be used for encapsulation of functional ingredients. Physical and chemical characteristics of the core and the elements that control the interfacial and assembling properties of the matrix materials is required to decide on suitable encapsulation process for designing encapsulated ingredients (Augustin and Hemar, 2009).

3.1 Spray Drying

Spray drying is a well-established and commonly used encapsulation process in food industry since it is more cost-effective than others. It is 30–50 times cheaper than freeze drying. Dried ingredients are valuable because of their availability and shelf life stability. To produce a spray dried encapsulated ingredient, the basic process includes dissolving the core in a dispersion of the encapsulant matrix. The dispersion is atomized and exposed heated air for rapid drying of droplets. The powder particles are then separated from the drying air (Augustin and Hemar, 2009; Gharsallaoui et al., 2007).

Encapsulant matrix needs high solubility in water. Desirable features are forming high solids at low viscosity, and good film and emulsifying capacity. Hydrophobic core materials are usually dissolved in oil phase and oil-in-water emulsions while hydrophilic cores are dispersed in aqueous phase comprising water-soluble encapsulant wall material (Table 7.1). Hydrophobic core material is coated by interfacial membrane after drying and it can be released by disruption the integrity of this membrane. Spray drying is used for generation of many encapsulated ingredients such as vitamins, minerals, flavors, enzymes, polyunsaturated oils, and probiotic microorganisms. It may be used for heat sensitive and volatile ingredients (Gharsallaoui et al., 2007; Augustin and Hemar, 2009).

3.2 Spray Cooling

Active core dispersed in liquefied matrix is atomized into a cool environment such as cool air. Generally high melting fats are used as the matrix material. Fat solidifies during cooling and the core is immobilized. Water-soluble functional ingredients such as enzymes, flavors, and food acids are incorporated into solid fat particles (lipospheres) for delaying the release of core material. These ingredients are applied in dry products (Augustin and Hemar, 2009).

3.3 Freeze Drying

Freeze drying can be used for ingredients which are very sensitive to heating processes. Commercial application of freeze drying is limited for high-value ingredients such as probiotics. Spray-drying process is cheaper and faster than freeze drying (Augustin and Hemar, 2009). Recently used wall materials for encapsulation by freeze drying were listed in Table 7.2.

3.4 Fluidized Bed Coating

This technique was originally developed as a pharmaceutical technique and now it is applied in the food industry to increase efficiency of functional ingredients and additives. Food

Table 7.1 Recently Used Wall Materials for Encapsulation of Different Food Ingredients by Spray Drying

Encapsulated Ingredient	Wall Materials and Stabilizers	References
Oregano, citronella, and marjoram flavors	Whey proteins/milk proteins	Baranauskiene et al. (2006)
Cardamom oleoresin	Gum arabic/modified starch/maltodextrin	Krishnan et al. (2005)
Bixin	Gum arabic/maltodextrin/sucrose	Barbosa et al. (2005)
D-Limonene	Gum arabic/maltodextrin/modified starch	Soottitantawat et al. (2005)
Cumin oleoresin	Gum arabic/maltodextrin/modified starch	Kanakdande et al. (2007)
Paprika oleoresin	Gum arabic/soy protein isolate	Rascon et al. (2011)
Black pepper oleoresin	Gum arabic/modified starch	Shaikh et al. (2006)
Short chain fatty acid	Gum arabic/maltodextrin	Teixeira et al. (2004)
Extra virgin olive oils	Gum arabic/maltodextrin/sodium caseinate/starch/lactose/gelatin	Calvo et al. (2010)
Fish oil	Maltodextrin combined with whey protein concentrate or modified starch	Jafari et al. (2008)
	Caseinate glycated with glucose, glucose syrup, or dextrans	Drusch et al. (2009)
	Barley protein	Wang et al. (2011)
Conjugated linoleic acid	Whey protein concentrate/whey protein isolate/maltodextrin/Maillard reaction products of whey protein isolate and maltodextrin	Choi et al. (2010b)
Flaxseed oil	Gum arabic	Tonon et al. (2011)
	Maltodextrin/gum arabic/whey protein concentrate/modified starches	Carneiro et al. (2013)
Polyphenols and anthocyanins from juice and ethanolic extracts of pomegranate	Soybean protein isolates/maltodextrin	Robert et al. (2010)
Probiotic (*Lactobacillus rhamnosus* GG) preparations	Whey protein isolate/maltodextrin/glucose/inulin	Ying et al. (2012)
Probiotic (*Lactobacillus plantarum*)	Regular and denatured whey protein isolates with sodium alginate	Rajam et al. (2012)
Probiotic (*Bifidobacterium* Bb-12)	Whey protein	De Castro-Cislaghi et al. (2012)

Table 7.1 Recently Used Wall Materials for Encapsulation of Different Food Ingredients by Spray Drying (*cont.*)

Encapsulated Ingredient	Wall Materials and Stabilizers	References
Chia essential oils	Whey protein concentrate/mesquite gum/gum arabic	Rodea-Gonzalez et al. (2012)
Cinnamon oleoresin	Gum arabic/Maltodextrin/Modified starch	Vaidya et al. (2006)
Phospholipid	Maltodextrin blended with sodium caseinate, gelatin, and soy protein	Yu et al. (2007)
Lycopene	β-Cyclodextrin	Nunes and Mercadante (2007)
Thymol	Whey protein Isolate/maltodextrin conjugate	Shah et al. (2012b)
Eugenol	Whey protein isolate/maltodextrin conjugate	Shah et al. (2012c)
Lipase enzyme isolated from endophytic fungus *Cercospora kikuchii*	Maltodextrin/β-cyclodextrin	Costa-Silva et al. (2010)
Elderberry (*Sambucus nigra* L.) juice	Maltodextrin/gum acacia/soya milk powder/ soya protein powder/isolated soya protein	Murugesan and Orsat (2011)
Fresh ginger oil	Maltodextrin/whey protein	Toure et al. (2011)
Thymol, geraniol	β-Cyclodextrin/modified starch	Mourtzinos et al. (2008)

Source: Adapted from Gharsallaoui et al. (2007) and updated with recently made researches.

technologists should consider more about the cost of this expensive technology. Top spray film coating of food ingredients seems more feasible than other fluidized bed coating methods (Dewettinck and Huyghebaert, 1999). This process contains spraying a coating solution into a fluidized bed of solid particles. A continuous coating film formed after a few cycles of wetting–drying processes. After several cycles of wetting–drying, a continuous film is formed (Guignon et al., 2002). The solid particles are suspended in air and matrix material for coating core materials is sprayed onto particles. The matrix material can be dispersion or concentrated solution, a hot melt or an emulsion. This technique can be used to provide extra protection for spray-dried powders (Augustin and Hemar, 2009). This technology is one of the few advanced technologies have the capacity to coat solid particles with any type of shell material such as polysaccharides, proteins, emulsifiers, fats,

Table 7.2 Recently Used Wall Materials for Encapsulation of Different Food Ingredients by Freeze Drying

Encapsulated Ingredient	Wall Materials and Stabilizers	References
Turmeric oleoresin	Maltodextrin/gelatin	Malacrida et al. (2013)
Fennel oleoresin	Gum arabic mixtures	Chranioti and Tzia (2013)
Biofilm-like *Lactobacillus rhamnosus* probiotics	Chitosan/alginate/carrageenan	Cheow and Hadinoto (2013)
Garcinia fruit extract	Whey protein isolate/maltodextrin	Ezhilarasi et al. (2013)
Potassium norbixinate and curcumin	Maltodextrin	Sousdaleff et al. (2012)
Limonene	Gum arabic/sucrose/gelatin	Kaushik and Roos (2007)
Argentine red wine (Cabernet Sauvignon) polyphenols	Maltodextrin	Sanchez et al. (2011)
Lycopene extract from tomato pulp	Gelatin/poly (γ-glutamic acid)	Chiu et al. (2007)
Flavourzyme enzyme	Chitosan/alginate/poly L-lysine	Anjani et al. (2007)
Fish oil	β-Cyclodextrin/polycaprolactone	Choi et al. (2010a)
Curcumin	Hydrophobically modified starch (HMS)	Yu and Huang (2010)
Thymol, geraniol	β-Cyclodextrin/modified starch	Mourtzinos et al. (2008)
Vacuum-dried pineapple pulp powder	Maltodextrin/gum arabic	Gabas et al. (2007)
Curcumin	Pluronic triblock copolymer micelles	Sahu et al. (2011)
	Baker's yeast (*Saccharomyces cerevisiae*) cells/β-cyclodextrin/modified starch	Paramera et al. (2011)
Probiotic lactic acid bacteria (*Lactobacillus* spp.)	Bacterial cellulose (*nata*)/skim milk/calcium alginate	Jagannath et al. (2010)
Probiotics (*Lactobacillus*, *Bifidobacterium*, and *Lactococcus* species	Alginate/lecithin/starch	Donthidi et al. (2010)
Prebiotic and probiotics (*Lactobacillus gasseri* and *Bifidobacterium bifidum*)	Chitosan/calcium alginate beads	Chavarri et al. (2010)
Cinnamaldehyde, thymol	β-Cyclodextrin	Cevallos et al. (2010)
Methanolic extract of *Hypericum perforatum* (St. John's wort)	β-Cyclodextrin	Kalogeropoulos et al. (2010)
Eugenol	β-Cyclodextrin	Seo et al. (2010)

Table 7.2 Recently Used Wall Materials for Encapsulation of Different Food Ingredients by Freeze Drying (*cont.*)

Encapsulated Ingredient	Wall Materials and Stabilizers	References
Olive leaf extract	β-Cyclodextrin	Mourtzinos et al. (2007)
β-Carotene	Mannitol matrix	Sutter et al. (2007)
	Regular and modified pinhão starch/gelatin	Spada et al. (2012)
Trehalose	Thermally responsive pluronic nanocapsules	Zhang et al. (2009)
Egg yolk immunoglobulin (IgY)	Chitosan/alginate	Li et al. (2007)

complex formulations, enteric coating, powder coatings, yeast cell extract, and so forth. Thus, the controlled release possibilities are remarkably more variable than any other encapsulation methods (Poshadri and Kuna, 2010). Fluidized bed coating is used to supply encapsulated versions of vitamin C, B vitamins, ferrous sulfate, ferrous fumarate, sodium ascorbate, potassium chloride, and variety of vitamin/mineral premixes. For bakery applications, products such as vitamin C, acetic acid, lactic acid, potassium sorbate, sorbic acid, calcium propionate, and salt are encapsulated by using fluidized bed coating method. Encapsulated vitamin C is used as effective potassium bromated replacer. Several food acids have been fluid-bed encapsulated to develop color and flavor systems in meat industry. These acids are used to obtain a reproducible pH in cured meat products and decrease the processing time. To prevent rancidity, fluid-bed encapsulated salt is used in meats (Dewettinck and Huyghebaert, 1999).

3.5 Extrusion

Extrusion microencapsulation has been used especially for the encapsulation of volatile and unstable flavors in glassy carbohydrate matrices. This process provides a very long shelf life to oxidation-prone flavor compounds like citrus oils. Shelf life can be up to 5 years for extruded flavor oils while it is 1 year for spray-dried flavors and a few months for unencapsulated citrus oils. Carbohydrate matrices in glassy states form a very good barrier and the extrusion method is very convenient for encapsulation of such flavors in carbohydrate matrices. Extrusion can be used for encapsulating nutraceuticals as well. Extrusion is the most suitable process for glassy carbohydrates as shell materials (Poshadri and Kuna, 2010). Syringe extrusion is used for production of alginate beads and for the formation of alginate beads an alginate solution is extruded as droplets into a calcium chloride solution at lower temperatures. In another extrusion method called spinning disk, core materials suspended in a matrix material solution are passed over a rotating cylinder to form microparticles. Double capillary is used in centrifugal extrusion method. Core material is in the inner capillary and matrix material is on the outer capillary (Augustin and Hemar, 2009).

3.6 Microencapsulation Based on Supercritical Fluids

In microencapsulation processes, the core material is dissolved in matrix material solubilized in a supercritical fluid such as carbon dioxide. The core material is encapsulated in matrix material by removing the carbon dioxide. The dispersion is sprayed through nozzle and a particulate material containing the core material is produced in a process called rapid expansion of supercritical solutions (RESS) (Augustin and Hemar, 2009). Romo-Hualde et al. (2012) obtained bioactive natural vitamins from red pepper (*Capsicum annuum* L.) by-products using supercritical fluid extraction and stabilized them with microencapsulation method.

3.7 Coacervation

Often called phase separation, coacervation is accepted as a real microencapsulation method and involves the precipitation or separation of colloidal phase from an aqueous phase. Water soluble or a nonsolvent polymer is used in simple coacervation and this polymer competes for solubilization of gelatin solution with hydrophobic interactions. The capsule is produced by ionic interactions between oppositely charged polymers in the complex coacervation method. In complex coacervation usually protein molecules and anionic macromolecules such as gelatin and gum arabic carry positive charges. Coacervation comprises the separation of liquid phase of coating material from a polymeric solution right after coating of this phase as a uniform layer around suspended core material particles (Poshadri and Kuna, 2010).

A complex is formed when the opposite charge of biopolymer solutions is mixed. Biopolymer type, pH, ionic strength, concentration, and the ratio of biopolymers have impacts on the affinity of biopolymers and the structure of the complex formed. One phase or two phases can occur depending on biopolymers involved and different conditions. In the two-phase system, one of the phases has reduced both of the biopolymers whereas the other phase is enriched in both biopolymers in a precipitated form or complex coacervate. One of the restrictions in using coacervates is their sensitivity against pH and ionic strength. To allow the permanence of coacervates, they may be cross-linked. Enzymatic cross-linkers such as transglutaminase are more acceptable than glutaraldehyde in the food industry. Plant polyphenols have been recently used to cross-link gelatin-based coacervates (Augustin and Hemar, 2009).

Heat-resistant nanocapsules of jasmine essential oil were obtained via a gelatin and gum arabic mixture using complex coacervation encapsulation method (Lv et al., 2014). Jun-Xia et al. (2011) microencapsulated sweet orange oil by complex coacervation with soybean protein isolate and gum arabic. Leimann et al. (2009) use simple coacervation for microencapsulation of lemongrass (*Cymbopogom citrates*) EO, known because of its broad-spectrum antimicrobial activity. The size of microcapsules was in the range of 10–250 μm. No agglomeration was observed in microcapsules when SDS (sodium dodecyl sulfate) at 0.03 wt% was used. Complex coacervation with transglutaminase as a hardening agent was used to prepare the gelatin/gum arabic microcapsules encapsulating peppermint oil. During the 40 days of storage of microcapsules in cold water only about 7% of peppermint oil was released, showing excellent storage stability (Dong et al., 2011).

Nori et al. (2011) encapsulated propolis extract by complex coacervation using isolated soy protein and pectin as encapsulant agents. They were successful to obtain the encapsulated propolis extract in a powdered form, alcohol free, stable protecting its antioxidant and antimicrobial activities and with the possibility of controlled release in foods. Eugenol was microencapsulated using gelatin/sodium alginate complex coacervation. Eugenol microcapsules had good flow properties and therefore improved handling. Treatment with dehydrating agent caused reduction in loading and percent encapsulation efficiency of eugenol microcapsules (Shinde and Nagarsenker, 2011). Martins et al. (2009) developed a novel coacervation method to produce microcapsules of polyactide (PLA) to encapsulate thyme oil. The new approach in their method was dissolving PLA in dimethylformamide (DMF), which is good solvent for PLA and highly soluble in water. The homogenous solution of PLA in DMF stimulates the precipitation of PLA around the thyme oil core. They achieved to form microcapsules with a mean particle size of 40 μm, 5 μm wall thickness, spherical shape, and rough surface. To optimize and control the kinetics of gelatin–acacia complex, coacervation method is used for encapsulation of emulsified oil droplets. This study showed the formation of nanomicroparticles could be highly related to kinetics of freezing process and particle properties can be adjusted by playing with the freezing operation (Nakagawa and Nagao, 2012).

Silva et al. (2012) encapsulated oily dispersion of lycopene by complex coacervation using gelatin and pectin and characterized the properties of microcapsules. Although they formed microcapsules with high values of encapsulation efficiency, they could not stop the degradation of the lycopene during storage period. Gum arabic–chitosan microcapsules containing commercially product Miglyol 812N (blend of triglycerides) as core phase were synthesized by complex coacervation. For the emulsion step, the optimum phase volume ratio chosen was 0.10 and an emulsion time of 15 min was selected. The results showed that the optimum formation of these complexes appears at pH 3.6 and a weight ratio of chitosan to gum arabic mixtures of 0.25 (Butstraen and Salaun, 2014).

Ozyildiz et al. (2013) showed antimicrobial action of red pepper seed oil (RPSO) after ozonation and they microencapsulated it for the preparation of functional textile material as a first reported material in this field. They ozonated and microencapsulated RPSO by the complex coacervation method using gelatin and gum arabic as wall materials.

3.8 Liposomes

Liposomes are spherical bilayer vesicles, which are formed as polar lipid dispersions in aqueous media. They have been widely used in the pharmaceutical industry in the delivery of drugs, vaccines, hormones, enzymes, and vitamins to their targets in the body but they have limited applications in the food industry. Phospholipids have commonly been used to prepare unilamellar or multilamellar liposomes. Liposomes can range from 25 nm to several micrometers in diameter, are easy to make, and can be stored by freeze-drying. They are carriers for both lipophilic and hydrophilic molecules. The entrapped active materials are stabilized against environmental factors such as pH, temperature, and ionic strength. The core material can be released when the gel to liquid transition temperature of phospholipids is reached. Cholesterol can be added to improve the resistance of liposomal system to degradation when it is exposed to in vitro and in vivo conditions. The composition of phospholipids determines the number of layers in liposomal systems. Fatty acids can also be used to form liposomes. Small lipophilic molecules can permeate through liposomal bilayer although sugars and large polar molecules cannot permeate (Augustin and Hemar, 2009; Poshadri and Kuna, 2010).

Liolios et al. (2009) incorporated carvacrol and thymol isolated from EO of *Origanum dictamnus* L. in phosphatidyl choline-based liposomes and were able to increase antimicrobial activity of these compounds after encapsulation against four Gram-positive bacteria: *Staphylococcus aureus* (ATCC 25923), *Staphylococcus epidermidis* (ATCC 12228), *Staphylococcus mutans* (ATCC 31989), and *Staphylococcus viridans* (ATCC 19952); four Gram-negative bacteria: *Pseudomonas aeruginosa* (ATCC 27853), *Escherichia coli* (ATCC 25922), *Enterobacter cloacae* (ATCC 13047), and *Klebsiella pneumoniae* (ATCC 13883); three human pathogenic fungi: *Candida albicans* (ATCC 10231), *Candida tropicalis* (ATCC 13801), and *Candida glabrata* (ATCC 28838); as well as against the food-pathogen, *Listeria monocytogenes* (11994). The antioxidant and antimicrobial activity of *O. dictamnus* methanol or dichloromethane extracts was increased after encapsulation in liposomes (Gortzi et al., 2007). Varona et al. (2011) studied the corporation of lavandin (*Lavandula hybrida*) EO in liposomes formed with lecithins and cholesterol. They used thin-film method and particles from gas-saturated solutions (PGSS) drying process to produce liposomes. Liposomes produced by a thin-film method were multivesicular or unilamellar/multilamellar with a mean diameter of 0.4–1.3 μm and encapsulation of EO was carried out at 66% efficiency. The EO of Brazilian cherry (*Eugenia uniflora* L.) leaves was incorporated into multilamellar liposomes by dry film hydration without causing phase separation in the membrane structure (Yoshida et al., 2010).

3.9 Molecular Inclusion Complexes with Cyclodextrins

This technique is performed at a molecular level. Cyclic oligosaccharides of α-(1, 4) linked glucopyranose units, cyclodextrins, are used as a microencapsulation method due to their capacity to form molecular inclusion complexes. The interior part of cyclodextrins is hydrophobic whereas their exterior side is hydrophilic. The size of hydrophobic internal cavity that carries hydrophobic molecules is determined based on the number of glucose units. The α-cyclodextrin is a cyclic derivative of starch made up of seven glucopyranose units. They are prepared from partially hydrolyzed starch by an enzymatic process. The internal cavity of about 0.65 nm diameter allows the inclusion of EO compounds and can take up one or more flavor volatile molecules. In this encapsulation method, the flavor compounds are entrapped inside the center of cyclodextrin molecule. Complex formation with a hydrophobic molecule is driven due to displacement of water from the interior of cyclodextrin is energetically favorable. The formation of inclusion complexes increases the solubility of hydrophobic molecules in water and core ingredient is protected against degradation. Unexpected odors and flavors can be masked. The hydrophobic molecule is usually displaced by heating (Augustin and Hemar, 2009; Poshadri and Kuna, 2010).

4 Encapsulation of Lipophilic Antimicrobials

Functional ingredients (such as drugs, antimicrobials, vitamins) are rarely used in their pure form—they are usually incorporated into delivery systems. The roles of delivery systems are (1) they serve as a vehicle for carrying the functional ingredient to the desired site of action; (2) they protect it from chemical or biological degradation during processing, storage, and utilization—this helps to keep it in active phase; (3) they control the release of functional ingredient; and (4) they have to be compatible with the other components in the system, as well as being compatible with qualitative and physicochemical attributes (Weiss et al., 2006). Finally the delivery system should be prepared with food-grade ingredients using simple, cost-effective processing operations if it will be used in the food industry (McClements et al., 2007).

The delivery system characteristics influence the efficiency of functional ingredients in industrial products. Different types of delivery systems has been developed for encapsulating functional ingredients including simple solutions, association colloids, emulsions, biopolymer matrices and so on (Weiss et al., 2006). Delivery systems have advantages and disadvantages for encapsulation, protection, and delivery of functional agents in terms of their cost, regulatory status, ease of use, biodegradability, biocompatibility, and so on (McClements et al., 2007).

5 Nano- Versus Microencapsulation

The majority of research has been conducted on the use of emulsion-based delivery systems for encapsulation of lipophilic compounds such as flavors, fat-soluble nutrients, and antimicrobial EOs (McClements et al., 2007; Shah, 2011). Glenn et al. (2010) encapsulated thymol, clove, origanum, and camphor white EOs in high-amylose starch-based porous microspheres. They used spray-drying method oil-in-water emulsions to produce capsules of the size of 5 μm. The encapsulation properties of skimmed milk powder (SMP) and whey protein concentrate (WPC) for the coating of EO of oregano and aroma extracts of citronella and sweet marjoram by spray-drying method were evaluated. SMP-coated products had a smoother surface while WPC contains more dents and wrinkles on the surface of capsules. Particle sizes ranged from 6 μm to 280 μm and 2 μm to 556 μm for SMP and WPC, respectively, for microencapsulated products (Baranauskiene et al., 2006). Farhang (2009) used maltodextrin at different ratios for microencapsulation of elemi and peppermint oils by spray drying of emulsion in a laboratory dryer. Donsì et al. (2012) used algae oil in oil-in-water emulsions and stabilized with whey protein isolate. The oxidative stability of emulsion was increased by using EDTA at concentrations ≥ 1 μM at both pH 3.0 and 7.0. Biopolymer materials such as chitosan, sodium alginate, and carrageenan have been used to produce stabilized and multilayered emulsions for additional protection of EOs (Perez et al., 2009; Shah, 2011; Wang and Lucey, 2003). Elshafei et al. (2010) made environmentally friendly pesticides using water-in-oil-in-water (W/O/W)–type emulsions of EOs from eucalyptus, linalool, and marjoram stabilized with xanthan gum. Solid lipid microparticles were designed to reduce volatility of the antimicrobial compound extracted from juniper berries (Gavini et al., 2005).

5.1 Milk Proteins and Carbohydrates as Emulsifying Agents

Different carbohydrate-based matrices such as starch, maltodextrin, gum arabic, and milk protein-based matrices were used to encapsulate several lipophilic ingredients like EOs and flavor volatiles. Whey protein–maltodextrin conjugates are recently used as emulsifying agents. Because of their tendency to be adsorbed at fluid interfaces, these amphiphilic whey proteins can be used as perfect emulsifiers during encapsulation of lipophilic compounds. Bovine serum albumin, α-lactalbumin, and β-lactoglobulin form about 60% of total whey proteins, whereas lactoferrine and immunoglobulins are some of the minor ingredients. Because of their low cost and nutritional value, whey protein isolate and whey protein concentrate gained more interest for encapsulation. Their applications in the food industry are limited by environmental conditions such as thermal processing, ionic strength, and pH. These proteins are denatured, unfolded, and aggregated during heating above 60°C. Negatively charged carboxyl and positively charged amine groups are present in proteins and these charged groups are important during interparticle repulsion. It is more effective around pH 5.0, which is close to their isoelectric point (pI) and the presence of high ionic concentrations. Interparticle affinity is increased at the isoelectric point of proteins and this causes protein precipitation resulting in increased turbidity and solid gel formation in the system. These effects are usually seen during storage and can negatively impact acceptability by consumer. Whey proteins have some unique benefits useful for their functional properties in food such as heat resistance and increased emulsifying capacity when they are processed together with food additives such as polysaccharides. However polysaccharides do not exhibit surface activities but they are extremely hydrophilic. Thus, they contribute to stabilizing and jellifying properties and rheology of the aqueous phase. For dispersion and foam stabilizing methods, heat-induced protein-polysaccharide conjugates have been suggested. Proteins are conjugated with carbohydrates by dry heating mixtures to stimulate covalent coupling of proteins and carbohydrates also known as Maillard reaction (Akhtar and Dickinson, 2003, 2007; Dickinson, 2003; Kato, 2002; Shah, 2011).

Condensation occurs between unprotonated amine group from lysine residues of proteins and carbonyl group of carbohydrates without chemical catalysis at the early stage of Maillard reaction (Wannes et al., 2010). The functional properties of whey proteins such as solubility, formation of soluble aggregates, and interfacial properties can be improved by conjugating them with alginates, carrageenans, xanthan, and pectins. Functional properties of conjugates are dependent upon hydrophilic property, viscosity, and binding number of polysaccharides. The suitable properties of conjugates can be adjusted by protein: polysaccharide ratio, size and conformation of polysaccharides and proteins based on their applications in food industry (Shah, 2011). Whey protein isolate has been conjugated with hydrolyzed polysaccharides such as maltodextrin and used for encapsulating lipophilic ingredients comprising orange oil, ginger oil, avocado oil, and conjugated linoleic acid to make them resistant to oxidative stress, volatilization, and improve their bioactivity (Akhtar and Dickinson, 2003; Bae and Lee, 2008; Choi et al., 2010b).

5.2 Nanoencapsulation of EOs for Enhanced Antimicrobial Effectiveness

The antimicrobial activity of EO is remarkably lowered in foods because of the poor solubility in water that causes nonhomogenous dispersion in food matrices (Gutierrez et al., 2008). They have high affinity to lipids, proteins, and other food compounds because of their hydrophobicity and chemical structure. This tendency reduces their antimicrobial activity (Burt, 2004).

To assure targeted antimicrobial efficacy is the same as in growth media, EOs need to be used at higher concentrations in foods and this causes changes in sensory properties of foods and decreases acceptance by consumers (Weiss et al., 2009). For this reason, EOs are encapsulated with stabilizing materials such as gum arabic and starches (Kanakdande et al., 2007), and proteins (Gbassi et al., 2009). As one of the emulsion-based systems, nanoemulsions are dispersions of nanoscale droplets less than 100 nm formed with high shear application. The separation of these droplets may be carried out by microfluidization or ultrasonication. Nanoemulsion systems require greater amount of surfactant when compared with microemulsions. Nanoemulsion of triglyceride oils can be achieved by using microfluidizers, low molecular weight surfactants and cosolvent for stability of the oil phase. Nanoemulsions are transparent systems whereas microemulsions scatter light. This property is important in food applications because lipophilic antimicrobials can be delivered in clear emulsions. Droplets have a smaller size and potential to increase the bioavailability of core material due to increased surface area when compared with conventional emulsion systems (Augustin and Hemar, 2009). Nanoencapsulation of thymol and carvacrol in liposomes (Liolios et al., 2009), eugenol (Shah et al., 2012b), thymol (Shah et al., 2012c), terpenes mixture, and D-limonene (Donsi et al., 2011) in nanocapsules was performed to enhance their antimicrobial activities. These processes need high energy use, expensive materials, or involve a lot of steps. Microemulsions nanoscale systems require more surfactants and cosurfactants, which are very expensive and under regulation; therefore their applications in food industry are limited (Shah, 2011).

The principals behind using emulsion-evaporation system to obtain food grade nanocapsules with encapsulated antimicrobials involve dissolving natural lipophilic antimicrobial in a volatile organic solvent to form an oil phase with its emulsification into aqueous phase containing dissolved whey protein isolate (WPI)–maltodextrin (MD) conjugates, followed by spray drying to remove hexane. Spray drying is widely used in food industry to stabilize volatile ingredients such as EOs during processing and storage periods and to increase bioactivity of encapsulated lipophilic compounds in food matrices because it is a simple, fast, and low-cost method. Additionally, products in a powdered form are compatible for long-term storage, handling, and transportation. WPI–MD conjugates are selected as emulsifiers. Amphiphilic whey proteins adsorb onto the lipophilic EO phase and hydrophilic maltodextrin moiety can be solved in aqueous phase and this will improve dispersion capacity and reduce aggregation of particles because of steric hindrance (Shah, 2011).

5.3 Physicochemical Properties

Shah et al. (2012c) chose the thymol as a model lipophilic antimicrobial due to its strong antibacterial and antifungal activity. They used the conjugate preparation method of Akhtar and Dickinson (2007) but modified heating temperature, drying method, MD chain lengths, and WPI:MD ratios. Emulsions created with different WPI:MD ratios were spray dried. Spray-dried powders were loaded with thymol. Nanocapsules were hydrated at 5% w/v powder concentration in deionized water at room temperature (20°C) for 14 h and adjusted to pH 3.0, 5.0, and 7.0 for characterization. They showed that optimization of emulsion preparation conditions and conjugation of WPI and MD can help minimize losses and improve the encapsulation efficiency after spray drying. When spray-dried capsules were hydrated, an overall 0.525% w/v overall thymol concentration was obtained for some nanodispersions, which is much higher than the solubility limit of thymol.

They could obtain some clear dispersion with an increase in visual clarity after heating at both salt concentrations. Samples made with a WPI:MD mass ratio of 1:2 produced dispersions with higher clarity after heating. They showed that formulations need to be optimized to achieve visual clarity, improved dispersion properties, and heat resistance in dispersions. Most of the formulations used to prepare nanodispersions gave less than 100 nm capsules after heating at pH 3.0 and 7.0, at both salt concentrations. The improved dispersibility and thermal stability of these dispersions are highly relevant to conjugates to possibly form core-shell structures where the MD shell allows for steric hindrance stabilizing particles. The presented scalable technology in their research may be used to disperse different types of lipophilic antimicrobial compounds in clear liquid forms, making an important contribution to functional beverage industry (Shah et al., 2012a).

Shah et al. (2012b) used eugenol as a model compound for encapsulation in WPI:MD40 with 1:2 ratios. Emulsifying the oil phase of eugenol in hexane into an aqueous phase with dissolved conjugates and spray drying are the two main steps of encapsulation process. Following the hydration of spray-dried powders, several samples containing eugenol above its solubility limit formed transparent dispersions at ph 3.0 and 7.0 after heating at 80°C for 15 min that corresponds to smaller than c. 100 nm mean diameters of particles.

5.4 Antimicrobial Activity in Growth Media

Shah (2011) dispersed thymol in nanocapsules obtained with WPI: MD180 conjugates at 1:2 ratios. The antimicrobial effects of nanodispersed (ND) and free thymol were tested against *E. coli* O157:H7, *Salmonella typhimurium*, *L. monocytogenes*, and *S. aureus* in TSB adjusted to pH at 35°C for *E. coli* and 32°C for *L. monocytogenes*. They showed that the MIC for ND and free thymol against all tested strains at pH 6.8 was 0.5 g/L, except for *E. coli* ATCC 43889 and *Lm* strains Scott A, which were inhibited by 0.3 g/L free oil. *Lm* was completely inhibited at this concentration at pH 5.5 after 3 h, whereas *E. coli* showed 1–3 log CFU/mL reduction after 48 h. The effect of temperature was not statistically significant. They were able to produce clear nanodispersions of thymol with promising antimicrobial activity against foodborne pathogens.

Shah et al. (2013) investigated the antimicrobial activities of ND eugenol (NE) and free eugenol (FE) against *E. coli* O157:H7 and *Lm* strains in TSB at 35 or 32°C. The antimicrobial efficacy of NE against all tested strains was not improved with MIC and MBC values being 0.25 g/L higher than those of FE. FE performed better in TSB because there was no interfering compound and the MIC and MBC were below solubility of eugenol.

5.5 Antimicrobial Activity in Food Systems

Shah et al. (2012b) compared the antimicrobial activities of free and ND thymol (NT) against *E. coli* O157:H7 and *L. monocytogenes* strains in apple cider and 2% reduced fat milk. Overall, 0.5 and 1.0 g/L thymol in nanodispersions and free thymol were inhibitory and bactericidal respectively, against all tested strains at all different pH and thymol concentrations. NT at 0.5 g/L concentration has bacteriostatic activity against *Lm* and *E. coli* for up to 48 h at pH 5.5. At pH 3.5, *Lm* controls could not survive after 12 h but *E. coli* survived and was inhibited by 0.5 g/L NT after 12 h and 48 h in apple cider. *E. coli* strains were sensitive to 4°C and pH 3.5. In 2% reduced milk at 35 or 32°C NT and free thymol inhibited tested bacterial strains at 4.5 g/L concentration.

Shah et al. (2013) investigated the antimicrobial activities of ND eugenol (NE) and free eugenol (FE) against *E. coli* O157:H7 and Lm strains in milk with different fat levels (whole, 2% reduced fat, and skim milk) at 35 or 32°C. NE was more effective than FE, with MIC and MBC values above the solubility of eugenol. The NE was thought to be equally dispersed in food matrices at concentrations above the solubility limit of eugenol and supplied the eugenol at specific locations when the binding to food compounds caused eugenol level decrease less than required amount for inhibition.

6 Conclusions

Plant-derived natural products contain active compounds that have antimicrobial properties. However, natural compounds are very sensitive to light and heat and many food processing conditions. As a result, these the compounds have poor stability in food systems. To overcome these challanges, different types of encapsualtion methods as described in this chapter have been proven to be effective when applied in food systems for the control and prevention of foodborne pathogens. By using nanoencapsulation technology, active ingredeints can be easily incorporate that can be delivered effectvely. In addition, nanocapsulation methods also shown to mask unwanted flavor as well as to improve the bioavailability of the active ingredeints present in the food systems. Therefore, use of enacapsulation techniques offer potential advantages over traditional food preservation methods.

References

Akhtar, M., Dickinson, E., 2003. Emulsifying properties of whey protein–dextran conjugates at low pH and different salt concentrations. Colloid. Surf. B 31, 125–132.

Akhtar, M., Dickinson, E., 2007. Whey protein–maltodextrin conjugates as emulsifying agents: an alternative to gum arabic. Food Hydrocoll. 21, 607–616.

Anjani, K., Kailasapathy, K., Phillips, M., 2007. Microencapsulation of enzymes for potential application in acceleration of cheese ripening. Int. Dairy J. 17, 79–86.

Augustin, M.A., Hemar, Y., 2009. Nano- and micro-structured assemblies for encapsulation of food ingredients. Chem. Soc. Rev. 38, 902–912.

Bae, E.K., Lee, S.J., 2008. Microencapsulation of avocado oil by spray drying using whey protein and maltodextrin. J. Microencapsul. 25, 549–560.

Baranauskiene, R., Venskutonis, P.R., Dewettinck, K., Verhe, R., 2006. Properties of oregano (*Origanum vulgare* L.), citronella (*Cymbopogon nardus* G.) and marjoram (*Majorana hortensis* L.) flavors encapsulated into milk protein-based matrices. Food Res. Int. 39, 413–425.

Barbosa, M.I.M.J., Borsarelli, C.D., Mercadante, A.Z., 2005. Light stability of spray-dried bixin encapsulated with different edible polysaccharide preparations. Food Res. Int. 38, 989–994.

Burt, S., 2004. Essential oils: their antibacterial properties and potential applications in foods—a review. Int. J. Food Microbiol. 94, 223–253.

Butstraen, C., Salaun, F., 2014. Preparation of microcapsules by complex coacervation of gum arabic and chitosan. Carbohydr. Polym. 99, 608–616.

Calvo, P., Hernandez, T., Lozano, M., Gonzalez-Gomez, D., 2010. Microencapsulation of extra-virgin olive oil by spray-drying: influence of wall material and olive quality. Eur. J. Lipid Sci. Technol. 112, 852–858.

Carneiro, H.C.F., Tonon, R.V., Grosso, C.R.F., Hubinger, M.D., 2013. Encapsulation efficiency and oxidative stability of flaxseed oil microencapsulated by spray drying using different combinations of wall materials. J. Food Eng. 115, 443–451.

Cevallos, P.a.P., Buera, M.P., Elizalde, B.E., 2010. Encapsulation of cinnamon and thyme essential oils components (cinnamaldehyde and thymol) in β-cyclodextrin—effect of interactions with water on complex stability. J. Food Eng. 99, 70–75.

Chavarri, M., Maranon, I., Ares, R., Ibanez, F.C., Marzo, F., Villaran, M.D.C., 2010. Microencapsulation of a probiotic and prebiotic in alginate-chitosan capsules improves survival in simulated gastro-intestinal conditions. Int. J. Food Microbiol. 142, 185–189.

Cheow, W.S., Hadinoto, K., 2013. Biofilm-like *Lactobacillus rhamnosus* probiotics encapsulated in alginate and carrageenan microcapsules exhibiting enhanced thermotolerance and freeze-drying resistance. Biomacromolecules 14, 3214–3222.

Chiu, Y.T., Chiu, C.P., Chien, J.T., Ho, G.H., Yang, J., Chen, B.H., 2007. Encapsulation of lycopene extract from tomato pulp waste with gelatin and poly(γ-glutamic acid) as carrier. J. Agric. Food Chem. 55, 5123–5130.

Choi, M.-J., Ruktanonchai, U., Min, S.-G., Chun, J.-Y., Soottitantawat, A., 2010a. Physical characteristics of fish oil encapsulated by β-cyclodextrin using an aggregation method or polycaprolactone using an emulsion–diffusion method. Food Chem. 119, 1694–1703.

Choi, K.-O., Ryu, J., Kwak, H.-S., Ko, S., 2010b. Spray-dried conjugated linoleic acid encapsulated with Maillard reaction products of whey proteins and maltodextrin. Food Sci. Biotechnol. 19, 957–965.

Chranioti, C., Tzia, C., 2013. Arabic gum mixtures as encapsulating agents of freeze-dried fennel oleoresin products. Food Bioprocess Technol. 7, 1057–1065.

Costa-Silva, T.A., Said, S., Souza, C.R.F., Oliveira, W.P., 2010. Stabilization of endophytic fungus *Cercospora kikuchii* lipase by spray drying in the presence of maltodextrin and β-cyclodextrin. Dry. Technol. 28, 1245–1254.

De Castro-Cislaghi, F.P., Silva, C.D.R.E., Fritzen-Freire, C.B., Lorenz, J.G., Sant'anna, E.S., 2012. Bifidobacterium bb-12 microencapsulated by spray drying with whey: survival under simulated gastrointestinal conditions, tolerance to nacl, and viability during storage. J. Food Eng. 113, 186–193.

Dewettinck, K., Huyghebaert, A., 1999. Fluidized bed coating in food technology. Trends Food Sci. Technol. 10, 163–168.

Dickinson, E., 2003. Hydrocolloids at interfaces and the influence on the properties of dispersed systems. Food Hydrocoll. 17, 25–39.

Dong, Z., Ma, Y., Hayat, K., Jia, C., Xia, S., Zhang, X., 2011. Morphology and release profile of microcapsules encapsulating peppermint oil by complex coacervation. J. Food Eng. 104, 455–460.

Donsi, F., Annunziata, M., Sessa, M., Ferrari, G., 2011. Nanoencapsulation of essential oils to enhance their antimicrobial activity in foods. LWT Food Sci. Technol. 44, 1908–1914.

Donsì, F., Annunziata, M., Vincensi, M., Ferrari, G., 2012. Design of nanoemulsion-based delivery systems of natural antimicrobials: effect of the emulsifier. J. Biotechnol. 159, 342–350.

Donthidi, A.R., Tester, R.F., Aidoo, K.E., 2010. Effect of lecithin and starch on alginate-encapsulated probiotic bacteria. J. Microencapsul. 27, 67–77.

Drusch, S., Berg, S., Scampicchio, M., Serfert, Y., Somoza, V., Mannino, S., Schwarz, K., 2009. Role of glycated caseinate in stabilisation of microencapsulated lipophilic functional ingredients. Food Hydrocoll. 23, 942–948.

Elshafei, G., El-Said, M.M., Attia, H.A.E., Mohammed, T.G.M., 2010. Environmentally friendly pesticides: essential oil-based w/o/w multiple emulsions for anti-fungal formulations. Ind. Crop. Prod. 31, 99–106.

Ezhilarasi, P.N., Indrani, D., Jena, B.S., Anandharamakrishnan, C., 2013. Freeze-drying technique for microencapsulation of Garcinia fruit extract and its effect on bread quality. J. Food Eng. 117, 513–520.

Farhang, B., 2009. Nanotechnology and applications in food safety. In: Gustavo, B.-C., Alan, M., David, L., Walter, S., Ken, B., Paul, C. (Eds.), Global Issues in Food Science and Technology. Academic Press, San Diego, CA.

Gabas, A.L., Telis, V.R.N., Sobral, P.J.A., Telis-Romero, J., 2007. Effect of maltodextrin and arabic gum in water vapor sorption thermodynamic properties of vacuum dried pineapple pulp powder. J. Food Eng. 82, 246–252.

Gavini, E., Sanna, V., Sharma, R., Juliano, C., Usai, M., Marchetti, M., Karlsen, J., Giunchedi, P., 2005. Solid lipid microparticles (SLM) containing juniper oil as anti-acne topical carriers: preliminary studies. Pharm. Dev. Technol. 10, 479–487.

Gbassi, G.K., Vandamme, T., Ennahar, S., Marchioni, E., 2009. Microencapsulation of *Lactobacillus plantarum* spp. in an alginate matrix coated with whey proteins. Int. J. Food Microb. 129, 103–105.

Gharsallaoui, A., Roudaut, G., Chambin, O., Voilley, A., Saurel, R., 2007. Applications of spray-drying in microencapsulation of food ingredients: an overview. Food Res. Int. 40, 1107–1121.

Glenn, G.M., Klamczynski, A.P., Woods, D.F., Chiou, B., Orts, W.J., Imam, S.H., 2010. Encapsulation of plant oils in porous starch microspheres. J. Agric. Food Chem. 58, 4180–4184.

Gortzi, O., Lala, S., Chinou, I., Tsaknis, J., 2007. Evaluation of the antimicrobial and antioxidant activities of *Origanum dictamnus* extracts before and after encapsulation in liposomes. Molecules 12, 932–945.

Guignon, B., Duquenoy, A., Dumoulin, E.D., 2002. Fluid bed encapsulation of particles: principles and practice. Dry. Technol. 20, 419–447.

Gutierrez, J., Barry-Ryan, C., Bourke, P., 2008. The antimicrobial efficacy of plant essential oil combinations and interactions with food ingredients. Int. J. Food Microb. 124, 91–97.

Gyawali, R., Hayek, S.A., Ibrahim, S.A., 2014a. Plant extracts as antimicrobials in food products: types. Handbook of Natural Antimicrobials for Food Safety and QualityWoodhead Publishing, Cambridge, UK, p. 31.

Gyawali, R., Hayek, S.A., Ibrahim, S.A., 2014b. Plant extracts as antimicrobials in food products: mechanisms of action, extraction methods, and applications. Handbook of Natural Antimicrobials for Food Safety and QualityWoodhead Publishing, Cambridge, UK, p. 49.

Gyawali, R., Ibrahim, S.A., 2012. Impact of plant derivatives on the growth of foodborne pathogens and the functionality of probiotics. Appl. Microbiol. Biotechnol. 95 (1), 29–45.

Gyawali, R., Ibrahim, S.A., 2014. Natural products as antimicrobial agents. Food Control 46, 412–429.

Gyawali, R., Ibrahim, S.A., Hasfa, S.H.A., Smqadri, S.Q., Haik, Y., 2011. Antimicrobial activity of copper alone and in combination with lactic acid against *Escherichia coli* O157:H7 in laboratory medium and on the surface of lettuce and tomatoes. J. Pathog. 2011, 650968.

Jafari, S.M., Assadpoor, E., Bhandari, B., He, Y., 2008. Encapsulation efficiency of food flavours and oils during spray drying. Dry. Technol. 26, 816–835.

Jagannath, A., Raju, P.S., Bawa, A.S., 2010. Comparative evaluation of bacterial cellulose (nata) as a cryoprotectant and carrier support during the freeze drying process of probiotic lactic acid bacteria. LWT Food Sci. Technol. 43, 1197–1203.

Jun-Xia, X., Hai-Yan, Y., Jian, Y., 2011. Microencapsulation of sweet orange oil by complex coacervation with soybean protein isolate/gum Arabic. Food Chem. 125, 1267–1272.

Kalogeropoulos, N., Yannakopoulou, K., Gioxari, A., Chiou, A., Makris, D.P., 2010. Polyphenol characterization and encapsulation in β-cyclodextrin of a flavonoid-rich *Hypericum perforatum* (St. John's wort) extract. LWT Food Sci. Technol. 43, 882–889.

Kanakdande, D., Bhosale, R., Singhal, R.S., 2007. Stability of cumin oleoresin microencapsulated in different combination of gum arabic, maltodextrin and modified starch. Carbohydr. Polym. 67, 536–541.

Kato, A., 2002. Industrial applications of Maillard-type protein-polysaccharide conjugates. Food Sci. Technol. Res. 8, 193–199.
Kaushik, V., Roos, Y.H., 2007. Limonene encapsulation in freeze-drying of gum Arabic–sucrose–gelatin systems. LWT Food Sci. Technol. 40, 1381–1391.
Krishnan, S., Bhosale, R., Singhal, R.S., 2005. Microencapsulation of cardamom oleoresin: evaluation of blends of gum arabic, maltodextrin and a modified starch as wall materials. Carbohydr. Polym. 61, 95–102.
Leimann, F.V., Gonçalves, O.H., Machado, R.A.F., Bolzan, A., 2009. Antimicrobial activity of microencapsulated lemongrass essential oil and the effect of experimental parameters on microcapsules size and morphology. Mater. Sci. Eng. C 29, 430–436.
Li, X.-Y., Jin, L.-J., Mcallister, T.A., Stanford, K., Xu, J.-Y., Lu, Y.-N., Zhen, Y.-H., Sun, Y.-X., Xu, Y.-P., 2007. Chitosan–alginate microcapsules for oral delivery of egg yolk immunoglobulin (IgY). J. Agric. Food Chem. 55, 2911–2917.
Liolios, C.C., Gortzi, O., Lalas, S., Tsaknis, J., Chinou, I., 2009. Liposomal incorporation of carvacrol and thymol isolated from the essential oil of *Origanum dictamnus* L. and in vitro antimicrobial activity. Food Chem. 112, 77–83.
Lv, Y., Yang, F., Li, X., Zhang, X., Abbas, S., 2014. Formation of heat-resistant nanocapsules of jasmine essential oil via gelatin/gum arabic based complex coacervation. Food Hydrocoll. 35, 305–314.
Malacrida, C.R., Ferreira, S., Telis, N., Vania, R., 2013. Stability at different temperatures of turmeric oleoresin encapsulated in maltodextrin/gelatin matrices by freeze-drying. J. Colloid Sci. Biotechnol. 2, 100–105.
Martins, I.M., Rodrigues, S.N., Barreiro, F., Rodrigues, A.E., 2009. Microencapsulation of thyme oil by coacervation. J. Microencapsul. 26, 667–675.
McClements, D.J., Decker, E.A., Weiss, J., 2007. Emulsion-based delivery systems for lipophilic bioactive components. J. Food Sci. 72, R109–R124.
Mourtzinos, I., Kalogeropoulos, N., Papadakis, S.E., Konstantinou, K., Karathanos, V.T., 2008. Encapsulation of nutraceutical monoterpenes in β-cyclodextrin and modified starch. J. Food Sci. 73, S89–S94.
Mourtzinos, I., Salta, F., Yannakopoulou, K., Chiou, A., Karathanos, V.T., 2007. Encapsulation of olive leaf extract in β-cyclodextrin. J. Agric. Food Chem. 55, 8088–8094.
Murugesan, R., Orsat, V., 2011. Spray drying of elderberry (*Sambucus nigra* L.) juice to maintain its phenolic content. Dry. Technol. 29, 1729–1740.
Nakagawa, K., Nagao, H., 2012. Microencapsulation of oil droplets using freezing-induced gelatin–acacia complex coacervation. Colloids Surf. A 411, 129–139.
Nori, M.P., Favaro-Trindade, C.S., De Alencar, S.M., Thomazini, M., De Camargo Balieiro, J.C., Castillo, C.J.C., 2011. Microencapsulation of propolis extract by complex coacervation. LWT Food Sci. Technol. 44, 429–435.
Nunes, I.L., Mercadante, A.Z., 2007. Encapsulation of lycopene using spray-drying and molecular inclusion processes. Braz. Arch. Biol. Technol. 50, 893–900.
Ozyildiz, F., Karagonlu, S., Basal, G., Uzel, A., Bayraktar, O., 2013. Micro-encapsulation of ozonated red pepper seed oil with antimicrobial activity and application to nonwoven fabric. Lett. Appl. Microbiol. 56, 168–179.
Paramera, E.I., Konteles, S.J., Karathanos, V.T., 2011. Stability and release properties of curcumin encapsulated in *Saccharomyces cerevisiae* β-cyclodextrin and modified starch. Food Chem. 125, 913–922.
Perez, A.A., Carrara, C.R., Sanchez, C.C., Santiago, L.G., Patino, J.M.R., 2009. Interfacial dynamic properties of whey protein concentrate/polysaccharide mixtures at neutral pH. Food Hydrocoll. 23, 1253–1262.
Poshadri, A., Kuna, A., 2010. Microencapsulation technology: a review. J. Res. Angrau 38, 86–102.

Rajam, R., Karthik, P., Parthasarathi, S., Joseph, G.S., Anandharamakrishnan, C., 2012. Effect of whey protein—alginate wall systems on survival of microencapsulated *Lactobacillus plantarum* in simulated gastrointestinal conditions. J. Funct. Foods 4, 891–898.

Rascon, M.P., Beristain, C.I., García, H.S., Salgado, M.A., 2011. Carotenoid retention and storage stability of spray-dried encapsulated paprika oleoresin using gum arabic and soy protein isolate as wall materials. LWT Food Sci. Technol. 44, 549–557.

Robert, P., Gorena, T., Romero, N., Sepulveda, E., Chavez, J., Saenz, C., 2010. Encapsulation of polyphenols and anthocyanins from pomegranate (*Punica granatum*) by spray drying. Int. J. Food Sci. Technol. 45, 1386–1394.

Rodea-Gonzalez, D.A., Cruz-Olivares, J., Roman-Guerrero, A., Rodriguez-Huezo, M.E., Vernon-Carter, E.J., Perez-Alonso, C., 2012. Spray-dried encapsulation of chia essential oil (*Salvia hispanica* L.) in whey protein concentrate-polysaccharide matrices. J. Food Eng. 111, 102–109.

Romo-Hualde, A., Yetano-Cunchillos, A.I., González-Ferrero, C., Saiz-Abajo, M.J., Gonzalez-Navarro, C.J., 2012. Supercritical fluid extraction and microencapsulation of bioactive compounds from red pepper (*Capsicum annum* L.) by-products. Food Chem. 133, 1045–1049.

Sahu, A., Kasoju, N., Goswami, P., Bora, U., 2011. Encapsulation of curcumin in pluronic block copolymer micelles for drug delivery applications. J. Biomater. Appl. 25, 619–639.

Sanchez, V., Baeza, R., Galmarini, M.V., Zamora, M.C., Chirife, J., 2011. Freeze-drying encapsulation of red wine polyphenols in an amorphous matrix of maltodextrin. Food Bioprocess Technol. 6, 1350–1354.

Seo, E.-J., Min, S.-G., Choi, M.-J., 2010. Release characteristics of freeze-dried eugenol encapsulated with β-cyclodextrin by molecular inclusion method. J. Microencapsul. 27, 496–505.

Shah, B.D., 2011. Nanodispersing lipophilic antimicrobials for improved food safety. PhD Dissertation. The University of Tennessee, Knoxville, TN.

Shah, B., Davidson, P.M., Zhong, Q., 2012a. Nanocapsular dispersion of thymol for enhanced dispersibility and increased antimicrobial effectiveness against *Escherichia coli* O157:H7 and *Listeria monocytogenes* in model food systems. Appl. Environ. Microbiol. 78, 8448–8453.

Shah, B.D., Davidson, P.M., Zhong, Q., 2012b. Encapsulation of eugenol using Maillard-type conjugates to form transparent and heat stable nanoscale dispersions. LWT Food Sci. Technol. 49, 139–148.

Shah, B.D., Davidson, P.M., Zhong, Q., 2013. Nanodispersed eugenol has improved antimicrobial activity against *Escherichia coli* O157:H7 and *Listeria monocytogenes* in bovine milk. Int. J. Food Microb. 161, 53–59.

Shah, B., Ikeda, S., Davidson, P.M., Zhong, Q., 2012c. Nanodispersing thymol in whey protein isolate-maltodextrin conjugate capsules produced using the emulsion–evaporation technique. J. Food Eng. 113, 79–86.

Shaikh, J., Bhosale, R., Singhal, R.S., 2006. Microencapsulation of black pepper oleoresin. Food Chem. 94, 105–110.

Shinde, U., Nagarsenker, M., 2011. Microencapsulation of eugenol by gelatin-sodium alginate complex coacervation. Indian J. Pharm. Sci. 73, 311–315.

Silva, D.F., Favaro-Trindade, C.S., Rocha, G.A., Thomazini, M., 2012. Microencapsulation of lycopene by gelatin–pectin complex coacervation. J. Food Process. Preserv. 36, 185–190.

Soottitantawat, A., Bigeard, F., Yoshii, H., Furuta, T., Ohkawara, M., Linko, P., 2005. Influence of emulsion and powder size on the stability of encapsulated d-limonene by spray drying. Innov. Food Sci. Emerg. Technol. 6, 107–114.

Sousdaleff, M., Baesso, M.L., Neto, A.M., Nogueira, A.C., Marcolino, V.A., Matioli, G., 2012. Microencapsulation by freeze-drying of potassium norbixinate and curcumin with maltodextrin: stability, solubility, and food application. J. Agric. Food Chem. 61, 955–965.

Spada, J.C., Marczak, L.D.F., Tessaro, I.C., Noreña, C.P.Z., 2012. Microencapsulation of β-carotene using native pinhão starch, modified pinhão starch and gelatin by freeze-drying. Int. J. Food Sci. Technol. 47, 186–194.

Sutter, S.C., Buera, M.P., Elizalde, B.E., 2007. β-Carotene encapsulation in a mannitol matrix as affected by divalent cations and phosphate anion. Int. J. Pharm. 332, 45–54.

Tassou, C.C., Drosinos, E.H., Nychas, G.J.E., 1995. Effects of essential oil from mint (*Mentha piperita*) on *Salmonella enteritidis* and *Listeria monocytogenes* in model food systems at 4° and 10°C. J. Appl. Microbiol. 78, 593–600.

Teixeira, M.I., Andrade, L.R., Farina, M., Rocha-Leao, M.H.M., 2004. Characterization of short chain fatty acid microcapsules produced by spray drying. Mater. Sci. Eng. C 24, 653–658.

Tonon, R.V., Grosso, C.R.F., Hubinger, M.D., 2011. Influence of emulsion composition and inlet air temperature on the microencapsulation of flaxseed oil by spray drying. Food Res. Int. 44, 282–289.

Toure, A., Lu, H.B., Zhang, X., Xueming, X., 2011. Microencapsulation of ginger oil in 18DE maltodextrin/whey protein isolate. J. Herbs Spices Med. Plants 17, 183–195.

Vaidya, S., Bhosale, R., Singhal, R.S., 2006. Microencapsulation of cinnamon oleoresin by spray drying using different wall materials. Dry. Technol. 24, 983–992.

Varona, S., Martin, A., Cocero, M.J., 2011. Liposomal incorporation of lavandin essential oil by a thin-film hydration method and by particles from gas-saturated solutions. Ind. Eng. Chem. Res. 50, 2088–2097.

Wang, T., Lucey, J.A., 2003. Use of multi-angle laser light scattering and size-exclusion chromatography to characterize the molecular weight and types of aggregates present in commercial whey protein products. J. Dairy Sci. 86, 3090–3101.

Wang, R., Tian, Z., Chen, L., 2011. A novel process for microencapsulation of fish oil with barley protein. Food Res. Int. 44, 2735–2741.

Wannes, W.A., Mhamdi, B., Sriti, J., Ben Jemia, M., Ouchikh, O., Hamdaoui, G., Kchouk, M.E., Marzouk, B., 2010. Antioxidant activities of the essential oils and methanol extracts from myrtle (*Myrtus communis* var. *italica* L.) leaf, stem and flower. Food Chem. Toxicol. 48, 1362–1370.

Weiss, J., Gaysinsky, S., Davidson, M., McClements, J., 2009. Nanostructured encapsulation systems: food antimicrobials. In: Gustavo, B.-C., Alan, M., David, L., Walter, S., Ken, B., Paul, C. (Eds.), Global Issues in Food Science and Technology. Academic Press, San Diego, CA, (Chapter 24).

Weiss, J., Takhistov, P., McClements, D.J., 2006. Functional materials in food nanotechnology. J. Food Sci. 71, R107–R116.

Ying, D.Y., Sun, J., Sanguansri, L., Weerakkody, R., Augustin, M.A., 2012. Enhanced survival of spray-dried microencapsulated *Lactobacillus rhamnosus* GG in the presence of glucose. J. Food Eng. 109, 597–602.

Yoshida, P.A., Yokota, D., Foglio, M.A., Rodrigues, R.A.F., Pinho, S.C., 2010. Liposomes incorporating essential oil of Brazilian cherry (*Eugenia uniflora* L.): characterization of aqueous dispersions and lyophilized formulations. J. Microencapsul. 27, 416–425.

Yu, H., Huang, Q., 2010. Enhanced in vitro anticancer activity of curcumin encapsulated in hydrophobically modified starch. Food Chem. 119, 669–674.

Yu, C., Wang, W., Yao, H., Liu, H., 2007. Preparation of phospholipid microcapsule by spray drying. Dry. Technol. 25, 695–702.

Zhang, W., Rong, J., Wang, Q.I.N., He, X., 2009. The encapsulation and intracellular delivery of trehalose using a thermally responsive nanocapsule. Nanotechnology 20, 275101.

8

SUPRAMOLECULAR STRATEGY OF THE ENCAPSULATION OF LOW-MOLECULAR-WEIGHT FOOD INGREDIENTS

Lucia Zakharova*,, Alla Mirgorodskaya*,**, Gulnara Gaynanova*,**, Ruslan Kashapov*,**, Tatiana Pashirova*,**, Elmira Vasilieva*,**, Yuriy Zuev[†], Oleg Synyashin*,****

**A.E. Arbuzov Institute of Organic and Physical Chemistry, Kazan Scientific Center, Russian Academy of Sciences, Kazan, Russian Federation; **Kazan National Research Technological University, Kazan, Russian Federation; [†]Kazan Institute of Biochemistry and Biophysics, Kazan Scientific Center, Russian Academy of Sciences, Kazan, Russian Federation*

1 Introduction

Due to increasing concern about human health, modern food technologies usually involve complicated formulations, with various bioactive ingredients used (Sagalowicz and Leser, 2010; Gutierrez et al., 2013; Maswal and Dar, 2014; McClements, 2013; Davidov-Pardo and McClements, 2014; Fathi et al., 2014). This is needed for stabilizing food products and improving their organoleptic characteristics. However, a large number of additives (stabilizers, preservatives, vitamins, and nutraceuticals) are hydrophobic substances poorly mixing with aqueous food dispersion and unstable against light, temperature, pH changing, and oxygen. Therefore, one of the key challenges of the food industry is to design effective delivery systems with controlled binding/release functions toward active components. While many stable biocompatible delivery systems with low toxicity, high loading capacity, and controlled release behavior have been designed, the majority of them are oriented toward pharma practice. To meet the criteria of the food industry, additional factors should be taken into account. In particular, (1) food products have to be prepared for daily consumption, and therefore all ingredients

Encapsulations. http://dx.doi.org/10.1016/B978-0-12-804307-3.00008-9

must be Generally Recognized as Safe (GRAS); (2) all the carrier components should be inert with regard to their influence on the transparency, color, flavor, and food value of products; (3) delivery systems need to show long-term stability including periods of food processing, transport and storage, thereby preserving their protective function toward food product. For example, these formulations should enhance the stability of sensitive compounds, for example, vitamins (Ziani et al., 2012; de Britto et al., 2012), decrease evaporation and degradation of volatile bioactives, for example, aromas (Maswal and Dar, 2014; Fadida et al., 2015), mask unpleasant tastes (Gharsallaoui et al., 2012), or limit exposure to oxygen, water, or light (Nedovic et al., 2011).

This chapter focuses on three lines of investigations, that is, (1) amphiphile-based formulations (micelles, micro- and nanoemulsions, liposomes, solid lipid nanoparticles); (2) polymer-based strategies for the encapsulation of food ingredients including aromas and flavors; and (3) molecular complexes based on the guest–host interactions. The survey of recent literature demonstrates that the first direction has received much attention, since the use of amphiphiles makes it possible to fabricate soft formulations through noncovalent methods with controlled size, morphology, and binding capacity. Therefore, this theme is addressed in a separate section, in which the application of lipid formulations in food industry is the main subject. Besides, the design of novel amphiphilic molecules, the elucidation of their self-assembly, and the structure-activity relations are discussed as well. The next section is devoted to the polymer-based strategies, with natural polymers emphasized. Milk proteins (casein), polysaccharides (chitosan), and others are discussed, including coating techniques through layer-by-layer deposition with the use of synthetic and particularly edible polyelectrolytes. The final part of the chapter concentrates on the involvement of macrocyclic compounds (calixarenes, cyclodextrines) for binding of small guest molecules via selective guest–host interactions; as well as on the binding/release behavior of loads, including both trigger-like mechanisms and sustained release.

2 Self-Assembled Delivery Systems Based on Amphiphilic Compounds

2.1 Structure-Activity Correlation as a Basis for the Design of Nanocontainers

The surfactant-based formulations are widely used in food industry, which is mainly due to their ability of self-assembling in solutions. Surfactant aggregates formed in aqueous media, the so-called direct micelles, have nonpolar interior responsible for their solubilization ability toward hydrophobic guests. Our scientific activity focuses on the design of amphiphilic compounds and elucidation of the structure-activity correlation (Zakharova et al., 2004, 2010b, 2012c; Zakharova and Konovalov, 2012). In particular, the correlation between chemical structure, morphology of aggregates and their functional activity (solubilization power, catalytic and antimicrobial activities, and drug delivery efficacy) has been confirmed. Despite the fact that cationic surfactants show higher toxicity compared to nonionic counterparts, they possess marked advantages due to their high affinity toward different biospecies, for example, DNA, cell surfaces, negatively charged proteins, and so forth. Besides, their head group is easily functionalized, which provides the possibility for the control of their morphological behavior. Therefore, systematic investigations involving homological series of cationic surfactants make it possible to elucidate the role of hydrophobicity, the structure of head group, and the spacer length in the case of dimeric (gemini) surfactants.

The series of cationic surfactants with alkyltrimethylammonium (TMA), triphenylphosphonium (TPP) (Gainanova et al., 2012; Vagapova et al., 2013a,b), morpholinium (Mirgorodskaya et al., 2014b; Yackevich et al., 2014), and 1,4-diazabicyclo[2.2.2]octane (DABCO) (Zhiltsova et al., 2012; Zakharova et al., 2012a; Karpichev et al., 2014; Pashirova et al., 2015) head groups were studied including mono- and dicationic analogs (Mirgorodskaya et al., 2012a,b, 2014c) (Fig. 8.1). Apart from single surfactant solutions, mixed systems with cosurfactants (Mirgorodskaya et al., 2014a) and polymers (Vasilieva et al., 2014a; Zakharova et al., 2010a; Mirgorodskaya et al., 2012c) were investigated as well. While the key role of hydrophobicity in surfactant aggregation is reliably documented, effect of the modification of head group structure is not predictable. Therefore, the structure-activity relations may provide an effective tool for the control of the aggregation behavior and functional activity of supramolecular assemblies, including their use as delivery systems. This may be exemplified by the replacement of methyl by hydroxyethyl fragment in the head group of TMA surfactants, which resulted in enhancement of solubilization capacity and catalytic effect of the systems (Mirgorodskaya et al., 2012a,b). This effect increases upon transition from mono- to dicationic analogs and is probably due to the additional involvement of specific interactions in the case of hydroxyethylated surfactants. Interestingly, catalytic (or inhibition) activity of amphiphiles in combination with their influence

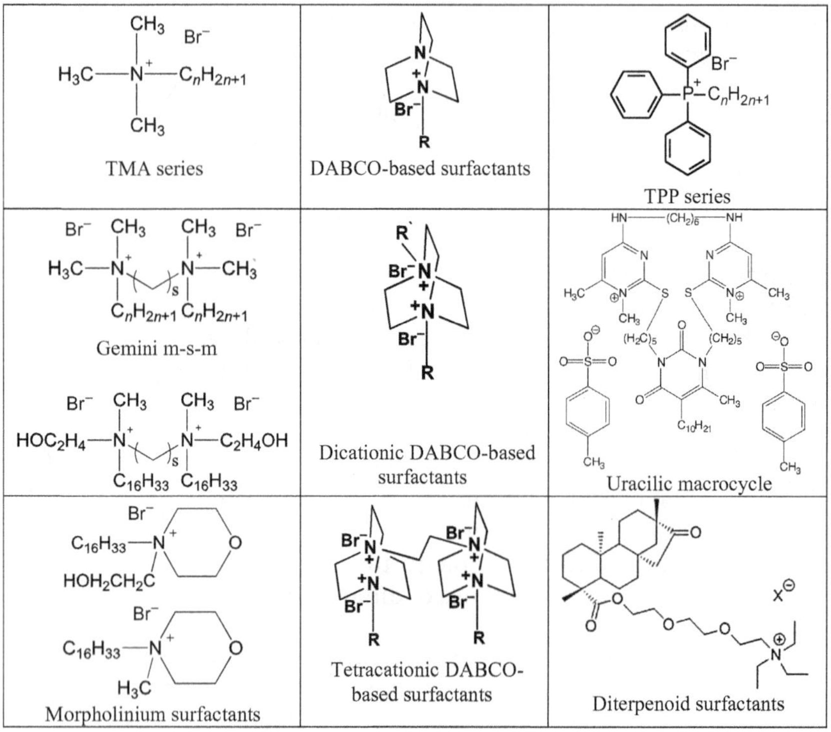

Figure 8.1. Structural formulas of cationic amphiphiles.

on the acid-based equilibria may be of particular interest from the viewpoint of modifying the stability of active components or degradation of pollutants in medicine, food, and analytical and ecological applications.

An even stronger effect is observed upon the variation of the scaffold of polar fragment or the nature of the charged atom, as it occurs in the series TMA–morpholinium–DABCO–TPP surfactants. While critical micelle concentration (cmc) shows little changes with the transition from TMA to morpholinium and DABCO head group, their aggregation behavior markedly differs. In particular, sharp growth of micelles occurs in aqueous solution of DABCO-16 micelles around the cmc (Zhiltsova et al., 2012; Pashirova et al., 2010), with aggregate size exceeding 100 nm. Mono- and diquaternized DABCO derivatives show superior solubilization, catalytic, and antimicrobial activities over those of TMA surfactants. Increase in alkyl chain length typically enhances aggregation and catalytic properties, while a maximal antimicrobial effect corresponds to DABCO-14 or DABCO-16 derivatives (Fig. 8.2).

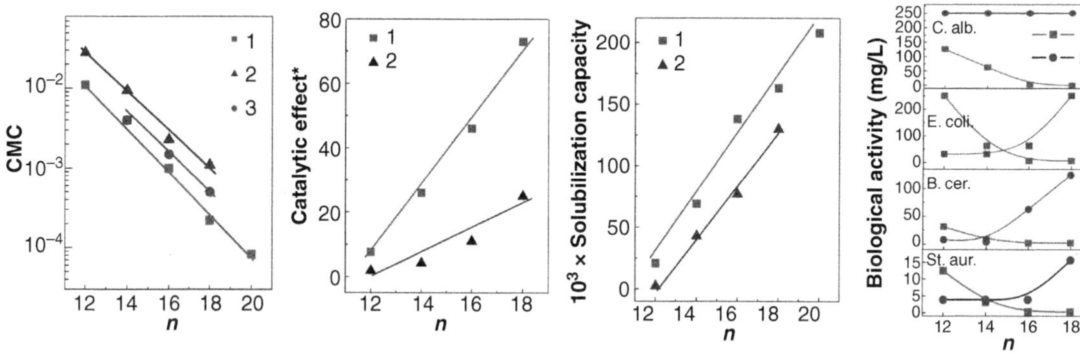

Figure 8.2. From left to right: cmc; the catalytic effect in reaction of hydrolysis of phosphorus acid esters (4-nitrophenyl ethylchloromethylphosphonate, triangles and 4-nitrophenyl butylchloromethylphosphonate, squares); the solubilization capacity toward hydrophobic dye Sudan I and the biological activity toward the collection of test-strains of microorganisms [*Staphylococcus aureus*-209 P *(St.aur.)*; *Escherichia coli* F50 *(E. coli)*; *Bacillus cereus* 8035 *(B. cer.)* and *Candida albicans (C. alb.)*] as function of number of carbon atoms in mono- (1), di-(2), and tetra- (3) quaternized derivatives of 1,4-diazabicyclo[2.2.2]octanes shown in Fig. 8.1.

The variation of charged atom in cationic surfactants from nitrogen to phosphorus results in a marked change in the interaction of charged species in the Stern layer, including their solvation and degree of counterion binding. For TPP series, a 10-fold lower cmc and markedly lower Krafft temperatures are observed compared to TMA surfactants (Gainanova et al., 2012), which is followed by increase in solubilization capacity and catalytic effect toward alkali degradation of organophosphorus ecotoxicants (Fig. 8.3).

Additional tools of the control of structural behavior and functional activity can be provided by the use of binary or polycomponent formulations (Zakharova et al., 2004, 2010a,b; Zakharova and Konovalov, 2012; Vagapova et al., 2013a,b; Mirgorodskaya et al., 2012c, 2014a; Vasilieva et al., 2014a). Addition of cosurfactants (Mirgorodskaya et al., 2014a), uncharged polymers (Vagapova et al., 2013a,b; Mirgorodskaya et al., 2012c) or polyelectrolytes (Vasilieva et al., 2014a; Zakharova et al., 2010a) results in synergetic decrease in cmc, as well as enhancement in solubilization capacity and catalytic activity. It is noteworthy that size behavior undergoes the most considerable modification upon transition from single surfactant solution to mixed systems, including polymer-colloid complexes, which makes it possible to precisely tune the size, morphology, and functional properties of the assembles (Vagapova et al., 2013a,b; Vasilieva et al., 2014a; Zakharova et al., 2010a; Mirgorodskaya et al., 2012c). Marked effect on properties of supramolecular systems can be achieved upon their structural and phase transitions, for example, reverse (water/oil)

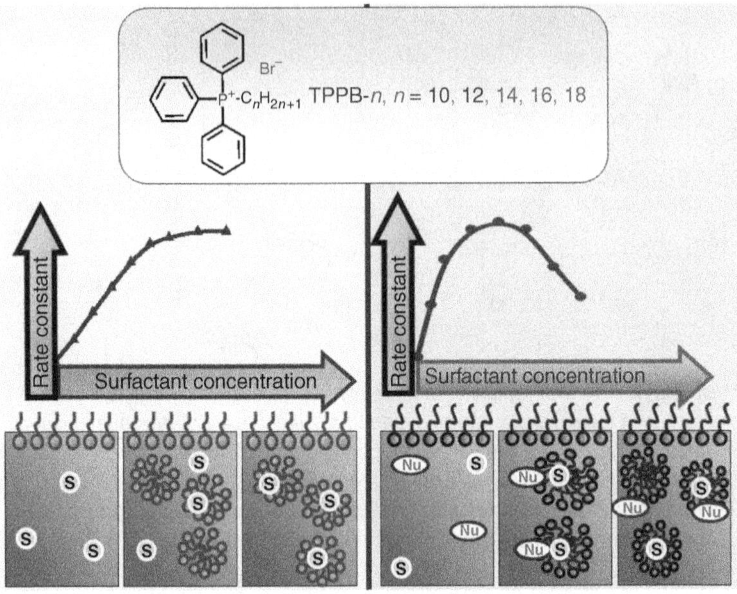

Figure 8.3. Schematic representation of solubilization and catalytic effect of TPP micelles toward bioactive substrates of lower (on the left) and higher (on the right) hydrophobicity; here S denotes substrate and Nu denotes nucleophile.

to bicontinuous to direct (oil/water) microemulsions (Fig. 8.4) (Zuev et al., 2004); sphere-rod micellar transition (Zakharova et al., 2004, 2010b), temperature induced percolation in reverse micelles (Zakharova et al., 2004, 2010b), micelle-to-vesicle (Voronin et al., 2013; Gabdrakhmanov et al., 2013), micelle-to-liquid crystal (Zakharova et al., 2004), and so forth.

To decrease the toxicity and enhance the biocompatibility of the systems, cationic surfactants with natural fragments—uracilic (Voronin et al., 2011; Zakharova et al., 2012d, 2014; Kharlamov et al., 2013b), diterpenoid (Voronin et al., 2013; Gabdrakhmanov et al., 2013; Kataev et al., 2014), glucamine (Gabdrakhmanov et al., 2015), and nonionic amphiphiles—were involved as well. For the series of dicationic uracilic (both of open-chain and macrocyclic) surfactants aggregation, solubilization and functional activities have been examined and compared with those of monocationic analog and TMA series. Their solution behavior was shown to depend on hydrophobicity, especially on the presence of long chain alkyl group in the uracil fragment (Zakharova and Konovalov, 2012). The latter markedly enhances the aggregation ability and functional activity. Due to the shielding effect of uracilic moiety localized around head groups, low binding degree of counterions and hence a rather high surface charge of aggregates

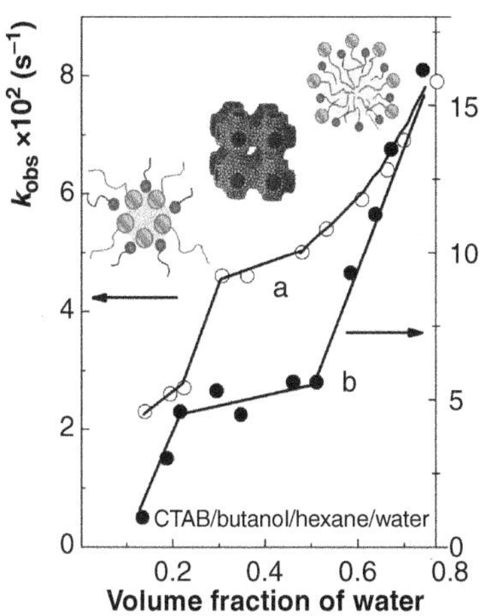

Figure 8.4. Dependences of observed rate constant of basic hydrolysis of 4-nitrophenyl acetate *(a)* and 4-nitrophenyl caprylate *(b)* on volume fraction of water in quaternary microemulsions (NaOH concentration in aqueous phase is 0.01 mol/L).

occurs. This in turn causes the polarization and even the dissociation of water in solvate shells, thereby generating the OH^-/OH_3^+ pair. While OH^- ions bind with nitrogen cations, H^+ ions acidify the bulk solution. For mixed systems based on uracilic macrocycle (Fig. 8.1) and nonionic surfactant, Triton X-100 stimuli-responsive behavior was revealed, with transition from cationic mixed micelles to nonionic aggregates occurring at the variation of pH (Kharlamov et al., 2013b). This can be used for the control of binding capacity of the system toward charged solutes.

2.2 Application of Micelles and Microemulsions in the Food Industry

The classification of food additives based on their technological function divides them on five categories, that is, agents (1) improving the color, flavor, and taste of products; (2) modifying the texture and appearance; (3) enhancing stability and shelf life of products; (4) facilitating the technological processes; (5) auxiliary substances. As supplements, natural compounds and synthetic counterparts with similar structures and properties are typically used, as well as novel nontoxic synthetic materials (Fig. 8.5). The use of these additives is often limited by their poor mixing with food, which can be overcome by introducing the dispersing additives. Micellizable surfactants favoring the formation of stable polycomponent colloid systems (microdispersions) are widely

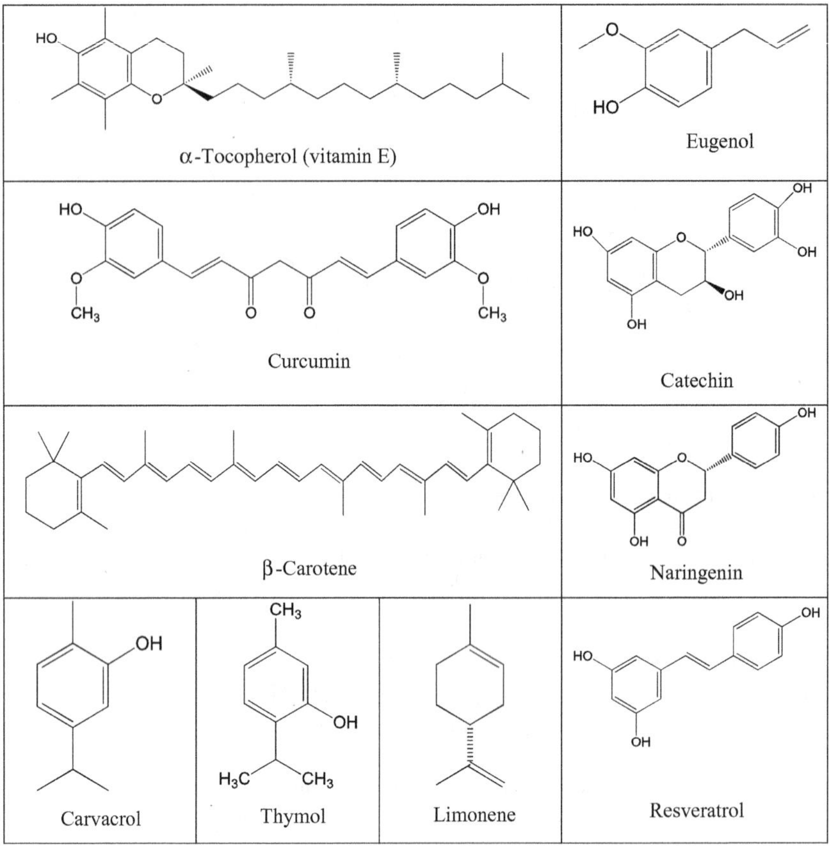

Figure 8.5. Structural formulas of selected active ingredients used for the modification and stabilization of food products.

used as dispersing agents. Among them, solubilizers promoting the formation of transparent thermodynamically stable colloid systems, that is, micellar solutions and microemulsions, are of particular interest. The use of solubilizers makes it possible to introduce hydrophilic solutes into oil media, and vice versa to mix hydrophobic compounds with aqueous solution, thereby increasing their biocompatibility and environmental protection (Garti and Aserin, 2012; Flanagan and Singh, 2006). As mentioned previously, solubilization ability of surfactant solution is caused by their self-assembly. Beyond cmc, direct micelles with nonpolar interior are formed in aqueous solution, while reverse micelles with polar head groups inside occur in oil media (Mittal, 1977). Solubilization capacity of surfactants strongly depends on their structure and increases with alkyl chain length (Zakharova et al., 2004, 2010b, 2012c; Zakharova and Konovalov, 2012; Gainanova et al., 2012; Zhiltsova et al., 2015). For cationic surfactants, the essential role of head group structure was revealed (Zakharova and Konovalov, 2012; Gainanova et al., 2012); in particular, solubilization capacity toward hydrophobic dye increases in the series TMA < TPP < DABCO head group.

A variety of publications are devoted to the application of surfactants in analytical assays for the quality control of food products, with micellar liquid chromatography (MLC) attracting the growing interest (El-Shaheny et al., 2015). In MLC technique, surfactant micellar solution is used as a micellar mobile phase, with following surfactant properties involved: (1) ability of solubilizing and concentrating of substances that are insoluble in dispersion media; (2) selective solubilization effect contributed by hydrophobic, electrostatic, donor–acceptor, and stacking interactions allowing to differentiate and separate components of dispersions; (3) ability of surfactant monomer adsorbing on the sorbent surface, thereby dynamically modifying the stationary phase. Substantial advantages are offered by MLC over other analytical techniques in terms of retention and selectivity, time saving, direct injection of the sample, high reproducibility, simultaneous separation of ionic and nonionic species, rapid gradient capability, and so forth. MLC technique is widely used for the control of harmful impurities in food products, for example, drugs (Nasr et al., 2014), toxic biogenic amines (cadaverine, 2-phenylethylamine, histamine, and spermidine) in fish sauce, antioxidants in meat products (Chin-Chen et al., 2013), and so forth.

In the food industry, micellar extraction is also widely applied for the recovery of single components or purification of edible raw materials, for example, purification of pectinase enzyme from psidium guajava (Amid et al., 2015b), lipase from pumpkin (Amid et al., 2015a), with the valuable enzyme recovered. Reverse micelles of anionic surfactants sodium dodecyl sulfate (SDS) and aerosol OT (AOT) were shown to be effective for the extraction of peanut protein from full-fat peanut powder (Guo et al., 2015). Extraction rate of around 80% was achieved by the optimization of conditions, such as pH, ionic force, and temperature. Reverse micelle extraction makes it possible to determine antibiotics in edible products (Chuo et al., 2014), to separate phenolic contaminants from olive mill wastewater (El-Abbassi et al., 2014), and so forth. Surfactant micellar solutions are of practical importance in analytical techniques of qualitative and quantitative analyses of food colorants (Sha et al., 2015) and their precursors, for example, for the detection of melamine (Nascimento et al., 2015) added to animal feed to overmark protein content.

Investigation of micellar systems is of interest from the viewpoint of modeling the processes occurring at the biointerface, which may provide information on their potential application in food processing and predict the possible results of interaction between food ingredients and solubilizers. In particular, a paper from Bhardwaj et al. (2014) discusses the interaction of cationic surfactant cetyltrimethylammonium bromide (CTAB) with oxidation inhibitors that can influence the antioxidant protection efficacy.

For chemically unstable additives, for example, flavoring agent citral, which is poorly soluble in water, the influence of surfactant on their decomposition should be taken into account (Park et al., 2015), especially in acidic conditions. This is exemplified by the kinetic study of the degradation of citral in nonionic micelles of Brij surfactants with different hydrophilic/lipophilic length ratio. The surfactant with a large polar fragment is shown to be more effective in the stabilization of the flavor, while the alkyl chain length slightly affects the destruction rate (Park et al., 2015). When surfactants are used as solubilizers of colorants, the surfactant influence on spectral characteristics, that is, color of solution should be controlled. In (Piyarat et al., 2014) the effect of surfactants of different types on the spectral behavior of red dyes from natural sources, roselle and lac, was studied. The former appeared to discolor in the presence of the surfactants, especially Tween-80, while the latter was rather stable.

Gemini surfactants with two bridged head groups bearing alkyl chains are of particular interest from the viewpoint of green chemistry criteria. Extremely low cmc values (within micromolar range) make it possible to decrease markedly their concentrations. Besides, these surfactants are characterized by high solubilization capacity, selectivity, and diverse morphological behavior, which give them advantages over single-chain analogs (Menger and Keiper, 2000). Interaction of dicationic surfactants of *m-s-m* type (here *m* and *s* are the number of carbon atoms in alkyl chains and polymethylene spacer, respectively) (Fig. 8.1) with food dye tartrazine measured by tensiometry, dynamic light scattering, transmission electron microscopy, and UV-Vis techniques was reported by Shahir et al. (2011), which can be used for analytical control of the colorant. Biodegradable gemini surfactants/xanthine oxidase complexation were testified by tensiometric, spectroscopic, microscopic, and molecular modeling techniques that may be of particular interest for the understanding of protein/amphiphile interactions involved in food processing (Akram et al., 2015). Upon the application of gemini surfactants with biologically active compounds, special attention should be paid to their marked influence on such properties of solutes as pK_a, oxidation, and hydrolytic stability, since micelles show ability to shift acid–base equilibria and accelerate (or retard) chemical processes. It is notable that geminis exert higher effect at lower concentrations compared to single-chain analogs, with even slight variation in surfactant structure inducing marked impact on the solubilization behavior. Such fine effects of geminis can be exemplified by the control of hydrolytic stability of fat acid esters (Mirgorodskaya et al., 2012a) and acid–base behavior of curcumin (Ke et al., 2014).

One of the most practically important carrier families based on surfactants includes microemulsions (ME). They are widely used in nano- and biotechnology including food industry (Djekic et al., 2013). MEs are self-assembling, thermodynamically stable, and macroscopically homogeneous disperse systems composed of water and oil phases, separated/stabilized by the surfactant layer (ternary ME), generally including cosurfactants, that is, C4—C6 alcohols and amines (quaternary ME) (Friberg and Bothorel, 1987). While the boundary between micelle (swollen micelle) and ME is quite relative, MEs are characterized by larger droplet size (up to 100 nm), higher surfactant concentration, larger fraction of disperse phase, and markedly higher solubilization power. The use of ME makes it possible to overcome the problem of solubility and compatibility of components, to control the reactivity and stability of systems, and to solve the tasks connected with preparing the ultradisperse materials required in nanotechnologies. MEs were reported to be used as nanoreactors for different processes (Tiwary, 2013; Holmberg, 2007; Zuev et al., 2004; Mirgorodskaya et al., 2013), including polymerization, oxidation, condensation, hydrolysis, esterification, and so forth. In biotechnologies ME-based techniques are widely involved for the enzymatic (lipase) hydrolysis of ester bonds and protein bioseparation processes (Das et al., 2014; Sintra et al., 2014). Application of MEs in food industry is based on their high ability of solubilizing dyes, flavors, taste substances, and vitamins. Tween-80 based bicontinuous ME was reported to increase markedly the solubility of carotenoids in aqueous media, thereby improving their bioavailability upon passing through gastrointestinal tract (Chow et al., 2015). This formulation provides a 17% enhancement of the carotenoids level in the serum and up to 30% color intensification compared to control protocol. Roohinejad et al. (2014, 2015); Chen and Zhong (2015) report on the optimization of the ME-based formulations for the encapsulation of β-carotene with the purpose of fabrication of particles of desirable sizes, with high loading capacity toward β-carotene and low cytotoxicity. A lot of publications are devoted to the solubilization of the natural polyphenolic dye curcumin, which is used for yellow coloring of food product and shows a marked antioxidant effect, antimicrobial, antiphlogistic, hepatoprotective, and other kinds of biological activities (Naksuriya et al., 2014). Many ME formulations based on synthetic surfactants and lipids were tested to increase the solubility and hence bioavailability of curcumin, with the oil/surfactant nature and ratio varied (eg, soybean oil, olive oil, wheat germ oil, vitamin E as a bulk phase, and Cremophor EL, Tween 20, Tween 80, or lecithin as an amphiphile) (Lin et al., 2014; Bergonzi et al., 2014).

Water-based MEs are successfully used for the solubility enhancement of food-grade flavor ingredient carvacrol showing antioxidant and antimicrobial activity (Shaaban and Edris, 2015). Factor affecting the efficacy of formulations based on nonionic and cationic surfactants including cosurfactants were tested to fabricate carvacrol microemulsion system. Rather high surfactant/carvacrol ratio of 9:1 was found to be required to solubilize 1.0 wt% carvacrol.

Oil/water MEs based on nonionic surfactants Tween-80 and Span-20 were formulated to increase the bioavailability of bioflavonoids exemplified by sour cherry kernel extract (Salimi et al., 2014). These MEs are nontoxic and show therapeutic effects (improving the cardiac function recovery) within the food formulation doze. Tween-80 MEs are shown to be nontoxic and make it possible to reach a 1000-fold enhancement of the solubility of the flavonoid norartocarpetin exhibiting high tyrosinase inhibitory activity (Zheng et al., 2015). These formulations are stable within 8 weeks under wide surfactant/oil/water ratio and may be of practical interest for preventing the browning effect in fresh-cut fruits.

Stable dilutable ME formulations based on Tween-20/lecithin mixture with droplet size of ≤12 nm were fabricated by Chen et al. (2015a) for the dissolution of peppermint oil (food, medicine, and cosmetic component). The lipid additives were shown to improve the stability of formulation (MEs underwent up to 500-fold water dilution and were stable over 70 days) and its loading capacity toward coenzyme Q10, as well as prevent its light degradation. In (Tu et al., 2013) octacosanol-loaded soybean oil/water microemulsion was prepared, with lecithin and ethanol used as a cosurfactant/stabilizer. The active component, fat alcohol octacosanol, shows antioxidant activity and increases human endurance; therefore overcoming the problem of low water solubility may give rise to its application in sports beverage. The extension of stable ME region was reported to depend on lipid/alcohol ratio, with 1:6 (w/w) ratio being optimal.

MEs show high technological potentiality toward different antioxidants from both natural and synthetic sources, for example, polyphenols, gallic acid, *p*-hydroxybenzoic acid, protocatechuic acid, and tyrosol (Ohara and Ohyama, 2014; Chatzidaki et al., 2015). The encapsulation in reverse ME is documented to prevent their degradation in the digestive tract, thereby maintaining high radical scavenging activity.

Separate research direction is the application of ME in technological processes connected with synthesis of food components. This can be exemplified by work (Sutili et al., 2015), in which continuous-flow esterification of protected fructose is reported, with the lipase immobilized in microemulsion used as a catalyst. Ultralow surface tension and superior solubilization capacity of ME make it possible to develop the technological protocols for extraction of vegetable oil from oil plants (Gadhave and Waghmare, 2014), which is a beneficial alternative compared to routine techniques based on organic solutions.

It is significant that wide application of the surfactant-based formulations, for example, ME in food technologies, poses problems of such criteria as low toxicity, microbial resistance, odor-free, colorless, high solubilizing efficacy, long-term stability, wetting and washing effect, and so forth. This strongly limits the range of surfactants that are certified for food production. Nonionic surfactants (Tween, Span, mono- and diglicerides of fat acids, phospholipids) have many advantages from this viewpoint due to their low toxicity in combination with high effectivity. Fabrication of MEs based on these surfactants usually assumes preliminary fundamental studies on the optimization of the composition, size, and morphology of droplets (Zheng et al., 2015).

2.3 Nanoemulsion Formulations

It would be improper to disregard a specific type of nanoplatform that has recently received increasing application, that is, nanoemulsion (NE) (McClements, 2013; Adjonu et al., 2014; Guttoff et al., 2015). NEs have obvious benefits over other emulsions due to the lower size of droplets, ranging from 20 to 200 nm. This modifies many bulk properties, such as viscosity, optical transparency, stability toward aggregation, and gravitational phase separation, and improved bioavailability. While MEs are thermodynamically stable, NEs are nonequilibrium systems. Due to their metastability NEs can undergo dilution with water without changing their droplet size. Unlike spontaneously formed MEs, NEs need to use the energy-consuming methods for their preparation (Fathi et al., 2012), which is usually classified into mechanical and nonmechanical techniques. The former assumes the use of high-energy treatment (homogenization, microfluidization, and sonication) of dispersions. Low-energy (nonmechanical) techniques can be roughly divided between isothermal and thermal methods (Komaiko and McClements, 2015b). The isothermal strategies involve spontaneous emulsification and emulsion phase inversion methods, which are of special interest due to their simplicity and low cost. Guttoff et al. (2015) report on a simple low-energy method for encapsulation of low polar vitamins E and D by spontaneous emulsification of oil phase containing surfactant and active component with water. Similarly, spontaneous titration method was developed for the preparation of carvacrol nanoemulsion showing antimicrobial activity (Landry et al., 2014). Currently, oil-in-water NEs are very attractive carriers for food and beverage supplements, and therefore a large number of publications are devoted to different aspects covering their fabrication (Komaiko and McClements, 2015a; Rebolleda et al., 2015), size and rheological behavior (Komaiko and McClements, 2015a,b), biological fate (McClements, 2013) and practical applications for encapsulation of antioxidants, for example, tocopherol (Plaza-Oliver et al., 2015; Rebolleda et al., 2015), vitamins (Guttoff et al., 2015; Mehmood, 2015), flavors (Zahi et al., 2015), carvacrol (Landry et al., 2014), and so forth.

2.4 Liposomes as Delivery Systems for Bioactive Supplements

Liposomes are the most widely used delivery systems in pharma, cosmetics, and food industry (McClements, 2015; Borel and Sabliov, 2014). They are especially attractive as edible carriers, since they may encapsulate substances of different polarity (hydrophilic, hydrophobic, amphiphilic) due to their structure composed of water pool surrounded by phospholipid bilayer. Importantly, liposomes can be obtained from natural building blocks; therefore they answer the criteria of biocompatibility, low toxicity, and can be easily certified for use in the food technology. Lecithin is known to be used for a long time as emulsifier and textural modifier and regarded as safe. Phospholipid components of liposomes show a number of benefits for human health; for example, soybean-derived phosphatidylserine is found to be safe, to improve brain functions and prevent memory loss due to aging (Richter et al., 2013; Moré et al., 2014). Sphingolipids are not only constituents of biomembranes, but also control key metabolic processes (Romero-Guevara et al., 2015; Kawabori et al., 2013). Noteworthy, liposomes not merely protect active ingredients but also enhance their efficacy. At first, liposome formulations were solely based on natural lipids, while both natural and synthetic amphiphiles, lipids, and surfactants, are used at present. Liposomes may be classified by (1) method of preparation; (2) size (small, medium, and large); and (3) lamellarity (mono, oligo- and multilamellar vesicles). There are various techniques for preparing the liposomes (Dua et al., 2012), with the choice between them determined by several factors, that is, physico-chemical properties of both liposome constituents and loads; toxicity and concentration of the cargo; nature of disperse media; way of administration; biological half-life; cost and opportunity of scaling (Wagner and Vorauer-Uhl, 2011; Bozzuto and Molinari, 2015). One of the most popular methods of liposome manufacturing is the thin lipid film hydration proposed by Bangham. It consists of the homogenization of lipids or surfactants in organic media, evaporation of the solvent, and dispergation of the film obtained in aqueous medium. However this technique is characterized by low loading efficacy of around 5–15%. Besides, liposomes manufactured through this method are large and multilamellar, and therefore sonication procedure or extrusion are needed to obtain small homogeneous liposomes.

Reverse phase evaporation method and solvent injection method are used to obtain water suspension of mono- and lamellar vesicles, respectively. The procedures involve the solubilization of lipids in organic solvent with its following removal from the system. The latter technique provides more effective encapsulation efficacy compared to the former. Detergent removal method makes it possible to fabricate large unilamellar vesicle through the fusion of the surfactant micelles enriched by lipids. This method is highly effective and characterized by narrow size distribution of fabricated particles. The limitation of the technique connected with the admixture of detergent in liposomes can be overcome by the use of dialysis and gel chromatography or other procedures for detergent removing. To obtain liposomes of desirable size by the aforementioned methods, further modifying is required, such as ultrasonication, extrusion, and high-pressure homogenization. Sonication is used to decrease liposomes in size; small unilamellar vesicles obtained are further cleaned by ultracentrifugation. Extrusion through membrane filters of a definite pore size results in a decrease in liposome dimension. High-pressure homogenization assumes a mechanical process breaking the vesicles into smaller particles in a homogenizing valve. Since the liposomes came into common use, many modified techniques of their manufacture were developed including spray drying, freeze drying, super critical reverse phase evaporation, and so forth (Laouini et al., 2012). Supercritical fluids are known to be an excellent organic solvent alternative, with such beneficial characteristics as nonflammable, low toxicity, and economy priced. In ref. Lesoin et al. (2011) different mechanisms of liposome formation are discussed involving the supercritical carbon dioxide, with the model considerations developed. According to one of protocols reported in Huang et al. (2014) lipids are extracted in supercritical phase of carbon dioxide and then precipitate in form of fine particles, from which liposomes are formed upon the water addition. Liposomes obtained by this way are identical to the so-called Bangham liposomes in their size, stability, and loading capacity. Dual asymmetric centrifugation method is suitable for viscose solutions. It assumes the rotation of vials around of both main rotation axis and own sample axis (Huang et al., 2014). Membrane contactors technique is widely used for effective mixing of two phases upon the fabrication of disperse systems, precipitates, polymeric or lipid particles, and liposomes. This method makes it possible to obtain ca.100 nm liposomes with 98% loading coefficient (Huang et al., 2014; Laouini et al., 2011).

2.4.1 Types and Characteristics of Vesicular Systems

Morphology, size, surface characteristics, and lamellarity strongly affect the biological behavior of liposomes. Multilamellar liposomes composed of two or more layers of 1 to 5 nm preferably incorporate lipophilic substances, while unilamellar vesicles possessing large water core are excellent carriers for hydrophilic substrates. Release of loads is directly connected with the lamellarity of liposomes, for example, substrate release from unilamellar vesicle with diameter of 130 nm occurs much faster compared to that from 250 nm multilamellar vesicle (Akbarzadeh et al., 2013). At present, one of main challenges in the field of practical application of liposomes in food and beverage production is low physical and chemical stability. The liposome formulations can be improved by decorating with biopolymers or surfactants. Besides, synthetic liposomes are manufactured based on ionic or nonionic surfactants. Among beneficial alternatives to natural vesicles, niosomes should be mentioned (Abdelkader et al., 2014; Marianecci et al., 2014), which are based on nonionic surfactants in combination with various excipients, for example, cholesterol. Unlike liposomes, niosomes are more stable and not subjected to hydrolysis and oxidation. As a helper lipid ionic amphiphile can be used that may markedly modify the properties of carriers. Charged liposomes are more effective and selective. Anionic liposomes have negatively charged surface formed by lipids dimyristoylphosphatidylglycerol or dipalmitoylphosphatidylglycerol and anionic surfactant, for example, sodium desoxycholate. For the first time, cationic liposomes were described in 1987. Usually, mixed cationic vesicles are used with addition of uncharged phospholipids, such as dioleylphosphatidylcholin or dioleylphosphatidylethanolamine. Liposomes admixed with cationic surfactants are known to exhibit improved transdermal delivery of some biologically active agents including drugs. Negatively charged cutaneous surface favors the penetration of positively charged carriers, which enhance transdermal delivery of loads. Cationic liposomes effectively vectorize different drugs including gene material (Ojeda et al., 2015). It was found that positively charged liposomes are selectively cumulated in tumor endothelial cells. Cationic liposomes provide advantages from the viewpoint of penetration through brain blood barrier as well (Schnyder and Huwyler, 2005), which makes it possible to reach therapeutic effect in the case of central nervous system afflictions. Another kind of carriers is proniosome, that is, liquid crystalline hybrid of noisome possessing higher stability and resistance to fusion and aggregation (Yasam et al., 2014). Liposomes decorated with chitosan often referred as chitosome show enhanced stability to pH/temperature/ionic force stresses and mucoadhesiveness compared to conventional liposome. Chitosan coated liposomes find applications as delivery systems in nanomedicine (Gonçalves et al., 2012) and food (Wang et al., 2015) industry. Structural aspects of chitosomes and chitosan/lipid binding affinity are detailed in Chiappisi and Gradzielski (2015). To modify stability of liposomes, layer-by-layer technique can be applied, which makes it possible to protect liposome surface with polyelectrolyte shell. This can be exemplified by Karadag et al. (2013) and Gibis et al. (2012), in which chitosan, maltodextrins (Karadag et al., 2013), and citrus pectin (Gibis et al., 2012) were used as coating materials, while polyphenolic grape seed extract was used as a load (Gibis et al., 2012). Four-layered liposomes with extract loaded were of ca.100 nm in diameter and show long-term storage stability at the variation of solution pH. Liposomes fabricated from unsaturated fatty acid, the so-called ufasomes, are promising delivery systems for drug and food industry. Work (Fan et al., 2014) exemplifies the formation of stable 20–50 nm ufasomes by intravesicle crosslinking of self-assembling conjugated linoleic acid. Sustainable release of the model guest of 5-fluorouracil from fabricated ufasomes was reported.

2.4.2 Application of Liposomes in Food Industry

Until recently, the application of liposomes in food formulations was limited in spite of their widespread use as delivery systems in cosmetics and pharmaceutics. At the present time liposomes are successfully used for the encapsulation of enzymes, proteins, vitamins, aromas, antioxidants, and antimicrobial agents, which is further illustrated by recent studies. Liposomes based on soy phosphatidylcholine with food ingredients (omega-3, omega-6, vitamin E) were fabricated in Marsanasco et al. (2015) for potential use in chocolate milk production. Micrometer-size liposomes stabilized with stearic acid and calcium stearate were loaded with folic acid; their size and rheological behaviors were tested for different formulations before and after the pasteurization. Use of liposomes for the enhancement of bioavailability of carotenoids is reported in Tan et al. (2014). It was shown that size distribution, morphology, and release of loads in a model gastrointestinal tract strongly affect by the structure of carotenoids, with a lutein > carotene > lycopene > canthaxanthin order observed in changing of their bioavailability. The thin-film hydration method followed by sonication and homogenization was applied in Machado et al. (2014) to fabricate soybean phosphatidylcholine-based liposomes loaded with a cyanobacterium *Spirulina platensis*. Encapsulation of spirulina considered as GRAS microorganism makes it possible to mask its odor, prevent the loss in food value, and favor the release directly in the digestive tract. Solvent- and surfactant-free heating method was developed in (Jahadi et al., 2015) for encapsulation of Flavourzyme used for cheese production. Different factors influencing the stability, loading capacity, and functionality of core material were evaluated, such as lipid/helper lipid and lipid/core material ratio, pH, temperature, stirring characteristics, with the formulation and protocol optimized. The liposomal delivery of antimicrobial peptides, bacteriocines, were reported in Arthur et al. (2014), including two peptides, nisin, and pediocin PA-1, classified as GRAS components for food industry. The possibility of encapsulation of tea polyphenol extract into liposomes based on milk phospholipids (as an alternative to soy phospholipids) was studied by Gülseren et al. (2012), with structural characteristics and bioefficacy evaluated. Separated direction involving the liposome based protocols is connected with analytical applications, including monitoring of safety of food products and ecological situation (Edwards et al., 2012). Apart from other carriers liposomes can be easily marked with different signal molecules that enable their detection through the biorecognition principles. The study by Chu and Wen (2013) focuses on the design of the antibody-tagged liposomal formulation for the detection of a common food allergen gliadin. Commercial personal glucose meters of high sensitivity can be constructed by the glucose encapsulation in the antibody-labeled liposomes (Zhao et al., 2015). Colorimetric detection of pathogenic bacteria was proposed for their routine analysis in food industry (Oliveira et al., 2013), in particular the chicken meat industry (Oliveira et al., 2015), which was mediated by the use of polydiacetylene(PDA)/lipid/lysine vesicles undergoing blue–red shift due to PDA associations with lysine. Beloglazova et al. (2013) reports on liposomes loaded with quantum dots, which is one of the first examples of the use of such kind of formulations as labels for the determination of food contaminants. Use of liposomes loaded with quantum dots instead of conventional fluorescence labeled immunosorbent assay makes it possible to amplify analytical signal, thereby enhancing the sensitivity of analyses.

2.5 Solid Lipid Nanoparticles and Nanostructured Lipid Carriers

Solid lipid nanoparticles (SLN) were designed in order to combine benefits of polymeric carriers, liposomes, and emulsions (Tamjidi et al., 2013; Müller et al., 1995; Shegokar et al., 2011). They are prepared from single solid lipid or lipid mixture (Müller et al., 1995). Due to low diffusion of bioactive substrate, its prolonged release from SLN can be achieved. SLN demonstrate a number of advanced features compared to other carriers (Shegokar et al., 2011; Martins et al., 2007; Müller et al., 2011; Kunzmann et al., 2011), namely (1) probability of controlled release and targeted delivery of substrates; (2) high encapsulation efficacy, (3) solvent free protocols, (4) possibility of encapsulating of both lipophilic and hydrophilic substances, (5) low toxicity, (6) high physical stability, (7) possibility of large-scale production. On the other hand, such shortcoming as poor loading capacity should be noted. Ratcharin et al. (2012) describes SLN loaded with ginger extract that were prepared through microemulsion technique with stearic acid used as solid lipids and Cremophor RH 40 as stabilizing surfactant. Ergocalciferol (vitamin D_2) was successfully encapsulated in tripalmitin SLN with Tween 20 used as surfactants (Patel and Martin-Gonzalez, 2012). Vitamin-loaded SLN were both spherical and rod shape and show increased transparence with an increase in ergocalciferol amount.

Nanostructured lipid carriers (NLC) are modified SLN in which both liquid and solid lipids occur in liquid phase (Tamjidi et al., 2013). It is assumed that the presence of oil domains in the NLS core enhances their loading capacity compared to SLN. The use of NLS based on hempseed oil or a blend of amaranth and hempseed oils is exemplified in Lacatusu et al. (2014). Nanoparticles loaded with the natural antioxidant, carotenoid enriched plant extract, have size of ca.100 nm and bear negative surface potential. SLN and NLC carriers for the encapsulation of hydrophilic active components such as didanosine (Kasongo et al., 2011) or epigallocatechin gallate (Fangueiro et al., 2014) were fabricated to prevent their oxidative destruction. Novel type of carriers, multiple lipid particles, for the encapsulation of water soluble ingredient is described in Zha et al. (2015), which combines advantages of SLN and multiple emulsions. Long-term stability and high-loading capacity of formulations toward both active component, coenzyme Q10 and tea polyphenols, was demonstrated.

3 Polymer-Based Formulations as Delivery Systems

Polymeric soft materials attract much interest of nanotechnologists in different spheres including food production (Joye and McClements, 2014; Sağlam et al., 2014; Maswal and Dar, 2014; Delavari et al., 2015; Hu and McClements, 2015; Arroyo-Maya and McClements, 2015; Kutyreva et al., 2015; Sultanova et al., 2015). Due to hard safety standards, special attention in food industry is paid to biopolymers, such as polysaccharides and proteins that can be used in both their natural and modified forms. Protein polymers are divided into two main groups, that is, animal-derived and plant-derived proteins (Joye and McClements, 2014). First group involves water-soluble globular protein albumin, milk protein casein, gelatin, fibroin, and whey protein. Plant-derived proteins have advantages over the former due to the lower risk of infection. Gliadin and zein are two typical plant-derived proteins with poor solubility in water. Another class of biopolymers includes saccharides, namely agar, alginate, carrageenan, cellulose, inulin, chitosan, pectin, and others. All the biopolymers listed find increasing technological applications in pharma and food industry. A variety of nanoparticulate formulations, morphology forms and protocols for their fabrication are designed. Large family of colloid systems based on hydrophilic biopolymers, the so-called hydrocolloids has received increased interest (Maswal and Dar, 2014; Delavari et al., 2015; Hu and McClements, 2015). Herein, we focus on two biopolymers presenting each of the classes, casein and chitosan.

3.1 β-Casein in Encapsulation of Low-Molecular-Weight Ingredients

Proteins are significant ingredients of food because they provide essential amino acids for human health. Besides, they are widely used in food industry and pharmacology to form different functional network structures and to stabilize emulsions or foams. Milk proteins have special physiological and functional properties, which for a long time are widely exploited in the food industry. Milk contains 30–35 g of protein per liter. Milk from the commercially important species, namely cow, goat, sheep, buffalo, camel, yak, horse, pig, and so forth. are quite well characterized (Evans, 1959; Jenness and Sloan, 1970; Park and Haenlein, 2006; Thompson et al., 2009).

Casein, accounting for about 80% of milk protein, is organized in micelles. In nature, casein micelles concentrate, stabilize, and transport essential nutrients, mainly calcium and protein, for the neonate, kids, and adults (DeKruif and Holt, 2003). Practically, casein micelles are natural nano-delivery systems, which are used in encapsulation of different ingredients in nanotechnological applications. The micelles are spherical colloids about 50–500 nm in diameter (Fox, 2003), composed of four caseins: two alpha (α_{s1} and α_{s2}), beta (β), and kappa (κ) in molar ratio 4:1:4:1, respectively (Wong et al., 1996). The caseins are held together in the micelles by hydrophobic interactions, electrostatic repulsions, and by bridging of calcium–phosphate nanoclusters bound to serine–phosphate residues of the caseins. The structure of casein micelles is important for their biological functions in living organisms, for their stability in milk and during processing of milk into various products, as well as for assimilation of the nutrients comprising the micelles. The micelles are very stable during processing and retain their basic structural identity through most of these processes.

β-Casein (β-CN) is one of the members of the casein family and is the most abundant milk protein (80%). Amphiphilic character and high surface activity of β-casein molecule convey good foaming and emulsifying properties (Dalgleish, 1997). The β-CN molecules have a strong tendency to self-associate in the aqueous environment, forming globe-shaped, surfactant-like micelles (Swaisgood, 1992; Gangnard et al., 2007). β-CN is the representative of intrinsically unstructured proteins (Van der Lee et al., 2014), and therefore its secondary (probably, also tertiary) structure and specific properties of its colloidal micelles are strongly dependent on the properties of its environment, such as the solvent structure and interactions with low- and high-molecular ligands including different proteins. It is the most abundant for engineering of micellar nanocontainers and encapsulation of different functional ingredients.

The balance between two main driving forces, the attraction of hydrophobic regions, and the electrostatic repulsion of hydrophilic charged regions is at the origin of the micellization of β-CN and other caseins. The association of β-CN is reversible,

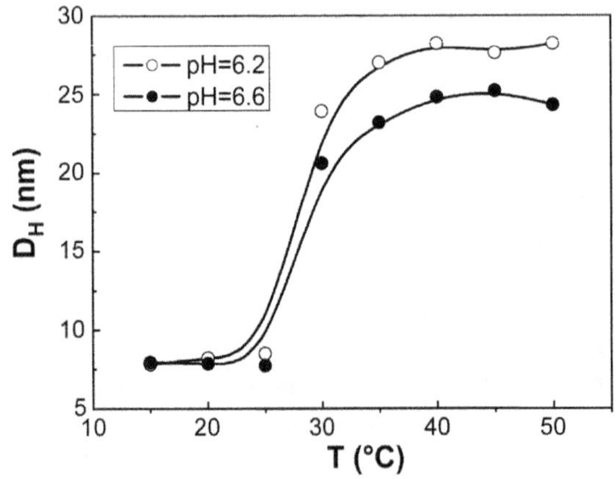

Figure 8.6. β-CN monomer-micelle transition at different pH, detected as the changes in hydrodynamic diameter of particles by dynamic light scattering.

dependent on temperature, pH, ionic strength, and protein concentration. There are two alternative models of β-CN micellization. In the first one the micellization process follows the "all or nothing" law (Evans et al., 1979), in the second "the shell model" represents the consecutive micellization (Kegeles, 1979). The molecules of β-CN form micelles with the size 25–30 nm under the increase of temperature up to 25–30°C (Fig. 8.6). The β-CN molecule is very convenient for artificial engineering of its primary structure thus changing its hydrophilic–hydrophobic balance, to induce new aggregation and solubilizing properties of micelles (Gaudin et al., 2009). This results from the lack of cysteine residues, intra- and intermolecular disulfide bonds in the native form of β-CN. The insertion of one or two cysteines in the peptide chain of protein, obtained by heterologous expression in the prokaryotic host *Escherichia coli*, results in the recombinant forms of β-CN. These molecules can be dimerized yielding oriented biamphiphilic opposite and somehow palindromic structures.

Mutant forms of β-CN retain their micellization properties but with some new characteristics. Dimeric β-CN displays significant changes in the temperature of monomer–micelle transition or its absence in comparison with native protein (Fig. 8.7a). The additional instrument to modify β-CN micellization process is the presence of reducer. For example, its presence can reduce the stability of micelles at high temperatures for mutant forms of β-CN with cysteine in the N-terminal part (Fig. 8.7b).

Modified forms of β-CN change their abilities to interact with other protein species and solubilize low molecular weight ingredients, what can be used in different applications. For example, they

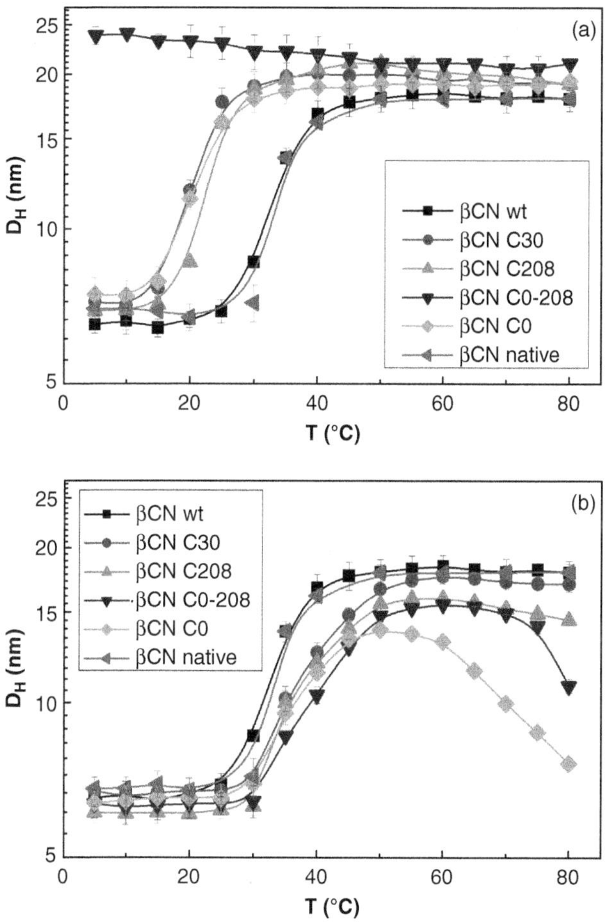

Figure 8.7. Hydrodynamic diameter of β-CN particles in water solution; oligomeric (a) and monomeric (b) forms. Legend depicts the location of cysteine in the protein polypeptide chain.

can be used to create special sorbents for the study of protein–protein interactions (Stroylova et al., 2013), based on the fixation of cysteine-inserted β-CN mutants with thiol-Sepharose resin. An amphiphilic β-CN molecule enables one to exploit its capacity to engage in different types of intermolecular interactions. Two different casein-Sepharose sorbents incorporating either C-4 or C-208 protein mutants bound to thiol-Sepharose were produced, exposing the hydrophobic domain in the case of the C-4 and the hydrophilic domain in the case of the C-208 mutant, respectively. The results obtained using the proposed sorbents with native β-CN, another partially unfolded protein prion, and an oligomeric globular glyceraldehyde-3-phosphate dehydrogenase showed that Sepharose modified with various proteins is suitable for isolation of proteins interacting with the chromatographic phase bound partners from multicomponent systems such as milk.

The occurrence of self-assembling molecular nanostructures is the important feature in nanotechnology. Foods naturally contain the composition of such systems, examples being the casein micelle in milk. The potential application of nanotechnology to encapsulation properties of casein micelles is likely to become an important area. Nano self-assemble properties of β-CN is having a considerable impact in different applications, in part through the use of new improved protein engineering becoming available to support nanotechnology research.

Numerous applications of casein-based formulations may be exemplified by high efficacy of casein-based carriers demonstrated with vitamin A, carotinoids, resveratrol, curcumin, and other active food ingredients. In a recent study, Bourassa et al. (2013) characterized the binding sites of retinol and retinoic acid with α- and β-caseins, with higher binding constant of retinol observed compared to retinoic acid. Immobilization of curcumin by hydrophobic domain of casein, as well as sodium caseinate/resveratrol complexation were testified in (Yazdi and Corredig, 2012; Acharya et al., 2013) by fluorescence spectroscopy.

Another application of β-CN micelles is discovered in the nanomedicine being convenient for drug delivery due to their biodegradability, biocompatibility, low toxicity, easy metabolism, abundant and unlimited sources, ease of manufacturing (Husseini and Pitt, 2008; MaHam et al., 2009; Elzoghby et al., 2013). There are modern examples of drug-loaded micelle design, both in the bulk solution and in dry form and even freeze-dried without any cryo-protectants, kept as a powder for at least 6 months (Turovsky et al., 2015; Bachar et al., 2012).

3.2 Natural Polymer Chitosan, a Representative of Polysaccharide Family

Chitosan is a partially deacetylated derivative of chitin. Chitosan possesses a number of unique physical and chemical properties, that is, biocompatibility, biodegradability, antimicrobial activity, and the ability of chemical modification for yielding various derivatives. The mucoadhesive nature of chitosan makes it a good candidate for prolonging the residence time and increasing cargo release time in the gastrointestinal tract (Fathi et al., 2014; Joye and McClements, 2014). Importantly, chitin and chitosan might be useful as a therapy for food allergies (Bae et al., 2013).

Chitosan particles are especially useful for encapsulation of hydrophilic macromolecules. These molecules associate with chitosan through electrostatic interaction or hydrogen bonding. Furthermore, chitosan is available in different molecular weights and acetylation degrees, which enables fine tuning of particle characteristics (Joye and McClements, 2014). The application of chitosan as a food preservative and for other uses has been limited by its insolubility at neutral and higher pH. Improvement of its solubility could favor its application as a food preservative (Gyawali and Ibrahim, 2014). The antibacterial activity of chitosan was examined against several gram-negative and gram-positive bacteria and reviewed in (Gyawali and Ibrahim, 2014). Chitosan has been applied mainly as an antimicrobial packaging films and coatings (Dasgupta et al., 2015; Gyawali and Ibrahim, 2014), thereby increasing the shelf life of foods and beverages. Chitosan based formulation shows antioxidant effect, which makes it possible to avoid a use of synthetic antioxidants. This is exemplified by the encapsulation and release of yerba mate extracts in alginate and in alginate–chitosan beads (Deladino et al., 2008).

Controlled release and stabilization of resveratrol (natural phytoalexin exhibiting antitumor and antiinflammatory activities) were achieved through incorporation into crosslinked chitosan microspheres by vanillin (Peng et al., 2010). The encapsulation efficiency of resveratrol within microspheres was up to 93.68%. Resveratrol contained within microspheres was stronger protected from light and heat compared with the free resveratrol.

Chitosan-based formulations are effective for the prolongation of shelf life of the cut fruits, for example, chitosan coating alone and in combination with citric acid pretreatment were found to be beneficial in reducing weight loss and resulted in minimal changes in pH, titrable acidity, and total soluble solid content and exhibited lower respiration rate on storage of shredded carrots (Pushkala et al., 2012). Chitosan-coating treatments effectively retarded enzymatic browning on minimally processed apples during storage (Qi et al., 2011).

There are publications on the use of chitosan as carriers of essential oil and aroma compounds. Essential oils encapsulation led to numerous new formulations with new applications. This ensures the protection of the fragile oil and controlled release (Asbahani et al., 2015).

The encapsulation and delivery techniques of citral mostly based on colloidal systems have been reviewed in detail (Maswal and Dar, 2014). McClements and coworkers showed (Djordjevic et al., 2007) that when oil droplets are first stabilized with an anionic surfactant, such as a phospholipid, with further addition of positively charged polymer, such as chitosan, the droplets get coated with a surfactant-polymer membrane, making globules positively charged. Such an emulsion protects essential oils (citral and limonene) from oxidation more efficiently than the ordinary emulsions stabilized by a single surfactant or amphiphilic layer (Djordjevic et al., 2007).

The effect of pomegranate juice (PJ) and chitosan (CH) coating enriched with *Zataria multiflora* essential oil (Z) on the shelf-life of chicken breast meat during refrigerated storage was investigated (Bazargani-Gilani et al., 2015). Treatments examined were the following: Control, PJ, PJ–CH, PJ–CH–Z 1%, and PJ–CH–Z 2%. The samples were stored at 4°C for 20 days and analyzed at 5-day intervals. All of treatments significantly decreased total viable counts, *Pseudomonas* spp., lactic acid bacteria, *Enterobacteriaceae*, *Psychrotrophic bacteria* as compared to control during the storage period.

The results obtained by Dima et al. (2014) prove the possibility of using the *Pimenta dioica* essential oil encapsulated in chitosan and chitosan/k-carrageenan microspheres for the preparation of some food products, mainly for meat products. Microencapsulated essential oil exhibited antimicrobial activity against *Candida utilis*, *Bacillus cereus*, and *Bacillus subtilis*.

The evaluation of the antibacterial effects of modified chitosan-based coating incorporating mandarin essential oil nanoemulsion in combination with three nonthermal treatments against *Listeria innocua* inoculated in green beans during storage at 4°C was carried out in Severino et al. (2014).

Chitosan microcapsules containing limonene essential oil as active ingredient were prepared by coacervation using three different concentrations of NaOH and fixed concentrations of chitosan and surfactant of 0.50 wt%. This technique shows that tuning NaOH concentration makes it possible to efficiently control the release rate of encapsulated active agents demonstrating great potential as insect repellent for textiles (Souza et al., 2014).

Currently, increased attention is paid to the design of chitosan-based formulations for the encapsulation of volatile aromas, with their release controlled. A new type of controlled-release system was prepared by the covalent attachment of a volatile antifungal agent, hexanal, to the biodegradable polymer chitosan. The antifungal activity of *N*-hexylimine-chitosan film was demonstrated on harvested wheat. Acidic stimulation resulted in release of hexanal in the grain storage container and a significant (up to 10-fold) decrease in the mold occurrence on the grain. The presented controlled-release system is based on dynamic covalent bonding. The formation of a stable covalent bond completely prevents the escape of active compound and can, however, be easily hydrolyzed (Fadida et al., 2015).

Chitosan incorporating hydroxypropyl-β-cyclodextrins and glycerol films capable of modulating loading capacity and release of active aroma compound carvacrol have been developed. Release of carvacrol from the films was greatly affected by relative humidity. The films showed antimicrobial activity against *Staphylococcus aureus* and *E. coli* after 20 days of storage at 25°C and 43% environmental relative humidity (Higueras et al., 2015).

B. pseudocatenulatum G4 was effectively encapsulated by alginate and coated by chitosan to obtain encapsulated Bifidobacterium with a high viability through the human GIT (Abdollahi et al., 2012). Furthermore, alginate–chitosan capsule was more resistant than noncoated alginate capsules to simulated intestinal fluid. More details of microorganisms protection and delivery of them into the gut were discussed in the review by Martín et al. (2015).

The thymol-loaded zein nanoparticles were successfully prepared with sodium caseinate and chitosan hydrochloride as electrosteric stabilizers using a simple and low-energy liquid–liquid dispersion method. Thymol encapsulated in nanoparticles had much stronger antimicrobial activity against *S. aureus* under the experimental conditions (Zhang et al., 2014).

The modified starch–chitosan and modified starch–maltodextrin–chitosan systems were used as agents for beetroot and saffron coloring-extracts microencapsulation by freeze drying (Chranioti et al., 2015).

A novel flavor microcapsule containing vanilla oil (VO) was developed using complex coacervation approach, aimed at control release of VO and enhancement of its thermostability for spice application in food industry. Viscosity of chitosan and VO/chitosan ratio were optimized for fabrication of microcapsules. Moreover, VO could remain at about 60% in the microcapsules after release for 30 days, which demonstrated that flavor microcapsules had good potential to serve as a high-quality food spice with long residual action and high thermostability (Yang et al., 2014).

Traditionally, active compounds are incorporated into initial food formulations. A limitation of this traditional method is that once the active compounds are consumed in reaction, protection ceases and the quality of food degrades at an increased rate. Another limitation is its inability to selectively target the food surface where most spoilage reactions occur; as a result, an extra amount of active compound is also unnecessarily added inside the food product. Controlled release packaging can overcome these two limitations by continuously replenishing active compounds to the food surface, compensating for the consumption or degradation of active compounds, so that a predetermined concentration of active compound is maintained in the food to achieve a desired shelf life (Hannon et al., 2015; Mastromatteo et al., 2010; Siripatrawan and Noipha, 2012; Sung et al., 2013). Montmorillonite nanoclay and rosemary essential oil were incorporated into chitosan film to improve its physical and mechanical properties as well as antimicrobial and antioxidant behavior (Abdollahi et al., 2012). Chitosan nanoparticles loaded with cinnamon essential oil (CE-NPs) exhibited the excellent antimicrobial and antioxidant property for the pork during refrigerated storage. The fresh pork was wrapped with the low density polyethylene films and the active films coated with different CE-NPs and displayed at 4°C for 15 days. The active films with 527 nm CE-NPs led to a significant decrease of microbial growth, pH, peroxide value, 2-thiobarbituric acid, and sensory scores of the pork ($P < 0.05$) than the other treatments at the end of storage (Hu et al., 2015).

In food industry, nanolayered coatings can be used as a strategy for the shelf life extension. The layer-by-layer (LbL) deposition of k-carrageenan and chitosan is used for encapsulation of the model substrate Methylene Blue (Pinheiro et al., 2012). The monitoring of loading and release of substrate from polyelectrolyte capsules were conducted under isothermal conditions via different mathematical models. The advantages of the LbL formulations will be discussed in the next section in more details.

3.3 Layer-by-Layer Strategy in Food Technologies

The LbL strategy assumes the encapsulation of an active component into the cavity formed by alternating deposition of polycations and polyanions (Chen et al., 1999), with electrostatic forces mainly contributed to the coating process. At the same time, van der Waals interactions, hydrogen bonding, and guest–host complexation may play important roles as well (Mauser et al., 2006). Different protocols were developed involving the use of sacrificial template with its subsequent removing (Chen et al., 1999). As a simple promising alternative, template free technique was proposed, with substrate particles used as a matrix (Zakharova et al., 2012b, 2015; Vasilieva et al., 2014b) (Fig. 8.8). Importantly, polyelectrolyte capsulation provides many advantages over other techniques, for example, high loading capacity, wide choice of natural polyelectrolytes, the possibility to tailor capsule properties, that is, wall thickness, permeability, and integrity by simple variation of shell material, and so forth. This is of particular importance for pharma and food technologies, where rigid criteria occur toward the building blocks. To meet these criteria, flavorless edible polymers compatible with food components are usually used as coating agents in food processing (Davidov-Pardo

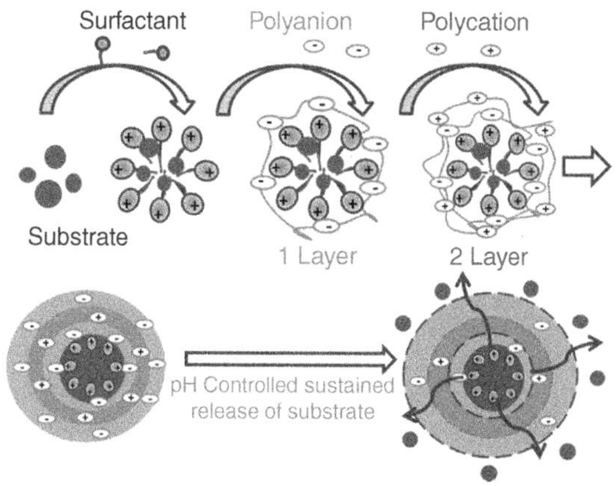

Figure 8.8. Schematic representation of the LbL technique for fabrication of core-shell nanoparticles with sustained release of cargo material.

and McClements, 2014), for example, enzymes, sodium alginate and poly-L-lysine, milk protein α-lactalbumin, chitosan, casein, carrageenan, pectin, carboxymethylcellulose, and so forth (Lvov et al., 1995; Borodina et al., 2008; Davidov-Pardo and McClements, 2014; Delavari et al., 2015; Chen et al., 2015b; Trabelsi et al., 2013; Bernela et al., 2014; Nakagawa et al., 2013).

Mild conditions of capsule fabrication make it possible to incorporate the variety of biomolecules preserving their biological activity. This can be exemplified by the LbL encapsulation of chymotrypsin, with 86% of the activity retained in the immobilized form (Borodina et al., 2008). In Karimi et al. (2014) the inulinase (the source of fructose syrup) immobilization technique was worked out involving the protein bearing negatively charged amino acid moieties with cationic polyelectrolyte poly-D-lysine. Polyelectrolyte capsulation markedly increases physical and thermal stability as well as temperature diapason of functional activity, which is of special importance for food processing assuming different kinds of treatment. Modified LbL protocol was developed in Sharipova et al. (2015) for the encapsulation of vitamin E. Vitamin E (α-tocopherol) is known as an effective antioxidant that plays a key role in the treatment of diseases of nervous system. The encapsulation makes it possible to prevent it from oxygen, temperature, and light degradation. The technique involved the formation of sodium polystyrene sulfonate/dodecyl trimethyl ammonium bromide complexes stabilizing the vitamin E loaded emulsion, which was further used as a charged template for alternative polystyrene sulfonate/chitosan deposition. The release profile obtained testifies the sustained discharge of the vitamin E within around 80 h. In ref. Bernela et al.'s (2014) polyelectrolyte-based formulation was optimized for the encapsulation of food preservative nisin, as a GRAS ingredient. Roughly spherical capsules of about 200 nm were engineered, with biocompatible polymers chitosan, sodium alginate, and pluronic F68 used as shell material. A 2-week monitoring of release profile reveals the transition from an initial burst nisin liberation to the sustained release, with prolonged functional activity observed. A large number of publications are devoted to the encapsulation of the water-insoluble food-grade dye curcumin, possessing a wide range of biological activities (Ariga et al., 2011; Bielska et al., 2013; Manju and Sreenivasan, 2011; Yang et al., 2015). Positively charged hyaluronic acid/chitosan nanoparticles of ca. 200 nm were fabricated for the curcuminoid delivery (Yang et al., 2015). Formulation was optimized by controlling such factors as the order of polyelectrolyte deposition, their ratios and initial concentrations, and solution pH, which made it possible to achieve the wide range of thermal stability, high encapsulation efficacy of around 90% and loading capacity of 6.5%. The temperature-responsive nanoparticulate formulations were prepared in Bielska et al. (2013) through alternate deposition of the oppositely charged derivatives of hydroxypropyl cellulose. The spherical nanosized (≤250 nm) capsules loaded with curcumin demonstrated the temperature-dependent size behavior and release of the guest molecule. In Manju and Sreenivasan (2011) hollow microcapsules loaded with curcumin were fabricated by poly(sodium 4-styrene sulfonic acid)/poly(ethylene imine) deposition onto the melamine formaldehyde template. An in vitro release study shows the fast-to-sustained release transition, with prolonged biological activity observed. The 10-layered chitosan/fucoidan nanocapsules were formulated as described in Pinheiro et al. (2015) with polystyrene nanoparticles used as sacrificial template. The encapsulation of bioactive compound poly-L-lysine was studied therein, with pH-controlled binding/release behavior observed. Until recently, thermal treatment is widely used for the preservation of food products, despite the strong negative effect on organoleptic characteristics and the texture of the edible product. This is especially important for fresh produce and fresh-cut fruits and vegetables. Synthetic coating agents based on organic solvents and surfactants sometimes used in practice are not suitable based on ecological and healthy criteria. The application of edible coatings to minimally processed fruits faces some technical problems related to the difficult adhesion of materials to the hydrophilic surface of the cut fruit. In response to the challenges, novel nonthermal techniques are currently developed as a natural alternative to synthetic waxes, which is especially effective in combination with antimicrobial effect (Donsì et al., 2015; Thakhiew et al., 2010). In

Poverenov et al. (2014) oppositely charged polysaccharide pair alginate/chitosan was used to prepare edible protective shell for fresh cut melon. The LbL-fabricated coating demonstrated advanced stability, adhesion, and antimicrobial characteristics over single-layered and uncoated formulation. Polysaccharide coatings based on modified cellulose and chitosan were formulated in Arnon et al. (2015) to prevent the deterioration of mandarin. The pectin/chitosan multilayered coating through direct polyelectrolyte self-assembly onto the fruit skin was used for the control of permeability of the film toward water vapor, oxygen, and carbon dioxide spoiling the quality of fresh-cut mangoes (Medeiros et al., 2010). The coated mangoes are found to perform better than uncoated mangoes due to the combination of antimicrobial and gas barrier effects. Antimicrobial edible coating loaded with β-cyclodextrin and *trans*-cinnamaldehyde complex was formulated in Brasil et al. (2012) by LbL technique for the protection of fresh-cut papaya. The comparison of the protected and uncoated fruits reveals improved physical and antimicrobial characteristics of the former.

4 Molecular Complexes Based on the Guest–Host Interactions

Application of inclusion complexes of host–guest type with cyclodextrins and calixarenes aims at improving stability, controlled delivery, and convenient handling of active ingredients. The host–guest complexation events based on cyclodextrin and calixarene occur commonly in aqueous media, and therefore the use of macrocyclic receptors is particularly advantageous in constructing water-soluble supramolecular architectures (Guo and Liu, 2012; Polyakov and Kispert, 2015). Moreover, such macrocycles have been shown to be very biocompatible (Polyakov and Kispert, 2015; Perret and Coleman, 2011). Consequently, construction of supramolecular complexes from these macrocycles is of fundamental interest for applications in food industry and biotechnology.

Cyclodextrins (CDs) are a family of cyclic oligosaccharides typically containing six (α-CD), seven (β-CD), or eight (γ-CD) 1,4-linked D-glucose units. α-CD and β-CD have found widespread use in food industry; both have obtained GRAS status on the Food and Drug Administration's list for use as food additives (Kurkov and Loftsson, 2013). α-CD is a good candidate for use as a food additive owing to its superior solubility over the larger β-CD. γ-CD is also highly soluble; however, it is easily degraded by the human digestive system, contributing as a high calorie additive when used for food products. α-CD is highly resistant to enzymatic degradation and its dietary fiber properties contribute to its desirable use as a food additive (Li et al., 2014a). Only β-CD and its derivatives have been widely used in food industry due to the proper inside cavity and lower production cost (Pinho et al., 2014). Unmodified or unsubstituted β-CD has poor water solubility and is parenterally unsafe due to its nephrotoxicity. Therefore, several synthetically modified and relatively safe β-CDs have been made and used in parenteral formulations, such as methyl-β-cyclodextrin (M-β-CD), hydroxypropyl-β-cyclodextrin (HP-β-CD), 2,6-di-O-methyl-β-cyclodextrin (DM-β-CD), 2,3,6-tri-O-methyl-β-cyclodextrin (TM-β-CD) sulfobutyl ether-β-cyclodextrin (SBE-β-CD), randomly methylated β-CD (RM-β-CD), and the so-called branched CDs such as maltosyl-β-CD.

Calixarenes are versatile macrocyclic oligomers made up of benzene units as CDs are made up of glucose units. They present a hydrophobic core sandwiched between two functionalizable rims. Since calixarenes possess a cylindrical architecture similar to CDs, they are expected to form inclusion complexes. In contrast to the CDs few reports are available on the application of calixarenes in food. However, the advantage of the calixarenes consisting of the modification at the upper and lower rim of the appropriate macrocyclic platform allows varieties of the receptor properties of these macrocycles over a wide range (Gutsche, 2008). Therefore, calixarene molecules may find application in different fields of food industry.

4.1 Effect of the Addition of CDs on Organoleptic Properties

Although the CD molecules are composed of glucose units, α-CD and β-CD do not taste sweet at all, while γ-CD has only a slightly sweet taste. Nevertheless, the bitterness of foods is readily masked by CDs due to the formation of an inclusion complex. Ono et al. (2011) measured the effects of α-CD, β-CD, and γ-CD as well as a derivatized β-CD on the bitterness of a range of antihistaminic drugs. They showed that the level of bitterness suppression was correlated with the binding coefficient. A higher binding coefficient would mean a lower free bitterant concentration and hence less bitter taste. However, there have been some exceptions reported to the expected relationship between binding and bitterness reduction. For example, Gaudette and Pickering did not observe any suppression of caffeine bitterness by β-CD (Gaudette and Pickering, 2012a).

In protein and peptide chemistry, the ability of CD to form inclusion complexes, altering the physicochemical properties of the substrate, has attracted much attention (Serno et al., 2011; Aachmann et al., 2012). Formation of peptide inclusion complexes with CDs has shown altered aggregation and stabilization properties (Wahlstrom et al., 2012). The incredible sweetness of the protein such as thaumatin compared to sugars is attributed to allosteric agonist activity of the sweet-taste receptor, a C-class G protein-coupled receptor (GPCR) (Ohta et al., 2011; Masuda et al., 2013). α-CD may interact with the active binding site on thaumatin, suggesting that a-cyclodextrin could be used to modify the interaction of thaumatin with GPCRs and affect the taste properties of thaumatin in food products (Healey et al., 2015).

CDs were capable of partially suppressing the bitterness of whey protein hydrosylates (Yang et al., 2012) and soybean antioxidant hydrosylates (Hou et al., 2013). In other food studies, Gaudette and Pickering (2012b) showed β-CD suppressed, but did not eliminate the bitter taste of catechin, and only in the presence of sucrose or a bitter-blocking compound. CDs were used in tea to mask the bitter taste caused by catechins. NMR experiments suggested that ß-CD binds theaflavin-3,3′-di-O-gallate, a major constituent in black tea, more tightly than α- or γ-CD (Nishizawa et al., 2014). β-CD also masks goaty flavor in yogurt, and it could be used in commercial goat yogurts and similar products, so the real or perceived nutritional advantages of goat milk are not lost to goaty flavor (Young et al., 2012).

The effect of the addition of CDs on others organoleptic properties has been reported in few works. CD addition has both positive and negative effects on the composition and quality of the mandarin juice (Andreu-Sevilla et al., 2011a). The treatment with β-CD caused higher intensities of typical aroma (fresh mandarin). However, addition of HP-β-CD resulted in the highest color intensity, vitamin C content, retinol equivalents, antioxidant activity, and overall quality even at the end of the cold storage period. Moreover, HP-β-CD improved the quality of mandarin juice more efficiently than β-CD (Navarro et al., 2011). Comparing the effects of the addition of α-, β-, and γ-CDs on pear juice shows that α-CD considerably improves the overall quality of pear juice and reduces its browning without producing significant decreases in aroma (Andreu-Sevilla et al., 2011b).

The addition of CDs prior to enzymatic clarification provides an alternative solution to the depletion of carotene content in clear blended carrot-orange juice. The addition of 3% HP-β-CD in homogenized blended carrot-orange juice improves carotene content, though reduces slightly its clarity (Karangwa et al., 2012).

The ability of β-CD to stabilize aronia juice anthocyanins was described in Howard et al. (2013). The addition of up to 3% β-CD to aronia juice prevented anthocyanin loss and resulted in excellent protection of anthocyanins after 8 months. Encapsulation of anthocyanins may also prove useful in improving their bioavailability in the gastrointestinal tract, which may improve health benefits (Oidtmann et al., 2012).

4.2 Extraction of Components

One of the main applications of β-CDs in food industry is the formation of inclusion complexes for the extraction of an undesired component, practically, cholesterol (Fauziah et al., 2013). The potential use of the composites developed in the reduction of the cholesterol content of foods was explored, specifically, in milk (López-de-Dicastillo et al., 2011). Crosslinked β-CD was found to be highly effective in cholesterol removal from cream cheese (Mei et al., 2012) and Gouda cheese (Jung et al., 2013) without any significant change in the flavor, texture, and sensory attributes of the cream cheese. Addition of crosslinked β-CD into food system such as milk and dairy products in order to lower cholesterol levels could not cause any adverse health effects (Park et al., 2011). The highest cholesterol removal rate of nearly 99% was obtained from fresh egg yolk in Su et al. (2015).

CDs can solubilize, stabilize, mask unpleasant odors and increase the bioavailability of food components (García-Padial et al., 2013). However, the complexes formed by the interaction of lipophilic compounds with CDs have limited aqueous solubility, which results in precipitation of solid CD complexes from water and other aqueous systems. Therefore, there is a need for a delivery system that can solubilize CDs and at the same time act as a host toward cholesterol. Research by Kaur (2014) demonstrates the use of nanoemulsions as a medium in food matrices, instead of water, for hosting cholesterol in β-CD. In addition, there is a report by Dos Santos et al. (2011) where water is used as a medium in such studies.

Magnetic mesoporous silica is functionalized with β-CD for selective separation of cholesterol from milk or a cellular environment (Sinha et al., 2015). The colloidal form of this silica provides effective interaction with cholesterol of any form, while magnetic property facilitates separation of bound cholesterol.

β-CD can also extract different protein-bound carbonyl compounds, free fatty acids, and phospholipids from soy protein isolate (Arora and Damodaran, 2011). A further very interesting application of β-CD-polyurethane polymer is to remove ochratoxin

A from wines (Appell and Jackson, 2012). The chemically modified quaternary ammonium CDs results in about 200-fold higher complex stability with ochratoxin A than was observed in the case of β-CD (Poór et al., 2015). In addition, further potential application of quaternary ammonium CDs was studied (Sebestyén et al., 2012; Wang et al., 2012). In another investigation, Ratanasooriya and Rupasinghe (2012) showed that the cyclodextrin technology was more effective in removal of polyphenols from grape than organic solvent-based extractions.

An antioxidant powder from spent eggplant prepickling solution exhibits low water-solubility and strong acetic acid odor. Treatment with γ-CD alleviates significantly these undesirable characteristics. In addition, the complex exhibited high antioxidant activity, equivalent to 56.2% of the activity of the delphinidin standard (Itoh et al., 2015). In another study, β-CD/ATP sorbent composite for dispersive solid-phase extraction prepared by bonding β-CD to modified attapulgite via silane coupling was used to determine the concentrations of four (fluoro)quinolones in honey samples (Cui et al., 2015).

In recent years an increasing interest in calixarenes as potential complexing agents for metals, which together with food and potable water can penetrate into the human body, is observed. It is supported by several reviews (Mokhtari et al., 2011a,b; Schühle et al., 2011; Kiegiel et al., 2013). For example, the pyridinyl hydrazone derivatives of thiacalix[4]arene may be successfully applied for separation of certain transition and heavy metals (Podyachev et al., 2011). The introduction of electron-donor substituents in phenyl fragments of hydrazone groups improves the affinity of these compounds toward Ag^{+} and Hg^{2+} ions (Podyachev et al., 2014).

The measurement of naturally occurring radionuclides in drinking water is important to assess their health impact (Persson and Holm, 2011). For this purpose the use of calixarenes is of great interest to extract radioelements. The calix[6]arene-derivatives columns were developed to extract and separate selectively the actinides from urines samples (Bouvier-Capely et al., 2014). The calixarene column was also successfully employed for the analysis of the illegal additive of melamine in milk product (Hu et al., 2013).

4.3 Enantioseparation

Enantioseparation is widely used for resolution and preparation of foods. Mixtures of chiral compounds in natural products can be separated by the use of chiral selectors, and CDs have been used for this purpose. The enantiomers of ethyl

2-hydroxy-4-methylpentanoate in wines from various vintages and origins were separated by chiral gas chromatography analysis on a γ-CD phase (Lytra et al., 2012). A wide array of CDs (native α-CD, β-CD, γ-CD, hydroxypropylated, methylated, alkyl-carboxylated, sulfated, and alkyl-sulfated CDs) has been screened for chiral resolution of racemic antimalarial drugs (Németh et al., 2011).

The position of carboxymethyl group on monosubstituted α-CD plays a very important role in the separation and the chiral recognition (Rezanka et al., 2012). α-CD derivative with carboxymethyl group attached at position 3 provided significantly better resolution than the native α-CD and its other individual monosubstituted carboxymethyl derivatives including their mixture. Different columns coated with modified CDs were tested to resolve racemic 4-alkyl-branched fatty acid methyl ester standards (Kaffarnik et al., 2015).

Calixarene derivatives were also applied for separation and determination of enantiomers. Study in (Memon and Memon, 2013) demonstrates the differential recognition of L-alanine amino acid by 5,11,17,26-tetrakis-[(*N*,*N*-dimethylamino) methyl]-25,26,27,28-tetrahydroxy-calix[4]arene. In other work (Mokhtari and Pourabdollah, 2012), *p*-tert-calix[4]-1,2-crown-3, -crown-4, -crown-5, and -crown-6 were used to prepare bonded phases for detection and quantitation of salbutamol enantiomers in six samples of livestock meat (pork, pork casing, beef, beef casing, mutton, and mutton casing).

4.4 Complexes with Antioxidants

The formation of cyclodextrin inclusion complexes with highly hydrophobic molecules has allowed large improvements not only in their aqueous solubility but also their stability and bioavailability, thus ensuring the wide application of CDs in food industrial areas to encapsulate antioxidants (López-Nicolás et al., 2014).

Coenzyme CoQ10 is an antioxidant whose activity is particularly important in regenerating vitamin E. Uekaji et al. (2011) reported the enhancement of the stability and bioavailability of coenzyme CoQ10 oxidized form by γ-CD complexation. In a series of studies, the authors investigated an easy and economical conversion of coenzyme CoQ10 oxidized form to its reduced form in complex powder, using inexpensive vitamin C as the reductant.

In order to enhance the cost-effectiveness practicability of enzymes in food technological processes, there is great need to immobilize them onto solid supports. There are a few studies on the use of calix[4]arene 1,3-distal glutaraldehyde derivative as a cross linker reagent for alpha amylase immobilization (Veesar et al., 2013; Ozyilmaz et al., 2014). The efficient immobilization of α-amylase from *Saccharomyces cerevisiae* was developed by using the surface functionalization of calix[4]arene as support (Veesar et al., 2015).

The novel reversed phase HPLC method was developed for the determination of pterostilbene in food samples (Rodríguez-Bonilla et al., 2011a). This method is based on the addition of CDs to the mobile phase where the 1:1 complexation of pterostilbene with CDs is carried out. The inclusion complexes formed by the interaction between HP-β-CD and both the pterostilbene *trans*-dehydromer pterostilbene and *cis*-dehydromer products obtained by the oxidation of pterostilbene by peroxidase may show higher solubility than these molecules in the absence of HP-β-CD and slow down the detoxification process (Rodríguez-Bonilla et al., 2011b).

β-CD was added to the formulations, in order to obtain inclusion complexes with the essential oils (Răileanu et al., 2013). It was clearly demonstrated that CDs (particularly HP-β-CD) have a significant positive impact on the antioxidant activity of essential oils from three Lamiaceae species in aqueous systems. The microencapsulation of essential oils with β-CD was applied in gastronomy to improve the sensorial properties of food (García-Segovia et al., 2011).

In food science CDs have been used for the controlled release of garlic essential oil (Wang et al., 2011b). The stoichiometry of the garlic oil/β-CD inclusion complex was 1:1, and the calculated apparent stability constant of garlic oil/β-CD complex was 1141 M^{-1}. The water solubility of garlic oil was significantly improved by the phase solubility study. Furthermore, the release of garlic oil from the inclusion complex was determined at a temperature range from 25 to 50°C and in an acidic dissolution medium (pH 1.5).

The complexation between CDs and *trans*-anethole (TA), major component in essential oil of several plants, was investigated (Ciobanu et al., 2013; Kfoury et al., 2014). Results showed that the encapsulation of TA inside the CD cavity was an efficient way to increase stability and solubility of TA, and β-CD displayed higher stability constants than α-CD and γ-CD. The release study in Zhang et al. (2015) suggests that β-CD provides the protection for AT against evaporation.

The inclusion of essential oils with a very high incorporation rate was achieved into β-CD (Hill et al., 2013). A study in Haloci et al. (2014) justifies the use of β-CD as complexion agent for essential oil of such aromatic plants as *S. Montana* in the food industry.

The development of antimicrobial materials and their application in the design of active packaging is creating considerable expectation in the food industry, since food safety is an area of great concern. Incorporation of low amounts of complex of β-CD and essential oils in the chitosan film significantly improves its antimicrobial properties (Higueras et al., 2013; Sun et al., 2014). Therefore, chitosan films containing complex of β-CD and essential oils could be used as active food-packaging material.

The catechin and epigallocatechin-gallate are molecules representative of the flavon-3-ols subgroup. These compounds present antidiabetic and antiobesity properties, besides the antioxidant action (Haidong et al., 2011). If catechin isolated from grape seed was successfully complexed with β-CD with a 1:1 stoichiometry (Krishnaswamy et al., 2012), DM-β-CD was the most suitable CD for the complexation of epigallocatechin (Folch-Cano et al., 2013). The authors also described that regardless of how all the inclusion complexes had similar geometries, the flavonoid antioxidant rings position inside the CD cavity was different.

In general, the coordination to the Al(III) ion lowers the oxidation potential of such flavonoids as catechin and quercetin, making them more susceptible to suffering oxidation processes. In fact, both compounds present the ability to scavenge 1,1-diphenyl-2-picrylhydrazyl radicals efficiently. However, their potential use as antioxidants is severely imparted by their low water solubility. With this regard, it was also shown that their inclusion into the β-CD cavity increases their solubility in aqueous media, and that this chemical modification does not compromise their antioxidant ability. Actually, the ability of the inclusion compounds to react with radicals is maintained, with only a slight decrease of the EC50 values (Dias et al., 2011).

The inclusion complexes of SBE-β-CD, HP-β-CD, and M-β-CD, with insoluble quercetin in tris–HCl buffer solutions at pH 7.40, were investigated through phase solubility analysis and spectroscopic methods. The results suggested that quercetin formed 1:1 inclusion complexes with three substituted CDs (Liu et al., 2013a), and hydrophilic SBE-β-CD is a more effective solubilizing agent for quercetin due to stronger ion–dipole interactions between the anions of the outer surface of SBE-β-CD and the hydroxyl groups of quercetin (Dong et al., 2013). The authors of another study reported the formation of inclusion complexes of 7-diethylaminocoumarin-3-carboxylic acid with β-CD and 2-HP-β-CD in buffer solution (pH 7.4) and in the solid state (Tablet et al., 2012). The solubility and stability of 7-hydroxy-4-methylcoumarin can be improved through the complexation with SBE-β-CD. It is beneficial for 7-hydroxy-4-methylcoumarin to release from the cavity of SBE-β-CD with the ascent of pH value (Liu et al., 2015a).

There are several investigations on encapsulation of flavonoids in various CDs for improving solubility (Lucas-Abellán et al., 2011), quality (Karangwa et al., 2012), and physico-chemical stability (Chao et al., 2012). The most successful formulations were the RM-β-CD complexes of various carotenoids (capsanthin, capsorubin, and lutein). The major drawback of these nanocapsulated carotenoids is that they contain a relatively high percent (95%) of CD, which is required to maintain their complexation ability and the water solubility of the complexes. The aqueous solutions of these complexes are stable over months, so no aggregation can be observed, and the complexation is not pH dependent. The RM-β-CD–lutein complex has been recently found to facilitate the incorporation of lutein in neurons (Horvath et al., 2012).

Complexation with CD protects the carotenoid during storage and transportation to the target. In vivo and in vitro experimental data demonstrated that CDs stabilize carotenoids and allow efficient cellular uptake (Yuan et al., 2013; Gharibzahedi et al., 2014). RM-β-CD has shown higher solubilization behaviors of poorly water soluble molecules in comparison with other types of CDs (Mazzaferro et al., 2011; Pradines et al., 2014).

β-CD–carvacrol complexes could be useful antimicrobial delivery systems, for example, for application in a variety of food systems where foodborne pathogens could present a risk (Santos et al., 2015). According to Wang et al. (2011c), since the primary sites of action of essential oils are at the membrane and inside the cytoplasm of bacteria, the improvement for the antimicrobial activity of the carvacrol is probably because β-CD may have enhanced carvacrol access to these regions by increasing carvacrol aqueous solubility. Since it is possible to use less concentration of carvacrol, these results shows that encapsulation with β-CD is able to improve delivery of these antimicrobials to the site where they can be active.

Self-assembly aggregation of CDs has been extensively investigated from the aggregation of native CDs to high-order complex aggregates. The aggregation of β-CD due to its self-assembling behavior influenced the release characteristics of eugenol (Chun et al., 2012). After 8 h of complexation in solution, β-CD–eugenol molecular inclusion complex started to precipitate, leading to self-aggregation, which retarded the release of eugenol.

The study in (Paczkowska et al., 2015) confirmed that it is possible to form inclusion complexes of rutin, citrus flavonoid glycoside, with β-CD, which may qualify as effective delivery systems. The CD complex protects rutin from thermal and UV degradation and, also, increases this phenolic antioxidant capacity (Nguyen et al., 2013).

The naringenin is a flavanone with a similar structure of the rutin, with good antioxidant capacity and capable of reduce the cholesterol plasma level. The formation of inclusion complex between naringenin and β-CD and its derivatives (HP-β-CD, DM-β-CD, M-β-CD, and TM-β-CD) was analyzed by several teams (Shulman et al., 2011; Yang et al., 2013). For the CDs mentioned previously the stoichiometry of the inclusion complex was 1:1, and complex with HP-β-CD was with higher stability constant. Shulman et al. (2011) proved that the solubility of the flavonoid was increased 400 times when complexed with HP-β-CD. Yang et al. (2013) demonstrated that water solubility and thermal stability of naringenin was improved when encapsulated by β-CD, DM-β-CD or TM-β-CD. In fact, the inclusion complexes remained stable when exposed to temperatures near 225°C. The solubility of the flavonoid was increased 400 times when complexed with the HP-β-CD.

The curcumin encapsulation by the native CDs was described by Patro et al. (2013). They were able to improve the solubility and oral availability with all three inclusion complexes; however the α-CD was the one that showed higher binding constant (1124 M^{-1}). López-Tobar et al. (2012) also applied to β-CD and γ-CD as curcumin carriers. The large cavity of γ-CD was more efficient in the complex formation. Moreover, both CDs were able to form 2:1 inclusion complexes with this phenolic compound, and the molecular interaction proposed was that the aromatic rings and the hydrogen bonds were involved and a change occurs from the curcumin planar ketoenol form to nonplanar form. The chemical stability and bioavailability may be upgraded by this conformational alteration. The same stoichiometry of the CD–curcumin complexes was observed in Rahman et al. (2012). The solubility of curcumin was described as well as the same molecular interaction, as referred previously. Dandawate et al. (2012) used a synthetic form of curcumin but accomplished the same stoichiometry and solubility. In this work, the anticarcinogenic, systemic bioavailability, and tissue distribution of the β-CD–synthetic curcumin complex were compared with the synthetic curcumin alone and concluded that they have been improved by the encapsulation.

CD derivatives were also tested as drug carriers of curcumin, in order to overcome the difficulties of its application as anticarcinogenic agent. For instance, two molecules of HP-β-CD formed stable complex with two molecule of curcumin (Ghanghoria et al., 2012; Mohan et al., 2012). The transdermal capacity of curcumin was raised by the complexation with HP-β-CD as well as the decrease of skin irritation. Besides, HP-β-CD, HP-α-CD and HP-γ-CD were used with the same goal. Mohan et al. compared the encapsulation of the three CD derivatives and reported that the encapsulation may occur both in 1:1 and 2:1 stoichiometry and the HP-γ-CD has a better complexion capacity. This HP-γ-CD–curcumin complex was capable of reducing cell proliferation and increases the apoptosis of cancer cells by interfering in the protein production (Rocks et al., 2012). M-β-CD was used to increase the solubility of the major compound of curcumin present in turmeric rhizome oleoresin (Hadi et al., 2015).

Complex between resveratrol and HP-γ-CD could be considered by the food and packaging industries as a new antibiofilm agent and/or quorum sensing inhibitor to enhance shelf life and increase food safety, meeting consumer expectations to have pathogen-free food without the use of chemical additives or preservatives (Duarte et al., 2015). Resveratrol plays a hydrophobic behavior, and is also extremely affected by exposure to oxygen, light, and oxidative enzymes, reducing its bioactivity. The use of CD to protect resveratrol and to increase its solubility, stability, and bioactivity was applied in several studies. The native α-CD and β-CD and two derivatives (HP-β-CD and DM-β-CD) were used to increase the concentration of resveratrol on solution and its stability. It was observed that the complex formation with native CD was only capable of complexing with part of the resveratrol molecule and that the HP-β-CD offered a cavity with a better fit to the bioactive molecule (Kumpugdee-Vollrath et al., 2012).

The biological properties of resveratrol (antioxidant and anticarcinogenic) were also enhanced by its encapsulation. For instance, Lu et al. (2011) used β-CD and HP-β-CD as resveratrol carrier agents and described the inhibition of the lipid peroxidation activity and the cytotoxicity to cancer cells without harming the healthy ones. Results have shown that, besides being able to significantly increase resveratrol aqueous solubility, CDs also improved or maintained the antioxidant activity of this compound (Davidov-Pardo and McClements, 2014), while being capable of protecting resveratrol from the external environmental factors, such as temperature, light, and pH, by entrapping it inside their cavities.

Hydroxycinnamic acids (such as caffeic, ferulic, *p*-coumaric, and sinapic acids) are a group of compounds highly abundant in food that may account for about one-third of the phenolic compounds in our diet (Teixeira et al., 2013). They have been shown to exhibit a broad range of biological activities (Lou et al., 2011; El-Medany et al., 2011; Shi et al., 2013). Coffee is one of the plants that accumulate hydroxycinnamic acids in quantities sufficient to have physiological effects. Green coffee beans are rich source of these acids. They contain 4–10% of these compounds (Budryn et al., 2014). Coffee bean extracts could be added to various kinds of food products, including those rich in proteins (Budryn et al., 2013). This kind of combination caused partial degradation of essential amino acids and reduced the susceptibility to proteolytic digestion (Bandyopadhyay et al., 2012). Therefore, it could be more beneficial to encapsulate hydroxycinnamic acids by forming inclusion complexes with β-CD and to add them in this form to foods (Nasirullah Kumar and Shariff, 2011). Inclusion of these acids with β-CD presumably does not limit their bioavailability and antioxidant activity. Budryn et al. (2015a) demonstrated to a lesser extent interactions of food protein with hydroxycinnamic acids from coffee encapsulated in β-CD compared with free hydroxycinnamic acids that caused higher availability of the polyphenols for the digestive tract. Taking into account high bioavailability of hydroxycinnamic acids supplied in the form of inclusion complexes with β-CD, such encapsulation could be a promising solution for reducing unfavorable protein–polyphenol interactions during processing of foods enriched with these acids (Budryn et al., 2015b).

The encapsulation of ferulic acid with HP-β-CD was also studied (Wang et al., 2011a). This complex obtained had lower stability. Nevertheless, the solubility and protection against decomposition caused by irradiation with UV light was enhanced by the complexation of the ferulic acid with this CD. The cutaneous permeation and distribution through skin of the complex of ferulic acid with α-CD was assessed by Monti et al. (2011). They concluded that this complex prevented the formation of the less active *cis*-isomer of ferulic acid and its degradation by UV light. The ferulic acid–α-CD complex presented lower penetration on the skin which enlarges the skin protection against UV damages, since the ferulic acid remains at the skin surface.

There are other examples of complexes of β-CD derivatives. Liu et al. (2012) successfully prepared the inclusion complex of glucosyl-β-CD with puerarin and luteolin (Liu et al., 2013b) with good yield via the freeze-drying method. An et al. (2013) reported the inclusion behavior of HP-β-CD with polydatin by fluorescence method. 6-O-α-Maltosyl-β-CD, as one of new β-CD derivatives, was used to improve poor aqueous solubility of polydatin by forming the inclusion complex, which can be used to develop new functional food (Liu et al., 2015b).

Most vitamins are well-known natural antioxidant agents that can be usefully employed for the preservation of foods to increase their shelf life. Vitamin C– and/or E–based chitosan nanoparticles were formulated following the ionic gelation technique and using SBE-β-CD as a crosslinking agent (Aresta et al., 2013). In vitro release studies showed a slow and continuous vitamin(s) release from these nanoparticles during 7 days of monitoring, suggesting the efficacy of the investigated nanosystems as preservative agents in foods.

Fat-soluble vitamin A, also known as retinol, is soluble in water due to its inclusion in the cavity of β-CDs (Vilanova and Solans, 2015). Importantly, the vitamin A within the cyclodextrin complex has higher stability against temperature and oxygen. In addition, this inclusion of vitamin A partially hinders its *trans–cis* isomerization upon UV light and as a consequence the formation of further toxic photodegradation products. Compared to other encapsulating species such as micelles, simple emulsions, solid–lipid nanoparticles, or gels, β-CDs provide better protection of vitamin A; namely the encapsulated form of vitamin A degrades more slowly. Therefore, β-CDs are promising vehicles for increasing water solubility, stability, and thereby the bioavailability of vitamin A in food fortification to treat vitamin A deficiency. Similar to CDs, tetrabutyl ether derivatives of *p*-sulfonatocalix[4]arene and ascorbic acid form a complex, and its inclusion complexes show potential for biological and medical applications (Zhou et al., 2011).

4.5 Sensors

Biosensor detecting techniques have attracted much attention in the content determination of food additives. The hydrophobic cavities of CDs were used to develop different sensors (Ncube et al., 2011). The technology to connect β-CD with invertase that can convert sucrose to glucose for quantification to fabricate the signal tag is crucial question. To do this, gold nanoparticles were used to coimmobilize invertase and mercaptol-β-CD composite owing to its excellent biocompatibility and easy immobilization of proteins (Li et al., 2014b). The sandwich-type strategy was developed on the basis of magnetic molecularly imprinted probenanoparticles and a β-CD/invertase-functionalized signal tag that could recognize the nitrobenzene segment of chloramphenicol through host–guest interaction (Chen et al., 2015). This method overcomes the drawbacks of conventional antibody-based immunoassays, and was successfully applied in the detection of chloramphenicol in animal-derived foods.

Detecting volatile amines is a significant topic in quality control of food diagnosis. Supramolecular strategy for selectively sensing aniline based on the aggregation of perylene–CD conjugate was demonstrated in (Jiang et al., 2011). β-CD/carbon nanotubes combination was widely applied for fabricating electrochemical sensors for some electroactive substrates (Yang et al., 2011; Rahemi et al., 2012; Gaichore and Srivastava, 2013).

As a result of the known host–guest interaction between β-CD and azo-compound (Venkatesh et al., 2013), an electrochemical voltammetric sensor for coloring matter Sudan I was fabricated for the first time by modifying glassy carbon electrode with single-walled carbon nanotubes–CD conjugate (Cao et al., 2015). Self-assembling nanoparticles formed with viologen–resorcin[4]arene cavitands were also applied for detection of this dye, and Sudan sensitivity depends on tail lengths on the lower rim of macrocycles (Kashapov et al., 2014).

Quantum dots (QDs) could be powerful tools for biosensors, bioanalytical methods, and for sensitive detection in relation to food. β-CD modified CdTe QDs were used as a nanosensor for acetylsalicylic acid and its metabolites (Algarra et al., 2012), and β-CD modified CdSe QDs as a recognition system for tyrosine enantiomers (Cao et al., 2013). The optical sensor for vanillin determination based on the selective supramolecular recognition of vanillin with β-CD modified CdSe/ZnS QDs was developed (Durán et al., 2015).

A CdSe/ZnS quantum dot–based fluorescence method for the fast and selective determination of quercetin aglycone in onion was obtained. Fluorescence quenching of the quantum dots is due to the formation of host–guest complexes between quercetin aglycone and β-CD. Besides, the developed technique is to be universal and capable of determining this flavonoid in methanolic extracts obtained from onion in the presence of quercetin glucosides and other phenolic compounds (Dwiecki et al., 2015).

Calixarenes are also applied as sensors for various analytes. The cysteine complex of piperidine-calix[4]arene acts as a convenient and effective dual-signal responsive switch for mercury (II) ion (Zhang et al., 2013), the recognition of which is essential because of its extreme toxicity in the food. In other work, the method for the determination of safranine T in food samples with satisfactory results was developed based on its interaction with calix[4]arene (4,10,16,22-tetramethoxylresorcinarene carboxylic acid derivatives) (Wang et al., 2013). The fluorescence intensity of this calixarene could be quenched by safranine T, and the fluorescence quenching was sensitized in CTAB. The quantum yield of safranine T was approximately 2.0 times higher in the presence of CTAB than that in the absence of CTAB. The mixed systems based on ionic calix[4]arenes and oppositely charged surfactants were also applied for detection and release of such food colorants as Sudan I and Orange OT in (Kashapov et al., 2011; Kharlamov et al., 2013a). As shown in Fig. 8.9, the solubilization study with these hydrophobic dyes demonstrated that only mixed system with calix[4]arene as minor component is capable of binding the organic probe, while mixed aggregates enriched by calix[4]arene show no binding capacity toward dyes.

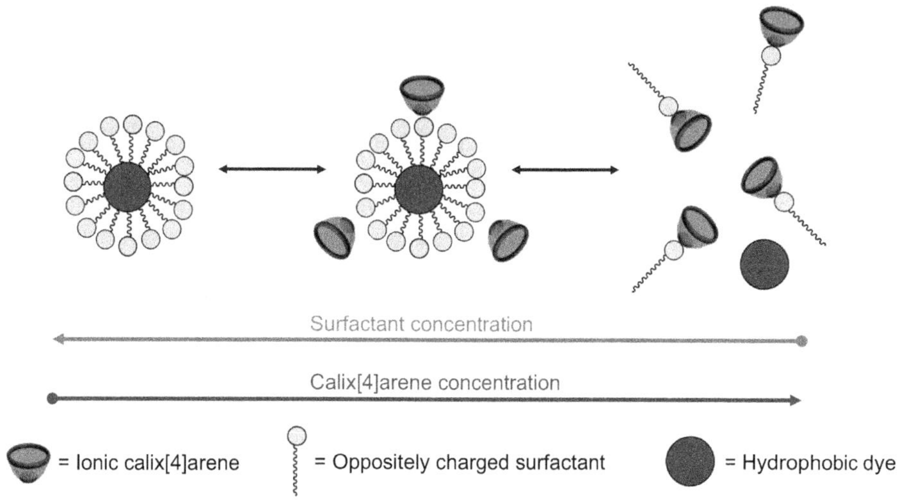

Figure 8.9. Changes in the supramolecular architecture with the variation in ionic calix[4]arene–oppositely charged surfactant ratio.

5 Conclusions

The review presented in the chapter is not intended to be a comprehensive analysis of publications covering delivery and storage systems for food industry. It is rather oriented on recent work illustrating trends outlined in the field. As before, the green chemistry approach is the mainstream of the research activity in the pharma and food technologies. Therefore, natural and low toxic synthetic materials are extremely attractive. Food formulations tend to be polycomponent and polyfunctional, with the low-cost and scaling-up possibility emphasized.

Many advantages are provided by the use of soft colloidal formulations, that is, micelles, microemulsions, and nanoemulsions that possess high solubilization capacity for a variety of water-insoluble supplements. Low droplet sizes, high surface-to-volume ratio, possibility of precisely tuning the properties of carriers and therefore their functional activity, that is, viscosity of formulation, physical, chemical and biological stability of active components, controlled release of cargo material, and so forth, are aspects of promising formulations. Numerous publications focusing on structure-activity relation of amphiphile-based systems make it possible to optimize colloidal formulations in the light of this information. Lipid-based formulations are traditional carriers for the food industry due to high efficacy and low toxicity. A survey of the literature confirms the steadily increased research activity focusing on liposomal preparations, including their optimization in composition and stability as well as application in analytical control of food production. Solid lipid nanoparticles that have received much attention in biomedicine are demonstrated to need thorough optimization for wider use in the food industry.

Polymeric nanoparticles are beneficial from the viewpoint of the use of covalently bound building blocks, which guarantees the stability of composition in time. The wide range of biopolymers and biocompatible synthetic analogs makes them attractive candidates for utilization in food technologies. This is exemplified herein by two natural biopolymers, casein and chitosan, presenting the key families of natural edible polymers, that is, proteins and polysaccharides. Casein is widely involved in fabrication of nanoparticulate structures due to its nutrition value and ability of self-assembling. Chitosan is shown to be a very promising coating agent in food industry both as a single component and in complexes with other building blocks. Due to bioadhesive properties and high affinity toward mucosal membranes, chitosan possesses high protective ability against uncontrolled biodegradation of food ingredients in the gastrointestinal tract. The layer-by-layer technique essentially enhances the polymer-based strategies due to high performance, simplicity, and variability of protocols, and the wide spectrum of edible polianion/polycation pairs that were shown to meet safety standards.

Cyclodextrins and calixarenes are reported to have diverse, important applications in food chemistry. These applications are based on the ability of these macrocycles to form molecular complexes with low-molecular compounds through the molecular recognition mechanism.

Acknowledgment

This work is supported by the Russian Scientific Foundation, grant no. 14-23-00073.

References

Aachmann, F.L., Larsen, K.L., Wimmer, R.J., 2012. Interactions of cyclodextrins with aromatic amino acids: a basis for protein interactions. Incl. Phenom. Macrocycl. Chem. 73, 349–357.

Abdelkader, H., Alani, A.W., Alany, R.G., 2014. Recent advances in nonionic surfactant vesicles (niosomes): self-assembly, fabrication, characterization, drug delivery applications and limitations. Drug Deliv. 21 (2), 87–100.

Abdollahi, M., Rezaei, M., Farzi, G., 2012. A novel active bionanocomposite film incorporating rosemary essential oil and nanoclay into chitosan. J. Food Eng. 111, 343–350.

Acharya, D.P., Sanguansri, L., Augustin, M.A., 2013. Binding of resveratrol with sodium caseinate in aqueous solutions. Food Chem. 141, 1050–1054.

Adjonu, R., Doran, G., Torley, P., Agboola, S., 2014. Whey protein peptides as components of nanoemulsions: a review of emulsifying and biological functionalities. J. Food Eng. 122, 15–27.

Akbarzadeh, A., Rezaei-Sadabady, R., Davaran, S., Joo, S.W., Zarghami, N., Hanifehpour, Y., Samiei, M., Kouhi, M., Nejati-Koshki, K., 2013. Liposome: classification, preparation, and applications. Nanoscale Res. Lett. 8, 102–110.

Akram, M., Bhat, I.A., Kabir-ud-Din, 2015. Interaction of a green ester-bonded gemini surfactant with xanthine oxidase: biophysical perspective. Int. J. Biol. Macromol. 78, 62–71.

Algarra, M., Campos, B.B., Aguiar, F.R., Rodriguez-Borges, J.E., Esteves da Silva, J.C.G., 2012. Novel β-cyclodextrin modified CdTe quantum dots as fluorescence nanosensor for acetylsalicylic acid and metabolites. Mater. Sci. Eng. C 32, 799–803.

Amid, M., Manap, M.Y., Hussin, M., Mustafa, S., 2015a. A novel aqueous two-phase system composed of surfactant and xylitol for the purification of lipase from pumpkin (*Cucurbita moschata*) seeds and recycling of phase components. Molecules 20, 11184–11201.

Amid, M., Murshid, F.S., Manap, M.Y., Hussin, M., 2015b. A novel aqueous micellar two-phase system composed of surfactant and sorbitol for purification of pectinase enzyme from psidium guajava and recycling phase components. Biomed. Res. Int., 1–8, Article ID 815413.

An, S., He, J., Sun, L., Ren, D., Ban, Y., 2013. Investigation of the inclusion behavior of HP-(-cyclodextrin with polydatin in solution and its analytical application. J. Mol. Struct. 1037, 9–14.

Andreu-Sevilla, A.J., Carbonell-Barrachina, A., Lopez-Nicolas, J.M., Garcıa-Carmona, F., 2011a. Sensory quality, volatile compounds and color of pear juice treated with β-cyclodextrin. J. Incl. Phenom. Macrocycl. Chem. 70, 453–460.

Andreu-Sevilla, A.J., López-Nicolás, J.M., Carbonell-Barrachina, A.A., García-Carmona, F., 2011b. Comparative effect of the addition of α-, β-, or γ-cyclodextrin on main sensory and physico-chemical parameters. J. Food Sci. 76, S347–S353.

Appell, M., Jackson, M.A., 2012. Sorption of ochratoxin A from aqueous solutions using β-cyclodextrin-polyurethane polymer. Toxins 4, 98–109.

Aresta, A., Calvano, C.D., Trapani, A., Cellamare, S., Zambonin, C.G., De Giglio, E., 2013. Development and analytical characterization of vitamin(s)-loaded chitosan nanoparticles for potential food packaging applications. J. Nanopart. Res. 15, 1592.

Ariga, K., Lvov, Y.M., Kawakamia, K., Jia, Q., Hilla, J.P., 2011. Layer-by-layer self-assembled shells for drug delivery. Adv. Drug Deliv. Rev. 63, 762–771.

Arnon, H., Granit, R., Porat, R., Poverenov, E., 2015. Development of polysaccharides-based edible coatings for citrus fruits: a layer-by-layer approach. Food Chem. 166, 465–472.

Arora, A., Damodaran, S., 2011. Removal of soy protein–bound phospholipids by a combination of sonication, β-cyclodextrin, and phospholipase A2 treatments. Food Chem. 127, 1007–1013.

Arroyo-Maya, I.J., McClements, D.J., 2015. Biopolymer nanoparticles as potential delivery systems for anthocyanins: fabrication and properties. Food Res. Int. 69, 1–8.

Arthur, T.D., Cavera, V.L., Chikindas, M.L., 2014. On bacteriocin delivery systems and potential applications. Future Microbiol. 9, 235–248.

Asbahani, A.E., Miladi, K., Badri, W., Sala, M., Addi, E.H.A., Casabianca, H., Mousadik, A.E., Hartmann, D., Jilale, A., Renaud, F.N.R., Elaissari, A., 2015. Essential oils: from extraction to encapsulation. Int. J. Pharm. 483, 220–243.

Bachar, M., Mandelbaum, A., Portnaya, I., Perlstein, H., Even-Chen, S., Barenholz, Y., Danino, D., 2012. Development and characterization of a novel drug nanocarrier for oral delivery, based on self-assembled β-casein micelles. J. Control. Rel. 160, 164–171.

Bae, M.-J., Shin, H.S., Kim, E.-K., Kim, J., Shon, D.-H., 2013. Oral administration of chitin and chitosan prevents peanut-induced anaphylaxis in a murine food allergy model. Int. J. Biol. Macromol. 61, 164–168.

Bandyopadhyay, P., Ghosh, A.K., Ghosh, C., 2012. Recent developments on polyphenol–protein interactions: effects on tea and coffee taste, antioxidant properties and the digestive system. Food Func. 3, 592–605.

Bazargani-Gilani, B., Aliakbarlu, J., Tajik, H., 2015. Effect of pomegranate juice dipping and chitosan coating enriched with *Zataria multiflora* Boiss. essential oil on the shelf-life of chicken meat during refrigerated storage. Innov. Food Sci. Emerg. Technol. 29, 280–287.

Beloglazova, N.V., Shmelin, P.S., Speranskaya, E.S., Lucas, B., Helmbrecht, C., Knopp, D., Niessner, R., Saeger, S., Goryacheva, I.Y., 2013. Quantum dot loaded liposomes as fluorescent labels for immunoassay. Anal. Chem. 85, 7197–7204.

Bergonzi, M.C., Hamdouch, R., Mazzacuva, F., Isacchi, B., Bilia, A.R., 2014. Optimization, characterization and in vitro evaluation of curcumin microemulsions. LWT Food Sci. Technol. 59, 148–155.

Bernela, M., Kaur, P., Chopra, M., Thakur, R., 2014. Synthesis, characterization of nisin loaded alginate chitosane pluronic composite nanoparticles and evaluation against microbes. LWT Food Sci. Technol. 59, 1093–1099.

Bhardwaj, V., Sharma, K., Chauhan, S., Sharma, P., 2014. Intermolecular interactions of CTAB and potential oxidation inhibitors: physico-chemical controlled approach for food/pharmaceutical function. RSC Adv. 4, 49400–49414.

Bielska, D., Karewicz, A., Kaminski, K., Kiełkowicz, I., Lachowicz, T., Szczubiałka, K., Nowakowska, M., 2013. Self-organized thermo-responsive hydroxypropyl cellulose nanoparticles for curcumin delivery. Eur. Polym. J. 49, 2485–2494.

Borel, T., Sabliov, C.M., 2014. Nanodelivery of bioactive components for food applications: types of delivery systems, properties, and their effect on ADME profiles and toxicity of nanoparticles. Ann. Rev. Food Sci. Technol 5, 197–213.

Borodina, T.N., Rumsh, L.D., Kunizhev, S.M., Sukhorukov, G.B., 2008. Polyelectrolyte microcapsules as the systems for delivery of biologically active substances. Biomed. Chem. 2, 88–93.

Bourassa, P., N'soukpoé-Kossi, C.N., Tajmir-Riahi, H.A., 2013. Binding of vitamin A with milk α- and β-caseins. Food Chem. 138, 444–453.

Bouvier-Capely, C., Bonthonneau, J.P., Dadache, E., Rebière, F., 2014. An alternative procedure for uranium analysis in drinking water using AQUALIX columns: application to varied French bottled waters. Talanta 118, 180–185.

Bozzuto, G., Molinari, A., 2015. Liposomes as nanomedical devices. Int. J. Nanomed. 10, 975–999.

Brasil, I.M., Gomes, C., Puerta-Gomez, A., Castell-Perez, M.E., Moreira, R.G., 2012. Polysaccharide-based multilayered antimicrobial edible coating enhances quality of fresh-cut papaya. LWT Food Sci. Technol. 47, 39–45.

Budryn, G., Nebesny, E., Rachwał-Rosiak, D., Oracz, J., 2013. Fatty acids, essential amino acids, and chlorogenic acids profiles, in vitro protein digestibility and antioxidant activity of food products containing green coffee extract. Int. Food Res. J. 20, 2133–2144.

Budryn, G., Nebesny, E., Rachwał-Rosiak, D., Pałecz, B., Hodurek, P., Miśkiewicz, K., Oracz, J., Zyzelewicz, D., 2014. Inclusion complexes of β-cyclodextrin with chlorogenic acids (CHAs) from crude and purified aqueous extracts of green Robusta coffee beans (*Coffea canephora* L.). Food Res. Int. 61, 202–213.

Budryn, G., Pałecz, B., Rachwał-Rosiak, D., Oracz, J., Zaczyńska, D., Belica, S., Navarro-González, I., Vegara Meseguer, J.M., Pérez-Sánchez, H., 2015a. Effect of inclusion of hydroxycinnamic and chlorogenic acids from green coffee bean in β-cyclodextrin on their interactions with whey, egg white, and soy protein isolates. Food Chem. 168, 276–287.
Budryn, G., Zaczyńska, D., Rachwał-Rosiak, D., Oracz, J., 2015b. Changes in properties of food proteins after interaction with free and β-cyclodextrin encapsulated hydroxycinnamic acids. Eur. Food Res. Technol. 240, 1157–1166.
Cao, Y., Wu, S., Liang, Y., Yu, Y., 2013. The molecular recognition of (-cyclodextrin modified CdSe quantum dots with tyrosine enantiomers: theoretical calculation and experimental study. J. Mol. Struct. 1031, 9–13.
Cao, Y., Fang, Z., Yang, D., Gao, Y., Li, H., 2015. Voltammetric sensor for Sudan I based on glassy carbon electrode modified by SWCNT/β-cyclodextrin conjugate. Nano 10, 1550026, (9 pages).
Chao, J., Wang, H., Zhao, W., Zhang, M., Zhang, L., 2012. Investigation of the inclusion behavior of chlorogenic acid with hydroxypropylcyclodextrin. Int. J. Biol. Macromol. 50, 277–282.
Chatzidaki, M.D., Mitsou, E., Yaghmur, A., Xenakis, A., Papadimitriou, V., 2015. Formulation and characterization of food-grade microemulsions as carriers of natural phenolic antioxidants. Colloids Surf. A 483, 130–136.
Chen, H., Zhong, Q., 2015. Thermal and UV stability of β-carotene dissolved in peppermint oil microemulsified by sunflower lecithin and Tween 20 blend. Food Chem. 174, 630–636.
Chen, J., Huang, L., Ying, L., Luo, G., Zhao, X., Cao, W., 1999. Self-assembly ultrathin films based on diazoresins. Langmuir 15, 7208–7212.
Chen, S., Gan, N., Zhang, H., Hu, F., Li, T., Cui, H., Cao, Y., Jiang, Q., 2015. A portable and antibody-free sandwich assay for determination of chloramphenicol in food based on a personal glucose meter. Anal. Bioanal. Chem. 407, 2499–2507.
Chen, H., Guan, Y., Zhong, Q., 2015a. Microemulsions based on a sunflower lecithin–Tween 20 blend have high capacity for dissolving peppermint oil and stabilizing coenzyme Q10. J. Agric. Food Chem. 63, 983–989.
Chen, H., Zhang, Y., Zhong, Q., 2015b. Physical and antimicrobial properties of spray-dried zein–casein nanocapsules with coencapsulated eugenol and thymol. J. Food Eng. 144, 93–102.
Chiappisi, L., Gradzielski, M., 2015. Coassembly in chitosan–surfactant mixtures: thermodynamics, structures, interfacial properties and applications. Adv. Colloid Interf. Sci. 220, 92–107.
Chin-Chen, M.L., Carda-Broch, S., Peris-Vicente, J., Rambla-Alegre, M., Esteve-Romero, J., Marco-Peiro, S., 2013. Evaluation of biogenic amines in fish sauce by derivatization with 3,5-dinitrobenzoyl chloride and micellar liquid chromatography. J. Food Compos. Anal. 29, 32–36.
Chow, P.Y., Gue, S.Z., Leow, S.K., Goh, L.B., 2015. Solid self-microemulsifying system (S-SMECS) for enhanced bioavailability and pigmentation of highly lipophilic bioactive carotenoid. Powder Technol. 274, 199–204.
Chranioti, C., Nikoloudaki, A., Tzia, C., 2015. Saffron and beetroot extracts encapsulated in maltodextrin, gum Arabic, modified starch, and chitosan: incorporation in a chewing gum system. Carbohyd. Polym. 127, 252–263.
Chu, P.-T., Wen, H.-W., 2013. Sensitive detection and quantification of gliadin contamination in gluten-free food with immunomagnetic beads based liposomal fluorescence immunoassay. Anal. Chim. Acta 787, 246–253.
Chun, J.-Y., You, S.-K., Lee, M.-Y., Choi, M.J., Min, S.G., 2012. Characterization of beta-cyclodextrin self-aggregates for eugenol encapsulation. Int. J. Food Eng. 8, 1–19.

Chuo, S.C., Ahmad, A., Mohd-Setapar, S.H., Ripin, A., 2014. Reverse micelle extraction—an alternative for recovering antibiotics. Pharm. Chem. 6, 37–44.

Ciobanu, A., Landy, D., Fourmentin, S., 2013. Complexation efficiency of cyclodextrins for volatile flavor compounds. Food Res. Int. 53, 110–114.

Cui, X., Zhang, P., Yang, X., Yang, M., Zhou, W., Zhang, S., Gao, H., Lu, R., 2015. β-CD/ATP composite materials for use in dispersive solidphase extraction to measure (fluoro)quinolone antibiotics in honey samples. Anal. Chim. Acta 878, 131–139.

Dalgleish, D.G., 1997. Structure-function relationships of caseins. In: Srinivasan Damodaran, A.P. (Ed.), Food Proteins and Their Applications. Marcel Dekker, New York, pp. 199–223.

Dandawate, P.R., Vyas, A., Ahmad, A., Banerjee, S., Deshpande, J., Swamy, K.V., Jamadar, A., Dumhe-Klaire, A.C., Padhye, S., Sarkar, F.H., 2012. Inclusion complex of novel curcumin analogue CDF and (-cyclodextrin (1:2) and its enhanced in vivo anticancer activity against pancreatic cancer. Pharm. Res. 29, 1775–1786.

Das, K., Maiti, S., Das, P.K., 2014. Probing enzyme location in water-in-oil microemulsion using enzyme-carbon dot conjugates. Langmuir 30, 2448–2459.

Dasgupta, N., Ranjan, S., Mundekkad, D., Ramalingam, C., Shanker, R., Kumar, A., 2015. Nanotechnology in agro-food: from field to plate. Food Res. Int. 69, 381–400.

Davidov-Pardo, G., McClements, D.J., 2014. Resveratrol encapsulation: designing delivery systems to overcome solubility, stability, and bioavailability issues. Trends Food Sci. Technol. 38, 88–103.

de Britto, D., de Moura, M.R., Aouada, F.A., Mattoso, L.H.C., Assis, O.B.G., 2012. *N, N, N*-trimethyl chitosan nanoparticles as a vitamin carrier system. Food Hydrocoll. 27, 487–493.

DeKruif, C.G., Holt, C., 2003. Casein micelle structure, functions and interactions. In: Fox, P.F., McSweeney, P.L.H. (Eds.), Advanced Dairy Chemistry Proteins. Part A Kluwer Academic/Plenum Publishers, New York, pp. 233–276.

Deladino, L., Anbinder, P.S., Navarro, A.S., Martino, M.N., 2008. Encapsulation of natural antioxidants extracted from Ilex paraguariensis. Carbohyd. Polym. 71, 126–134.

Delavari, B., Saboury, A.A., Atri, M.S., Ghasemi, A., Bigdeli, B., Khammari, A., Maghami, P., Moosavi-Movahedi, A.A., Haertle, T., Goliaei, B., 2015. Alpha-lactalbumin: a new carrier for vitamin D_3 food enrichment. Food Hydrocoll. 45, 124–131.

Dias, K., Nikolaou, S., Giovani, W.F., 2011. The in vitro antioxidant properties of the Al-quercetin/CD and Al-catechin/CD inclusion compounds, rationalized in terms of their electrochemical behaviour. Med. Chem. Res. 21, 2920–2925.

Dima, C., Cotârlet, M., Alexe, P., Dima, S., 2014. Microencapsulation of essential oil of pimento [*Pimenta dioica* (L) Merr.] by chitosan/k-carrageenan complex coacervation method. Innov. Food Sci. Emerg. Technol. 22, 203–211.

Djekic, L., Cirkovic, V., Heleta, M., Krajisnik, D., Primorac, M., 2013. Water-dilutable biocompatible microemulsion systems: design and characteriation. Tenside Surfact. Det. 50, 409–413.

Djordjevic, D., Cercaci, L., Alamed, J., McClements, D.J., Decker, E.A., 2007. Chemical and physical stability of citral and limonene in sodium dodecyl sulfate-chitosan and gum arabic-stabilized oil-in-water emulsions. J. Agric. Food Chem. 55, 3585–3591.

Dong, L., Liu, M., Chen, A., Wang, Y., Sun, D., 2013. Solubilities of quercetin in three β-cyclodextrin derivative solutions at different temperatures. J. Mol. Liq. 177, 204–208.

Donsì, F., Marchese, E., Maresca, P., Pataro, G., Vu, K.D., Salmieri, S., Lacroix, M., Ferrari, G., 2015. Green beans preservation by combination of a modified chitosan based-coating containing nanoemulsion of mandarin essential oil with high pressure or pulsed light processing. Postharvest Biol. Technol. 106, 21–32.

Dos Santos, C., Buera, M.P., Mazzobre, M.F., 2011. Phase solubility studies and stability of cholesterol/β-cyclodextrin inclusion complexes. J. Sci. Food Agric. 91, 2551–2557.

Dua, J.S., Rana, A.C., Bhandari, A.K., 2012. Liposome: methods of preparation and applications. Int. J. Pharm. Stud. Res. 3, 14–20.

Duarte, A., Alves, A.C., Ferreira, S., Silva, F., Domingues, F.C., 2015. Resveratrol inclusion complexes: antibacterial and anti-biofilm activity against *Campylobacter* spp. and *Arcobacter butzleri*. Food Res. Int. 77, 244–250.

Durán, G.M., Contento, A.M., Ríos, Á., 2015. β-Cyclodextrin coated CdSe/ZnS quantum dots for vanillin sensoring in food samples. Talanta 131, 286–291.

Durga, P.A., Sanguansri, L., Augustin, M.F., 2013. Binding of resveratrol with sodium caseinate in aqueous solutions. Food Chem. 141, 1050–1054.

Dwiecki, K., Kwiatkowska, P., Siger, A., Staniek, H., Nogala-Kałucka, M., Polewski, K., 2015. Determination of quercetin in onion (*Allium cepa*) using β-cyclodextrin-coated CdSe/ZnS quantum dot-based fluorescence spectroscopic technique. Int. J. Food Sci. Technol. 50, 1366–1373.

Edwards, K.A., Bolduc, O.R., Baeumner, A.J., 2012. Miniaturized bioanalytical systems: enhanced performance through liposomes. Curr. Opin. Chem. Biol. 16, 444–452.

El-Abbassi, A., Kiai, H., Raiti, J., Hafidi, A., 2014. Cloud point extraction of phenolic compounds from pretreated olive mill wastewater. J. Environ. Chem. Eng. 2, 1480–1486.

El-Medany, A., Bassiouni, Y., Khattab, M., Mahesar, A., 2011. Chlorogenic acid as potential anti-inflammatory analgesic agent: an investigation of the possible role of nitrogen-based radicals in rats. Int. J. Pharmacol. Toxicol. Sci. 1, 24–33.

El-Shaheny, R.N., El-Maghrabey, M.H., Belal, F.F., 2015. Micellar liquid chromatography from green analysis perspective. Open Chem. 13, 877–892.

Elzoghby, A.O., Helmy, M.W., Samy, W.M., Elgindy, N.A., 2013. Novel ionically crosslinked casein nanoparticles for flutamide delivery: formulation, characterization, and in vivo pharmacokinetics. Int. J. Nanomed. 8, 1721–1732.

Evans, D.E., 1959. Milk composition of mammals whose milk is not normally used for human consumption. Dairy Sci. Abstr. 21, 277–288.

Evans, M.T.A., Phillips, M.C., Jones, M.N., 1979. The conformation and aggregation of bovine b-casein II. Thermodynamics of thermal association and the effects of changes in polar and apolar interactions on micellization. Biopolym. 18, 1123–1140.

Fadida, T., Selilat-Weiss, A., Poverenov, E., 2015. *N*-hexylimine-chitosan, a biodegradable and covalently stabilized source of volatile, antimicrobial hexanal. Next generation controlled release system. Food Hydrocoll. 48, 213–219.

Fan, Y., Fang, Y., Ma, L., 2014. The self-crosslinked ufasome of conjugated linoleic acid: investigation of morphology, bilayer membrane and stability. Colloids Surf. B 123, 8–14.

Fangueiro, J.F., Andreani, T., Fernandes, L., Garcia, M.L., Egea, M.A., Silva, A.M., Souto, E.B., 2014. Physicochemical characterization of epigallocatechin gallate lipid nanoparticles (EGCG-LNs) for ocular instillation. Colloids Surf. B 123, 452–460.

Fathi, M., Mozafari, M.R., Mohebbi, M., 2012. Nanoencapsulation of food ingredients using lipid-based delivery systems. Trends Food Sci. Technol. 23, 13–27.

Fathi, M., Martin, A., McClements, D.J., 2014. Nanoencapsulation of food ingredients using carbohydrate based delivery systems. Trends Food Sci. Technol. 39, 18–39.

Fauziah, C.I., Zaibunnisa, A.H., Osman, H., Wan Aida, W.M., 2013. Thermal analysis and surface morphology study of cholesterol: β-cyclodextrin inclusion complex. Adv. Mat. Res. 812, 221–225.

Flanagan, J., Singh, H., 2006. Microemulsions: a potential delivery system for bioactives in food. Food Sci. Nutr. 46, 221–237.

Folch-Cano, C., Guerrero, J., Speisky, H., Jullian, C., Olea-Azar, C., 2013. NMR and molecular fluorescence spectroscopic study of the structure and thermodynamic parameters of EGCG/β-cyclodextrin inclusion complexes with potential antioxidant activity. J. Incl. Phenom. Macrocycl. Chem. 78, 287–298.

Fox, P.F., 2003. Milk proteins: general and historical aspects. In: Fox, P.F., McSweeney, P.L.H. (Eds.), Advanced Dairy Chemistry Proteins. Part A. Kluwer Academic/Plenum Publishers, New York, pp. 1–48.

Friberg, S.E., Bothorel, P., 1987. Microemulsions: Structure and Dynamics. CRC Press, Boca Raton.

Gabdrakhmanov, D.R., Voronin, M.A., Zakharova, L.Y., Konovalov, A.I., Khaybullin, R.N., Strobykina, I.Y., Kataev, V.E., Faizullin, D.A., Gogoleva, N.E., Konnova, T.A., Salnikov, V.V., Zuev, Y.F., 2013. Supramolecular design of biocompatible nanocontainers based on amphiphilic derivatives of a natural compound isosteviol. Phys. Chem. Chem. Phys. 15, 16725–16735.

Gabdrakhmanov, D., Valeeva, F., Syakaev, V., Lukashenko, S., Zakharov, S., Kuryashov, D., Bashkirtseva, N., Zakharova, L., Latypov, S., Sinyashin, O., 2015. Novel supramolecular system based on cationic amphiphile bearing glucamine fragment: structural behavior and binding the hydrophobic probe. Mendeleev Commun. 25, 174–176.

Gadhave, A.D., Waghmare, J.T., 2014. A short review on microemulsion and its application in extraction of vegetable oil. Int. J. Res. Eng. Technol. 3, 148–158.

Gaichore, R.R., Srivastava, A.K., 2013. Voltammetric determination of nifedipine using a β-cyclodextrin modified multi-walled carbon nanotube paste electrode. Sens. Actuators B 188, 1328–1337.

Gainanova, G.A., Vagapova, G.I., Ibragimova, A.R., Valeeva, F.G., Syakaev, V.V., Tudriy, E.V., Galkina, I.V., Kataeva, O.N., Zakharova, L.Y., Latypov, S.K., Konovalov, A.I., 2012. Self-assembling systems based on amphiphilic alkyltriphenylphosphonium bromides. Elucidation of the role of head group. J. Colloid Interf. Sci. 367, 327–336.

Gangnard, S., Zuev, Y., Gaudin, J.-C., Fedotov, V., Choiset, Y., Axelos, M.A.V., Chobert, J.-M, Haertle, T., 2007. Modifications of the charges at the N-terminus of bovine beta-case: Consequences on its structure its micellisation. Food Hydrocoll. 21, 180–190.

García-Padial, M., Martínez-Ohárriz, M.C., Navarro-Blasco, I., Zornoza, A., 2013. The role of cyclodextrins in ORAC-fluorescence assays. Antioxidant capacity of tyrosol and caffeic acid with hydroxypropyl-β-cyclodextrin. J. Agric. Food Chem. 61, 12260–12264.

García-Segovia, P., Barreto-Palacios, V., Bretón, J., Martínez-Monzó, J., 2011. Microencapsulation of essential oils using β-cyclodextrin: applications in gastronomy. J. Culinary Sci. Technol. 9, 150–157.

Garti, N., Aserin, A., 2012. Micelles and microemulsions as food ingredient and nutraceutical delivery systems. In: Garti, N., McClements, D.J. (Eds.), Encapsulation Technologies and Delivery Systems for Food Ingredients and Nutraceuticals. Woodhead Publishing, Cambridge, UK, pp. 211–251.

Gaudette, N.J., Pickering, G.J., 2012a. The efficacy of bitter blockers on health-relevant bitterants. J. Funct. Foods 4, 177–184.

Gaudette, N.J., Pickering, G.J., 2012b. Optimizing the orosensory properties of model functional beverages: the influence of novel sweeteners, odorants, bitter blockers, and their mixtures on (+)-catechin. J. Food Sci. 77, S226–S232.

Gaudin, J.-C., Le Parc, A., Castrec, B., Ropers, M.-H., Choiset, Y., Shchutskaya, Y., Yousefi, R., Muronetz, V.I., Zuev, Y., Chobert, J.-M., Haertlé, T., 2009. Engineering of caseins and modulation of their structure and interactions. Biotechnol. Adv. 27, 1124–1131.

Ghanghoria, R., Kesharwani, P., Agashe, H.B., Jain, N.K., 2012. Transdermal delivery of cyclodextrin-solubilized curcumin. Drug Deliv. Transl. Res. 3, 272–285.

Gharibzahedi, S.M.T., Razavi, S.H., Mousavi, M., 2014. Characterizing the natural canthaxanthin/2-hydroxypropyl-beta-cyclodextrin inclusion complex. Carbohyd. Polym. 101, 1147–1153.

Gharsallaoui, A., Roudaut, G., Beney, L., Chambin, O., Voilley, A., Saurel, R., 2012. Properties of spray-dried food flavors microencapsulated with two-layered membranes: roles of interfacial interactions and water. Food Chem. 132, 1713–1720.

Gibis, M., Vogta, E., Weiss, J., 2012. Encapsulation of polyphenolic grape seed extract in polymer-coated liposomes. Food Funct. 3, 246–254.

Gonçalves, M.C.F., Mertins, O., Pohlmann, A.R., Silveira, N.P., Guterres, S.S., 2012. Chitosan-coated liposomes as an innovative nanocarrier for drugs. J. Biomed. Nanotechnol. 8, 240–250.

Gülseren, İ., Guri, A., Corredig, M., 2012. Encapsulation of tea polyphenols in nanoliposomes prepared with milk phospholipids and their effect on the viability of HT-29 human carcinoma cells. Food Dig. 3, 36–45.

Guo, D.-S., Liu, Y., 2012. Calixarene-based supramolecular polymerization in solution. Chem. Soc. Rev. 41, 5907–5921.

Guo, Z., Chen, F., Yang, H., Liu, K., Zhang, L., 2015. Kinetics of protein extraction in reverse micelle. Int. J. Food Prop. 18, 1707–1718.

Gutierrez, F.J., Albillos, S.M., Casas-Sanz, E., Cruz, Z., Garcıa-Estrada, C., Garcıa-Guerra, A., Garcıa-Reverter, J., Garcıa-Suarez, M., Gaton, P., Gonzalez-Ferrero, C., Olabarrieta, I., Olasagasti, M., Rainieri, S., Rivera-Patino, D., Rojo, R., Romo-Hualde, A., Saiz-Abajo, M.J., Mussons, M.-L., 2013. Methods for the nanoencapsulation of β-carotene in the food sector. Trends Food Sci. Technol. 32, 73–83.

Gutsche, C.D., 2008. Calixarenes: An Introduction, second ed. The Royal Society of Chemistry, Cambridge.

Guttoff, M., Saberi, A.H., McClements, D.J., 2015. Formation of vitamin D nanoemulsion-based delivery systems by spontaneous emulsification: factors affecting particle size and stability. Food Chem. 171, 117–122.

Gyawali, R., Ibrahim, S.A., 2014. Natural products as antimicrobial agents. Food Control 46, 412–429.

Hadi, B.J., Sanagi, M.M., Aboul-Enein, H.Y., Ibrahim, W.A.W., Jamil, S., Mu'azu, M.A., 2015. Microwave-assisted extraction of methyl β-cyclodextrin-complexed curcumin from turmeric rhizome oleoresin. Food Anal. Methods 8, 2447–2456.

Haidong, L., Fang, Y., Zhihong, T., Changle, R., 2011. Study on preparation of β-cyclodextrin encapsulation tea extract. Int. J. Biol. Macromol. 49, 561–566.

Haloci, E., Toska, V., Shkreli, R., Goci, E., Vertuani, S., Manfredini, S., 2014. Encapsulation of *Satureja montana* essential oil in β-cyclodextrin. J. Incl. Phenom. Macrocycl. Chem. 80, 147–153.

Hannon, J.C., Kerry, J., Cruz-Romero, M., Morris, M., Cummins, E., 2015. Advances and challenges for the use of engineered nanoparticles in food contact materials. Trends Food Sci. Technol. 43, 43–62.

Healey, R.D., Prasad, S., Rajendram, V., Thordarson, P., 2015. Unravelling the interaction between α-cyclodextrin with the thaumatin protein and a peptide mimic. Supramol. Chem. 27, 414–419.

Higueras, L., Lopez-Carballo, G., Cerisuelo, J.P., Gavara, R., Hernandez-Munoz, P., 2013. Preparation and characterization of chitosan/HP-β-cyclodextrins composites with high sorption capacity forcarvacrol. Carbohyd. Polym. 97, 262–268.

Higueras, L., Lopez-Carballo, G., Gavara, R., Hernandez-Munoz, P., 2015. Incorporation of hydroxypropyl-β-cyclodextrins into chitosan films to tailor loading capacity for active aroma compound carvacrol. Food Hydrocoll. 43, 603–611.

Hill, L.E., Gomes, C., Taylor, T.M., 2013. Characterization of β-cyclodextrin inclusion complexes containing essential oils (transcinnamaldehyde, eugenol, cinnamon bark, and clove bud extracts) for antimicrobial delivery applications. LWT Food Sci. Technol. 51, 86–93.

Holmberg, K., 2007. Organic reactions in microemulsions. Eur. J. Org. Chem. 2007, 731–742.

Horvath, G., Szoke, E., Kemeny, A., Bagoly, T., Deli, J., Szente, L., Pal, S., Sandor, K., Szolcsanyi, J., Helyes, Z., 2012. Lutein inhibits the function of the transient receptor potential A1 ion channel in different in vitro and in vivo models. J. Mol. Neurosci. 46, 1–9.

Hou, L., Wang, J., Zhang, D., 2013. Optimization of debittering of soybean antioxidant hydrolysates with β-cyclodextrins using response surface methodology. J. Food. Sci. Technol. 50, 521–527.

Howard, L.R., Brownmiller, C., Prior, R.L., Mauromoustakos, A., 2013. Improved stability of chokeberry juice anthocyanins by beta-cyclodextrin addition and refrigeration. J. Agric. Food Chem. 61, 693–699.

Hu, K., McClements, D.J., 2015. Fabrication of biopolymer nanoparticles by antisolvent precipitation and electrostatic deposition: zein-alginate core/shell nanoparticles. Food Hydrocoll. 44, 101–108.

Hu, K., Zhang, Y., Liu, J., Chen, K., Zhao, W., Zhu, W., Song, Z., Ye, B., Zhang, S., 2013. Development and application of a new 25,27-bis(l-phenylalaninemethylester-*N*-carbonylmethoxy)-26,28-dihydroxy-*para*-tert-butylcalix[4]arene stationary phase. J. Sep. Sci. 36, 445–453.

Hu, J., Wang, X., Xiao, Z., Bi, W., 2015. Effect of chitosan nanoparticles loaded with cinnamon essential oil on the quality of chilled pork. LWT Food Sci. Technol. 63, 519–526.

Huang, Z., Li, X., Zhang, T., Song, Y., She, Z., Li, J., Deng, Y., 2014. Progress involving new techniques for liposome preparation. Asian J. Pharm. Sci. 9, 176–182.

Husseini, G.A., Pitt, W.G., 2008. Micelles and nanoparticles for ultrasonic drug and gene delivery. Adv. Drug Deliv. Rev. 60, 1137–1152.

Itoh, K., Akutsu, S., Watanabe, T., Yoshizawa, F., Uda, Y., 2015. Improved water-solubility and decreased acetate-odor of a cyclodextrin-treated nasunin-containing powder obtained from spent eggplant pre-pickling solution. Nippon Shokuhin Kagaku Kogaku Kaishi 62, 201–206.

Jahadi, M., Khosravi-Darani, K., Ehsani, M.R., Mozafari, M.R., Saboury, A.A., Pourhosseini, P.S., 2015. The encapsulation of flavor zyme in nanoliposome by heating method. J. Food Sci. Technol. 52, 2063–2072.

Jenness, R., Sloan, R.E., 1970. The composition of milk of various species: a review. Dairy Sci. Abstr. 32, 599–612.

Jiang, B.-P., Guo, D.-S., Liu, Y., 2011. Reversible and selective sensing of aniline vapor by perylene-bridged bis(cyclodextrins) assembly. J. Org. Chem. 76, 6101–6107.

Joye, I.J., McClements, D.J., 2014. Biopolymer-based nanoparticles and microparticles: fabrication, characterization, and application. Curr. Opin. Colloid Interf. Sci. 19, 417–427.

Jung, H.J., Ganesan, P., Lee, S.J., Kwak, H.S., 2013. Comparative study of flavor in cholesterol-removed Gouda cheese and Gouda cheese during ripening. J. Dairy Sci. 96, 1972–1983.

Kaffarnik, S., Heid, C., Kayademir, Y., Eibler, D., Vetter, W., 2015. High enantiomeric excess of the flavor relevant 4 alkyl-branched fatty acids in milk fat and subcutaneous adipose tissue of sheep and goat. J. Agric. Food Chem. 63, 469–475.

Karadag, A., Özçelik, B., Sramek, M., Gibis, M., Kohlus, R., Weiss, J., 2013. Presence of electrostatically adsorbed polysaccharides improves spray drying of liposomes. J. Food Sci. 78 (2), E206–E221.

Karangwa, E., Hayat, K., Rao, L., Nshimiyimana, D., Foh, M.B.K., Li, L., Ntwali, J., Raymond, L.V., Xia, S., Zhang, X., 2012. Improving blended carrot-orange juice quality by the addition of cyclodextrins during enzymatic clarification. Food Bioprocess. Technol. 5, 2612–2617.

Karimi, M., Habibi-Rezaei, M., Safari, M., Moosavi-Movahedi, A.A., Sayyah, M., Sadeghi, R., Kokini, J., 2014. Immobilization of endo-inulinase on poly-D-lysine coated $CaCO_3$ micro-particles. Food Res. Int. 66, 485–492.

Karpichev, E.A., Zakharova, L.Y., Gaisin, N.K., Gnezdilov, O.I., Zhil'tsova, E.P., Pashirova, T.N., Lukashenko, S.S., Anikeev, A.V., Gorbań, O.A., Konovalov, A.I., Popov, A.F., 2014. Self-assembly of symmetrical and dissymmetrical dicationic surfactants in the solid phase and in solution. Russ. Chem. Bull. 63, 68–75.

Kashapov, R.R., Pashirova, T.N., Kharlamov, S.V., Ziganshina, A.Y., Ziltsova, E.P., Lukashenko, S.S., Zakharova, L.Y., Habicher, W.D., Latypov, S.K., Konovalov, A.I., 2011. Novel self-assembling system based on resorcinarene and cationic surfactant. Phys. Chem. Chem. Phys. 13, 15891–15898.

Kashapov, R.R., Kharlamov, S.V., Sultanova, E.D., Mukhitova, R.K., Kudryashova, Y.R., Zakharova, L.Y., Ziganshina, A.Y., Konovalov, A.I., 2014. Controlling the size and morphology of supramolecular assemblies of viologen–resorcin[4] arene cavitands. Chem. Eur. J. 20, 14018–14025.

Kasongo, K.W., Pardeike, J., Müller, R.H., Walker, R.B., 2011. Selection and characterization of suitable lipid excipients for use in the manufacture of Didanosine-loaded solid lipid nanoparticles and nanostructured lipid carriers. J. Pharm. Sci. 100, 5185–5196.

Kataev, V.E., Strobykina, I.Y., Zakharova, L.Y., 2014. Quaternary ammonium derivatives of natural terpenoids: synthesis and properties. Russ. Chem. Bull. 63, 1884–1900.

Kaur, K., 2014. Nanoemulsions as an effective medium for encapsulation and stabilization of cholesterol/β-cyclodextrin inclusion complex. J. Sci. Food Agric. 95, 2718–2728.

Kawabori, M., Kacimi, R., Karliner, J.S., Yenari, M.A., 2013. Sphingolipids in cardiovascular and cerebrovascular systems: pathological implications and potential therapeutic targets. World J. Cardiol. 5, 75–86.

Ke, D., Wu, Y., Wang, X., 2014. Shift of acid–base equilibrium of curcumin in its complexes withgemini surfactant hexamethylene-1,6-bis-(dodecyldimethylammonium bromide). Colloids Surf. A 443, 481–487.

Kegeles, G., 1979. A shell model for size distribution in micelles. J. Phys. Chem. 83, 1728–1732.

Kfoury, M., Auezova, L., Greige-Gerges, H., Ruellan, S., Fourmentin, S., 2014. Cyclodextrin, an efficient tool for *trans*-anethole encapsulation: chromatographic, spectroscopic, thermal and structural studies. Food Chem. 164, 454–461.

Kharlamov, S.V., Kashapov, R.R., Pashirova, T.N., Zhiltsova, E.P., Lukashenko, S.S., Ziganshina, A.Y., Gubaidullin, A.T., Zakharova, L.Y., Gruner, M., Habicher, W.D., Konovalov, A.I., 2013a. A supramolecular amphiphile based on calix[4] resorcinarene and cationic surfactant for controlled self-assembly. J. Phys. Chem. C 117, 20280–20288.

Kharlamov, S.V., Voronin, M.A., Semenov, V.E., Gabdrakhmanov, D.R., Nikolaev, A.E., Reznik, V.S., Zakharova, L.Y., Konovalov, A.I., 2013b. Tunable biomimetic systems based on a novel amphiphilic pyrimidinophane and a helper nonionic surfactant. Colloids Surf. B 111, 218–223.

Kiegiel, K., Steczek, L., Zakrzewska-Trznadel, G., 2013. Application of calixarenes as macrocyclic ligands for uranium(VI): a review. J. Chem. 2013 (762819), 1–16.

Komaiko, J., McClements, D.J., 2015a. Food-grade nanoemulsion filled hydrogels formed by spontaneous emulsification and gelation: optical properties, rheology, and stability. Food Hydrocoll. 46, 67–75.

Komaiko, J., McClements, D.J., 2015b. Low-energy formation of edible nanoemulsions by spontaneous emulsification: factors influencing particle size. J. Food Eng. 146, 122–128.

Krishnaswamy, K., Orsat, V., Thangavel, K., 2012. Synthesis and characterization of nano-encapsulated catechin by molecular inclusion with beta-cyclodextrin. J. Food Eng. 111, 255–264.

Kumpugdee-Vollrath, M., Ibold, Y., Sriamornsak, P., 2012. Solid state characterization of *trans* resveratrol complexes with different cyclodextrins. J. Asian Assoc. Schools of Pharm. 1, 125–136.

Kunzmann, A., Andersson, B., Turnherr, T., Krug, H., Scheynius, A., Fadeel, B., 2011. Toxicology of engineered nanomaterials: focus on biocompatibility, biodistribution and biodegradation. Biochim. Biophys. Acta 1810, 361–373.

Kurkov, S.V., Loftsson, T., 2013. Cyclodextrins. Int. J. Pharm. 453, 167–180.

Kutyreva, M.P., Khannanov, A., Zakharova, L.Y., Ulakhovich, N.A., Kutyrev, G.A., Gabdrakhmanov, D.R., 2015. Self-organization and solubilization in binary systems based on hyperbranched polyesters polyols. Colloids Surf. A 468, 40–48.

Lacatusu, I., Badea, N., Niculae, G., Bordei, N., Stan, R., Meghea, A., 2014. Lipid nanocarriers based on natural compounds: an evolving role in plant extract delivery. Eur. J. Lipid Sci. Technol. 116, 1708–1717.

Landry, K.S., Chang, Y., McClements, D.J., McLandsborough, L., 2014. Effectiveness of a novel spontaneous carvacrol nanoemulsion against *Salmonella enterica* Enteritidis and *Escherichia coli* O157:H7 on contaminated mung bean and alfalfa seeds. Int. J. Food Microbiol. 187, 15–21.

Laouini, A., Jaafar-Maalej, C., Sfar, S., 2011. Liposome preparation using a hollow fiber membrane contactord: application to spironolactone encapsulation. Int. J. Pharm. 415, 53–61.

Laouini, A., Jaafar-Maalej, C., Limayem-Blouza, I., Sfar, S., Charcosset, C., Fessi, H., 2012. Preparation, characterization and applications of liposomes: state of the art. J. Colloid. Sci. Biotechnol. 1, 147–168.

Lesoin, L., Boutin, O., Crampon, C., 2011. CO_2/water/surfactant ternary systems and liposome formation using supercritical CO_2: a review. Colloids Surf. A 377, 1–14.

Li, Z., Chen, S., Gu, Z., Chen, J., Wu, J., 2014a. Alpha-cyclodextrin: enzymatic production and food applications. Trends Food Sci. Technol. 35, 151–160.

Li, Y., Yuan, R., Chai, Y., Zhuo, Y., Su, H., Zhang, Y., 2014b. Horseradish peroxidase-loaded nanospheres attached to hollow gold nanoparticles as signal enhancers in an ultrasensitive immunoassay for alpha-fetoprotein. Microchim. Acta 181, 679–685.

Lin, C.-C., Lin, H.-Y., Chi, M.-H., Shen, C.-M., Chen, H.-W., Yang, W.-J., Lee, M.-H., 2014. Preparation of curcumin microemulsions with food-grade soybean oil/lecithin and their cytotoxicity on the HepG2 cell line. Food Chem. 154, 282–290.

Liu, B., Zhao, J., Liu, Y., Zhu, X., Zeng, J., 2012. Physicochemical properties of the inclusion complex of puerarin and glucosyl-β-cyclodextrin. J. Agric. Food Chem. 60, 12501–12507.

Liu, M., Dong, L.N., Chen, A.J., Zheng, Y., Sun, D.Z., Wang, X., Wang, B., 2013a. Inclusion complexes of quercetin with three β-cyclodextrins derivatives at physiological pH: spectroscopic study and antioxidant activity. Spectrochim. Acta A 115, 854–860.

Liu, B., Li, W., Zhao, J., Liu, Y., Zhu, X., Liang, G., 2013b. Physicochemical characterisation of the supramolecular structure of luteolin/cyclodextrin inclusion complex. Food Chem. 141, 900–906.

Liu, M., Chen, A., Wang, Y., Wang, C., Wang, B., Sun, D., 2015a. Improved solubility and stability of 7-hydroxy-4-methylcoumarin at different temperatures and pH values through complexation with sulfobutyl ether-β-cyclodextrin. Food Chem. 168, 270–275.

Liu, B., Li, Y., Xiao, H., Liu, Y., Mo, H., Ma, H., Liang, G., 2015b. Characterization of the supermolecular structure of polydatin/6-O-α-maltosyl-β-cyclodextrin inclusion complex. J. Food Sci. 80, 1156–1161.

López-de-Dicastillo, C., Jordá, M., Catalá, R., Gavara, R., Hernández-Muñoz, P., 2011. Development of active polyvinyl alcohol/β-cyclodextrin composites to scavenge undesirable food components. J. Agric. Food Chem. 59, 11026–11033.

López-Nicolás, J.M., Rodríguez-Bonilla, P., García-Carmona, F., 2014. Cyclodextrins and antioxidants. Crit. Rev. Food Sci. Nutr. 54, 251–276.

López-Tobar, E., Blanch, G.P., Ruiz del Castillo, M.L., Sanchez-Cortesa, S., 2012. Encapsulation and isomerization of curcumin with cyclodextrins characterized by electronic and vibrational spectroscopy. Vibrat. Spec. 62, 292–298.

Lou, Z., Wang, H., Zhu, S., Ma, C., Wang, Z., 2011. Antibacterial activity and mechanism of action of chlorogenic acid. J. Food Sci. 76, 398–403.

Lu, Z., Chen, R., Fu, R., Xiong, J., Hu, Y., 2011. Cytotoxicity and inhibition of lipid peroxidation activity of resveratrol/cyclodextrin inclusion complexes. J. Incl. Phenom. Macrocycl. Chem. 73, 313–320.

Lucas-Abellán, C., Mercader-Ros, M.T., Zafrilla, M.P., Gabaldón, J.A., Núñez-Delicado, E., 2011. Comparative study of different methods to measure antioxidant activity of resveratrol in the presence of cyclodextrins. Food. Chem. Toxicol. 49, 1255–1260.

Lvov, Y., Ariga, K., Ichinose, I., Kunitake, K., 1995. Assembly of multicomponent protein films by means of electrostatic layer-by-layer adsorption. J. Am. Chem. Soc. 117, 6117–6123.

Lytra, G., Tempere, S., de Revel, G., Barbe, J.-C., 2012. Distribution and organoleptic impact of ethyl 2-hydroxy-4-methylpentanoate enantiomers in wine. J. Agric. Food Chem. 60, 1503–1509.

Machado, A.R., Assis, L.M., Costa, J.A.V., Badiale-Furlong, E., Motta, A.S., Micheletto, Y.M.S., Souza-Soares, L.A., 2014. Application of sonication and mixing for nanoencapsulation of the cyanobacterium *Spirulina platensis* in liposomes. Int. Food Res. J. 21, 2201–2206.

MaHam, A., Tang, Z., Wu, H., Wang, J., Lin, Y., 2009. Protein-based nanomedicine platforms for drug delivery. Small 5, 1706–1721.

Manju, S., Sreenivasan, K., 2011. Hollow microcapsules built by layer by layer assembly for the encapsulation and sustained release of curcumin. Colloids Surf. B 82, 588–593.

Marianecci, C., Di Marzio, L., Rinaldi, F., Celia, C., Paolino, D., Alhaique, F., Esposito, S., Carafa, M., 2014. Niosomes from 80s to present: the state of the art. Adv. Colloid Interf. Sci. 205, 187–206.

Marsanasco, M., Márquez, A.L., Wagner, J.R., Chiaramoni, N.S., del, V., Alonso, S., 2015. Bioactive compounds as functional food ingredients: characterization in model system and sensory evaluation in chocolate milk. J. Food Eng. 166, 55–63.

Martín, M.J., Lara-Villoslada, F., Ruiz, M.A., Morales, M.E., 2015. Microencapsulation of bacteria: a review of different technologies and their impact on the probiotic effects. Innov. Food Sci. Emerg. Technol. 27, 15–25.

Martins, S., Sarmento, B., Ferreira, D.C., Souto, E.B., 2007. Lipid-based colloidal carriers for peptide and protein delivery—liposomes versus lipid nanoparticles. Int. J. Nanomed. 2, 595–607.

Mastromatteo, M., Mastromatteo, M., Conte, A., Del Nobile, M.A., 2010. Advances in controlled release devices for food packaging applications. Trends Food Sci. Technol. 21, 591–598.

Masuda, T., Taguchi, W., Sano, A., Ohta, K., Kitabatake, N., Tani, F., 2013. Five amino acid residues in cysteine-rich domain of human T1R3 were involved in the response for sweet-tasting protein, thaumatin. Biochimie 95, 1502–1505.

Maswal, M., Dar, A.A., 2014. Formulation challenges in encapsulation and delivery of citral for improved food quality. Food Hydrocoll. 37, 182–195.

Mauser, T., Déjugnat, C., Sukhorukov, G.B., 2006. Balance of hydrophobic and electrostatic forces in the pH response of weak polyelectrolyte capsules. J. Phys. Chem. B 110, 20246–20253.

Mazzaferro, S., Bouchemal, K., Gallard, J.-F., Iorga, B.I., Cheron, M., Gueutin, C., Steinmesse, C., Ponchel, G., 2011. Bivalent sequential binding of docetaxel to methyl-β-cyclodextrin. Int. J. Pharm. 416, 171–180.

McClements, D.J., 2013. Edible lipid nanoparticles: digestion, absorption, and potential toxicity. Prog. Lipid Res. 52, 409–423.

McClements, D.J., 2015. Encapsulation, protection, and release of hydrophilic active components: potential and limitations of colloidal delivery systems. Adv. Colloid Interf. Sci. 219, 27–53.

Medeiros, B.G.S., Pinheiro, A.C., Carneiro-da-Cunha, M.G., Vicente, A.A., 2010. Development and characterization of a nanomultilayer coating of pectin and chitosan—Evaluation of its gas barrier properties and application on 'Tommy Atkins' mangoes. J. Food Eng. 110, 457–464.

Mehmood, T., 2015. Optimization of the canola oil based vitamin E nanoemulsions stabilized. Food Chem. 183, 1–7.

Mei, J., Guo, Q., Wu, Y., Li, Y., Yu, H., 2012. Study of proteolysis, lipolysis, and volatile compounds of a Camembert-type cheese manufactured using a freeze-dried Tibetan kefir coculture during ripening. Food Sci. Biotechnol. 21, 159–165.

Memon, F.N., Memon, S., 2013. Differential recognition of d and l-alanine by calix[4]arene amino derivative. J. Incl. Phenom. Macrocycl. Chem. 77, 413–420.

Menger, F.M., Keiper, J.S., 2000. Gemini surfactants. Angew. Chem. Int. Ed. 112, 1906–1920.

Mirgorodskaya, A.B., Yackevich, E.I., Lukashenko, S.S., Zakharova, L.Y., Konovalov, A.I., 2012a. Solubilization and catalytic behavior of micellar system based on gemini surfactant with hydroxyalkylated head group. J. Mol. Liq. 169, 106–109.

Mirgorodskaya, A., Yackevich, E., Syakaev, V., Zakharova, L., Latypov, S., Konovalov, A., 2012b. Micellization and catalytic properties of cationic surfactants with head groups functionalized with a hydroxyalkyl fragment. J. Chem. Eng. Data 57, 3153–3163.

Mirgorodskaya, A.B., Yatskevich, E.I., Zakharova, L.Y., Konovalov, A.I., 2012c. Geminal surfactant–nonionic polymer mixed micellar systems. Colloid J. 74, 91–98.

Mirgorodskaya, A.B., Yackevich, E.I., Zakharova, L.Y., Konovalov, A.I., 2013. Microemulsions based on cationic surfactant with hydroxyalkyl fragment in the head group. Chem. Phys. Lett. 567, 18–22.
Mirgorodskaya, A.B., Karpichev, Y.A., Zakharova, L.Y., Yackevich, E.I., Kapitanov, I.V., Lukashenko, S.S., Popov, A.F., Konovalov, A.I., 2014a. Aggregation behavior and interface properties of mixed surfactant systems gemini 14-s-14/CTABr. Colloids Surf. A 457, 425–432.
Mirgorodskaya, A.B., Lukashenko, S.S., Yatskevich, E.I., Kulik, N., Voloshina, A.D., Kudryavtsev, D.B., Panteleeva, A.R., Zobov, V.V., Zakharova, L.Y., Konovalov, A.I., 2014b. Aggregation behavior, anticorrosion effect, and antimicrobial activity of alkylmethylmorpholinium bromides. Prot. Met. Phys. Chem. 50, 530–534.
Mirgorodskaya, A.B., Yackevich, E.I., Valeeva, F.G., Pankratov, V.A., Zakharova, L.Y., 2014c. Solubilizing and catalytic properties of supramolecular systems based on gemini surfactants. Russ. Chem. Bull. 63, 82–87.
Mittal, K.L., 1977. Micellization. Solubilization and MicroemulsionsPlenum Press, New York.
Mohan, P.R.K., Sreelakshmi, G., Muraleedharan, C.V., Joseph, R., 2012. Water soluble complexes of curcumin with cyclodextrins: characterization by FT-Raman spectroscopy. Vibrat. Spec. 62, 77–84.
Mokhtari, B., Pourabdollah, K., 2012. Chromatographic separation of clenbuterol by bonded phases bearing nano-baskets of *p*-tert-calix[4]-1,2-crown-3, -crown-4, -crown-5 and -crown-6. J. Sci. Food Agric. 92, 2679–2688.
Mokhtari, B., Pourabdollah, K., Dalali, N., 2011a. Applications of nano-baskets of calixarenes in chromatography. Chromatographia 73, 829–847.
Mokhtari, B., Pourabdollah, K., Dallali, N., 2011b. A review of calixarene applications in nuclear industries. J. Radioanal. Nucl. Chem. 287, 921–934.
Monti, D., Tampucci, S., Chetoni, P., Burgalassi, S., Saino, V., Centini, M., Staltari, L., Anselmi, C., 2011. Permeation and distribution of ferulic acid and its a-cyclodextrin complex from different formulations in hairless rat skin. AAPS PharmSciTech 12, 514–520.
Moré, M.I., Freitas, U., Rutenberg, D., 2014. Positive effects of soy lecithin-derived phosphatidylserine plus phosphatidic acid on memory, cognition, daily functioning, and mood in elderly patients with Alzheimer's disease and dementia. Adv. Ther. 31, 1247–1262.
Müller, R.H., Mehnert, W., Lucks, J.S., Schwarz, C., Mühlen, A., Weyhers, H., Freitas, C., Rühl, D., 1995. Solid lipid nanoparticles (SLN)—an alternative colloidal carrier system for controlled drug deliver. Eur. J. Pharm. Biopharm. 41, 62–69.
Müller, R.H., Gohla, S., Keck, C.M., 2011. State of the art of nanocrystals—special features, production, nanotoxicology aspects and intracellular delivery. Eur. J. Pharm. Biopharm. 78, 1–9.
Nakagawa, K., Sowasod, N., Tanthapanichakoon, W., Charinpanitkul, T., 2013. Hydrogel based oil encapsulation for controlled release of curcumin by using a ternary system of chitosan, kappa-carrageenan, and carboxymethylcellulose sodium salt. LWT Food Sci. Technol. 54, 600–605.
Naksuriya, O., Okonogi, S., Schiffelers, R.M., Hennink, W.E., 2014. Curcumin nanoformulations: a review of pharmaceutical properties and preclinical studies and clinical data related to cancer treatment. Biomaterials 35, 3365–3383.
Nascimento, C.F., Rocha, D.L., Rocha, F.R.P., 2015. A fast and environmental friendly analytical procedure for determination of melamine in milk exploiting fluorescence quenching. Food Chem. 169, 314–319.
Nasirullah Kumar, P., Shariff, R., 2011. Development of nutraceutical carriers for functional food applications. Nutr. Food Sci. 41, 34–43.
Nasr, J.J., Shalan, S., Belal, F., 2014. Simultaneous determination of tylosin and josamycin residues in muscles, liver, eggs and milk by MLC with a monolithic

column and time-programmed UV detection: application to baby food and formulae. Chem. Cent. J. 8, 1–9.

Navarro, P., Nicolas, T.S., Gabaldon, J.A., Mercader-Ros, M.T., Calín-Sánchez, A., Carbonell-Barrachina, A.A., Pérez-López, A.J., 2011. Effects of cyclodextrin type on vitamin C, antioxidant activity, and sensory attributes of a mandarin juice enriched with pomegranate and goji berries. J. Food Sci. 76, S319–S324.

Ncube, P., Krause, R.W., Mamba, B.B., 2011. Fluorescent sensing of chlorophenols in water using an azo dye modified β-cyclodextrin polymer. Sensors 11, 4598–4608.

Nedovic, V., Kalusevic, A., Manojlovic, V., Levic, S., Bugarski, B., 2011. An overview of encapsulation technologies for food applications. Procedia Food Sci. 1, 1806–1815.

Németh, K., Tárkányi, G., Varga, E., Imre, T., Mizsei, R., Iványi, R., Visy, J., Szemán, J., Jicsinszky, L., Szente, L., Simonyi, M., 2011. Enantiomeric separation of antimalarial drugs by capillary electrophoresis using neutral and negatively charged cyclodextrins. J. Pharm. Biomed. Anal. 54, 475–481.

Nguyen, T.A., Liu, B., Zhao, J., Thomas, D.S., Hook, J.M., 2013. An investigation into the supramolecular structure, solubility, stability and antioxidant activity of rutin/cyclodextrin inclusion complex. Food Chem. 136, 186–192.

Nishizawa, M., Hosoya, T., Hirokawa, T., Shinya, K., Kumazawa, S., 2014. NMR spectroscopic characterization of inclusion complexes of theaflavin digallate and cyclodextrins. Food Sci. Technol. Res. 20, 663–669.

Ohara, M., Ohyama, Y., 2014. Delivery and application of dietary polyphenols to target organs, tissues and intracellular organelles. Curr. Drug. Metab. 15, 37–47.

Ohta, K., Masuda, T., Tani, F., Kitabatake, N., 2011. The cysteine-rich domain of human T1R3 is necessary for the interaction between human T1R2-T1R3 sweet receptors and a sweet-tasting protein, thaumatin. Biochem. Biophys. Res. Commun. 406, 435–438.

Oidtmann, J., Schantz, M., Mäder, K., Baum, M., Berg, S., Betz, M., Kulozik, U., Leick, S., Rehage, H., Schwarz, K., 2012. Preparation and comparative release characteristics of three anthocyanin encapsulation systems. J. Agric. Food Chem. 60, 844–851.

Ojeda, E., Puras, G., Agirre, M., Zárate, J., Grijalvo, S., Pons, R., Eritja, R., Martinez-Navarrete, G., Soto-Sanchez, C., Fernández, E., Pedraz, J.L., 2015. Niosomes based on synthetic cationic lipids for gene delivery: the influence of polar head-groups on the transfection efficiency in HEK-293, ARPE-19 and MSC-D1 cells. Org. Biomol. Chem. 13, 1068–1081.

Oliveira, T.V., Soares, N.F.F., Silva, D.J., Andrade, N.J., Medeiros, E.A.A., Badaró, A.T., 2013. Development of PDA/Phospholipids/Lysine vesicles to detectpathogenic bacteria. Sens. Actuators B 188, 385–392.

Oliveira, T.V., Soares, N.F.F., Andrade, N.J., Silva, D.J., Medeiros, E.A.A., Badaró, A.T., 2015. Application of PCDA/SPH/CHO/Lysine vesicles to detect pathogenic bacteria in chicken. Food Chem. 172, 428–432.

Ono, N., Miyamoto, Y., Ishiguro, T., Motoyama, K., Hirayama, F., Iohara, D., Seo, H., Tsuruta, S., Arima, H., Uekama, K., 2011. Reduction of bitterness of antihistaminic drugs by complexation with β-cyclodextrins. J. Pharm. Sci. 100, 1935–1943.

Ozyilmaz, E., Sayin, S., Arslan, M., Yilmaz, M., 2014. Improving catalytic hydrolysis reaction efficiency of sol–gel-encapsulated *Candida rugosa* lipase with magnetic β-cyclodextrin nanoparticles. Colloids Surf. B. 113, 182–189.

Paczkowska, M., Mizera, M., Piotrowska, H., Szymanowska-Powałowska, D., Lewandowska, K., Goscianska, J., Pietrzak, R., Bednarski, W., Majka, Z., Cielecka-Piontek, J., 2015. Complex of rutin with β-cyclodextrin as potential delivery system. PLoS One 10, e0120858.

Park, Y.W., Haenlein, G.F.W., 2006. Handbook of Milk of Non-Bovine Mammals. Blackwell Publishing, Oxford.
Park, J.-H., Choi, K.-H., Kwak, H.-S., 2011. Single- and 14-day repeat-dose toxicity of cross-linked β-cyclodextrin in rats. Int. J. Toxicol. 6, 700–706.
Park, S.J., Hong, C.R., Choi, S.J., 2015. Citral degradation in micellar structures formed with polyoxyethylene-type surfactants. Food Chem. 170, 443–447.
Pashirova, T.N., Zhiltsova, E.P., Kashapov, R.R., Lukashenko, S.S., Litvinov, A.I., Kadirov, M.K., Zakharova, L.Y., Konovalov, A.I., 2010. Supramolecular systems based on 1-alkyl-4-aza-1-azoniabicyclo[2.2.2]octane bromides. Russ. Chem. Bull. 59, 1745–1752.
Pashirova, T.N., Lukashenko, S.S., Zakharov, S.V., Voloshina, A.D., Zhiltsova, E.P., Zobov, V.V., Souto, E.B., Zakharova, L.Y., 2015. Self-assembling systems based on quaternized derivatives of 1,4-diazabicyclo[2.2.2]octane in nutrient broth as antimicrobial agents and carriers for hydrophobic drugs. Colloids Surf. B 127, 266–273.
Patel, M.R., Martin-Gonzalez, M.F.S., 2012. Characterization of ergocalciferol loaded solid lipid nanoparticles. J. Food Sci. 77, 8–13.
Patro, N.M., Sultana, A., Terao, K., Nakata, D., Jo, A., Urano, A., Ishida, Y., Gorantla, R.N., Pandit, V., Devi, K., Rohit, S., Grewal, B.K., Sophia, E.M., Suresh, A., Ekbote, V.K., Suresh, S., 2013. Comparison and correlation of in vitro, in vivo and in silico evaluations of alpha, beta and gamma cyclodextrin complexes of curcumin. J. Incl. Phenom. Macrocycl. Chem. 78, 471–483.
Peng, H., Xiong, H., Li, J., Xie, M., Liu, Y., Bai, C., Chen, L., 2010. Vanillin cross-linked chitosan microspheres for controlled release of resveratrol. Food Chem. 121, 23–28.
Perret, F., Coleman, A.W., 2011. Biochemistry of anionic calix[n]arenes. Chem. Commun. 47, 7303–7319.
Persson, B.R.R., Holm, E., 2011. Polonium-210 and lead-210 in the terrestrial environment: a historical review. J. Environ. Radioact. 102, 420–429.
Pinheiro, A.C., Bourbon, A.I., Quintas, M.A.C., Coimbra, M.A., Vicente, A.A., 2012. K-carrageenan/chitosan nanolayered coating for controlled release of a model bioactive compound. Innov. Food Sci. Emerg. Technol. 16, 227–232.
Pinheiro, A.C., Bourbon, A.I., Cerqueira, M.A., Maricato, É., Nunes, C., Coimbra, M.A., Vicent, A.A., 2015. Chitosan/fucoidan multilayer nanocapsules as a vehicle for controlledrelease of bioactive compounds. Carbohyd. Polym. 115, 1–9.
Pinho, E., Grootveld, M., Soares, G., Henriques, M., 2014. Cyclodextrin-based hydrogels toward improved wound dressings. Crit. Rev. Biotechnol. 34, 328–337.
Piyarat, K., Walaisiri, M., Pornpen, W., 2014. Comparison of stability of red colorants from natural sources, roselle and lac in micelles. Int. Food Res. J. 21, 325–330.
Plaza-Oliver, M., de Baranda, J.F.S., Robledo, V.R., Castro-Vázquez, L., Gonzalez-Fuentes, J., Marcos, P., Lozano, M.V., Santander-Ortega, M.J., Arroyo-Jimenez, M.M., 2015. Design of the interface of edible nanoemulsions to modulate the bioaccessibility of neuroprotective antioxidants. Int. J. Pharm. 490, 209–218.
Podyachev, S.N., Burmakina, N.E., Syakaev, V.V., Sudakova, S.N., Habicher, W.D., Konovalov, A.I., 2011. Pyridinyl hydrazone derivatives of thiacalix[4]arene as selective extractants of transition metal ions. J. Incl. Phenom. Macrocycl. Chem. 71, 161–168.
Podyachev, S.N., Kashapova, N.E., Syakaev, V.V., Sudakova, S.N., Zainullina, R.R., Gruner, M., Habicher, W.D., Barsukova, T.A., Yang, F., Konovalov, A.I., 2014. Mercury(II) and silver(I) receptors based on tetrathiacalix[4]arene hydrazones. J. Incl. Phenom. Macrocycl. Chem. 78, 371–380.
Polyakov, N.E., Kispert, L.D., 2015. Water soluble biocompatible vesicles based on polysaccharides and oligosaccharides inclusion complexes for carotenoid delivery. Carbohyd. Polym. 128, 207–219.

Poór, M., Kunsági-Máté, S., Szente, L., Matisz, G., Secenji, G., Czibulya, Z., Koszegi, T., 2015. Interaction of ochratoxin A with quaternary ammonium beta-cyclodextrin. Food Chem. 172, 143–149.

Poverenov, E., Danino, S., Horev, B., Granit, R., Vinokur, Y., Rodov, V., 2014. Layer-by-layer electrostatic deposition of edible coating on fresh cut melon model: anticipated and unexpected effects of alginate–chitosan combination. Food Bioprocess. Technol. 7, 1424–1432.

Pradines, B., Gallard, J.-F., Iorga, B.I., Gueutin, C., Loiseau, P.M., Ponchel, G., Bouchemal, K., 2014. Investigation of the complexation of albendazole with cyclodextrins for the design of new antiparasitic formulations. Carbohydr. Res. 398, 50–55.

Pushkala, R., Parvathy, K.R., Srividya, N., 2012. Chitosan powder coating, a novel simple technique for enhancement of shelf life quality of carrot shreds stored in macro perforated LDPE packs. Innov. Food Sci. Emerg. Technol. 16, 11–20.

Qi, H., Hu, W., Jiang, A., Tian, M., Li, Y., 2011. Extending shelf-life of fresh-cut 'Fuji' apples with chitosan-coatings. Innov. Food Sci. Emerg. Technol. 12, 62–66.

Rahemi, V., Vandamme, J.J., Garrido, J.M.P.J., Borges, F., Brett, C.M.A., Garrido, E.M.P.J., 2012. Enhanced host–guest electrochemical recognition of herbicide MCPA using a β-cyclodextrin carbon nanotube sensor. Talanta 99, 288–293.

Rahman, S., Cao, S., Steadman, K.J., Wei, M., Parekh, H.S., 2012. Native and β-cyclodextrin-enclosed curcumin: entrapment within liposomes and their in vitro cytotoxicity in lung and colon cancer. Drug Deliv. 19, 346–353.

Răileanu, M., Todan, L., Voicescu, M., Ciuculescu, C., Maganu, M., 2013. A way for improving the stability of the essential oils in an environmental friendly formulation. Mater. Sci. Eng. C 33, 3281–3288.

Ratanasooriya, C., Rupasinghe, H.P.V., 2012. Extraction of phenolic compounds from grapes and their pomace using β-cyclodextrin. Food Chem. 134, 625–631.

Ratcharin, N., Wongtrakul, P., Indranupakorn, R., 2012. Preparation of zingiber officinale extract loaded solid lipid nanoparticles. Adv. Mater. Res. 506, 389–392.

Rebolleda, S., Sanz, M.T., Benito, J.M., Beltran, S., Escudero, I., San-Jose, M.L.G., 2015. Formulation and characterisation of wheat bran oil-in-water nanoemulsions. Food Chem. 167, 16–23.

Rezanka, M., Rezanka, P., Sykora, D., Jindrich, J., Král, V., 2012. Impact of substituent position in monosubstituted α-cyclodextrins on enantioselectivity in capillary electrophoresis. J. Sep. Sci. 35, 811–815.

Richter, Y., Herzog, Y., Lifshitz, Y., Hayun, R., Zchut, S., 2013. The effect of soybean-derived phosphatidylserine on cognitive performance in elderly with subjective memory complaints: a pilot study. Clin. Interv. Aging 8, 557–563.

Rocks, N., Bekaert, S., Coia, I., Paulissen, G., Gueders, M., Evrard, B., Van Heugen, J.C., Chiap, P., Foidart, J.M., Noel, A., Cataldo, D., 2012. Curcumin-cyclodextrin complexes potentiate gemcitabine effects in an orthotopic mouse model of lung cancer. Br. J. Cancer 107, 1083–1092.

Rodríguez-Bonilla, P., López-Nicolás, J.M., Méndez-Cazorla, L., García-Carmona, F., 2011a. Development of a reversed phase high performance liquid chromatography method based on the use of cyclodextrins as mobile phase additives to determine pterostilbene in blueberries. J. Chromatogr. B 879, 1091–1097.

Rodríguez-Bonilla, P., Méndez-Cazorla, L., López-Nicolás, J.M., García-Carmona, F., 2011b. Kinetic mechanism and product characterization of the enzymatic peroxidation of pterostilbene as model of the detoxification process of stilbene-type phytoalexins. Phytochemistry 72, 100–108.

Romero-Guevara, R., Cencetti, F., Donati, C., Bruni, P., 2015. Sphingosine 1-phosphate signaling pathway in inner ear biology: new therapeutic strategies for hearing loss? Front. Aging. Neurosci. 7, 60–73.

Roohinejad, S., Middendorf, D., Burritt, D.J., Bindrich, U., Everett, D.W., Oey, I., 2014. Capacity of natural β-carotene loaded microemulsion to protect Caco-2 cells from oxidative damage caused by exposure to H_2O_2. Food Res. Int. 66, 469–477.

Roohinejad, S., Oey, I., Wen, J., Lee, S.J., Everett, D.W., Burritt, D.J., 2015. Formulation of oil-in-water β-carotene microemulsions: effect of oiltype and fatty acid chain length. Food Chem. 174, 270–278.

Sagalowicz, L., Leser, M.E., 2010. Delivery systems for liquid food products. Curr. Opin. Colloid Interf. Sci. 15, 61–72.

Sağlam, D., Venema, P., van der Linden, E., de Vries, R., 2014. Design, properties, and applications of protein micro- and nanoparticles. Curr. Opin. Colloid Interf. Sci. 19, 428–437.

Salimi, A., Motaharitabar, E., Goudarzi, M., Rezaie, A., Kalantari, H., 2014. Toxicity evaluation of microemulsion (nano size) of sour cherry kernel extract for the oral bioavailability enhancement jundishapur. J. Nat. Pharm. Prod. 9, 16–23.

Santos, E.H., Kamimura, J.A., Hill, L.E., Gomes, C.L., 2015. Characterization of carvacrol beta-cyclodextrin inclusion complexes as delivery systems for antibacterial and antioxidant applications. LWT Food Sci. Technol. 60, 583–592.

Schnyder, A., Huwyler, J., 2005. Drug transport to brain with targeted liposomes. NeuroRx 2, 99–107.

Schühle, D.T., Peters, J.A., Schatz, J., 2011. Metal binding calixarenes with potential biometic and biomedical applications. Coord. Chem. Rev. 255, 2727–2745.

Sebestyén, Z., Buvári-Barcza, Á., Rohonczy, J., 2012. pH-dependent complex formation of amino acids with β-cyclodextrin and quaternary ammonium bcyclodextrin. J. Incl. Phenom. Macrocycl. Chem. 73, 199–210.

Serno, T., Geidobler, R., Winter, G., 2011. Protein stabilization by cyclodextrins in the liquid and dried state. Adv. Drug Deliv. Rev. 63, 1086–1106.

Severino, R., Vu, K.D., Donsì, F., Salmieri, S., Ferrari, G., Lacroix, M., 2014. Antibacterial and physical effects of modified chitosan based-coating containing nanoemulsion of mandarin essential oil and three non-thermal treatments against Listeria innocua in green beans. Int. J. Food Microbiol. 191, 82–88.

Sha, O., Zhu, X., Feng, Y., Ma, W., 2015. Aqueous two-phase based on ionic liquid liquid–liquid microextraction for simultaneous determination of five synthetic food colourants in different food samples by high-performance liquid chromatography. Food Chem. 174, 380–386.

Shaaban, H.A., Edris, A.E., 2015. Factors affecting the phase behavior and antimicrobial activity of carvacrol microemulsions. J. Oleo Sci. 64, 393–404.

Shahir, A.A., Javadian, S., Razavizadeh, B.B.M., Gharibi, H., 2011. Comprehensive study of tartrazine/cationic surfactant interaction. J. Phys. Chem. B 115, 14435–14444.

Sharipova, A.A., Aidarova, S.B., Grigoriev, D., Mutalieva, B., Madibekova, G., Tleuova, A., Millerba, R., 2015. Polymer–surfactant complexes for microencapsulation of vitamin E and its release. Colloids Surf. B 137, 152–157.

Shegokar, R., Singh, K.K., Müller, R.H., 2011. Production & stability of stavudine solid lipid nanoparticles: from lab to industrial scale. Int. J. Pharm. 416, 461–470.

Shi, H., Dong, L., Dang, X., Liu, Y., Jiang, J., Wang, Y., Lu, X., Guo, X., 2013. Effect of chlorogenic acid on LPS-induced proinflammatory signaling in hepatic stellate cells. Inflamm. Res. 62, 581–587.

Shulman, M., Cohen, M., Soto-Gutierrez, A., Yagi, H., Wang, H., Goldwasser, J., 2011. Enhancement of naringenin bioavailability by complexation with hydroxypropoyl-β-cyclodextrin. PloS One 6, e18033.

Sinha, A., Basiruddin, S.K., Chakraborty, A., Jana, N.R., 2015. β-cyclodextrin functionalized magnetic mesoporous silica colloid for cholesterol separation. ACS Appl. Mater. Interf. 7, 1340–1347.

Sintra, T.E., Ventura, S.P.M., Coutinho, J.A.P., 2014. Superactivity induced by micellar systems as the key for boosting the yield of enzymatic reactions. J. Mol. Catal. B 107, 140–151.

Siripatrawan, U., Noipha, S., 2012. Active film from chitosan incorporating green tea extract for shelf life extension of pork sausages. Food Hydrocolloid. 27, 102–108.

Souza, J.M., Caldas, A.L., Tohidi, S.D., Molina, J., Souto, A.P., Fangueiro, R., Zille, A., 2014. Properties and controlled release of chitosan microencapsulated limonene oil. Rev. Bras. Farmacogn. 24, 691–698.

Stroylova, Y.Y., Konnova, T., Zuev, Y.F., Chobert, J.-M., Choiset, Y., Haertlé, T., Muronetz, V.I., 2013. Selective introduction of sulfhydryl groups into recombinant proteins for study of protein-protein interactions. Chromatographia 76, 621–628.

Su, Y., Tian, Y., Yan, R., Wang, C., Niu, F., Yang, Y., 2015. Study on a novel process for the separation of phospholipids, triacylglycerol and cholesterol from egg yolk. J. Food Sci. Technol. 52, 4586–4592.

Sultanova, E.D., Krasnova, E.G., Kharlamov, S.V., Nasybullina, G.R., Yanilkin, V.V., Nizameev, I.R., Kadirov, M.K., Mukhitova, R.K., Zakharova, L.Y., Ziganshina, A.Y., Konovalov, A.I., 2015. Thermoresponsive polymer nanoparticles based on viologen-cavitand. ChemPlusChem 80, 217–222.

Sun, X., Sui, S., Ference, C., Zhang, Y., Sun, S., Zhou, N., Zhu, W., Zhou, K., 2014. Antimicrobial and mechanical properties of β-cyclodextrin inclusion with essential oils in chitosan films. J. Agric. Food Chem. 62, 8914–8918.

Sung, S.-Y., Sin, L.T., Tee, T.-T., Bee, S.-T., Rahmat, A.R., Rahman, W.A.W.A., Tan, A.-C., Vikhraman, M., 2013. Antimicrobial agents for food packaging applications. Trends Food Sci. Technol. 33, 110–123.

Sutili, F.K., Nogueira, D.O., Leite, S.G.F., Miranda, L.S.M., de Souza, R.O.M.A., 2015. Lipase immobilized in microemulsion based organogels (MBGs) as an efficient catalyst for continuous-flow esterification of protected fructose. RSC Adv. 5, 37287–37291.

Swaisgood, H.E., 1992. Chemistry of the casein. In: Fox, P.F., (Ed.), Advanced Dairy Chemistry 1 Proteins. Springer, New York, pp. 63–111.

Tablet, C., Metei, I., Pincu, E., Meltzer, V., Hillebrand, M., 2012. Spectroscopic and thermodynamic studies of 7-diethylamino-coumarin-3-carboxylic acid in interaction with β- and 2-hydroxypropyl-β-cyclodextrins. J. Mol. Liq. 168, 47–53.

Tamjidi, F., Shahedi, M., Varshosaz, J., Nasirpour, A., 2013. Nanostructured lipid carriers (NLC): a potential delivery system for bioactive food molecules. Innov. Food Sci. Emer. Technol. 19, 29–43.

Tan, C., Zhang, Y., Abbas, S., Feng, B., Zhang, X., Xia, S., 2014. Modulation of the carotenoid bioaccessibility through liposomalencapsulation. Colloids Surf. B 123, 692–700.

Teixeira, J., Gaspar, A., Garrido, E.M., Garrido, J., Borges, F., 2013. Hydroxycinnamic acid antioxidants: an electrochemical overview. Biomed. Res Int. 2013, Article ID 251754 (11 pages).

Thakhiew, W., Devahastin, S., Soponronnarit, S., 2010. Effects of drying methods and plasticizer concentration on some physical and mechanical properties of edible chitosan films. J. Food Eng. 99, 216–224.

Thompson, A., Boland, M., Singh, H., 2009. Milk Proteins: From Expression to Food. Academic Press, New Zealand.

Tiwary, L.K., 2013. Reverse microemulsions as a novel reaction media. Orient. J. Chem. 29, 375–379.

Trabelsi, I., Bejar, W., Ayadi, D., Chouayekh, H., Kammoun, R., Bejar, S., 2013. Encapsulation in alginate and alginate coated chitosan improved the survival of newly probiotic in oxgall and gastric juice. Int. J. Biol. Macromol. 61, 36–42.

Tu, Z., Wen, Z., Liu, J., Wang, C., He, J., 2013. Preparation of octacosanol microemulsion and its application in sports beverage. J. Chin. Inst. Food Sci. Technol. 13, 108–112.

Turovsky, T., Khalfin, R., Kababya, S., Schmidt, A., Barenholz, Y., Danino, D., 2015. Celecoxib encapsulation in β-casein micelles: structure, interactions, and conformation. Langmuir 31, 7183–7192.

Uekaji, Y., Nakata, D., Shiga, H., Jo, A., Tachi, I., Fukumi, H., Urano, A., Terao, K., 2011. Formation of CoQ10 reduced form by mixing CoQ10 oxidized form γ-CD complex and vitamin C in powder. J. Incl. Phenom. Macrocycl. Chem. 70, 447–451.

Vagapova, G.I., Ibragimova, A.R., Zakharov, A.V., Dobrynin, A.B., Galkina, I.V., Zakharova, L.Y., Konovalov, A.I., 2013a. Novel biomimetic systems based on polyethylene glycols and amphiphilic phosphonium salt: self-organization and solubilization of hydrophobic guest. Eur. Polym. J. 49, 1031–1039.

Vagapova, G.I., Valeeva, F.G., Gainanova, G.A., Syakaev, V.V., Galkina, I.V., Zakharova, L.Y., Latypov, S.K., Konovalov, A.I., 2013b. Novel self-assembling systems based on amphiphilic phosphonium salt and polyethylene glycol. Kinetic arguments for synergetic aggregation behavior. Colloids Surf. A 419, 186–193.

Van der Lee, R., Buljan, M., Lang, B., Weatheritt, R.J., Daughdrill, G.W., Dunker, A.K., Fuxreiter, M., Gough, J., Gsponer, J., Jones, D.T., Kim, P.M., Kriwacki, R.W., Oldfield, C.J., Pappu, R.V., Tompa, P., Uversky, V.N., Wright, P.E., Babu, M.M., 2014. Classification of intrinsically disordered regions and proteins. Chem. Rev. 114, 6589–6631.

Vasilieva, E.A., Ibragimova, A.R., Lukashenko, S.S., Konovalov, A.I., Zakharova, L.Y., 2014a. Mixed self-assembly of polyacrylic acid and oppositely charged gemini surfactants differing in the structure of head group. Fluid Phase Equilib. 376, 172–180.

Vasilieva, E.A., Ibragimova, A.R., Mirgorodskaya, A.B., Yackevich, E.I., Dobrynin, A.B., Nizameev, I.R., Kadirov, M.K., Zakharova, L.Y., Zuev, Y.F., Konovalov, A.I., 2014b. Polyelectrolyte micro- and nanocapsules with varied shell permeability: controlling the rate of esters hydrolysis. Russ. Chem. Bull. 63, 232–238.

Veesar, I.A., Memon, S., Syed, M.N., 2013. Synthetic *p*-tetrasulphonatocalix[4] arene as novel excipient for lipase-complex. Biochem. Eng. J. 79, 71–76.

Veesar, I.A., Solangi, I.B., Memona, S., 2015. Immobilization of α-amylase onto a calix[4]arene derivative: evaluation of its enzymatic activity. Bioorg. Chem. 60, 58–63.

Venkatesh, G., Sivasankar, T., Karthick, M., Rajendiran, N., 2013. Inclusion complexes of sulphanilamide drugs and β-cyclodextrin: a theoretical approach. J. Incl. Phenom. Macrocycl. Chem. 77, 309–318.

Vilanova, N., Solans, C., 2015. Vitamin A palmitate–β-cyclodextrin inclusion complexes: characterization, protection and emulsification properties. Food Chem. 175, 529–535.

Voronin, M.A., Gabdrakhmanov, D.R., Semenov, V.E., Valeeva, F.G., Mikhailov, A.S., Nizameev, I.R., Kadirov, M.K., Zakharova, L.Y., Reznik, V.S., Konovalov, A.I., 2011. A novel bolaamphiphilic pyrimidinophane as building block for design of nanosized supramolecular systems with concentration-dependent structural behavior. ACS Appl. Mater. Inter. 3, 402–409.

Voronin, M.A., Gabdrakhmanov, D.R., Khaibullin, R.N., Strobykina, I.Y., Kataev, V.E., Idiyatullin, B.Z., Faizullin, D.A., Zuev, Y.F., Zakharova, L.Y., Konovalov, A.I., 2013. Novel biomimetic systems based on amphiphilic compounds with a diterpenoid fragment: role of counterions in self-assembly. J. Colloid Interf. Sci. 405, 125–133.

Wagner, A., Vorauer-Uhl, K., 2011. Liposome technology for industrial purposes. J. Drug Deliv. 2011, 1–9, Article ID 591325.

Wahlstrom, A., Cukalevski, R., Danielsson, J., Jarvet, J., Onagi, H., Rebek, J., Linse, S., Graslund, A., 2012. Specific binding of a β-cyclodextrin dimer to the amyloid β peptide modulates the peptide aggregation process. Biochemistry 51, 4280–4289.

Wang, J., Cao, Y., Sun, B., Wang, C., 2011a. Characterisation of inclusion complex of *trans*-ferulic acid and hydroxypropyl-β-cyclodextrin. Food Chem. 124, 1069–1075.

Wang, J., Cao, Y., Sun, B., Wang, C., 2011b. Physicochemical and release characterisation of garlic oil-β-cyclodextrin inclusion complexes. Food Chem. 127, 1680–1685.

Wang, T., Li, B., Si, H., Chen, L., 2011c. Release characterisitics and antibacterial activity of solid state eugenol/beta-cyclodextrin inclusion complex. J. Incl. Phenom. Macrocycl. Chem. 71, 207–213.

Wang, Y., Cohen, B., Jicsinszky, L., Douhal, A., 2012. Femtosecond to second studies of a water-soluble porphyrin derivative in chemical and biological nanocavities. Langmuir 28, 4363–4372.

Wang, W., Zhu, X., Yan, C., 2013. Determination of safranine T in food samples by CTAB sensitised fluorescence quenching method of the derivatives of calix[4] arene. Food Chem. 141, 2207–2212.

Wang, Q., Lv, S., Lu, J., Jiang, S., Lin, L., 2015. Characterization, stability, and in vitro release evaluation of carboxymethyl chitosan coated liposomes containing fish oil. J. Food Sci. 80, 1460–1467.

Wong, D.W.S., Camirand, W.M., Pavlath, A.E., 1996. Structures and functionalities of milk proteins. Crit. Rev. Food Sci. Nutr. 36, 807–844.

Yackevich, E.I., Mirgorodskaya, A.B., Lukashenko, S.S., Zakharova, L.Y., 2014. Polyfunctional supramolecular systems based on surfactants containing the hydroxyalkyl moiety in the head group. Russ. Chem. Bull. 63, 1801–1806.

Yang, L.Z., Xu, Y., Wang, X.H., Zhu, J., Zhang, R.Y., He, P.G., Fang, Y.Z., 2011. The application of β-cyclodextrin derivative functionalized aligned carbon nanotubes for electrochemically DNA sensing via host–guest recognition. Anal. Chim. Acta 689, 39–46.

Yang, S., Mao, X.-Y., Li, F.-F., Zhang, D., Leng, X.-J., Ren, F.-Z., Teng, G.-X., 2012. The improving effect of spray-drying encapsulation process on the bitter taste and stability of whey protein hydrolysate. Eur. Food Res. Technol. 235, 91–97.

Yang, L.-J., Ma, S.-X., Zhou, S.-Y., Chen, W., Yuan, M.-W., Yin, Y.-Q., Yang, X.D., 2013. Preparation and characterization of inclusion complexes of naringenin with β-cyclodextrin or its derivative. Carbohyd. Polym. 98, 861–869.

Yang, Z., Peng, Z., Li, J., Li, S., Kong, L., Li, P., Wang, Q., 2014. Development and evaluation of novel flavour microcapsules containing vanilla oil using complex coacervation approach. Food Chem. 145, 272–277.

Yang, L., Gao, S., Asghar, S., Liu, G., Song, J., Wang, X., Ping, Q., Zhang, C., Xiao, Y., 2015. Hyaluronic acid/chitosan nanoparticles for delivery of curcuminoidand its in vitro evaluation in glioma cells. Int. J. Biol. Macromol. 72, 1391–1401.

Yasam, V.R., Jakki, S.L., Natarajan, J., Kuppusamy, G., 2014. A review on novel vesicular drug delivery: proniosomes. Drug Deliv. 21 (4), 243–249.

Yazdi, S.R., Corredig, M., 2012. Heating of milk alters the binding of curcumin to casein micelles. A fluorescence spectroscopy study. Food Chem. 132, 1143–1149.

Young, O.A., Gupta, R.B., Sadooghy-Saraby, S., 2012. Effects of cyclodextrins on the flavor of goat milk and its yogurt. J. Food Sci. 77, S122–S127.

Yuan, C., Du, L., Jin, Z., Xu, X., 2013. Storage stability and antioxidant activity of complex of astaxanthin with hydroxypropyl-β-cyclodextrin. Carbohydr. Polym. 91, 385–389.

Zahi, M.R., Liang, H., Yuan, Q., 2015. Improving the antimicrobial activity of D-limonene using a novel organogel-based nanoemulsion. Food Control 50, 554–559.

Zakharova, L.Y., Konovalov, A.I., 2012. Supramolecular systems based on cationic surfactants and amphiphilic macrocycles. Colloid J. 74, 194–206.

Zakharova, L.Y., Mirgorodskaya, A.B., Zhiltsova, T.P., Kudryavtseva, L.A., Konovalov, A.I., 2004. Catalysis of nucleophilic substitution reactions in supramolecular systems. Russ. Chem. Bull. 53, 1385–1401.

Zakharova, L.Y., Mirgorodskaya, A.B., Yackevich, E.I., Yurina, A.V., Syakaev, V.V., Latypov, S.K., Konovalov, A.I., 2010a. The polyethyleneimine + cationic surfactant systems: the self-organization and reactivity study. J. Chem. Eng. Data 55, 5848–5855.

Zakharova, L.Y., Mirgorodskaya, A.B., Zhiltsova, T.P., Kudryavtseva, L.A., Konovalov, A.I., 2010b. Reactions in Supramolecular Systems in Molecular Encapsulation: Organic Reactions in Constrained Systems. Wiley, Hoboken, NJ, pp. 397–420.

Zakharova, L.Y., Gaysin, N.K., Gnezdilov, O.I., Bashirov, F.I., Kashapov, R.R., Zhiltsova, E.P., Pashirova, T.N., Lukashenko, S.S., 2012a. Micellization of alkylated 1.4-diazabicyclo[2.2.2]octane by nuclear magnetic resonance technique using pulsed gradient of static magnetic field. J. Mol. Liq. 167, 89–93.

Zakharova, L.Y., Ibragimova, A.R., Vasilieva, E.A., Mirgorodskaya, A.B., Yackevich, E.I., Nizameev, I.R., Kadirov, M.K., Zuev, Y.F., Konovalov, A.I., 2012b. Polyelectrolyte capsules with tunable shell behavior fabricated by the simple layer-by-layer technique for the control of the release and reactivity of small guests. J. Phys. Chem. C 116, 18865–18872.

Zakharova, L., Kashapov, R., Vagapova, G., Gabdrakhmanov, D., Vasilieva, E., 2012c. Comparative study of aqueous solutions of cationic surfactants: structure/activity relation in their aggregation and solubilization behavior and complexation with oligonucleotide. Chem. Lett. 41, 1226–1228.

Zakharova, L., Voronin, M., Gabdrakhmanov, D., Semenov, V., Giniyatullin, R., Syakaev, V., Latypov, S., Reznik, V., Konovalov, A., Zuev, Y., 2012d. Supramolecular systems based on mono- and dicationic pyrimidinic amphiphiles: self-organization and complexation with oligonucleothides. ChemPhysChem 13, 788–796.

Zakharova, L.Y., Semenov, V.E., Syakaev, V.V., Voronin, M.A., Gabdrakhmanov, D.R., Valeeva, F.G., Mikhailov, A.S., Voloshina, A.D., Reznik, V.S., Latypov, S.K., Konovalov, A.I., 2014. Amphiphilic macrocycles bearing biofragment: molecular design as factor controlling self-assembly. Mater. Sci. Eng. C 38, 143–150.

Zakharova, L.Y., Vasilieva, E.A., Gaynanova, G.A., Mirgorodskaya, A.B., Ibragimova, A.R., Salnikov, V.V., Uchegbu, I.F., Konovalov, A.I., Zuev, Y.F., 2015. The polyacrylic acid/modified chitosan capsules with tunable release of small hydrophobic probe and drug. Colloids Surf. A 471, 93–100.

Zha, G., Hu, D., Sun.F R., Ni, S., Xia, Q., 2015. Effect of emulsification process on multiple lipid particles encapsulating both coenzyme Q10 and tea polyphenols. J. Food Process Eng. 38, 144–154.

Zhang, X., Zhao, H., Cao, 2013. Hg^{2+} wettability and fluorescence dual-signal responsive switch based on a cysteine complex of piperidine-calix[4]arene. Org. Biomol. Chem. 11, 8262–8268.

Zhang, Y., Niu, Y., Luo, Y., Ge, M., Yang, T., Yu, L., Wang, Q., 2014. Fabrication, characterization and antimicrobial activities of thymol-loaded zein nanoparticles stabilized by sodium caseinate–chitosan hydrochloride double layers. Food Chem. 142, 269–275.

Zhang, W., Li, X., Yu, T., Yuan, L., Rao, G., Li, D., Mu, C., 2015. Preparation, physicochemical characterization and release behavior of the inclusion complex of *trans*-anethole and β-cyclodextrin. Food Res. Int 74, 55–62.

Zhao, Y., Du, D., Lin, Y., 2015. Glucose encapsulating liposome for signa lamplification forquantitative detection of biomarkers with glucometer readout. Biosens. Bioelectron. 72, 348–354.

Zheng, Z.-P., Dong, X., Yuan, K., Lan, S., Zhu, Q., Wang, M., Chen, J., 2015. Preparation, characterization, and preliminary antibrowning evaluations of norartocarpetin microemulsions. J. Agric. Food Chem. 63, 1615–1621.

Zhiltsova, E.P., Pashirova, T.N., Kashapov, R.R., Gaisin, N.K., Gnezdilov, O.I., Lukashenko, S.S., Voloshina, A.D., Kulik, N.V., Zobov, V.V., Zakharova, L.Y., Konovalov, A.I., 2012. Alkylated 1,4-diazabicyclo[2.2.2]octanes: self-association, catalytic properties, and biological activity. Russ. Chem. Bull. 61, 113–120.

Zhiltsova, E.P., Lukashenko, S.S., Pashirova, T.N., Valeeva, F.G., Zakharova, L.Y., 2015. Self-assembling systems based on diquaternized derivatives of 1,4-diazabicyclo[2.2.2]octane. J. Mol. Liq. 210, 136–142.

Zhou, Y., Ding, X., Fang, X., Li, T., Tang, D., Lu, Q., 2011. Studies on the inclusion behavior of amphiphilic *p*-sulfonatocalix[4]arene with ascorbic acid by spectrofluorometric titrations. Opt. Photonics J. 1, 59–64.

Ziani, K., Fang, Y., McClements, D.J., 2012. Encapsulation of functional lipophilic components in surfactant-based colloidal delivery systems: vitamin E, vitamin D, and lemon oil. Food Chem. 134, 1106–1112.

Zuev, Y.F., Mirgorodskaya, A.B., Idiatullin, B.Z., 2004. Structural properties of microheterogeneous surfactant-based catalytic system. Multicomponent self-diffusion NMR approach. Appl. Magn. Res. 27, 489–500.

9

NOVEL APPROACHES IN NANOENCAPSULATION OF AROMAS AND FLAVORS

Kata Trifković*, Verica Đorđević*, Bojana Balanč*, Ana Kalušević, Steva Lević**, Branko Bugarski*, Viktor Nedović****

**University of Belgrade, Department of Chemical Engineering, Faculty of Technology and Metallurgy, Belgrade, Serbia; **University of Belgrade, Department of Food Technology and Biochemistry, Faculty of Agriculture, Belgrade-Zemun, Serbia*

1 Introduction

Preservation of food flavor during production, storage, and controlled release of aroma compounds during consumption has attracted great attention in the past years. Flavor is one of the most important food properties that affect consumers' perception of food quality. Food flavor is a sensorial perception of food that is created during food consumption. Flavor (commonly known as aroma) could be further defined as a substance that causes the reaction of receptors in the nose (Zuidam and Heinrich, 2010).

Food aroma is usually a complex mixture of different organic chemical compounds. A huge number of these molecules has been isolated and identified mainly as hydrocarbons, esters, aldehydes, and so forth. Generally, these chemical compounds exhibit relatively low boiling points (faster evaporation compared with water) and could exist at room temperature as gas, liquid, or solid-state substances (eg, vanillin, camphor, and menthol are solid at room temperature). The chemical compounds that are frequently used in food industry as flavors are listed in Table 9.1. Flavors are used either as single flavor compounds, or as flavor mixtures, such as essential oils.

According to the origin, flavors are divided into two major groups: natural and artificial flavors. Natural flavors originate from natural sources (eg, spices, fruits, herbs, or animal products) and could be produced (extracted, purified, and modified)

Encapsulations. http://dx.doi.org/10.1016/B978-0-12-804307-3.00009-0

Table 9.1 List of Some Major Flavor Compounds

Compound	Source	Sensorial Characteristics	Application in Food Products
Vanillin	Chemical synthesis and natural sources	Vanilla odor	Beverages, candies, milk products, chewing gums, breakfast cereals
Limonene	Natural sources	Lemon odor	Beverages, candies, bakery products
Menthol	Chemical synthesis	Peppermint odor	Beverages, candies, chewing gums
Isoamyl acetate	Chemical synthesis	Banana odor	Beverages, candies, dairy products
n-Butyric acid	Natural sources—fermentation	Buttery odor	Milk products, bakery products, meat products
Geraniol	Natural sources	Rose odor	Candies, chewing gums, beverages
Diacetyl	Chemical synthesis and fermentation of glucose; identified in many milk products	Buttery odor	Bakery and dairy products, snack foods
γ-Lactone	Chemical synthesis	Peach odor	Beverages, chewing gums, puddings
Cinnamaldehyde	Chemical synthesis and natural sources	Cinnamon odor	Beverages, chewing gums, candies, puddings
Benzaldehyde	Chemical synthesis and natural sources	Bitter almond aroma	Fruit juices, candies, beverages
Ethyl maltol	Chemical synthesis	Fruit odor	Fruit juices, candies, beverages, breakfast cereals, jams, milk products
Allyl hexanoate	Chemical synthesis	Pineapple odor	Beverages, candies, chewing gums
Ethyl vanillin	Chemical synthesis	Vanilla odor	Beverages, candies, milk products, chewing gums, breakfast cereals
Furaneol	Natural sources	Strawberry odor	Ice cream, gelatin and puddings, candy, bakery products, beverages
Camphor	Natural sources	Camphor odor	Cooking, mainly for dessert dishes in Asia
Carvone	Natural sources	Caraway or spear mint odor	Chewing gums and candies
Anethole	Natural sources	Anise odor	Alcoholic drinks: ouzo and pernod

by different processes (eg, distillation, extraction, hydrolysis, etc.). Natural compounds extracted from plant materials have been used for centuries for medical treatments. Usually, these compounds are isolated in the form of essential oils using steam distillation or solvent extraction. Further, essential oils could be used as sources of specific compounds. For example, the mint essential oil consists of many organic molecules such as α-pinene, β-pinene, β-myrcene, limonene, menthone, isomenthone, menthyl acetate, neomenthol, β-caryophyllene, menthol, pulegone, and piperitone (Rohloff, 1999). The fermentation processes are also in use for production of some naturally derived flavors like *n*-butyric acid. Nevertheless, the sustainability of using natural sources for isolation of flavor compounds could be questioned since some isolation processes actually produce a waste that must be effectively treated. An example of this process is production of benzaldehyde from the pits of peaches. During this process waste cyanide is produced, while the yield of benzaldehyde is relatively low. In this case, chemical synthesis of benzaldehyde is far more sustainable (Burdock, 2009).

The artificial (synthetic) flavors are substances produced by chemical synthesis. Chemical synthesis of flavor compounds is usually based on conversion of organic precursor into the final product through one or several steps. For example, diacetyl that is used in many food products as butter flavor, is synthesized in several separated steps from methyl ethyl ketone. Some chemicals could be obtained using several different processes, for example, γ-lactone for peach flavor (Burdock, 2009; Gil et al., 2009).

Besides flavors, food regulations also recognize flavor enhancers which could be defined as substances that are added to enhance or modify food aroma (eg, sodium glutamate, inositol, maltol, or ethyl maltol). Synthetic compounds have been widely used in the food industry since their flavoring properties are sometimes superior compared to natural aromas; a good example is natural vanillin (3-methoxy 4-hydroxybenzaldehyde) versus synthetic vanilla aroma, that is, ethyl vanillin (3-ethoxy-4-hydroxybenzaldehyde). Ethyl vanillin was found to be an excellent replacement for natural vanilla flavor since having up to four times stronger flavor power compared to naturally occurring vanillin. On the other hand, the value of naturally occurring vanilla is estimated to be between \$80 and \$120 per kg (and depends on climate conditions) while ethyl vanillin is less expensive (\$12–19 per kg).

However, one should keep in mind that the flavor intensity could be so high that it becomes unpleasant at higher concentrations; in addition, in high concentrations flavor can have a negative impact on human health, by creating headaches, nausea, vomiting, and even allergies and intolerance (Taylor and Dormedy, 1998; Jinap and Hajeb, 2010). Therefore, the dosage of flavors in food products must be strictly controlled.

The effects of flavor compounds are not only limited to sensorial properties of food products. Thus, essential oils exhibit some medical and therapeutical effects depending on their chemical compositions. Also, some pure compounds were found to be a promising alternative for commercial antimicrobial agents. According to Fitzgerald et al. (2004), vanillin has certain antimicrobial activity toward bacterial cells such as *Escherichia coli*, *Lactobacillus plantarum*, and *Listeria innocua*. Yemiş et al. (2011) showed that ethyl vanillin, vanillin, and vanillic acid have antimicrobial activity against *Cronobacter* sp. D-Limonene has been identified as a powerful inhibitor against *Saccharomyces cerevisiae* that is cultivated in the medium based on citrus peel (Pourbafrani et al., 2007). Moreover, D-limonene showed some promising results as a substance for different tumor treatments (Uedo et al., 1999; Del Toro-Arreola et al., 2005).

One of the main concerns of food industry regarding food aroma is preserving of flavor compounds during food processing and storage. Conditions such as high temperatures or the presence of oxygen or light and high humidity can cause degradation and loss of aroma compounds. Briefly, the reasons for encapsulation of food flavors are: improving of flavor stability and handling, better safety of encapsulated aroma as well as possibility for controlled release. Encapsulation could be used for creation of specific visual effects in the final product as well as for masking of some sensorial unpleasant effects (ie, off taste) (Zuidam and Shimoni, 2010). Furthermore, in the form of solid encapsulates, liquid flavor could be easily manipulated and stored. Moreover, encapsulated flavors could be released under specific conditions, for example, at elevated temperatures or upon being triggered by mechanical forces. The preservation of flavors and their controlled release at elevated temperatures is desirable for real industrial processes that include thermal treatments. There are numerous studies showing that encapsulated flavors showed high stability compared to free compounds when they were submitted to heating (Kayaci and Uyar, 2012; Rocha et al., 2012; Lević et al., 2014, 2015; Sosa et al., 2014a).

Encapsulation of food flavors has been developed continuously for several decades. Among many techniques used for flavor encapsulation, spray drying is still the main industrial scale technique for encapsulation of large quantities of food flavor ingredients. Approximately 80–90% of all flavor encapsulates have been produced by spray drying. Other techniques such as spray chilling or melt extrusion are lused less on an industrial scale. Spray drying is a well-established and flexible technique with huge possibilities for optimization and production of particles in the broad range of sizes. This is main reason for its applications not only in the food processes (Zuidam and Shimoni, 2010; Zuidam and Heinrich, 2010) but also in other industries, especially in pharmacy (Paudel et al., 2013).

However, in recent years, many industries have adopted nanotechnologies. The nanoscale is defined by the National Nanotechnology Initiative as dimensions between roughly 1 and 100 nm (Tamjidi et al., 2013), but food nanotechnologies deal with the production, processing, and application of particles with sizes between 10 nm and 1 μm that serve as nanocarriers (Li et al., 2010; Ezhilarasi et al., 2013). This interpretation of "nanoscale" is derived from food technologists probably because of the following facts: the portion of particles in size 1–100 nm is contained even in the systems where the average particle size is, for example, 600 nm; up to 600 nm particles have the same effects in the gastrointestinal tract as the ones with 1–100 nm size; the particles are expected to be spherical and the dimension may be measured as the radius or the diameter (Tamjidi et al., 2013). According to the literature data, on the global market there are several hundred companies that already use nanotechnology. The nanofood market is expected to approach US$7 billion in 2015 and has the potential to grow to US$20.4 billion in 2020 (http://www.hkc22.com/Nanofood.html).

The nanoencapsulates, due to their very small size, are able to improve the macro scale characteristics of food by reducing the impact on the food texture, taste, and color, but also to improve stability during shelf life and processing (Ezhilarasi et al., 2013). The optical transparency is one of the most important attributes of some food products, such as beverages. Nanoemulsions are used in flavored clear beverages and fortified beverages as a means to incorporate water-insoluble flavors into aqueous beverages, and, at the same time, to avoid unwanted coloring or clouding of the product; the last issue cannot be achieved with any microscale system. Furthermore, nanoencapsulation has the potency to improve bioavailability (due to the greater surface area of nanoparticles per mass unit), prolong the release, but also to enable precision targeting of the flavor compounds at a higher level than microencapsulation (Mozafari et al., 2006). Another advantage of nanoencapsulation is the possibility to achieve the specific effect with a smaller quantity of the flavor compound compared to that of unencapsulated flavor (due to the nanosize and targetability of the carriers). Therefore, in recent years there has been a great interest in nanoencapsulation of flavors for food application. There

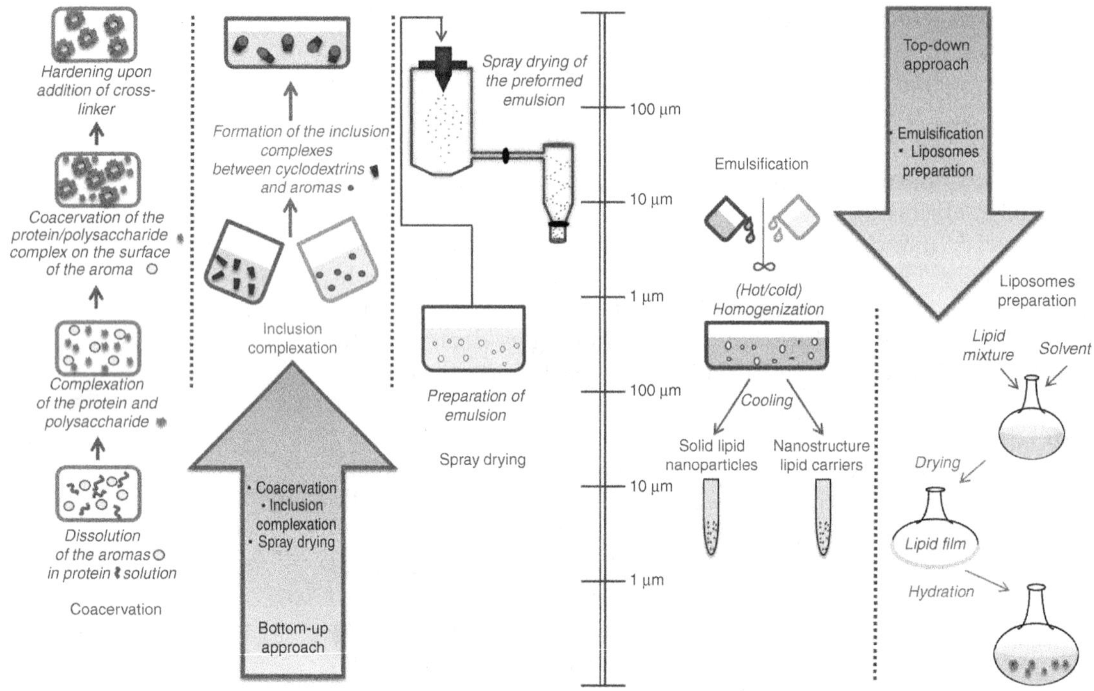

Figure 9.1. Bottom-up and top-down approaches in nanoencapsulation techniques.

are two main design principles used by nanoencapsulation techniques: top-down or bottom-up (Fig. 9.1). Top-down approaches involve tools which apply forces that induce particle formation and size reduction of the particles. In the bottom-up approach, particles are constructed by self-assembly and self-organization of molecules.

The following section aims to provide the current state-of-the-art of nanoencapsulation technology for flavors and aromas.

2 Flavor Nanoencapsulation by Spray Drying

Spray drying is widely used technique for encapsulation of various food ingredients such as flavors and aromas, antioxidants, vitamins, minerals, food-colorants, fats, and oils, in order to provide protection of these sensible compounds, as well as to prolong product shelf life during storage (Pillai et al., 2012). Numerous merits have been ascribed to encapsulation of the actives by the spray-drying process, especially when encapsulates are anticipated for the food industry. The equipment availability, low cost of the process, diversity of available carrier materials, stability of the final product, retention of volatile compounds, as well as possibility of continuous large-scale production, represent some of the advantages of the spray drying (Desobry and Debeaufort, 2011).

Nowadays, spray drying of the food flavors and aromas has been exploited well enough. Some of the authors report that more than 90% of encapsulated aromas have been prepared by spray drying (Gibbs et al., 1999; Madene et al., 2006; Vaidya et al., 2006). As most of the food aromas are in the liquid form at the room temperature, the goal of the spray drying is to enable easier handling by converting liquid into solid forms, while providing high loading efficiency upon the encapsulation process (Jafari et al., 2008a). The majority of aromas used in the food industry are of lipophilic nature; on the other hand, carrier material to be used in spray-drying process has to fulfill one major demand: to be water soluble. Thus, two immiscible phases, aqueous solution of carrier material and oil phase consisting of lipophilic aroma/flavor, present the input of spray-drying process. The most exploited carrier materials are: gum arabic, modified starch, maltodextrins, polysaccharides (eg, alginate, chitosan, carboxymethylcellulose and its derivatives, guar gum), and proteins (eg, whey proteins, soy proteins, sodium caseinate). Encapsulation by spray drying is attained via two steps: the first of which implies dissolution, emulsification, or dispersion of the core material in an aqueous solution of the carrier material (eg, preparation of O/W emulsion), while the second involves atomization and spraying of the core/carrier mixture into a hot chamber (Zuidam and Shimoni, 2010). As the outcome of spray-drying process, spherical particles are obtained, commonly of the micron size (10–100 μm) (Madene et al., 2006). However, preformed formulations are usually of nanodimensions, for example, nanosuspensions and nanoemulsions; therefore spray drying is considered a nanoencapsulation technique (Jafari et al., 2007, 2008a). In line with that, the influence of preapplied nanoencapsulation technique (eg, nanoemulsification) detrimentally affects the latter spray-drying process in terms of the finally achieved particle size and efficiency of encapsulation. On the other hand, it is possible to control powder particle size and morphology to some extent, by optimizing process parameters of the spray-drying process (Anandharamakrishnan et al., 2007, 2008).

2.1 Emulsions Processed by Spray Drying

The emulsification process plays an important role in the encapsulation of flavors and aromas as a first preparation step in the spray-drying process; hence, it has been extensively studied in the literature. The presence of a significant amount of a wall material gives some specific features to food emulsions. The slow kinetic motion of adsorption of surface-active wall material during high-energy emulsification process causes formation of the emulsions with the micron-sized droplets. Microemulsions generally are not desirable in food applications, not only because of the large particle size, but because of their low stability as well (Jafari et al., 2007). Lately, emulsions with droplet size in the nanorange have become very appealing and have been proposed as a means to enhance encapsulation efficiency during spray drying and to prevent loss of volatiles (Reineccius, 2001; Desai and Park, 2005; Madene et al., 2006). Encapsulation efficiency and active loading are the most significant attributes of the process, and they are related to the amount of unencapsulated oil at the surface of the spray-dried particles, as well as to the oil content directly entrapped in the particles (Baranauskiene et al., 2007). Jafari et al. (2007) studied the encapsulation of nanoparticles of D-limonene by spray-drying technique. They have examined different emulsifying devices (rotor-stator system, which served as a tool to achieve preemulsion, microfluidizer, and ultrasound probe) and their influence on the size of the emulsion droplets and accordingly, spray-dried particles. It seemed that emulsifying conditions did not affect significantly the emulsion droplet size (0.74 and 0.75 μm) but the size of the spray-dried powders, where the largest particles were obtained after microfluidization process (37.3 μm) (Jafari et al., 2007). All the examined techniques provided about the same retention of volatiles, for example, encapsulation efficiency, while microfluidized spray-dried emulsions exhibited the lowest unencapsulated oil amount on the particles surface. On the contrary, Fang et al. (2005) showed that particle size of spray-dried powders had detrimental influence on the encapsulation efficiency; the larger particles exhibited lower surface oil content, and consequently, higher encapsulation efficiency (ie, improved flavor retention). These opposing findings may be explained by the factor X, defined by Re and Liu (1996); this factor represents the correlation between the powder particle size and size of emulsion droplets; specifically [X = (powder size − emulsion size)/powder size]. These authors showed that by increasing the difference between the emulsion droplet size and powder particle size (eg, by increasing X from 0.2 to 0.8) it is possible to increase encapsulation efficiency. Since the difference in the emulsion particle size and powder particle size in the study of Jafari et al. (2007) was rather small, the resulting encapsulation efficiency (volatiles retention) was not improved regardless the increase in powder particle size.

2.2 Influence of Emulsion Droplet Size on Spray-Dried Powders

The influence of emulsion droplet size on retention of flavors upon spray drying, precisely the improvement of flavor retention upon reduction of emulsion droplet size, was confirmed by several authors (Liu et al., 2001a; Soottitantawat et al., 2005a; Baranauskiene et al., 2007; Penbunditkul et al., 2011). Soottitantawat et al. (2005a) spray-dried D-limonene with different kinds of matrices (gum arabic, maltodextrin, and modified starch), had highlighted the influence of decreasing emulsion droplet size on the increasing flavor retention, with practically no impact of the powder particle size. The evaporation of the volatiles is apparently easier from the large emulsion droplets in comparison to smaller ones (Soottitantawat et al., 2003). Similarly, Penbunditkul et al. (2011) confirmed increased retention of bergamot oil inside Hi-Cap 100 matrix upon decrement in emulsion droplet size; this in particular can be ascribed to hydrophobic flavors (Soottitantawat et al., 2003). The noticed effect of emulsion droplet size on retention of flavor was correlated to the emulsion viscosity. Namely, when emulsion viscosity was varied from 0.01 up to 1 Pa s (by changing concentration of Hi-Cap 100), the influence on emulsion droplet size (as well as on powder particle size) and consequently on flavor retention was evident (Penbunditkul et al., 2011). It was also observed that with the increase in emulsion viscosity (especially over 0.03 Pa s), the droplet size decreased, resulting in increased flavor retention; they have suggested the preferred droplet size of the emulsion below 1 μm in order to achieve nearly 100% of flavor retention. Tonon et al. (2011) confirmed this trend: with an increase in emulsion viscosity (by increase of solid content), the mean diameter of emulsions decreased. The reason behind this might be in the stabilization of the emulsions and evasion of the droplets' coalescence, via diminished rate of particle sedimentation upon viscosity increment (McClements, 2005). On the contrary, Janiszewska et al. (2015) found that a decrease in viscosity in lemon aroma emulsion led to formation of smaller emulsion droplet size, probably due to decreased adhesiveness of the aroma phase droplets. A similar finding was reported by Carmona et al. (2013), who encapsulated orange essential oil within a mixture of whey protein concentrate and maltodextrin (1:3). These opposite reports indicated that other features like wall material (type, concentration), as well as flavor properties (hydrophilic/hydrophobic nature, concentration) influenced the droplet formation process.

2.3 Influence of Wall Materials on Spray-Dried Powders

The influence of wall material utilized in the spray-drying process on the final features of produced powders (eg, flavor retention, surface oil content, shelf life, and release rate of encapsulated flavors) has been under scrutiny (Liu et al., 2001a; Alamilla-Beltran et al., 2005; Bruckner et al., 2007; Kanakdande et al., 2007; Penbunditkul et al., 2012; Sosa et al., 2014b). The wall material properties, such as emulsifying abilities and film-forming capabilities during spray drying of emulsion droplets, distinguish the superior materials among different candidates. Among various type of materials used for the spray-drying purposes, starch and its modified analogues (and/or starch-based materials such as maltodextrin) are of great interest. The capability of modified starches to act as emulsifying agents has been confirmed by numerous studies (Bao et al., 2003; Xiaoyan et al., 2006; Jafari et al., 2008b; Liu et al., 2008). For example, when a modified starch, HI-CAP 100 [octenyl succinic anhydride (OSA) waxy maize starch], was used for encapsulation of bergamot oil, it exhibited a favorable effect on emulsion droplet size in comparison to gum arabic (Penbunditkul et al., 2012). More specifically, the emulsion droplets obtained under the same homogenizing conditions, were smaller in size when HI-CAP 100 was used. This result is rather surprising, since the viscosity of both emulsions, with (40%) HI-CAP 100 and (10%) gum arabic, was about the same, which basically means that the effect of viscosity on the process of emulsification can be neglected. The explanation could be found in different emulsifying properties of the two wall materials, despite the fact that HI-CAP 100 contained higher solid content (Penbunditkul et al., 2012). Soottitantawat et al. (2005a) also gave a priority to Hi-CAP 100 compared to gum arabic in achieving fine emulsion droplets' size (eg, 670 and 800 nm, respectively) regardless of the fact that gum arabic exhibited slightly higher flavor retention (79 vs. 82%). In contrast to that, Krishnan et al. (2005b) showed that gum arabic ensured better retention of cardamom oleoresin flavoring compounds (precisely, 1,8-cineole, α-terpinyl acetate and volatiles) in comparison to maltodextrin and HI-CAP starch (30% of solid content was used). This effect was ascribed to gum arabic plasticity, which is rather higher than its glassines; the plasticity was proved to inhibit the rupturing of a matrix (Rosenberg et al., 1990; Sheu and Rosenberg, 1995; McNamee et al., 1998; Sankarikutty et al., 1988; Bertolini et al., 2001). In another study, Krishnan et al. (2005a) reported that with an increase in gum arabic content in gum arabic–maltodextrin and gum arabic–Hi-CAP 100 formulations, the entrapment efficiency of the cardamom constituents increased, as well as protection over them. The combination of all three carrier materials (in ratio gum arabic:maltodextrin:Hi-CAP 100 of 4/6:1/6:1/6) seemed to provide the superior protection over cardamom oleoresin, compared to all other mixtures, even compared to pure gum arabic as a carrier. Exactly the same result was obtained in study when cumin oleoresin was used as a core compound (Kanakdande et al., 2007).

Not only the composition of wall material, but also the saturation degree of polymer(s) solution plays important role in emulsification process. The influence of saturation of HI-CAP 100 solution (40%) on flavor retention and surface oil content was investigated by Penbunditkul et al. (2012). Higher retention of flavors was reported when saturated HI-CAP 100 solution was used, for example, up to 114% for linalool. In addition, the saturation of carrier solution affected the surface oil content of powders, in terms of decreasing the content roughly one order of magnitude (for particular flavor compounds from bergamot essential oil: linalool and linalyl acetate). The explanation for the previously mentioned effect can be found in the chemical structure of modified starch, such that OSA modified starch appears as an interfacial layer between water and oil phases (Nilsson and Bergenstahl, 2006). This indicates the necessity of providing enough time for polymer dissolution in order to assure solubilization of the polymer and a proper saturation of the solution, that is, evading residual granule structures (Yusoff and Murray, 2011).

Modified starches have been used with other carrier materials to form the polymer blends for spray-drying purposes. Sosa et al. (2014b) encapsulated orange essential oil within modified starch/maltodextrin polymer blends, with addition of disaccharides (sucrose or trehalose). As the emulsifying agents, two different modified starches have been utilized, HI-CAP and Capsul starch. The stability of encapsulated flavor during time, for example, the loss of the encapsulated orange essential oil, has been studied at different storage temperatures, 25 and 37°C. In the formulations where trehalose was used, higher flavor retention has been achieved with the HI-CAP than with Capsul, while for sucrose formulations the selected starch did not exhibit any influence on the retention behavior. At the same time, storage temperature decreased the retention of flavor upon 270 days (for HI-CAP/trehalose formulations, retention decreased from ~65% at 25°C to ~55% for 37°C), indicating the reduced storage stability of flavor formulations upon temperature increase. Moreover, HI-CAP was evaluated against whey protein concentrate (WPC) as a wall material for encapsulation of D-limonene (Jafari et al., 2007); the surface oil content of HI-CAP particles was lower (55.4 mg/100 g powder) in comparison to WPC ones (170.1 mg/100 g powder), while retention of D-limonene was higher (86.2 and 76.3%, respectively), proving the greater applicability of HI-CAP for encapsulation of flavor compounds. Similarly, Jafari et al. (2008a) confirmed the HI-CAP supremacy over WPC; this was not surprising, giving the fact that emulsion droplet size was found to be lower when HI-CAP was used, with a narrower unimodal size distribution in comparison to WPC (with a bimodal size distribution) (Risch and Reineccius, 1988; Liu et al., 2000a,b; Soottitantawat et al., 2003). In addition, the more substantial difference between particle size of spray-dried powders and emulsion droplets of HI-CAP samples (0.74 and 37.3 μm, respectively) in contrast to WPC samples (0.98 and 21.3 μm, respectively) have resulted in the higher encapsulation efficiency, that is, better oil retention inside the powder particles and reduced oil content on particle surface (Jafari et al., 2007), as a consequence of the previously described *X* factor (Re and Liu, 1996).

Gum arabic is one of the widely used materials for spray drying of food ingredients. It is well known that gum arabic is composed of the simple sugars (ie, galactose, arabinose, rhamnose, and glucuronic acids) (Anderson and Stoddart, 1966; Street and Anderson, 1983), but contains as well a component of protein nature (~2% w/w) that forms covalent bonds within the molecular arrangement (Anderson et al., 1985), consequently affecting the functional properties of gum arabic to a great extent (Randall et al., 1988). It is highly soluble in water and able to operate as an emulsifying agent in oil-in-water emulsions; hence, gum arabic is extensively used for encapsulation, that is, retention of volatile and flavor compounds (McNamee et al., 2001). For example, Fernandes et al. (2013) encapsulated rosemary essential oil within the gum arabic matrix of various concentrations (10–30%); it was shown that mediocre concentration of wall material (24%) has been proved as the best. However, gum arabic is produced in the area of world with unstable climatic and even more unpredictable political circumstances, so it is an expensive material with uncertain availability. In line with this, different materials have been evaluated as a partial replacement for gum arabic in spray-drying processes. McNamee et al. (2001) investigated the possibility of utilization of gum arabic and different carbohydrate materials (maltodextrin, corn starch, native maize starch, lactose, sucrose, and glucose) mixtures; it was found that in order to achieve the increased encapsulation efficiency, it is necessary to increase the solid content of the wall material. To accomplish this, it was possible to add some of the carbohydrate materials, instead of increasing the content of gum arabic (while keeping the ratio gum arabic/encapsulated oil at 2); in that way, process cost are kept at minimum, since carbohydrates are cheaper than gum arabic. Of the proposed carbohydrate materials, the best result for encapsulation efficiency (91.8%) was obtained when glucose (50% of glucose and 50% of gum arabic) was used, while addition of maize starch led to low encapsulation efficiency (30%) (McNamee et al., 2001). Regardless this high encapsulation efficiency, the increased sweetness and hygroscopicity of powders obtained with glucose often limits the application in food products; hence the gum arabic-maltodextrin powders have been chosen as the optimal ones (with the encapsulation efficiency ~72%), and maltodextrin has been highlighted as the most suitable substitute of gum arabic. The influence of maltodextrin concentration (10–30%) in maltodextrin-gum arabic system on retention of ethylbutyrate was reflected in retention increase (from almost zero up to ~40%) (Yoshii et al., 2001). The additional increase in flavor retention (up to ~80%) was recorded upon addition of 1% gelatin in the maltodextrin-gum arabic mixture, due to the fact that gelatin enhances formation of coating over droplet surface during spray drying. Recently, Sarkar et al. (2012) obtained differently modified (irradiated and enzymatically partially hydrolyzed) guar gum and arabic gum blends for encapsulation of mint oil. They showed that higher flavor retention upon 8 weeks of storage was obtained when mixtures of modified guar gums were used instead of sole gum arabic (eg, 65.78% as opposite to 58.12%). The overall conclusion is that polymer mixtures in comparison to single polymer materials in most cases exhibit superior features of the final spray-dried powders.

In classical emulsions, small molecular surfactants are usually used to act as a surface active agent. As the alternative to those, surface-active biopolymers are usually employed in the food emulsions. What's more, biopolymers proved supremacy over small molecular surfactants when final features of spray-dried particles are concerned. For example, Jafari et al. (2007) compared the properties of the powder particles obtained with Tween 20 as a surfactant versus HI-CAP and/or WPC. The surface oil content of particles with Tween 20 was very high—270.2 mg/100 g powder, which was nearly 2 and 5 times higher when compared to WPC and HI-CAP samples, respectively. As already discussed, smaller emulsion droplets is usually associated with a better aroma retention (Liu et al., 2000a, 2001b; Soottitantawat et al., 2003, 2005a; Alamilla-Beltran et al., 2005; Bruckner et al., 2007; Kanakdande et al., 2007; Penbunditkul et al., 2012; Sosa et al., 2014b). However, Jafari et al. (2007) obtained that despite the fact that droplet size was significantly smaller when Tween 20 was used (0.22 μm) in comparison with HI-CAP and WPC samples (0.74 and 0.98 μm, respectively), the flavor retention was better with HI-CAP (86.2%) than with Tween 20 (76.9%). Two possible effects could be behind these results: (1) inferior stability of emulsions containing Tween 20 compared to biopolymer emulsions, and/or (2) poor film-forming abilities of Tween 20. Some previous studies (Klinkesorn et al., 2004; Damodaran, 2005) showed that free biopolymer (wall material) residues and surfactant molecules, could possible interfere, provoking the "depletion flocculation" and "bridging flocculation," which further may induce coalescence of emulsions. In addition, the film (membrane) formed around oil droplet by biopolymer is much more protective (stronger) than the one formed by small surfactants, which prevents the coalescence of oil droplets (Bibette et al., 1999; McClements, 2005) The previously stated justifies the favoring of surface-active biopolymers utilization in food emulsions over small molecule surfactants.

2.4 Influence of Aroma Type and Concentration on Spray-Dried Powders

Another parameter that plays important role in emulsification process is aroma concentration. For instance, it was shown that increase in aroma concentration (from 2 to 10%) led to increase of emulsion viscosity (from 31.1 to 40.9 mPa s), when high-shear homogenization method was applied (Janiszewska et al., 2015). This can be elucidated by the presence of air within the emulsions (due to the process of mixing) and the fact that the higher content of air in emulsions causes the higher emulsion viscosity (McClements, 2005). Further, the increase in emulsion viscosity provokes additional increase in emulsion droplet size (from 2.40 to 4.09 μm), and consequently, in powder particle size (6.8–7.3 μm). On the contrary, if the method used for emulsion preparation was high-pressure homogenization instead of high-shear homogenization, the increase in aroma concentration had quite the opposite influence on emulsion viscosity—it decreased (from 15.6 to 13.4 mPa s, with increase of aroma content from 2 to 10%) (Janiszewska et al., 2015). The physical properties of the spray-dried particles were also affected by aroma concentration—the porosity of particles increased along with increase in aroma concentration (Janiszewska et al., 2015). The same authors observed the increase in emulsion droplet size upon increase in aroma concentration, regardless of the homogenization method. Regardless, the influence of homogenization method on water content of powders, as well as powders' particle size, was evident: powder particle size was not changed linearly with increase in aroma content (in the case of high-pressure homogenization method), while the water content in powders was lower for high-pressure homogenized samples in comparison to high-shear homogenized samples (Janiszewska et al., 2015). Similar conclusions were reached by Jafari et al. (2007); this could be due to lower viscosity of emulsions obtained by high-pressure homogenization that could promote evaporation of water. Quite the reverse, the high water content for powders obtained by the high-shear homogenization method can be ascribed to higher viscosity of emulsions and bigger emulsion droplets (when compared to the high-pressure method), that consequently resulted in hampered water evaporation (Turchiuli et al., 2014). It is evident that in complex systems like emulsions, the influence of individual parameters that plays a role in emulsification process cannot be elucidated for itself, but in segregation of several factors.

Since natural aromas and flavors often come in the form of extracts/essential oils, which presents multicomponent flavor systems, it is interesting to examine the influence of multicomponent flavors as opposite to the each individual flavor compound on the emulsification and/or spray-drying process, as well as on the features of the produced powder particles. Such studies have been performed on encapsulation of bergamot oil (and its individual components: D-limonene, linalool, and linalyl acetate) (Penbunditkul et al., 2011, 2012), oregano (*Origanum vulgare* L.), citronella (*Cymbopogon nardus* G.), and marjoram (*Majorana hortensis*) essential oils (and the corresponding isolated flavoring compounds: *p*-cymene, limonene, linalool, thymol, carvacrol) (Baranauskiene et al., 2006), as well as peppermint (*Mentha piperita* L.) essential oil and its constituents (Baranauskiene et al., 2007). The challenge of multicompound systems lays in diverse solubility of individual flavors (for example, D-limonene is practically insoluble in water, while linalool and linalyl acetate exhibit water solubility to some extent) (Penbunditkul et al., 2011). When emulsion stability is concerned, it was noticed that upon reconstitution pure D-limonene droplet size was smaller than the initial D-limonene emulsion size, while the reconstituted droplet size of multicomponent bergamot oil emulsion did not alter, indicating greater stability of multiflavor emulsion systems when compared to single flavor compounds (Penbunditkul et al., 2011). In addition, the emulsion droplet size of bergamot oil was smaller; this (when keeping in mind the fact that with a decrease in emulsion size, flavor retention increases) led to the conclusion that flavor retention of single flavor systems was poorer than of the multiflavor bergamot oil system. Knowing that flavor components form complexes by mutual interactions, the improved retention of flavors in multicomponent system was expectable (Penbunditkul et al., 2011). The content of different essential oil constituents differed between the particle surface and bulk essential oil. For example, when compared oregano essential oil and spray-dried particles containing oregano essential oil, the content of *p*-cymene on the surface of spray-dried powders was up to 26.1% while the content in the bulk oregano essential oil was 30.3% (Baranauskiene et al., 2006). Similarly, the content of menthone on the powder particle surface (up to 14.9%) and in the pure peppermint essential oil (19.5%) revealed significant difference (under the same process conditions). These variations could be elucidated by the evaporation of the volatile compounds during the spray drying. On the other hand, the amount of less volatile compounds, such as menthol, in the surface oil content of powder particles was considerably higher (up to 62.9%) compared to percentage (47.5%) of the bulk peppermint essential oil (Baranauskiene et al., 2007). Penbunditkul et al. (2012) related the flavor nature/solubility to its retention upon spray drying. Namely, when investigating the flavor retention of individual flavors from multicomponent flavor system such as bergamot oil, it was noticed that the linalool retention was higher than 100%. In an attempt to explain this, it can be assumed that transformation of linalyl acetate to linalool occurred, either by biotransformation (Zhu and Lockwood, 2000) or catalytic reaction (Ramishvili et al., 2007). In addition, the flavors' nature (eg, functional groups of flavors) influences the retention, as it is verified that flavoring alcohols (eg, linalool) are better retained in comparison to other compounds (Le Thanh et al., 1992; Goubet et al., 1998). Furthermore, water solubility of flavors influenced the surface oil retention as well; poorly water soluble D-limonene exhibited low surface oil content (up to 0.14%) in comparison to highly water-soluble linalool (10.5%) (Penbunditkul et al., 2012). Water-soluble compounds diffuse at higher rate (along with water) through the shell of spray-dried particle, while this is impossible to occur with poorly soluble D-limonene.

2.5 Influence of Spray-Drying Conditions on the Powder Properties

The process parameters of the spray-drying affect the final features of powder particles. The most crucial influence has been ascribed to inlet air temperature. Upon temperature increase (from 135 to 195°C), the moisture content of powders decreased (from 3.16 to 0.26%), as shown by Fernandes et al. (2013). The noticed effect of inlet temperature has also been verified by other studies (Finney et al., 2002; Ersus and Yurdagel, 2007). In a similar manner, with increase of inlet air temperature (from 120 to 180°C), surface oil content of multiflavor aroma decreased, while the best results (no oil on the surface of powders) were achieved for inlet temperature of 160°C (Penbunditkul et al., 2012). This particular temperature proved to be best for flavor retention, too (for all the aroma compounds from multicomponent flavor system); nonetheless, the observed trend was that increase in inlet temperature (up to 180°C) led to increase in flavor retention (Penbunditkul et al., 2012). Low inlet temperature causes the low drying rate, which further induces the slow crust formation and consequently low retention of flavors and high content of surface oil; on the other hand, at high inlet temperature the drying rate is higher, but this promotes higher evaporation of flavors which finally results in low flavor retention and high surface oil content. Thus, the inlet air temperature of 160°C proved to be best when flavor (rosemary essential oil) retention and surface oil content were concerned (Penbunditkul et al., 2012). The inlet air temperature affects the hygroscopicity of powders as well; the decrease of inlet temperature led to reduction of hygroscopicity of spray-dried rosemary essential oil within the gum arabic (Fernandes et al., 2013), that is probably caused by high moisture content of the powders produced under such conditions, and subsequently minor gradient of water concentration between the powders and the surrounding atmosphere. Similarly, when coffee oil was spray dried with gum arabic, hygroscopicity increased along the inlet air temperature increase (Frascareli et al., 2012).

The behavior of spray-dried powders in contact with water, for example, their wettability, presents one of the most significant powder properties (Bae and Lee, 2008). The time necessary for powders to become wet depends mostly on inlet air temperature; the higher the temperature, the higher the wettability time (Fernandes et al., 2013). It is important to notice that not only the inlet air temperature itself, but in conjunction with the wall material concentration, impacts the wettability time of powders, where the best result can be achieved with a high inlet air temperature and a high wall material concentration (Fernandes et al., 2013). This is related to low moisture content of the particles produced under these conditions. There have been studies that emphasize the influence of powders caking on wettability of particles, where it was shown that upon caking the liquid diffuses more easily into powder pores (Buffo et al., 2002). Since the caking is related to powders with high moisture content, the high wettability time for powders with low moisture content is then expected.

The physical features of powders, such as powders bulk and tapped density and particle density, were also influenced by inlet air temperature. The bulk density of the spray-dried powders with rosemary essential oil changed (from 0.25 to 0.36 g/mL) within the temperature range of 135–195°C (Fernandes et al., 2013); similar bulk densities were noticed when oregano essential oil was spray dried (0.34–0.45 g/mL) (Botrel et al., 2012). This demonstrated that with an increase of inlet air temperature, the bulk density decreased; this was also confirmed by other studies (Goula and Adamopoulos, 2008; Souza et al., 2009). As the increase in the temperature accelerates the evaporation and promotes the pore formation, it is reflected in decreased density (Walton, 2000). This particular effect can be associated also to decrease in tapped density upon inlet air temperature increase (Fernandes et al., 2013). The parameter of tapped density is important from the packaging aspect, since high-density products require smaller packaging vessels (Quispe-Condori et al., 2011). The increase in inlet temperature leads to increase of drying rate, which further causes the greater droplets expansion and finally lower density of spray-dried powders (Walton, 2000). Numerous authors confirmed this effect (Cai and Corke, 2000; Finney et al., 2002; Chegini and Ghobadian, 2007; Bae and Lee, 2008; Fernandes et al., 2013). Regarding the particle density of spray-dried powders, it was proven that it depends not only on particle size and moisture of the powders, but also on the parameters of spray-drying process (Botrel et al., 2012). The inlet air temperature influenced the particle density in negative manner—with an increase in inlet temperature, particle density decreased (Fernandes et al., 2013). It could be due to steam formation during the drying process, which induced the enlargement of particles (Finney et al., 2002).

Besides the inlet temperature, emulsion feed rate also has to some extent an impact on the features of spray-dried powders. With higher emulsion feed rate, higher flavor retention, and lower surface oil content were achieved (Penbunditkul et al., 2012). This can be associated with the drying rate, where changes in emulsion feed rates influence the drying rate of every emulsion droplet during atomization, which further affects the formation of protective crust over droplets. It is proved that this crust acts as a protective membrane that hinders flavors loss (Lee et al., 2006). In respect to this, when optimizing the emulsion feed rate, it is necessary to provide drying rate high enough for instantaneous formation of crust that will disable the flavor loss (Penbunditkul et al., 2012). Flow rate of feed emulsion influences the moisture content of spray-dried powders, where the high feed rates lead to increased amount of water in powders (Fernandes et al., 2013); this appears as a consequence of the large volume of emulsion to be dried and a shortened contact time with the hot air that is not sufficient for evaporation of higher amount of water. The consequent effect of flow rate increase on the hygroscopicity of powders is reflected in its reduction.

2.6 Encapsulated Aroma Stability and Release

The stability of encapsulated aroma is a very important property from the aspect of encapsulates applications. The spray drying should enhance stability of flavor during processing and storage and in real food. It means that is necessary to ensure stability of encapsulated aroma formulations against oxidation; the stability of flavor is usually quantified by determination of oxidation rate and release rate (Soottitantawat et al., 2004). The release of encapsulated aroma is a complex function of different factors: (1) the environmental conditions (such as relative humidity, temperature); (2) the wall material features, for example, glass transition temperature; and/or (3) type of encapsulated aroma. It was shown that relative humidity affects the release rate of encapsulated aroma in terms of a release increment with an increase in relative humidity; this was proved when release of ethyl butyrate was studied (Yoshii et al., 2001). However, the mentioned dependency was not universal, since the noticed trend was not applicable to all examined relative humidity conditions (Soottitantawat et al., 2004). The release rate of encapsulated aroma, for example, D-limonene, was dependent on two mechanisms: diffusion of the aroma through the matrix and loss of the aroma during storage. In particular for D-limonene, it has been shown that the diffusion plays the crucial role in release rate (Whorton, 1995; Soottitantawat et al., 2004). Additionally, when comparing release rates of D-limonene (Soottitantawat et al., 2004) and ethyl butyrate (Yoshii et al., 2001), it was noted that release of D-limonene was somewhat slower, proving that this kind of encapsulated aroma affects the release rate to some extent.

The storage temperature is another parameter of environmental conditions that affects the release rate of aroma compounds. An increase in temperature of storage led to increase of L-menthol release rate (Soottitantawat et al., 2005b). As mentioned previously, the rate of release is a complex function of various factors; hence the influence of temperature in correlation with the relative humidity is more distinguished and thereof as such discussed. When determining the influence of temperature on release rate, the Arrhenius equation could be applied (Soottitantawat et al., 2005b). If the relative humidity was higher (83% vs 75%), the influence of increased storage temperature on increase of release rate was more evident. This is once again confirmation that release of encapsulated aroma is facilitated in the higher humidity conditions.

Release rate depends on several factors, one of them being the type of wall material used and its features. Baranauskiene et al. (2007) examined the encapsulation of peppermint essential oil within the different modified starches (HI-CAP 100, N-LOK, CAPSUL, ENCAPSUL 855, CRYSTAL TEX 627, CIEmCap 12633, CIEmCap 12634, CIEmCap 12635) and the consequent influence on flavor release. The release of all the peppermint oil components was most pronounced in the first hour of storage with the obvious influence of the wall material on release (retention) rate. The highest release of aroma compounds was emitted when HI-CAP 100 was used (Baranauskiene et al., 2007). Furthermore, the influence of storage relative humidity was evident confirming the trend that with an increase in humidity, the release rate increased, regardless the type of starch utilized for the production of the spray-dried powders. However, the impact of relative humidity was dependent on type of wall material: the less pronounced effect was reported when HI-CAP 100 was used (Baranauskiene et al., 2007). The water activity, for example, relative humidity appears to have the overriding effect on aroma release, since the modifications of the wall material structure caused by changes in humidity determinate the final release features of the powders. Rosenberg et al. (1990) examined the flavor retention at various humidity conditions and showed that up to 64% of relative humidity no structural changes in matrix (gum arabic) structure were noticed. Nevertheless, the powder particles began to swell and agglomerate, and any further increase in relative humidity led to decomposition of wall material (at humidity 75–92%), where at 95% of relative humidity the wall material was destroyed. Similar was confirmed by other authors, highlighting the following: as far as the wall material structure is undisturbed, the aroma retention is high (Whorton and Reineccius, 1995); as soon as the particle structure is altered by water diffusion, the aroma release is intensified (Soottitantawat et al., 2005b).

The specific characteristic of wall material, its glass transition temperature, plays an important role in the release of encapsulated aroma. It is known that at lower water content the release of aroma is low (given that the wall material is in the glassy state and the mobility of aroma compounds is reduced), while the increase in water content expectedly leads to an increase in release rate, since the matrix alternates to the more plasticized state where the mobility of aroma compounds is higher (Baranauskiene et al., 2007). Labrousse et al. (1992) and later Soottitantawat et al. (2004) correlated the parameter ($T - T_g$) to the release rate, where T represented storage temperature and T_g glass transition temperature; it was evident that with increase in the T_g nearly to the point where $(T - T_g)$ becomes zero, the release rate of aroma increased. This also agrees with the theory that increased mobility of aroma compounds occurs near the temperature of glass transition, for example, when the wall material begins to plasticize. In the $0 < (T - T_g) < 50$ range, when powder particles are in the rubbery state, aroma release is low, due to the closed pore gaps and the reduced surface area for the aroma evaporation (Soottitantawat et al., 2004). A similar conclusion has arisen when L-menthol was encapsulated within the various matrixes: gum arabic, HI-CAP 100, and CAPSUL (Soottitantawat et al., 2005b); at a low relative humidity, the release of aroma compounds was decreased.

Oxidation of aromas has been frequently examined, as the process which deteriorates aroma stability. It is known that D-limonene forms various products upon reaction of oxidation (Bertolini et al., 2001). Soottitantawat et al. (2004) examined formation of limonene oxide (limonene-1, 2-epoxide) and carvone as products of parallel reactions that occurred during the oxidation of D-limonene (Anandaraman and Reineccius, 1986). The conditions of high relative humidity promoted the oxidation of aroma to the greatest extent, particularly at the beginning of the storage. What was interesting is the fact that during a longer period of storage, increased humidity influenced the aroma oxidation in an opposite manner: with an increase in relative humidity, the oxidation of D-limonene decreased (Soottitantawat et al., 2004). This phenomenon could be related to the rapid aroma degradation resulted in appearance of other oxidation products and since the screening was based on limonene oxide and carvone, it seems that oxidation was lowered. In addition to that, the decreased oxygen content at the increased water content (at higher relative humidity), due to the low oxygen solubility, led to the reduction of aroma oxidation (Soottitantawat et al., 2004). Another parameter of storage, storage temperature, affects the stability of encapsulated aroma in terms of oxidative stability. When coffee oil was encapsulated, the oxidative stability was determined via assessment of peroxide value upon 8 weeks of storage, and under two different storage temperatures: 25 and 60°C (Frascareli et al., 2012). It is worth emphasizing, peroxide values of unencapsulated and encapsulated coffee oil upon storage of two weeks, indicated the protective effect of the encapsulation against oxidation of coffee oil, since the peroxide values of unencapsulated oil was not higher than those related to spray-dried coffee oil. In addition, when spray-dried powders were stored 8 weeks at 60°C, the oxidation of aroma was more pronounced, given that the peroxide value for the gum arabic encapsulated coffee was 4.4 meq peroxide/kg oil, while the peroxide value after 8 weeks of storage at 25°C was ~1 meq peroxide/kg oil. The relative high stability of coffee oil against oxidation could be attributed to the presence of polyphenolic compounds and products of Maillard reaction that exhibited high antioxidant capacity (Anese et al., 2006).

In general, it can be concluded that two main factors influence and restraint the release of aroma compounds from encapsulates: (1) the loss (evaporation) of the aroma compounds in the food matrix (known as "thermodynamic factor") and (2) the resistance to mass transfer of powders to air (so-called "kinetic factor") (de Roos, 2000). In addition, the release mechanism of aroma compounds from powder particles might be dependent on different solvent criterion: melting, diffusion, particle degradation, or fracture (Madene et al., 2006). Some researches highlighted that release of aroma compounds from spray-dried powders is governed mostly by diffusion (Soottitantawat et al., 2004, 2005a,b; Baranauskiene et al., 2007). Usually, to describe the release mechanism of aromas from spray-dried powders and to fit experimental data, Avrami's equation is used: $R = \exp[-(kt)^n]$, where R stands for retention of D-limonene, t is the storage time, and k is the release rate constant. The n parameter of the fitting equation indicates the kind of release mechanism: when $n = 1$ the release can be describes as first order mechanism; when $n < 1$ (in theory $n = 0.54$), the diffusion of the aroma compounds is limiting factor of release rate; finally, when $n > 1$, the release is rapid, with an initial induction period (Soottitantawat et al., 2004). For example, when fitting the release of ethyl butyrate from the gum arabic powder particles, the value for n was 0.55, 0.79, and 0.74 (depending on relative humidity: 75, 60, and 45%, respectively), and the authors Yoshii et al. (2001) concluded that release mechanism was diffusion, for example, first-order mode. Similarly, the release of L-menthol proved to follow the diffusion mechanism, with the n values from 0.10 to 1.00 (depending on the wall material used) (Soottitantawat et al., 2005b). The release of encapsulated aroma compounds is associated with the water adsorption and respective hydration of the spray-dried powders; upon diffusion of water the particle start to swell, and the particle surface starts to crack and to release the aroma compounds (Whorton and Reineccius, 1995).

Beside the effect of wall material on the release of encapsulated aromas, the composition of the food matrices, in which the spray-dried powders are implemented, also affects the release rate of aroma compounds. If the food matrix is rich in lipids, the release of aromas is decreased, except for hydrophilic components with a value for log P (where P is permeability of volatiles) around zero (Guichard, 2002). Further, if the salts content in the foods is high, the release of aroma compounds will be promoted (Druaux and

Voilley, 1997). Hence, to formulate the optimal dosage forms of the spray-dried aroma compounds, as well as to apply those in correct manner to the shrewdly chosen real food products, it is necessary to take different variables into account.

3 Flavors Encapsulated in Nanoemulsions

Conventional emulsions utilized in food are divided in two major categories: (1) the emulsion that consists of oil droplets dispersed within a watery phase is called oil-in-water (O/W) emulsion, while (2) the emulsion that contains water droplets dispersed in an oily phase is called water-in-oil (W/O) emulsion. These systems are generally prepared by homogenizing oil, water, and emulsifier together by a high-energy mixing device (colloid mill, sonicator, microfluidizer, or high shear mixer). The average size of emulsions droplets divides the emulsions on macroemulsions (0.1–100 μm), microemulsions (5–50 nm), and nanoemulsions (20–100 nm). Macroemulsions are thermodynamically unstable and optically opaque. On the other hand, there is often confusion between thermodynamically stable microemulsions and thermodynamically unstable (but kinetically stable) nanoemulsions. Microemulsions are thermodynamically stable, usually of white appearance, low viscous and isotropic dispersions. Typically microemulsions are easily prepared, but require higher concentrations of a surfactant (emulsifier) usually in conjunction with a cosurfactant (such as short- and medium-chain alcohols). The surfactants are added to stabilize the interfacial layer between the two phases which have been formed by the addition of energy to the system (Flanagan and Singh, 2006), while cosurfactants contribute to a decrease of interfacial tension between the phases (Madadlou et al., 2014). Unfortunately, the application of microemulsions in foods is still limited due to the toxicity of cosurfactants which are commonly used; therefore in the last few years a strong effort has been dedicated in order to produce the food-grade microemulsions (Augustin and Hemar, 2009).

Nanoemulsions are transparent, more resistant to oxidation than microemulsions (Maswal and Dar, 2014), but at the same time, more prone to growth in particle size (by Ostwald ripening) (Sagalowicz and Leser, 2010) and tend to break down during storage. Overcoming of Ostwald ripening, which is undesirable phenomenon in nanoemulsions, is one of the main challenges in preparation of stable nanoemulsions for practical applications. Decreasing of oil solubility in the aqueous phase is an effective method to delay or avoid Ostwald ripening in O/W

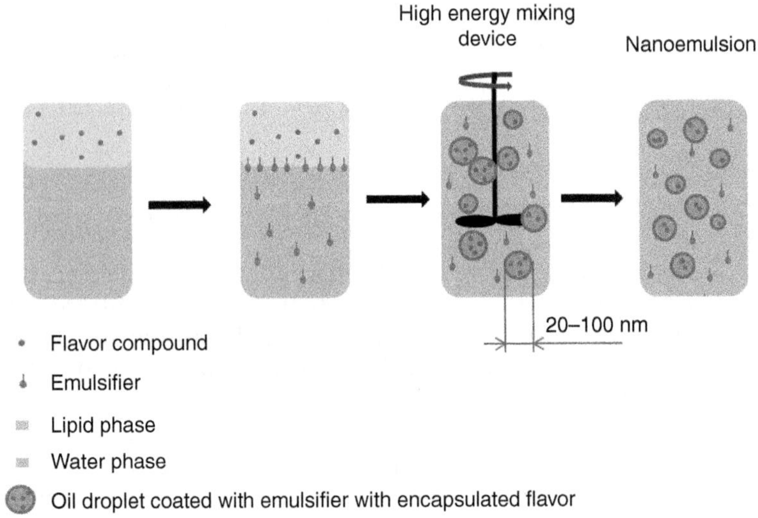

Figure 9.2. Simplification of the nanoemulsion preparation.

nanoemulsions. Ostwald ripening also depends on molecular weight of the oil phase so that higher molecular weight oil reduces Ostwald ripening rate (Wooster et al., 2008).

Nanoemulsions cannot be formed spontaneously; therefore an energy input, commonly from a mechanical device or from the chemical potential of the constituents, is required (Fig. 9.2). In general, nanoemulsions produced by ultrasonication displayed broader and bimodal particle size distributions compared to nanoemulsions obtained by using microfluidization. High-pressure homogenization is in industrial use for preparation of emulsions due to a possibility to control final size of droplets and size distribution by adjusting energy input. This preparation method allows the use of a variety of surfactants and cosurfactants (eg, proteins and GRAS polysaccharides) as well as different oils, but the drawbacks are very expensive equipment, low energetic efficiencies, and high production costs (Đorđević et al., 2015).

Furthermore, the amount of a surfactant which is required in order to stabilize nanoemulsions is usually high, but significantly less than that necessary for microemulsion stabilization (Mason et al., 2006). The market demand for low-fat or reduced-fat products imposes some limits on the amount of emulsions which can be added to food. For example, the flavor oils used in soft drinks and waters are often less than 0.1% of the final product; consequently the surfactant level in the final product is also low (Chang and McClements, 2014).

Additionally, nanoemulsions exhibit remarkably strong elasticity at low volume fractions of the dispersed phase, as well as enhanced shelf stability compared to microemulsions (Mason et al., 2006). These properties along with the optically transparency of nanoemulsions are of great interest for the food industry (but also personal care products and pharmaceuticals), as they enable the delivery of liposoluble flavors in clear emulsions. Nevertheless, the drawback of nanoemulsions is the fact that the droplets are very small, so most of the encapsulated flavors diffuse out of them in a very short period. Hence, different parameters should be taken into consideration in order to produce nanoemulsion with preferable properties (energy intensity, order of addition of the components, viscosity of phases, emulsifier properties) (McClements, 2010).

The choice of emulsifier reflects on the emulsion stability but also on flavor release. The most broadly used emulsifier for the stabilization of emulsions containing flavor oils (such as limonene, which is used in beverages) is gum arabic. Gum arabic is a complex polysaccharide which contains the small amount of proteins (~2.5% for gum from *A. senegal*). These proteins turn out to be important for its emulsification properties. The positive effect of proteins on emulsion stability provoked considerable interest in identifying other naturally occurring polysaccharide–protein complexes (such as pectin) (Siew and Williams, 2008; Klein et al., 2010; Evans et al., 2013). The main drawback of gum arabic is that it has to be used at quite high concentrations, for example, 20% w/w for stabilizing 10% w/w flavor oils (Reiner et al., 2010). Additionally, there are some frequent complications related to gum arabic usage, such as consistent quality, cost, and availability (Reiner et al., 2010). Therefore, there is a raising attention in science toward finding substitutes to gum arabic (such as modified starches and proteins).

Regarding the release of flavor, an interfacial film formed by the emulsifier retard mass transfer. Thus, commercial emulsifiers (monoglycerides) are able to form self-assembly structures for use in low-fat and fat-free food products, but also to retain the release of volatile aroma compounds. Phan et al. (2008) showed that the unsaturated monoglyceride mixed with the triglyceride in the ratio of 60:40 (as the oil phase) displayed a better retention for six aroma compounds compared to the water-in-oil emulsion having the same hydration value. The similar results were found by Mao et al. (2012), who investigated the introduction of limonene into the oil phase of emulsions. The authors observed the release of limonene by headspace analysis and indicated a delay in the release in the case of emulsions which contained 2% w/w monoglycerides.

In general, the concentration gradient of aroma in the food and the air above the food is the driving force of aroma release under dynamic conditions. The literature data give evidence that the release rate depends on compositional and structural properties of emulsions such as size of oil droplets, lipid fraction, and viscosity (Landy et al., 2007; Benjamin et al., 2012). Besides, properties of the flavor molecules (solubility, polarity, molecular size) but also the interactions between flavor molecules and other emulsion constituents (oils, emulsifiers, and thickening agents) have influence on flavor retention (van Ruth et al., 2002; Roberts et al., 2003; Karaiskou et al., 2008; Mao et al., 2012). Thus, Karaiskou et al. (2008) reported that the release rate of limonene was meaningfully reduced when oil droplet size was enlarged. The influence of hydrophobicity of the aroma on the release kinetic was described by Philippe et al. (2003) in a thermodynamic study, and the authors concluded that the partition of aromas was dependent on their hydrophobicity.

A novel trend in the field of emulsions designed for food application is preparation of organogel-based nanoemulsions which are easy for preparation and have high stability. Organogel-based nanoemulsions can be comprehended as thermoreversible semisolid systems formed by capturing of liquid oil in three-dimensional networks of structuring agent (organogelators), such as monoglycerides, fatty alcohols, and fatty acids (Zahi et al., 2014). This kind of nanoemulsion was used for delivery of D-limonene (Zahi et al., 2014). The authors concluded that organogelator type had an important impact on the formulations, in which stearic acid had provided the nanoemulsion with the smallest droplet size. It was also concluded that the kind of emulsifier and its concentration also had a considerable effect on droplet size and Tween 80, provided the smallest mean droplet diameter around 112 nm. The author suggested that this effect may derive from the molecular geometry of the surfactants. The similar effect of Tween 80 n the liposome size was found for the orange oil nanoemulsion formed by isothermal low-energy methods (Chang and McClements, 2014). The increase in Tween 80 concentration up to 20% decreased the mean droplet diameter down to 25 nm and the systems became optically transparent, which is preferred in commercial applications.

Interfacial engineering technology is an emerging field in food encapsulation. The main principle is modification of the interface of emulsion droplets with biopolymers. These modifications can change the droplets permeability, which leads to improved flavor retention and enhanced resistance to environmental stress (Gu et al., 2005). The emulsions formed using biopolymers can be described as emulsion droplets coated with nanolaminated interfacial layers. They can be prepared by utilization of layer-by-layer adsorption of oppositely charged polyelectrolytes onto a primary emulsion droplet (Moreau et al., 2003; Ogawa et al., 2004; Augustin and Hemar, 2009). The multilayer emulsions are characterized as more stable than conventional ones toward particle aggregation at high salt concentrations, thermal processing, freeze-thaw cycling, and high calcium concentrations, the conditions which are commonly present in foods (Ogawa et al., 2004; Gu et al., 2005; Mun et al., 2005). Anyway, pH and biopolymer concentration are crucial to avoid aggregation and to obtain the stable multilayered emulsions. The charge density of biopolymers can be adjusted through pH variation. On one hand, it is important to have sufficient biopolymer to completely saturate the oil–water interface, but on the other hand, the overmuch addition of biopolymer can promote depletion and flocculation (Guzey and McClements, 2006).

It is essential to stress that multilayer emulsions retain flavors during longer period of time compared to conventional ones. For example, when carvone (volatile flavor oil) was encapsulated in chitosan-coated emulsions the retention of the compound was significantly improved as compared to emulsions not containing chitosan (Kaasgaard and Keller, 2010). Đorđević et al. (2015) reported that the release of entrapped compounds from the inner phase of multilayer emulsions can occur by three mechanisms: (1) swelling or shear can cause the rupture or break down of the emulsion and can result in leaking of the inner phase content into the outer phase; (2) the phenomena of "thinning the liquid film" which occurs between the internal droplets and the outer interface, followed by its break; and (3) the emulsion remains intact but the entrapped compounds cross very thin lamella either by diffusion of the emulsifier or by reverse micelle transport. The release kinetic of flavors in these systems can be optimized by adjustable thickness and properties of the interfacial layer of the used biopolymer.

Unfortunately, there is a lack of literature introducing the preparation of entirely food-grade emulsions. Just a small number of surfactants are considered of food-grade (eg, sorbitans or polysorbates). Even though, in recent years there has been a great interest in identifying and characterizing new natural surfactants. Furthermore, the choice of edible oils needed for preparation of microemulsions or nanoemulsions (corn oil, fish oil, or soybean oil) is challenging since it is relatively difficult to prepare nanoscaled droplets using high-pressure homogenization due to the quite high viscosity and interfacial tension of the oils (Rao and McClements, 2012). On the other hand, a customer requirement for the healthier food is another criterion for selection of the oils, which should not contain a high amount of saturated fats and cholesterol (Lin et al., 2014). Rukmini et al. (2012) managed to prepare a microemulsion using virgin coconut oil, ternary food grade nonionic surfactant (Span 20, Span 60, and Tween 20), and deionized water. However, this study confirmed that the microemulsion was only suitable for food applications involving mild heating. Similarly, there are no successful nanoemulsion applications in beverages on the market, first and foremost because of a lack of functional edible and allowed emulsifiers (Zhang, 2011). Nevertheless, nanoemulsions are getting more attention in recent years as carriers for flavor molecules that are required to be released in the mouth during consumption.

They seem to be especially attractive for the beverage industry (eg, production of flavored clear beverages, fortified beverages, and mouthwashes) as a means to incorporate water-insoluble flavors into aqueous beverages.

4 Encapsulation of Flavors in Lipid-Based Nanoparticles

Solid lipid nanoparticles (SLNs) are very similar to nanoemulsions because they contain emulsifier-coated lipid droplets dispersed in an aqueous phase. The main difference is that lipid phase is either completely or partially solidified. The preparation procedure is carried out by homogenizing in the presence of an emulsifier at a temperature above the melting point of the lipid (hot homogenization procedure). The emulsion is further cooled under controlled conditions to solidify the lipids with incorporated flavor and to promote lipid crystallization. In this way the prepared SLNs have a quite small diameter, in the range between 50 nm and 1 m. Anyway, this procedure can be problematic for highly heat-sensitive lipophilic components. Therefore, cold homogenization has been applied instead (Fathi et al., 2012). Melted lipids with incorporated flavor are cooled and after solidification ground to powder by a mill. The obtained particles are further dispersed in a cold emulsifier solution, and this suspension is then subjected to homogenization (at room temperature or lower). During this procedure the special attention should be dedicated to the temperature increase during homogenization and milling.

The lipid phase used for preparation of SLNs can be composed of a variety of lipids, for example, mixtures of purified triglycerides, complex triglyceride mixtures, fatty acids, or even waxes (Wissing and Muller, 2002; Wissing et al., 2004; Augustin and Hemar, 2009).

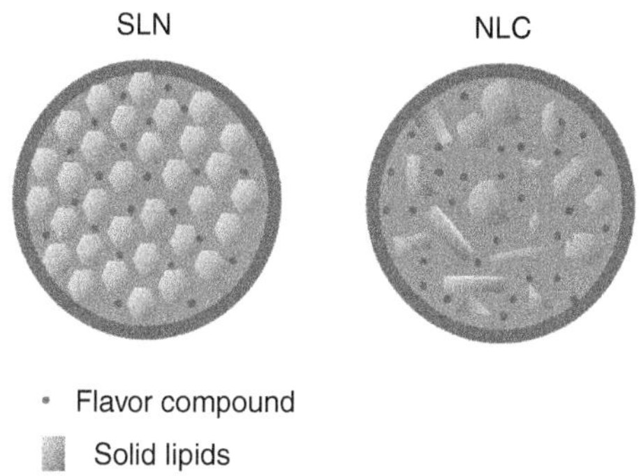

Figure 9.3. Morphological differences between SLN and NLC.

The choice of lipids, the lipid melting point, and the type of crystal network formed after the cooling step all have a huge influence on encapsulation, protection, stabilization, and delivery of lipophilic flavors. SLNs have some advantages in comparison to nanoemulsions and liposomes: they provide high encapsulation efficiency, the organic solvents are avoided during preparation procedure, there is a possibility of large-scale production as well as sterilization, the release kinetic (prolonged release) of encapsulated flavor can be optimized, and they offer higher targetability (Fathi et al., 2012; Đorđević et al., 2015). On the other hand SLNs have low encapsulation load, risk of burst during storage, and the lipid phase is usually highly saturated, which may have an adverse impact on human health.

Some of these limitations can be overcome by utilization of nanostructure lipid carriers (NLCs). NLCs usually contain the mix of different lipid molecules (oils and solid lipids) and the preparation methods are the same as for SLNs. In this way imperfect crystals are formed in NLCs while the SLNs have homogeneous crystal structure (Fig. 9.3). This difference provides a lower chance for burst phenomenon during storage and also enhanced encapsulation load because the flavor fits better into the more imperfect crystalline structure (McClements, 2010; Fathi et al., 2012, 2013). It is also reported in the literature that NLCs have smaller particle size in comparison to SLN (Fang et al., 2008).

Ghosh et al. (2008) conducted an in vitro study on kinetics of flavors release (ethyl butanoate, ethyl pentanoate, ethyl heptanoate, and ethyl octanoate) from solid and liquid lipid particles (of the same composition), and they found that the initial release trends were similar, but a much higher release was detected from the solid particles. This phenomenon was ascribed to differences in partitioning of the flavor to solid and liquid lipid phase. They also pointed out that melting of solid lipid particles in the mouth led to a sharp reduction in headspace flavor concentration. Finally, the literature suggests that burst release of flavor can be achieved by adjusting the physical state of core lipid, which is extremely suitable for increasing the release in ready-meals and fast foods during heating and consuming (Tamjidi et al., 2013).

5 Flavors Encapsulated in Liposomes

Liposomes are membrane-like vesicles mostly consisted of phospholipids and they are structured of one or more concentric spheres of lipid bilayers separated by water compartments. Liposomes are divided depending on their size and lamellarity: multilamellar vesicles (MLV), with diameter greater than 0.5 μm; large unilamellar vesicles (LUV), with diameter greater than 100 nm; and small unilamellar vesicles (SUV), with diameter between 20 and 100 nm (Sherry et al., 2013).

Liposomes are widely used for encapsulation, especially for pharmaceutical applications (Vemuri and Rhodes, 1995; Samad et al., 2007; Xu et al., 2012; Allen and Cullis, 2013), but nowadays liposomes are also attractive for food application (Marsanasco et al., 2011; Imran et al., 2015). This is due to the fact that liposomes are well characterized, easily prepared, very versatile in their carrier properties, biocompatible, biodegradable, and usually prepared from GRAS materials (such as egg, soy, or milk lipids) (Xia and Xu, 2005; Izadyari et al., 2014). Furthermore, liposome encapsulation enables the usage of smaller quantity of the active ingredient to achieve the specific effect compared to the quantity of unencapsulated actives needed for the same effect (due to the liposome nanosize and targetability) (Mozafari et al., 2006). One more fact which favors the encapsulation in liposomes is their capability to encapsulate both hydrophilic and lipophilic ingredients (such are most of the flavors), but also amphiphilic molecules (Yoshida et al., 2010). MLVs possess large lipidic phase and therefore they are more suitable for the encapsulation of lipophilic compounds. In contrast, unilamellar vesicles ensure more capacity for the encapsulation of hydrophilic compounds. Weiss et al. (2006) reported that this feature of liposomes can be utilized to enable incorporation of flavors into edible coatings or laminate films suitable for use in the food industry, with a goal to increase the shelf life and quality of coated foods.

The application of liposomes for flavor encapsulation is progressing research field. Since liposomes can be tailor-made, they possess unique characteristic to program complex flavor patterns in food products and to release the flavor at prearranged rates. The ability to adjust the liposome bilayer composition, from the standpoint of phase transition temperature, allows flavors and aromas to stay entrapped and protected during storage, but to be released in the mouth where the trigger is the increase in temperature up to physiological point (33.2–38.2°C) (Singh et al., 2012). Phase transition temperature depends on the following parameters: polar head group, length of acyl chain, degree of saturation of the hydrocarbon chains, ionic strength and nature, and of the suspension medium (Mozafari et al., 2008). It has been noticed that decrease in chain length, unsaturation of the acyl chains, likewise presence of branched chains and bulky head groups lead to lower values of phase transition temperature (Mozafari et al., 2008). Related to that, very interesting application of liposomes with encapsulated flavor was described by Lengerich et al. (1991). They illustrated the preparation of cookies having liposomes with encapsulated flavor and suggested that the liposomes may release the flavor over a period of time. The authors also indicated that the rate of the release depends on the phase transition temperature of the lipids, and on the chosen flavor, as well as on the concentration of the flavor. A special kind of liposomes (thermosensitive liposomes) can be prepared by addition of specific polymers to the lipid bilayer. The thermosensitive liposomes are destabilized above the polymer's critical temperature as a result of interaction between the liposome membrane and the hydrophobic polymer chains; in consequence, the release of the encapsulated flavor occurs. This kind of flavor release is convenient for the ready meals, which are subjected to increased cooking temperature during preparation (Kono et al., 1999).

Essential oils are usually considered as food flavors because they often have an odor. Thus, the use of liposomes for the encapsulation of essential oils is an attractive solution for the problems with physicochemical instability of essential oils and their reduced bioavailability (Detoni et al., 2012). Different methods have been used for encapsulation of flavors and essential oils in liposomes, from the most conventional Bangham method (Bangham, 1978) to more sophisticated approaches (Patil and Jadhav, 2014). Most of the methods provide MLV with wide size distribution. In order to obtain liposomes suspension with homogenous and reduced size, variety of treatment can be employed, but most common ones are sonication and extrusion. Sonication is frequently used for homogenization of liposomes with encapsulated essential oils (Valenti et al., 2001; Sinico et al., 2005; Detoni et al., 2012). Nevertheless, the resulting liposomes exhibit low encapsulation efficiency due to degradation of phospholipids and other materials during ultrasonic wave application (Patil and Jadhav, 2014). For the sake of comparison, MLVs prepared from enriched soy phosphatidylcholine exhibited encapsulation efficiency around 74%, while the sonicated SUVs displayed 66% of encapsulated essential oil (Sinico et al., 2005). Recently, Khatibi et al. (2015) used thin film evaporation method, ethanol injection method, and sonication after thin film evaporation in order to prepare nanoliposomes with encapsulated *Zataria multiflora* Boiss essential oil. The average particle size of the nanoliposomes encapsulating the essential oil was 395.3 nm for the thin film evaporation method, 239.7 nm for ethanol injection method, and for those treated by sonication the size was 99.9 nm. The authors found that incorporation of essential oil decreased the average vesicle size, but on the other hand the smaller size and unilamellar structure of liposomes produced by sonication provided the lowest encapsulation efficiency. Similar result was observed when essential oil of Brazilian cherry (*Eugenia uniflora* L.) was incorporated in liposomes obtained by thin film hydration method using hydrogenated soy lecithin (Yoshida et al., 2010). This result was explained by the ability of essential oils to cause higher cohesion packing among the apolar chains in the membrane vesicles (Valenti et al., 2001). However, the encapsulation efficiency also depends on the flavor type, physicochemical properties of each active (Detoni et al., 2012.) as well as on phospholipid concentration, phospholipid type, and ratio of flavor to phospholipid (Sherry et al., 2013). The influence of phospholipids type on entrapment efficiency of essential oil of *Anethi fructus* was investigated by Ortan et al. (2009). The authors suggested that the egg yolk phosphatidylcholine provided the highest encapsulation efficiency (98%) while dimiristroylphosphatidyl choline had the lowest one (55.2%).

Despite many obvious advantageous properties of liposomes, there are still certain limits for liposome application in food products. Difficulties in scaling up the preparation process, at adequate cost in use level, are still present to some extent. The constant researches on new methods such as microfluidization and discovery of low-priced alternatives for phospholipids (based on hydrophobic emulsifiers) are promising steps toward the broader use of liposomes in food industry (Isailović et al., 2013).

6 Flavors Encapsulated in Cyclodextrins

Cyclodextrins (CDs) are cyclic oligosaccharides composed of α-1,4-linked glucopyranose subunits. These inexpensive starch derivatives with truncated molecular structure are produced by enzymatic degradation. There is a progressive increase of publications and/or patents related to the studies and productions of CDs during the past four decades. Among CDs commonly used

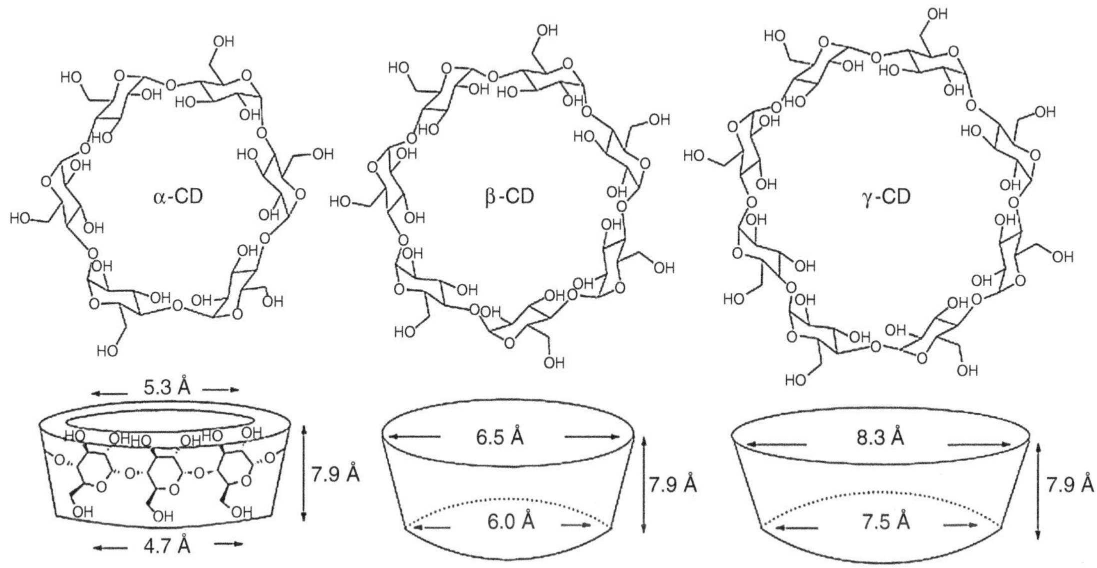

Figure 9.4. Flat chemical structure of a CD molecule and a cross-section of the toroid formed by a CD molecule. Reproduced from Freitag and Galoppini (2011) with permission of The Royal Society of Chemistry (RSC).

at technological level are: β-CDs (Nakamura and Horikoshi, 1977; Horikoshi, 1979), α-CDs (Flaschel et al., 1984), and γ-CDs (Bender, 1983). However, the purification of α- and γ-CDs increases significantly the cost of production, thus 97% of the used CDs are β-CDs (Astray et al., 2009). Those three mostly studied CDs are homogeneous, crystalline, and nonhygroscopic with previously mentioned torus-like macrorings of glucopyranose units (6, 7, and 8 of them in α- β-and γ-CD, respectively), which form cavities. Fig. 9.4 illustrates that the size of the cavity increases with the size of the CD oligomer. The presence of the hydroxyl groups on each glucopyranoside unit gives to these compounds a toroidal structure in aqueous solution. Compared to the linear oligosaccharides CDs, cyclic CDs show lower water solubility, even nine times lower in the case of β-CD (1.85 g/100 mL at room temperature) (Astray et al., 2009; Đorđević et al., 2015). Even δ-CD has better aqueous solubility than the β-CD, but still is less soluble than α- and γ-CD. The nine-membered δ-CD, isolated from the commercially available CD conversion mixture, is the least stable among the CDs known at this time. During the past 15 years larger CDs have been isolated and studied in order to have cavity diameter increased. However, the structures of larger CDs are not regular cylinder shaped, but collapsed, so their real cavity is even smaller than in the γ-CD (Astray et al., 2009).

Cyclodextrins are nontoxic, inexpensive, poorly absorbed in the upper gastrointestinal tract, and completely metabolized by the colon microflora; these characteristics provided them a very broad application in different areas, such as in the pharmaceutical, chemical, cosmetic, and food industries. To be more specific, the annual application of CDs has been increasing roughly 20–30%, and 80–90% of those applications were in food products (Szente and Szejtli, 2004). Still, the most important characteristic of CDs is the capability to form solid inclusion complexes (*host–guest* complexes) with a very wide range of liquid as well as solid and gaseous compounds (Eastburn and Tao, 1994). Upon molecular complexation with a CD, the properties of the component in the complex are modified considerably.

The lipophilic cavity of CD provides a microenvironment into which appropriately sized non or less polar molecules can be located. For this reason, CDs applications are primarily intended for the entrapment, delivery, and stabilization of small molecules. For example, flavor compounds are being nanoencapsulated into CDs for several benefits: better retention, protection from possible deterioration, and controlled delivery. In addition, CDs have been used for elimination of unpleasant tastes, contaminations and undesired compounds especially in the food, cosmetics, toiletry, and tobacco products, where the flavor is one of the most important characteristics (Singh et al., 2002; Martín del Valle, 2004).

6.1 Regulations for CDs Application in the Food Industry

Regulatory statuses of cyclodextrins application vary from one country to another. Generally, all three natural CDs are already used as food additives in the USA and Japan (the largest consumer of CDs in the world). In 2005 the US Environmental Protection Agency (EPA) proclaimed regulation that eliminates the need to establish a maximum permissible level for residues of α-, β-, and γ-CDs in various food commodities (US Federal Register, 2005), so they are on the GRAS (generally recognized as safe) list. In Japan the commercialization of CDs in the food sector is restricted only by considerations of purity. In Australia and New Zealand, α- and γ-cyclodextrins are identified as a novel food in 2004 and 2003 (Szente and Szejtli, 2004; Cravotto et al., 2006) In EU countries α- and γ-cyclodextrins represent a novel food since 2008 and 2001, respectively, while β-cyclodextrin is additive E-459 and a carrier for food additives (1 g/kg).

6.2 Inclusion Complexes of Cyclodextrins

An inclusion complex is a complex formed from two molecules: CDs as empty capsules of a determined molecular size (host) able to include a variety of molecules (guest) in the cavity (Fig. 9.5). The "host" molecule, totally or partly includes the "guest" molecules by physical forces (Astray et al., 2009). Complexation with CD is enabled through a noncovalent interaction between the molecule of interest and the cyclodextrin molecule often called "the molecular container." Complex formation is a dynamic process where the *guest* molecule associates and dissociates from the *host*.

If in this complex *host–guest* ratio is 1:1, the interaction can be simplified as follows:

$$\mathrm{CD} + \mathrm{G} \rightleftarrows \mathrm{CD\text{-}G} \quad K = k_{\mathrm{R}} / k_{\mathrm{D}}$$

where, CD, cyclodextrin; G, *guest* molecule; CD-G, inclusion complex; k_R, recombination constant; k_D, dissociation constant; and K, equilibrium constant.

The constant K is the central characteristic of this association. As the *guest* is larger, the formation and dissociation of the inclusion complex will be slower and ionization decreases the rate of complex formation and dissociation (Astray et al., 2009).

In order to increase the yield of this process, the affinity of the guest molecule to CD or solubility of the inclusion complex has to be increased. For this purpose, polymers, acids, or bases have been added. For acidic or basic ionizable *guest* compounds, adjusting of the pH is suitable in order to obtain the higher solubility of the *guest*, thus easier inclusion formation and better complexation

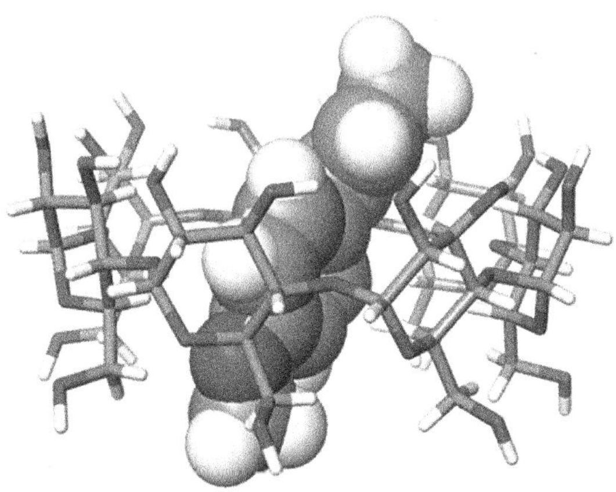

Figure 9.5. Inclusion complexes. Adapted from Kfoury et al. (2014).

efficiency. The dispersion of an active compound in water can be improved with supercritical carbon dioxide and consequently increasing process yield. The most important parameters of complexation process are temperature, stirring rate, and time, and centrifugation speed, while the hydrophilic–hydrophobic nature of the *guest* molecule has a large impact on encapsulation efficiency (Dias et al., 2010; Gomes et al., 2014; Đorđević et al., 2015). Encapsulation efficiency can be very high, but it depends on the technique applied (kneading, coprecipitation, dry mixing, sealing, slurry complexation, neutralization, spray drying, freeze-drying, and solvent evaporation) (Das et al., 2013).

One of the most important features of the inclusion complexes is stability constant (*K*). It represents a dynamic equilibrium between the free molecules and the complex in solution. The determination of the stability constant of inclusion complexes is crucial since it shows the ability of cyclodextrins to form inclusion complexes (Vilanova and Solans, 2015). Static headspace gas chromatography seems to be the most appropriate method to determine stability constant. Thus, Ciobanu et al. (2013) studied the interactions between 13 volatile flavor compounds and 6 CDs in inclusion complexes (1:1). They have found that α-CD and γ-CD gave lower stability constants than β-CDs and the complexation efficiency of native β-CD is close to the modified β-CDs (Ciobanu et al., 2013).

6.3 Electrospinning of Inclusion Complexes

Electrospinning is cost-effective and the most versatile technique for the production of functional nanofibers from various polymers, their blends and composites. This technique utilizes electrostatic forces to generate nanofibers. The device consists of a capillary through which the liquid to be electrospun is pumped, a source of high voltage which generates a charge in the solution, and a grounded collector, which is in contact with the counter electrode. Nanofibers produced by electrospinning have found applications in different areas due to the large surface-to-volume ratio and their ability to form a highly porous mesh (Ramakrishna et al., 2006).

Apart from high-molecular-weight polymers and high polymer concentrations, the inclusion complexes of CDs with *guest* molecules have been treated by electrospinning (Fig. 9.6). Thus, the electrospinning of methyl-β-cyclodextrin (MβCD) nanofibers without using a polymeric carrier matrix has been achieved (Celebioglu and Uyar, 2010). In addition, Celebioglu and Uyar (2011) have done the electrospinning of polymer-free nanofibers

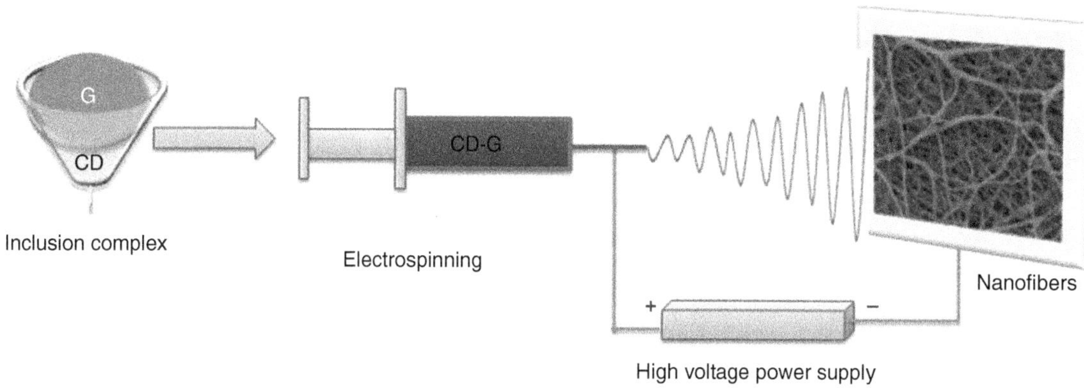

Figure 9.6. Electrospinning of inclusion complex.

from hydroxypropyl-β-cyclodextrin (HPβCD) and its inclusion complexes (IC) with triclosan (HPβCD/ triclosan-IC). They anticipated that the electrospinning of nanofibers from CD-ICs would be particularly attractive because of the exclusive properties obtained by combining the very large surface area of nanofibers/ nanowebs with specific functionality of the CD-ICs. It was found that having 1:1 host–guest complexation was optimal for HPβCD/ triclosan-IC nanofibers without any free guest molecules (Celebioglu and Uyar, 2011).

In addition, nanofibers of different CDs (α, β, γ) were produced using the electrospinning method in order to prolong the shelf life and high temperature stability of flavor compounds. For example, Kayaci and Uyar (2012) produced functional polyvinyl alcohol (PVA) nanowebs incorporating vanillin/cyclodextrin inclusion complex (vanillin/CD-IC). Their PVA/vanillin/CD-IC nanofibres were approximately 200 nm in diameter. Among different nanofiber formulations made from α-, β-, γ- CD, PVA/vanillin/γ-CD-IC was the most efficient in stabilization of vanillin, as well as for slow release. The result suggested that the interaction between vanillin and γ-CD was stronger in comparison to α- and β-CD (Kayaci and Uyar, 2012).

Mascheroni et al. (2013) have demonstrated possibility to encapsulate flavor compounds directly into an edible nanofibrous matrix by elecrospinning. In their study, a single step process has been used, based on electrospinning of the solution containing CD (β-cyclodextrin), polysaccharide (pullulan), and the flavor compound (perillaldehyde) with goal to achieve a humidity-triggered release. This nanofibrous structure ensured the aroma retention at ambient conditions (23°C and 55% RH) and even at high temperature (up to 230°C), but the release occurred beyond a given relative humidity threshold (90%). This result is very useful for further application in food packaging. The authors suggested that the load limit for the flavor compound is correlated with the solubility of the β-cyclodextrins and with their possibility to be electrospun.

Recently, Wen et al. (2016) produced polyvinyl alcohol/cinnamon essential oil/β-cyclodextrin (PVA/CEO/b-CD) antimicrobial nanofibrous film with average diameter 240 ± 40 nm. This film exhibited excellent antimicrobial activity and effectively prolong the shelf life of strawberry, indicating its potential for the application in active food packaging

6.4 Cyclodextrins and Flavors

As mentioned earlier, CDs have found an important place in the food industry due to their characteristics to protect, stabilize, solubilize, mask, and control flavor and their stability and release. Some examples of their applications in food processing are presented in Table 9.2.

The complexation with cyclodextrins provides a promising alternative to the conventional encapsulation technologies for flavor protection. The explanation lies in the unique property of the molecular encapsulation to provide a protection to basically any flavor component and to inhibit or to exclude molecular interactions between the different components present in a complex food system (Szente and Szejtli, 2004). Although this kind of encapsulation occurs at a molecular level, the final forms of encapsulates can be from nano to microsize. Namely, flavor-CD inclusion complexes are able to form large aggregates in water (He et al., 2008). The aggregates can have many geometric shapes (rhombic, trapezoidal, and parallelogram; Zhu et al., 2014).

CDs have a long tradition in removing or masking undesirable taste or smell, while the oldest applications appeared in the Far East. For example, CDs have been used for deodorizing soybean milk and soy proteins and for removing the specific fish odors (Sakakibara et al., 1985). Rice stored for longer period acquires a peculiar off-flavor, which can be removed by cooking in the presence of 0.01–0.4% β-CD (Takeda Chem. Ind. Ltd., 1981). Also, the bitter taste of milk casein hydrolysate can be eliminated by adding 10% β-CD to the protein hydrolysate (Suzuki, 1975).

Of all encapsulating techniques, molecular encapsulation of flavors with β-CD has been proven to be the most effective for providing protection against heat and evaporation (Pagington, 1986; Qi and Hedges, 1995). Jouquand et al. (2004) found that the addition of β-CD as stabilizing or thickening agents could retain some aroma compounds in food matrices during thermal processes (eg, cooking, pasteurization) (Jouquand et al., 2004). It is confirmed

Table 9.2 Practical Advantages of CDs Application in Flavor Nanoencapsulation

Aims of CDs Application	CD	Flavor or Food Product	References
Protection against degradations	HP-βCD, β-CD	L-menthol, naringin, sweet orange flavor	Liu et al. (2000b, 2013b); Zhu et al. (2014)
Antimicrobial and/or antioxidant activity	β-CD, HP-β-CD	Cinnamon essential oil, perillaldehyde, carvacrol, benzyl isothiocyanate, grape juice	Mascheroni et al. (2013); Kamimura et al. (2014); Li et al. (2015); Shao et al. (2014); Wen et al. (2016)
Retention and control release	β-CD	Essential oils, D-Limonene	Yuliani et al. (2006); Hădărugă et al. (2008); Kfoury et al. (2015)
Suppressing of unpleasant odours or taste	β-CD, α-CD, γ-CD	Goat milk, ginseng, naringin, caffeine	Tamamoto et al. (2010); Young et al. (2012)
Solubilization or solubility improvement	HP-β-CD	Vanillin, naringin	Karathanos et al. (2007); Mantegna et al. (2012); Liu et al. (2013b)
Stabilization	γ-CD, β-CD	Vanillin and coffee flavor, spices in sausages, fragrances, thymol, and cinnamaldehyde	Singh et al. (2002); Kayaci and Uyar (2012); Cevallos et al. (2010); Kfoury et al. (2015)

by Munoz-Botella et al. (1995) that β-CD, as a carrier, allows the flavor to be preserved to a higher degree and longer period of time compared to other encapsulants in food products (Munoz-Botella et al., 1995). Kfoury et al. (2014) showed that the degradation of *trans*-anethole (AN) induced by UVC irradiation was markedly reduced by the formation of CD inclusion complexes.

Stability of the dried CD-flavor complexes is governed by the water uptake, which means that relative humidity largely affects complex stability during storage. For example, Cevallos et al. (2010) showed that both β-CD complexes with thymol and cinnamaldehyde remained stable up to 75% relative humidity during long storage period.

There are many flavor/CDs complexes-based products available on market such as powdered flavors (eg, citrus fruits, vanilla, and apple), spices like ginger, garlic, mustard, horseradish, and other herbs such as peppermint or basil, and green tea. Granular sugar coated with maple and vanilla flavor included with CD, also

(Hashimoto, 2002). CDs have been also used in alcoholic beverages like whisky or beer (Singh et al., 2002).

A bitter taste is the main reason for the rejection of various food products, such as commercial citrus juices. The flavonoids and limonoids (naringin and limonin, respectively) are the classes of chemical compounds, responsible for a bitter taste of food products. Differential scanning calorimetry thermograms confirmed the ability of cross-linked CDs to form inclusion complexes with naringin and decrease the bitterness (Binello et al., 2004).

Hădărugă et al. (2008) studied nanoencapsulation of the unsaturated fatty acids and essential oils in α- and β-cyclodextrins. The results confirmed the protection capacity of these CDs, better thermal stability, and very good yields. In addition, their research showed that the complex of the garlic essential oil in β-CD provides nanoencapsulates with no "garlic" odor (Hădărugă et al., 2008).

There are many examples on CD complexes successfully implemented into final food products. Some of them are Cyroma-line, flavored sugar for backing produced in Hungary where CDs have a role of flavor preservation during heating; Flavono, chewing gum produced in Japan with CDs in order to stabilize flavor; Gymet, a Japanese commercial dietary fiber drink in which unpleasant taste has been masked by CD complexation; Stick Lemon, a Japanese instant tea drink in which CDs preserve flavor; FlavorAktiv standard kit, in which CDs are also used, to preserve beer flavor standards produced in Great Britain. CAVAMAX® W8 Curcumin produced by Wacker Chemie AG, Germany, is the trade name that represents the food supplement which is a complex of curcumin with γ-cyclodextrin (Szejtli and Szente, 2005; Das et al., 2013; Đorđević et al., 2015). Vivid® Baking Flavor Powder is designed for industrial production, too. Generally, this product is suitable for different kinds of bakery products. With a small added amount of this flavor powder, the final products are enabled to possess specific flavor with richness. Two most important characteristics of this product are resistance to high-temperature baking and pure fragrance. There are different kinds of Vivid Baking flavors on the market, like passion fruit, sabayon, tiramisu, cheese, green tea, milk, butter, milk, egg, orange, banana, chocolate, and so forth.

7 Flavors Encapsulated in Complex Coacervates

Complex coacervation was described for the first time by Bungenberg de Jong in the 1950s (Bungenberg de Jong, 1949). Essentially, it is a process of liquid–liquid phase separation, which is endorsed by different interactions between adversely charged polymers in a solution, for example, electrostatic and hydrophobic interactions, hydrogen bonding, as well as attractive interactions that usually occur upon polarization of polymers (Boral and Bohidar, 2010). The coacervates can be categorized into two types, simple and complex. The simple coacervation implies only one polymeric macromolecule that, under certain conditions (usually in the presence of self-charges, both positive and negative or upon addition of salts) reorganizes its intermolecular structure via self-charge neutralization (Mohanty and Bohidar, 2005). On the other hand, complex coacervates are formed when two adversely charged biopolymers correlate to establish complex, by separation of phase rich in polymer and aqueous phase (Gupta et al., 2012).

The process of complex coacervation can be divided into following steps: (1) preparation of the polymer solution that contains two different polymers, usually protein and polysaccharide (this is commonly done at the temperature that exceeds polysaccharide gelling point, and pH higher than protein isoelectric point); (2) emulsification/dispersion of aroma in the prepared polymers solution; (3) coacervation—pH-induced phase separation into two phases (insoluble phase rich in polymer and aqueous phase containing both polymers); (4) gelation—formation of the wall, by deposition of the polymers on the surface of the aroma droplets, provoked by lowering of the temperature below the protein isoelectric point; and (5) hardening—stabilizing the structure of the formed particles by addition of a cross-linking agent (Xiao et al., 2014). An alternative procedure of complex coacervation implies dissolution of the protein with the aroma compounds, then complexation of the protein and polysaccharide and coacervation to the aroma surface; the prepared particles then may be subjected to the hardening process, by addition of a cross-linker (Jincheng et al., 2010). The size of the obtained particles is dependent on the applied pH and ion concentration, ratio matrix-to-aroma, as well as on the biopolymer types (de Vos et al., 2010).

Protein–polysaccharide complexes are most widely used for the encapsulation purposes, while protein–protein coacervates (Elzoghby et al., 2012) and polysaccharide–polysaccharide coacervates (Fathi et al., 2014) have been used to a lesser extent. The well-known and most exploited protein–polysaccharide coacervate complex is certainly gelatin–gum arabic; it has been used for encapsulation of various aroma compounds and essential oils, for example allyl isothiocyanate (Zhang et al., 2011), limonene and menthol (Leclercq et al., 2009), peppermint oil (Dong et al., 2007, 2008, 2011), jasmine oil (Lv et al., 2012), lemon oil (Specos et al., 2010a), citronella oil (Specos et al., 2010b; Liu et al., 2013a), vetiver oil (Prata et al., 2008a,b), garlic oil (Siow and Ong, 2013), patchouli oil (Au et al., 2011), and so forth. In general, for complex coacervates, the following proteins can be used (apart from gelatin): whey proteins, casein, soy proteins, pea proteins, cereal proteins, silk proteins (Xiao et al., 2014). Among the complex-forming polysaccharides are pectin, chitosan, agar, alginate, and carrageenan, as well as sodium carboxymethyl cellulose (Xiao et al., 2014).

Coacervation has widely been used for the entrapment of aromatic compounds (Jun-xia et al., 2011; Ezhilarasi et al., 2013; Dima et al., 2014; Lv et al., 2014; Yang et al., 2014). It is a promising encapsulation technology, since high payloads can be achieved (up to 99%), and the produced particles acquire good controlled-release potential, on the basis of mechanical stress, temperature, and/or sustained release (Gouin, 2004). Particles produced by complex coacervation are commonly micron-sized; nevertheless, an increasing demand for nanodelivery systems led to development of complex coacervation at nanoscale. Although the complex coacervation has been extensively studied over the years, nanoencapsulation of aromatic compounds via complex coacervation is still not fully developed and potentials of this method have not been exploited enough. The average size of the coacervate nanocapsules depends on the cocervation process conditions as well as the drying technique used for the preparation, for example, oven drying, vacuum drying, freeze drying; commonly, it is in the range of 100–600 nm (Ezhilarasi et al., 2013). Several studies have been dedicated to nanoencapsulation of capsaicin (it is a natural compound isolated from chili peppers, used as a food additive due to its various biological activities, and as a food spice that adds piquancy) (Xing et al., 2004; Wang et al., 2008a,b; Jincheng et al., 2010). Wang et al. (2008a) encapsulated capsaicin via simple coacervation technique. The material used for coacervation was gelatin, which exhibited self-assembly into coacervates, with the addition of glutaraldehyde as a cross-linking agent. The dried coacervates (100 nm in size) provided improved thermal stability and a somewhat higher melting point of the aroma compound. The authors suggested that this result is caused by deposition of the cross-linked gelatin over the surface of the capsaicin. This paper also indicates that the thickness of a capsule wall could be optimized by manipulating with process conditions (eg, gelatin viscosity, shearing force, cross-linking time). Similarly, Xing et al. (2004) encapsulated capsaicin, but they employed complex coacervation of gelatin and gum arabic. The coacervation of protein and polysaccharide polymers were endorsed by addition of the tannins, and complex coacervates were hardened via cross-linking with glutaraldehyde. The tannins addition intensified the synergistic effects of hydrogen bonding and hydrophobic interactions within the coacervates; this resulted in improved particle morphology and distribution of the particles size. Jincheng et al. (2010) also used gelatin, gum arabic, and tannins for the formation of complex coacervates. They optimized parameters of the process, in terms of determining the optimal agitation rate (15,000 rpm), viscosity of the protein–polymer solution (5–20 cPs), sufficient time for the cross-linking reaction to occur (20–40 min), utilization of the tannins, as well as the addition of emulsifier molecules (Tween 60) to promote the formation of a stable emulsion that was further subjected to coacervation and hardening (via cross-linking with glutaraldehyde). The produced spherical nanocapsules (around 100 nm in size) exposed enhanced thermal properties which were attributed to the cross-linking reaction. Particles produced by complex coacervation expose a unique property—they are capable of enduring the exchange of solvent via diffusion through the coacervate layer, without affecting particle structure (Thies, 2007). Usually, a water-immiscible compound originally encapsulated in the particles of complex coacervate can be exchanged with some other liquid compound that has fine water miscibility. This property enables the loading of complex coacervate particles with liquids and flavors that cannot be added at the time of the particle formation. Thus, Soper et al. (2000) introduced the oil-loaded particles with water-swollen shell suitable for absorption of flavor from the gas phase. The purging of particles with gas containing the aroma (0.5–5 h depending on aroma type) the absorption of aroma was detected.

Since modern food regulation do not support utilization of glutaraldehyde in food, researchers constantly seek for alternatives, which will possess equally, if not even higher, cross-linking abilities. In that respect, utilization of the food-safe sodium tripolyphosphate (TPP) as a cross-liker was investigated. Additionally, in efforts to make the complex coacervates more suitable for food applications, the enzyme cross-linking has also been proposed as a method to avoid potentially toxic ingredients (ie, glutaraldehyde), for example utilization of transglutaminase as a cross-linker (Prata et al., 2008a,b) along with other food-grade cross-linkers, such as tannic acid (Zhang et al., 2011), or genipin (Maji and Hussain, 2009).

8 Conclusions

The flavoring industry is a pioneer in implementation of encapsulation technology in food sector. Still, it utilizes about three times more free flavors compared to encapsulated forms, and most of flavor formulations are of microsize. There are several reasons behind this fact: cost of nanoencapsulation is rather high, the potential risks of nanomaterials to human health are unknown and need to be explored and studied, the regulatory issues on nanofoods are still being developed, and the knowledge of release of encapsulated aroma when incorporated in real food products is rather poor. It is likely to expect that in future spray-dried encapsulation will again dominate the food industry. In fact, most of encapsulation techniques ultimately depend on suitable drying techniques to produce nanoencapsulates in powder form. Spray drying needs some modifications for retaining the nanoparticle size, with a special design of equipments needed to produce nanoencapsulates in powder form. Furthermore, each encapsulation technique has some unique operating factors, which affect the final outcome of nanoencapsulates, and those factors need to be investigated and optimized. Development of novel nanoencapsulated flavors is going to progress in line with development of more sophisticated dispersion technologies employing complex mixtures of biopolymers as well as low molecular weight surfactants, and novel multilayered interfacial structures.

Acknowledgment

This work was supported by the Ministry of Education, Science, and Technological Development, Republic of Serbia (Projects No. III46010 and III46001).

References

Alamilla-Beltran, L., Chanona-Perez, J.J., Jimenez-Aparicio, A.R., Gutierrez-Lopez, G.F., 2005. Description of morphological changes of particles along spray drying. J. Food Eng. 67, 179–184.

Allen, T.M., Cullis, P.R., 2013. Liposomal drug delivery systems: from concept to clinical applications. Adv. Drug Deliv. Rev. 65, 36–48.

Anandaraman, S., Reineccius, G.A., 1986. Stability of encapsulated orange peel oil. Food Technol. 40, 88–93.

Anandharamakrishnan, C., Rielly, C.D., Stapley, A.G.F., 2007. Effects of process variables on denaturation of whey protein during spray drying. Dry. Technol. 25, 799–807.

Anandharamakrishnan, C., Rielly, C.D., Stapley, A.G.F., 2008. Loss of solubility of α-lactalbumin and β-lactoglobulin during spray drying of whey proteins. LWT Food Sci. Technol. 41, 270–277.

Anderson, D.M.W., Stoddart, J.F., 1966. The use of molecular sieve chromatography on *Acacia senegal* gum. Carbohydr. Res. 2, 104–114.

Anderson, D.M.W., Howlett, J.F., McNab, C.G.A., 1985. The amino acid composition of the proteinaceous component of gum arabic (*Acacia senegal* (L.) Willd.). Food Addit. Contam. 2, 159–164.

Anese, M., Manzocco, L., Nicoli, M.C., 2006. Modeling the secondary shelf life of ground roasted coffee. J. Agric. Food Chem. 54, 5571–5576.

Astray, G., Gonzalez-Barreiro, C., Mejuto, J.C., Rial-Otero, R., Simal-Gándara, J., 2009. A review on the use of cyclodextrins in foods. Food Hydrocoll. 23, 1631–1640.

Au, W.M., Yang, Z.M., Wong, T.K., Li, L., Liang, G.Q., Zhong, H.Y., et al., 2011. Preparation of antibacterial cotton fabric containing patchouli oil microcapsules by chemical crosslinking method. Adv. Mater. Res. 221, 308–315.

Augustin, M.A., Hemar, Y., 2009. Nano- and micro-structured assemblies for encapsulation of food ingredients. Chem. Soc. Rev. 38, 902–912.

Bae, K.E., Lee, S.J., 2008. Microencapsulation of avocado oil by spray drying using whey protein and maltodextrin. J. Microencapsul. 25, 549–560.

Bangham, A.D., 1978. Properties and uses of lipid vesicles: an overview. Ann. NY Acad. Sci. 308, 2–7.

Bao, J., Xing, J., Phillips, D.L., Corke, H., 2003. Physical properties of octenyl succinic anhydride modified rice, wheat, and potato starches. J. Agric. Food. Chem. 51, 2283–2287.

Baranauskiene, R., Venskutonis, P.R., Dewettinck, K., Verhe, R., 2006. Properties of oregano (*Origanum vulgare* L.), citronella (*Cymbopogon nardus* G.) and marjoram (*Majorana hortensis* L.) flavours encapsulated into milk protein-based matrices. Food Res. Int. 39, 413–425.

Baranauskiene, R., Bylaite, E., Rate, J.H., Zukauskaite, J., Venskutonis, R.P., 2007. Flavour retention of peppermint (*Mentha piperita* L.) essential oil spray-dried in modified starches during encapsulation and storage. J. Agric. Food Chem. 55, 3027–3036.

Bender, H., 1983. An improved method for the preparation of cyclooctaamylose, using starches and the cyclodextrin glycosyltransferase of *Klebsiella peumoniae* M 5 al. Carbohydr. Res. 124, 225–233.

Benjamin, O., Silcock, P., Leus, M., Everett, D.W., 2012. Multilayer emulsions as delivery systems for controlled release of volatile compounds using pH and salt triggers. Food Hydrocoll. 27, 109–118.

Bertolini, A.C., Siani, A.C., Grosso, C.R., 2001. Stability of monoterpenes encapsulated in gum arabic by spray-drying. J. Agric. Food Chem. 49, 780–785.

Bibette, J., Calderon, F.L., Poulin, P., 1999. Emulsions: basic principles. Rep. Prog. Phys. 62, 969–1033.

Binello, A., Cravotto, G., Nano, G.M., Spagliardi, P., 2004. Synthesis of chitosan–cyclodextrin adducts and evaluation of their bitter-masking properties. Flavour Frag. J. 19, 394–400.

Boral, S., Bohidar, H.B., 2010. Effect of ionic strength on surface-selective patch binding-induced phase separation and coacervation in similarly charged gelatin–agar molecular systems. J. Phys. Chem. B 114, 12027–12035.

Botrel, D.A., Borges, S.V., Fernandes, R.V.B., Viana, A.D., Gomes, J.M.C., Marques, G.R., 2012. Evaluation of spray drying conditions on properties of microencapsulated oregano essential oil. Int. J. Food Sci. Technol. 47, 2289–2296.

Bruckner, M., Bade, M., Kunz, B., 2007. Investigations into the stabilization of a volatile aroma compound using a combined emulsification and spray drying process. Eur. Food Res. Technol. 226, 137–146.

Buffo, R.A., Probst, K., Zehentbauer, G., Luo, Z., Reineccius, G.A., 2002. Effects of agglomeration on the properties of spray-dried encapsulated flavours. Flavour Frag. J. 17, 292–299.

Bungenberg de Jong, H.G., 1949. Complex colloid systems. In: Kruyt, H.R. (Ed.), Colloid Science. Elsevier, New York, NY, pp. 280–283.

Burdock, A.G., 2009. Fenaroli's Handbook of Flavour Ingredients, sixth ed. CRC Press, Boca Raton, FL.

Cai, Y.Z., Corke, H., 2000. Production and properties of spray-dried *Amaranthus betacyanin* pigments. J. Food Sci. 65, 1248–1252.

Carmona, P.A.O., Tonon, R.V., da Cunha, R.L., Hubinger, M.D., 2013. Influence of emulsion properties on the microencapsulation of orange essential oil by spray drying. J. Colloid Sci. Biotechnol. 2, 130–139.

Celebioglu, A., Uyar, T., 2010. Cyclodextrin nanofibers by electrospinning. Chem. Commun. 46, 6903–6905.

Celebioglu, A., Uyar, T., 2011. Electrospinning of polymer-free nanofibers from cyclodextrin inclusion complexes. Langmuir 27, 6218–6226.

Cevallos, P.A.P., Buera, M.P., Elizalde, B.E., 2010. Encapsulation of cinnamon and thyme essential oils components (cinnamaldehyde and thymol) in β-cyclodextrin: effect of interactions with water on complex stability. J. Food Eng. 99, 70–75.

Chang, Y., McClements, D.J., 2014. Optimization of orange oil nanoemulsion formation by isothermal low-energy methods: influence of the oil phase, surfactant, and temperature. J. Agric. Food Chem. 62, 2306–2312.

Chegini, G.R., Ghobadian, B., 2007. Spray dryer parameters for fruit juice drying. World J. Agric. Sci. 3, 230–236.

Ciobanu, A., Landy, D., Fourmentin, S., 2013. Complexation efficiency of cyclodextrins for volatile flavor compounds. Food. Res. Int. 53, 110–114.

Cravotto, G., Binello, A., Baranelli, E., Carraro, P., Trotta, F., 2006. Cyclodextrins as food additives and in food processing. Curr. Nutr. Food Sci. 2, 343–350.

Damodaran, S., 2005. Protein stabilization of emulsions and foams. J. Food Sci. 70, R54–R66.

Das, S.K., Rajabalaya, R., David, S., Gani, N., Khanam, J., Nanda, A., 2013. Cyclodextrins—the molecular container. Res. J. Pharm. Biol. Chem. Sci. 4, 1694–1720.

de Roos, K.B., 2000. Physicochemical models of flavour release from foods. In: Roberts, D.D., Taylor, A.J. (Eds.), Flavour Release. American Chemical Society, Washington, DC, pp. 126–141.

de Vos, P., Faas, M.M., Spasojevic, M., Sikkema, J., 2010. Encapsulation for preservation of functionality and targeted delivery of bioactive food components. Int. Dairy J. 20, 292–302.

Del Toro-Arreola, S., Flores-Torales, E., Torres-Lozano, C., Del Toro-Arreola, A., Tostado-Pelayo, K., Ramirez-Dueñas, M.G., et al., 2005. Effect of D-limonene on immune response in BALB/c mice with lymphoma. Int. Immunopharmacol. 5, 829–838.

Desai, K.G.H., Park, H.J., 2005. Recent developments in microencapsulation of food ingredients. Dry. Technol. 23, 1361–1394.

Desobry, S., Debeaufort, F., 2011. Encapsulation of flavours, nutraceuticals, and antibacterials. In: Baldwin, E.A., Hagenmaier, R., Bai, J. (Eds.), Edible Coatings and Films to Improve Food Quality. second ed. CRC Press, Boca Raton, FL, pp. 333–373.

Detoni, C.B., de Oliveira, D.M., Santo, I.E., Pedro, A.S., El-Bacha, R., da Silva Velozo, E., et al., 2012. Evaluation of thermal-oxidative stability and antiglioma activity of *Zanthoxylum tingoassuiba* essential oil entrapped into multi- and unilamellar liposomes. J. Liposome Res. 22, 1–7.

Dias, H.M.A.M., Berbicz, F., Pedrochic, F., Baesso, M.L., Matioli, G., 2010. Butter cholesterol removal using different complexation methods with beta-cyclodextrin, and the contribution of photoacoustic spectroscopy to the evaluation of the complex. Food. Res. Int. 43, 1104–1110.

Dima, C., Cotârlet, M., Alexe, P., Dima, S., 2014. Microencapsulation of essential oil of pimento [*Pimenta dioica* (L) Merr.] by chitosan/k-carrageenan complex coacervation method. Innov. Food Sci. Emerg. Technol. 22, 203–211.

Dong, Z.J., Toure, A., Jia, C.S., Zhang, X.M., Xu, S.Y., 2007. Effect of processing parameters on the formation of spherical multinuclear microcapsules encapsulating peppermint oil by coacervation. J. Microencapsul. 24, 634–646.

Dong, Z.J., Xia, S.Q., Hua, S., Hayat, K., Zhang, X.M., Xu, S.Y., 2008. Optimization of cross-linking parameters during production of transglutaminase-hardened spherical multinuclear microcapsules by complex coacervation. Colloids Surf. B 63, 41–47.

Dong, Z.J., Ma, Y., Hayat, K., Jia, C., Xia, S., Zhang, X., 2011. Morphology and release profile of microcapsules encapsulating peppermint oil by complex coacervation. J. Food Eng. 104, 455–460.

Đorđević, V., Balanč, B., Belščak-Cvitanović, A., Lević, S., Trifković, K., Kalušević, A., et al., 2015. Trends in encapsulation technologies for delivery of food bioactive compounds. Food Eng. Rev. 7, 452–490.

Druaux, C., Voilley, A., 1997. Effect of food composition and microstructure on volatile flavour release. Trends Food Sci. Technol. 8, 364–368.

Eastburn, S.D., Tao, B.Y., 1994. Applications of modified cyclodextrins. Biotechnol. Adv. 12, 325–339.

Elzoghby, A.O., Samy, W.M., Elgindy, N.A., 2012. Protein-based nanocarriers as promising drug and gene delivery systems. J. Control. Release 161, 38–49.

Ersus, S., Yurdagel, U., 2007. Microencapsulation of anthocyanin pigments of black carrot (*Daucus carota* L.) by spray drier. J. Food Eng. 80, 805–812.

Evans, M., Ratcliffe, I., Williams, P.A., 2013. Emulsion stabilisation using polysaccharide–protein complexes. Curr. Opin. Colloid Interf. Sci. 18, 272–282.

Ezhilarasi, P.N., Karthik, P., Chhanwal, N., Anandharamakrishnan, C., 2013. Nanoencapsulation techniques for food bioactive components: a review. Food Bioprocess Technol. 6, 628–647.

Fang, X., Shima, M., Adachi, S., 2005. Effects of drying conditions on the oxidation of linoleic acid encapsulated with gum arabic by spray drying. Food Sci. Technol. Res. 11, 380–384.

Fang, J.-Y., Fang, C.-L., Liu, C.-H., Su, Y.-H., 2008. Lipid nanoparticles as vehicles for topical psoralen delivery: solid lipid nanoparticles (SLN) versus nanostructured lipid carriers (NLC). Eur. J. Pharm. Biopharm. 70, 633–640.

Fathi, M., Mozafari, M.R., Mohebbi, M., 2012. Nanoencapsulation of food ingredients using lipid-based delivery systems. Trends Food Sci. Technol. 23, 13–27.

Fathi, M., Varshosaz, J., Mohebbi, M., Shahidi, F., 2013. Hesperetin-loaded solid lipid nanoparticles and nanostructure lipid carriers for food fortification: preparation, characterization, and modeling. Food Bioprocess Technol. 6, 1464–1475.

Fathi, M., Martin, A., McClements, D.J., 2014. Nanoencapsulation of food ingredients using carbohydrate-based delivery systems. Trends Food Sci. Technol. 39, 18–39.

Fernandes, R.V.B., Borges, S.V., Botrel, D.A., 2013. Influence of spray drying operating conditions on microencapsulated rosemary essential oil properties. Food Sci. Technol. 33, 171–178.

Finney, J., Buffo, R., Reineccius, G.A., 2002. Effects of type of atomization and processing temperatures on the physical properties and stability of spray-dried flavours. J. Food Sci. 67, 1108–1114.

Fitzgerald, D.J., Stratford, M., Gasson, M.J., Ueckert, J., Bos, A., Narbad, A., 2004. Mode of antimicrobial action of vanillin against *Escherichia coli*, *Lactobacillus plantarum* and *Listeria innocua*. J. Appl. Microbiol. 97, 104–113.

Flanagan, J., Singh, H., 2006. Microemulsions: a potential delivery system for bioactives in food. Crit. Rev. Food Sci. Nutr. 46, 221–237.

Flaschel, E., Landert, J.P., Spiesser, D., Renken, A., 1984. The production of acyclodextrin by enzymatic degradation of starch. Ann. NY Acad. Sci. 434, 70–77.

Frascareli, E.C., Silva, V.M., Tonon, R.V., Hubinger, M.D., 2012. Effect of process conditions on the microencapsulation of coffee oil by spray drying. Food Bioprod. Process. 90, 413–424.

Freitag, M., Galoppini, E., 2011. Molecular host–guest complexes: shielding of guests on semiconductor surfaces. Energy Environ. 4, 2482–2494.

Ghosh, S., Peterson, D.G., Coupland, J.N., 2008. Temporal aroma release profile of solid and liquid droplet emulsions food biophysics. Food Biophys. 3, 335–343.

Gibbs, B.F., Kermasha, S., Alli, I., Mulligan, C.N., 1999. Encapsulation in the food industry: a review. Int. J. Food Sci. Nutr. 50, 213–224.

Gil, S., Parra, M., Rodriguez, P., Segura, J., 2009. Recent developments in γ-lactone synthesis. Mini-Rev. Org. Chem. 6, 345–358.

Gomes, L.M., Petito, N., Costa, V.G., Falcão, D.Q., de Lima Araujo, K.G., 2014. Inclusion complexes of red bell pepper pigments with β-cyclodextrin: preparation, characterisation and application as natural colorant in yogurt. Food. Chem. 148, 428–436.

Goubet, I., Le Quere, J.L., Voilley, A.J., 1998. Retention of aroma compounds by carbohydrates: influence of their physicochemical characteristics and of their physical state: a review. J. Agric. Food. Chem. 46, 1981–1990.

Gouin, S., 2004. Microencapsulation: industrial appraisal of existing technologies and trends. Trends Food Sci. Technol. 15, 330–347.

Goula, A.M., Adamopoulos, K.G., 2008. Effect of maltodextrin addition during spray drying of tomato pulp in dehumidified air: I. Powder properties. Dry. Technol. 26, 726–737.

Gu, Y.S., Regnier, L., McClements, D.J., 2005. Influence of environmental stresses on stability of oil-in-water emulsions containing droplets stabilized by beta-lactoglobulin-iota-carrageenan membranes. J. Colloid Interface Sci. 286, 551–558.

Guichard, E., 2002. Interactions between flavour compounds and food ingredients and their influence on flavour perception. Food Rev. Int. 18, 49–70.

Gupta, R., Basu, S., Shivhare, U.S., 2012. A review on thermodynamics and functional properties of complex coacervates. Int. J. Appl. Biol. Pharm. Technol. 3, 64–86.

Guzey, D., McClements, D.J., 2006. Formation, stability and properties of multilayer emulsions for application in the food industry. Adv. Colloid Interface Sci. 128–130, 227–248.

Hădărugă, N.G., Hădărugă, D.I., Bandur, G.N., Riviş, A., Pârvu, D., Lupea, A.X., 2008. Protection and controlled release of fatty acids and essential oils by nanoencapsulation in cyclodextrins (a review). J. Agroaliment. Proc. Technol. 14, 381–389.

Hashimoto, H., 2002. Present status of industrial application of cyclodextrins in Japan. J. Incl. Phenom. Macrocycl. Chem. 44, 57–62.

He, Y., Fu, P., Shen, X., 2008. Cyclodextrin-based aggregates and characterization by microscopy. Micron 29, 495–516.

Horikoshi, K., 1979. Production and industrial applications of beta cyclodextrin. Process Biochem. 14, 26–30.

Imran, M., Revol-Junelles, A.M., Paris, C., Guedon, E., Linder, M., Desobry, S., 2015. Liposomal nanodelivery systems using soy and marine lecithin to encapsulate food biopreservative nisin. LWT Food Sci. Technol. 62, 341–349.

Isailović, B., Kostić, I., Zvonar, A., Đorđević, V., Gašperlin, M., Nedović, V., et al., 2013. Resveratrol loaded liposomes produced by different techniques. Innov. Food Sci. Emerg. Technol. 19, 181–189.

Izadyari, A., Akbarzadeh, A., Vaziri, A., Attar, A., Alavi, A., 2014. Preparation of liposomal ferrous sulfate nanocapsules by reverse-phase evaporation method and nanocapsules structure analysis to apply in the food and medicinal industries. Bull. Appl. Res. Sci. 4, 140–145.

Jafari, S.M., He, Y., Bhandari, B., 2007. Encapsulation of nanopartricles of D-limonene by spray drying: role of emulsifiers and emulsifying agent. Dry. Technol. 25, 1079–1089.

Jafari, S.M., Assadpoor, E., Bhandari, B., He, Y., 2008a. Nano-particle encapsulation of fish oil by spray drying. Food Res. Int. 41, 172–183.

Jafari, S.M., Assadpoor, E., He, Y., Bhandari, B., 2008b. Encapsulation efficiency of food flavours and oils during spray drying. Dry. Technol. 26, 816–835.

Janiszewska, E., Jedlinska, A., Witrowa-Rajchert, D., 2015. Effect of homogenization parameters on selected physical properties of lemon aroma powder. Food Bioprod. Process. 94, 405–413.

Jinap, S., Hajeb, P., 2010. Glutamate: its applications in food and contribution to health. Appetite 55, 1–10.

Jincheng, W., Xiaoyu, Z., Siahao, C., 2010. Preparation and properties of nanoencapsulated capsaicin by complex coacervation method. Chem. Eng. Commun. 197, 919–933.

Jouquand, C., Ducruet, V., Giampaoli, P., 2004. Partition coefficients of aroma compounds in polysaccharide solutions by the phase ratio variation method. Food Chem. 85, 467–474.

Jun-xia, X., Hai-yan, Z., Jian, Y., 2011. Microencapsulation of sweet orange oil by complex coacervation with soybean protein isolate/gum arabic. Food Chem. 125, 1267–1272.

Kaasgaard, T., Keller, D., 2010. Chitosan coating improves retention and redispersibility of freeze-dried flavour oil emulsions. J. Agric. Food Chem. 58, 2446–2454.

Kamimura, J.A., Santos, E.H., Hill, L.E., Gomes, C.L., 2014. Antimicrobial and antioxidant activities of carvacrol microencapsulated in hydroxypropyl-beta-cyclodextrin. LWT Food Sci. Technol. 57, 701–709.

Kanakdande, D., Bhosale, R., Singhal, R.S., 2007. Stability of cumin oleoresin microencapsulated in different combination of gum arabic, maltodextrin and modified starch. Carbohydr. Polym. 67, 536–541.

Karaiskou, S., Blekas, G., Paraskevopoulou, A., 2008. Aroma release from gum arabic or egg yolk/xanthan-stabilized oil-in-water emulsions. Food Res. Int. 41, 637–645.

Karathanos, V.T., Mourtzinos, I., Yannakopoulou, K., Andrikopoulos, N.K., 2007. Study of the solubility, antioxidant activity, and structure of inclusion complex of vanillin with β-cyclodextrin. Food Chem. 101, 652–658.

Kayaci, F., Uyar, T., 2012. Encapsulation of vanillin/cyclodextrin inclusion complex in electrospun polyvinyl alcohol (PVA) nanowebs: prolonged shelf life and high temperature stability of vanillin. Food Chem. 133, 641–649.

Kfoury, M., Landy, D., Auezova, L., Greige-Gerges, H., Fourmentin, S., 2014. Effect of cyclodextrin complexation on phenylpropanoids' solubility and antioxidant activity. Beilstein J. Org. Chem. 10, 2322–2331.

Kfoury, M., Auezova, L., Greige-Gerges, H., Fourmentin, S., 2015. Promising applications of cyclodextrins in food: improvement of essential oils retention, controlled release and antiradical activity. Carbohydr. Polym. 131, 264–272.

Khatibi, S.A., Ali, M., Mir-Hassan, M., Ghasem, A., Afshin, A.B., 2015. Effect of preparation methods on the properties of *Zataria multiflora* Boiss. essential oil loaded nanoliposomes: characterization of size, encapsulation efficiency, and stability. Pharm. Sci. 20, 141–148.

Klein, M., Aserin, A., Svitov, I., Garti, N., 2010. Enhanced stabilization of cloudy emulsions with gum arabic and whey protein isolate. Colloids Surf. B 77, 75–81.

Klinkesorn, U., Sophanodora, P., Chinachoti, P., McClements, D.J., 2004. Stability and rheology of corn oil-in-water emulsions containing maltodextrin. Food Res. Int. 37, 851–859.

Kono, K., Nakai, R., Morimoto, K., Takagishi, T., 1999. Thermosensitive polymer-modified liposomes that release contents around physiological temperature. Biochim. Biophys. Acta 1416, 239–250.

Krishnan, S., Bhosale, R., Singhal, R.S., 2005a. Microencapsulation of cardamom oleoresin: evaluation of blends of gum arabic, maltodextrin, and a modified starch as wall materials. Carbohydr. Polym. 61, 95–102.

Krishnan, S., Kshirsagar, A.C., Singhal, R.S., 2005b. The use of gum arabic and modified starch in the microencapsulation of a food flavouring agent. Carbohydr. Polym. 62, 309–315.

Labrousse, S., Roos, Y., Karel, M., 1992. Collapse and crystallization in amorphous matrices with encapsulated compounds. Sci. Aliments 12, 757–769.

Landy, P., Pollien, P., Pytz, A., Leser, M.E., Sagalowicz, L., Blank, I., et al., 2007. Model studies on the release of aroma compounds from structured and nonstructured oil systems using proton-transfer reaction mass spectrometry. J. Agric. Food Chem. 55, 1915–1922.

Le Thanh, M., Thibeaudeau, P., Thibaut, M.A., Voilley, A., 1992. Interactions between volatile and nonvolatile compounds in the presence of water. Food Chem. 43, 129–135.

Leclercq, S., Harlander, K.R., Reineccius, G.A., 2009. Formation and characterization of microcapsules by complex coacervation with liquid or solid aroma cores. Flavour Frag. J. 24, 17–24.

Lee, S., Hwang, S., Lee, K., Ahn, I.-S., 2006. Microscopic analysis of ester hydrolysis reaction catalyzed by Candida rugosa lipase. Colloid. Surf. B 47, 78–84.

Lengerich, B.V., Haynes, L.C., Levine, H., Otterburn, M.S., Mathewson, P., Finley, J., 1991. United States 4,999,208A. Extrusion baking of cookies having liposome encapsulated ingredients.

Lević, S., Obradović, N., Pavlović, V., Isailović, B., Kostić, I., Mitrić, M., et al., 2014. Thermal, morphological, and mechanical properties of ethyl vanillin immobilized in polyvinyl alcohol by electrospinning process. J. Therm. Anal. Calorim. 118, 661–668.

Lević, S., Lijaković, I.P., Đorđević, V., Rac, V., Rakić, V., Knudsen, T.Š., et al., 2015. Characterization of sodium alginate/D-limonene emulsions and respective calcium alginate/D-limonene beads produced by electrostatic extrusion. Food Hydrocoll. 45, 111–123.

Li, X., Anton, N., Arpagaus, C., Belleteix, F., Vandamme, F.T., 2010. Nanoparticles by spray drying using innovative new technology: the Büchi Nano Spray Dryer B-90. J. Control. Release 147, 304–310.

Li, W., Liu, X., Yang, Q., Zhang, N., Du, Y., Zhu, H., 2015. Preparation and characterization of inclusion complex of benzyl isothiocyanate extracted from papaya seed with β-cyclodextrin. Food Chem. 184, 99–104.

Lin, C.C., Lin, H.Y., Chi, M.H., Shen, C.M., Chen, H.W., Yangd, W.J., et al., 2014. Preparation of curcumin microemulsions with foodgrade soybean oil/lecithin and their cytotoxicity on the HepG2 cell line. Food Chem. 154, 282–290.

Liu, X.D., Furuta, T., Yoshii, H., Linko, P., 2000a. Retention of emulsified flavour in a single droplet during drying. Food Sci. Technol. Res. 6, 335–339.

Liu, X.D., Furuta, T., Yoshii, H., Linko, P., Coumans, W.J., 2000b. Cyclodextrin encapsulation to prevent the loss of l-menthol and its retention during drying. Biosci. Biotechnol. Biochem. 64, 1608–1613.
Liu, X.-D., Atarashi, T., Furuta, T., Yoshii, H., Aishima, S., Ohkawara, M., et al., 2001a. Microencapsulation of emulsified hydrophobic flavours by spray drying. Dry. Technol. 19, 1361–1374.
Liu, X.D., Atarashi, T., Furuta, T., Yoshii, H., Aishima, S., Ohkawara, M., et al., 2001b. Microencapsulation of emulsified hydrophobic flavours by spray drying. Dry. Technol. 19, 1361–1374.
Liu, Z., Li, Y., Cui, F., Ping, L., Song, J., Ravee, Y., et al., 2008. Production of octenyl succinic anhydride-modified waxy corn starch and its characterization. J. Agric. Food. Chem. 56, 11499–11506.
Liu, C.H., Zhou, H.J., Liu, J.Y., Li, X.T., Fang, H., Yang, Z.H., 2013a. Preparation of antibacterial citronella oil microcapsules and their application in cotton fabrics. Adv. Mater. Res. 627, 271–274.
Liu, B., Zhu, X., Zeng, J., Zhao, J., 2013b. Preparation and physicochemical characterization of the supramolecular inclusion complex of naringin dihydrochalcone and hydroxypropyl-β-cyclodextrin. Food Res. Int. 54, 691–696.
Lv, Y., Zhang, X., Abbas, S., Karangwa, E., 2012. Simplified optimization for microcapsule preparation by complex coacervation based on the correlation between coacervates and the corresponding microcapsule. J. Food Eng. 111, 225–233.
Lv, Y., Yang, F., Li, X., Zhang, X., Abbas, S., 2014. Formation of heat-resistant nanocapsules of jasmine essential oil via gelatin/gum arabic based complex coacervation. Food Hydrocoll. 35, 305–314.
Madadlou, A., Jaberipour, S., Eskandari, M.H., 2014. Nanoparticulation of enzymatically cross-linked whey proteins to encapsulate caffeine via microemulsification/heat gelation procedure. LWT Food Sci. Technol. 57, 725–730.
Madene, A., Jacquot, M., Scher, J., Desobry, S., 2006. Flavour encapsulation and controlled releasea review. Int. J. Food Sci. Technol. 41, 1–21.
Maji, T.K., Hussain, M.R., 2009. Microencapsulation of *Zanthoxylum limonella* oil (ZLO) in genipin crosslinked chitosan–gelatin complex for mosquito repellent application. J. Appl. Polym. Sci. 111, 779–785.
Mantegna, S., Binello, A., Boffa, L., Giorgis, M., Cena, C., Cravotto, G., 2012. A one-pot ultrasound-assisted water extraction/cyclodextrin encapsulation of resveratrol from *Polygonum cuspidatum*. Food Chem. 130, 746–750.
Mao, L., O'Kennedy, B.T., Roos, Y.H., Hannon, J.A., Miao, S., 2012. Effect of monoglyceride self-assemble dstructure on emulsion properties and subsequent flavour release. Food Res. Int. 48, 233–240.
Marsanasco, M., Márquez, A.L., Wagner, J.R., del V. Alonso, S., Chiaramoni, N.S., 2011. Liposomes as vehicles for vitamins E and C: an alternative to fortify orange juice and offer vitamin C protection after heat treatment. Food Res. Int. 44, 3039–3046.
Martín del Valle, E.M., 2004. Cyclodextrins and their uses: a review. Process Biochem. 39, 1033–1046.
Mascheroni, E., Fuenmayor, C.A., Cosio, M.S., Di Silvestro, G., Piergiovanni, L., Mannino, S., Schiraldi, A., 2013. Encapsulation of volatiles in nanofibrous polysaccharide membranes for humidity-triggered release. Carbohydr. Polym. 98, 17–25.
Mason, T.G., Wilking, J.N., Meleson, K., Chang, C.B., Graves, S.M., 2006. Nanoemulsions: formation, structure, and physical properties. J. Phys. Condens. Matter 18, 635–666.

Maswal, M., Dar, A.A., 2014. Formulation challenges in encapsulation and delivery of citral for improved food quality. Food Hydrocoll. 37, 182–195.

McClements, D.J., 2005. Food Emulsions: Principles, Practice, and Techniques, second ed. CRC Press, Boca Raton, FL.

McClements, D.J., 2010. Emulsion design to improvet the delivery of functional lipophilic components. Annu. Rev. Food Sci. Technol. 1, 241–269.

McNamee, B.F., O'Riordan, E.D., O'Sullivan, M., 1998. Emulsification and microencapsulation properties of gum arabic. J. Agric. Food Chem. 46, 4551–4555.

McNamee, B.F., O'Riordan, E.D., O'Sullivan, M., 2001. Effect of partial replacement of gum arabic with carbohydrates on its microencapsulation properties. J. Agric. Food Chem. 49, 3385–3388.

Mohanty, B., Bohidar, H.B., 2005. Microscopic structure of gelatin coacervates. Int. J. Biol. Macromol. 36, 39–46.

Moreau, L., Kim, H.J., Decker, E.A., McClements, D.J., 2003. Production and characterization of oil-in-water emulsions containing droplets stabilized by beta-lactoglobulin-pectin membranes. J. Agric. Food. Chem. 51, 6612–6617.

Mozafari, M.R., Flanagan, J., Matia-Merino, L., Awati, A., Omri, A., Suntres, Z.E., et al., 2006. Recent trends in the lipid-based nanoencapsulation of antioxidants and their role in foods. J. Sci. Food Agric. 86, 2038–2045.

Mozafari, M.R., Johnson, C., Hatziantoniou, S., Demetzos, C., 2008. Nanoliposomes and their applications in food nanotechnology. J. Liposome Res. 18, 309–327.

Mun, S., Decker, E.A., McClements, D.J., 2005. Influence of droplet characteristics on the formation of oil-in-water emulsions stabilized by surfactant chitosan layers. Langmuir 21, 6228–6234.

Munoz-Botella, S., del Castillo, B., Martín, M.A., 1995. Cyclodetrin properties and applications of inclusion complex formation. Ars. Pharm. 36, 187–198.

Nakamura, N., Horikoshi, K., 1977. Production of Schardinger β-dextrin by soluble and immobilized cyclodextrin glycosyltransferase of an alkalophilic *Bacillus* sp. Biotechnol. Bioeng. 19, 87–99.

Nilsson, L., Bergenstahl, B., 2006. Adsorption of hydrophobically modified starch at oil/water interfaces during emulsification. Langmuir 22, 8770–8776.

Ogawa, S., Decker, E.A., McClements, D.J., 2004. Production and characterization of O/W emulsions containing droplets stabilized by lecithin-chitosan-pectin mutilayered membranes. J. Agric. Food. Chem. 52, 3595–3600.

Ortan, A., Campeanu, G., Dinu-Pirvu, C., Popescu, L., 2009. Studies concerning the entrapment of *Anethum graveolens* essential oil in liposome. Rom. Biotechnol. Lett. 14, 4411–4417.

Pagington, J.S., 1986. Beta-cyclodextrin and its uses in the flavour industry. In: Birch, G.G., Lindley, M.G. (Eds.), Developments in Food Flavours. Elsevier Applied Science, London, pp. 131–150.

Patil, Y.P., Jadhav, S., 2014. Novel methods for liposome preparation. Chem. Phys. Lipids 177, 8–18.

Paudel, A., Worku, A.Z., Meeus, J., Guns, S., Van den Mooter, G., 2013. Manufacturing of solid dispersions of poorly water-soluble drugs by spray drying: formulation and process considerations. Int. J. Pharm. 453, 253–284.

Penbunditkul, P., Yoshii, H., Ruktanonchai, U., Charinpanitkul, T., Soottitantawat, A., 2011. Effect of feed liquid viscosity on flavour retention of bergamot oil encapsulated in spray-dried modified starch powder. International Congress on Engineering and Food, Athens, Greece, pp. 1–6.

Penbunditkul, P., Yoshii, H., Ruktanonchai, U., Charinpanitkul, T., Assabumrungrat, S., Soottitantawat, A., 2012. The loss of OSA-modified starch emulsifier property during the high-pressure homogeniser and encapsulation of multi-flavour bergamot oil by spray drying. Int. J. Food Sci. Technol. 47, 2325–2333.

Phan, V.A., Liao, Y.C., Antille, N., Sagalowicz, L., 2008. Delayed volatile compound release properties of self-assembly structures in emulsions. J. Agric. Food Chem. 56, 1072–1077.
Philippe, E., Seuvre, A.M., Colas, B., Langendorff, V., Shippa, C., Voilley, A., 2003. Behavior of flavour compounds in model food systems: a thermodynamic study. J. Agric. Food Chem. 51, 1393–1398.
Pillai, D.S., Prabhasankar, P., Jena, B.S., Anandharamakrishnan, C., 2012. Microencapsulation of Garcinia cowa fruit extract and effect of its use on pasta process and quality. Int. J. Food Prop. 15, 590–604.
Pourbafrani, M., Talebnia, F., Niklasson, C., Taherzadeh, M.J., 2007. Protective effect of encapsulation in fermentation of limonene-contained media and orange peel hydrolyzate. Int. J. Mol. Sci. 8, 777–787.
Prata, A.S., Menut, C., Leydet, A., Trigo, J.R., Grosso, C.R.F., 2008a. Encapsulation and release of a fluorescent probe, khusimyl dansylate, obtained from vetiver oil by complex coacervation. Flavour Frag. J. 23, 7–15.
Prata, A.S., Zanin, M.H.A., Re, M.I., Grosso, C.R., 2008b. Release properties of chemical and enzymatic crosslinked gelatin–gum arabic microparticles containing a fluorescent probe plus vetiver essential oil. Colloids Surf. B 67, 171–178.
Qi, Z.H., Hedges, A.R., 1995. Use of cyclodextrins for flavours. In: Ho, C.T., Tan, C.T., Tong, C.H. (Eds.), Flavour Technology: Physical Chemistry, Modification and Process, ACS Symposium Series, vol. 610. American Chemical Society, Washington, DC, pp. 231–243.
Quispe-Condori, S., Saldaña, M.D.A., Temelli, F., 2011. Microencapsulation of flax oil with zein using spray and freeze drying. LWT Food Sci. Technol. 44, 1880–1887.
Ramakrishna, S., Fujihara, K., Teo, W.E., Yong, T., Ma, Z., Ramaseshan, R., 2006. Electrospun nanofibers: solving global issues. Mater. Today 9, 40–50.
Ramishvili, T., Yushchenko, V., Charkviani, M., 2007. Catalytic conversions of linalool and linalyl acetate over large-pore zeolites and mesoporous MCM-41. Moscow Univ. Chem. Bull. 62, 180–186.
Randall, R.C., Phillips, G.O., Williams, P.A., 1988. The role of the proteinaceous component on the emulsifying properties of gum arabic. Food Hydrocoll. 2, 131–140.
Rao, J., McClements, D.J., 2012. Food-grade microemulsions and nanoemulsions: role of oil phase composition on formation and stability. Food Hydrocoll. 29, 326–334.
Re, M.I., Liu, Y.J., 1996. Microencapsulation by spray drying: influence of wall systems on the retention of the volatile compounds. In: Proceedings of the 10th International Drying Symposium, Krakow, Poland, July 30–August 2, pp. 541–549.
Reineccius, G.A., 2001. Multiple-core encapsulation: the spray drying of food ingredients. In: Vilstrup, P. (Ed.), Microencapsulation of Food Ingredients. Leatherhead Food RA Publishing, London, pp. 151–185.
Reiner, S.J., Reineccius, G.A., Peppard, T.L., 2010. A comparison of the stability of beverage cloud emulsions formulated with different gum acacia- and starch-based emulsifiers. J. Food Sci. 75, 236–246.
Risch, S.J., Reineccius, G.A., 1988. Spray-dried orange oil—effect of emulsion size on flavour retention and shelf stability. ACS Symposium Series, vol. 370, pp. 67–77.
Roberts, D.D., Pollien, P., Watzke, B., 2003. Experimental and modeling studies showing the effect of lipid type and level on flavour release from milk-based liquid emulsions. J. Agric. Food. Chem. 51, 189–195.
Rocha, G.A., Fávaro-Trindade, C.S., Ferreira Grosso, C.R., 2012. Microencapsulation of lycopene by spray drying: characterization, stability, and application of microcapsules. Food Bioprod. Process. 90, 37–42.

Rohloff, J., 1999. Monoterpene composition of essential oil from peppermint (*Mentha × piperita* L.) with regard to leaf position using solid-phase microextraction and gas chromatography/mass spectrometry analysis. J. Agric. Food. Chem. 47, 3782–3786.

Rosenberg, M., Kopelman, I.J., Talmon, Y., 1990. Factors affecting retention in spray-drying microencapsulation of volatile materials. J. Agric. Food Chem. 38, 1288–1294.

Rukmini, A., Raharjo, S., Hastuti, P., Supriyadi, S., 2012. Formulation and stability of water-in-virgin coconut oil microemulsion using ternary food grade nonionic surfactants. Int. Food. Res. 9, 259–264.

Sagalowicz, L., Leser, M.E., 2010. Delivery for liquid food products. Curr. Opin. Colloid Interface Sci. 15, 61–72.

Sakakibara, S., Sugisawa, K., Matsui, F., Sengoku, K., 1985. Jpn. Pat., JP 851 248 075.

Samad, A., Sultana, Y., Aqil, M., 2007. Liposomal drug delivery systems: an update review. Curr. Drug Deliv. 4, 297–305.

Sankarikutty, B., Sreekumar, M.M., Narayanan, C.S., Mathew, A.G., 1988. Studies on microencapsulation of cardamom oil by spray drying technique. J. Food Sci. Technol. 25, 352–356.

Sarkar, S., Gupta, S., Variyar, P.S., Sharma, A., Singhal, R.S., 2012. Irradiation depolymerized guar gum as partial replacement of gum arabic for microencapsulation of mint oil. Carbohydr. Polym. 90, 1685–1694.

Shao, P., Zhang, J., Fang, Z., Sun, P., 2014. Complexing of chlorogenic acid with β-cyclodextrins: inclusion effects, antioxidative properties and potential application in grape juice. Food Hydrocoll. 41, 132–139.

Sherry, M., Charcosset, C., Fessi, H., Greige-Gerges, H., 2013. Essential oils encapsulated in liposomes: a review. J. Liposome Res. 232, 268–275.

Sheu, T.Y., Rosenberg, M., 1995. Microencapsulation by spray drying ethyl caprylate in whey protein and carbohydrate wall systems. J. Food Sci. 60, 90–103.

Siew, C.K., Williams, P.A., 2008. Role of protein and ferulic acid in the emulsification properties of sugar beet pectin. J. Agric. Food Chem. 56, 4164–4171.

Singh, M., Sharma, R., Banerjee, U.C., 2002. Biotechnological applications of cyclodextrins. Biotechnol. Adv. 20, 341–359.

Singh, H., Thompson, A., Liu, W., Corredig, M., 2012. Liposomes as food ingredients and nutraceutical delivery systems. In: Garti, N., McClements, D.J. (Eds.), Encapsulation Technologies and Delivery Systems for Food Ingredients and Nutraceuticals. Woodhead Publishing, Cambridge, UK, pp. 287–318.

Sinico, C., De Logu, A., Lai, F., Valenti, D., Manconi, M., Loy, G., et al., 2005. Liposomal incorporation of *Artemisia arborescens* L. essential oil and in vitro antiviral activity. Eur. J. Pharm. Biopharm. 1, 161–168.

Siow, L.F., Ong, C.S., 2013. Effect of pH on garlic oil encapsulation by complex coacervation. J. Food Process. Technol. 4, 1–9.

Soottitantawat, A., Yoshii, H., Furuta, T., Ohgawara, M., Linko, P., 2003. Microencapsulation by spray drying: influence of emulsion size on the retention of volatile compounds. J. Food Sci. 68, 2256–2262.

Soottitantawat, A., Yoshii, H., Furuta, T., Ohgawara, M., Forssell, P., Partanen, R., et al., 2004. Effect of water activity on the release characteristics and oxidative stability of D-limonene encapsulated by spray drying. J. Agric. Food Chem. 52, 1269–1276.

Soottitantawat, A., Bigeard, F., Yoshii, H., Furuta, T., Ohkawara, M., Linko, P., 2005a. Influence of emulsion and powder size on the stability of encapsulated D-limonene by spray drying. Innov. Food Sci. Emerg. Technol. 6, 107–114.

Soottitantawat, A., Takayama, K., Okamura, K., Muranaka, D., Yoshii, H., Furuta, T., et al., 2005b. Microencapsulation of L-menthol by spray drying and its release characteristics. Innov. Food Sci. Emerg. Technol. 6, 163–170.

Soper, J.C., Kim, Y.D., Tomas, M.D., 2000. United States Patent 6,045,835. Method of encapsulating flavours and fragrances by controlled water transport into microparticles.

Sosa, N., Schebor, C., Pérez, O.E., 2014a. Encapsulation of citral in formulations containing sucrose or trehalose: emulsions properties and stability. Food Bioprod. Process. 92, 266–274.

Sosa, N., Zamora, M.C., van Baren, C., Schebor, C., 2014b. New insights in the use of trehalose and modified starches for the encapsulation of orange essential oil. Food Bioprocess Technol. 7, 1745–1755.

Souza, A.S., Borges, S.V., Magalhães, N.F., Ricardo, H.V., Cereda, M.P., Daiuto, E.R., 2009. Influence of spray drying conditions on the physical properties of dried pulp tomato. Food Sci. Technol. 29, 291–294.

Specos, M.M.M., Escobar, G., Marino, P., Puggia, C., Tesoriero, M.V.D., Hermida, L., 2010a. Aroma finishing of cotton fabrics by means of microencapsulation techniques. J. Ind. Text. 40, 13–32.

Specos, M.M.M., Garcıa, J.J., Tornesello, J., Marino, P., Vecchia, M.D., Tesoriero, M.V.D., et al., 2010b. Microencapsulated citronella oil for mosquito repellent finishing of cotton textiles. Trans. R. Soc. Trop. Med. Hyg. 104, 653–658.

Street, C.A., Anderson, D.M.W., 1983. Refinement of structures previously proposed for gum arabic and other acacia gum exudates. Talanta 30, 878–893.

Suzuki, J., 1975. Japan Kokai, JP 7 569 100.

Szejtli, J., Szente, L., 2005. Elimination of bitter, disgusting tastes of drugs and foods by cyclodextrins. Eur. J. Pharm. Biopharm. 61, 115–125.

Szente, L., Szejtli, J., 2004. Cyclodextrins as food ingredients. Trends Food. Sci. Technol. 15, 137–142.

Takeda Chem. Ind. Ltd., 1981. Japan Kokai, JP 81 127 058.

Tamamoto, L.C., Schmidt, S.J., Lee, S.Y., 2010. Sensory properties of ginseng solutions modified by masking agents. J. Food Sci. 75, 341–347.

Tamjidi, F., Shahedi, M., Varshosaz, J., Nasirpour, A., 2013. Nanostructured lipid carriers (NLC): a potential delivery system for bioactive food molecules. Innov. Food Sci. Emerg. Technol. 19, 29–43.

Taylor, S.L., Dormedy, E.S., 1998. The role of flavouring substances in food allergy and intolerance. Adv. Food Nutr. Res. 42, 1–44.

Thies, C., 2007. Microencapsulation of flavors by complex coacervation. In: Lakkis, J.M. (Ed.), Encapsulation and Controlled Release Technologies in Food Systems. Blackwell Publishing, Ames, IA, pp. 149–170.

Tonon, R.V., Grosso, C.R.F., Hubinger, M.D., 2011. Influence of emulsion composition and inlet air temperature on the microencapsulation of flaxseed oil by spray drying. Food Res. Int. 44, 282–289.

Turchiuli, C., Jimenez Munguia, M.T., Hernandez Sanchez, M., Cortes Ferre, H., Dumoulin, E., 2014. Use of different supports for oil encapsulation in powder by spray drying. Powder Technol. 255, 103–108.

Uedo, N., Tatsuta, M., Iishi, H., Baba, M., Sakai, N., Yano, H., et al., 1999. Inhibition by D-limonene of gastric carcinogenesis induced by *N*-methyl-*N*'-nitro-*N*-nitrosoguanidine in Wistar rats. Cancer Lett. 137, 131–136.

US Federal Register, 2005. Alpha-cyclodextrin, beta-cyclodextrin, and gammacyclodextrin: exemption from the requirement of a tolerance. US Federal Register 70, 38780–38785.

Vaidya, S., Bhosale, R., Singhal, R.S., 2006. Microencapsulation of cinnamon oleoresin by spray drying using different wall materials. Dry. Technol. 24, 983–992.

Valenti, D., De Logu, A., Loy, G., Sinico, C., Bonsignore, L., Cottiglia, F., 2001. Liposome-incorporated *Santolina insularis* essential oil: preparation, characterization, and in vitro antiviral activity. J. Liposome Res. 11, 73–90.

van Ruth, S.M., King, C., Giannouli, P., 2002. Influence of lipid fraction, emulsifier fraction, and mean particle diameter of oil-in-water emulsions on the release of 20 aroma compounds. J. Agric. Food Chem. 50, 2365–2371.

Vemuri, S., Rhodes, C.S., 1995. Preparation and characterization of liposomes as therapeutic delivery systems: a review. Pharm. Acta Helv. 70, 95–111.

Vilanova, N., Solans, C., 2015. Vitamin A palmitate–β-cyclodextrin inclusion complexes: characterization, protection and emulsification properties. Food Chem. 175, 529–535.

Walton, D.E., 2000. The morphology of spray-dried particles a qualitative view. Dry. Technol. 18, 1943–1986.

Wang, J.C., Chen, S.H., Xu, Z.C., 2008a. Synthesis and properties research on the nanocapsulated capsaicin by simple coacervation method. J. Dispers. Sci. Technol. 29, 687–695.

Wang, X., Jiang, Y., Wang, Y.W., Huang, M.T., Hoa, C.T., Huang, Q., 2008b. Enhancing anti-inflammation activity of curcumin through O/W nanoemulsions. Food Chem. 108, 419–424.

Weiss, J., Takhistov, P., McClements, D.J., 2006. Functional materials in food nanotechnology. J. Food Sci. 71, 107–116.

Wen, P., Zhu, D.H., Wu, H., Zong, M.H., Jing, Y.R., Han, S.Y., 2016. Encapsulation of cinnamon essential oil in electrospun nanofibrous film for active food packaging. Food Control 59, 366–376.

Whorton, C., 1995. Factors influencing volatile release from encapsulation matrices. In: Risch, S.J., Reineccius, G.A. (Eds.), Encapsulation and Controlled Release of Food Ingredients. ACS Symposium Series, vol. 590. American Chemical Society, Washington, DC, pp. 134–142.

Whorton, C., Reineccius, G.A., 1995. Evaluation of the mechanisms associated with the release of encapsulated flavour materials from maltodextrin matrices. In: Risch, S.J., Reineccius, G.A. (Eds.), Encapsulation and Controlled Release of Food Ingredients. ACS Symposium Series, vol. 590. American Chemical Society, Washington, DC, pp. 143–160.

Wissing, S.A., Muller, R.G., 2002. The influence of the crystallinity of lipid nanoparticles on their occlusive properties. Int. J. Pharm. 242, 377–379.

Wissing, S.A., Kayser, O., Muller, R.H., 2004. Solid lipid nanoparticles for parenteral drug delivery. Adv. Drug Deliv. Rev. 56, 1257–1272.

Wooster, T.J., Golding, M., Sanguansri, P., 2008. Impact of oil type on nanoemulsion formation and Ostwald ripening stability. Langmir 24, 12758–12765.

Xia, S., Xu, S., 2005. Ferrous sulfate liposomes: preparation, stability, and application in fluid milk. Food Res. Int. 38, 289–296.

Xiao, Z., Liu, W., Zhu, G., Zhou, R., Niu, Y., 2014. A review of the preparation and application of flavour and essential oils microcapsules based on complex coacervation technology. J. Sci. Food Agric. 94, 1482–1494.

Xiaoyan, S., Guoqing, H., Hui, R., Qihe, C., 2006. Preparation and properties of octenyl succinic anhydride modified early indica rice starch. Starch 58, 109–117.

Xing, F., Cheng, G., Yi, K., Ma, L., 2004. Nanoencapsulation of capsaicin by complex coacervation of gelatin, acacia, and tannins. J. Appl. Polym. Sci. 96, 2225–2229.

Xu, X., Khan, M.A., Burgess, D.J., 2012. Predicting hydrophilic drug encapsulation inside unilamellar liposomes. Int. J. Pharm. 423, 410–418.

Yang, Z., Peng, Z., Li, J., Li, S., Kong, L., Li, P., Wang, Q., 2014. Development and evaluation of novel flavour microcapsules containing vanilla oil using complex coacervation approach. Food Chem. 145, 272–277.

Yemiş, G.P., Pagotto, F., Bach, S., Delaquis, P., 2011. Effect of vanillin, ethyl vanillin, and vanillic acid on the growth and heat resistance of *Cronobacter* species. J. Food Prot. 12, 2000–2228.

Yoshida, P.A., Yokota, D., Foglio, M.A., Rodrigues, R.A.F., Pinho, S.C., 2010. Liposomes incorporating essential oil of Brazilian cherry (*Eugenia uniflora* L.): characterization of aqueous dispersions and lyophilized formulations. J. Microencapsul. 27, 416–425.

Yoshii, H., Soottitantawat, A., Liu, X.D., Atarashi, T., Furuta, T., Aishima, S., et al., 2001. Flavour release from spray-dried maltodextringum arabic or soy matrices as a function of storage relative humidity. Innov. Food Sci. Emerg. Technol. 2, 55–61.

Young, O.A., Gupta, R.B., Sadooghy-Saraby, S., 2012. Effects of cyclodextrins on the flavor of goat milk and its yogurt. J. Food Sci. 77, 122–127.

Yuliani, S., Torley, P.J., D'Arcy, B., Nicholson, T., Bhandari, B., 2006. Extrusion of mixtures of starch and D-limonene encapsulated with β-cyclodextrin: flavour retention and physical properties. Food Res. Int. 39, 318–331.

Yusoff, A., Murray, B.S., 2011. Modified starch granules as particle-stabilizers of oil-in-water emulsions. Food Hydrocoll. 25, 42–55.

Zahi, M.R., Wan, P., Liang, H., Yuan, Q., 2014. Formation and stability of D-limonene organogel-based nanoemulsion prepared by a high-pressure homogenizer. J. Agric. Food Chem. 62, 12563–12569.

Zhang, J., 2011. Novel emulsion-based delivery systems. Dissertation submitted to the faculty of the graduate school of the University of Minnesota.

Zhang, Z.Q., Pan, C.H., Chung, D., 2011. Tannic acid cross-linked gelatin–gum arabic coacervate microspheres for sustained release of allyl isothiocyanate: characterization and in vitro release study. Food Res. Int. 44, 1000–1007.

Zhu, W., Lockwood, G.B., 2000. Enhanced biotransformation of terpenes in plant cell suspensions using controlled release polymer. Biotechnol. Lett. 22, 659–662.

Zhu, G., Xiao, Z., Zhou, R., Zhu, Y., 2014. Study of production and pyrolysis characteristics of sweet orange flavor-β-cyclodextrin inclusion complex. Carbohydr. Polym. 105, 75–80.

Zuidam, J.N., Heinrich, E., 2010. Encapsulation of aroma. In: Zuidam, N.J., Nedovic, V.A. (Eds.), Encapsulation Technologies for Food Active Ingredients and Food Processing. Springer, Dordrecht, pp. 127–160.

Zuidam, N.J., Shimoni, E., 2010. Overview of microencapsulation use in food products or processes and methods to make them. In: Zuidam, N.J., Nedovic, V.A. (Eds.), Encapsulation Technique for Active Food Ingredients and Food Processing. Springer, New York, NY, pp. 3–29.

10

NANOCOMPOSITE FOR FOOD ENCAPSULATION PACKAGING

Daibing Luo*,, Liangzhuan Wu*, Jinfang Zhi***

**Key Laboratory of Photochemical Conversion and Optoelectronic Materials, Technical Institute of Physics and Chemistry, Chinese Academy of Sciences, Beijing, P.R. China; **Sichuan University, Analytical & Testing Center, Chengdu, P.R. China*

1 Introduction

1.1 Food Nanotechnology

Health-promoting components can enhance the nutritional function of foods and contribute health benefits to the human body, reduce risk of chronic disease, boost energy, and enhance body building functions. However, many health-promoting components suffer from bioactive instability since they are easily degraded in the formulation, processing, and storage processes because most of them are sensitive to heat, pH, oxygen, ion, and light. Hence various kinds of technologies for food packages have been developed to avoid the decomposition of food (Fig. 10.1) (Pawlik and Norton, 2014; Aguilera and Lillford, 2008).

Encapsulating foods in packaging materials is necessary for transporting, protecting, labeling, and advertising. In recent years nanotechnology has found innumerable applications in food industry (Fathi et al., 2012; Duncan, 2011). Food encapsulation requires protection, tampering resistance, and special physical, chemical, or biological needs. The encapsulation packaging is significant in preserving the foods to make them safe and marketable. Innovations in food encapsulation packaging can lead to quality packaging and show consumers a friendly approach in determining the shelf life, biodegradable period, and other information (Abbas et al., 2009).

Until now nanotechnology and nanomaterials have already played important roles in the packaging industry. The history of food nanotechnology can be traced back to the pasteurization process introduced by Pasteur to kill spoilage bacteria, which

Encapsulations. http://dx.doi.org/10.1016/B978-0-12-804307-3.00010-7

Figure 10.1. Food encapsulation.

formed the first step of the revolution in food processing and improvement in quality of foods. Nanotechnology and nanomaterials can be used to create or improve a number of packaging characteristics that are advantageous for the packaged foods. Nanotechnology has been employed for preparation of stronger and lighter materials, improving biodegradability or recyclability, incorporating sensors or indicators for consumer information, or for traceability or authentication (product security to avoid fraud) (Chellaram et al., 2014; Chen et al., 2014). Nanomaterials are increasingly being used in the food packing industry due to the range of advanced functional properties to conventional materials. Materials in the nanoscale range may present different electronic properties, which in turn exert influence on their mechanical, catalytic, optical, and other reactive properties. Nanomaterials can be employed to reinforce biodegradable packaging, produce fewer amounts of wasted materials for recycling, and meanwhile reduce the permeability to gases. Functional nanomaterials can prolong shelf life, decrease the demand of preservative materials, and provide hygienic surfaces that are easy to clean and can inhibit microbe accumulation or formation. Antimicrobial packaging is the most common use of nanomaterials. As a simple passive barrier, antimicrobial packaging can reduce the growth of harmful microbes (Joseph and Morrison, 2006).

Incorporation of nanomaterials into capsulation packaging materials will yield lightweight, durable, and low gas-permeable nanocomposites contributing to food quality by extending shelf life, preserving flavor and aroma, and reducing contact with microorganisms.

1.2 Nanocomposite

Research on the use of nanocomposites for food encapsulation packaging started in the 1990s with the incorporation of montmorillonite clay to polyethylene, nylon, polyvinvyl chloride, and starch polymers. Nanocomposite has a multiphase where one of the phases has at least one phase of less than 100 nm, or structures with repeating distances in nanoscale between the different phases in the material (Kamigaito, 1991). In mechanical properties, nanocomposites show extinct difference from conventional composite materials because of the high ratio of surface to volume from the reinforcing phase and the high aspect ratio. The nanocomposite materials can be prepared from nanosized particles, sheets, rods, tubes, or fibers. When only one dimension is in the nanometer range, the composites are known as polymer-layered crystal nanocomposites (Alexandre and Dubois, 2000; Azeredo, 2009). Nanocomposites can be used for encapsulation packaging by interacting with foods, thus giving pathways for releasing beneficial materials such as antioxidant and antimicrobial agents, or depressing some unfavorable compounds generation such as water vapor and oxygen. Polymeric nanocomposites can be used in design and production of packaging and coating materials, microelectromechanical systems, sensors, thermal control materials, and so forth (Sozer and Kokini, 2012).

Different types of nanofillers and polymers have been combined to prepare nanocomposite materials, which mainly include clay, natural biopolymers, natural antimicrobial agents, metals, and metal oxides, and so forth (Lagaron, 2011). An important one among the numerous nanofillers is clay, because it is generated from the earth's crust, cost effective, and provides reinforcement effect and easy processability. Usually, addition of nanofillers of about 5% is enough to result in an improvement in the substrate polymer properties. The enhanced mechanical properties of nanocomposite materials may benefit from the nanofillers with high rigidity besides excellent affinity between polymers and nanofillers at the interface. Besides the effects of the reinforcements from nanofillers, an interphase composition of altered mobility around each nanostructure is induced by well-distributed nanofillers, thus forming a percolating interphase network structure in the composite, which plays an important role in improving the structural strength of nanocomposites (Arora and Padua, 2010). The formation of nanocomposites facilitates improving thermal, mechanical, and barrier properties of polymers, and has proven to be a promising strategy. Further research and investigation may be inspired by the various nanocomposites naturally formed in environments.

1.3 Functionality and Advantages

1. Improved packaging strength
   Introduction of nanomaterials into polymeric matrix can improve the gas barrier properties, as well as temperature and humidity resistance of the packaging.

2. Antimicrobial ability
   Use of antibacterial nanomaterials to interact directly with the food product or the environment provides better protection of the products. For example, silver nanoparticles and silver coatings can provide antimicrobial properties, with other materials being used as oxygen or UV scavengers.
3. Smart packaging: nanosensing
   Technology in terms of smart packaging has been explored for the possibility of preserving food for as long as possible. This novel concept is designed for biochemical sensing of microbial changes in the food, for example, detecting of specific pathogens developing in the food, or specific gases generated from food spoiling. Some smart packaging has been developed to be used as a monitoring device not only for food safety but also to avoid counterfeit. Food packaging containing smart materials in nanosized and flexible nanoelectronics could actively control or adjust the environmental parameters inside the packaging, and inform consumers when the packaged food has begun to deteriorate. In addition, dirt-repellent or self-cleaning coatings for food packages are also being developed.
4. Self-cooling packaging
   Self-cooling packaging, which makes use of a chemical or physical process, such as evaporation of gases, to keep the temperature inside the packaging cool, thus keeping the food fresh, has been developed by the assistance of nanotechnology. Furthermore, the microsized powered systems could make use of a flexible or thin-film photovoltaic cell for food cooling by using thermoelectric materials. This technology will reduce the need for large-scale and long-time refrigeration in the supply chain, although it may generate a higher cost in this case.
5. Warning tags
   Another interesting developmental direction is the use of nanoparticles to create color-changing packaging which will respond when food is going off. Chemical or physical methods can be used to achieve this effect. The chemical pathway is based on chemical indicators which change colors when there are gases generated by food oxidization. The physical pathway takes advantage of nanoparticles embedded in the polymer layers that can change their optical properties upon their relative position in the nanocomposite structure. These changes can be used to design warning tags in which an intense color is produced when the packaging stretches, creating obvious visional signals of gas-releasing decomposition.

6. Green packaging
   Nanocomposite materials which are antimicrobial and biodegradable are welcome and expected to be used in food packaging in the near future.

2 Nanocomposite for Food Encapsulation Packaging

2.1 Nanocomposite Thin Films for High Barrier and Flame Protection

Nanocomposites as barrier materials have been widely applied in food encapsulation packaging and protective coatings. Multilayered composite polymer films have been developed and studied for improving barrier properties. As shown in Fig. 10.2, by multilayering and annealing polymer composite films, improvements in barrier and fire properties, exceeding those of contemporary nanocomposites, can be achieved (http://www.usm.edu/nrg/research.html). The introduction of crystalline nanofillers is helpful to increase the tortuosity thereby slower diffusion processes and lower the permeability (Sanchez-Garcia et al., 2008) of gases and humidity. The barrier properties are desirable if the nanofillers are less permeable, and are well dispersed in the matrix with a high aspect ratio (Lagaron et al., 2004).

The enhanced barrier properties of most polymer-based nanocomposites take advantage of the improved tortuosity of the diffusion or permeation path for gases. These wall-like nanocomposites force gases to travel longer paths to diffuse through the coatings. The presence of nanoparticles with high aspect ratios

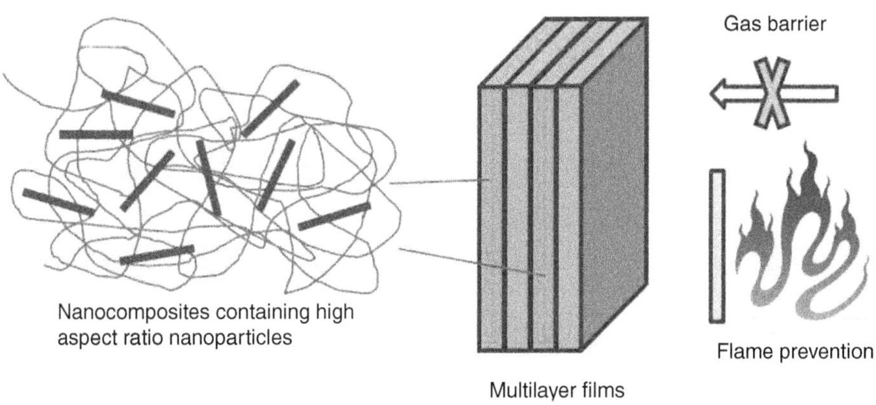

Figure 10.2. Nanocomposite for flame protection and gas barrier.

in the packaging dramatically decreases the transfer rate of gases such as oxygen, carbon dioxide, and water vapor crossing the packages (Farhoodi, 2015). The nanoparticles inside polymeric nanocomposites could also bring lots of active zones with better reinforcing effects. Furthermore, variety or change in the size and number of nanoparticles per unit volume of polymers will result a significant impact on the properties of the polymers (Dalmas et al., 2007; Jordan et al., 2005).

Some foods have to be sealed in a protective, oxygen-free environment. Many normal packaging coatings made from flexible plastics are somehow slightly permeable to oxygen and other gases. Over time, this shortage allows the protective atmosphere slowly leak out, meanwhile oxygen and water vapor can leak in, thus resulting damage to the packaged foods. A coating made of metal or glass is totally impermeable to gases and would prevent leakage. However this kind of encapsulation is obviously impractical since it would reduce flexibility of the coating, and would be more expensive than single plastic packaging. An encapsulation coating a few nanometers thick is enough to create an impermeable layer, even without losing of flexibility or increasing much to the cost. Layer-by-layer (LBL) assembly (Rudra et al., 2006) is an effective strategy by which a multilayered films of nanosized layers can be prepared by sequential reaction or adsorption of polymers and metals on a solid support. For a mechanical durability of the metal nanoparticles deposited onto the polymer substrate, protective layers are deposited onto the nanocomposite layer in the present study. Recently RUSNANO Group has produced a flexible packaging based on a nanocomposite for food encapsulation which can replace some traditional materials (Fig. 10.3) (http://en.rusnano.com/press-centre/news/88595).

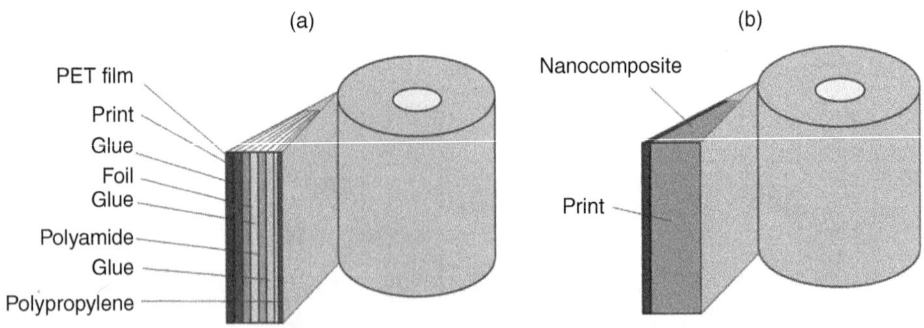

Figure 10.3. Schematic of a nanostructured, multilayer barrier film for use in food packaging. The polyamide and metal nanofilms provide high impermeability to gases and moisture, preserving the encapsulated fresh foods for a longer time, while retaining the flexibility of the multilayer films (a). As an alternative product, a new packaging material based on a nanocomposite extends shelf life, reduces the amount of preservatives in foodstuff, while simplifying the composition of packaging materials (b).

The most common methods to improve barrier efficiency are the use of polymer blends and coating articles with high barrier properties, and the employment of multilayered films with high barrier layers. Between the surfaces of these layers, the imbalance caused by the negative surface charge may be balanced by certain exchangeable cations (eg, Na^+ and Ca^{2+}). The parallel layers in the packaging materials are well linked together by weak electrostatic forces between each other (Tan et al., 2008). For constant nanofiller content, particles in smaller size increases their numbers, resulting them closer to each other; thus the interface layers from adjacent particles overlap to a certain degree, significantly altering the macroscopic properties (Qiao and Brinson, 2009). Owing to the high barrier effect to gases and humidity, bionanocomposite materials have found applications in food packaging because they could enhance shelf life of food products.

In addition, some reactive nanofillers incorporated in polymers can act as sensors. This kind of nanocomposite materials could respond to the environmental changes such as humidity, temperature, and oxygen, which bring contamination and degradation to food products (Azeredo, 2009; Bouwmeester et al., 2009). Moreover, if there are sealing defects in food packaging sometimes not found, the nanosensors will indicate the undesirable changes in the food and to alert consumers if the food has gone bad. Carbon nanotube (CNT) is the most common nanofiller that exhibits sensing properties. Recently Georgia Technology Institution has developed a biosensor based on multiwall carbon nanotube (MWNT) and polyamide to monitor toxic proteins, microorganisms, and potential spoilage of foods and drinking (Han et al., 2011).

Some kinds of nanofillers show a scavenging ability in reaction to oxygen. Introduction of these nanofillers into food encapsulation packaging can facilitate reducing and maintaining oxygen level, which is helpful for food preservation since existence of high concentration of oxygen may result in deterioration of many foods (Azeredo, 2009). Packaging materials with oxygen scavenger properties for foods can decrease oxidation to foods and favorable in keeping food freshness. For example, TiO_2 is one of the functional nanofillers with good oxygen scavenger properties upon exposure to UV light radiation (Xiao et al., 2004).

While using flavor or odor absorbent materials in packaging nanocomposites, it is important to consider the off-odors and off-flavors produced by hazardous microorganisms, since in many cases they are used as early warnings of deterioration by the consumers.

2.2 Nanofiller Materials or Nanostructures

The large contact area of nanofillers affords greater reinforcement to the relatively weaker polymer matrix to strain due to their high moduli. In general, low amounts (<10%) of nanofillers incorporation to polymeric systems were found to effectively improve both the mechanical and barrier properties.

The antibacterial mechanisms of the nanofillers are varied from cell wall–damaging abrasiveness, generation of metal ions to limit reactions of certain oxidative enzymes, interference with DNA/RNA replication or denaturation of proteins, and so forth (Chen et al., 2014; Percival et al., 2005). To enhance the antibacterial activities, a great number of efforts have been made in designing the ideal antibacterial nanostructures. The extremely high specific surface area and their effectiveness at very low loading levels make certain nanofillers excellent candidates for improvement of material properties of polymers (Sozer and Kokini, 2012). More importantly, the incorporation of nanofillers enhances the mechanical and physical properties, and reduces the weight and gas diffusivity of the coatings.

Representative antibacterial materials relate to the use of biocides or antibacterial metallic ions, such as Ag^+ ions, or killing by highly active reagents, such as hydrogen peroxide, hydroxyl radical, and superoxide species generated in the photocatalysis of TiO_2 (Zhang et al., 2013). Chemical compatibility between the nanofiller and the matrix plays a critical role in the nanofiller dispersion within the matrix and in the adhesion between both phases. Clays, carbon nanotubes, metals, silica, and zirconia are some of the commercially available nanofillers which can be used to tailor material properties of polymers when used in combination (Vaia and Maguire, 2007). Optimization of the properties, such as reduced permeability, enhanced optical clarity, and resistance to oxidation, could be possible by controlling the arrangement and distribution of the nanofillers (Vaia and Maguire, 2007). The alignment of nanofillers in the polymer depends on the type of the flow which occurs during the mechanical processing of the nanocomposites. The processing includes either extensional, shear, or mixed, which is similar to that in complex fluids (Sozer and Kokini, 2012). Nanofiller superstructures can be formed by self-assembly that is categorized in two main groups: static self-assembly and dynamic self-assembly.

Traditionally, only the materials that can induce the death or depress the growth of bacteria have been referred to as antibacterial materials. Metallic nanoparticles have earned interest in food packaging application due to unique properties such as high surface area, excellent catalytic, optical, and electrical properties. The most common metal used in nanocomposites is silver, due to the antimicrobial properties as well as being stable and displaying low volatility at high temperature (Heo et al., 2015). Silver nanoparticles or silver ions have been widely used in the food industry for their antibacterial activity. A typical example for the packaging industry is the use of nanosilver. Because of its excellent antimicrobial properties, nanosilver has been used in encapsulation packaging materials and inner surfaces of fridges and dishwashers, as well as being incorporated into plastic food containers. Silver nanoparticles show toxicity to a wide range of microorganisms and can kill both Gram-positive and Gram-negative bacteria, such as *Staphylococcus aureus* (*S. aureus*) and *Escherichia coli* (*E. coli*) (Kong and Jang, 2008; Vimala et al., 2010).

Silver nanoparticle-based nanocomposites present high transmittance, excellent antibacterial activity, no fingerprint marks, and a good mechanical endurance for touching and swiping by fingers, which make them ideal encapsulation packaging materials. Various strategies have been reported for producing silver nanoparticles such as chemical reduction, hydrothermal, photochemical, laser, and sol–gel methods (Poortavasoly et al., 2014). Interaction of silver nanoparticles with bacteria membrane and intracellular proteins interferes with cell division and causes cell death. In the presence of silver ions, bacteria DNA conglomeration defense mechanisms will protect bodies from the toxic environment, but this will weaken the replication ability of bacteria (Morones et al., 2005).

Besides silver as the oldest and most popular reagent, some other metals, such as molybdenum and copper, are now used as antibacterial species to cause the death of adhered bacteria by their metallic ions. These metal nanoparticles exhibit antimicrobial activity and have been incorporated into polymers as matrix fillers to prepare antimicrobial packaging materials (Cushen et al., 2014).

Silica nanoparticles have been reported to enhance the mechanical and barrier properties of several polymeric matrices. Addition of SiO_2 nanoparticles into polypropylene (PP) matrix improves the tensile properties including strength, modulus, and elongation of the material. The mechanical properties of the silica nanoparticles–doped nanocomposite are improved by the formation of a percolating network and strong filler–matrix interaction. It was found that SiO_2 can form a twisting path for gases when used as nanofillers in food packaging. The SiO_2 nanofillers can also improve the tensile property of nanocomposite films. The most used silicates in nanocomposite materials are montmorillonite, hectorite, and saponite. Silicates can be modified by surface engineering to improve wetting property with the polymer matrix (Kornmann et al., 2001a).

TiO_2 attracts increasing attention because it is chemical inert, antibacterial from its photoirradiation effect, corrosion resistant, high hardness, and high refractive index. Moreover, TiO_2 is inexpensive and can be produced in the food industry at a large scale. TiO_2 has three crystalline structures known as rutile, brookite, and anatase, respectively. TiO_2 nanoparticles have surface activity for reaction with most biological molecules, such as some unspecific binding with DNA molecules. However, the antibacterial ability of nano-TiO_2 is mainly dependent upon UV irradiation, which limits its application in food packaging (Naicker et al., 2005). Currently the research of TiO_2 mainly activated on visible-light photocatalysis is still under development. The surface energy of TiO_2 nanoparticles plays an important role in the interface interaction between polymers and nanofillers. It was reported that the surface energy of a TiO_2 nanoparticle increases and was close to a constant value as the particle size gets larger. Besides, TiO_2 nanoparticles can also be used as an indicator to detect oxygen-exposure levels. As oxygen concentration within the packaging increases, the TiO_2 nanoparticles within the nanocomposites will bleach due to photosensitive redox dyes generated in the polymer medium (Mills and Hazafy, 2009). DuPont Company has announced the production and commercialization of a new kind of nano-TiO_2 plastic additive, which could decrease UV damage to foods packaged in transparent encapsulation (ElAmin, 2007a).

ZnO is popularly used as a metal oxide due to its deodorizing and antibacterial properties. ZnO nanoparticles become more antibacterial as their particle size gets smaller. The addition of ZnO nanofiller has been found to improve the color adhesion, humidity, hydrophobicity, UV barrier, elongation, and thermal stability of the nanocomposite coatings, as well as decrease the tensile, strength, moisture permeability, and elastic modulus. When using bovine gelatin and sago starch together as the polymer substrates with ZnO nanofillers, researchers observed an increase in both mechanical and heal insulation properties and a notable decrease in oxygen permeability of the packaging films (Espitia et al., 2012).

Cellulose nanomaterials are now attracting attention in food encapsulation because of their low cost, lightweight, and high strength. Cellulose nanofibers are environmentally friendly and easy to recycle by combustion, and require low energy in manufacturing. Cellulose chains are synthesized from living organisms as microfibers, which are formed by bundles of elongated molecules linked and stabilized by hydrogen bonds. A nanocomposite based on chitosan film with 15% cellulose nanofibers and plasticizer constituents with 18% glycerol is comparable to most synthetic polymers with regard to stiffness and strength, but its percentage of elongation and moisture barrier are not good (Azeredo et al., 2010). It was found that cellulose nanoparticles are of great importance in improving the mechanical properties of natural polymers. Recent study shows cellulose nanofibers with ribbon-shaped structures can be used to reinforce polymers. The cellulose nanofiber-reinforced resins can improve the elastic modulus to a value of 28 GPa (Nakagaito et al., 2005). Cellulose nanofibers are often added to nanocomposite matrix to improve tensile properties, water vapor permeability, and glass transition temperature (Arora and Padua, 2010).

Natural clays have been introduced into polymeric substrates of packaging to improve characteristics such as strength, coating, and barrier properties. Nanocomposites made of polymer-clay have been developed for encapsulation packaging of foods since 1990. The intercalated nanocomposites are generally prepared based on the penetration of polymer phases into the intersecting layer region in the clay, which generates alternating polymer/inorganic layers at a repeated distance of a few nanometers in an ordered multilayer structure (Weiss et al., 2006). The nanocomposites in exfoliated structures involve extensive polymer penetration, containing the clay layered and randomly dispersed in the polymer substrate. Exfoliated nanocomposites show excellent properties due to the strong and optimal interaction between clay and polymers (Adame and Beall, 2009). Montmorillonite, with aluminum hydroxide between two silica tetrahedral layers, is the popularly used type of clay filler. The surface negative charges can be exchanged with other cations, so the parallel layers in clay are able to link together by weak electrostatic force. Nanoclays can be incorporated into plastic packaging of bottles for drinks, which prevent oxygen from migrating through the plastic bottle walls and destabilizing the drink and therefore extending the shelf life of food products. Besides, clay as a common nanoabsorber has the ability to lower the volatile concentration of some flavor and odor compounds that are products of a result of biochemical reactions such as fermentation or ripening of foods (Brody et al., 2008). Recently it has been found that nanoclays can be used with natural polymers to produce green nanocomposites that are nontoxic, biodegradable, and biocompatible.

Carbon nanotubes also bring antimicrobial/antibacterial effects to the encapsulation materials. Multiwalled carbon nanotubes which are less toxic to human and animal subjects inactivate *E. coli* cells by direct contact (Kang et al., 2008). Encapsulation packaging involving carbon nanotubes is currently being developed which has the ability to pump out oxygen or carbon dioxide in the packages to prevent food or beverage deterioration (foodqualitynews.com, 2005). In addition to their outstanding mechanical properties, carbon nanotubes are good electrical conductive material that can be used in smart packaging for nanosensing application. When dispersed in an insulating matrix, they allow the material to be conductive (Dalmas et al., 2007).

Recently researchers have focused on graphene to exploit a sp^2-hybrid carbon network. In particular, graphene is considered as an ideal two-dimensional reinforcing component for composite materials possessing superior carrier transport, high mechanical stiffness, extremely large surface area, and fine thermal/chemical stability (Liu et al., 2011). Graphene nanoplates are novel, highly promising carbon-based nanosized fillers (Arora and Padua, 2010). The new graphene-based hybrids with metal nanoparticles such as Pt, Au, and Ag have shown potential applications in the areas of optics, electronics, catalysis, and sensors.

2.3 Polymers

Polymers have found wide applications in the textile industry, packaging, filtration, automotive, biomedical, and many other fields. Polymers are always used in the antibacterial nanocomposites as excellent substrates. Various modification methods have been developed to improve their characteristics without altering its bulk properties. Recently an interesting method named aminolysis has been reported to modify the polymer surface and enhance the hydrophilicity and biocompatibility without affecting the bulk properties of the polymer (Poortavasoly et al., 2014; Dalmas et al., 2007). The effective role of aminolysis on polymers has advantages including a fabric with improved wettability, comfort, handling, and dyeability (Textor et al., 2010). Various kinds of nanoparticles can be anchored into various polymer substrates by a facile two-phase assembly method (Liu et al., 2011). Nanotechnology application to these polymers may provide possible methods of improving both the encapsulation properties and the cost–price–efficiency (Sorrentino et al., 2007).

Some cationic polymers show antibacterial activity. Encapsulation coatings with antimicrobial cationic polymers can greatly inhibit the growth of various bacteria and microbes without releasing toxic chemicals in low molecular weight into the environment. Antimicrobial polymers usually contain polycationic structures, such as substituted quaternary ammonium compounds, phosphonium salts, *N*-alkyl pyridinium salts, and rhodamine derivatives (Milović et al., 2005; Qiao et al., 2012; Lv et al., 2010; Liu et al., 2009; Hwang et al., 2007). Unlike common polymer films with antimicrobial properties, cationic polymeric nanoparticles based on self-assembled method can form secondary structures before interacting with the microbial membranes, and thus are expected to have better antimicrobial properties (Nederberg et al., 2011).

Polymers with self-healing properties have been developed for some years, and until now some products have been commercialized in the market (Hager et al., 2010). These materials used as outer encapsulation layers in food packaging allow small damages (eg, punctures and tears) in the wrapping, thus reducing the cost caused by wastage due to damaged packaging. Nowadays, antifingerprint polymers have been developed by some enterprises (Aguilar De Armas and Román, 2014). For an antifingerprint of the packaging materials using flexible substrate, fluoride films can be used to deposit on flexible substrates (Heo et al., 2015). Currently in the encapsulation field many reports on the formation and properties of biopolymer films are focused on their application as edible films.

Addition of nanofillers to a polymer matrix can enhance its performance, often dramatically, by simply capitalizing on the nature and properties of the nanofillers. Introduction of these nanostructures into the polymer matrix leads significant improvements in the compressive and flexural mechanical properties of polymeric nanocomposites. Dispersion of nanofillers or controlled nanostructures in nanocomposites can offer novel physical properties and behaviors that are not found in the corresponding unfilled substrates, which changes or modifies the essence of the original matrices. These nanocomposites have some enhanced properties including fire resistance and accelerated biodegradability.

Biopolymer-based gels have the ability to trap molecules, provide protection to the entrapped active cores and to reduce the diffusion rate of the active until an external stimulus is applied to weaken the gel network. To date, numerous biopolymers have been exploited to develop biodegradable food encapsulation materials (Othman, 2014). Biopolymers are one of the favorable alternatives to be exploited and developed into eco-friendly food packaging materials due to its biodegradability. Biopolymers based on monomeric units are linked by covalent bonds, forming chain-like molecular structures. Biopolymers can be generally degraded thoroughly through the natural reactions on organisms, generating decomposed products such as H_2O and CO_2, which are harmless toward the environment. Starch, cellulose, agar, gelatin, gluten, alginate, whey protein, and collagen are natural biopolymers that are commonly used in food encapsulation. Nanocomposite encapsulation can be established as a promising route to enhance both mechanical and barrier properties of biopolymers. The nanofillers play an important role in structural reinforcement of the biopolymeric nanocomposites and improving the mechanical and barrier properties of the matrix. In this case, the matrix tension is delivered to the nanofillers via the boundary between them (Arfat et al., 2014; Azeredo, 2009). The development of nanocomposite based on biopolymers for food encapsulation is important not only to reduce environmental problems but also to improve the functions of food packaging materials.

Among the biopolymers, in the case of cost, the most common type that has been studied to prepare bionanocomposite materials for food packaging applications is starch and its derivatives. Silicate, clay, and TiO_2 are excellent nanofillers to incorporate into biopolymers, which can improve the biopolymers' mechanical and barrier properties. Besides, these nanofillers have found applications in food packaging that can be used as biosensors, antimicrobial agents, and oxygen scavengers (Rhim et al., 2013; Farhoodi, 2015). As important biopolymers, proteins have been used for nanocomposite synthesis in food industry. Industrial application of various proteins with film-forming ability has been employed for a long time. Proteins derived from animal tissues that are mainly used in commercial applications are casein, collagen, whey protein, fish myofibrillar protein, and egg white (Zhao et al., 2008). Plant-sourced proteins that are used in the food industry mainly include soybean protein, wheat gluten, and zein (corn protein) (Lee et al., 2005a). It was found that nanofibers made from lobster shells or organic corns are both antimicrobial and biodegradable.

Nowadays, synthetic polymers, which include polylactic acid (PLA), polycaprolactone (PCL), polyglycolic acid (PGA), polyvinyl alcohol (PVA), polyvinyl chloride (PVC), polyethylene, polypropylene, polybutylene, polystyrene, and polyethylene terephthalate (PET), have found wide applications in the food industry. The most common synthetic polymer used in food encapsulation is PLA. In addition, polyethylene terephthalate, polyethylene, and polystyrene are also commonly used in food packaging systems (Farhoodi, 2015). Advantages of the synthetic biopolymers include the enhancement in various properties such as durability, flexibility, high gloss, clarity, and tensile strength. For instance, high-density polyethylene can be used in packaging for bottles and bags while low-density polyethylene can be employed for trays and general-purpose containers. Polypropylene, with excellent chemical resistance, lowest density of the plastics, a high melting point, is suitable for hot-fill liquids in films and microwavable containers (Michaels, 1995). PET films with clearing, tough, and excellent gas barrier properties, are widely used for soft drink bottle packaging. Furthermore, PET has good resistance to heat, solvents, mineral oils, and acids. Therefore, PET-based encapsulation technology has been gradually applied for packaging of many kinds of foods, especially beverage and mineral water products. The demand for PET films to produce plastic bottles for carbonated drinks is rapidly growing in recent years (van Willige et al., 2002).

Conductive polymers with their conjugated π electron backbones, which can be synthesized by chemical or electrochemical oxidation methods, are very important for nanosensor preparation because of their electrical, electronic, magnetic, and optical properties (Ahuja et al., 2007). Nanoparticles treated by electrochemically polymerized methods have a notable ability to switch between conducting oxidization and insulating reduction state, which is the foundation of many particular applications (Rajesh et al., 2004). Conductive polymer constituents embedded in nanocomposite matrix can be used to detect gases produced by food spoilage microoganisms. These polymers are excellent templates for immobilization of biorecognition elements such as enzymes, antibodies, or DNA to be used in biosensor applications (Arshak et al., 2009).

3 Preparation Methods

Various methods and technologies may be used to produce encapsulated ingredients. Many of these have been adapted in the chemical and pharmaceutical industries (Augustin and Heman, 2009). The use of low-cost inorganic substances in the food industry is common since they improve mechanical and thermal properties of polymers and polymeric composites. It was found that the best technique for nanofiller synthesis is to use polymer-assisted fabrication of these nanomaterials. In these processes, nanoparticles interact with each other by van der Waals force, hydrophobic and solvation forces of colloidal, as well as electrical double layer and steric interactions. The reacted nanoparticles can be stabilized against aggregation by using mechanochemical approaches such as ultrasound sonication (Rozenberg and Tenne, 2008).

The nanocomposite preparation can be realized by dispersion in polymeric matrix by various methods including sol–gel technology, intercalation of polymer from solution, intercalative polymerization, and melt intercalation (Sozer and Kokini, 2012). Among these methods melt intercalation and intercalation of polymer from solution are widely used for food encapsulation. Most of the future progress in nanotechnology relies on polymer science, which can help in the development of cost-effective, environmentally friendly, and multifunctional ergonomic products.

1. Dispersion

   Spray drying is a well-established process in many sectors of the food industry. It is a commonly used method for encapsulation because it is more cost-effective than other techniques. In addition, there are several recent articles that describe methods of nanocomposites' preparation in detail. The basic process for production of a spray-dried encapsulated ingredient involves dissolving the core in a dispersion of the matrix material. The dispersion is atomized into heated air to facilitate the rapid removal of water as the droplets are mixed with the hot air in the drying chamber. The powder particles are then separated from the drying air at the outlet at lower temperatures (Augustin and Heman, 2009). Only aqueous-based dispersions can be used in this technology. Hence the matrix material requires good solubility in water. The ability to achieve high solids at low viscosity and good film and emulsifying properties are desirable.

   Recently complementary studies focusing on both the antimicrobial and optical properties of the chitosan materials have been investigated (Pinto et al., 2012). Nanocomposite composed of Ag nanoparticles can be dispersed in distinct chitosan-containing matrices for antibacterial packaging. The antimicrobial properties of chitosan-based materials have been thoroughly described. The dispersion of Ag nanoparticles in the chitosan matrix leads to macroscopically homogeneous films with variable optical transparency, depending both on the amount of Ag and average particle size.

   Spray drying is always used for the production of many encapsulated food ingredients—vitamins, minerals, flavors, polyunsaturated oil, enzymes, and probiotic microorganisms. It is notable that spray drying may be used for heat sensitive and volatile ingredients as the wall material protects the core and limits losing of volatile components.

2. Pad-dry-cure method (Lee and Jeong, 2004)

   Metal nanoparticles with antiracial property can be coated on polymers with an ultrasonic-assisted electrolysis method creating high conductivity. For example, triethanolamine has been used as a reducing agent in the synthesis of silver nanoparticles through chemical reduction. Then the polymer fabric is treated with silver nitrate and sodium hydroxide to form a silver nanolayer on the fabric surface after posttreatment with ammonia. Later, a step of simultaneous aminohydrolysis of polymer fabric and reduction of silver nitrate into silver nanoparticles has been developed (Poortavasoly et al., 2014).

Chemical reduction is the most common method used for preparation of Ag-nanoparticles as stable and colloidal dispersions. Silver nitrate ionized to Ag^{+} and NO_3^- in water, and silver ions were reduced to silver metal (Ag^0) during the oxidation of the $—CH_2CH_2OH$ groups in TEA to $—CH_2CHO$. Ag nanoparticles coordinate with hydroxyl groups on polymer fabric and substitute with hydrogen (Rad et al., 2011). The synthesis is often performed in the presence of stabilizers in order to avoid undesirable agglomeration of colloids. Ag-nanoparticles with smaller size and larger surface area available for interaction with microbial cells result in better bactericidal effect than larger ones (An et al., 2008).

Preparation of silver nanoparticles with triethanolamine by pad-dry-cure method (Poortavasoly et al., 2014) is indicated as follows:

$$ROH + Ag^0 \rightarrow ROAg + 1/2\, H_2 \quad (10.1)$$

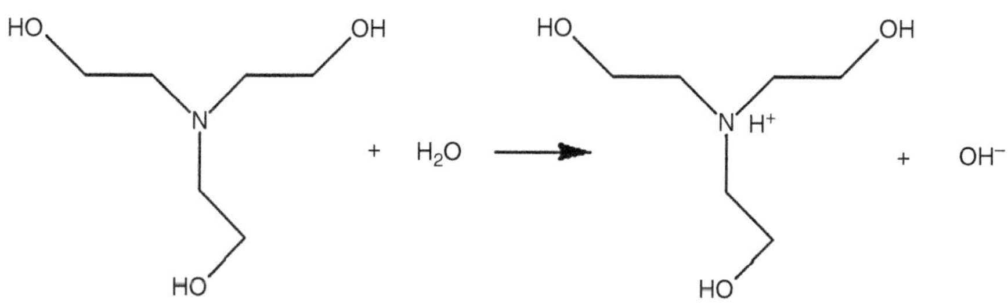

(10.2)

$$AgNO_3 \rightarrow Ag^+ + NO_3^- \quad (10.3)$$

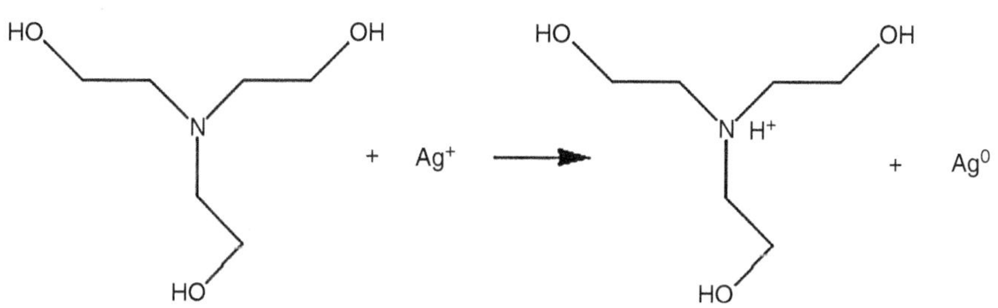

(10.4)

3. PTC
A novel chemical-plating-like solution deposition approach for preparation of large-area flexible thin films of metal oxides with d^0 configuration has been reported. This method was triggered at a low temperature (about 80°C) through the redox reaction between peroxo-metal complex (PMC) and molecules with π-conjugated structures. Flexible thin films with a large area of crystalline metal oxides, including TiO_2, V_2O_5, and MoO_3, were successfully obtained by this approach (Li et al., 2013; Wu et al., 2014, 2015). This kind of synthesis method could be employed as a versatile technology for preparation of nanocomposite materials for food encapsulation.
4. Exfoliation–adsorption
In this method, a layered filler material (eg, silicate) is first exfoliated into single layers, in which the target polymer or monomer is soluble. Due to the weak forces that stack the filler layers together, the layers can be easily dispersed in adequate solvent. When the solvent is evaporated, the target polymer then adsorb onto the delaminated sheet, thus the sheets reassemble the polymer to sandwich form. In that case, an ordered multistructure is obtained (Alexandre and Dubois, 2000). Exfoliated nanofillers can effectively enhance gas barrier properties of polymeric packages, which form maze-like structures that present tortuous pathways to diffusive gases, dramatically slowing their permeation rate (Arora and Padua, 2010).
5. In situ polymerization
This method takes advantage of layered silicates swelling by absorption of a monomer solution or a liquid monomer. In this process, the monomer molecules diffuse into the interval space or galleries of the layered silicate, where the polymerization takes place within the intercalated sheets. The polymerization can be initiated by heat, radiation, organic initiator, or diffusion of a suitable initiator (Alexandre and Dubois, 2000; LeBaron et al., 1999).
6. Melt processing
In this process, firstly the layered filler material (eg, silicate) is added into the polymer substrate in the molten state without using any solvent. Then it is mechanically mixed with a thermoplastic polymer by conventional methods including extrusion and injection moulding. Under this circumstance, nanocomposites could be formed through the polymer chains' intercalation or exfoliation. This method is suitable for the polymers that are difficult for adsorption or in situ polymerization (Kornmann et al., 2001b).

4 Surface Adhesion

Recently, a new idea with respect to superhydrophobicity has attracted more attention, especially for its ability to reduce bacterial adhesion on the surface. Developing antibacterial materials based on superhydrophobicity is a new strategy that has been recently developed (Zhang et al., 2013). It is more and more acceptable to create superhydrophobic surfaces; in this case they reduce bacterial adhesion rather than killing them directly. Superhydrophobicity is able to dramatically decrease the adhesion force between bacteria and solid surfaces, which enables the easy removal of bacteria before thick biofilms are formed on the surfaces (Yan et al., 2011). However, the adhesion and proliferation of bacteria on abiotic surfaces and the subsequent biofilm formation are still challenges in both health-care and industrial applications.

In general, bacteria are structurally and chemically complex and dynamic to environmental changes. Classically, the DLVO (Derjaguin, Landau, Verwey, Overbeek) theory has been widely employed to describe the bacterial adhesion to a flat surface. Bacterial cells have been treated as inert particles, and the particle adhesion is determined by van der Waals interactions generally with attractive effect and repulsive interactions from the electrical double layer of the cell and the surface (Bhattacharjee et al., 1998).

A good relationship between wettability and bacterial adhesion has been found. Past studies have also shown that a hydrophilic surface can effectively attract bacteria such as *E. coli* and *S. aureus*, while a hydrophobic surface facilitates attracting *Taiwanensis*, *Pseudoxanthomonas*, and *Staphylococcus epidermidis* and reducing the adherence of bacteria including *Deinococcus geothermalis*, *Meiothermus silvanus*, *S. aureus*, and *Streptococcus mutans* (Zhang et al., 2013).

It was found that irregularities in polymeric surfaces promote bacterial adhesion and biofilm formation, while ultra-smooth surfaces do not favor bacterial adhesion and biofilm formation. This may be explained by the fact that a rough surface possesses a greater surface area with more favorable sites for bacterial colonization (Zhang et al., 2013).

In addition, charged polymers with nanobrushes are of great interest for antibacterial applications (Lee et al., 2011). Polymer brushes impart surfaces with bacterial resistance. The number of adherent bacteria can be greatly decreased since the smaller size and curly structure has the ability to resist bacterial colonization.

5 New Concepts in Food Encapsulation

1. Nanosensor

Nanosensors can assist in determining the environment for foods. Food products are continuously affected by oxygen content, temperature, and pathogens; hence indicators or sensors are needed for proper alarming. Consumers are concerned about the expiration dates of foods, which are related to distribution and storage conditions and the period of time that the food product is predicted to be exposed to these environments. Sensors can be used to detect the physical qualities of substances and convert the information into readable signals. Nanosensors in the nanocomposite encapsulation can respond to environmental changes (eg, temperature, humidity, and oxygen level), degradation of products, and microbial contamination (Bouwmeester et al., 2009). Recent progress shows that the current smart packaging segment is dominated by oxygen scavengers, moisture absorbers, and barrier packaging products, accounting for about 80% of the market (Chellaram et al., 2014). Nanosensors can assist in the case of temperature increases, or in the presence of micropores or sealing defects in packaging systems that can expose food products to a unexpected levels of oxygen, which will result in undesirable changes. In fact due to the short quality guarantee period, bakery and meat products have used the most nano-enabled smart packaging technology to date.

Currently nanobased sensors embedded in encapsulation packaging to detect spoilage, chemical contaminants, pathogens, product tampering, or to monitor ingredients of foods through the processing chain sensing are under development, and some products have been commercialized in the market (Nachay, 2007). Several kinds of gas sensors have been developed for food encapsulation based on chemical reaction routes, which transforms chemical reactions between particles taking place on the surfaces into response signals. One of the most popular types of sensing materials is based on metal oxides due to their high sensitivity and stability (Šetkus, 2002). The conventional systems take several days to confirm the presence of pathogens in food; however, new super sensors will be able to detect pathogens immediately.

There has been a growing interest to prepare irreversible and nontoxic oxygen sensors embedded in films to prevent the leakage of oxygen into oxygen-free food packaging systems, which can be realized by packaging in vacuum or nitrogen. For example, TiO_2 nanoparticles are generally used to monitor the reduction of methylene blue (MB) by triethanolamine in polymer substance by photosensibility upon UVA light irradiation (Lee et al., 2005b). This nanosensor system will turn to its original blue color when exposed to oxygen, while upon UV irradiation it becomes colorless. The rate of the color recovery is proportional to the level of oxygen exposure. This technique could be used in oxygen indicator packaging systems for protecting a variety of oxygen-sensitive foods. Nanocrystalline $SnO_{2,}$ as another example, when combined with glycerol as an electron donor, methylene blue as a redox dye, and hydroxyethyl cellulose as an encapsulation polymer together, can be also used as an O_2 photosensitizer (Mills and Hazafy, 2009). The color of these detectors gradually changes in response to a small amount of oxygen.

Food spoilage is often caused by intrusive microorganisms, whose metabolism can produce some gases that could be detected by certain metal oxides or conducting polymer nanocomposites (CPC). These materials are useful for quantification or identification of microorganisms aiming at their gas emissions. Sensors based on the CPC materials contain conducting particles embedded in insulating polymeric matrices. The sensors will respond to the gases from microorganisms by resistance changes (Arshak et al., 2007). A typical example is the use of gold nanoparticles that incorporated enzymes for microbe's detection (Chellaram et al., 2014). Nanofibrils of perylene-based fluorophores have the ability to indicate fish and meat spoilage by detection of gaseous amines. ZnO and TiO_2 nanocomposites can also be used for the detection of volatile organic compounds (Chellaram et al., 2014).

Temperature increasing may endanger food safety in the presence of pathogens or toxins. Recently several kinds of time-temperature indicators have been made to give the rate of change from visual indication (Smolander et al., 2004). The working mechanisms of this kind of indicator may be based on enzymatic reaction, polymerization, melting point, and diffusion of a substrate (Poças et al., 2008).

Humidity is also a threat to food safety. Porous metal films with carbon coating, a silicon tenside, and a polymeric wetting agent can be used to prepare humidity sensors which can be employed for food handing, storage, and transport. Exposure to water or vapor induces an exceptional sensitivity with optical shifts in the visible light range on these sensors (Luechinger et al., 2007).

Smart packaging containing immobilized enzymes such as lactase or cholesterol reductase can be employed for designing food products that require certain enzyme treatments for customers suffering from high cholesterol levels or lactose intolerance (Sozer and Kokini, 2012).

A new concept of smart packaging known as BioSwitch has been proposed by TNO (Zeist, The Netherlands). In this kind of package, an antimicrobial agent will be released when triggered by environmental changes such as pH, temperature, and light. This will afford protection to packaged foods in specific conditions. For example, antimicrobial-encapsulated polysaccharides can be embedded in encapsulation materials as the trigger agents. The majority of bacteria utilize polysaccharides for their growth. As the amount of bacteria increases within the package, the antimicrobial agents will be released to hinder further bacterial growth in these types of packaging materials (De Jong et al., 2005; Huff, 2009).

Nanosensors can be designed within the food packaging for rapid detection of toxins and food poisoning microorganisms upon changes in color, mass, and temperature (Lee et al., 2009). They also can be used for tracking and monitoring foods during transportation and storage.

2. Edible coating
Edible coatings have been designed and developed to protect foods from browning reactions, flavor changes, microorganisms, and textual deteriorations. They can be used as antimicrobial and antibrowning agents, texture enhancers, and nutraceuticals to ensure not only food safety and quality but also food functionality. The new tendency for edible coatings involves the use of nanoencapsulation and nanocomposite to enhance their physical properties and enable them to act as delivery systems (Rojas-Graü et al., 2009). Whey protein has received significant attention as edible film and coating material. But the effect of active ingredients incorporated into food package needs to be evaluated for their functional, mechanical, and sensory properties or possible potential risk. United States Company Sono-Tek Corporation in 2007 developed an edible antibacterial nano-coating which can be applied directly to bakery goods (ElAmin, 2007b).

3. Self-heating, cooling, healing, and filtering
The exothermic chemical reactions of lime occur in the presence of water, or hydrolysis of calcium chloride can be employed in the self-heating packages. The military requires ready-to-eat packages, which are important in combat and rescue actions. Food packaging meeting this goal uses the

exothermic reactions of magnesium oxidation and potassium permanganate and glycerine to heat highly viscous liquid and solid products. The same principle can be used for self-cooling packaging, which could be applied for drinks such as beverages for conventional fast cooling (www.idspackaging.com). Recently fullerene nanotubes have been found to improve the self-cooling efficiency. Carbon dioxide and nitrogen can be used as the refrigerants held by fullerene nanotubes at a pressure slightly higher than atmospheric pressure. Self-cooling beverage and food container have adopted this technology conditioned by fullerene nanotubes (World patent number 0073718). Nowadays self-healing polymers for treatment of fractures, tears, and punctures have been under research since 1990. Self-healing materials have the ability to automatically recover from damage (Robinson and Morrison, 2010). Self-healing nanocomposites involve reformation of polymer bonds relating nanoparticle migration within the composite to the site of damage by the micropower of repulsive forces between the polymer matrix and nanofillers. Another interesting profile of today's encapsulation is nanofilter function. Nanocomposites with nanofilter systems are able to purify water supplies form biological contamination and salt (Gelman and Wolf, 2010). Nanofiltration has been proving its worth for the filtration and separation aspect of food processing.

Encapsulation packaging that incorporates nanomaterials can be smart, which means that it can respond to environmental conditions or repair itself to alert consumers of contamination or the presence of pathogens, humidity, and oxygen. However, the potential use of these smart packages in food industry will be dependent on the final cost and feasibility to apply to food products.

6 Industry, Enterprise, and Market

Until 2014 the global food market revolving nanotechnology was about US$4.13 billion (Duncan, 2011). In terms of nanotechnology research and development (R&D) numbers, the USA leads, followed by Japan, China, and the European Union. Some important companies in food nanotechnology, most of which are located in USA, have been working on the nanocomposite for food encapsulation for many years and have led the developing tendency in this field. Some famous giants of food enterprises, such as Heinz, Nestlé, Unilever, and Kraft have cost funding to support specific research programs to share the profit from the technology and occupy the future market.

In the nanoclay field companies such as Nanocor, Inc., and Southern Clay Products, Inc., have made efforts to import montmorillonite (MMT) into nanocomposites, which makes packaging films much lighter, stronger, and more heat-resistant, and improves gas barrier properties. A successful process based on Nylon-6 has been developed by the enterprise. Nylon-6 has been incorporated into clays because it is fluid and able to penetrate the interspace between the clay layers easily. When extruded, the Nylon-6 platelets rearrange their direction parallel to the surface, which is very useful in improving barrier properties. Nanocomposite materials containing Nylon-6 can lower the transmission rate of oxygen about 4 times that of single Nylon-6 material (Brody, 2003). Some food companies, including Nanocor and Mitsubishi Gas Chemical, have developed a series of nanocomposites with nylon MXD6 named Imperm, which has shown much enhanced barrier properties. Imperm can be widely used in various types of packaging for encapsulation of bottles for drinking (Brody, 2006). Until now nanocomposites involving Imperm have been commercialized as effective oxygen barrier layers used on bottles for beer, dairy foods, fruit juices, and carbonated drinks. The shelf life of a variety of foods such as processed meats, cereals, confectionery, cheese, and boil-in-bag products can also be greatly enhanced by multilayer films with Imperm embedded in them (Brody, 2007).

Bayer Corporation produces nanocomposite packaging films and coatings with nanoclay dispersion to block oxygen, carbon dioxide, and moisture from touching foods. One of the products developed by Bayer named KU2-2601 packaging film, which is lighter, stronger, and more heat resistant than those currently available in the market. This film is known as a hybrid system that is enhanced by a large number of silicate nanoparticles, thus greatly reducing the entry pathways of gases and moisture into the encapsulated foods (nanoparticles make Durethan films airtight and glossy, Bayer polymers).

Nanocor Company produces nanocomposite-reinforced plastic beer bottles to address oxidation and flavor problems in the beer-packaging industry (Brody et al., 2008).

A key enterprise for military ration service, known as the U.S. Army Natick Soldier Center (Natick, Massachusetts), has made efforts to prepare alternatives to laminations to greatly extend the shelf life of shelf-stable foods. They also develop new technologies for food packaging for fast heating in microwave ovens, meanwhile lowering waste production from package materials. Their products are mainly intended to serve US forces in military actions. They have successfully incorporated nanoclay into plastic matrices to improve mechanical strength, thermal resistance, and barrier properties. Recently Natick's research has focused on preparation of PE, PET, and ethylene vinyl alcohol (EVOH) with 1–5% nanoclay platelet weight. They have successfully dispersed the clay platelets to maximize orientation, which plays an important role in producing tortuous pathways for outside gases. This increases almost 80% in thermal resistance and about 100% in mechanical strength.

Kraft Foods Company has developed an electronic microdevice like a tongue that can be embedded in food packages. This novel device can change color to indicate whether the food has deteriorated because of spoiling microorganisms, with an array of nanosensors sensitive to gases released from these microorganisms (Joseph and Morrison, 2006).

Research on polymer-layered silicate was once boosted by the Toyota research group. They found that polymer-clay nanocomposites exhibited superior strength modulus, gas, and humidity barrier properties as compared to the original polymers (Usuki et al., 1995).

The Nordic Innovation Center has developed a product named ENZY-COAT by bionanotechnology to design food encapsulation from enzymes using nanofiller dispersed coating method. They demonstrated the use of enzymes as oxygen scavengers to prevent foods from oxidation discoloration, slime formation, textural changes, off-odor, and off-flavor developments (Järnström, 2008).

The use of aluminum nanolayers in most snack, beverage, and milk packages has become one of the leaders in the polyester film market. This kind of packaging uses higher barrier coating by vermiculite nanoplatelets that bind to positively charged aroma and flavor molecules. This can retain volatile oils up to 25–50% and is better than the traditional packages (www.plastesmart.com) (Sozer and Kokini, 2012).

Kodak started the research in the packaging market since 2005. It has produced antimicrobial food and medical packaging materials with its especially flexible nanotechnology. They focus on the use of antimicrobial materials such as silver ions, nisin, magnesium oxide, and zinc oxide incorporated into packaging nanocomposites. They also produce active products such as oxygen absorbers and biolayers for food packaging (Sozer and Kokini, 2012).

Some companies have made efforts in developing smart packaging in the self-heating and self-cooling fields. Nestlé has focused its research on coffee cans that self-heat by simply shaking. Caldo Caldo, an Italian branch, is pursuing similar technology for products such as coffee, cappuccino, chocolate, and tea. Self-cooling technology has been successfully used in the market for cooling beer kegs by zeolite heat pumps and endothermic reactions between sodiumthiosulfate pentahydrate and water (www.idspackaging.com).

The food packaging industry recently has started to use radio frequency identification (RFID) to trace foods from raw materials, which takes advantage of radio waves for food traceability. RFID systems have been widely used due to an increasing demand to control and monitor the agricultural food production line in terms of quality, health, and safety issues (www.foodlife.com). RFID intelligent tags allow real-time monitoring of both ingredients and end products of agricultural foods during various steps of growth, processing, packaging, transportation, distribution, and storage (Wang et al., 2006).

The level of funding for nanotechnology in foods until now in developing countries may be comparatively lower in developing countries, but this has not hindered the impact of some countries or districts on the global stage. China, India, South Korea, Iran, and Thailand are catching up with a focus on applications specific to the economic growth and demands of their countries (Joseph and Morrison, 2006).

7 Limitations and Shortcomings

Even though great positive results have been achieved, the development of nanotechnology in the food industry has been slow and faced severe challenges in recent years. Many food companies are reluctant to fund continued research into food nanotechnology due to limitations caused by increased costs, restrictive legislation, and negative feedback from consumers.

The rapid adoption of nanocomposite materials for food packaging in a wide range has also raised concerns regarding their safety, environmental impacts, policies, and regulation. Some shortcomings to using nanocomposites as food packaging have now attracted increasing attention. As a typical example, the particle aggregation problem and nanomaterial recovery are two big challenges when using Ag nanoparticles in applications (Liu et al., 2011). Studies have shown that the toxicity of Ag nanoparticles is size dependent and the smaller sized nanoparticles exhibit higher antibacterial activity due to higher specific surface area and ease of cell penetration (Cushen et al., 2014; Marambio-Jones and Hoek, 2010). Some results indicated that both the antimicrobial properties and the detrimental effect of Ag nanomaterials or ions on biological cells should be carefully considered when using them in food packaging (Carlson et al., 2008).

The fact that some compounds behave distinctly differently at the nanoscale should be noted. The extremely small-sized particles and increased surface area might let nanofillers easily pass through the skin and respiratory system and also into the environmental cycle (Sozer and Kokini, 2012). In terms of consumers' safety, it is important to evaluate the potential migration of packing constituents into foods and to assess their potential hazard for a comprehensive risk assessment. There are growing concerns about the extent to which nanomaterials will migrate into foods from packaging films, and how they perform on the health of consumers. So far the benefits are highly evident since some nanocomposites for food packaging have already been on the market, and the advantages they offer to prolong the shelf life of foods and simplify the manufacture, processing, and management of products are tangible.

Even so, some researchers pointed out that there is risk of migration of nanocomposites into foods. For instance, their results showed that silver presented the highest level of migration into food simulants and acidic food. In addition, heating was found to increase the migration rate. Two possible pathways may cause the migration: the detachment of silver nanoparticles from the composite or the oxidative dissolution of silver ions. Researchers showed that the percentage of the nanofillers in the composites plays an important role in the migration rate. Studies indicate that normally less than 10 mg/dm^2 of the nanocomposites migrate into water and the levels of metals measured were below the permitted values (Marambio-Jones and Hoek, 2010).

Another potential health risk may occur from the migration of by-products of the fermentation process: materials with incomplete reaction or in an intermediate state. In this case the final product should be risk assessed and authorized before its use in the manufacture of nanocomposite material and articles.

The majority of materials used for food encapsulation are nonbiodegradable that does not satisfy increasing demands for the theme of sustainability and the environmental safety of society (Othman, 2014). Some types of materials are derived from petroleum products and cause the problem in waste disposal. Most materials used for food packaging are practically undegradable, which result in a serious global environmental problem. But the popular use of biodegradable and edible polymers has been restricted due to problems related to negative effects such as brittleness and poor gas barrier properties.

Even usage of biopolymers has some potential disadvantages. A distinct drawback of using biopolymers is that, compared to the conventional nonbiodegradable materials, especially those made from petroleum, most food packaging materials have poor mechanical and barrier properties. Biopolymers of food-packaging grade are generally easy to crack, show distortion at lower temperatures, and show less resistance to long process operations.

For an encapsulated ingredient to be successful in the marketplace, it has to meet several criteria. The functionality of an encapsulated ingredient has to be tested in the final food product, taking into account the storage stability of the encapsulated ingredient, its compatibility with the food matrix, the processing stresses it has to withstand during food manufacture when it is in intimate contact with other ingredients, and how it breaks down when consumed (Augustin and Heman, 2009).

Consumers with the potential risk of nanomaterials in the encapsulation packaging must be informed of the dangers by usage of a new labeling or tag system that includes new safety symbols. If a nanomaterial to be used in food encapsulation is produced from the start, it is strongly suggested a comparison should be carefully made in the properties of the bulk substance, to determine if there has been a change. Safety data indicators or sheets must also be made available to users and consumers. Furthermore, it is advised that toxicity data should be open to the public after complete characterization of nanomaterials both in bulk and surface properties (Mills and Hazafy, 2009; Elzey et al., 2009). Regulatory guidance and consumer education may be conducted upon the acceptance of commercialization of nanotechnology for foods.

8 Conclusions

This chapter clearly shows that nanoenhanced encapsulation has much to offer the food industry. The impact of nanotechnology is focused mostly in food packaging technology with the improvement of biodegradable packaging coatings and sensors incorporated into intelligent packages. The benefits of nanocomposites range from stronger and more flexible films, to smart packaging which can vastly simplify stock management and monitor food condition. The addition of nanofillers to the polymer substrates has found new potential to prepare innovative nanocomposite materials with enhanced sealing properties and performance. Nanofillers can be used in food packaging as antimicrobial agents, biosensors, gas sensors, temperature sensors, and oxygen scavengers. Nanofillers embedded in the nanocomposites can also be engineered to trigger an electrical or chemical signal in the presence of a contaminant such as bacteria. However, with the wide use of nanotechnology in food industry, there are some problems and limitations which must be concerned. Until now it has not yet become evident to what extent nanoparticles should be embedded in packaging films. Further, the effects of exposure to nanomaterials on consumer health is not yet clearly understood. A potential risk arises from nanofillers' entry to the biosystemic cycle in the form of areoles or liquid suspensions. Bioavailability is another concern which might influence their toxicological impact. Multilayered thin films based on nanocomposites show excellent properties, but considerable amounts of them are likely difficult to reuse or recycle. In addition, some limited raw materials may be used, and manufacturing processes require huge energy consumption and produce large amounts of waste materials. In that case, the benefits versus the negative effects of nanocomposites should be carefully compared and evaluated in a well-balanced manner. In addition, the sustainability of nanoenhanced packaging must also be considered. Even these issues are still under exploration, with the viewpoint that there is no doubt that in the coming years there will be important developments in nanotechnology for food encapsulation packaging. The increase in the application of nanotechnology should be accelerated by the accomplishment of regulatory issues. Scientists, industry, governments, and the public should join together in the scientific and informational fields to make strong progress in food nanotechnology.

References

Abbas, K.A., Saleh, A.M., Mohamed, A., Mohdazhan, N., 2009. The recent advances in the nanotechnology and its applications in food processing: a review. J. Food Agric. Environ. 7, 14–17.

Adame, D., Beall, G.W., 2009. Direct measurement of the constrained polymer region in polyamide/clay nanocomposites and the implications for gas diffusion. Appl. Clay Sci. 42, 545–552.

Aguilar De Armas, M.R., Román, J.S., 2014. Smart Polymers and Their Applications. Woodhead Publishing Limited, Cambridge, UK.

2008. Structure–property relationship in foods. In: Aguilera, J.M., Lillford, P.J. (Eds.), Food Materials Science: Principles and Practice. Springer, New York, NY, pp. 229–254.

Ahuja, T., Mir, I.A., Kumar, D., Rajesh, 2007. Biomolecular immobilization on conducting polymers for biosensing applications. Biomaterials 28, 791–805.

Alexandre, M., Dubois, P., 2000. Polymer-layered silicate nanocomposites: preparation, properties and uses of a new class of materials. Mater. Sci. Eng. R 28 (1–2), 1–63.

An, J., Zhang, M., Wang, S., Tang, J., 2008. Physical, chemical and microbiological changes in stored green asparagus spears as affected by coating of silver nanoparticles—PVP. LWT Food Sci. Technol. 41 (6), 1100–1107.

Arfat, Y.A., Benjakul, S., Prodpran, T., Sumpavapol, P., Songtipya, P., 2014. Properties and antimicrobial activity of fish protein isolate/fish skin gelatin film containing basil leaf essential oil and zinc oxide nanoparticles. Food Hydrocoll. 41, 265–273.

Arora, A., Padua, G.W., 2010. Review: nanocomposites in food packaging. J. Food Sci. 75, R43–R49.

Arshak, K., Adley, C., Moore, E., Cunniffe, C., Campion, M., Harris, J., 2007. Characterization of polymer nanocomposite sensors for quantification of bacterial cultures. Sensor. Actuat. B 126, 226–231.

Arshak, K., Velusamy, V., Korostynska, O., Oliwa-Stasiak, K., Adley, C., 2009. Conducting polymers and their applications to biosensors: emphasizing on food borne pathogen detection. IEEE Sens. J. 9 (12), 1942–1951.

Augustin, M.A., Heman, Y., 2009. Nano- and micro-structured assemblies for encapsulation of food ingredients. Chem. Soc. Rev. 38, 902–912.

Azeredo, H.M.C.D., 2009. Nanocomposites for food packaging applications. Food Res. Int. 42, 1240–1253.

Azeredo, H., Mattoso, L.H.C., Avena-Bustillos, R.J., Munford, M.L., Wood, D., McHugh, T.H., 2010. Nanocellulose reinforced chitosan composite films as affected by nanofiller loading and plasticizer content. J. Food Sci. 75, N1–N7.

Bhattacharjee, S., Elimelech, M., Michal, B., 1998. DLVO interaction between colloidal particles: beyond Derjaguin's approximation. Croat. Chim. Acta 7, 883–903.

Bouwmeester, H., Dekkers, S., Noordam, M.Y., Hagens, W.I., Bulder, A.S., De Heer, C., 2009. Review of health safety aspects of nanotechnologies in food production. Regul. Toxicol. Pharm. 53, 52–62.

Brody, A.L., 2003. "Nano, nano" food packaging technology. Food Technol. 57 (12), 52–54.

Brody, A.L., 2006. Nano and food packaging technologies converge. Food Technol. 60 (3), 92–94.

Brody, A.L., 2007. Nanocomposite technology in food packaging. Food Technol. 61 (10), 80–83.

Brody, A.L., Bugusu, B., Han, J.H., Sand, C.K., McHugh, T.H., 2008. Innovative food packaging solutions. J. Food Sci. 73 (8), R107–R116.

Carlson, C., Hussain, S., Schrand, A., Braydich-Stolle, L., Hess, K., Jones, R., Schlager, J., 2008. Unique cellular interaction of silver nanoparticles: size-dependent generation of reactive oxygen species. J. Phys. Chem. B 112, 13608–13619.

Chellaram, C., Murugaboopathi, G., John, A.A., Rivakumar, R., Ganesan, S., Krithika, S., Priya, G., 2014. Significance of nanotechnology in food industry. APCBEE Procedia 8, 109–113.

Chen, J., Wang, F.Y.K., Liu, Q.M., Du, J.Z., 2014. Antibacterial polymeric nanostructures for biomedical applications. Chem. Commun. 50, 14482–14493.

Cushen, M., Kerry, J., Morris, M., Cruz-Remero, M., Cummins, E., 2014. Evaluation and simulation of silver and copper nanoparticle migration from polyethylene nanocomposites to food and an associated exposure assessment. J. Agric. Food. Chem. 62, 1403–1411.

Dalmas, F., Cavaillé, J.-Y., Gauthier, C., Chazeau, L., Dendievel, R., 2007. Viscoelastic behavior and electrical properties of flexible nanofiber-filled polymer nanocomposites. Influence of processing conditions. Compos. Sci. Technol. 67, 829–839.

De Jong, A.R., Boumans, H., Slaghek, T., Van Veen, J., Rijk, R., Van Zandvoort, M., 2005. Active and intelligent packaging for food: is it the future? Food Addit. Contam. A 22, 975–979.

Duncan, T.V., 2011. Applications of nanotechnology in food packaging and food safety: barrier materials, antimicrobials and sensors. J. Colloid Interface Sci. 363, 1–24.

ElAmin, A., 2007a. Nanoscale particles designed to block UV light. FoodProductionDaily.com, October 18, 2007. Available from: http://foodproductiondaily.com/news/ng.asp?id=80676

ElAmin, A. 2007b. Nanoscale coating process developed for baking sector. FoodProductionDaily.com, February 28, 2007. Available from: http://www.foodproductiondaily.com/news/ng.asp?id=74584

Elzey, S., Larsen, R.G., Howe, C., Grassian, V.H., 2009. Nanoscience and nanotechnology: environmental and health impacts. In: Klabunde, K.J., Richards, R.M. (Eds.), Nanoscale Materials in Chemistry. second ed. John Wiley & Sons, Inc., Hoboken, NJ, pp. 681–727.

Espitia, P.J.P., Soares, N.F.F., Coimbra, J.S.R., de Andrade, N.J., Cruz, R.S., Medeiros, E.A.A., 2012. Zinc oxide nanoparticles: synthesis, antimicrobial activity, and food packaging applications. Food Bioprocess Technol. 5, 1447–1464.

Farhoodi, M., 2015. Nanocomposite materials for food packaging applications: characterization and safety evaluation. Food Eng. Rev. 8, 35–51.

Fathi, M., Mozafari, M.R., Mohebbi, M., 2012. Nanocapsulation of food ingredients using lipid-based delivery systems. Trends Food Sci. Technol. 23, 13–27.

Gelman, D.L., Wolf, D.G., 2010. Applications of nanotechnology to the food & water infrastructure. J. Homeland Secur. Manag. 6 (1), 1–10, Article 19.

Hager, M.D., Greil, P., Leyens, C., van der Zwaag, S., Schubert, U.S., 2010. Self-healing materials. Adv. Mater. 22, 5424–5430.

Han, W., Yu, Y.J., Li, N.T., Wang, L.B., 2011. Application and safety assessment for nano-composite materials in food packaging. Chinese Sci. Bull. 56, 1216–1225.

Heo, S.-Y., Choi, H.-J., Park, B.-J., Um, J.-H., Jung, H.-J., Jeong, J.-R., Yoon, S.-G., 2015. Effect of protective layer on enhanced transmittance, mechanical durability, anti-fingerprint, and antibacterial activity of the silver nanoparticles deposited on flexible substrate. Sensor. Actuat. A. 221, 131–138.

Huff, K., 2009. Active and intelligent packaging: innovations for the future. In: IEEE International Conference on Microwaves, Communications, Antennas and Electronic Systems, 2008, COMCAS 2008, Tel Aviv, Israel, pp. 1–8. Available from: http://www.iopp.org/files/public/VirginiaTechKarleighHuff.pdf

Hwang, J.M., Yeom, S.H., Jung, K.Y., 2007. Synthesis of oxazolidinone phosphonates as antibacterial agents. J. Ind. Eng. Chem. 13, 474–479.

Järnström, L., 2008. Oxygen scavenging and aroma affecting enzymes embedded in barrier coatings. In: Järnström, L. (Ed.), Nordic Innovation Centre Project 06085, Final report, January 30, 2008. Available from: www.nordicinnovation.net

Jordan, J., Jacob, K.I., Tannenbaum, R., Sharaf, M.A., Jasiuk, I., 2005. Experimental trends in polymer nanocomposites—a review. Mater. Sci. Eng. A 393, 1–11.

Joseph, T., Morrison, M., 2006. Nanotechnology in agriculture and food. A Nanoforum report. Available from: www.nanoforum.org

Kamigaito, O., 1991. What can be improved by nanometer composites? J. Jpn. Soc. Powder Powder Metall. 38 (3), 315–321.

Kang, S., Herzberg, M., Rodrigues, D.F., Elimelech, M., 2008. Antibacterial effects of carbon nanotubes: size does matter! Langmuir 24, 6409–6413.

Kong, H., Jang, J., 2008. Antibacterial properties of novel poly(methyl methacrylate) nanofiber containing silver nanoparticles. Langmuir 24, 2051–2056.

Kornmann, X., Lindberg, H., Berglund, L.A., 2001a. Synthesis of epoxy–clay nanocomposites. Influence of the nature of the curing agent on structure. Polymer 42, 4493–4499.

Kornmann, X., Lindberg, H., Berglund, L.A., 2001b. Synthesis of epoxy–clay nanocomposites: influence of the nature of the clay on structure. Polymer 42 (4), 1303–1310.
Lagaron, J.-M., 2011. Multifunctional and Nanoreinforced Polymers for Food Packaging. Woodhead Publishing Limited, Cambridge, UK.
Lagaron, J.M., Catala, R., Gavara, R., 2004. Structural characteristics defining high-barrier polymeric materials. Mater. Sci. Technol. 20, 1–7.
LeBaron, P.C., Wang, Z., Pinnavaia, T.J., 1999. Polymer-layered silicate nanocomposites: an overview. Appl. Clay Sci. 15, 11–29.
Lee, H.J., Jeong, S.H., 2004. Bacteriostasis of nanosized colloidal silver on polyester nonwovens. Text. Res. J. 74, 442–447.
Lee, C., Scheufele, D.A., Lewenstein, B.V., 2005a. Public attitudes toward emerging technologies: examining the interactive effects of cognitions and affect on public support for nanotechnology. Sci. Commun. 27 (2), 240–267.
Lee, S.K., Sheridan, M., Mills, A., 2005b. Novel UV-activated colorimetric oxygen indicator. Chem. Mater. 17 (10), 2744–2751.
Lee, J.B., Roh, Y.H., Um, S.H., Funabashi, H., Cheng, W., Cha, J.J., Kiatwuthinon, P., Muller, D.A., Luo, D., 2009. Multifunctional nanoarchitectures from DNA-based ABC monomers. Nat. Nanotechnol. 4, 430–436.
Lee, H.-S., Eckmann, D.M., Lee, D., Hickok, N.J., Composto, R.J., 2011. Symmetric pH-dependent swelling and antibacterial properties of chitosan brushes. Langmuir 27, 12458–12465.
Li, Y.Z., Yu, Y., Wu, L.Z., Zhi, J.F., 2013. Processable polyaniline/titania nanocomposites with good photocatalytic and conductivity properties prepared via peroxo-titanium complex catalyzed emulsion polymerization. Appl. Surf. Sci. 273, 135–143.
Liu, L., Xu, K., Wang, H., Jeremy Tan, P.K., Fan, W., Venkatraman, S.S., Li, L., Yang, Y.-Y., 2009. Self-assembled cationic peptide nanoparticles as an efficient antimicrobial agent. Nat. Nanotechnol. 4, 457–463.
Liu, L., Liu, J.C., Wang, Y.J., Yan, X.L., Sun, D.D., 2011. Facile synthesis of nanodispersed silver nanoparticles on grapheme oxide sheets with enhanced antibacterial activity. New J. Chem. 35, 1418–1423.
Luechinger, N.A., Loher, S., Athanassiou, E.K., Grass, R.N., Stark, W.J., 2007. Highly sensitive optical detection of humidity on polymer/metal nanoparticle hybrid films. Langmuir 23, 3473–3477.
Lv, M., Su, S., He, Y., Huang, Q., Hu, W., Li, D., Fan, C., Lee, S.-T., 2010. Long-term antimicrobial effect of silicon nanowires decorated with silver nanoparticles. Adv. Mater. 22, 5463–5467.
Marambio-Jones, C., Hoek, E.M.V., 2010. A review of the antibacterial effects of silver nanomaterials and potential implications for human health and the environment. J. Nanopart. Res. 12, 1531–1551.
Michaels, F., 1995. Modern Plastics Encyclopedia. McGraw-Hill, New York, NY.
Mills, A., Hazafy, D., 2009. Nanocrystalline SnO_2-based, UVB-activated, colourimetric oxygen indicator. Sensor. Actuat. B 136 (2), 344–349.
Milović, N.M., Wang, J., Lewis, K., Klianov, A.M., 2005. Immobilized *N*-alkylated polyethylenimine avidly kills bacteria by rupturing cell membranes with no resistance developed. Biotechnol. Bioeng. 90, 715–722.
Morones, J.R., Elechiguerra, J.L., Camacho, A., Holt, K., Kouri, J.B., Ramirez, J.T., Yacaman, M.J., 2005. The bactericidal effect of silver nanoparticles. Nanotechnology 16, 2346–2353.
Nachay, K., 2007. Analyzing nanotechnology. Food Technol. 61 (1), 34–36.
Naicker, P.K., Cummings, P.T., Zhang, H., Banfield, J.F., 2005. Characterization of titanium dioxide nanoparticles using molecular dynamics simulations. J. Phys. Chem. B 109, 15243–15249.

Nakagaito, A.N., Iwamoto, S., Yano, H., 2005. Bacterial cellulose: the ultimate nano-scalar cellulose morphology for the production of high-strength composites. Appl. Phys. A 80 (1), 93–97.

Nederberg, F., Zhang, Y., Tan, J.P.K., Xu, K., Wang, H., Yang, C., Gao, S., Guo, X.D., Fukushima, K., Li, L., Hedrick, J.L., Yang, Y.-Y., 2011. Biodegradable nanostructures with selective lysis of microbial membranes. Nat. Chem. 3, 409–414.

Othman, S.H., 2014. Bio-nanocomposite materials for food packaging applications: type of biopolymer and nano-sized filler. Agric. Agric. Sci. Procedia 2, 296–303.

Pawlik, A.K., Norton, I.T., 2014. Bridging benchtop research and industrial processed foods: structuring of model food emulsions. Food Struct. 1, 24–38.

Percival, S.L., Bowler, P.G., Russell, D., 2005. Bacterial resistance to silver in wound care. J. Hosp. Infect. 60, 1–7.

Pinto, R.J.B., Fernandes, S.C.M., Freire, C.S.R., Sadocco, P., Causio, J., Neto, C.P., Trindade, T., 2012. Antibacterial activity of optically transparent nanocomposite films based on chitosan or its derivatives and silver nanoparticles. Carbohydr. Res. 348, 77–83.

Poças, M., Delgado, T.F., Oliveira, F.A.R., 2008. Smart packaging technologies for fruits and vegetables smart packaging technologies. In: Kerry, J., Butler, P. (Eds.), Smart Packaging Technologies for Fast Moving Consumer Goods. Wiley, Chichester, UK, pp. 151–166.

Poortavasoly, H., Montazer, M., Harifi, T., 2014. Simultaneous synthesis of nano silver and activation of polyester producing higher tensile strength aminohydroxylated fiber with antibacterial and hydrophilic properties. RSC Adv. 4, 46250–46256.

Qiao, R., Brinson, L.C., 2009. Simulation of interphase percolation and gradients in polymer nanocomposites. Compos. Sci. Technol. 69 (3–4), 491–499.

Qiao, Y., Yang, C., Coady, D.J., Ong, Z.Y., Hedrick, J.L., Yang, Y.-Y., 2012. Highly dynamic biodegradable micelles capable of lysing Gram-positive and Gram-negative bacterial membrane. Biomaterials 33, 1146–1153.

Rad, P.S., Montazer, M., Rahimi, M.K., 2011. Simultaneous antimicrobial and dying of wool: a facial method. J. Appl. Polym. Sci. 122, 1405–1411.

Rajesh, Takashima, W., Kaneto, K., 2004. Amperometric phenol biosensor based on covalent immobilization of tyrosinase onto an electrochemically prepared novel copolymer poly(*N*-3-aminopropyl pyrrole-copolymer) film. Sensor. Actuat. B 102, 271–277.

Rhim, J.-W., Park, H.-M., Ha, C.-S., 2013. Bio-nanocomposites for food packaging applications. Prog. Polym. Sci. 38, 1629–1652.

Robinson, D.K.R., Morrison, M.J., 2010. Nanotechnologies for food packaging: reporting the science and technology research trends. Report for the Observatory Nano. August 2010. Available from: www.observatorynano.eu

Rojas-Graü, M.A., Soliva-Fortuny, R., Martín-Belloso, O., 2009. Edible coatings to incorporate active ingredients to fresh-cut fruits: a review. Trends Food Sci. Technol. 20, 438–447.

Rozenberg, B.A., Tenne, R., 2008. Polymer-assisted fabrication of nanoparticles and nanocomposites. Prog. Polym. Sci. 33 (1), 40–112.

Rudra, J.S., Dave, K., Haynie, D.T., 2006. Antimicrobial polypeptide multilayer nanocoatings. J. Biomater. Sci. Polym. Ed. 17 (11), 1301–1315.

Sanchez-Garcia, M.D., Gimenez, E., Lagaron, J.M., 2008. Morphology and barrier properties of solvent cast composites of thermoplastic biopolymers and purified cellulose fibers. Carbohydr. Polym. 71, 235–244.

Šetkus, A., 2002. Heterogeneous reaction rate based description of the response kinetics in metal oxide gas sensors. Sensor. Actuat. B 87, 346–357.

Smolander, M., Alakomi, H.L., Ritvanen, T., Vainionpää, J., Ahvenainen, R., 2004. Monitoring of the quality of modified atmosphere packaged broiler chicken cuts stored in different temperature conditions. A. Time-temperature indicators as quality-indicating tools. Food Control 15, 217–229.

Sorrentino, A., Gorrasi, G., Vittoria, V., 2007. Potential perspectives of bionanocomposites for food packaging applications. Trends Food Sci. Technol. 18 (2), 84–95.

Sozer, N., Kokini, J., 2012. The Applications of Nanotechnology, first ed. Elsvier Inc. USA.

Tan, W., Zhang, Y., Szeto, Y., Liao, L., 2008. A novel method to prepare chitosan/montmorillonite nanocomposites in the presence of hydroxy-aluminum oligomeric cations. Compos. Sci. Technol. 68, 2917–2921.

Textor, T., Fouda, M.M.G., Mahltig, B., 2010. Deposition of durable thin silver layer onto polyamides employing a heterogeneous Tollens' reaction. Appl. Surf. Sci. 256, 2337–2342.

Usuki, A., Koiwai, A., Kojima, Y., Kawasumi, M., Okada, A., Kurauchi, T., Kamigaito, O., 1995. Interaction of nylon 6-clay surface and mechanical properties of nylon 6-clay hybrid. J. Appl. Polym. Sci. 55 (1), 119–123.

Vaia, R.A., Maguire, J.F., 2007. Polymer nanocomposites with prescribed morphology: going beyond nanoparticle-filled polymers. Chem. Mater. 19 (11), 2736–2751.

van Willige, R.W.G., Linssen, J.P.H., Meinders, M.B.J., Van Der Steger, H.J., Voragen, A.G.J., 2002. Influence of flavor absorption on oxygen permeation through LDPE, PP, PC and PET plastics food packaging. Food Addit. Contam. 19 (3), 303–313.

Vimala, K., Mohan, Y.M., Sivudu, K.S., Varaprasad, K., Ravindra, S., Reddy, N.N., Padma, Y., Sreedhar, B., MohanaRaju, K., 2010. Fabrication of porous chitosan films impregnated with silver nanoparticles: a facile approach for superior antibacterial application. Colloids Surf. B 76, 248–258.

Wang, N., Zhang, N., Wang, M., 2006. Wireless sensors in agriculture and food industry: recent development and future perspective. Comput. Electron. Agric. 50 (1), 1–14.

Weiss, J., Takhistov, P., McClements, D.J., 2006. Functional materials in food nanotechnology. J. Food Sci. 71 (9), R107–R116.

Wu, L.Z., Yu, Y., Han, X.Y., Zhang, Y., Zhang, Y., Li, Y.Z., Zhi, J.F., 2014. An electroless-plating-like solution deposition approach for large-area flexible thin films of transition metal oxide nanocrystals. J. Mater. Chem. C 2, 2266–2271.

Wu, L.Z., Yu, Y., Zhi, J.F., 2015. Low cost and large-area fabrication of self-cleaning coating on polymeric surface based on electroless-plating-like solution deposition approach. RSC Adv. 5, 10159–10164.

Xiao, L., Green, A.N.M., Haque, S.A., Mills, A., Durrant, J.R., 2004. Light-driven oxygen scavenging by titania/polymer nanocomposite films. J. Photochem. Photobiol. A 162, 253–259.

Yan, Y.Y., Gao, N., Barthlott, W., 2011. Mimicking natural superhydrophobic surfaces and grasping the wetting process: a review on recent progress in preparing superhydrophobic surfaces. Adv. Colloid Interface Sci. 169, 80–105.

Zhang, X.X., Wang, L., Levänen, E., 2013. Superhydrophobic surfaces for the reduction of bacterial adhesion. RSC Adv. 3, 12003–12020.

Zhao, R.X., Torley, P., Halley, P.J., 2008. Emerging biodegradable materials: starch- and protein-based bionanocomposites. J. Mater. Sci. 43, 3058–3071.

11

MICROENCAPSULATED BIOACTIVE COMPONENTS AS A SOURCE OF HEALTH

Leslie Violeta Vidal Jiménez
University of Concepcion, Faculty Agricultural Engineering, Chillan, Chile

1 Introduction

The development of multifunctional ingredients that can lead to different beneficial physiological effects for the human organism represents an important challenge to meet the expectations of consumers with specific needs. An alternative to these requirements is incorporating biologically active compounds into food, such as vitamins, minerals, dietetic fiber, and flavonoids. Flavonoids are chemical compounds exhibiting antioxidant and antimicrobial activity and functionality to be incorporated in different food matrices; they can be extracted from fruit and vegetables. Consuming foods rich in polyphenols is associated with bitterness, specifically with the oral sensation of astringency that is negatively perceived, thus forcing food producers to mask their incorporation in food products. Likewise, highly reactive groups of these polyphenols are degraded in a short time when extracted and lose their health-promoting properties; the need arises to protect them from degradation and make them bioavailable in the digestive route. In this respect, microencapsulation is presented as an alternative technology for stabilizing compounds of interest, which allow responding to the competitive demands of flavor and health arising from bioactive principles and food additives or ingredients. These components can be safeguarded by protective matrices that allow them to be incorporated in food and increase their useful life and functionality.

Therefore, the objective of this research was to encapsulate polyphenols derived from a maqui, *Aristotelia chilensis* (Mol.) Stuntz. (Elaecarpaceae), leaf extract to obtain microcapsules with antioxidant activity and evaluate the effect of encapsulant and surfactant concentrations in maintaining antioxidant capacity

Encapsulations. http://dx.doi.org/10.1016/B978-0-12-804307-3.00011-9

to avoid the astringent perception and low bioavailability when glycosylated in the stomach and affecting polyphenol efficacy.

The microencapsulation process was carried out based on the preparation of water oil (W/O) emulsion, which consisted in working with an aqueous phase of maqui leaf extract and gum arabic along with an oil phase of liquid vaseline. The antioxidant capacity of the maqui leaf extract was determined by measuring the inhibition percentage of the DPPH*(517 nm) and ABTS*(730 nm) radical; this capacity was maintained at stable levels of 95–99% for five months. The chromatographic profile of the maqui leaf extract allowed identifying the phenolic components and their relative quantities of phenolic acids (54.36%), flavonoids (42.10%), and stilbenes (3.55%).

Incorporating gum arabic to the extract in 5 and 15% concentrations did not produce complexation, nor interfered with the antioxidant activity (ABTS*) of the extract, and remained at levels close to 95%.

Mean microencapsulation yield of the maqui leaf extract with 5% gum arabic varied between 38% and 48%, while yield was maintained at 39% with the 15% gum arabic rate. Both yields responded in a similar way to the changes in gum arabic concentrations (5 and 15%) in the aqueous phase of the emulsion, but the 5% concentration had values ranging from 1.0 to 10 μm for microcapsule size.

It was also determined that the antioxidant capacity (as inhibition of ABTS radical) of the extract, once the microcapsules were formed, varied between 30% and 35%, compared with 94% in the aqueous phase of the emulsion. Similarly, the retention of the phenolic components of the extract in the microcapsules decreases when gum arabic content increases in the emulsion.

Finally, this study has shown that microencapsulation of maqui, *Aristotelia chilensis* (Mol.) Stuntz. (Elaecarpaceae) leaf extract antioxidants with gum arabic could help to improve the liberation, absorption, and/or protection of these active ingredients, that is, phenolic components. This would allow the compound to be beneficial in the right place and thus increase the bioavailability and organoleptic quality of food by avoiding undesirable sensory properties.

2 Microencapsulation

Microencapsulation technology defined as "micropackaging" that enables encapsulating liquid, solid, or gaseous functional agents. The mechanism of microencapsulation is to catch droplets or particles of a sensitive bioactive material or thin films of a

Figure 11.1. Microcapsules types. Lupo et al., 2012.

coating material (Fuchs et al., 2006). The structure formed by the microencapsulating agent around the microencapsulated substance (core) is called wall: it protects the core from deterioration and allows its release under desired conditions. A microcapsule consists of a semipermeable irregular and wiry membrane, or spherical (Fig. 11.1) around a solid/liquid center, with a diameter ranging from a few micrometers to 1,000 μm being influenced by the physico-chemical properties of the active ingredient and coating material and the chosen technological process (Desai and Park, 2005).

The core comprises the microcapsule inner phase is also called active ingredient or while the membrane can be named outer layer or matrix. Microcapsules or microspheres are defined as the product of the microencapsulation process whichever is their morphology and internal structure. The microcapsules are differentiated mainly microsphere distribution of the active ingredient. In the first case the core may be solid or liquid in nature including a kind of reservoir or coated with a film of the material, while the microspheres of the active ingredient are highly dispersed as particles or molecules in a matrix. Obtaining one structure type or another depends on the physicochemical properties of the active substance and the matrix and the technique employed for their preparation (Desplanques et al., 2012).

2.1 Methods of Microencapsulation

There are several methods to consider when choosing or developing an encapsulated active ingredients (Soottitantawata et al., 2005). Different techniques like spray drying (Saénz et al., 2009); air suspension coating, extruding, spray cooling, and spray chilling, coacervation, inclusion complexes, and interfacial polymerization (Desai and Park, 2005) are available to encapsulate active agents. The encapsulation method selection will depend on cost, the properties of the material to be encapsulated, the desired microcapsule size, system application, and release mechanisms of the active ingredient required (Fig. 11.2).

Emulsification shows itself as a technique effective for microencapsulated functional compounds (Gaysinsky et al., 2008). It has been shown that microencapsulation of these components

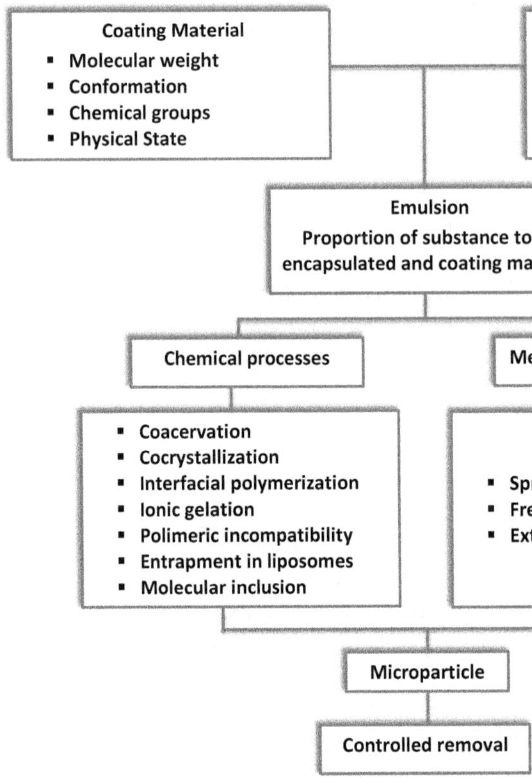

Figure 11.2. Schematic illustration of the different processes of microencapsulation. Madene et al., 2006.

provides protection against degradative reactions, prevents loss of volatile aromatics, and allows processing fluids to manipulate solids to allow for incorporation in various food matrices.

The emulsion via encapsulation technique has been defined as the process of dispersion of a liquid in another immiscible liquid where the dispersed phase consists of the matrix that includes the encapsulated component. Essentially it consists of a mixture of liquid stirring with a mechanical system that generates turbulence; this process can form a dispersion (multiphase system in which one phase is dispersed in a continuous phase within another). Emulsions are thermodinamically unstable dispersions of a least two inmiscibles liquids or partially miscible liquids. Lowering of interfacial tension is one way in which the increase surface free energy can be reduced, leading to a more stable emulsion, due to the formation of smaller emulsion droplets with narrower droplets size distribution and greater kinetic stability (Golemanov et al., 2006). One generally corresponds to a liquid aqueous phase and the other an oil phase: the abbreviations for water (W) and oil (O) are used for these phases. We speak of a normal emulsion when the emulsion contains oil droplets (O) dispersed in water (W) and is called oil/water emulsion (O/W), while if we have a dispersed aqueous phase in a medium of oil phase, there is talk of a water/oil (W/O) emulsion. They can also generate double emulsions, where two types of liquids of different nature are dispersed within a continuous phase. There may also be more complex systems, for example, drops of oil emulsion of a multiple water/oil/water (W/O/W) type.

Currently, there has been a growing interest in the development of water-in-oil in water (W/O/W) since they present a number of advantages over conventional O/W emulsions, such as controlled or triggered release, reducing the content fat, and protection of labile ingredients (Surh et al., 2007). However, there are difficulties associated with the preparation of multiple emulsions in use in the food industry due to internal problems coalescence and diffusion expulsion of water molecules from the internal aqueous phase to external aqueous phase.

Once it has formed the emulsion, it is required that the system remains stable for as long as possible. The emulsion stability as well as ease of their formation requires the application of surfactants.

2.2 Surfactants

Surfactant is a term used to designate the abbreviated "interface-active compounds." Chemically, surfactants are characterized by a molecular structure having a group with limited attraction for the solvent, nominated liofhobic group, adjacent to another group with strong attraction for the solvent, called lipophilic group. If the solvent is water, these groups are known as the hydrophobic and hydrophilic portions of the surfactant. Usually, the hydrophobic group is a straight or branched carbon chain, while the hydrophilic portion is a group with some polar character.

Surfactants show a strong tendency to migrate to interfaces, in such a way that its polar group is oriented toward the water and a polar group is oriented toward an organic solvent or in the surface. This causes changes in interfacial interactions leading to a decrease in surface tension. Therefore, they are known as surface agents or surfactants. Furthermore, these compounds have the ability to form micellar aggregates of different characteristics.

When surfactants are found in low concentrations in a system, they are adsorbed in the surfaces or interfaces, substantially modifying the superficial or interfacial energy of such systems. When a surfactant is dissolved in water adsorption of surfactant molecules it occurs on the water surface, reducing its surface tension, which causes a decrease in the surface free energy of the system. If this is a system with oil/water/surfactant, the surfactant is adsorbed in the oil-water interface, decreasing the interfacial free energy.

The emulsion stability implies that they exhibit resistance to coalescence of the droplets in the dispersed phase. The process of destabilization of an emulsion involves different mechanisms (sedimentation, flocculation, coalescence, and Ostwald ripening) that can happen simultaneously or consecutively (Mirhosseini et al., 2008). Stability evaluation is influenced by formulation variables (temperature, nature of the emulsifying agent of the water, and oil phase); composition (relative proportions of water and oil and emulsifier concentration), and fluomechanical factors (equipment used, stirring intensity, and procedure). However, the single most reliable measurement of the stability of an emulsion is the variation of number of drops with time (Mirhosseini and Bahareh, 2012). Therefore, the stability is generally related to the volume of the separated phases. After a time, the system is typically separated into three zones: a central zone containing a cream or emulsion of high internal phase and two separate phases: internal (coalesced) and outer (clarified).

Relevant phenomena involved in the stability of emulsions are:

- Physical nature of the interfacial film: Maximum resistance committed constituted by the surfactant molecules with strong intermolecular interactions (Golemanov et al., 2006). Emulsifying agents with ideal characteristics to ensure high stability, are mixtures of two or more surfactants, the most recommended combination, constituted by a hydrophilic surfactant, and one lipophilic.
- Viscosity of the continuous phase: Determines the frequency of collisions between droplets of the dispersed component, through diffusion coefficient. An increase in viscosity reduces the coefficient of diffusion and coalescence slows accordingly. This is one reason for the incorporation of the external phase emulsifying agents.
- Droplet size distribution: The smaller the droplet size the more stable the emulsions and the narrower the distribution of those sizes.
- Phase volume ratio (WOR): Refers to the relative ratio of the volume of dispersed phase and continuous phase volume known as WOR (water-oil ratio). When the internal phase is less than 30% of the total volume, individual droplets do not interfere with each other and the physical properties of the system are primarily determined by the nature of the continuous phase.
- Temperature: The temperature changes greatly affect the stability of emulsions, through its effect on interfacial tension, viscosity and interfacial nature film, partition coefficient of the

surfactant, viscosity and thermal agitation phases of the dispersed particles.

2.3 Microencapsulation Applications

Microencapsulation has been applied in different systems in order to stabilize different components of interest to industry. Chanamai and McClements (2002), studied the microencapsulation process soybean oil by forming an emulsion. Stabilizing oil with modified starch, gum arabic and whey proteins, depending on the concentration of calcium chloride, pH and temperature were compared. Resulting whey proteins as emulsifier better due to electrostatic repulsion mechanism compared to the steric repulsion of gum arabic and modified starch. Furthermore, Soottitantawata et al. (2005) used as encapsulating agents in the formation of emulsions, mixtures of gum arabic (10% w/w), maltodextrin (30% w/w), modified starches and soluble soy protein in water (10% w/w) for microencapsulating d-limonene, ethyl butyrate, and ethyl propionate, and subsequent controlled release of these compounds in preventing degradation during spray drying to obtain powdered flavors. The results show that the kinetics of release of the flavor is determined by the type of encapsulating material used and the size of the generated microcapsules. The increase in average droplet size produced a decrease in flavor retention.

The behavior of emulsions stabilized with whey protein in the presence of mixtures of polysaccharides, acacia-chitosan at various concentrations were studied in terms of their stability, microstructure, and functional properties by Moschakis et al. (2010). Consequently, the behavior of the emulsion and the microstructure are dependent on the polysaccharide concentration. The relationship between the molecular structure of the polysaccharide and its ability to enhance the viscosity of the aqueous phase and to induce depletion floculation, controls the kinetics of the overall phase separation process, both concentration-dependent variations of electrostatic forces polysaccharides and positively charged protein whose mechanism responsible. It is the depletion flocculation.

Also Desplanques et al. (2012) studied the effect of the chemical structure of xanthan gum–arabic gum mixture on the stability of oil in water emulsions (O/W). These authors concluded that the content of arabinogalactan protein (AGP) present in the gum arabic and pyruvate xanthan gum groups allowed for better stabilization of the interface O/W, indicating that this partnership in the emulsion, recognizes the structure of xanthan gum while the droplet size depends on the content of arabinogalactan protein (AGP) present in the gum arabic. Although the molecular interactions contributed stabilizing gums, the highest content of AGP in the fraction of the emulsion stability was observed when intervened by only gum arabic.

Mixtures of maltodextrins and gum arabic were used in cardamom oil microencapsulation by spray drying (Madene et al., 2006) as solid carriers, providing adequate viscosity to stabilize the emulsion.

In the past decade, research has shown a growing interest in the encapsulation of natural extracts with high polyphenol content from medicinal plants, herbs, flowers, and seeds. This has emerged with the aim of reaching new nutritional proposals, allowing the use of antioxidants as food additives without changing the taste, aroma, and color of the original products and to increase the use of its protective benefits against free radicals and degenerative diseases. Various studies indicate that the large hydrophilic character may be offset by applying special treatments in the microencapsulation process. In this regard, Zhang et al. (2007) evaluated the matrix of maltodextrin (60%) and gum arabic (40%) in the microencapsulation of procyanidins from grape seeds. These were unchanged during the critical drying step. The encapsulation efficiency was around 85%, significantly improving stability. Within the same carbohydrate matrix, epigallocatechin gallate (EGCG) from green tea was encapsulated, yielding the same encapsulation efficiency of 85%. These particles were able to inhibit tumorigenesis process steps (Rocha et al., 2011). Also, chitosan was used as a coating material for encapsulating olive leaf extract (OLE) (Kosaraju et al., 2006). OLE loaded microspheres (27%) with a smooth surface on the microsphere pattern, indicating the influence of structural interactions polyphenols present in the extract and the polymer matrix was obtained.

Recently, a soy extract rich in polyphenols was immobilized on a composite matrix of maltodextrin, starch, and silica (Tixosil® 333) (Georgetti et al., 2008). The results show that the Tixosil 333 reduces degradation of the encapsulated polyphenol and protects its antioxidant activity. Adding this carrier during the drying step ensures the stability and effectiveness of the final product. Studies by Krishnaiah et al. (2009) showed that the carrageenan is an efficient encapsulating agent to maintain the antioxidant activity of various natural polyphenolic extracts.

Encapsulation of polyphenolic compounds in protein–lipid matrices shows significant efficacy. The grape seed extract, and apple and olive leaves rich in oleoropeína, were immobilized within a matrix of sodium caseinate and soya lecithin (Kosaraju et al., 2008). Electronic microscopic observations and sieve analysis revealed the presence of spherical particles with a uniform size

(80% particle/6–60 microns). These results favor the retention of polyphenols after encapsulation by spray drying, which can be used for implementing method nutraceuticals. The encapsulation of an extract of oak (Quercus resin), with high polyphenol content, was performed by high-pressure homogenization (Rocha et al., 2010). This extract showed instability, bad taste, and strong astringency, which makes encapsulation necessary before incorporation into food products. Therefore, it was encapsulated in a matrix of sodium caseinate and lactose, having a high antioxidant activity even at very low concentrations of phenolic compounds.

Xiong et al. (2006) encapsulated four structures of anthocyanins from black currant extract (delphinidin-3-O-glucoside, delphinidin-3-O-rutinoside, cyanidin-3-O-glucoside, and cyanidin 3-O rutinósido-) using β-glucan as encapsulating agent, spherical and cubic forms based on the effect of temperature, pH, and the presence of ferric–ferrous ions in antioxidant activity and stability of the encapsulation process these structures. The investigation was conclusive in establishing that the release of encapsulated anthocyanin was most significant in the cubic forms of the encapsulant, and more stable than the free anthocyanins antioxidant activity.

Catechins are powerful natural antioxidants, but unstable in biological fluids under alkaline conditions and in some experimental protocols. To avoid the problems mentioned above, catechin and (–)-epigallocatechin were immobilized on chitosan nanoparticles tripolyphosphate (Dube et al., 2010). The results showed that the antioxidant activity after 24 h was 88.3% and 73.4%, respectively. Where 50% of the encapsulated catechin was degraded, while 8 h were sufficient to degrade the same amount of free catechin. Epigallocatechin was unstable, because after 40 min over 50% denatured.

Lyophilized extract of yerba mate (*Ilex paraguariensis*) containing 62.11 ± 1.16 mg gallic acid/g, was encapsulated by ionic gelling (calcium alginate), and complex coacervation in a matrix of calcium alginate and chitosan (Deladino et al., 2008). Both types of packages were resistant oven-dried and lyophilized. The antioxidant activity was greater than 85%, of the phenolic compound immobilized in alginate, compared to 50% obtained with the alginate–chitosan matrix. Gallic acid release revealed the influence of the nature of the encapsulating material with respect to the release of natural antioxidants present in yerba mate. Similar results obtained Anbinder et al. (2011) encapsulation of extracts of yerba mate (*Ilex paraguariensis*) in calcium alginate and an alginate–chitosan matrix. While both systems showed a high content of polyphenols released in the simulated gastric fluid, covering with chitosan capsules allows greater release of polyphenols. This was attributed to the protective barrier it presents chitosan and the strong interaction of the complex formed by yerba mate extract and chitosan.

Levic et al. (2011) tested the encapsulation of d-limonene in a matrix of calcium alginate and polyvinyl alcohol (20:80) by "freezing–thawing" method to ensure formation of cryogel structure polivilílico alcohol. The results of thermal analysis show that the release of d-limonene extends over a wide temperature range compared to free flavor, indicating improved stability of the encapsulated flavor compared to free flavor. However, leakage of fragrance at higher concentrations (10% w/w) appears to be the problem due to the hydrophobic nature of aroma, which leads to separation of the polymer/flavor mixture into two layers. The combined homogenization with the appropriate polymer composition is critical to achieve high retention of limonene during sample preparation prior to encapsulation.

Propolis, a mixture rich in polyphenols collected by bees from certain plants, has known therapeutic properties. However, its use as a food additive is limited because it is solubilized in alcohol and has a strong taste. Propolis encapsulation by complex coacervation with pectin and soy protein appears to be an interesting alternative (Nori et al., 2011). The result is a powder that is readily dispersible in solvents other than alcohol, with unquestionable antioxidant and antimicrobial properties, and the release of the active material can be controlled.

Grape proanthocyanidin microparticles were encapsulated by interfacial polycondensation considering the polyphenolic compound as membrane material, and the cross-linking reaction of the molecule as a stabilizer, while the radical scavenging activity is mantienía. The cross-linking reaction grape proanthocyanidin terephthaloyl dichloride involved with phenolic hydroxy groups leading to the establishment of ester linkages that were detected by infrared spectroscopy (Munin and Edwards, 2011). The microcapsules obtained cross-linked and pH 9 and 11, were smaller than 10 microns and were stable for more than five months at 45°C in an aqueous medium. The microcapsules were slowly degraded in plasma and showed antioxidant activity of interest, although slightly lower than the initial grape proanthocyanidin. The method of preparation of these microcapsules by interfacial cross-linking of polyphenols is patented.

Microencapsules were prepared from compounds and extracts obtained from a large number of fruits and vegetables, for example, vegetable juices such as tomato, cucumber, carrot, lettuce, beets, spinach, celery, and parsley (García et al., 2004). Also, volatile substances such as orange oils, cinnamic aldehyde, ethyl buturato, etilpropionato, have been encapsulated using arabic and maltodextrins rubber, limiting the degradation of said compounds during processing and storage (Madene et al., 2006).

2.4 Encapsulating Materials

The substance coating the active agent is basically a filmogenic capacity material, which can be selected from a wide variety of natural or synthetic polymers, depending on the physicochemical nature of the material to be coated, and the characteristics desired in the final microstructure. The coating composition is the main determinant of the functional properties of the microcapsule and the method to be used to improve protection of a particular ingredient. Effective coating material must have good rheological properties at high levels and should be easy to handle during processing or storage.

There are a variety of materials used in the microencapsulation process, such as the following.

- Polymers: The following are the main polymeric encapsulants.
  - Derivatives of cellulose: These feature prominently as encapsulants (Burey et al., 2009). Cellulose is the most abundant of all organic materials, fibrous part of plant tissues, is water soluble, and its solubility increased by treatment with alkali to swell the structure, followed by reaction with trichloroacetic acid chloride methyl or propylene oxide producing carboxymethylcellulose (CMC), hydroxypropylmethylcellulose (HPMC) or hydroxypropylcellulose (HPC). The anionic and nonanionic CMC MC, HPMC, and HPC have excellent film-forming characteristics. Derived HPC cellulose is a thermoplastic polymer that can be injection molded or extruded; it is edible and biodegradable. The MC is less hydrophilic, cellulose ethers do not offer a good moisture barrier, but have excellent barrier to the migration of fats and oils. Aqueous solutions of MC form relatively strong gels to review about 50° C. HPMC solutions are low thermally induced gels force between 50–85°C. HPC solutions form gels that are not hot, but precipitate at temperatures of 40°C. The HPC is a polymeric surfactant soluble in cold water. It is a good thickener that can be surface-treated to permit direct addition of water at pH 7.5 or less, and has good electrolyte tolerance, being stable in pH ranges from 3 to 11. Some types of HPC are also dispersible alcohols and glycols.
  - It described the existence of two major mechanisms in the formation of the coating film of HPC in a microcapsule:

The first in which the polymers have dominated the forces of interpenetration and a second mechanism in which cohesion and adhesion forces dominated. Depending on the conditions, the film may have an intermediate to the above mechanism, depending upon the degree of ionization of the polymer. Both mechanisms are conducted in solution, which facilitates the interaction between the polymer chains with the active agent, protecting the functional groups of the polymer having a high degree of interpenetration, and forms a more stable three-dimensional network molecular level interactions. Following this, the formation of a tough laminate film occurs. Clearly, the properties of films formed for encapsulating depend largely on the nature of the polymer. Thus, the bioadhesive characteristics of cellulose polymers like HPMC, HPC, and CMC can improve encapsulation efficiency and retention of the active agents, as such are hydrophilic molecules, which have groups capable of forming hydrogen bonds.

- Alcohol–Polivinil (Leiman et al., 2009): hydrophilic polymer can be used as capsule wall-forming material as nylon membranes have also been used to encapsulate and entrap enzymes such as pepsin, pectin esterase for the clarification of juices, invertase for sucrose inversion.
- Quitosano: Polymer used in the microencapsulation and in the food industry for its nontoxic, biodegradable, biocompatible, and mucoadhesive properties. It stands as an antimicrobial and antioxidant compound (Alishahi and Alder, 2012).
- Alginato: Polymer extracted from algae and used as a nontoxic, biocompatible, and soluble encapsulant (Nazzaro et al., 2009). Calcium alginate has been widely used in the immobilization of lactic acid bacteria (LAB), for its ease of use, nontoxic nature, and low cost (Bastos et al., 2009). Immobilization of bacteria in biodegradable microcapsules creates a suitable environment for their survival, providing temporarily immobilized bacteria protection from the environment, soil, competitors, and predators.
- Eudragit: Refers to a group of polymers derived from methacrylic acid that are available in different ionic forms. They are highly soluble due to its alkaline pH value, and by neutralizing the carboxyl groups with formation of the corresponding salt, and therefore exhibit the character of anionic polyelectrolyte in solution. Different types of Eudragit were used in preparing microparticles, allowing the release of active substances in the intestine, preventing inactivation of

drugs in the stomach, for example, in the preparation of microparticles that allow oral administration of peptides and proteins (Villamizar and Martinez, 2008).

- Lipids: The main lipid encapsulating agents include: milk fat, lecithin, waxes, stearic acid, monoglycerides, diglycerides, waxes, hydrogenated oils such as palm oil, cotton, and soybeans. These are excellent film formers, capable of covering the individual particles and providing a uniform encapsulation (Yañez et al., 2002).
- Carbohydrates: Widely used in encapsulation, specifically spray drying technique for food ingredients and encapsulation support, this broad group includes starches, maltodextrins, and gums (Murúa et al., 2009).
- Starches: They are widely used in the food industry (Madene et al., 2006; Murúa et al., 2009) starch-based ingredients (modified starches, maltodextrins, β-cyclodextrins). Among the most important stand the starch potato (*Solanum tuberosum*), corn (*Zea mays*), wheat (*Triticum aestivum*), rice (*Oryza sativa*), tapioca (*Manihot esculenta*) (Fuchs et al., 2006; Yañez et al., 2002) and inulin (Saénz et al., 2009). The native and modified tapioca starch and maltodextrin has been investigated for its ability to be used as wall material for encapsulation of β-carotene (Li et al., 2009).
- Maltodextrins: Obtained by acid or enzymatic hydrolysis of starches. In selecting wall materials to encapsulate, maltodextrin is a good solution because of its cost and effectiveness. Exhibits low viscosity, high proportion of solids. At high concentrations, they are odorless, colorless, and also allow the formation of free-flowing powders without masking the original taste. They are commercially available at different molecular weights and are widely used in the food industry (Madene et al., 2006; Saénz et al., 2009).
- Proteins: Several proteins have been widely used as microencapsulating agents, including highlights sodium caseinate, whey protein, soy protein isolates (Madene et al., 2006; Murúa et al., 2009), waxes (Fuchs et al., 2006), gluten, gelatin (Yañez et al., 2002), casein, soy, wheat (Saénz et al., 2009) and gelatin, the latter used their good emulsifying, film formation, water solubility, and biodegradability.
- Gums: Obtained by chemical modification of native polysaccharides, high molecular weight, hydrophilic and hydrophobic characteristics, have colloidal properties, form gels or viscous solutions when combined with the appropriate solvent. They possess characteristics of thickening agents, gelling agents, emulsifiers, and emulsion stabilizers. Natural gums, are

defined as carbohydrate polymers, highly hydrophilic, insoluble in alcohol and other organic solvents, and are very versatile for most encapsulation methods (Madene et al., 2006; Murúa et al., 2009). Among them are locust bean gum, gum arabic, guar gum, tamarind gum, gellan, and xanthan gum; applicability has been seen to immobilize bacterial cells, for which were used alginates and carrageenan.

Among the most widely used encapsulating materials in the food industry is gum arabic (Alishahi and Alder, 2012). This has been considered the quintessential emulsifying agent. It is a product obtained from the acacia tree and is defined as a complex and slightly acidic polysaccharide, belonging to the group of natural hydrocolloids. It stands out for its versatile functional properties as filmogenic agent colloidal achieving maximum protection, and as an encapsulating agent of essential oils (Fuchs et al., 2006). It has also been used as an emulsifier and stabilizing agent in emulsions (O/W, W/O) and within a wide range of pH.

Its multimolecular structure consists of three fractions (Fig. 11.3). The first fraction consists of arabinogalactan (AG) polysaccharide chains, low molecular weight, and low protein content (0.5%), this fraction representing 90% of the molecule. The second fraction corresponds to arabinogalactan protein (AGP) complex of low molecular weight and higher protein content (10%), representing 10% of the molecule, and the third fraction is a glycoprotein (GP) of low molecular weight and high protein (50%) content, representing 1% of the molecule. Its structure corresponds to polysaccharide arabinogalactan is highly branched acids, consisting of: D-galactose (44%), L-arabinose (24%), D-glucuronic acid (14.5%), L-rhamnose (13%), acid-4-O-methyl-D-glucuronide (1.5%). Major brands are formed by units β-D-galactopyranosyl units attached by glycosidic linkages (1 → 3), with side chains of

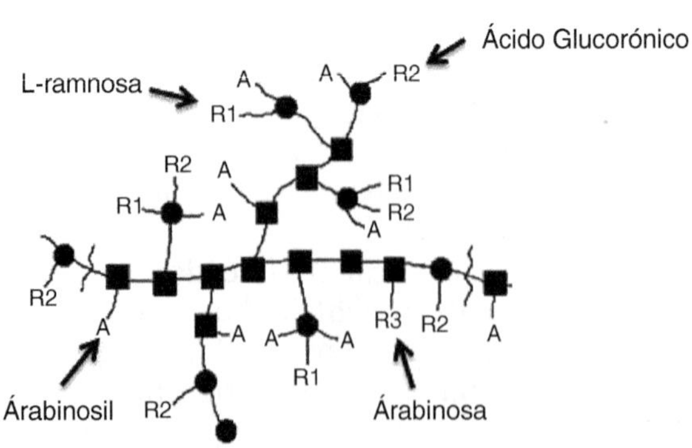

Figure 11.3. Molecular structure of the gum arabic. Mirhosseini and Bahareh, 2012.

two to four units of the same sugar and link type, connected thereto by glycosidic linkages (1 $\rightarrow$ 6).

Fractions containing a high content of protein (AGP and GP), give this polymer the emulsifying property and acting at the interface between the oil droplets and the water, where it is believed that the predominant amino acids in the polypeptide (hydroxyproline and serine), hydrophobic character of the molecule attached to the surface of the droplets. The ability to form a protective film is given by the AG fraction, which has low viscosity and high solubility in water by its highly branched structure, decreasing interchain interactions and facilitating solvation, while it blocks arabinogalactan hydrophilic character extending in the aqueous phase, providing stability to the emulsion (Chanamai and McClements, 2002).

Also, glucuronic acids of the molecule are partially filled, developing negative charges around the oil droplets in the emulsion. The fraction of higher viscosity AGP delivery system slows the attraction of the droplets in the emulsion and thus may help prevent agglomeration of the droplets.

2.5 Characterization of the Microcapsules

The microcapsules obtained by any of the procedures described above must be characterized according to various tests to ensure quality, uniformity, and bioavailable behavior. The characteristic tests that are usually performed are: morphology, particle size and internal structure, production yield, encapsulation efficiency and active ingredient content, study drug release, and fitness and polymer-active substance interactions

Depending on the conformation there are different types of microcapsules, which have a wide variety of structures: spherical or irregular, simple, one wall with a cover membrane, a multiwall structure with walls of some or several or numerous compositions cores inside a wall structure, which gives differentiated properties and applicability.

Moreover they can be of the matrix type, that is, the active material is highly dispersed in the polymer matrix. These may have a foamed structure in which the active material is distributed throughout the microcapsule and the cover, which either remains intact or in an open netlike structure. Microcapsules can be found with active material dispersed in the matrix covering the entire sphere and the periphery.

The morphological characteristics of microcapsules are analyzed by optical microscopy and scanning electron microscopy (SEM). The latter allows to detect the possible aggregation, of particles, characterize the internal structure, determine the internal compaction and size of the microcapsules.

The size distribution of the microcapsules is determined by using microscopic techniques, sieving, sedimentation, by laser diffraction or Coulter Counter equipment.

2.6 Release Mechanisms of Encapsulated Compounds

The release behavior of bioactive agents is the result of the phenomenon of diffusion and the polymer mass transfer restrictions at the interface polymer/liquid. Thus, modifying or designing a controlled release system requires prior knowledge of the mechanism of solute diffusion through the polymer material (Champagne and Fustier, 2007).

The release mechanisms of the active agents of the microcapsules can be carried out by different methods, such as dissolution in water, application of shear, temperature, chemical, or enzymatic reactions, and changes in osmotic pressure. The release of components of a capsule can be controlled by diffusion of the capsule wall or membrane that covers the wall (Betz et al., 2012). Permeability through the matrix and the solubility of component capsule wall influence the diffusion rate. The compound to be spread must be soluble in the matrix. Although the vapor pressure of volatile substances on either side of the matrix can be determined by the spreading force. Selecting a matrix or membrane is important, so their chemical nature, morphology and the glass transition temperature, the degree of swelling and cross-linking also influence the diffusion of the active agent through the membrane.

The release of active ingredient from within the microcapsules is governed by a number of factors that are dependent of the polymer (solubility pH-dependent, molecular weight, crystalline state), the active ingredient (solubility, molecular weight) of the microparticle itself (type internal structure, theoretical content of the active ingredient to polymer).

For the study of release, researchers can use the procedure specified by the USP (United States Pharmacopeial) and methods of flow, agitation vial, and dialysis membranes, among others.

The release rate is a more abrupt rupture of the microcapsule cover by applying a force. The force required to rupture the capsule wall is determined by the material forming the wall and by the thickness thereof, and both parameters can be controlled during the microencapsulation process. Another method of release is the dissolution of the microcapsule coating in between. This can occur by melting, action of a solvent, enzymatic attack, hydrolysis, slow or disintegration chemical or photochemical reaction. Both the breakdown and dissolution of cover systems are immediate release of the microencapsulated agent.

If, however, it is desired to obtain a gradual and sustained release thereof, in which the internal phase is to be released over time, the mechanism of diffusion through a porous coating is the best choice. The encapsulated ingredient will slowly diffuse through the cover. The rate of diffusion is controlled by the permeability of the same, as well as the size and shape of the encapsulated substance.

Another alternative for sustained release over time is the use of coatings able to change their morphology to certain stimuli, such as changes in pH, ionic strength, temperature, and light. In this case, control over the release profile can be accomplished by modifying the environmental conditions.

Regardless of the delivery system, the diffusion coefficient of the bioactive agent through the polymer depends on the structural and morphological parameters thereof, as well as the solute concentration. With this in mind, it can be said that one of the most laborious tasks in the field of technology of controlled release, is the development of formulations of polymers (matrix type) able to release active substances at a constant speed for a predetermined time. One approach is the use of hydrophilic polymers having the ability to swell in aqueous media without dissolving and releasing the dissolved or dispersed active principle, providing a substantially constant speed.

Migration of the active ingredient to the aqueous medium involves a process of absorption of water or biological fluid, and a simultaneous process of desorption active ingredient, by a diffusion mechanism, controlled by swelling the polymeric material presented.

The analysis of the dynamic behavior of polymer swelling and its effect on solute diffusion, therefore requires a knowledge of the thermodynamics of polymer medium/solution system and the system solute/polymer/dissolution medium. The solvent into a polymer is in a glassy state that produces a considerable increase in the macromolecular mobility, which implies a decrease in glass transition temperature, T_g. If the solvent is very compatible with the polymer, the decrease in T_g is not enough for the polymer reaches its rubbery state. Therefore, when the system reaches thermodynamic equilibrium, the polymer is in the glassy state and, under these conditions, the release of any active ingredient becomes very slow and has limited application. Conversely, if the solvent is thermodynamically good, the probability of reaching the elastomeric state is high and the solute is capable of smoothly spreading the swollen regions from the external environment.

3 Antioxidant Properties of Phenolic Compounds

Since the scientific and agribusiness view the antioxidant properties of phenolic compounds, has generated great interest for its abundance in the diet, and its preventive role associated with oxidative stress (Robert et al., 2010; Munin and Edwards, 2011).

Polyphenols are major classes of secondary metabolites of plants, where they play different physiological functions involved in growth, reproduction, and defense against pathogenic processes, predators, or ultraviolet radiation. The levels of these compounds may vary considerably within the same plant species, and even between varieties due to genetic and environmental factors influencing the germination, growth, and quality of crops (Ordoñez et al., 2006).

In foods, the phenolic compounds are typically present conjugated to sugars such as glucose, galactose, arabinose, rhamnose, xylose, and glucuronic or galacturonic acids. They can also join carboxylic acids, organic acids, amines, and lipids (Manach et al., 2004).

The antioxidant activity of the phenolic compounds is attributed to their ease to yield hydrogen atoms of an aromatic hydroxyl group to a free radical and the possibility of relocation of loads (resonant effect) in the system of double bonds of the aromatic ring (Apak et al., 2007).

Phenolic compounds also possess a chemical structure ideal for capturing metal ions (mainly iron and copper) and inhibit free radical formation through the Fenton reaction. The type of compound, the degree of methoxylation, and the number of hydroxyl groups are among the parameters that determine the antioxidant activity.

Many of these phenolic compounds are also responsible for the organoleptic properties of plant foods and therefore directly affect the quality of some of them. Phenolic compounds are pigments such as anthocyanins, responsible for the red, blue, and purple characteristics of many fruits, vegetables, and red wine tones; or flavonols, present mainly in fruits and vegetables, that give a creamy-yellow color. Some polyphenols such as citrus flavanones (naringin) provide a bitter taste, whereas other fruits confer astringency as hydrolysable tannins.

Depending on their structure, polyphenols can be classified into two groups: flavonoids (flavanones, flavones, isoflavones, flavonols, flavan-3-ols, proanthocyanidins, and anthocyanins) and nonflavonoids (hydroxycinnamic acids, hydrolysable tannins, hydroxybenzoic acids, and stilbenes) (Heinonen, 2007; Gaysinsky et al., 2008).

The proposed explanations of the beneficial health effects exerted by flavonoids' mechanisms are antioxidant effect, metal chelation, enzyme inhibition, and gene regulation (Neto, 2007).

Flavonoids can exert their antioxidant activity in many biological systems, but their distribution depends on their relative hydrophilicity/hydrophobicity and their interactions with certain macromolecules (Koksal et al., 2011). These factors determine the local concentration of flavonoids, which affects their ability to regulate certain cellular phenomena. The protective effect exerted by flavonoids such as catechins present in tea, has been attributed to its ability to neutralize or sequester free radicals. Several studies have shown the importance of these compounds with special emphasis on structure-activity relationships and mechanisms underlying the biological activity of flavonoids, which regulate the activity of certain cellular enzymes and that part of this regulation, it would be related to the flavonoids ability to alter the structure of the plasma membrane (Lu et al., 2010). In this effect, flavonoids would act on various cellular processes closely related to the plasma membrane, as cell signaling, cell cycle, arachidonic acid metabolism, cell proliferation, apoptosis, and mitochondrial functionality (Fuentealba et al., 2012). This sorts the pharmacological actions of flavonoids into the following groups: vasoprotectors modify levels of cholesterol and lipids, antiplatelet agents, enzyme modifiers, estrogenic activity, anticancer activity, antibacterial and antifungal activity, antiurémica activity, spasmolytic activity, antiallergic activity, anti-inflammatory, and antiviral activity (Schreckinger et al., 2010).

Free radicals and reactive oxygen species are generated during normal metabolism of oxygen or are induced by exogenous factors, and represent a potential risk to cells and tissues. Flavonoids exert a protective effect, as part of exogenous antioxidant defense systems of the body, that is, those defenses that are generated through diet. Three types of mechanisms that may account for the antioxidant activity of these defenses are as follows: (1) electron transfer determines that the antioxidant becomes a radical active molecule, (2) electron transfer causes the formation of an antioxidant molecule to be stable or inactive, (3) small molecules act as antioxidant enzymes (Nohynek et al., 2006).

It is believed that the mechanism by which flavonoids exert their antioxidant activity is based on electron transfer, which involves the emergence of a radical active molecule and the ability of these compounds to chelate metals. But flavonoids must meet two additional requirements to be considered antioxidant molecules: (1) at low concentrations must protect the compounds from oxidation or damage induced by free radicals, and (2) formed flavonoid radical (radical call aroxilo) It must be stable enough to be effective antioxidant function. The unstable nature of aroxilo radical can cause prooxidant effect as shown by some flavonoids, however collaboration antioxidant molecules, promotes recovery aroxilo radical by other antioxidants, such as ascorbate. Flavonoids have antioxidant capacity that varies depending on the number and position of their hydroxyl groups attached to the ring structures (Pastene, 2009).

The nature of flavonoid compounds present a series of structural features that allow us to assess its possible antioxidant properties: (1) the presence of a catechol group (3', 4'-dihydroxy) in ring B; (2) the presence of an unsaturated double bond between C2 and C3 in the C ring; (3) the presence of a hydroxyl group at C3 in ring C. Phenols with a catechol structure in the B ring, despite having increased antioxidant potential are metabolized more easily.

To understand the potential of flavonoids to act as antioxidants in vivo, it is necessary to consider other factors such as bioavailability and interactions in the gastrointestinal tract, as well as the influence of conjugation and metabolism. Polyphenols have a rather low bioavailability—that is, can be easily destroyed during digestion—and can have unpleasant flavors or cause astringency. Astringency is a feeling of dryness or constriction in the mouth that develops and dissipates slowly.

The molecular basis of this phenomenon are interactions between polyphenols and glycoproteins and mucopolysaccharides forming saliva, resulting in insoluble aggregates that precipitate and clog palate lubrication, giving the rough feeling of the particular astringency. Both astringency and bitter taste are generally perceived as negative attributes in beverages and soy products, milk, and juices. Because of this, the food industry continually removes polyphenols and other compounds present in plant products. However, these bioactive phytochemicals or phytonutrients remain promising for the creation of designer foods for the prevention of chronic diseases (Howell et al., 2005; Puupponen et al., 2005). These competing demands for taste and health pose a dilemma for the food industry.

One solution to this problem is to try to mask the astringency of such compounds. Satisfactory results have been reported using food-grade polysaccharides (guar, xanthan gum, and carboxymethyl cellulose and rubber) because the increasing viscosity may change the perception of astringency and the release of volatile and nonvolatile compounds (Troszynska et al., 2010).

Therefore, the application of phenolic compounds requires the development of a protection matrix capable of maintaining the structural integrity of the polyphenol during consumption or administration, masking taste, improving water solubility and bioavailability, and transporting it to a physiological target precisely.

Among existing methods of stabilization, microencapsulation technology is significantly positioned (Pasin et al., 2012; Onwulata, 2013) to permit stabilization and/or protection of active ingredients or substances against oxidation (Parra, 2010), photosensitivity, volatility, or reaction with other compounds present in the food; allowing in turn an improvement in dosage; masking the disagreeable tastes and odors of some of these compounds; controlling the release of active substances to a determined location or at a certain speed; and improvement in the bioavailability of certain nutrients.

Therefore, in this research it is proposed to encapsulate polyphenols derived from an extract of the leaves of Maqui—*A. chilensis* (Mol.) Stuntz. (Elaecarpaceae)—to obtain microcapsules with the antioxidant activity of *A. chilensis* and to evaluate the effect of the concentrations of the encapsulating agent and surfactant in maintaining the antioxidant capacity, to avoid the astringent perception and low bioavailability in the stomach deglycosylated, affecting the effectiveness of polyphenols. The microencapsules thus obtained will safeguard the extraction of antioxidants from natural and native sources, masking its astringent and bitter taste, which limits its use in food. These compounds will be protected by the microencapsulation process and thus be incorporated as active ingredients in different food products, which will contribute to creating foods with high added value, since their production by chemical synthesis is difficult because of its structural complexity.

3.1 Phenolic Compounds

Polyphenols are secondary metabolites present in all vascular plants, and constitute a large family of substances ubiquitous and varied, from simple molecules to complex structures. These natural substances have in common the presence of one or more benzene rings that support one or more hydroxyl groups and are derived from the metabolism of shikimic acid and/or polyacetal (Xiong et al., 2006). To date, numerous plant-derived polyphenolic compounds have been characterized and classified, with differences in the chemical structure affecting essentially the oxidation, hydroxylation, methylation, glycosylation, and possible connections to other molecules (such as primary metabolites carbohydrates, lipids, proteins, or phenolic secondary metabolites, etc.).

In the adaptive evolution of living species, oxygen was achieved by enzyme systems that not only facilitated consumption but also detoxification of highly reactive metabolites, reactive oxygen species (ROS). When the capacity of an organism is exceeded for detoxifying ROS occurs, oxidative stress as a result of a pronounced imbalance between pro-oxidants and antioxidants. (Heinonen, 2007). Today, it has amply demonstrated the implication of this phenomenon of cellular aggression in numerous pathologies.

Free radical damage seems to be partially limited by the action of natural antioxidants present in food compounds called polyphenols. In addition to the specific properties of some kinds, two fundamental properties are involved in the antioxidant capacity: protein interaction or ions, and free radical scavenging activity. Polyphenols can act using different modes of action, such as formation of molecular complexes with pro-oxidant proteins, chelation of potentially pro-oxidant metal ions (Fe^{3+}, Al^{3+}, Cu^{2+}), or direct impingement of ROS (Leopoldini et al., 2011).

Among its properties, the antioxidant power of polyphenols is probably the most documented (Velioglu et al., 2006; Ruiz et al., 2010; Koksal et al., 2011). Numerous in vitro studies have shown that polyphenolic compounds can sequester directly molecular active oxygen species such as superoxide radical ($O_2^{\bullet}$), hydrogen peroxide (H_2O_2), hydroxyl radical ($HO^{\bullet}$), singlet oxygen ($^1O_2$) or peroxyl radicals ($RO_2^{\bullet}$). Polyphenols have structural characteristics ideal for its antioxidant action, mainly due to its ability to donate hydrogen atoms (11.1) or electron (11.2).

$$\text{H-atom transfer (HAT) } X^{\bullet} + ArOH \rightarrow XH + ArO^{\bullet} \quad (11.1)$$

$$\text{Singlet electron transfer (SET) } X^{\bullet} + ArOH \rightarrow X^{-} + ArOH^{\bullet} + \quad (11.2)$$

In the transfer mechanism of the hydrogen atom (HAT), the phenolic antioxidant (ArOH) reacts with the free radical ($X^{\bullet}$) and becomes a free radical ($ArO^{\bullet}$) by transfer of a hydrogen atom by cleavage homolytic O-H bond. The formation and stability of $ArO^{\bullet}$ is dependent on the structural characteristics of the compound ArOH. Determinant factors imply the presence, number, and relative position of additional phenolic hydroxyl, groups participating in the formation of intramolecular hydrogen bonds, and the dependent conformation the possibility of electron delocalization of most of the molecule. All these factors affect the dissociation energy (BDE) of phenolic OH bond: the weaker the OH bond is, the easier the transfer of atom H.

The second mechanism is the singlet electron transfer (SET) from ArOH of the free radical $X^{\bullet}$ with formation of a stable radical cation $ArOH^{\bullet+}$. The ionization potential (IP) of physicochemical ArOH is an important parameter for evaluating the effectiveness of antioxidant plant polyphenols: IP decreased, facilitates the transfer of an electron. Basic physicochemical parameters IP and

BDE polyphenols can be used to determine the potential effectiveness of each mechanism. Stabilizing the resulting phenoxy radical, $ArOH^{\bullet+}$ and ArO, is due to the delocalization of the unpaired electron on the aromatic ring by resonance or hyperconjugation effects.

The high tendency of polyphenols to chelate metal ions may contribute to their antioxidant activity to prevent redox active transition metal from the catalytic formation of free radicals (Leopoldini et al., 2011). Polyphenols may be inactivated by chelating iron ions and consequently suppress the Fenton reaction that produces superoxide release, which gives rise to harmful ROS. It has been widely reported that flavonoids can chelate metals and possible metal binding sites have been identified. Polyphenolic compounds include hydroxyl and carboxyl groups capable of binding metal ions having strong positive, such as iron (III) and copper (II) loads. In the chelating bidentate ligands are powerful sequestrants of metal cations compared with monodentate ligands. Phenolic groups protonated ligands are poor metal cations, because once deprotonated, a central oxygen having a high charge density is generated. Furthermore, the metal chelating ability of polyphenols could be related to the high nucleophilicity of the aromatic rings instead of specific chelating groups within the molecule.

However, many health benefits reported in the scientific literature also result from the ability of the polyphenols to interact with proteins (enzymes, membrane receptors, tissue proteins), specifically allowing protection or modulatation of proteins' activity (Howell et al., 2005; Nohynek et al., 2006; Neto, 2007; Céspedes et al., 2008).

Polyphenols act as potent inhibitors of ROS generated by enzymes such as xanthine oxidase, cyclooxygenase, and lipoxygenase, by complexation of the protein. Polyphenolic complexation process is directly influenced by the characteristics of the proteins (solubility, molecular weight, hydrodynamic volume, isoelectric point, and amino acid composition) and polyphenol characteristics (molecular weight, structure conformational flexibility, water solubility) (Richard et al., 2006). The physico-chemical conditions (pH, nature of the solvent, temperature, ionic strength, presence of other organic molecules, such as polysaccharides) are also considered. The main types of interactions involved in the complexation mechanism correspond to the noncovalent bonding and hydrophobic interactions.

The literature shows that, in vitro and/or in vivo, polyphenols are able to reduce the inflammation by inhibiting edema, stopping tumor growth, presenting proapoptotic and antiangiogenic actions, modulating the immune system, preventing bone disturbances incriminated in osteoporosis, increasing capillary resistance, by actions on the components of the blood vessels, protecting the cardiovascular system, protecting the retina, and limiting overweight.

The chemical criterion for the antioxidant capacity of flavonoids is as follows:

- Presence of o-dihydroxy structure in the B ring; this gives more stability to the form of the radical and participates in the delocalization of the electrons.
- Double link in conjunction with function 4-oxo ring C.
- Groups 3- and 5- OH with 4-oxo function in rings A and C are required to exert maximum antioxidant potential.

Among the major phenolic compounds present in berries (Fig. 11.4), it is common to find derivatives of hydroxybenzoic and hydroxycinnamic acids, anthocyanins, flavanols, catechin, and condensed and hydrolyzable tannins, which are mainly characterized by the number of carbon atoms of basic molecular skeleton, represented mainly by: cinnamic acids (C6-C3), benzoic acids (C6-C1-C6 or C2), flavonoids (C6-C3-C6), proanthocyanidins or condensed tannins ((C6-C3-C6) n), stilbenes (C6-C2-C6), coumarin (C6-C3), lignans (C3-C6-C3-C6), and lignin ((C6-C3) n).

3.2 Native Species as Potential Sources of Antioxidants

The *A. chilensis* (Mol.) Stuntz. (Elaecarpaceae) Chilean native species is a small tree (up to 4 m), evergreen, which is distributed from the IV to the XI region of Chile. The leaves of the species have been used by the Mapuche Indians as an infusion to treat inflammation of the throat. Its fruits (maqui) are edible and are used as antidiarrheal. This is supported by findings of phytochemicals describing the presence of polyphenols, such as flavonoids in *A. chilensis* sheets (Suwalsky et al., 2008).

Phenolic compounds constitute a large family of secondary metabolites with different chemical characteristics and biological properties, however, share some of them, being one of the most important to neutralize the action of free radicals, preventing or delaying lipoperoxidation processes, and consequently cell damage (Delporte, 2007; Céspedes et al., 2010). Why, for some time it has been studying the impact on the health of the population, the consumption of foods and supplements that contain them, with a decrease in mortality and morbidity due to degenerative diseases, especially cardiovascular level (Howell et al., 2005; Nohynek et al., 2006; Neto, 2007; Céspedes et al., 2008).

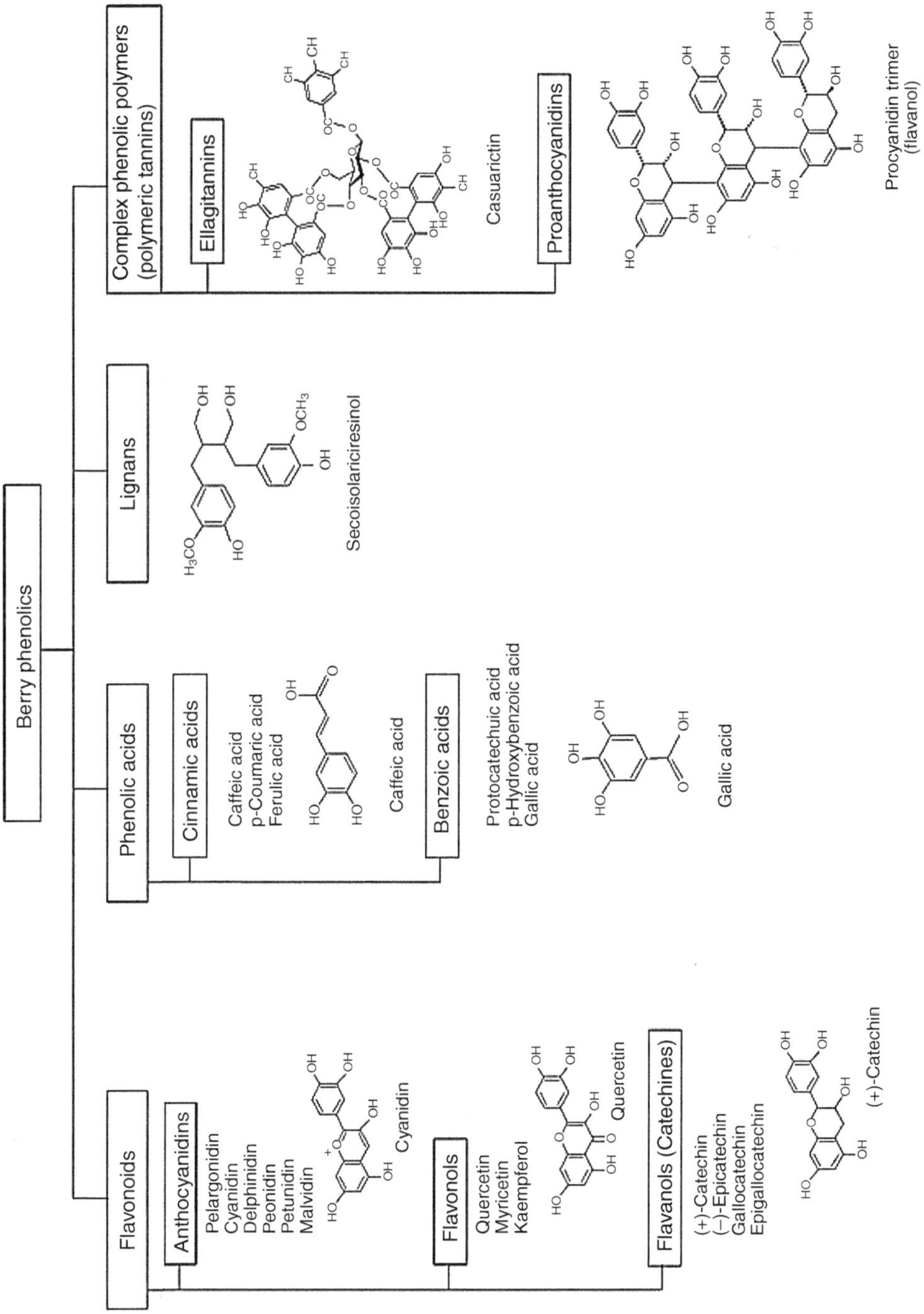

Figure 11.4. Chemical structure of berries' main phenolics. Puupponen et al., 2005.

These observations are based on records that indicate that a decrease in plasma antioxidants promotes lipid peroxidation, contributing to this endothelial dysfunction in hypertension and hypercholesterolemia, this being the first step of atherosclerotic disease, and studies that show the correlation between the incidence of degenerative diseases and low concentrations of plasma antioxidants (Manach et al., 2005).

It has been estimated that about 2% of the oxygen consumed by a normal body goes to the formation of reactive oxygen species (ROS) of which several are free radicals. When the generation of ROS exceeds the antioxidant defenses of the body, whatever the mechanism (UV radiation, pollution, strenuous physical exercise or other) damage is caused by chemical injury of biological structures, a process called oxidative stress, which is involved in the development of many diseases such as atherosclerosis and cancer (Seeram, 2008). In quantity, the major reactive oxygen species generated during cell respiration are superoxide anion ($O2^{\bullet -}$) and hydrogen peroxide (H_2O_2), which gives rise to hydroxyl radical ($OH^{\bullet}$). Superoxide anion and hydroxyl radical species are highly reactive because they possess one or more unpaired electrons being able to set off chain reactions and cause damage to cell membranes that may be irreversible, reaching even to cell death (Nohynek et al., 2006). The hydroxyl radical is perhaps most harmful, due to its high reactivity and short half life, so it is capable of reacting with all types of biomolecules, the most vulnerable polyunsaturated fatty acids (PUFAs), suffering the loss of hydrogen atoms from its methylene groups, also becoming lipoperoxides radicals, which, if not neutralized, will continue the damage. The production of hydroxyl radicals by the Fenton reaction is facilitated by transition metals such as Cu^{+2} and $Fe^{+2,}$ which catalyze this reaction.

Because of this, it is of particular interest to determine the contribution that regular consumption of these products may represent, and their potential for obtaining standardized applications in the phyto-pharmaceutical, cosmetic, and food extracts.

The determination of the antioxidant capacity of native Chilean berries was performed using different methods: FRAP (ferric reducing activity power), TRAP (radical-trapping antioxidant parameter Total), TAR (total antioxidant reactivity), TBARS (thiobarbituric acid reactive substances), radical DPPH (2,2-diphenyl-1-picrilhidracil) radical ABTS (2,2'-azino-bis-(3-ethylbenzothiazoline) -6-sulfonic acid), ORAC (Oxygen radical absorbance capacity). The FRAP method measures the reduction of ferric ion (Fe^{+3}) to ferrous (Fe^{+2}) in the presence of antioxidants, while TRAP methods, TAR, TBARS, radical DPPH, radical ABTS, and ORAC measure the ability of antioxidants to catch different types of free radicals (Kuskoski et al., 2005; Apak et al., 2007; Céspedes et al., 2008).

Preliminary studies by Miranda-Rottmann et al. (2002) compared the total polyphenol content and antioxidant capacity (TRAP and TAR methods) of different juice concentrates (blueberry, raspberry, strawberry, cranberry, blackberry, and maqui) and red wine. The antioxidant capacity was directly proportional to the total polyphenol content of the juice of maqui standing in LDL protective effect exerted on the plasma and intracellular oxidative stress of endothelial cells, suggesting that the antiatherogenic properties of *A. chilensis* presents the rest of concentrated juice and red wine analyzed.

Early studies (Céspedes et al., 2008) performed on this species show that the leaves have a hydrocarbon of 29 carbon atoms n-nonacosane, a steroid β-sitosterol, the flavonoid quercetin, and a complex mixture of alkaloids. It has also been determined that the fruits are anthocyanins (Escribano-Bailón et al., 2005), which would be responsible for the characteristic purple pigmentation of the fruits of maqui. Escribano-Bailón et al. (2005) determined the presence of nine of these compounds: 3,5-diglucoside of cyanidin; 3,7-diglucoside delphinidin; Delphinidin 3-glucoside acylated with p-coumaric acid; 3 delphinidin glucoside; 3,5-malvidin diglucoside; 3,5-diglucoside malvidinaacilada with p-coumaric acid; 3,5-diglucoside petunidin, and petunidin 3-glucoside. Subsequent work with stems and leaves of *A. chilensis* allowed researchers to isolate and identify indole alkaloids type: aristotelina, aristotelona, aristotelinina, and sadder, all in very small quantities.

Araya et al. (2006) determined antioxidant capacity (FRAP method) of a number of fruits and vegetables consumed in Chile. The maqui results were highlighted significantly over the rest of the species analyzed. Fruits values were between 0.02 mMFe/100 g for cucumber to 12.32 mM Fe/100 g for maqui; the value of other berries such as strawberry and blackberry (3.10 and 3.55 mM Fe/100 g). In the intermediate zone were fruits like lemon (0.25 and 0.23 mM Fe/100 g), and the lowest values were apple (var. Fuji) and peaches. Given these results, the maqui stands out for its specific antioxidant capacity in vitro.

Céspedes et al. (2008) determined that the methanol extract of fruits maqui presented antioxidant and cardioprotective activity on ischemia/reperfusion in acute heart disease of rats in vivo. Moreover, this extract was able to prevent these adverse events in the heart of the animal, by decreasing lipid oxidation and reducing the concentration of TBARS. In addition, these authors determined the antioxidant capacity of methanol extract of maqui by

DPPH* (IC50), ORAC, FRAP, and TBARS. The methanol extract exhibited an IC50 of 1.62 ug/mL and TBARS a value of 2.51 mg/mL, compared with maqui juice, which submitted an IC50 of 12.1 mg/mL and TBARS a value of 9.58 mg/mL. The antioxidant capacity of the methanol extract of maqui was strongly correlated with the total polyphenol content (12.97 PM at EC/g), which was observed by higher ORAC values (29.69 PM at EC/g) and FRAP (EC at 25 PM/g). These results demonstrated that these fruits may be useful as sources of antioxidants, cardioprotectors, and nutraceuticals.

Avello et al. (2008) evaluated the antioxidant capacity (BARS method) plasma before and after intake of infusions of maqui sheet (1%). The content of total phenols was 0.074 infusion mMEAG. In this study, volunteers with healthy nonsmoking habit and body mass index within the normal range drank this tea twice a day for three days. The results showed an average increase of antioxidant capacity observed at 24 h through TBARS (30.27%).

The *A.chilensis* species has an exceptionally high phenolic content with high antioxidant capacity, which has a protective effect against oxidation in human low-density lipoprotein, and also protects endothelial cells from intracellular oxidative stress, suggesting that *A. chilensis* could be regarded as having the property of being atherogenic (Céspedes et al., 2008).

Recently, Schreckinger et al. (2010) presented in vitro evidence suggesting an antiinflammatory and inhibitory adipogenesis of polyphenols maqui activity. In addition, the polyphenols in maqui could also promote action on digestive enzymes of interest. Indeed, recent in vitro studies by Rubilar et al. (2011) report that crude extracts of maqui may be capable of inhibiting the activity of alpha-glucosidase and alpha-amylase enzymes responsible for breakdown of carbohydrates into glucose. Such activity could be a potential of some polyphenols in maqui for "modular" postprandial glycemia.

4 Microencapsulation via Emulsion of Chilean Blackberry

4.1 *Aristotelia Chilensis* (Elaeocarpaceae), Maqui Leaf Extracts

Chile's economy has been increased in the agro-food field by the production of berries, with these crops being valued in world markets for quality, value, and opportunity in innovation. The progress of production, purification, and extraction technologies must be compatible with the economic and technical aspects of the product, to increase scientific and commercial interest (Onwulata, 2013).

Though, to settle their productive principles, it is very important to transfer the production of the products in the market (Parra, 2010; Pasin et al., 2012). An alternative method is to ensure the quality of food additives or ingredients and bioactive principles is microencapsulation. The matrices protect these compounds and increase the functionality and life span of foods (Banjare and Ghillare, 2012). Nowadays, people demand healthy foods, which do not contain any synthetic preservatives that can damage their physiological functions (Manach et al., 2005; Ah-Hen et al., 2012). To protect the production of food, an alternative can be the inclusion of natural active agents, like antimicrobials or antioxidants, which can also have an effect on human health (Burris et al., 2012). Thus, interest developed in natural compounds, which possess antimicrobial and antioxidant capacities. Phenolic compounds present important antimicrobial and antioxidant properties, making them a good candidate in agro-industry. They are able to avoid or delay lipid peroxidation process, to neutralize the action exercised by free radicals, and to prevent cell damage (Delporte, 2007; Céspedes et al., 2010). It is estimated that only 2% of the total amount of oxygen consumed by an organism helps in the formation of reactive oxygen species (ROS), or free radicals. In the moment when ROS generation can surpass the protection of an organism's antioxidant, in a manner independent of the mechanism (environmental pollution, UV radiations, intense physical activity, etc.), biological structures support damages caused by chemical injuries in an oxidative stress process, implicated in reduced plasmatic antioxidant concentrations and degenerative pathologies (Seeram, 2008). Red fruits and berries are important antioxidants, possessing a high concentration of phenolic compounds, especially flavonoids, characterized by their significant properties, like antiatherogenic, antioxidant, anticarcinogenic, antimicrobial, and antiinflamatory activities (Howell et al., 2005; Nohynek et al., 2006).

Aristotelia chilensis ([Molina] Stuntz, Elaecarpaceae), known as maqui, is a plant famous for its fruit, and can be found in the southern regions of Chile. The residents of this area use its leaves to mace an infusion tea, which have beneficial effects against mouth sores, diarrhea, pharyngitis or tonsillitis, acting as an febrifuge and analgesic. Their use is upholded by its phytochemical properties, which describe the presence of polyphenols (flavonoids) in the leaves. Therefore, a growing interest is granted to natural maqui products, in which can be found increased concentrations of phenolic compounds, carrying antimicrobial and antioxidant activities (Delporte, 2007; Avello et al., 2009; Céspedes et al., 2010). Through the encapsulation method (Cartes et al., 2009), these properties can be conserved over time, and can be used as active ingredients in the agro-food chain. The aim of this study was the investigation of conditions suitable in the obtaining process of microcapsules with antioxidant properties, extracted from maqui leaves by emulsification and subsequent retention after microencapsulation.

4.2 Phenolic Characterization and Extraction

Maqui leaves were collected in April 2007 from Universidad de Concepción, Biobío Region, Chile, and were dried at 35°C. 50 g of sifted (EasySieve Retsch, Haan, Germany) leaves powder (Ultra Centrifugal Mill ZM 200 Retsch, Haan, Germany) were macerated in a mortar and crumbled at room temperature for 30 minutes with an ethanol and water solution (40% v/v) in a 6:1 solvent solid ratio. Afterward, the extract was concentrated in a rotary evaporator at the temperature of 35°C and lyophilized for 24 h. The total amount of flavonoid could be determined by the method described by Kumazawa et al. (2004), and the results were titled as quercetin equivalent. Folin-Ciocalteu method (Rubilar et al., 2011) helped in the obtaining of polyphenols, and the results were expressed as gallic acid equivalents (GAE). Total tannins were determined by the Folin-Ciocalteu technique (Velioglu et al., 2006). The amount of alkaloids from the maqui leaves extract were calculated by alkaline titration (Ordoñez et al., 2006), the results being expressed as milligrams of hyoscyamine.

4.3 Antioxidant Capacity and Extract Stability

The antioxidant capability was determined by two methods: by inhibiting the 2,2'-azino-bis(3-ethylbenzothiazoline-6-sulfonic acid) (ABTS) free radical (Kuskoski et al., 2005), and by inhibiting 1,1-diphenyl-2-picrylhydrazyl (DPPH) free radical (Koksal et al., 2011). The oxidation processes are prevented by the use of chromogenic compounds (DPPH and ABTS), which can capture free radicals. Quantifying the total phenols for 5 mo at 30-d intervals in refrigerated samples, the stability of the maqui leaf extract could be determined.

4.4 Identification and Quantification of Phenolic Compounds in the Extract

The determination of phenolic components were identified with a high performance liquid chromatography (HPLC) (Series

1050, Hewlett-Packard, Hopkins, Minnesota, USA) with a LiChrospher 100 RP-18 (5 μm, 125-4 mm) column, and elution solvent mixtures A (96% ultrapure water; 3% acetic acid, and 1% acetonitrile) and B (72% ultrapure water; 3% acetic acid, and 25% acetonitrile). The gradient was realized beginning with 100% A, and ending with 100% B (0–10 min), and maintained for 40 min at 100% B, after which the column was set up at the initial 100% A. The flow was 0.8 mL min^{-1} detected in UV at 280 nm, by absorbance. The standards corresponded to flavonoids phenolic acids and grade HPLC (Merck, Darmstadt, Germany).

4.5 Microencapsulation by Extract Emulsion

The maqui leaf extract microcapsules were obtained using the water-in-oil emulsions (W/O) (Fig. 11.5). The aqueous phase emulsion was prepared by dissolving 0.25% maqui leaf extract in ethanol solution and 5 and 15% gum arabic in distilled water. After 10–15 min, when the gum arabic was dissolved, the extract was added. The oil phase consisted of 1 and 2% nonionic surfactant (Tween® 80) and pure liquid Vaseline. Both phases were homogenized at 60°C in a thermostatic bath, and then stirred for 5 min. Water-in-oil emulsions were prepared by dissolving 20% aqueous phase with 80% oil phase. The emulsification process was performed by progressively dispersing aqueous phase into oil phase in a magnetic stirrer agitator at the temperature of 60°C for 5 min. After that, an Ultra-Turrax T18 homogenizer (Fisher Scientific, Waltham, Massachusetts, USA) was used to mix them, increasing the speed from 0 to 10,000 rpm every 20 s for 3 min. Subsequently, emulsions were cooled to room temperature (~23°C) for later analysis.

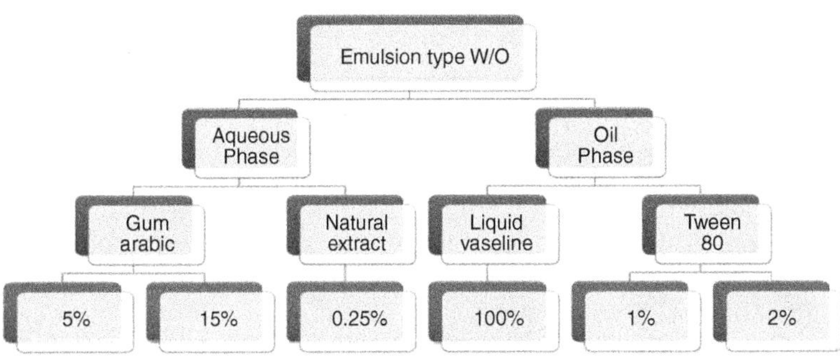

Figure 11.5. Aqueous and oil phase concentrations used to obtain microcapsules.

4.6 Determination of Microcapsule Yield, Size, and Morphology

The overall percentage yield (OY%) was counted by the ratio of microcapsule mass as a function of the mass of dry solids to encapsulate. From the total microcapsule mass could be calculated the microencapsulation yield (MY%), as a function of the total emulsion mass. Using an optical microscope (Carl Zeiss Standard 25 ICS) with a 100× objective, the microcapsules could be directly observed. In order to find the average size, the microcapsules' diameter was measured on two fields of 20 randomly selected units. Both the external and internal structural particularities were observed with a scanning electron microscope (SEM, JEOL JSM-6380 LV, Tachikawa, Tokyo, Japan). The microcapsules were then dried at ambiental temperature for 1 mo. After that, the microcapsules were sprinkled in samples of 10 × 10 mm, and a double-sided adhesive tape was added on the surface, to provide adhesion. After this process, they were put into a sputter coater (Edwards S150, England) metallizer and covered with gold, and viewed at 30 kV on a Scanning Electron Microscope, at 3,700× magnification. The images were provided using a digital camera, with the microscope operative system, in order to obtain the micrographs.

4.7 Determination of Microcapsule Antioxidant Capacity

About 1 g of microcapsules was washed in 10 mL water-isopropanol (6:1) solution and stirred in a regulated vortex mixer (ZX3, Velp Scientifica, Usmate, Italy) for 5 min. The microcapsules were then frozen for 90 min at –18°C, and thawed at room temperature. The probes were stirred for 2 min in a vortex mixer, and centrifuged for 15 min at 6000 rpm, at room temperature. After consecutive washings with isopropanol-water (1:6), the aqueous phase was extracted with syringes, evaluating the antioxidant capacity using the ABTS method.

4.8 Statistical Analysis

The influence of gum arabic concentrations (5 and 15%) were performed at 5% significance level, in the aqueous phase of the W/O emulsion. With the Duncan's test ($P < 0.05$), there could be comparison with the means. The normality was checked in accordance with Modified Shapiro Wilks and the homogeneity of variances in conformity with Bartlett. There was performed Kruskal-Wallis nonparametric ANOVA and a Conover (1999) contrast test in order to establish the surfactant influence on the antioxidant activity of the extract released from the microcapsules. Using an InfoStat Professional version 2008 software (Di Rienzo et al., 2008), there were achieved the statistical analyses.

4.9 Characterization of Maqui Leaf Extract

The qualitative screening and chemical characterization of maqui leaf ethanolic extract at 1% mostly shows flavonoids at a concentration of 0.061 ± 0.01 mg mL^{-1}, phenolic compounds acting as colorings, antioxidants, and providing flavor to the species that contain them. Tannins and alkaloids were also identified with concentrations of 0.615± 0.02 mg mL^{-1} and 1.447 ± 0.01 mg mL^{-1}, respectively.

The general phytochemical analysis of *A. chilensis* coincides with Pastene (2009) regarding the presence of flavonoids. These authors point out that flavonoids would be present as heterosides since genins are mostly soluble in nonpolar solvents, while tannins have a lower degree of polymerization. Extracts of the flavor to the species that contain them. Tannins and alkaloids were also identified with concentrations of 0.615 ± 0.02 mg mL^{-1} and 1.447 ± 0.01 mg mL^{-1}, respectively.

The general phytochemical analysis of *A. chilensis* coincides with Pastene (2009) regarding the presence of flavonoids. These authors point out that flavonoids would be present as heterosides since genins are mostly soluble in nonpolar solvents, while tannins have a lower degree of polymerization. Extracts of the Chilean *A. chilensis* species have shown interesting results both for antioxidant action and antibacterial capacity due to the phenolic composition of leaves (Suwalsky et al., 2008; Rubilar et al., 2011).

Flavonols and phenolic components of flavonols found in the analyzed extract coincide with those reported by Céspedes et al. (2010), who determined the presence of ferulic acid, rutin, quercetin, and mirecetin by HPLC in the *A. chilensis* methanolic extract and aqueous extract, as well as isoquercetin and α-catechin in the above fractions.

Quantifying maqui leaf extract phenolic components by HPLC (Table 11.1) confirmed the preponderance of phenolic acids (54.36%), flavonoids (42.10%), and stilbenes (3.55%). Identifying the components (Fig. 11.6) and their relative quantities (%) in the maqui leaf ethanolic extract showed higher percentages of phenolic acids (47.55% gallic acid), flavonols (21.75% catechin), and anthocyanins (14.45% pelargonidin). These have a general structure consisting of two aromatic rings linked by three carbons forming an oxygenated heterocyclic ring substituted with hydroxyl groups whose basic structural formula is made up of anthocyanidins.

Table 11.1 Phenolic Compounds of Maqui *(Aristotelia chilensis)* Leaf Extract

Phenolic Compounds		μM	%
Phenolic acids			
	Hydroxybenzoic acid	399.18	47.55
	Hydroxybenzoic acid	57.14	6.81
Flavonoids			
Flavonols	Quercetin	10.92	1.30
	isoquercetin	2.91	0.35
	mirecetin	18.95	2.26
	rutin	15.05	1.79
Anthocyanins	Pelargonidin	121.29 1.71	14.45
	peonidin		0.20
Flavonols	catechin	182.59	21.75
Stilbenes	resveratrol	29.79	3.55

4.10 Determination of Extract Antioxidant Capacity

The DPPH method determined that the antioxidant capacity of maqui leaves ethanolic extract is 158.15 ± 3.36 mM EAG (extract at 0.1%) and 189.50 ± 10.25 mM EAG in the case of 1% extract. For 0.1%, the ABTS method showed 165.08 ± 7.12 mM EAG, while for 1% extract, the results were 165.650 ± 9.94. The results obtained are higher than those obtained by Rubilar et al. (2011), who got values of 47.03 ± 0.1 mM. EAG using the DPPH method, where a total polyphenols content in crude maqui leafs extracts of 69.0 ± 0.9 mg EAG g^{-1}. The increased antioxidant capacity obtained in the present study is the result of working with a hydroalcoholic solution (60:40), with a 1:6 solid-solvent ratio, being previously studied in comparison with the quantification of total phenols (40 ± 0.57 mM EAG) in different solvents.

Rubilar et al. (2011), demonstrated that in comparison with the myrtle leaf extract, the maqui leaf extract has a higher antioxidant capacity. Using ethanol water (50:50) as a solvent and a 1:5 solid-solvent ratio, the polyphenol contents were 32.5 ± 3.1 mg EAG g^{-1}. This fact confirms the correction in extraction capacity, in the case when the percentage of ethanol is higher. During refrigerated storage for 5 months, the stability of the extract antioxidant capacity was 100% inhibition of DPPH radical at 1% and 95% inhibition at 0.1%.

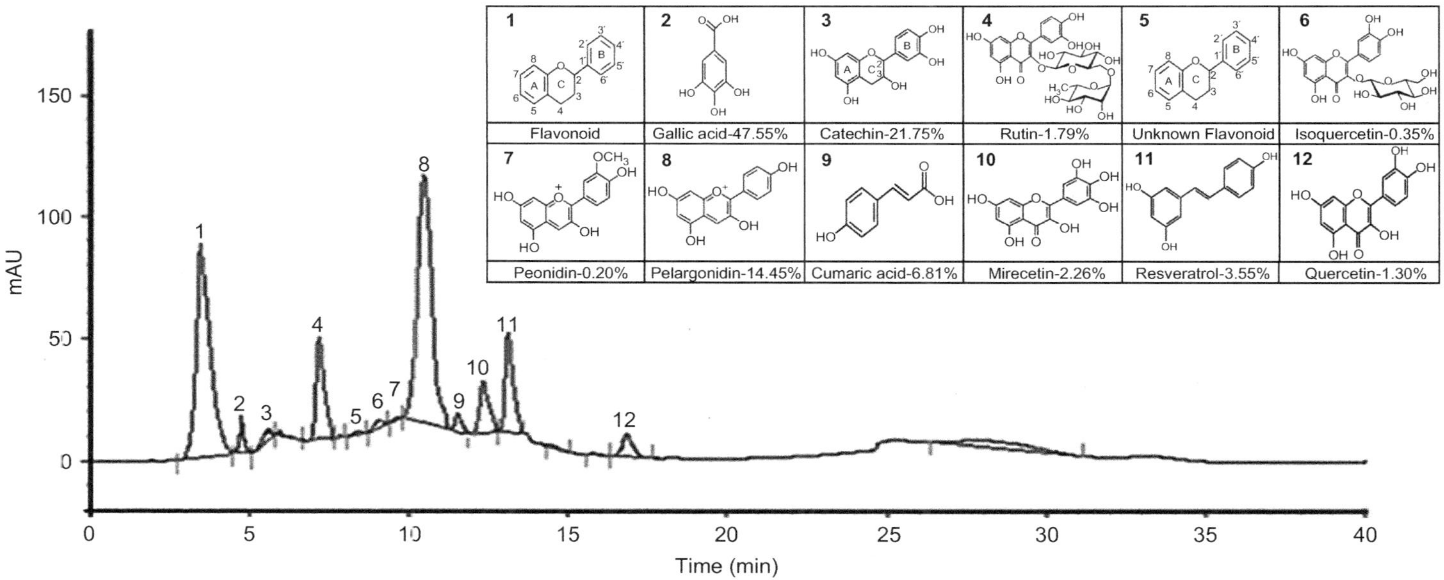

Figure 11.6. HPLC chromatogram of maqui leaf extract (detection in nanometers). *(1) Flavonoid; (2) Gallic acid; (3) Catechin; (4) Rutin; (5) Unknown Flavonoid; (6) Isoquercetin; (7) Peonidin; (8) Pelargonidin ; (9) Coumaric acid; (10) Mirecetin; (11) Resveratrol; (12) Quercetin.*

Using the Folin-Ciocalteu technique, the total amount of polyphenol of the maqui leaf extract was 78.5 ± 0.43 mM EAG, decreasing with only 10% after 5-mo storage. The results are similar with those obtained by Avello et al. (2009), who obtain a high value of phenol content in the *A. chilensis* hydroalcoholic extract, of 40.00 ± 0.1 mM EAG. In a similar manner, the research on maqui leaf hydroalcoholic extract (60:40) at 1%, shows 100% inhibition of the DPPH radical, and an increased antioxidant capacity, between 6.0 and 0.6 μM EAG.

4.11 Encapsulation of Maqui (*Aristotelia Chilensis*) Leaf Extract

Because the maqui leaf extract is soluble in water, there was selected a W/O emulsion. By combining the gum arabic which possess important properties like low viscosity, high solubility in water, mild flavor, and also good superficial activity, with the extract, there was obtained the aqueous phase. The oil phase was realized using liquid Vaseline and 1% and 2% concentrations of surfactant (Tween 80) with a reduced hydrophilic-lipophilic balance (HLB 3.3), selected for the capacity of forming stable W/O emulsions. Vaseline type Tween 80 (HLB 15) was soluble in liquid Vaseline (HLB 4).

The emulsion stability increases with the decreases of the size, against gravitational separation. There was not observed a creaming process in time; this fact suggests that there are resistant interfacial forces which, due to the repulsive forces occurred between the microcapsules for steric stabilization mechanisms, prevent the coalescence process and produce a stable emulsion (Marcuzzo et al., 2010).

An important advantage was the small diameter with dimensions less than 100 μm, preventing the side effects on the sensorial characteristics when they are used as ingredients in incorporations.

4.12 Emulsion Yield

The best microcapsule yields were obtained with emulsions containing 5 and 15% gum arabic. An increase in gum arabic concentration provoked a significant decrease in encapsulation yield since gum arabic viscosity increases when its concentration increases; at 20°C, a 30% gum arabic solution has a viscosity of 361.5 cP, 16.7 cP at 15%, and only 7.6 cP at 5% (Table 11.2). Guarda et al. (2011) used gum arabic concentrations of 15 and 30% to obtain microcapsules of active compounds with antimicrobial capacity, for O/W emulsions; this reaffirms that gum arabic viscosity prevented good solubility of maqui leaf extract at concentrations higher than 15%.

Table 11.2 Gum Arabic Viscosity as a Function of Temperature

	Viscosity Concentration				
	5%	10%	15%	20%	30%
Temperature (°C)	GA	GA	GA	GA	GA
			cP		
20.0	7.6	12.3	16.7	195.3	361.5
29.5	6.2	10.3	14.6	128.3	238.6
39.2	5.3	8.73	13.7	95.9	159.2
50.0	4.5	7.62	12.2	68.7	108.4
60.0	4.2	6.82	10.3	48.6	76.6

GA, Gum Arabic

4.13 Characterization of Microcapsules

Table 11.3 shows the overall and microcapsule yields with the mean droplet size diameter. Yields are within the range reported by Saénz et al. (2009) for phenolic compounds with variations between 20% and 80%.

Size was expressed as the mean diameter value of a total of 30 droplets observed with an optical microscope (Fig. 11.7). Emulsion mean droplet diameter was 2.73 ± 0.37 and 2.91 ± 0.47 μm for

Table 11.3 Mean Size and Yield of Microcapsules Prepared with Maqui Leaf Extract

Relationship Gumr: Tween 80	Microcapsule Yield	Overall Yield	Mean Size
	%		μm
5:1	37.69 ± 6.7	43.44 ± 8.0	2.93 ± 0.4
5:2	48.05 ± 16.8	55.38 ± 15.5	2.56 ± 0.3
15:1	39.15 ± 3.7	45.13 ± 4.3	2.80 ± 0.7
15:2	39.10 ± 0.6	45.07 ± 0.6	3.36 ± 0.2

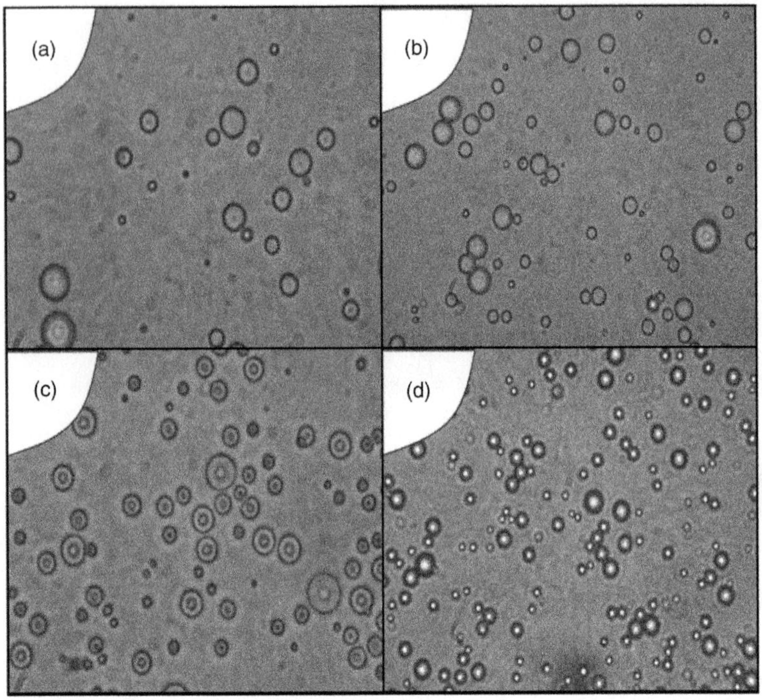

Figure 11.7. Optical microscopy (100×) of microcapsules prepared in w/o emulsion containing 0.25% maqui leaf extract, gum arabic (GA), and Tween 80 (T80). (a) 5% GA, 1% T80, (b) 5% GA, 2% T80; (c) 15% GA, 1% T80; (d) 15% GA, 2% T80

microcapsules with 5 and 15% gum arabic, respectively. Size distribution varied in accordance with the percentage of gum arabic.

Microcapsules with 5% gum arabic resulted in a bimodal size distribution with two maxima, the first at 2 μm and the second at approximately 5 μm, while microcapsules with 15% gum arabic resulted in a monodisperse size distribution with a maximum of approximately 2 μm. The values are consistent with the range of 0.2–5 μm defined by Medonça et al. (2009). The droplet size normal distribution is shown in (Fig. 11.8). The interval of droplet size is between 1 and 10 μm with a distribution between 2 and 6 μm. As size decreases, emulsion stability increases against gravitational separation. A creaming process was not observed over time; this suggests that resistant interfacial forces occur, which produce a stable emulsion over time and prevent the coalescence process due to the repulsive forces between the microcapsules for steric stabilization mechanisms (Marcuzzo et al., 2010). The microcapsules diameters were less than 100 μm, which is an advantage to prevent adverse effects on the sensory characteristics when they are incorporated as ingredients.

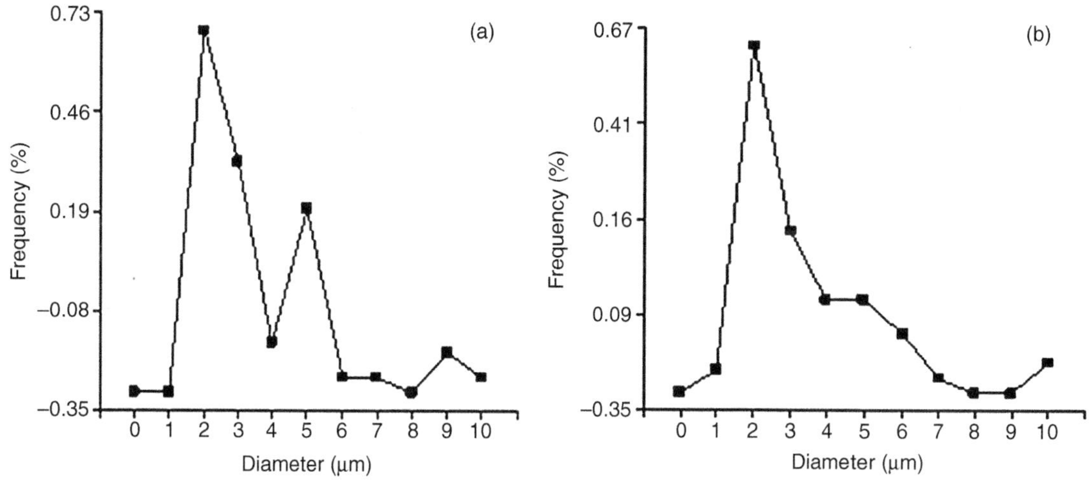

Figure 11.8. Mean normal distribution of W/O droplet particle size measured by optical microscopy in samples of 5% (a) and 15% (b) gum arabic.

4.14 Morphology of Microcapsules with SEM

The morphology of microcapsules can be studied using the scanning electron microscope, showing that the emulsions contain a microcapsule population forming droplets with a spherical shape, which doesn't present any damage or pores on their surface (Fig. 11.9), an increased degree of integrity, and also a reduced presence of nicks on the surface structure (McClements, 2012). This is due to the method used, in accordance with the results obtained by Saénz et al. (2009), who used the spray-drying process, showing a patchy morphology of the surface, because it shrinks during the process of drying. Fig. 11.9 (1–8) illustrates the external topography of the microcapsules, with a concentration of 5% and 15% of gum arabic. In both concentration of surfactant (1% and 2%), there can be observed spherical capsules having with distinct compaction in the oil phase. The properties of gum arabic, the concentration of the surfactant, and the cross-linking from the water-in-oil emulsion give the size distribution. The agglomerates appear in a higher number when the concentration of the surfactant increases, the dilution in an aqueous medium being necessary. The choice of surfactant concentrations under study is therefore validated.

4.15 Study of Microencapsulated Maqui Leaf Extract

The protective capacity of maqui leaves ethanolic extract is 99.66%, of the radicals determined through ABTS technique, and

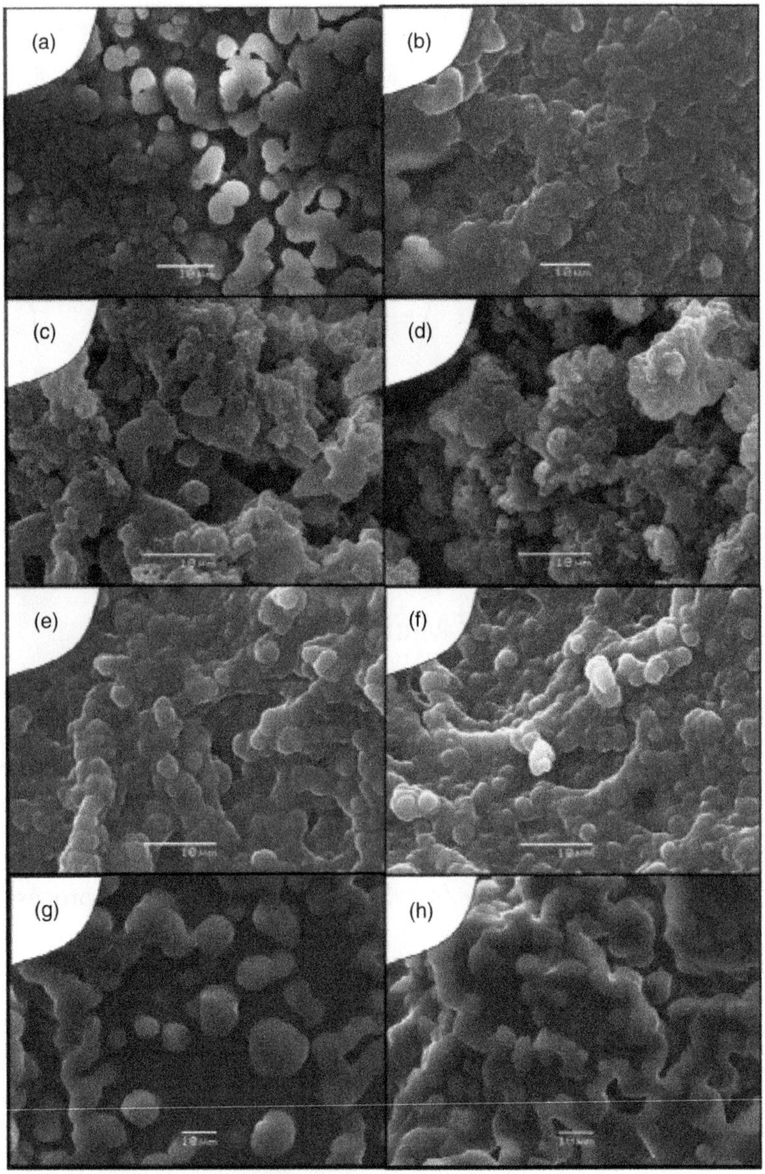

Figure 11.9. Micrographs (3700×) of microcapsules from W/O emulsion with 0.25% maqui leaf extract, gum arabic (GA), and Tween 80 (T80). (a–b) 5% GA, 1%T80, (c–d) 5% GA, 2%T80; (e,f) 15% GA, 1% T80; (g–h) 15% GA, 2%T80.

96.68% in the case of using the DPPH method, the extract being considered a mixture of compounds that possess a high antioxidant capacity. Another effect produced by the incorporation of gum arabic is the decreasing of the antioxidant capacity of the extract, the inhibition being 94% using 5% gum arabic, and 93% in the case of 15% gum arabic. This fact can be attributed to the complex formed by gum arabic protein with the extract, this being also shown in other research study (Su et al., 2008), in which it was demonstrated that the aminoacid composition of gum arabic manages the high molecular weight protein fraction to absorb the oil-water interface. The emulsifying property of gum arabic is dependent by the absorption capacity of the protein component, a distribution of the low and high molecular weight parts of the protein and the N content. Thereby, the complex formed by gum arabic microparticles with maqui leaf extract may produce a powerful cross-linking effect by forming a three-dimensional network bonded by various polymer chains, reducing the capacity of swelling of the cross-linked microparticles, and also the mobility of the macromolecular chains.

The type and content of aminoacid can be affected by the addition of gum arabic protein, defining the interfacial cross-linking (McClements, 2012) If the treatment means for the percentage antioxidant capacity of the maqui extract is compared with the gum arabic concentration (5% and 15%) and surfactant (1% and 2%) (Table 11.4), it will be shown that the extract presents a high percentage of antioxidant capacity 165.08 mM EAG because it is dissolved in the aqueous phase.

The antioxidant capacity varies from the liberated microcapsule antioxidant capacity, decreasing from 94% to 30%. This decrease implies the liberation of the active principle from the internal aqueous phase by the external oil phase. Isopropanol:water solution was used to wash this phase, to reduce the phenolic

Table 11.4 Mean Antioxidant Capacity of Prepared Microcapsules Compared with Maqui Leaf Extract

	Samples				
Variables	Extract	5% GA/1%T80	5% GA/2%T80	15% GA/1%T80	15% GA/2%T80
ABTS radical inhibition, %	99.66e	30.49 a	30.42 a	35.62 c	31.24b
Antioxidant capacity, mM EAG	165.08d	50.26 a	50.07 a	59.59 c	51.63 b

Nonparametric Kruskal Wallis ANOVA. Different letters in the rows indicate differences: ($P < 0.05$) according to Conover. GA, Gum arabic compound polarity from the initial extract, in comparison with those kept in the microcapsule aqueous phase (Surh et al., 2007). Studies are oriented to avoid flavonoid deterioration, and try to find effective methods to prevent its degradation. This new technology of microencapsulation is proposed as an alternative for applying, using, and conserving bioactive agents. The gum arabic used as an encapsulating agent in the aqueous phase, develops a protective matrix of the phenolic components found in the maqui leaves extract. In order to obtain microcapsules (Onwulata, 2013), the W/O emulsion is stabilized by the surfactant action (Tween 80) in the droplet interface.

5 Conclusions

According to the properties of the component that will be encapsulated, the surfactant concentration, microcapsule size, and the composition of gum arabic, the encapsulation technology by water-oil (W/O) emulsion can be applied to maqui leaf extract in a ratio of 80% oil phase and 20% aqueous phase. Maqui leaf extract antioxidant activity was 99.66% and the results for the aqueous phase of the emulsion were 94.38% for 5% gum arabic and 93.06% for 15% gum arabic. This verifies the capacity of gum arabic concentrations as an encapsulating material to reach the highest antioxidant capacity of the extract in the aqueous phase of the emulsion. The average antioxidant activity of the maqui leaves extract microcapsules was 30% compared to 94% in the aqueous phase of the emulsion. This gradient is given by the loss of polarity of the extract's phenolic components in the aqueous phase after the external oil phase was removed by washing it in an isopropanol-water mixture. This study determined the influence of 80% concentration of the external oil phase as an important factor in the measuring of antioxidant capacity in the internal aqueous phase of the microcapsule. The obtained results show that W/O microcapsules are an alternative for the natural preservative systems, alongside with their synthetic counterparts.

References

Ah-Hen, K., Fuenzalida, C., Hess, S., Contreras, A., Vega, A., Lemus, R., 2012. Antioxidant capacity and total phenolic compounds of twelve selected potato landrace clones grown in Southern Chile. Chilean J. Agric. Res. 72, 3–9.

Alishahi, A., Alder, M., 2012. Applications of chitosan in the seafood industry and aquaculture: a review. Food Bioproc. Technol. 5, 817–830.

Anbinder, P., Deladino, L., Navarro, A., Arnalovy, J., Martino, M., 2011. Yerba Mate Extract Encapsulation with Alginate and Chitosan Systems: Interactions between Active Compound Encapsulation Polymers. J. Encapsul. Adsorp. Sci. 1, 80–87.

Apak, R., Güçlü, K., Demirata, B., Özyürek, M., Çelik, S., Bektaşoğlu, B., Berker, K., Özyurt, D., 2007. Review: comparative evaluation of various total antioxidant capacity assays applied to phenolic compounds with the CUPRAC assay. Molecules 12, 1496–1547.

Araya, H., Clavijo, C., Herrera, C.l., 2006. Capacidad antioxidante de frutas y verduras cultivados en Chile. Archivos Latinoamericanos de Nutrición. 56 (4), 361–365.

Avello, M., Valladares, R., Ordoñez, J., 2008. Capacidad antioxidante de *Aristotelia chilensis* (molina) stuntz. Rev. Cubana Plant. Med. 13 (4), 1–7.

Avello, M., Valdivia, R., Sanzana, R., Mondaca, M., Mennickent, S., Aeschliman, V., Bittner, M., Becerra, J., 2009. Extractos antioxidantes y antimicrobianos de *Aristotelia chilensis* y *Ugni milinae* y sus aplicaciones como preservantes en productos cosméticos. BLACPMA 8 (6), 479–486.

Banjare, L., Ghillare, N., 2012. Development of biocompatible nanoparticles for sustained topical delivery of Rutin. Int. J. Pharma. Biolog. Arch. 3 (2), 326–332.

Bastos, D., Araujo, K., Leao, M., 2009. Ascorbic acid retaining using a new calcium alginate-capsule-based edible film. J. Microencapsul. 26 (2), 97–103.

Betz, M., Steiner, B., Schantz, M., Oidtmann, J., Mader, K., Richling, E., Kulozik, U., 2012. Antioxidant capacity of bilberry extract microencapsulated in whey protein hydrogels. Food Res. Int. 47, 51–57.

Burey, P., Bhandari, B., Howes, T., Gidley, M., 2009. Gel particles from spray-dried disordered polysaccharides. Carbohyd. Polym. 76 (2), 206–213.

Burris, K., Harte, F., Davidson, M., Stewart, N., Zivanovic, S., 2012. Composition and bioactive properties of yerba mate (*Ilex paraguariensis* A.St. -Hill.): A review. Chilean J. Agric. Res, 72, 268–274.

Cartes, P., Castellanos, H., Ríos, D., Sáez, K., Spierccolli, S., Sánchez, M., 2009. Encapsulated somatic embryos and zygotic embryos for obtaining artificial seeds of rauli-beech (*Nothofagus alpine* (Poepp. & Endl.) Oerst). Chilean J. Agric. Res. 69, 107–111.

Céspedes, C., Alarcón, J., Gavila, G., Nieto, A., 2010. Antiinflammatory activity of *Aristotelia chilensis* Mol. (Stuntz) (Elaeocarpaceae). Boletín Latinoamericano y del Caribe de Plantas Medicinales y Aromáticas 9, 127–135.

Céspedes, C., El-Hafidi, M., Pavon, N., Alarcón, J., 2008. Antioxidant and cardioprotective activities of phenolic extracts from fruits of Chilean blackberry *Aristotelia chilensis* (Elaeocarpaceae). Maqui Food Chem. 107, 820–829.

Céspedes, C., Valdez, M., Avila, J., El-Hafidi, Alarcón, J., Paredes, O., 2010. Phytochemical profile and the antioxidant of Chilean wild blackberry fruits, *Aristotelia chilensis* (Mol) Stuntz (Elaeocarpaceae). Food Chem. 119, 886–895.

Chanamai, R., McClements, D., 2002. Comparison of gum arabic, modified starch, and whey protein isolate as emulsifiers: influence of pH, $CaCl_2$ and temperatures. J. Food Sci. Food Chem. Toxicol. 67 (1), 120–125.

Champagne, Cl., Fustier, P., 2007. Microencapsulation for the improved delivery of bioactive compounds into foods. Curr. Opin. Biotechnol. 18 (2), 181–190.

Conover, W., 1999. Practical Nonparametric Statistics. John Wiley & Sons, New York.

Deladino, L., Anbinder, P., Navarro, A., Martino, M., 2008. Encapsulation of natural antioxidants extracted from *Ilex paraguariensis*. Carbohyd. Polym. 71 (1), 126–134.

Delporte, C., 2007. Determinación de las actividades antiinflamatorias, analgésicas, antioxidantes y antimicrobianas de las hojas de *Aristotelia chilensis* (maqui). Identificación de los compuestos activos. Boletín Latinoamericano y del Caribe de Plantas Medicinales y Aromáticas 5 (6), 136.

Desai, K., Park, H., 2005. Recent developments in microencapsulation of food ingredients. Dry. Tech. 23, 1361–1394.

Desplanques, S., Renou, F., Grisel, M., Malhiac, C., 2012. Impact of chemical composition of xanthan and acacia gums on the emulsification and stability of oil-in-water emulsions. Food Hydrocolloid. 27, 401–440.
Di Rienzo, J., Casanoves, F., Balzarini, M., González, L., Tablada, M., Robledo, C., 2008. InfoStat, versión 2008. Grupo InfoStat, Facultad de Ciencias Agrarias, Universidad Nacional de Córdoba, Córdoba, Argentina.
Dube, A., Ng, K., Nicolazzo, J., Larson, I., 2010. Effective use of reducing agents and nanoparticle encapsulation in stabilizing catechins in alkaline solution. Food Chem. 122, 662–667.
Escribano-Bailón, M., Alcalde-Eon, C., Muñoz, O., Rivas, J., Santos, C., 2005. Anthocyanins in berries of maqui [*Aristotelia chilensis* (Mol.) Stuntz]. Phytochem. Anal. 17 (1), 8–14.
Fuchs, M., Turchiuli, C., Bohin, M., Cuvelier, M., Ordonnaud, C., Peyrat, M., Dumoulin, E., 2006. Encapsulation of oil in powder using spray drying and fluidized bed agglomeration. J. Food Eng. 75, 27–35.
Fuentealba, J., Dibarrart, A., Sáez, F., Fuentes, M., Oyanedel, C., Guzmán, J., Pérez, Cl., Becerra, J., Aguayo, L., 2012. Synaptic silencing and plasma membrane dyshomeostasis induced by amyloid-β peptide are prevented by *Aristotelia chilensis* enriched extract. J. Alzheimer s Dis. 31, 879–889.
García, G., González, M., Ochoa, M., Medrano, H., 2004. Microencapsulación del jugo de cebada verde mediante secado por aspersión. Revista Ciencia y Tecnología Alimentaria. 4 (4), 262–266.
Gaysinsky, S., Davidson, P., McClements, J., Weiss, J., 2008. Formulation and characterization of phytophenol-crrying Antimicrobial Microemulsions. Food Biophys. 3, 54–65.
Georgetti, S., Casagrande, R., Souza, C., Oliveira, W., Fonseca, M., 2008. Spray drying of the soybean extract: effects on chemical properties and antioxidant activity. LWT—Food Sci. Technol. 41, 1521–1527.
Golemanov, K., Tcholakova, S., Denkov, N., Gurkov, T., 2006. Selection of surfactans for stable paraffin in water dispersions, undergoing solid-liquid: Transition of the dispersed particles. Langmuir 22, 3560–3569.
Guarda, A., Rubilar, J., Miltz, J., Galotto, M., 2011. The antimicrobial activity of microencapsulated thymol and carvacrol. Int. J. Food Microbiol. 146 (2), 144–150.
Heinonen, M., 2007. Review: antioxidant activity and antimicrobial effect of berry phenolics—a Finnish perspective. Mol. Nutr. Food Res. 51, 684–691.
Howell, A., Reed, J., Krueger, C., Winterbottom, R., Cunningham, D., Leahy, M., 2005. A-type cranberry proanthocyanidins and uropathogenic bacterial anti-adhesion activity. Phytochemistry 66, 2281–2291.
Koksal, E., Bursal, E., Dikici, E., Tozoglu, F., Gulcin, I., 2011. Antioxidant activity of *Melissa officinalis* leaves. J. Med. Plant. Res. 5 (2), 217–222.
Kosaraju, S., D'ath, L., Lawrence, A., 2006. Preparation and characterisation of chitosan microspheres for antioxidant delivery. Carbohyd. Polym. 64, 163–167.
Kosaraju, S., Labbett, D., Emin, M., Konczak, I., Lundin, L., 2008. Delivering polyphenols for healthy ageing. Nutr. Diet. 65, S48–S52.
Krishnaiah, D., Sarbatly, R., Mohan Rao, S., Nithyanand, R., 2009. Optimal operating conditions of spray-dried noni fruit extract using κ-carrageenan as adjuvant. J. Appl. Sci. 9, 3062–3067.
Kumazawa, S., Hamazaka, T., Nakayama, T., 2004. Antioxidant activity of propolis of various geographic origins. Food Chem. 84, 329–339.
Kuskoski, M., Asuero, A., Troncoso, A., Mancini, J., Fett, R., 2005. Aplicación de diversos métodos químicos para determinar actividad antioxidante en pulpa de frutos Ciênc. Tecnol. Aliment. Campinas 25 (4), 726–732.

Leiman, F., Goncalves, O., Machado, R., Bolzan, A., 2009. Antimicrobial activity of microencapsulated lemongrass essential oil and the effect of experimental parameters on microcapsules size and morphology. Mater. Sci. Eng. 29 (2), 430–436.
Leopoldini, M., Russo, N., Toscano, M., 2011. The molecular basis of working mechanism of natural polyphenolic antioxidants. Food Chem. 125, 288–306.
Levic, S., Rac, V., Manojlovi, V., Raki, V., Bugarski, B., Flock, T., Krcyczmonik, K., Nedovi, V., 2011. Limonene encapsulation in alginate/poly (vinyl alcohol). Procedia Food Sci. 1, 1816–1820.
Li, B., Wang, L., Li, D., Bhandari, B., Jun, S., Lan, Y., Chen, X., Mao, Z., 2009. Fabrication of starch-based microparticles by an emulsification-cross-linking method. J. Food Eng. 92 (3), 250–254.
Lu, J., Lin, P., Yao, Q., Chen, C.h., 2010. Chemical and molecular mechanisms of antioxidants: experimental approaches and model systems. J. Cell Mol. Med. 14 (4), 840–860.
Lupo, B., González, C., Maestro, A., 2012. Microencapsulation in alginate for food: technologies and applications. Rev. Venez. Cienc. Tecnol. Aliment. 3 (1), 130–151.
Manach, C., Scalbert, A., Morand, C., Rémésy, C., Jiménez, L., 2004. Polyphenols: food sources and bioavailability. Am. J. Clin. Nutr. 79, 727–747.
Manach, C., Williamson, G., Morand, C., Scalbert, A., Remesy, C., 2005. Bioavailability and bioefficacy of polyphenols in humans. I. Review of 97 bioavailability studies. Am. J. Clin. Nutr. 81, 230S–242S.
Madene, A., Scher, J., Desobry, S., 2006. Flavour encapsulation and controlled release—a review. Int. J. Food Sci.Technol. 4 (1), 1–21, 2006.
Marcuzzo, E., Sensidoni, A., Debeaufort, F., Voilley, A., 2010. Encapsulation of aroma compounds in biopolymeric emulsion-based edible films to control flavor release. Carbohyd. Polym. 80 (3), 984–988.
McClements, D., 2012. Crystals and crystallization in oil-in-water emulsions: Implications for emulsion-based delivery systems. Adv. Colloid Interf. Sci. 174, 1–30.
Medonça, C., Silva, Y., Bockel, W., Simó-Alfonso, E., Ramis-Ramos, G., Piatnicki, C., Bica, C., 2009. Role of the surfactant nature in soybean w/o microemulsions. J. Colloid. Interf. Sci. 337, 579–585.
Miranda-Rottmann, S., Aspillaga, A., Pearez, D., Vasquez, L., Martínez, A., Leighton, L., 2002. Juice and phenolic fractions of the berry *Aristotelia chilensis* inhibit LDL oxidation *in vitro* and protect human endothelial cells against oxidative stress. J. Agric. Food Chem. 50, 7542–7547.
Mirhosseini, H., Tan, C., Hamid, N., Yusof, S., 2008. Optimization of the contents of arabic gum, xanthan gum, and orange oil affecting turbidity, average particle size, polydispersity index and density in orange beverage emulsion. Food Hydrocolloid. 22, 1212–1223.
Mirhosseini, H., Bahareh, A., 2012. A review study on chemical composition and molecular structure of new plant gum exudates and seed gums. Food Res. Int. 46, 387–398.
Moschakis, T., Murray, B., Biliaderis, C., 2010. Modifications in stability and structure of whey protein-coated o/w emulsions by interacting chitosan and gum arabic mixed dispersions. Food Hydrocolloid. 24, 8–17.
Munin, A., Edwards, F., 2011. Encapsulation of natural polyphenolic compounds: a review. Pharmaceutics 3, 793–829.
Murúa, B., Beristain, C., Martínez, F., 2009. Preparation of starch derivatives using reactive extrusion and evaluation of modified starches as shell materials for encapsulation of flavoring agents by spray drying. J. Food Eng. 91 (3), 380–386.

Nazzaro, F., Fratianni, F., Coppola, R., Sada, A., Orlando, P., 2009. Fermentative ability of alginate–prebiotic encapsulated *Lactobacillus acidophilus* and survival under simulated gastrointestinal conditions. J. Funct. Foods 1 (3), 319–323.

Neto, C., 2007. Review: cranberry and blueberry: evidence for protective effects against cancer and vascular diseases. Mol. Nutr. Food Res. 51, 652–664.

Nohynek, L., Alakomi, H., Kähkönen, M., Heinonen, M., Helander, I., Oksman, K., Puupponen, R., 2006. Berry phenolics: antimicrobial properties and mechanisms of action against severe human pathogens. Nutr. Cancer. 54 (1), 18–32.

Nori, M., Favaro-Trindade, C., de Alencar, S., Thomazini, M., de Camargo, J., Contreras, C., 2011. Microencapsulation of propolis extract by complex coacervation. LWT—Food Sci. Technol. 44, 429–435.

Onwulata, C., 2013. Microencapsulation and functional bioactive foods. J. Food Process. Preserv. 37 (5), 510–532.

Ordoñez, P., Vega, M., Malagón, O., 2006. Phytochemical study of native plant species used in traditional medicine in Loja Province. Lyonia: J. Ecol. Appl. 10 (2), 65–71.

Parra, R., 2010. Revisión Microencapsulación de Alimentos. Rev. Fac. Agr. Medellín. 63 (2), 5669–5684.

Pasin, B., González, C., Maestro, A., 2012. Microencapsulación con alginato en alimentos. Técnicas y aplicaciones. Rev. Venez. Cienc. Tecnol. Aliment. 3 (1), 130–151.

Pastene, E., 2009. Estado actual de la búsqueda de plantas con actividad antioxidante. BLACPMA 8 (6), 449–455.

Puupponen, R., Nohynek, L., Hartmann, S., Kahkonen, M., Heinonen, M., Maatta, K., Oksman, K., 2005. Berry phenolics selectively inhibit the growth of intestinal pathogens. J. Appl. Microbiol. 98, 991–1000.

Richard, T., Lefeuvre, D., Descendit, A., Quideau, S., Monti, J., 2006. Recognition characters in peptide-polyphenol complex formation. Biochem. Biophys. Acta 1760, 951–958.

Robert, P., Gorena, T., Romero, N., Sepúlveda, E., Chávez, J., Sáenz, C., 2010. Encapsulation of polyphenols and anthocyanins from pomegranate (*Punica granatum*) by spray drying. Int. J. Food Sci. Technol. 45, 1386–1394.

Rocha, N., Gallegos, J., González, R., Harte, F., Medina, L., Ochoa, L., Soto, M., 2010. Effect of high-pressure homogenization on the physical and antioxidant properties of quercus resinosa infusions encapsulated by spray-drying. J. Food Sci. 75, 57–61.

Rocha, S., Generalov, R., Pereira, M., Peres, I., Juzenas, P., Coelho, M., 2011. Epigallocatechin gallate-loaded polysaccharide nanoparticles for prostate cancer chemoprevention. Nanomedicine 6, 79–87.

Rubilar, M., Jara, Cl., Poo, Y., Acevedo, F., Gutierrez, C., Sineiro, J., Shene, C., 2011. Extracts of maqui (*Aristotelia chilensis*) and murta (*Ugni molinae Turcz.*): Sources of antioxidants compounds and α-glucosidase/α-amylase inhibitors. J. Agric. Food Chem. 59, 1630–1637.

Ruiz, A., Hermosín, I., Mardones, Cl., Vergara, C., Herlitz, E., Vega, M., Dorau, C., Winterhalter, P., Dietrich von Baer, 2010. Polyphenols and antioxidant activity of calafate (*Berberis microphylla*) fruits and other native berries from Southern Chile. J. Agric. Food Chem. 58, 6081–6089.

Saénz, C., Tapia, S., Chávez, J., Robert, P., 2009. Microencapsulation by spray drying of bioactive compounds from cactus pear (*Opuntia ficus-indica*). Food Chem. 114, 616–622.

Schreckinger, M., Wang, J., Yousef, G., Lila, M., Gonzalez, E., 2010. Antioxidant capacity and in vitro inhibition of adipogenesis and inflammation by phenolic extracts of *Vaccinium floribundum* and *Aristotelia chilensis*. J. Agric. Food Chem. 58, 8966–8976.

Seeram, N., 2008. Berry fruits for cancer prevention: current Status and Future Prospects. J. Agric. Food Chem. 56 (3), 630–635.
Soottitantawata, A., Takayama, K., Okamura, K., Muranaka, D., Yoshii, H., Furuta, T., Ohkawara, M., Linko, P., 2005. Microencapsulation of l-menthol by spray drying and its release characteristics. Innov. Food Sci. Emerg. Technol. 6, 163–170.
Su, J., Flanagan, J., Singh, H., 2008. Improving encapsulation efficiency and stability of wáter-in-oil-i-water emulsions using a modified gum arabic (*Acacia*(sen) SUPER GUM™). Food Hydrocolloid. 22, 112–120.
Surh, J., Vladisavljevic, G., Mun, S., McClements, J., 2007. Preparation and characterization of water/oil and water/oil/water emulsions containing biopolymer-gelled water droplets. J. Agric. Food Chem. 55, 175–184.
Suwalsky, M., Vargas, P., Avello, M., Villena, F., Sotomayor, C., 2008. Human erythrocytes are affected in vitro by flavonoids of *Aristotelia chilensis* (maqui) leaves. Int. J. Pharmaceut. 363 (12), 85–90.
Troszynska, A., Narolewska, O., Robredo, S., Estrella, I., Hernández, T., Lamparski, G., Amarowicz, R., 2010. The effect of polysaccharides on the astringency induced by phenolic compounds. Food Qual. Pref. 21 (5), 463–469.
Velioglu, Y., Mazza, G., Gao, L., Oomah, B., 2006. Antioxidant activity andtotal phenolics in selected fruits, vegetables, and grain products. J. Agric. Food Chem. 46, 4113–4117.
Villamizar, L., Martínez, F., 2008. Determinación de las condiciones de microencapsulación de un baculovirus entomopatógeno mediante coacervación con Eudragit S100®. Revista Vitae. 15 (1), 123–131.
Xiong, S., Melton, L., Easteal, A., Siew, D., 2006. Stability and antioxidant activity of black currant anthocyanins in solution and encapsulated in glucan gel. J. Agric. Food Chem. 54, 6201–6208.
Yañez, J., Salazar, J., Chaires, L., Jimenez, J., Marquez, M., Ramos, E., 2002. Aplicaciones biotecnológicas de la microencapsulación. Revista Avance y Perspectiva 21, 313–319.
Zhang, L., Mou, D., Du, Y., 2007. Procyanidins: extraction and micro-encapsulation. J. Sci. Food Agric. 87, 2192–2197.

12

BIOCOMPATIBLE MICROEMULSIONS FOR THE NANOENCAPSULATION OF ESSENTIAL OILS AND NUTRACEUTICALS

Sergio Enrique Flores-Villaseñor, René Darío Peralta-Rodríguez*, Jorge Carlos Ramirez-Contreras*, Gladis Yakeline Cortes-Mazatán*, Alba Nidia Estrada-Ramírez****

**Department of Polymerization Processes, Center for Research in Applied Chemistry, Saltillo, Coahuila, Mexico; **Doctoral Program in Polymer Technology, Center for Research in Applied Chemistry, Saltillo, Coahuila, Mexico*

1 Introduction

Oil/water (O/W), bicontinuous, and water/oil (W/O) microemulsions are at least three-component (oil, water, surfactant) systems that form spontaneously provided the adequate temperature and proportions of the ingredients for their formation are met. However, frequently a cosurfactant such as a low-molecular weight alcohol is used in O/W microemulsions. These isotropic, optically transparent and thermodynamically stable systems can be used to incorporate lipophilic water-insoluble materials (O/W microemulsions), such as beta-carotene, into food and beverage compositions, whereas W/O microemulsions are used to incorporate water-soluble materials. Because of their small droplet size (5–100 nm in diameter), the visible light is not scattered and thus microemulsions appear as clear or transparent solutions. To our knowledge, microemulsions were first introduced by Hoar and Schulman (1943) at Cambridge University. Hoar and Schulman came up with a transparent homogeneous solution by titrating a turbid emulsion with hexanol. They prepared the first reported microemulsion by dispersing oil in an aqueous amphiphiles solution and incorporating an alcohol as cosurfactant, to produce a transparent stable formulation. Schulman

Encapsulations. http://dx.doi.org/10.1016/B978-0-12-804307-3.00012-0

and coworkers subsequently termed these systems as *microemulsions* (Schulman et al., 1959). Alternative names for these systems are often used, such as transparent emulsion, swollen micelle, micellar solution, and solubilized oil. However, some of these terms are misleading and the term *microemulsion* is preferred, even though characteristic dimensions of microemulsions are in the nanometer scale (*micro-* means 10^{-6} and *nano-* means 10^{-9}). Since their introduction, microemulsions have been used in many fields: fuels, detergents, agrochemicals, soil remediation, foods, pharmaceutics, and cosmetics (Eccleston, 1988; Paul and Moulik, 1997, 2001; Lawrence and Rees, 2000; Zheng et al., 2011; Peltola et al., 2003; Valenta and Schultz, 2004), and have been reviewed many times (Sharma and Shah, 1985; Paul and Moulik, 2001; Salager et al., 2005; Flanagan and Singh, 2006; Boonme, 2007; Garti and Yuli-Amar, 2008; Li et al., 2014). Additionally, a number of books have been published on microemulsions, some of the most recent are the books by Najjar (2012); Fanun (2008), and Stubenrauch (2008).

Attractive characteristics of microemulsions for food, pharmaceutical, and cosmeceutical applications are their ultra-low interfacial tension (close to water) between the immiscible phases, their simplicity of manufacture (mixing of components with gentle agitation), and their capacity to solubilize hydrophobic and hydrophilic compounds (Fanun, 2012). Further, microemulsions are nonviscous liquids and possess the not shear-dependent flow behavior of Newtonian type and they have been found to be suitable for applications in beverages and drug solubilization (Spernath and Aserin, 2006). An interesting type of microemulsions are the U-type, which consist of a single isotropic region which upon dilution shows a continuous transition from W/O microemulsion to an O/W microemulsion with no phase separation. In general, most microemulsions for pharmaceutical and cosmetic applications are prepared with nonfood grade surfactants and oils. Health concerns in the development of food systems for human consumption are stricter than for pharmaceutical and cosmetic uses, regarding the type of surfactant and oil. Thus, it is more difficult the formulation of U-type, or any other class, microemulsions for food applications (Garti et al., 2003). As an example of a successful formulation of a microemulsion, Spernath et al. (2003) demonstrated the use of U-type food-grade microemulsions as vehicles for solubilizing phytosterols in amounts up to 12 times more than their solubility in R-(+)-limonene. Phytosterols are steroid alcohols naturally occurring in plants (plant sterols): administration of phytosterols to human subjects reduces the total plasma cholesterol and LDL (low-density lipoprotein) cholesterol levels (Pelletier et al., 1995; Jones et al., 1997).

Some food and beverage products need to have an appearance that is either transparent or slightly cloudy, such as fortified waters and beverages. Thus, flavor or aroma delivery systems must not affect these optical attributes so that consumer attraction for these goods is unaffected. Microemulsion systems are ideally suited to carry out this function by encapsulating aromas and flavors (Sanguansri and Augustin, 2006; Garti, 2003). Aromas and flavors have already been encapsulated in microemulsions, for instance, Rao and McClements (2011b) demonstrated the preparation of flavor oil (lemon oil) microemulsions using the GRAS surfactant Tween 80. Their research provides useful information into the preparation and stability of different types of food grade colloid dispersions (emulsions, nanoemulsions, and microemulsions) containing flavor oils. Lin et al. (2014) report microemulsions prepared with completely food grade components (surfactants: soybean lecithin and Tween 80; soybean oil as oil phase) for the encapsulation of curcumin and pinpoint that their results have the capability to provide cutting-edge applied techniques in the area of nutraceuticals and functional foods. Food applications of nanoencapsulation for targeted delivery of bioactives (eg, vitamins, carotenes, polyphenols, omega-3 fatty acids, coenzyme Q10, etc.) are just appearing. A basic strategy to deliver nutrients, proteins, and antioxidants to the body more effectively through food products is nanoencapsulation (Sanguansri and Augustin, 2006). Encapsulation can also reduce the evaporation rate of volatile species such as flavors and aromas. However, there is still need in the area of target release of bioactives in the body via encapsulation technologies using exclusively food-grade encapsulants. Further, as pointed out by Sanguansri and Augustin (2006), in the near future, food will not only need to be a good source of nutrients with good sensory appeal, but also contribute to the well-being and health of individuals. Almost a decade after this statement, nanoencapsulation of food ingredients is still a pending issue for development.

Self-microemulsifying delivery systems (SMEDS) represent an alternate approach to nanoencapsulation of flavors, aromas, drugs, nutraceuticals, and cosmeceuticals. In a recent paper, Chu et al. (2014) report the use of SMEDS and SEDS (self-emulsifying delivery systems) to enhance the absorption of lipophilic drugs. SEDS and SMEDS are mixtures of oils and surfactants, ideally isotropic and sometimes containing cosolvents, which emulsify spontaneously to produce fine O/W emulsions or microemulsions when introduced into an aqueous phase under gentle agitation (Gursoy and Benita, 2004; Ritesh et al., 2008). The advantages of some formulations using the SMEDS and SEDS approach as well as commercial products are summarized by Patel et al. (2008). These systems are well suited for encapsulation of essential oils and nutraceuticals. To our knowledge there are no reports on the use of the SMEDS techniques for encapsulation of food components.

1.1 Surfactants

The term *surfactant* (surface-active-agent) designates a substance which exhibits some superficial or interfacial activity. Surfactants are amphiphiles although not all of them display surface activity; in effect, only the amphiphiles with more or less equilibrated hydrophilic and lipophilic tendencies are likely to migrate to the surface or interface. All surfactants have a polar or ionic portion, hydrophilic head, attached to a nonpolar hydrophobic tail (a hydrocarbon or fluorocarbon chain containing 8–18 carbon atoms) (Tadros, 2005). Depending on the nature of the hydrophilic head, surfactants can be classified as cationic, anionic, zwitterionic, and nonionic. Cationic surfactants have a positive charge on their polar group (linear alkyl-amines and alkyl-ammoniums, etc.), whereas anionic surfactants have a negative charge (carboxylates, sulfates, sulfonates, etc.). Zwitterionic surfactants have both positive and negative charges (eg, trimethylammonium carboxylates, commonly known as betaines), depending on the environment in which they are placed. Nonionic surfactants have no charge on their hydrophilic group (ethoxylated alcohols and alkylphenols, fatty acid esters, etc.) (Sekhon, 2013). Fig. 12.1 shows a schematic representation of this classification.

Surfactant chains may be saturated or unsaturated, linear or branched, aliphatic and/or aromatic, but most food grade surfactants have either one or two linear aliphatic chains, which may be saturated or unsaturated. Not all surfactants perform the same and care should be taken in their selection. To select a surfactant for a given application, the molecular structure and conditions in which it will be used have to be accounted for. The chosen

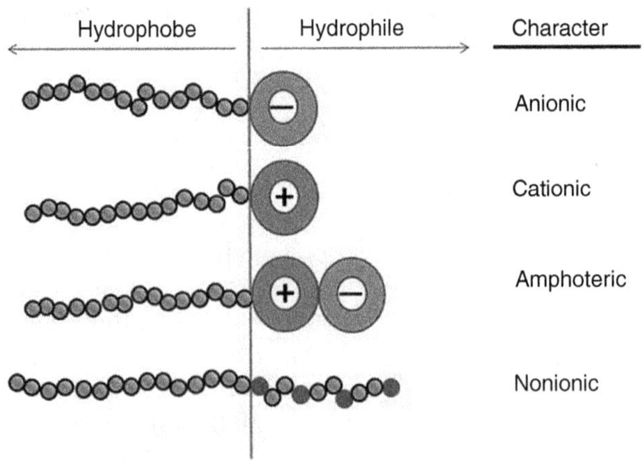

Figure 12.1. Different types of surfactants.

surfactant must meet besides technical aspects (physicochemical performance, usage levels, ingredient compatibility, stability, and ease of utilization), legal status, and costs (McClements et al., 2009).

Kralova and Sjöblom (2009) review the surfactants used in the food industry and critically analyze the main parameters for the hazard assessment of food surfactants: acute toxicity, subacute repeated studies, allergy, reproductive toxicity, long-term studies, and mutagenicity tests. They also include a detailed discussion of biosurfactants and point to the potential applications of these surfactants as multipurpose ingredients or additives because of their combination of specific features such as emulsifying, antiadhesives, and antimicrobial activities.

The role of surfactants in microemulsion formulation is to lower the interfacial tension which facilitates the dispersion process and provides the flexibility of the droplets. The surfactant should have the appropriate lipophilic character to give the correct curvature at the interfacial region. The emulsifying capability of an agent may be classified according to the hydrophile-lipophile balance (HLB) in its molecules. This is a function of the weight percentage of hydrophilic groups in the molecule (Kralova and Sjöblom, 2009). HLB values for commercial emulsifying agents range from 1 to 20. Generally, low HLB surfactants form W/O microemulsions (HLB < 12), high HLB (> 12) are suitable for O/W microemulsions. Commonly used surfactants are aerosol OT, and mainly, nonionic surfactants, such as polysorbates, alkyl polyethers, and sorbitan monoesters (Salager, 2002).

In general, microemulsions require higher surfactants/cosurfactant concentration than their counterparts, emulsions and nanoemulsions, to reduce the surface tension between oil phase and water phase which leads to increased toxicity (Lin et al., 2014). Several surfactants are used in food and pharmaceutical applications. In food applications surfactants have been used as dispersing agents, emulsifiers, foamers, stabilizers, and so forth. (Kralova and Sjöblom, 2009), while in pharmaceutical applications, for solubilization of hydrophobic drugs in aqueous media, as components of emulsions, as surfactant self-assembly vehicles for oral and transdermal drug delivery, as plasticizers in semisolid delivery systems, and as agents to improve drug absorption and penetration (Sekhon, 2013; Kumar et al., 2011). Two series of commercial nonionic surfactants widely used in foods and pharmaceutics formulations are sorbitan esters (Span brand or equivalent) and their ethoxylated derivatives (Tween brand or equivalent). The lipophilic group in these surfactants ranges from one monolaureate unit (C12) to three oleate units (three C18).

These molecules although of some complexity, are easy to prepare from commonly available natural raw materials (eg, fat), which make them biologically compatible for food and pharmaceutical applications (Salager, 2002). In the case of cosmeceuticals, mild surfactants in the vehicle have been selected, typically anionic surfactants, although more recently nonionic and amphoteric surfactants, which are now considered milder than anionic types, are preferred (Epstein, 2009). Table 12.1 shows a compilation of biocompatible surfactants that have been used in food, nutraceutical, and pharmaceutical applications. Table 12.1 is not exhaustive but gives an idea of the surfactants available to the food technologist to formulate products. Among all, nonionic surfactants predominate, mainly the sorbitan oleates (Span series) and the polyoxyethylene sorbitan oleates (Tween series).

1.1.1 Synthetic Surfactants

The demand for synthetic surfactants is one of the highest and more widespread among all chemicals (Kulapina et al., 2013) because of their many industrial applications. The use of surfactants in food manufacture goes back to the decade of 1930: the margarine industry introduced the use of mono- and diglycerides (Csáki, 2011). However, the breakthrough in the use of surfactants in the food industry occurred during the 1960s in the United Kingdom with the invention of a process to reduce the fermentation periods in breadmaking by the use of a dough preparation and maturation high-speed mixing machine (Cauvain, 2003). This process makes use of chemical improvers, including surfactants. In 2008 the worldwide production of 20 different types of food surfactants was close to 500,000 tons, and 50% of these were used by the baked foods industry. In the food products industry, six groups of synthetic surfactants are more widely used as follows (Csáki, 2011).

1.1.1.1 Mono- and Diglycerides of Fatty Acids

These surfactants are the most popularly used emulsifiers in the food industry. They are nonionic surfactants made from natural raw materials. Their use is in bakery products, frozen desserts, icings toppings, and peanut butter.

1.1.1.2 Diacetyltartaric Acid Esters of Mono- and Diglycerides

Anionic surfactants are more hydrophilic than its constituents (mono- and diglycerides). The use of these surfactants is as dough

Table 12.1 Biocompatible Surfactants: Characteristics and Some Applications

Generic, Chemical, or Trademark Name	Surfactant Type	HLB	Uses	References
Synthetic surfactants				
Sorbiton monolaurate (Span 20), sorbitan monooleate (Span 80), sorbitan sesquioleate (Span 83), and sorbitan trioleate (Span 85)	Nonionic	Span 20: 8.6, Span 80: 4.3, Span 83: 3.7, Span 85: 1.8	Suspension aerosols	Mishra et al. (2009)
Sorbitan monopalmitate (Span 40), sorbitan trioleate (Span 85), cetyl trimethyl ammonium bromide (CTAB)	Nonionic and cationic	Span 40: 6.7, Span 85: 1.8, CTAB: 10	Transdermal drug delivery systems as permeation enhancers	Pang and Han (2014)
Polysorbate 20 (Tween 20)	Nonionic	16.7	Clove bud oil and eugenol microemulsions were used as antioxidant and antimicrobial activities	Hamed et al. (2012), cited by Xue (2015)
			Transdermal penetration drug	Mishra et al. (2009)
			Stabilizer in poly(alkylcyanoacrylate) nanoparticles (biodegradable polymer). Tissue adhesives in surgery (well tolerated in vivo)	Couvreur et al. (1979), cited by Soppimath et al. (2001)
Polysorbate 60 (Tween 60)	Nonionic	14.9	Stabilizer in order to study the solubilization of phytosterols and cholesterol in R(+)-limonene as oil phase	Spernath et al. (2003), cited by Spernath and Aserin (2006)

(*Continued*)

Table 12.1 Biocompatible Surfactants: Characteristics and Some Applications (*cont.*)

Generic, Chemical, or Trademark Name	Surfactant Type	HLB	Uses	References
			Preparation of essential oil microemulsions	Ma and Zhong (2015)
			Elaboration of nanoemulsion using cinnamon oil as antimicrobial agent	Ghosh et al. (2013b)
Polysorbate 80 or polyoxyethylene sorbitan monooleate (Tween 80)	Nonionic	15	Parenteral phospholipid-based microemulsion containing etoposide as anticancer drug	Jain et al. (2010)
			Parenteral products (polysorbates) to prevent aggregation; intramuscular administration	Broadhead and Gibson (2009)
			Ophthalmic suspension	Gibson (2009)
			U-type microemulsion systems used for the solubilization of nutraceuticals	Garti et al. (2006), cited by Spernath and Aserin (2006)
			U-type microemulsion systems using R(+)-limonene/ethanol (1:2 w/w) as oil phase for the solubilization of lutein	Garti (2003), cited by Spernath and Aserin (2006)
Tweens 20, 40, 60, and 80	Nonionic	Tween 20: 16.7, Tween 40: 15.6, Tween 60: 14.9, Tween 80: 15	Stabilizer and solubilization of lycopene (essential oil) via microemulsions	Spernath et al. (2002), cited by Spernath and Aserin (2006)

Table 12.1 Biocompatible Surfactants: Characteristics and Some Applications (*cont.*)

Generic, Chemical, or Trademark Name	Surfactant Type	HLB	Uses	References
Nonionic: fatty acid esters of sorbitan (spans: 20, 40, 60, 65, 80) and ethoxylated derivatives (Tweens: 20, 40, 60, 65, 80, and 85) and poloxamer 188 more used; zwitterionic: phospholipids as phosphatidylcholine (lecithin); cationic: quaternary ammonium chloride also known as quats; anionic: sodium lauryl sulfate (SDS), dioctyl sodium sulfosuccinate, alkyl benzene sulfonates, alkyl ether phosphates, sodium lauryl sulfate	Nonionic, zwitterionic, cationic, and anionic	Span 20: 8.6, Span 40: 6.7, Span 60: 4.7, Span 65: 2.1, Span 80: 4.3, Tween 20: 16.7, Tween 40: 15.6, Tween 60: 14.9, Tween 65: 10.5, Tween: 80: 15, Tween 85: 11, SDS: 40	Nonionic: emulsifier, wetter, solubilizer and dispersant in pharmaceutical industry; zwitterionic: excellent dermatological properties; cationic: bactericidal activity; anionic: sodium lauryl sulfate is used pharmaceutically as a preoperative skin cleaner, having bacteriostatic action against gram-positive bacteria, and also in medicated shampoos	Sekhon (2013)
Polyoxyethylene sorbitan monooleate (Tween 80) and mixture of sucrose monopalmitate (SMP)	Nonionic	Tween 80: 15	Lemon oil nanoemulsions using these stabilizers in order to improve the acidity stability of oil	Rao and McClements (2012), cited by Xue (2015)
Tween 80, egg yolk phospholipid, and cholesterol	Nonionic and zwitterionic	Tween 80: 15	Nanoliposomes as delivery system using coenzyme Q10 as nutraceutical	Xia et al. (2007), cited by Quintanilla-Carvajal et al. (2010)
Poloxamers and lecithin	Nonionic and zwitterionic	Lecithin: 4–9	Surfactants/emulsifying agents used as amphiphilic excipients in the elaboration of gelatin capsules	Davies (2009)

(*Continued*)

Table 12.1 Biocompatible Surfactants: Characteristics and Some Applications (*cont.*)

Generic, Chemical, or Trademark Name	Surfactant Type	HLB	Uses	References
Methylcellulose (MC), hydroxypropyl methylcellulose (HPMC), carboxymethylcellulose sodium, polyvinyl alcohol, povidone, polyethylene glycol 400 (PEG-400), and poloxamer 407		MC: 10–12, HPMC: 10–12, poloxamer 407: 18–23 PEG-400: 11.6	Surfactants/emulsifiers used as ophthalmic viscosity-enhancing agents	Gibson (2009)
β-Dodecyl maltoside (β-C12G2), an alkyl polyglycoside (APG), sucrose 6-monolaurin, saccharide-fatty acid ester, polysorbate (ethoxylated sorbitan-oleic acid ester), 1-monolaurin (a monoacylglycerol, or MAG), α-tocopheryl polyethylene glycol succinate (TPGS)	Nonionic	TPGS: 13	Pharmaceutical preparation	Sandeep et al. (2013)

conditioners for yeast-leavened baked products (eg, white bread, flour mixes for convenience foods). They also find applications in dairy products and are approved to be used in special infant formulae.

1.1.1.3 Sodium Stearoyl-2-Lactylate (SSL) and Calcium Stearoyl-2-Lactylate

The sodium and calcium stearoyllactylates are obtained from reacting fatty acids, lactic acid, with a suitable sodium or calcium source (Lauridsen, 1976). These anionic surfactants are used primarily to prepare bread for better dough gas retention and dough stability; they also produce a finer cell structure in the product.

1.1.1.4 Sucrose Esters of Fatty Acids

Nonionic surfactants obtained by esterification of fatty acids with sucrose. Depending on the ester composition sucrose esters can have a very wide HLB range (1–18) and consequently a wide variety of applications (dressing sauces, light mayonnaise, infant foods, dairy products, and ice creams). Sucrose fatty acid esters (SEs) are recognized as very functional emulsifiers. Otomo (2009) reports on the basic properties and applications of a group of SEs commercialized under the trade name of Ryoto Sugar Ester® (Mitsubishi–Kagaku Foods Corporation, Tokyo). The HLB values of Ryoto's stearic series of SEs is from 1 to 16, that is, cover a wide range of needs in the food and nonfood industries to prepare O/W to W/O emulsion systems (emulsions, nanoemulsions, and microemulsions).

1.1.1.5 Polyglycerol Esters of Fatty Acids (PGE)

This is another series of nonionic surfactants obtained by the esterification of vegetable long aliphatic chain acids (even number of carbon atoms, C10–C18) and polyglycerol. The HLB range of these surfactants can be between 3 and 18, and depends on the degree of glycerol polymerization (typically, 2–10 units long) (Hepworth, 2006), and the fatty acid/polyglycerol ratio (Csáki, 2011). Polyglycerol esters are applied in the formulation of low-fat margarines, spreads, butter creams, and breakfast cereals.

1.1.1.6 Sorbitan Esters of Fatty Acids (Span Series, HLB = 2–9) and Their Ethoxylated Derivatives Polysorbates (Tween Series, HLB = 10–17)

These chemicals are nonionic surfactants widely used in the food industry because of their excellent emulsifying properties, and also find applications as aerating agents and lubricants in cakes, toppings, cookies, and crackers. As an example, polysorbate 60 (a Tween surfactant) is used as dough strengthener coemulsifier in bakery products. Sorbitan esters of fatty acids and polysorbates have been and are used in surfactant mixtures. As far back as 1980, the use of mixtures of Span 20/Tween 80 or Span 60/Tween 60 to enhance physical stability of aqueous emulsions of the essential oil of hops was reported (Chilton and Laws, 1980). More recently, Losada-Barreiro et al. (2013) evaluated the effects of the HLB of mixtures of four nonionic amphiphiles (Tween 20, 40, 80, and Span 20) on the partition, between the aqueous and oil phases, plus the interface, of gallic acid, propyl gallate, and alpha-tocopherol (antioxidants) in edible emulsions formulated with corn oil, acidic water, and a mixture. These antioxidants are frequently added in the industry to retard or inhibit lipid oxidation in food products (an important chemical reaction that may occur during food processing and storage) causing loss of food quality.

1.1.2 Natural Surfactants (Biopolymers)

Functional macromolecules such as polar lipids, proteins, glycolipids, polysaccharides, and phospholipids, have been known as good emulsifying agents and stabilizers in natural food colloids and in man-made food products (Bouyer et al., 2012; Garti, 1999). These natural polymers are already used for pharmaceutical applications (Rowe et al., 2006). In the case of proteins, they are used as emulsifiers in the formation and stabilization of fat emulsions, typically, sausages, bologna, soup, and cakes (Cabra et al., 2008).

1.1.2.1 Proteins

Many proteins can act as emulsifiers because of their ability to adsorb at the oil–water interface (Bouyer et al., 2012) increasing emulsion stability. Many proteins are too hydrophobic, or too hydrophilic, and thus it is convenient to modify them by either chemical or enzymatic routes so that they become more surface active (Kralova and Sjöblom, 2009). Examples of native emulsifiers used in dairy products like milk, ice cream are lactoglobulins, lysozymes, and ovalbumins (Kralova and Sjöblom, 2009). Chemically modified products, in some cases, perform even better than the native proteins; they appear in the market as modified soy proteins, egg proteins, and whey proteins.

1.1.2.2 Polysaccharides

Polysaccharides are rigid, water-soluble, and they are not considered as classical emulsifiers. Polysaccharides make good emulsion stabilizing agents because of their hydrophilicity, high molecular weight and oil–water interfaces forming an extended network in the continuous phase which becomes highly viscous and can even form a gel (Bouyer et al., 2012). Only few polysaccharide derivatives possess surface properties that enable their adsorption at the oil–water interface (Laplante et al., 2005; Dickinson, 2009; Benna-Zayani et al., 2008; Dickinson, 2003).

1.1.2.3 Phospholipids

Phospholipids are natural, highly surface-active compounds widely used to prepare food emulsions (Garti, 1999). Frequently used phospholipid emulsifiers include phosphatidylcholine, phosphatidylethanolamine, phosphatidylinositol, and phosphatidic acid (Bergenståhl, 2008). Lecithin is a very common natural emulsifier with many health benefits that is widely accepted by consumers and legislators (Oke et al., 2010). The main source of vegetable lecithin is soybeans: soy lecithin is extracted from the by-product of soybean oil (Wu and Wang, 2003).

1.1.3 Mixtures of Surfactants

Small molecule surfactants and polysaccharides are two groups of amphiphilic materials that have been studied for the stabilization of emulsions in food and pharmaceutical applications (Yang et al., 2013). Commonly employed polysaccharides include xanthan, alginate, chitosan, carrageenan, pectin, rhamsan, and dextran, while small molecule surfactants frequently used are sodium dodecyl sulfate (SDS), mono- and diglycerides, sorbitan esters, and phospholipids (Bais et al., 2005).

Proteins and polysaccharides can be combined under the proper proportions, temperature, pH, and ionic strength for improving the stability of emulsions (Bouyer et al., 2012). Biopolymer particles, which may be soluble complexes, fluid droplets, or hydrogel particles, can be formed from a biopolymer solution (polysaccharides and proteins used either individually or in combination) and used to encapsulate and deliver nutraceuticals and functional food components (McClements et al., 2009). Nonionic small molecule surfactants of the sorbitan esters series (monolaurate, monooleate, or triooleate) and chitosan were combined and studied by Grant et al. (2006) for preparation of emulsions. The chitosan-sorbitan monooleate complex produced a cream that may be used for the development of stable emulsions for applications in the food and pharmaceuticals industries.

1.1.4 Biosurfactants

A recent development in surfactants for biological applications (foods, pharmaceutics, cosmetics) is the use of microbial surfactants, so-called biosurfactants (Banat et al., 2010). Biosurfactants are produced by a variety of microorganisms from different renewable resources like vegetable oils and carbohydrates (Kitamoto et al., 2009). The term *biosurfactant* refers to any usable and isolated compound obtained from microorganisms that has some influence on interfaces (Kralova and Sjöblom, 2009). Some examples of biosurfactants are rhamnolipids (from *Pseudomonas aeruginosa*), sophorolipids (obtained from *Candida bombicola*), emulsan (from *Acinetobacter calcoaceticus*), and surfactin (from *Bacillus subtilis*) (Nitschke and Costa, 2007). Rodrigues (2015) reviews the fundamentals and applications of microbial surfactants in the formulation of drug delivery nanodevices. Microbial surfactants are synthesized by microorganisms and are generally classified by their chemical composition and molecular weight, as low (eg, glycolipids and lipopeptides) and high molecular weight (eg, polysaccharides, proteins, and lipoproteins) surfactants. High-molecular-weight biosurfactants, also called bioemulsifiers, are more effective in stabilizing O/W emulsions (Banat et al., 2010). Microbial surfactants are considered as possessing advantages over their chemical counterparts and their application in the food, cosmetics, and bioremediation industries is very promising.

Most work on biosurfactants applications has been focusing on bioremediation of pollutants (Mulligan, 2005); however, these diverse microbial compounds (biodegradable, low toxicity) exhibit a variety of useful properties for the food industry, especially as emulsifiers, foaming, wetting, solubilizers (Banat et al., 2000), antiadhesive, and antimicrobial agents (Singh and Cameotra, 2004). For example, biosurfactants have been used to control consistency, retard staling and solubilize flavor oils in bakery and ice cream formulations; also, they have been used as fat stabilizers and antispattering agents during cooking of oil and fats (Kosaric, 2001). Another potential application in food industry is as antiadhesive agents that could be used to protect against bacterial biofilms on food surfaces (Kralova and Sjöblom, 2009). Biosurfactants have also been used to prepare microemulsions, as reported by Kitamoto et al. (2009) and Nguyen and Sabatini (2011).

1.2 Nutraceuticals

Nutraceuticals are diet supplements that deliver a concentrated form of a presumed bioactive agent from a food, presented in a nonfood matrix, and used with the purpose of enhancing health in dosages that exceed those that could be obtained from normal foods (Zeisel, 1999). This isolated component concept of nutraceuticals clearly distinguishes them from functional foods, defined by the American Dietetic Association as "...whole, fortified, enriched, or enhanced foods which have a potentially beneficial effect on health when consumed as part of a varied diet on a regular basis at effective levels." (Hasler and Brown, 2009; Hasler, 2003). The hybrid term between nutrients and pharmaceuticals, *nutraceuticals*, has been coined to designate phytochemicals that are present in the diet and have been associated with health benefits (Gul et al., 2015). It is common to use the terms *nutraceutical* and *functional foods* interchangeably. In their recent review (Gul et al., 2015) and in a recent book (Boye, 2015), details and examples of the differences between these two concepts are given. The emergence of nutraceuticals with health benefits provides an excellent opportunity to improve public health. Nutraceuticals are sold in presentations similar to drugs: pills, extracts, tablets, and so forth. The Food and Drug Administration (FDA; http://vm.cfsan.fda.gov) regulates dietary supplements under a different set of regulations than those covering conventional foods and drug products. In Europe, no specific regulation exists to control nutraceuticals. In this chapter, nutraceuticals will be considered in line with Gul et al. (2015) based on Zeisel's approach (1999) as dietary supplements. The importance of nutraceuticals and other natural health products has been well recognized in connection with health promotion, disease risk reduction, and reduction in health care costs (Shahidi, 2009).

1.3 Cosmeceuticals

Although the purpose of this review is on the use of microemulsions for nanoencapsulation of flavors and aromas for food and nutraceuticals applications, it is considered convenient to give a brief account on the use of microemulsions in cosmeceuticals. The composite term *cosmeceuticals* was originally created by R.E. Reed in 1962, and A.M. Kligman repopularized the term in 1993 (Saint-Leger, 2012). The word cosmeceuticals represents a linguistic fusion of cosmetics and pharmaceutics, and indicates the expected bioactivity of the cosmetic product (Magdassi, 1997). In a few words, cosmeceuticals are cosmetics formulated with pharmaceutical-type ingredients (Patravale and Mandawgade, 2008). These authors, further state that nutraceutical ingredients formulated in cosmetic delivery systems constitute nutracosmetics. Some ingredients, for example, vitamins, can be used in either nutraceuticals, nutracosmetics, and in cosmeceuticals applications, although the end function is generally different. Microemulsions formulated for food applications could also be directed for nutraceuticals and cosmeceuticals uses. The application of microemulsions to encapsulate active ingredients in cosmeceuticals has been reviewed by several authors and the interested reader is addressed to the references (Edris and Malone, 2012; Patravale and Mandawgade, 2008; Magdassi, 1997; Higgins and Wesley, 2015; Newburger, 2009).

1.4 Oil Phase (Essential Oils, Vitamin E, *Trans*-anethole, Cinnamon Oil, Thyme Oil, Peppermint Oil)

In antiquity, the Greeks used spices for their perfume, medicinal and preservative properties and to impart aroma and flavor to food. The first distillation of essential oils (EO) appeared in the East (India and Persia) more than 2000 years ago and was improved in the 9th century by the Arabs (Bilia et al., 2014). Since the middle ages, EOs have been widely used for bactericidal, virucidal, fungicidal, antiparasitical, insecticidal, medicinal, and cosmetic applications, and nowadays in pharmaceutical, sanitary, cosmetic, agricultural, and food industries. Because of the mode of extraction, mostly by distillation from aromatic plants, EOs contain volatile molecules such as terpenes and terpenoids, phenol-derived aromatic components, and aliphatic components (Bakkali et al., 2008). At present, around 3000 essential oils are known, 300 of which have commercial importance (Bilia et al., 2014) for the pharmaceutical, agronomic, food, sanitary, cosmetic, and perfume industries. EOs that have found applications in foods are, for example, mint (*Mentha spicata* L., Lamiaceae), ginger (*Zingiber officinale* Rosc., Zingiberaceae), lemon (*Citrus limon* Burm.f., Rutaceae), grapefruit (*Citrus paradisi* Macf., Rutaceae), chamomile (*Matricaria chamomilla* L., Compositae), thyme (*Thymus vulgaris* L., Lamiaceae), cinnamon (*Cinnamomum zeylanicum* N. Lauraceae), and many others.

Vitamin E is a fat and oil-soluble vitamin that protects human cells against damage from free radicals. Further, it can also be used in cancer treatment, although its efficacy is questioned and additional research is needed (http://www.mayoclinic.org/drugs-supplements/vitamin-e/evidence/hrb-20060476). Nevertheless, vitamin E is administered to patients with deficiencies to compensate this nutrient. Vitamin E supplementation is carried out with capsules that have the disadvantage of low bioavailability. Thus, microemulsions with vitamin E as the oil phase in an O/W system represents a viable and convenient alternative to capsules. Feng et al. (2009) report a study on food-grade vitamin E microemulsions formulated with poloxyl 35 (EL-35, Cremophor EL) noninonic surfactant and as food-grade cosurfactants, ethanol (EtOH) and propylene glycol (PG). Food-grade oils used were ethyl butyrate, ethyl caprylate, ethyl oleate, and n-octane. Feng et al. concluded that the microemulsion of water/vitamin E/ethyl butyrate/EL-35/ethanol can form a large microemulsion region and this system behaves as a U-type microemulsion, that is, upon dilution of a W/O microemulsion it can transform into an O/W microemulsion without separating into phases. Further, they also conclude that this system may further be used in the food industry. A major drawback of this study is the use of the Cremophor EL surfactant that has been associated with undesirable side effects when used as a vehicle for poorly water-soluble drugs (Gelderblom et al., 2001).

Trans-anethole is an alkoxy-propenylbenzene derivative with important use and great commercial interest as a flavoring substance in baked goods, candy, ice cream, chewing gum, and alcoholic beverages. *Trans*-anethole [1-methoxy-4-(1-propenyl) benzene] is found in spices and herbs such as anise, sweet and bitter fennel, anise myrtle, lemon balm, coriander, and so forth. (Zafeiropoulou et al., 2010). Additionally, *trans*-anethole inhibits pathogens that contaminate food and therefore can be used for food preservation (Kfoury et al., 2014). In spite of these beneficial properties, the use of *trans*-anethole in foods has been limited because of its volatility and very low aqueous solubility that hinder the contact with pathogens in high moisture foods. In as much as *trans*-anethole can be easily oxidized, decomposed, or isomerized when exposed to air, light, or heat during food processing and storage (Elgendy and Khayyat, 2008), it is imperative to enhance its solubility and stability. This represents an important challenge and nanoencapsulation in microemulsions could be used to alleviate these problems. Newberne et al. (1999) report that based on results of extensive studies, and under conditions of its intended use as a flavoring agent, *trans*-anethole is considered as nongenotoxic, noncarcinogenic and generally recognized as safe (GRAS).

Cinnamon, which has the scientific name *Cinnamomum zeylanicum*, originated in tropical Asia. used nowadays for the treatment of a variety of health disorders including respiratory problems, skin infections, blood impurity, menstrual problems, and various heart disorders. Cinnamon oil has been proven to show several biological functions such as analgesic, antithrombotic, antioxidant, antimicrobial, antidiabetic, antipyretic, antispastic, antiulcerogenic, anxiolytic, antiulcerous, and antiallergic effects (Ghosh et al., 2013a; https://www.organicfacts.net/health-benefits/essential-oils/health-benefits-of-cinnamon-oil.html). The most important part of cinnamon is its bark, which can be used in a variety of ways. Cinnamon has the ability to control blood sugar (Ghosh et al., 2013b), so diabetics find it very useful because cinnamon aids them in using less insulin helping patients with diabetes type 2. The health benefits of cinnamon can be attributed to its antibacterial, antifungal, antimicrobial, astringent, anticlotting, and antioxidant properties (https://www.organicfacts.net/health-benefits/essential-oils/health-benefits-of-cinnamon-oil.html; Perdones et al., 2014).

Cinnamon is rich in essential minerals such as manganese, iron, and calcium, while also having a high content of fiber (https://www.organicfacts.net/health-benefits/essential-oils/health-benefits-of-cinnamon-oil.html). The main compound of cinnamon leaf essential oil is eugenol (70–95%), followed by cinnamaldehyde (an excellent flavoring agent) which can be present in a proportion of 1–5% (Vangalapati et al., 2012; Perdones et al., 2014).

Essential oil of cinnamon (CM) has been widely studied for biological activity in several fields. Lu et al. (2010) reported the bacterial-inhibiting effect of CM in alginate–calcium coating. Maqbool et al. (2011) revealed that 0.4% CM combined with 10% gum arabic was efficient in controlling postharvest anthracnose of banana and papaya Wang et al. (2014). Also, Özcan and Arslan (2011) found that 2.5 and 5.0 g/kg CM showed a strong antioxidant effect. Although CM is a GRAS material, the main disadvantages center on the balance between antifungal activity and odor acceptability. In other works, Bullerman et al. (1977), Montes-Belmont and Carvajal (1998), and Sinha et al. (1993) found that cinnamon and clove oils inhibited the growth of *Aspergillus flavus* and the production of aflatoxin. Velluti et al. (2003) reported that cinnamon, clove, lemongrass, oregano, and palmarose essential oils have inhibitory effect on the growth and Fumonisin B1 (FB1) production by *F. proliferatum* in maize grains. According to Marasas et al. (2004), FB1 is the most frequent toxic fumonisin affecting living organisms (humans and animals). FB1 is teratogenic and carcinogenic (Gelderblom et al., 2001). Further, epidemiological studies have demonstrated that human exposure to FB1 causes esophageal cancer (Marasas et al., 1988; Voss et al., 2002) and neural tube defects in newborn infants (Hendricks et al., 1999; Moore et al., 1997; Sydenham et al., 1990).

Thyme essential oil is extracted from an aromatic plant in North America, Europe, and North Africa (*Thymus vulgaris*) by hydrodistillation (Ziani et al., 2011; Martins et al., 2011). This oil is widely used in the manufacture of cosmetics, perfumes, and in the flavorings in food. (Martins et al., 2011). This oil is a mixture of many compounds, but the antimicrobial activity against various bacteria and yeasts is primarily attributed to carvacrol, cinnamaldehyde. Thymol [5-methyl-2-(1-methylethyl) phenol] is a monoterpene present in certain Lamiaceae families (oreganos and screws) and exhibit antimicrobial activity against several bacteria and fungi (Ponce et al., 2010).

Peppermint is one of the many flavors currently used widely within the food industry. It is a GRAS essential oil and very popular flavor used in a wide range of foods (chocolate fillings, chewing gums, sugar confectioneries, etc.), personal care (toothpastes, mouthwashes), pharmaceuticals (antacids), and liqueur products; it is also used in the fragrance, and cleaning industries (Gañán et al., 2015; Roy et al., 1996; Andersen and Jensen, 1984; Carretto et al., 2010). Additionally, peppermint oil is employed as an antiseptic, stimulant, carminative agent throughout the world (Mahboubi and Kazempour, 2014; Alankar, 2009; Hussain et al., 2010; Işcan et al., 2002). Peppermint oil (yellowish color, intense agreeable odor, and taste followed by a cool feeling) is obtained from the leaves of the perennial herb *Mentha x piperita* L., originated in the Mediterranean region (now cultivated worldwide) and of a group of plants belonging to the Labiateae family. This plant is a perennial smooth leaves herb with strong, pepper-like, pungent odor. The oil is generally extracted by steam distillation and also by supercritical CO_2 extraction with advantage of the latter since the degradation of temperature-sensitive products is avoided. Distillation could be followed by rectification and fractionation before use (Mahboubi and Kazempour, 2014; Alankar, 2009). The major constituents of the oil include the terpenes (−)-menthol (30–55%) and (−)-menthone (14–32%) (Grigoleit and Grigoleit, 2005). Fiocco et al. (2011) investigated the use of peppermint oil as an antimelanogenic agent as an alternative, natural tyrosinase inhibitor for applications in foods (antifood-browning) and cosmetics (skin-whitening). Ingredients from natural origin have better acceptance by consumers because they are usually considered safer than synthetic ones. Fiocco et al. concluded that peppermint essential oil has a potential for these intended uses.

1.5 Emulsions, Microemulsions, and Nanoemulsions

1.5.1 Nanotechnology in Food and Nutraceuticals Industries

Nanotechnology is defined as the study of manipulating matter on an atomic and molecular scale which allows the creation of new techniques and devices with sizes of 1–100 nm in scale, and their applications range from mechanics to medicine. Specifically, in the food industry, nanotechnology was born in the pasteurization process introduced by Pasteur in an effort to eliminate spoilage bacteria, which have a size close to 1000 nm. With this event, food processing was revolutionized, improving the quality of the products. In later years, the model of DNA (2.5 nm in size) introduced by Watson and Crick generated new research and improvements in biotechnological, biomedical, agricultural, and productive processes (Chellarama et al., 2014).

1.5.2 Food Processing

Technological advances have made possible better food processing operations since important aspects such as the elimination of toxins, prevention of pathogens, preservation, and improved consistency can be accounted for in more detail, generating processed foods that are less susceptible to decay than fresh foods, an important point for transportation and storage. All these aspects of food processing are more effectively achieved by the incorporation of nanotechnology. Nanocapsules delivery systems play an important role in the food processing sector, maintaining functional properties by encapsulating simple solutions, colloids, emulsions, and other biopolymers (Chellarama et al., 2014).

The use of colloidal dispersions has become of great interest in various industries such as foods, because they contain very small particles (radius, $r < 100$ nm), enhancing their use in applications such as encapsulation, systems protection, and delivery of bioactive lipophilic components as nutraceuticals, pharmaceuticals, vitamins, antifungal agents, antioxidants, and others (McClements, 2012a). One of the advantages offered by the colloidal emulsion systems (emulsion, microemulsion, nanoemulsion) is easy processing (mixing, cutting, and homogenization) (Rao and McClements, 2011a).

1.5.3 Differences Between Emulsions, Nano and Microemulsions

1.5.3.1 Emulsions

Emulsions are dispersions of one liquid phase in fine droplets in the immiscible liquid phase (eg, oil in water), thermodynamically unstable systems, consisting of oil, surfactant, and water. Particle size is greater than 300 nm, and their appearance tends to be cloudy or opaque (Rao and McClements, 2011b). Some typical food emulsions are soft cream, ice cream, butter, margarine, salad dressings, and meat emulsions (Barbosa-Cánovas et al., 1996; Rao and McClements, 2011a; Tabilo-Munizaga and Barbosa-Cánovas, 2005).

1.5.3.2 Nanoemulsions

Nanoemulsions are thermodynamically unstable systems that typically consist of oil, surfactant, and water (Mason et al., 2006; Sonneville-Aubrun et al., 2004; Tadros et al., 2004) with small particle sizes ($r < 100$ nm), and so they tend to be either transparent or slightly turbid (Rao and McClements, 2011b). Like conventional emulsions (with sizes > μm), nanoemulsions are in a nonequilibrium thermodynamic state. However, the kinetics of destabilization of nanoemulsions is so slow (months) that they are considered kinetically stable. This is mainly due to their very small size, resulting in the prevention of droplet flocculation and coalescence: Ostwald ripening, Brownian motion, applied shear or gravitational forces govern the destabilizing process (Anton and Vandamme, 2011; McClements, 2012b).

Nanoemulsions are generally formulated through the high-energy methods, using specific devices (ultrasound generators or high pressure homogenizers) able to supply enough energy to increase the W/O interfacial area for generating submicronic droplets (Anton and Vandamme, 2011). This low-energy emulsification is in fact an efficient method enabling the formation of kinetically stable and potentially concentrated emulsion droplets ranging in size from 10 to 100 nm (Rao and McClements, 2011b).

1.5.3.3 Microemulsions

Microemulsions are thermodynamically stable systems prepared from three basic ingredients: oil, surfactant (sometimes cosurfactant/cosolvent), and water (Flanagan and Singh, 2006; Garti et al., 2001; Leser et al., 2006; Spernath and Aserin, 2006). Microemulsions can be of three types: O/W (oil: dispersed phase; water: continuous phase); W/O (water: dispersed phase; oil: continuous phase); and bicontinuous (oil and water as continuous phases). O/W and W/O microemulsions consist of small spherical particles with sizes lower than 100 nm presenting transparent or slightly turbid apparancies (Lesmes and McClements, 2009; Rao and McClements, 2011b).

The formation of microemulsions is spontaneous, but depends mainly on temperature and composition. They can exhibit many kinds of structures involving the formation of one, two or three phases in equilibrium such as wormlike, bicontinuous spongelike, liquid crystalline, or hexagonal, spherical swollen micelles (Anton and Vandamme, 2011; McClements, 2012b). Microemulsions are usually easier to prepare than nanoemulsions and emulsions, but they generally require much higher surfactant concentrations (Rao and McClements, 2011b). As this chapter is directed to microemulsions, in section 1.6 they will be addressed more thoroughly.

1.5.4 Nano Versus Microemulsions

Microemulsions and nanoemulsions have similar size and geometric structures that are exclusively in the nanometer range, giving them a bluish and translucent appearance (Anton and Vandamme, 2011). However, they are clearly distinct types of colloidal dispersions: a microemulsion is thermodynamically stable, while a nanoemulsion is not. A recent review by McClements (2012b) gives detailed aspects of differences and similarities between these two systems and only a few features will be discussed here.

An important confusion between microemulsion and nanoemulsion is due to the prefix used, as micro refers to 10^{-6} m, while nano means 10^{-9} m, so being strict and following the same, a nanoemulsion contains smaller particles than a microemulsion, which is totally contrary to what occurs in practice, as the particles found in the microemulsions are lower than those found in the nanoemulsions. This confusion of terms is due to the historical development in the science of colloids (McClements, 2012b).

1.5.4.1 Terminology

Microemulsions The term *microemulsion* is generally used to refer to thermodynamically stable isotropic liquids formed by mixing oil, water, and surfactants together. Mixtures of oil, water, and surfactant can form a variety of different systems depending on their composition and the environmental conditions (particularly temperature) (McClements, 2012b).

Nanoemulsions A nanoemulsion is a thermodynamically unstable colloidal dispersion consisting of two immiscible liquids, using water, surfactant, oil and, in some cases cosurfactant, one of the liquids being dispersed as small spherical droplets ($d < 200$ nm) in the other liquid and could be considered to be a conventional emulsion that contains very small particles (McClements, 2012b).

1.5.4.2 Composition

Generally microemulsions and nanoemulsions require the same ingredients for processing, that is, water, oil, surfactant, and cosurfactant (in some cases). The big difference between these colloidal systems is the surfactant-oil ratios used, being greater in the microemulsions than in the nanoemulsions (McClements, 2012b).

1.5.4.3 Optical Properties

Usually the physical appearance of nanoemulsion and microemulsion colloidal systems tends to be translucent and/or transparent as long as the particle size is less than 60 nm, so it becomes a little more difficult to distinguish when the optical properties are only analyzed, but all is not lost, as one of the possible means to distinguish them is the shape of the distribution of particle sizes: in microemulsions a single narrow peak is generally observed, while nanoemulsions can have several peaks that may be broad or narrow (McClements, 2012b).

1.5.4.4 Particle Structure

The particles in microemulsions can be both spherical (S) and nonspherical (NS), and also they could form sponge-like structures in bicontinuous systems. The shape, S or NS, is a function of the optimum curvature that the surfactant monolayer can acquire and the amount of oil incorporated into the system. For nanoemulsions, the particles are generally spherical because the interfacial tension (γ, dynes/centimeter) is relatively high and the radius (r, cm) of the particle is relatively low, so it has high Laplace pressure ($\Delta P_L = 2\gamma/r$) favoring the reduction of interfacial area (McClements, 2012b).

1.5.4.5 Practical Methods of Distinguishing Nanoemulsions and Microemulsions

One of the practical methods for distinguishing the two systems is the fact that a microemulsion can be subjected to cooling, heating and mechanical agitation but when the perturbation is removed, the microemulsion returns to its original condition; in nanoemulsions, that is not the case, when they are subjected to similar perturbations and then returned to the original condition, their properties may change from its original condition. All the differences described earlier are summarized in Table 12.2.

1.6 Microemulsions

In 1943, Hoar and Schulman reported that addition of medium chain alcohols to soap emulsions produced microemulsion phases (Hoar and Schulman, 1943). The term *microemulsion* first entered the scientific literature in 1959, through the work of Jack H. Schulman and others (Schulman et al., 1959) at Columbia University (Acharya and Hartley, 2012). In the decade of 1970, because of the oil crisis, research and application of microemulsions presented a significant growth with their use for the enhanced oil recovery (Eastoe, 2010). These colloidal systems are of particular interest in food, medicine, pharmaceutical, and other industries because they can be easily fabricated using relatively simple processing operations such as mixing (McClements, 2012a).

Table 12.2 Principal Differences Between Emulsion, Microemulsion, and Nanoemulsion

System	Composition	Stability	Droplet Diameter (nm)	Formation	Appearance	Droplet Shape
Emulsion	Water, oil, surfactant	Kinetic	>500	High energy process	Turbid	Spherical
Miniemulsion	Water, oil, surfactant, hydrophobe	Kinetic	30–500	High energy process	Turbid	Spherical
Microemulsion	Water, oil, surfactant, cosurfactant	Thermodynamic	<100	Spontaneous	Transparent	Spherical, lamellar, cylindrical
Nanoemulsion	Water, oil, surfactant, cosurfactant	Kinetic	<200	High energy process	Transparent (D_p < 30 nm), turbid (D_p > 30 nm)	Spherical

Microemulsions are very simply to prepare, they present smaller diameter sizes (<100 nm) than emulsions (>100 nm) or nanoemulsions (<200 nm) (Rao and McClements, 2011b). Nowadays, the application of microemulsions in food and beverage products is limited because of the low number of food-grade surfactants available for preparing and stabilizing these systems (Kralova and Sjöblom, 2009; Rao and McClements, 2011b).

Depending on the ratios between the components there are three main types of microemulsions: at high water content, microemulsions consist of small oil droplets surrounded by an interfacial film comprised of both surfactant and cosurfactant molecules dispersed in the aqueous phase (O/W microemulsions), whereas at lower water concentrations the situation is reversed and the system consists of water droplets dispersed in oil (W/O microemulsions) (McClements, 2012b), when a gradual transition from O/W to W/O microemulsion exists the systems are known as bicontinuous microemulsions.

Microemulsions can be formed by two different methods; the first one is named phase titration method. In this method, the objective is to construct a ternary or pseudoternary diagram in which each corner represents 100% of a particular component, as shown in Fig. 12.1. The region can be separated into W/O or O/W microemulsion by simply considering the composition that whether it is oil rich or water rich. The other method used for microemulsion preparation is the phase inversion method (Saito, 1969) and consists in increasing temperature to cause loss of water of the headgroups shrinking the head group area of the nonionic surfactant causing a phase inversion from O/W to W/O microemulsions in two-phase mixtures with either excess oil or excess water. The temperature at which phase inversion takes place is known as the phase inversion temperature (PIT) (Robb, 1997). Alternatively, the addition of excess of the dispersed phase will cause the inversion of the system. During phase inversion drastic physical changes occur including particle sizes.

1.6.1 Thermodynamics of Microemulsion Formation

The free energy of microemulsion formation can be considered to depend on the extent to which surfactant lowers the surface tension of the oil-water system and change in entropy of the system such that

$$\Delta G = \Upsilon \Delta A - T \Delta S \tag{12.1}$$

where ΔG, is the free energy of formation, Υ is the surface tension of the oil–water interface, ΔS is the change in entropy of the system, and T is the temperature (McClements, 2012b).

This explains the thermodynamic stability of the microemulsions. Thus, the main driving force for microemulsion formation is the ultra low interfacial tension, which is usually achieved by the use of two or more emulsifiers, one predominantly water soluble and other predominantly oil soluble called cosurfactant, which reduce the interfacial tension (γ) to the order of $<10^{-2}$ mN/m generally required for microemulsion formation (Khar et al., 2010). The thermodynamics and physicochemical characteristics of microemulsion formation are discussed elsewhere (Rosano et al., 1988; Moulik and Rakshit, 2006) and the interested reader is referred to these and other publications.

1.6.2 Characterization

The characterization of microemulsions is a difficult task due to their complexity, variety of structures, and components involved in these systems. The starting point for the characterization of microemulsions is the ternary phase diagram, which is made up of oil, water, and surfactant-cosurfactant mixtures to identify the compositions which form a single phase (Acharya and Hartley, 2012). The basic components in a physcochemical characterization of microemulsion systems are (Sajal et al., 2011):

- Phase stability and phase behavior.
- Microstructure, dimension.
- Shape, surface, and distribution.
- Interaction and dynamics.

The points mentioned earlier can be evaluated by different techniques, as follows.

1.6.2.1 Electronic Microscopy

Transmission electronic microscopy (TEM) is the most important technique for the study of microstructure of microemulsions because it directly produces images at high resolution (<5 nm) and it captures any coexistent contrasting structures and microstructural transitions generated due to the passage of electrons through the sample (Sajal et al., 2011; Acharya and Hartley, 2012).

Scanning electron microscopy (SEM) allows the direct mapping of surface features of microemulsions, but, due to the problems in sample preparation, this technique has been used less frequently than TEM (Sajal et al., 2011; Acharya and Hartley, 2012).

1.6.2.2 Scattering Techniques

Scattering techniques such as dynamic light scattering (DLS), small angle X-ray scattering (SAXS) and small angle neutron scattering (SANS) are used to study microemulsions structure, specially size, shape and dynamics of their components (Sajal et al., 2011; Acharya and Hartley, 2012). The basic principle of these techniques involves applying an incident beam of radiation to the sample, and recording the intensity and angle of the scattered beam. The scattering arises from the interaction of the radiation with regions of different refractive index (DLS), electron density (SAXS) or nuclear composition (SANS). In SAXS to reach the scale of interest for microemulsios (>10 nm) scattering angles smaller than 1 degree must be used (Acharya and Hartley, 2012).

1.6.2.3 Rheology

The rheological properties of microemulsions depend on the composition, type, shape, and number density of aggregates present, as well as the interactions between these aggregates. Bicontinous microemulsions exhibit a Newtonian behavior (constant viscosity) at low to medium shear rates, probably due to fragmentation of the bicontinous structure (Sharma et al., 2010; Acharya and Hartley, 2012).

1.6.2.4 Conductivity

The conductivity in microemulsions reveals whether an aqueous or oil phase or both phases are continuous. This technique has been used to determine the type of microemulsion and to estimate the phase boundaries resulting from changes in composition or temperature (Lutz et al., 2007; Li et al., 2010; Acharya and Hartley, 2012).

1.6.3 Applications

Microemulsions have many applications for example in fuels, cosmetics, detergency, food, agrochemicals, analytical applications, lubricating, and recently as pharmaceutical devices due to their easy of formation, thermodynamic stability, transparency, nanometric size, and other properties Flores et al. (2016).

2 Examples of Components and Characterization Procedures

2.1 Chemicals

The following oils are examples that can be used in food applications: (±) tocopherol (vitamin E), *trans*-anethole (from anis essential oil), thyme oil, cinnamon bark oil, and commercial peppermint oil. An example of a surfactant–cosurfactant system could be D-α-tocopherol polyethylenglycol 1000 succinate (TPGS-1000) and 2-methyl-1-butanol; also Tween 80/PEG-400 are included as another example using a common surfactant in foods, Tween 80, combined with the cosurfactant PEG-400. Bidistillated water is commonly used in microemulsion preparation.

2.2 Microemulsion Formation

The water tritation method is a common procedure to locate microemulsion zones and pseudoternary diagrams are usually constructed to examine the formation of O/W microemulsions using four components (water, surfactant, cosurfactant, oil). A weight ratio of surfactant to cosurfactant varies but it is frequent to use 50/50 in weight. In the examples that follow the ratio was 51/49% w/w and the ratios of water to surfactant-cosurfactant were from 99/1 to 75/25% W/W. The oil phase is added drop to drop under constant stirring at selected temperatures (eg, 25 and 37°C). Transparent liquids are recorded as microemulsions. The data acquired are plotted in phase diagrams with appropriate software. Selected microemulsion compositions are then used for the rest of the investigation.

2.3 Stability Testing

The stability of microemulsions can be determined by different methods, three of them are briefly described. The first method consists in thermally stressing the microemulsion: heating and freezing cycles where the microemulsions undergo +40°C during 24 h (average diameter, D_p is determined) followed by freezing to –40°C for 24 h (D_p is determined) during more or less, 21 days. The second method used is centrifugation at, for example, 1000 g during 15 min; in this test the turbidity index can be determined before and after centrifugation. The third method is dilution, the microemulsion is diluted within a given interval, for example, from 1:100 to 1:1000 v/v and D_p is determined at each dilution. In all methods the physical appearance is also monitored (visual changes, phase separation, etc.).

2.4 Characterization Techniques

2.4.1 *Particle Size Measurements*

The droplet size (D_p) of microemulsions can be determined by several commonly available techniques. A very used technique is dynamic light scattering (DLS). Additionally, D_p and shape of droplets of microemulsions can be analyzed by the SAXS technique.

2.4.2 pH Measurement

The pH of microemulsions can be measured easily and is usually available in any laboratory. Usually, the pH is determined at 25°C and could also be desirable to have these values recorded at higher temperatures, for example at 40°C. This is a nondestructive measurement and is carried out in undiluted samples.

2.4.3 Viscosity and Rheological Behavior

The microemulsions viscosity is preferably evaluated in a rheometer, since this technique not only provides the viscosity value at the selected temperature(s) but also the rheological behavior. The measurements require small amounts of samples, as low as 5 mL, without any dilution.

2.4.4 Conductivity

The measurements of conductivity of microemulsions are a complementary tool to effectively elucidate the microstructural changes. It can be used to obtain important information concerning the phase transition upon changes in different parameters affecting microemulsions, such as ionic strength, pH, temperature, dilution, and composition (Zhang et al., 2013).

2.4.5 Transmission Electronic Microscopy (TEM)

The TEM analysis of microemulsions provides information on the shape and size of the samples. Samples are usually stained with phosphotungstic acid at 2%.

2.4.6 Density

Density of microemulsions is useful from the point of view of processing and it is usually determined at 25°C and any other convenient temperature. It is an easy to carry out technique and there are simple and sophisticated apparatuses to realize the measurements. The apparatus has to be calibrated with distilled, deionized, and degassed water before use (Buchacher and Herbst, 1999).

2.4.7 Zeta Potential

The zeta potential of a dispersed system indicates the surface charge in the particles. It is also an indication of the contribution of each component to the net surface charge. For nonionic microemulsions it is expected to find a zeta potential close to zero. There are different commercial apparatuses that measure zeta potential as well as particle diameter.

3 Discussions

3.1 Physical Properties of Surfactants

A very important component in microemulsions is the surfactant. It is thus convenient to know several characteristics of surfactants used for microemulsion formation. The critical micelle concentration (CMC), aggregation number (N_{agg}), and hydrophilic–lipophilic balance (HLB) values of several surfactants and cosurfactants that have been used in our group to prepare microemulsions, are reported in Table 12.3. For isobutanol, the HLB was calculated according to Griffin (1954) and assuming that the alcohol is a surfactant. For the mixed surfactants systems (TPGS-1000 + iso-BuOH and Tween 80 + PEG-400), the CMC values were calculated from,

$$\mathrm{CMC_M} = y_1\mathrm{CMC_1} + y_2\mathrm{CMC_2} \qquad (12.2)$$

Assuming an ideal mixture (activity coefficients of free surfactant monomers for each surfactant type in the mixture are equal to unity), where y_1 and y_2 are the mole fraction of surfactants 1 and 2 in solution on a surfactant base, and $\mathrm{CMC_1}$ and $\mathrm{CMC_2}$ are the critical micelle concentrations of pure surfactants 1 and 2, respectively. The HLB was calculated from a similar equation:

$$\mathrm{HLB_M} = x_1\mathrm{HLB_1} + x_2\mathrm{HLB_2} \qquad (12.3)$$

where $\mathrm{HLB_1}$ and $\mathrm{HLB_2}$ are the HLB values of surfactants 1 and 2, x_1, and x_2 the weight fractions in the mixture, and $\mathrm{HLB_M}$, the HLB of the surfactant mixture.

The reported data shown in Table 12.3 for TPGS-1000 differ considerably from the results determined by surface tension measurements in a rising bubble tensiometer found by our group. Further, although in one case the temperatures are similar (25 vs 22°C) and in the other there is a difference of 15°C, temperature alone does not explain the substantial differences between the values. Another source of variation when comparing results of CMC is the technique used in the determination. Sadoqi et al. (2009) determined the CMC of TPGS-1000 using the technique of steady-state fluorescence and found the value reported in Table 12.3 (equivalent to ~0. 003% W/W), and compared with literature values: surface tension and solubility; the values found by these techniques were 0.02% and 0.1–0.2 mg/mL, respectively. Moreover, the continuous medium also influences the value of CMC. Contradictory CMC values for the same surfactant are frequent in the literature. For instance,

Table 12.3 Properties of TPGS-1000, Mixture of TPGS-1000 With Isobutanol, Tween 80, and Tween 80 with PEG-400

Surfactant	T(°C)	CMC (mM) Reported	CMC (mM) Experimental	N_{agg}	HLB
TPGS-1000	25	0.02	0.7730	10	13
	40		0.5790		
iso-BuOH	—	—	—	—	4.6
TPGS-1000 + iso-BuOH	25	—	1.558	—	8.9
	40		1.426		
Tween 80	25	0.02	0.0197	60	15
	40		0.0206		
Tween 80 + PEG-400	25	—	0.0233	—	13.3
	40		0.0209		11.6

for sodium dodecyl sulfate (SDS), Fuguet et al. (2005) measured the CMC of this anionic surfactant at 25°C by four different methodologies and varied the electrolyte concentration (phosphate buffer, pH 7). In one technique, conductometric measurements, these authors found that the CMC of SDS varied from 8.08 mM (0 mM electrolyte concentration) to 1.99 mM (50 mM electrolyte concentration). Thus, when reporting values of CMC for a surfactant, medium composition, and temperature must be specified as well as the technique used for the determinations.

3.2 Partial Phase Diagrams

It is well known that the solubilization capability of a nonionic surfactant reaches its maximum at a particular temperature called the HLB temperature, at which a microemulsion phase coexists with excess water and oil. The HLB temperature is fixed and different in each system depending on the types of surfactants and/or oils. Since microemulsions have a bicontinuous structure at the HLB temperature, surfactant molecules are distributed among water domains, oil domains, and the water-oil interfaces inside micro emulsion (Kunieda et al., 1995). It's very important to know this distribution in order to estimate the net solubilization capability of a given surfactant. The hydrophile-lipophile property of nonionic surfactants, especially polyoxyethylene type, is dramatically changed with increasing temperature due to the conformation change of the hydrophilic chain becoming more lipophilic at high temperature (Kunieda et al., 1995; Mitra and Paul, 2005; Warisnoicharoen et al., 2000).

The phase diagrams constructed by the titration method for the oils and surfactants reviewed here at 25 and 40°C are presented in Fig. 12.2. The area outside the frame indicates a turbid region. It could be noted that the area of microemulsion region

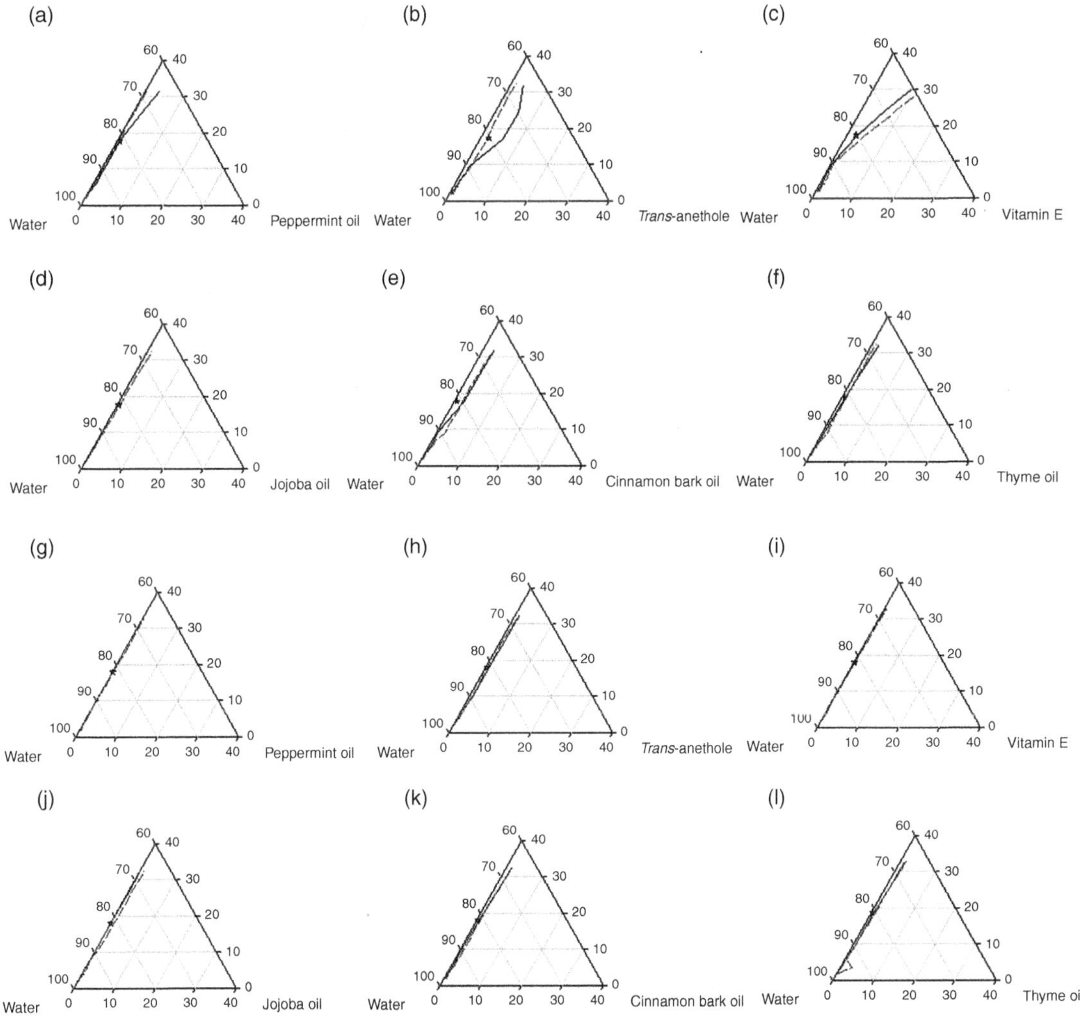

Figure 12.2. Partial phase diagrams for the oil in water microemulsions prepared with TPGS-1000/isobutanol (a–f) and with Tween 80/PEG-400 (g–l) using different oils. *Full lines* correspond to diagrams at 25°C and *dotted lines* to diagrams at 40°C. The *dot* marked in the diagrams corresponds to the composition of the microemulsion used.

was smaller when Tween 80/PEG-400 was used than for the combination TPGS-1000/isobutanol. The dots shown in each diagram correspond to the microemulsions compositions that were selected and are summarized in Table 12.4.

In the partial phase diagrams two different behaviors were observed: the first one was that for the peppermint, *trans*-anethole, vitamin E and thyme oils systems using TPGS-1000 as surfactant and isobutanol as cosurfactant: a reduction was noted in the microemulsion region, which coincides with reports in the literature and it is mainly attributable to the fact that nonionic surfactants become lipophilic at elevated temperatures due to the dehydration of the oxyethylene group of the chains thus causing a reduction in the amount of oil incorporated at a temperature of 40°C (Mitra and Paul, 2005; Kunieda et al., 1995). The second observed behavior was that opposite to the TPGS-1000 systems described earlier, when Tween 80 was the surfactant and PEG-400 the cosurfactant, it can be said that this surfactant mixture has an HLB value less sensible to temperature since no significant changes in the microemulsion region were observed (Mitra and Paul, 2005).

3.3 Microemulsions

Twelve microemulsion systems were considered to determine the adequability to produce o/w microemulsion regions. The systems and their compositions are described in Table 12.4.

The physicochemical data of selected microemulsions prepared by our group are presented in Table 12.5. Jojoba oil microemulsions were not considered in the following discussions.

Variation in pH influences microemulsions stabilized with pH sensitive amphiphiles and it is more pronounced when acidic or alkaline surfactants are the stabilizers. An important point in the case of carboxylic acids and amines is the change in the phase behavior from W/O to O/W by increasing the pH. It can be noticed that all microemulsions presented an acid pH that changed little as the temperature rose from 25 to 40°C. At 25°C the pH of microemulsions varied between 4.06 (M2CL-4, *trans*-anethole) and 5.61 (M2CL-2, thyme oil) whereas at 40°C the range was 4.06 (M2CL-8, thyme oil, Tween 80 + PEG-400) to 5.16 (M2CL-2, thyme oil, TPGS-1000 + iso-BuOH). Anjali et al. (2010), reported pH values for sunflower oil microemulsions stabilized with Tween 20 determined at 20°C and they vary from 4.53 to 4.93; they observed that when oil and surfactant concentration was increased, the pH of the system decreased. Liu and Chang (2011) investigated eucalyptol (10–40 wt%), water (10–40 wt%), polysorbate 80 (12.5 wt%), and ethanol (37.5 wt%) microemulsions unloaded

Table 12.4 Identification and Composition of Microemulsion Systems*

Sample Identification	Surfactant	Co-surfactant	Oil
	TPGS-1000	Isobutanol	
MET	8.90	8.55	2.03 (Transanethole)
MEM	9.03	8.67	0.70 (Peppermint oil)
MEV	8.88	8.53	2.24 (Vitamin E)
MEJ	9.05	8.70	0.035 (Jojoba oil)
M2CL-1	9.05	8.87	0.660 (Cinnamon bark oil)
M2CL-2	9.05	8.87	0.660 (Thyme oil)
	Tween 80	PEG-400	
M2CL-3	9.10	8.91	0.110 (Peppermint oil)
M2CL-4	9.09	8.90	0.240 (Transanethole)
M2CL-5	9.09	8.91	0.210 (Vitamin E)
M2CL-6	9.10	8.92	0.060 (Jojoba oil)
M2CL-7	9.08	8.90	0.026 (Cinnamon bark oil)
M2CL-8	8.87	8.69	2.65 (Thyme oil)

*Composition in weight percent, balance is water

and loaded with curcumin (diferuloylmethane, a phenolic compound), the major pigment in turmeric (extracted from the powdered rhizome of *Curcuma longa Linn*) that has been used for imparting color and flavor to foods, for transdermal topic delivery of curcumin. These microemulsions were characterized by pH, conductivity, viscosity, size, and surface tension. The pH value remained at 7.7 for both unloaded and curcumin-loaded microemulsions.

The zeta potential of liquid-liquid and solid-liquid dispersed systems is a measure of the surface charge of the particles. Knowledge of the values of this parameter could be useful in food formulation as a guide to expected stability of the finished product. Most values were positive (*trans*-anethole, cinnamon bark oil, vitamin E) and a few of them negative (peppermint and thyme essential oils). Absolute values of zeta potential were between 0.3 and 0.8 mV and reflect the nature of the nonionic surfactants used to stabilize these microemulsions, as well as the oil phase, that is, the surface charge is expected to be zero. Basheer et al. (2013) report values of pH between 6.76 and 7.80 and zeta potential between –32 and –75 mV in O/W microemulsions of Tween 80 and *n*-butanol in a weight ratio of 3:1 using isopropyl palmitate as the oil phase. In a recent study, Roohinejad et al. (2015) prepared microemulsions with different amounts of Tween 80 as surfactant, medium chain monoglycerides (Capmul MCM) as the oil phase and phosphate buffer (0.01 M, pH 6.8) as the aqueous phase. The microemulsions were loaded with β-carotene. pH values for unloaded microemulsions ranged from 5.1 to 7.1 and for loaded microemulsions the values were 5.3–7.1. Zeta potentials were all negative and ranged from −13.1 to −36.7 mV in unloaded microemulsions and from −16.1 to −39.9 mV in loaded microemulsions. In another recent research, Ariviani et al. (2015) prepared also β-carotene-loaded microemulsions by spontaneous emulsification method using different oil phases (virgin coconut oil, VCO, and palm oil) and β-carotene levels. The pH values reported did not show significant difference and varied from 6.63 to 6.70. The results of Ariviani et al. for zeta potential also were all negative and ranged from −11.9 to −14.9 mV for the 0.05% and 0.025% β-carotene loaded microemulsions in VCO. For palm-oil microemulsions with similar β-carotene loads the values were −17.0 and −25.0 mV, respectively. In the pharmaceutical area, Biruss and Valenta (2008) report values of zeta potential for tributyrin microemulsions stabilized with the nonionic surfactant polyoxyethylene-10-dodecyl ether (commercialized as Brij(R) 35, and Laureth-10 and others), and found a value of −9.34 ± 0.19 mV. On the other hand, Paolino et al. (2002), in a study of lecithin microemulsions for the topical administration of ketoprofen, found a value of −19.7 ± 1.2 mV for an unloaded microemulsion stabilized with soybean lecithin; for a lecithin microemulsion containing oleic acid as penetration enhancer, the value was −39.5 ± 2.7 mV. When both microemulsions were loaded with ketoprofen, the values were −24.1 ± 1.6 mV and −38.4 ± 1.9 mV, respectively. Thus, from the values of zeta potentials found in our group and other researchers, it is evident that the surface charge of microemulsions strongly depends on the chemical nature of the individual components that make up the formulation.

The average diameters of microemulsions prepared with TPGS-100 + iso-BuOH behaved differently from microemulsions stabilized with the mixture of Tween 80 + PEG-400. In the first case, D_p were between 6.35 and 13.49 nm at both temperatures and in the second case D_p were between 14.4 and 27.44 nm. Thus, microemulsions prepared with the mixture of the two nonionic surfactants yield bigger droplets because of the influence of the polyoxiethylene chains. As an example of this behavior, the average particle diameter of the MEV microemulsion at 25°C was 8.81 nm versus 16.59 nm for the M2CL-5, as can be seen in Table 12.5. Basheer et al. (2013) report 26 and 147 nm, Roohinejad et al. (2015) 10.9–64.7 nm

Table 12.5 Physical Properties of Selected Biocompatible Microemulsions

Sample	Particle Size (nm)		Z Potential (mV)		Conductivity (mS/cm)		pH		Viscosity (mPa•s)		Density (g/cm³)		Surface Tension (mN/m)	
Temperature (°C)														
	25	40	25	40	25	40	25	40	25	40	25	40	25	40
MET	10.16	12.45	0.6	0.4	12	16	4.69	4.76	4.78	6.07	0.994800	0.987050	30.73	34.07
MEM	6.35	8.65	0.6	−0.3	14	21	4.79	4.90	3.43	2.15	0.995772	0.989007	27.93	31.47
MEV	8.81	11.32	0.5	0.3	17	22	4.76	4.82	5.6	7.69	0.994152	0.987021	27.80	31.60
MEJ	6.89	9.02	0.7	0.4	16	23	4.93	5.05	3.41	2.27	0.995523	0.988707	33.47	33.60
M2CL-1	8.83	10.46	0.6	0.4	98	106	4.92	4.71	2.93	4.33	0.995885	0.989349	29.2	29.88
M2CL-2	9.77	13.49	0.8	−0.4	14	17	5.61	5.16	3.31	2.75	0.993578	0.987381	28.87	29.03
M2CL-3	14.48	16.98	0.6	−0.4	98	107	4.15	4.24	2.56	2.18	1.021466	1.014663	47.5	42.35
M2CL-4	16.4	20.4	0.6	0.4	95	100	4.06	4.10	3.04	2.53	1.020602	1.014137	44.7	42.33
M2CL-5	16.59	20.22	0.6	0.4	97	105	4.15	4.21	2.84	2.58	1.020750	1.014172	45.5	42.60
M2CL-6	15.56	16.67	−50.8	−45.2	99	107	4.19	4.06	2.87	2.19	1.023130	1.016183	46.03	41.55
M2CL-7	14.4	20.15	0.6	0.4	94	102	4.14	4.15	5.26	2.17	1.019888	1.015018	46.25	40.8
M2CL-8	14.8	27.44	0.6	−0.4	93	109	4.07	4.06	8.96	10.67	1.020753	1.014012	43.8	41.65

for unloaded microemulsions and 12.8–101.0 nm for loaded microemulsions, and Ariviani et al. (2015) found 20 and 22.6 nm for VCO microemulsions loaded with β-carotene 0.025 and 0.05%wt, respectively, 23.0 and 22.9 nm for palm oil microemulsions loaded with β-carotene, 0.025 and 0.05% wt, respectively. Biruss and Valenta (2008) determined an average particle diameter of 5.41 ± 0.02 nm for their cosurfactant-free o/w microemulsion (polyoxyethylene-10-dodecyl ether, tributyrin, distilled water). The droplet size of the microemulsions with various oil/water ratios reported by Liu and Chang (2011) remained in the 2–5-nm range. Thus, our results are typical of what has been reported for microemulsions. A TEM micrograph of the TPGS-1000/isobutanol/*trans*-anethole microemulsion is presented in Fig. 12.3. It can be observed that microemulsion droplets are spherical and of small size.

3.3.1 Electrical Conductivity

Electrical conductivity of selected microemulsions shown in Table 12.5 could be considered high values, since they are O/W microemulsions (Biruss and Valenta, 2008). Roohinejad et al. (2015) found values from 12.15 to 520 μS/cm whereas Ariviani et al. (2015) and Basheer et al. (2013) do not report these determinations. However, Biruss and Valenta (2008) report that the electrical conductivity of their system was 0.33 ± 0.002 mS/cm, a very low value compared with the

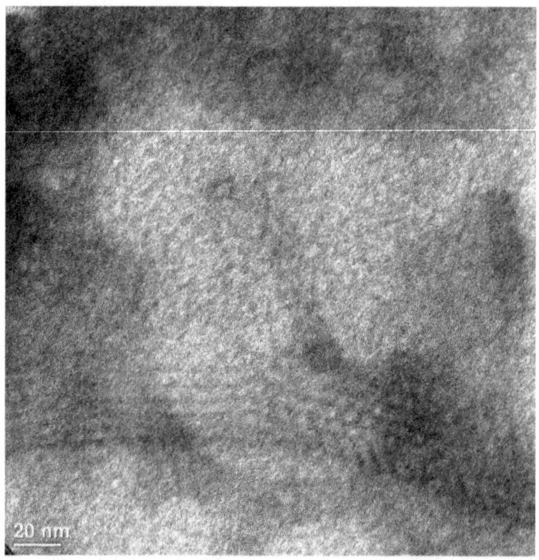

Figure 12.3. TEM micrography. TEM micrograph of microemulsion prepared with TPGS-1000/isobutanol/*trans*-anethole (MET). The microemulsion was stained with phosphotungstic acid at 2%.

data in Table 12.5. This value reflects the nonconductive properties of the components. In their study, Liu and Chang (2011) found that the conductivity of the microemulsions increased from 12 to 30 μS/cm when the water content increased from 10 to 40%. Conductivity is a useful technique to characterize microemulsion systems. Electrical conductivity shows inflection points when the water content increases from a W/O microemulsion to a bicontinuous and to an O/W microemulsion. Kalaitzaki et al. (2015) studied the transitions at 25°C from an oil-rich microemulsion to a water rich microemulsion and detected the transitions using electrical conductivity measurements. Their microemulsions were stabilized with polyoxyethylene sorbitan monolaurate (Tween 20) and polyoxyethylene sorbitan monopalmitate (Tween 40) and mixtures thereof. The oil phase was composed of (R)-(+)-limonene and 1,2-propanediol (propylene glycol). The values of electrical conductivities were up to about 18 mS·cm^{-1}. In general, high water content microemulsions present higher conductivity values than low water content ones. It is known that O/W and bicontinuous microemulsions present higher conductivity values than W/O microemulsions (Bumajdad and Eastoe, 2004). It is also observed that conductivity increases with temperature for all cases presented in Table 12.5. From the data presented here for electrical conductivity of nonionic microemulsions, it is clear that this property depends on the nature of the components.

3.3.2 Viscosity

Taking into account that the viscosity of water at 25°C is 0.8902 mPa·s and at 40°C is reduced to 0.6532 mPa·s (Kestin et al., 1978), viscosity results observed in Table 12.5 can be considered low at both temperatures. No noticeable changes on viscosity occur by the different surfactants, cosurfactants and oil phases. In general, viscosity decreases with increasing temperature with four exceptions: the MET, MEV, M2CL-1 and M2CL-8 microemulsions showed increased viscosity at 40°C. Basheer et al. (2013) report that their microemulsions behaved as Newtonian fluids at 25°C. Roohinejad et al. (2015) report values from 3.9 to 3887.8 mPa·s. Ariviani et al. found values as follows: for VCO, 0.025% and 0.050% β-carotene loaded, 5.90 and 6.05 mPa·s, respectively; for palm oil, 0.025% and 0.050% β-carotene loaded, 6.08 and 5.98 mPa·s, respectively. The values reported by Ariviani et al. (2015) are not significantly different from each other, according to these authors. Liu and Chang (2011) report the viscosity range of their microemulsions determined at 32°C, from 6 to 14 mPa·s with an increase in the water content from 10 to 40%, whereas Biruss and Valenta (2008) report a viscosity of 450 mPa·s for their polyoxyethylene-10-dodecyl ether microemulsion of tributyrin at 20°C. Thus, the microemulsions reported in Table 12.5 are low viscosity systems, which is a desirable feature in food process operations. Furthermore, when comparing the viscosities at 25 and 40°C, it appears that in most cases both the use of TPGS-1000 and of Tween 80 tend to decrease viscosity with increasing temperature. This effect occurs because the cohesive forces linking molecules diminish with temperature by increasing the distance between molecules.

3.3.3 Density

The density of the microemulsions is also shown in Table 12.5. It is evident that the properties of the two groups of microemulsions (different surfactant-cosurfactant combination) behave differently. When TPGS-1000–iso-BuOH is used, densities at 25 and 40°C remain below 1. This is not the case for the group in which the surfactant-cosurfactant pair is Tween 80 and PEG-400, where density values are above 1. This behavior could be explained by the interactions between water molecules with the surfactant-cosurfactant combination. For the TPGS-1000–iso-BuOH combination, repulsive forces between the vitamin E fraction of TPGS-1000 and the hydrophobic iso-BuOH with water predominate, increasing the volume of the mixture and thus reducing the density. On the other hand, for the Tween 80 and PEG-400 combination, a more hydrophilic surfactant-cosurfactant system, attractive forces predominate, reducing the volume and increasing the density of the system. Another point that can be seen in Table 12.5 is the fact that increasing the temperature from 25 to 40°C, in all systems, density decreased, that is, with more energy in the system because of the higher temperature, the distance between molecules is increased, thus the number of molecules per unit volume (liquid density) decreases. There are scarce reports in the literature on density measurements of microemulsions for food applications. Ariviani et al. (2015) present specific gravity data for their β-carotene-loaded microemulsions and found no significant difference between their tested formulations with values in the range between 1.010 and 1.011, at 20°C. In terms of density, these values correspond to 1.0098–1.0108 g/cm^3, which are similar to the values reported in Table 12.5.

3.3.4 Surface Tension

One of the characteristics of microemulsions is their low interfacial tension with values typically between 20 and 25 mN/m. The surface tension at air–water interfaces is usually of the order of 72 mN/m. The interfaces oil-microemulsion could be around 10^{-3} to 10^{-6} mN/m (Bera et al., 2014). The values of surface tension of the different microemulsions reported in Table 12.5 are between

27.80 and 34.07 mN/m for the TPGS-1000/isobutanol systems and from 40.80 to 47.5 mN/m for the Tween 80/PEG-400 systems. When Tween 80/PEG-400 was used as surfactant-cosurfactant we observed that there is a decrease in surface tension values when the temperature increases from 25 to 40°C, this tendency could be explained because the molecular interaction between the liquid molecules becomes weaker since the hydrogen bond is weak and it is the main factor for association between molecules. On other hand, when TPGS-1000/isobutanol was used as the surfactant-cosurfactant mixture, the tendency was reversed, namely, an increase of surface tension was observed when the temperature increased. The interfacial tension of microemulsions reported by Liu and Chang (2011) slightly increased from 26 to 28 mN/m with an increase in the water content.

3.3.5 Stability

To assess any change in stability of microemulsions (phase separation, phase inversion, aggregation, creaming, and cracking) they are subjected to freeze–thaw cycles. For this test, the MET, MEM, and MEV systems were subjected to thermal stress and centrifugation and the results did not show any significant changes since only a small increase from 2 to 3 nm in particle size were observed after 21 days under the freezing and heating cycles. Further, in terms of changes in the turbidity index, these were not observed given that the turbidity index before and after the centrifugation test at 10,000g had the same value for all systems, the transparency was not lost for any microemulsion. These tests confirm one of the most important properties of microemulsions, their thermodynamic stability.

4 Conclusions and Outlook

The 10 food-grade microemulsions exemplified here were clear, low viscosity (2.17–10.67 mPa·s), low surface tension (27.80–46.25 mN·m) with densities close to water. Average particle diameters were from 6.35 to 27.44 nm. The pH of these microemulsions was slightly acid (4.06–5.61) and conductivity was between 12 and 109 mS/cm. Zeta potentials were near zero, reflecting the nature of the nonionic surfactants used to stabilize the microemulsions. Although the microemulsions reported here are particular cases of the systems considered and is not expected that any of them would be used in product formulation, confirm that these systems can be used with advantage as carriers of essential oils, nutraceuticals, and other hydrophobic liquid components (oil phase) of interest in foods, pharmaceuticals and cosmetics. Further, these oils could actually be carriers of oil-soluble substances (solid or liquid) with specific activities such as flavors, aromas, vitamins, drugs, and others.

Future work could be oriented toward the development of fully food-grade surfactants with no side effects and from natural origin as much as possible (biosurfactants). Microemulsion formulation must be aimed to meet very specific goals for carefully established applications.

Acknowledgments

Flores-Villaseñor and Estrada-Ramírez acknowledge a fellowship from CONACyT (Mexico) for PhD studies at CIQA. Peralta-Rodríguez acknowledges CONACyT for grant CB-2011/168472.

References

Abrahamsson, B., Johansson, D., Torstensson, A., Wingstrand, K., 1994. Evaluation of solubilizers in the drug release testing of hydrophilic matrix extended-release tablets of felodipine. Pharm. Res. 11, 1093–1097.

Abrahamsson, B., Ungell, A.L., 2009. Biopharmaceutical support in formulation development. In: Gibson, M. (Ed.), Pharmaceutical Preformulation and Formulation: A Practical Guide from Candidate Drug Selection to Commercial Dosage Form. second ed. Informa Healthcare, New York, NY.

Acharya, D.P., Hartley, P.G., 2012. Progress in microemulsion characterization. Curr. Opin. Colloid Interface Sci. 17, 274–280.

Alankar, S., 2009. A review on peppermint oil. Asian J. Pharm. Clin. Res. 2, 27–33.

Andersen, P.H., Jensen, N.J., 1984. Mutagenic investigation of peppermint oil in the salmonella/mammalian salmonella test. Mutat. Res. 138, 1720.

Anjali, C.H., Madhusmita, D., Chandrasekaran, N., Mukherjee, A., 2010. Antibacterial activity of sunflower oil microemulsion. Int. J. Pharm. Pharm. Sci. 2 (Suppl. 1), 123–128.

Anton, N., Vandamme, T.F., 2011. Nano-emulsions and micro-emulsions: clarifications of the critical differences. Pharm. Res. 28, 978–985.

Ariviani, S., Anggrahini, S., Naruki, S., Raharjo, S., 2015. Characterization and chemical stability evaluation of β-carotene microemulsions prepared by spontaneous emulsification method using VCO and palm oil as oil phase. Int. Food Res. J. 22, 2432–2439.

Attwood, D., Florence, A.T., 2012. Physical Pharmacy, second ed. Pharmaceutical Press, London.

Bakkali, F., Averbeck, S., Averbeck, D., Idaomar, M., 2008. Biological effects of essential oils—a review. Food Chem. Toxicol. 46, 446–475.

Bais, D., Trevisan, A., Lapasin, R., Partal, P., Gallegos, C., 2005. Rheological characterization of polysaccharide–surfactant matrices for cosmetic O/W emulsions. J. Colloid Interf. Sci. 290, 546–556.

Banat, I.M., Franzetti, A., Gandolfi, I., Bestetti, G., Martinotti, G.M., Fracchia, L., Smyth, T.J., Marchant, R., 2010. Microbial biosurfactants production, applications and future potential. Appl. Microbiol. Biotechnol. 87, 427–444.

Banat, I.M., Makkar, R.S., Cameotra, S.S., 2000. Potential commercial applications of microbial surfactans. Appl. Microbiol. Biotechnol. 53, 495–508.

Barbosa-Cánovas, G.V., Kokini, J.L., Ma, L., Ibarz, A., 1996. The rheology of semiliquid foods. Adv. Food Nutr. Res. 39, 1–69.

Basheer, H.S., Noordin, M.I., Ghareeb, M.M., 2013. Characterization of microemulsions prepared using isopropyl palmitate with various surfactants and cosurfactants. Trop. J. Pharm. Res. 12, 305–310.

Benna-Zayani, M., Kbir-Ariguib, N., Trabelsi-Ayadi, M., Grossiord, J.L., 2008. Stabilization of W/O/W double emulsion by polysaccharides as weak gels. Colloids Surf. A 316, 46–54.

Bera, A., Mandal, A., Kumar, T., 2014. Physicochemical Characterization of Anionic and Cationic Microemulsions: Water Solubilization, Particle Size Distribution, Surface Tension, and Structural Parameters. J. Chem. Eng. Data. 59, 2490–2498.

Bergenståhl, B., 2008. Physicochemical aspects of an emulsifier functionality. In: Hasenhuettl, G.L., Hartel, R.W. (Eds.), Food Emulsifiers and Their Applications. Springer, New York, NY, pp. 173–194.

Bilia, A.R., Guccione, C., Isacchi, B., Righeschi, C., Firenzuoli, F., Bergonzi, M.C., 2014. Essential oils loaded in nanosystems: a developing strategy for a successful therapeutic approach. Evid. Based Complement. Alternat. Med. 651593, 14.

Biruss, B., Valenta, C., 2008. The advantage of polymer addition to a nonionic oil in water microemulsion for the dermal delivery to progesterone. Int. J. Pharm. 349, 269–273.

Boonme, P., 2007. Applications of microemulsions in cosmetics. J. Cosmet. Dermatol. 6, 223–228.

Bouyer, E., Mekhloufi, G., Rosilio, V., Grossiord, J.L., Agnely, F., 2012. Proteins, polysaccharides, and their complexes used as stabilizers for emulsions: alternatives to synthetic surfactants in the pharmaceutical field. Int. J. Pharm. 436, 359–378.

Boye, J.I. (Ed.), 2015. Nutraceutical and Functional Food Processing Technology. IFT Advances in Food Science. Wiley Blackwell, Chichester, UK.

Broadhead, J., Gibson, M., 2009. Parenteral dosage forms. In: Gibson, M. (Ed.), Pharmaceutical Preformulation and Formulation: A Practical Guide from Candidate Drug Selection to Commercial Dosage Form. second ed. Informa Healthcare, New York, NY.

Buchacher, P., Herbst, K., 1999. DMA 4500/5000. Instruction: HandbookAnton Paar GmbH, Austria.

Bullerman, L.B., Lieu, F.Y., Seire, A.S., 1977. Inhibition of growth and aflatoxin production by cinnamon and clove oils, cinnamic aldehyde and eugenol. J. Food Sci. 42, 1107–1116.

Bumajdad, A., Eastoe, J., 2004. Conductivity of water-in-oil microemulsions stabilized by mixed surfactants. J. Colloid. Interface Sci. 274, 268–276.

Cabra, V., Arreguin, R., Farres, A., 2008. Emulsifying properties of proteins. Bol. Soc. Quím. Méx. 2, 80–89.

Cabra, V., Arreguin, R., Vazquez-Duhalt, R., Farres, A., 2007. Effect of alkaline deamidation on the structure, surface hydrophobicity, and emulsifying properties of the Z19 alpha-zein. J. Agric. Food Chem. 55, 439–445.

Carretto, C.F.P., Almeida, R.B.A., Furlan, M.R., Jorge, A.O.C., Junqueira, J.C., 2010. Antimicrobial activity of *Mentha piperita* L. against *Candida* spp. Braz. Dent. J. 13, 4–9.

Cauvain, S.P. (Ed.), 2003. Bread Making: Improving Quality. Woodhead Publishing Limited, Cambridge, UK.

Chellaram, C., Murugaboopathi, G., John, A.A., Sivakumar, R., Ganesan, S., Krithika, S., Priya, G., 2014. Significance of nanotechnology in food industry. APCBEE Procedia 8, 109–113.

Chilton, H.M., Laws, D.R.J., 1980. Stability of aqueous emulsions of the essential oil of hops. J. Inst. Brew. 86 (3), 126–130.

Chu, J., Cheng, Y.L., Rao, A.V., Nouraei, M., Zarate-Muñoz, S., Acosta, E.J., 2014. Lecithin-linker formulations for self-emulsifying delivery of nutraceuticals. Int. J. Pharm. 471, 92–102.

Couvreur, P., Kante, B., Roland, M., Goit, P., Bauduin, P., Speiser, P., 1979. Polycyanoacrylate nanocapsules as potential lysosomotropic carriers: preparation, morphology and sorptive properties. J. Pharm. Pharmacol. 31, 331–332.

Csáki, K.F., 2011. Synthetic surfactant food additives can cause intestinal barrier dysfunction. Med. Hypotheses 76, 676–681.

Cui, S.W., 2005. Food Carbohydrates: Chemistry, Physical Properties and Applications. Taylor & Francis, Boca Raton, FL.

Davies, P., 2009. Oral solid dosage forms. In: Gibson, M. (Ed.), Pharmaceutical Preformulation and Formulation: A Practical Guide from Candidate Drug Selection to Commercial Dosage Form. second ed. Informa Healthcare, New York, NY.

Day, N., 2009. Aqueous nasal dosage forms. In: Gibson, M. (Ed.), Pharmaceutical Preformulation and Formulation: A Practical Guide from Candidate Drug Selection to Commercial Dosage Form. second ed. Informa Healthcare, New York, NY, pp. 1–555.

Dickinson, E., 2003. Hydrocolloids at interfaces and the influence on the properties of dispersed systems. Food Hydrocoll. 17, 25–39.

Dickinson, E., 2009. Hydrocolloids as emulsifiers and emulsion stabilizers. Food Hydrocoll. 23, 1473–1482.

Eastoe, J., 2010. Microemulsions. In: Cosgrove, T. (Ed.), Colloid Science: Principles, Methods, and Applications. second ed. John Wiley & Sons, New York, NY, pp. 91–116.

Eccleston, G.M., 1988. Microemulsions. In: Boylan, J.C. (Ed.), Encyclopedia of Pharmaceutical Technology. Marcel Dekker, New York, NY, pp. 375–421.

Edris, A.E., Malone, C.F.R., 2012. Preferential solubilization behaviours and stability of some phenolic-bearing essential oils formulated in different microemulsion systems. Int. J. Cosmet. Sci. 34, 441–450.

Elgendy, E.M., Khayyat, S.A., 2008. Oxidation reactions of some natural volatile aromatic compounds: anethole and eugenol. Russ. J. Org. Chem. 44, 823–829.

Epstein, H., 2009. Cosmeceutical vehicles. Clin. Dermatol. 27, 453–460.

Fanun, M., 2008. A study of the properties of mixed nonionic surfactants microemulsions by NMR, SAXS, viscosity and conductivity. J. Mol. Liq. 142, 103–110.

Fanun, M., 2012. Microemulsions as delivery systems. Curr. Opin. Colloid Interface Sci. 17, 306–313.

Feng, J.L., WuWang, Z., Zhang, J., NiWang, Z., Liu, F., 2009. Study on food-grade vitamin E microemulsions based on nonionic emulsifiers. Colloids Surf. A. 339, 1–6.

Fiocco, D., Fiorentino, D., Frabboni, L., Benvenuti, S., Orlandini, G., Pellati, F., Gallone, A., 2011. Lavender and peppermint essential oils as effective mushroom tyrosinase inhibitors: a basic study. Flavour Frag. J. 26, 441–446.

Flanagan, J., Singh, H., 2006. Microemulsions: a potential delivery systems for bioactives in food. Crit. Rev. Food Sci. 46, 221–237.

Flores, S.E., Rial-Hermida, M.I., Ramirez, J.C., Pazos, A., Concheiro, A., Alvarez-Lorenzo, C.I., Peralta, R.D., 2016. Microemulsions for colorectal cáncer treatments. Considerations and formulation of methotrexate. Minirev. Med. Chem. 16, 498–508.

Fuguet, E., Rafols, C., Roses, M., Bosch, E., 2005. Critical micelle concentration of surfactants in aqueous buffered and unbuffered systems. Anal. Chem. Acta 548, 95–100.

Gañán, N.A., Dambolena, J.S., Martini, R.E., Bottini, S.B., 2015. Supercritical carbon dioxide fractionation of peppermint oil with low menthol content—experimental study and simulation analysis for the recovery of piperitenone. J. Supercrit. Fluids 98, 1–11.
Garti, N., 1999. What can nature offer from an emulsifier point of view: trends and progress? Colloids Surf. A. 152, 125–146.
Garti, N., 2003. Microemulsions as microreactors for food applications. Curr. Opin. Colloid Interface Sci. 8, 197–211.
Garti, N., Amar, I., Yaghmur, A., Spernath, A., Aserin, A., 2003. Interfacial modification and structural transitions induced by guest molecules solubilized in U-type nonionic microemulsions. J. Disp. Sci. Technol. 24, 397–410.
Garti, N., Avrahami, M., Aserin, A., 2006. Improved solubilization of celecoxib in U-type nonionic microemulsions and their structural transitions with progressive aqueous dilution. J. Colloid Interface Sci. 299, 352–365.
Garti, N., Yaghmur, A., Leser, M.E., Clement, V., Watzke, H.J., 2001. Improved oil solubilization in oil/water food grade microemulsions in the presence of polyols and ethanol. J. Agric. Food Chem. 49, 2552–2562.
Garti, N., Yuli-Amar, I., 2008. Micro- and nano-emulsions for delivery of funtional food ingredients. In: Garti, N. (Ed.), Delivery and Controlled Release of Bioactives in Foods and Nutraceuticals. CRC Press, New York, NY, pp. 149–183.
Gaysinsky, S., Davidson, P.M., Bruce, B.D., Weiss, J., 2005. Growth inhibition of *Escherichia coli* O157: H7 and *Listeria monocytogenes* by carvacrol and eugenol encapsulated in surfactant micelles. J. Food Protect. 68, 2559–2566.
Gelderblom, W.C., Abel, S., Smuts, C.M., Marnewick, J., Marasas, W.F., Lemmer, E.R., Ramljak, D., 2001. Fumonisin-induced hepatocarcinogenesis: mechanisms related to cancer initiation and promotion. Environ. Health Perspect. 109 (Suppl. 2), 291–300.
Ghosh, V., Saranya, S., Mukherjee, A., Chandrasekaran, N., 2013a. Antibacterial microemulsion prevents sepsis and triggers healing of wound in wistar rats. Colloids Surf. B. 105, 152–157.
Ghosh, V., Saranya, S., Mukherjee, A., Chandrasekaran, N., 2013b. Cinnamon oil nanoemulsion formulation by ultrasonic emulsification: investigation of its bactericidal activity. J. Nanosci. Nanotechnol. 13, 114–122.
Gibson, M., 2009. Ophthalmic dosage forms. In: Gibson, M. (Ed.), Pharmaceutical Preformulation and Formulation: A Practical Guide from Candidate Drug Selection to Commercial Dosage Form. second ed. Informa Healthcare, New York, NY.
Grant, J., Cho, J., Allen, C., 2006. Self-Assembly and Physicochemical and Rheological Properties of a Polysaccharide-Surfactant System Formed from the Cationic Biopolymer Chitosan and Nonionic Sorbitan Esters. Langmuir 22, 4327–4335.
Griffin, W.C., 1954. Calculation of HLB values of nonionic surfactants. J. Soc. Cosmet. Chem. 5, 249–256.
Grigoleit, H.G., Grigoleit, P., 2005. Pharmacology and preclinical pharmacokinetics of peppermint oil. Phytomedicine 12, 612–616.
Gul, K., Singh, A.K., Jabeen, R., 2015. Nutraceuticals and functional foods: the foods for future world. Crit. Rev. Food Sci. Nutr. DOI:10.1080/10408398.2014.903384.
Gursoy, R.N., Benita, S., 2004. Self-emulsifying drug delivery systems (SEDDS) for improved oral delivery of lipophilic drugs. Biomed. Pharmacother. 58, 173–182.
Hadi, J.N., Norazian, M.H., Kausar, A., 2011. Natural surfactants for pharmaceutical emulsions. In: Ahmed, Q.U. (Ed.), Current Issues in Pharmacy. IIUM printing sdn. Bhd., Selangor, Darul Ehsan, pp. 178–195.
Hamed, S.F., Sadek, Z., Edris, A., 2012. Antioxidant and antimicrobial activities of clove bud essential oil and eugenol nanoparticles in alcohol-free microemulsion. J. Oleo Sci. 61, 641–648.

Hasler, C.M., 2003. Book review. In: Rapport, L., Lockwood, B. (Eds.), Nutraceuticals. Pharmaceutical Press, London, pp. 996–997.

Hasler, C.M., Brown, A.C., 2009. Position of the American Dietetic Association: functional foods. J. Am. Diet Assoc. 109, 735–746.

Hendricks, K.A., Simpson, J.S., Larsen, R.D., 1999. Neural tube defects along the Texas Mexico border, 1993–1995. Am. J. Epidemiol. 149, 1119–1127.

Hepworth, P., 2006. Non-ionic surfactants. In: Farn, R.J. (Ed.), Chemistry and Technology of Surfactants. Blackwell Publishing Ltd, Oxford, pp. 133–151.

Higgins, S., Wesley, N.O., 2015. Topical retinoids and cosmeceuticals: where is the scientific evidence to recommend products to patients? Curr. Derm. Rep. 4, 56–62.

Hoar, T.P., Schulman, J.H., 1943. Transparent water-in-oil dispersions: the oleophatic hydro-micelle. Nature 152, 102–103.

Hu, Z.P., Tawa, R., Konishi, T., Shibata, N., Takada, K., 2001. A novel emulsifier, Labrasol, enhances gastrointestinal absorption of gentamicin. Life Sci. 69, 2899–2910.

Hussain, A.I., Anwar, F., Nigam, P.S., Ashraf, M., Gilani, A.H., 2010. Seasonal variation in content, chemical composition and antimicrobial and cytotoxic activities of essential oils from four *Mentha* species. J. Sci. Food Agric. 90, 1827–1836.

Işcan, G., Kirimer, N., Kürkcüoğlu, M., Başer, K.H., Demirci, F., 2002. Antimicrobial screening of *Mentha piperita* essential oils. J. Agric. Food Chem. 50, 3943–3946.

Jain, J., Fernandes, C., Patravale, V., 2010. Formulation development of parenteral phospholipid-based microemulsion of etoposide. AAPS PharmSciTech 11 (2), 826–831.

Jones, P.J.H., MacDougall, D.E., Ntanios, F., Vanstone, C.A., 1997. Dietary phytosterols as cholesterol-lowering agents in humans. Can. J. Physiol. Pharm. 75, 217–227.

Kalaitzaki, A., Xenakis, A., Papadimitriou, V., 2015. Highly water dilutable microemulsions: a structural study. Colloid. Polym. Sci. 293, 1111–1119.

Kang, B.K., Lee, J.S., Chon, S.K., Jeong, S.Y., Yuk, S.H., Khang, G., Lee, H.B., Cho, S.H., 2004. Development of self-microemulsifying drug delivery systems (SMEDDS) for oral bioavailability enhancement of simvastatin in beagle dogs. Int. J. Pharm. 274, 65–73.

Kestin, J., Solokov, M., Wakeham, W.A., 1978. Viscosity of liquid water in the range 8°C to 150°C. J. Phys. Chem. Ref. Data 7, 941–948.

Kfoury, M., Auezova, L., Greige-Gerges, H., Ruellan, S., Fourmentin, S., 2014. Cyclodextrin, an efficient tool for *trans*-anethole encapsulation: chromatographic, spectroscopic, thermal, and structural studies. Food Chem. 164, 454–461.

Khar, R.K., Pathan, S.A., Jain, G.K., Ahmad, S.A.F.J., 2010. Microemlusion: practical applications and concepts. Pharm. Student 25, 32–40.

Kim, S.K., Lee, E.H., Vaishali, B., Lee, S., Lee, Y.K., Kim, C.Y., Moon, H.T., Byun, Y., 2005. Tricaprylin microemulsion for oral delivery of low molecular weight heparin conjugates. J. Control. Release 105, 32–42.

Kitamoto, D., Morita, T., Fukuoka, T., Masa-aki, K., Tomohiro, I., 2009. Self-assembling properties of glycolipid biosurfactants and their potential applications. Curr. Opin. Colloid Interface Sci. 14, 315–328.

Kommuru, T.R., Gurley, B., Khan, M.A., Reddy, I.K., 2001. Self-emulsifying drug delivery systems (SEDDS) of coenzyme Q10: formulation development and bioavailability assessment. Int. J. Pharm. 212, 233–246.

Kosaric, N., 2001. Biosurfactants and their application for soil bioremediation. Food Technol. Biotechnol. 39, 295–304.

Kralova, I., Sjöblom, J., 2009. Surfactants used in food industry: a review. J. Disp. Sci. Technol. 30, 1363–1383.
Kulapina, E.G., Chernova, R.K., Makarova, N.M., Pogorelova, E.S., 2013. Methods for determining synthetic surfactants. Rev. J. Chem. 3, 323–362.
Kumar, K.S., Dhachinamoorthi, D., Saravanan, R., Gopal, U., Shanmugam, V., 2011. Microemulsions as carrier for novel drug delivery: a review. Int. J. Pharm. Sci. Rev. Res. 10, 37–45.
Kunieda, H., Nakano, A., Pes, M.A., 1995. Effect of oil on the solubilization in microemulsion systems including nonionic surfactant mixtures. Langmuir 11, 3302–3306.
Laplante, S., Turgeon, S.L., Paquin, P., 2005. Emulsion stabilizing properties of various chitosans in the presence of whey protein isolate. Carbohydr. Polym. 59, 425–434.
Lauridsen, J.B., 1976. Food Emulsifiers: Surface Activity, Edibility, Manufacture, Composition, and Application. J. Am. Oil Chemists' Soc. 53, 400–407.
Lawrence, M.J., Rees, G.D., 2000. Microemulsion-based media as novel drug delivery systems. Adv. Drug Deliv. Rev. 45, 89–121.
Leser, M.E., Sagalowicz, L., Michel, M., Watzke, H.J., 2006. Self-assembly of polar food lipids. Adv. Colloid Interface Sci. 123, 125–136.
Lesmes, U., McClements, D.J., 2009. Structure-function relationships to guide rational design and fabrication of particulate food delivery systems. Trends Food Sci. Technol. 20, 448–457.
Li, P., Ghosh, A., Wagner, R.F., Krill, S., Joshi, Y.M., Serajuddin, A.T.M., 2005. Effect of combined use of nonionic surfactant on formation of oil-in-water microemulsions. Int. J. Pharm. 288, 27–34.
Li, W., Xu, P., Zhou, H.C., Yang, L.R., Liu, H.Z., 2014. Physicochemical Characterization of Anionic and Cationic Microemulsions: Water Solubilization, Particle Size Distribution, Surface Tension, and Structural Parameters. J. Chem. Eng. Data. 59, 2490–2498.
Li, X.C., He, G.H., Zheng, W.J., Xiao, G.K., 2010. Study on conductivity property and microstructure of TritonX-100/alkanol/n-heptane/water microemulsion. Colloids Surf. A 360, 150–158.
Lin, C.C., Lin, H.Y., Ming-Hung Chi, M.H., Shen, C.M., Chen, H.W., Yang, W.J., Lee, M.H., 2014. Preparation of curcumin microemulsions with food-grade soybean oil/lecithin and their cytotoxicity on the HepG2 cell line. Food Chem. 154, 282–290.
Liu, Ch., Chang, F.Y., 2011. Development and characterization of eucalyptol microemulsions for topic delivery of curcumin. Chem. Pharm. Bull. 59, 172–178.
Losada-Barreiro, S., Sánchez-Paz, V., Bravo-Díaz, C., 2013. Effects of emulsifier hydrophile–lipophile balance and emulsifier concentration on the distributions of gallic acid, propyl gallate, and α-tocopherol in corn oil emulsions. J. Colloid Interface Sci. 389, 1–9.
Lu, F., Ding, Y., Ye, X., Liu, D., 2010. Cinnanom and nisin in alginate-calcium coating maintain quality of fresh northern snakehead fish filets. LWT Food Sci. Technol. 43, 1331–1335.
Lutz, R., Aserin, A., Wachtel, E.J., Ben-Shoshan, E., Danino, D., Garti, N., 2007. A study of the emulsified microemulsion by SAXS, Cryo-TEM, SD-NMR, and electrical conductivity. J. Disp. Sci. Technol. 28, 1149–1157.
Ma, Q., Zhong, Q., 2015. Incorporation of soybean oil improves the dilutability of essential oil microemulsions. Food Res. Int. 71, 118–125.
Magdassi, S., 1997. Delivery systems in cosmetics. Colloids Surf. A 123–124, 671–679.
Mahboubi, M., Kazempour, N., 2014. Chemical composition and antimicrobial activity of peppermint (*Mentha piperita* L.) essential oil. J. Sci. Technol. 36, 83–87.

Maqbool, M., Ali, A., Alderson, P.G., Mohamed, M.T.M., Siddiqui, Y., Zahid, N., 2011. Postharvest application of gum arabic and essential oils for controlling anthracnose and quality of banana and papaya during cold storage. Postharvest Biol. Technol. 62, 71–76.

Marasas, W., Jaskiewicz, K., Venter, F., Van-Schalkwyk, D., 1988. *Fusarium moniliforme* contamination of maize in oesophageal cancer areas in Transkei. S. Afr. Med. J. 74, 110–114.

Marasas, W.F.O., Riley, R.T., Hendricks, K.A., Stevens, V.L., Sadler, T.W., Gelineau-Van Waes, J., Missmer, S.A., Cabrera, J., Torres, O., Gelderblom, W.C.A., Allegood, J., Martínez, C., Maddox, J., Miller, J.D., Starr, L., Sullards, M.C., Roman, A.V., Voss, K.A., Wang, E., Merril, Jr., A.H., 2004. Fumonisins disrupt sphingolipid metabolism, folate transport, and neural tube development in embryo culture and in vivo: a potential risk factor for human neural tube defects among populations consuming fumonisin contaminated maize. J. Nutr. 134, 711–716.

Martins, I.M., Rodrigues, S.N., Barreiro, M.F., Rodrigues, A.E., 2011. Release of thyme oil from polylactide microcapsules. Ind. Eng. Chem. Res. 50, 13752–13761.

Mason, T.G., Wilking, J.N., Meleson, K., Chang, C.B., Graves, S.M., 2006. Nanoemulsions: formation, structure, and physical properties. J. Phys. Condens. Matter 18, 635–666.

McClements, D.J., 2004. Protein-stabilized emulsions. Curr. Opin. Colloid Interface Sci. 9, 305–313.

McClements, D.J., 2012a. Edible delivery systems for nutraceuticals: designing functional foods for improved health. Ther. Deliv. 3, 801–803.

McClements, D.J., 2012b. Nanoemulsions versus microemulsions: terminology, differences, and similarities. Soft Matter 8, 1719–1729.

McClements, D.J., Decker, E.A., Park, Y., Weiss, J., 2009. Structural design principles for delivery of bioactive components in nutraceuticals and functional foods. Crit. Rev. Food Sci. 49, 577–606.

Mishra, M., Muthuprasanna, P., Surya prabha, K., Sobhita Rani, P., Satish Babu, I.A., Sarath Chandiran, I., Arunachalam, G., Shalini, S., 2009. Basics and potential applications of surfactants: a review. Int. J. Pharm. Technol. Res. 1, 1354–1365.

Mitra, R.K., Paul, B.K., 2005. Effect of temperature and salt on the phase behavior of nonionic and mixed nonionic-ionic microemulsions with fish-tail diagrams. J. Colloid Interface Sci. 291, 550–559.

Montes-Belmont, R., Carvajal, M., 1998. Control of *Aspergillus flavus* in maize with plant essential oils and their components. J. Food Protect. 61, 616–619.

Moore, C.A., Li, S., Li, Z., Hong, S.X., Gu, H.Q., Berry, R.J., Mulinare, J., Erickson, J.D., 1997. Elevated rates of severe neural tube defects in a high-prevalence area in northern China. Am. J. Med. Genet. 73, 113–118.

Moulik, S.P., Rakshit, A.K., 2006. Physicochemistry and applications of ME. J. Surf. Sci. Technol. 22, 159–186.

Mulligan, C.N., 2005. Environmental applications for biosurfactants. Environ. Pollut. 133, 183–198.

Najjar, R. (Ed.), 2012. Microemulsions: An Introduction to Properties and Applications. InTech Publisher, Rijeka, Croatia, p. 262.

Newberne, P., Smith, R.L., Doull, J., Goodman, J.I., Munro, I.C., Portoghese, P.S., Wagner, B.M., Weil, C.S., Woods, L.A., Adams, T.B., Lucas, C.D., Ford, R.A., 1999. The FEMA GRAS assessment of *trans*-anethole used as a flavouring substance. Food Chem. Toxicol. 37, 789–811.

Newburger, A.E., 2009. Cosmeceuticals: myths and misconceptions. Clin. Dermatol. 27, 446–452.

Nguyen, T.T., Sabatini, D.A., 2011. Characterization and emulsification properties of rhamnolipid and sophorolipid biosurfactants and their applications. Int. J. Mol. Sci. 12, 1232–1244.
Nitschke, M., Costa, S.G.V.A.O., 2007. Biosurfactants in food industry. Trends Food Sci. Technol. 18, 252–259.
Oke, M., Jacob, J.K., Paliyath, G., 2010. Effect of soy lecithin in enhancing fruit juice/sauce quality. Food Res. Int. 43, 232–240.
Otomo, N., 2009. Basic properties of sucrose fatty acid esters and their applications. In: Hayes, D.G., Kitamoto, D., Solaiman, D.K.Y., Ashby, R.D. (Eds.), Biobased Surfactants and Detergents: Synthesis, Properties, and Applications. AOCS Press, Champaign, IL, pp. 275–298.
Özcan, M.M., Arslan, D., 2011. Antioxidant effect of essential oils of rosemary, clove, and cinnamon on hazelnut and poppy oils. Food Chem. 129 (1), 171–174.
Pang, Z., Han, C., 2014. Review on transdermal drug delivery systems. J. Pharm. Drug Dev. 2, 1–10.
Paolino, D., Ventura, C.A., Nisticò, S., Puglisi, G., Fresta, M., 2002. Lecithin microemulsions for the topical administration of ketoprofen: percutaneous adsorption through human skin and in vivo human skin tolerability. Int. J. Pharm. 244 (1–2), 21–31.
Patel, P.A., Chaulang, G.M., Akolkotkar, A., Mutha, S.S., Hardikar, S.R., Bhosale, A.V., 2008. Self-emulsifying drug delivery system: a review. Res. J. Pharm. Technol. 1, 313–323.
Patravale, V.B., Mandawgade, S.D., 2008. Novel cosmetic delivery systems: an application update. Int. J. Cosmet. Sci. 30, 19–33.
Paul, B.K., Moulik, S.P., 1997. Microemulsions: an overview. J. Disp. Sci. Technol. 18, 301–367.
Paul, B.K., Moulik, S.P., 2001. Uses and applications of microemulsions. Curr. Sci. India 80, 990–1001.
Pelletier, X., Belbraouet, S., Mirabel, D., Mordret, F., Perrin, J.L., Pages, X., Debry, G., 1995. A diet moderately enriched in phytosterols lowers plasma cholesterol concentrations in normocholesterolemic humans. Ann. Nutr. Metab. 39, 291–295.
Peltola, S., Saarinen-Savolainen, P., Kiesvaara, J., Suhonen, T.M., Urtti, A., 2003. Microemulsions for topical delivery of estradiol. Int. J. Pharm. 254, 99–107.
Perdones, A., Vargas, M., Atarés, L., Chiralt, A., 2014. Physical, antioxidant, and antimicrobial properties of chitosane cinnamon leaf oil films as affected by oleic acid. Food Hydrocoll. 36, 256–264.
Ponce, A., Buera, M.P., Elizalde, B.E., 2010. Encapsulation of cinnamon and thyme essential oils components (cinnamaldehyde and thymol) in β-cyclodextrin: effect of interactions with water on complex stability. J. Food Eng. 99, 70–75.
Quintanilla-Carvajal, M.X., Camacho-Díaz, B.H., Meraz-Torres, L.S., Chanona-Pérez, J.J., Alamilla-Beltrán, L., Jimenéz-Aparicio, A., Gutiérrez-López, G.F., 2010. Nanoencapsulation: a new trend in food engineering processing. Food Eng. Rev. 2, 39–50.
Rao, J., McClements, D.J., 2011a. Food-grade microemulsions, nanoemulsions and emulsions: fabrication from sucrose monopalmitate and lemon oil. Food Hydrocoll. 25, 1413–1423.
Rao, J., McClements, D.J., 2011b. Formation of flavor oil microemulsions, nanoemulsions and emulsions: influence of composition and preparation method. J. Agric. Food Chem. 59, 5026–5035.
Rao, J., McClements, D.J., 2012. Lemon oil solubilization in mixed surfactant solutions: rationalizing microemulsion & nanoemulsion formation. Food Hydrocoll. 26, 268–276.

Ritesh, B.P., Rakesh, P.P., Madhabhai, M.P., 2008. Self-emulsifying drug delivery systems. PharmTech.com, http://www.pharmtech.com/self-emulsifying-drug-delivery-systems.

Robb, I.D., 1997. Specialist Surfactants, first ed. Blackie Academic and Professional, London.

Rodrigues, L.R., 2015. Microbial surfactants: fundamentals and applicability in the formulation of nano-sized drug delivery vectors. J. Colloid Interface Sci. 449, 304–316.

Roohinejad, S., Oey, I., Wen, J., Lee, S.J., Everett, D.W., Burritt, D.J., 2015. Formulation of oil-in-water β-carotene microemulsions: effect of oil type and fatty acid chain length. Food Chem. 174, 270–278.

Rosano, H.L., Cavallo, J.L., Chang, D.L., Whittam, J.H., 1988. Microemulsions: a commentary on their preparation. J. Soc. Cosmet. Chem. 39, 201–209.

Rowe, R.C., Sheskey, P.J., Owen, S.C., 2006. Handbook of Pharmaceutical Excipients, fifth ed. Pharmaceutical Press and the American Pharmacists Association, London.

Roy, B.C., Goto, M., Kodama, A., Hirose, T., 1996. Supercritical CO_2 extraction of essential oils and cuticular waxes from peppermint leaves. J. Chem. Technol. Biotechnol. 67, 21–26.

Sadoqi, M., Lau-Cam, C.A., Wu, S.H., 2009. Investigation of the micellar properties of the tocopheryl polyethylene glycol succinate surfactants TPGS 400 and TPGS 1000 by steady state fluorometry. J. Colloid Interface Sci. 333, 585–589.

Saint-Leger, D., 2012. 'Cosmeceuticals'. Of men, science and laws. Int. J. Cosmet. Sci. 34, 396–401.

Saito, H.K.S., 1969. The Stability of O/W type emulsions as functions of temperature and the HLB of emulsifiers: the emulsification by PIT-method. J. Colloid Interface Sci. 30, 258–263.

Sajal, K.J., Roopa, K., Venkatesh, D.P., Geethalakshami, A., 2011. Formulation development and characterization of microemulsion drug delivery systems containing antiulcer drug. Int. J. Drug Dev. Res. 3, 336–343.

Salager, J.-L., 2002. Surfactants Types and Uses. FIRP Booklet # E300-A. Universidad de los Andes, Facultad de Ingeniería, Escuela de Ingeniería Química. Mérida-Venezuela.

Salager, J.-L., Antón, R.E., Sabatini, D.A., Harwell, J.H., Acosta, E.J., Tolosa, L.I., 2005. Enhancing solubilization in microemulsions—state of the art and current trends. J. Surfactants Deterg. 8, 3–21.

Sandeep, S.B., Narendra, M., Karunakar, S., Parihar, M.S., 2013. Biosurfactants: a new pharmaceutical additive for solubility enhancement and pharmaceutical development. Biochem. Pharmacol. 2, 1–5.

Sanguansri, S., Augustin, M.A., 2006. Nanoscale materials development—a food industry perspective. Trends Food Sci. Technol. 17, 547–556.

Schulman, J.H., Stoeckenius, W., Prince, L.M., 1959. Mechanism of formation and structure of microemulsions by electron microscopy. J. Phys. Chem. 63, 1677–1680.

Sekhon, B.S., 2013. Surfactants: pharmaceutical and medicinal aspects. J. Pharm. Technol. Res. Manag. 1, 11–36.

Semo, E., Kesselman, E., Danino, D., Livney, Y.D., 2007. Casein micelle as a natural nano-capsular vehicle for nutraceuticals. Food Hydrocoll. 21, 936–942.

Shah, V.P., Konecny, J.J., Everett, R.L., McCullough, B., Nooriazadeh, A.C., Skelly, J.P., 1989. In vitro dissolution profile of water insoluble drug dosage forms in the presence of solubilizers. Pharm. Res. 6, 162–168.

Shahidi, F., 2009. Nutraceuticals and functional foods: whole versus processed foods. Trends Food Sci. Technol. 20, 376–387.

Sharma, M.K., Shah, D.O., 1985. Introduction to macro- and microemulsions. In: Shah, D.O. (Ed.), Macro and Microemulsions, Theory and Applications. ACS Symposium Series. American Chemical Society, Washington, DC, p. 272.

Sharma, G., Wilson, K., Van der Walle, C.F., Sattar, N., Petrie, J.R., Kumar, M., 2010. Microemulsions for oral delivery of insulin: design, development, and evaluation in streptozotocin induced diabetic rats. Eur. J. Pharm. Biopharm. 76, 159–169.

Siddalingappa, B., Nekkanti, V., Betageri, G.V., 2013. Insoluble drug delivery technologies: review of health benefits and business potentials. OA Drug Des. Deliv. 1, 1–5.

Singh, P., Cameotra, S.S., 2004. Potential applications of microbial surfactants in biomedical sciences. Trends Biotechnol. 22, 142–146.

Sinha, K.K., Sinha, A.K., Prasad, G., 1993. The effect of clove and cinnamon oils on growth of and aflatoxin production by *Aspergillus flavus*. Lett. Appl. Microbiol. 16, 114–117.

Sonneville-Aubrun, O., Simonnet, J.T., L'Alloret, F., 2004. Nanoemulsions: a new vehicle for skincare products. Adv. Colloid Interface Sci. 108, 145–149.

Soppimath, K.S., Aminabhavi, T.M., Kulkarni, A.R., Rudzinski, W.E., 2001. Biodegradable polymeric nanoparticles as drug delivery devices. J. Control. Release 70, 1–20.

Spernath, A., Aserin, A., 2006. Microemulsions as carriers for drugs and nutraceuticals. Adv. Colloid Interface Sci. 128–130, 47–64.

Spernath, A., Yaghmur, A., Aserin, A., Hoffman, R.E., Garti, N., 2002. Food grade microemulsions based on nonionic emulsifiers: media to enhance lycopene solubilization. J. Agric. Food Chem. 50, 6917–6922.

Spernath, A., Yaghmur, A., Aserin, A., Hoffman, R.E., Garti, N., 2003. Self-diffusion nuclear magnetic resonance, microstructure transitions, and solubilization capacity of phytosterols and cholesterol in Winsor IV food-grade microemulsions. J. Agric. Food Chem. 51, 2359–2364.

Strickley, R.G., 1999. Parenteral formulations of small molecule therapeutics marketed in the United States Part I. PDA J. Pharm. Sci. Technol. 53, 324–349.

Stubenrauch, C., 2008. Microemulsions Background, New Concepts, Applications, Perspectives. John Wiley & Sons, Hoboken, NJ.

Sydenham, E.W., Thiel, P.G., Marasas, W.F.O., Shephard, G.S., Van Schalkwyk, D.J., Koch, K.R., 1990. Natural occurrence of some *Fusarium* mycotoxins in corn from low and high-risk esophageal cancer prevalence areas of the Transkei. J. Agric. Food Chem. 38, 1900–1903.

Tabilo-Munizaga, G., Barbosa-Cánovas, G.V., 2005. Rheology for the food industry. J. Food Eng. 67, 147–156.

Tadros, F.T., 2005. Applied Surfactants: Principles and Applications. Wiley-VCH, Weinheim, Germany.

Tadros, F.T., Izquierdo, R., Esquena, J., Solans, C., 2004. Formation and stability of nano-emulsions. Adv. Colloid Interface Sci. 108–109, 303–318.

Ungell, A.L., Abrahamsson, B., 2009. Biopharmaceutical support in candidate drug selection. In: Gibson, M. (Ed.), Pharmaceutical Preformulation and Formulation: A Practical Guide From Candidate Drug Selection to Commercial Dosage Form. second ed. Informa Healthcare, New York, NY.

Valenta, C., Schultz, K., 2004. Influence of carrageenan on the rheology and skin permeation of microemulsion formulations. J. Control. Release 95, 257–265.

Vangalapati, M., Satya, S.N., Prakash, D.V.S., Avanigadda, S., 2012. A review on pharmacological activities and clinical effects of cinnamon species. Res. J. Pharm. Biol. Chem. Sci. 3, 653–663.

Velluti, A., Sanchis, V., Ramos, A.J., Egido, J., Marin, S., 2003. Inhibitory effect of cinnamon, clove, lemongrass, oregano, and palmarose essential oils on growth and fumonisin B1 production by *Fusarium proliferatum* in maize grain. Int. J. Food Microbiol. 89, 145–154.

Voss, K.A., Howard, P.C., Riley, R.T., Sharma, R.P., Bucci, T.J., Lorentzen, R.J., 2002. Carcinogenicity and mechanism of action of fumonisin B1: a mycotoxin produced by *Fusarium moniliforme* (= *F. verticillioides*). Cancer Detect. Prev. 26, 1–9.

Walters, K.A., Brain, K.R., 2009. Topical and transdermal delivery. In: Gibson, M. (Ed.), Pharmaceutical Preformulation and Formulation: A Practical Guide from Candidate Drug Selection to Commercial Dosage. Informa Healthcare, New York, NY, pp. 475–516.

Wang, Y., Zhao, R., Yu, L., Zhang, Y., He, Y., Yao, J., 2014. Evaluation of cinnamon essential oil microemulsion and its vapor phase for controlling postharvest gray mold of pears (*Pyrus pyrifolia*). J. Sci. Food Agric. 94, 1000–1004.

Warisnoicharoen, W., Lansley, A.B., Lawrence, M.J., 2000. Nonionic oil-in-water microemulsions: the effect of oil type on phase behavior. Int. J. Pharm. 198, 7–27.

Wu, Y., Luo, Y., Wang, Q., 2012. Antioxidant and antimicrobial properties of essential oils encapsulated in zein nanoparticles prepared by liquid-liquid dispersion method. LWT Food Sci. Technol. 48, 283–290.

Wu, Y., Wang, T., 2003. Soybean Lecithin Fractionation and Functionality. J. Am. Oil Chem. Soc. 80, 319–326.

Xia, S., Xu, S., Zhang, X., Zhong, F., 2007. Effect of coenzyme Q10 incorporation on the characteristics of nanoliposomes. J. Phys. Chem. B 111, 2200–2207.

Xue, J., 2015. Essential oil nanoemulsions prepared with natural emulsifiers for improved food safety. Doctoral Dissertations. The University of Tennessee, Knoxville, TN.

Yang, Y., Leser, M.E., Sher, A.A., McClements, D.J., 2013. Formation and stability of emulsions using a natural small molecule surfactant: Quillaja saponin (Q-Naturale®). Food Hydrocolloids 30, 589–596.

Zafeiropoulou, T., Evageliou, V., Gardeli, C., Yanniotis, S., Komaitis, M., 2010. Retention of *trans*-anethole by gelatin and starch matrices. Food Chem. 123, 364–368.

Zainol, N.A., Ming, T.S., Darwis, Y., 2015. Development and charaterization of cinnamon leaf oil nanocream for topical application. Indian J. Pharm. Sci. 77, 422–433.

Zeisel, S.H., 1999. Regulation of "nutraceuticals". Science 285, 1853–1855.

Zhang, H., Taxipalati, M., Que, F., Feng, F., 2013. Microstructure characterization of a food-grade U-type microemulsion system by differential scanning calorimetry and electrical conductivity techniques. Food Chem. 141, 3050–3055.

Zheng, G., Zhao, Z., Wong, J.W.C., 2011. Role of non-ionic surfactants and plant oils on the solubilization of organochlorine pesticides by oil-in-water microemulsions. Environ. Technol. 32, 269–279.

Ziani, K., Chang, Y., McLandsborough, L., McClements, D.J., 2011. Influence of surfactant charge on antimicrobial efficacy of surfactant-stabilized thyme oil nanoemulsions. J. Agric. Food Chem. 59, 6247–6255.

Zuidam, N.J., Nedović, V.A., 2010. Encapsulation Technologies for Active Food Ingredients and Food Processing. Springer, New York, NY.

13

NANOENCAPSULATION STRATEGIES APPLIED TO MAXIMIZE TARGET DELIVERY OF INTACT POLYPHENOLS

Sandra Pimentel-Moral*,, Vito Verardo†, Paz Robert‡, Antonio Segura-Carretero*,**, Antonio Martínez-Férez**,§**

**University of Granada, Department of Analytical Chemistry, Granada, Spain; **Research and Development Centre for Functional Food (CIDAF), Health Science Technological Park, Granada, Spain; †University of Almería, Agrifood Campus of International Excellence, Department of Chemistry and Physics (Analytical Chemistry Area), Research Centre for Agricultural and Food Biotechnology (BITAL), Almería, Spain; ‡University of Chile, Department of Food Science and Chemical Technology, Santiago, Región Metropolitana, Chile; §University of Granada, Chemical Engineering Department Granada, Spain*

1 Introduction

Polyphenols are secondary metabolites found in all vascular plants, and constitute a big family of varied substances. All these natural compounds have one or more benzenic cycles with hydroxyl functions and derive from the metabolism of shikimic acid and/or polyacetate. In this sense, the most important characteristic of polyphenols is their ability to scavenge reactive oxygen species (ROS). However, polyphenolic compounds exhibit many physiological properties such as anti-allergenic, antiatherogenic, antiinflammatory, antimicrobial, antioxidant, cardioprotective, and vasodilatory effects. Due to this broad bioactivity, polyphenols have become very interesting functional compounds and great interest has developed in establishing new strategies for incorporation into the diet.

But these bioactivities are very short term due to rapid oxidation or insufficient gastric residence time, low permeability and/or solubility within the gut. Indeed, polyphenols are very unstable under typical food processing and storage conditions

Encapsulations. http://dx.doi.org/10.1016/B978-0-12-804307-3.00013-2

(eg, temperature, oxygen, light exposure, etc.) or through the gastrointestinal tract (eg, pH, enzymes, interaction with other nutrients, etc.), provoking a partial or even total loss that restricts their application in products for human consumption.

At this point, it is important to remind ourselves that the effectiveness of nutraceutical products to prevent a disease depends on preserving the bioavailability of the active compound. Therefore, oral administration of polyphenols needs an additional protection to keep their structural integrity and increase their bioavailability and physiological target (Fang and Bhandari, 2010; Munin and Edwards-Lévy, 2011; Quideau et al., 2011; Đorđević et al., 2014), which makes encapsulation, in particular nanoencapsulation, the target technology.

In general terms, encapsulation may be defined as a process to capture one substance (active ingredient) within another substance (cover material). It is a technology in which there is a physical barrier to protect the bioactive agents of reactions against the environment during storage, or against other components when added to a food matrix. It also presents an economic advantage in its ability to control the rate of release of the encapsulated ingredient, allowing the food industry to increase the effectiveness of their bioactive ingredients by not having to dispense them in excess. As a consequence, by means of encapsulation, the food industry has expanded the range of available bioactive products, since encapsulated materials can be protected from moisture, heat, or other extreme conditions, thereby enhancing their stability and maintaining viability (Gibbs et al., 1999).

The encapsulated substance is called the core, active, internal, or payload phase. The substance that is encapsulating can be called the coating, carrier material, capsule, membrane, shell, or matrix, and can be composed of different biopolymers such as sugars, gums, proteins, natural and modified polysaccharides, lipids and synthetic polymers, and so forth. The functions of these micro and nanocapsules include to protect an unstable compound from the environment, to avoid side-effects of the encapsulated ingredient in the consumer, to isolate two incompatible compounds that needs to coexist in the same medium, or to control the release of the encapsulated compound (Fang and Bhandari, 2010; Munin and Edwards-Lévy, 2011).

2 Encapsulation Technologies

The cost limitations for raw materials in the food industry are stricter than in other industries, like in the pharmaceutical or cosmetic domains. Therefore, selection of the microencapsulation procedure for food applications is governed not only by the physical and chemical characteristics of the core and wall materials and the intended application of the selected food compound, but also by economic issues. To date, different wall materials and techniques have been developed and applied for food nanoencapsulation applications (Desai and Jin Park, 2005).

The encapsulation technologies for delivery of bioactive compounds like polyphenols are classified under chemical and mechanical techniques. Chemical methods include liposomes, coacervation, molecular inclusion, yeast encapsulation, emulsion, and ionic gelation. Others, like spray-drying, freeze-drying, spray-cooling/chilling, fluidized bed drying, extrusion, and spinning discs, are based on mechanical principles.

2.1 Mechanical Technologies

2.1.1 Spray Drying

Spray-drying is a technique used in the food and pharmaceutical industries where the solvent evaporates quickly from the droplet and allows the encapsulation of bioactive materials within a protective matrix. This protective matrix is inert to the material being encapsulated through a brief exposure to high temperatures in a spray chamber (Mahdavi et al., 2014). Although most often considered a dehydration process, spray-drying is well established for producing dried powdered products and microencapsulation of polyphenols and other heat labile compounds (Ishwarya et al., 2015). In the food industry, spray-drying is the most applied encapsulation technique. However, it is considered an immobilization method rather than a true encapsulation method because some bioactive components may be exposed superficially on the microparticles (De Vos et al., 2010).

The target of this technique is the dispersion of the component to be encapsulated in a carrier material, followed by atomization and spraying of the mixture into a hot chamber (Fig. 13.1) (Madene et al., 2006). The first phase consists in the elaboration of feed liquid (solution, dispersion, or emulsion) composed by the coating agent and the bioactive component. This mixture is atomized with the formation of droplets. Then, droplets are dehydrated by hot drying medium (usually air), where the solvent (mainly water) is rapidly evaporated in a heated chamber. The dehydrated powdered particles are then transported to a cyclone separator for recovery. During the drying process, a layer is formed at the droplet surface and the concentrations of ingredients in the drying droplet amplify. This process takes a few seconds and feed liquid is converted into powdered particles without any appreciable

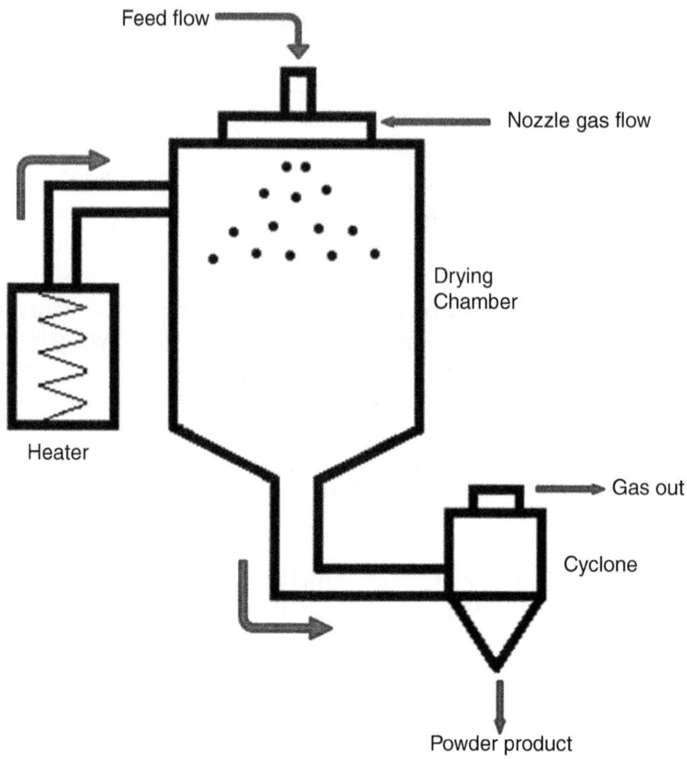

Figure 13.1. Schematic view of spray-drying process.

loss (Gouin, 2004; Mahdavi et al., 2014; Munin and Edwards-Lévy, 2011; Đorđević et al., 2014).

The spray-dried products have important properties, such as increased thermal stability, regular particle shape and size, and reduced moisture content. The final product will depend on process parameters (ie, inlet and outlet air temperatures and feed temperature), on formulation parameters (ie, core material/encapsulating agent ratio, solid content, and viscosity) and equipment; therefore the process can be modified to control the properties of spray-dried powders. The characterization of powders generally involved the determination of total phenolic compounds, superficial polyphenols, encapsulation efficiency, recovery of polyphenols, and powder yield (Gharsallaoui et al., 2007).

Also, the spray drying has been successfully used for the encapsulation of a number of polyphenols rich materials in the last years, including *Hibiscus sabdariffa* L. extract (Chiou and Langrish, 2007; Idham et al., 2012), *Paeoniarockii* extract (Sansone et al., 2014), Cactus pear (Ruiz-Gutiérrez et al., 2014; Sáenz et al., 2009), *Gypsophila* extract (Ozdikicierler et al., 2014), *Averrhoacarambola pomace* (Saikia et al., 2015), bayberry *procyanidins* (Fang and Bhandari, 2011, 2012); black currant pomace extract (Bakowska-Barczak and Kolodziejczyk, 2011); bilberry pomace extract (Baum et al., 2014; Oidtmann et al., 2012); blackberry pulp (Ferrari et al., 2012); black mulberry fruit juice (Fazaeli et al., 2012); blueberry extract (Flores et al., 2014); blueberry fruit extract (Jimenez-Aguilar et al., 2011); black carrot extract (Ersus and Yurdagel, 2007); Andes berry fruit extract (Villacrez et al., 2014); apple skin extract (Kosaraju et al., 2008), grape seed extract (Kosaraju et al., 2008); olive leaf extract (Kosaraju et al., 2008; Kosaraju, 2005); artichoke extract (Gavini et al., 2005); carozo fruit extract (Osorio et al., 2010); jabuticaba peel extract (Silva et al., 2013); kokum fruit extract (Nayak and Rastogi, 2010); mate tea leaf (Yatsu et al., 2011); noni fruit (Krishnaiah, 2009); pomegranate fruit juice (Robert et al., 2010), and white oak leaf infusion (Rocha-Guzmán et al., 2010).

In those works a variety of wall materials have been successfully used for encapsulation of polyphenols, such as: natural gums (arabic gum and mesquite gum), proteins (whey protein isolate, soybean protein isolate), carbohydrates (starch of different sources, maltodextrins of different dextrose equivalents, inulin and modified starches), and chitosan and carrageenan.

In this spray-drying process, the most important steps are the selection of wall materials, the bioactive component, and the release system. In this sense, the expected features of a good carrier material are appropriate dissolution characteristics, low viscosity at great concentrations, and excellent emulsifying capabilities (unbroken emulsions, if possible) (Đorđević et al., 2014).

Carbohydrates, proteins, and new biopolymers are the most appropriate wall materials for spray-drying encapsulation. Gum arabic is the most used carbohydrate due to its low viscosity, solubility, emulsification characteristics, and high capacity to withhold volatile compounds, developing the role of active surface agent and matrix dryer (Fang and Bhandari, 2010; Đorđević et al., 2014). However, the increasing prices of these gums limit enormously their applications in the food industry and they are often replaced by less expensive maltodextrins. Maltodextrins are frequently used for encapsulation of flavors and polyphenols. For example, different wall materials (eg, gum arabic, whey protein concentrate, maltodextrin egg albumin, and starch sodium octenyl succinate) were studied in persimmon powders obtained by spray-drying. Maltodextrin, gum arabic, and starch sodium octenyl succinate showed higher total polyphenol retention and better reconstitution. In addition, persimmon powders made with maltodextrin and gum arabic showed a continuous and smooth surface and a more spherical physical form compared to those produced with starch sodium octenyl succinate, whey protein concentrate, and egg albumin that presented wrinkled surfaces (Jing et al., 2014).

This technique has also been applied to obtain nanocapsules. Bioactive compounds like folic acid have been encapsulated using a novel nanospray-drying device able to obtain smaller encapsulation structures. The polymer used for folic acid encapsulation was a whey protein concentrate and a commercial resistant starch (Pérez-Masiá et al., 2015).

In 2011, bayberry juice was dried by spray-drying with maltodextrins and then stored under different temperatures and water activities (a w). The retention of the total phenolic content and total anthocyanins during the drying process was about 96% and 94%, respectively, implying that spray-drying was a good technique for drying heat sensitive polyphenols (Fang and Bhandari, 2011). More recently, an extract of star fruit (*Averrhoa carambola*) pomace was encapsulated with maltodextrin by spray-drying and freeze-drying. The methods were compared and the highest encapsulating efficiency was obtained in freeze-dried encapsulates (78–97%) (Saikia et al., 2015).

Regarding the mixture of carriers, gum arabic and maltodextrins have arisen as good solid carriers. Apintanapong and Noomhorm (2003) analyzed different ratios of gum arabic and maltodextrins as wall materials for encapsulation by spray-drying a flavor component from aromatic rice. Ratios 70:30 of gum arabic/maltodextrin showed better results in quality capsules.

Chitosan has also been used as a carrier material for *Paeonia rokii* (PPR) extracts. The formulation 1:1 chitosan/PPR had the highest polyphenol content (Sansone et al., 2014). Other wall materials, such as soluble fiber for encapsulating cactus pear, have developed a final powder with good characteristics after encapsulation at 160°C and 22.5% soluble fiber (Ruiz-Gutiérrez et al., 2014). The encapsulation of polyphenols obtained by the spray-drying method permits a good entrapping of these bioactive compounds and, because of that, it can be used for nutraceutical application.

The major advantages of this technique are the production of micro and nanocapsules in a simple and continuous operation, low operating cost, small size, easy handling, rapid solubility of the capsules, high-quality capsules in a good yield, good retention of volatiles, and good stability of the final product (Madene et al., 2006; Mahdavi et al., 2014; Đorđević et al., 2014). However, this technique has some disadvantages, such as the production of very fine powders that need further processing, or limitation in the choice of wall materials (Ishwarya et al., 2015). Due to the

Table 13.1 Advantages and Disadvantages of Spray-Drying

Advantages	Disadvantages
• Ease scalability • Moderate operating cost • Capsules of high quality and payload • Capsule solubility can be modified • New approaches make possible particle sizes up to approx. 100 nm • High stability capsules • Process easy to control	• Nonhomogeneous microcapsules can arise • Limitation in the choice of wall material • Produces very fine powder which needs further processing • Is necessary to optimize the process for heat-sensitive material

thermosensitivity of some antioxidants, it is necessary to optimize the spray-drying process at lower temperatures and to make a good choice of cover agents to preserve the antioxidants during storage (Davidov-Pardo et al., 2012). The principal advantages and disadvantages of spray-drying encapsulation are summarized in Table 13.1.

2.1.2 Spray-Cooling/Spray-Chilling

Spray-cooling and spray-chilling represent encapsulation techniques used to reach heat stability and delay the release of the encapsulated compound transforming powders. The active ingredient might be soluble in the lipids, or be present as dry particles or flowing emulsions. These technologies have a mechanism similar to the spray-drying operation, but the principle of these methods is contrary to spray-drying, because the carrier is cooled instead of evaporated. The bioactive ingredients are fused with lipids coating and atomized in droplets. After this, the droplets are blended with a cooling medium and a result is achieved in powder form. The choice of either method will depend on the melting point of the lipids used for the coating. In spray-chilling, the melting temperature is in a range of 34–42°C, while for spray-cooling, it is higher than spray-chilling.

Spray-chilling is generally applied to decelerate volatilization and increase the protection in the heated processing of sensitive active agents with water-soluble properties such as vitamins, minerals, and flavors, because the wall material is a vegetable oil with a melting point of 32–42°C. In this technique, the lipids coating is melted and atomized into a container that includes a carbon dioxide ice bath as a hot-melt fluidized bed and the droplets are solidified to constitute a cover layer. Spray-chilled products have applications in bakery products, dry soup mixes, and foods containing a high level of fat (Madene et al., 2006; Onwulata, 2012).

Spray-cooling is a technique similar to spray-chilling but the temperature of the reactor, where the wall material is sprayed, is different.

The disadvantage of spray-chilling and spray-cooling is that they need special handling and storage conditions.

2.1.3 Freeze-Drying

Freeze-drying is a technique used for the dehydration of heat sensitive components. It is one of the most effective processes for drying thermosensitive substances that are unstable in aqueous solutions.

This process starts with a decrease of the sample temperature (freezing) of the bioactive ingredient controlling the ice crystal size; a subsequent reduction of pressure and increase of heat is needed to sublimate the water present in the extract containing the bioactive component. The results reach a porous form that preserves the nutritional properties. In the last stage, a second drying is applied to absorb the water attached to the porous matrix, with an increasing temperature resulting in a product moisture level of around 0.5% (Fig. 13.2) (Fang and Bhandari, 2010; Lopez-Quiroga et al., 2012; Madene et al., 2006).

Drying conditions influence the antioxidant capacity and phenolic profile. Vacuum freeze-drying (VFD) decreases deterioration, but increases the cost. Atmospheric freeze-drying (AFD) can be used with an alternative and combined with infrared radiation. Reyes et al. (2011) studied the effects of particle size and type of freeze-drying (vacuum or atmospheric) on the antioxidant capacity and total polyphenols of blueberries. This study concluded that the antioxidant activity of freeze-dried blueberries did not diverge much from that of the fresh fruits. In order to reduce the deterioration of the nutritional properties of freeze-dried blueberries, it is

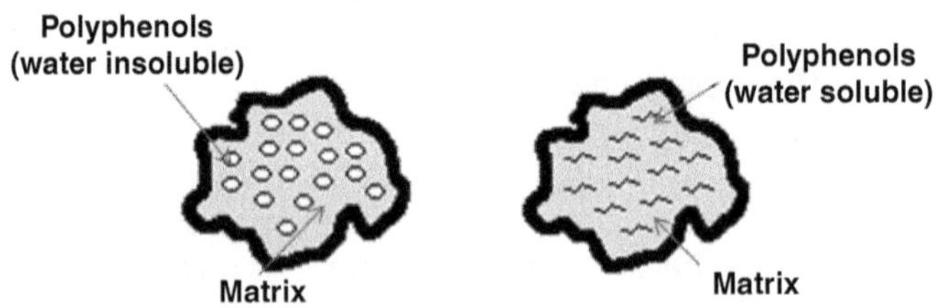

Figure 13.2. Polyphenolic capsules produced by freeze-drying.

recommended to apply vacuum freeze-drying with infrared radiation and small particle size.

This technology is applied to obtain micro and nanoparticles of polyphenols. Guzman-Villanueva et al. (2013) encapsulated curcumin in micro-nanoparticles, synthesized by ionotropic gelation. The obtained microparticles were fractioned and freeze-dried. The average particle size of the curcumin nanoparticles was 480 ± 70 nm. In addition, the in vitro release profile showed up to 95% release of curcumin in these nanoparticles after 9 h in PBS at pH 7.4.

Others encapsulation applications of polyphenols by freeze-drying have been studied. Laine et al. (2008) evaluated storage stability, phenolic content, and antioxidant activity of phenolic-rich cloudberry extract capsules by freeze-drying with maltodextrin DE 18.5 and maltodextrin DE 5-8, and of unencapsulated cloudberry extract. Encapsulated cloudberry extract provided better phenolic protection and antioxidant capacity during storage than the unencapsulated sample.

In other study, a solution of Argentinian red wine and maltodextrin DE 10 was freeze-dried. In this process, there was very little loss of total polyphenols. This "wine powder" showed 3.7 times higher content of polyphenols than red wine (Sanchez et al., 2013). Gurak et al. (2013) studied the freeze-drying process in pre-concentrated grape juice: the results reported that polyphenolic concentration did not vary during 120 d of storage, because the freeze-drying process increased the stability and shelf life of this product.

The major disadvantages of freeze-drying are high energy use and long processing time because of the low temperature. This makes for high capital and operating costs. Besides, an open porous structure is obtained, which is in general not a very good barrier between the active component and its surroundings. Compared to spray-drying, freeze-drying is up to 30–50 times more expensive (Gharsallaoui et al., 2007).

2.1.4 Fluid Bed Drying

Fluidized bed-coating is an encapsulation method based on extra coating and wall material being used on powder particles in a continuous set-up or in a batch processor. First, the bioactive agents, which will be encapsulated, are suspended in a hot atmosphere in a drying chamber with a predefined temperature. Then, the coating agent is atomized over the active particles and a film is created. Each particle will be progressively coated every time it is in the spraying zone. The water is evaporated and the wall material is united with the bioactive agent. This evaporation is controlled

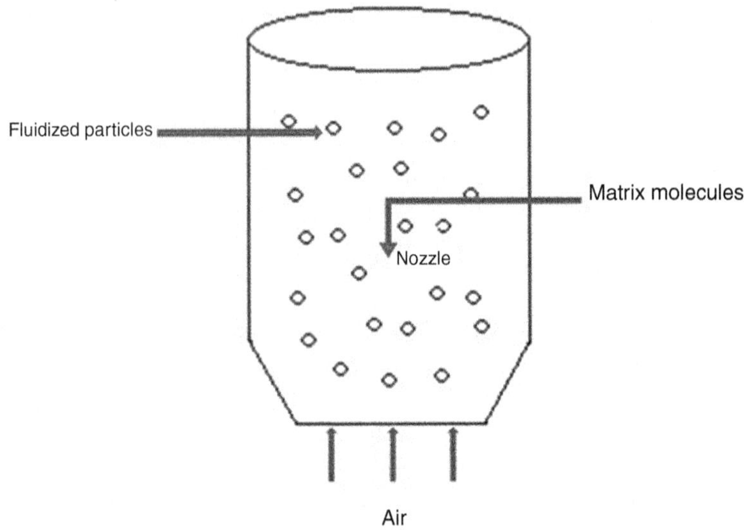

Figure 13.3. Schematic view of fluid bed drying process.

by different ratios such as spraying rate, water content, humidity, and temperature, among others (Fig. 13.3).

The difference between spray-drying and fluidized bed drying is that the fluidized bed method can be achieve at lower temperatures than spray-drying, because there is a excellent mass and heat transport and equal temperature distribution. In addition, the fluid bed core material is smooth, spherical, and in powdered form to minimize the amount of fluid needed to coat it and to reduce the likelihood of uneven coating along the jagged edges, which limits coating efficiency and functionality. The integrity and solubility of the wall material is essential to get a good fluidized bed-coating procedure and to obtain a controlled release of bioactive components (Onwulata, 2012; Đorđević et al., 2014).

Therefore, Sun et al. (2013) studied different coating materials to get a controlled delivery of fluidized bed-coated menthol powder. Gelatin coating was the best material (60% of the menthol powder was delivered after 11 min in water at 37°C).

Generally, the coating material is solid and it might be proteins, gums, or starch derivatives. The fluid bed particles should be spherical and dense, and should have a narrow particle size distribution and good flowability. Spherical particles have the lowest possible surface area and require less coating material for the same shell thickness than nonspherical ones. In general, coating can be used to make the powder more resistant to humidity. This technique may be applied to give a second coating to spray-dried products or alone on products with a sensitive core containing aromas, flavors, polyphenols, or oils. Fluidized bed-coating has also been used to create an additional layer of molecules for targeted release in the gut (De Vos et al., 2010; Dewettinck and Huyghebaert, 1999; Zuidam and Shimoni, 2010). The problem in this technique is the coalescence of coating material and agglomeration of particles. After solvent evaporation, the liquid bridges become solidified, thus forming agglomerates. This occurs when the moisture content during the process is too high, due generally to insufficient drying or the use of a temperature above glass transition of the polymer solution (Saleh et al., 2003). Many variables such as air temperature, atomization pressure, and coating solution concentration should be controlled in order to diminish agglomerations (Jiménez et al., 2006; Prata et al., 2012).

Another disadvantage of the fluidized bed is the high cost. The food industry is obliged to cut production costs and should therefore adopt a somewhat different approach to this rather expensive technology. However, this technique is used in the meat industry and in nutritional supplements like encapsulated vitamins and minerals. In fact, fluid bed encapsulated salt is used in meats to prevent development of rancidity (Dewettinck and Huyghebaert, 1999).

Fluid bed technology is also applied in polyphenols coating. Oehme et al. (2011) encapsulated anthocyanins to obtain their delivery in the colon. They developed the capsules by ionotropic gelation and subsequent fluid bed drying to decrease the water content and bead size.

2.1.5 Extrusion and Spinning Disc

The extrusion technique has been widely applied in the food industry from 1957, when it was patented by Swisher. The fundament of this method is the extrusion of a liquid combination of wall material and active compound through an orifice and to achieve the formation of droplets. This encapsulation technique is divided into two phases. In the first phase, an O/W emulsion is prepared, where the oil phase is involved in the aqueous phase. In the second phase, the emulsion is submitted to an extrusion process, which consists of small drop size formation using a syringe or atomization based on simple dripping, vibrating jet/nozzle, coaxial airflow, electrostatic extrusion, jet cutting, and spinning disc. The drops formed fall into a gelling solution or they are subjected to a physical process like cooling or heating (Dima et al., 2015; Đorđević et al., 2014).

Melt injection and melt extrusion are techniques to encapsulate an active ingredient in a carbohydrate melt. In melt injection, the melt is pressured through a filter and then is made inactive by a dehydrating dissolvent like isopropanol or liquid nitrogen. This involves hardening of the wall material and the active component is encapsulated. Melt extrusion and melt injection are very similar methods, but in case of melt extrusion, extrudates are not surface washed. In addition, melt injection uses screws in a vertical position, while in melt extrusion, screws are in the horizontal position (Yanniotis and Stoforos, 2014).

Simple dripping uses low liquid velocity and extruded liquid sticks. When the gravitational force is higher than surface tension, a drop is formed. If the liquid velocity grows, droplet coalescence will be produced. Moreover, the size of the droplet, which depends on the orifice diameter, is another important parameter to consider. The main disadvantages of this method are the low quantity of droplets produced and their large diameter.

Increasing the liquid velocity with an uninterrupted liquid jet in a continuous stream can result in breakage into droplets due to vibrations/forces and surface tension. If the used force is electrostatic force, this method is referred to as electrostatic extrusion, also called electrospraying. This technique is based on the high voltage spinning of a polymer solution to produce nanofibers/nanowebs. The electrostatic extrusion technique permits the production of small particles and regular size distribution. Another method that allows the production of particles with uniform shape and size is the coaxial airflow technique. In this technique, a stream of compressed air is applied to pull the liquid droplets from the nozzle at a faster rate compared to normal gravitational force. However, the main drawback is low production rate.

Jet cutting, vibrating jet/nozzle, and spinning or rotating disc are techniques that have a higher production capacity due to a controllable liquid jet break-up. The differences between them are the mechanism of the jet break-up, by cutting wire (jet cutting), by vibrating a nozzle (vibrating jet/nozzle), or by rotating discs (spinning disc) (Đorđević et al., 2014).

In addition, rotating disc extrusion can be combined with a facility for recycling of the excess coating liquid. The active agent is dispersed in the carrier material; when the suspension is extruded through a rotating disc, the excess coating fluid is atomized and separated from coated particles. Excess coating fluid is then recycled, while the resulting capsules are hardened by cooling or solvent extraction (Madene et al., 2006).

The main advantage of extrusion techniques is that they do not need severe conditions, either in terms of temperature or solvents. In addition, extruders can be manipulated to obtain the temperatures and speeds desired. Thus, it is possible to control the screw design, feed rate, and moisture content. However, among all the extrusion techniques, only spinning disc atomization has been shown to be easily scalable, with production capacities of tons/day using a multidisc system. Another advantage of the spinning disc is the small amounts of water that are required to change the state of carbohydrates in the extruder, so subsequent drying is not necessary, making this technique more economical.

The disadvantages of this technique are the large particle size obtained and the narrow range of coating materials. Among the ingredients that can be applied, polysaccharides are one of the more attractive, due to their technological characteristics and because they are recognized as safe (GRAS). The principal polysaccharide derivatives used in this technique are chitosan, alginate, kappa-carrageenan, and gellan gum. These compounds need be dissolved in a water o organic suspension and then electrospinning is applied by adapting the procedure and/or modifying the suspension characteristics with the provision of proper additives (eg, whey protein). Another problem is the processing of viscous polymer solutions. When these solutions are used as coating materials, a decrease in viscosity is obtained when the temperature increases. Thus, an apparatus heating nozzle/pulsating head has been developed for control of viscosity of biopolymers through the controllable temperatures when they pierce the pulsation chamber, before extrusion and break-up. Use of such a device should enable encapsulation into viscous materials like highly concentrated biopolymer solutions, gelatins, and gums (Guevara, 2008; Đorđević et al., 2014).

In a recent work, a natural extract (rich in phenolic compounds) obtained from a brewery waste stream was incorporated into a polymer film to develop an active packaging film by extrusion. Several nanoclays of the natural extract with different polymeric matrices (ethylene vinyl acetate and low-density polyethylene) were prepared with a corotating twin-screw extruder. The mixtures were processed at a speed of 150 rpm at 150°C for 6 min and the antioxidant effectiveness of these functional nanoclays has been verified in meat samples: the meat samples containing the nanoclays of the natural extract show a retarding of oxidation of around 60% (Barbosa-Pereira et al., 2014). Polyphenols from bilberry pomace extract were encapsulated by extrusion with amidated pectin (Baum et al., 2014; Kropat et al., 2013; Oidtmann et al., 2012).

2.2 Chemical Technologies

2.2.1 Emulsions and Nanoemulsions

Emulsion is a technology used to encapsulate active agents in aqueous solutions, which can either be used directly in their liquid state or can be dried to form powders after emulsification by spray-drying, freeze-drying, or extrusion. Therefore, the emulsions can be a part of the encapsulation process and have a number of advantages as delivery systems for nutraceutical and functional food components.

This technique consists of at least two immiscible liquids, with one liquid being dispersed as small spherical droplets in the other. Emulsions can be classified according to the spatial distribution of oil and water phases. Oil droplets dispersed in an aqueous phase is an oil-in-water (O/W) emulsion, whereas water droplets dispersed in an oil phase is a water-in-oil (W/O) emulsion. In conventional emulsion (O/W), the oil droplets are surrounded by a thin interfacial layer consisting of emulsifier molecules. All emulsions are formed through a nonpolar phase (the oil), a polar phase (the aqueous), and an amphiphilic phase (the interfacial layer). It is therefore possible to incorporate active polar, nonpolar, and amphiphilic ingredients within the same delivery system.

In addition to the simple O/W or W/O systems, there are various types of multiple emulsion like oil-in-water-in-oil (O/W/O) or water-in-oil-in-water (W/O/W) emulsions (Fang and Bhandari, 2010; McClements, 2015).

Nanoemulsions and emulsions can be differentiated based on their particle radius: $r < 100$ nm for nanoemulsions, $r > 100$ nm for emulsions. Moreover, conventional emulsions are more turbid because the drops have the same order dimension as the light wavelength, and this causes their dispersion. On the contrary, nanoemulsions are less turbid, because they spread less light.

The instability of conventional emulsions is reached when they are exposed to adverse environmental conditions, such as heating, drying, freezing, high mineral concentration solutions, chilling, and pH extremes. In addition, a limited number of emulsifiers can be used to form the interfacial layers which surround the oil droplets.

In the water phase of O/W emulsions, a lot of hydrophilic components like vitamins, proteins, polysaccharides, colorings, flavorings, etc.) are dissolved. These contribute to the stability and definition of the functional attributes of emulsions and interact either with solvent molecules or with molecules adsorbed to the O/W interface. Polysaccharides contribute to emulsion stability by increasing the aqueous phase viscosity or gel formation.

Currently, the emulsions are being used for encapsulation of polyphenols and other labile compounds. Emulsions O/W were reported to encapsulate polyphenols from bayberry fruit extract (Zheng et al., 2011) and from bilberry frit extract (Betz et al., 2012), in both works polymers (ethyl cellulose and whey protein isolate, respectively) were used in order to stabilize the emulsion.

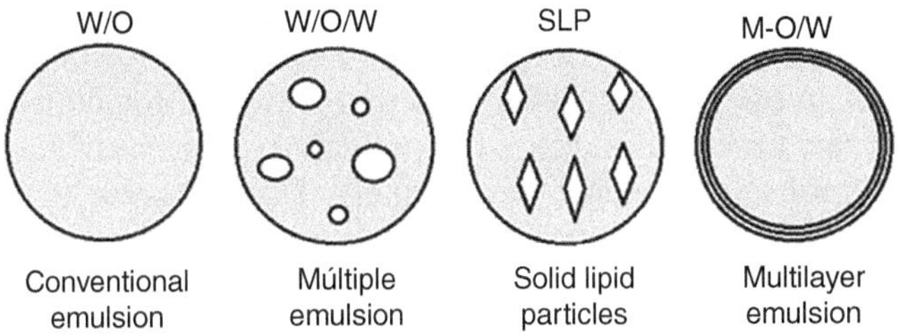

Figure 13.4. Types of emulsion.

The emulsion's functional characteristics can be controlled by monitoring the concentration and size distribution of the lipid particles, as well as the composition, thickness, and electric characteristics of the interfacial layer (Chung and McClements, 2014; Davidov-Pardo et al., 2012; Dima et al., 2015).

Fig. 13.4 shows the different emulsion types that can be used in the food industry.

Nanoemulsions of phenolic compounds, extracted from grape marc by high pressure, were formulated with sunflower oil (liquid), or palm oil (solid) like lipid phase, as well as the combination of emulsifier. These nanoemulsions were produced by high-pressure homogenization. The results obtained with sunflower oil-based nanoemulsions were the most stable because the droplet size had no significant variation. Moreover, the antioxidant activity was greater for the encapsulated grape marc polyphenols than for unencapsulated. Therefore, the nanoemulsions improved the delivery of polyphenols through the biological membranes (Sessa et al., 2013).

2.2.1.1 Multiple Emulsions

In water-in-oil-in-water ($W_1/O/W_2$) emulsions, small water droplets contained within larger oil droplets are dispersed within an aqueous medium. Multiple emulsions can be used to encapsulate hydrophilic components like polyphenols within the inner water phase (W_1), which may be useful for taste masking of bitter bioactive ingredients or if a system contains two hydrophilic components that would interact with one another.

These multiple emulsions are generally obtained using two stages. In the first stage, a W_1/O emulsion is formed by homogenizing water, oil, and an oil-soluble emulsifier together; in the second stage, a $W_1/O/W_2$ emulsion is obtained by homogenizing the W_1/O emulsion with water containing a water-soluble emulsifier. The secondary homogenization step is usually carried out using lower energy intensity than the primary step so as not to break the initial W_1/O emulsion.

W/O/W emulsions have several advantages over simple O/W emulsions, as delivery systems like water-soluble active food compounds can be trapped within the inner water phase, which may have advantages for a number of applications like controlled rate or in response to specific environmental triggers, for example, in the mouth, stomach, or small intestine. In addition, bioactive agents like polyphenols could be protected from chemical degradation by isolating them from other water-soluble ingredients. In addition, a W/O/W emulsion can be produced that has the same overall dispersed phase volume fraction and droplet size distribution as a conventional O/W emulsion, but with a reduced fat content; therefore, it should be possible to obtain reduced-fat products with similar physicochemical and sensory characteristics to full-fat products, including appearance, texture, mouth feel, and flavor.

Multiple emulsions are thermodynamically unstable systems that break down over time and therefore they must be carefully designed to give them sufficient kinetic stability for practical applications. $W_1/O/W_2$ emulsions break down due to the same mechanisms as conventional emulsions, plus some additional mechanisms, such as water diffusion and droplet expulsion from the oil droplets (Chung and McClements, 2014; McClements, 2012, 2015).

Other researchers studied polyphenols encapsulation by multiple emulsions. Encapsulation of polyphenols from bilberry fruit extract was performed by double emulsion (W/O/W), using PGPR as emulsifier and pectin (calcium chloride) as emulsion stabilizing agent (Frank et al., 2012). Hemar et al., 2010, showed that resveratrol could be encapsulated in W/O/W emulsions, and that it did not adversely affect the physicochemical properties of the emulsions, and that less than 10% of it was released into the external aqueous phase during storage. In other study (Matos et al., 2014) resveratrol was encapsulated in W/O/W emulsions formed using polyglycerol polyricinoleate (PGPR) as an lipophilic surfactant to form the primary W/O emulsion, and a combination Tween 20 and sodium carboxymethylcellulose as hydrophilic surfactants to form the W/O/W emulsions. The results of these multiple emulsions showed that they are physically stable for 30 days with $<10\%$ of resveratrol being released into the external aqueous phase.

Multiple nanoemulsions were formulated to encapsulate olive leaf extract in soybean oil. The primary W/O nanoemulsion had an average droplet size of 6.16 nm. Multiple emulsions were stabilized by WPC alone and WPC-pectin with average droplet size of 675 and 1443 nm. Olive leaf extract nanocapsules showed more antioxidant activity than unencapsulated olive leaf extract (Mohammadi et al., 2016).

2.2.1.2 Multilayer Emulsions

O/W emulsion attributes can be modified by building laminated coatings around the oil droplets using the layer-by-layer (LbL) deposition technique.

This method consists of the adsorption of a charged polyelectrolyte onto an oppositely charged surface through an electrostatic attraction. Charge reversal of surface is produced because the total number of charges on the adsorbed polyelectrolyte molecules is greater than the number of charges on the surface. Multilayer emulsion preparations formed by the LbL method are carried out in various steps. The first stage is the preparation of an O/W emulsion by homogenization. The droplets of this primary emulsion are electrically charged due to the ionic hydrophilic emulsifier adsorption.

An aqueous dissolution of an oppositely charged polyelectrolyte is added to the primary emulsion, obtaining a secondary emulsion. The oil droplets in the secondary emulsion are coated with a polyelectrolyte layer bound through electrical interaction to an ionic surfactant molecule layer. Repetition of these adsorption steps leads to the formation of multilayer emulsions. These layers are maintained together by electrostatic attraction, and may therefore dissociate if pH or ionic strength is changed. Dissociation can be prevented by covalently cross-linking the adsorbed layers after they have been produced around a lipid droplet—eg, using enzymes, chemicals, or heating.

Bioactive compounds like polyphenols could be incorporated within the oil droplet core or within the laminated shell surrounding the droplets. For example, a lipophilic active ingredient could be trapped into the oil phase prior to homogenization, whereas a hydrophilic ionic active ingredient could be applied to make up one of the layers surrounding the oil droplets. Release of a bioactive compound incorporated in the core of the emulsion could be controlled by designing the response of the shell to the medium. Therefore, these systems ensure the release of bioactive ingredients in different parts of the gastrointestinal tract.

Another important advantage of this type of emulsions is that interfacial layers produce better emulsion stability and thus more protection for active ingredient incorporated within the core. (Chung and McClements, 2014; Dima et al., 2015).

In the work of Berendsen et al. (2015), a procyanidin-rich extract was encapsulated in a water-in-oil-in-water emulsion produced by premix membrane emulsification. This emulsion was stable for up to 14 days, where the procyanidin release was measured. A thicker interfacial WPI-Chi layer (4.2 nm versus 2.2 and 3.9 nm for WPI-CMC and WPI-GA, respectively) resulted in a lower procyanidin release.

2.2.1.3 Solid Lipid Particles

Solid lipid particle (SLP) emulsions consist of emulsified (partially) coated solid lipid particles dispersed in an aqueous continuous phase. Due to crystallizing in the lipid phase, it is often possible to slow down molecular diffusion processes, thereby increasing the stability of chemically labile ingredients and controlling the physical location of a lipophilic agent. SLP emulsions are generally formed using a "hot homogenization" process that involves homogenizing an oil and water phase with a hydrophilic surfactant at a temperature above the melting point of the lipid phase. The emulsion is then cooled so that some or all of the lipids within the droplets crystallize. Therefore, it is very important that the temperature of the emulsion remains substantially above the crystallization temperature of the highest melting lipid to prevent any fat solidification during homogenization.

The first process in encapsulating a bioactive agent is to dissolve the compound in the melted lipid carrier: this mixture is dispersed in a water-soluble emulsifier solution that has been heated to the same temperature as the melted lipid. Then, the emulsion is formed by a high-pressure homogenizer.

In these emulsions, the droplet size should be between 60–120 nm approximately and all lipid droplets should be covered by the emulsifier. Next, the emulsion is cooled down to solidify the lipid droplets to form solid lipid particles. The stability of the lipid particles depends mainly on cooling speed. The aggregation of SLP due to partial coalescence may rapidly occur, in addition to expulsion of the bioactive ingredient from the lipid matrix if the cooling speed is not controlled and the size of the emulsion droplets is too large. Also, a lot of lipid emulsifiers are applied to stabilize the initial emulsion and the cooling conditions.

For heat labile bioactive agents that may degrade when kept at an elevated temperature during the process, an alternative technique exists, known as "cold homogenization." In this method, the lipid is melted and mixed with the bioactive compound, but is then rapidly solidified using dry ice or liquid nitrogen. The high cooling rates favor a homogenous distribution of the functional component within the lipid matrix. Then, this solidified lipid mixture is milled in ball or mortar mills to produce particles with a size of between 50 and 100 μm. Next, the milled particles are suspended in a surfactant solution and homogenized at or below room temperature.

Solid lipid particles present more advantages than conventional lipid emulsions. These particles increase the chemical stability of the loaded active compound more than liposomes, by protecting the ingredient from processes like hydrolysis, oxidation, and photodegradation, while also enhancing bioavailability. Therefore, if chemical stability of the functional ingredient is a concern, solid lipid particle emulsions may prove a good choice and it is possible to obtain more precise control over the release kinetics of the bioactive ingredient.

It should be noted that SLP emulsions could also be used in multiple emulsions (by crystallizing the oil phase in a W/O/W emulsion) or in multilayer emulsions (by coating the lipid droplets with biopolymers) (McClements, 2015; Summerlin et al., 2015).

Neves et al. (2013) developed solid lipid nanoparticle (SLN) and nanostructured lipid carriers (NLCs) loaded with resveratrol to improve resveratrol oral bioavailability. These particles were formed by a modified hot homogenization method. The results obtained demonstrated both types of lipid nanoparticle as being stable during two months' storage and most of the resveratrol was released after incubation in simulated gastric and intestinal fluids. In addition, resveratrol did not change the overall physicochemical characteristics of either system, but it did cause a decrease in the crystalline order within the lipid phase of the nanoparticles. In another study, resveratrol was encapsulated by glyceryl behenate SLN to investigate the possibility of brain targeting. SLNs were prepared by a solvent evaporation technique employing high-speed homogenization followed by ultrasonication. These SLNs were formed with varying drug-lipid ratios (1:5–1:15) and the particle size and the encapsulation efficiency (EE) increased when varying the drug-lipid ratio from 1:5 to 1:15. The resveratrol SLN exhibited a sustained release in a phosphate buffer. In addition, the cytotoxicity assay showed that SLNs were as effective as free resveratrol as an antitumor agent. In addition, in vivo biodistribution was examined in rats and demonstrated that SLN significantly increased brain concentrations of resveratrol when compared with free resveratrol, 17.3 mg/g and 3.5 mg/g (Jose et al., 2014). It is also possible to combine the advantages of water-in-oil-in-water (W/O/W) multiple emulsions and solid lipid nanoparticles (SLN) with multiple solid particles (MLPs). Zhao et al. (2015) synthesized MLPs to encapsulate both coenzyme Q10 and tea polyphenols (CT-MLPs). CT-MLPs were prepared using the modified two-step emulsification process. The obtained CT-MLPs also showed high encapsulation efficiency, retention ratio, and stability during a 60-day stability study (Zhao et al., 2015).

2.2.2 Liposomes

Liposomes are spherical vesicular particles generally formed of concentric phospholipid bilayers dispersed in an aqueous medium (Fig. 13.5). These particles were described and synthesized for

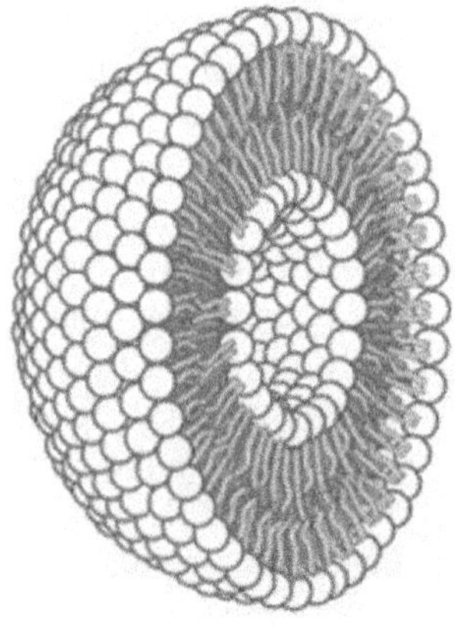

Figure 13.5. Liposome structure.

the first time in 1965 for targeted drug delivery, but interest in their application in the food industry has only recently emerged. The majority of microencapsulation techniques currently used in the food industry are based on biopolymer matrices composed of biopolymers like gums, proteins, sugars, dextrins, pectins, or other biopolymers, but liposomes have recently begun to gain in importance (Đorđević et al., 2014; Taylor et al., 2005).

The possibilities of application of this technique in the food industry have a lot of advantages. With this method, it is possible to capture water-soluble, lipid-soluble and amphiphilic ingredients. Also, liposomes can be synthesized using natural compounds such as egg, soy, dairy, or sunflower lecithin and other natural ingredients, enabling faster and easier implementation of the liposomes in food systems, overcoming regulatory barriers.

Although there has been little use of liposomes in this industry due to the interaction of liposomes with food ingredients being quite poorly understood, and manufacturing of liposomes being expensive due to the costs of raw materials and manufacturing procedures, nowadays, with the increasing understanding of the functional characteristics of liposomes—such as their physicochemical properties, their interplay with food components or lower costs of raw materials—it has become practicable to use liposomes to deliver bioactive ingredients such as polyphenols (Taylor et al., 2007).

Polyphenols are very varied molecules in which the number of rings and hydroxyl groups will considerably influence the polyphenol solubility in surfactants. So, the liposomal stability will be optimal depending on the polyphenol itself. In addition, the literature describing liposomal forms of polyphenol has been detailed for the food industry due to their antioxidant properties (Mignet et al., 2013).

The process for the formation of liposomes consists mainly of the hydrophilic–hydrophobic interactions between phospholipids and water molecules.

Bioactive compounds can be captured within their aqueous medium at a low yield, or within or connected to the membrane at a high yield. Depending upon the method of synthesis, liposomes can be formed in a range of sizes. Based on the size and the number of bilayers, liposomes can be identified as multilamellar (MLV: >1 bilayer), giant unilamellar (GUV: >0.5 mm), large unilamellar (LUV: >0.1 mm), or small unilamellar (SUV: <0.1 mm) vesicles. The liposomes formulated from nonionic synthesis surfactants are called niosomes and are formed in cheaper conditions. The unilamellar liposomes may involve only water-soluble ingredients in their molecule. The multilamellar liposomes may involve both lipophilic compounds inside the bilayer structure and hydrophilic compounds in the aqueous medium of liposome inside. Amphiphilic compounds may also be involved in liposomes. Liposomes and niosomes are formed by various methods such as lipid layer hydration, reversed phase evaporation, transmembrane pH gradient method, ether injection, microfluidization method, sonication, and extrusion. A number of factors should be taken into account when selecting the liposome formulation technique, such as the type of liposome formed, the phospholipid characteristics, the relationship between the phospholipids and the dispersion medium, the nature of the encapsulated bioactive molecules, and so forth. For example, the encapsulating efficiency of liposomes is highest when they are formed by freezing–thawing, followed by thin-film evaporation and then reverse phase evaporation, while melting and sonication has the lowest efficiency (Dima et al., 2015; Summerlin et al., 2015; Zuidam and Shimoni, 2010).

For example, salidroside (rhodioloside) is a glucoside of tyrosol found in the plant *Rhodiola rosea* and it is one of the compounds responsible for the antidepressive and anxiolytic actions of this plant. This compound was encapsulated by liposomal systems prepared by using various methods. The encapsulating efficiency of liposomes was highest when prepared by freezing–thawing, followed by thin film evaporation, then reverse phase evaporation: the lowest efficiency was with melting and sonication. Salidroside liposomes show a slower increase in particle size than liposomes without salidroside, suggesting salidroside plays an important role in preventing the aggregation and fusion of liposomes. These differences might come from the different morphologies of liposomes prepared by different methods (Fan et al., 2007).

Evidence of liposomes enhancing the bioactivity and bioavailability of polyphenols has been reported by a number of researchers. One demonstration that liposomes improve the bioactivity and the bioavailability of polyphenols is reported in this work, where curcumin was encapsulated in lecithin liposomes (LEC) by using a microfluidizer (Takahashi et al., 2009). Curcumin is an unstable compound and is poorly absorbed by the human gastrointestinal tract after oral administration. Liposome encapsulated curcumin (LEC) can be prepared from commercially available lecithins (SLP-PC70) and curcumin. The encapsulation efficiency for curcumin was 68.0% and was composed of small unilamellar liposomes with a diameter of approximately 263 nm. The most important factor in this work is that oral administration of the LECs favored the intestinal absorption of curcumin, leading to an increase in the plasma antioxidant activity.

2.2.3 Coacervation

Coacervation is a modified emulsification technique. It is a relatively simple technology, balancing the electrostatic interaction between the two compounds of the encapsulation emulsion to create water- and heat-resistant nanocapsules. Coacervation is an aqueous-phase separation procedure, where a complex is formed when a solution of a bioactive ingredient is combined with a matrix molecule which leads to deposition of the newly formed coacervate phase around the bioactive component suspended or emulsified in the same reaction medium.

There are two types of coacervation: simple or complex coacervation. Simple coacervation involves only one type of polymer, with the addition of a strongly water-soluble agent to the colloidal solution. For complex coacervation, two or more types of polymer are applied. The first stage in a typical complex coacervation procedure is the suspension or emulsification of the core material in either gelatin or gum arabic solution. When a solution of the core agent is mixed with an oppositely charged encapsulating material, a complex, resulting in phase segregation and associative complexation, is formed; the charges must be large enough to induce interaction, but not too large to avoid the precipitation (Fig. 13.6). Thus, the core agent used in coacervation must be compatible with the recipient polymer and be insoluble in the coacervation medium. The coacervates are generally further stabilized by thermal treatment, cross-linking or desolvation methods. The procedure of complex coacervation is divided into three stages: formation of three immiscible phases, deposition of the coating and solidification of the coating.

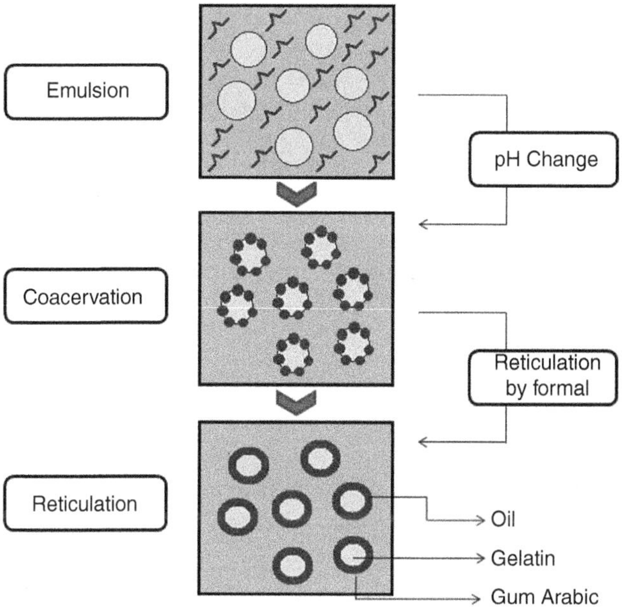

Figure 13.6. Schematic view of coacervation process.

By changing the temperature, type of matrix, ion concentration or ratio of bioactive ingredient, it is possible to obtain capsules with different characteristics and sizes (Fang and Bhandari, 2010; Madene et al., 2006; Munin and Edwards-Lévy, 2011; Onwulata, 2012; Đorđević et al., 2014).

In most cases, coacervates are protein/polysaccharide components, but protein–protein mixtures were also under scrutiny. The most common and usually used pair in complex coacervation is gelatin and gum arabic. Two oppositely charged biopolymer compounds are combined at a pH under the protein isoelectric point (pI), producing a separation of phase rich in biopolymers and a posterior complex, which precipitates and the bioactive ingredient is maintained in the coacervate phase. For example, to form stable suspension of hydrated gelatin and gum arabic, is needed a decrease of pH until 4.3 to have opposite charges that permit the step of separation. So, the whole coacervation procedure relies on protein–polysaccharide interactions that are governed by pH change.

The micro and nanocapsules formed by coacervation are slightly soluble in water, are temperature and mechanical shock resistant, have high retention efficiency and a good active component release.

Finally, these coacervated capsules can be dried by drying procedures like spray-drying or freeze-drying and are involved in the final characteristics of these micro- or nanocapusles whose final forms may change and not be round (that is the normal form of these particles). (Dima et al., 2015; Fang & Bhandari, 2010; Đorđević et al., 2014).

This technology can be applied to increase the potential benefits of labile functional compounds, such as the encapsulation of polyphenols. For example, yerba mate extract (rich in antioxidant components) was encapsulated with calcium alginate and calcium alginate-chitosan. This study shows the influence of coating materials on the release of the polyphenols of yerba mate extract, because the release in water was achieved in a shorter time for chitosan coated beads than with the alginate beads (Deladino et al., 2008).

Also, tea polyphenols were nanoencapsulated by complex coacervation for efficient delivery of these compounds. The first stage was formulating a conjugate gelatin–dextran obtained by Maillard reaction. This conjugate was mixed with tea polyphenols and core micelles of complex coacervation (C3Ms) were formed. The average diameter was 86 nm and the polyphenol release was sustained (Zhou et al., 2012).

Despite the obvious advantages of complex coacervation, until recently this technology has not been commonly used in the food industry, due to it being a complex and expensive technique. The high price of coacervation encapsulation might be accepted if the complex coacervates provide unique characteristics that are not attainable without them.

There are several factors, such as wall material concentration, emulsification process, and coacervation procedure, that should be optimized to increase the yield of micro- and nanocapsules. In addition, this technology is more suited to hydrophobic components as core materials than encapsulation of hydrophilic components. To encapsulate hydrophilic components, it is necessary to make some changes to the technique. Other limitations of coacervation are the evaporation of volatile compounds or dissolution of the bioactive ingredient into the processing solvent due to residual core ingredients sometime clinging to the exterior of the capsule (Madene et al., 2006; Đorđević et al., 2014).

2.2.4 Molecular Inclusion

Molecular inclusion is a technology that uses cyclodextrins as wall materials. Cyclodextrins are a type of cyclic oligosaccharide derived from starch, with six, seven, or eight glucose residues linked by a 1-4 glycosidic bond in a cylinder-shaped structure.

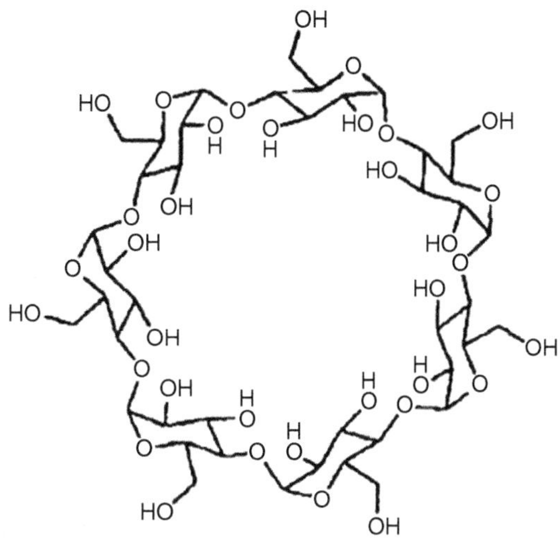

Figure 13.7. Cyclodextrin.

Three types of cyclodextrin exist: α-cyclodextrin, β-cyclodextrin, and γ-cyclodextrin. These compounds are absorbed in the upper gastrointestinal tract and are metabolized by the colon microflora. β-cyclodextrin is the most often used because the purification of α- and γ-cyclodextrins significantly increases their cost of production. In addition, only β-cyclodextrins are accepted globally with a maximum level in foods of 5 mg/kg per day according to Joint FAO/WHO Expert Committee on Food Additives (JECFA) (Fig. 13.7).

The inclusion complexes are synthesized by interactions between components in which a smaller guest compound fits into and is surrounded by the lattice of cyclodextrins.

The inner cavity of β-cyclodextrin creates a relatively hydrophobic environment, whereas its external surface has a hydrophilic property. This structure determines the physicochemical attributes of cyclodextrins. Thus, these molecules permit total or partial inclusion of bioactive ingredients like polyphenols. For the retention of polyphenols, the steric hindrance, polarity and volatility of the wall material, chemical functionality, weight, and shape can be influential.

For complexing β-cyclodextrin with polyphenol components, the techniques most often applied are to mix a cyclodextrin with an active compound in an aqueous medium and filter off the precipitated complex; or to blend solid cyclodextrin with guest molecules in a powerful mixer, making a solution of cyclodextrin; or to knead the active component with the cyclodextrin–water paste (Fang & Bhandari, 2010; Madene et al., 2006).

One important disadvantage of cyclodextrins is their lower water solubility compared to other oligosaccharides. β-cyclodextrins are less soluble than other cyclodextrins like α- and γ-cyclodextrins (1.85 g/100 mL for β-cyclodextrins and 14.5 g/100 mL and 23.2 g/100 mL for α- and γ-CDs, respectively). A possible solution to this limitation is the development of highly water-soluble cyclodextrin derivatives. These derivatives are methylated, sulfobutylated and hydroxypropylated cyclodextrins. However, inclusion of polyphenols improves their water solubility, notably for the less water-soluble phytochemicals compounds (Đorđević et al., 2014).

Curcumin was encapsulated by cyclodextrin complex and the anti-inflammatory and antiproliferative effects were observed. The results obtained were that cyclodextrins of curcumin are more active than free curcumin in the inhibiting of inflammatory factors. Also, this encapsulated curcumin was more effective than free curcumin in inducing apoptosis of leukemic cells (Yadav et al., 2010).

In another study, complexes of β-cyclodextrin with olive and olive oil polyphenols were synthesized and investigated by NMR spectroscopy and thermodynamical molecular dynamic studies to verify the molecular inclusion. This study shows that polyphenol compounds are captured with β-cyclodextrin; this encapsulation decreases the bitter taste and protects the polyphenols against environmental reactions during storage (Rescifina et al., 2010).

Rutin, phloridzin, and chlorogenic acid are the most important polyphenols found in apples and their products. Ramirez et al. (2014) studied the encapsulation of these polyphenols by β-cyclodextrin nanosponges and the bioavailability increased.

2.2.5 Ionic Gelation

Ionic gelation technology is based on extruding polymer solution in water through a syringe needle or a nozzle. These droplets, with the bioactive ingredient, are dissolved in a dispersant phase and spherical gel particles are formed.

Chitosan nanoparticles (carboxymethyl and chitosan hydrochloride), formed by this method, have been synthesized to capture a tea polyphenol extract. Carboxymethyl chitosan and chitosan hydrochloride are two different water-soluble chitosans, anionic and cationic respectively, which form nanoparticles through ionic gelation between the carboxyl groups of carboxymethyl chitosan and the amine groups of chitosan hydrochloride in a water solution. The results obtained show that in in vitro studies, sustained and controlled release of tea polyphenols was achieved, and they can be a highly promising cancer drug carrier system, where tea polyphenols have shown interesting antitumor characteristics

(Liang et al., 2011). Also, *Elsholtzia splendens* extract was encapsulated by ionic gelation. *E. splendens* is a plant genus in the Lamiaceae (mint family) and is mainly distributed across eastern Asia. *E. splendens* extract-loaded chitosan nanoparticles were prepared by ionic gelation between the protonated amino groups of chitosan and anions of tripolyphosphate (TPP). The antioxidant activity of free *E. splendens* extract and *E. splendens* extract-loaded chitosan nanoparticles was measured, the latter showing a more effective inhibitory activity on lipid peroxidation. Thus, ionic gelation is a potential method for improving the antioxidant activity of *E. splendens* extract (Lee et al., 2010). The same method was used more recently to encapsulate naringenin, a citrus flavonone and a potential anti-inflammatory agent. The nanoparticles were synthesized by chitosan and tripolyphosphate (TPP) as a cross-linker. This study showed the better performance of chitosan encapsulated naringenin over free naringenin, and suggested an efficient system for delivering this flavonone with antioxidant and anticancer attributes (Kumar et al., 2015).

Besides chitosan, other materials are used for encapsulation of polyphenols by ionic gelation. For example, extract of polyphenols of pomegranate peels was encapsulated by sodium alginate. It is an anionic polymer that can be easily cross-linked with calcium chloride; this is because the calcium ions are bound to carboxylate residues of both mannuronic acid and guluronic acid, which are components of sodium alginate. The study showed that this technique allowed the encapsulation of 43.90% of the total extracted polyphenols and 46.34% of the total extracted proanthocyanidins. Therefore, calcium alginate nanocapsules also could be a good carrier for polyphenols (Zam et al., 2014).

2.2.6 Yeast Encapsulation

This technology uses yeast cells (*Saccharomyces cerevisiae*) to encapsulate bioactive ingredients like polyphenols. The main advantage of this method is the cost effectiveness of the encapsulation procedure. This yeast strain occurs as a by-product of fermentation procedures and can be readily produced on a large scale. Therefore, *S. cerevisiae* holds a great interest for commercial exploitation of this biomass.

For the yeast encapsulation procedure, apart from yeast cells, water, and an active agent are necessary. However, encapsulation in yeast requires that the bioactive agent crosses the cover of the yeast. Although the yeast envelope is a structure to protect the bioactive compound, it is also inconvenient because the food bioactive ingredients should cross through the plasma membrane without irreversible changes occurring in the yeast cell.

A new way to capture the bioactive agent in yeast cells is through osmoporation of yeast cells by water glycerol solution, although the bioactive component is hydrophilic (Pedrini et al., 2013). This method could be used for the encapsulation of polyphenol molecules because osmoporation involves a permeabilization in the rehydration phase by water–glycerol solution (Đorđević et al., 2014).

There is a high number of examples of polyphenols encapsulated by yeast encapsulation. For example, photosensitive resveratrol was encapsulated in yeast cells. The storage stability and the release in simulated gastric fluid were investigated. In addition, the scavenging capacity of encapsulated resveratrol versus DPPH radical was compared with that of unencapsulated resveratrol. The results obtained showed that resveratrol encapsulated by yeast presented high bioavailability due to the increasing solubility of resveratrol, and higher radical-scavenging activity than unencapsulated resveratrol (Shi et al., 2008).

In other work, curcumin was encapsulated in baker's yeast (*S. cerevisiae*) cells, β-cyclodextrin (β-CD) and modified starch (MS). These three methods were compared in terms of storage stability and the release in simulated gastric and pancreatic fluid. Yeast encapsulation was the better encapsulation technique in this study. In yeast microcapsules, a slow and prolonged release occurred, while the other encapsulation methods observed a rapid dissolution of curcumin. Also, this study showed that the envelope of the cell that surrounds the plasma membrane in the yeast cells permitted more protection of the curcumin than other methods, such as, β-cyclodextrin (β-CD) and modified starch (MS), against deleterious photochemical reactions and against heat degradation following isothermal (inert or oxidative) heating at 20°C (Paramera et al., 2011).

Different encapsulation methods described above could also be combined in order to design micro and nanoparticles with a specific application.

3 Controlled Release of Polyphenols in the Gut

As already mentioned, polyphenols are compounds with functional activities, such as anti-cancer, antioxidant, cardio and neuroprotection; but these therapeutic potentials are limited due to their slow bioavailability and low concentrations at the target site.

After ingestion, loaded nanoparticles reach the gastrointestinal tract and remain in contact with the intestinal epithelium for approximately 4–8 h. Conditions such as the pH, presence of enzymes, interactions with other nutrients, or insufficient gastric time, low permeability or insolubility within the gut, may affect the digestibility and availability of the encapsulated bioactive agents, or polyphenols (Armstrong and Bharali, 2013; Onwulata, 2012).

For target release of polyphenol compound in the gut and to facilitate the dissolution of capsules in specific parts of the gastrointestinal tract, it is necessary to consider three factors. The first is the strong peristaltic waves in the colon. In the upper part of the gastrointestinal tract, the pressure is lower due to the fluids in the stomach and the small intestine. By increasing the mechanical resistance of nanocapsules so that they can withstand the pressure in the stomach and small intestine, we can facilitate the release of active agents in the lower part of the gut. Moreover, other mechanisms such as pH changes and the time of transport in the gastrointestinal tract should be useful to obtain a good release of bioactive compounds. The pH in the stomach is acidic and grows in the small and large intestine. The use of pH sensitive polymers that remain intact in the stomach and will not be attacked by digestive enzymes can facilitate the release of bioactive compounds in specific parts of the small or large intestine. The third method is the activity of enzyme systems produced by the microbiota. The bacterial number in the gut and the associated enzyme activities are specific for different parts of the gut and allow for precise delivery of bioactive ingredients (De Vos et al., 2010).

The design of novel colon-targeted delivery systems based on natural biodegradable polymers has recently gained importance in the therapy of colon-based diseases. Several formulations are designed to resist the release of the bioactive ingredient in the stomach, with an additional nondisintegration or lag phase included in the formulation so that the release of the drug takes place in the colon. However, the polyphenols could also have a local action in different sites of gastrointestinal tract, acting as antioxidants and therefore preventing the oxidation of lipids and proteins.

3.1 Materials to Enable Target Delivery in the Colon

A great number of polysaccharides have been studied for their characteristics as colon-specific bioactive ingredient carrier systems, such as chitosan, sodium alginate, pectin, chondroitin sulfate, cyclodextrin, dextran, guar gum, inulin, amylose, and locust-bean gum. These are broken down by the colonic microflora to simple saccharides. However, their high water solubility is a problem in the use of these polysaccharides. A wonderful possibility is to modify the solubility while still retaining their biodegradability.

For example, chitosan is soluble at low pH ranges. For good use in colon-specific delivery, an enteric layer is needed over the chitosan that will protect it from the acidity of the stomach. When the formulation arrives in the intestine, the pH increases and the enteric layer dissolves, releasing the chitosan-coated core. These cores are acted upon by the microflora of the colon, degrading the chitosan and releasing the bioactive compound.

Another water-soluble polysaccharide is pectin. This polysaccharide is not able to shield its bioactive ingredient load effectively during its passage through the stomach and small intestine. So, less water-soluble derivatives of pectin have been developed that are degradable by the colonic microflora. Most of the pectin/modified pectins gave better colon specific release with hydrophobic bioactive agents (Kosaraju, 2005; Sinha and Kumria, 2001).

In addition, to overcome the bioavailability problems, advanced nanoparticulate carriers are designed to provide localized or targeted delivery of these agents, and may represent a more viable therapeutic option. Several carriers—, for example, liposomes, polymeric nanoparticles, solid lipid nanoparticles, emulsions systems, and others described earlier—have shown noteworthy improvements in the preventive/therapeutic activities of many antioxidants by increasing their bioavailability and target ability.

3.2 Examples of Controlled Delivery of Polyphenols

Anthocyanins are flavonoid pigments found in red/purplish fruits and vegetables. These compounds have demonstrated activity against colon cancer and inflammatory gut diseases. However, most polyphenols are low in vivo bioavailability, limiting their application in cancer prevention and treatment. Therefore, several techniques are applied to encapsulate these bioactive components. For example, a study was conducted to encapsulate these bioactive agents in shellac and shellac/hydroxypropyl methylcellulose (HPMC) by ionotropic gelation and fluid bed drying. Simulated gastric fluid, ileostomy fluid, and colostomy fluid revealed a retardation of anthocyanins during simulated passage through the stomach and ileum as well as the desired release of pigments in the colon (Oehme et al., 2011). In another study, water-in-oil-in-water emulsions were applied to capture bilberry extract rich in anthocyanins in the inner water phase, and the stability of anthocyanins could be shown after an in vitro gastrointestinal passage. In fact, the exterior (O/W) emulsifier played an important role in the stability of multiple emulsions in gastrointestinal conditions and the location of release (Frank et al., 2012). Microencapsulated blueberry extract anthocyanins with whey protein isolate or gum arabic by spray-drying were studied in vitro digestion. The results showed that the site of release depends on the polymer nature. Microparticles with gum arabic had high release rate in gastric digestion site; whereas those microparticles with whey protein had gradual anthocyanins release rate and sustained antioxidant activity throughout the digestion tract (Flores et al., 2014).

Oidtmann et al. (2012) compared three bilberry extract (BE) anthocyanin encapsulation systems: pectin amide with calcium chloride by extrusion (1800 μm); whey protein isolate by emulsification/heat gelation, resulting in W/O emulsion (180 μm) and pectin amide with maltodextrin by spray-drying and coating with shellac (250–500 μm) with nanoencapsulated BE. All systems were not able to prevent the release of anthocyanins from encapsulated BE in gastric fluid (SFG) and intestinal fluid (FeSSIF). However, anthocyanins remained constant in SFG for 120 min, and they were degraded in FeSSIF. These results show that the design of the particles, mainly encapsulation method and properties of coating material, determine the applicability of the particles in the gastrointestinal tract (localized action and/or absorption).

Trans-resveratrol was nanoencapsulated for prostate cancer treatment. Nanoparticles were formulated with a blend of poly(epsilon-caprolactone) (PCL) and poly(D, L-lactic-co-glycolic acid)-poly(ethylene glycol) conjugate (PLGA-PEG-COOH) by a nanoprecipitation technique. Trans-resveratrol nanocapsules were able to control the bioactive ingredient release at pH 6.5 and 7.4. In addition, in gastrointestinal simulated fluids, nanocapsules released about 55% of resveratrol in the first 2 h in an acidic medium, and their total bioactive agent content within the subsequent 5 h at pH 7.4. Therefore, these nanoparticles have a potential use for chemoprevention or chemotherapy of prostate cancer (Sanna et al., 2013).

4 Conclusions

The use of nanoencapsulated food ingredients for controlled-release applications is a promising alternative to solve the major problem of unstable components like polyphenols in the food industry. The challenges are to select the appropriate nanoencapsulation technology and coating material. Despite the wide range of encapsulating ingredients that have been developed, manufactured, and successfully marketed in the pharmaceutical and cosmetic industries, nanoencapsulation has found a comparatively much smaller market in the food industry (Desai and Jin Park, 2005).

Presently, technologies with higher effectiveness with regard to preservation of stability, bioactivity and bioavailability of polyphenols are spray-drying, freeze-drying, emulsions, inclusion complexes and liposomes, unlike fluid bed drying, spinning disc and spray-cooling/-chilling technologies. In this sense, spraying technologies are already well established on an industrial scale and are expected to have a dominant role in the near future. Others, like yeast encapsulation, are low-cost processes, but today are not widely used for encapsulation of polyphenols. However, complex systems based on both methods are being developed for controlled release of bioactive compounds. Indeed, the food industry expects encapsulates to fulfill many demands: therefore two or more encapsulation processes to obtain encapsulates with superior functional characteristics are being explored.

References

Apintanapong, M., Noomhorm, A., 2003. The use of spray drying to microencapsulate 2-acetyl-1-pyrroline, a major flavour component of aromatic rise. Int. J. Food Sci. Technol. 38 (2), 95–102.

Armstrong, D., Bharali, D., 2013. Oxidative Stress and Nanotechnology. Springer, New York, 1028.

Bakowska-Barczak, A.M., Kolodziejczyk, P.P., 2011. Black currant polyphenols: their storage stability and microencapsulation. Ind. Crop. Prod. 34 (2), 1301–1309.

Barbosa-Pereira, L., Angulo, I., Lagarón, J.M., Paseiro-Losada, P., Cruz, J.M., 2014. Development of new active packaging films containing bioactive nanocomposites. Innov. Food Sci. Emerg. Technol. 26, 310–318.

Baum, M., Schantz, M., Leick, S., Berg, S., Betz, M., Frank, K., et al., 2014. Is the antioxidative effectiveness of a bilberry extract influenced by encapsulation? J. Sci. Food Agric. 94, 2301–2307.

Berendsen, R., Güell, C., Ferrando, M., 2015. A procyanidin-rich extract encapsulated in water-in-oil-in-water emulsions produced by premix membrane emulsification. Food Hydrocolloid. 43, 636–648.

Betz, M., Steiner, B., Schantz, M., Oidtmann, J., Mäder, K., Richling, E., et al., 2012. Antioxidant capacity of bilberry extract microencapsulated in whey protein hydrogels. Food Res. Int. 47, 51–57.

Chiou, D., Langrish, T.A.G., 2007. Development and characterization of novel nutraceuticals with spray drying technology. J. Food Eng. 82 (1), 84–91.

Chung, C., McClements, D.J., 2014. Structure-function relationships in food emulsions: improving food quality and sensory perception. Food Struct. 1 (2), 106–126.

Da Silva Pedrini, M.R., Dupont, S., De Anchieta Câmara, A., Beney, L., Gervais, P., 2014. Osmoporation: a simple way to inernalize hydrophilic molecules into yeast. Appl. Microbiol. Biotechnol. 98 (3), 1271–1280.

Davidov-Pardo, G., Moreno, M., Arozarena, I., Marín-Arroyo, M.R., Bleibaum, R.N., Bruhn, C.M., 2012. Sensory and consumer perception of the addition of grape seed extracts in cookies. J. Food Sci. 77 (12), S430–S438.

De Vos, P., Faas, M.M., Spasojevic, M., Sikkema, J., 2010. Encapsulation for preservation of functionality and targeted delivery of bioactive food components. Int. Dairy J. 20 (4), 292–302.
Deladino, L., Anbinder, P.S., Navarro, A.S., Martino, M.N., 2008. Encapsulation of natural antioxidants extracted from *Ilex paraguariensis*. Carbohyd. Polym. 71 (1), 126–134.
Desai, K.G.H., Jin Park, H., 2005. Recent developments in microencapsulation of food ingredients. Dry. Tech. 23, 1361–1394.
Dewettinck, K., Huyghebaert, A., 1999. Fluidized bed coating in food technology. Trends Food Sci. Tech. 10 (4–5), 163–168.
Dima, Ş., Dima, C., Iordăchescu, G., 2015. Encapsulation of functional lipophilic food and drug biocomponents. Food Eng. Rev. 7 (4), 417–438.
Đorđević, V., Balanč, B., Belščak-Cvitanović, A., Lević, S., Trifković, K., Kalušević, A., Nedović, V., 2014. Trends in encapsulation technologies for delivery of food bioactive compounds. Food Eng. Rev. 7(4), 452–490.
Ersus, S., Yurdagel, U., 2007. Microencapsulation of anthocyanin pigments of black carrot (*Daucus carota* L.) by spray drier. J. Food Eng. 80, 805–812.
Fan, M., Xu, S., Xia, S., Zhang, X., 2007. Effect of different preparation methods on physicochemical properties of salidroside liposomes. J. Agric. Food Chem. 55 (8), 3089–3095.
Fang, Z., Bhandari, B., 2010. Encapsulation of polyphenols—a review. Trends Food Sci.Tech. 21 (10), 510–523.
Fang, Z., Bhandari, B., 2011. Effect of spray drying and storage on the stability of bayberry polyphenols. Food Chem. 129 (3), 1139–1147.
Fang, Z., Bhandari, B., 2012. Comparing the efficiency of protein and maltodextrin on spray drying of bayberry juice. Food Res. Int. 48, 478–483.
Fazaeli, M., Emam-Djomeh, Z., Ashtari, A.K., Omid, M., 2012. Food and bioproducts processing effect of spray-drying conditions and feed composition on the physical properties of black mulberry juice powder. Food Bioprod. Process 90, 667–675.
Ferrari, C.C., Pimentel, S., Germer, M., de Aguirre, J.M., 2012. Effects of spray-drying conditions on the physicochemical properties of blackberry powder effects of spray-drying conditions on the physicochemical properties of blackberry powder. Dry. Tech. 30, 154–163.
Flores, F.P., Singh, R.K., Kerr, W.L., Pegg, R.B., Kong, F., 2014. Total phenolics content and antioxidant capacities of microencapsulated blueberry anthocyanins during in vitro digestion. Food Chem. 153, 272–278.
Frank, K., Walz, E., Gräf, V., Greiner, R., Köhler, K., Schuchmann, H.P., 2012. Stability of anthocyanin-rich W/O/W-emulsions designed for intestinal release in gastrointestinal environment. J. Food Sci. 77 (12), 50–57.
Gavini, E., Alamanni, M.C., Cossu, M., Giunchedi, P., 2005. Tabletted microspheres containing *Cynara scolymus* (var. Spinoso sardo) extract for the preparation of controlled release nutraceutical matrices. J. Microencapsul. 22 (5), 487–499.
Gharsallaoui, A., Roudaut, G., Chambin, O., Voilley, A., Saurel, R., 2007. Applications of spray-drying in microencapsulation of food ingredients: an overview. Food Res. Int. 40 (9), 1107–1121.
Gibbs, B.F., Kermasha, S., Alli, I., Mulligan, C.N., 1999. Encapsulation in the food industry: a review. Int. J. Food Sci. Nutr. 50 (3), 213–224.
Gouin, S., 2004. Microencapsulation: industrial appraisal of existing technologies and trends. Trends Food Sci. Technol. 15 (7–8), 330–347.
Guevara, B., 2008. Encapsulación técnicas y aplicaciones en la Ind alimentos 2008. pdf. Temas Selectos Ingeniería de Alimentos/Universidad de Las Américas.

Gurak, P.D., Cabral, L.M.C., Rocha-Leão, M.H., 2013. Production of grape juice powder obtained by freeze drying after concentration by reverse osmosis. Brazilian Arch. Biol. Technol. 56 (6), 1011–1017.

Guzman-Villanueva, D., El-Sherbiny, I.M., Herrera-Ruiz, D., Smyth, H.D.C., 2013. Design and in vitro evaluation of a new nano-microparticulate system for enhanced aqueous-phase solubility of curcumin. BioMed. Res. Int. 2013 (1), 724763.

Hemar, Y., Cheng, L.J., Oliver, C.M., Sanguansri, L., Augustin, M., 2010. Encapsulation of resveratrol using water-in-oil-in-water double emulsions. Food Biophys. 5 (2), 120–127.

Idham, Z., Muhamad, I.I., Sarmidi, M.R., 2012. Degradation kinetics and color stability of spray-dried encapsulated anthocyanins from *Hibiscus sabdariffa* L. J. Food Process Eng. 35 (4), 522–542.

Ishwarya, S.P., Anandharamakrishnan, C., Stapley, A.G.F., 2015. Spray-freeze-drying: A novel process for the drying of foods and bioproducts. Trends Food Sci. Technol. 41 (2), 161–181.

Jiménez, T., Turchiuli, C., Dumoulin, E., 2006. Particles agglomeration in a conical fluidized bed in relation with air temperature profiles. Chem. Eng. Sci. 61 (18), 5954–5961.

Jiménez-Aguilar, D.M., Ortega-Regules, A.E., Lozada-Ramírez, J.D., Pérez-Pérez, M.C.I., Vernon-Carter, E.J., Welti-Chanes, J., 2011. Color and chemical stability of spray-dried blueberry extract using mesquite gum as wall material. J. Food Compos. Anal. 24, 889–894.

Jing, D., Zhen-Zhen, G., Ze, X., Bo, Z., Zhang, Y., Chun-Mei, L., 2014. Comparison of the efficiency of five different drying carriers on the spray drying of persimmon pulp powders. Dry. Tech. 32 (10), 1157–1166.

Jose, S., Anju, S.S., Cinu, T.a., Aleykutty, N.a., Thomas, S., Souto, E.B., 2014. In vivo pharmacokinetics and biodistribution of resveratrol-loaded solid lipid nanoparticles for brain delivery. Int. J. Pharmaceut. 474 (1–2), 6–13.

Kosaraju, S.L., 2005. Colon-targeted delivery systems: review of polysaccharides for encapsulation and delivery. Crit. Rev. Food Sci. 45 (4), 251–258.

Kosaraju, S.L., Labbett, D., Emin, M., Konczak, I., Lundin, L., 2008. Delivering polyphenols for healthy aging. Nutr. Dietetics 65, S48–S52.

Krishnaiah, D., 2009. Optimal operating conditions of spray dried noni fruit extract using k–carrageenan as adjuvant. J. Appl. Sci. 9, 3062–3067.

Kropat, C., Betz, M., Kulozik, U., Leick, S., Rehage, H., Boettler, U., et al., 2013. Effect of microformulation on the bioactivity of an anthocyanin-rich bilberry pomace extract (*Vaccinium myrtillus* L.) in vitro. J. Agric. Food Chem. 61, 4873–4881.

Kumar, S.P., Birundha, K., Kaveri, K., Devi, K.T.R., 2015. Antioxidant studies of chitosan nanoparticles containing naringenin and their cytotoxicity effects in lung cancer cells. Int. J. Biol. Macromol. 78, 87–95.

Laine, P., Kylli, P., Heinonen, M., Jouppila, K., 2008. Storage stability of microencapsulated cloudberry (*Rubus chamaemorus*) phenolics. J. Agric. Food Chem. 56 (23), 11251–11261.

Lee, J.S., Kim, G.H., Lee, H.G., 2010. Characteristics and antioxidant activity of *Elsholtzia splendens* extract-loaded nanoparticles. J. Agric. Food Chem. 58 (6), 3316–3321.

Liang, J., Li, F., Fang, Y., Yang, W., An, X., Zhao, L., … Hu, Q., 2011. Synthesis, characterization and cytotoxicity studies of chitosan-coated tea polyphenols nanoparticles. Colloid. Surf. B, 82(2), 297–301.

Lopez-Quiroga, E., Antelo, L.T., Alonso, A.a., 2012. Time-scale modeling and optimal control of freeze-drying. J. Food Eng. 111 (4), 655–666.

Madene, A., Jacquot, M., Scher, J., Desobry, S., 2006. Flavour encapsulation and controlled release—a review. Int. J. Food Sci. Technol. 41 (1), 1–21.

Mahdavi, S.A., Jafari, S.M., Ghorbani, M., Assadpoor, E., 2014. Spray-drying microencapsulation of anthocyanins by natural biopolymers: a review. Dry. Tech. 32 (5), 509–518.
Matos, M., Gutiérrez, G., Coca, J., Pazos, C., 2014. Preparation of water-in-oil-in-water (W1/O/W2) double emulsions containing trans-resveratrol. Colloid. Surf. A 442, 69–79.
McClements, D.J., 2012. Edible delivery systems for nutraceuticals: designing functional foods for improved health. Ther. Deliv. 3 (7), 801–803.
McClements, D.J., 2015. Encapsulation, protection, and release of hydrophilic active components: potential and limitations of colloidal delivery systems. Adv. Colloid Interf. Sci. 219, 27–53.
Mignet, N., Seguin, J., Chabot, G.G., 2013. Bioavailability of polyphenol liposomes: a challenge ahead. Pharmaceutics 5 (3), 457–471.
Mohammadi, A., Jafari, S.M., Esfanjani, A.F., Akhavan, S., 2016. Application of nano-encapsulated olive leaf extract in controlling the oxidative stability of soybean oil. Food Chem. 190, 513–519.
Munin, A., Edwards-Lévy, F., 2011. Encapsulation of natural polyphenolic compounds: a review. Pharmaceutics 3, 793–829.
Nayak, C.A., Rastogi, N.K., 2010. Effect of selected additives on microencapsulation of anthocyanin by spray drying. Dry. Tech. 28, 1396–1404.
Neves, A.R., Lúcio, M., Martins, S., Lima, J.L.C., Reis, S., 2013. Novel resveratrol nanodelivery systems based on lipid nanoparticles to enhance its oral bioavailability. Int. J. Nanomed. 8, 177–187.
Oehme, A., Valotis, A., Krammer, G., Zimmermann, I., Schreier, P., 2011. Preparation and characterization of shellac-coated anthocyanin pectin beads as dietary colonic delivery system. Mole. Nutr. Food Res. 55 (Suppl. 1), 75–85.
Oidtmann, J., Schantz, M., Mäder, K., Baum, M., Berg, S., Betz, M., et al., 2012. Preparation and comparative release characteristics of three anthocyanin encapsulation systems. J. Agric. Food Chem. 60, 844–851.
Onwulata, C.I., 2012. Encapsulation of new active ingredients. Annu. Rev. Food. Sci. Technol. 3, 183–202.
Osorio, C., Acevedo, B., Hillebrand, S., Carriazo, J., Winterhalter, P., Morales, A.L., 2010. Microencapsulation by spray-drying of anthocyanin pigments from corozo (*Bactris guineensis*) fruit. J. Agric. Food Chem. 58, 6977–6985.
Ozdikicierler, O., Dirim, S.N., Pazir, F., 2014. The effects of spray drying process parameters on the characteristic process indices and rheological powder properties of microencapsulated plant (*Gypsophila*) extract powder. Powder Technol. 253, 474–480.
Paramera, E.I., Konteles, S.J., Karathanos, V.T., 2011. Stability and release properties of curcumin encapsulated in *Saccharomyces cerevisiae*, β-cyclodextrin and modified starch. Food Chem. 125 (3), 913–922.
Pérez-Masiá, R., López-Nicolás, R., Periago, M.J., Ros, G., Lagaron, J.M., López-Rubio, A., 2015. Encapsulation of folic acid in food hydrocolloids through nanospray drying and electrospraying for nutraceutical applications. Food Chem. 168, 124–133.
Prata, A.S., Maudhuit, A., Boillereaux, L., Poncelet, D., 2012. Development of a control system to anticipate agglomeration in fluidized bed coating. Powder Technol. 224, 168–174.
Quideau, S., Deffieux, D., Douat-Casassus, C., Pouységu, L., 2011. Plant polyphenols: Chemical properties, biological activities, and synthesis. Angew. Chem. Int. Ed. 50 (3), 586–621.
Ramirez-Ambrosi, M., Caldera, F., Trotta, F., Berrueta, L.Á., Gallo, B., 2014. Encapsulation of apple polyphenols in b-CD nanosponges. J. Incl. Phenom. Macro. 80, 85–92.

Rescifina, A., Chiacchio, U., Iannazzo, D., Piperno, A., Romeo, G., 2010. β-cyclodextrin and caffeine complexes with natural polyphenols from olive and olive oils: NMR, thermodynamic, and molecular modeling studies. J. Agric. Food Chem. 58 (22), 11876–11882.

Reyes, A., Evseev, A., Mahn, A., Bubnovich, V., Bustos, R., Scheuermann, E., 2011. Effect of operating conditions in freeze-drying on the nutritional properties of blueberries. Int. J. Food Sci. Nutr. 62 (3), 303–306.

Robert, P., Gorena, T., Romero, N., Sepúlveda, E., Chávez, J., Sáenz, C., 2010. Encapsulation of polyphenols and anthocyanins from pomegranate (*Punica granatum*) by spray drying. Int. J. Food Sci. Technol. 45, 1386–1394.

Rocha-Guzmán, N.E., Gallegos-Infante, J.A., González-Laredo, R.F., Harte, F., Medina-Torres, L., Ochoa-Martínez, L.A., Soto-García, M., 2010. Effect of high-pressure homogenization on the physical and antioxidant properties of *Quercus resinosa* infusions encapsulated by spray-drying. J. Food Sci. 75, N57–N61.

Ruiz-Gutiérrez, M.G., Amaya-Guerra, C.A., Quintero-Ramos, A., de Jesús Ruiz-Anchondo, T., Gutiérrez-Uribe, J.A., Baez-González, J.G., Campos-Venegas, K., 2014. Effect of soluble fiber on the physicochemical properties of cactus pear (Opuntia ficus indica) encapsulated using spray drying. Food Sci. Biotechnol., 23(3), 755–763.

Sáenz, C., Tapia, S., Chávez, J., Robert, P., 2009. Microencapsulation by spray drying of bioactive compounds from cactus pear (*Opuntia ficus-indica*). Food Chem. 114, 616–622.

Saikia, S., Mahnot, N.K., Mahanta, C.L., 2015. Optimisation of phenolic extraction from *Averrhoa carambola* pomace by response surface methodology and its microencapsulation by spray and freeze drying. Food Chem. 171, 144–152.

Saleh, K., Steinmetz, D., Hemati, M., 2003. Experimental study and modeling of fluidized bed coating and agglomeration. Powder Technol. 130 (1–3), 116–123.

Sanchez, V., Baeza, R., Galmarini, M.V., Zamora, M.C., Chirife, J., 2013. Freeze-drying encapsulation of red wine polyphenols in an amorphous matrix of maltodextrin. Food Bioprocess Technol. 6 (5), 1350–1354.

Sanna, V., Siddiqui, I.A., Sechi, M., Mukhtar, H., 2013. Resveratrol-loaded nanoparticles based on poly (epsiloncaprolactone) and poly (D, L-lactic-*co*-glycolic acid)-poly (ethylene glycol) blend for prostate cancer treatment. Mol. Pharm. 10 (10), 3871–3881.

Sansone, F., Picerno, P., Mencherini, T., Porta, A., Lauro, M.R., Russo, P., Aquino, R.P., 2014. Technological properties and enhancement of antifungal activity of a *Paeonia rockii* extract encapsulated in a chitosan-based matrix. J. Food Eng. 120 (1), 260–267.

Sessa, M., Casazza, A.a., Perego, P., Tsao, R., Ferrari, G., Donsì, F., 2013. Exploitation of polyphenolic extracts from grape marc as natural antioxidants by encapsulation in lipid-based nanodelivery systems. Food Bioprocess Technol. 6 (10), 2609–2620.

Shi, G., Rao, L., Yu, H., Xiang, H., Yang, H., Ji, R., 2008. Stabilization and encapsulation of photosensitive resveratrol within yeast cell. Int. J. Pharmaceut. 349 (1–2), 83–93.

Silva, P.I., Stringheta, P.C., Teófilo, R.F., de Oliveira, I.R.N., 2013. Parameter optimization for spray-drying microencapsulation of jaboticaba (*Myrciaria jaboticaba*) peel extracts using simultaneous analysis of responses. J. Food Eng. 117, 538–544.

Sinha, V.R., Kumria, R., 2001. Polysaccharides in colon-specific drug delivery. Int. J. Pharmaceut. 224 (1–2), 19–38.

Summerlin, N., Soo, E., Thakur, S., Qu, Z., Jambhrunkar, S., Popat, A., 2015. Resveratrol nanoformulations: challenges and opportunities. Int. J. Pharmaceut. 479 (2), 282–290.

Sun, P., Zeng, M., He, Z., Qin, F., Chen, J., 2013. Controlled release of fluidized bed-coated menthol powder with a gelatin coating. Dry. Technol. 31 (13–14), 1619–1626.

Takahashi, M., Uechi, S., Takara, K., Asikin, Y., Wada, K., 2009. Evaluation of an oral carrier system in rats: bioavailability and antioxidant properties of liposome-encapsulated curcumin. J. Agric. Food Chem. 57 (19), 9141–9146.

Taylor, T.M., Davidson, P.M., Bruce, B.D., Weiss, J., 2005. Liposomal nanocapsules in food science and agriculture. Crit. Rev. Food Sci. Nutr. 45 (7–8), 587–605.

Taylor, T.M., Gaysinsky, S., Davidson, P.M., Bruce, B.D., Weiss, J., 2007. Characterization of antimicrobial-bearing liposomes by ζ-potential, vesicle size, and encapsulation efficiency. Food Biophys. 2 (1), 1–9.

Villacrez, J.L., Carriazo, J.G., Osorio, C., 2014. Microencapsulation of Andes berry (*Rubus glaucus* Benth.) aqueous extract by spray drying. Food Bioprocess Tech., 1445–1456.

Yadav, V.R., Prasad, S., Kannappan, R., Ravindran, J., Chaturvedi, M.M., Vaahtera, L., Aggarwal, B.B., 2010. Cyclodextrin-complexed curcumin exhibits anti-inflammatory and antiproliferative activities superior to those of curcumin through higher cellular uptake. Biochem. Pharma., 80(7), 1021–1032.

Yanniotis, S., Stoforos, N.G., 2014. Modelling food processing operations with computational fluid dynamics: a review. Scientia Agriculturae Bohemica 45 (1), 1–10.

Yatsu, F.K.J., Borghetti, G.S., Bassani, V.L., 2011. Technological characterization and stability of *Ilex paraguariensis* St. Hil. Aquifoliaceae (Maté) spray-dried powder. J. Med. Food 14, 413–419.

Zam, W., Bashour, G., Abdelwahed, W., Khayata, W., 2014. Alginate-pomegranate peels' polyphenols beads: effects of formulation parameters on loading efficiency. Braz. J. Pharmaceu. Sci. 50 (4), 741–748.

Zhao, G., Hu, C., Sun, R., Ni, S., Li, Q., Xia, Q., 2015. Development of novel composite antioxidant multiple lipid particles from combination of W/O/W multiple emulsions and solid lipid nanoparticles. Eur. J. Lipid Sci. Technol. 117 (7), 1056–1065.

Zheng, L., Ding, Z., Zhang, M., Sun, J., 2011. Microencapsulation of bayberry polyphenols by ethyl cellulose: preparation and characterization. J. Food Eng. 104, 89–95.

Zhou, H., Sun, X., Zhang, L., Zhang, P., Li, J., Liu, Y.N., 2012. Fabrication of biopolymeric complex coacervation core micelles for efficient tea polyphenol delivery via a green process. Langmuir 28 (41), 14553–14561.

Zuidam, N.J., Shimoni, E., 2010. Cap.: 2 overview of microencapsulation for use in food products or process and methods to make them. Encapsulation Technologies for Active Food Ingredients and Food ProcessingSpringer, New York, 3–30.

14

NANOENCAPSULATION TECHNOLOGY TO CONTROL RELEASE AND ENHANCE BIOACTIVITY OF ESSENTIAL OILS

Jayamanti Pandit, Mohd. Aqil, Yasmin Sultana
Jamia Hamdard, Department of Pharmaceutics, Faculty of Pharmacy, New Delhi, India

1 Introduction

Essential oils, also called volatile or ethereal oils, are aromatic liquids obtained from various plant materials (Burt, 2004). They are useful in perfume and flavor industries and are currently extensively explored for uses in pharmaceutical, cosmetic, and textile industries. In food and pharmaceutical industries their main uses are as flavoring, preservatives, antioxidants, and insecticides. Recently essential oils have been evaluated for their antimicrobial, analgesic, sedative, antiinflammatory, spasmolytic, and local anesthetic properties in formulations or alone (Silva et al., 2003). Essential oils (EOs) are very sensitive to environment and easily degrade upon exposure to oxygen, light, and heat. They are hydrophobic and therefore insoluble in water, which reduces their bioavailability and absorption in the body. Therefore, a method or technique that takes into account these issues and makes an adequate formulation of EOs that is able to protect their useful properties and activities. The common goal of such a technique or methodology is to stabilize, to mask disagreeable taste and color, to protect bioactivity, and to improve the release or to achieve controlled release of the substance.

Encapsulation may be defined as a process to entrap one substance within another substance, thereby producing a delivery carrier with a diameter of a few nm to a few mm. The substance that is encapsulated may be called the core material, the active agent, internal phase, or payload phase. The material that is encapsulating may be called the coating, shell, carrier material, wall material, external phase, or matrix (Zuidam and Shimoni, 2010). Many scientists

Encapsulations. http://dx.doi.org/10.1016/B978-0-12-804307-3.00014-4

are working on these formulations of essential oil to improve the efficacy by formulating emulsions, micelles, liposomes, and other lipid-based delivery and microencapsulation (Augustin et al., 2001).

A few years back scientists realized the need to control the release of oils to give sustained biological effect of the essential oil.

Nanoencapsulation technique is one such approach to overcome the problem of stability with a sustained release effect. It is the process to encapsulate the active agent in the nanometric size and modify the physicochemical and bioactivity of bioactive agents. This chapter presents the state of art or insight of the present technique of nanoencapsulation application for EOs. It also includes their protection and release mechanism with the potential advantage and limitation. Work done so for in the nanoencapsulation of essential oils has been included.

2 Chemical Composition of EOs

The extensive biological application of EOs can be defined due to the complexity and variability of their chemical compositions. Many factors can affect the final composition of these EOs or oils. They can be obtained from any part of plants, that is, leaves, stems, buds, flowers, twigs, seeds, fruits, roots, wood, or bark. Constituents obtained are highly volatile and lipophilic and a molecular weight of below 300 (Sell, 2010).

Most of the EOs are complex mixtures of volatile organic compounds produced as secondary metabolites in plants; this include hydrocarbons, that is, terpenes and sesquiterpenes) and oxygenated compounds, that is, alcohols, esters, ethers, aldehydes, ketones, lactones, phenols, and phenol ethers (Guenther, 1972).

Generally these oils contain about 20–60 components up to more than 100 single substances, at different concentrations from which two or three components are present at fairly high concentrations (20–70%) compared to other components that remain in trace amounts. For example, the major components of the *Origanum* essential oils are carvacrol (30%) and thymol (27%) (Bilia et al., 2014). The biological properties of the EOs are decided by these major components.

3 Limits and Challenges for the Use of Essential Oils for Their Biological Activities

Traditionally, essential oils have been used for many biological properties like bactericidal, virucidal, fungicidal, antiparasitical, insecticidal, and other medicinal properties such as analgesic, sedative, antiinflammatory, spasmolytic, and local anesthetic (Adorjan and Buchbauer, 2010; Bakkali et al., 2008; Buchbauer et al., 1993). Recently they have been explored as antioxidants and preservatives in foods (Tiwari et al., 2009) and protectants for crops (Adorjan and Buchbauer, 2010). They are also incorporated into packaging materials for foodstuff (Kuorwel et al., 2011).

The preparations based on EOs have certain limitations that reduce their utility in pharmaceutical and food industries. Because of the potential for allergic reactions, EOs are not recommended for direct application on the skin. The presence of unsaturated carbon chains, which are the main constituents of EOs, leads to susceptibility to oxidation mediated by light or heat (Neumann and Garcia, 1992). The oxidation products of terpenes, that is, oxidated sesquiterpenes with lacone rings, terpenoids, and other plant metabolites have been shown to possess high allergenic activity (Vigan, 2010; Hammer et al., 2006; Sköld et al., 2002).

Furthermore, the high volatility of these oils discourages their free use, that is, without a pharmaceutical vehicle. The low aqueous solubility is another hurdle that limits application of EOs for pharmacological activity. The neglectable or very low solubility of EOs in biological fluids impedes their absorption, leading to a very low bioavailability. All these factors limit the utility of EOs as candidates for pharmacotherapeutic treatments, which need solutions for pharmaceutical technology studies.

4 Nanoencapsulation Technology

In recent years, encapsulation has offered an approach for effective protection of functional properties of EOs. Encapsulation aims to preserve stability of the bioactive compounds during processing and storage and to prevent undesirable interactions with food matrix. Most of the bioactive food compounds are characterized by rapid inactivation. The encapsulation procedure slows down the degradation processes (eg, oxidation or hydrolysis) or prevents degradation until the product is delivered at the desired sites (McClements and Lesmes, 2009).

Thus, the bioactive component would be kept as fully functional. Encapsulation of bioactive EOs may act as barriers between sensitive bioactive materials and the environment, and helps to mask the bad taste and aroma and produce stable food ingredients with enhanced bioavailability.

In addition to the previously mentioned advantages, EOs can be applied for modification of physical state of the original material in order to (1) allow easier handling, (2) help separate the components of the mixture that would otherwise react with one another, and (3) to provide an adequate concentration and uniform dispersion of an active agent (Desai and Park, 2005).

A significantly large part of literature on the encapsulation of EOs focuses on microencapsulation, which are used for the protection of the active compounds against environmental factors, for example, oxygen, light, moisture, and pH, to decrease oil volatility and to transform the oil into a powder. Encapsulation in nanometric particles is an alternative for overcoming these problems with an additional advantage due to the nano size (≤500 nm), which may increase the cellular absorption mechanisms and increasing bioefficacy.

A nanosystem as a formulation is accepted for every route of administration. Nanocarriers applied to the skin to facilitate local therapies though the mechanisms of penetration through skin are still part of discussion for many scientists. Topical drug delivery with nanoparticles delivers the active agent into the deeper layers of skin but generally they do not reach the viable epidermis. The use of nanocarriers provides a sustained and slow release of the active constituents, where these can act as a reservoir (Schneider et al., 2009; Prow et al., 2011).

The other routes of administration of EOs are by oral intake and inhalation. Within these routes the nanoencapsulated delivery systems encounter the mucosal barrier present in the nasal, lung, stomach, gut, and oral (sublingual and buccal) cavity. Further, they can improve the stability of EOs against enzymatic degradation and attain the desired therapeutic levels in target tissues with a lower number of doses for the required duration.

The thick, viscous, and sticky mucus present over all mucosal tissues as part of the protective barrier of the body can rapidly trap and remove foreign particles and hydrophobic molecules. To overcome this barrier, mucoadhesive nanocarriers are specially designed which have the ability to adhere to the mucus, leading to retention of the carrier system that enhances absorption and bioavailability of the active constituent. The interaction between mucosal lining and the nanocarrier is the property of the polymers, which can form hydrogen bonding and hydrophobic or electrostatic interactions with mucin (Roger et al., 2010; Thanki et al., 2013; Lai et al., 2009; Kushwaha et al., 2011; Singh et al., 2011).

Particle size, shape, and surface properties of the nanoencapsulated carrier also play a crucial role in deciding the uptake of nanosized delivery systems across the mucosal membrane. The nanocarriers with a size range of 50–300 nm with positive zeta potential and hydrophobic surface were preferentially taken up by these cells compared to their counterparts (Roger et al., 2010).

The two absorption mechanisms for nanocarriers are the paracellular route, which is slow and passive, and transport through a lipoidal route, that is, the transcellular process that is responsible for the transport of lipophilic drugs (Singh et al., 2011). The other way of enhanced absorption of nanocarriers is by receptor-mediated endocytosis and transcytosis, phagocytosis via specialized microfold cells (M cells) of the Peyer's patches, and other mucosa associated lymphoid tissues (MALT), and lymphatic absorption via chylomicron uptake mechanism (Thanki et al., 2013).

5 Nanoencapsulated Delivery of EOs

Nanoencapsulated delivery carriers of EOs can possess a number of desirable features for therapeutic application, which include: (1) sustained and controlled release of EOs locally, (2) enhanced tissue penetration due to the nanometric size, (3) cellular uptake and subcellular trafficking, and (4) protection of encapsulated EO therapeutics at both extracellular and intracellular levels (Chaudhry et al., 2008).

A lot of substances are being used to coat or encapsulate solids, liquids, or gases of different types and properties. However, due to the rigid regulations for food additives, many of the widely accepted coating materials for drug encapsulation are not been approved for food. Many of these substances have not been certified for food applications as "generally recognized as safe" (GRAS) materials (Wandrey et al., 2009).

The selection of materials for design of protective shell of encapsulates must be food-grade, biodegradable, and able to form a barrier between the internal phase and its surroundings. The materials used for encapsulation in the food sector are mainly natural biomolecules that have to provide protection of the active material against environmental conditions and also preserve the activity within capsules' structure during processing or storage under various conditions. It should not react or bind with the food material or EOs and must have good rheological characteristics at high concentration. The most widely used materials for encapsulation in food applications are polysaccharides. Starch and their derivates such as amylose, amylopectin, dextrins, maltodextrins, polydextrose, syrups, and cellulose and their derivatives are commonly used. Plant exudates and extracts, such as gum arabic, gum tragacanth, gum karaya, mesquite gum, galactomannans, pectins, and soluble soybean polysaccharides, are employed, too. Subsequently, marine extracts such as carrageenans and alginate are also present in foods. Some polysaccharides of microbial and animal origin, like dextran, chitosan, xanthan, and gellan, are also exploited. Apart from natural and modified polysaccharides, proteins and lipids are also accessed for encapsulation, such as milk and whey proteins, gelatin, and gluten. Among lipid materials a few fatty acids and fatty alcohols, waxes (beeswax, carnauba-wax, candellia wax), glycerides, and phospholipids are suitable for encapsulation. Polyvinylpyrrolidone (PVP), paraffin, shellac, and a few inorganic materials are also employed (Wandrey et al., 2009; Nedovic et al., 2011).

The selection of coating material depends on the type of active constituents and its characteristics, and the site of application of the encapsulated active agents. The cost constraint is a key factor that decides the most appropriate materials. All properties of potential wall material must be analyzed in order to predict their behavior under conditions present in food formulations (Wandrey et al., 2009).

The nanoencapsulation techniques for the encapsulation of EOs can be broadly classified as polymer-based nanoparticles and lipid-based nanoparticles. Apart from this, molecular complexes such as inclusion complexes with cyclodextrins are also reported. A schematic representation of different nanoencapsulation techniques for EOs is reported in Fig. 14.1.

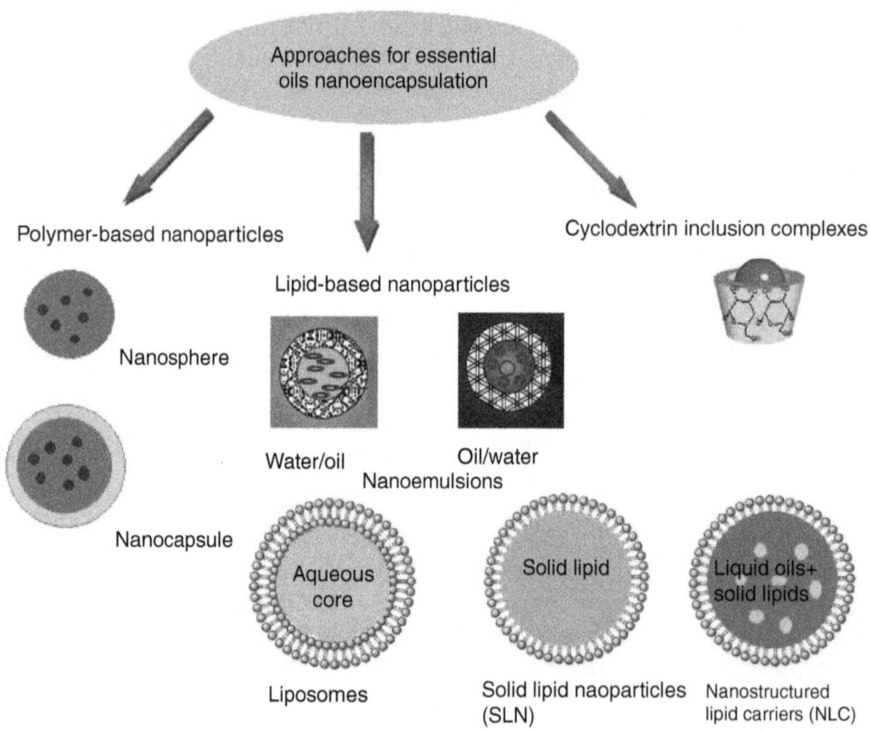

Figure 14.1. Illustration of various nanoencapsulation delivery systems.

5.1 Methods of Nanoencapsulation

Various techniques have been developed and used for nanoencapsulation purposes. They are emulsification, coacervation, emulsification–solvent evaporation, nanoprecipitation, supercritical fluid, and inclusion complexation technique and they can produce capsules in the nanometer range varying from 10 to 1000 nm (Ezhilarasi et al., 2013). The particle size of the nanocarrier is crucial, so these methods are classified as top-down or bottom-up approaches. Top-down approach involves the use of precise tools that allow size reduction and structure shaping for desired application of the nanocarrier being developed. In the bottom-up approach, nanocarriers are constructed by self-assembly and self-organization of molecules, which were influenced by many factors including pH, temperature, concentration, and ionic strength (Augustin and Sanguansri, 2009). Fig. 14.2 shows the two approaches of nanoencapsulation techniques with respective sizes (Sanguansri and Augustin, 2006; Mishra et al. 2010). Among these emulsification, coacervation, and supercritical fluid technique are used for encapsulation of both hydrophilic and lipophilic compounds (McClements et al., 2009; Chong et al., 2009) and the inclusion complexation, emulsification–solvent evaporation, and

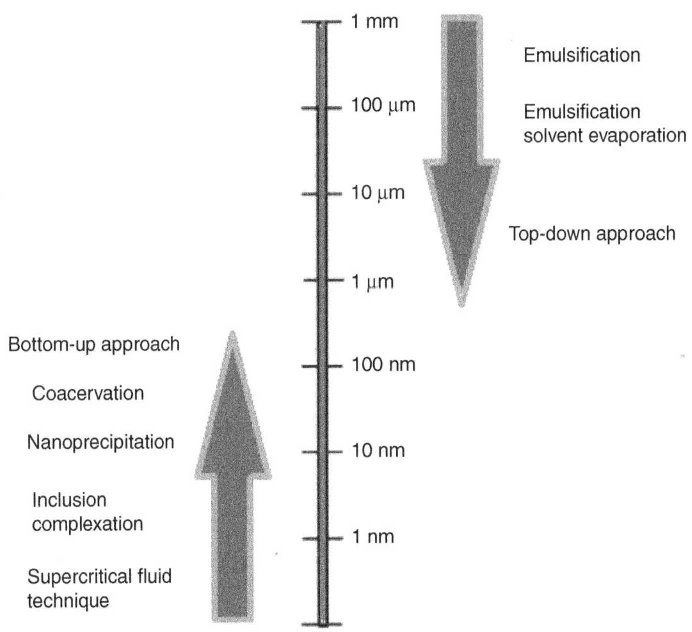

Figure 14.2. Illustration of various nanoencapsulation techniques based on the size.

nanoprecipitation techniques are mostly used for lipophilic compounds (Reis et al., 2006).

5.2 Release Mechanism of Encapsulated EO

Encapsulation can be distinguished as reservoir type and matrix type. The reservoir type has a shell surrounding the active agent. This type is also called capsule, single core, mono-core, or core-shell type. The encapsulated material comes out with the application of pressure and releases its contents. The poly- or multiple-core type of encapsulates with several reservoir chambers in one particle also exist.

If the active agent is dispersed in the matrix of polymer or carrier material it is called matrix type. Matrix type encapsulation can form relatively smaller droplets with more homogenous distribution of encapsulate. Active agents in the matrix type system are in general also present at the surface, in contrast to those in the reservoir type. A modified system where the matrix type is further surrounded with a shell is also known. It can be called as coated matrix type. Fig. 14.3 illustrates all types of encapsulation with spherical shapes, although they can also be cylindrical, oval, or irregular shaped (Zuidam and Shimoni, 2010).

Release of the encapsulated active agent from any of these types of carrier depends upon solubility, diffusion, and biodegradation of the shell or matrix materials. The release of the active agent can be modified by the choice of polymer or coating material. The release also depends upon the loading efficiency of active agent and size of encapsulate. Larger particles have a smaller initial burst release than smaller particles. The loading capacity is directly proportional to the burst and release rate of the nanoencapsulated molecules (Kumari et al., 2010).

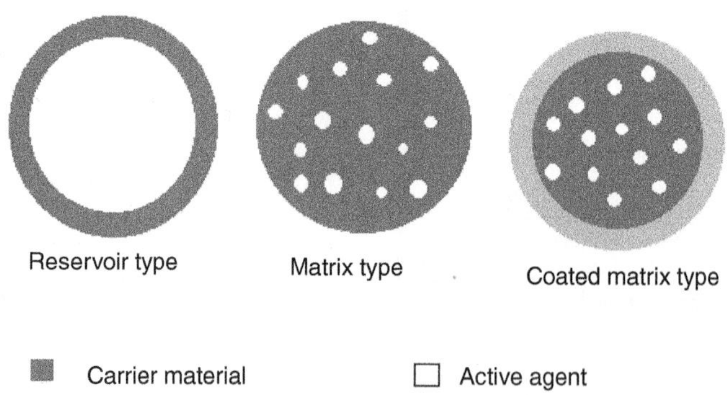

Figure 14.3. Schematic view of different types of nanoencapsulation.

In the case of matrix type, where the active agent is uniformly distributed in the polymer phase, the release occurs by diffusion or erosion of the matrix under sink conditions. Either diffusion or erosion can be rate limiting. If the diffusion of an active agent is faster than matrix erosion, the mechanism of release is largely controlled by a diffusion process. In this case the rapid initial release or burst is mainly due to weakly bound or adsorbed active agents to the surface of the encapsulate (Soppimath et al., 2001).

Diffusion is the possible mechanism of the EO release from matrix and is controlled by the solubility of an EO in the matrix (which must establish a concentration gradient in the matrix, which drives diffusion) and the permeability of the EO through the matrix. It is the dominant mechanism of controlled release from encapsulation matrices (Cussler, 1997). The vapor pressure of a volatile substance is also a driving force that can influence diffusion through the matrix type encapsulation (Gibbs et al., 1999a).

The principal steps involved in the release of EOs from the matrix system are: diffusion of the active agent to the surface of the matrix, partition of the volatile component between the matrix and outside media, and transport away from the matrix surface (Fan and Singh, 1989).

In cases where the matrix undergoes degradation, the release of an active agent may be controlled by diffusion, erosion, or a combination of both. This degradation or erosion may be homogeneous or heterogeneous. Heterogeneous erosion occurs when degradation is confined to a thin layer at the surface of the matrix system, whereas homogenous erosion is a result of degradation occurring at a uniform rate throughout the polymer matrix (Pothakamury and Barbosa-Canovas, 1995).

Sometimes the release of the active agent, which is dissolved or dispersed in a polymeric matrix, is controlled by the swelling of the matrix. When such a system is placed in a thermodynamically compatible medium, the polymer swells because of absorption of fluid from the medium. The EO in the swollen part of the matrix then diffuses out (Fan and Singh, 1989). The degree of swelling is controlled by water absorption of surrounding medium or solvents such as water, glycerine, or propylene glycerol (Gibbs et al., 1999b).

6 Polymer-Based Nanoparticles

Polymeric nanoparticles have been classified as nanospheres and nanocapsules. Nanocapsules are defined as two distinguished compartments, a polymeric shell and a core. Active agents that are encapsulated may present inside the core or adsorbed on the surface. Nanospheres are matrix type systems in which the active agent is homogeneously dispersed. The polymer may be selected based on the use and properties desired. Biocompatible polymers from synthetic or natural origin are more preferred in food industries Uhrich et al. (1999).

These NPs are prepared by different methods depending on the solubility of active agent. The most commonly used method is based on an antisolvent procedure, known as nanoprecipitation or solvent displacement (Reis et al., 2006). The encapsulated EO in NP has several advantages. Apart from controlled release, these NPs also enhance apparent water solubility and the activity of the EO (Li et al., 2012; Iannitelli et al., 2011). They also reduce the undesirable affect or cytotoxicity of the entrapped active agent (Keawchaoon and Yoksan, 2011; Chen et al., 2009). Essential oils encapsulated polymeric particles prepared from different polymers are listed in Table 14.1.

Eugenol is widely used in food, pharmaceutical, cosmetics, and active packaging applications, owing to its effective antimicrobial and antioxidant properties (Devi et al., 2010). However, the activity is reduced during processing and storage due to the volatile nature and sensitivity toward atmosphere, that is, oxygen, light, and heat (Choi et al., 2009).

Woranuch and Yoksan (2013) worked to improve the thermal stability of eugenol by encapsulating into chitosan nanoparticles. The influences of the initial eugenol content and tripolyphosphate (TPP) concentration on the loading capacity (LC), encapsulation efficiency (EE), morphology, and surface charge of the eugenol-loaded chitosan nanoparticles was determined. Nanoparticles were prepared via an emulsion–ionic gelation crosslinking method and the thermal stability of eugenol was investigated. This was verified through its extrusion at 155°C with model plastic, that is, thermoplastic flour (TPF). TPF-containing encapsulated eugenol showed an eightfold higher retention of eugenol content and 2.7-fold greater radical scavenging activity than that containing naked eugenol (Woranuch and Yoksan, 2013).

In another work, oregano essential oil (OEO) has been encapsulated in chitosan nanoparticles evaluated in vitro. The NP exhibited a size range of 40–80 nm with regular distribution and spherical shape. The encapsulation efficiency (EE) and loading capacity (LC) were about 21–47% and 3–8%, respectively. In vitro release studies showed biphasic release of EO (Hosseini et al., 2013).

Carvacrol-loaded chitosan nanoparticles were prepared by ionic gelation of chitosan with pentasodium tripolyphosphate. The NPs has encapsulation efficiency (EE) and loading capacity (LC) in the ranges of 14–31% and 3–21%, respectively. The particles

Table 14.1 Encapsulation of Essential Oils in Polymeric Particles

Nanoformulation	Essential Oil	Inferences	References
Chitosan nanoparticles	(1) Eugenol (2) Oregano (3) Carvacrol (4) Eugenol and carvacrol	CS-NP was thermally stable and could be used as antioxidant in thermal processing It shows two-phase release profile of the antioxidant oil Enhanced antimicrobial activity Reduced cytotoxicity of EOs as compared to free ESO	Woranuch and Yoksan (2013) Hosseini et al. (2013) Keawchaoon and Yoksan (2011) Chen et al. (2009)
PCL nanoparticles	Eugenol Tea tree oil Tea tree oil	Enhanced stability against light oxidation Enhanced activity against fungal infection Nanocapsules showed higher protection against volatilization	Choi et al. (2009) Flores et al. (2013) Flores et al. (2011)
Alginate/cashew gum NP Cashew gum NP	*Lippia sidoides* EOs (rich in thymol) *Eucalyptus staigeriana* essential oil (ESO)	A good EE of 50% with tailored release rate following Korsmeyer–Peppas mechanism Potential for use as a natural food preservative	de Oliveira et al. (2014) Herculano et al. (2015)
Chitosan/cashew gum NP	*Lippia sidoides* EOs (rich in thymol)	Higher loading and EE of EO with slow and sustained release was obtained	Abreu et al. (2012)
Gelatin and arabic gum nanoparticle	Jasmine EOs	Heat-resistant nanocapsules of EO were obtained	Lv et al. (2014)
Zein NP	Thymol and carvacrol	Enhanced antimicrobial activity and aqueous solubility	Wu et al. (2012)
Zein NP coated with sodium caseinate (SC)	Thymol	Effectively suppress gram-positive bacteria for longer time than the unencapsulated EO	Zhang et al. (2014)
PLGA NP	Carvacrol Eugenol or *trans*-cinnamaldehyde	Enhanced antimicrobial activity with biphasic release profile Presented efficient antimicrobial activity with two phase release profile	Iannitelli et al. (2011) Gomes et al. (2011)
MC/EC NP	Thymol	Effective microbial suppression by NP as well as when this was used in cream and gel formulation	Wattanasatcha et al. (2012)
PEG-coated nanoparticles	Garlic essential oil	Slow and persistent release of the active components to control the pests for longer duration	Yang et al. (2009)

NP, nanoparticle; PCL, polycaprolactone; PLGA, poly(lactic-*co*-glycolic acid); MC, methyl cellulose; EC, ethylcellulose; and PEG, polyethylene glycol.

were spherical shaped with an average diameter of 40–80 nm, and a zeta potential value of 25–29 mV. Carvacrol-loaded chitosan nanoparticles showed antimicrobial activity against *Staphylococcus aureus*, *Bacillus cereus*, and *Escherichia coli* with an MIC of 0.257 mg/mL. The release of carvacrol from chitosan nanoparticles followed a Fickian behavior and superior release rate in an acidic medium to either alkaline or neutral media, respectively (Keawchaoon and Yoksan, 2011).

In another study, nanoparticles with two components of essential oils, eugenol, and carvacrol, were prepared by grafting onto chitosan nanoparticles. For this, free aldehyde groups were introduced in eugenol and carvacrol and then grafted to chitosan nanoparticles via the Schiff base reaction. The antioxidant activities were assayed with diphenylpicrylhydrazyl (DPPH) and carried out with *Escherichia coli* (*E. coli*) and *Staphylococcus aureus* (*S. aureus*). The antioxidant activity of carvacrol- and eugenol-grafted chitosan nanoparticles was compared to the chitosan nanoparticles. Results showed that the chitosan nanoparticles grafted with EOs component achieved an antibacterial activity equivalent to or better than that of the unmodified chitosan nanoparticles. Cytotoxicity assay on 3T3 mouse fibroblast showed a significantly lower toxicity of EOs encapsulated nanocapsules than those of the pure EOs (Chen et al., 2009).

Choi et al. (2009) studied the stability of eugenol by inclusion with β-cyclodextrin (β-CD) and 2-hydroxypropyl-β-cyclodextrin (2-HP-β-CD) and by encapsulation in polycaprolactone (PCL). The formulations that were evaluated for the type of complex, size, zeta-potential, and thermal properties were determined using differential scanning calorimetry (DSC). Other studies such as thermogravimetric analysis (TGA), transmission electron microscopy (TEM), scanning electron microscopy (SEM), and atomic force microscopy (AFM) were also done. TGA analysis showed the EE of PCL, β-CD eugenol, and 2-HP-β-CD eugenol inclusion complexes were 100, 90.9, and 89.1%, respectively. The oxidation stability study revealed the PCL nanoencapsulation was more efficient than the molecular inclusion method and resulting high stability of encapsulated eugenol. The eugenol encapsulation protects from light oxidation during storage time due to complete wrapping of eugenol by PCL layer (Choi et al., 2009).

Flores et al. (2013) have developed the nanocapsules and nanoemulsions containing *Melaleuca alternifolia* essential oil (tea tree oil) and evaluated the antifungal efficacy in an onychomycosis model. The antifungal activity was evaluated against *Trichophyton rubrum* in two different in vitro models of dermatophyte nail infection. First, nail powder was infected with *T. rubrum* in a 96-well plate and then treated with the formulations. After 7 and 14 days, the cell viability was verified and the plate counts for the samples were 2.37, 1.45, and 1.0 log CFU/mL for the emulsion, nanoemulsion, and nanocapsules containing tea tree oil, respectively. In the second model, nail fragments were infected with the microorganism and treated with the formulations. The diameter of the fungal colony was measured and found to be 2.88 ± 2.08 mm^2, 14.59 ± 2.01 mm^2, and 40.98 ± 2.76 mm^2 for the nanocapsules, nanoemulsion, and emulsion containing tea tree oil, respectively, and 38.72 ± 1.22 mm^2 for the untreated nail. The result of the study demonstrated that the encapsulation of oil in nanocapsules was most efficient to reduce *T. rubrum* growth in both the nail infection models (Flores et al., 2013).

In another study, the feasibility of tea tree oil was evaluated as oil phase for preparing nanocapsules and nanoemulsions aiming to protect its volatilization. The nanostructure was found in the size range 160–220 nm with a polydispersity index below 0.25 and negative zeta potential. The oil content was 96% after preparation and after 60 min; the oil content was 30, 33, and 54% in relation to the initial values for emulsion, NE and NC, respectively. After heating for 30 min, the oil content in NC was about 67%, while lower values of 42 and 40% were obtained for NE and emulsion, respectively. Results show that the inclusion of tea tree oil in nanocapsules showed higher protection against volatilization (Flores et al., 2011).

A thymol-rich *Lippia sidoides* essential oil was encapsulated in alginate/cashew gum nanoparticles via spray drying and investigated for fungicide and bactericide activities. The NPs' sizes ranged from 223 to 399 nm, and zeta potential values ranged from −30 to −36 mV. The encapsulation efficiency was up to 55% with the in vitro release of oil between 45 and 95% within 30–50 h. The cashew gum and alginate in combination has proven to enhance the hydrophilic character of the polymer matrices, and therefore quicker release of oil from the matrix. The oil release profile also revealed that the use of alginate in synergy with cashew gum presents a potential delivery system with tailored release rate, loading, and encapsulation efficacy (de Oliveira et al., 2014).

Eucalyptus staigeriana essential oil (ESO) was encapsulated using cashew gum (CG) as wall material. The nanoparticles were evaluated for antimicrobial activity against *Listeria monocytogenes* (gram-positive) and *Salmonella* enteritidis (gram-negative) by determining their minimum bactericidal concentration (MBC). The data from MBC showed the greater activity of NPs against gram-positive bacteria, due to a likely synergistic effect between the CG and ESO. It was suggested that ESO NPs have potential for use as a natural food preservative (Herculano et al., 2015).

The EO from *Lippia sidoides* nanoparticles made of chitosan and cashew gum aimed to improve essential oil loading and release profiles. The developed NPs showed high loading (11.8%) and encapsulation efficiency (70%), with size ranges from 335–558 nm. In vitro release profiles showed slow and sustained release of EO from NPs. The nanocarriers presented efficacy against *St. aegypti* larvae, where the mortality rate was related to the loading values and gum:chitosan ratios. A ratio of gum:chitosan of 1:1 and gum:chitosan of 1:10 showed 87 and 75% of mortality, respectively after 48 h (Abreu et al., 2012).

Encapsulation of jasmine essential oil was achieved by gelatin and arabic gum, which gives a heat-resistant property to nanocapsules. The heat-resistance capability of developed NPs against 80°C was evaluated by both structural characteristics (size, polydispersity index, and zeta potential) and flavor analysis. The results showed that these encapsulated flavor nanocapsules were stable at 80°C for 7 h, even the gas chromatography–mass spectroscopy (GC–MS) analysis revealed that jasmine essential oil began to deteriorate after 5 h (Lv et al., 2014).

It was demonstrated that encapsulating EOs, that is, thymol and carvacrol, in zein nanoparticles can enhance their solubility up to 14-fold without hindering their ability to scavenge free radicals or to control *E. coli* growth. Results from this study support the use of nanoencapsulation to facilitate the application of EOs in food preservation (Wu et al., 2012).

Thymol-loaded zein nanoparticles stabilized with sodium caseinate (SC) and chitosan hydrochloride (CHC) were prepared and evaluated for their antibacterial effect. The SC-stabilized nanoparticles had well-defined size range and negatively charged surface. CHC-coated SC-stabilized zein nanoparticles showed increased particle size, and reversal of zeta potential value from negative to positive with improved encapsulation efficiency. Both the thymol-loaded zein nanoparticles and SC stabilized zein nanoparticles had showed spherical shape and smooth surface, while the surfaces of CHC-SC stabilized zein nanoparticles seemed rough with a few clumps. The study concluded that the encapsulated thymol was more effective in suppressing gram-positive bacterium than unencapsulated thymol for a longer time period (Zhang et al., 2014).

Another group of scientists worked on carvacrol-loaded PLGA nanocapsules and characterized it for antimicrobial activity. Prepared NPs showed a size of about 209.8 nm with polydispersity and zeta potential of 0.260 and −18.99, respectively. The loading efficiency and encapsulation efficiency were found to be 21 and 26%, respectively. In vitro release profile showed an initial "burst" release followed by a slower release. The antimicrobial activity of the encapsulated carvacrol EO was enhanced and it was illustrated that the nanoparticles significantly altered rheological characteristics of bacterial biofilms and facilitated the action of carvacrol (Iannitelli et al., 2011).

Eugenol- or *trans*-cinnamaldehyde-loaded polylactic glycolic acid (PLGA) nanocapsules showed a two-phase EO release profile. The first phase was rapid (under 30 min) and approximately 20% of the EO was estimated; the second phase showed prolonged and consistent release and about 64% of eugenol and 87% of *trans*-cinnamaldehyde were detected after 72 h. In case of PLGA-based NPs, the release was governed mostly by diffusion, with a possible influence of polymer swelling and bulk erosion (Gomes et al., 2011).

Thymol-loaded methyl cellulose/ethyl cellulose–based polymeric nanoparticles have shown relatively high loading, that is 43.53% thymol (weight of encapsulated thymol to weight of the thymol-loaded NPs), and were able to reduce the levels of *E. coli* in an oil/water lotion and in a hydrophilic gel, of *P. aeruginosa* in an oil/water lotion and of *S. aureus* in an oil/water lotion and in a water/oil cream. The encapsulated thymol interestingly, more efficiently preserve these formulations compared to free thymol (Wattanasatcha et al., 2012).

Polyethylene glycol (PEG)–coated NPs loaded with garlic EO was prepared by using melt–dispersion method and evaluated for their insecticidal activity against adult *Tribolium castaneum*. The NPs showed a size of less than 240 nm with an oil-loading efficiency of 80% at an optimal ratio of EO to PEG (10%). The control efficacy against adult *T. castaneum* remained over 80% even after 5 months. This is attributed due to the slow and consistent release of the EO from the NPs. Free garlic EOs was found only 11% effective in controlling the insect. This suggests the feasibility of PEG-coated NPs loaded with garlic EO as an effective carrier to control store products from pests (Yang et al., 2009).

7 Lipid-Based Nanoencapsulation

Lipid-based nanoencapsulation systems are emerging as the most promising rapidly developing encapsulation technologies employed in the field of nanotechnology (Mozafari et al., 2006). Lipid is used as the oil phase in these formulations, either as internal or external phases. Compared with other encapsulation strategies such as polymeric nanoparticles, lipid-based nanoencapsulation systems have many advantages. Being prepared from natural ingredients, they are able to entrap material with different solubilities (Bummer, 2004; Yurdugul and Mozafari, 2004). These lipid-based carriers have the ability to protect an ingredient from free radicals, metal ions, pH, and enzymes that might cause degradation of the sensitive food ingredient. They can accommodate water-soluble material as well as lipid-soluble agents, in the same formulation if required, hence providing a synergistic effect (Suntres and Shek, 1996). Lipid-based nanocarriers may be targeted to the desired site inside the body via active (eg, by incorporation of antibodies) and passive (eg, targeting based on particle size) mechanisms (Mozafari and Mortazavi, 2005). These include nanometric-scaled emulsions and lipid nanoparticles, roughly divided in liposomes, solid lipid nanoparticles (SLN), and nanostructured lipid carriers (NLC). The aims of association of EO with lipid-based nanocarrier are to enhance stability and aqueous solubility of EO, maintaining or enhancing biological activity with the controlled targeting effect. The main lipid-based nanoencapsulation systems that can be used for encapsulation of EO for their protection and delivery in particular are discussed next.

7.1 Liposomes

Liposomes are self-assembled spherical vesicles, comprising one or several concentric phospholipidic bilayers with an internal aqueous phase. Accordingly, they can classified as (1) unilamellar vesicles (ULV), which contains one bilayer; (2) multilamellar vesicles (MLV), which have several concentric bilayers; or (3) multivesicular vesicles (MVV), which consist of nonconcentric bilayers. The size of liposomes can vary from the small size of 20 to the larger, which can exceed 1 μm. With a hydrophilic compartment and lipophilic palisade, they can be used as carriers for both lipophilic and hydrophilic molecules (Musthaba et al., 2009; Mozafari et al., 2008). They are biodegradable, nontoxic, nonimmunogenic, and biocompatible (Anweker et al., 2011). They are an efficient carrier for incorporating natural compounds, such as essential oil (EO) components, by improving their solubility and stability (Coimbra et al., 2011; Sherry et al., 2013). A list of essential oil incorporated in liposomes is shown in Table 14.2.

The antiherpetic activity of *Santolina insularis* EO was investigated after encapsulation into liposomes. Vesicles were made from hydrogenated soy phosphatidylcholine and cholesterol, and their stability was examined for over one year. During the storage period, drug leakage from vesicles and the average size distribution were estimated. The stability of the incorporated oil was checked by evaluating its quali-quantitative composition. The in vitro antiviral activity was studied against herpes simplex virus type 1 (HSV-1) by plaque reduction and yield reduction assays. Results showed efficient entrapment of *Santolina insularis* EO in

Table 14.2 Essential Oils Encapsulated in Liposomes

Nanoformulation	Essential Oil	Inferences	References
Hydrogenated soy PC liposomes	*Santolina insularis* EO	Reduced degradation and nontoxic in concentration range tested	Valenti et al. (2001)
Hydrogenated and nonhydrogenated PC, liposomes	*Artemisia arborescens* L. EO	Enhanced antiherpetic activity with hydrogenated PC	Sinico et al. (2005)
Liposomes	*Thymus* species (*boissieri, longicaulis, leucospermus*, and *ocheus*) extracts	Superior antioxidant as well as antimicrobial activities of encapsulated extract	Gortzi et al. (2006)
Liposomes	*Origanum dictamnus* extracts	Both antimicrobial and antioxidant activity was higher than the extracts in pure form.	Gortzi et al. (2007)
Liposomes	*O. dictamnus* L. EO	Enhanced antimicrobial activities after the encapsulation	Liolios et al. (2009)
Liposomes	*Zanthoxylum tingoassuiba* EO	Enhance essential oil targeting to cells	Detoni et al. (2009)
Liposomes	Caffeic acid (derivatives), carvacrol (derivatives), thymol, pterostilbene (derivatives), and *N*-(3-oxo-dodecanoyl)-L-homoserine lactone	Improved solubility and stability	Coimbra et al. (2011)
Liposomal gel	*Eucalyptus camaldulensis* EO	Enhanced stability as well as antifungal activity	Moghimipour et al. (2012)
Liposomes	*Z. tinguassuiba* EO	Enhanced thermal-oxidative stability with significant apoptotic-inducing activity for glioma cells	Detoni et al. (2012)
Liposomes	Bergamot essential oil (BEO)	Increased water solubility and anticancer activity under in vitro	Celia et al. (2013)
Penetration enhancer containing vesicles (PEVs)	*S. insularis* essential oil	Enhanced skin penetration	Castangia et al. (2015)
Liposomes	Tea tree oil	Enhanced inhibition and bactericidal effect on the TTO-tolerant strain	Ge and Ge (2016)
Nanoliposomes	EO of *Zataria multiflora* Bioss.	Reduced degradation during storage	Khatibi et al. (2014)
Liposomes	Clove essential oil	Improved the stability of clove essential oil	Sebaaly et al. (2015)
Liposomes	Clove essential oil	Exhibited efficient antimicrobial activity for *S. aureus* in tofu	Cui et al. (2015)

the prepared liposomes, which successfully enhanced its stability. Stability studies showed that the vesicle dispersions were stable for at least one year without any oil leakage or vesicle size alteration throughout this period. Antiviral activity assays demonstrated that *Santolina insularis* EO is effective in inactivating HSV-1; however, free EO proved to be more effective than liposomal oil, whereas liposomal Santolina EO was found nontoxic in the range of the concentration tested (Valenti et al., 2001).

Similarly, antiherpetic activity of *Artemisia arborescens* L. EO-encapsulated liposomes was investigated under in vitro condition. To investigate the influence of vesicle structure and composition on the antiviral activity of the encapsulated oil, two different liposomal formulations [multilamellar (MLV) and unilamellar (SUV)] were prepared. Hydrogenated (P90H) and nonhydrogenated (P90) soy phosphatidylcholine were used for the formulation. They were examined for the stability and antiviral activity against herpes simplex virus type 1 (HSV-1) by tetrazolium-based colorimetric method. Results showed that a good amount of *Artemisia* EO can be encapsulated, which was stable for at least 6 months. The liposomal encapsulated *A. arborescens* EO showed enhanced antiherpetic activity, which was more in case of P90H. Both the free and SUV-incorporated oil showed nonsignificant difference in antiviral activity, while the P90H MLV showed a higher activity than P90MLV. They observed EC50 values of 18.3 and 43.6 g/mL for P90H MLV and P90MLV, respectively. Thus, incorporation of *A. arborescens* EO in multilamellar liposomes greatly improved its activity against intracellular HSV-1 (Sinico et al., 2005).

The antioxidant and antimicrobial activity of methanol or dichloromethane extracts of *O. dictamnus* were determined using Rancimat and malondialdehyde (MDA) by HPLC methods. The antioxidant action of extract was further compared with that of some common commercial antioxidants such as butylated hydroxytoluene (BHT) and α-tocopherol. Extracts with high antioxidant activity were selected and encapsulated in liposomes. The antioxidant action of the encapsulated EO was determined using differential scanning calorimetry (DSC). All the extract showed antioxidant and antimicrobial activities, which were reported superior to α-tocopherol. The highest antioxidant activity was reported for the methanol extract of *O. dictamnus* (240 ppm), which is higher than BHT. Results of the liposomal formulation showed higher antioxidant as well as antimicrobial activities compared with the same extracts in pure form (Gortzi et al., 2007).

The major components of *Origanum dictamnus* L. EO are carvacrol, thymol, *p*-cymene, and c-terpinene. The antioxidant and antimicrobial activities of these components after encapsulation in phosphatidyl choline-based liposomes were tested against gram positive and gram-negative bacterial strains. Their activity against three human pathogenic fungi and a food-borne pathogen, *Listeria monocytogenes*, was also evaluated. The possibility of synergistic or antagonistic effect between carvacrol/thymol and carvacrol/c-terpinene was also investigated by determining the antimicrobial activities of the mixtures before and after encapsulation in liposomes. The results showed enhanced antimicrobial activities of all tested compounds after the encapsulation (Liolios et al., 2009).

Zanthoxylum tingoassuiba EO consists of a mixture of organic compounds among which methyl-*N*-methylanthranilate and sesquiterpene alcohol alpha-bisabolol are the main compounds. They have shown significant activity against *S. aureus*, *S. aureus* isolated multiresistant, and the dermatophyte fungi. When this oil was loaded into dipalmitoyl-phosphatidylcholine (DPPC) multilamellar liposomes (MLV) using a thin film hydration method, liposomes showed a mean diameter of 9.37 ± 4.69 μm and found to have more spherical and narrower size distribution than empty liposomes. Results showed entrapment of appreciable amounts ($43.7 \pm 6.0\%$) of *Zanthoxylum tingoassuiba* EO in the prepared vesicular dispersions. The incomplete and consisted release of EO from liposomes suggested that EO-loaded liposomes will be useful in pharmaceutical applications to enhance essential oil delivery in the target cells (Detoni et al., 2009).

The EO extract from *Atractylodes macrocephala* Koidz was encapsulated in liposomes using the rapid expansion of supercritical solutions (RESS) technique. In this method, both the liposomal materials and EO were dissolved in the mixture of supercritical carbon dioxide (SC-CO_2)/ethanol and then the solution was sprayed into an aqueous medium through a coaxial nozzle to form liposomal suspension. Under the optimum conditions of a pressure of 30 MPa, a temperature of 338 K, and an ethanol mole fraction in SC–CO_2 [$x(CH_3CH_2OH)$] of 15%, the developed liposomes showed an entrapment efficiency, drug loading, and average particle size of 82.18%, 5.18%, and 173 nm, respectively. The liposomes appeared as double-layered colloidal spheres with a uniform and narrow particle size distribution. These results indicated that the modified RESS technique is an innovative way to form liposomes' incorporation of multicomponents extract from plant sources (Wen et al., 2010).

In a study, a few natural component having strong antiinflammatory properties were selected and encapsulated into liposome to overcome poor water solubility or chemical instability of these compounds. The caffeic acid (derivatives), carvacrol (derivatives), thymol, pterostilbene (derivatives), and *N*-(3-oxo-dodecanoyl)-L-homoserine lactone were used in the experiment. Results showed that lipophilic 3-oxo-C_{12}-homoserine lactone and stilbene derivatives were efficiently loaded into liposomal lipid bilayer (EE reported as 50–70%). The liposomes solubilize these compounds, which allow intravenous administration without use of solvents. In the case of carvacrol and thymol, it is not possible to load the EOs into the lipid bilayer or they are rapidly extracted from the liposomes in the presence of serum albumin (3-oxo-C_{12}-homoserine lactone and pterostilbene derivatives). Derivatization of the compound into a water-soluble prodrug was shown to improve loading efficiency and stability. The phosphate forms of carvacrol and pterostilbene were loaded into the aqueous interior of the liposomes and the encapsulation was not affected by the presence of serum albumin. Chemical instability of resveratrol was also improved after encapsulation in liposome, which prevents inactivation of *cis–trans* isomerization. Caffeic acid, by derivatization into a phenyl ester, was encapsulated in liposomes without chemical degradation. The liposomal formulation of 3-oxo-C_{12}-homoserine lactone and resveratrol, showed an inhibition of tumor growth of approximately 70% in a murine tumor model after intravenous administration. This was due to the simple solubilization affect of liposomes (Coimbra et al., 2011).

The essential oil of *Eucalyptus camaldulensis* was investigated for its effectiveness on dermatophytes' growth after encapsulation into liposomal gel. Liposomes were prepared by the freeze-thaw method and characterized for particle size and size distribution. Further liposomes were dispersed into gel hydroxethyl cellulose (HEC) to form liposomal gel. The antifungal activity of the essential oil and liposomal gel was determined against *Microsporum canis, M. gypseum, Trichophyton rubrum,* and *T. verrucosum,* using the well diffusion method. Results showed the particle size of liposomes range from 40.5 to 298 nm for different formulations with sufficient entrapment efficiency (95 ± 0.57%); 0.125 mL of EO was reported as the minimum inhibitory volume for antifungal activity. Phenol, 1, 8 cineole, limonene, alcohol, pinene, and terpinen were detected as the main constituents of the EO by GC–MS. They have reported that gel formulation of liposomal-loaded essential oil may lead to improved antifungal activity (Moghimipour et al., 2012).

In another study, Detoni and coworkers evaluated the antiglioma potential of a component, α-bisabolol, present in *Zanthoxylum tinguassuiba* essential oil (ZtEO). They have prepared ZtEO-loaded liposomes and evaluated the oxidative stability of ZtEO and the ability to reduce glioblastoma cell viability. Results of thermal analysis indicated that the thermal-oxidative stability of the liposomal ZtEO was enhanced compared to its free form. A significant apoptotic-inducing activity for glioma cells was observed for Liposomal ZtEO. Therefore, liposomal systems carrying ZtEO may be a potential alternative for gliobastoma treatment (Detoni et al., 2012).

Bergamot essential oil (BEO) and its fractions have been shown to exhibit anticancer efficacy but cannot be used in cancer therapy due to its poor water solubility, low stability, and limited bioavailability. To alleviate these drawbacks, BEO liposomes were prepared and reported to improve water solubility and enhance anticancer activity of the phytocomponents under in vitro conditions against human SH SY5Y neuroblastoma cells (Celia et al., 2013).

A modified form of liposomes called penetration enhancer containing vesicles (PEVs), loaded with *Santolina insularis* essential oil was formulated and evaluated for its physicochemical features and stability, and ability to deliver the oil to the skin. *S. insularis* essential oil is predominantly composed of terpenes of which the most abundant are phellandrene (22.6%), myrcene (11.4%), and curcumenes (12.1%). They were prepared using phosphatidylcholine as lipid phase and ethylene or propylene glycol was added to the water phase [10% (v/v)] to improve vesicle performances as delivery systems. The results showed polyhedral, faceted, unilamellar vesicles of 115 nm in diameter, where the presence of the glycols improved vesicle stability under accelerated aging conditions, without changes in size or migration phenomena (eg, sedimentation and creaming). The confocal laser scanning microscopy images of the pig-skin treated with EO formulation displayed a penetration ability of PEVs greater than that of control liposomes. All the formulations showed a marked in vitro biocompatibility in human keratinocytes. Thus, it was suggested that PEVs enhance the delivery of *S. insularis* essential oil to the skin (Castangia et al., 2015).

The EO containing the antibacterial and antifungal constituents of *Zataria multiflora* Bioss. can be used as a substitute for synthetic drugs and food preservatives. This can be possible after developing a suitable carrier that can offer stability to EOs. The EO was encapsulated into nanoliposomes by three different methods including thin film evaporation, ethanol injection, and sonication methods. The effect of different methods on liposomes and their physical properties was studied by means of particle size, polydispersity index, zeta potential, and encapsulation efficiency. Results showed that the sonication method produces liposomes with the smallest mean size of 99 nm and better dispersivity. The encapsulation efficiency of nanoliposomes containing EO was in the following order: thin film evaporation > ethanol injection > sonication.

MLV (multilamellar vesicles) liposomes prepared by the thin layer evaporation method showed better stability after 1 month of storage at 4 ± 1°C. It was concluded that the preparation methods have an effect on the physical properties and storage stability of liposomes (Khatibi et al., 2014).

The antibacterial activities of clove oil and liposome-encapsulated clove oil were investigated. It was demonstrated that the clove oil exhibited favorable antimicrobial activity for both *Escherichia coli* (*E. coli*) and *Staphylococcus aureus* (*S. aureus*) but have low chemical stability. After incorporation into a liposome formulation to increase its stability, the optimal polydispersity index (PDI), zeta potential, and entrapment efficiency of liposome was found to be 0.196, −24.5 mV, and 20.41% respectively at the concentration of clove oil to 5.0 mg/mL. It was observed that the pore-forming toxins (PFTs) secreted by *S. aureus* was utilized to activate release of clove oil from liposome. The encapsulated clove oil has no effect on *E. coli* that doesn't secrete PFTs, therefore the antimicrobial component does not reach bacteria. Gas chromatography (GC) assay proved that when liposome comes into contact with *S. aureus* that secrete PFTs, PFTs would insert into the lipid bilayers of liposomes and form pores through which the encapsulated clove oil was released (Cui et al., 2015).

In another study, natural soybean phospholipid-based liposomes were developed to improve the stability of clove EO and its main component, eugenol. Various lipids such as those using a saturated (Phospholipon 80H, Phospholipon 90H) and unsaturated soybean (Lipoid S100), in combination with cholesterol, were used to prepare liposomes at various eugenol and clove oil concentrations using ethanol injection method. They were characterized and compared for their size and surface morphology, polydispersity index, zeta potential, loading rate, and encapsulation efficiency. Stability of these formulations was checked after storing them for 2 months at 4°C. It was reported that liposomes exhibited nanometric oligolamellar and spherical-shaped vesicles and protected eugenol from degradation induced by UV exposure; they also maintained the DPPH scavenging activity of free eugenol (Sebaaly et al., 2015).

7.2 Lipid Nanoparticles

Lipidic nanoparticles can be discussed under two headings, according to the chemical composition of the lipidic phase: solid lipid nanoparticles (SLNs) and nanostructured lipid carriers (NLCs). In case of SLN, the lipophilic component is dissolved in a solid lipid or in a solid lipid mixture while the lipidic phase is a solid lipid and liquid lipid (oil) mixture at room temperature for NLCs (Dima et al., 2015). The release of active agent from SLNs is due to the transition phase of the lipids in the cooling process. During the cooling, the predominant lipid polymorph, that is, α get transforms into a polymorphous form β and β'. Further transition of β to β' leads to expulsion of the molecules from the lipophilic component of the lipid NPs.

Compared to the other liquid lipid-based systems—emulsions, nanoemulsions, microemulsions, self-nano emulsifying drug delivery systems (SNEDDS), self-micro emulsifying drug delivery systems (SMEDDS), liposomes—lipid nanoparticles have many advantages. They have increased stability and entrap more active agents, can protect the lipophilic components and showed controlled and targeted release of encapsulated components; they improve the bioavailability of lipophilic components with reduced toxicity risk because of absence of organic solvents; they can be produced and processed on an industrial scale at a lower cost (Tamjidi et al., 2012). Both SLNs and NLCs are mainly prepared by two techniques: hot homogenization and cold homogenization. Homogenization may be achieved by means of different types of equipment such as a high-pressure homogenizer (HPH), a high-intensity ultrasonic probe/jet/bath, or a microfluidizer. Many researchers studied the lipid NPs for encapsulating essential oils and other food nutrients (Tamjidi et al., 2013; Fathi et al., 2012). Various essential oils encapsulated lipid nanoparticles are discussed in Table 14.3.

Solid lipid nanoparticles loaded with *Artemisia arborescens* essential oil was evaluated for pest control. Two SLN formulations were developed by high-pressure homogenization technique where Compritol 888 ATO was taken as lipid and Poloxamer 188 or Miranol Ultra C32 used as surfactants. The results of the study illustrated a high physical stability for both formulations at various storage temperatures for a period of 2 months. The average diameter of *EO*-loaded SLN did not vary during this period and only a slight increase in size was observed after spraying the SLN dispersions. In vitro release study showed that SLN were able to reduce the volatility of essential oil if compared with the reference emulsions. This showed that the *Artemisia arborescens*–loaded SLN formulations can be a suitable carriers in agriculture (Lai et al., 2006).

In another study, *Artemisia arborescens* EO incorporated in SLN has been investigated for in vitro antiherpetic activity on transdermal delivery. SLN formulations were prepared by hot-pressure homogenization technique using Compritol 888 ATO as lipid. The surfactants, Poloxamer 188 and Miranol Ultra C32, were used in SLN1 and SLN 2, respectively. After a day, the SLN

Table 14.3 Essential Oils Encapsulated in Lipid Nanoparticles

Formulation	Essential Oil	Inferences	References
Compritol 888 ATO, SLN	*Artemisia arborescens* EO	Reduced volatility	Lai et al. (2006)
Compritol 888 ATO, SLN	*A. arborescens* EO	Improved the oil accumulation into the skin	Lai et al. (2007)
SLN	*Nigella Sativa* EO	High physical stability	Alhaj et al. (2010)
SLN	Essential oil of *Zataria multiflora*	Potential carrier for EO delivery	Moghimipour et al. (2013)
SLN	Frankincense and myrrh essential oils (FMO)	Reduced evaporation loss and increased antitumor efficacy of active constituents	Shi et al. (2012)
SLN	Clove (*Syzygium aromaticum*) extract	Enhanced stability	Cortés-Rojas et al. (2014)
NLC	*Zingiber zerumbet* oil	Increased stability	Rosli et al. (2015)

1 showed a size of 223 nm with a polydispersion index of 0.243, while the particle size of SLN 2 was found to be 219 nm with 0.301 polydispersion index (PI). Both formulations had high physical stability with very slight increase in the particle size and PI after two years. The antiviral activity of free and SLN-incorporated EO was tested in vitro against Herpes Simplex Virus-1 (HSV-1), which showed that entrapment does not affect the antiherpetic activity of EO. In vitro diffusion experiments through newborn pig skin were performed to determine the permeation and accumulation of EO-incorporated SLN into the skin strata, where an almond oil *Artemisia essential* oil solution was taken as a control. Results showed that skin permeation of EO occurred only when the oil was delivered from the control solution, while a significant amount of EO released by SLN formulations was found into the skin strata. SLN seems to adhere to the skin surface that increase skin hydration and promote penetration of active compounds as well as the carrier into the stratum corneum. The SLN could not improve essential oil diffusion through the inner, more hydrophilic skin layers. On the contrary, the EO in almond oil promotes permeation of EO through skin layers, probably by increasing skin hydration that in this case favors oil solution to diffuse through the skin. This indicated SLN as a good carrier to incorporate the essential oil with a good yield and long-term stability (Lai et al., 2007).

Nigella sativa essential oil–encapsulated SLN were developed by Alhaj and coworkers. Formulation showed high physical stability at various storage temperatures during 3 months of storage. The effect on average diameter of *N. sativa* essential oil–loaded SLN was minimum, which strongly recommends SLN as a suitable carrier for food and pharmaceutical industry (Alhaj et al., 2010).

The essential oils obtained from frankincense and myrrh oil (FMO) exhibit a broad spectrum of biological activities such as antimicrobial, antiinflammatory, and antitumor activities. The instability and poor water solubility of FMO result in poor oral bioavailability, which limits its clinical application. A study demonstrated frankincense and myrrh essential oils (FMO) incorporated solid lipid nanoparticles were developed for oral delivery using Compritol 888 ATO as the solid lipid and soybean lecithin and Tween 80 as the surfactants. The obtained SLNs showed a mean size of 113.3 nm with a zeta potential of −16.8 mV, and an encapsulation efficiency of 80.60%. Results showed an increase in the antitumor efficacy of FMO entrapped in SLNs in H22-bearing Kunming mice. The component, Compritol 888 ATO showed reasonable FMO solubilization capacity (Shi et al., 2012).

Solid lipid nanoparticles (SLNs) of essential oil of *Zataria multiflora* was formulated by two methods: precipitation technique and hot homogenization method for achieving the best encapsulation. They were characterized by differential scanning calorimetry (DSC), transmission electron microscopy (TEM), particle size analysis, and essential oil release was evaluated by using a dialysis membrane method. Results showed spherical-shaped particles of a mean size of 650 nm with the encapsulation efficiency of 38.66%. Results of particle size determination showed a mean size of 650 nm and SLNs were spherical as shown by TEM. About 93.2% of the EO was released after 24 h, indicating the SLNs as the potential carrier system of *Z. multiflora* essential oil (Moghimipour et al., 2013).

In this study, lipid formulations containing clove extract were spray dried to encapsulate the volatile and poor water-soluble compounds—eugenol and eugenyl acetate—aiming to obtain solid redispersible powders. Five formulations were prepared to test two different solid lipids (stearic acid and Compritol), two surfactants (Poloxamer 188 and Polysorbate 80), and three drying carriers (maltodextrin, lactose, and arabic gum mixture). All five formulations after drying were characterized by the eugenol and eugenyl acetate retention, in vitro antioxidant activity and other physical properties. The formulation containing glyceryl behenate, Poloxamer 188, and Maltodextrin DE10 presented better retention of bioactive compounds and good antioxidant activity. Further, the influence of the dispersion methods such as high-shear mixing ultraTurrax, ultrasonication and high-pressure homogenization and the drying technique, that is, spray- and freeze-drying. They reported that the freeze-dried samples presented significantly higher retention of eugenol and eugenyl acetate than the spray-dried ones (Cortés-Rojas et al., 2014).

In a study, nanostructured lipid carrier (NLC) encapsulated *Zingiber zerumbet* oil (NLC-ZZ) was prepared by ultrasonication technique. Particle size, polydispersity index (PDI), zeta-potential, encapsulation efficiency, and physical morphology were determined. Results showed that the NLC-ZZ exhibited nanometer size (96.59 nm), stable polydispersity index (0.192) with a high zeta potential charge (–39.88 mV). The encapsulation efficiency of lipid carrier to encapsulate Zingiber zerumbet oil was found above 80%. The nanosize and stability of NLC-ZZ, makes it suitable for topical and transdermal delivery since it can enhance the penetration to the deeper layer of skin (Rosli et al., 2015).

7.3 Nanoemulsions

Nanoemulsions are stable colloidal systems with nanometric size range (≤100 nm). It is prepared by dispersing one liquid in another immiscible liquid with aid of emulsifiers (Burguera and Burguera, 2012). Compared with microemulsions, nanoemulsions are optically transparent and possess a relatively high kinetic stability even for several years, due to their very small size (Fathi et al., 2012; Blanco-Padilla et al., 2014). Nanoemulsions can be formulated either by high-energy methods (high-pressure homogenization, microfluidization, and ultrasonication) or by low-energy methods (solvent diffusion). Choice of method depends on the material and stability of active agent (Donsì et al., 2011).

Nanoemulsions can be (1) the oil in water (O/W) where the oil phase is dispersed in the form of nano droplets in the external aqueous phase with the use of stabilizer or emulsifiers; (2) the multiple emulsions can be of two types, oil-in-water-in-oil (O/W/O) and water-in-oil-in-water (W/O/W), where, for example, in case of W/O/W type, the water droplets were entrapped within large oil droplets that were finally dispersed within an external aqueous phase; and (3) the multilayer emulsions which carry nanometric oil droplets surrounded by different polyelectrolytes layers (Weiss et al., 2006).

Most of the O/W nanoemulsions were used to encapsulate and deliver poorly water-soluble essential oils with different properties for improving the physical stability of the active compound as well as its bioactivity. For example, the antibacterial activity of certain oil was increased after encapsulation in nanoemulsion because the nano size droplets carrying oil has increased interaction of EO with the bacterial cell membrane and further disturbs the structural integrity of membrane leading to leakage of bacterial intracellular constituents. This nonspecific action of entrapped EO is able to decrease the development of resistant microbial strains (Maswal and Dar, 2014). Essential oil encapsulated in various nanoemulsions is shown in Table 14.4. The results of entrapped EOs demonstrated that nanoemulsion is an efficient approach to increase physical stability and provide an easy way to transfer the active molecules to the sites of action.

The antimicrobial activity after encapsulation of D-Limonene and a mixture of terpenes extracted from *Melaleuca alternifolia* into nanoemulsions based on essential oils of sunflower and palm oil was developed by high-pressure homogenization. The effect of encapsulation on the antimicrobial activity of terpenes was investigated by determining the minimum inhibitory concentration (MIC) and minimum bactericidal concentration (MBC) for three different classes of microorganisms (*Lactobacillus delbrueckii, Saccharomyces cerevisiae,* and *Escherichia coli*). Results showed that the increase in the antimicrobial activity depends on the formulation and mean diameter of the nano droplets as well as on the microorganism's class. The antimicrobial activity was tested in pear and orange juices inoculated with *L. delbrueckii* and these encapsulated terpenes in nanosize showed higher antimicrobial properties. Therefore, lower antimicrobial concentrations are required for a bactericidal action under accelerated aging at 32°C, without any visual and organoleptic changes in the juice (Donsì et al., 2011).

Buranasuksombat et al. (2011) has investigated the effect of oil droplet size on antimicrobial activity. O/W emulsions were prepared from lemon myrtle oil (LMO) possessing antimicrobial activity and soyabean oil, which has no antimicrobial properties. The antimicrobial properties were determined against five bacteria. Results showed that all emulsions made from LMO had the same level of antimicrobial effects against the five bacteria whereas all soybean oil emulsions had no antimicrobial effect. It was concluded that the antimicrobial property of nanoemulsions was due to the active agent encapsulated in the emulsions and not from high surface tensions and cell wall diffusion activity of nanosized droplet (Buranasuksombat et al., 2011).

Eugenol-incorporated O/W nanoemulsion was prepared by ultrasound cavitation method using sesame oil, Tween 80, and water, which was stable for more than 1 month. Encapsulated eugenol exhibited antibacterial activity against *Staphylococcus*

Table 14.4 Essential Oils Encapsulated in Nanoemulsions

Formulation	Essential Oil	Inferences	References
Nanoemulsion	Terpenes from melaleuca, sunflower oil, palm oil, and D-limonene	Enhanced antimicrobial activity without affecting organoleptic properties of the juice	Donsì et al. (2011)
Nanoemulsion	Lemon myrtle oil	Enhanced antimicrobial properties	Buranasuksombat et al. (2011)
Nanoemulsion	Basil oil (*Ocimum basilicum*)	Nanoemulsion exhibited considerable antibacterial activity even after diluting 10-, 100- and 1000-fold with water	Ghosh et al. (2012)
Nanoemulsion	Peppermint oil (PO) with medium-chain triacylglycerol	Prolonged antibacterial activities with extended shelf life	Liang et al. (2012)
Nanodispersion	Thymol	Enhanced antimicrobial activity	Shah et al. (2012)
	Carvacrol, limonene, and cinnamaldehyde		Donsì et al. (2012)
Nanoemulsion	Eucalyptus oil	Enhanced antibacterial activity	Saranya et al. (2012)
Nanoemulsion	Essential oils of andiroba (*Carapa guaianensis*) and aroeira (*Schinus molle*)	Increased activity against *T. evansi* in vitro	Baldissera et al. (2013)
Nanoemulsion	Lemongrass, clove, tea tree, thyme, geranium, marjoram, palmarosa, rosewood, sage, or mint	Enhanced bactericidal activity	Salvia-Trujillo et al. (2015)
Nanoemulsion	*Thymus daenensis* EO	Amplified antibacterial activity	Moghimi et al. (2016)
SNEDDS	Zedoary turmeric oil (ZTO)	1.7- and 2.5-fold increase in AUC and C_{max}, respectively compared with the nonencapsulated ZTO	Zhao et al. (2010)

aureus and reported to reduced microorganism population 3 log (CFU/mL) after 120 min due to changes in bacterial membrane permeability (Ghosh et al., 2012).

Peppermint oil (PO)–loaded nanoemulsion was prepared using medium-chain triacylglycerol and a food-grade biopolymer that is, modified starch as stabilizer. The developed PO nanoemulsions showed a mean diameters of <200 nm with high stability over at least 30 days of storage time. The antimicrobial properties of PO have been evaluated by two assays, the minimum inhibitory concentration (MIC) and time-kill dynamic processes, against bacterial strains of *Listeria monocytogenes* Scott A and *Staphylococcus aureus* ATCC 25923. Compared with PO, the PO nanoemulsions showed prolonged antibacterial activities. The results illustrated that the nanoemulsion technology can provide novel applications of essential oils in extending the shelf life of aqueous food products (Liang et al., 2012).

In a study, free and nanodispersed (ND) thymol were compared in terms of their antimicrobial efficacies against strains of *Escherichia coli* and *Listeria monocytogenes* in apple cider and 2% reduced-fat milk. The antimicrobial assay was performed at pH 5.5 and 3.5 in apple cider at 0.3-, 0.5-, 0.75-, and 1.0-g/L thymol concentrations at 35, 32, 25, and 4°C. It was found that 0.5 and 1.0 g/L thymol either in free form or in nanodispersion were inhibitory and bactericidal, respectively, against bacterial strains under all treatment conditions. In the case of 2% reduced-fat milk at 35 or 32°C, ND and free thymol demonstrated the inhibition of bacteria at 4.5 g/L. Results showed promising and novel, enabling thymol-containing nanodispersions as an effective carrier to target pathogens in food, especially in clear beverages (Shah et al., 2012).

Sunflower oil–based nanoemulsion was developed as a carrier for carvacrol, limonene, and cinnamaldehyde by high-pressure homogenization and stabilized by different emulsifiers. Antimicrobial activities of these formulations were assayed against *Escherichia coli, Lactobacillus delbrueckii,* and *Saccharomyces cerevisiae.* The antimicrobial activity was found to be dependent on the concentration of active agent in the aqueous phase, which was further dependent on their emulsifier capability to solubilize them. Sugar esters and glycerol monooleate were reported to have high antimicrobial activity due to their ability to solubilize high concentration of the essential oil in the aqueous phase. The complete inactivation was achieved in case of *E.coli* and *L. delbrueckii* while a reduced population of *S. cerevisiae* (2 log) observed by carvacrol nanoemulsion after 2 h. Thereafter, complete inactivation of *S. cerevisiae* was also attained after 24 h. However, a less effective antimicrobial activity was exhibited in case of limonene and cinnamaldehyde nanoemulsions. It was found that complete inactivation of *E. coli, L. delbrueckii,* and *S. cerevisiae* was obtained after 24 h at more concentrated EO nanoemulsions. This was due to the easy availability of the EO from the nanoemulsion, which has significantly enhanced bactericidal effect over shorter time scales (Donsì et al., 2012).

The O/W based nanoemulsion was formulated using eucalyptus oil, Tween 20, and ethanol by ultrasonication and the mean droplet size was observed 20.17 nm by dynamic light scattering. The antibacterial activity of the EO-loaded nanoemulsion was tested against *P. mirabilis*. Results showed that EO-loaded nanoemulsions were highly stable, transparent, and found to be effective bactericidal against tested pathogen. One hundred percent growth inhibition was obtained with nanoemulsion, which was confirmed by dilution plate count and antibacterial susceptibility method (Saranya et al., 2012).

Nanoemulsions containing carvacrol and eugenol was prepared by high-pressure homogenization and ultrasonication using triacylglyceride (Miglyol 812N) or Tween 80. Antimicrobial activity of the developed emulsion was evaluated against *Escherichia coli* C 600 and *Listeria innocua*.

Carvacrol emulsions at a concentration of 800 ppm with a droplet size of 3000 nm reported to completely inhibit *L. innocua*, while nanoemulsion (mean droplet size of 80 nm) was only able to delay the growth of microbes. The results was attributed to the fact that macroemulsions were more effective as antimicrobial carrier compared to the nanoemulsions because of an increased sequestering of antimicrobials in emulsion interfaces and reduced solubilization in excess of Tween 80 micelles (Terjung et al., 2012).

Joe and coworkers have developed a sunflower oil–surfactin–based O/W nanoemulsion, where biosurfactant surfactin is a cyclic lipopeptide antibiotic obtained from *B. subtilis*. The nanoemulsion was shown to have higher antibacterial activity against *S. typhi*, *L. monocytogenes*, and *S. aureus* compared with streptomycin (positive control) at a concentration of 100 mg/L. High fungicidal activity was reported against *Aspergillus niger*, *Rhizopus nigricans*, and *Penicillium* sp. compared with sodium benzoate (positive control) with sporicidal activity against *Bacillus cereus* and *Bacillus circulans*, which was 3 times higher than the positive control. They concluded that sunflower oil–surfactin nanoemulsion can be used in food products such as raw chicken, apple juice, milk, and mixed vegetable to reduce the native cultivable bacterial and fungal populations (Joe et al., 2012).

Essential oil obtained from Basil (*Ocimum basilicum*) which contains 88% of estragole was encapsulated in nanoemulsion prepared by ultrasonic emulsification using Tween 80 and water. The antibacterial activity of encapsulated EO against *E. coli* was determined after different dilution. Results showed that the nanoemulsion were sufficient to inactivate the bacterial strain even after dilution. Complete inactivation of *E. coli* was achieved with a 10-fold and 100-fold diluted nanoemulsion after 45 min, while a 40% reduction was shown after 60 min incubation period with the 1,000-fold diluted sample. The alteration in the bacterial cell membrane was the possible mechanism shown by fluorescence microscopy and FTIR (Ghosh et al., 2013).

The susceptibility of *Trypanosoma evansi* to the essential oils of andiroba (*Carapa guaianensis*) and *aroeira* (*Schinus molle*) under in vitro condition was investigated in their conventional and nanoencapsulated forms. For this pure oils in different concentrations (0.5, 1.0, and 2.0%) were used. An untreated (negative control) and 0.5% diminazene aceturate treated (positive control) samples were used as comparative parameters. Later, the nanoemulsions oils at concentrations of 0.5 and 1.0% were tested in the same way. All the tests were performed in triplicates and the numbers of parasites in each group were quantified after 1, 3, and 6 h from onset of the study. It was observed that the reduction in the number of parasites after 1 h is dose-dependent phenomenon. A significant reduction in the parasites concentration was achieved at low concentrations after 3 h, as well as at 6 h with no live parasites in case of the essential oils–treated sample. The results indicate that oils of *andiroba* and *aroeira* in their conventional as well as nanoemulsion forms were active against *T. evansi* under in vitro test. This can be attributed to the fact that these oils can be used as an alternative for the treatment of *T. evansi* infections (Baldissera et al., 2013).

A W/O/W multiple nanoemulsions loaded with lactoferrin was prepared by homogenization using lecithin and poloxamers. Where, lactoferrin is an iron-binding protein that can inhibit the growth of or kills iron-dependent pathogenic bacteria. Both free and encapsulated lactoferrin showed a minimum inhibitory concentration (MIC) of 2000 mg/mL for *S. aureus* and *L. innocua* and 200 mg/mL for *Candida albicans*. The antibacterial activity remained the same for the free and encapsulated lactoferrin but these multiple nanoemulsions can be employed to formulate oral elixir and beverages (Balcão et al., 2013).

Lemongrass oil (LO)–encapsulated nanoemulsions were prepared using carnauba–shellac wax (CSW) by high-pressure homogenization (Kim et al., 2013; Jo et al., 2014) and alginate by ultrasonication and microfluidization (Salvia-Trujillo et al., 2014). The nanoemulsion with CSW was showed a decrease by 8.18 log CFU/g the total population of *E. coli* and *L. monocytogenes* after 2 h. Stability of apples coated with unloaded and LO-loaded CSW nanoemulsions was studied. It was reported that a decrease of 0.8 and 1.4 log CFU of aerobic bacteria was obtained with unloaded and LO-loaded CSW nanoemulsions, respectively after 5 months of storage. The coatings inhibited the development of yeast and mold and even the growth of *Salmonella typhimurium* and *E. coli* O157:H7 was inhibited on apples and plums, respectively. This property of EO-loaded nanoemulsions could be used to preserve various physicochemical qualities of fruits. In case of LO-alginate nanoemulsions, the antibacterial effect against *E. coli* was dependent on the nanoemulsion production process. The microfluidization enhanced antimicrobial activity, while ultrasounds diminished the activity (Kim et al., 2013; Jo et al., 2014).

Nanoemulsions containing a range of EOs were prepared by high-shear homogenization using Tween 80 and sodium alginate as stabilizers. EOs such as lemongrass, clove, tea tree, thyme, geranium, marjoram, palmarosa, rosewood, sage, or mint were considered for the study. Size of the developed nanoemulsion was reduced by microfluidization except of palmarosa and rosewood oil emulsions, which were already in the nanorange. The ζ-potential of the emulsions were more than −30 mV, which indicates a strong electrostatic repulsion of the dispersed oil droplets in the aqueous phase. In vitro bactericidal action against *Escherichia coli* was performed and observed a 4.1, 3.6, 2.8, or 3.9 log-reductions with lemongrass, clove, thyme, or palmarosa-loaded nanoemulsions, respectively, after 30 min of contact time. This showed a faster and enhanced inactivation kinetic was achieved in case of nanoemulsions containing lemongrass or clove essential oils in comparison with their respective coarse emulsions. This illustrated the promising advantages of using nanoemulsions as carriers of flavoring and preservative agents in the food industry (Salvia-Trujillo et al., 2015).

Thymus daenensis–loaded water-dispersible nanoemulsion was formulated using high-intensity ultrasound with a mean diameter of 143 nm. The antibacterial activity of the free EO and EO-loaded nanoemulsion was measured against *Escherichia coli*. The minimum inhibitory concentration (MIC) and minimum bactericidal concentration (MBC) were measured. It was reported that nanoemulsions enhanced the bactericidal affect of the EO and this was due to easier access of the essential oils to the bacterial cells in the form of nanoemulsion (Moghimi et al., 2016).

A self-nanoemulsifying drug delivery system (SNEDDS) was developed for the oral delivery of zedoary turmeric oil (ZTO). With the help of pseudoternary phase diagrams the regions of the efficient self-emulsification were identified. ZTO was used with a second oil phase to enhance drug-loading efficiency. An optimized formulation consisting of ZTO, ethyl oleate, Tween 80, transcutol P (30.8:7.7:40.5:21, w/w), with 30% drug loading was prepared. The active components were stable in the optimized SNEDDS during the storage period of 12 months at 25°C. A 1.7-fold and 2.5-fold enhanced in AUC and C_{max} respectively was shown of germacrone (GM) after oral administration of ZTO-SNEDDS in rats compared with the unformulated ZTO (Zhao et al., 2010).

8 Molecular Complexes

A molecular complex is referred to the physical association between a host and a guest (active ingredient) molecule, which can improve solubility and stability of the active agent. In case of EOs, the molecular inclusion is achieved by using cyclodextrins (CDs) as the encapsulating materials. CDs are a group of naturally occurring cyclic oligosaccharides obtained from starch, with six, seven, or eight glucose residues linked by a (1–4) glycosidic bonds in a cylinder-shaped structure. These are classified into three types as α-, β-, and γ-cyclodextrins, corresponding to 6, 7, and 8 glucopyranose units linked by α-(1,4) bonds, respectively. The dimension of the internal cavity is crucial for the "encapsulation" of guest molecules (Duchêne et al., 1999; Pagington, 1986). In the hydrophobic cavity they are able to enclose highly hydrophobic molecules in a molecular ratio and constituting a true molecular encapsulation (Dodziuk, 2006).

The major advantages of CD-complexation in pharmaceutical, food, cosmetics, and toiletries industries are (1) protection of the active ingredients against oxidation, light-induced decomposition, and thermal decomposition; (2) evaporation and sublimation losses; (3) elimination or reduction of undesired tastes/odors; (4) to reduce gastric-intestinal irritation or ocular disturbances; (5) to prevent interactions with other additive and many more advantage such as to convert oils and liquid active agents into microcrystalline or amorphous powders and to reduce hygroscopicity (Rubistein, 1989). The majority work that has been done on EOs inclusion complex are concerning the encapsulation of essential oils with β-CD and its derivatives, that is, methylated-β-cyclodextrin, hydroxypropyl-β-cyclodextrin, and low methylated-β-cyclodextrin.

Thymol and cinnamaldehyde complexes with β-CD were investigated and the release of encapsulated compounds was determined. The inclusion complex with both of the components is formed in a 1:1 molar ratio. Water sorption for cyclodextrin and complex was evaluated. They were stored at different relative humidity; vary from 22 to 97%, at 25°C. The release of encapsulated active agent was determined following the melting enthalpy of each guest. Both β-CD and the complexes exhibited similar water sorption isotherms and it was observed that water sorption occurred in β-CD up to RH 50%, which remained constant till RH

75% and beyond this value water sorption increased rapidly and achieved maximum value at RH 97%. While water sorbed by the complexes at each RH was smaller than that found for β-CD, no thymol or cinnamaldehyde release was detected till RH < 84%; beyond this it increased abruptly, coincidentally with the abrupt increase of absorbed water.

Water sorption significantly affects the stability of β-CD complexes, which is thus governed by the shape of the water sorption isotherm. The stability studies showed that both the inclusion complexes remain stable up to 75% RH during long storage times. The release of thymol and cinnamaldehyde from the β-CD complexes were detectable after RH 84%, at which a sharp increase of water content occurred. Therefore, by selecting an appropriate storage conditions for encapsulated EO in β-CD, it is easy to predict the shelf life of functional products formulated with nanoencapsulated compounds (Cevallos et al., 2010).

β-Caryophyllene (BCP), a natural sesquiterpene present in the EO of many plants, has exhibited many biological activities such as antimicrobial, antioxidant, antiinflammatory, anticarcinogenic, anxiolytic-like, and local anesthetic effects. The volatility and poor water solubility of BCP, limits its use in the pharmaceutical field. Liu and coworkers prepared BCP inclusion complex with β-CD and investigated its oral bioavailability and the pharmacokinetics after a single oral dose of 50 mg/kg in rats, which was further compared with free BCP (Liu et al., 2013). It was found that BCP was rapidly released from IC and the in vivo data showed that BCP/β-CD IC achieved an earlier T_{max}, higher C_{max}, and the $AUC_{0-12\ h}$ showed approximately 2.6 times larger than those of free BCP. Thus, it showed that β-CD has significantly improved the oral bioavailability of the drug in rats compared to free BCP (Liu et al., 2013).

The essential oil of *Chamomilla recutita* (L.) contains up to 50% (−)-α-bisabolol, which contributes to the antiinflammatory properties of camomile oil. Bisabolol is a highly lipophilic component and on exposure to atmosphere it oxidizes and decreases antiinflammatory activity. Fifty percent (−)-α-Bisabolol was able to form an inclusion complex with β-CD in solution as well as in the solid state. The molecular associations of β-CD with pure (−)-α-bisabolol or (−)-α-bisabolol as a component of camomile EO was investigated by Waleczek and coworkers. They studied the phase solubility and reported the complex constant 273 M^{-1} and 304 M^{-1} for the pure (−)-α-bisabolol and (−)-α-bisabolol as a constituent of the EO, respectively. They observed significant difference in intrinsic solubility of pure (−)-α-bisabolol (4.85×10^{-4} M) and (−)-α-bisabolol as a component of the EO

(1.82×10^{-4} M). Computer simulation showed that an inclusion complex is possible in a stoichiometric composition of 2:1 (β-CD:drug) (Waleczek et al., 2003).

Thymol and cinnamaldehyde are frequently used as flavors in food industries but they also have antimicrobial, antioxidant, and antiseptic properties, making them more useful as preservatives in food and pharmaceutical industries. Thymol is a monoterpene obtained from Lamiaceae plants, especially oreganos and thymes while 65–75% Cinnamaldehyde (3-phenyl-2-propenal) is present in the cinnamon EO. As they exist in EO, they are very sensitive to the effects of light, oxygen, humidity, and high temperatures. Hill and coworkers elucidated the affect of essential oil complexation with β-cyclodextrin (EO-β-CD) on the physicochemical characteristics of essential oils and their resulting antimicrobial activity. Cinnamon bark extract (contain *trans*-cinnamaldehyde), clove bud extract (contain eugenol), and a 2:1 (*trans*-cinnamaldehyde:eugenol) mixture were microencapsulated by the freeze-drying method. Furthermore, the complexes were characterized for particle size, morphology, polydispersity index (PDI), entrapment efficiency, and phase solubility. All particles were morphologically smooth and spherical in shape with a nonsignificant difference in size distribution, and tendency to form agglomerate. The entrapment efficiencies vary from 41.7 to 84.7%, where the pure constituents were higher than extracts.

The antimicrobial activity were analyzed for the oil and their cyclodextrin complex, against *Salmonella enterica* serovar *typhimurium* LT2 and *Listeria innocua*. Results showed that EO-β-CD complexes were able to inhibit both bacterial strains at lower concentrations than free oils. This can be attributed to their increased water solubility, which determined an increased contact between pathogens and essential oils. Thus, EO inclusion complexes could be useful as antimicrobial delivery systems with a broad spectrum of application in food systems, where gram-positive and gram-negative bacteria could present a risk (Hill et al., 2013).

The extract from garlic (*Allium sativum* L.) is called garlic oil (GO) and mainly consists of diallyl disulfide, diallyl trisulfide, allyl propyl disulfide, a small quantity of disulfide, and probably diallyl polysulfide (Pranoto et al., 2005). GO is reported to be more potent than aqueous extracts of garlic and exhibits a wide range of pharmacological properties such as antimicrobial, antidiabetic, antimutagenic, and anticarcinogenic effects (Agarwal, 1996). However, due to its volatility, strong odor, insolubility in water, and low physicochemical stability, its application is limited. The inclusion complex with of β-CD was prepared by the coprecipitation method in a molar ratio of 1:1. Furthermore, this was investigated for the stability and water solubility. The apparent stability constant of inclusion complex was 1141 M^{-1}, with the improved water solubility (Wang et al., 2011).

Isothiocyanates (ITCs), a naturally occurring antimicrobial compound rich in cruciferous vegetables, such as broccoli and cabbage, is obtained by mechanical disruption of cruciferous plant tissues (Delaquis and Mazza, 1995). ITCs, in their allyl isothiocyanate (AITC) form, have been extensively studied for their application in food preservation. The inclusion complex between these isothiocyanates (ITCs), namely, allyl isothiocyanate (AITC) and phenyl isothiocyanate (PITC), and randomly methylated β-cyclodextrin (RM-β-CD) were studied. They have reported that RM-β-CD have a strong solubilizing effect on the poorly water-soluble AITC and PITC in the aqueous phase. Both of these compounds were reported to form inclusion complexes with RM-β-CD at guest to host ratios of 1:1 and 1:2 in the aqueous phase (Neoh et al., 2012).

Ciobanu and coworkers have investigated the novel controlled release systems for the delivery of few EOs used as ambient odors. They used static headspace gas chromatography (SH-GC) to study the interactions of cyclodextrins (CDs) and β-CD polymers with linalool and camphor in *Lavandula angustifolia* essential oil. The retention capacity of α-CD, β-CD, γ-CD, hydroxypropyl-β-cyclodextrin (HPBCD), randomly methylated-β-cyclodextrin (RAMEB), a low methylated-β-cyclodextrin (CRYSMEB) and crosslinked β-CD polymers for linalool and camphor two major components of *Lavandula angustifolia* essential oil were investigated. All CDs and CD polymers reported to form stable inclusion complexes in the molar ratio of 1:1 and reduced the volatility of the aroma. Static experiments were used to determine the retention capacity of the CD derivatives. The RAMEB showed the higher complex formation constant both for the standard compounds (833 M^{-1} for linalool and 1194 M^{-1} for camphor) and for the compounds in the in *Lavandula angustifolia* essential oil (1074 M^{-1} for linalool and 2963 M^{-1} for camphor). It was concluded that β-CD polymers can be used as novel delivery system which allows the controlled release of aroma compounds (Ciobanu et al., 2012).

Inclusion complexes of basil and tarragon essential oils (EOs) and estragole (ES) as pure compound was formulated with α-cyclodextrin (α-CD), β-cyclodextrin (β-CD), hydroxypropyl-β-cyclodextrin(HP-β-CD), randomly methylated-β-cyclodextrin (RAMEB), a low methylated-β-cyclodextrin (CRYS-MEB), and γ-cyclodextrin (γ-CD) were characterized by differential scanning calorimetry (DSC) and Fourier transform infrared spectroscopy

(FT-IR). The inclusion complexes were prepared by the freeze-drying method for different CD:ES molar ratios and their formation constants (K_f) were determined in aqueous solution by nonlinear regression analysis using static headspace gas chromatography and ultraviolet–visible spectroscopy. All inclusion complexes showed the controlled release of ES as well as increased DPPH radical scavenging activity and photostability of ES and ES-containing EOs (ESEOs). These findings suggested that encapsulation with CDs could be an efficient tool to improve the use of essential oils in cosmetic and food fields (Kfoury et al., 2015).

9 Conclusions

Essential oils, due to their wider application, have captured attention in the food, drug, and cosmetic commercial arena. However, the drawbacks, such as low water solubility and stability, strong flavor, and aroma, along with the side effects associated with their use, have limited their application at larger level.

Emergence of nanotechnology and its application in food science advances the suitability and durability of these food related products, that is, essential oils. Nanoencapsulation of EOs in the form of liposomes, SLN, nanoemulsion, and polymeric nanoparticles offers a potential approach to alleviate all the issues related to chemical stability and solubility. Furthermore, the nanocarriers facilitate safe and easy handling of the EO by changing the state, masking the undesirable taste, and showing controlled and sustained release with reduced toxic and side effects. These are able to retain the chemical and biological activity of EOs for longer periods of time, and in most of the cases the activity was enhanced after encapsulation of EO in suitable carrier system.

References

Abreu, F.O., Oliveira, E.F., Paula, H.C., de Paula, R.C., 2012. Chitosan/cashew gum nanogels for essential oil encapsulation. Carbohydr. Polym. 89 (4), 1277–1282.

Adorjan, B., Buchbauer, G., 2010. Biological properties of essential oils: an updated review. Flavour Frag. J. 25 (6), 407–426.

Agarwal, K.C., 1996. Therapeutic actions of garlic constituents. Med. Res. Rev. 16 (1), 111–124.

Alhaj, N.A., Shamsudin, M.N., Mohamed Alipiah, N., Zamri, H.F., Abdul, A.B., Ibrahim, S., Abdullah, R., 2010. Characterization of *Nigella sativa* L. essential oil-loaded solid lipid nanoparticles. Am. J. Pharmacol. Toxicol. 5 (1), 52–57.

Anweker, H., Patel, S., Singhai, A.K., 2011. Liposome as drug carriers. Int. J. Pharm. Life Sci. 2, 945–951.

Augustin, M.A., Sanguansri, P., 2009. Nanostructured materials in the food industry. Adv. Food Nutr. Res. 58 (4), 183–213.

Augustin, M.A., Sanguansri, L., Margetts, C., Young, B., 2001. Microencapsulation of food ingredients. Food Aust. 53, 220–223.

Bakkali, F., Averbeck, S., Averbeck, D., Idaomar, M., 2008. Biological effects of essential oils—a review. Food Chem. Toxicol. 46 (2), 446–475.

Balcão, V.M., Costa, C.I., Matos, C.M., Moutinho, C.G., Amorim, M., Pintado, M.E., Gomes, A.P., Vila, M.M., Teixeira, J.A., 2013. Nanoencapsulation of bovine lactoferrin for food and biopharmaceutical applications. Food Hydrocoll. 32 (2), 425–431.

Baldissera, M.D., Da Silva, A.S., Oliveira, C.B., Zimmermann, C.E., Vaucher, R.A., Santos, R.C., Rech, V.C., Tonin, A.A., Giongo, J.L., Mattos, C.B., Koester, L., Santurio, J.M., Monteiro, S.G., 2013. Trypanocidal activity of the essential oils in their conventional and nanoemulsion forms: in vitro tests. Exp. Parasitol. 134 (3), 356–361.

Bilia, A.R., Guccione, C., Isacchi, B., Righeschi, C., Firenzuoli, F., Bergonzi, M.C., 2014. Essential oils loaded in nanosystems: a developing strategy for a successful therapeutic approach. Evid. Based Complement. Alternat. Med. 2014, 651593.

Blanco-Padilla, A., Soto, K.M., Hernández Iturriaga, M., Mendoza, S., 2014. Food antimicrobials nanocarriers. Scientific World Journal 2014, 837215.

Buchbauer, G., Jirovetz, L., Jäger, W., Plank, C., Dietrich, H., 1993. Fragrance compounds and essential oils with sedative effects upon inhalation. J. Pharm. Sci. 82 (6), 660–664.

Bummer, P.M., 2004. Physical chemical considerations of lipid-based oral drug delivery solid lipid nanoparticles. Crit. Rev. Ther. Drug Carrier Syst. 21, 1–20.

Buranasuksombat, U., Kwon, Y., Turner, M., Bhandari, B., 2011. Influence of emulsion droplet size on antimicrobial properties. Food Sci. Biotechnol. 20, 793–800.

Burguera, J.L., Burguera, M., 2012. Analytical applications of emulsions and microemulsions. Talanta 96, 11–20.

Burt, S., 2004. Essential oils: their antibacterial properties and potential applications in foods—a review. Int. J. Food Microbiol. 94, 223–253.

Castangia, I., Manca, M.L., Caddeo, C., Maxia, A., Murgia, S., Pons, R., Demurtas, D., Pando, D., Falconieri, D., Peris, J.E., Fadda, A.M., Manconi, M., 2015. Faceted phospholipid vesicles tailored for the delivery of *Santolina insularis* essential oil to the skin. Colloids Surf. B Biointerfaces 132, 185–193.

Celia, C., Trapasso, E., Locatelli, M., Navarra, M., Ventura, C.A., Wolfram, J., Carafa, M., Morittu, V.M., Britti, D., Di Marzio, L., Paolino, D., 2013. Anticancer activity of liposomal bergamot essential oil (BEO) on human neuroblastoma cells. Colloids Surf. B 112, 548–553.

Cevallos, P.A.P., Buera, M.P., Elizalde, B.E., 2010. Encapsulation of cinnamon and thyme essential oils components (cinnamaldehyde and thymol) in β-cyclodextrin: effect of interactions with water on complex stability. J. Food Eng. 99 (1), 70–75.

Chaudhry, Q., Scotter, M., Blackburn, J., Ross, B., Boxall, A., Castle, L., Watkins, R., 2008. Applications and implications of nanotechnologies for the food sector. Food Addit. Contam. Part A 25 (3), 241–258.

Chen, F., Shi, Z., Neoh, K.G., Kang, E.T., 2009. Antioxidant and antibacterial activities of eugenol and carvacrol-grafted chitosan nanoparticles. Biotechnol. Bioeng. 104, 30–39.

Choi, M.J., Soottitantawat, A., Nuchuchua, O., Min, S.G., Ruktanonchai, U., 2009. Physical and light oxidative properties of eugenol encapsulated by molecular inclusion and emulsion diffusion method. Food Res. Int. 42 (1), 148–156.

Chong, G.H., Yunus, R., Abdullah, N., Choong, T.S.Y., Spotar, S., 2009. Coating and encapsulation of nanoparticles using supercritical antisolvent. Am. J. Appl. Sci. 6, 1352–1358.

Ciobanu, A., Mallard, I., Landy, D., Brabie, G., Nistor, D., Fourmentin, S., 2012. Inclusion interactions of cyclodextrins and crosslinked cyclodextrin polymers with linalool and camphor in Lavandula angustifolia essential oil. Carbohydr. Polym. 87 (3), 1963–1970.
Coimbra, M., Isacchi, B., Van Bloois, L., Torano, J.S., Ket, A., Wu, X., Broere, F., Metselaar, J.M., Rijcken, C.J., Storm, G., Bilia, R., Schiffelers, R.M., 2011. Improving solubility and chemical stability of natural compounds for medicinal use by incorporation into liposomes. Int. J. Pharm. 416, 433–442.
Cortés-Rojas, D.F., Souza, C.R., Oliveira, W.P., 2014. Encapsulation of eugenol rich clove extract in solid lipid carriers. J. Food Eng. 127, 34–42.
Cui, H., Zhao, C., Lin, L., 2015. The specific antibacterial activity of liposome-encapsulated clove oil and its application in tofu. Food Control. 56, 128–134.
Cussler, E.L., 1997. Diffusion, Mass Transfer in Fluid Systems, second ed. Cambridge University Press, Cambridge, UK.
de Oliveira, E.F., Paula, H.C., de Paula, R.C., 2014. Alginate/cashew gum nanoparticles for essential oil encapsulation. Colloids Surf. B 113, 146–151.
Delaquis, P.J., Mazza, G., 1995. Antimicrobial properties of isothiocyanates in food preservation. Food Technol. 49 (11), 73–84.
Desai, K.G.H., Park, H.J., 2005. Recent developments in microencapsulation of food ingredients. Dry. Technol. 23, 1361–1394.
Detoni, C.B., Cabral-Albuquerque, E.C.M., Hohlemweger, S.V.A., Sampaio, C., Barros, T.F., Velozo, E.S., 2009. Essential oil from *Zanthoxylum tingoassuiba* loaded into multilamellar liposomes useful as antimicrobial agents. J. Microencapsul. 26 (8), 684–691.
Detoni, C.B., de Oliveira, D.M., Santo, I.E., Pedro, A.S., El-Bacha, R., da Silva Velozo, E., Ferreira, D., Sarmento, B., de Magalhães Cabral-Albuquerque, E.C., 2012. Evaluation of thermal-oxidative stability and antiglioma activity of *Zanthoxylum tingoassuiba* essential oil entrapped into multi- and unilamellar liposomes. J. Liposome Res. 22 (1), 1–7.
Devi, K.P., Nisha, S.A., Sakthivel, R., Pandian, S.K., 2010. Eugenol (an essential oil of clove) acts as an antibacterial agent against *Salmonella typhi* by disrupting the cellular membrane. J. Ethnopharmacol. 130 (1), 107–115.
Dima, S., Dima, C., Iordachescu, G., 2015. Encapsulation of functional lipophilic food and drug biocomponents. Food Eng. Rev. 7, 1–22.
Dodziuk, H., 2006. Cyclodextrins and Their Complexes. Wiley-VCH, GmbH & KGaA, Weinheim, Germany.
Donsì, F., Sessa, M., Mediouni, H., Mgaidi, A., Ferrari, G., 2011. Encapsulation of bioactive compounds in nanoemulsion-based delivery systems. Proc. Food Sci. 1, 1666–1671.
Donsì, F., Annunziata, M., Vincensi, M., Ferrari, G., 2012. Design of nanoemulsion-based delivery systems of natural antimicrobials: effect of the emulsifier. J. Biotechnol. 159 (4), 342–350.
Duchêne, D., Wouessidjewe, D., Ponchel, G., 1999. Cyclodextrins and carrier systems. J. Control. Release 62 (1), 263–268.
Ezhilarasi, P.N., Karthik, P., Chhanwal, N., Anandharamakrishnan, C., 2013. Nanoencapsulation techniques for food bioactive components: a review. Food Bioprocess Technol. 6, 628–647.
Fan, L.T., Singh, S.K., 1989. Controlled Release: A Quantitative Treatment. Springer-Verlag, Berlin, Germany.
Fathi, M., Mozafari, M.R., Mohebbi, M., 2012. Nanoencapsulation of food ingredients using lipid-based delivery systems. Trends Food Sci. Technol. 23 (1), 13–27.
Flores, F.C., Ribeiro, R.F., Ourique, A.F., Rolim, C.M.B., Silva, C.D.B.D., Pohlmann, A.R., Ruy, C.R.B., Guterres, S.S., 2011. Nanostructured systems containing an essential oil: protection against volatilization. Química Nova 34 (6), 968–972.

Flores, F.C., De Lima, J.A., Ribeiro, R.F., Alves, S.H., Rolim, C.M.B., Beck, R.C.R., Da Silva, C.B., 2013. Antifungal activity of nanocapsule suspensions containing tea tree oil on the growth of *Trichophyton rubrum*. Mycopathologia 175 (3–4), 281–286.

Ge, Y., Ge, M., 2016. Distribution of *Melaleuca alternifolia* essential oil in liposomes with Tween 80 addition and enhancement of in vitro antimicrobial effect. J. Exp. Nanosci. 11 (5), 345–358.

Ghosh, V., Mukherjee, A., Chandrasekaran, N., 2012. Ultrasonic emulsification of food-grade nanoemulsion formulation and evaluation of its bactericidal activity. Ultrason. Sonochem. 4, 497–500.

Ghosh, V., Mukherjee, A., Chandrasekaran, N., 2013. Ultrasonic emulsification of food-grade nanoemulsion formulation and evaluation of its bactericidal activity. Ultrason. Sonochem. 20 (1), 338–344.

Gibbs, B.F., Kermasha, S., Alli, I., Mulligran, C.N., 1999a. Pressure-and heat-induced gelation of mixed beta-lactoglobulin/polysaccharide solutions: scanning electron microscopy of gels. Food Hydrocoll. 13, 339–351.

Gibbs, B.F., Kermasha, S., Alli, I., Mulligan, C.N., 1999b. Encapsulation in the food industry. Int. J. Food Sci. Nutr. 50, 213–224.

Gomes, C., Moreira, R.G., Castell-Perez, E., 2011. Poly (DL-lactide-*co*-glycolide) (PLGA) nanoparticles with entrapped *trans*-cinnamaldehyde and eugenol for antimicrobial delivery applications. J. Food Sci. 76 (2), N16–N24.

Gortzi, O., Lalas, S., Chinou, I., Tsaknis, J., 2006. Reevaluation of antimicrobial and antioxidant activity of Thymus spp. exctracts before and after encapsulation in liposomes. J. Food Protec. 69, 2998–3005.

Gortzi, O., Lalas, S., Chinou, I., Tsaknis, J., 2007. Evaluation of the antimicrobial and antioxidant activities of *Origanum dictamnus* extracts before and after encapsulation in liposomes. Molecules 12, 932–945.

Guenther, E., 1972. The Essential Oils. Krieger Publishing Company, Malabar, FL.

Hammer, K.A., Carson, C.F., Riley, T.V., Nielsen, J.B., 2006. A review of the toxicity of *Melaleuca alternifolia* (tea tree) oil. Food Chem. Toxicol. 44, 616–625.

Herculano, E.D., de Paula, H.C., de Figueiredo, E.A., Dias, F.G., Pereira, V.D.A., 2015. Physicochemical and antimicrobial properties of nanoencapsulated *Eucalyptus staigeriana* essential oil. LWT Food Sci. Technol. 61 (2), 484–491.

Hill, L.E., Gomes, C., Taylor, T.M., 2013. Characterization of beta-cyclodextrin inclusion complexes containing essential oils (*trans*-cinnamaldehyde, eugenol, cinnamon bark, and clove bud extracts) for antimicrobial delivery applications. LWT Food Sci. Technol. 51 (1), 86–93.

Hosseini, S.F., Zandi, M., Rezaei, M., Farahmandghavi, F., 2013. Two-step method for encapsulation of oregano essential oil in chitosan nanoparticles: preparation, characterization, and in vitro release study. Carbohydr. Polym. 95 (1), 50–56.

Iannitelli, A., Grande, R., Stefano, A.D., Giulio, M.D., Sozio, P., Bessa, L.J., Cellini, L., 2011. Potential antibacterial activity of carvacrol-loaded poly (DL-lactide-*co*-glycolide) (PLGA) nanoparticles against microbial biofilm. Int. J. Mol. Sci. 12 (8), 5039–5051.

Jo, W.S., Song, H.Y., Song, N.B., Lee, J.H., Min, S.C., Song, K.B., 2014. Quality and microbial safety of "Fuji" apples coated with carnauba-shellac wax containing lemongrass oil. LWT Food Sci. Technol. 55 (2), 490–497.

Joe, M.M., Bradeeba, K., Parthasarathi, R., Sivakumaar, P.K., Chauhan, P.S., Tipayno, S., Benson, A., Sa, T., 2012. Development of surfactin based nanoemulsion formulation from selected cooking oils: evaluation for antimicrobial activity against selected food associated microorganisms. J. Taiwan Inst. Chem. Eng. 43 (2), 172–180.

Keawchaoon, L., Yoksan, R., 2011. Preparation, characterization, and in vitro release study of carvacrol-loaded chitosan nanoparticles. Colloids Surf. B 84, 163–171.
Kfoury, M., Auezova, L., Ruellan, S., Greige-Gerges, H., Fourmentin, S., 2015. Complexation of estragole as pure compound and as main component of basil and tarragon essential oils with cyclodextrins. Carbohydr. Polym. 118, 156–164.
Khatibi, S.A., Misaghi, A., Moosavy, M.H., Amoabediny, G., Basti, A.A., 2014. Effect of preparation methods on the properties of *Zataria multiflora* Boiss. essential oil loaded nanoliposomes: characterization of size, encapsulation efficiency, and stability. J. Pharm. Sci. 20 (4), 141–148.
Kim, I.H., Lee, H., Kim, J.E., Song, K.B., Lee, Y.S., Chung, D.S., Min, S.C., 2013. Plum coatings of lemongrass oil-incorporating carnauba wax-based nanoemulsion. J. Food Sci. 78 (10), E1551–E1559.
Kumari, A., Yadav, S.K., Yadav, S.C., 2010. Biodegradable polymeric nanoparticles based drug delivery systems. Colloids Surf. B 75, 1–18.
Kuorwel, K.K., Cran, M.J., Sonneveld, K., Miltz, J., Bigger, S.W., 2011. Essential oils and their principal constituents as antimicrobial agents for synthetic packaging films. J. Food Sci. 76 (9), R164–R177.
Kushwaha, S.K.S., Keshari, R.K., Rai, A.K., 2011. Advances in nasal trans-mucosal drug delivery. J Appl. Pharm. Sci. 1 (7), 21–28.
Lai, F., Wissing, S.A., Müller, R.H., Fadda, A.M., 2006. *Artemisia arborescens* L. essential oil-loaded solid lipid nanoparticles for potential agricultural application: preparation and characterization. AAPS Pharm. Sci. Technol. 7 (1), E10–E18.
Lai, F., Sinico, C., De Logu, A., Zaru, M., Müller, R.H., Fadda, A.M., 2007. SLN as a topical delivery system for *Artemisia arborescens* essential oil: in vitro antiviral activity and skin permeation study. Int. J. Nanomed. 2 (3), 419–425.
Lai, S.K., Wang, Y.Y., Hanes, J., 2009. Mucus-penetrating nanoparticles for drug and gene delivery to mucosal tissues. Adv. Drug Deliv. Rev. 61 (2), 158–171.
Li, K.K., Yin, S.W., Yang, X.Q., Tang, C.H., Wei, Z.H., 2012. Fabrication and characterization of novel antimicrobial films derived from thymol-loaded zein–sodium caseinate (SC) nanoparticles. J. Agric. Food Chem. 60 (46), 11592–11600.
Liang, R., Xu, S., Shoemaker, C.F., Li, Y., Zhong, F., Huang, Q., 2012. Physical and antimicrobial properties of peppermint oil nanoemulsions. J. Agric. Food Chem. 60, 7548–7555.
Liolios, C.C., Gortzi, O., Lalas, S., Tsaknis, J., Chinou, I., 2009. Liposomal incorporation of carvacrol and thymol isolated from the essential oil of *Origanum dictamnus* L. and in vitro antimicrobial activity. Food Chem. 112 (1), 77–83.
Liu, H., Yang, G., Tang, Y., Cao, D., Qi, T., Qi, Y., Fan, G., 2013. Physicochemical characterization and pharmacokinetics evaluation of β-caryophyllene/β-cyclodextrin inclusion complex. Int. J. Pharm. 450 (1), 304–310.
Lv, Y., Yang, F., Li, X., Zhang, X., Abbas, S., 2014. Formation of heat-resistant nanocapsules of jasmine essential oil via gelatin/gum arabic based complex coacervation. Food Hydrocoll. 35, 305–314.
Maswal, M., Dar, A.A., 2014. Formulation challenges in encapsulation and delivery of citral for improved food quality. Food Hydrocoll. 37, 182–195.
McClements, D., Lesmes, U., 2009. Structure-function relationships to guide rational design and fabrication of particulate food delivery systems. Trends Food Sci. Technol. 20, 448–457.
McClements, D.J., Decker, E.A., Park, Y., Weiss, J., 2009. Structural design principles for delivery of bioactive components in nutraceuticals and functional foods. Crit. Rev. Food Sci. Nutr. 49 (6), 577–606.
Mishra, B., Patel, B.B., Tiwari, S., 2010. Colloidal nanocarriers: a review on formulation technology, types and applications toward targeted drug delivery. Nanomed. Nanotechnol. Bio. Med. 6, 9–24.

Moghimi, R., Ghaderi, L., Rafati, H., Aliahmadi, A., McClements, D.J., 2016. Superior antibacterial activity of nanoemulsion of *Thymus daenensis* essential oil against *E. coli*. Food Chem. 194, 410–415.

Moghimipour, E., Aghel, N., Mahmoudabadi, A.Z., Ramezani, Z., Handali, S., 2012. Preparation and characterization of liposomes containing essential oil of *Eucalyptus camaldulensis* leaf. Jundishapur J. Nat. Pharm. Prod. 7 (3), 117–122.

Moghimipour, E., Ramezani, Z., Handali, S., 2013. Solid lipid nanoparticles as a delivery system for *Zataria multiflora* essential oil: formulation and characterization. Curr. Drug Deliv. 10 (2), 151–157.

Mozafari, M.R., Mortazavi, S.M., 2005. Nanoliposomes: From Fundamentals to Recent Developments. Trafford, Oxford, UK.

Mozafari, M.R., Flanagan, J., Matia-Merino, L., Awati, A., Omri, A., Suntres, Z.E., Singh, H., 2006. Recent trends in the lipid-based nanoencapsulation of antioxidants and their role in foods. J. Sci. Food Agric. 86 (13), 2038–2045.

Mozafari, M.R., Johnson, C., Hatziantoniou, S., Demetzos, C., 2008. Nanoliposomes and their applications in food nanotechnology. J. Liposome Res. 18 (4), 309–327.

Musthaba, S.M., Baboota, S., Ahmed, S., Ahuja, A., Ali, J., 2009. Status of novel drug delivery technology for phytotherapeutics. Exp. Opin. Drug Deliv. 6 (6), 625–637.

Nedovic, V., Kalusevic, A., Manojlovic, V., Levic, S., Bugarski, B., 2011. An overview of encapsulation technologies for food applications. Procedia Food Sci. 1, 1806–1815.

Neoh, T.L., Yamamoto, C., Ikefuji, S., Furuta, T., Yoshii, H., 2012. Heat stability of allyl isothiocyanate and phenyl isothiocyanate complexed with randomly methylated β-cyclodextrin. Food Chem. 131 (4), 1123–1131.

Neumann, M., Garcia, N., 1992. Kinetics and mechanic of the light-induced deterioration of lemon oil. J. Agric. Food Chem. 40, 957–960.

Pagington, J.S., 1986. β-Cyclodextrin and its uses in the flavour industry. In: Birch, G.G., Lindley, M.G. (Eds.), Developments in Food Flavours. Elsevier Applied Science, London.

Pothakamury, U.R., Barbosa-Canovas, G.V., 1995. Fundamental aspects of controlled release in foods. Trends Food Sci. Technol. 6, 397–406.

Pranoto, Y., Salokhe, V.M., Rakshit, S.K., 2005. Physical and antibacterial properties of alginate-based edible film incorporated with garlic oil. Food Res. Int. 38 (3), 267–272.

Prow, T.W., Grice, J.E., Lin, L.L., Faye, R., Butler, M., Becker, W., Roberts, M.S., 2011. Nanoparticles and microparticles for skin drug delivery. Adv. Drug Deliv. Rev. 63 (6), 470–491.

Reis, C.P., Neufeld, R.J., Ribeiro, A.J., Veiga, F., 2006. Nanoencapsulation I. Methods for preparation of drug-loaded polymeric nanoparticles. Nanomedicine 2, 8–21.

Roger, E., Lagarce, F., Garcion, E., Benoit, J.P., 2010. Biopharmaceutical parameters to consider in order to alter the fate of nanocarriers after oral delivery. Nanomedicine 5 (2), 287–306.

Rosli, N.A., Hasham, R., Azizc, A.A., Azizd, R., 2015. Formulation and characterization of nanostructured lipid carrier encapsulated *Zingiber zerumbet* oil using ultrasonication technique. J. Adv. Res. 11 (1), 16–23.

Rubistein, M.H., 1989. Pharmaceutical Technology, Drug Stability. Ellis Horwood, Chichester, UK, (Chapter 1).

Salvia-Trujillo, L., Rojas-Graü, M.A., Soliva-Fortuny, R., Martín-Belloso, O., 2014. Impact of microfluidization or ultrasound processing on the antimicrobial activity against *Escherichia coli* of lemongrass oil-loaded nanoemulsions. Food Control 37, 292–297.

Salvia-Trujillo, L., Rojas-Graü, A., Soliva-Fortuny, R., Martín-Belloso, O., 2015. Physicochemical characterization and antimicrobial activity of food-grade emulsions and nanoemulsions incorporating essential oils. Food Hydrocoll. 43, 547–556.
Sanguansri, P., Augustin, M.A., 2006. Nanoscale materials development—a food industry perspective. Trends Food Sci. Technol. 17 (10), 547–556.
Saranya, S., Chandrasekaran, N., Mukherjee, A., 2012. Antibacterial activity of eucalyptus oil nanoemulsion against *Proteus mirabilis*. Int. J. Pharm. Pharm. Sci. 4 (3), 668–671.
Schneider, M., Stracke, F., Hansen, S., Schaefer, U.F., 2009. Nanoparticles and their interactions with the dermal barrier. Dermatoendocrinology 1 (4), 197–206.
Sebaaly, C., Jraij, A., Fessi, H., Charcosset, C., Greige-Gerges, H., 2015. Preparation and characterization of clove essential oil-loaded liposomes. Food Chem. 178, 52–62.
Sell, C., 2010. Chemistry of essential oils. In: Baser, K.H., Buchbauer, G. (Eds.), Handbook of Essential Oils. Science, Technology, and Applications. CRC Press, Boca Raton, FL, pp. 121–150.
Shah, B., Davidson, P.M., Zhong, Q., 2012. Nanocapsular dispersion of thymol for enhanced dispersibility and increased antimicrobial effectiveness against *Escherichia coli* O157:H7 and *Listeria monocytogenes* in model food systems. Appl. Environ. Microbiol. 78, 8448–8453.
Sherry, M., Charcosset, C., Fessi, H., Greige-Gerges, H., 2013. Essential oils encapsulated in liposomes: a review. J. Liposome Res. 23, 268–275.
Shi, F., Zhao, J.-H., Liu, Y., Wang, Z., Zhang, Y.-T., Feng, N.-P., 2012. Preparation and characterization of solid lipid nanoparticles loaded with frankincense and myrrh oil. Int. J. Nanomed. 7, 2033–2043.
Silva, J., Abebe, W., Sousa, S.M., Duarte, V.G., Machado, M.I.L., Matos, F.J.A., 2003. Analgesic and anti-inflammatory effects of essential oils of *Eucalyptus*. J. Ethnopharmacol. 89, 277–283.
Singh, S.G., Singh, R.P., Gupta, S.K., Kalyanwat, R., Yadav, S., 2011. Buccalmucosa as a route for drug delivery: mechanism, design and evaluation. Res. J. Pharm. Bio. Chem. Sci. 2 (3), 358–372.
Sinico, C., De Logu, A., Lai, F., Valenti, D., Manconi, M., Loy, G., Leonardo, B., Fadda, A.M., 2005. Liposomal incorporation of *Artemisia arborescens* L. essential oil and in vitro antiviral activity. Eur. J. Pharm. Biopharm. 59 (1), 161–168.
Sköld, M., Börje, A., Matur, M., Karlberg, A.T., 2002. Studies on the autoxidation and sensitizing capacity of the fragance chemical linalool, identifying a linalool hydroperoxide. Contact Dematitis 46, 267–272.
Soppimath, K.S., Aminabhavi, T.M., Kulkarni, A.R., Rudzinski, W.E., 2001. Biodegradable polymeric nanoparticles as drug delivery devices. J. Control. Release 70 (1), 1–20.
Suntres, Z.E., Shek, P.N., 1996. Alleviation of paraquat-induced lung injury by pretreatment with bifunctional liposomes containing α-tocopherol and glutathione. Biochem. Pharmacol. 52, 1515–1520.
Tamjidi, F., Nasirpour, A., Shahedi, M., 2012. Physicochemical and sensory properties of yogurt enriched with microencapsulated fish oil. Food Sci. Technol. Int. 18 (4), 381–390.
Tamjidi, F., Shahedi, M., Varshosaz, J., Nasirpour, A., 2013. Nanostructured lipid carriers (NLC): a potential delivery system for bioactive food molecules. Innovat. Food Sci. Emerg. Technol. 19, 29–43.
Terjung, N., Löffler, M., Gibis, M., Hinrichs, J., Weiss, J., 2012. Influence of droplet size on the efficacy of oil-in-water emulsions loaded with phenolic antimicrobials. Food Funct. 3 (3), 290–301.
Thanki, K., Gangwal, R.P., Sangamwar, A.T., Jain, S., 2013. Oral delivery of anticancer drugs: challenges and opportunities. J. Control. Release 170 (1), 15–40.

Tiwari, B.K., Valdramidis, V.P., O'Donnell, C.P., Muthukumarappan, K., Bourke, P., Cullen, P.J., 2009. Application of natural antimicrobials for food preservation. J. Agric. Food. Chem. 57 (14), 5987–6000.

Uhrich, K.E., Cannizzaro, S.M., Langer, R.S., Shakesheff, K.M., 1999. Polymeric systems for controlled drug release. Chem. Rev. 99 (11), 3181–3198.

Valenti, D., De Logu, A., Loy, G., Sinico, C., Bonsignore, L., Cottiglia, F., Donatella, G., Fadda, A.M., 2001. Liposome-incorporated *Santolina insularis* essential oil: preparation, characterization, and in vitro antiviral activity. J. Liposome Res. 11 (1), 73–90.

Vigan, M., 2010. Essential oils: renewal of interest and toxicity. Eur. J. Dermatol. 20, 685–692.

Waleczek, K.J., Marques, H.C., Hempel, B., Schmidt, P.C., 2003. Phase solubility studies of pure (−)-α-bisabolol and camomile essential oil with β-cyclodextrin. Eur. J. Pharm. Biopharm. 55 (2), 247–251.

Wandrey, C., Bartkowiak, A., Harding, S.E., 2009. Materials for encapsulation. In: Zuidam, N.J., Nedovic, V.A. (Eds.), Encapsulation Technologies for Food Active Ingredients and Food Processing. Springer, Dordrecht, The Netherlands, pp. 31–100.

Wang, J., Cao, Y., Sun, B., Wang, C., 2011. Physicochemical and release characterisation of garlic oil-β-cyclodextrin inclusion complexes. Food Chem. 127 (4), 1680–1685.

Wattanasatcha, A., Rengpipat, S., Wanichwecharungruang, S., 2012. Thymol nanospheres as an effective anti-bacterial agent. Int. J. Pharm. 434 (1), 360–365.

Weiss, J., Takhistov, P., McClements, D.J., 2006. Functional materials in food nanotechnology. J. Food Sci. 71 (9), R107–R116.

Wen, Z., Liu, B., Zheng, Z., You, X., Pu, Y., Li, Q., 2010. Preparation of liposomes entrapping essential oil from *Atractylodes macrocephala* Koidz by modified RESS technique. Chem. Eng. Res. Des. 88 (8), 1102–1107.

Woranuch, S., Yoksan, R., 2013. Eugenol-loaded chitosan nanoparticles: II. Application in bio-based plastics for active packaging. Carbohydr. Polym. 96 (2), 586–592.

Wu, Y., Luo, Y., Wang, Q., 2012. Antioxidant and antimicrobial properties of essential oils encapsulated in zein nanoparticles prepared by liquid–liquid dispersion method. LWT Food Sci. Technol. 48, 283–290.

Yang, F.L., Li, X.G., Zhu, F., Lei, C.L., 2009. Structural characterization of nanoparticles loaded with garlic essential oil and their insecticidal activity against *Tribolium castaneum* (Herbst) (Coleoptera: Tenebrionidae). J. Agric. Food Chem. 57 (21), 10156–10162.

Yurdugul, S., Mozafari, M.R., 2004. Recent advances in micro- and nano-encapsulation of food ingredients. Cell. Mol. Biol. Lett. 9 (Suppl. 2), 64–65.

Zhang, Y., Niu, Y., Luo, Y., Ge, M., Yang, T., Yu, L.L., Wang, Q., 2014. Fabrication, characterization and antimicrobial activities of thymol-loaded zein nanoparticles stabilized by sodium caseinate–chitosan hydrochloride double layers. Food Chem. 142, 269–275.

Zhao, Y., Wang, C., Chow, A.H., Ren, K., Gong, T., Zhang, Z., Zheng, Y., 2010. Self-nanoemulsifying drug delivery system (SNEDDS) for oral delivery of Zedoary essential oil: formulation and bioavailability studies. Int. J. Pharm. 383 (1), 170–177.

Zuidam, N.J., Shimoni, E., 2010. Overview of microencapsulates for use in food products or processes and methods to make them. In: Zuidam, N.J., Nedović, V.A. (Eds.), Encapsulation Technologies for Active Food Ingredients and Food Processing. Springer Science + Business Media, LLC, Dordrecht, The Netherlands, pp. 3–29.

15

NANOENCAPSULATION OF ESSENTIAL OILS FOR SUSTAINED RELEASE: APPLICATION AS THERAPEUTICS AND ANTIMICROBIALS

Sumit Gupta, Prasad S. Variyar

Food Technology Division, Bhabha Atomic Research Centre, Mumbai, Maharashtra, India

1 Introduction

Nanotechnology is an emerging area of technology dealing with production, processing, and application of materials with size less than 1000 nm (Bagchi et al., 2013). The British Standards Institution has defined nanotechnology as the design, characterization, production, and application of structures, devices, and systems by controlling shape and size at the nanoscale that is, 10^{-9} m. (Bawa et al., 2005). Encapsulation is a technology with many potential applications in areas such as the pharmaceutical and food industries. Encapsulating bioactive molecules at nanoscale (10^{-9} m) is known as nanoencapsulation (Quintanilla-Carvajal et al., 2010). It is mainly used for providing protection to bioactive compounds such as polyphenols, micronutrients, enzymes, antioxidants, and nutraceuticals from harsh environmental conditions during handling and processing and for controlled release at targeted site (Gouin, 2004; Ezhilarasi et al., 2013). Controlled and targeted release results in enhanced effectiveness of micronutrients, ensuring optimal dosage and thereby improving cost effectiveness of the product (Mozafari, 2006).

With regards to flavor encapsulation most of the reported literature deals with microencapsulation of flavor extracts. Microcapsules are particles having a diameter between 1 and 5000 μm (Bilia et al., 2014). Essential oils (EOs) are prone to degradation due to volatilization and chemical reaction, mainly oxidation, of

Encapsulations. http://dx.doi.org/10.1016/B978-0-12-804307-3.00015-6

their constituents. Control of the volatilization rate and degradation, thereby prolonging the sensory characteristics of aroma extracts, is the main aim of present microencapsulation strategies. Microencapsulation provides both stabilization and a controlled release of the entrapped materials. It also results in conversion of volatile liquid EOs into stable solid encapsulated products, provides protection to the active compounds against environmental factors (eg, oxygen, light, moisture, and pH), reduced flammability and increased water dispensability, thereby improving ease of handling, safety, and applicability to various products, especially in water-based formulations (Sansukcharearnpon et al., 2010).

Although microcapsules may guarantee excellent protection of EOs against degradation or evaporation, they in general do not affect bioefficacy. In contrast, encapsulation in nanometric size delivery systems, apart from providing other benefits (stabilization and protection), due to the subcellular size, reduce mass transfer resistances thus increasing the passive cellular absorption mechanisms and in effect bioefficacy (Weiss et al., 2009). Proof of this concept was given by the improvement of the antimicrobial activity of EOs when encapsulated into liposomal delivery systems (Gortzi et al., 2007; Liolios et al., 2009). The encapsulation of eugenol and carvacrol into nanometric surfactant micelles also resulted in enhanced antimicrobial activity (Gaysinsky et al., 2005). Oral administration of EO nanoemulsion of *Curcuma zedoaria* resulted in up to 2.5 times increased bioavailability of the active component germacrone, as compared to unformulated oil (Zhao et al., 2010). Antiviral activity of *Artemisia arborescens* EO increased when it was nanoencapsulated in liposomes (Sinico et al., 2005). Oral bioavailability of β-caryophyllene having several activities such as antimicrobial, antiinflammatory, antioxidant, anticarcinogenic, and local anesthetic, increased 2.6 times when it was nanoencapsulated by formation of inclusion complex using β-cyclodextrin (Liu et al., 2013).

These studies clearly demonstrate that nanoencapsulation of EOs can increase their bio efficacies. Major strategies employed for encapsulating EOs in nanometric size are polymer-based nanocapsules, nanoemulsions, liposomes, solid lipid nanocarriers and molecular complexes. The present review focuses on methodology of each kind of nanocapsules, its advantages, and their effect on the biological properties of EOs.

2 Chemical Composition of Essential Oils

EOs (also called volatile or ethereal oils) are oily liquids characterized by a strong odor and are formed by aromatic plants as secondary metabolites. The term *essential oil* was used for the first time in the 16th century by Paracelsus von Hohenheim, who named the effective component of a drug "Quinta essential" (Guenther, 1950). In plants essentials oils are synthesized in different plant organs such as flowers, leaves, seeds, stems, wood, fruits, and roots and are stored in epidermic cells, glandular trichomes, canals, cavities, or secretory cells and can be extracted using several different techniques (Bilia et al., 2014; Bakkali et al., 2008). Steam distillation is the most common method used for commercial production of EOs. Several other techniques such as fermentation, solvent extraction, expression under pressure, supercritical fluid, and subcritical water extractions are also reported (Burt, 2004). EOs were extracted using distillation for more than 2000 years in the East (India and Persia); however, the methodology was improved by Arabs in the 9th century (Guenther, 1972; Bauer et al., 2001). The first authentic written description of distillation of EOs is credited to Villanova (c. 1235–1311), a Catalan physician (Guenther, 1948). Although by the 13th century, several pharmacological effects of EOs were described in pharmacopoeias and were being prepared by pharmacies, it was not until the 16th century that their use was widespread in Europe (Bauer et al., 2001). Some of the most common EOs used were turpentine, juniper wood, rosemary, lavender, clove, mace, nutmeg, anise, and cinnamon. However, by the middle of the 20th century, the role of EOs declined in pharmaceutical preparations with their application almost entirely focused on perfumes, cosmetics, and food flavorings (Bilia et al., 2014). Presently, about 3000 EOs are known, of which approximately 10%, that is, 300, are commercially important and are mainly used for the flavors and fragrances market (Braak van de and Leijten, 1999).

In nature, EOs provide protection to the plants from attack of bacteria, virus, fungi, and insects and also give defense against herbivores. They also aid the plant in attracting insects to facilitate dispersion of pollens and seeds or to repel undesirable ones (Bakkali et al., 2008). EOs have been mainly employed for their antibacterial, antifungal, and insecticidal activities. While antimicrobial properties of EOs have been long recognized and extensively reviewed in the past (Shelef, 1983; Nychas, 1995), there is a renewed interest in their various biological activities owing to the growing interest in nature-derived products by consumers (Burt, 2004). In recent years various other biological activities of EOs were studied and there are several comprehensive reviews describing antibacterial, antimutagenic, anticancer, antiviral, antidiabetic, antiarthrosclerosis, and antioxigenic effect of EOs (Bakkali et al., 2008; Burt, 2004; Bilia et al., 2014). Apart from these activities, EOs or their components have been shown to exhibit antimycotic

(Azzouz and Bullerman, 1982; Akgul and Kivanç, 1988; Jayashree and Subramanyam, 1999; Mari et al., 2003), antiparasitic (Pandey et al., 2000; Pessoa et al., 2002), and insecticidal (Karpouhtsis et al., 1998; Konstantopoulou et al., 1992) properties.

The main components of EOs are secondary metabolites produced by plants comprising of hydrocarbons such as terpenes and sesquiterpenes and various oxygenated compounds including phenols, lactones, ketones, aldehydes, ethers, esters, alcohols and phenol ethers (Guenther, 1972). Each EO can contain a minimum of 20–60 components in varying concentrations. The majority of the oils are characterized by two or three major components present in fairly high concentrations (20–70%), with other components being present in trace amounts (Bakkali et al., 2008). For example, menthol (59%) and menthone (19%) are major components of *Mentha piperita* EO and linalool (68%) of the *Coriandrum sativum* EO. Generally, major compounds present define the biological properties of EOs (Bakkali et al., 2008). Generally all chemical constituents of EOs are characterized by low molecular weight of less than 200 Da. Chemically, components of EOs can be divided into two groups of distinct biosynthetic origin. Terpenes and terpenoids constitute major group while aromatic and aliphatic compounds constitute the second group (Fig. 15.1).

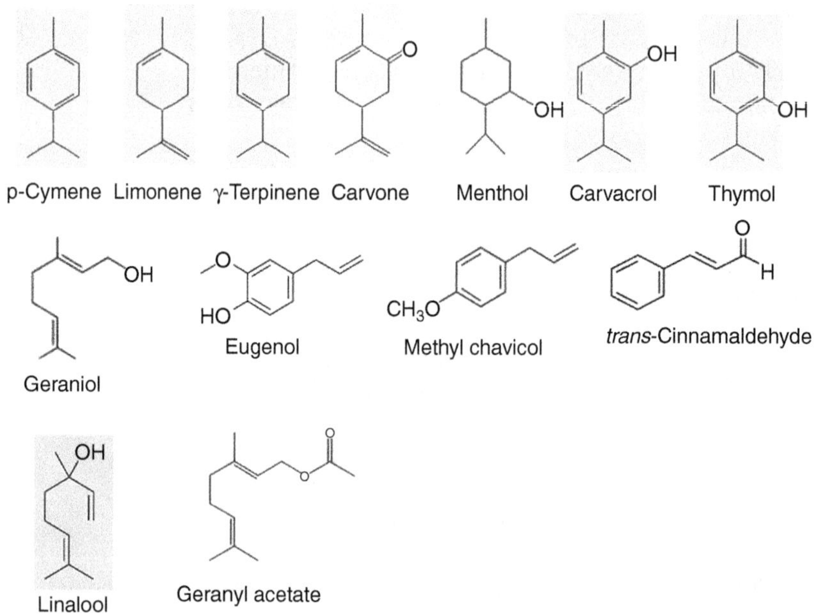

Figure 15.1. Major components of essential oils.

Terpenes are group of structurally and functionally diverse compounds and are made up of combinations of several 5 carbon base (C5) isoprene units. Formation of the isopentenyl diphosphate (IPP) precursor is the initial step in terpene biosynthesis. Repetitive addition of IPPs then form prenyldiphosphate precursor. Terpene specific synthetases then modify allylic prenyldiphosphate to form the terpene skeleton. Finally, secondary enzymatic modification by redox reaction of the terpene skeleton results in synthesis of different terpenes. C5-building blocks isopentenyl diphosphate (IPP) and dimethylallyl diphosphate (DMAPP) which is a isomer of IPP, mainly result in formation of monoterpenes (C10) and sesquiterpenes (C15); however, hemiterpenes (C5), diterpenes (C20), triterpenes (C30), and tetraterpenes (C40) are also observed in nature (Bakkali et al., 2008). Terpenes containing oxygen are called terpenoids.

The monoterpenes are formed from the addition of two isoprene units (C10) and are observed in a variety of structures in nature. They are the most widely present molecules constituting more than 90% of the EOs. Several examples of monoterpenes are monocyclic hydrocarbons (limonene, terpinenes, p-cymene, and phellandrenes); acyclic hydrocarbons (myrcene and ocimene); bicyclic hydrocarbons (pinenes, camphene, and sabinene); monocyclic alcohols (menthol, α-terpineol, and carveol); bicyclic alcohols (borneol, fenchol, chrysanthenol, and thuyan-3-ol) and acyclic alcohols (geraniol, linalool, citronellol, lavandulol, and nerol); Monoterpenes are also present as carbonyl compounds such as acyclic aldehydes (geranial, neral, and citronellal); monocyclic ketones (menthones, carvone, pulegone, and piperitone) and bicyclic ketones (camphor, fenchone, thuyone, and pinocarvone). Examples of monoterpene acyclic esters are linalyl acetate or propionate and citronellyl acetate while monocyclic esters are represented by menthyl or α-terpinyl acetate. Presence of bicyclic esters such as isobornyl acetate is also reported. Apart from these compounds monoterpenes are also reported to be present as ethers (1, 8-cineole and menthofuran), phenols (thymol, carvacrol), and peroxides (ascaridole) (Bilia et al., 2014; Bakkali et al., 2008).

The sesquiterpenes are formed from three isoprene units (C15). The increase in chain length increases the number of cyclizations, thus allowing a large number of structures. Sesquiterpenes are similar to monoterpenes in structure and function. Example of sesquiterpene hydrocarbons are β-bisabolene, cadinene, azulene, β-caryophyllene, farnesene, and zingiberene. Several sesquiterpenes alcohols reported in EOs are β-nerolidol, farnesol, bisabolol, β-santalol, and patchoulol. Germacrone, β-vetinone, and turmerones are some of the examples of sesquiterpene ketones while caryophyllene oxide and humulene epoxides are sesquiterpene epoxides reported in EOs (Bilia et al., 2014).

Aromatic compounds are another class of compounds present in EO and are formed via the shikimic acid pathway leading to phenylalanine (Pichersky et al., 2006) and occur less frequently than the terpenes. Alcohols such as cinnamic alcohol, aldehydes such as cinnamaldehyde, phenols comprising chavicol and eugenol, methoxy derivatives (eg, anethole, estragole, and methyl eugenol) and methylenedioxy compounds (eg, apiole, myristicin, and safrole) are some of the examples of aromatic compounds reported to be present in EOs. Apart from terpenes and aromatic compounds some other secondary metabolites can also be present in EO. The presence of nitrogen or sulfur-containing compounds such as isothiocyanate derivatives are reported in garlic and mustard oils. In addition, the presence of certain photoactive molecules such as coumarins and furocoumarins and short-chain aliphatic substances, for example 3-octanone and methyl nonyl ketone, is also demonstrated.

3 Nanoencapsulation Strategies of Essential Oils for Various Biological Activities

EOs can undergo significant chemical changes during storage and handling due to high volatility and decomposition owing to direct exposure to heat, humidity, light, or oxygen. Isomerization, cyclization, dehydrogenation, or oxidation reactions triggered either enzymatically or chemically can cause degradation of EOs constituents (Scott, 2005). Conditions during processing and storage of the plant material, during distillation while extraction of EO, and its subsequent handling and storage strongly influence the rate of degrading reactions (Schweiggert et al., 2007). Specific knowledge of the chemical composition and properties of EO is fundamental for an adequate use (Turek and Stintzing, 2013). Encapsulation increases the physical stability of EOs and therefore protects them from interactions with the environment. Encapsulation also results in decreased volatility and toxicity, increased bioactivity, controlled release and improved convenience (Ravi Kumar, 2000). Nanoencapsulation of EOs also helps in sustained and controlled release, deep tissue penetration, and cellular uptake due to nanometric size and protection at both extracellular and intracellular levels. Nanocapsules can be prepared with a large variety of materials and designs. In the present review three types of nanoencapsulates, namely polymer-based nanocapsules, lipid-based nanocarriers, and finally molecular inclusion complexes are discussed (Fig. 15.2). General advantages of nanoencapsulation are demonstrated in Fig. 15.3

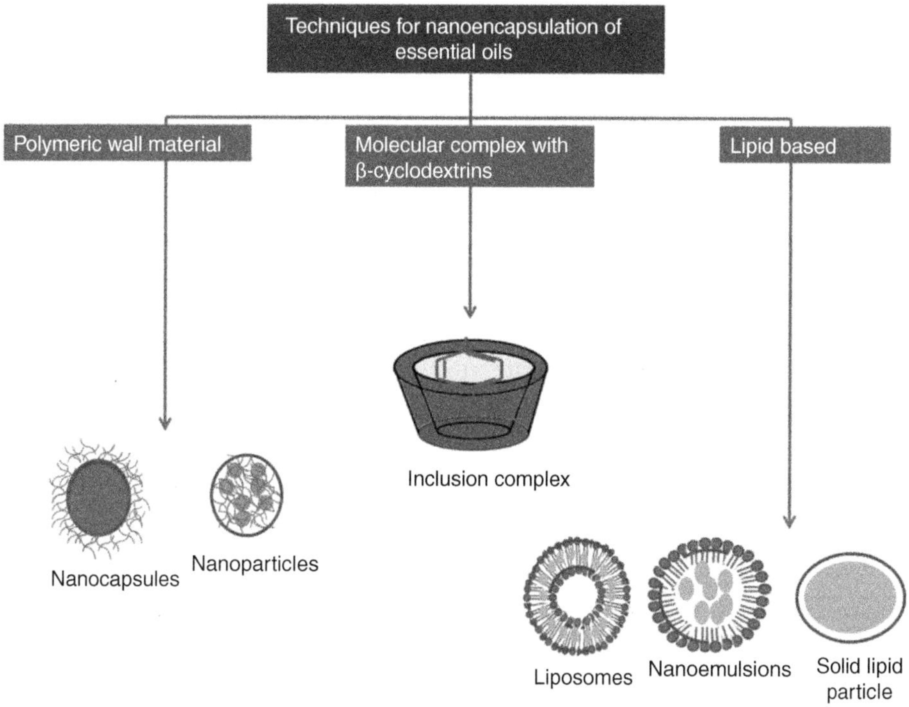

Figure 15.2. Different types of carriers available for nanoencapsulation of essential oils.

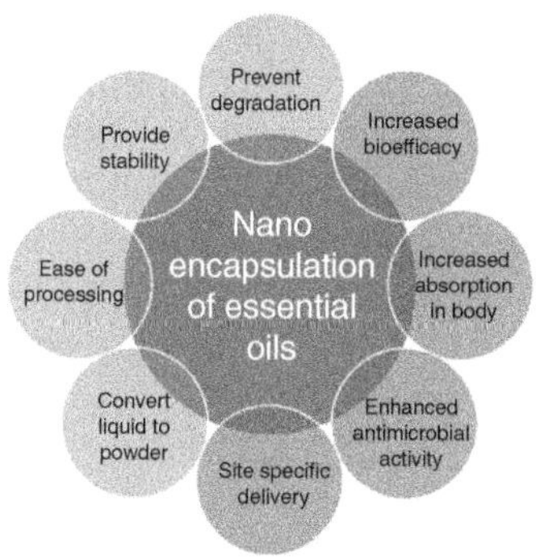

Figure 15.3. Various advantages provided by nanoencapsulation.

3.1 Polymer-Based Nanocapsules

Polymers are widely used as wall material for encapsulation of flavor components. Poly-α-cyanoacrylate alkyl esters, polyvinyl alcohol, polylactic acid, polyglycolic acid, and polylactic glycolic acid are some examples of biocompatible but synthetic polymers used for nanoencapsulation of flavors. Apart from synthetic origin, use of several biopolymers is also well reported. Biopolymers can be divided into two classes: polysaccharides and proteins. Use of several polysaccharides of plant origin—for example, pectin, cellulose, and its derivatives—starch and its derivatives, arabic gum, carrageenan, and alginate and from microbial (eg, xanthan gum) or animal origin (eg, chitosan) are reported. Various proteins such as albumin, gelatine, soy proteins, and casein are also used for preparation of nanocapsules for flavor encapsulation. Several different techniques have been used to fabricate nanocapsules such as complex coacervation (Dong and Bodmeier, 2006), freeze-drying (Kaushik and Roos, 2007), spray drying (Stulzer et al., 2009), and polyelectrolyte complexation (Schatz et al., 2005).

Polymeric nanocarriers are classified as nanocapsules and nanospheres. In nanocapsules an oily EO core is surrounded by a polymeric wall while in nanospheres, which are matrix systems, EOs are conjugated with the polymer matrix (Bilia et al., 2014). Various processes by which EOs are released from nanocarriers are diffusion through the matrix, dissolution of matrix, desorption of the surface-bound/adsorbed EO components, matrix erosion by enzyme degradation, or a combination of these processes (Soppimath et al., 2001). Among the wall materials used, polysaccharides are most widely employed as wall materials due to their stability, safe and nontoxic nature, hydrophilic character, and biodegradability. Furthermore, polysaccharides are abundantly present in nature and have a low processing cost.

Due to the aforementioned advantages, nanocarriers having polysaccharide-based wall materials have promising properties to deliver and protect the physiological properties of EOs and have been successfully applied as delivery systems (Liu et al., 2008). Recently, chitosan (CS) has attracted a great attention in the encapsulation of bioactive compounds because of its generally recognized as safe (GRAS) status. Its various other biological properties such as nontoxicity, biocompatibility, and biodegradability, along with its ability to form membranes, gels, beads, fibers, and particles provides it a distinct advantage. It has solubility in aqueous medium due to which it circumvents the need for organic solvents. Chitosan is prepared by deacetylation of chitin (chemically D-glucosamine and *N*-acetyl-D-glucosamine linked by beta (1–4) linkages), which is the main component of the crustacean exoskeleton and the cell wall of some fungi. Several studies have successfully demonstrated use of chitosan as wall material for encapsulation of EOs or their components employing various encapsulation techniques. Positive charge of chitosan attracts negatively charged lipid molecules, resulting in their strong binding. Furthermore, due to high hydrophilic–lipophilic balance of chitosan, oil-in-water emulsions are easily stabilized.

Eugenol was encapsulated into chitosan nanoparticles formed via emulsion–ionic gelation cross-linking method. Tripolyphosphate (TPP) when added slowly in emulsion containing eugenol and chitosan resulted in ionic gelation of chitosan to form nanoparticles. Nanoparticles formed showed an average size of less than 100 nm with loading capacity and encapsulation efficiency of 12 and 20%, respectively. The particles were found to have positively charged surface and demonstrated a zeta potential value ranging from +16.2 to +33.5 mV. The eugenol-loaded chitosan nanoparticles had improved thermal stability as compared to unencapsulated standard eugenol, when evaluated using extrusion at 155°C (Woranuch and Yoksan, 2013). Subsequently, these eugenol-loaded nanocapsules were used for preparation of extruded active films from thermoplastic starch. Although the addition of eugenol-loaded chitosan nanoparticles caused decreased mechanical properties and increased oxygen permeability of films, it provided enhanced antioxidant activity and decreased water vapor permeability. Films containing encapsulated eugenol had higher antioxidant capacity compared to films prepared with unencapsulated standard eugenol.

Oregano EO, which is known for its excellent antioxidant and antimicrobial activity has also been encapsulated in chitosan nano particles. The method of preparing nanoparticles was formation of oil-in-water emulsions using chitosan followed by gelation of emulsion droplets using TPP. Nano particles obtained had a spherical shape and size range of 40–80 nm with regular distribution. Nanoparticles demonstrated encapsulation efficiency of 21–47% and loading capacity of 3–8% when 0.1–0.8 EO was used per gram of chitosan. In vitro release studies showed an initial burst effect and followed by a slow release. Thus, it was demonstrated that nanoparticulation could be used for controlling the release behavior of EO (Hosseini et al., 2013).

EO from *Lippia sidoides* was encapsulated in nanoparticles made of chitosan and cashew gum. *L. sidoides* is a plant native to Brazil and EO extracted from the leaves of this plant is rich in thymol, a compound well known for bactericidal and fungicidal activities. Nanoparticles were obtained by spray drying of emulsions containing EO, chitosan, and cashew gum. Tween was used as an emulsifier. Ratio of chitosan to cashew gum and wall material to oil used was optimized to obtain high EO loading and improved release profiles. Samples having matrix-to-oil ratio of 10:2 with gum:chitosan 1:1 and 5% gum concentration, demonstrated high loading (11.8%), and encapsulation efficiency (70%). Nanocapsules showed an average size of 335–558 nm and presented slower and sustained release during in vitro studies. Efficacy of nanocarriers against *St. aegypti* larvae was evaluated and compared with pure EO. It was observed that mortality rate of larvae was related to loading of oil in nanocapsules; however, no effect of size of nanoparticles could be observed. It was observed that samples prepared with increasing chitosan content in the matrix presented increasing mortality rates. In particular a sample prepared with chitosan to gum ratio of 10 with matrix to oil of 10:2 presented the highest mortality rate of 90%. It was inferred that high chitosan content resulted in sustained release of EO and better biological activity (Abreu et al., 2012).

Carvacrol is a major component of the EOs derived from oregano, thyme, marjoram, and summer savory and is generally recognized as a safe food additive. Although it has been used for flavoring and as an antimicrobial and antioxidant in food products, it is a volatile compound and decomposes easily during processing. Encapsulation can be used as a tool for providing protection as well as sustained release of this compound. Carvacrol-loaded chitosan nanoparticles were prepared using ionic gelation technique. Nanoparticles prepared using the initial carvacrol content of 0.25–1.25 g/g of chitosan demonstrated encapsulation efficiency (EE) and loading capacity (LC) in the ranges of 14–31% and 3–21%, respectively. The nanoparticles had spherical shape with an average diameter of 40–80 nm. Zeta potential was found to be 25–29 mV, thus confirming a positive charged surface. Nanoparticles further showed antimicrobial activity against various microorganisms such as *Staphylococcus aureus, Bacillus cereus,* and *Escherichia coli* with comparable minimum inhibitory concentration to that of free carvacrol. The release of carvacrol from chitosan nanoparticles followed a Fickian behavior and was found to be relatively quick in an acidic solution, followed by alkaline and neutral media, respectively. The amount of liberated carvacrol was 53, 23, and 33% in buffer solutions with pH of 3, 7, and 11, respectively, on day 30. Nanoparticles were found to be effective for sustained slow release of active ingredient (Keawchaoon and Yoksan, 2011). Similarly, EO from *Eucalyptus citriodora* having citronellal as a main constituent and possessing antimicrobial, antioxidant, anti-inflammatory, antifungal, analgesic, insect repellant, and insecticide activities was encapsulated in chitosan. Nanoemulsion was formed by adding EO emulsified in Tween and 1% acidic solution of chitosan. Nanoemulsion particles demonstrated a mean size of 232 nm with unimodal distribution and polydispersity (Ribeiro et al., 2014). Nanoencapsulated EO was evaluated for its efficacy against gastrointestinal nematodes of sheep. Although in in vitro tests demonstrated better efficacy of nanoencapsulates for preventing larval hatching as compared to free EO, however, no significant effect could be observed under in vivo conditions. *Zataria multiflora* EO was also encapsulated in chitosan using ionic gelation technique. Chitosan nanoparticles were formed using 0.25 g EO per g of wall material with average size of 125–175 nm and entrapment efficiency of 45%. Prepared nanoparticles were then evaluated for antifungal activity against *Botrytis cinerea*. Significantly better antifungal activity and sustained release was obtained for encapsulated EO as compared to free EO in in vitro tests. The in vivo experiment also showed that the encapsulated oils at 1500 ppm concentration significantly decreased both disease severity and incidence of *B. cinerea* inoculated strawberries during 7 days of storage at 4°C (Mohammadi et al., 2015).

Apart from chitosan, zein is another GRAS biopolymer that has received increased attention as wall material for encapsulation of bioactive compounds. Zein is a prolamin (alcohol soluble protein) derived from corn and has up to 50%hydrophobic amino acid residues on the surface and therefore is not water soluble. Zein is only soluble in 55–90% aqueous alcohol and easily precipitates as nanoparticles after mixing with water to an overall alcohol concentration incapable of dissolving zein. This antisolvent precipitation property has been used to encapsulate several bioactive compounds such as fish oil, a-tocopherol, vitamin D_3, daidzin, and curcumin in zein nanoparticles (Chen et al., 2015). Three EOs—oregano, red thyme, and cassia (100% pure oil)—were encapsulated by phase separation into zein nanospheres. Ethanol solution containing EO and zein was dispersed in water along with high-speed mixing. The resulting solution was then freeze dried to obtain dried nanoparticles. Topographical images indicated that the powders were made up of irregularly shaped particles (50 μm) containing close-packed nanospheres. In vitro release profile of EOs was studied both at acidic (3.5 pH) and basic pH (7.4). Results indicated that under acidic conditions complete dissolution of nanoparticles was obtained after 52 h using pepsin digestion; however, rapid release was observed under basic conditions with 60% of oil released in fewer than 4 h. From results of in vitro studies it was concluded that particles have limited digestibility in the stomach, slow release in the small intestine, and more rapid release in the large intestine. They could be useful for oral or injectable administration of biological materials intended for colon delivery (Parris et al., 2005). Similarly, thymol and carvacrol were encapsulated in zein particles using liquid–liquid dispersion method. Ethanolic solutions of zein and EO components were dispersed in deionized water using high-speed mixer. Ethanol was then removed by purging nitrogen and dried nano particles were obtained by freeze-drying. Particle sizes formed were in the range of 430–740 nm with 50%entrapment efficiency. Encapsulating in zein nanoparticles enhanced solubility of thymol and carvacrol up to 14-fold without hindering their ability to scavenge free radicals or to control *E. coli* growth (Wu et al., 2012).

Zein, a protein, has an isoelectric point at around 6.2, which may result in poor physical stability and redispersibility of freeze-dried nanoparticles at a neutral pH in aqueous systems. This limits its application as delivery systems in food and pharmaceutical industries. Therefore, its stabilization with other water-soluble and amphiphilic biopolymers has been attempted. Components such as sodium caseinate can be used to form complexes with zein nanoparticles to create repulsive steric and electrostatic forces (Chen et al., 2015). Eugenol and thymol were coencapsulated in zein/casein nanoparticles at pH 6.0–8.0. Hot ethanolic solution of zein, sodium caseinate, EO components was sheared into buffered water. The resulting solution was spray dried to obtain dry nanoparticles. Particle size obtained for spray-dried zein/casein nanoparticles was smaller than 200 nm. Aqueous dispersions of nanoparticles prepared were formed easily and were stable. In aqueous dispersions, controlled release was observed for both EO components in 24 h, with higher rate of release shown by eugenol as compared to thymol. Spray-dried nanoparticles containing these compounds were also analyzed for bactericidal and bacteriostatic effects against *E. coli* O157:H7 and *Listeria monocytogenes* in milk and were found to be effective in inhibiting the growth of both of these microorganisms.

In another study thymol-loaded zein nanoparticles were stabilized using sodium caseinate (SC) and chitosan hydrochloride. Zein nanoparticles were prepared by mixing ethanolic solution of zein and thymol with deionized water with continuous mixing. To stabilize nanoparticles using SC, ethanolic solution was added in deionized water containing SC. After mixing, ethanol was removed by purging nitrogen. Zein nanoparticles when stabilized using SC showed a shift of isoelectric point from 6.18 to 5.05, and had a desirable redispersibility in water at neutral pH after lyophilization. SC-coated nanoparticles were further coated with chitosan hydrochloride (CHC) by adding aqueous solution of CHC in a solution containing SC stabilized nanoparticles with continuous stirring. This resulted in increased particle size, reversal of zeta potential value from negative to positive, and improved encapsulation efficiency. CHC stabilized nanoparticles demonstrated encapsulation efficiency of 80%. When tested for antimicrobial activity, stabilized nanoparticles demonstrated enhanced antimicrobial activity against *S. aureus* as compared to free thymol (Zhang et al., 2014).

Recently, thymol-loaded zein–SC nanoparticles were prepared using an antisolvent technique, with the average particle size and zeta potential of about 200 ± 20 nm and −40 mV, respectively. Ethanolic solutions of thymol and zein were added into aqueous solutions of SC with stirring to form nanoparticles that were then spray dried. Prepared particles demonstrated significant antioxidant activity and antimicrobial activity against *E. coli* and *Salmonella*. Nanoparticles also delayed growth of *S. aureus* in liquid broth medium (Li et al., 2013). In another study carried out by the same group final solutions containing zein-SC-thymol nanoparticles were casted and air dried to form antimicrobial films. Nano-particle-based films were then compared with only SC or zein films. Zein–SC nanoparticle-based films exhibited higher mechanical resistance and water barrier capacity than the SC films and good extensibility as compared with zein films. Films were observed to have antimicrobial activity against *E. coli* and *Salmonella* as well as DPPH radical scavenging activity. The release kinetic profile of thymol from zein/SC nanoparticles-based films followed a two-step biphasic process, that is, an initial burst effect followed by subsequent slower release. It was proposed that owing to their antimicrobial and antioxidant activity, these films could possibly be used in inner lining of food packaging (Li et al., 2012).

Apart from zein, other proteins used for nanoencapsulation include whey proteins and casein. Milk proteins are among the best natural surfactants for food applications (Ly et al., 2008). Furthermore, it has been demonstrated earlier that by conjugation with more hydrophilic oligosaccharides and polysaccharides, interfacial properties of milk proteins, in particular of whey proteins, can be further improved (Kato, 2002). The conjugation of oligo- or polysaccharide moiety results in improved emulsion stability because they provide steric hindrance against droplet aggregation. Whey proteins were conjugated by Maillard reaction to maltodextrins by heating at 90°C for 2 h. Conjugates were then employed to form nanodispersions of antimicrobial compounds thymol and eugenol. Oil in water emulsions were prepared for EO components using conjugated proteins and emulsions were then spray dried to form nanoparticles. Dried nanoparticles were redispersed in water to form clear nanodispersions. It was demonstrated that nanodispersions prepared from conjugated whey proteins were stable even at pH 5.0, which is closer to isoelectric point for whey proteins. Formation of nanodispersions resulted in improved solubility of EO components in water. Nanodispersions of thymol were tested for antimicrobial activity against *E. coli* and *L. monocytogenes* in apple cider and reduced fat milk. Although there was no significant difference between antimicrobial activity of free and encapsulated thymol, it was observed that free thymol tend to form crystals while nanodispersions result in improved solubility and clear solutions. Therefore, nanodispersions could provide a better alternative for adding this natural antimicrobial in clear beverages (Shah et al., 2012). Similarly, regarding eugenol nanodispersions, it was observed that antimicrobial activity was similar for both free and nanodispersed eugenol when tested in tryptic soy broth. However, nanodispersed eugenol demonstrated significantly better antimicrobial activity when tested in a bovine milk sample. It was due to the fact that in milk free eugenol reacted with food components such as proteins thus lowering its efficacy (Shah et al., 2013). In another study, thymol was encapsulated in sodium caseinate using high shear homogenization. Encapsulation efficiency was observed to be approximately 90% with particle size of 110 nm. Dispersion formed was transparent at neutral pH and was stable for 30 days at room temperature, as determined by dynamic light scattering and atomic force microscopy. When molecular binding was studied by fluorescence spectroscopy, thymol was observed to bind with tyrosine and possibly other amino acid residues away from tryptophan of caseins. At pH 4.6 (isoelectric point of caseins), the stabilization of thymol nanoparticles against aggregation was carried out using soluble soybean polysaccharide (SSPS). SSPS could effectively prevent aggregation of thymol nanoparticles due to combined electrostatic and steric repulsions. The encapsulated thymol showed the significantly improved antilisterial activity in milk with different fat levels when compared to thymol crystals, resulting from the quicker mixing and increased solubility in the milk serum. Thus, the transparent thymol nanodispersions have promising applications to improve microbiological safety and quality of foods (Pan et al., 2014).

Alginate is an anionic polysaccharide derived from cell wall of several brown algae. Chemically alginates are linear copolymers containing blocks of (1,4)-linked β-D-mannuronate (M) and α-L-guluronate (G) residues. The blocks are composed of consecutive G residues (GGGGGG), consecutive M residues (MMMMMM), and alternating M and G residues (GMGMGM) (Lee and Mooney, 2012). It is widely used in food systems as a gelling and thickening agent due to its biocompatibility, low toxicity, relatively low cost, and mild gelation by addition of divalent cations such as Ca^{2+}. Cashew gum is a biopolymer extracted from the exudate of *Anacardium occidentale*, a common tree of Brazil's Northeastern region. The gum main chain is composed of galactose (72%), with side-chains of arabinose (4.6%), glucose (14%), rhamnose (3.2%), and uronic acid (4.7%). *L. sidoides* EO, which has significant fungicidal and bactericidal activities was encapsulated in Alginate/cashew gum nanoparticles prepared via spray-drying.

Nanoparticles formed were in the average size range of 223–399 nm having negative surface charge with zeta potential values ranging from −30 to −36 mV. Encapsulated oil levels varied from 1.9 to 4.4% with an encapsulation efficiency of up to 55%. Nanoparticles were able to release between 45 and 95% of oil within 30–50 h, as evaluated by the in vitro studies. Cashew gum helped in maximizing the hydrophilic character of the polymer matrices thus allowing a quicker release at satisfactory oil loading. These results showed that alginate in combination with cashew gum acts synergistically for EO encapsulation and could provide tailored release rate, loading, and encapsulation efficacy (de Oliveira et al., 2014). Gelatin and arabic gum were used to form heat-resistant flavor nanocapsules of jasmine EO. Gelatin and gum Arabic nanocapsules containing jasmine oil were prepared using high-speed homogenization. Span-80 and Tween-80 were used as surfactants. After homogenization, nanoparticles formed were hardened using transglutaminase. Nanoparticles were formed in size range of 50–100 nm. Heat-resistance capability of nanoparticles was evaluated at 80°C by both structural characteristics (size, polydispersity index, and zeta potential) and flavor analysis. The results showed that the nanocapsules structurally were stable at 80°C for 7 h; however, GCMS analysis revealed that encapsulated jasmine oil began to degrade after 5 h of heating (Lv et al., 2014). Results suggest that heat-stable nanoparticles will be useful for flavor delivery in the food and pharmaceutical industry.

Polymeric nanoparticles containing thymol with methyl cellulose/ethyl cellulose as wall materials were synthesized using solvent displacement method. To prepare nanoparticles water was slowly added into ethanolic solution of thymol with methyl cellulose/ethyl cellulose. Prepared nanoparticles demonstrated a loading capacity of 43.53% (weight of encapsulated thymol to weight of the thymol-loaded spheres) with an average size of 420 nm. Nanoparticles were evaluated for their efficacy against three microorganisms *E. coli, S. aureus,* and *Pseudomonas aeruginosa* in three cosmetic preparations. Nanocapsules could significantly reduce microbiologic counts levels in all three cosmetic preparations for a longer period of 3 months as compared to free thymol, which showed antimicrobial activity for 2–4 weeks. Study also indicated that the encapsulated thymol was an effective preservative, as good as the traditional methylparaben, even when used at 12-fold to 52-fold lower concentrations (by mass or molarity) (Wattanasatcha et al., 2012). In another study, the six chemical components of EOs, that is, camphor, citronellal, eucalyptol, limonene, menthol, and 4-*tert*-butylcyclohexyl acetate, having different chemical functionalities were encapsulated using a polymer-blend of ethylcellulose (EC), hydroxypropyl methylcellulose (HPMC), and poly (vinyl alcohol) [PV(OH)] as wall material by solvent displacement technique. Nanoparticles prepared demonstrated ≥40% loading capacity with ≥80% encapsulation with particle size of less than 450 nm. The release profiles of the encapsulated fragrances were evaluated. Limonene showed the fastest release with essentially no retention by the nanoparticles, while eucalyptol and menthol showed the slowest release (Sansukcharearnpon et al., 2010).

Poly lactic glycolic acid (PLGA) is a copolymer of poly lactic acid (PLA) and poly glycolic acid (PGA). It is widely used for sustained drug-delivery applications because it is biocompatible and biodegradable, exhibits a wide range of erosion times, has tunable mechanical properties, and is an FDA approved polymer (Makadia and Siegel, 2011). PLGA nanoparticles containing eugenol and *trans*-cinnamaldehyde having loading efficiency of 90% and particle size of 180 nm were prepared. To prepare nanocapsules PLGA along with EO component was dissolved in dichloromethane. In this mixture aqueous solution of polyvinylalcohol (PVA) was added and resulting mixture was homogenized at high speeds. Polylactic glycolic acid nanocapsules presented a two-phase release for both the components. In the first rapid phase approximately 20% of the EO load was detected in less than 30 min; however, after that a prolonged slow release was observed and after 72 h 64% of eugenol and 87% of transcinnamaldehyde was detected. Since PLGA is a stable polymer with a low degradation rate, the observed release of EO components was mainly due to diffusion with a possible influence of polymer swelling and bulk erosion. In the first rapid release phase molecules adsorbed on polymeric wall are released while in slow release phase EO components present in the core of the nanocapsules diffuses through the polymeric wall. The antimicrobial efficacy of these nanoparticles was evaluated against *Salmonella* spp. (gram-negative bacterium) and *Listeria* spp. (gram-positive bacterium). MICs ranged from 10 to 20 mg/mL, respectively for both nanoparticles, demonstrating a broad spectrum of application in food systems (Gomes et al., 2011). Polylactic glycolic acid (PLGA) nanocapsules containing carvacrol have also been prepared with size about 209.8 nm, polydispersity index of 0.260, zeta potential of −18.99, drug loading of 21, and 26 % encapsulation efficiency. Nanocapsules were prepared by solvent displacement technique in which solution of PLGA and carvacrol in acetone was mixed with aqueous solution of pluronic under constant stirring. In vitro release profile showed an initial phase characterized by rapid release followed by a second phase of much slower release due to the concentration gradient. Approximately 60 and 95% of carvacrol was released from nanoparticles after 3 and 24 h, respectively. It was also observed that nanoparticles significantly altered rheological characteristic of bacterial biofilms, thus facilitating the action of carvacrol resulting in enhanced antimicrobial activity (Giulio et al., 2011).

Apart from natural polymers, use of synthetic biodegradable polymers such as poly-caprolactone, a biodegradable polyester, is also demonstrated. Use of polycaprolactone was demonstrated to prepare nanoparticles encapsulating eugenol. Solution of PLGA and eugenol in ethyl acetate was slowly mixed with aqueous solution of pluronic and resulting mixture was homogenized at high speed to form nanocapsules. Particles had size range of 320 nm with 100% encapsulation efficiency. It was reported that encapsulation of eugenol into polycaprolactone nanoparticles could enhance its stability against light oxidation (Choi et al., 2009). *Achyrocline satureioides* EO nanoencapsulated in poly-caprolactone were evaluated for antiparasitic activity against *Trypanosoma evansi* in rats. Nanocapsules were formed by dissolving polycaprolactone along with EO in acetone. In this organic phase aqueous solution of polysorbate was added dropwise. Acetone was removed using a rotary evaporator to obtain solution of nanoparticles, which was further evaluated. Although the treatment with EO did not eliminate the parasites from the bloodstream, it reduced the number of trypanosomes. Nanoencapsulated form was found to be more effective in reducing the number of parasites in blood stream as compared with free EO. Treatment with *A. satureioides* EO in nanoencapsulated form more efficiently reduced the histological damage in the liver samples and provided better consumption of ROS generated by *T. evansi* infection as compared to free EO (Baldissera et al., 2014).

3.2 Lipid-Based Nanocarriers

Although carbohydrate- and protein-based nanoparticles have several advantages, due to application of different complicated heat or chemical treatments during production their industrial full scale-up is rather difficult. On the other hand, lipid-based nanocarriers have possibility of industrial production and have the advantage of higher encapsulation efficiency and low toxicity (Fathi et al., 2012). Further, since these nanoparticles are composed of lipids and/or phospholipids, they can interact with several cell types. So, liposomes could possibly provide an alternative for treatment of microbial infections, due to their capacity of interaction with infected cells. Encapsulation of EOs within lipid-based nanoparticles can have different objectives such as enhancement in stability, increased solubility in water-based systems, improved biological activity, and target drug delivery. In this review lipid-based nanocarriers including nanoemulsions, liposomes, and solid lipid nanoparticles are discussed.

3.2.1 Nanoemulsions

Nanoemulsions (also known as miniemulsions or submicron emulsions) are nanoscale droplets of multiphase colloidal dispersions formed by dispersing of one liquid in another immiscible liquid by physical sheer-induced rupturing. Different size ranges have been reported in the literature for nano emulsions; for example, less than 10 nm, 10–100 nm, and 100–500 nm (Fathi et al., 2012). Emulsions having particle size greater than 500 nm are classified as microemulsions. Preparation of emulsions in nano size confers several distinct advantages over micro sized emulsions. Formation of microemulsions require large amount of surfactants, which can cause toxicity when used in pharmaceutical applications. In contrast nanoemulsions can be prepared utilizing relatively lesser amount of surfactants. Microemulsions cause multiple scattering of visible light and therefore generally have white milky appearance, whereas nanoemulsions having much lesser droplet size appear optically transparent. This is a very favorable feature of nanoemulsions for applying them as the nutrient carriers in beverages. Another interesting property of nanoemulsions is that it exhibits stability over gravity as compared to microemulsions due to Brownian motion of nanosized particles. Nanoemulsions are also metastable and can be diluted with water without change in the particle size distribution (McClements and Li, 2010; Shakeel and Ramadan, 2010; Gutieırrez et al., 2008). Nanoemulsions are usually prepared by either mechanical or nonmechanical techniques. Mechanical (high-energy) methods include high-pressure homogenization, microfluidization and ultrasonication while nonmechanical methods include solvent diffusion technique.

Sunflower oil was used for encapsulation of carvacrol, limonene, and cinnamaldehyde. Nanoemulsions were prepared by high-pressure homogenization and then stabilized using four different stabilizers that is, lecithin, pea proteins, sugar ester, and a combination of Tween 20 and glycerol monooleate. Particle size of prepared nano emulsions were found to be in the range of 160–300 nm. The antimicrobial activity of prepared nano capsules was evaluated against three different microorganisms that is, *E. coli*, *Lactobacillus delbrueckii*, and *Saccharomyces cerevisiae*. It was observed that the nanoemulsions based on sugar esters or on a combination of Tween 20 and glycerol monooleate increased the equilibrium concentration of the EO components in aqueous phase well above their water solubility. Increased concentration in aqueous phase resulted in enhanced bactericidal activity over shorter time scales (2 h) due to immediate availability of antimicrobial compounds. In contrast, nanoemulsions based on lecithin or pea proteins promoted only slightly the aqueous-phase concentration of the active molecules, causing the antimicrobial activity to be evident only over a longer time scale (24 h). These results demonstrate that nanoemulsion systems could be designed to obtain antimicrobial activity over desired time scale (Donsì et al., 2012). A terpene mixture and D-limonene were encapsulated into nanoemulsions based on food-grade ingredients, prepared by high-pressure homogenization at 300 MPa. For preparation of nanoemulsions, sunflower oil and palm oil were used as organic phases while glycerol monooleate, soy lecithin, and modified starch were used as emulsifying agents. Different formulations of wall materials were tried and particle size obtained was in the range of 150–300 nm. Antimicrobial activities of nanoemulsions formed were tested against three different microorganisms (*L. delbrueckii, S. cerevisiae,* and *E. coli*). Nanoemulsions were also evaluated for their efficacy in pear and orange juices inoculated with *L. delbrueckii.* In vitro nanoemulsions demonstrated better antimicrobial efficacy as compared to free EO components and furthermore when tested in fruit juices, nanoemulsions could effectively inhibit the microbial growth without altering the organoleptic properties (Donsì et al., 2011). In another study, zedoary turmeric oil, which is an EO extracted from the dry rhizome of *C. zedoaria* was encapsulated in self-nanoemulsifying system designed for oral delivery of EO. The optimized formulation consisting of EO, ethyl oleate, Tween 80, Transcutol P (30.8: 7.7: 40.5: 21, w/w), and loaded with 30% drug was prepared. The nanoemulsion had a mean particle size of 68.3 ± 1.6 nm and zeta potential of −41.2 ± 1.3 mV. The active components remained stable in the optimized formulation stored at 25°C for entire duration of 12 months. Following oral administration in rats, both AUC (area under the curve: in plot of concentration of drug in blood serum versus time) and C_{max} (maximum concentration reached by drug after administration) of germacrone, a representative bioactive marker of zedoary turmeric oil, increased by 1.7-fold and 2.5-fold, respectively, compared with the unformulated zedoary turmeric oil (Zhao et al., 2010).

Although nanoemulsions demonstrate stability toward gravitational separation, flocculation, and coalescence but they are prone to destabilization due to an effect known as Ostwald ripening (OR). OR is associated with growth of large oil droplets in an emulsion at an expense of smaller droplets. This phenomenon is mainly seen in emulsions prepared with oils having appreciable water solubility. Due to water solubility oil from smaller droplets diffuse toward larger droplets through an intermittent aqueous phase because of greater solubility at surface of larger droplets (Chang et al., 2012). Rate of OR is generally directly proportional to solubility of oil phase in water. EO nanoemulsions are relatively prone to OR due to their appreciable solubility in water. High amounts of water insoluble oils could be added in nanoemulsions to prevent OR. In a study carried out on formation of thyme oil nanoemulsions for potential antimicrobial activity it was demonstrated that nanoemulsions with good physical stability could be prepared using organic OR inhibitors such as corn oil and medium chain triglycerides (MCT) in the lipid phase. OR inhibitors were mixed with the thyme oil prior to addition in aqueous phase containing Tween 80 as emulsifier. Nanoemulsions were prepared by high-pressure homogenization carried out at 10 kPa. Prepared emulsions demonstrated mean particle diameter of 160 nm and stability up to a storage period of 3 days. However, results of antimicrobial activity against acid-resistant spoilage yeast, *Zygosaccharomyces bailii* (ZB), suggested that the oil phase composition (ripening inhibitor type and concentration) had an appreciable influence on the antimicrobial activity. In general, with increasing ripening inhibitor levels in the lipid phase a reduction in the antimicrobial efficacy of nanoemulsions was observed. For example, it was observed that for nanoemulsions containing 60% corn oil (w/w) in the lipid phase as a ripening inhibitor, the minimum inhibitory concentration (MIC) of the thyme oil required to inhibit ZB growth was 375 μg/mL. However, no inhibition in ZB growth was observed even at thyme oil concentrations of 6000 μg/mL when a corn oil concentration was increased to 90% (w/w) in the lipid phase. Antimicrobial activity of nanoemulsions was also found to be dependent on ripening inhibitor types used. When used at the same concentration as ripening inhibitor in lipid phase MCT resulted in greater reduction in antimicrobial efficacy of thyme oil as compared to corn oil. At concentration of 70% w/w ripening inhibitor in the lipid phase, the MICs of thyme oil for nanoemulsions containing corn oil and MCT were 750 and 3000 μg/mL, respectively (Chang et al., 2012).

Increased antimicrobial properties of nanoemulsion as compared to free EO is due to small size of oil particles having high surface tension which can fuse and subsequently disrupt the membrane of bacteria, fungi, and viruses, but did not have an affect on eukaryotic cells of higher organisms. Since nanoemulsion result in increased antimicrobial activity it could reduce required amount of active substances for killing microorganisms as compared to conventional methods.

3.2.2 Liposomes

Liposomes are one of the most widely studied colloidal delivery systems and were first developed for drug delivery purposes as early as 1970s (Gregoriadis, 2006; Musthaba et al., 2009). Liposomes consist of vesicular self-assembled system comprising of one or more bilayers, usually formed using a phospholipid, surrounding an aqueous core. Polar head groups of phospholipids are subjected to the aqueous phases of the inner and outer media, and the hydrophobic hydrocarbon tails are associated into the bilayer and spherical core shell structures are formed (Goyal et al., 2005; Jesorka & Orwar, 2008). Liposomes can contain (1) one bilayer forming unilamellar vesicles (ULV), (2) several concentric bilayers forming multilamellar vesicles, or (3) nonconcentric bilayers forming multivesicular vesicles (MVV). Liposomes can be prepared from a very small size of order of 20 nm to sizes exceeding 1 μM. Liposomes have several advantages, such as possibility of large-scale production using natural ingredients and entrapment and release of water-soluble, lipid-soluble, and amphiphilic materials as well as targetability (Fathi et al., 2012). Similar to nanoemulsions liposomes are kinetically stable. Different procedures have been proposed to produce nanosized liposomes. Mechanical procedures include sonification, extrusion, high-pressure homogenization, microfluidization, and colloid mill whereas nonmechanical procedures to form liposomes are depletion of mixed detergent-lipid micelles and reverse-phase evaporation.

Liposomal vesicles were prepared for *Santolina insularis* EO using hydrogenated soya phosphatydilcholine and cholesterol. Prepared liposomes were stable and demonstrated neither oil leakage nor size alteration during a storage period of 1 year. Liposomes were further evaluated and compared with free EOs for antiviral activity against herpes simplex virus type 1 (HSV-1). It was observed that free *S. insularis* EO presented significant antiviral activity mainly due to direct virucidal effects. However, significantly lower activity was observed when EO encapsulated with liposomes was evaluated. It was also reported that when cells were preincubated with EO prior to virus adsorption, the ED (50) values were significantly lower. These results suggest an intracellular mechanism in the antiviral activity of *S. insularis* (Valenti et al., 2001). Positively charged multilamellar (MLV) and unilamellar (ULV) liposomes of *A. arborescens* L. EO were prepared using hydrogenated (P90H) and nonhydrogenated (P90) soy phosphatidylcholine. Vesicles were stable up to a storage period of 6 months but storage for longer duration (1 year) resulted in vesicle fusion. Liposomes were further evaluated and compared with free EO for antiviral activity against herpes simplex virus type 1 (HSV-1). Results obtained demonstrated that the nanoencapsulation of *A. arborescens* EO in liposomes especially when vesicles were made with P90H significantly enhanced its in vitro antiherpetic activity in comparison to free EO. However, no such improvement in antiviral activity was observed when liposomes were prepared with P90. Multilamellar vesicles demonstrated a higher activity when prepared with P90H (EC50 values of 18.3 μg/mL) as compared to vesicles prepared with nonhydrogenated P90 having EC50 value of 43.6 μg/mL. These results indicated that incorporation of *A. arborescens* EO in multilamellar liposomes greatly improved its activity against intracellular HSV-1 (Sinico et al., 2005).

Liposomes have also been prepared and evaluated for their antimicrobial efficacy. Well-known antimicrobial compounds such as carvacrol, thymol, p-cymene, and γ-terpinene were encapsulated in phosphatidyl choline-based liposomes. Prepared liposomes were evaluated for possible improvement of their antioxidant and antimicrobial activities against four gram positive and four gram-negative bacteria and three human pathogenic fungi, as well as the food-borne pathogen, *L. monocytogenes*. Possible synergistic or antagonistic effect between EO components was also evaluated by analyzing antimicrobial activities of carvacrol/thymol and carvacrol/γ-terpinene mixtures before and after encapsulation in liposomes. Enhanced antimicrobial activities for all the compounds were observed after the encapsulation as compared to respective free forms (Liolios et al., 2009). Another study indicated enhanced stability to UV light and humidity when liposomes encapsulated with carvacrol derivatives and thymol was used (Coimbra et al., 2011).

3.2.3 Solid-Lipid Nanoparticles

Solid lipid nanoparticles (SLN) are nanoscale-size particles prepared using lipids that remain solid at room temperature (or/and body temperature). Generally, particle sizes reported for SLN are in the range between 50 nm and 1 μm. The lipid component may comprise of a broad range of lipid and lipid-like molecules such as triacylglycerols or waxes (Mehnert and Mader, 2001) with active ingredients solubilized homogeneously either in the core of the SLNs or in the outside part (McClements et al., 2007). SLN have attracted increase research interest in recent years due to several advantages over nanoemulsions and liposomes. SLN have high encapsulation efficiency with possibility of large-scale production and sterilization. Use of organic solvents can be avoided in preparation of SLN. Further, solid matrix provides better prevention to encapsulated core against chemical degradation and offers more flexible rate of release for bioactive components. Hot homogenization and cold homogenization are two basic techniques widely used for large-scale production of SLNs (Fathi et al., 2012).

Aqueous dispersions of solid lipid nanoparticles (SLNs) of frankincense and myrrh EOs (FMO) were successfully prepared employing high-pressure homogenization method. Solid lipid used for preparing SLNs was Compritol 888 ATO while soybean lecithin and Tween 80 were used as the surfactants. SLNs prepared were of round shape with a mean size of 113.3 ± 3.6 nm. Zeta potential and encapsulation efficiency was observed to be −16.8 ± 0.4 mV and 80.60% ± 1.11%, respectively. It was observed that nanoparticles had significantly better antitumor activity when orally administered in mice as compared to free EO dispersion due to increased solubility (Shi et al., 2012). In another study *A. arborescens* EO was encapsulated into SLN prepared by high-pressure homogenization using Compritol 888 ATO as wall material. Particles with average size of 200–250 nm with up to 90% entrapment efficiency were obtained and were stable for a storage period of 2 years after preparation. SLN were not significantly different in their antiviral activity when compared with free EO. Nevertheless, they greatly improved skin accumulation of the EO and thus proved to be a good carrier for the cutaneous delivery of the antiviral *A. arborescens* EO (Lai et al., 2007). Solid lipid nanoparticles (SLNs) have also been prepared of EO of *Z. multiflora*. Spherical particles with mean particle size and encapsulation efficiency of 650 nm and 38.66% were obtained. During in vitro studies 93.2% of the EO was released after 24 h. Furthermore, from DSC studies it was concluded that EO can interact with the lipid matrix during the preparation of SLNs. The results of characterization indicated the suitability of application of SLN as carrier system for EO of *Z. multiflora* (Moghimipour et al., 2013).

4 Inclusion Complexes

Inclusion complexes are entities comprising two or more molecules. One of the molecule, the "host," accepts, totally or partly, the "guest" molecules by physical forces. Nanoencapsulation inclusion complexes of EO are reported with cyclodextrins. Cyclodextrins (CDs) are cyclic oligomers of α-D-glucopyranose that can be produced due to the transformation of starch by certain bacteria such as *Bacillus macerans* (Astray et al., 2009). Cyclodextrins have toroid-shaped structures with rigid lipophilic cavities and a hydrophilic outer surface insuring good dissolution of the complex in an aqueous environment. They are able to enclose highly hydrophobic molecules inside their hydrophobic cavity, constituting a true molecular encapsulation (Dodziuk, 2006). There are three main types of CDs: α-, β-, and γ-cyclodextrins, corresponding to 6, 7, and 8 glucopyranose units linked by α-(1,4) bonds, respectively. The dimensions of the internal cavity are 0.5–0.8 nm and are crucial for the encapsulation of guest molecules. The purification of α- and γ-cyclodextrins increases considerably the cost of production, so that 97% of the cyclodextrins used in the market are β-cyclodextrins. The water solubility of CDs is unusual. β-cyclodextrin is at least nine times less soluble (1.85 g/100 mL at room temperature) than the other cyclodextrins (14.5 g/100 mL and 23.2 g/100 mL for α- and γ-cyclodextrins, respectively). The inclusion of a guest in a CD cavity consists basically of a substitution of the included water molecules by the less polar guest. The process is energetically favored by the interactions of the guest molecule with the solvated hydrophobic cavity of the host. Generally the steps involved are: (1) Substitution of the energetically unfavored polar–apolar interactions (between the included water and the CD cavity on the one hand, and between water and the guest on the other) by the more favored apolar–apolar interaction (between the guest and the cavity), and the polar–polar interaction (between bulk water and the released cavity-water molecules). (2) CD-ring strain release on complexation and (3) Van der Waals interactions and hydrogen bonds between host and guest (Astray et al., 2009). The major advantages of the use of CD-complexation in foods, pharmaceutical and cosmetics include typically stable and standardized compositions, simple dosing and handling of dry powders, with the consequent reductions of packing and storage costs, conversion of liquid EOs into crystalline powder form, rendering such materials suitable for the manufacture of powders, granules, tablets, and improve handling. It also results in improvement of the molecular stability such as physical stability by the retardation of the crystal growth and chemical stability by the deceleration or even suppression of chemical reactivity, such as volatility, photodegradation, dehydration, hydrolysis, sublimation, oxidation, thermal decomposition, stereochemical transformations, and isomerization. Inclusion complexes can also increase water solubility, dissolution, and release rates of hydrophobic EOs and consequently increasing bioavailability and bioefficacy (Bilia et al., 2014; Astray et al., 2009; Marques, 2010). The preparation method most often used for a complex consists of stirring or shaking the aqueous solution (cold or warm, neutral, or acidic) of CD together with the guest molecule or its solution. After equilibrium has been attained, water is eliminated by freeze-drying, spray-drying, or by any other convenient method. However, most frequently, the microcrystalline product is separated by filtration (Marques, 2010).

β-Caryophyllene (BCP) is a natural sesquiterpene having several biological activities such as antimicrobial, anticarcinogenic, antiinflammatory, antioxidant, anxiolytic like, and local anesthetic effects. However, its volatility and poor water solubility limit its application in pharmaceutical field. It was demonstrated that after single oral dose in rats' inclusion complex displayed earlier Tmax (time after administration of drug when maximum plasma concentration is reached), higher Cmax and the AUC0-12 h showed approximately 2.6 times higher increase as compared to free EO component. The β-CD significantly increased the oral bioavailability of the drug in rats than free BCP (Liu et al., 2013).

In another study, extracts of cinnamon bark and clove bud and pure EO components such as *trans*-cinnamaldehyde, eugenol, and a 2:1 (transcinnamaldehyde : eugenol) mixture were microencapsulated in β-CD by the freeze-drying method. All particles showed a spherical shape, smooth surface, no significant differences in size distribution, and strong tendency to agglomerate. In general, entrapment efficiencies observed were in range of 41.7–84.7%. Pure compounds demonstrated significantly higher ($P < 0.05$) entrapment efficiency as compared to extracts. β-CD complexes were further evaluated and compared with free EOs for their antimicrobial activity against *Salmonella enterica* serovar Typhimurium LT2 and *Listeria innocua*. All the samples effectively inhibited bacterial growth within the concentration range tested, except free eugenol. The EO-β-CD complexes inhibited both bacterial strains at significantly lower concentrations as compared to free oils. Increased antimicrobial activity is attributed to the increased water solubility due to inclusion complexation leading to increased contact between pathogens and EO. Although the cinnamon bark and clove bud oils/β-CD complexes showed the lowest entrapment efficiency but surprisingly demonstrated most significant antimicrobial activities. These results indicate usefulness of EO inclusion complexes as an effective antimicrobial delivery systems. Inclusion complexes could provide broad spectrum of application in food systems where gram-positive and -negative bacteria could present a risk (Hill et al., 2013). Although garlic (*Allium sativum* L.) EO is well known for various pharmaceutical properties including antimicrobial, antidiabetic, antimutagenic, and anticarcinogenic effects but find limited applicability due to its low water solubility, strong odor, volatility, and physiochemical instability. Garlic EO was efficiently complexed with β-CD to form an inclusion complex by the coprecipitation method in a molar ratio of 1:1. The aqueous solubility and stability of EO were significantly increased by inclusion in β-CD (Wang et al., 2011). Another study reported inclusion complexation of isothiocyanates (ITC). ITC are hydrolysis products of glucosinolates and naturally occur in cruciferous vegetables such as broccoli and cabbage. ITCs, particularly allylisothiocyanate are well studied and reported for its antibacterial activity. However, due to its low water solubility its actual use in food systems is difficult. Inclusion complex of allylisothiocyanate and phenyl isothiocyanate was formed with randomly methylated β-cyclodextrin. Significantly improved solubility of both the compounds was observed due to complex formation. Possibility of formation of active packaging films using encapsulated ITCs was proposed (Neoh et al., 2012). Efficacy of inclusion complex for controlled release of aroma components has also been demonstrated. *Lavandula angustifolia* (Ciobanu et al., 2012) and *M. piperita* (Ciobanu et al., 2013) EOs were encapsulated in β-clyclodextrin and cross-linked cyclodextrin polymers. Although it was observed that β-clyclodextrin is a more versatile molecule for encapsulation of EOs as compared to cross-linked cyclodextrin polymers, it was observed that cross-linked cyclodextrin polymers allowed better control of release of aroma components.

5 Future Prospects

EOs are important antimicrobials due to the synergism of their components and their capability of modulating antibiotic resistance. However, due to their low water solubility, strong organoleptic characteristics (flavor and aroma), and low stability together with the high volatility they find a little application in medicine. Most of these drawbacks can be overcome by nanoencapsulating EOs. Nanoencapsulation can provide chemical stability from oxidation, light-induced reactions, moisture and high temperatures, and various other factors that can lead to rapid degradation of the active components. In addition, nanocarriers convert liquid EO into solid powders, thus ensuring their safer and easier handling, improve water solubility, enhance bioavailability and bioefficacy, and improve water solubility. Nanoencapsulation of EOs in liposomes, solid lipid nanoparticles, nano- and microemulsions, and polymeric nanoparticles represent a promising strategy for overcoming EOs limitations, lowering their dose and increasing long-term safety of these constituents. Although a number of different types of delivery systems have been developed, there is still a relatively poor understanding of the major factors governing the rational design of these systems for particular applications.

Several modified carbohydrate delivery systems are reported that demonstrate great potential but future studies should be carried out on toxicity and the biological fate of modified systems during digestion, absorption, and excretion. Furthermore, future studies should focus on physiochemical interactions of nanocarriers with food systems. There should be more emphasis on analyzing the effect of addition of nanocapsules on sensory quality of food products. More research efforts should be directed to prepare active packaging films using nanocapsules.

References

Abreu, F.O., Oliveira, E.F., Paula, H.C., de Paula, R.C., 2012. Chitosan/cashew gum nanogels for essential oil encapsulation. Carbohyd. Polym. 89 (4), 1277–1282.

Akgul, A., Kivanç, M., 1988. Inhibitory effects of selected Turkish spices and oregano components on some foodborne fungi. Int. J. Food Microbiol. 6 (3), 263–268.

Astray, G., Gonzalez-Barreiro, C., Mejuto, J.C., Rial-Otero, R., Simal-Gándara, J., 2009. A review on the use of cyclodextrins in foods. Food Hydrocolloid. 23 (7), 1631–1640.

Azzouz, M.A., Bullerman, L.B., 1982. Comparative antimycotic effects of selected herbs, spices, plant components and commercial antifungal agents. J. Food Protect 45 (14), 1298–1301.

Bagchi, D., Bagchi, M., Moriyama, H., Shahidi, F., 2013. Bio-Nanotechnology: A Revolution in Food, Biomedical and Health Sciences, first ed. Wiley-Blackwell, Sussex, U.K.

Bakkali, F., Averbeck, S., Averbeck, D., Idaomar, M., 2008. Biological effects of essential oils—a review. Food Chem. Toxicol. 46 (2), 446–475.

Baldissera, M.D., Oliveira, C.B., Rech, V.C., Rezer, J.F.P., Sagrillo, M.R., Alves, M.P., da Silva Ana, P.T., Leal, D.B.R., Boligon, A.A., Athayde, M.L., Da Silva, A.S., Mendes, R.E., Monteiro, S.G., 2014. Treatment with essential oil of *Achyrocline satureioides* in rats infected with *Trypanosoma evansi*: relationship between protective effect and tissue damage. Pathol. Res. Pract. 210 (12), 1068–1074.

Bauer, K., Garbe, D., Surburg, H., 2001. Common Fragrance and Flavor Materials: Preparation, Properties and Uses, fourth ed. Wiley-VCH, Weinheim, Germany.

Bawa, R., Bawa, S.R., Maebius, S.B., Flynn, T., Wei, C., 2005. Protecting new ideas and inventions in nanomedicine with patents. Nanomed.-Nanotechnol. 12, 150–158.

Bilia, A.R., Guccione, C., Isacchi, B., Righeschi, C., Firenzuoli, F., Bergonzi, M.C., 2014. Essential oils loaded in nanosystems: a developing strategy for a successful therapeutic approach. J. Evid.-Based Complementary Altern. Med. 2014, 651593.

Braak van de, S., Leijten, G., 1999. Essential Oils and Oleoresins: A Survey in the Netherlands and Other Major Markets in the European Union. CBI, Centre for the Promotion of Imports from Developing Countries, Rotterdam.

Burt, S., 2004. Essential oils: their antibacterial properties and potential applications in foods—a review. Int. J. Food Microbiol. 94 (3), 223–253.

Chang, Y., McLandsborough, L., McClements, D.J., 2012. Physical properties and antimicrobial efficacy of thyme oil nanoemulsions: influence of ripening inhibitors. J. Agric. Food. Chem. 60 (48), 12056–12063.

Chen, H., Zhang, Y., Zhong, Q., 2015. Physical and antimicrobial properties of spray-dried zein–casein nanocapsules with coencapsulated eugenol and thymol. J. Food Eng. 144, 93–102.

Choi, M.J., Soottitantawat, A., Nuchuchua, O., Min, S.G., Ruktanonchai, U., 2009. Physical and light oxidative properties of eugenol encapsulated by molecular inclusion and emulsion—diffusion method. Food Res. Int. 42 (1), 148–156.

Ciobanu, A., Mallard, I., Landy, D., Brabie, G., Nistor, D., Fourmentin, S., 2012. Inclusion interactions of cyclodextrins and crosslinked cyclodextrin polymers with linalool and camphor in *Lavandula angustifolia* essential oil. Carbohyd. Polym. 87 (3), 1963–1970.

Ciobanu, A., Mallard, I., Landy, D., Brabie, G., Nistor, D., Fourmentin, S., 2013. Retention of aroma compounds from *Mentha piperita* essential oil by cyclodextrins and cross-linked cyclodextrin polymers. Food Chem. 138 (1), 291–297.

Coimbra, M., Isacchi, B., Bloois, L.V., Torano, J.S., Ket, A., Wu, X., Broere, F., Metselaar, J.M., Rijcken, C.J.F., Storm, G., Bilia, R., Schiffelers, R.M., 2011. Improving solubility and chemical stability of natural compounds for medicinal use by incorporation into liposomes. Int. J. Pharm. 416 (2), 433–442.

de Oliveira, E.F., Paula, H.C., de Paula, R.C., 2014. Alginate/cashew gum nanoparticles for essential oil encapsulation. Colloid. Surf. B. 113, 146–151.

Dodziuk, H., 2006. Cyclodextrins and Their Complexes. Wiley-VCH, GmbH & KGaA, Weinheim, Germany.

Dong, W., Bodmeier, R., 2006. Encapsulation of lipophilic drugs within enteric microparticles by a novel coacervation method. Int. J. Pharm. 326 (1), 128–138.

Donsì, F., Annunziata, M., Sessa, M., Ferrari, G., 2011. Nanoencapsulation of essential oils to enhance their antimicrobial activity in foods. LWT – Food Sci. Technol. 44 (9), 1908–1914.

Donsì, F., Annunziata, M., Vincensi, M., Ferrari, G., 2012. Design of nanoemulsion-based delivery systems of natural antimicrobials: effect of the emulsifier. J. Biotechnol. 159 (4), 342–350.

Ezhilarasi, P.N., Karthik, P., Chhanwal, N., Anandharamakrishnan, C., 2013. Nanoencapsulation techniques for food bioactive components: a review. Food Bioprocess Tech. 6 (3), 628–647.

Fathi, M., Mozafari, M.R., Mohebbi, M., 2012. Nanoencapsulation of food ingredients using lipid-based delivery systems. Trends Food Sci. Technol. 23 (1), 13–27.

Gaysinsky, S., Davidson, P.M., Bruce, B.D., Weiss, J., 2005. Growth inhibition of *Escherichia coli* O157: H7 and *Listeria monocytogenes* by carvacrol and eugenol encapsulated in surfactant micelles. J. Food Protect 68, 2559–2566.

Giulio, M.D., Sozio, P., Bessa, L.J., Laserra, S., Paolini, C., Protasi, F., Cellini, L., Iannitelli, A., Grande, R., Stefano, D.A., Giulio, M.D., Sozio, P., Bessa, L.J., Laserra, S., Paolini, C., Protasi, F., Cellini, L., 2011. Potential antibacterial activity of carvacrol-loaded poly(DL-lactide-*co*-glycolide) (PLGA) nanoparticles against microbial biofilm. Int. J. Mol. Sci. 12 (8), 5039–5051.

Gomes, C., Moreira, R.G., Castell-Perez, E., 2011. Poly (DL- lactide-*co*-glycolide) (PLGA) nanoparticles with entrapped trans-cinnamaldehyde and eugenol for antimicrobial delivery applications. J. Food Sci. 76 (2), N16–N24.

Gortzi, O., Lalas, S., Tsaknis, J., Chinou, I., 2007. Enhanced bioactivity of *Citrus limon* (Lemon Greek cultivar) extracts, essential oil, and isolated compounds before and after encapsulation in liposomes. Planta Med. 73 (9), 184.

Gouin, S., 2004. Microencapsulation: industrial appraisal of existing technologies and trends. Trends Food Sci. Tech. 15 (7–8), 330–347.

Goyal, P., Goyal, K., Kumar, S.G.V., Singh, A., Katare, O.P., Mishra, D.N., 2005. Liposomal drug delivery systems—clinical applications. Actapharmaceutica. 55, 1–25.

Gregoriadis, G., 2006, third ed. Liposome Technology: Interactions of Liposomes with the Biological MilieuVolume IIICRC Press, New York.

Guenther, E., 1948. The Essential Oils. D. Van Nostrand, New York.

Guenther, E., 1950. In the Essential Oilvol. IVD. VanNostrand, New York.

Guenther, E., 1972. The Essential Oils. Krieger Publishing Company, Malabar, FL.

Gutieırrez, J.M., Gonzaılez, M., Sole, I., Pey, C.M., Nolla, J., 2008. Nano-emulsions: new applications and optimization of their preparation. Curr. Opin. Colloid Interface Sci. 13, 245–251.

Hill, L.E., Gomes, C., Taylor, T.M., 2013. Characterization of beta-cyclodextrin inclusion complexes containing essential oils (*trans*-cinnamaldehyde, eugenol, cinnamon bark, and clove bud extracts) for antimicrobial delivery applications. Food Sci. Technol. 51 (1), 86–93.

Hosseini, S.F., Zandi, M., Rezaei, M., Farahmandghavi, F., 2013. Two-step method for encapsulation of oregano essential oil in chitosan nanoparticles: preparation, characterization and in vitro release study. Carbohyd. Polym. 95 (1), 50–56.

Jayashree, T., Subramanyam, C., 1999. Antiaflatoxigenic activity of eugenol is due to inhibition of lipid peroxidation. Lett. Appl. Microbiol. 283, 179–183.

Jesorka, A., Orwar, O., 2008. Liposomes: technologies and analytical applications. Annu. Rev. Anal. Chem. 1, 801–832.

Karpouhtsis, I., Pardali, E., Feggou, E., Kokkini, S., Scouras, Z.G., Mavragani-Tsipidou, P., 1998. Insecticidal and genotoxic activities of oregano essential oils. J. Agric. Food. Chem. 46 (3), 1111–1115.

Kato, A.K.I.O., 2002. Industrial applications of Maillard-type protein-polysaccharide conjugates. Food Sci. Technol. Res. 8 (3), 193–199.

Kaushik, V., Roos, Y.H., 2007. Limonene encapsulation in freeze-drying of gum arabic–sucrose–gelatin systems. LWT – Food Sci. Technol. 40 (8), 1381–1391.

Keawchaoon, L., Yoksan, R., 2011. Preparation, characterization and in vitro release study of carvacrol-loaded chitosan nanoparticles. Colloid. Surf. B. 84 (1), 163–171.

Konstantopoulou, I., Vassilopoulou, L., Mavragani-Tsipidou, P., Scouras, Z.G., 1992. Insecticidal effects of essential oils: a study of the effects of essential oils extracted from eleven Greekaromatic plants on *Drosophila auraria*. Experientia. 48 (6), 616–619.

Lai, F., Sinico, C., De Logu, A., Zaru, M., Müller, R.H., Fadda, A.M., 2007. SLN as a topical delivery system for *Artemisia arborescens* essential oil: in vitro antiviral activity and skin permeation study. Int. J. Nanomed. 2 (3), 419.

Lee, K.Y., Mooney, D.J., 2012. Alginate: properties and biomedical applications. Prog. Polym. Sci. 37 (1), 106–126.

Li, K.K., Yin, S.W., Yang, X.Q., Tang, C.H., Wei, Z.H., 2012. Fabrication and characterization of novel antimicrobial films derived from thymol-loaded zein–sodium caseinate (SC) nanoparticles. J. Agric. Food. Chem. 60 (46), 11592–11600.

Li, K.K., Yin, S.W., Yin, Y.C., Tang, C.H., Yang, X.Q., Wen, S.H., 2013. Preparation of water-soluble antimicrobial zein nanoparticles by a modified antisolvent approach and their characterization. J. Food Eng. 119 (2), 343–352.

Liolios, C.C., Gortzi, O., Lalas, S., Tsaknis, J., Chinou, I., 2009. Liposomal incorporation of carvacrol and thymol isolated from the essential oil of *Origanum dictamnus* L. and in vitro antimicrobial activity. Food Chem. 112 (1), 77–83.

Liu, Z., Jiao, Y., Wang, Y., Zhou, C., Zhang, Z., 2008. Polysaccharides based nanoparticles as drug delivery systems. Adv. Drug Delivery Rev. 60 (15), 1650–1662.

Liu, H., Yang, G., Tang, Y., Cao, D., Qi, T., Qi, Y., Fan, G., 2013. Physicochemical characterization and pharmacokinetics evaluation of β-caryophyllene/β-cyclodextrin inclusion complex. Int. J. Pharm. 450 (1), 304–310.

Lv, Y., Yang, F., Li, X., Zhang, X., Abbas, S., 2014. Formation of heat resistant nanocapsules of jasmine essential oil via gelatin/gum arabic based complex coacervation. Food Hydrocolloid. 35, 305–314.

Ly, M.H., Aguedo, M., Goudot, S., Le, M.L., Cayot, P., Teixeira, J.A., Le, T.M., Belin, J.M., Waché, Y., 2008. Interactions between bacterial surfaces and milk proteins, impact on food emulsions stability. Food Hydrocolloid. 22 (5), 742–751.

Makadia, H.K., Siegel, S.J., 2011. Poly lactic-*co*-glycolic acid (PLGA) as biodegradable controlled drug-delivery carrier. Polymers 3 (3), 1377–1397.

Mari, M., Bertolini, P., Pratella, G.C., 2003. Nonconventional methods for the control of post-harvest pear diseases. J. Appl. Microbiol. 94 (5), 761–766.

Marques, H.M.C., 2010. A review on cyclodextrin encapsulation of essential oils and volatiles. Flavor Fragance J. 25 (5), 313–326.

McClements, D.J., Li, Y., 2010. Structured emulsion-based delivery systems: controlling the digestion and release of lipophilic food components. Adv. Colloid Interf. Sci. 59 (2), 213–228.

McClements, D.J., Decker, E.A., Weiss, J., 2007. Emulsion based delivery systems for lipophilic bioactive components. J. Food Sci. 72 (8), R109–R124.

Mehnert, W., Mader, K., 2001. Solid lipid nanoparticles: production, characterization and applications. Adv. Drug Delivery Rev. 47 (2–3), 165–196.

Moghimipour, E., Ramezani, Z., Handali, S., 2013. Solid lipid nanoparticles as a delivery system for *Zataria multiflora* essential oil: formulation and characterization. Curr. Drug Deliv. 10 (2), 151–157.

Mohammadi, A., Hashemi, M., Hosseini, S.M., 2015. Nanoencapsulation of *Zataria multiflora* essential oil preparation and characterization with enhanced antifungal activity for controlling *Botrytis cinerea*, the causal agent of gray mould disease. Innovative Food Sci. Emerg. Technol. 28, 73–80.

Mozafari, M.R., 2006. Bioactive entrapment and targeting using nanocarrier technologies: an introduction. In: Mozafari, M.R. (Ed.), Nanocarrier Technologies: Frontiers of Nanotherapy. Springer, Netherlands, pp. 1–16.

Musthaba, M.S., Baboota, S., Ahmed, S., Ahuja, A., Ali, J., 2009. Status of novel drug delivery technology for phytotherapeutics. Expert Opin. Drug Dis. 6, 625–637.

Neoh, T.L., Yamamoto, C., Ikefuji, S., Furuta, T., Yoshii, H., 2012. Heat stability of allylisothiocyanate and phenyl isothiocyanate complexed with randomly methylated β-cyclodextrin. Food Chem. 131 (4), 1123–1131.

Nychas, G.J.E., 1995. Natural antimicrobials from plants. In: Gould, G.W. (Ed.), New Methods of Food Preservation. Blackie Academic and Professional, London, pp. 58–89.

Pan, K., Chen, H., Davidson, P.M., Zhong, Q., 2014. Thymol nanoencapsulated by sodium caseinate: physical and antilisterial properties. J. Agric. Food. Chem. 62 (7), 1649–1657.

Pandey, R., Kalra, A., Tandon, S., Mehrotra, N., Singh, H.N., Kumar, S., 2000. Essential oil compounds as potent source of nematicidal compounds. J. Phytopathol. 148 (7–8), 501–502.

Parris, N., Cooke, P.H., Hicks, K.B., 2005. Encapsulation of essential oils in zein nanospherical particles. J. Agric. Food. Chem. 53 (12), 4788–4792.

Pessoa, L.M., Morais, S.M., Bevilaqua, C.M.L., Luciano, J.H.S., 2002. Anthelmintic activity of essential oil of *Ocimum gratissimum* Linn. and eugenol against *Haemonchus contortus*. Vet. Parasitol. 109 (1–2), 59–63.

Pichersky, E., Noel, J.P., Dudareva, N., 2006. Biosynthesis of plant volatiles: nature's diversity and ingenuity. Science 311 (5762), 808–811.

Quintanilla-Carvajal, M.X., Camacho-Díaz, B.H., Meraz-Torres, L.S., Chanona-Pérez, J.J., Alamilla-Beltrán, L., Jimenéz-Aparicio, A., Gutiérrez-López, G.F., 2010. Nanoencapsulation: a new trend in food engineering processing. Food Eng. Rev. 2 (1), 39–50.

Ravi Kumar, N.M., 2000. Nano and microparticles as controlled drug delivery devices. J. Pharm. Pharm. Sci. 3 (2), 234–258.

Ribeiro, J.C., Ribeiro, W.L.C., Camurça-Vasconcelos, A.L.F., Macedo, I.T.F., Santos, J.M.L., Paula, H.C.B., Bevilaqua, C.M.L., 2014. Efficacy of free and nanoencapsulated *Eucalyptus citriodora* essential oils on sheep gastrointestinal nematodes and toxicity for mice. Vet. Parasitol. 204 (3), 243–248.

Sansukcharearnpon, A., Wanichwecharungruang, S., Leepipatpaiboon, N., Kerdcharoen, T., Arayachukeat, S., 2010. High loading fragrance encapsulation based on a polymer-blend: preparation and release behavior. Int. J. Pharm. 391 (1), 267–273.

Schatz, C., Bionaz, A., Lucas, J.M., Pichot, C., Viton, C., Domard, A., Delair, T., 2005. Formation of polyelectrolyte complex particles from self-complexation of N-sulfated chitosan. Biomacromolecules. 6 (3), 1642–1647.

Schweiggert, U., Carle, R., Schieber, A., 2007. Conventional and alternative processes for spice production—a review. Trends Food Sci. Technol. 18 (5), 260–268.

Scott, P.W.R., 2005. Essential oils. In: Worsfold, P., Townshend, A., Poole, C. (Eds.), Encyclopedia of Analytical Science. second ed. Elsevier, London, UK, pp. 554–561.

Shah, B., Davidson, P.M., Zhong, Q., 2012. Nanocapsular dispersion of thymol for enhanced dispersibility and increased antimicrobial effectiveness against *Escherichia coli* O157: H7 and *Listeria monocytogenes* in model food systems. Appl. Environ. Microbiol. 78 (23), 8448–8453.

Shah, B., Davidson, P.M., Zhong, Q., 2013. Nanodispersed eugenol has improved antimicrobial activity against *Escherichia coli* O157: H7 and *Listeria monocytogenes* in bovine milk. Int. J. Food Microbiol. 161 (1), 53–59.

Shakeel, F., Ramadan, W., 2010. Transdermal delivery of anticancer drug caffeine from water-in-oil nanoemulsions. Colloid. Surf. B. 75, 356–362.

Shelef, L.A., 1983. Antimicrobial effects of spices. J. Food Saf. 6 (1), 29–44.

Shi, F., Zhao, J.H., Liu, Y., Wang, Z., Zhang, Y.T., Feng, N.P., 2012. Preparation and characterization of solid lipid nanoparticles loaded with frankincense and myrrh oil. Int. J. Nanomed. 7, 2033.

Sinico, C., De Logu, A., Lai, F., Valenti, D., Manconi, M., Loy, G., Bonsignore, L., Fadda, A.M., 2005. Liposomal incorporation of Artemisia arborescens L. essential oil and in vitro antiviral activity. Eur. J. Pharm. Biopharm. 59 (1), 161–168.

Soppimath, K.S., Aminabhavi, T.M., Kulkarni, A.R., Rudzinski, W.E., 2001. Biodegradable polymeric nanoparticles as drug delivery devices. J. Control. Rel. 70 (1–2), 1–20.

Stulzer, H.K., Tagliari, M.P., Parize, A.L., Silva, M.A.S., Laranjeira, M.C.M., 2009. Evaluation of cross-linked chitosan microparticles containing acyclovir obtained by spray-drying. Mater. Sci. Eng. C. 29 (2), 387–392.

Turek, C., Stintzing, F.C., 2013. Stability of essential oils: a review. Compr. Rev. Food Sci. Food Saf. 12 (1), 40–53.

Valenti, D., DeLogu, A., Loy, G., Sinico, C., Bonsignore, L., Cottiglia, F., Donatella, G., Fadda, A.M., 2001. Liposome-incorporated *Santolina insularis* essential oil: preparation, characterization and in vitro antiviral activity. J. Liposome Res. 11 (1), 73–90.

Wang, J., Cao, Y., Sun, B., Wang, C., 2011. Physicochemical and release characterisation of garlic oil-β-cyclodextrin inclusion complexes. Food Chem. 127 (4), 1680–1685.

Wattanasatcha, A., Rengpipat, S., Wanichwecharungruang, S., 2012. Thymol nanospheres as an effective antibacterial agent. Int. J. Pharm. 434 (1–2), 360–365.

Weiss, J., Gaysinksy, S., Davidson, M., McClements, J., 2009. Nanostructured encapsulation systems: food antimicrobials. In: Barbosa-Cánovas, G.V., Mortimer, A., Lineback, D., Spiess, W., Buckle, K. (Eds.), IUFoST World Congress Book: Global Issues in Food Science and Technology. Elsevier, Amsterdam, pp. 425–479.

Woranuch, S., Yoksan, R., 2013. Eugenol-loaded chitosan nanoparticles: thermal stability improvement of eugenol through encapsulation. Carbohyd. Polym. 96 (2), 578–585.

Wu, Y., Luo, Y., Wang, Q., 2012. Antioxidant and antimicrobial properties of essential oils encapsulated in zein nanoparticles prepared by liquid–liquid dispersion method. LWT – Food Sci. Technol. 48 (2), 283–290.

Zhang, Y., Niu, Y., Luo, Y., Ge, M., Yang, T., Yu, L.L., Wang, Q., 2014. Fabrication, characterization and antimicrobial activities of thymol-loaded zein nanoparticles stabilized by sodium caseinate–chitosan hydrochloride double layers. Food Chem. 142, 269–275.

Zhao, Y., Wang, C., Chow, A.H., Ren, K., Gong, T., Zhang, Z., Zheng, Y., 2010. Self-nanoemulsifying drug delivery system (SNEDDS) for oral delivery of Zedoary essential oil: formulation and bioavailability studies. Int. J. Pharm. 383 (1), 170–177.

16

NANOENCAPSULATION AND NANOCONTAINER BASED DELIVERY SYSTEMS FOR DRUGS, FLAVORS, AND AROMAS

Shailesh Ghodke*, Bharat Bhanvase, Shirish Sonawane[†], Satyendra Mishra*, Kalpana Joshi[‡]**

**North Maharashtra University, University Institute of Chemical Technology, Jalgaon, Maharashtra, India; **Chemical Engineering Department, Laxminarayan Institute of Technology, Rashtrasant Tukadoji Maharaj Nagpur University, Nagpur, Maharashtra, India; [†]National Institute of Technology, Warangal, Telangana, India; [‡]Department of Biotechnology, Sinhgad College of Engineering, Pune, Maharashtra, India*

1 Introduction

With the rapid changes in the fields of science, engineering, and technology, the use of application-specific materials is in vogue. A major impact had been shown by various developments of nanoscience/nanotechnology wherein materials on nanoscale are synthesized that have ability to change their surface properties in their release behavior. This behavior is seen when a small change in external environment of such materials is being used. This change is because of rearrangement of surface properties such as structure, aggregation state, interaction with encapsulated material, and external environment (Skirtach and Kreft, 2009; Moghimi et al., 2005). Such materials are especially being introduced in a variety of fields such as modern medicine, food, cosmetics, material science, and biochemistry for one of the primary reasons, that is, delivery systems (Chilkoti et al., 2002; Min-Hui and Keller, 2009). For these delivery systems various material chemists have been interested in the development of small (and usually variable) size and low toxicity, and the controlled permeability of the micro/nano-container wall. The interest has been in spatially

Encapsulations. http://dx.doi.org/10.1016/B978-0-12-804307-3.00016-8

confined, engineered micro- and nanosystems containing various species (eg, catalysts, DNA, drugs, enzymes) that are applicable in diagnostics, catalysis, therapy, and bioengineering in both aqueous and nonaqueous media (Shchukin et al., 2005). Some more applications consist of delivery of aroma chemicals, flavors, and fragrances in consumer items such as those produced by the food packaging, textile, and detergent industries (Fisk et al., 2011). The protection of food flavors from loss and degradative reactions, like oxidation, can be easily achieved through aroma compound encapsulations (Marcuzzo et al., 2010). Application of nanocontainers in the food products can increase food's spreadability and stability, and can support in developing healthier low-fat food products (Anandharamakrishnan, 2014).

Nanoencapsulation is found to be the efficient method for fabricating such nanocontainers. Nanoencapsulation is defined as "methodology of entrapping core of active molecules within a transporting shell material and forming a closed container within a nanometric dimension"; the carrier formed is typically in the range of 0.1 nm to 1 μm (1000 nm). Most successful examples of such micro- and nanoencapsulation are hollow polymeric nanospheres (Li and Szoka, 2007; Gill and Ballesteros, 1998; Lee et al., 2001; Lu et al., 2009), lipid bilayers (Shenton et al., 2001), micelles and microemulsions (Ball and Haymet, 2001; Seregina et al., 1997), thermoplastic aliphatic polyester (polylactic acid) (Hosseinkhani et al., 2015), liposomes (Chekhonin et al., 2012), inorganic hosts (Shengnan et al., 2010), and so forth. For this purpose not only polymers are being used; there are some other techniques to (self-) assemble them into functional materials have been used under vigorous research (Stuart et al., 2010). In order to have prolonged release characteristics it becomes necessary to apply a protective covering on to the nanocontainer's surface. It is common knowledge that changes in solvent quality may strongly affect stability of the particles' dispersions or adsorption of the particles at interface, due to the responsive properties of the polymer molecules. An increase in complexity of the particle and the polymer structures and immersion of this particle in a complex solvent will result in a broad variety of multifunctional systems with important applications.

For any delivery system the responsive properties of the particle may originate from the polymeric shell, polymer core, and both the core and the shell. Fig. 16.1 shows the commonly used nanocontainers for transport of active molecules. Fig. 16.1a represents a typical single or multiwall nanocontainer able to encapsulate single/multicore. Layer-by-layer assemblies with more than two polyelectrolyte layers are able to encapsulate active molecules

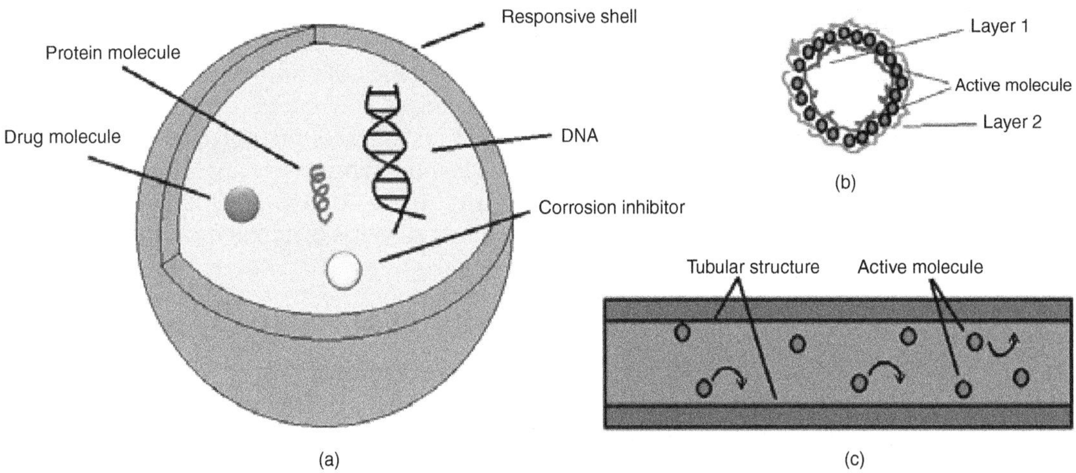

Figure 16.1. Responsive nanocontainers for delivery of active molecules. (a) Core–shell assembly; (b) layer–layer assembly; (c) inorganic clay mineral (nanotubules).

(as shown in Fig. 16.1b). Use of tubular structures is also possible for to encapsulate active molecule within hollow lumen (as depicted in Fig. 16.1c). The external stimuli can be used to tune uptake or release of small molecules in and out of the pores; this is because the stimuli change the permeability of the shell material. The general concept of responsive release can be divided mainly into two major types according to the nature of the interaction between the active molecule and the nanocontainer. In the complexation approach, the active molecule is entrapped within the nanocontainer, the release can be initiated by structural change within the shell material (eg, charging of functional groups, carrier degradation, cleavage of shell, etc.), while in the nanocontainer-conjugate approach, the molecule release is the result of splitting of the link between the nanocontainer and the active molecule (Fleige et al., 2012). In this chapter, the delivery of active molecules such as drug, fragrances, and aroma delivery will be discussed. In the description of nanocontainers for delivery system the constructional aspects of nanocontainers are discussed in detail. This detailed section depicts the various facets of nanocontainers, their formation, and delivery efficiency in detail. All these shells are necessarily responsive and compatible with the kind of environment in which these are applied (Fu et al., 2011; Gil and Hudson, 2004).

In the present review we have studied nanoencapsulation and various types of nanocontainers frequently used for delivery of active molecules like drug, flavor, and aroma. Mechanism of response and various types of stimuli used for prolonged delivery of active molecule. The various types of stimuli studied thoroughly are chemical (pH, ionic strength, solvent, electrochemical stimuli), physical (temperature, laser, ultrasound, mag field, mechanical deformation), and biological (enzyme triggering, receptor implementing triggering). Efforts are made to present the state of art of nancontainer-based delivery systems of active molecules.

2 Nanocontainers for Delivery System

Development of functional nanocontainers for delivery of active material is a real challenge for a delivery system. The primary task of these materials is to protect the content from external influence/environment and release it only in response to certain environmental conditions at the desired destination. Materials used to fabricate nanocontainers play the most crucial role for stimuli-responsive modality. They are responsible for the release of active materials either temporally or spatially within the structure in response to the trigger.

Various methods can be employed to encapsulate the active cargo within the nanocontainer shell. Delivery of such materials can be achieved through the use of specific stimuli causing change in permeability of container wall. Hence in order to regulate the transport of cargo a nanocontainer having appropriate permeability characteristics must be carefully selected. In order to achieve maximum efficiency of nanocontainers certain characteristics like mechanical permeability, stability, elasticity, morphology, biocompatibility, and surface characteristics can be adjusted accordingly. Over the past few years, there were several reports on various types of nanocontainers with the capability to encapsulate and release active materials. Following are the various nanocontainers reported for successful delivery of active materials.

2.1 Polymeric Nanocontainers

Polymeric nanocontainers have been class of highly versatile and diverse materials for the delivery systems. With the wide range of properties and functions of polymeric materials it became easier to synthesize nanocontainers of desired property and applications. Potential use of polymeric nanocontainers are delivery of drugs (Mu et al., 2009), proteins (Lee et al., 2001), genes (Li et al., 2012), catalysts (Morris et al., 1999; Lee et al., 2001), and so forth. Polymeric nanocontainers are mainly of two types: hollow spherical nanocontainers and polymeric nanotubes. The spherical nanocontainers are usually prepared by suspension polymerization, emulsion polymerization, self-assembly, core-shell precursors and dendrimers. The polymeric nanotubes are prepared by emulsion polymerization, self-assembly macromolecules, template-directed synthesis, and electrospinning.

The expediency in polymerization process, uniform product size and simplicity in handling are the major advantages of suspension polymerization process (Islam et al., 2002). Suspension polymerization is a process in which the monomer, or mixture of monomers, is dispersed by mechanical agitation in a liquid phase, usually water, in which the monomer droplets are polymerized while they are dispersed by continuous agitation. Okubo et al. (2002, 2003, 2004) have reported various nanostructure formations and their applications. The nanocontainers formed by this method generally tend to reach in micrometric dimensions. Emulsion polymerization is a type of radical polymerization that usually starts with an emulsion incorporating water, monomer, and surfactant. The most common type of emulsion polymerization is an oil-in-water emulsion, in which droplets of monomer (the oil) are emulsified (with surfactants) in a continuous phase of water. Emulsion polymerization is a rather complex process because nucleation, growth, and stabilization of polymer particles are controlled by the free radical polymerization mechanisms in combination with various colloidal phenomena. Various recent studies have been reported for the synthesis of polymeric hollow nano/microspheres (Xiaokun et al., 2008; Pollert et al., 2006; Zhao et al., 2009). Herein, often there is segregation of free radicals among the discrete monomer-swollen polymer particles. Hence there is reduction in the probability of bimolecular termination of free radicals resulting in a faster polymerization rate and polymer with a higher molecular weight (Chern, 2006).

The high level of control over the dendritic architecture (size, branching density, surface functionality) makes dendrimers ideal carriers in these applications (Svenson, 2009). A dendrimer is a macroscale; growth of the structure starts from a central core and builds on a repetitive structural arrangements. Unlike traditional polymers, such structures are found to be well defined, having low size polydispersity yet with a large molecular size. Some studies have reported the formation of hollow polymeric nanocontainers via the synthesis of dendrimers and removal of the internal core (Sunder et al., 1999). Nevertheless, complex methodologies involved in the preparation of these containers have limited their widespread application (Patri et al., 2005).

Polyamide nanocapsules containing *Aloe vera* L. by an emulsion diffusion technique with in vivo studies were reported by

Esmaeili and Ebrahimzadeh (2015). Ratio of polymer to oil, the concentration of polymers, and the plant extract were found to be key parameters for the nanocontainer diameters. Among the existing strategies to overcome these drawbacks, inclusion of hydrophobic drugs into polymeric micelles is one of the most attractive alternatives. Amphiphilic poly(ethylene oxide)–poly(propylene oxide) block copolymers are thermoresponsive materials that display unique aggregation properties in aqueous medium (Chiappetta and Sosnik, 2007). Harada and Kataoka have given an excellent review based on supramolecular assemblies of block copolymers in aqueous media as nanocontainers relevant to biological applications (Harada and Kataoka, 2006). The use of block copolymers in corrosion protection under self-repairing protective coating containing active nanoreservoirs has been reviewed by Shchukin and Mçhwald (Shchukin and Mçhwald, 2007). Later, use of PEO-*b*-PPO-*b*-PEO block copolymer for fragrance delivery in ethanol-water mixture was reported by Berthier et al. (2010). Mainly the study focused on the influence of block copolymers on the evaporation of volatile molecules in ethanolic solution. Swelling studies were performed to analyze the capacity of volatile molecules in aggregates of copolymers.

Hofmeister et al. (2014) reported the pH and temperature sensitive nancontainers for fragrance encapsulation and controlled release. α-pinene was used as hydrophobic model compound for the encapsulation process to produce nanocontainers with 200 nm diameter having ≥90% encapsulation efficiency. The method used was miniemulsion-analogous free radical polymerization. The release studies postulated that, upon deprotonation the nanocontainer absorbs water, resulting in increasing chain segment mobility, reduced diffusion barrier resulting into triggered release. Monomer ratio and acid functionalization was found to be responsible for higher encapsulation efficiency. Another temperature induced fragrance delivery was reported by Theisinger et al. (2009). Poly(methyl methacrylate), polystyrene, or acrylic copolymer were used to form 100 nm nanocontainer encapsulating a hydrophobic fragrance. The study revealed that the release behavior can be tuned by the temperature in relation to the T_g (glass transition temperatures) of the polymer, which makes these nanocontainers interesting candidates for temperature dependent delivery systems. Poly(ethylene oxide)-4-methoxycinnamoylphthaloylchitosan (PCPLC) was also found to be able to form nanocontainers in the range of 300–320 nm by solvent displacement method (Tachaprutinun et al., 2009). The study concluded that nanoencapsulated astaxanthin showed minimal heat degradation of olefinic functionality in contrast to that of the unencapsulated pigment molecules, which were almost completely destroyed.

2.2 Layer-by-Layer Assemblies

With the flexibility and versatility of synthesis the layer-by-layer assembly process has been extensively practiced in the variety of applications. Some of the other desirable characteristics of layer-by-layer are mechanical stability, elasticity, morphology, biocompatibility, permeability, and surface characteristics can be adjusted accordingly. The most significant advantages of layer-by-layer assemblies are multifunctionality and availability of various stimuli to affect and control their properties; also, the permeability is determined by the balance of electrostatic interactions within the multilayer (Pomorska et al., 2011). The building blocks of layer-by-layer assemblies are the polyelectrolytes (class of polymers that carry charged functional groups) essential to form a container. In order to assemble polyelectrolyte multilayers complexation of polymers is required which results due to opposite charges of polymers. The alternative depositions can be made to form a spherical or substrate template.

Encapsulation of cargo is possible either during synthesis of assemblies (ie, incorporation) or after the formation of assemblies (ie, adsorption or physical interaction) (Pomorska et al., 2011). The release mechanism of layer-by-layer assembly initiates with the ionization of weak polyelectrolyte of the functional groups, which tends to increase repulsion between the uncompensated charges (Andreeva et al., 2010). In order to balance these charges small counter ions penetrate the layered structure. Further, the higher ionic concentration inside the assembly increases the osomotic pressure over the nanocontainers shell. Finally when the surrounding solute/water enters into the assembly, it leads to swelling followed by pore opening into the surrounding area (Kozlovskaya et al., 2006). Many of the polymers widely used in deposition at surfaces, such as polystyrene sulfonate (PSS) and poly(allylamine hydrochloride) (PAH), show extremely slow equilibration times in solution, and chain desorption from surfaces is kinetically frozen and usually not observed (Sukhishvili, 2005).

In the extensive review of responsive layer-by-layer materials for drug delivery, Wohl and Engbersen described the use of layer-by-layer (LbL) assemblies for drug and gene delivery (Wohl and Engbersen, 2012). The study reports the formation of LbL assemblies can be possible wherein therapeutic payload can be encapsulated both in the shell and the interior, and the composition of the capsules can be tailored. Apart from other properties of LbL system nanoscale dimensions of nanocontainers enables easy circulation during in vivo applications. Recently Parekh et al. (2014) reported nanoencapsulation of camptothecin with improved activity using LbL approach. The study attempted to reduce the hydrolysis of Camptothecin (CPT) nanocapsules to an inactive carboxylic form at neutral and alkaline conditions. The result claimed synthesis of 160 nm nanocontainer of heparin and block copolymer (polyethylene glycol plus L-lysine). Stability and prolonged delivery of nanocontainer was obtained with addition of polyethylene glycol (molecular weight: 5 or 20 kDa). The layer-by-layer approach was successful in obtaining 7–8 polyelectrolyte bilayers. Gu et al. demonstrated a LbL assembly of PADH (3-dimethylaminopropyl and hydrazide grafted PAsp) and PACA (carboxyl and aldehyde grafted PAsp) employing hydrazone for crosslinking for protein delivery (Gu et al., 2013). Poon et al. (2011) illustrated a promising approach for systemic tumor targeting using LbL nanoparticles. The idea was the use of charged species to aid or inhibit their cellular uptake and extend this idea to enable tumor targeting in vivo by incorporating a pH-responsive layer that exposes the underlying charged surface when localized in an acidic tumor microenvironment. The nanocontainers functionalized with charged species were not more than 90 nm, which significantly indicates the application for in vivo studies. The formulation of a novel delivery system for ellagic acid formulated via layer-by-layer (LbL) electrostatic deposition of biopolymers onto soybean lecithin liposomes was achieved by Madrigal-Carballo et al. (2010).

2.3 Silica-Based Delivery System

Mesoporous silica nanoparticles as a promising drug carrier have become the new area of interest in the field of biomedical application in recent years because of their unique characteristics and abilities to efficiently and specifically entrap cargo molecules (Zhu et al., 2014). Mesoporous silicas are inorganic materials synthesized in the presence of surfactants as templates for the polycondensation of silica species, originating from different sources of silica [sodium silicate, alkoxydes like tetraethylortosilicate –(TEOS) and tetramethylortosilicate (TMOS)]. Synthesis conditions such as source of silica, type of surfactant, ionic strength, pH, and composition of the reaction mixture, temperature, and duration of synthesis affect the surfactant micellar conformation, the silica–surfactant interactions and the degree of silica polycondensation (Renzo et al., 1999). High acidic environment of stomach often tend to decompose the pharmaceutical or nutriceutical cargo of enzymes, DNAs, and RNAs. This demands a drug carrier that would not leak or degrade until it delivers the molecule successfully. Silica is one of the carriers most often used for delivery of drug and corrosion inhibitors (Dalmoro et al., 2012). Some studies also indicate the delivery of antibiotics using silica (Meseguer-Olmo et al., 2006; Radin et al., 2004).

It is found that MSNs can deliver anticancer drugs to tumors by accumulating in tumor xenografts and improve the anticancer drug efficiency. The promotion of human malignant melanoma growth by mesoporous silica nanoparticles through decreased reactive oxygen species was studied by Huang et al. (2010). The experimental results indicated that the silica nanoparticles were not having any toxic effects. Citronelleal fragrant molecule delivery was made possible with the use of poly[propyl-4 methoxycinnamamide silsesquioxane] nanocontainers from triethoxysilane monomers containing the chromophoric 2,4-dimethoxycinnamoyl and 4-methoxycinnamoyl moieties, using the sol–gel process (Kidsaneepoiboon et al., 2011). The hydrophobic interaction between citronellal molecules and the cinnamoyl moieties of the silica network structure probably helped retard the release of the volatile citronellal molecules, resulting in an obvious slower release of citronellal from nanocontainer. Silica nanocontainers (140–220 nm) containing fragrances through the miniemulsion technique were reported by Cao et al. (2015). The hydrophobic liquid droplets consisting of TEOS (tetraethoxysilane) and fragrance worked as templates to form container morphology. Further an attempt was made to encapsulate hydrophilic perfume into the mesoporous silica followed by layer-by-layer deposition of poly(diallyldimethylammonium chloride) and poly(sodium 4-styrenesulfonate) (Wang et al., 2008). The study concluded that the absorption of perfume into nanocontainers as well as the polyelectrolyte layers contributed to retention and prolonged release of perfume during release.

2.4 Halloysite

Haollysite nanotubes are aluminosilicate clays mined from natural deposits. Chemically Halloysites are similar to kaolin clays and are two-layered (1:1) aluminosilicate. The only difference between Kaolin clay and Halloysite is the morphology of crystals. Halloysite nanotube consists of a negative charge on external surface and positive charge on internal surface in pH 2~8 aqueous solutions (Vergaro et al., 2010; Xing et al., 2012). Two major advantages of Halloysite are that they have hollow nanotubular structure in the submicrometer range and a large specific surface area (Zhai et al., 2010). The hollow multilayer tubule formation is a result of neighboring alumina. During basic structure formation, the presence of water of hydration causes the silica layers to form curves and create the rolled statures forming multilayered tubular halloysite structures. This provides the advantage of formation of cylindrical halloysite nanotubes with very small inner diameter and hence can be used as nanocontainers for delivery of molecules for various applications.

Levis and Deasy (2002) have reported the use of halloysite for delivery of drug molecules. The surface charge was predominantly negative over most of the physiologically relevant pH range (>2) and the specific surface area of the material was very large (~57 m^2/g), indicating that the material has significant potential for extensive binding of cationic drugs. Further investigation of equilibrium, kinetics, release kinetics, and thermodynamic aspects for drug (5-aminosalicyclic acid) delivery were studied in detail (Viseras et al., 2008; Aguzzi et al., 2013). For prolonged release and higher drug loading various efforts have been made. Encapsulation/loading of molecules is the result of electorstatic attraction of cargo molecules to the inner surface of the tubules. Recently, Tan et al. (2013) reported loading and in vitro release of Ibuprofen in tubular halloysite. Ibuprofen molecule first tends to form a weak bond with –OH. Release of such particles is found to be rapid. In order to restrict the molecules to strong bond with the drug molecule, 3-aminopropyltriethoxysilane (APTES) was introduced to build a strong affinity through electrostatic attraction, between the carboxyl groups of IBU and the aminopropyl groups of the grafted APTES. As a result the APTES-modified halloysite nanocontainers were found to have the largest IBU loading, which was 25.4% greater than that in unmodified halloysite.

Higher loading efficiency is one of the desirable criteria for the use of nanocontainers. In this light, some studies necessarily focus on increasing the lumen diameter (Abdullayev et al., 2012; Zhang et al., 2012). This is predominantly performed by acid etching. During etching there is first diffusion of hydrogen ions into hollow lumen followed by the interaction of hydrogen with alumina and then diffusion of reaction products outside the halloysite nanotubules. As a result there is successful removal of alumina from the inner lumen of halloysite. Lately, Wang et al. have studied alkali activation of halloysite for adsorption and release of ofloxacin (OFL) (Wang et al., 2013). Herein, alkali activation on the physicochemical properties, structure, and morphology of halloysite nanotube were performed. Afterward, the adsorption and in vitro release properties of halloysite for cationic OFL were evaluated. The results indicate that alkali activation dissolves amorphous aluminosilicate, free silica, and alumina, which results in the increase in pore volume and pore size. Considering the ability of halloysite for responsive delivery of active molecules, Ghodke et al. (2015) depicted the successful attempts of using halloysite nanocontainers for encapsulation of fragrant molecules. The loaded nanocontainers were analyzed for responsive release studies. The release studies were conducted using various pH solutions. It is established that the Korsmeyer–Peppas release model is best among the six models used to predict the release from Halloysite nanocontainers.

2.5 Ultrasonic Technique

One of the most recent techniques used to form nanocontainers is ultrasound, in the frequencies from 20 kHz to 1 GHz. This technique involves formation of microbubble as template on whose surface an organic or inorganic shell is formed from monomers, precursors, and nanoparticles adsorbed at the cavitation interface (Suslick and Crum, 1997). Bubble collapse in liquids results in an enormous concentration of energy from the conversion of the surface energy and kinetic energy of the liquid motion into heat or chemical energy (Shchukin and Mçhwald, 2007). The high local temperatures (5000–7000 K inside the microbubble) and pressures combined with rapid cooling provide unique conditions for forming micro- and nanocontainers (Rae et al., 2005). In some of earlier works related to container formation, Suslick et al. used high-intensity ultrasound air and oil-filled protein microspheres (Suslick and Grinstaff, 1990; Suslick et al., 1994). Emulsification and cavitation are responsible for formation of microcontainers wherein disulfide cross-linking of cysteine residues between protein molecules results in the formation of bubbles. These containers were found to be much more stable, with high loading capacity. The approach described in this discussion is depicted in Fig. 16.2.

The characteristics of the ultrasonically obtained containers can be changed to increase their stability and loading capacity. This can be done by decorating the surface of the nanocontainers with suitable hydrophilic and hydrophobic moieties functioning as selective ligands (such as L-cysteine, L-lysine, chitosan, and β-cyclodextrin) (Cavalieri et al., 2006; Shchukin and Mçhwald, 2007). Shchukin et al. (2005) described the application of polyelectrolyte multilayers for air encapsulation and the formation of polyelectrolyte microcapsules with a gaseous interior. The work further explained the layer-by-layer approach for encapsulation of active molecules. Poly(allylamine hydrochloride) (PAH) and poly(styrenesulfonate) (PSS) layers were adsorbed on containers as forms of positive and negative layers, respectively.

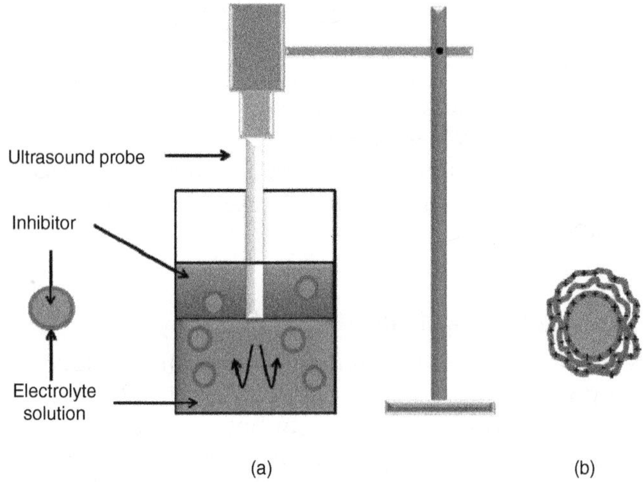

Figure 16.2. Method for encapsulation of active molecule for the use as nanocontainers delivery system. (a) Formation of polyelectrolytes nanocontainer; (b) coating of nanocontainers by oppositely charged polyectrolytes using layer-by-layer approach for triggered release.

Teng et al. (2007) reported a novel approach for encapsulation of hydrophobic materials into a hydrophilic multifunctional shell, based on combining an ultrasonic technique and a layer-by-layer protocol. Uniform, stable, and monodisperse polyglutamate/PEI/PAA nanocontainers of about 600 nm were obtained. This provides an opportunity to increase loading efficiency. The results also found that, using sodium dodecyl sulfate as surfactant, the amount of nanocontainers, their monodispersity, and stability can be increased dramatically. However the methods of forming organic nanocontainers suffer some disadvantages (Suslick and Grinstaff, 1990; Suslick et al., 1994; Cavalieri et al., 2006; Teng et al., 2007): first, it is difficult to modify the protein shell by introducing a desired material in order to attain additional functionality. Second, the method yields large size nanocontainers (almost microspheres) with higher polydispersity. Third and last the ultrasonic technology used for the preparation of the microcapsules is limited to proteins possessing at least one cysteine residue or sulfide residue to obtain stable microcapsules.

The characteristics of the product obtained in the synthesis of nanocontainers using ultrasound provides a platform of responsive releases for various applications. Ease of operation also plays an important role in the synthesis of nanocontainers. A future prospect for this method is fabrication of nanocontainers using polyelectrolyte followed by a layer-by-layer approach for responsive release. Anionic biopolymer–alginate was used to produce turmeric-oil-encapsulated nanocontainers (Lertsutthiwong et al., 2008). Alginate was selected because of some favorable properties such as good biocompatibility, biodegradability, nontoxicity, mucoadhesion, gelation, and film formation properties. The lowest nanocontainer size was found to be 95 nm ethanol as a solvent and Tween 80 as surfactant. Study concluded that the uniform size of the nanocontainer was possible because of use of ultrasound. And the size of the nanocontainer could be controlled by the time of sonication. Esmaeili et al. (2013) reported the synthesis of oil-filled nanocontainers under ultrasound. *Crataegus azarolus* L. was used as model fragrant molecule to be encapsulated in the triblock copolymer (PEG–PBA–PEG). Results indicated that with an increase in amount of polymer added, the average size of the nanocontainer increased. However, the size of nanocontainer decreased with increase in model fragrant. This is because the interaction of model fragrance and polymer causes an aggregation of polymer in the nanocapsules structure. The degree to which this occurs depends on the concentration of the extract; when the amount of extract in the nanocapsules increases, the extract acts as a clog factor, causing an increase in the particle size and decreasing the stability.

3 Mechanism of Response

Many studies are reported for environmentally sensitive nanocontainers that can respond to elusive stimuli such as pH, temperature, ionic strength light, magnetic field, biosensing, chemical separation, catalysis, and biomaterials applications (Sukhorukov et al., 2005). The mechanism of response is decided either by nature of nanocontainers material or the nature of response desired in the particular application. LbL systems are generally considered inherently responsive, as changes in pH or ionic strength of the surrounding media obviously influence the layer interactions (Wohl and Engbersen, 2012). Electrostatic interactions between oppositely charged polyelectrolytes are responsible for keeping alternating layers of electrolytes in forming LbL structures of nanocontainers. Also multilayers are stabilized by inter layer hydrogen bonding interactions. At high salt concentrations, electrostatically assembled films are destabilized; this is because of the fact that the ions shield the charges of the polyelectrolytes. In LbL structures the protonation/deprotonation of the charged group takes place as a result of pH alteration. Further protonation leads to stronger repulsion, causing swellings in the nanocontainers that result in increased permeability. And deprotonation decreases the polymer interactions, leads to shrinking of nanocontainers core. Shrinking of nanocontainers decreases permeability (Delcea et al., 2011).

For a drug molecule, release from polymeric nanocontainers is due to destabilization of nanocontainers itself or by decomposition of the pH-sensitive linking unit that connects the drug to the container (Fleige et al., 2012). In one of the studies of demonstration of mesoporous silica nanoparticles as nanocontainers, the release of corrosion inhibitor (1H-benzotriazole, BTA) is the effect of electrostatic repulsion (Borisova et al., 2011). At pH values different from neutral, both the silica particles as well as the inhibitor molecules have the same charge (positive at pH < 6 and negative at pH > 6). This leads to larger electrostatic repulsion forces and faster releases (Steitz et al., 2002). These results are very much favorable for the subsequent application of the loaded silica nanocontainers in anticorrosive active coatings, as the corrosion is usually followed by alkaline or acidic pH shift. Thus, a release of the inhibitor in response to a pH change in the local environment is provided without the need for an additional polyelectrolyte shell as was employed before (Delcea et al., 2011; Ahrens et al., 2004). In case of LbL assemblies it is a well-known fact that the temperature increase can provide enough thermal energy to surpass the barrier necessary for polymeric film rearrangements. Polyelectrolyte multilayer films deposited on flat substrates exhibit only negligible changes in thickness upon heating. However, they shrink noticeably if heated at 100% humidity, indicating that water desorption takes place (Borisova et al., 2011; Steitz et al., 2002; Ahrens et al., 2004). In designing thermoresponsive nanocontainers the temperature has to be raised above the glass transition temperature (T_g) of the polymer complex. Fig. 16.3 indicates a typical nanocontainer and the frequent responses used in delivery studies for various applications. It shows the mechanism of leaching out/transport of encapsulated material from the nanoconainer. External stimuli may induce partial disintegration of nanocontainer walls, thus obtaining the release of their contents. Physical methods induce disintegration of the wall by mechanical (pressure) or thermal (heat) stimulation. Chemical methods induce on the other hand the rupture of specific chemical bonds inside the nanocontainer walls, thus leading to the collapse of the whole structure. Light may be used both to disrupt a nanocontainer wall through physical routes for example, local thermal heating (Radt et al., 2004; Cui et al., 2011), or to destroy specific bonds in the structure (Sarti and Bordi, 2013). Table 16.1 indicates the stimuli for nanocontainers loading and release.

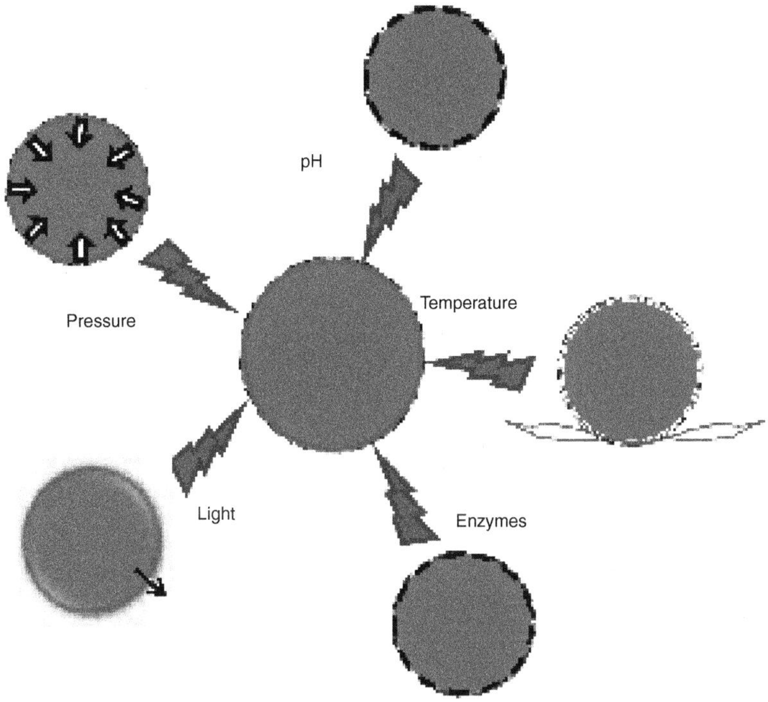

Figure 16.3. Schematic illustration of various stimuli and mechanism of cargo release through nanocontainers.

Table 16.1 Stimuli for Nanocontainers Loading and Release

Type of Stimuli	Factor	References
Physical	Temperature	(Chen et al., 2004; Wei et al., 2007; Guo et al., 2010; Baier et al., 2013)
	Ultrasound	(Fomina et al., 2012; Husseini and Pitt, 2008; Schlicher et al., 2006; Nyborg, 2001; Aw and Losic, 2013; Rapoport et al., 2009)
	Magnetic	(Banerjee and Chen, 2008; Franchini et al., 2010; Ding et al., 2012; Hua et al., 2011)
	Mechanical (impact)	(Tedim et al., 2010; Hodoroaba et al., 2014; Zheludkevich et al., 2005; Bhanvase et al., 2014; Jafari et al., 2010)
	Light	(Li et al., 2012; Swaminathan et al., 2014; Fomina et al., 2012)
Chemical	pH	(Shchukin and Mçhwald, 2007; Pomorska et al., 2011; Shu et al., 2010)
	Ionic strength	(Yuxi et al., 2012; Antipov et al., 2003)
	Electrochemical	(Hodoroaba et al., 2014)
	solvent	(Okubo et al., 2004; Teng et al., 2007)
Biological	Enzyme	(Zhai et al., 2010)
	receptor	(Park et al., 2009; Yang et al., 2012)

4 Active Molecules to be Delivered

4.1 Drugs

In the past decade molecular assemblies with nanometric range have shown tremendous potential as diagnostic and therapeutic tools as genuine nanocontainers, able to interact with biological systems at molecular levels and with a high degree of specificity. In case of delivery of active molecules of drug/protein, it has a therapeutic concentration range, above which it is toxic and below which it is ineffective. When overdosed, the protein could potentially cause severe side effects, while the variation of the protein concentration in the bloodstream could possibly complicate the therapeutic effects. Ideally, the concentration of the therapeutic protein should be maintained continuously within the therapeutic range for a prolonged time period to achieve the optimal therapeutic effects without toxicity and unfavorable side effects. In this respect, a controlled delivery system may be an appropriate choice for protein administration (Levis and Deasy, 2003). For this purpose nanocontainers can be realized over a much wider range of sizes and shell materials, allowing for fine tuning of their properties. Synthesis of such a nanocontainer is focused so as to be able to target specific cells or tissues.

A major issue of synthesizing or selecting a nanocontainer is the biocompatibility and biodegradability of a material. One wide class fulfilling the requirement is aliphatic polyester polymers. Polymers and copolymers of poly-lacticacid (PLA), polyglycolicacid (PGA), and poly-lactic-*co*-glycolicacid (PLGA) are used very often for vaccine and drug delivery. Poly ε-caprolactone (PCL) possesses several interesting properties, including its high permeability to small drug molecules, and an exceptional ability to form blends with other polymers (Anderson and Shive, 1997; Hillaireau and Couvreur, 2009). Moreover, the slower degradation rate of the PCL homopolymer as compared to PLGA and polyglycolic acid-*co*-lactic acid makes it more suitable for long-term delivery systems, extending to a period of more than 1 year (Sarti and Bordi, 2013). Chitosan (CS) is a natural polysaccharide similar to cellulose, possessing several interesting properties for microsphere realization. Among others, it is mucoadhesive, thus favoring the interaction with tissues, and may induce a relaxation of tight junctions between the cells of epi/endothelium tissues, thus favoring drug absorption. Finally, it is charged, and can be neutralized by increasing the pH, leading to a transition from soluble to insoluble. Based on these characteristics, an innovative method to produce chitosan/DNA nanospheres aimed to gene delivery has been recently proposed (Masotti et al., 2008).

Some of the recent studies depict that the aliphatic polyester possesses hydrophobic core used for encapsulation of lipophilic drugs. Also the residence time of such nanocontainers in the blood system after intravenous injections are increased by providing hydrophilic coatings. Core-shell polymeric nanoparticles like D,L-poly (lactic acid) are found to be advantageous in encapsulating lipophilic drugs; this may be due to their hydrophobic behavior. Also after intravenous injections in the blood system, coating nanoparticles with hydrophilic polymers was found to increase their residence time (Knop et al., 2010; Martín del Valle et al., 2009; Alexis et al., 2008). Doughnut- or torus-shaped β-cyclodextrin (β-CD) units are found to be such hydrophilic polymers, which can be used for coating nanoparticles (Fagui and Amiel, 2012).

In one of the studies, Fagui et al. (2011) described well-defined core–shell nanoparticles containing cyclodextrin in the shell. The investigations showed that the Poly-β-CD is securely adsorbed at the surface and the shell can be viewed as a monolayer of poly-β-CD coils. This ensures formation of core-shell nanoparticles with hydrophilic coating with the ability to encapsulate hydrophobic core.

Another challenge in delivery system synthesis is successful loading of large protein and drug molecules within the hollow lumen of nanocontainers. Weda et al. (2008) demonstrated that this can be done by maintaining sufficiently mild conditions and encapsulating the large molecules during synthesis of hollow shell material. The study elaborated the use of phase separation technique for synthesis of poly(nisopropylacrylamide) (PNIPAM) nanocontainer. Controlled heating was provided in order to make use of lower critical solution temperature (LCST). Dimethacrylate (PEG–DMA) and sodium dodecyl sulfate were used as cross-linking agent and surfactant, respectively. The container formation and release was found to depend on molar mass and the mass distribution of the polymer of PNIPAM. During drug release the shell acts as a membrane while the shell thickness or density will specify and mark the molar mass range of the core polymer to be released. He et al. (2015) has given an excellent review on core-shell particles for controllable release of drug. The study reports two important factors in drug release: (1) the encapsulating shells that are responsive to various stimulus such as temperature, light, enzyme, or other stimuli enhance the degradation of core or/and shell, and thus increase the release of the drugs encapsulated within. (2) Drugs encapsulated into core or shell are mostly via chemical bonding or ionic bonding and the release is affected by the specific particle form (shape), polymer compositions of core and shell, loading approach, and encapsulating locations, and the specific characteristics of the drug.

The limitations of shape, size, and surface charge of loading material leads to use of a shell material that will not be affected by any of these. A certain delivery system requires tailoring different functionalities impregnating inorganic and organic substances both inside the capsule volume and in the shell. Also certain lipid-based drugs require versatility and multifunctionality of delivery system (eg, encapsulating several drugs in one capsule, magnetic, ultrasonic delivery, etc.) (Shchukina and Shchukin, 2011). Drug molecules when delivered to the region other than the target site cause adverse reactions hence selective permeability is also one of the important parameters for developing a delivery system. In overcoming these disadvantages LbL techniques have attracted much attention for drug delivery because of their unique advantages, especially in terms of their multifunctionality and responsiveness to various stimuli. Some prominent properties of these nanocontainers are controlled permeability, morphology, and surface charges. This is due to the flexibility available in tailoring the nanocontainer assemblies during formation. Also biocompatibility, mechanical stability, and elasticity are some characteristics offered by LbL approach. Layer-by-layer deposition is as a result of electrostatic interactions between oppositely charged polyectrolytes. The intrinsic advantage of the LbL fabrication method is unmet by any other technique, as it offers the potential of entrapping simultaneously drugs, fluorescent probes, or colloid nanoparticles (eg, quantum dots or magnetic particles) with tunable functionalities into the biodegradable multilayers of one unique hollow capsule (post loading method) (Skirtach et al., 2011; de Villiers et al., 2011)

Recently Shu et al. (2010) reported hollow and degradable polyelectrolyte nanocapsules for protein drug delivery. The study employed bovine serum albumin (BSA) as a model protein for encapsulation purpose. Physical interaction as well as electrostatic attraction was used to encapsulate the protein molecules. In the final step of nanocontainer formation silica core was removed at acidic pH (pH 5.4) with the help of 2:8 M solution of HF/NH_4F solution for 5 min. During drug release studies the initial burst release was decreased to less than 10% (pH 1.4, simulating pH in the stomach), the release increased subsequently for about 60% in the first 2 h (pH 7.4, simulating pH in the bloodstream). At pH 7.4 there is significant weak electrostatic interaction between chitosan and dextran sulfate. This is due to the presence of amino and sulfate groups in the form of $-NH_2$ and $-SO_3$ groups. This causes a swelling in the nanocontainer wall, resulting in increasing BSA release.

4.2 Flavor and Aroma Delivery

Uses of nanocontainers have provided advantages like higher stability, lower evaporation, and unmasking unpleasant tastes for their use in food industries (Jafari et al., 2008; Ezhilarasi et al., 2013; Fang and Bhandari, 2010). The idea in developing such nanocontainers is that, since encapsulated material is protected from other components in the food and from the environment, the use of encapsulation can improve the nutritional content of food without affecting the taste, aroma, or texture of food, mask off-flavors, and enhance the shelf-life and stability of the ingredient and the finished food product (Augustin and Hemar, 2009). Microcapsules are reported to be ineffective in offering antimicrobial activity; this is because the nanometric dimensions of nanocontainers being subcellular sized may increase passive cellular absorption, causing reduced mass transfer resistance (Donsi et al., 2011). Most of the flavoring agents are found to be liquids at ambient temperature. Most of these food oils exhibit considerable sensitivity to air, light, irradiation and elevated temperature (Jafari et al., 2008; Quintanilla-Carvajal et al., 2010; Bejrapha et al., 2010).

This leads to the urgent need of synthesizing potential nanoencapsulating candidates, which are more bioavailable and stimuli specific as compared to microencapsulation (Ezhilarasi et al., 2013; Mozafari et al., 2006). As discussed previously α-pinene loaded polymeric nanocontainer of 200 nm average size were found to be stable for several months (Hofmeister et al., 2014). Nanocontainer shell polarity represented by the monomer ratio between MMA and BMA and more significantly by the degree of acid functionalization, was identified as a main factor to obtain surprisingly higher α-pinene encapsulation efficiencies. During nanocontainer batch synthesis pH sensitive monomer–methyl methacrylate (MMA) served as the main monomer, accompanied by the more hydrophobic butyl methacrylate (BMA) in lower amounts, which was systematically varied in content to gradually lower the hydrophilicity and the T_g of the corresponding polymeric shells. Another study of encapsulating turmeric oil in alginate nanocontainer was found to be producing 100 nm nanocontainers stable for about 120 days, carrying negative charge (Lertsutthiwong et al., 2008). In another successful attempt of encapsulating different fragrant molecules, Sansukcharearnpon studied the formation of nanocontainer from a polymer-blend of ethylcellulose (EC), hydroxypropyl methylcellulose (HPMC), and poly(vinyl alcohol) (PV(OH)) (Sansukcharearnpon et al. 2010). During nanocontainer formation, the chains of PV(OH) and HPMC were found to be useful for creating stable structures. They must be present because of good water-solubility characteristics. During drying these left-out polymers also cover up the surface of the nanocontainers, making them readily dispersible in water. Due to some desirable properties like nontoxicity, biocompatibility, and antibacterial activity, chitosan was used to fabricate nanocontainer (Xiao et al., 2014). It was observed that with the weight ratio of chitosan:tripolyphosphate as 5:1 of 1.5 mg/mL chitosan and 100% (w/w) tuberose fragrance, 174 nm and 20.8 eV TC-NP were obtained. It was found that with increase in fragrance loading the size of nanocontainer also increased; this was the result of aggregation and adhesion of nanocontainers caused by reduction of surface charge.

Mesoporous silica has been reported to be an efficient template for fragrance delivery. It is well known to obtain increased loading onto silica nanospheres the size pore openings are to be increased (Wang et al., 2008; Wang et al., 2011). Use of silica in fabrication of perfumed nanocontainers proves to be advantageous due to its nonsticky nature (Kidsaneepoiboon et al., 2011). Higher surface area (190 m^2/g) and mean pore diameter (36 nm) were found to be useful for successful encapsulation of perfume molecules (Cao et al., 2015). Some other studies are reported in literature highlighting the use of nanocontainers on cotton fiber surfaces. This is because even if fragrant molecules are added during textile finishing, the activity of these molecules tends to diminish during storage and transport. This disadvantage can be avoided by using responsive fragrance-loaded nanocontainers. A model fragrance, *Osmanthus*, was encapsulated in chitosan–sodium tripolyphosphate nanocontainers of 130 nm average size (Hu et al., 2012). The nanocontainers were simply deposited onto the cotton fabric by immersing them into 0.4% solution of nanocontainer. The study concluded that SEM displayed that the cotton fabrics treated by *Osmanthus* fragrance-loaded chitosan nanocontainers (OF-NPs) had an excellent washing resistance. This was attributed to the hydroxyl group interaction between the cotton fabrics and the OF-NPs. In another attempt of using chitosan for nanocontainer formation, the methanol extracts of *O. sanctum* having antimicrobial effect were loaded inside the sodium alginate chitosan nanoparticles (Rajendran et al., 2013). About 35 nm sized nanocontainers were loaded onto cotton fabric by pad dry-cure method. The study highlighted that the cotton fabrics finished with the methanol extract of *O. sanctum*–loaded nanoparticles possessed remarkable antibacterial activities with excellent wash durability.

5 Stimuli for Controlled Release

All the encapsulation and release modalities discussed previously are having a typical purpose to release the cargo on stimuli (change within or surrounding of a nanocontainer). The typical stimuli used are of chemical, physical, or biochemical origin, which tends to modify the structural composition/conformation of nanocontainers. Some nanocontainers are relatively large and abrupt physical and chemical changes in sharp response to applied stimuli (Fleige et al., 2012). Hence it is very important not only two select a suitable nanocontainer but also an appropriate stimuli.

5.1 Chemical Stimuli for Permeability Changes: pH, Ionic Strength, Solvent, Electrochemical Stimuli

Recently various nanocontainers have been developed wherein deposition of oppositely charged weak polyelectrolytes on a desired substrate creates a well-defined system with regulated storage/release properties (Pomorska et al., 2011; Shchukin et al., 2006) in which the permeability of the container shell is determined by the balance of electrostatic interactions within the multilayer. During the corrosion process, change in pH or ionic strength are followed by ionization of the weak polyelectrolytes of the functional groups, which results in increased repulsion between uncompensated charges (Andreeva et al., 2010). Also the influence of pH and salt concentrations can be both reversible (ie, increased permeability, which is interesting for LbL capsule loading) or irreversible (ie, LbL film disassembly), depending on the film composition and presence of crosslinks (Wohl and Engbersen, 2012).

The release of cargo from a polyelectrolyte nanocontainer is a result of change in permeability at a particular pH. As a typical observation it is permeable at pH 3 and impermeable at pH 7. This is due to polyelectrolyte interactions in the shell wall. During nanocontainer formation the charge densities on both polyelectrolytes determine their stoichiometric ratio during adsorption. Since the polymers are irreversibly adsorbed in the shell wall, a pH decrease does not induce polymer desorption. However, charging of one of the polyelectrolytes may occur, which would induce positive (negative) charge into the shell wall. This may alter the shell-wall morphology by enhancing the repulsion, which could lead to defects in the polyelectrolyte shell (Shchukin and Mçhwald, 2007). Shu et al. (2010) developed hollow and degradable polyelectrolyte nanocapsules for protein drug delivery. The results concluded that during responsive release, controlled release was possible by increasing or decreasing the thickness of the LbL membrane. The drug delivery was found to be the result of fickian diffusion. The penetration of water into nanocontainer surface resulted in rubbery matrix and finally the protein diffused out of the nanocontainer. This phenomenon was responsible for obtaining initial slow release, which became consistent for an extended period of time. QCM (quartz crystal microbalance) study of the adsorption of polyelectrolyte-covered mesoporous TiO_2 nanocontainers on SAM (self-assembled monolayer)–modified Au surfaces was carried out by Pomorska et al. (2011). The study confirmed that the polyelectrolyte termination layer has a profound influence on the surface charge o f the nanocontainers and their subsequent adsorption kinetics on SAM-modified gold-coated quartz substrates. Recently, a dual responding with pH and ionic strength were developed by Yuxi et al. (2012). Herein, formation and characterization of natural polysaccharide hollow nanocapsules via template layer-by-layer self-assembly was demonstrated successfully. Screening of electrostatic interaction between the oppositely charged polyelectrolyte polymers by salt ions causes change in the ionic strength of LbL nancontainers. The responsive release is the effect of reduction of the electrostatic attractions inside the multilayers or forming defects or cavities in the multilayer network (Delcea et al., 2011; Antipov et al., 2003). Most of the cargo release studies are in water; however, it is reported that the organic solvents induce permeation through the LbL-assembled nanocontainers (Lvov et al., 2001). The presence of ethanol influences the activity of urease, which was found to be lower than the activity of free urease in bulk solution. The hydration water between the polyelectrolytes might have been removed by the organic solvent resulting in separation of the polyion network and the formation of pores (Delcea et al., 2011; Lvov et al., 2001; Shchukin and Mçhwald, 2007). Huang et al. (2008) reported that unique hollow polypyrrole nanostructured arrays with a conical shape have been produced by a stepwise electropolymerization process. The fabricated conical nanocontainers showed a reversible switchable behavior between open and closed states. This was possible with the movement of counter ions during electrically controlled reversible oxidation and reduction processes. Völker et al. (2008) reported layer-by-layer self-assembled redox polyelectrolytes on passive steel. Electrochemical experiments have shown that the oxidation reduction of the osmium sites lies in the passivity potential interval and the rate of electron transfer from the underlying metal to osmium sites in the polymer over layer is significantly hindered as compared to thiolated gold electrodes.

5.2 Physical Stimuli for Affecting Permeability: Temperature, Light, Ultrasound, Magnitude Field, Mechanical Deformation

As discussed in an earlier section it is found that most of the polymer nanocontainers reported until now load and release guest molecules from their interior only by diffusion. It is therefore rather difficult to control the loading or releasing process. Also some applications require external stimulus for delivery of active molecule. This leads to development of nanocontainers wherein physical stimuli is required for controlled delivery application. Typical physical stimuli are temperature, laser, ultrasound, magnitude field, mechanical deformation. In this context one may design a nanocontainer that can undergo reversible structural transitions from a closed to an open state with the help of external stimuli (Chen et al., 2004). This would result in a targeted release of encapsulated material between nanocontainer and the target environment. Thermo-responsive polymers are the important building blocks of any temperature responsive nanocontainer. Two distinct classes of thermos-responsive polymers are available for nanocontainer formation. Both the classes report the points at which the polymers and solvents are completely miscible (Meng et al., 2009). And hence these classes of polymers represent lower critical solution temperature (LCST) as well as upper critical solution temperature (UCST). Practically when the surrounding temperature reaches above LCST the polymer solution becomes turbid whereas the clear polymer solution is obtained when surrounding temperature is below LCST. The temperature responsive release may be because of the balance between segment–segment interactions and segment–solvent intermolecular interactions can be shifted by temperature changes (Motornov et al., 2010).

The most common polymer used to design temperature responsive nanocontainers is poly(*N*-isopropylacrylamide) (PNIPAAm). All biomedical applications require a nanocontainer suitable to the body temperature (typically 37°C). PNIPAAm is found to have the LCST value at 32°C and hence is found to be useful for temperature responsive delivery systems pertaining to biomedical delivery application (Lehner et al., 2012; Wei et al., 2007). The typical thermoresponsive polymers used in delivery system are listed in Table 16.2. Poly(*N*,*N*-diethylacrylamide) is typically used for synthesis of hydrogels for drug delivery applications (Chen et al., 2009; Panayiotou and Freitag, 2005). Niu et al. recently studied crosslinkable PEO-PPO-PEO triblocks as building blocks of thermo-responsive nanoshells (Niu et al., 2011). During synthesis amphiphilic block copolymers based on PEO-PPO-PEO

Table 16.2 Thermoresponsive Polymers Used in Designing Nanocontainers for Delivery System

Polymer	LCST Range (°C)	Active Molecule	References
Poly(*N*-isopropylacrylamide) (PNIPAAm)	32	Prednisone acetate, aminophylline	(Wei et al., 2007; Chen et al., 2009)
Poly(*N*,*N*-diethylacrylamide) (PDEAAm)	25–32	Aminophylline, insulin	(Panayiotou and Freitag, 2005; Vihola et al., 2008)
Poly(*N*-vinylcaprolactam) (PVCL)	25–35	Nadolol, propranolol, ketoprofen, salicylic acid	(Guo et al., 2010)
Poly[2-(dimethylamino)ethyl methacrylate] (PDMAEMA)	50	Paclitaxel	(Guo et al., 2010)
Poly(ethylene oxide) (PEO)	85	Paclitaxel	(Shahin and Lavasanifar, 2010)
Poly(propylene glycol)	27–31	Methyl orange	(Allı and Hazer, 2008)

poly(ethylene oxide)-poly-(propylene oxide)-poly(ethylene oxide) block copolymer (Pluronic) and poly (ε-caprolactone) (PCL) were synthesized by ring-opening polymerization of ε-caprolactone in the presence of PEO-PPO-PEO block copolymer having hydroxyl groups at two ends of chains. Stannous octoate acts as a catalyst. Dialysis process in deionized water was used to form indomethacin (IMC)-loaded and unloaded nanospheres using pluronic/PCL block copolymer with different composition. The size of pluronic/PCL block copolymeric nanospheres increased with increase in temperature. This was justified with a reason of strong chain–chain aggregation due to the increase of intermolecular and intramolecular interaction due to change of the solubility and hydrophilicity of the PPO block. Moreover, this change of size exhibited a reversible tendency according to the repetitive thermal cycles. Skirtach et al. (2004) reported remote activation of capsules containing Ag nanoparticles and IR dye by laser light. Ag nanoparticles or IR dye was introduced into the PAH/PSS capsules. A near-infrared continuous-wave laser diode was used during release studies. It was found that the release properties were dependent on the intensity of the laser beam, absorption properties of the materials constituting the shell of the capsules, and their composition. A novel and versatile method of forming cross-linked self-assembled structures using a combination of ring-opening and RAFT (radical) polymerization was given by Hales et al. (2004). The process encompasses addition-fragmentation equilibria superimposed on a normal free radical chain polymerization. This result into formation of polymers with thiocarbonylthio end-groups and a narrow molecular weight distribution (Barner-Kowollik et al., 2003). The synthesized nanocontainer was comprised of a polylactide core, a cross-linked shell, and a thermosensitive corona. During turbidity studies via UV spectroscopy, significant changes to the thermal transition behavior of the aggregates was observed because of cross-linking (hexandiol diacrylate was used as a cross linker). One of the recent studies depicted cross-linkable PEO-PPO-PEO triblocks as building blocks of thermo-responsive nanoshells (Geest et al., 2007). This was made possible by shell-cross-linking poly(ethylene oxide)-poly(propylene oxide)-poly(ethylene oxide) (PEO-PPO-PEO) dimethacrylate triblocks.

The near–infrared (IR) laser light (800–1200 nm) is less harmful and has a much deeper penetration depth in tissues than visible light, and light-responsive capsules have potential for in vivo drug delivery. This laser light can then induce structural changes in drug-containing vesicles injected in tissues located at the surface of the body (Angelatos et al., 2005). During stimuli generation the short pulse of IR light is irradiated on the noble metal particles (like gold nanoparticles). The irradiation causes energy absorption by metal nanoparticles, which transform it into heat, which locally disturbs the integrity of the polyelectrolyte capsules. Li et al. (2012) reported the photo and pH dual-responsive polydiacetylene smart nanocontainer. For this study photo-responsive azobenzene derivative/cyclodextrin (AzoeCD) supramolecular complex showed the photo activity. Polydiacetylene (PDA) vesicles matrix showed the pH responsiveness.

Recently Swaminathan et al. published a review article on photoresponsive polymer nanocarriers with multifunctional cargo (Swaminathan et al., 2014). It demonstrates that encapsulation of multiple photoresponsive species with distinct functions within the very same polymer nanocontainer/nanocarrier is possible. This is because of the solvophobic interactions between the hydrophobic domains of the macromolecular components and their guests are responsible for these supramolecular events. In another review it was reported that in the case of gold nanoparticles used for photothermal effects, thermal stability of the potential cargo must be considered due to local heat generation (Fomina et al., 2012). Also, there is a requirement to create systems that degrade into small molecules upon irradiation or completion of function. These fully degradable systems would be desirable for a variety of biomedical applications.

Another physical phenomenon practiced very regularly is use of ultrasound triggered drug release. Ultrasound (US) consists of pressure waves having frequencies of 20 kHz or greater and is generated by piezoelectric transducers. Change in voltage applied to a transducer results into mechanical displacement of surface that is in contact of media (water, gel, etc.). During stimuli generation the US transducers needs to be in direct contact with the tissue or skin. The air present surrounding should be excluded through the application of a fluid such as water or ultrasonic gel. The movement/mechanical displacement creates physical push and stress in the cell walls and tissues. The magnitude is not so strong to disrupt the cell membrane unless the gas bubbles are present (Husseini and Pitt, 2008; Schlicher et al., 2006; Nyborg, 2001). Recently, Aw and Losic (2013) studied US-enhanced release of therapeutics from drug-releasing implants based on titania nanotube arrays. A pulsating sonication probe was used to generate stimuli in phosphate buffered saline (PBS) at pH 7.2 as the medium. The system was found to be advantages as stimulated release can be obtained throughout the lifespan of TNT (Titania nanotube). With such a study, it is observed that such dug release implant systems can be utilized for practical applications of feasible and tenable releases. In one of the application of chemotherapy ultrasound activated nanoemulsions/microbubbles were used for controlled drug delivery (Rapoport et al., 2009). The strategy used during application was to inject drug-loaded nanoemulsion that convert into microbubble in-situ under the action of therapeutic ultrasound.

Magnetic materials and their oxides can be incorporated within the nancontainer shells so as to respond to the magnetic field for drug delivery applications. When external magnetic field is applied to the magnetic nanoparticles, the particles could be retained within the targeted organ for a specific period. The container encapsulating these magnetic particles plays a vital role in delivery systems (Banerjee and Chen, 2008). Heated in high frequency the magnetic nanocontainer trigger the drug release. One of the studies reports a potential theranostic approach in cancer treatment of bovine serum albumin (BSA)–based magnetic nanocarrier for MRI diagnosis and hyperthermic therapy (Franchini et al., 2010). The study reported synthesis of novel BSA-based nanocarrier containing $CoFe_2O_4$ NPs as a promising nanosystem to perform theranostic treatment. The results were monitored using MRI in brain and liver of normal rats and hyperthermic treatments in standard human tumor cell line hela cells. For delivery of hydrophobic drugs Ding et al. (2012) reported a double-targeted magnetic nanocarrier. Solution evaporation technique was used to fabricate the nanocontainer (20–60 nm) wherein magnetite (Fe_3O_4) nanoparticles encapsulated in self-assembled micelles of the amphiphilic copolymer MPEG–PLGA [methoxy poly(ethylene glycol)-poly(D,L-lactide-coglycolide)]. The nanocontainer was found to be highly stable and displayed superparamagnetic behavior at room temperature. Nanocontainers were also found to be noncytotoxic even after 24 h incubation at a concentration of 400 μg/mL. Superhigh-magnetization nanocarrier as a doxorubicin delivery platform for magnetic targeting therapy was studied by Hua et al. (2011). During nanocontainer formation an aqueous stable self-doped poly[*N*-(1-onebutyric acid)]aniline (SPAnH) was used as a nanocontainer shell in order to house a magnetic Fe_3O_4 core. Emphasis on the possibility of obtaining maximum magnetization. In the magnetization property the value (89.7 emu/g) was found to be more (73.7 emu/g) typical value is than the commercial grade contrast agent used for magnetic resonance imaging application. Covalent bonding between the $-NH_2$ of DOX and the –COOH of nanocontainers was found to be responsible for immobilization of doxorubicin (DOX) to enhance the thermal stability of doxorubicin (DOX) and magnetic targeting (MT) efficiency.

One or more causes for failure of a typical coating are wrong mechanical damage, choice of coating, misapplication of coating, defective coating, and exposure to unwanted/unanticipated environment (Samadzadeh et al., 2010). Once corrosion starts the coating no longer protects the defective zone and paint creepage is initiated at the defect and it becomes necessary to provide coatings with ability to repair the defects in an autonomous way and provide long-term protection (Tedim et al., 2010). Shchukin and Mçhwald (2007) presented a review on self-repairing coatings containing active nanoreservoirs. The concept of such nanocontainer-based systems is based on incorporating corrosion inhibitor, which is subsequently released upon mechanical shock/stress/damage. Fig. 16.4 shows the mechanism where the coating has undergone a mechanical deformation. A scratched or impacted coating produces a small rupture on the core of a nanocontainer, causing leakage of corrosion inhibitor. Seepage of inhibitor then fills the micro cracks produced because of deformation. Zheludkevich et al. (2005) studied oxide nanoparticle reservoirs for storage and prolonged release of the corrosion inhibitors. During the study oxide nanoparticles were used as a reinforcement of the hybrid sol–gel matrix and also as a reservoir for corrosion

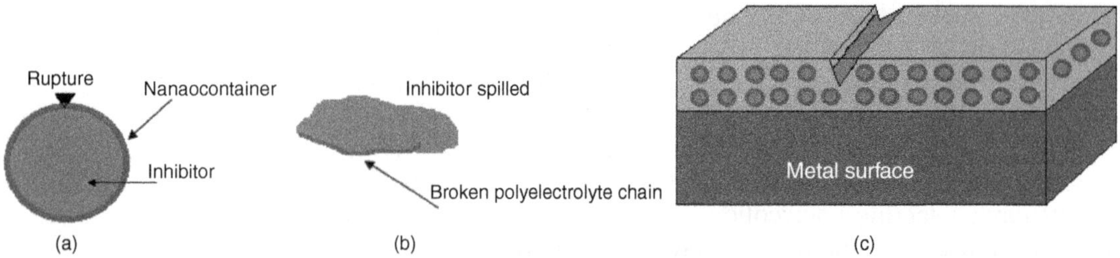

Figure 16.4. Self-healing coating mechanism of corrosion inhibitor form nanocontainers: nanocontainer loaded with inhibitor and the point of rupture due to mechanical shock. (a) Broken nanocontainer spilling the loaded inhibitor in the region of microcracks; (b) inhibitor layer formation onto the metal surface (c).

inhibitors. Zirconia was used to absorb the inhibitor ions during the preparation procedure and then slowly release them in contact with moisture. The nanocontainers were found to have an effective inhibiting effect on the corrosion processes on aluminum alloys, suppressing the cathodic reactions. Enhancement of active corrosion protection via combination of inhibitor-loaded nanocontainers was studied by some authors (Tedim et al., 2010; Hodoroaba et al., 2014). Zheludkevich et al. (2005) studied polymer-based polyepoxide (PE) capsules and SiO_2 based inorganic capsules filled with a corrosion inhibitor. The nanocontainers formed were synthesized from oil-in-water emulsions as products of polymerization (PE) or polycondensation (SiO_2) reactions, wherein the corrosion inhibitor (2-methylbenzothiazole) is the oil phase. Recently Bhanvase et al. (2014) reported kinetic properties of layer-by-layer-assembled cerium zinc molybdate nanocontainers during corrosion inhibition. Cerium zinc molybdate (CZM) nanoparticles were used as a core material in the CZM nanocontainer. CZM nanocontainer was prepared by LbL by depositing polyaniline (PANI), imidazole, and polyacrylic acid (PAA) layers on CZM nanoparticles in the presence of US irradiation. The study also showed the presence of US irradiation can accelerate the nucleation rate resulting in reduction of CZM nanoparticles size and which was found to be 26 nm. Jafari et al. (2010) reported that the polyelectrolyte-coated assemblies such as SiO_2 nanoparticles, halloysite nanotubes, and polyelectrolyte nanocapsules are able to successfully transport corrosion inhibitor such as benzotriazole in coating matrix. The release of benzotriazole (BTA) was found to depend on the polyelectrolyte shell components and the quantity of BTA. Corrosion inhibitor release rate followed the sequence of PAH/PSS + PMA > PAH/PSS > PAH/PMA.

5.3 Biological Stimuli for Release and Targeting: Enzyme Triggering, Receptor Implementing Triggering

In case of disease or injury, a particular biological system secretes a specific enzyme. Such secretion can then be used to trigger the drug release in the biological system. A biologically relevant molecule or phenomenon can be used in order to produce biological stimuli. The enzyme secretion is generally spatial and temporal and this can be used as a stimulus for the nanocontainer. Some studies report use of hydrogel for spatial delivery of proteins (Williams et al., 2011). Wen et al. presented the controlled protein delivery based on enzyme-responsive nanocapsules (Wen et al., 2011). During study, bovine serum albumin (BSA) and vascular endothelial growth factor (VEGF) were used as the model proteins. During release studies it was found that, in the presence of plasmin, nVEGF (protein nanocontainer) degradation leads to release of the encapsulated VEGF. In one more study, hyaluronic acid (HA)-based nanocapsules containing the antimicrobial agent polyhexanide were specifically cleaved in the presence of hyaluronidase (Baier et al., 2013). For release study, a model dye and the antimicrobial polyhexanide was monitored using fluorescence and UV spectroscopy. Mesoporous silica nanoparticles coated with molecular valves having an ester-linked stopper and the other with an amide-linked stopper, were used to deliver a tracer dye (Rhodamine B) (Patel et al., 2008). Subsequently, Park et al. (2009) studied enzyme responsive nanocontainers with cyclodextrin gatekeepers and synergistic effects in release of guests. Stimuli-responsive gatekeepers were introduced onto the stimuli-responsive silica nanoparticles. Cyclodextrin (CD) gatekeepers and two enzymes (α-amylase and lipase) were used during the study. First the silica nanoparticles (~60 nm) were functionalized with amine groups by treatment with 3-aminopropyltriethoxysilane to obtain Si-MP-NH_2, which was then allowed to react with propargyl bromide to provide Si-MP-alkyne. Further Yang et al. (2012) reported the enzyme-responsive nanocontainer as an intelligent signal-amplification platform for a multiple proteases assay. It was found the addition of protease to the substrate solution induces pore opening with the subsequent release of the entrapped dyes. The results seem promising in order to use in the diagnosis of protease-related disease and screening of potential drugs with high sensitivity in a high throughput way.

6 Case Study

Various fields such as cosmetics, foods, medicine, tobacco, textiles, leather, paper making, and so on primarily employ fragrance and aromas at large. These fragrant/perfume molecules are obtained from natural sources and are volatile in nature. Most volatile fragrance materials are easily lost during the manufacture, storage, and use of the perfumes or the perfumed consumer products (Xiao et al., 2014). We recently attempted to encapsulate and deliver the fragrant molecules within the hollow lumen of a halloysite nanocontainer (Ghodke et al., 2015). The purpose was to safeguard the volatile fragrant molecule and provide pH responsive delivery system. The attempts were successful in employing hollow lumen for housing the fragrance molecules and the polyelctrolyte coating provided adequate sensitivity within various pH environments. The nanocontainers were found to be pH responsive and the responsive nature was analyzed in the range of pH 3–7. The structural analysis of nanocontainer indicated that the container material is essentially aluminosilicate clay with surface morphology indicating outside diameter in the range of 30–50 nm, 15 nm lumen, and length equal to 800 ± 300 nm.

The fragrance release studies were obtained in a batch mode. In a typical batch 1 g of loaded nanocontainers were added to deionized water of desired pH. Fig. 16.5 release rate of perfume molecules from halloysite nanocontainers. The release from these containers was found to be consistent. The results indicate that the release rate of perfume molecules increased with respect to time and gets stabilized. This behavior is because of decrease in

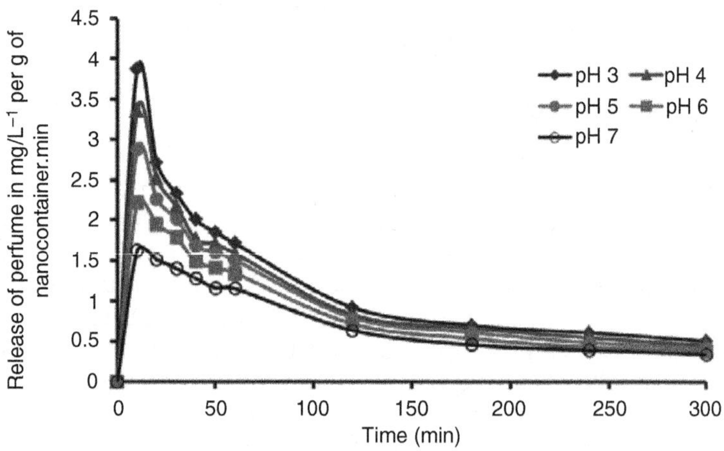

Figure 16.5. Release rate of perfume molecules from halloysite nanocontainers.

diffusion rate of perfume with decrease in concentration gradient with respect to exposure time. The release rate of perfume was found to increase from 1.62 to 3.86 mg/L per g of nanocontainer. min with decrease in pH from pH value from 7 to 3. Also at lower values of pH increased repulsion between moieties was produced in the inner lumen causing increase in the amount of fragrance release.

7 Future Prospects

A wide variety of stimuli responsive nanocontainers are available for regulating the permeability, attachment and targeting, encapsulation, and release of active molecules. Efforts are made to avoid unwanted environment by protecting vulnerable active molecules from degradation by light or by enzymatic attack in their passage through the digestive tract. According to Barratt (2000), nanocontainers can increase the therapeutic efficacy of active molecules because their biodistribution follows that of the carrier, rather than depending on the physicochemical properties of the active molecule itself. The nanocontainers formed in the range of 250–500 nm can be injected directly into the systemic circulation without the risk of blocking blood vessels (Barratt, 2000; Mora-Huertas et al., 2010; Letchford and Burt, 2007). Formation of polymeric nanocontainers is relatively simple and easy but the stability and structural integrity is to be controlled by varying the degree of cross-linking. For these polymeric nanocontainers, biocompatibility and biodegradability has been analyzed as an imminent challenge for drug delivery systems. The hollow spherical polymer nanocontainers are found to have greater ability to encapsulate large quantities of guest molecules or large-sized guests within the "empty" core domain as compared to polymer microspheres or micelles (Chen et al., 2004). During polymeric nanocontainer formation the covalent or ionic interactions provides sufficient mechanical strength as compared to polymer vesicles.

Many times the strategy to design a nanocontainer depends upon the physicochemical characteristics of cargo, particularly its solubility and the therapeutic objective of nanocontainer administration, for example, the route chosen and drug release profile (Mora-Huertas et al., 2010). Another major challenge in nanocontainer synthesis is the identification of formulation strategies that enhance the circulation time of cargo, thus allowing them adequate time to reach targeted sites. Also many biological delivery applications require capsules of size less than 50 nm with a narrow size distribution (Motornov et al., 2010). One of the most promising candidates in this regard is LbL assembly. They have been found to be responsive to a large range of different stimuli, both chemical and physical. Substantial efforts have been taken in development of multilayers with responsiveness to multiple stimuli. However in vivo drug release studies has only tentatively been explored but has recently received more and more concern (Wohl and Engbersen, 2012). The nanocontainer designed has particular material requirements, for example, specific charge densities. The template based LbL methodology is another method of forming nanocontainer but has a shortcoming of getting the final container size in the micrometer scale. An increasing amount of research is ongoing for using halloysite nanocontainer for delivery systems. Halloysite is found to possess negative electrical potential of c. –50 mV, which allows halloysite good dispersibility and colloidal stability in water and is nontoxic up to concentrations of 75 μg/mL (Vergaro et al., 2010). For using halloysite for encapsulation, one fact to be noted is that the surface hydroxyl groups and external surface siloxane (Si–O–Si) groups only have weak interactions with guest molecules through hydrogen bonding or Van der Waals forces, which limits the loading of guest molecules (Tan et al., 2013; Aguzzi et al., 2007). Also it is practically possible to modify halloysite surface with specific functional groups. Some efforts have been put forward to increase the hollo lumen of halloysite by changing the physicochemical properties, structure, and morphology (Abdullayev et al., 2012). However sufficient research on surface area increment and host-guest interaction is yet to come to offer different examples.

Silica-based nanocontainers are reported to have a large surface area (>900 m^2/g) tuned from 50 to 300 nm allowing a facile endocytosis by living animal and plant cells without any significant cytotoxicity (Slowing et al., 2008). More and more in vivo studies related to circulation properties in blood, possible immunogenicity, and accumulation are necessary to come to valuable conclusions. Use of ultrasound not only produces emulsions but also induces the cross-linking of protein molecules adsorbed at the surface of oil. Ultrasound improves the mobility of droplets in the emulsion; this causes increased chances of bigger droplets to being close to the sound-emitting surface of the ultrasonic probe wherein big droplets are easily broken to small (nanoscale) droplets (Han et al., 2010). Ultrasonic protein nanocontainers proved to be having good biocompatibility having features such as higher loading capacity and multistage controlled release. Plenty of opportunities in these kinds are available. One may look over the possibility of use of inorganic substrate to develop the bottom approach for building responsive nanocontainer. However, this becomes difficult because of the absence of functional groups for selective binding.

Photo-triggered nanocontainers are also of interest because of ability to apply at with extremely high spatial and temporal precision. These nanocontainers are designed to respond to UV or visible light. However, the application is found to have limited use, such as in the field of cosmetics (topical applications) where stimulus penetration is not necessary (Swaminathan et al., 2014). One should also be aware of detrimental effects of short length-high energy irradiations. This can be overcome by a system that is sensitive to NIR (near infrared region) light. More efforts in design and synthesis of such systems are needed. Another method that is harmless is use of ultrasound to trigger the release. It has the advantages such as the absence of ionizing radiations, and the facile regulation of tissue penetration depth by tuning frequency, duty cycles, and time of exposure (Mura et al., 2013). Ultrasound-triggered drug delivery systems have strong potential; however, many of the systems reported in literature are still in the early stage of development. Encapsulation of functional food is still performed at lower levels (1–5%); this is because of the encapsulation data and feasible encapsulation scale-up techniques (Verica ðordevic, 2014).

8 Conclusions

In a variety of industrial applications the need for smart and innovative nanocontainer delivery system has become critical. From the present literature analysis, we find that a wide range of nanocontainer modalities for responsive release have been developed. It is very much necessary to understand the principles and underlying mechanisms of these stimuli so as to develop responsive nanocontainers for delivery of active molecules. Container materials and the release phenomenon are the two major aspects that pose a real challenge to the researchers. The present study highlights the various materials used for fabrication of the nanocontainer and the mechanisms that play a role in bringing about the responsive release. Although various nanocontainer assemblies are reported with great flexibility in the design, most of the studies report in vitro release studies and a very few report in vivo releases. Sometimes, the complexity of designing a nanocontainer makes it difficult to scaling up and hence has a limited practical use. Hence simpler the architecture of the system better the chances of scale up and practical use. With this one should look after the good mechanical stability, reproducibility and responsiveness to various stimuli so as to get the prolonged release. All the results and discussions reported in the literature are promising and encouraging but still needs more systematic and fundamental investigations. The nanocontainer for delivery system could find potential applications in corrosion protection, personal care, drug delivery, intelligent catalyst, and so forth.

References

Abdullayev, E., Joshi, A., Wei, W.B., Zhao, Y.F., Lvov, Y., 2012. Enlargement of halloysite clay nanotube lumen by selective etching of aluminum oxide. ACS Nano 6, 7216–7226.

Aguzzi, C., Cerezo, P., Viseras, C., Caramella, C., 2007. Use of clays as drug delivery systems: possibilities and limitations. Appl. Clay Sci. 36, 22–36.

Aguzzi, C., Viseras, C., Cerezo, P., Salcedo, I., Sánchez-Espejo, R., Valenzuela, C., 2013. Release kinetics of 5-aminosalicylic acid from halloysite. Colloid. Surf. B 105, 75–80.

Ahrens, H., Büscher, K., Eck, D., Förster, S., Luap, C., Papastavrou, F., Schmitt, J., Steitz, R., Helm, C.A., 2004. Poly(styrene sulfonate) adsorption: electrostatic and secondary interactions. Macromol. Symp. 211, 93–106.

Alexis, F., Pridgen, E., Molnar, L.K., Farokhzad, O.C., 2008. Factors affecting the clearance and biodistribution of polymeric nanoparticles. Mole. Pharm. 5 (4), 505–515.

Allı, A., Hazer, B., 2008. Poly(N-isopropylacrylamide) thermoresponsive cross-linked conjugates containing polymeric soybean oil and/or polypropylene glycol. Eur. Polym. J. 44 (6), 1701–1713.

Anandharamakrishnan, C., 2014. Techniques for nanoencapsulation of food ingredients. In: Hartel, R.W. (Ed.), Springer Briefs in Food, Health, and Nutrition. Madison, New York, pp. 29–41.

Anderson, J.M., Shive, M.S., 1997. Biodegradation and biocompatibility of PLA and PLGA microspheres. Adv. Drug Deliv. Rev 28, 5–24.

Andreeva, D.V., Skorb, E.V., Shchukin, D.G., 2010. Layer-by-layer polyelectrolyte/inhibitor nanostructures for metal corrosion protection. Appl. Mat. Interf. 2 (7), 1954–1962.

Angelatos, A.S., Radt, B., Caruso, F., 2005. Light-responsive polyelectrolyte/gold nanoparticle microcapsules. J. Phys. Chem. B. 109, 3071–3076.

Antipov, A.A., Sukhorukov, G.B., Möhwald, H., 2003. Influence of the ionic strength on the polyelectrolyte multilayers' permeability. Langmuir 19, 2444–2448.

Augustin, M.A., Hemar, Y., 2009. Nano- and micro-structured assemblies for encapsulation of food ingredients. Chem. Soc. Rev. 38, 902–912.

Aw, M.S., Losic, D., 2013. Ultrasound enhanced release of therapeutics from drug-releasing implants based on titania nanotube arrays. Int. J. Pharm. 443, 154–162.

Baier, G., Cavallaro, A., Vasilev, K., Mailänder, V., Musyanovych, A., Landfester, K., 2013. Enzyme responsive hyaluronic acid nanocapsules containing polyhexanide and their exposure to bacteria to prevent infection. Biomacromolecules 14, 1103–1112.

Ball, R., Haymet, A.D.J., 2001. Bistability and hysteresis in self-assembling micelle systems: phenomenology and deterministic dynamics. Phys. Chem. Chem. Phys. 3, 4753–4761.

Banerjee, S.S., Chen, D.H., 2008. Cyclodextrin conjugated magnetic colloidal nanoparticles as a nanocarrier for targeted anticancer drug delivery. Nanotechnology 19, 265602.

Barner-Kowollik, C., Davis, T.P., Heuts, J.P.A., Stenzel, M.H., Vana, P., Whittaker, M.J., 2003. RAFTing down under: tales of missing radicals, fancy architectures, and mysterious holes. Polym. Sci. A Polym. Chem. 41 (3), 365–375.

Barratt, G.M., 2000. Therapeutic applications of colloidal drug carriers. Pharm. Sci. Technol. Today 3, 163–171.

Bejrapha, P., Min, S.G., Surassmo, S., Choi, M.J., 2010. Physicothermal properties of freeze-dried fish oil nanocapsules frozen under different conditions. Dry. Tech. 28, 481–489.

Berthier, D.L., Schmidt, I., Fieber, W., Schatz, C., Furrer, A., Wong, K., Lecommandoux, S., 2010. Controlled release of volatile fragrance molecules from PEO-b-PPO-b-PEO block copolymer micelles in ethanol-water mixtures. Langmuir 26 (11), 7953–7961.

Bhanvase, B.A., Patel, M.A., Sonawane, S.H., 2014. Kinetic properties of layer-by-layer assembled cerium zinc molybdate nanocontainers during corrosion inhibition. Corros. Sci. 88, 170–177.

Borisova, D., Mohwald, H., Shchukin, D.G., 2011. Mesoporous silica nanoparticles for active corrosion protection. ACS Nano 5 (3), 1939–1946.

Cao, Z., Xu, C., Ding, X., Zhu, S., Chen, H., Qi, D., 2015. Synthesis of fragrance/silica nanocapsules through a sol–gel process in miniemulsions and their application as aromatic finishing agents. Colloid. Polym. Sci. 293, 1129–1139.

Cavalieri, F., El-Hamassi, A., Chiessi, E., Paradossi, G., Villa, R., Zaffaroni, N., 2006. Tethering functional ligands onto shell of ultrasound active polymeric microbubbles. Biomacromolecules 7, 604–611.

Chekhonin, V.P., Baklaushev, V.P., Yusubalieva, G.M., Belorusova, A.E., Gulyaev, M.V., Tsitrin, E.B., Grinenko, N.F., Gurina, O.I., Pirogov, Y.A., 2012. Targeted delivery of liposomal nanocontainers to the peritumoral zone of glioma by means of monoclonal antibodies against GFAP and the extracellular loop of Cx43. Nanomed: Nanotechnol. Biol. Med. 8 (1), 63–70.

Chen, J., Liu, M., Liu, H., Ma, L., 2009. Synthesis, swelling and drug release behavior of poly(*N,N*-diethylacrylamide-*co*-N-hydroxymethyl acrylamide) hydrogel. Mater. Sci. Eng. C 29, 2116–2123.

Chen, X., Ding, X., Zheng, Z., Peng, Y., 2004. A self-assembly approach to temperature-responsive polymer nanocontainers. Macromol. Rapid Commun. 25, 1575–1578.

Chern, C.S., 2006. Emulsion polymerization mechanisms and kinetics. Prog. Polym. Sci. 31, 443–486.

Chiappetta, D.A., Sosnik, A., 2007. Poly(ethylene oxide)–poly(propylene oxide) block copolymer micelles as drug delivery agents: improved hydrosolubility, stability, and bioavailability of drugs. Eur. J. Pharm. Biopharm. 66, 303–317.

Chilkoti, A., Dreher, M.R., Meyer, D.E., Raucher, D., 2002. Targeted drug delivery by thermally responsive polymers. Adv. Drug Delivery Rev. 54, 613–630.

Cui, W., Lu, X., Cui, K., Wu, J., Wei, Y., Lu, Q., 2011. Photosensitive nanoparticles of chitosan complex for controlled release of dye molecules. Nanotechnology 22, 065702.

Dalmoro, V., Santos, J.H.Z.dos., Armelin, E., Alemán, C., Azambuja, D.S., 2012. Phosphonic acid/silica-based films: a potential treatment for corrosion protection. Corros. Sci. 60, 173–180.

de Villiers, M.M., Otto, D.P., Strydom, S.J., Lvov, Y.M., 2011. Introduction to nanocoatings produced by layer-by-layer (LbL) self-assembly. Adv. Drug Deliv. Rev. 63, 701–715.

Delcea, M., Möhwald, H., Skirtach, A.G., 2011. Stimuli-responsive LbL capsules and nanoshells for drug delivery. Adv. Drug Deliv. Rev. 63, 730–747.

Ding, G., Guo, Y., Lv, Y., Liu, X., Xu, L., Zhang, X., 2012. A double-targeted magnetic nanocarrier with potential application in hydrophobic drug delivery. Colloid. Surf. B 91, 68–76.

Donsi, F., Annunziata, M., Sessa, M., Ferrari, G., 2011. Nanoencapsulation of essential oils to enhance their antimicrobial activity in foods. LWT—Food Sci. Technol. 44, 1908–1914.

Esmaeili, A., Ebrahimzadeh, M., 2015. Preparation of polyamide nanocapsules of aloe vera l. delivery with in vivo studies. AAPS PharmSciTech. 16 (2), 242–249.
Esmaeili, A., Rahnamoun, S., Sharifnia, F., 2013. Effect of O/W process parameters on *Crataegus azarolus* L. nanocapsule properties. J. Nanobiotechn. 1, 11–16.
Ezhilarasi, P.N., Karthik, P., Chhanwal, N., Anandharamakrishnan, C., 2013. Nanoencapsulation techniques for food bioactive components: a review. Food Bioprocess Technol. 6, 628–647.
Fagui, A.E., Amiel, C., 2012. PLA nanoparticles coated with a β-cyclodextrin polymer shell: preparation, characterization and release kinetics of a hydrophobic compound. Int. J. Pharm. 436, 644–651.
Fagui, A.E., Dalmas, F., Lorthioir, C., Wintgens, V., Volet, G., Amiel, C., 2011. Well-defined core–shell nanoparticles containing cyclodextrin in the shell: a comprehensive study. Polymer 52, 3752–3761.
Fang, Z., Bhandari, B., 2010. Encapsulation of polyphenols—a review. Trends Food Sci. Technol. 21, 510–523.
Fisk, I.D., Linforth, R.S.T., Taylor, A.J., Gray, D.A., 2011. Aroma encapsulation and aroma delivery by oil body suspensions derived from sunflower seeds (*Helianthus annus*). Eur. Food Res. Technol. 232, 905–910.
Fleige, E., Quadir, M.A., Haag, R., 2012. Stimuli-responsive polymeric nanocarriers for the controlled transport of active compounds: concepts and applications. Adv. Drug Deliv. Rev. 64, 866–884.
Fomina, N., Sankaranarayanan, J., Almutairi, A., 2012. Photochemical mechanisms of light-triggered release from nanocarriers. Adv. Drug Deliv. Rev. 64, 1005–1020.
Franchini, M.C., Baldi, G., Bonacchi, D., Gentili, D., Giudetti, G., Lascialfari, A., Corti, M., Marmorato, P., Ponti, J., Micotti, E., Guerrini, U., Sironi, L., Gelosa, P., Ravagli, C., Ricci, A., 2010. Bovine serum albumin-based magnetic nanocarrier for MRI diagnosis and hyperthermic therapy: a potential theranostic approach against cancer. ACS Small 6 (3), 366–370.
Fu, G.D., Li, G.L., Neoh, K.N., Kang, E.T., 2011. Hollow polymeric nanostructures—synthesis, morphology. and function. Prog. Polym. Sci. 36, 127–167.
Geest, B.G.D., Sanders, N.N., Sukhorukov, G.B., Demeester, J., Smedt, S.C.D., 2007. Release mechanisms for polyelectrolyte capsules. Chem. Soc. Rev. 36, 636–649.
Ghodke, S.A., Sonawane, S.H., Bhanvase, B.A., Mishra, S., Joshi, K.S., 2015. Studies on fragrance delivery from inorganic nanocontainers: encapsulation, release and modeling studies. J. Inst. Eng. India Ser. E. 96 (1), 45–53.
Gil, E.S., Hudson, S.A., 2004. Stimuli-responsive polymers and their bioconjugates. Prog. Polym. Sci. 29, 1173–1222.
Gill, I., Ballesteros, A., 1998. Encapsulation of biologicals within silicate, siloxane, and hybrid sol–gel polymers: an efficient and generic approach. J. Am. Chem. Soc. 12, 8587–8598.
Gu, X., Wang, J., Wang, Y., Wang, Y., Gao, H., Wu, G., 2013. Layer-by-layer assembled polyaspartamide nanocapsules for pH-responsive protein delivery. Colloid. Surf. B. 108, 205–211.
Guo, S., Qiao, Y., Wang, W., He, H., Deng, L., Xing, J., Xu, J., Liangd, X.J., Dong, A., 2010. Poly(3-caprolactone)-graft-poly(2-(*N*, *N*-dimethylamino) ethyl methacrylate) nanoparticles: pH dependent thermo-sensitive multifunctional carriers for gene and drug delivery. J. Mater. Chem. 20, 6935–6941.
Hales, M., Barner-Kowollik, C., Davis, T.P., Stenzel, M.H., 2004. Shell-cross-linked vesicles synthesized from block copolymers of poly(D,L-lactide) and poly(*N*-isopropyl acrylamide) as thermoresponsive nanocontainers. Langmuir 20, 10809–10817.

Han, Y., Shchukin, D., Yang, J., Simon, C.R., Fuchs, H., Mohwald, H., 2010. Biocompatible protein nanocontainers for controlled drug release. ACS Nano 4 (5), 2838–2844.
Harada, A., Kataoka, K., 2006. Supramolecular assemblies of block copolymers in aqueous media as nanocontainers relevant to biological applications. Prog. Polym. Sci. 31, 949–982.
He, D., Wang, S., Lei, L., Hou, Z., Shang, P., He, X., Nie, H., 2015. Core–shell particles for controllable release of drug. Chem. Eng. Sci. 125, 108–120.
Hillaireau, H., Couvreur, P., 2009. Nanocarriers' entry into the cell: relevance to drug delivery. Cell Mol Life Sci. 66, 2873–2896.
Hodoroaba, V.D., Akcakayiran, D., Grigoriev, D.O., Shchukin, D.G., 2014. Characterization of micro- and nanocapsules for self-healing anti-corrosion coatings by high resolution SEM with coupled transmission mode and EDX. Analyst 139, 2004–2010.
Hofmeister, I., Landfester, K., Taden, A., 2014. pH-sensitive nanocapsules with barrier properties: fragrance encapsulation and controlled release. Macromolecules 47, 5768–5773.
Hosseinkhani, B., Callewaert, C., Vanbeveren, N., Boon, N., 2015. Novel biocompatible nanocapsules for slow release of fragrances on the human skin. New Biotechnol. 32 (1), 40–46.
Hu, J., Xiao, Z.B., Ma, S.S., Zhou, R.J., Wang, M.X., Li, Z., 2012. Properties of *Osmanthus* fragrance-loaded chitosan–sodium tripolyphosphate nanoparticles delivered through cotton fabrics. J. Appl. Poly. Sci. 123, 3748–3754.
Hua, M.Y., Yang, H.W., Liu, H.L., Tsai, R.W., Pang, S.T., Chuang, K.L., Chang, Y.S., Hwang, T.L., Chang, Y.H., Chuang, H.C., Chuang, C.K., 2011. Superhigh-magnetization nanocarrier as a doxorubicin delivery platform for magnetic targeting therapy. Biomaterials 32, 8999–9010.
Huang, J., Quan, B., Liu, M., Wei, Z., Jiang, L., 2008. Conducting polypyrrole conical nanocontainers: formation mechanism and voltage switchable property. Macromol. Rapid Commun. 29, 1335–1340.
Huang, X., Zhuang, J., Teng, X., Li, L., Chen, D., Yan, X., Tang, F., 2010. The promotion of human malignant melanoma growth by mesoporous silica nanoparticles through decreased reactive oxygen species. Biomater. 31, 6142–6153.
Husseini, G.A., Pitt, W.G., 2008. Micelles and nanoparticles for ultrasonic drug and gene delivery. Adv. Drug Deliv. Rev. 60, 1137–1152.
Islam, M.S., Yeum, J.H., Das, A.K., 2002. Synthesis of poly(vinyl acetate–methyl methacrylate) copolymer microspheres using suspension polymerization. J. Colloid Interf. Sci. 368 (1), 400–405.
Jafari, A.H., Hosseinib, S.M.A., Jamalizadeh, E., 2010. Investigation of smart nanocapsules containing inhibitors for corrosion protection of copper. Electrochim. Acta. 55, 9004–9009.
Jafari, S.M., Assadpoor, E., He, Y., Bhandar, B., 2008. Encapsulation efficiency of food flavors and oils during spray drying. Dry. Tech. 26, 816–835.
Kidsaneepoiboon, P., Wanichwecharungruang, S.P., Chooppawa, T., Deephum, R., Panyathanmaporn, T., 2011. Organic–inorganic hybrid polysilsesquioxane nanospheres as UVA/UVB absorber and fragrance carrier. J. Mater. Chem. 21, 7922–7930.
Knop, K., Hoogenboom, R., Fischer, D., Schubert, U.S., 2010. Poly(ethylene glycol) in drug delivery: pros and cons as well as potential alternatives. Angew. Chem. Int. Ed. 49, 6288–6308.
Kozlovskaya, V., Kharlampieva, E., Mansfield, M.L., Sukhishvili, S.A., 2006. Poly(methacrylic acid) hydrogel films and capsules: response to pH and ionic strength, and encapsulation of macromolecules. Chem. Mater. 18, 328–336.

Lee, J.M., Bermudez, H., Discher, B.M., Sheehan, M.A., Won, Y.Y., Bates, F.S., 2001. Preparation, stability, and in vitro performance of vesicles made with diblock copolymers. Biotechnol. Bioeng. 73, 135–145.

Lehner, R., Wang, X., Wolf, M., Hunziker, P., 2012. Designing switchable nanosystems for medical application. J. Control. Release 161, 307–316.

Lertsutthiwong, P., Noomun, K., Jongaroonngamsang, N., Rojsitthisak, P., Nimmannit, U., 2008. Preparation of alginate nanocapsules containing turmeric oil. Carbohyd. Polym. 74, 209–214.

Letchford, K., Burt, H., 2007. A review of the formation and classification of amphiphilic block copolymer nanoparticulate structures: micelles, nanospheres, nanocapsules and polymersomes. Eur. J. Pharm. Biopharm. 65, 259–269.

Levis, S.R., Deasy, P.B., 2002. Characterization of halloysite for use as a microtubular drug delivery system. Int. J. Pharm. 243, 125–134.

Levis, S.R., Deasy, P.B., 2003. Use of coated microtubular halloysite for the sustained release of diltiazem hydrochloride and propranolol hydrochloride. Int. J. Pharm. 253, 145–157.

Li, J., Yu, Z., Jiang, H., Zou, G., Zhang, Q., 2012. Photo and pH dual-responsive polydiacetylene smart nanocontainer. Mater. Chem. Phys. 136, 219–224.

Li, W., Szoka, F.C., 2007. Lipid-based nanoparticles for nucleic acid delivery. Pharm. Res. 24, 438–449.

Lu, Y., Proch, S., Schrinner, M., Drechsler, M., Kempe, R., Ballauff, M., 2009. Thermosensitive core–shell microgel as nanoreactor for catalytic active mental nanoparticles. J. Mater. Chem. 19, 3955–3961.

Lvov, Y., Antipov, A., Mamedov, A., Möhwald, H., Sukhorukov, G.B., 2001. Urease encapsulation in nanoorganized microshells. Nano Lett. 1, 125–128.

Madrigal-Carballo, S., Lim, S., Rodriguez, G., Vila, A.O., Krueger, C.G., Gunasekaran, S., Reed, J.D., 2010. Biopolymer coating of soybean lecithin liposomes via layer-by-layer self-assembly as novel delivery system for ellagic acid. J. Fun. Foods. 2 (2), 99–106.

Marcuzzo, E., Sensidoni, A., Debeaufort, F., Voilley, A., 2010. Encapsulation of aroma compounds in biopolymeric emulsion based edible films to control flavor release. Carbhy. Polym. 80, 984–988.

Martín del Valle, E.M., Galan, M.A., Carbonell, R.G., 2009. Drug delivery technologies: the way forward in the new decade. Ind. Eng. Chem. Res. 48, 2475–2486.

Masotti, A., Bordi, F., Ortaggi, G., Marino, F., Palocci, C., 2008. A novel method to obtain chitosan/DNA nanospheres and a study of their release properties. Nanotechnology 19, 055302.

Meng, F., Zhong, Z., Feijen, J., 2009. Stimuli-responsive polymersomes for programmed drug delivery. Biomacromolecules 10, 197–209.

Meseguer-Olmo, L., Ros-Nicolas, M.J., Vicente-Ortega, V., Alcaraz-Banos, M., Clavel-Sainz, M., Arcos, D., Ragel, C.V., Vallet-Regi, M., Meseguer-Ortiz, C., 2006. A bioactive sol–gel glass implant for in vivo gentamicin release: experimental model in rabbit. J. Orthop. Res. 24, 454–460.

Min-Hui, L., Keller, P., 2009. Stimuli-responsive polymer vesicles. Soft Matter 5, 927–937.

Moghimi, S.M., Hunter, A.C., Murray, J.C., 2005. Nanomedicine: current status and future prospects. J. Fed. Amer. Soc. Exp. Biol. 19, 311–330.

Mora-Huertas, C.E., Fessi, H., Elaissari, A., 2010. Polymer-based nanocapsules for drug delivery. Int. J. Pharm. 385, 113–142.

Morris, C.A., Anderson, M.L., Stroud, R.M., Merzbache, C.I., Rolison, D.R., 1999. Silica sol as a nanoglue: flexible synthesis of composite aerogels. Science 284, 622–624.

Motornov, M., Roiter, Y., Tokarev, I., Minko, S., 2010. Stimuli-responsive nanoparticles, nanogels, and capsules for integrated multifunctional intelligent systems. Prog. Polym. Sci. 35, 174–211.

Mozafari, M.R., Flanagan, J., Matia-Merino, L., Awati, A., Omri, A., Suntres, Z.E., Singh, H., 2006. Recent trends in the lipidbased nanoencapsulation of antioxidants and their role in foods. J. Sci. Food Agric. 86 (13), 2038–2045.

Mu, B., Shen, R.P., Liu, P., 2009. Crosslinked polymeric nanocapsules from polymer brushes grafted silica nanoparticles via surface initiated atom transfer radical polymerization. Colloid. Surf. B. 74, 511–515.

Mura, S., Nicolas, J., Couvreur, P., 2013. Stimuli-responsive nanocarriers for drug delivery. Nat. Mater. 12, 991–1003.

Niu, G., Djaoui, A.B., Cohn, D., 2011. Crosslinkable PEO-PPO-PEO triblocks as building blocks of thermo-responsive nanoshells. Polymer 52, 2524–2530.

Nyborg, W.L., 2001. Biological effects of ultrasound: development of safety guidelines. Part II: General review. Ultrasound Med. Biol. 27, 301–333.

Okubo, M., Konishi, Y., Minami, H., 2004. Production of hollow particles by suspension polymerization of divinylbenzene with nonsolvent. Prog. Colloid Polym. Sci. 124, 54–59.

Okubo, M., Konishi, Y., Inohara, T., Minami, H., 2002. Production of hollow polymer particles by suspension polymerizations for ethylene glycol dimethacrylate/toluene droplets dissolving styrene–methyl methacrylate copolymers. J. Appl. Polym. Sci. 86, 1087–1091.

Okubo, M., Konishi, Y., Inohara, T., Minami, H., 2003. Size effect of monomer droplets on the production of hollow polymer particles by suspension polymerization. Colloid Polym. Sci. 281, 302–307.

Panayiotou, M., Freitag, R., 2005. Synthesis and characterisation of stimuli-responsive poly(N,N′-diethylacrylamide) hydrogels. Polymer 46, 615–621.

Parekh, G., Pattekari, P., Joshi, C., Shutava, T., Levchenko, T., Torchilin, V., Lvov, Y., DeCoster, M., 2014. Layer-by-layer nanoencapsulation of camptothecin with improved activity. Int. J. Pharm. 465, 218–227.

Park, C., Kim, H., Kim, S., Kim, C., 2009. Enzyme responsive nanocontainers with cyclodextrin gatekeepers and synergistic effects in release of guests. J. Am. Chem. Soc. 131, 16614–16615.

Patel, K., Angelos, S., Dichtel, W.R., Coskun, A., Yang, Y.W., Zink, J.I., Stoddart, F., 2008. Enzyme-responsive snap-top covered silica nanocontainers. J. Am. Chem. Soc. 130, 2382–2383.

Patri, A.K., Kukowska-Latallo, J.F., Baker, Jr., J.R., 2005. Targeted drug delivery with dendrimers: comparison of the release kinetics of covalently conjugated drug and noncovalent drug inclusion complex. Adv. Drug Deliv. Rev. 57 (15), 2203–2214.

Pollert, E., Knížek, K., Maryško, M., Závěta, K., Lančok, A., Boháček, J., Horák, D., Babič, M., 2006. Magnetic poly(glycidyl methacrylate) microspheres containing maghemite prepared by emulsion polymerization. J. Magn. Magn. Mater. 306 (2), 241–247.

Pomorska, A., Yliniemi, K., Wilson, B.P., Shchukin, D., Johannsmann, D., Grundmeier, G., 2011. QCM study of the adsorption of polyelectrolyte covered mesoporous TiO_2 nanocontainers on SAM modified Au surfaces. J. Colloid Interf. Sci. 362, 180–187.

Poon, Z., Chang, D., Zhao, X., Hammond, P.T., 2011. Layer-by-layer nanoparticles with a pH-sheddable layer for in vivo targeting of tumor hypoxia. ACS Nano 5 (6), 4284–4292.

Quintanilla-Carvajal, M.X., Camacho-Diaz, B.H., Meraz-Torres, L.S., Chanona-Perez, J.J., Alamilla-Beltran, L., Jimenez-Aparicio, A., Gutierrez-Lopez, G.F., 2010. Nanoencapsulation: a new trend in food engineering processing. Food Eng. Rev. 2, 39–50.

Radin, S., El-Bassyouni, G., Vresilovic, E.J., Schepers, E., Ducheyne, P., 2004. In vivo tissue response to resorbable silica xerogels as controlled-release materials. Biomaterials 26, 1043–1052.
Radt, B., Smith, T., Caruso, F., 2004. Optically addressable nanostructured capsules. Adv. Mater. 16, 2184–2189.
Rae, J., Ashokkumar, M., Eulaerts, O., Sonntag, C.V., Reisse, J., Grieser, F., 2005. Estimation of ultrasound induced cavitation bubble temperatures in aqueous solutions. Ultrason. Sonochem. 12, 325.
Rajendran, R., Radhai, R., Kotresh, T.M., Csiszar, E., 2013. Development of antimicrobial cotton fabrics using herb-loaded nanoparticles. Carbohyd. Polym. 91, 613–617.
Rapoport, N.Y., Kennedy, A.M., Shea, J.E., Scaife, C.L., Nam, K.H., 2009. Controlled and targeted tumor chemotherapy by ultrasound-activated nanoemulsions/microbubbles. J. Control. Release 138, 268–276.
Renzo, F.D., Testa, F., Chen, J.D., Cambon, H., Galarneau, A., Plee, D., Fajula, F., 1999. Textural control of micelle-templated mesoporous silicates: the effects of cosurfactants and alkalinity. Micropor. Mesopor. Mater. 28, 437–446.
Samadzadeh, M., Hatami Boura, S., Peikari, M., Kasirih, S.M., Ashrafi, A., 2010. A review on self-healing coatings based on micro/nanocapsules. Prog. Org. Coat. 68, 159–164.
Sansukcharearnpon, A., Wanichwecharungruang, S., Leepipatpaiboon, N., Kerdcharoen, T., Arayachukeat, S., 2010. High loading fragrance encapsulation based on a polymer-blend: preparation and release behavior. Int. J. Pharm. 391, 267–273.
Sarti, S., Bordi, F., 2013. Polymeric hollow micro and nanospheres for biotechnological applications: a focused review. Mater. Lett. 109, 134–139.
Schlicher, R.K., Radhakrishna, H., Tolentino, T.P., Apkarian, R.P., Zarnitsyn, V., Prausnitz, M.R., 2006. Mechanism of intracellular delivery by acoustic cavitation. Ultrasound Med. Biol. 32, 915–924.
Seregina, M.V., Bronstein, L.M., Antonietti, M., 1997. Preparation of noble-metal colloids in block copolymer micelles and their catalytic properties in hydrogenation. Chem. Mater. 9, 923–931.
Shahin, M., Lavasanifar, A., 2010. Novel self-associating poly(ethylene oxide)-b-poly(ε-caprolactone)-based drug conjugates and nano-containers for paclitaxel delivery. Int. J. Pharm 389 (1–2), 213–222.
Shchukin, D.G., Mçhwald, H., 2007. Self-repairing coatings containing active nanoreservoirs. ACS Small 3 (6), 926–943.
Shchukin, D.G., Kohler, K., Mohwald, H., Sukhorukov, G.B., 2005. Gas-filled polyelectrolyte capsules. Angew. Chem. Int. Ed. 44, 3310–3314.
Shchukin, D.G., Zheludkevich, M., Yasakau, K., Lamaka, S., Ferreira, M.G.S., Möhwald, H., 2006. Layer-by-layer assembled nanocontainers for self-healing corrosion protection. Adv. Mater. 18, 1672–1678.
Shchukina, E.M., Shchukin, D.G., 2011. LbL-coated microcapsules for delivering lipid-based drugs. Adv. Drug Deliv. Rev. 63, 837–846.
Shengnan, W., Zhang, M., Zhong, L., Zhang, W., 2010. A strategy to immobilize noble metal nanoparticles on silica microspheres. J. Mol. Catal. A Chem. 327, 92–100.
Shenton, W., Mann, S., Clfen, H., Bacher, A., Fisher, M., 2001. Synthesis of nanophase iron oxide in lumazine synthase capsids. Angew. Chem. Int. Ed. 113, 456–459.
Shu, S., Sun, C., Zhang, X., Wu, Z., Wang, Z., Li, C., 2010. Hollow and degradable polyelectrolyte nanocapsules for protein drug delivery. Acta Biomater. 6, 210–217.

Skirtach, A.G., Yashchenok, A.M., Möhwald, H., 2011. Encapsulation, release, and applications of LbL polyelectrolyte multilayer capsules. Chem. Commun. 47, 12736–12746.

Skirtach, A.G., Antipov, A.A., Shchukin, D.G., Sukhorukov, G.B., 2004. Remote activation of capsules containing Ag nanoparticles and IR dye by laser light. Langmuir 20, 6988–6992.

Skirtach, A.G., Kreft, O., 2009. Stimuli sensitive nanotechnology for drug delivery. In: Devilliers, M.M., et al. (Eds.), Nanotechnology in Drug Delivery. Springer, New York, 545–578.

Slowing, I.I., Vivero-Escoto, J.L., Wu, C.W., Lin, V.S.Y., 2008. Mesoporous silica nanoparticles as controlled release drug delivery and gene transfection carriers. Adv. Drug Deliv. Rev. 60, 1278–1288.

Steitz, R., Leiner, V., Tauer, K., Khrenov, V., Klitzing, R.V., 2002. Temperature-induced changes in polyelectrolyte films at the solid–liquid interface. Appl. Phys. A Mater. Sci. Process. 74, 519–521.

Stuart, M.A.C., Huck, W.T.S., Genzer, J., Müller, M., Ober, C., Stamm, M., Sukhorukov, G.B., Szleifer, I., Tsukruk, V.V., Urban, M., Winnik, F., Zauscher, S., Luzinov, I., Minko, S., 2010. Emerging applications of stimuli-responsive polymer materials. Nat. Mater. 9, 101–113.

Sukhishvili, S.A., 2005. Responsive polymer films and capsules via layer-by-layer assembly. Curr. Opin. Colloid Interf. Sci. 10, 37–44.

Sukhorukov, G., Fery, A., Mohwald, H., 2005. Intelligent micro- and nanocapsules. Prog. Polym. Sci. 30, 885–897.

Sunder, A., Kramer, M., Hanselman, R., Muhlhaupt, R., Frey, H., 1999. Molecular nanocapsules based on amphiphilic hyperbranched polyglycerols. Angew. Chem. Int. Ed. 38, 3552–3555.

Suslick, K.S., Crum, L.A., 1997. Sonochemistry and sonoluminescence. In: Encyclopedia of Acoustics. Wiley-Interscience, New York, pp. 271–282.

Suslick, K.S., Grinstaff, M.W., 1990. Protein microencapsulation of nonaqueous liquids. J. Am. Chem. Soc. 112, 7807–7809.

Suslick, K.S., Grinstaff, M.W., Kolbeck, K.J., Wong, M., 1994. Characterization of sonochemically prepared proteinaceous microspheres. Ultrason. Sonochem. 1 (1), 65–68.

Svenson, S., 2009. Dendrimers as versatile platform in drug delivery applications. Eur. J. Pharm. Biopharm. 71 (3), 445–462.

Swaminathan, S., Garcia-Amoros, J., Fraix, A., Kandoth, N., Sortino, S., Raymo, F.M., 2014. Photoresponsive polymer nanocarriers with multifunctional cargo. Chem. Soc. Rev. 43, 4167.

Tachaprutinun, A., Udomsup, T., Luadthong, C., Wanichwecharungruang, S., 2009. Preventing the thermal degradation of astaxanthin through nanoencapsulation. Int. J. Pharm. 374, 119–124.

Tan, D., Yuan, P., Annabi-Bergaya, F., Liu, D., Wang, L., Liu, H., He, H., 2013. Loading and in vitro release of ibuprofen in tubular halloysite. Micropor. Mesopor. Mater. 179, 89–98.

Tedim, J., Poznyak, S.K., Kuznetsova, A., Raps, D., Hack, T., Zheludkevich, M.L., Ferreira, M.G.S., 2010. Enhancement of active corrosion protection via combination of inhibitor-loaded nanocontainers. ACS Appl. Mater. Interf. 2 (5), 1528–1535.

Teng, X., Shchukin, D.G., Möhwald, H., 2007. Encapsulation of water-immiscible solvents in polyglutamate/polyelectrolyte nanocontainers. Adv. Funct. Mater. 17 (8), 1273–1278.

Theisinger, S., Schoeller, K., Osborn, B., Sarkar, M., Landfester, K., 2009. Encapsulation of a fragrance via miniemulsion polymerization for temperature-controlled release. Macromol. Chem. Phys. 210, 411–420.

Vergaro, V., Abdullayev, E., Cingolani, R., Lvov, Y., Leporatti, S., 2010. Cytocompatibility and uptake of halloysite clay nanotubes. Biomacromolecules 11, 820–828.

Verica ðordevic, 2014. Trends in encapsulation technologies for delivery of food bioactive compounds. Food Eng. Rev., DOI 10.1007/s12393-014-9106-7, 452–490.

Vihola, H., Laukkanen, A., Tenhu, H., Hirvonen, J., 2008. Drug release characteristics of physically cross-linked thermosensitive poly(N-vinylcaprolactam) hydrogel particles. J. Pharm. Sci. 97, 4783–4793.

Viseras, M.T., Aguzzi, C., Cerezo, P., Visera, C., Valenzuela, C., 2008. Equilibrium and kinetics of 5-aminosalicylic acid adsorption by halloysite. Micropor. Mesopor. Mater. 108, 112–116.

Völker, E., Calvo, E.J., Williams, F.J., 2008. Layer-by-layer self-assembled redox polyelectrolytes on passive steel. Israel J. Chem. 48, 305–312.

Wang, P., Zhu, Y., Yang, X., Chen, A., 2008. Prolonged-release performance of perfume encapsulated by tailoring mesoporous silica spheres. Flavour Frag. J. 23, 29–34.

Wang, Q., Zhang, J., Wang, A., 2013. Alkali activation of halloysite for adsorption and release of of loxacin. Appl. Surf. Sci. 287, 54–61.

Wang, S., Zhang, M., Wang, D., Zhang, W., Liu, S., 2011. Synthesis of hollow mesoporous silica microspheres through surface sol–gel process on polystyrene-*co*-poly(4-vinylpyridine) core–shell microspheres. Micropor. Mesopor. Mater. 139, 1–7.

Weda, P., Trzebicka, B., Dworak, A., Tsvetanov, Ch.B., 2008. Thermosensitive nanospheres of low-density core: an approach to hollow nanoparticles. Polymer 49, 1467–1474.

Wei, H., Zhang, X., Cheng, C., Cheng, S.X., Zhuo, R.X., 2007. Self-assembled, thermosensitive micelles of a star block copolymer based on PMMA and PNIPAAm for controlled drug delivery. Biomaterials 28, 99–107.

Wen, J., Anderson, S.M., Du, J., Yan, M., Wang, J., Shen, M., Lu, Y., Segura, T., 2011. Controlled protein delivery based on enzyme-responsive nanocapsules. Adv. Mater. 23, 4549–4553.

Williams, R.J., Hall, T.E., Glattauer, V., White, J., Pasic, P.J., Sorensen, A.B., Waddington, L., McLean, K.M., Currie, P.D., Hartley, P.G., 2011. The in vivo performance of an enzyme-assisted self-assembled peptide/protein hydrogel. Biomaterials 32 (22), 5304–5310.

Wohl, B.M., Engbersen, J.F.J., 2012. Responsive layer-by-layer materials for drug delivery. J. Control. Release. 158, 2–14.

Xiao, Z., Tian, T., Hu, J., Wang, M., Zhou, R., 2014. Preparation and characterization of chitosan nanoparticles as the delivery system for tuberose fragrance. Flavour Fragr. J. 29, 22–34.

Xiaokun, M., Zhou, B., Deng, Y., Sheng, Y., Wang, C., Pan, Y., Wang, Z., 2008. Study on $CaCO_3$/PMMA nanocomposite microspheres by soapless emulsion polymerization. Colloid. Surf. A 312 (2–3), 190–194.

Xing, W.N., Ni, L., Huo, P.W., Lu, Z.Y., Liu, X.L., Luo, Y.Y., Yan, Y.S., 2012. Preparation high photocatalytic activity of CdS/halloysite nanotubes (HNTs) nanocomposites with hydrothermal method. Appl. Surf. Sci. 259, 698–704.

Yang, X., Pu, F., Chen, C., Ren, J., Qu, X., 2012. An enzyme-responsive nanocontainer as an intelligent signal-amplification platform for a multiple proteases assay. Chem. Commun. 48, 11133–11135.

Yuxi, L., Jing, Y., Ziqi, Z., Junjie, Rui, Z., Fanglian, Y., 2012. Formation and characterization of natural polysaccharide hollow nanocapsules via template layer-by-layer self-assembly. J. Colloid. Interf. Sci. 379, 130–140.

Zhai, R., Zhang, B., Liu, L., Xie, Y., 2010. Haoqin Zhang, Jindun Liu, Immobilization of enzyme biocatalyst on natural halloysite nanotubes. Catal. Commun. 12, 259–263.
Zhang, A.B., Pan, L., Zhang, H.Y., Liu, S.T., Ye, Y., Xia, M.S., Chen, X.G., 2012. Effects of acid treatment on the physico-chemical and pore characteristics of halloysite. Colloid. Surf. A. 396, 182–188.
Zhao, C.S., Liu, X.L., Yang, M., Fang, J.Y., Zhang, J.J., Liu, F.Q., 2009. The preparation of copolymerized fluorescent microspheres of styrene using detergent-free emulsion polymerization. Dyes Pigm. 82 (2), 134–141.
Zheludkevich, M.L., Serra, R., Montemor, M.F., Ferreira, M.G.S., 2005. Oxide nanoparticle reservoirs for storage and prolonged release of the corrosion inhibitors. Electrochem. Commun. 7, 836–840.
Zhu, C.L., Wang, X.W., Lin, Z.Z., Xie, Z.H., Wang, X.R., 2014. Cell microenvironment stimuli-responsive controlled-release delivery systems based on mesoporous silica nanoparticles. J. Food Drug Anal. 22 (1), 18–28.

17

CYCLODEXTRINS-BASED NANOCOMPLEXES FOR ENCAPSULATION OF BIOACTIVE COMPOUNDS IN FOOD, COSMETICS, AND PHARMACEUTICAL PRODUCTS: PRINCIPLES OF SUPRAMOLECULAR COMPLEXES FORMATION, THEIR INFLUENCE ON THE ANTIOXIDATIVE PROPERTIES OF TARGET CHEMICALS, AND RECENT ADVANCES IN SELECTED INDUSTRIAL APPLICATIONS

Paweł K. Zarzycki*, Bożena Fenert*, Bronisław K. Głód**
**Koszalin University of Technology, Section of Toxicology and Bioanalytics, Koszalin, Poland; **Siedlce University of Natural Sciences and Humanities, Department of Analytical Chemistry, Institute of Chemistry, Faculty of Science, Siedlce, Poland*

Encapsulations. http://dx.doi.org/10.1016/B978-0-12-804307-3.00017-X

1 Introduction

Cyclodextrins (CDs) belong to a wide group of low-molecular mass organic compounds classified as macrocycles or donut-like compounds, due to their three-dimensional shape. In 1891 Villiers published the paper describing methodology of starch digestion by *Bacillus amylobacter* culture and isolation of crystalline substance (3 g of pure target compound from 1 kg of starch) (Villiers, 1891). He discovered that this substance, which was characterized by molecular composition $(C_6H_{10}O_5)_2 \times 3H_2O$, was stable under acid hydrolysis conditions and did not posses reducing properties. He named this substance cellulosine. Moreover, he described two distinct crystalline forms of this substance, that to our present knowledge were very likely alpha and beta-cyclodextrins (α-, β-CD), composed of six and seven glucopyranose units. Fundamentals of cyclodextrins chemistry were laid down a few years later, mainly by Schardinger, which described the analytical protocol for distinguishing of α-dextrin from β-dextrin using iodine reaction resulting with different colors of thin layers containing α- or β-dextrin/iodine complexes (Schardinger, 1911; Szejtli, 1998). Cyclic structure of pure dextrins fractions composed of maltose units containing only α-1,4-glycosidic linkages was hypothesized in 1936 (Freudenberg et al., 1936). A few years later the chemical structure of native dextrins named α-, β- cyclodextrin was finally recognized and next the molecule (γ-cyclodextrin) was discovered (Freudenberg and Cramer, 1948; French, 1957). Chemical structure of congeneric series of native cyclodextrins and general scheme of their biosynthesis involving cyclodextrin glucosyltransferase (an enzyme that is present in several microorganisms including *Bacillus macerans, Klebsiella oxytoca, Bacillus circulans, or Alkalophyllic bacillus;* Szejtli, 1998) from the starch natural polymer are presented in Figs. 17.1 and 17.2, respectively.

With the beginning of the 1950s the inclusion properties of cyclodextrins were recognized and extensively investigated (Cramer, 1954). This research, which is based on host–guest supramolecular complexes formations involving native cyclodextrins or CDs dertivatives (acting as the host molecules) and plethora of low-molecular mass organic chemicals (guest molecules) is still dominating all the aspects of cyclodextrins chemistry, physics as well as analytical, medical, and industrial applications.

From a practical point of view three native cyclodextrins (α-, β-, γ-CD) and their derivatives are most extensively investigated. They are homogonous, can be easily extracted resulting with high purity crystalline products and relatively nonexpensive (particularly β-cyclodextrin and its hydroxypropyl derivatives). Cyclodextrins

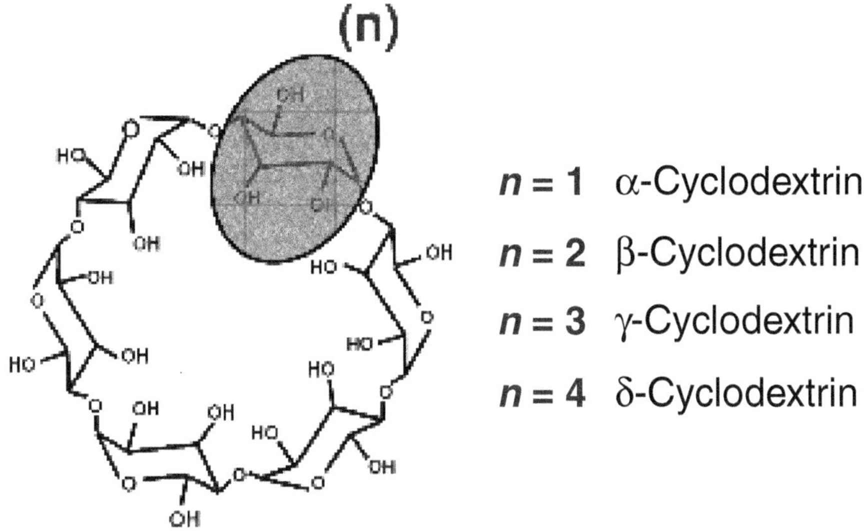

Figure 17.1. Chemical structure of native cyclodextrins: congeneric series composed of 6–9 maltose units.

are water soluble and due to the polar hydroxyl groups location (outside of the donut surface) the internal cavity is relatively nonpolar in comparison to different water nonsoluble macrocycles, for example, calixarenes (Fig. 17.3).

Dimensions of the cyclodextrin cavity increase with the number of glucose units in the macrocyclic ring (0.47–0.53, 0.6–0.65,

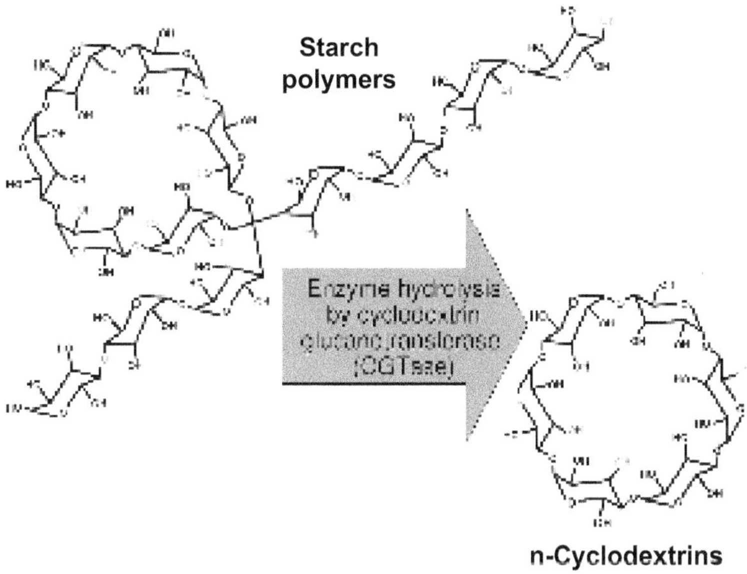

Figure 17.2. Enzymatic synthesis of cyclodextrin from starch polymer.

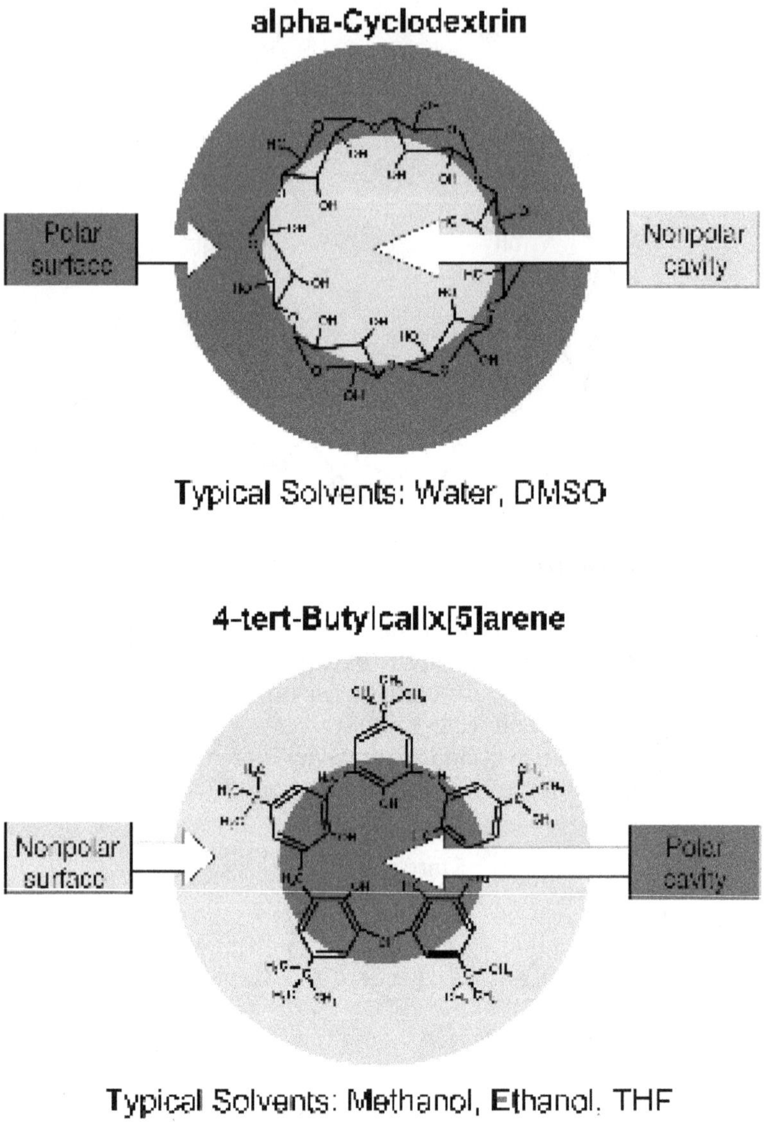

Figure 17.3. Polarity distribution within cyclodextrin and calixalene structures.

and 0.75–0.83 nm for α-, β- and γ-CD, respectively). However, it is noteworthy to say that the height of the cavity remains the same for all of them. Generally, the ability of macrocycles to form a host–guest (inclusion) complex with external (guest) molecule is a function of two critical factors. The first key factor is steric and depends on relative size/shape of the cyclodextrin cavity to the size/shape of interacting guest molecule. Without particular size and shape of the guest molecule, the proper fit into the cyclodextrin cavity is not possible. The second key factor involves the thermodynamic interactions between components of the cyclodextrin-guest-solvent system. Successful complex formation requires favorable net of electrostatic driving force that pulls the guest molecule into cyclodextrin cavity and to remove from this space already included guest molecules, especially solvent molecules (Lamparczyk, 1985; Lamparczyk et al., 1987; Lehn, 1995).

The key physicochemical parameter limiting number of analytical, medical and industrial applications is CDs solubility in water. Particularly, α- and γ-cyclodextrin is one order of magnitude more soluble than β-cyclodextrin. Solubility of cyclodextrins is strongly affected by the presence of organic cosolvents (eg, decreasing with methanol addition or increasing with several organic liquids at given concentration range like ethanol, acetonitrile or solid additives including urea) as well as temperature (Fig. 17.4) (Chatjigakis et al., 1992; Zarzycki et al., 2006).

The great interest in cyclodextrins applications, which is observed over the last decades in analytical chemistry, medicine, pharmacy, cosmetology, and the food industry, particularly in comparison to other macrocyclic compounds, is due to several subjects that were highlighted in a number of books and experimental as well as review papers (Szejtli, 1982; Szejtli, 1983; Duchene, 1987; Lehn, 1995; Szejtli, 1998; Hedges, 1998; Uekama et al., 1998):

1. Cyclodextrins can be produced in large amounts from natural and nonexpensive materials like starch via simple green chemistry protocols.
2. Host–guest complexes involving CDs can significantly modify the physicochemical properties of initial guest molecules, for example, increasing solubility and bioavailability that is the key issue in food and pharmaceutical industry.
3. Stereoselective interaction (CDs are chiral molecules) with target substances can be conveniently controlled by simple factors including pH, temperature, and the presence of low-molecular mass additives. This enables a number of analytical applications of these macrocycles and their derivatives.
4. Cyclodextrins are basically nontoxic if delivered *per os* in reasonable amounts and therefore, they can be consumed by the humans as food or can be used as cosmetics ingredients.

Generally, the modern applications of cyclodextrins are as follows: (1) in analytical chemistry, as selective and chiral complexation agents, improving fluorescence, absorbance, electrochemical reactions and chromatographic separation; (2) in cosmetology, as protective active ingredients in perfumes; (3) in food industry, as, for example, the medium for removing cholesterol from egg

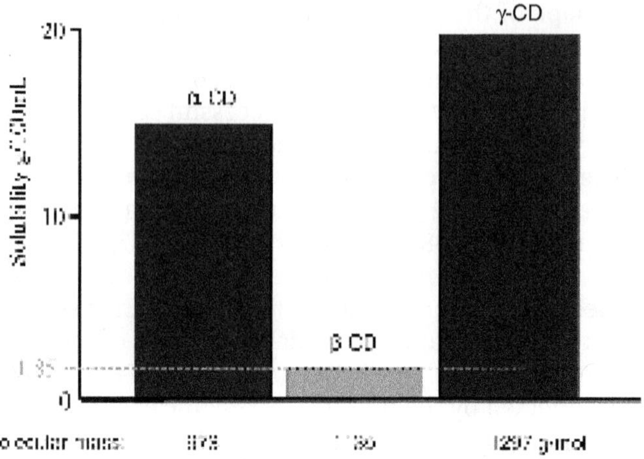

	β-Cyclodextrin		2-Hydroxypropyl β-CD	
Temperature	Methanol/water (mM)	ACN/water (mM)	Methanol/water (mM)	ACN/water (mM)
+30°C	8.9	25.4	>717	>717
−10°C	1.63	7.9	>437	>437

Figure 17.4. Solubility of various cyclodextrins in water and binary solvents (mole fraction Xs=0.16) at different temperatures. *Source: Data published by Chatjigakis et al. (1992), Zarzycki et al. (2006).*

and dairy products; and (4) in medicinal and pharmaceutical chemistry, enabling molecular encapsulation as well as increasing solubility of target chemicals and drugs in water. According to the literature search through the Web of Science databases, the number of research papers focusing on industrial applications of cyclodextrins associated with pharmaceuticals, food products, and cosmetics is systematically increasing (Fig. 17.5).

2 Cyclodextrins and Their Complexes

As is evident in the picture presented in Fig. 17.3, the interior of cyclodextrin cavity is relatively nonpolar. It is formed by a circular configuration of hydrogen atoms and glucoside oxygen atoms, while all the hydroxyl groups are located outside of the macrocyclic structure. This particular 3D structure strongly affects solubility of

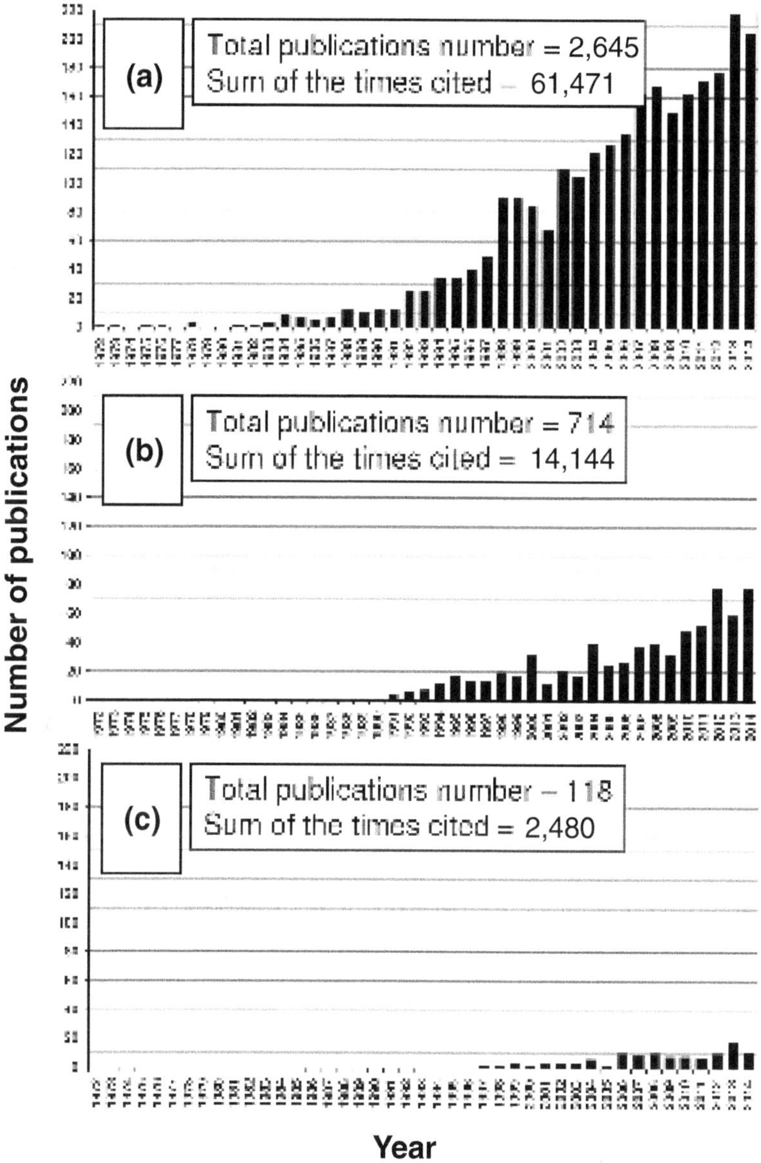

Figure 17.5. Comparison of the manuscripts number associated with topics: cyclodextrin in (a) pharmaceuticals, (b) food, and (c) cosmetics, respectively, that were published up to 2014 (according to Web of Science database).

their inclusion complexes, and usually CDs complexes even with nonpolar guest organic compounds are more soluble in water. The main advantage of CD complexation is that this process is highly stereoselective. Cyclodextrins molecules working as active component of chromatographic stationary or mobile phases can selectively recognize given low-molecular mass enantiomers, for example, polycyclic aromatic hydrocarbons, steroids, and their derivatives (Sybilska et al., 1993; Lamparczyk et al., 1994; Zarzycki et al., 2008).

The development of food, cosmetic, or pharmaceutical industries, where very often volatile, slightly soluble (eg, in human body fluids) compounds are used, demands application of newly invented substances or their mixtures that increase durability of edible products and availability of bioactive substances from pharmaceutical formulations. Cyclodextrins were recognized as nontoxic encapsulation candidates that may complex hydrophobic compounds, causing an increase of their solubility in hydrophilic solvents. Many compounds (eg, pesticides) in a humid environment are released from the CD torus, hence their use in the agronomy (Ramos-Campos et al., 2015). It should be highlighted that despite of long-term extensive research, the detailed mechanism of the host–guest supramolecular complexes formation based on cyclodextrins guest molecules, particularly in multicomponent liquid phase environment, as well as properties of such complexes are still not exactly known (Zarzycki and Lamparczyk, 1998; Głód et al., 2000). As can be seen from the pictures presented within Fig. 17.6, even considering very simple system (water solution of β-cyclodextrin at room temperature) the results of supramolecular interaction between CD and inert gas (*n*-hexane) molecules can be surprising. Under such conditions within a few seconds after gas

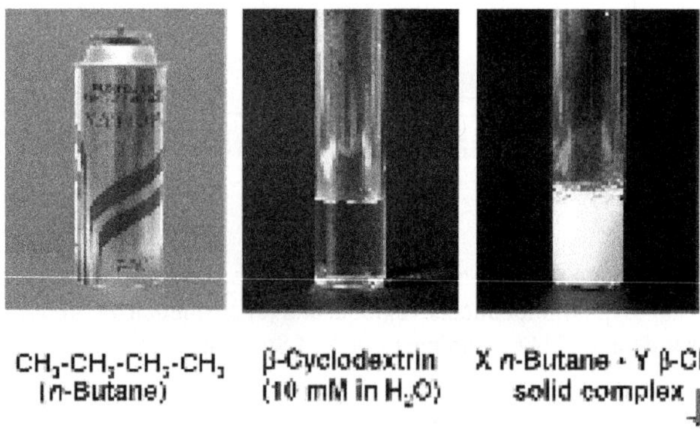

Figure 17.6. Solid host–guest complex formation between *n*-butane and β-cyclodextrin molecules in water at room temperature. Experimental data provided by P.K. Zarzycki.

injection through the small pipe located within the solution, the nonsoluble complex of β-cyclodextrin and *n*-butane molecules can be observed and isolated. By replacing native β-cyclodextrin with its hydroxypropyl derivative a solid complex cannot be generated. There is a similar result if *n*-alkane molecules are replaced by their *n*-alcohol derivative.

For many industrial applications the cyclodextrins as encapsulation agents must act in a more complex environment, including a number of host–guest competitive substances: mainly low-molecular mass organic additives. Therefore, prediction of host–guest complexes stability with target molecules can be difficult. As an example we may consider a still relatively simple chromatographic system, which is composed of octadecylsilane (C18 *n*-alkane) stationary phase (chromatographic plates) and binary mobile phases consisting of water and organic additive (acetonitrile, methanol). Native β-cyclodextrin, its methyl derivative and very well soluble in water hydroxypropyl derivative are selected as the chromatographic analytes. According to planar chromatography principles the strong interaction of chromatographed compounds with stationary phase results with R_F values less than 0.1 (in such a case an analyte spot is located close to the start line) and weak interaction is observed if analyte migrates close to the mobile phase front ($R_F > 0.9$). As can be seen from the retention profiles presented in Fig. 17.7, the composition of mobile phase is spectacularly changing the cyclodextrins retention, particularly

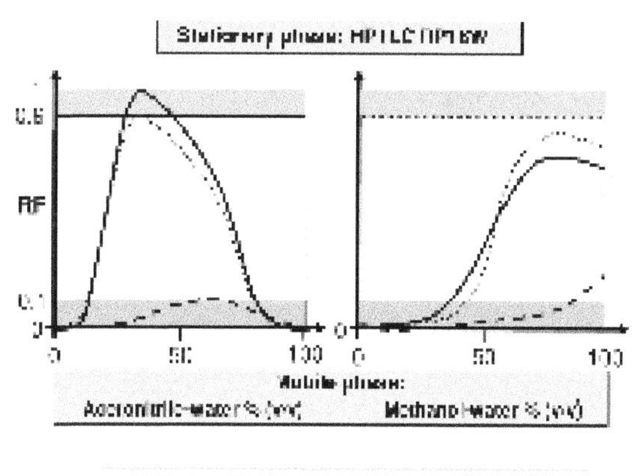

Figure 17.7. Retention profiles of β-cyclodextrin and its methyl as well as hydroxypropyl derivatives using *n*-alkane stationary phase (low carbon load octadecylsilica–water tolerable layer) and whole range (0–100%) binary mobile phases acetonitrile/water and methanol/water, according to data published by Zarzycki et al. (1995, 1997, 1999).

if acetonitrile is used as the mobile phase additive. Interestingly, there is no massive difference in retention between low soluble native β-cyclodextrin and its very well soluble in water hydroxypropyl derivative. In contrast, methyl derivative shows completely different chromatographic behavior despite of mobile phase composition, except pure water and pure acetonitrile, where all CDs are strongly retarded by stationary phase, despite their solubility. This illustrates the fact that host–guest complexes formation based on cyclodextrins can be significantly affected by a number of factors and prediction of complex stability may be fairly difficult for samples composed of multicomponent liquid matrix, which is typical for food products or pharmaceutical formulations.

In spite of sample composition problems some of the cyclodextrins-based complexes have been recognized as the temperature supersensitive objects (Zarzycki and Lamparczyk, 1996). This phenomenon is illustrated in Fig. 17.8, where thermochromic properties of alkaline solution of β-cyclodextrin and phenolphthalein (PP) are shown. At subambient temperature, CD-PP complex is almost colorless (association constant $K_{1:1}$ = 152,000, calculated at 0°C), while at elevated temperature, addition of cyclodextrin

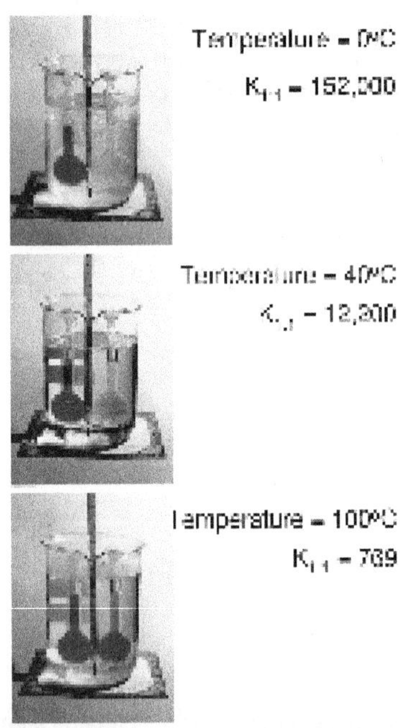

Figure 17.8. Termochromic solution composed of β-cyclodextrin-phenolphthalein host–guest complex (right bottles) and reference solution consisting of alkaline phenolphthalein solution without CD additive (left bottles) at temperatures 0°C (top), 40°C (middle), and 100°C (bottom).

does not strongly affect the phenolphthalein purple color intensity ($K_{1:1}$ = 769 for 100°C) (Zarzycki and Lamparczyk, 1998). Reported above stoichiometric proportion and binding constants K were calculated from spectrophotometric data using Scott's approach involving Eq. (17.1) (Scott, 1956; Ikeda et al., 1975; Connors, 1987):

$$C_{\beta\text{-CD}}/\Delta A = C_{\beta\text{-CD}}/(C_{PP} \cdot \Delta\varepsilon) + 1/(C_{PP} \cdot \Delta\varepsilon \cdot K) \quad (17.1)$$

where: $C_{\beta\text{-CD}}$ and C_{PP} are the total molar concentrations of β-cyclodextrin and phenolphthalein, respectively, while ΔA is the change in absorbance after addition of β-cyclodextrin ($\Delta A = A_{PP}$ $A_{\beta\text{-CD}\bullet PP}$), K is the binding constant and $\Delta\varepsilon$ is the difference of the molar absorbance for free and complexed phenolphthalein. If resulting plot of $C_{\beta\text{-CD}}/\Delta A$ against $C_{\beta\text{-CD}}$ yields a straight line, the 1:1 complexation stoichiometry can be expected. According to Fig. 17.8 the phenolphthalein alkaline solution without CD additive remains purple in all temperatures investigated.

Described phenomenon can be strongly disturbed by competitive interaction with low-molecular mass molecule like tetrahydrofuran (THF). In a water environment, a small amount of THF addition may permanently block the cyclodextrin cavity for other host molecules and CD-PP complex cannot be formed (Zarzycki and Lamparczyk, 1998). Competitive interaction with cyclodextrin cavity is particularly significant for molecules containing long *n*-alkanes chains, that was previously demonstrated using chromatographic experiment involving C-18 stationary phase, consisting of long *n*-alkanes structures (Fig. 17.7).

Based on the previously mentioned phenomenon number of chromatographic protocols for efficient multiple separation of steroids stereoisomers from complex biological materials including blood and plasma as well as environmental samples derived from surface waters, treated and untreated sewage waters, or activated sludge extracts were established (Zarzycki and Smith, 2001; Clifton et al., 2007; Zarzycki et al., 2009). Principles of temperature-dependent inclusion chromatography can be illustrated through the scheme within Fig. 17.9, where retention profiles of (I) target analyte (chromatographed using plain binary mobile phase without cyclodextrin), (II) target analyte interacting with cyclodextrin acting as the host additive and (III) free complexing agent (under conditions, in which CD is not retarded by stationary phase due to proper selection of binary mobile phase components for example, using optimization graph from Fig. 17.7) are visualized at different temperatures (Lamparczyk et al., 1994; Zarzycki and Smith, 2001). It has been discovered that the chromatographic retention of a

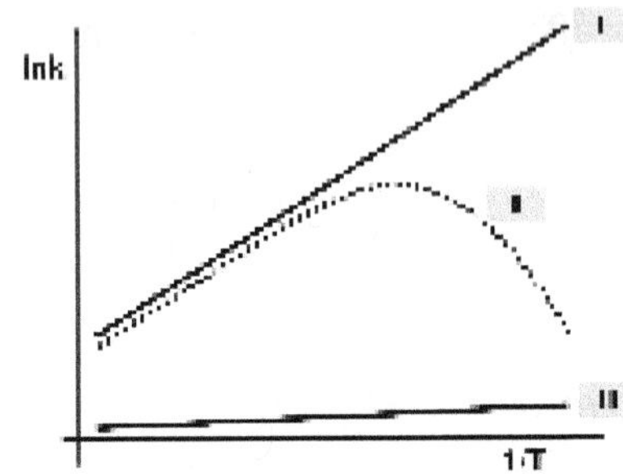

Figure 17.9. General scheme illustrating principles of temperature-dependent liquid chromatography. The lines and curve represent retention profiles observed under column liquid chromatography conditions for target analyte in different temperatures using binary mobile phase (I), analyte interacting with cyclodextrin applied as complexation agent (II), as well as cyclodextrin that is nonretarded by stationary phase under selected mobile phase conditions (III), based on steroids and β-cyclodextrin. Source: Data published by Zarzycki and Smith (2001).

number of steroids and related compounds is strongly influenced by temperature when the mobile phase is modified with cyclodextrin. At a wide range of subambient temperatures investigated the retention of cyclodextrin is significantly lower than the retention of target analytes chromatographed using an unmodified mobile phase. The experimental data indicates that retention of inclusion complexes can be varied between two lines formed by the Van't Hoff plot of the cyclodextrin and the Van't Hoff plot of the uncomplexed solute. The deviation magnitude from the line I presented in Fig. 17.9 and temperature in which the deviation begins depends on the target molecule stereochemistry and cyclodextrin cavity shape/size. It is noteworthy to say that at subambient temperature the retention of target analyte can be shorter than retention of uncomplexed analyte at elevated temperature, which is uncommon for classical chromatographic system involving binary mobile phases.

The widespread use of cyclodextrin has led to a number of methods that are used to study the properties of their complexes including high-performance liquid chromatography (Lamparczyk and Zarzycki, 1995; Zarzycki et al., 2008; Wantusiak and Głód, 2012), UV-Vis spectrophotometry (Zarzycki and Lamparczyk, 1996; Zarzycki and Lamparczyk, 1998; Filipský et al., 2013), IR, 1H, and 13C NMR spectroscopy (Chandrasekaran et al., 2015) as well as fluorescence, circular dichroism (Soumitra and Suresh, 2015), differential scanning calorimetry, laser Raman spectroscopy, scanning electron microscopy (Krishnaswamy et al., 2012), X-ray diffraction (Poorghorban et al., 2015), AFM (Ramos-Campos et al., 2015), conductometry, viscosimetry (Ali et al., 2015), cyclic and differential pulse voltammetry or polarography (Ferreira et al., 2010; Głód et al., 2014). These analytical techniques based on different working principles allow to determine the key physicochemical parameter reflecting supramolecular interaction—the complexation constants—and to examine the effect of such interaction on the properties of the formed complexes. Worthy of note, the electrochemical techniques are used when a "guest" molecule has no chromophoric groups.

During the examination of complex compounds the two types of stability constants of complexes, kinetic and thermodynamic, are considered. Thermodynamic stability determines the compound's tendency to stay in the thermodynamic stability state. The thermodynamic stability constant, K, depends on the enthalpy and entropy of the system. The higher its value the more stable is the complex. The kinetic stability is expressed by the rate of complexation reactions.

Thermodynamic stability constant can be determined using, for example, the absorption spectroscopy (Naseri et al., 2007). The complexation constant can be calculated from the Benesi-Hildebrand equation (17.2). The equation was derived for the complexes of the 1:1 stoichiometry, assuming that one of the reactants is present in large excess (Benesi and Hildebrand, 1949; Filipský et al., 2013; Liu et al., 2013). The measurement of the absorption are performed twice, before and after the complexing reaction. In such case the K value can be computed based on the following equation:

$$\frac{1}{\Delta A}=\frac{1}{\Delta\varepsilon[G]_0}+\frac{1}{K[G]_0\Delta\varepsilon[H]_0} \tag{17.2}$$

where: ΔA, difference between the absorbance of the "guest" molecule compound and its complex with the "host" (cyclodextrin)l; $[G]_0$, concentration of included compound; $[H]_0$, concentration of cyclodextrin; K, complexation constant; $\Delta\varepsilon$, difference between the absorption coefficients.

Complexation constant can be also determined from the NMR chemical shifts, using a modified Benesi-Hildebrand equation (17.3), assuming the fact that the magnetic properties of compounds result from the presence of unpaired electrons or magnetic nuclei in their molecules:

$$\frac{1}{\Delta\delta}=\frac{1}{K\cdot\Delta\delta_C}\cdot\frac{1}{[CD]_0}+\frac{1}{\Delta\delta_C} \tag{17.3}$$

where: $\Delta\delta$, difference between chemical shifts of the tested compound and complex; $\Delta\delta_C$, difference between chemical shifts of the test compound and cyclodextrin; [*CD*], concentration of cyclodextrin. Advantage of this approach is that the changes in proton chemical shifts of guest and host molecules allow researchers to determine the structure of the complex created (Dias et al., 2008).

3 General Information about Free Radicals

The human body is exposed to free radicals generated under the influence of many internal or external (environmental) factors. Recently, the increased interest in free radicals is observed. Thus, the compounds neutralizing their action, namely the free radical scavengers, usually identified with antioxidants, are also intensively investigated (Głód et al., 2009). Compounds characterized by antioxidative properties are used as food additives and dietary supplements.

The term "free radicals" refers to molecules, atoms, or ions, which are capable of independent existence and have at least one unpaired electron. The last parameter determines paramagnetic properties of radicals. Free radicals, in general, are characterized by high reactivity. Their reactivity is caused by their tendency to join or get rid of the excess electron (Głód et al., 2000; Halliwell, 2012).

Approximately one quarter of the Earth's mass falls on oxygen. The original Earth's atmosphere was a mixture of gases characterized by reducing properties. The free oxygen in the molecular form first appeared on the Earth with the development of photosynthetic organisms, around 2 billion years ago. At the end of the Precambrian interval (560 million years ago, approximately) the increased concentration of oxygen caused the extinction of more than 99% of living organisms. Oxygen dissolved in water was necessary for the appearance and further development of higher organisms (eukaryotes and multicellular). During oxygen reduction, the so-called reactive oxygen species (ROS) were produced. Besides oxygen-free radicals ROS include, among others, singlet oxygen or hydrogen peroxide. As the source of reactive oxygen species may be external or internal factors. The internal factors include, among others, mitochondrial respiratory chain, some of enzymatic reactions or immune cells. External factors include, for example, ionizing or UV radiation, ultrasounds or toxic compounds, like cigarette smoke. Free radicals are involved in some functions of a living organism, such as the regulation of metabolism or supporting of the immune system. However, above certain concentrations radicals can damage the cell membranes, DNA molecules and contribute to the carcinoma formation (Głód et al., 2009).

4 Antioxidants

Some of free radicals adversely affect the human body. They are removed from biological systems and living organisms by the free radical scavengers or antioxidants. Antioxidants are substances that in low concentrations, compared with the oxidized compound, inhibit or retard the oxidation of this compound. Concentration of the antioxidants decreases with the organism age or as a result of various diseases. This deficiency can be supplemented, for example, with food. The most important antioxidants include ascorbic acid (vitamin C), α-tocopherol (vitamin E), melatonin, beta-carotene and the polyphenolic compounds, in particular, flavonoids.

The flavonoids comprise a large group of polyphenolic compounds characterized by a 15-carbon skeleton. They consist of two aryl groups and heterocyclic ring with the oxygen atom (Sokolová et al., 2012). There are six main classes of flavonoids: flavones, flavonols, flavanones, flavanols, isoflavones, and anthocyanidins. They are the plant pigments that are mainly found in flowers, leaves, fruits, and stems. Mostly, they impart yellow color, while anthocyanins impart hues between red and blue. These colors depend on the pH and the presence and concentration of metal ions. They are produced in response to damage to the plant stress or UV radiation. Such substances are commonly presented in a number of food products like grapes, green tea, cabbage, tomatoes, and onions.

Many flavonoids are strong antioxidants. Their antioxidant activities are mainly due to the presence of hydroxyl groups in compound. They depend on their number and location. In particular, the two hydroxyl groups in *ortho* or *para* positions determine the extremely strong antioxidative properties of the compound. The antioxidant activities (reduction of free radicals) of the flavonoids are based on the cleavage of hydrogen atom from the hydroxyl group. Flavanoids are oxidized to phenoxy radicals, which are stabilized by resonance structures. The greater the number of these structures, the more stable the radical that is formed; flavonoid is characterized by the stronger antioxidant properties. Thus, double bonds, hydroxyl, and carbonyl groups increase the number of mesomeric structures and increase the antioxidant properties of the compound. Phenolic compounds can also transfer the electrons and chelate the transition metals. Moreover, they enhance the effects of other antioxidants and protect them.

Antioxidative properties of flavonoids are caused mainly by the reductive properties of hydroxyl groups (in particular, a catechol group) in the ring B. Another mechanism involves the complexation of transition metals, which prevents the production of free radicals (De Souza and De Giovani, 2005). It was shown that the complexed flavonoids were better DPPH radical scavengers than the corresponding flavonoids (Dias et al., 2008). Similar results were obtained using cyclic voltammetry (CV). The CV data show a considerable decrease of the oxidation potentials (De Souza and De Giovani, 2005) of the complexes in comparison to the flavonoids. It means the flavonoid-metal complexes are more effective antioxidants than the flavonoids.

5 Analysis of Antioxidants

Antioxidants can be determined chromatographically (by HPLC, high-performance liquid chromatography, or by TLC, thin-layer chromatography), electroanalytically (potentiometrically or using one of voltammetric techniques) spectrophotometrically, and so forth (De Rijke et al., 2006). The samples often contain a lot of antioxidants of different powers and concentrations. The weak antioxidants are frequently present in high concentrations. They can react with each other or act synergistically. Finally, the sample composition is often unknown (eg, a newly discovered herb or medicines or blood samples of a patient with an unusual illness). It turned out that in such cases much more information is obtained by measuring the total antioxidant potential (TAP). In the literature there are a number of other names, like total antioxidant capacity, activity, reactivity, and so forth (Głód et al., 2000). The measure of TAP is the sum of all antioxidants in a given sample, of the products of their concentrations (c_i) and the reaction rate constants with the corresponding radical (k_i) (Głód et al., 2009):

$$\mathrm{TAP} = \sum_i \prod_i k_i c_i \tag{17.4}$$

From the point of view of the measurement mechanism TAP can be divided into the method (1) based on the transfer of hydrogen (HAT), (2) a single electron (SET) or (3) the intermediate methods.

The TAP assay techniques can be divided into direct (eg, potentiometric or voltammetric) and indirect. In the indirect ones, the sample reacts with a stable or produced by enzymatic, redox, photo-, thermo-, electro- or radio-chemical reaction radicals (Malinka et al., 2003). The reaction can be monitored by analyzing the change of the radicals' concentration. If it is not possible, then so-called sensor (spin trap) is added to the reaction mixture (Głód et al., 2011; Głód et al., 2012a; Piszcz et al., 2014a). Then, the changes of concentration of the reaction product of sensor with radicals (with and without the sample) are determined. The measurements can be carried out either after a specified time of reaction or during the reaction (kinetic measurements). The literature describes a number of related methods, such as the determination of the total concentration of polyphenols, flavonoids, -SH groups, color or acidity of lipids.

The measure of the antioxidative properties is the redox potential. Alternatively, it may be a potential at the maximum of the voltammetric peak height or surface area of the anode part of the voltammetric curve. The stronger the antioxidant is, the lower the oxidized potential. The concentration of antioxidant in turn influences a peak height and area. From the other side the peak area is independent on the oxidation potential.

The determination of TAP of pure compound (antioxidant) is relatively simple (eg, the redox potential). The problem occurs when the sample contains many various antioxidants. Their voltammetric peaks overlap. It was proposed (Głód et al., 2014) that in this case the TAP's measure is the surface area under the voltammetric curve in which the potential is related to the reduction potential of the hydroxyl radical (strongest naturally occurring in biological systems radical; 2.08 V vs Ag/AgCl), and is presented in a exponential scale (Głód et al., 2014):

$$CPA = \int_{e(2{,}08-E_1)}^{e(2{,}08-E_2)} \left(I(E) - I(E)_{\min}\right) de^{(2{,}08-E)} \quad (17.5)$$

where I denotes current and $I_{\min}$ baseline current.

The disadvantage of the voltammetric measurements is their low sensitivity and high detection limit. These parameters can be improved by eliminating the capacitive current. This current hardly arises in the flowing systems (eg, in the electrochemical detection in HPLC). A small volume of the detector cell (causing a large linear flow rate of liquid) and stable flow of liquid results in a very large convection current. As a result, the detection limit is more than a trillion times smaller in comparison with the classical voltammetry (Piszcz et al., 2014b). In the case of complex samples (a mixture of compounds) the TAP measure is the sum of the surface areas of all peaks recorded at the anode potential of the working electrode (Wantusiak and Głód, 2012).

6 Encapsulation of Selected Antioxidants by Cyclodextrins

Low-molecular mass antioxidants as well as the transition metal ions may act as the host molecules and can interact with macrocyclic structures including cyclodextrins. The formation of the cyclodextrin inclusion complexes changes the redox potentials of included compounds. Complexation of the depolarizer increases its size, in general decreasing the diffusion current. Change of the formal potential (as well as the half-wave potential and/or the potential of voltammetric peak maximum) during the complexation is described by the de Ford and Hume equation (De Ford and Hume, 1951; Georgieva et al., 2010):

$$\Delta E = RT/nF \ln(1 + K \cdot c_{\downarrow} CD). \qquad (17.6)$$

Sign of this change depends on whether the stronger the substrate or the product of the electrode reaction is complexated.

Cyclodextrins may encapsulate a number of different low-molecular mass antioxidants, in particular flavonoids and/or phenolic acids (Carlotti et al., 2011; Vilanova and Solans, 2015). CDs as food additives may affect the antioxidant and antibacterial properties and particularly decrease the rates of degradation of antioxidants (Dominguez-Canedo et al., 2015). Encapsulated antioxidants can be released under the pH change (acidification) or the increase of temperature (Ferreira et al., 2007). It has been reported that target compounds can be complexed by CDs to increase their resistance to high temperatures (Milanovic et al., 2010), moisture, oxygen, visible, and UV light (Ferreira et al., 2007; Quirós-Sauceda et al., 2014), solubility in water (Kolanowski et al., 2004), controlled transfer and release (Soottitantawat et al., 2004), strengthening antimicrobial properties, stabilizing smelt (Chun et al., 2015). They are used, inter alia, for the preparation of nanoparticles based on iron, using environmentally friendly plant extracts (Zhuang et al., 2015) and the selective electrodes (Sheng et al., 2012).

Using voltametric technique, the interaction of cyclodextrins and selected antioxidants that are present in edible products can be investigated. Particularly, at the voltammetric curves of quercetin two peaks are observed. Cyclodextrins slightly decrease the voltammetric peak heights, due to the decreasing of diffusion coefficient during complexation. This confirms the formation of inclusion complexes (proves the complexation of quercetin by cyclodextrins). Small peak shifts toward positive values mean decrease of antioxidative activity of the sample. Similar results were obtained for other flavonoids (Piszcz, 2014b).

Measurements of TAP related to the strong radicals (eg, hydroxyl ones) are disturbed by reaction of these radicals with cyclodextrins. In turn, larger in size, weak radicals (eg, DPPH; 2,2-diphenyl-1-picrylhydrazyl) can be complexed by cyclodextrins. In practice, cyclodextrins may increase or decrease the antioxidative properties of target compounds encapsulated (Folch-Cano et al., 2010), depending on which part of molecule "enter" into the cyclodextrin torus. If the hydroxyl groups of the flavonoids B ring are in the CD torus then decrease of antioxidant activity is observed. Cyclodextrins usually decrease TAP values measured voltammetrically. The complexation of flavonoids by cyclodextrin may increase as it was described earlier for quercetin (Liu et al., 2013) or decrease the TAP value measured in relation to the DPPH radicals in case of different target oxidant for example, catechin (Folch-Cano et al., 2010).

Flavonoids can be complexed by cyclodextrins. On the other hand, flavonoids can complex transition metal ions (De Souza and De Giovani, 2005; Dias et al., 2008; Ebrahimzadeh et al., 2009). These complexed ions are not able to catalyze Fenton reaction. In this sense, cyclodextrins are antioxidants. Complexing in this case affects the antioxidant properties of both metals and the flavonoids. It turned out that the antioxidative properties (determined using Folin-Ciocalteau or DPPH methods) of complexes are smaller than the combined antioxidant properties of the transition metal ion and the ligand. The flavonoid–metal complex may be complexed by cyclodextrin. In this case the effect of cyclodextrin on the antioxidative properties of flavonoids is relatively small too. It turned out that TAP of such double complex dos not depend on the order of complexation.

7 Recent Advances of Cyclodextrins Application in Pharmaceutics, Food, and Cosmetics Products

According to data concerning literature search, which was presented in Fig. 17.5, an increasing interest in cyclodextrins research, especially investigating their encapsulation properties, is still recorded. There are a number of experimental and review papers that were published over the past few years, highlighting the unique properties of cyclodextrins based on supramolecular complexes and dealing with industrial applications of these nontoxic macrocycles. Data gathered in Tables 17.1–17.3 consist of several typical examples of CD/target compounds inclusion complexes, which in our opinion may give the general picture of the modern applications of native cyclodextrins and their derivatives as well as the studies trends within three major areas including: pharmaceutical formulations (Table 17.1), food (Table 17.2), and cosmetics (Table 17.3).

Table 17.1 Examples of Cyclodextrins-Drugs Related Preliminary Studies, Applications in Pharmaceuticals and Drug Supplements

No.	Product Category	Cyclodextrin Type	Research Objects, Target Molecules, and Macrocyclic Additive Application	References
1.	Peptide–cyclodextrin conjugates	3-monoamino-β-CD	The study of a LHRH (luteinizing hormone-releasing hormone) analogue conjugated with modified β-cyclodextrin revealed the existence of intramolecular interactions that could lead to an improved drug delivery. Particularly, both the phenyl group of tyrosine (Tyr) as well as the indole group of tryptophan (Trp) can be encapsulated inside the cyclodextrin cavity.	Kordopati et al. (2015)
2.	Cordycepine inclusion complexes	α-, β-, and γ-CD	In this study, the CDs were used to enhance the water solubility, dissolution rate, and bioavailability of cordycepin, a natural herbal medicine. Successful cyclodextrins encapsulation of target molecule may be considered as an important step in the design of novel formulations of cordycepin for herbal medicine or healthcare products.	Zhang et al. (2015)
3.	Carriers for drug delivery and optical imaging	β-CD	In this research two soluble cyanine/beta-cyclodextrin derivatives have been synthesized under simultaneous ultrasound/microwave irradiation conditions. The potential use of investigated supramolecular systems as a versatile carriers for drug delivery and optical imaging has been evaluated via in vitro experiments on HeLa cells and by monitoring cell entrance using confocal laser scanning microscopy.	Carmona et al. (2015)
4.	Complexes of norfloxacin with cyclodextrin for oral administration	β-CD	This article describes the research focusing on norfloxacin and beta-cyclodextrin complex formation increasing drug solubility and maximizing the oral drug absorption. Guest-host interactions were investigated through physicochemical characterization and the dissolution as well as microbiological activity study. Provided data demonstrated that the complex could represent an efficient drug delivery system.	Mendes et al. (2015)

Table 17.1 Examples of Cyclodextrins-Drugs Related Preliminary Studies, Applications in Pharmaceuticals and Drug Supplements (*cont.*)

No.	Product Category	Cyclodextrin Type	Research Objects, Target Molecules, and Macrocyclic Additive Application	References
5.	Acetazolamide encapsulated by HP-β-CD	Hydroxypropyl-β-CD	The article reports encapsulation of acetazolamide inside hydroxypropyl-β-cyclodextrin, which leads to more convenient drug administration and improved treatment efficacy with decreased side effects. In the authors' opinion results obtained could be useful for the design and development of novel ACZ formulations that can reduce gastrointestinal toxicity, while still maintaining their essential therapeutic efficacies. Presented research involved animal (rat) studies.	Mora et al. (2015)
6.	Salmon calcitonin particles for nasal delivery	β-CD	Research focusing on preparation of a nasal powder formulation of salmon calcitonin (sCT) using number of absorption enhancers including β-cyclodextrin to improve its bioavailability. Reported research involved in vivo pharmacokinetic study in rats.	Cho et al. (2015)
7	Injectable supramolecular hydrogel with γ-cyclodextrin	γ-CD	Study of injectable hydrogel formed by γ-cyclodextrins and polymers used as a biocompatible, biodegradable, and controllable drug delivery system. This research demonstrated the formation of an insulin/γ-CD SMGel of PCL-PEG-PCL, based on the self-assembly of PEG threaded with γ-CD without any covalent interactions. Reported data suggested the potential use of the prepared hydrogel as an injectable sustained-release system for insulin.	Khodavedi et al. (2014)

(*Continued*)

Table 17.1 Examples of Cyclodextrins-Drugs Related Preliminary Studies, Applications in Pharmaceuticals and Drug Supplements (*cont.*)

No.	Product Category	Cyclodextrin Type	Research Objects, Target Molecules, and Macrocyclic Additive Application	References
8.	Nanoplatform composed of cyclodextrin/multiwalled carbon nanotubes nanohybrid for guanine-based drugs entrapment and delivery	β-CD	Cyclodextrin/multiwalled carbon nanotubes (CD-MWCNT) nanohybrid platform for the entrapment and delivery of guanine-based drugs was synthesized by click chemistry approach and characterized by several complementary techniques. The preliminary antiviral data indicated that the Acyclovir loaded into β-CD-MWCNT nanoplatform interferes with HSV-1 replication and the antireplicative effect was higher than the free drug. The release studies showed a sustained delivery of Acy without initial burst effect confirming a strong interaction of drug with the nanoplatform sites.	Innazzo et al. (2014)
9.	Various pharmaceutical formulations including oral, parenteral, nasal, pulmonary, and skin delivery of drugs	α-, β- and γ-CD	In this review publication the application of cyclodextrins was described particularly for oral, parenteral, nasal, pulmonary, and skin delivery of drugs. Reported experimental and clinical data suggest that CDs can be used not only as excipiens for centrally acting marked drugs like antiepileptics, but also as active pharmaceutical ingredients to treat neurological disease (Niemann-Pick type C disease, stroke, neuroinfections, and brain tumors).	Vecernyes et al. (2014)
10.	Drug delivery systems to the skin	β-CD	Skin permeation studies highlighted that β-NS-PYRO (β-CD nanosponges cross-linked with pyromellitic dianhydride) in gels and cream-gels containing diclofenac significantly decreased the amount of drug permeated through the skin while increasing its amount in stratum corneum and viable epidermis. The authors concluded that CD contained nanosponges seems to be a multifunctional coingredient with potential in topical monophasic and biphasic formulations to stabilize light-sensitive drugs and to localize the action of highly penetrating drugs in the external layers of skin.	Conte et al. (2014)

Table 17.1 Examples of Cyclodextrins-Drugs Related Preliminary Studies, Applications in Pharmaceuticals and Drug Supplements (*cont.*)

No.	Product Category	Cyclodextrin Type	Research Objects, Target Molecules, and Macrocyclic Additive Application	References
11.	Electrospun nanofibrous materials containing naproxen-cyclodextrin complex	β-CD HP- β-CD	Naproxen (NAP) and its β-CD, HP-β-CD inclusion complexes were prepared by freeze-drying process. Prepared NAP/β-CD and NAP/HP-β-CD inclusion complexes were loaded to electrospun thermoplastic polyurethane fibers separately and their release characteristics were compared with pure NAP loaded fibres. This research indicates that NAP–inclusion complex loaded into electrospun nanofibres could be used as drug delivery systems for acute pain treatments since they possess a highly porous structure that can release the drug immediately.	Akaduman et al. (2014)
12.	Sildenafil pressurized metered dose inhalers	HP-β-CD, γ-CD, α-CD	Sildenafil citrate was complexed with cyclodextrins to enhance its water solubility prior to development as an inhaled preparation for pulmonary hypertension using pressurized metered dose inhalers (pMDI). It has been documented that the available sildenafil in the blood vessels of chicken egg embryos after spraying sildenafil-CDs pMDIs was within the range of 751–825 ng/mL, which was much higher than that of a sildenafil only pMDI.	Sawatdee et al. (2014)
13.	Multifunctional bioreducible targeted and synergistic codelivery system for anticancer drug paclitaxel (PTX)	γ-CD	For potential cancer therapy a multifunctional bioredycible targeted and synergistic codelivery system for anticancer drug paclitaxel (PTX) and p53 gene supramolecular self-assembling inclusion complex (IC) was prepared. This system was based on PTX and star-shaped cationic polymer containing γ-CD and multiple oligoathlenimine (OEI) arms with folic acid (FA) conjugated via a disulfide linker. The IC, termed as γ-CD-OEI-SS-FA/PTX complex was formed between PTX and the hydrophobic cavity of γ-CD core of the star polymer. The authors state that multifunctional self-assembly system with the synergistic effects of redox-sensitive folic-acid-targeted and PTX-enhanced p53 gene delivery may be promising for cancer therapeutic application.	Zhao et al. (2014)

(*Continued*)

Table 17.1 Examples of Cyclodextrins-Drugs Related Preliminary Studies, Applications in Pharmaceuticals and Drug Supplements (*cont.*)

No.	Product Category	Cyclodextrin Type	Research Objects, Target Molecules, and Macrocyclic Additive Application	References
14.	Gel ointments containing tranilast nanoparticles and additives including HP-β-CD	2-hydroxypropyl-β-CD	The authors reported that tranilast nanoparticles can be applied to the formulation of a transdermal system, and that a transdermal formulation using drug nanoparticles and several additives including 2-hydroxypropyl-β-CD might be a delivery option for the clinical treatment of rheumatoid arthritis. This study involved animals (rats) research. In this study, they prepared a gel ointment containing Tranilast (antiallergic agent, clinically used in the treatment of bronchial asthma) nanoparticles and investigated its usefulness.	Nagai and Ito (2014)
15.	Bioresorbable chlorhexidine delivery systems based on modified paper points	β-CD	This research reports application of adsorbent points transformed into bioresorbable chlorhexidine delivery systems for the treatment of the periodontal pocket by preventing its recolonization by the subgingival microflora. The paper points (PPs) were first oxidized to promote their resorption, then grafted with β-cyclodextrin (CD) or maltodextrin (MD) in order to achieve sustained delivery of chlorhexidine. The optimized oxidized-dextrin-grafted PPs responded to the authors' initial specifications in terms of resorption and chlorhexidine digluconate (digCHX) release rates and therefore could be adopted as a reliable complementary periodontal therapy. The biological evaluation proved the nontoxicity of modified material itself, in contrast to the digCHX loaded material, which showed important in vitro cytotoxicity comparable to that of PerioChip®.	Tabary et al. (2014)
16.	R-α lipoic acid cyclodextrin (RALA-CD) complex	γ-CD	The aim of this study was to determine the effect of RALA-CD on energy expenditure and underlying molecular targets in female laboratory mice. The authors' conclusion was that RALA-CD is a regulator of energy expenditure in laboratory mice	Nikolai et al. (2014)

Table 17.1 Examples of Cyclodextrins-Drugs Related Preliminary Studies, Applications in Pharmaceuticals and Drug Supplements (*cont.*)

No.	Product Category	Cyclodextrin Type	Research Objects, Target Molecules, and Macrocyclic Additive Application	References
17.	Multifunctional cotton containing silver nanoparticles and cyclodextrin	β-CD, monochlorotriazinyl-β-cyclodextrin	The protocols for functionalization of cotton fabrics against antibacterial was reported. The fabrics were reacted with cyclodextrin polyacrylic acid copolymer, which was newly synthesized. It has been found that the use of the copolymer for reduction of silver ions to silver nanoparticles was a better selection than sodium borohydride, which represents the conventional reducing agents. The new products displayed superior antibacterial activity along with good fabric stabilization.	Hebeish et al. (2014)
18.	Liposomal gels containing glutathione/cyclodextrins complexes for cutaneous administration	β-CD, HP-β-CD, Methyl-β-CD	The aim of this work was to develop and characterize a formulation intended for the cutaneous administration of glutathione (γ-glutamylcysteinylglycine, GSH), potentially useful for cellular defense against UV-induced damage. Particularly, liposomes containing GSH or GSH/cyclodextrins (CDs) inclusion complexes as well as liposomes dispersed within a hydrophilic gel, were evaluated. These formulations were designed in order to obtain a system combining the advantages of liposomes as vehicles for topical drug delivery with those of CDs as penetration enhancers, which can be potentially useful for cutaneous administration	Cutrignelli et al. (2014)

(*Continued*)

Table 17.1 Examples of Cyclodextrins-Drugs Related Preliminary Studies, Applications in Pharmaceuticals and Drug Supplements (*cont.*)

No.	Product Category	Cyclodextrin Type	Research Objects, Target Molecules, and Macrocyclic Additive Application	References
19.	Microbeads containing cod-liver oil encircled with natural cyclodextrins	γ-CD, α-CD, β-CD, HP-α-CD, HP-β-CD, HP-γ-CD	The authors evaluated ability of cyclodextrins to solubilize cod-liver oil in aqueous solutions. It has been found that only native (unmodified with hydroxypropyl groups) cyclodextrins were able to fully disperse 10% (v/v) cod-liver oil in aqueous solutions. Results of this research revealed that encapsulating cod-liveroil with γ-CD delays oxidative degradation when oxygen is present, but does not significantly decrease or increase the long term stability of cod-liver oil under anaerobic conditions. Cod-liver oil/γ-CD microbeads could be compressed into tablets without decreasing the integrity of encapsulation, therefore such product might be pharmaceutical industry interest as a efficient carrier for lipophilic drugs.	Konrádsdóttir et al. (2014)
20.	Emulgel formulation containing inclusion complex of calcipotriol with cyclodextrin	Anionically charged CD (Captisol®)	This study focusing on development a new topical drug delivery system of calcipotriol in order to improve the solubility and dissolution characteristic of the drug and reduce the undesirable side effects. This delivery system was developed for psoriasis treatment. Dissolution profiles of calcipotriol from emulgel formulations were compared with a commercial product of the drug. The drug release was significantly increased with the emulgel formulations compared to the commercial cream product.	Badilli et al. (2014)

Table 17.1 Examples of Cyclodextrins-Drugs Related Preliminary Studies, Applications in Pharmaceuticals and Drug Supplements (*cont.*)

No.	Product Category	Cyclodextrin Type	Research Objects, Target Molecules, and Macrocyclic Additive Application	References
21.	Fast-dissolving tacrolimus solid dispersion-loaded prolonged release tablet	HP-β-CD	The goal of this research was to develop a novel prolonged release tablet bioequivalent to the commercial sustained release capsule. The tacrolimus-loaded fast-dissolving solid dispersions formulated with HP-β-CD and dioctyl sulphosuccinate with the solvent evaporation method remarkably raised the solubility and dissolution of poorly water-soluble tacrolimus. The tacrolimus loaded fast drying solid dispersion was prepared and such pharmaceutical formulation was intended to use in the prophylaxis of organ rejection. It has been found that tacrolimus-loaded prolonged release tablet might be bioequivalent to the tacrolimus-loaded commercial capsule.	Cho et al. (2014)
22.	Photostable gel formulation protecting diclofenac from photodegradation	β-CD	The authors invented the gel containing the light-absorbers such as octisilate, octyl methoxycinnamate, and a combination thereof. The best protection from light was obtained by incorporating of drug (diclofenac) in β-cyclodextrin. New gel formulations were designed to increase the photostability of diclofenac by incorporating chemical light-absorbers or entrapping the drug into cyclodextrin.	Ioele et al. (2014)
23.	Ocular inserts containing lodocaine HCl/β-CD complex	β-CD	Ocular insert containing lidocaine HCl/β-CD complex was designed to give a sufficient level of anesthetic. The results of in vivo study showed that the addition of β-cyclodextrin may significantly increase the drug content in the aqueous humor when compared with ocuserts containing lidocaine HCl alone.	Shukr (2014)

(*Continued*)

Table 17.1 Examples of Cyclodextrins-Drugs Related Preliminary Studies, Applications in Pharmaceuticals and Drug Supplements (*cont.*)

No.	Product Category	Cyclodextrin Type	Research Objects, Target Molecules, and Macrocyclic Additive Application	References
24.	Cyclodextrins derivatives as the antioxidant agents and metal-induced antiaggregants	6A-deoxy-6A-[(8-hydroxyquinolyl)-2-methylamino]-β-cyclodextrin	The authors synthesized and evaluated new cyclodextrin derivatives, their potential as multifunctional drug-like molecules, the metal-binding ability and antioxidant activity. Moreover, new ligands and their metal complexes were tested. In particular, the metal-induced antiaggregants property has been applied in the development of an effective assay that exploits the formation of amyloid fibrils when β-lactoglobulin A is heated in the presence of metal ions.	Oliveri et al. (2014)
25.	Transdermal delivery of the in situ hydrogels of curcumin and its inclusion complexes of HP-β-CD for melanoma treatment	Hydroxypropyl-β-cyclodextrin	The increasing solubility of curcumin (Cur) was achieved due to CD encapsulation and the photochemical stability of Cur was improved. The in situ hydrogels (ISGs) of Cur and its inclusion complexes were prepared using poloxamers 407 and 188 as the matrix. It has been demonstrated that the cytotoxicity of Cur on melanoma cells was related to blocking of cellular proliferation in the G_2/M stage followed by cellular apoptosis. The authors concluded that the ISGs of Cur inclusion complexes are a promising formulation for melanoma treatment.	Sun et al. (2014)
26.	Encapsulation agent involving β-cyclodextrin-based dendrimers	β-CD based glycodendrimer	The encapsulation behavior of the β-CD-dendrimer was examined using naproxen and naltrexone as the guest molecules. The structure of the designed β-CD-dendrimer allowed two types of possible sites for encapsulation of the guest: in cavities of the dendritic structure and in hydrophobic cavities of β-CDs. Investigated dendrimer demonstrated the ability to encapsulate both small and large model drug compounds, with sufficiently high drug loading, which could be potentially useful in drug delivery systems.	Namazi and Heydari (2014)

Table 17.1 Examples of Cyclodextrins-Drugs Related Preliminary Studies, Applications in Pharmaceuticals and Drug Supplements (*cont.*)

No.	Product Category	Cyclodextrin Type	Research Objects, Target Molecules, and Macrocyclic Additive Application	References
27.	Bioadhesive tablets containing cyclodextrin complex of itraconazole for the treatment of vaginal candidiasis	HP- β-CD, γ-CD, Met-β-CD, SBE-β-CD (sulphobutyl ether)	Bioadhesive tablets containing CD complex of itraconazole (ITR) for treatment of vaginal candidiasis were investigated. The antifungal activity of the complexes was analyzed on *Candida albicans* strains. The swelling, matrix erosion, and bioadhesion properties of formulations and the drug release rate of these tablets were analyzed, and the most therapeutically effective vaginal formulation was determined. The mucoadhesive tablet containing ITR-SBE7-β-CD inclusion complexes prolonged the residence time and the drug release on the vaginal mucosa. This formulation could be an alternative to commercially available ITR formulations.	Cevher et al. (2014)
28.	Dry powder for inhalation containing ciclesonide encapsulated within CD	Metylated β-CD	The aim of this work was to verify if it is possible to obtain a dry powder for inhalation containing ciclsenide (corticosteroid hormone) included within methylated β-CD by applying the simple and inexpensive technique of kneading. It was concluded that the kneading method is feasible for production of a ciclesonide-CD inclusion complex with a reasonable inclusion efficiency and a good aerodynamic behavior, suitable for lung administration via inhalation.	Arriagas and Cabral-Marques (2014)
29.	Orally disintegrating tablets of type-2 antidiabetes agent glimepiride with cyclodextrin additive	γ-CD, α-CD, β-CD, HP-β-CD, HB-β-CD	The direct compression method was employed for the preparation of glimepiride tablets, containing CDs and Sangelose® (hydroxypropylmethylcellulose stearoxy ether). Obtained results suggested that α-CD and β-CD can be particularly useful for the Sangelose®-based orally disintegrating tablets formulation, compared to γ-CD, HP-β-CD and HB-β-CD because of the short disintegration time of the tablets containing α-CyD and β-CyD, their shear-thinning effect on Sangelose® solutions and their solubility enhancing effect on the drug.	Aldawsari et al. (2014)

(*Continued*)

Table 17.1 Examples of Cyclodextrins-Drugs Related Preliminary Studies, Applications in Pharmaceuticals and Drug Supplements (*cont.*)

No.	Product Category	Cyclodextrin Type	Research Objects, Target Molecules, and Macrocyclic Additive Application	References
30.	Transdermal drug delivery of zaltoprofen	HP-β-CD	In this study, zaltoprofen (nonsterodial antiinflammatory drug) gels were prepared using carbomer with mixture solution of PEG 400, Tween 80, and HP-β-CD. It has been demonstrated that zaltoprofen transdermal delivery did not cause dermal irritations in the animals and such mixture might be used as a universal vehicle for development of transdermal formulations.	Baek et al. (2013)
31.	Doxorubicin loaded multifunctional vesicles	β-CD	The authors developed the gold nanoparticle (AuNP) in which polyethylene glycol (PEG) and poly(*N*-isopropylacrylamide) (PNIPAM) were attached on the surface of a gold nanocrystal through the host–guest inclusion between adamantane groups (ADA) and β-cyclodextrin (β-CD). It has been demonstrated that developed vesicle can effectively load doxorubicyn anticancer drug.	Ha et al. (2013)
32.	Cardiovascular effects of HP-β-CD	HP-β-CD	HP-β-CD was infused intravenously over 10 min. in anesthetized dogs in nondenervated animals and at 40% in denervated animals. CD increased renal arteriolar resistance and decreased renal blood flow at all doses, almost immediately after infusion start, more drastically in females. Results of studies performed suggest that autonomous nervous feedback loops are functional in normal animals and that HP-βC-D has no direct chronotropic effect. It was concluded that systemic and renal hemodynamic changes should be considered as potential background effects at 200–400 mg/kg. At higher doses (800 mg/kg), changes are more pronounced and could mask/exacerbate hemodynamic response of drug candidate; such doses should be avoided in nonclinical safety studies.	Rosseels et al. (2013)

Table 17.1 Examples of Cyclodextrins-Drugs Related Preliminary Studies, Applications in Pharmaceuticals and Drug Supplements (*cont.*)

No.	Product Category	Cyclodextrin Type	Research Objects, Target Molecules, and Macrocyclic Additive Application	References
33.	Ofloxacin supramolecular complexes with native cyclodextrins and their derivatives	α-, β- and γ-CD and their hydroxypropyl, methyl, trimethyl, carboxymethyl, carboxyethyl, sulfopropyl, sulfobutyl, succinyl, and 6-monoamino-6-monodeoxy derivatives	In this study the enantiomer-specific characterization of ofloxacin–cyclodextrin complexes was carried out by a set of complementary analytical techniques. It has been concluded that the α-cyclodextrin cavity can accommodate the oxazine ring only, whereas the whole tricyclic moiety can enter the β- and γ-cyclodextrin cavities. Molecular modeling studies provided evidence that the substitution pattern plays an important role in the complex stability. These equilibrium and structural information offer molecular basis for improved drug formulation.	Tóth et al. (2013)
34.	Inclusion complexes of sulphanilamide drugs and β-cyclodextrin	β-CD	Theoretical approach was applied to access and compare the relative stability of inclusion complexes formed by number of sulphanilamide drugs including sulphadiazene, sulphisomidine, sulphamethazine, and sulphanilamide with β-cyclodextrin. The main goal of this study was to predict that (1) the heterocyclic ring is encapsulated in the hydrophobic part and aniline ring is present in the hydrophilic part of the β-CD cavity and (2) intermolecular hydrogen bonds were formed between host and guest molecules.	Venkatesh et al. (2013)
35.	Inclusion complex of Rifabutin with β-cyclodextrin	β-CD	The effect of cyclodextrin additive on the improvement of solubility and antimicrobial activity of poorly water-soluble drug Rifabutin was investigated. The in vitro antimicrobial and antibiofilm activity of rifabutin sensible microorganisms was significantly increased by on inclusion complexation process.	Priya et al. (2013)

(*Continued*)

Table 17.1 Examples of Cyclodextrins-Drugs Related Preliminary Studies, Applications in Pharmaceuticals and Drug Supplements (*cont.*)

No.	Product Category	Cyclodextrin Type	Research Objects, Target Molecules, and Macrocyclic Additive Application	References
36.	Benznidazole/ cyclodextrin complexes for oral administration	β-CD, HP-β-CD, Me β-CD	The aim of this study was to evaluate the impact of stoichiometric and nonstoichiometric complex between benznidazole (BZL) and various CDs. The authors state that the inclusion complexes were found to improve the dissolution rate of BZL by 4.3-fold in comparison with BZL alone. Complexation of BZL with CDs derivatives increased its plasma concentrations when fed to rats, with AUC_{0-5} values increasing up to 3.7-fold and C_{max} increasing 2.5-fold in comparison with BZL alone. Remarkable increase in these parameters was observed in the case of the nonstoichiometric complexes. Thus, these CDs complexes may be used to efficiently deliver BZL in patients suffering from Chagas' disease.	Leonardi et al. (2013)
37.	Inclusion complexes of propafenone with α- and β-cyclodextrins	α-CD, β-CD	Host–guest inclusion complexes of cyclodextrins with a potential cardiovascular drug propafenone hydrochloride (PFO), were prepared and characterized using number of spectroscopic techniques. Investigations of thermodynamic and electronic properties confirmed the stability of the inclusion complexes.	Siva et al. (2013)
38.	Ternary system of dihydroar-temisinin with hydroxypropyl-β-cyclodextrin and lecithin	HP-β-CD	Main purpose of this study was to simultaneously improve the solubility and stability of dihydroartemisinin (DHA) in aqueous solutions by a ternary cyclodextrin system comprised of DHA, hydroxypropyl-β- cyclodextrin and a third auxiliary substance. The authors concluded that the ternary system investigated enhance DHA solubility and stability in aqueous solutions, and such mixture might have an important pharmaceutical potential in the development of a better oral formulation of DHA.	Wang et al. (2013)

Table 17.1 Examples of Cyclodextrins-Drugs Related Preliminary Studies, Applications in Pharmaceuticals and Drug Supplements (*cont.*)

No.	Product Category	Cyclodextrin Type	Research Objects, Target Molecules, and Macrocyclic Additive Application	References
39.	β-Cyclodextrin hydrogels for the ocular release of antibacterial thiosemicarbazones	HP-β-CD	The work investigated the development of soft contact lenses functionalized with CDs and loaded with thiosemicarbazone active substance that displays a broad antibiotic spectrum. Microbiological tests against *P. aeruginosa* and *S. aureus* confirmed the ability of TSC-loaded pHEMA-*co*-β-CD (poly(2-hydroxyethyl methacrylate)) network to inhibit bacterial growth.	Glisoni et al. (2013)
40.	Dexamethasone eye drops containing γ-cyclodextrin-based nanogels	2-hydroxypropyl-γ-CD	The authors designed and tested in vivo the aqueous eye drops of dexamethasone, based on 2-hydroxypropyl-γ-CD nanogels. The eye drops consisted of an aqueous solution of dexamethasone in 2-hydroxypropyl-γ-cyclodextrin medium containing γ-CD nanogels. The nanogel eye drops (containing 25 mg dexamethasone per mL) were tested in rabbits and compared to the commercially available product Maxidex® (suspension with 1 mg dexamethasone per mL). It has been documented that the dexamethasone nanogel eye drops were well tolerated with no macroscopic signs of irritation, redness or other toxic effects.	Moya-Ortega and Alves (2013)
41.	Solid forms of norfloxacin and β-cyclodextrin	β-CD	In this study the solid-state properties of inclusion complexes of β-CD and two different solid forms of norfloxacin were investigated to obtain promising candidates for the preparation of alternative matrices used in pharmaceutical oral formulations. The authors highlighted that investigated complexes can be valid tools for the preparation of pharmaceutical dosage products. Moreover, they represent an option to promote the pharmaceutical design for the improvement of physicochemical properties.	Chattah et al. (2013)

(*Continued*)

Table 17.1 Examples of Cyclodextrins-Drugs Related Preliminary Studies, Applications in Pharmaceuticals and Drug Supplements (*cont.*)

No.	Product Category	Cyclodextrin Type	Research Objects, Target Molecules, and Macrocyclic Additive Application	References
42.	Cyclodextrin nanocomposites of telmisartan for oral drug delivery	β-CD	Presented research explores a particle engineering approach which synergistically coalesce two principally different solubility enhancement strategies namely ternary β-cyclodextrin complexation and top-down nanonization in a unit process. As a result of this study improvement in in vitro dissolution characteristics in multimedia and biorelevant media was observed, in comparison with plain drug and marketed formulation.	Sangwai and Vavia (2013)
43.	Baicalein and HP-γ-CD complex in poloxamer thermal sensitive hydrogel for vaginal administration	Hydroxypropyl-γ-CD	This study aimed to prepare a chemically and physically stable formulation of baicalein (Ba) in an in situ thermally sensitive hydrogel for vaginal administration. An inclusion complex of Ba and HP-γ-CD was developed to increase the stability and solubility of Ba in an aqueous solution. The authors documented that in animal study (female Sprague–Dawley rats), treatment by using Ba/HP-γ-CD thermosensitive hydrogel could produce a restoration of damaged tissues. Moreover, thermosensitive hydrogel formulation of the Ba/HP-γ-CD complex appears to be a promising treatment for cervicitis.	Zhou et al. (2013)

In general, the main focus of pharmaceutics-related latest research is to design the effective and specific carriers for drugs delivery (Carmona et al., 2015), improving active substances stability, including photostability of pharmaceutical formulations (Ioele et al., 2014; Sun et al., 2014) and enhancing drugs bioavailability (Zhang et al., 2015). Supramolecular complexes are composed of common native cyclodextrins (α-, β-, γ-CD), their more soluble derivatives (for example, hydroxypropyl-CDs), but also involving more complex encapsulation structures like β-cyclodextrin-based dendrimers (Namazi and Heydari, 2014). Inclusion complexes are extensively applied for a number of pharmaceutical formulations

Table 17.2 Typical Applications of Cyclodextrins in Food Products, Food Supplements and Packaging Technology

No.	Product Category	Cyclodextrin Type	Research Objects, Target Molecules and Macrocyclic Additive Application	References
1.	Active packaging involving nanocrystal-reinforced soy protein films and cyclodextrins	β-CD	This research demonstrated that β-CD-containing soy protein isolate films were able to sequester cholesterol when brought into contact with cholesterol-rich food such as milk. This effect was more marked as the amount of CD into the films increased. Developed protocols allowed yielding biodegradable films with optimized physical and mechanical properties, which were assayed as active food coating.	Gonzales and Igardzabal (2015)
2.	Vitamin A palmitate/β-cyclodextrin inclusion complex	β-CD	In this paper the formation of water-soluble inclusion complexes of Vitamin A Palmitate with β-cyclodextrins, without the use of organic solvents, is described. The objective was to increase the water solubility of Vitamin A palmitate and its stability against different external factors to eventually enrich aqueous-based products. The stability of target compound in the complexes toward temperature, oxygen, and UV light was investigated. The results showed a notably increase of Vitamin A palmitate water solubility and stability in front of those variables when encapsulated. Moreover, β-CDs encapsulation agent seem to be a promising vehicle to increase the water solubility, stability and thereby the bioavailability of Vitamin A palmitate in food fortification to treat Vitamin A deficiency.	Vilanova and Solans (2015)
3.	Interaction of ochratoxin A with quaternary ammonium β-CD	β-CD, 2-hydroxy-3-*N,N,N*-trimethylamino)propyl-β-CD chloride	Interaction of ochratoxin A (OTA) quaternary ammonium beta-CD (QABCD) was investigated. The authors hypothesized, that QABCD may be a suitable tool for the decontamination of different OTA-contaminated drinks; furthermore, for alleviation of the toxic effects of OTA, such complex formation may reduce its absorption from the intestine.	Poór et al. (2015)

(*Continued*)

Table 17.2 Typical Applications of Cyclodextrins in Food Products, Food Supplements and Packaging Technology (*cont.*)

No.	Product Category	Cyclodextrin Type	Research Objects, Target Molecules and Macrocyclic Additive Application	References
4.	α-Cyclodextrin enzymatic production and food applications	α-CD, β-CD, γ-CD	This review focuses on the properties, enzymatic production, and food applications of α-cyclodextrin, as well as its differences with β- and γ-CDs. Compared with other native cylodextrins, α-CD has the smallest internal cavity and highest resistance to enzymatic hydrolysis. It was documented that this macrocyclic molecule has special applications in food industry, especially as a natural, soluble dietary fiber.	Li et al. (2014)
5.	Improved solubility and stability of stilbene antioxidants involving cyclodextrins encapsulation	HP-β-CD, HP-γ-CD	Aiming at the development of an active food packaging, the goal of this study was to increase stilbenes (resveratrol, pterostilbene, pinosylvin) aqueous solubility and stability using HP-CDs and bile salts. This work showed that stable stilbene solutions can be achieved via CD encapsulation, contributing for their incorporation in several materials for the food and pharmaceutical industries. Besides improving stilbene solubility, CDs also increased their photostability, which can avoid stilbene reactivity and encourage their application not only in the food industry but also in the pharmaceutical industry.	Silva et al. (2014)
6.	Antimicrobial packaging of fresh chicken breast fillets	HP-β-CD	Chitosan/cyclodextrin films (CS:CD) incorporating carvacrol were proposed as antimicrobial packaging of fresh chicken breast fillets. Active films were used to inhibit microbial growth in packaged chicken breast fillets. A general microbial inhibition was observed, increasing with the size of the active device.	Higueras et al. (2014)

Table 17.2 Typical Applications of Cyclodextrins in Food Products, Food Supplements and Packaging Technology (*cont.*)

No.	Product Category	Cyclodextrin Type	Research Objects, Target Molecules and Macrocyclic Additive Application	References
7.	Natural colorant in yogurt based on inclusion complexes of red bell pepper pigments with β-CD	β-CD	The aim of this study was to prepare inclusion complexes between red bell pepper pigments and CD using two different procedures as well as to characterize the prepared inclusion complex and to evaluate the color stability of host–guest complex added to yogurt. The final product was characterized by a higher color stability during storage.	Gomes et al. (2014)
8.	Resveratrol production involving cyclodextrin additive	Dimethyl-β-CD, Sulfo-β-CD, 2-hydroxypropyl-β-CD	The authors reviewing experimental papers dealing with resveratrol production at large scale using plant cell suspensions improved by complexation with cyclodextrins and several elicitors. The main compounds used for resveratrol elicitation were methyljasmonate, CDs, being the two latter often combined for their synergistic action, and chitosan.	Jeandet et al. (2014)
9.	Complexation of chlorpropham with HP-β-CD and its application in potato sprout inhibition	HP-β-CD	The authors investigated the effect of HP-β-CD on the improvement of chlorpropham (CIPC) as a potato sprout inhibitor. The inclusion complex was found to significantly improve the water solubility, thermal stability, and dissolution rate of CIPC. Moreover, supramolecular complex displayed a better effect on sprout inhibition.	Huang et al. (2014)
10.	Canthaxanthin/HP-β-CD inclusion complex	HP-β-CD	Phase-solubility, storage stability and antioxidant activity of the inclusion complex of 2-HP-β-CD with canthaxanthin (CTX) was investigated CTX was biosynthesized by *Dietzia natronolimnaea* HS-1 in a continuous bioreactor. The studies revealed that the CTX/CD complex showed higher scavenging capacity than native CTX against DPPH, ABTS and hydroxyl radicals. The authors suggested that complex formation can be used as a promising strategy to improve the food application of CTX.	Gharibzahedi et al. (2014)

(*Continued*)

Table 17.2 Typical Applications of Cyclodextrins in Food Products, Food Supplements and Packaging Technology (*cont.*)

No.	Product Category	Cyclodextrin Type	Research Objects, Target Molecules and Macrocyclic Additive Application	References
11.	Chitosan/cyclodextrin membrane for gallic acid encapsulation and controlled release	2-Hydroxypropyl-β-CD	A new type of chitosan/2-hydroxypropyl-β-cyclodextrin composite membrane have been developed for the encapsulation and controlled release of gallic acid. The composite membrane based with gallic acid/CD supramolecular complex showed improved antioxidant capacities compared to plain chitosan membrane. The authors highlighted that presented experimental data will facilitate the design and preparation of composite membrane based on chitosan and could open a wide range of applications, particularly its use as an antioxidant in food, food packaging, biomedical (biodegradable soft porous scaffolds for enhance the surrounding tissue regeneration), pharmaceutical and cosmetics industries.	Paun et al. (2014)
12.	Nanoen-capsulated black pepper oleoresin using HP-β-CD for antioxidant and antimicrobial applications		Hydroxypropyl-β-cyclodextrin inclusion complex with black pepper olereosin was synthesized using the kneading method and characterized for its physico-chemical properties, antioxidant and antimicrobial activities. The authors reported that black pepper oleoresin encapsulated within CD was able to inhibit *Salmonella* at lower ($P < 0.05$) concentrations than its corresponding free extract. Therefore, black pepper oleoresin/CD nanocapsules could have important applications in the food industry as antimicrobial and antioxidant system.	Teixeira et al. (2013)

Table 17.2 Typical Applications of Cyclodextrins in Food Products, Food Supplements and Packaging Technology (*cont.*)

No.	Product Category	Cyclodextrin Type	Research Objects, Target Molecules and Macrocyclic Additive Application	References
13.	Immobilization of β-CD on glass for cholesterol reduction from milk	β-CD	β-Cyclodextrin was converted into β-CD-N_3 by chemical modification and subsequently attached covalently on glass surface by click reaction. β-CD on solid surface was used to make a complexation with cholesterol to remove it from milk. It has been demonstrated that modified surface was used repeatedly for eight cycles and maintained its efficiency, with 68 ± 2% cholesterol reduction.	Tahir and Lee (2013)
14.	Encapsulation of volatiles in nanofibrous polysaccharide membranes for food packaging applications	β-CD	In this research a single-step electrospinning process was applied to a blend of edible carbohydrate polymers (pullulan and β-cyclodextrin) to encapsulate bioactive aroma compounds for a humidity-triggered release. According to reported data the membrane hosts small and homogeneously dispersed crystals of cyclodextrin–aroma complexes that are formed during the electrospinning. With this type of structure, the release of aroma compound is negligible at ambient conditions (23°C and 55% UR) even at high temperature (up to 230°C), and it occurs beyond a given relative humidity threshold (90%), therefore the system demonstrates promising features for the food packaging industry.	Mascheroni et al. (2013)

Table 17.3 Examples of Inclusion Complexes Applications in Cosmetics and Related Products

No.	Product Category	Cyclodextrin Type	Research Objects, Target Molecules and Macrocyclic Additive Application	References
1.	Complexation of estragole with cyclodextrins	α-CD, β-CD, HP-β-CD, randomly methylated-β-CD), low randomly methylated-β-CD, γ-CD	This work was aimed on characterization of inclusion complexes of estragole (as pure compound and as main component of basil and tarragon essential oils) with a number of cyclodextrins. Inclusion complexes formation allowed the controlled release of estragole. Moreover, increased DPPH radical scavenging activity and photostability of target compounds were observed in the presence of macrocycles. These findings suggest that encapsulation with CDs could be an efficient tool to improve the use of estragole and related compounds in aromatherapy, cosmetic, and food fields.	Kfoury et al. (2015)
2.	Monoterpenes complexated with HP-β-CD	HP-β-CD	The results of this research indicated that comlexation of HP-β-CD and eight monoterpenes including eucalyptol, geraniol, limonene, α- and β-pinene, pulegone. and thymol could be a promising strategy to enlarge the application of monoterpens in cosmetic, pharmaceutical, and food industries.	Kfoury et al. (2014)
3.	Cyclodextrins as encapsulation agents for plant bioactive compounds	Native cyclodextrins and their derivatives	This review paper highlighted the use of cyclodextrins as encapsulating agents for bioactive plant molecules in the pharmaceutical and cosmetology fields.	Pinho et al. (2014)
4.	Thermogelling hydrogels of cyclodextrin/poloxamer polypseudorotaxanes as aqueous-based nail lacquers	Methylated-β-CD	This work investigated the use of in situ gelling hydrogels based on polypseudorotaxanes of Pluronic F-127 and partially methylated β-cyclodextrin as aqueous nail lacquers. Particularly, the formulations were tested for their ability to deliver ciclopirox and triamcinolone across human nail plate and bovine hoof. It has been documented that the new polypseudorotaxanes formulation delivered more ciclopirox across human nail than a marketed organic lacquer. This supports the growing hypothesis that aqueous-based nail lacquers represent a superior formulation strategy in nail topical delivery.	Nogueiras-Nieto et al. (2013)

Table 17.3 Examples of Inclusion Complexes Applications in Cosmetics and Related Products (*cont.*)

No.	Product Category	Cyclodextrin Type	Research Objects, Target Molecules and Macrocyclic Additive Application	References
5.	Improved stability of the essential oils components involving CD encapsulation	β-CD	The authors combined molecular encapsulation of the lavender and mint essential oils volatile components by cyclodextrins and sol-gel process. Protected from both the cyclodextrin and silica matrix, the essential oils components became more resistant versus the effects of the environment factors. The authors concluded that the resulted powders can find applications in domains as agriculture, food industries, cosmetics, pharmaceutical and medicine.	Raileanu et al. (2013)
6.	Cyclodextrin stabilized emulsions and cyclodextrinosomes	β-CD, α-CD	This report deals with preparation of o/w emulsions stabilised by microcrystals of cyclodextrin–oil inclusion complexes. The inclusion complexes are formed by threading cyclodextrins from the aqueous phase on*n*-tetradecane or silicone oil molecules from the emulsion drop surface which grow further into microrods and microplatelets depending on the type of cyclodextrin (CD) used. The authors concluded that the CD-stabilized emulsions can be applied in a range of surfactant-free formulations with possible applications in cosmetics, home, and personal care. Cyclodextrinosomes could find applications in pharmaceutical formulations as microencapsulation and drug delivery vehicles.	Mathapa and Paunov (2013)
7.	Application of cyclodextrins for delivery systems in cosmetics	Native cyclodextrins	In this review paper the authors identify and discuss some delivery systems used in consumer health products, also based on cyclodextrins encapsulation for example, to improve the penetration and action of vitamin C into the skin to inhibit melanin formation.	Hougeeir and Kircik (2012)
8.	Cyclodextrin-grafted viscose loaded with aescin formulations for a cosmeto-textile materials	β-CD	The authors developed an efficient synthetic procedure for the preparation of β-cyclodextrin (β-CD)-grafted viscose by means of a two-step ultrasound-assisted reaction. This cosmeto-textile applications can be used in the treatment of chronic venous insufficiency in legs by means of elastic bandages loaded with natural products which possess flebotonic properties (eg, aescin, menthol, *Centella asiatica* and *Ginkgo biloba*). According to presented data the invented cosmeto-textile can be easily recharged and has the strong base characteristics needed for possible industrial production.	Cravotto et al. (2011)

designed for nasal (Cho et al., 2015), parenteral (Vecernyes et al., 2014), skin (Conte et al., 2014) and pulmonary (Sawatdee et al., 2014) drugs delivery as well as oral administration (Mendes et al., 2015). In spite of typical water-based CDs solutions that are used for improving solubility, stability, and drugs bioavailability, there is an increasing interest in designing more sophisticated and complex systems, also involving common solids or new discovered nanomaterials, for example:

- nanoplatforms composed of cyclodextrin/multiwalled carbon nanotubes nanohybrid for guanine-based drugs entrapment and delivery (Innazzo et al., 2014)
- electrospun nanofibrous materials containing naproxen-cyclodextrin complex (Akaduman et al., 2014)
- bioresorbable chlorhexidine delivery systems based on modified paper points (Tabary et al., 2014)
- multifunctional cotton containing silver nanoparticles and cyclodextrin (Hebeish et al., 2014)
- liposomal gels containing glutathione/cyclodextrins complexes for cutaneous administration (Cutrignelli et al., 2014)
- microbeads containing cod-liver oil encircled with natural cyclodextrins (Konrádsdóttir et al., 2014)
- emulgel formulation containing inclusion complex of calcipotriol with cyclodextrin (Badilli et al., 2014)
- fast-dissolving tacrolimus solid dispersion-loaded prolonged release tablet (Cho et al., 2014)
- ocular inserts containing lodocaine HCl/β-CD complex (Shukr, 2014)
- bioadhesive tablets containing cyclodextrin complex of itraconazole for the treatment of vaginal candidiasis (Cevher et al., 2014)

Applications of cyclodextrins driven host–guest complexes in food technology (Table 17.2) are mainly focusing on the packaging techniques involving “smart membranes,” which are designed to protect the food products from the oxygen or for selective extraction of low molecular mass impurities as well as for improving the food properties like color, flavor and overall stability. Particularly, the latest applications demonstrate:

- active packaging involving nanocrystal-reinforced soy protein films and cyclodextrins (Gonzales and Igardzabal, 2015)
- antimicrobial packaging of fresh chicken breast fillets (Higueras et al., 2014)
- preserving natural colorant in yogurt based on inclusion complexes of red bell pepper pigments with β-CD (Gomes et al., 2014)
- complexation of chlorpropham with HP-β-CD and its application in potato sprout inhibition (Huang et al., 2014)

- production of chitosan/cyclodextrin membrane for gallic acid encapsulation and controlled release (Paun et al., 2014)
- inventing of nano-encapsulated black pepper oleoresin using HP-β-CD for antioxidant and antimicrobial applications (Teixeira et al., 2013)
- immobilisation of β-CD on glassfor cholesterol reduction from milk (Tahir and Lee, 2013)
- encapsulation of volatiles in nanofibrous polysaccharide membranes for food packaging applications (Mascheroni et al., 2013)

A relatively low number of experimental papers was recently devoted to cosmetics applications; however, this topic is closely related to pharmaceutical applications and therefore also within this group several novel applications were investigated (Table 17.3). They were mainly focused on the stability improvement of the volatile target components derived from the essential oils. Additionally, applications of inclusion complexes for aqueous-based lacquers, emulsions, and cosmeto-textile materials used in the treatment of chronic venous insufficiency in legs by means of elastic bandages loaded with natural products possessing flebotonic properties were also reported, particularly:

- cyclodextrins as encapsulation agents for plant bioactive compounds (Pinho et al., 2014)
- thermogelling hydrogels of cyclodextrin/poloxamer polypseudorotaxanes as aqueous-based nail lacquers (Nogueiras-Nieto et al., 2013)
- improved stability of the essential oils components involving CD encapsulation (Raileanu et al., 2013)
- cyclodextrin stabilised emulsions and cyclodextrinosomes (Mathapa and Paunov, 2013)
- cyclodextrin-grafted viscose loaded with aescin formulations for cosmeto-textile materials (Cravotto et al., 2011)

8 Conclusions

Although cyclodextrins were discovered over 100 years ago, these relatively simple macrocyclic compounds are still of interest for a number of researchers' communities, particularly: analytical chemists, pharmacists, cosmetologists, and food engineers. This is mainly due to water solubility, lack of toxicity, high thermal stability, and unique physicochemical properties, which allow researchers to modify the solubility, stability, bioavailability, and antioxidant properties of the active components of interest via simple electrostatic interactions. Cyclodextrins-involved host–guest complexation can be fairly well controlled by competitive interactions with a number of low-molecular mass compounds as well as physicochemical parameters like pH, humidity, or temperature. There are an increasing number of nontrivial applications involving smart or active packaging and nanomaterials-driven carriers for drugs delivery through newly invented structures, which are based on cellulose and chitosan polymers in the form of nanocapsules, electrospinning nanofibers, or carbon nanotubes. Researchers are also focusing on the large three-dimensional structures composed of a battery of CDs units, including dendrimers that enable selective encapsulation of chiral active substances, particularly nonpolar chemicals. For that reason cyclodextrins may very likely dominate the future macrocycles applications in the food, pharmaceutics, and cosmetics industries for the foreseeable future.

References

Akaduman, C., Ozguney, I., Kumbasar, E.P.A., 2014. Electrospun thermoplastic polyurethane mats containing naproxen-cyclodextrin inclusion complex. Autex. Res. J. 14 (4), 239–246.

Aldawsari, H., Altaf, A., Banjar, Z., Okubo, M., et al., 2014. Combined use of cyclodextrins and hydroxypropylmethycellulose stearoxy ether (Sangelose®) for the preparation of orally disintegrating tablets of type-2 antidiabetes agent glimepiride. J. Incl. Phenom. Macrocycl. Chem. 80, 61–67.

Ali, M.S., Rub, M.A., Khan, F., Al-Lohedan, H.A., Kabir-ud-Din, 2015. beta-Cyclodextrin-promazine hydrochloride interaction: conductometric and viscometric studies. J. Saudi Chem. Soc. 19, 83–87.

Arriagas, N., Cabral-Marques, H., 2014. Obtetion and characterization of a ciclesonide: methyl-β cyclodextrin complex. J. Incl. Phenom. Macrocycl. Chem. 80, 125–132.

Badilli, U., Amasya, G., Sen, T., Tarmici, S., 2014. Topical emulgel formulation containing inclusion complex of calcipotriol with cyclodextrin. J. Incl. Phenom. Macrocycl. Chem. 78, 249–255.

Baek, J.S., Lim, J.H., Kang, J.S., Shin, S.C., Jung, S.H., Cho, C.W., 2013. Enhanced transdermal drug delivery of zaltoprofen using a novel formulation. Int. J. Pharm. 453, 358–362.

Benesi, H., Hildebrand, J., 1949. A spectrophotometric investigation of the interaction of iodine with aromatic hydrocarbons. J. Am. Chem. Soc. 71, 2703–2707.

Carlotti, M.E., Sapino, S., Ugazio, E., Caron, G., 2011. On the complexation of quercetin with methyl-β-cyclodextrin: photostability and antioxidant studies. J. Incl. Phenom. Macrocycl. Chem. 70, 81–90.

Carmona, T., Marcelo, G., Rinaldi, L., et al., 2015. Soluble cyanine dye/β-cyclodextrin derivatives: potential carries for drug delivery and optical imaging. Dyes Pigments 114, 204–214.

Cevher, E., Acma, A., Sinani, G., et al., 2014. Bioadhesive tablets containing cyclodextrin complex of itraconazole for the treatment of vaginal candidiasia. Int. J. Biol. Macromol. 69, 124–136.

Chandrasekaran, S., Sameena, Y., Enoch, I.V.M.V., 2015. Modulation of the interaction of Coumarin with DNA by b-cyclodextrin. J. Incl. Phenom. Macrocycl. Chem. 81, 225–236.

Chatjigakis, A.K., Donze, C., Coleman, A.W., 1992. Solubility behavior of beta-cyclodextrin in water/cosolvent mixtures. Anal. Chem. 64, 1632–1634.
Chattah, A.K., Mroue, K.H., Pfund, L.Y., et al., 2013. Insights into novel supramolecular complexes of two solid forms of Norfloxacin and β-Cyclodextrin. J. Pharm. Sci. 102, 3717–3724.
Cho, J.H., Kim, Y., Kim, D., et al., 2014. Devolopment of novel fast-dissolving tacrolimus solid dispersion–loaded prolonged release tablet. Eur. J. Pharm. Sci. 54, 1–7.
Cho, W., Kim, M.-S., Jung, M.-S., et al., 2015. Design of salomon calcitonin particles for nasal delivery using spray-drying and novel supercritical fluid-assisted spray-drying processes. Int. J. Pharm. 478, 288–296.
Chun, J.-Y., Jo, Y.-J., Bjrapha, P., Choi, M.-J., Min, S.-G., 2015. Antimicrobial effect of α- or β-cyclodextrin complexes with trans-cinnamaldehyde against *Staphylococcus aureus* and *Escherichia coli*. Dry. Techn. Int. J. 33, 377–383.
Clifton, V.L., Bisits, A., Zarzycki, P.K., 2007. Characterization of human fetal cord blood steroid profiles in relation to fetal sex and mode of delivery using temperature-dependent inclusion chromatography and principal component analysis (PCA). J. Chromatogr. B. 855, 249–254.
Connors, A.K., 1987. Binding Constants. Wiley, New York.
Conte, C., Caldera, F., Catanzano, O., et al., 2014. β-cyclodextrin nanosponges as multifunctional ingredient in water-containing semisolid formulations for skin delivery. J. Pharm. Sci. 103, 3941–3949.
Cramer, F., 1954. Eischlussverbindungen (Inclusion Compounds). Springer-Verlag, Berlin.
Cravotto, G., Beltramo, L., Sapino, S., Binello, A., Carlotti, M.E., 2011. A new cyclodextrin-grafted viscose loaded with aescin formulations for a cosmeto-textile approach to chronic venous insufficiency. J. Mater. Sci: Mater. Med. 22, 2387–2395.
Cutrignelli, A., Lopedota, A., Denora, N., Laquintana, V., et al., 2014. Characterization and release studies of liposomal geles containing glutathione/cyclodextins complexes potentially useful for cutaneous administration. J. Pharm. Sci. 103, 1246–1254.
De Ford, D.D., Hume, D.N., 1951. The determination of consecutive formation constants of complex ions from polarographic data. J. Am. Chem. Soc. 73, 5321–5322.
De Rijke, E., Out, P., Niessen, W.M.A., Ariese, F., Gooijer, C., Brinkman, U.A.Th., 2006. Analytical separation and detection methods for flavonoids. J. Chromatogr. A 1112, 31–63.
De Souza, R.F.V., De Giovani, W.F., 2005. Synthesis spectral and electrochemical properties of Al(III) and Zn(II) complexes with flavonoids. Spectrochim. Acta Part A 61, 1985–1990.
Dias, K., Nikolaou, S., De Giovani, W.F., 2008. Synthesis and spectral investigation of Al(III) catechin/β-cyclodextrin and Al(III) quercetin/β-cyclodextrin inclusion compounds. Spectrochim. Acta A 70, 154–161.
Dominguez-Canedo, I.L., Beristain-Guevara, C.I., Diaz-Sobac, R., Vázquez-Luna, A., 2015. Degradation of carotenoids and capsaicin in the molecular inclusion complex of habanero chili oleoresin (*Capsicum chinense*) in beta-cyclodextrin. CyTA J. Food 13, 151–158.
Duchene, D., 1987. Cyclodextrins and their industrial uses. Editions de Sante, Paris.
Ebrahimzadeh, M.A., Nabavi, S.M., Nabavi, S.F., 2009. Correlation between the in vitro iron chelating activity and polyphenol and flavonoid contens of some medicinal plants. Pakistan J. Biol. Sci. 12, 934–938.
Ferreira, F.R., da Silva, E.G., De Leob, L.P.M., Calvob, E.J., Bentoa, E.S., Goularta, M.O.F., Abreua, F.C., 2010. Electrochemical investigations into host–guest interactions of a natural antioxidant compound with β-cyclodextrin. Electrochim. Acta 56, 797–803.

Ferreira, I., Rocha, S., Coelho, M., 2007. Encapsulation of antioxidants by spray drying. Chem. Eng. Trans. 11, 713–717.
Filipský, T., Říha, M., Hrdina, R., Vávrova, K., Mladěnka, P., 2013. Mathematical calculations of iron complex stechiometry by direct UV-Vis spectrophotometry. Bioorg. Chem. 49, 1–8.
Folch-Cano, C., Jullian, C., Speisky, H., Olea-Azar, C., 2010. Antioxidant activity of inclusion complexes of tea catechins with β-cyclodextrins by ORAC assays. Food Res. Int. 43, 2039–2044.
Freudenberg, K., Blomquist, G., Ewald, L., So, K., 1936. Hydrolyse und Acetolyse der arke und der Schardinger-Dextrine. Ber. Dtsch. Chem. Ges. 69, 1258–1266.
Freudenberg, K., Cramer, F., 1948. Die Konstitution der Schardinger-Dextrine α, β und γ. Z. Naturforsch. 3b, 464.
French, D., 1957. The Schardinger dextrins. Adv. Carbohydr. Chem. 12, 189–260.
Georgieva, S., Nedeltcheva, T., Nikolova, L., 2010. Study of labile complexes by differential pulse polarography. J. Uni. Chem. Technol. Metall. 45, 201–206.
Gharibzahedi, S.M.T., Razavi, S.H., Mousavi, M., 2014. Characterizing the natural canthaxanthin/2-hydroxypropyl-β-cyclodextrin inclusion complex. Carbohyd. Polym. 101, 1147–1153.
Glisoni, R.J., Garía-Fernández, M.J., Pino, M., Moglioni, G.G.A.G., et al., 2013. β-Cyclodextrin hydrogels for the ocular release of antibacterial thiosemicarbazones. Carbohyd. Polym. 93, 449–457.
Głód, B.K., Czapski, G.A., Haddad, P.R., 2000. Application of high-performance liquid chromatography to the investigation of free radical reactions in biological systems, TRAC. Trends Anal. Chem. 19, 492–497.
Głód, B.K., Kiersztyn, I., Piszcz, P., 2014. Total antioxidant potential assay with cyclic voltammetry and/or differential pulse voltammetry measurements. J. Electroanal. Chem. 719C, 24–29.
Głód, B.K., Piszcz, P., Kiersztyn, I., Lamert, A., Zarzycki, P., 2009. Application of HPLC to the determination of free radicals, antioxidants, and total antioxidant potential. Cam. Sep. 1, 41–66.
Głód, B.K., Piszcz, P., Czajka, K., Zarzycki, P.K., 2011. A new total antioxidant potential measurements using RP-HPLC assay with fluorescence detection. J. Chromatogr. Sci. 49, 401–404.
Głód, B.K., Piszcz, P., Czajka, J., Zarzycki, P.K., 2012a. Evaluation of total antioxidant potential of selected biogenic polyamines, nonalcoholic drinks, and alcoholic beverages using improved RP-HPLC assay involving fluorescence detection. Food Chem. 131, 1026–1029.
Głód, B.K., Piszcz, P., Beta, A., Zarzycki, P.K., 2012b. RP-HPLC, with fluorescence detection, assay for the determination of total antioxidant potential (TAP). J. Liquid Chromatogr. Rel. Technolog. 35, 1194–1201.
Gomes, L.M.M., Petito, N., Costa, V.G., et al., 2014. Inclusion complexes of red bell pepper pigments with β-cyclodextrin: preparation, characterization, and application as natural colorant in yogurt. Food Chem. 148, 428–436.
Gonzales, A., Igardzabal, C.I.A., 2015. Nanocrystal-reinforced soy protein films and their application as active packaging. Food Hydrocolloid. 43, 777–784.
Ha, W., Kang, Y., Peng, S.L., Ding, L.S., Zhang, S., Li, B.J., 2013. Vesicular gold assemblies based on host–guest inclusion and its controllable release of doxorubicin. Nanotechnology 24, 1–8.
Halliwell, B., 2012. Free radicals and antioxidants: updating a personal view. Nutr. Rev. 70, 257–265.
Hebeish, A., El-Shafei, S., Sharaf, S., Zaghloul, 2014. In situ formation of silver nanoparticles for multifunctional cotton containing cyclodextrin. Carbohyd. Polym. 103, 442–447.
Hedges, A.R., 1998. Industrial applications of cyclodextrins. Chem. Rev. 98, 2035–2044.

Higueras, L., López-Carballo, G., Hernández-Munoz, P., Catalá, R., et al., 2014. Antimicrobial packaging of chicken fillets based on the release of carvacrol from chitosan/cyclodextrin films. Int. J. Food Microbiol. 188, 53–59.
Hougeeir, F.G., Kircik, L., 2012. A revive of delivery systems in cosmetics. Dermatol. Ther. 25, 234–237.
Huang, Z., Tian, S., Ge, X., et al., 2014. Complexation of chlorpropham with hydroxypropyl-β-cyclodextrin and its application in potato sprout inhibition. Carbohyd. Polym. 107, 241–246.
Ikeda, K., Uekama, K., Otagiri, M., 1975. Inclusion complexes of β-cyclodextrin with antiinflammatory drugs fenamates in aqueous solution. Chem. Pharm. Bull. 23, 201–208.
Innazzo, D., Mazzagilia, A., Scala, A., et al., 2014. β-cyclodextrin-grafted on multiwalled carbon nanotubes as versatile nanoplatform for entrapment of guanine-based drugs. Colloid. Surf. B 123, 264–270.
Ioele, G., De Luca, J., Tavano, L., Ragno, G., 2014. The difficulties for a photolabile drug in topical formulatios: the case of diclofenac. Int. J. Pharm. 465, 284–290.
Jeandet, P., Clement, C., Courot, E., 2014. Resveratrol production at large scale using plant cell suspension. Eng. Life Sci. 14, 622–632.
Kfoury, M., Auezova, L., et al., 2014. Investigation of monoterpens complexation with hydroxypropyl-β-cyclodextrin. J. Incl. Phenom. Macrocycl. Chem. 80, 51–60.
Kfoury, M., Azezova, L., Ruellan, S., Greige-Gerges, H., Fourmentin, S., 2015. Complexation of estragole as pure compound and main component of basil and tarragon essential oils with cyclodextrins. Carbohyd. Polym. 118, 156–164.
Khodavedi, E., Heidari, Z., et al., 2014. Injectable supramolecular hydrogel from insulin-loaded triblock PCL-PEG-PCL copolymer and γ-cyclodextrin with sustained-release property. AAPS Pharm. Sci. Tech. 16 (1), 140–149.
Kordopati, G.G., Tselios, T.V., Kellici, T., et al.,2015. A novel synthetic luteinizing hormone-releasing hormone (LHRH) analogue coupled with modified (-cyclodextrin: insight into its intramolecular interactions. Biochim. Biophys. Acta. 1850, 159–168.
Krishnaswamy, K., Orsat, V., Thangavel, K., 2012. Synthesis and characterization of nanoencapsulated catechin by molecular inclusion with beta-cyclodextrin. J. Food Eng. 111, 255–264.
Kolanowski, W., Laufenberg, G., Kunz, B., 2004. Fish oil stabilisation by microencapsulation with modified cellulose. Int. J. Food Sci. Nutr. 55, 333–343.
Konrádsdóttir, F., Geirsson, T., Halldórsson, A., et al., 2014. Physicochemical characterization and stability of microbeades containing cod liver oil encircled with natural cyclodextrins. J. Incl. Phenom. Macrocycl. Chem. 78, 485–499.
Lamparczyk, H., 1985. The role of electrostatic interactions in the retention index concept: universal interaction indices for GLC, HPLC and TLC. Chromatographia 20, 283–288.
Lamparczyk, H., Atomura, M., Jinno, K., 1987. Qualitative description of dispersive and inductive electrostatic interactions in reversed-phase liquid chromatography. Chromatographia 23, 752–759.
Lamparczyk, H., Zarzycki, P.K., Nowakowska, J., 1994. Effect of temperature on separation of norgestrel enantiomers by high-performance liquid chromatography. J. Chromatogr. A 668, 413–417.
Lamparczyk, H., Zarzycki, P.K., 1995. Effect of temperature on separation of estradiol stereoisomers and equilin by liquid chromatography using mobile phases modified with beta-cyclodextrin. J. Pharm. Biomed. Anal. 13, 543–549.
Lehn, J.M., 1995. Supramolecular Chemistry: Concepts and Perspectives. VCH, Weinheim.

Leonardi, D., Bombardiere, M.E., Salomon, C.J., 2013. Effects of benznidazole: cyclodextrin complexes on the drug bioavailability upon oral administration to rats. Int. J. Biol. Macromol. 62, 543–548.

Li, Z., Chen, S., Gu, Z., et al., 2014. Alpha-cyclodextrin: enzymatic production and food applications. Trends Food Sci. Tech. 35, 151–160.

Liu, M., Dong, L., Chen, A., Zheng, Y., Sun, D., Wang, X., Wang, B., 2013. Inclusion complexes of qercetin with three β-cyclodextrins derivatives at physiological pH: spectroscopic study and antioxidant activity. Spectrochim. Acta A 115, 854–860.

Malinka, W., Kaczmarz, M., Filipek, B., Sapa, J., Glod, B.K., 2003. Preparation of novel derivatives of pyridothiazine-1,1-dioxide and their CNS and antioxidant properties. Farmaco 57, 737–746.

Mascheroni, E., Fuenmayor, C.A., Cosio, M.S., Di Silvetro, G., Piergiovanni, L., Mannino, S., Schiraldi, A., 2013. Encapsulation of volatiles in nanofibrous polysaccharide membranes for humidity-triggered release. Carbohyd. Polym. 98, 17–25.

Mathapa, B.G., Paunov, V.N., 2013. Cyclodextrin stabilized emulsions and cyclodextrinosomes. Phys. Chem. Chem. Phys. 15, 17903–17914.

Mendes, C., Wiemes, B.P., Buttchevitz, A., et al., 2015. Investigation of beta-cyclodextrin-norfloxacin inclusion complexes. Part 1. Preparation, physiochemical and microbiological characterization. Expert Rev. Anti-Infe. 13 (1), 119–129.

Milanovic, J., Manojlovic, V., Levic, S., Rajic, N., Nedovic, V., Bugarski, B., 2010. Microencapsulation of flavors in carnauba wax. Sensor 10, 901–912.

Mora, M.J., Petiti, J.P., Longhi, M.R., et al., 2015. Intestinal uptake and toxicity evaluation of acetazolamide and its multicomponent complexes with hydroxypropyl-β-cyclodextrin in rats. Int. J. Pharm. 478, 258–267.

Moya-Ortega, M.D., Alves, T.F.G., Alvarez-Lorenzo, C., et al., 2013. Dexamethasone eye drops containing γ-cyclodextrin-based nanogels. Int. J. Pharm. 441, 507–515.

Nagai, N., Ito, Y., 2014. Therapeutic effects of gel ointments containing tranilast nanoparticles on paw edema in adjuvant-induced arthritis rats. Biol. Pharm. Bull. 37, 96–104.

Namazi, H., Heydari, A., 2014. Synthesis of β-cyclodextrin-based dendrimer as a novel encapsulation agent. Polym Int. 63, 1447–1455.

Naseri, G., Ashnagar, A., Husseini, F., 2007. Study of the inclusion complexation of piroxicam-β-cyclodextrin and determination of the stability constant (K) by UV-visible spectroscopy. Sci. Iran. 14, 308–315.

Nikolai, S., Huebbe, P., Metges, C., Schloesser, A., 2014. R-α lipoic acid γ-cyclodextrin complex increases energy expenditure: a 4-month feeding study in mice. Nutrition 30, 228–233.

Nogueiras-Nieto, L., Delgado-Charro, M.B., Otero-Espinar, F.J., 2013. Thermogelling hydrogels of cyclodextrin/poloxamer polypseudorotaxanes as aqueous-based nail lacquers: application to the delivery of triamcinolone acetonide and ciclopirox olamine. Eur. J. Pharm. Biopharm. 83, 370–377.

Oliveri, V., Attanasio, F., Puglisi, A., et al., 2014. Multifunctional 8-hydroxyquinoline-appended cyclodextrins as new inhibitors of metal-induced protein aggregation. Chem. Eur. J. 20, 8954–8964.

Paun, G., Neagu, E., Tache, A., Radu, G.L., 2014. New type of chitosan/2-hydroxypropyl-β-cyclodextrin composide membrane for gallic acid encapsulation and controlled release. Acta Chim. Slov. 61, 27–33.

Pinho, E., Grootveld, M., Soares, G., Henriques, M., 2014. Cyclodextrins as encapsulation agents for plant bioactive compounds. Carbohyd. Polym. 101, 121–135.

Piszcz, P., Woźniak, M., Asztemborska, M., Głód, B.K., 2014a. Comparative analysis of antioxidative activity of flavonoids using HPLC-ED and photometric assays. Food Anal. Methods 7, 1474–1480.

Piszcz, P., Żurawski, K., Głód, B.K., 2014b. Application of HPLC to study the reaction of free radicals with antioxidants and/or toxins. J. Chem. Article ID: 385908, 6 pages, doi:10.1155/2014/385908.
Poorghorban, M., Das, U., Alaidi, O., Chitanda, J.M., Michel, D., Dimmock, J., Verrall, R., Grochulski, P., Badea, I., 2015. Characterization of the host–guest complex of a curcumin analog with beta-cyclodextrin and beta-cyclodextrin-gemini surfactant and evaluation of its anticancer activity. Int. J. Nanomed. 10, 503–515.
Poór, M., Kunsagi-Mate, S., et al., 2015. Interaction of ochratoxin A with quaternary ammonium beta-cyclodextrin. Food Chem. 172, 143–149.
Priya, A.S., Sivakamavalli, J., Vaseeharan, B., Stalin, T., 2013. Improvement on dissolution rate of inclusion complex of Rifabutin drug with β-cyclodextrin. Int. J. Biol. Macromol. 62, 472–480.
Quirós-Sauceda, A.E., Ayala-Zavala, J.F., Olivas, G.I., González-Aguilar, G.A., 2014. Edible coatings as encapsulating matrices for bioactive compounds: a review. J. Food Sci. Technol. 51, 1674–1685.
Raileanu, M., Todan, L., Voicescu, M., et al., 2013. A way for improving the stability of the essential oils in an environmental friendly formulation. Mater. Sci. Eng. C. 33, 3281–3288.
Ramos-Campos, E.V., de Oliveira, J.L., Fraceto, L.F., Singh, B., 2015. Polysaccharides as safer release systems for agrochemicals. Agron. Sustain. Dev. 35, 47–66.
Rosseels, M.L., Delaunois, A.G., Hanon, E., Guillaume, P., et al., 2013. Hydroxypropyl-β-cyclodextrin impacts renal and systemic hemodynamics in the anesthetized dog. Regul. Toxicol. Pharm. 67, 351–359.
Sangwai, M., Vavia, P., 2013. Amorphous ternary cyclodextrin nanocomposites of telmisartan for oral drug delivery: improved solubility and reduced pharmacokinetic variability. Int. J. Pharm. 453, 423–432.
Sawatdee, S., Hiranphan, P., Laphanayos, K., Srichana, T., 2014. Evaluation of sildenafil pressurized metered dose inhalers as a vasodilator in umbilical blood vessels of chicken egg embryos. Eur. J. Pharm. Biopharm. 86, 90–97.
Schardinger, F., 1911. Bildung kristallisierter Polysaccharide (Dextrine) aus Starke kleisterdurch Mikrobien. Zentralbl. Bakteriol. Parasintenk. 29, 188–197.
Scott, R.L., 1956. Some comments on the Benesi-Hildebrand equation. Recueil 75, 787–789.
Sheng, Q.-L., Luo, K., Liu, R.-X., Zheng, J.-B., 2012. Direct electrochemistry of glucose oxidase immobilized at a novel composite material beta-cyclodextrin/poly(4-aminothiophenol)/Au nanoparticle modified electrode. J. Chinese Chem. Soc. 59, 154–160.
Shukr, M., 2014. Formulation, in vitro and in vivo evaluation of lidocaine HCl ocular inserts for topical ocular anesthesia. Arch. Pharm. Res. 37, 882–889.
Siva, S., Thulasidhasan, J., Rajendiran, N., 2013. Host–guest inclusion complex of propafenone hydrochloride with α- and β-cyclodextrins: spectral and molecular modeling studies. Spectrochim. Acta A-M. 115, 559–567.
Silva, F., Figueiras, A., Gallardo, E., et al., 2014. Strategies to improve the solubility and stability of stilbene antioxidant: a comparative study between cyclodextrins and bile acids. Food Chem. 145, 115–125.
Sokolová, R., Ramešová, Š., Degano, I., Hromado, M., Gál, M., Žabka, J., 2012. The oxidation of natural flavonoid quercetin. Chem. Commun. 48, 3433–3435.
Soottitantawat, A., Yoshii, H., Furuta, T., Ohgawara, M., Forssell, P., Partanen, R., Poutanen, K., Linko, P., 2004. Effect of water activity on the release characteristics and oxidative stability of d-limonene encapsulated by spray drying. J. Agric. Food Chem. 52, 1269–1276.
Soumitra, H., Suresh, K.G., 2015. Physicochemical properties of inclusion complexes of sanguinarine with natural cyclodextrins: spectroscopy, calorimetry, and NMR studies. RSC Adv. 5, 1873–1882.

Sun, Y., Du, L., Liu, Y., Li, X., et al., 2014. Transdermal delivery of the in situ hydrogels of curcumin and its inclusion complexes of hydroxypropyl-β-cyclodextrin for melanoma treatment. Int. J. Pharm. 469, 31–39.

Sybilska, D., Asztemborska, M., Bielejewska, A., Kowalczyk, J., Dodziuk, H., Duszczyk, K., Lamparczyk, H., Zarzycki, P., 1993. Chromatographic studies on inclusion isomeric dimethylnaphthalenes by beta- and gamma-cyclodextrin. Chromatographia 35, 637–642.

Szejtli, J., 1982. Cyclodextrins and Their Inclusion Complexes. Akademiai Kiado, Budapest.

Szejtli, J., 1983. Dimethyl-beta-cyclodextrin as parenteral drug carrier. J. Inclusion Phenom. 1, 135–150.

Szejtli, J., 1998. Introduction and general overview of cyclodextrin chemistry. Chem. Rev. 98, 1743–1753.

Tabary, N., Chai, F., Blanchemain, N., Neut, C., et al., 2014. A chlorhexidine-loaded biodegradable cellulosic device for periodontal pockets treatment. Acta Biomater. 10, 318–329.

Tahir, M.N., Lee, Y., 2013. Immobilisation of β-cyclodextrin on glass: characterization and application for cholesterol reduction from milk. Food Chem. 139, 475–481.

Teixeira, B.N., Ozdemir, N., Hill, L.E., Gomes, C.L., 2013. Synthesis and characterization of nano-encapsulated black pepper oleoresin using hydroxypropyl beta-cyclodextrin fof antioxidant and antimicrobial applications. J. Food Sci. 78 (12), 1913–1920.

Tóth, G., Mohácsi, R., Rácz, A., et al., 2013. Equilibrium and structural characterization of ofloxacin-cyclodextrin complexation. J. Incl. Phenom. Macrocycl. Chem. 77, 291–300.

Uekama, K., Hirayama, F., Irie, T., 1998. Cyclodextrin drug carrier systems. Chem. Rev. 98, 2045–2076.

Vecernyes, M., Fenyvesi, F., et al., 2014. Cyclodextrins, blood-brain barrier, and treatment of neurological disease. Arch. Med. Res. 45, 711–729.

Venkatesh, G., Sivasankar, T., Karthick, M., Rajendiran, N., 2013. Inclusion complexes of sulphanilamide drugs and β-cyclodextrin: a theoretical approach. J. Incl. Phenom. Macrocycl. Chem. 77, 309–318.

Vilanova, N., Solans, C., 2015. Vitamin A palmitate-beta-cyclodextrin inclusion complexes: characterization, protection, and emulsification properties. Food Chem. 175, 529–535.

Villiers, A., 1891. Fermentation of starch by the butyric ferment. Compt. Rend. 112, 536–538.

Wang, D., Li, H., Gu, J., Guo, T., Yang, S., Guo, Z., Zhang, X., Zhu, W., Zhang, J., 2013. Ternary system of dihydroartemisinin with hydroxypropyl-β-cyclodextrin and lecithin: simultaneous enhancement of drug solubility and stability in aqueous solutions. J. Pharmaceut. Biomed. 83, 141–148.

Wantusiak, P., Głód, B.K., 2012. Application of UV detection in HPLC in the total antioxidant potential assay. Cent. Eur. J. Chem. 10, 1786–1790.

Zarzycki, P.K., Nowakowska, J., Chmielewska, A., Lamparczyk, H., 1995. Retention properties of cyclodextrins in RP-HPTLC. J. Planar. Chromatogr. 8, 227–231.

Zarzycki, P.K., Lamparczyk, H., 1996. A simple experiment demonstrating the temperature effect in supramolecular chemistry. J. Chem. Educ. 73, 459–460.

Zarzycki, P.K., Nowakowska, J., Chmielewska, A., Wierzbowska, M., Lamparczyk, H., 1997. Thermodynamic study of retention of selected macrocycles using RP-HPTLC plates and methanol/water mobile phases. J. Chromatogr A. 787, 227–233.

Zarzycki, P.K., Lamparczyk, H., 1998. The equilibrium constant of beta-cyclodextrin-phenolphthalein complex: influence of temperature and tetrahydrofuran addition. J. Pharm. Biomed. Anal. 18, 165–170.

Zarzycki, P.K., Wierzbowska, M., Nowakowska, J., Chmielewska, A., Lamparczyk, H., 1999. Interactions between native cyclodextrins and n-alcohols studied using thermostated thin-layer chromatography. J. Chromatogr. A 839, 149–156.
Zarzycki, P.K., Smith, R., 2001. Separation of steroids using temperature-dependent inclusion chromatography. J. Chromatogr. A 912, 45–52.
Zarzycki, P.K., Włodarczyk, E., Lou, D.W., Jinno, K., 2006. Evaluation of methanol-water and acetonitrile-water binary mixtures as the elements for temperature-dependent inclusion chromatography. Anal. Sci. 22, 453–456.
Zarzycki, P.K., Ohta, H., Saito, Y., Jinno, K., 2008. Interaction of native α-cyclodextrin, β-cyclodextrin and γ-cyclodextrin and their hydroxypropyl derivatives with selected organic low molecular mass compounds at elevated and subambient temperature under RP-HPLC conditions. Anal. Bioanal. Chem. 391, 2793–2801.
Zarzycki, P.K., Włodarczyk, E., Baran, M.J., 2009. Determination of endocrine disrupting compounds using temperature-dependent inclusion chromatography I. Optimization of separation protocol. J. Chromatogr. A 1216, 7602–7611.
Zhang, J.-Q., Wu, D., Jiang, K.-M., et al., 2015. Preparation, spectroscopy, and molecular modeling studies of the inclusion complex of cordycepin with cyclodextrin. Carbohyd. Res. 406, 55–64.
Zhao, F., Yin, H., Li, J., 2014. Supramolecular self-assembly forming a multifunctional synergistic system for targeted codelivery of gene and drug. Biomaterials 35, 1050–1062.
Zhou, Q., Zhong, L., Wei, X., Chou, G., Wang, Z., 2013. Baicalein and hydroxypropyl-γ-cyclodextrincomplex in poloxamer thermal sensitive hydrogel for vaginal administration. Int. J. Pharm. 454, 125–134.
Zhuang, Z., Huang, L., Chen, Z., 2015. Effects of cyclodextrin on the morphology and reactivity of iron-based nanoparticles using Eucalyptus leaf extract. Ind. Crop. Prod. 69, 308–313.

18

NANOENCAPSULATION OF FLAVORS AND AROMAS BY CYCLODEXTRINS

Eva Fenyvesi, Lajos Szente

CycloLab Cyclodextrin Research & Development Laboratory Ltd., Budapest, Hungary

1 Introduction

Cyclodextrins (CDs) are cyclic oligosaccharides produced from starch by fermentation. The mixture of dextrins obtained by hydrolysis of starch is treated by a special enzyme, CD glycosyl transferase (CGTase) able to catalyze the cyclization of linear maltooligosacharides to yield CDs (Szejtli, 1982; Schmid, 1996). The enzymatic conversion results in a so-called "conversion mixture" (CMx) containing cyclic and acyclic dextrins. The acyclic dextrins are called also linear dextrins and maltodextrins. They are glucose-1,4 oligomers of various length, consisting of usually 3–20 glucose units. The cyclodextrins are separated from this mixture by selective precipitation/complexation (Fig. 18.1).

Cyclodextrins are natural compounds produced by some soil living bacteria (eg, *B. macerans, B. subtilis, B. coagulans*) having CGTase enzyme. For these microorganisms the degradation of CDs is one of the consecutive reactions in the metabolism of starch: the formation, uptake, and intracellular degradation of CDs is a beneficial starch-degradation pathway for bacteria possessing both CD-forming and CD-degrading enzymes (Usanov et al., 1990; Pocsi, 1999).

CDs are glucose oligomers similar to amylose in starch and contain at least 6 glucopyranose units. The three major CDs, α-, β-, and γ-CDs (ACD, BCD, and GCD) are the smallest members of this family and consist of 6, 7, and 8 glucopyranose units, respectively (Fig. 18.2). These CDs are produced in thousands of tons yearly while the large-ring CDs are still of academic interest and are prepared in small quantities as fine chemicals.

Encapsulations. http://dx.doi.org/10.1016/B978-0-12-804307-3.00018-1

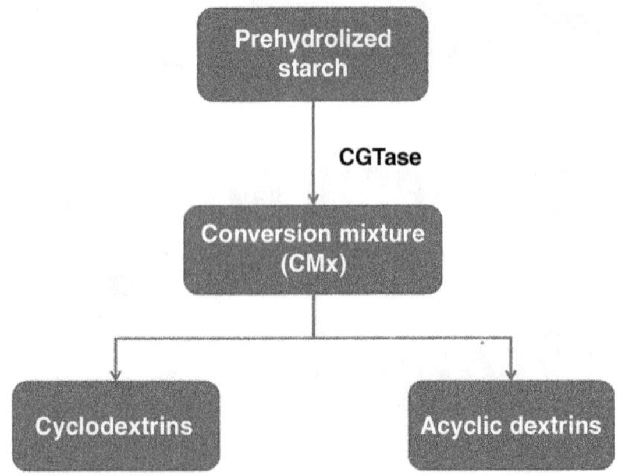

Figure 18.1. Flow diagram of cyclodextrin production.

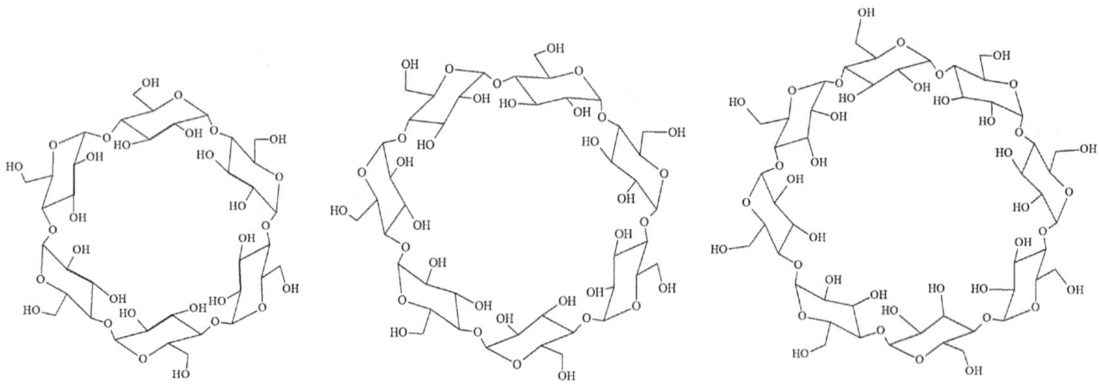

Figure 18.2. Chemical structure of α-, β- and γ-cyclodextrins.

The arrangement of the glucose units gives conical shape to the CDs, forming a cavity. All the hydroxyl moieties are located at both edges facing outside (the primary hydroxyls at the narrower primary face, the secondary hydroxyls at the broader secondary face) and provide a polar surface to the molecule while the cavity is less hydrophilic because of the glucosidic bonds. Owing to the peculiar structure of CDs (hydrophilic outside, less hydrophilic inside the cavity) they are able to include various organic molecules and form inclusion complexes. The driving force is the replacement of high-energy water molecules in the cavity of the host CD with a hydrophobic "guest." The prerequisite of complex formation is the geometrical fit similarly to key and hole. ACD, BCD, and GCD differ in their size (0.5–0.9 nm in diameter), which gives a choice to select the most proper host for a given guest. The stoichiometry of the complex is usually 1:1; that is why this process is called also molecular encapsulation. The larger guest molecules can interact with two or more CDs, and it can happen that two of the smaller guests are included into one (larger) cavity. As both the empty CDs and the complexes can aggregate different stoichiometries can be also observed (Loftsson et al., 2002).

The included guest usually shows enhanced solubility because of the hydrophilic carbohydrate shell. It is protected against hydrolysis, and light- or heat-induced degradation if the sensitive part of the molecule is included. The opposite effects (catalytic decomposition) may occur when the sensitive part of the molecule is protruding from the cavity. Also the volatility of the compounds is decreased. By complexation the bitter taste and malodor of compounds can be reduced. The versatile applications including pharmaceutical, food, cosmetic, environmental, agricultural, and other fields are based on these effects of inclusion complex formation. These effects are utilized in the flavor encapsulation by cyclodextrins, too.

2 History of Flavor Encapsulation by Cyclodextrins

Cyclodextrins were discovered by chance by the French pharmacist Villiers (1854–1932), who studied the bacterial decomposition (putrefaction) of starch (Crini, 2014). It was Franz Shardinger (1853–1920) working in the Food Research Institute in Vienna who could isolate them in a yield enough for detailed studies. He isolated also *Bacillus macerans*, the microbes producing the crystalline α- and β-dextrins, discovered the complex forming ability and postulated their cyclic structure. As his findings were fundamental on these materials they were called Schardinger dextrins in his honor.

The most important milestones of cyclodextrin research were the discovery of γ-dextrin (Karl Freudenberg, 1886–1983), proving the cyclic structure (Dexter French, 1918–1981), and recognizing the main effects of the inclusion complex formation (Friedrich Cramer, 1923–2003). At the beginning of the 1970s József Szejtli (1931–2004) proved that the CDs were not toxic and initiated the industrial production. In a few years the production was started not only in Hungary, but also in Germany, France, and Japan.

The global market of CDs is continuously increasing. The production was estimated to reach 250,000 tons per year with Japan sharing the 40% of it (Zhiyuan Bio-tech, 2013). Although this estimation seems to be exaggerating, it should be around several hundred thousand tons (Fenyvesi et al., 2015). In Asia, China, South Korea, and India also produce CDs. In China there are around 50 companies producing various CDs; 20 of them contribute with more than 1000 tons per year. In Japan nearly 90% of CDs were used for foods in 1988 (Hashimoto, 1988), in the United States 14% of CD production was utilized by the food industry in 2008 (Frost and Sullivan, 2008).

Rogers and Whaley patented stabilization of fruit aromas by a mixture of cyclic and acyclic dextrins Rogers and Whaley (1962). According to the claims, starch is hydrolyzed by α-amylase to get a mixture of acyclic dextrins that, however, similarly to amylose have helical structure and are able to include smaller components of aromas forming inclusion complexes as the CDs do. A cyclic dextrin (α-, β-, γ-, δ- and/or ε-cyclodextrin) is added and a very versatile host mixture is obtained: the various components of food flavors and aromas can find their proper counterpart easily. In this way the taste of orange concentrate was improved by orange aroma encapsulated by ACD and hydrolyzed starch to get a drink which closely resembles the taste of fresh orange juice. D-limonene, menthol, and shiitake flavor, respectively, were stabilized with a mixture of BCD and maltodextrins (Furuta et al., 1994; Liu et al., 2000; Shiga et al., 2004).

Also amylose (unlike to amylopectin) can bind some flavor components (volatile aliphatic alcohols, aldehydes, ketones, esters, amines of some aromatic and heteroaromatic compounds) via H-bonding and inclusion (Kuge and Takeo, 1968; Maier and Bauer, 1972; Wulff et al., 2005). Enhanced solubility of flavor compounds was obtained with high-amylose maize starch via inclusion complex formation (Tapanapunnitkul et al., 2008).

Most flavors are sensitive to oxygen, heat, and light. Complexation can protect them from these harmful effects. Szejtli in his early review showed that 90% less oxygen was used by anethole (one of the main components of anise and fennel oils) in complex form compared to the uncomplexed compound (Szejtli, 1977). Also the volatility is reduced significantly: while the uncomplexed anethole is fast evaporated, no detectable anethole can be found in the head space of the complex as long as it is stored in solid, dry state. The unpleasant odor of sweet potato juice was reduced by applying maltosyl CDs that bind the flavor components and reduces their volatility (Tamaki et al., 2007). Odorless garlic powder was prepared (Kawashima, 1986).

The extracts of spices, such as garlic, onion, dill, and caraway were stabilized by CD complexation (Lindner et al., 1981). The aroma content hardly decreased after 10 years of storage under ambient conditions (Szente and Szejtli, 1988b).

Fig. 18.3 shows the storage stability of some essential oil/BCD complexes after 14 years of storage (Szente and Szejtli, 2004).

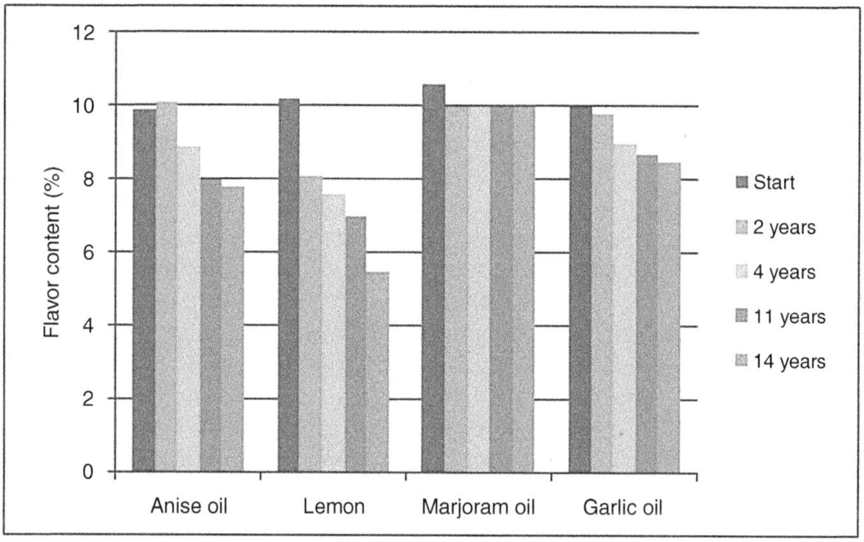

Figure 18.3. Essential oil content in BCD complexes stored for years under ambient conditions.

The flavors are usually multicomponent systems. For instance, about 950 individual components in coffee flavor and aroma have been identified (Yeretzian et al., 2003). Natural and synthetic coffee flavors were successfully stabilized against heat and light by BCD (Szente and Szejtli, 1986). Upon contact with water the flavor substances were released immediately. The transformation of greasy, oily, or liquid coffee aroma concentrates into a microcrystalline stable inclusion complex may have practical importance as an additive to enhance the sensory properties and quality of instant coffee products. GCD is suitable for stabilization of several coffee flavor components (Schmid et al., 1995). In Japan instant coffee with additional coffee flavor stabilized by CD complexation has been already on the market (Hashimoto, 2002).

Not only the flavors but also the antioxidants in coffee can be complexed by CD resulting in enhanced antioxidant effect and decreased bitter taste (Zhao et al., 2010, 2011; Szejtli and Szente, 2005).

Gray and Roberts (1970) studied the volatile food aromas in the presence of various aroma-adsorbing materials including BCD. Among the aldehydes, ketones, alcohols, amines, and sulfides from various substrates especially the amines (triethylamine and ethylamine) showed preferential sorption on BCD.

The volatility decreasing effect of complexation was reported using terpenes as model flavors by Reineccius and Risch (1986). Later on these authors published several books on flavor encapsulation comparing various techniques: such as extrusion, coacervation, microencapsulation, and molecular inclusion by CDs (Risch and Reineccius, 1988, 1995).

The stability of essential oils, such as lemon, orange, hop, and chamomile oil was improved by CD complexation (Thoss et al., 1994). Candies with maltosyl CD-stabilized strawberry flavor were patented (Takaku et al., 1989).

The hygroscopicity of freeze-dried pineapple juices can be reduced by applying BCD as additive (Phanindrakumar et al., 2005).

Szejtli (1982) summarized the advantages of molecular encapsulation of flavors and aromas in various fields of utilization:

- In households the simpler processing (no need to bother with plants, for example, with chopping onion), extended shelf life and reliable quality have to be mentioned. Furthermore, the aroma powders or tablets don't require large room for storing.
- In the catering, trade work and storage room can be saved, transport is simple, loss on storage is low, the choice is high independent of the season and location.
- In dietetics and hospitals, the main advantage is that the taste can be provided without the fibrous plant components that might irritate the digestion system of the patients. Delicious dishes can be prepared even for those on diets.
- In canned meat and food, the standard composition and low risk of microbial contamination make the encapsulated flavors and aromas an attractive choice.
- Flavoring animal feed and fodder helps maintain consistent feed consumption, maximizing feed intake and growth of the animals. Doing this with molecularly encapsulated flavors, which are powders with long shelf life, instead of solid or liquid plant extracts is not only more convenient but also easier to standardize.

The exponential increase of the publications related to cyclodextrin encapsulation of flavors and aromas for food industry shows the high potential of this technology (Fig. 18.4). The main advantages of the complexation: decreased volatility, improved stability against oxygen, light and hydrolysis, elongated shelf-life, controlled release, and so forth have been shown in various reviews (Szejtli et al., 1979; Szente and Szejtli, 1988a; Szente et al., 1988b; Hedges et al., 1995; Bhandari et al., 2001; Furuta et al., 2001; Szente and Szejtli, 2004; Cravotto et al., 2006; Hashimoto, 2008; Astray et al., 2009; Marques, 2010; Ciobanu et al., 2013; Martina et al., 2013, Martina and Cravotto, 2015).

In spite of the abundant literature on flavor and aroma complexation the studies are still continued with various compounds and various techniques. For instance, a recent paper gives results

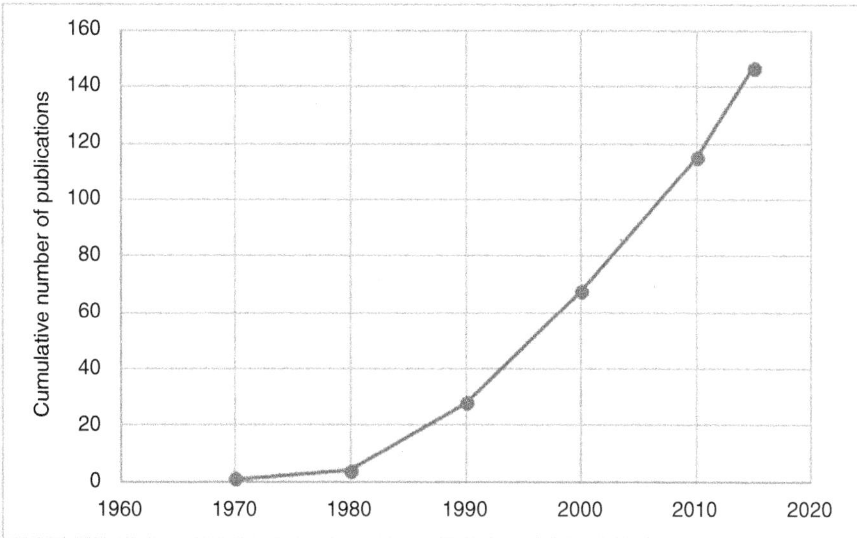

Figure 18.4. Cumulative number of scientific papers related to cyclodextrin-encapsulation of flavors, aromas, and spices used for food industry in CyclodextrinNews.

on the salting-in (volatility decreasing) effect of CD complexation on two model fruit aroma compounds: ethyl butanoate and butyl ethanoate (Baránková and Dohnal, 2015). This effect was contrasted with the salting-out (volatility increasing) effect of inorganic salts, such as sodium chloride and sodium hydrogen carbonate.

Mixtures of ACD and highly branched cyclic dextrins (BrCD) were used to improve the flavor retention in spray-dried rice flavor oil (Kawakami et al., 2009).

More recently, a single-step electrospinning process was applied to encapsulate flavors: a blend of pullulan and BCD was mixed with aroma compounds and the resulting membrane released the aroma only in humid conditions (Mascheroni et al., 2013). Electrospinning technique was used to produce poly(vinyl alcohol) nanowebs containing vanillin complexed by CD and poly (methyl methacrylate) nanofibers containing menthol complex. The inclusion complexes enhanced the stability of vanillin, reduced the volatility of menthol and provided slow release of both flavors (Uyar et al., 2009; Kayaci and Uyar, 2012).

Some of the latest inventions on utilizing CDs for flavors in the Espacenet Patent database include fragrant and crispy sliced garlic stabilized with BCD (Chao and Li, 2015), yellow-head catfish sausage with BCD-stabilized flavor (Li et al., 2015), powder seasoning for dressing water-containing food material, capable of maintaining the flavor by using BCD (Ando and Tanimoto, 2015), red wine powder prepared by spray drying red wine together with BCD, which ensures long storage stability for red wine flavor and functional components (Wang et al., 2015), cucumber powder (Shi, 2014), mulberry fruit, and raspberry combined health drink (Wu and Wu, 2014), and so forth.

3 Approval Status of Cyclodextrins

Although it is conceivable that the mankind has been consuming cyclodextrins for thousands of years as CDs, at least their branched variations (glucosylated and maltosylated) can be detected in beer and bread and other food products containing enzyme- and heat-processed starch (Szente et al., 2006), it was not easy to get approval for food application.

The authorization process was the fastest in Japan where CDs were declared to be enzymatically modified starch and, therefore, their use in food products has been permitted since 1978. All CDs produced by methods not using any organic solvents are approved as natural food additives (Horikoshi, 1988). In addition to the parent ACD, BCD, and GCD also the enzymatically modified branched CDs: glucosyl CD and maltosyl CDs, as well as the conversion mixture containing the three parent CDs and some residual dextrins are authorized in food (Tanada and Kawasaki, 2012).

In Hungary, the Ministry of Health approved the use of BCD for stabilization of natural flavors (flavor/BCD complexes) in 1983. In France, S.A.L. International in cooperation with Chinoin (Hungary) received a limited approval for the use of BCD as a flavor carrier in 1986.

In the Netherlands, and later on in the Benelux countries (Belgium, Luxemburg, and the Netherlands) CDs were approved as enzymatically modified starch products. In 1987, the Spanish authorities also approved the utilization of BCD in foods.

Based on the toxicological data JECFA (Joint FAO/WHO Expert Committee on Food Additives) classified both ACD and GCD as "ADI not specified" (ADI = Allowed Daily Intakes), which means both CDs can be used in food at any concentration and quantity (JECFA, 1999, 2001). GRAS (generally recognized as safe in a wide range of intended uses in food) approvals were obtained in the United States and novel food applications have been filed in Europe (FDA, 2000, 2001, 2004). For BCD ADI of 0–5 mg/kg body weight (bw) was allocated based on the NOEL (no effect level) of 1.25% in the diet (equal to 470 mg/kg bw/day) in a 1-year study in dogs and a safety factor of 100 (JECFA, 1995). It has been classified as GRAS for the use as a flavor protectant in human food

(Wacker Chemie, 2013). In Australia and New Zealand ACD and GCDs are regarded as Novel Food.

The three parent CDs, ACD, BCD, and GCD are registered in the Codex Alimentarius of Joint FAO/WHO Expert Committee on Food Additives (JECFA, 2013) with INS (International Numbering System) No. 457, 459, and 458, respectively, among the General Food Standard Additives (GFSA). BCD is registered in EU as E-459 additive (Commission Directive 2003/95/E, 2003).

CDs are allowed to be used in bread, cracker, soft drinks, beverages, fruit juices, instant coffee and tea, coffee whitener, dairy products, candy, chewing gum, spices, and seasonings, carrier for vitamins and polyunsaturated fatty acids (PUFA), nutritional supplements, and so forth.

Recently a Japanese company, Ezaki Gliko Ltd. (Osaka, Japan) received approval for branched cyclic dextrin (BrCD) (FDA, 2012). This product is a mixture of branched oligosaccharides composed of α-D-glucose monomers with at least 80% of cyclic dextrin molecules with molecular weights ranging from 30,000 to 1,000,000. Cyclic dextrin is a partially degraded maize starch containing short linear chains that are composed of α-(1,4)-linked glucose units with branching through α-(1,6) glucosidic bonds. The metabolic fate of cyclic dextrin is similar to that of GCD.

As ACD and BCD are not digestible and they are metabolized only by the colon microflora, these CDs are considered as soluble fibers. On the other hand, GCD is easily digestable similarly to starch. The recently discovered antiobesity and antidiabetic effect makes these flavor-encapsulants bioactive food supplements (Asp et al., 2006; Comerford et al., 2011; Jarosz et al., 2013).

4 Formulation of Flavors with Cyclodextrins: Methods of Preparation, and Analysis

The most frequently applied methods for complex formation are coprecipitation, freeze-drying or spray drying of common solutions or suspensions, and kneading. There are some examples on the preparation of the complex by mechanochemical activation that is by grinding (cogrinding) using ball milling or vibrating mill. In the case of volatile guest molecules, such as most of the flavors, the sealed heating method works as well.

In the case of coprecipitation the solution of the guest compound is added drop wise to the aqueous solution of CD at ambient or at slightly enhanced temperature. After cooling the forming complex precipitates and the crystals are filtered out and dried.

There is a special variation of this method when supercritical carbon dioxide is used as solvent (Locci et al., 2004; He and Li, 2009).

In the suspension method CD is suspended in water and the guest is added as it is or diluted with/dissolved in ethanol or aqueous ethanol. The solid complex is obtained by freeze-drying (Maier et al., 1987; Karathanos et al., 2007) or spray-drying. A variety of flavor compounds were encapsulated by the three natural CDs using spray-drying and the following general conclusions were drawn (Reineccius et al., 2002):

- The relative retention of flavor is decreasing with increasing flavor/CD ratio (CD excess is favorable in this respect);
- Initial flavor retention was the highest with GCD, but the losses upon storage at enhanced humidity were also the highest with GCD compared to ACD and BCD;
- The release of flavor from the complex into aqueous ethanol solution varies among flavors and CDs and depends on the temperature as well. Thus, the flavor profile and accordingly the perception of flavor may be altered by complexation.

Another possibility for preparing the flavor/CD complexes is kneading: CD and the guest compound are mixed in stoichiometric ratio and a small amount of water is added. The flavor can be dissolved in/diluted with ethanol. During kneading the water molecules in the cavity are replaced by the guest molecules. Having finished the kneading (the consistency of the dense suspension is changed) the solvent is removed by drying. Some examples: lemon oil was microencapsulated by kneading with BCD and the resulting paste samples of the complex were vacuum- or spray-dried. Ten selected lemon oil flavor volatiles (α- and β-pinene, sabinene, β-myrcene, limonene, γ-terpinene, terpinolene, linalool, neral, and geranial) complexed with BCD by kneading. The optimum mixing time (15 min) ensures the maximum encapsulation of lemon oil (97.7 mg/g of BCD) (Bhandari et al., 1999). Preparing limonene/CD complexes by kneading a minimum amount of water (~1%) is necessary; the completely dry CD does not include the flavor (Furuta et al., 1993). Inclusion complex formation is not hindered by ethanol for BCD and GCD/ limonene systems, particularly at lower moisture content but is inhibited for ACD as this will complex ethanol, too. Comparing various alcohols, the smaller (in molecular size) the alcohol is, the more enhanced is the inclusion of D-limonene to BCD.

The grinding technology is similar to kneading but no solvent is used. Mandelic acid cogrinded with CDs was only partially complexed; 10–20% uncomplexed mandelic acid remained in the product (Fodor et al., 1997). This method has been very popular in the pharmaceutical industry when both the host and the guest are crystalline and an amorphous complex of improved solubility is obtained. There are only a few examples when flavors or essential oils are encapsulated by cogrinding, for example, citrus reticulata oil by BCD (Cai et al., 1995), borneol, linalool, and vanillin by BCD (Song et al., 2000; Zhao and Sheng, 2007; Gao et al., 2006).

As CD contains some water even in dry state, the complex formation with sealed heating method is also possible. The components are weighed in a container and sealed. After heating for a while the water gets replaced with the guest and the complex is obtained without drying. The complex of borneol/methyl-beta-cyclodextrin was prepared by this method and a well soluble, amorphous powder was obtained (He and Li, 2009).

4.1 Analysis of Flavor Complexes

Before starting the preparation of a complex it is advisable to perform an interaction study to see if the compound to be included in the CD cavity has affinity to the specific CD. The simplest and most frequently used method is the phase solubility study: the solubility of the guest compound is measured in aqueous solutions of increasing CD concentration. If there is an interaction usually an enhancement in solubility can be observed. The solubility versus CD concentration curves (solubility isotherms) are classified according to Higuchi and Connors (1965) into A and B type. The type A isotherms show continuous increase while the type B reach a plateau at a given CD concentration. The affinity for complex formation is characterized by the complex association constant, which can be calculated from the initial slope of the isotherms and the intrinsic solubility of the guest compound. As the latter is often not easy to determine because of practical insolubility, the term *complexation efficiency* (CE) was introduced (Loftsson et al., 2005, 2007). It expresses the ratio of CD in complex and in free form. The method can be used for one component at a time when the concentration of the flavor is measured by UV photometry. Such solubility study was reported by Ikeda et al. (1982) who studied the interactions of 12 compounds from different essential oils, such as anethole, d-camphor, linalool, and so forth with parent ACD and BCD using the phase solubility method. In this way also the apparent complex stability constants were determined and it was found that binding constants of BCD complexes were greater than those obtained for ACD complexes. These complex stability differences were due to the steric (host-guest fitting) and hydrophobic factors of the studied terpenoids.

When non-UV active component is studied, a competition method with methyl orange can be used (Decock et al., 2006).

For the multicomponent systems chromatographic methods (HPLC, gas chromatography) or capillary electrophoresis (CE) are suggested.

The decreased concentration of the flavor in the headspace of aqueous solutions of increasing CD concentration gives the possibility to compare the affinity of flavors to various CDs and calculate the complex association constants (Decock et al., 2008; Ciobanu et al., 2013). As an example, the interaction of 13 aroma components with various CDs (ACD, BCD, GCD and BCD derivatives) was compared by head-space gas chromatography (HSGC) to establish the order of complexation affinity.

The CE method is simple but needs a special apparatus. The method can be used only if either the flavor or the CD is ionizable in the buffer used for the measurement. The mobility of the charged flavor compound(s) in the electric field is changed when complexed. From this change as a function of the CD concentration in the background electrolyte the complex association constant can be calculated (Rundlett and Armstrong, 1996).

CE is useful also for quantitative determination of the ionizable flavor compounds in the complexes. In the case of racemic mixtures, the enantiomer ratio can be measured, too (Varga et al., 2015).

The flavor content of the complexes depends on the molecular mass of the components. The molecular mass of ACD, BCD, and GCD is 972, 1135, and 1297 g/mol, respectively. The molecular mass of the majority of the flavor components, mono- and sesquiterpenoids, and phenylpropane derivatives falls in the range of 120–160 g/mol. As usually the molar ratio of included flavor to the CD is 1:1, the flavor content in the complex is 9–16%.

The flavor content for the UV active flavors can be measured by UV photometry, HPLC, CE (if ionizable), or HSGC. For the non-UV active components HSGC is the most suitable technique.

While phase solubility, HSGC, and CE use solutions, the solid complexes can be characterized by thermoanalytical methods (Giordano et al., 2001). Comparing the thermal behavior of the single components with their physical mixture and complex gives information on the extent of inclusion and on the stoichiometry of complex. The high thermostability of CDs (they start to decompose above 220–250°C) makes possible to differentiate between the strongly bound (included into the cavity) and only surface-bound fraction of less stable guests. In thermogravimetry (TGA) or differential thermogravimetry (DTG) the mass loss is measured as a function of increasing temperature. It can be coupled to flame ionization detector and the release of the volatile organic components from the complexes can be followed by this evolved gas analysis (EGA) method (Novak et al., 2006). In differential scanning calorimetry (DSC) the heat flow is detected. The absence of melting, evaporation, or sublimation peaks or their shifting toward higher temperature is the sign of complex formation. Thermogravimetry coupled with mass spectrometry detector is useful for studying the complexation of multicomponent essential oils (Fernandes et al., 2009).

Also HSGC proves the enhanced thermal stability of the encapsulated flavors (Vincieri et al., 1986).

Earlier also the pyrolysis thin layer chromatography (TAS) was used, which demonstrated the difference in volatility of the complexed and uncomplexed components (Szente and Szejtli, 1988b). As this technique needs special apparatus and gives rather qualitative results nowadays it is not applied.

When crystalline components are used for complex formation the change in crystallinity (amorphization) can be a sign of inclusion complexation (Szejtli et al., 1979; Furuta et al., 1994; Tapanapunnitkul et al., 2008).

The structure of the flavors inclusion complexes can be characterized by various spectroscopic methods, such as NMR (Mulinacci et al., 1996; Divakar and Maheswaran, 1997; Karathanos et al., 2007; Pirnau et al., 2009), Raman (Moreira da Silva et al., 1995), fluorescence (Ishikawa et al., 2007), infrared (Kayaci and Uyar, 2011), and UV spectroscopy (Astray et al., 2010).

Dynamic HSGC can be applied for characterizing the dynamic release of the components from essential oils and their complexes (Kfoury et al., 2015). The headspace above the sample is extracted several times to be injected into the GC. The continuous decrease of the concentration of the flavor components in the head space allows researchers to calculate the retention efficacy of each component.

5 Flavor Complexes in Food Processing

Tea aromatizing essential oils were encapsulated by BCD and mixed to Ceylon black tea to get a long-lasting aroma retention (Szente et al., 1988a). The aroma components were immediately released in hot water.

Retention of aroma compounds during heat treatment in the presence and absence of BCD was measured by HSGC (Jouquand et al., 2004). There was no general rule: the retention of the hydrophobic flavors (2-heptanone, ethyl butanoate, 2-octanone, and hexanal) did not change with heating, that of 1,2-hexanal and 2-hexanone increased, while of 2-butanone and 1-hexanol decreased with increasing temperature. The equilibrium between the inclusion complex formation and H-bonding with the water molecules is influenced by temperature depending on the structure of the less hydrophobic flavor molecules.

CDs have been widely used to improve the color of different fruit juices. The enzymatic browning of fruits and vegetables (apple and potato) is controlled by combining ascorbic acid with cinnamate and BCD (Sapers and Hicks, 1989) or ascorbate triphosphate with BCD (Gacche et al., 2003). The mechanism is the encapsulation of the substrates of polyphenol oxidase (Irwin et al., 1994), such as complexation of catechol (Ohnishi and Matsubara, 1996) and chlorogenic acid (de la Rosa et al., 2010).

The effects of the addition of ACD, BCD, and GCD on pear juice were compared and ACD was found the most beneficial significantly increasing the global quality of the pear juice by reducing its browning but without producing a significant reduction in the aroma quality (Andreu-Sevilla et al., 2011). Among the samples of freeze-dried pear juice containing dissolved ACD, BCD, or GCD, the juice with ACD retained the greatest amount of aliphatic acetate esters characteristic compounds of La France pear (Tobitsuka et al., 2005). The ACD concentration necessary for the complete inhibition of browning (90 mM) decreased the sensory quality, so an optimum concentration (15 mM) of ACD was used to get a pear juice with an acceptable color but intensive fruity and pear-like odors and aromas (Lopez-Nicolas et al., 2009).

Application of lemon oil/BCD complex in juices and juice-based beverages can eliminate the off-note development, maintain the lemon profile, minimize packaging interaction in PET, and ensure high photostability (Strassburger et al., 2010).

Apricot powder prepared by drying the juice aromatized with BCD-encapsulated natural extracts or natural-identical model mixture encapsulated with BCD showed higher volatile compounds retention than that produced without BCD (Di Cesare et al., 1996). The sensorial properties (color, odor, aroma, and taste) of the reconstituted juices were acceptable.

6 Cyclodextrins in Aroma Preserving and Antibiotic Active Food Packaging

Traditional food packages are passive barriers designed to delay the adverse effects of the environment on the food product. Active packaging, however, allows packages to interact with food and the environment and play a dynamic role in food preservation. The inhibition of the release of the aroma components through the packaging would be desirable. On the other hand, nowadays there is a growing interest in the application of essential oils as natural antioxidants and antibiotics instead of synthetic preservatives both in food and food packaging. CDs, being stable to >200°C are suitable encapsulating agents for these volatile flavor components. They are not decomposed and protect the encapsulated flavor during the processing of packaging materials.

Allylisothiocyanate (mustard flavor with antibiotic properties) is used as CD complex incorporated into the material of packaging plastic films, and used for packaging of meat, or in the material of the sandwich boxes in Japan (Hashimoto, 2002). Essential oils complexed with BCD in chitosan film have long lasting antimicrobial properties (Sun et al., 2014).

Allylisothiocyanate (AITC) and menthol complexed by CDs and incorporated into polyethylene films showed controlled release of these flavors (Balogh et al., 2008). CDs in PVC packaging films inhibited the release of plasticizers (Fenyvesi et al., 2007) as another advantage of their application.

The high volatility, strong odor, and poor water solubility of AITC can be improved by complexation. Incorporating its inclusion complex with ACD and BCD into polylactide-co-polycaprolactone films antibacterial packaging for cheese was developed (Plackett et al., 2006, 2007). The concept was to avoid preservatives added to the food by placing them into the packaging. The controlled release not only lengthens the shelf-life of the product packaged but the consumption of the preservatives can also be avoided. Another advantage is the biodegradability of CDs and their complexes. Even the AITC complex of BCD was fast biodegraded (Verstichel et al., 2004).

Another biodegradable packaging material was prepared by electrospinning of AITC/BCD complex mixed in soy protein isolate/poly(ethylene oxide) blend and poly(lactic acid) (PLA) (Vega-Lugo and Lim, 2009). The complexation of AITC with ACD but not with BCD suppressed the decomposition of AICD (Ohta et al., 2000; Jiang et al., 2006). The release was accelerated with increased relative humidity (Li et al., 2007) as it is generally observed in the case of CD inclusion complexes. This behavior makes this kind of encapsulation so attractive in food packaging: the moisture-triggered release is exactly what is required. The antimicrobial activity of PLA films with AITC/BCD and carvacrol/BCD ensured complete inhibition of *B. cinerea* growth during 10 days (Raouche et al., 2011).

Various other packaging systems were also elaborated

- BCD inclusion complexes with flavor volatiles, such as D-limonene, alpha-pinene, and 2-methoxy-3-methylpyrazine, have been incorporated into low-density polyethylene (LDPE) powder by dry mixing and then thermally pressed into films providing extended food shelf-life (Koontz and Marcy, 2007).

- BCD complex of trans-cinnamaldehyde was embedded into edible coating consisting of chitosan and pectin and this coating was applied to fresh cut fruits, such as papaya (Brasil et al., 2012).
- Films produced by extrusion of ethylene-vinyl alcohol copolymer with 10–40% BCD were prepared for aroma-preserving food packaging (Lopez-de-Dicastillo et al., 2010). The presence of BCD slightly increased the fragility and crystallinity of the polymer. Terpene flavors were preferentially absorbed by the film.
- Chitosan/hydroxypropyl-BCD composite films showed high retention of carvacrol (Higueras et al., 2013).
- Joo et al. (2011) made a masterbatch first by mixing BCD with PLA, extruding and pelletizing, then this masterbatch with 30% BCD content was used for casting films. This technique increased the compatibility between BCD and PLA. Films containing *trans*-2-hexanal, a naturally occurring plant flavor with antibiotic properties reduced the microbial growth (Joo et al., 2012). On the other hand, incorporation of the complex into the film resulted in impaired physico-chemical characteristics, reduced tensile strength, decreased elongation at break, and increased permeability.
- Grafting of CDs derivatives to cellulose surface, as in coffee filter paper, flavor components are stabilized and released in a controlled way (Bergamasco et al., 2006). Tissue paper modified with CDs (by using monochlorotriazinyl β-CD) is useful for tea bags and coffee bags to preserve the aroma and improve freshness (Wintersgill, 2004). Cellulosic web coated with a layer containing CD can be also used for this purpose (Wood and Beaverson, 1999).
- Cellulose acetate films modified with CDs were used for packaging of ground, roasted coffee. Both the organoleptic properties and the HSGC chromatograms showed that CD-modified cellophane preserves the aroma and hinders oxidation better than the unmodified cellophane (Asslisi, 2011).

7 Conclusions

A thorough literature survey on encapsulation of food flavors and aromas covering the history, approval status, production, and analysis as well as on application in food processing and packaging is given. The main advantages of flavor encapsulation by CDs are as follows:

- Enhanced stability (decreased volatility, enhanced heat- and photostability, inhibited hydrolysis, etc.)
- Moisture-triggered release (controlled release)

- Biodegradable encapsulating agents derived from plants (sustainable raw materials)
- Approved as safe food ingredient (GRAS, novel food)
- Have beneficial bioactive effect on lipid and carbohydrate metabolism

References

Ando, S., Tanimoto, S., 2015. Powder seasoning and dressed food product. JP2014226084. http://worldwide.espacenet.com.

Andreu-Sevilla, A.J., Lopez-Nicolas, J.M., Carbonell-Barrachina, A.A., Garcıa-Carmona, F., 2011. Comparative effect of the addition of α-, β-, or γ-cyclodextrin on main sensory and physico–chemical parameters. J. Food Sci. 76, S347–S353.

Asp, M.L., Hertzler, S.R., Chow, J., Wolf, B.W., 2006. Gamma-cyclodextrin lowers postprandial glycemia and insulinemia without carbohydrates malabsorption in healthy adults. J. Am. Coll. Nutr. 25, 49–55.

Asslisi, C., 2011. Active packaging of ground coffee using cyclodextrin-containing cellophane films. Dissertation. Corvinus University Budapest and CycloLab.

Astray, G., Gonzalez-Barreiro, C., Mejuto, J.C., Rial-Otero, R., Simal-Gandara, J., 2009. A review on the use of cyclodextrins in foods. Food Hydrocoll. 23 (7), 1631–1640.

Astray, G., Mejuto, J.C., Morales, J., Rial-Otero, R., Simal- Gandara, J., 2010. Factors controlling flavors binding constants to cyclodextrins and their applications in foods. Food Res. Int. 43 (4), 1212–1218.

Balogh, K., Fenyvesi, É., Makk, J., Márialigeti, K., M., Sényi, J., Siró, I., Orgoványi, J., Otta, K., Szente, L., 2008. Cyclodextrin complexes of natural antimicrobial compounds in active packaging. In: Proceedings of Fourteenth International Cyclodextrins Symposium, Kyoto, Japan, May 8–11, 2008, pp. 320–323.

Baránková, E., Dohnal, V., 2015. Effect of additives on volatility of aroma compounds from dilute aqueous solutions. Fluid Phase Equilibria 407, 217–223.

Bergamasco, R. de C., Zanin, G.M., de Moraes, F.F., 2006. Grafting of cyclodextrins onto filter paper. Thirteenth International Cyclodextrin Symposium, Torino, May 14–17.

Bhandari, B.R., D'Arcy, B.R., Padukka, I., 1999. Encapsulation of lemon oil by paste method using β-cyclodextrin: encapsulation efficiency and profile of oil volatiles. J. Agric. Food Chem. 47 (12), 5194–5197.

Bhandari, B., D'Arcy, B., Young, G., 2001. Flavor retention during high temperature short time extrusion cooking process: a review. Int. J. Food Sci. Technol. 36 (5), 453–461.

Brasil, I.M., Gomes, C., Puerta-Gomez, A., Castell-Perez, M.E., Moreira, R.G., 2012. Polysaccharide-based multilayered antimicrobial edible coating enhances quality of fresh-cut papaya. LWT—Food Sci. Technol. 47 (1), 39–45.

Cai, Z., Wu, L., Zhong, G., Gao, S., 1995. Effects of inclusion of *Citrus reticulata* volatile oil by β-cyclodextrin. Zhongguo Zhongyao Zazhi 20 (10), 603–604, (Chinese) (Chem. Abstr.: 124:241719).

Chao, P., Li, Y., 2015. Processing process of fragrant and crispy sliced garlic. CN104366338. http://worldwide.espacenet.com.

Ciobanu, A., Landy, D., Fourmentin, S., 2013. Complexation efficiency of cyclodextrins for volatile flavor compounds. Food Res. Int. 53 (1), 110–114.

Comerford, K.B., Artiss, J.D., Jen, K.L.C., Karakas, S.E., 2011. The beneficial effects alpha-cyclodextrin on blood lipids and weight loss in healthy humans. Obesity 19, 1200–1204.

Commission Directive 2003/95/E, 2003. Available from: http://faolex.fao.org/docs/pdf/eur40476.pdf.

Cravotto, G., Binello, A., Baranelli, E., Carraro, P., Trotta, F., 2006. Cyclodextrins as food additives and in food processing. Curr. Nutr. Food Sci. 2 (4), 343–350.

Crini, G., 2014. A history of cyclodextrins. Chem. Rev. 114, 10940–10975.

Decock, G., Fourmentin, S., Surpateanu, G.G., Landy, D., DeCock, P., Surpateanu, G., 2006. Experimental and theoretical study on the inclusion compounds of aroma components with beta-cyclodextrins. Supramol. Chem. 18 (6), 477–482.

Decock, G., Landy, D., Surpateanu, G., Fourmentin, S., 2008. Study of the retention of aroma components by cyclodextrins by static head-space gas chromatography. J. Incl. Phenom. Macrocycl. Chem. 62 (3–4), 297–302.

de la Rosa, L.A., Mercado-Mercado, G., Rodrigo-Garcia, J., Gonzalez-Aguilar, G.A., Alvarez-Parrilla, E., 2010. Peach polyphenol oxidase inhibition by beta-cyclodextrin and 4-hexylresorcinol is substrate dependent. CyTA—J. Food 8 (2), 87–93.

Di Cesare, L.F., Nani, R., Mariani, N., D'Angelo, V., 1996. Influence of maltodextrin, cyclodextrin, and palm oil on the aroma retention of apricot powders. Ind. Bevande 25 (142), 101–107.

Divakar, S., Maheswaran, M.M., 1997. Structural studies on inclusion compounds of β-cyclodextrin with some substituted phenols. J. Incl. Phenom. Mol. Recognit. Chem. 27 (2), 113–126.

FDA, 2000. Agency Response Letter GRAS Notice No. GRN 000046 http://www.accessdata.fda.gov/scripts/fdcc/?set=GRASNotices&id=46.

FDA, 2001. Agency Response Letter GRAS Notice No. GRN 000074 http://www.accessdata.fda.gov/scripts/fdcc/?set=GRASNotices&id=74.

FDA, 2004. Agency Response Letter GRAS Notice No. GRN 000155 http://www.accessdata.fda.gov/scripts/fdcc/?set=GRASNotices&id=155.

FDA, 2012. Agency Response Letter GRAS Notice No. GRN 000404 http://www.fda.gov/food/ingredientspackaginglabeling/gras/noticeinventory/ucm327929.htm.

Fenyvesi, E., Balogh, K., Siro, I., Orgovanyi, J., Senyi, J.M., Otta, K., Szente, L., 2007. Permeability and release properties of cyclodextrin-containing poly(vinyl chloride) and polyethylene films. J. Incl. Phenom. Macrocycl. Chem. 57 (1–4), 371–374.

Fenyvesi, E., Vikmon, A.M., Szente, L., 2015. Cyclodextrins in food technology and human nutrition: benefits and limitations in 2012. Crit. Rev. Food Sci. Nutr., 2015 Mar 12:0. [Epub ahead of print] doi:10.1080/10408398.2013.809513.

Fernandes, L.P., Oliveira, W.P., Sztatisz, J., Szilagyi, I.M., Novak, Cs., 2009. Solid state studies on molecular inclusions of *Lippia sidoides* essential oil obtained by spray drying. J. Therm. Anal. Calorim. 95 (3), 855–863.

Fodor, M., Novak, Cs., Rakosa, R., Tomor, K., Pokol, G., Gal, S., 1997. Solid and liquid state investigations of mandelic acid cyclodextrin complexes. J. Therm. Anal. 48 (3), 515–525.

Frost and Sullivan, 2008. U.S. Starch Market. N107-01. http://www.frost.com/prod/servlet/research.pag.

Furuta, T., Yoshii, H., Miyamoto, A., Yasunishi, A., Hirano, H., 1993. Effects of water and alcohol on the formation of inclusion complexes of d-limonene and cyclodextrins. Supramol. Chem. 1 (3–4), 321–325.

Furuta, T., Yoshii, H., Kobayashi, T., Nishitarumi, T., Yasunishi, A., 1994. Powdery encapsulation of d-limonene by kneading with mixed powders of β-cyclodextrin and maltodextrin at low water content. Biosci. Biotechnol. Biochem. 58 (5), 847–850.

Furuta, T., Yoshii, H., Shiga, H., 2001. Microencapsulation of food flavorings by inclusion in cyclodextrin and by spray drying and its release characteristics. Foods Food Ingr. J. Jpn. 191, 23–32.
Gacche, R.N., Zore, G.B., Ghole, V.S., 2003. Kinetics of inhibition of polyphenol oxidase mediated browning in apple juice by beta-cyclodextrin and l-ascorbate-2-triphosphate. J. Enzyme Inhib. Med. Chem. 18 (1), 1–5.
Gao, J., Ren, Y., Wang, X., 2006. Study on clathrate of vanillin and cyclodextrin. Zhongguo Xinyao Zazhi 15 (21), 1858–1861, (Chinese) (Chem. Abstr.:147:38434).
Giordano, F., Novak, C., Moyano, J.R., 2001. Thermal analysis of cyclodextrins and their inclusion compounds. Thermochim. Acta 380 (2), 123–151.
Gray, J.I., Roberts, D.G., 1970. Retention and release of volatile food flavor compounds. J. Food Technol. 5, 231.
Hashimoto, H., 1988. Application of cyclodextrins to foods, toiletries, and other products in Japan. In: Huber, O., Szejtli, J. (Eds.), In: Proc. Int. Symp. Cyclodextrins. 4th Kluwer, Dordrecht, pp. 533–543.
Hashimoto, H., 2002. Present status of industrial application of cyclodextrins in Japan. Eleventh International Cyclodextrin Symposium, Reykjavik, Iceland, 5–8 May 2002.
Hashimoto, H., 2008. CyD applications in food, cosmetic, toiletry, textile, and wrapping material fields. In: Dodziuk, H. (Ed.), Cyclodextrins and Their Complexes. Wiley-VCH Verlag GmbH & Co, Weinheim, pp. 452–459.
He, J., Li, W., 2009. Preparation of borneol-methyl-beta-cyclodextrin inclusion complex by supercritical carbon dioxide processing. J. Incl. Phenom. Macrocycl. Chem. 65 (3–4), 249–256.
Hedges A. R., Shieh W. J., Sikorski C. T., 1995. Use of cyclodextrin for encapsulation in the use and treatment of food products. In: ACS., Symp., Ser., 590: Encapsulation, Controlled Release of Food Ingredients (206th National Meeting of the American Chemical Society, Chicago, Illinois, August 22–27, 1993), pp. 60–73.
Higuchi, T., Connors, K.A., 1965. Phase-solubility techniques. Adv. Anal. Chem. Instr. 4, 117–212.
Higueras, L., Lopez-Carballo, G., Cerisuelo, J.P., Gavara, R., Hernandez-Munoz, P., 2013. Preparation and characterization of chitosan/HP-beta-cyclodextrins composites with high sorption capacity for carvacrol. Carbohyd. Polym. 97, 262–268.
Horikoshi, K., 1988. Enzymology and molecular genetics of CD-forming enzymes. In: Huber, O., Szejtli, J. (Eds.), In: Proc. Int. Symp. Cyclodextrins. fourth ed., 7–17 Kluwer, Dordrecht, pp. 7–18.
Ikeda, Y., Matsumoto, K., Kunihiro, K., Fuwa, T., Uekama, K., 1982. Inclusion complexation of essential oils with α- and β-cyclodextrins. Yakugaku Zasshi 102 (1), 83–88, (Chem. Abstr.: 96:168511).
Irwin, P.L., Pfeffer, P.E., Doner, L.W., Sapers, G.M., Brewster, J.D., Nagahashi, G., Hicks, K.B., 1994. Binding geometry, stoichiometry, and thermodynamics of cyclomalto-oligosaccharide (cyclodextrin) inclusion complex formation with chlorogenic acid, the major substrate of apple polyphenol oxidase. Carbohyd. Res. 256 (1), 13–27.
Ishikawa, H., Kuwano, A., Matsumoto, K., 2007. Complexation of vanillin and ethylvanillin with alpha-, beta-, and gamma-cyclodextrin. J. Fac. Agricult., Kyushu Univ. 52 (1), 87–90.
Jarosz, P.A., Fletcher, E., Elserafy, E., Artiss, J.D., Jen, K.-L.C., 2013. The effect of gamma-cyclodextrin on postprandial lipid and glycemic responses to a fat-containing meal. Metab. Clin. Exp. 62 (10), 1443–1447.

JECFA, 1995. Summary of Evaluations Performed by the Joint FAO/WHO Expert Committee on Food Additives: Beta-cyclodextrin. http://www.inchem.org/documents/jecfa/jeceval/jec_465.htm.

JECFA, 1999. Safety Evaluation of Certain Food Additives: Gamma-cyclodextrin. http://www.inchem.org/documents/jecfa/jecmono/v042je11.htm.

JECFA, 2001. Summary of Evaluations Performed by the Joint FAO/WHO Expert Committee on Food Additives: Alpha-cyclodextrin http://www.inchem.org/documents/jecfa/jeceval/jec_464.htm.

JECFA, 2013. Codex Alimentarius Gamma-cyclodextrin. http://www.codexalimentarius.net/gsfaonline/additives/details.html?id=355.

Jiang, Z., Zhang, Q., Tian, H., Li, R., 2006. The reaction of allyl isothiocyanate with hydroxyl/water and beta-cyclodextrin using ultraviolet spectrometry. Food Technol. Biotechnol. 44 (3), 423–427.

Joo, M.J., Auras, R., Almenar, E., 2011. Preparation and characterization of blends made of poly(L-lactic acid) and beta-cyclodextrin: Improvement of the blend properties by using a masterbatch. Carbohyd. Polym. 86 (2), 1022–1030.

Joo, M.J., Merkel, C., Auras, R., Almenar, E., 2012. Development and characterization of antimicrobial poly(L-lactic acid) containing trans-2-hexenal trapped in cyclodextrins. Int. J. Food Microbiol. 153 (3), 297–305.

Jouquand, C., Ducruet, V., Giampaoli, P., 2004. Partition coefficients of aroma compounds in polysaccharide solutions by the phase ratio variation method. Food Chem. 85, 467–474.

Karathanos, V.T., Mourtzinos, I., Yannakopoulou, K., Andrikopoulos, N.K., 2007. Study of the solubility, antioxidant activity and structure of inclusion complex of vanillin with beta-cyclodextrin. Food Chem. 101 (2), 652–658.

Kawakami, K., Fujita, A., Mikami, T., Yoshii, H., Paramita, V., Neoh, T.L., Furuta, T., 2009. Formation of rice flavor powder with β-cyclodextrin by spray drying. Food Res. Technol. 229, 239–245.

Kawashima, Z., 1986. Manufacture of odorless garlic powder. JP 61091128. (Chem. Abstr.: 105:113943).

Kayaci, F., Uyar, T., 2011. Solid inclusion complexes of vanillin with cyclodextrins: their formation, characterization, and high-temperature stability. J. Agric. Food Chem. 59 (21), 11772–11778.

Kayaci, F., Uyar, T., 2012. Encapsulation of vanillin/cyclodextrin inclusion complex in electrospun polyvinyl alcohol (PVA) nanowebs: Prolonged shelflife and high temperature stability of vanillin. Food Chem. 133, 641–649.

Kfoury, M., Auezova, L., Greige-Gerges, H., Fourmentin, S., 2015. Promising applications of cyclodextrins in food: improvement of essential oils retention, controlled release, and antiradical activity. Carbohyd. Polym. 131, 264–272.

Koontz, J.L., Marcy, J.E., 2007. Controlled release of active ingredients from polymer food packaging by molecular encapsulation with cyclodextrins. Polym. Preprints (Am. Chem. Soc., Div. Polym. Chem.) 48 (2), 742.

Kuge, T., Takeo, K., 1968. Complexes of starchy materials with organic compounds: Part I. Affinity observed by gas chromatography. Agric. Biol. Chem. 32, 753–758.

Li, C., Dai, X., Wang, N., Shang, X., Chen, D., Duan, C., 2015. Flavor enhanced type fermented yellow-head catfish sausage and preparation method thereof. CN104305349. http://worldwide.espacenet.com.

Li, X., Jin, Z., Wang, J., 2007. Complexation of allyl isothiocyanate by alpha- and beta-cyclodextrin and its controlled release characteristics. Food Chem. 103 (2), 461–466.

Lindner, K., Szente, L., Szejtli, J., 1981. Food flavoring with β-cyclodextrin-complexed flavor substances. Acta Aliment. 10 (3), 175–186.

Liu, X.D., Furuta, T., Yoshii, H., Linko, P., Coumans, W.J., 2000. Cyclodextrin encapsulation to prevent the loss of l-menthol and its retention during drying. Biosci. Biotechnol. Biochem. 64 (8), 1608–1613.

Locci, E., Lai, S., Piras, A., Marongiu, B., Lai, A., 2004. ^{13}C-CPMAS and ^{1}H-NMR study of the inclusion complexes of beta-cyclodextrin with carvacrol, thymol, and eugenol prepared in supercritical carbon dioxide. Chem. Biodivers. 1 (9), 1354–1366.

Loftsson, T., Hreinsdóttir, D., Márson, M., 2005. Evaluation of cyclodextrin solubilization of drugs. Int. J. Pharm. 302, 18–28.

Loftsson, T., Hreinsdóttir, D., Márson, M., 2007. The complexation efficiency. J. Incl. Phenom. Macrocycl. Chem. 57, 545–552.

Loftsson, T., Magnusdottir, A., Masson, M., Sigurjonsdottir, J.F., 2002. Self-association and cyclodextrin solubilization of drugs. J. Pharm. Sci. 91 (11), 2307–2316.

Lopez-de-Dicastillo, C., Gallur, M., Catala, R., Gavara, R., Hernandez-Munoz, P., 2010. Immobilization of beta-cyclodextrin in ethylene-vinyl alcohol copolymer for active food packaging applications. J. Membr. Sci. 353 (1–2), 184–191.

Lopez-Nicolas, J.M., Andreu-Sevilla, A.J., Carbonell- Barrachina, A.A., Garcia-Carmona, F., 2009. Effects of addition of alpha-cyclodextrin on the sensory quality, volatile compounds, and color parameters of fresh pear juice. J. Agric. Food Chem. 57 (20), 9668–9675.

Maier, H.G., Bauer, A., 1972. Binding of volatile aroma components on starch. Starch/Staerke 24, 101–107.

Maier, H.G., Moritz, K., Ruemmler, U., 1987. Thermostable binding of aroma compounds to starch. Part I: Binding by freeze-drying. Starch/Staerke 39, 126.

Marques, H.M.C., 2010. A review on cyclodextrin encapsulation of essential oils and volatiles. Flavour Frag. J. 25 (5), 313–326.

Martina, K., Binello, A., Lawson, D., Jicsinszky, L., Cravotto, G., 2013. Recent applications of cyclodextrins as food additives and in food processing. Curr. Nutr. Food Sci. 9 (4), 167–179.

Martina, K., Cravotto, G., 2015. Cyclodextrin as food additives in food processing. In: Functional Polymers in Food Science: From Technology to Biology (vol. II, Chapter 12) Ed. Scrivener Publishing LLC, Salem, MA. Published Online: 27 Mar 2015. doi:10.1002/9781119108580.ch12.

Mascheroni, E., Fuenmayor, C.A., Cosio, M.S., Di Silvestro, G., Piergiovanni, L., Mannino, S., Schiraldi, A., 2013. Encapsulation of volatiles in nanofibrous polysaccharide membranes for humidity-triggered release. Carbohyd. Polym. 98, 17–25.

Moreira da Silva, A.M., Amado, A.M., Ribeiro-Claro, P.J.A., Empis, J., Teixeira-Dias, J.J.C., 1995. β-Cyclodextrin complexes of benzaldehyde, vanillin, and cinnamaldehyde: a Raman spectroscopic study. J. Carbohydr. Chem. 14 (4&5), 677–684.

Mulinacci, N., Melani, F., Vincieri, F.F., Mazzi, G., Romani, A., 1996. 1H-NMR NOE and molecular modeling to characterize thymol and carvacrol β-cyclodextrin complexes. Int. J. Pharm. 128 (1,2), 81–88.

Novak, Cs., Ehen, Zs., Fodor, M., Jicsinszky, L., Orgovanyi, J., 2006. Application of combined thermo-analytical techniques in the investigation of cyclodextrin inclusion complexes. J. Therm. Anal. Calorim. 84 (3), 693–701.

Ohnishi, M., Matsubara, T., 1996. Complex formation between sugar ligands and catechol. Studies on the fluorescence titration with TNS and its effect on the polyphenol oxidase-catalyzed browning reaction. Starch/Staerke 48 (6), 223–238.

Ohta, Y., Takatani, K., Kawakishi, S., 2000. Kinetic and thermodynamic analyses of the cyclodextrin-allyl isothiocyanate inclusion complex in an aqueous solution. Biosci. Biotechnol. Biochem. 64 (1), 190–193.

Phanindrakumar, H.S., Radhakrishna, K., Mahesh, S., Jagannath, J.H., Bawa, A.S., 2005. Effect of pretreatments and additives on the thermal behavior and hygroscopicity of freeze-dried pineapple juice powder. J. Food Process Preserv. 29 (5/6), 307–318.

Pirnau, A., Bogdan, M., Floare, C.G., 2009. NMR spectroscopic characterization of beta-cyclodextrin inclusion complex with vanillin. J. Physics: Conf. Series 182.

Plackett, D.V., Holm, V.K., Johansen, P., Ndoni, S., Nielsen, P.V., Sipilainen-Malm, T., Soedergaard, A., Verstichel, S., 2006. Characterization of L-polylactide and L-polylactide-polycaprolactone copolymer films for use in cheese-packaging applications. Packag. Technol. Sci. 19 (1), 1–24.

Plackett, D., Ghanbari-Siahkali, A., Szente, L., 2007. Behavior of alpha- and beta-cyclodextrin-encapsulated allyl isothiocyanate as slow-release additives in polylactide-*co*-polycaprolactone films. J. Appl. Polym. Sci. 105 (5), 2850–2857.

Pocsi, I., 1999. Physiological and ecological evaluation of bacterial cyclodextrin glycosyltransferases (CGTases). Biologia (Bratislava) 54 (6), 603–616.

Raouche, S., Mauricio-Iglesias, M., Peyron, S., Guillard, V., Gontard, N., 2011. Combined effect of high pressure treatment and antimicrobial bio-sourced materials on microorganisms' growth in model food during storage. Innov. Food Sci. Emerg. Technol. 12 (4), 426–434.

Reineccius, G.A., Risch, S.J., 1986. Encapsulation of artificial flavors by beta-cyclodextrin. Perfume Flavor 11 (1), 3–6.

Reineccius, T.A., Reineccius, G.A., Peppard, T.L., 2002. Encapsulation of flavors using cyclodextrins: comparison of flavor retention in alpha, beta, and gamma types. J. Food Sci. 67 (9), 3271–3279.

Risch, S.J., Reineccius, G.A., (Eds.), 1988. Flavor Encapsulation. ACS Symposium Series No. 370, American Chemical Society, Washington, DC.

Risch, S.J., Reineccius, G.A., (Eds.), 1995. Encapsulation and Controlled Release of Food Ingredients. ACS Symposium Series No. 590, American Chemical Society, Washington, DC.

Rogers, W.I., Whaley, W.M., 1962. US Patent 3,061,444.

Rundlett, K.L., Armstrong, D.W., 1996. Examination of the origin, variation, and proper use of expressions for the estimation of association constants by capillary electrophoresis. J. Chromatogr. A 721, 173–186.

Sapers G.M., Hicks K.B., 1989. Inhibition of Enzymatic Browning in Fruits and Vegetables. ACS Symposium Series No. 405, Chapter 3, 29–43.

Schmid, G., 1996. Enzymology of cyclodextrins. Szejtli, J., Osa, T. (Eds.), Comprehensive Supramolecular Chemistry, vol. 3, Elsevier, Oxford, UK, pp. 615–626.

Schmid, G., Regiert, M., Antlsperger, G., 1995. Applicability of gamma-cyclodextrin in food. 209th ACS National Meeting, American Chemical Society, Anaheim, California, April 2–6.

Shi, A., 2014. Cucumber powder and processing method thereof. CN104146220. http://worldwide.espacenet.com.

Shiga, H., Yoshii, H., Ohe, H., Yasuda, M., Furuta, T., Kuwahara, H., Ohkawara, M., Linko, P., 2004. Encapsulation of shiitake (*Lenthinus edodes*) flavors by spray drying. Biosci. Biotechn. Biochem. 68 (1), 66–71.

Song, H., Guo, T., Qin, D., Zhao, M., Zhang, R., 2000. Technology of preparing borneol-β-cyclodextrin inclusion complexes. Shenyang Yaoke Daxue Xuebao 17 (3), 170–173, (Chinese) (Chem. Abstr.: 133:48807).

Strassburger, K., Startup, W., Levey, V., Mattingly, T., Briggs, J., Harrison, J., Wilson, T., 2010. Enhanced stability of citral in juice beverages by applying cyclodextrin micro emulsion technology. ACS Symposium Series, 1036 (Flavor in Noncarbonated Beverages), 143–158.

Sun, X., Sui, S., Ference, C., Zhang, Y., Sun, S., Zhou, N., et al., 2014. Antimicrobial and mechanical properties of beta-cyclodextrin inclusion with essential oils in chitosan films. J. Agric. Food Chem. 62 (35), 8914–8918.

Szejtli, J., 1977. Some possible applications of cyclodextrins in the pharmaceutical industry. Starch/Staerke 29 (1), 26–33.
Szejtli, J., 1982. Cyclodextrins and their inclusion complexes. Akadémiai Kiadó, Budapest.
Szejtli, J., Szente, L., Banky-Elod, E., 1979. Molecular encapsulation of volatile, easily oxidizable labile flavor substances by cyclodextrins. Acta Chim. Acad. Sci. Hung. 101 (1–2), 27–46.
Szejtli, J., Szente, L., 2005. Elimination of bitter, disgusting tastes of drugs and foods by cyclodextrins. Eur. J. Pharm. Biopharm. 61 (3), 115–125.
Szente, L., Gal-Fuzy, M., Szejtli, J., 1988a. Tea aromatization with beta-cyclodextrin complexed flavors. Acta Aliment. 17, 193.
Szente, L., Harangi, J., Greiner, M., Mandel, F., 2006. Cyclodextrins found in enzyme- and heat-processed starch-containing foods. Chem. Biodivers. 3 (9), 1004–1014.
Szente, L., Harangi, J., Szejtli, J., 1988b. Long-term storage stability studies on flavor-beta-cyclodextrin complexes. In: Huber, O., Szejtli, J. (Eds.), In: Proc. Int. Symp. Cyclodextrins. 4th Kluwer, Dordrecht, pp. 545–549.
Szente, L., Szejtli, J., 1986. Molecular encapsulation of natural and synthetic coffee flavor with β-cyclodextrin. J. Food Sci. 51 (4), 1024–1027.
Szente, L., Szejtli, J., 1988a. New results on the molecular encapsulation of flavors by cyclodextrins. Abh. Akad. Wiss. Ddr, abt. Math., Naturwiss. Tech. (2), 101.
Szente, L., Szejtli, J., 1988b. Stabilization of flavors by cyclodextrins. ACS Symp. Ser. 370, 148.
Szente, L., Szejtli, J., 2004. Cyclodextrins as food ingredients. Trends Food Sci. Tech. 15 (3–4), 137–142.
Takaku H., Kogure Y., Kuwabara N., Oku S., 1989. Candies containing maltosyl cyclodextrin. Jpn. Kokai JP 01 05,453. (Chem. Abstr. 110:191530).
Tamaki, K., Tamaki, T., Suzuki, Y., 2007. Deodorization of off-odour during sweet potato juice production by employing physical and chemical deodorants. Food Chem. 105 (2), 454–461.
Tanada, S., Kawasaki, N., 2012. Cyclodextrin interfacial behavior. Somsundaran, P., Hubbard, A.T. (Eds.), Encyclopedia of Surface Colloid Science, vol. 3, second ed. Taylor and Francis, Milton Park, Abingdon-on-Thames, U.K., pp. 1653–1663.
Tapanapunnitkul, O., Caiseri, S., Peterson, D.G., Thompson, D.B., 2008. Water solubility of flavor compounds influences formation of flavor inclusion complexes from dispersed high-amylose maize starch. J. Agric. Food Chem. 56 (1), 220–226.
Thoss, M., Schwabe, L., Froemming, K.H., 1994. Storage and photostability of cyclodextrin inclusion compounds of lemon, orange, hop, and chamomile oil. Pharmazie 49 (4), 252–257.
Tobitsuka, K., Miura, M., Kobayashi, S., 2005. Interaction of cyclodextrins with aliphatic acetate esters and aroma components of La France pear. J. Agric. Food Chem. 53 (13), 5402–5406.
Usanov, N.G., Loginov, O.N., Melent'ev, A.I., 1990. Synthesis of cyclodextrin glucanotransferases by microorganisms utilizing cyclodextrins as the only source of carbon. Dokl. Akad. Nauk SSSR 310, 1489–1492.
Uyar, T., Nur, Y., Hacaloglu, J., Besenbacher, F., 2009. Electrospinning of functional poly (methyl methacrylate) nanofibers containing cyclodextrin-menthol inclusion complexes. Nanotechnology 20, 125703/1-10.
Varga, E., Sohajda, T., Juvancz, Z., Kendrovics, R.B., Szekely, E., Bansaghi, G., 2015. Development of electrophoretic methods for simultaneous determination of enantiomeric ratio and composition of diastereomeric salt mixtures. Chromatographia 78 (13–14), 881–888.
Vega-Lugo, A.-C., Lim, L.-T., 2009. Controlled release of allyl isothiocyanate using soy protein and poly(lactic acid) electrospun fibers. Food Res. Intern. 42 (8), 933–940.

Verstichel, S., De Wilde, B., Fenyvesi, E., Szejtli, J., 2004. Investigation of the aerobic biodegradability of several types of cyclodextrins in a laboratory-controlled composting test. J. Polym. Environ. 12 (2), 47–55.

Vincieri, F.F., Mazzi, G., Papini, P., Gelsomini, N., 1986. Monitoring by headspace gas chromatography (HSGC) of formation and thermal stability of terpene β-cyclodextrin complexes. Farmaco, Ed. Prat. 41 (3), 98–106.

Wacker Chemie, 2013. Cyclodextrins in food and nutraceutical applications. http://www.wacker.com/cms/media/publications/downloads/6088_EN.pdf.

Wang, J., Jin, Z., Ge, Y., Li, L., Wang, L., Zhang, X., Cai, C., Xu, X., Zhou, X., Jiao, A., Tian, Y., Yang, N., 2015. CN104312884. http://worldwide.espacenet.com.

Wintersgill, S., 2004. Treated fibers for making a beverage infusion package with improved freshness and a beverage infusion package made therefrom. WO 2004067841.

Wood, W.E., Beaverson, N.J., 1999. Packaging system comprising cellulosic web with a permeant barrier or contaminant trap. US 5,776,842.

Wu, H., Wu, G., 2014. Mulberry fruit and raspberry combined health drink and preparation technology thereof. CN104116096. http://worldwide.espacenet.com.

Wulff, G., Avgenaki, G., Guzmann, M.S.P., 2005. Molecular encapsulation of flavours as helical inclusion complexes of amylose. J. Cereal Sci. 41, 239–249.

Yeretzian, C., Jordan, A., Lindinger, W., 2003. Analysing the headspace of coffee by proton-transfer-reaction mass-spectrometry. Int. J. Mass Spectr. 223-224, 115–139.

Zhao, W., Chao, J., Du, R., Huang, S., 2011. Spectroscopic studies on the inclusion behavior between caffenic acid and gamma-cyclodextrin. J. Incl. Phenom. Macrocycl. Chem. 71 (1–2), 25–34.

Zhao, X., Sheng, J., 2007. Preparation of inclusion compound of linalool and beta-cyclodextrin. Riyong Huaxue Gongye 37 (2), 97–98, 127 (Chinese) (Chem. Abstr.: 148:586049).

Zhao, M., Wang, H., Yang, B., Tao, H., 2010. Identification of cyclodextrin inclusion complex of chlorogenic acid and its antimicrobial activity. Food Chem. 120 (4), 1138–1142.

Zhiyuan Bio-tech, 2013. http://www.bzzysw.com/en/main.asp?newsid=12&id=4.

19

NATURAL BIOPOLYMERS AS NANOCARRIERS FOR BIOACTIVE INGREDIENTS USED IN FOOD INDUSTRIES

Mohamed H. Abd El-Salam, Safinaz El-Shibiny
Dairy Department, National Research Centre, Cairo, Egypt

1 Introduction

There is a growing recognition about the role of the functional foods and nutraceuticals in improving health and quality of life. This has been mostly derived from the scientific evidences to support the concept of health-promoting ingredients, health awareness, and the proactive role of consumers in their health management. The potential health effects of numerous food bioactive ingredients have been described including phytochemicals, vitamins, bioactive peptides, antioxidants, essential fatty acids, and so forth. The characteristic properties of free forms of many micronutrients are responsible for the encountered difficulties in their incorporation in commercial food products. The observed constraints can be summarized in the following:

1. Many micronutrients have low solubility in oil and/or water.
2. The susceptibility of many micronutrients to degradation by physical treatments, or chemical and enzymatic reactions during food processing, transport, storage, or preparation
3. The reduced acceptability of food products fortified with micronutrients because of distinct off-flavors. Therefore, masking the off-flavors of these micronutrients is needed before their addition to foods.
4. Some micronutrients can interact with other food constituents. They may adversely affect the bioavailability of the micronutrient and/or the product stability. In order to avoid these effects, added micronutrients should be isolated from other food constituents.

Encapsulations. http://dx.doi.org/10.1016/B978-0-12-804307-3.00019-3

5. Some micronutrients have an inherently low or variable oral bioavailability. It is noteworthy that the enhancement of the bioavailability is desirable for many bioactive food ingredients since their effective bioactivity can be increased and a lower amount might be needed (Acosta, 2009).
6. The small micronutrients quantities needed for fortification make it difficult to incorporate them homogenously in the food matrix.

Therefore, it is necessary to protect these bioactive ingredients from harsh environmental conditions, and to mask their disagreeable sensory properties to ensure the retention of their health benefits without affecting the quality of the carrier food. This can best achieved by encapsulation of the bioactive ingredients before their inclusion in foods. One of most important applications of food nanotechnology has been the development of nanodelivery systems for micronutrients. Colloidal nanoparticles (NPs) are among these systems that can protect and release the encapsulated micronutrients (McClements et al., 2009). Nanoparticles include nanospheres in which the encapsulated material is uniformly dispersed in the matrix and nanocapsules in which the entrapped material is confined in a cavity surrounded by a layer of the used polymer.

Polymer-based NPs can be fabricated easily into variable designs, which can offer effective stability of the encapsulated bioactive ingredient against adverse environmental conditions. Also NPs allows for sustained and controlled release of the entrapped bioactives in the target site. Various functional materials have been developed for micro/nanoencapsulation of bioactive ingredients. Smaller particles provide a larger surface area for the diffusion path of the entrapped material and in turn their bioavailability. These particles may be prepared from (1) a single polymer species by self-assembly or controlled aggregation or (2) from mixed polymer system by inducing phase separation based on aggregative or segregative interactions. The most common shape of NPs is spheroid but they may have a variety of internal structures such as homogeneous, core-shell, dispersion, or cluster structures. The importance of the particle size can be understood from its effects on the optical clarity, physicochemical stability, encapsulation, and release characteristics, and bioavailability of encapsulated bioactive ingredients. In nanotechnology NPs are defined as particles of radius less than 100 nm (Weiss et al., 2006). However, in many studies, the reported particle sizes of the developed colloidal delivery systems for food bioactives exceeded this limit (up to 500 nm) while still described as NPs.

Table 19.1 Advantages and Limitations of Polysaccharides and Proteins as Nano Carriers for Bioactive Food Ingredients

Polysaccharides	Proteins
1. They are safe, biocompatible, biodegradable	1. They are safe, biocompatible, biodegradable
2. Can be modified to achieve the required properties	2. Have a wide range of functional properties
3. Versatile carriers to bind and entrap a verity of hydrophilic and hydrophobic bioactive food ingredients	3. Consistent composition and properties
4. They are considered as a suitable shell under high temperature processes	4. Can interact with wide range of hydrophilic and hydrophobic bioactive compounds
5. Resistant of gastric and intestinal conditions	5. Can be easily restructured in nano forms by several physical and chemical treatments
6. Slightly affected by pH and ionic strength of solution	6. Liberates on digestions several bioactive peptides
7. Their composition and properties are greatly affected by the source and method of extraction	7. Easily affected by changes in pH and ionic strength
	8. Poor resistance to intestinal conditions
	9. May develop allergic reactions

One of the main criteria for selecting the encapsulant materials is their safety and that they are of food grade, that is, generally recognized as safe (GRAS). This limits the number of functional materials that can be used in the delivery of bioactive food ingredients. In addition, the choice will depend on the ability of the biopolymer to self-assemble into particles with defined functional properties (eg, size, morphology, charge, permeability, and environmental stability). Also, the cost of the raw materials, ease of production, and regulatory status should be considered. The fabrication of the delivery systems should be easily scaled to industrial production. Natural biopolymers can be considered as the main inventory for suitable materials to be used as nanocarriers for the bioactive food ingredients. Polysaccharides and proteins are the two categories of natural biopolymers that can be used in nanoencapsulation and each category has its advantages and limitations (Table 19.1).

The past decade has witnessed major advances in the development of food-grade colloidal delivery systems based on proteins, polysaccharides, and their conjugates. These systems have been designed to encapsulate, protect, and release many food bioactive components. Although the natural polymers have been used in other nanodelivery systems for nutraceuticals such as nanoemulsions and liposomes, they are used mainly in delivery systems based on NPs.

Therefore, the present chapter has been devoted to the preparation of nano carriers based on natural biopolymers NPs and micelles for the nanodelivery of bioactive food ingredients.

2 Natural Biopolymers Used in Nanoencapsulation

Natural biopolymers are the most suitable materials for the formation of NPs that can be incorporated in foods. Depending on their characteristics and the conditions of aggregations, polymers can self-assemble in various particle structures. In addition, the electrical properties of the obtained NPs are important to keep them stable in the medium. Proteins and polysaccharides are the two classes of the natural polymers that have been used or have the potential to be used in nanoencapsulation of bioactive food ingredients.

2.1 Proteins

Proteins are a class of natural polymers that have unique functionalities and potentialities in food, medical, and material applications. They are ideal materials for nanoparticle preparation because of their amphiphilicity, which allows them to interact well with both the entrapped bioactive ingredients and the medium. Nanoparticles derived from natural proteins are biodegradable and metabolizable. Proteins are classified in groups of defined characteristics (Fig. 19.1).

2.1.1 Globular Proteins

Globular proteins are characterized by their ordered structures, which variably change into unordered structures (denatured) and subsequent formation of heat-set gels when subjected to heat treatments under controlled ionic strength and pH. Also, they can form cold-set gels by thermal denaturation followed by gelation in the presence of a coagulating agent. Further, the amphiphilic properties of proteins have been used as a driving force for self-assembly into supramolecular structures. The size and shape of formed aggregates can be determined by controlling the heat denaturation conditions (Chen et al., 2006). Both the native and denatured forms of globular proteins have been used as nanocarriers of bioactive food ingredients.

2.1.1.1 Whey Proteins

Whey proteins are mixtures of globular proteins (β-lactoglobulin (β-Lg), α-lactalbumin (α-La), blood serum albumin

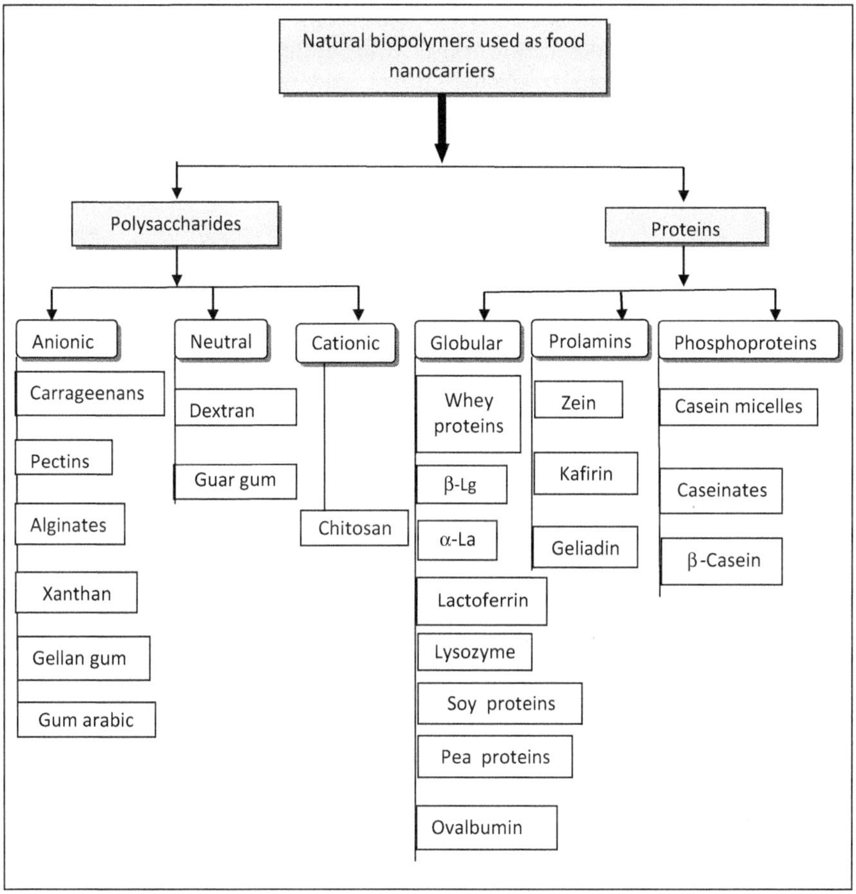

Figure 19.1. Natural biopolymers used as nanocarriers.

(BSA), lactoferrin (Lf), immunoglobulins (Ig) that differ greatly in their composition and functional properties (McSweeney and Fox, 2013). Whey proteins are generally resistant to some proteolytic enzymes and therefore can possibly serve for enteric delivery. Whey proteins retain their native structures in their commercial preparations namely: whey protein concentrates (WPC), which have variable protein contents (35–80%) and whey protein isolates (WPI) of protein content >90%. These preparations have been used extensively as nanodelivery vehicles for many bioactive food ingredients. Beta-lactoglobulin (β-Lg), the major whey protein, is able to bind small hydrophobic molecules in three binding sites namely: its internal cavity, the outer surface near Trp19-Arg124, and the monomer–monomer interface of the β-Lg dimer (McSweeney and Fox, 2013). Thermal treatments induce denaturation of bovine β-Lg through an initial dissociation from the native dimer to monomer β-Lg, followed by a change in the polypeptide chain conformation and subsequent aggregation. Denatured bovine β-Lg can form particulate, fibrous, or amyloid fibrile materials depending upon the precise heating conditions. α-La is the second major whey protein, which can apparently bind retinol and palmitic acid. α-La is more resistant to thermal denaturation than β-Lg. Partial hydrolysis of α-La with microbial enzyme from *Bacillus licheniforms* resulted in the formation of nanotubes (Gravel and-Bikker and de Kruif, 2006). BSA, a natural carrier of small molecules in the blood, contains three domains specified for metal ion, lipid, and nucleotide binding respectively. Lf is an iron binding protein that has several beneficial health effects including its antimicrobial effect of harmful microorganisms, regulating the immno system. Mechanical treatment such as high hydrostatic pressure treatment (HHP) changes markedly the conformation of heat denatured whey proteins and their interaction with lipophilic bioactive compounds (Relkin et al., 2014). Increasing the time-intensity of HHP decreased the particle size and increased the α-tocopherol–protein interaction.

2.1.1.2 Ovalbumin

Ovalbumin (OVA) is the main protein of egg white proteins (EWP), which has good functionality, particularly foaming and gelling properties. OVA is a globular protein of 43 kDa molecular weight and containing 385 amino acid residues, 1 disulphide bond, and 4 free sulphydryl groups. As a globular protein OVA can be modified through different treatments such as heating, high pressure, enzymatic reaction, shaking, and so forth. Environmental conditions under which denaturation takes place can lead to OVA aggregation driven by molecular interactions between exposed hydrophobic patches and disulfide bonds formation. Moreover, under controlled ionic strength and pH conditions, heated OVA dispersions can produce protein aggregates with different sizes and morphologies (Nyemb et al., 2014), which affect its binding to hydrophobic molecules such as the polyunsaturated fatty acids, PUFA (Sponton et al., 2015)

2.1.1.3 Pea Proteins

Dry pea seeds contain about 20–25% crude protein, among them 70% are salt-soluble globulins that are composed of two fractions, namely: legumin 11S and vicilin/convicilin 7S. These are constituted of heterogeneous subunits assembled into high molecular weight oligomers. Pea proteins heat-set gels exhibit lower textural features than those obtained with their soy counterparts processed under the same conditions. In addition, concentrated pea proteins suspensions and high salt concentrations are required to induce heat gelation of pea proteins. Moreover, slow heating/cooling rates applied would be required for pea protein molecules to form gel network of enhanced elasticity (Mession et al., 2015).

2.1.1.4 Soy Proteins

Glycinin and β-conglycinin are the two major components of soy protein. Glycinin is a hexameric protein composed of alternating (1) acidic and (2) basic units around the ring and linked by disulfide bridges, except for the acidic polypeptide A4 (Chen et al., 2014a). Three different β-conglycinin subunits are known as α′, α, and β, which are associated via hydrophobic interactions. Above the isoelectric point, the onset denaturation temperature of glycinin is around 80–90°C, and β-conglycinin starts to denature around 60–75°C Soy protein isolate (SPI), is the most commercially available soy protein product (Liu and Tang, 2013). SPI is unique in its high content (over 0.6 mol/mol protein) of hydrophobic amino acids, which enables a strong hydrophobic interaction with encapsulated compounds. In addition, SPI contains a considerable percentage of polar and charged residues, which leads to good water solubility and facilitates association with bioactive compounds through electrostatic attraction and hydrogen bonding (Teng et al., 2013b).

2.1.2 Phosphoproteins

Bovine milk contains four different caseins, namely: αs_1-, αs_2-, β- and κ-caseins (CN), present as naturally self-assembled micelles (McSweeney and Fox, 2013). The structures of caseins are quite unique. They are open-structured rheomorphic phosphoproteins, which have distinct hydrophobic and hydrophilic domains. The hydrophobic and charged residues are not uniformly distributed along the polypeptide chains of the four caseins. Therefore, caseins have a distinctly amphipathic character that allow them to form self-assembled aggregates of different sizes. The casein micelles are spherical colloidal particles with an average diameter of 150 nm in which the different caseins are held together mainly by hydrophobic interactions, and by calcium phosphate nanoclusters, bridging between their serine-phosphate residues. Casein micelles can be reassembled from acid casein. The reassembled casein micelles have almost similar sizes and properties to the natural micelles. Also, individual caseins can self-assemble to form micelles of variable sizes. In the case of β-casein, it forms nanosized micelles while the other caseins form micelles of larger and uncontrolled sizes. Therefore, β-casein has been used as nanocarriers for several bioactive ingredients. Also, sodium caseinate (SCN) has been successfully used in combination with other biopolymer as wall material for encapsulation of several bioactive food ingredients.

2.1.3 Prolamines

Prolamines are a group of cereal storage proteins characterized by their solubility in aqueous ethanol solutions but insoluble in water. Zein is the major storage protein of maize (*Zea mays*). It consists of four major (α-β-γ- and ô-zein) hydrophobic, alcohol-soluble proteins (prolamines), while the α-component accounts for almost 85% of the whole protein (Rishi and Munir, 2001). About two thirds of the amino acid residues of zein are hydrophobic, which allows for the solubilzation of the protein in 60–90% aqueous ethanol solution, but not in water (Dong et al., 2013). The characteristic solubility of zein in ethanol solution, in addition to being biodegradable and biocompatible makes it suitable material for encapsulation of hydrophobic bioactive ingredients. Zein has relatively low surface charge and in turns a weak electrostatic repulsion between its particles, leading to the formation of large aggregates. Zein is a very good candidate to generate colloidal particles by the simple antisolvent method (Patel et al., 2010). NPs of sizes that range from 100 to 200 nm can be easily fabricated by shearing an ethanolic solution (55–90%) of zein in deionized water (Zhong and Jin, 2009). Smaller particles can be formed at high shear rate, high ethanol, and low zein concentrations. Zein can interact with other charged biopolymers, such as caseinate and CS to form hybrid particles (Patel et al., 2010; Luo et al., 2011). Also, zein particles may exhibit mucoadhesive properties (Joye and McClements, 2014) increasing their residence time in the gastrointestinal tract and the bioavailability of the entrapped active ingredients. Gliadin is one of the two wheat storage proteins. It is a monomeric protein rich in glutamine and proline residues and a very low content of charged amino acids (Duclairoir et al., 2002). Similar to zein, gliadin is soluble in 70% aqueous ethanol solution, has low surface charge and is biocompatible and biodegradable. The relatively high level of neutral and lipophilic amino acids of gliadin promotes its interaction with mucosa to increase the bioavailability of the encapsulated bioactive ingredients (Joye et al., 2015b). Kafirin is the prolamine protein found in sorghum grain. It is one of the structurally tunable and commercially viable biomaterials suitable for encapsulation of bioactive food ingredients. Its unique properties, such as water insolubility, biodegradability, low toxicity, high hydrophobicity, nonallergenicity, and reduced susceptibility to proteolysis support its use in nanoencapsulation. Compared to zein, kafirin is relatively more hydrophobic and less digestible (Xiao et al., 2015), which enables it to be a more effective encapsulation vehicle with stronger protective function.

2.2 Polysaccarides

Polysaccharides are natural polymers of monosaccharides that vary in the type, number, distribution, and bonding of the monomers in the chain. They can be classified according to their charge into neutral, acidic, and cationic (Fig. 19.1). Due to their massive molecular structure and their ability to self-assemble, many polysaccharides can be used to build delivery systems for bioactive ingredients.

2.2.1 Chitosan

CS is a linear polysaccharide, composed of glucosamine and *N*-acetyl glucosamine units linked via β(1→4) linkage. Depending on the method of preparation these two monomers are randomly or block distributed throughout the biopolymer chain. CS is the only natural polysaccharide that has a surface of high-density positive charge. Therefore, it has the ability to dissolve in acidic medium. CS prepared from shrimps has been approved as GRAS (FDA, 2005) (www.fda.gov/ucm/groups/fdagov-public/@fdagov../ucm337459.pdf). CS has several nutritional and health properties, such as hypocholesterolemic, immunity-enhancing, anticancer effects, and acceleration of mineral absorption (Xia et al., 2011). Also, CS has a wide spectrum of antimicrobial properties against foodborne microorganisms (Rabea et al., 2003). On the basis of this property and its film-forming property, food industry has used CS singly or in combination with other antimicrobial agents in coating fresh cuts to extend their shelf life. Also, CS and its Maillard conjugates have been widely used to stabilize and improve overall quality of emulsions. In addition, CS has many favorable properties as a candidate for encapsulation of food bioactives, such as nontoxicity, biocompatibility and biodegradability (Kean and Thanou, 2010). CS can be modified by chemical and physical treatments. Chemically modified CS can be prepared by N-alkylation, hydroxyalkaylation, carboxyalkaylation, and so forth (Luo and Wang, 2013). Also, CS can be modified by physical treatments such as electromagnetic radiation and sonication. The modified CSs are characterized by improved solubility in aqueous solutions at different pHs, modulated surface charges, and higher absorption efficiency. Modified CS can be tailored for special nanoencapsulation purposes.

2.2.2 Pectin

Pectin is a complex mixture of linear polysaccharides of variable composition depending on the source and the conditions applied during its isolation. Pectin consists mainly of D-galacturonic acid (GalA) units, joined in chains by means of α-(1→4) glycosidic linkage. Part of the carboxyl groups of the GalA is naturally present as methyl esters. Pectin contains from a few hundred to about 1,000 saccharide units, which corresponds to average molecular weights from about 50,000 to 150,000 Da. Pectin is classified on the basis of the degree of estrification (DE) to high methoxyl (HM) pectin (DE 60–75%), and low methoxyl (LM) pectin (DE 20–40%). These two groups of pectin form gels by different mechanisms. HM pectin forms gels at pH around 3.0 using a minimum amount of soluble sugars. HM-pectin gels are thermally reversible. LM-pectin produces gels at wide range of pH and in the presence of a controlled amount of calcium or other divalent cations (Sundar Raj et al., 2012).

2.2.3 Alginates

Alginates are natural anionic polysaccharides isolated from brown marine algae (*Phaeophyceae*). They are linear copolymers of (1→4)-β-D-mannuronopyranosyl and (1→4)-α-L-guluronopyranosyl units in homopolymeric sequences. Alginates can form gels in the presence of divalent cations, such as Ca^2. Thus alginate beads can be fabricated by dropwise addition of sodium alginate solution into $CaCl_2$ solution to induce cross linking of the polymer chain to form an egg-box like configuration. The size of the formed particles depends on solution concentration and the dimensions of the initial extruded droplets. At low pH values, alginate beads are insoluble, which makes them favorable candidates for delivery of acid sensitive bioactive ingredients in the intestine or to retard the release of bioactive compounds in acidic foods.

2.2.4 Gum Arabic

Gum arabic (GA) is the exudates gum of *Acacia* trees. It is a complex mixture of arabinogalactan oligosaccharides, polysaccharides, and glycoproteins. It is a branched, neutral, or slightly acidic substance. Several factors affect the chemical composition of GA including the source, climate, season, age of trees, rainfall, and time of exudation. The backbone of GA has been identified to consist of α-(1→3)-linked D-galactopyranosyl units. The side chains are composed of 2–5 β-(1→3)-linked D-galactopyranosyl units, joined to the main chain by 1,6-linkages. Both the main and the side chain contain units of α-larabinofuranosyl, α-l-rhamnopyranosyl, β-D-glucuronopyranosyl, and 4-O-methyl β-D-glucuronopyranosyl. GA has a unique ability to create a strong protective film around oil droplets due to its highly branched arabinogalactan-protein structure (Wandrey et al., 2010). The hydrophobic polypeptides anchor the polysaccharide onto the surface of the oil droplet and the hydrophilic carbohydrate chains prevent the aggregation by forming a thick charged layer. GA is compatible with most other plant hydrocolloids, proteins, carbohydrates, and modified starches. GA acts as a useful prebiotic, which promotes beneficial intestinal physiological effects (Wandrey et al., 2010).

2.2.5 Carrageenans

Carrageenans are a family of high molecular weight anionic polysaccharides obtained from marine algae (Wandrey et al., 2010). They have variable structures depending on the source and conditions of their preparation. Generally the polymer chains comprise alternating (1→3)-linked β-D-galactopyranosyl and (1→4)-linked α-D-galactopyranosyl units. Three types of carrageenans are commercially available namely: κ- (kappa), ι- (iota), and λ- (lambda), which contains one, two, and three sulfated groups/repeating dimer unit of the polymer, respectively. Viscous solutions or thermally reversible gels can be obtained from carrageenans depending on their types. In addition the texture and rheological properties of the obtained gels differ markedly from soft to elastic and from firm to brittle. κ- and ι -carrageenans have the ability to form elastic gels in the presence of certain cations such as K^+ and Ca^{2+}. Synergistic effects can be obtained when combining carrageenans with other gum types.

2.2.6 Xanthan

Xanthan is microbial gum produced by *Xanthomonas campestris*. It is a high molar mass anionic polyelectrolyte that occurs as a mixed salt of sodium, potassium, and calcium. Its backbone consists of β-(1→4)-D-glucopyranosyl units with every second unit having a trisaccharide side chain attached at the C-3 position, one D-glucuronosyl unit between two D-mannosyl units (Katzbauer, 1998). Approximately 40–50% of the terminal mannosyl units are 4,6-pyruvated. Xanthan is a nongelling gum but forms a high viscous solution. The secondary structure of xanthan gum consists of a fivefold helical structure. The xanthan gum molecule undergoes a thermally induced order–disorder conformational transition. The disordered form is favored by low salt concentrations and high temperatures, whereas salt stabilizes the ordered conformation.

2.2.7 Gellan

Gellan is an anionic polysaccharide of high molecular weight commercially produced by the microorganism *Sphingomonas elodea* in two forms, namely: low acyl (deacetylated) and high acyl gellan. It consists of tetrasaccharide repeating unit (one rhamnose, one glucuronic acid, and two glucose units). Low acyl gellan forms thermo-reversible gel upon cooling of its solutions in the presence of gelling cations while the high acyl gellan form gels without the need for cations. The resulting gels have variable textures depending on the degree of acetylation and concentration of the divalent cations present.

2.2.8 Dextrans

Dextrans are mainly linear neutral polymers of α-D-glucose linked by α-(1→6) glycosidic bonds, which can have variable amounts of α-(1→3) branches. They are microbial polysaccharides obtained by the fermentation of sucrose. The exact structure of each type of dextran depends on its specific producing microbial strain. Dextrans can be modified by the formation of covalent attachment between the hydroxyl groups of the polysaccharides and various organic functional groups especially hydrophobic compounds. Modification of dextrans changes their properties depending on the degree and type of substitution.

2.2.9 Cyclodextrins

They are a family of cyclic oligosaccharides prepared degradation and cyclyzation of the degradation products of amylase,or amylopectin by treatment with dilute acid or by enzymes from *Bacillus macerans*. Typical cyclodextrins contain six (α), seven (β) and eight (γ) glucose monomer units connected by α-(1→4) glucosidic bonds and forming cone-shaped ring structure with a cavity depth of 0.7–0.8 nm. The central cavity is lipophilic while the outside surface is hydrophilic. Therefore, hydrophobic food bioactives can be entrapped in their inner cavity through the formation of inclusion complexes.

2.3 Protein–Polysaccharides Complexes

Interactions between natural biopolymers under specific conditions (eg, pH, temperature, heating time, ionic strength, and concentration) can promote the formation of hydrogels with different functional properties in comparison with those of single polymer species. In addition to their role as protective carrier for bioactive food ingredients, they are designed to exhibit the required functional attributes within the final product (eg, optical properties, rheological properties, release characteristics, encapsulation properties, and physicochemical stability) (Jones and McClements, 2011). Hybrid biopolymer complexes prepared from proteins and polysaccharides may have new functional properties that combine the advantages of both building materials for use as nanocarriers for bioactive food ingredients. The protein–polysaccharides can be prepared in two different ways:

1. Protein–polysaccharide interaction via physical bonding such as van der Waals, electrostatic, hydrophobic, and hydrogen bonding. Proteins and polysaccharides of opposite charge can interact via electrostatic interaction. For example, anionic polysaccharide can interact with cationic groups on the protein surface, which may vary substantially depending primarily on environmental conditions such as pH, ionic strength, and temperature. Two approaches can be followed in preparing protein–polysaccharide complexation. In the first method, denatured forms of globular proteins are prepared by heating at specific conditions of temperature/time and pH for each protein followed by the formation of their complexes with ionic polysaccharides through hydrophobic interactions. In the second method, a mixture of the protein and polysaccharides solutions is heated at a temperature above the thermal denaturation of the protein at a specific pH. The particle size formed by the first method depends mainly on the size of the protein particle obtained after the heating step, while variable factors control the size of particles formed by the second method (Jones and McClements, 2011).
2. By covalent bonding as in the case of Maillard-type protein-polysaccharide conjugates. The reaction can be done by controlled dry heating of protein–polysaccharide mixtures where the ε-amino groups in the protein and the reducing end carbonyl group in polysaccharide interact to form protein-polysaccharide conjugates (Kato, 2002). The reaction rate seems to depend on the protein conformation being faster in the case of unfolded proteins like caseins. The folded proteins attach only one or two polysaccharides, while unfolded proteins attach several polysaccharide units. The protein-polysaccharide conjugates exhibit excellent emulsifying properties. In addition they had excellent heat stability, and antimicrobial activities. Also, conjugation of the allergen protein with polysaccharides may be effective in reducing their allergenicity.

An unlimited number of protein–polysaccharides can be prepared via the two approaches. The diversity and sensitivity of protein–polysaccharide complexes to external environmental conditions make them very versatile as building blocks for the development of stimuli-sensitive carriers for controlled delivery of encapsulated compounds, to specific sites of the gastrointestinal (GI) tract and at a specified rate of their release (Semenova et al., 2014)

3 Methods of Fabrication of Nanoparticles from Natural Polymers

A wide range of different methods have been described in the literature for the fabrication of carbohydrate- and protein-based nanoparticles (Jones and McClements, 2011; Joye and McClements, 2014). Table 19.2 summarizes the principles, advantages, and limitations of some of the mostly widely used techniques for encapsulating food bioactives.

4 Nanoencapsulation of Bioactive Food Ingredients

4.1 Nanoencapsulation of Phytochemicals

Phytochemicals are defined as natural chemical compounds found in "plants" which possess biological activities beneficial to human health. The prefix "phyto" in phytochemicals originates from a Greek word meaning "plant." They include several groups of chemicals such as polyphenols (flavonoids and nonflavonoid polyphenols), phytosterols, and carotenoids. Their potentials for health promotion and disease prevention are due to their multiple biological activities, such as antioxidant, antiinflammatory, anticancer, antiviral, and antibacterial properties, their stimulation of the immune system, the modulation of enzyme activities, gene expression, regulation of cholesterol synthesis, and blood pressure (Liu, 2003).

Many phytochemicals are insoluble in an aqueous solution and have poor solubility in oil. They are chemically unstable and degrade rapidly when exposed to the external environmental conditions, such as oxygen, temperature, light, pH, and some other reactive substances. For these reasons, the production of phytochemicals as food ingredients and their utilization in the development of functional foods are limited. It has been shown that these problems can be overcome by using encapsulation technologies.

Table 19.2 Some Methods Extensively Used in the Preparation of Nanoparticles for the Delivery of Nutraceuticals (Jones and McClements, 2011; Joye and McClements, 2014)

Method	Principles	Advantages/limitations
Antisolvent/desolvation	Addition of a nonsolvent to a solution to induce solute precipitation. The main driving force is the imbalance of molecular interactions between solute, solvent, and antisolvent.	1. Simple/improved control over particle properties 2. Widely used in pharmaceutical industry to produce NPs 3. The need for the use of organic solvents
Extrusion/ion-gelation	Inducing gelation of dilute aqueous solution of charged polyelectrolyte by extrusion in solution containing ions of opposite charge	1. Rapid method can be used on an industrial scale to produce biopolymer particle 2. The particle size is dependent on the way and conditions of extrusion 3. Applying of ultrasonic during gelation can assist in reducing the particle size
Coacrevation	Addition of two oppositely charged biopolymers to interact with each other through electrostatic forces forming a precipitate.	1. One of the most easily implemented method for fabrication of NPs from natural biopolymers 2. Difficult to control particle size and to prevent particle agglomeration 3. The stability of the NPs depends on the biopolymer material used
Spray drying	Atomization of a solutions or suspensions of bioactives and biopolymers to fine droplets into an opposite.stream of hot air, which induces the quick evaporation of solvent from the droplets and formation of nano/microspheres	1. Water-based dispersions are used and so avoids the use of organic solvents 2. A versatile single-step technique 3. Simple, economical, fast, and easy to scale-up 4. Reduces storage and transport costs and capacity 5. Large particle sizes
Inclusion complexation	Molecular encapsulation of a bioactive guest molecule into a cavity-bearing host molecule	1. Suitable for masking of odors or flavors or preservation of aromas
Fluid gel formation	Applying shear forces on a biopolymer solution undergoing gelation	1. Formation of tightly packed suspensions of gelled particles. 2. The particles produced can be easily deformed and have an irregular shape 3. The parameters should be carefully controlled

4.1.1 Polyphenols

Polyphenols are secondary metabolites in plants. They consist of a large number of diverse groups of compounds ranging from simple molecules of low molecular weight up to very complex molecules of high molecular weights. Polyphenols have several potential health effects, but they suffer from low solubility, stability, and bioavailability in their free forms. In addition, they can interact with food constituents (fat, carbohydrate, proteins), which may have significant effects on their biological activities (Jakobek, 2015). The interactions between polyphenol and proteins occur mainly through hydrophobic interactions while the formation of hydrogen bonds play the main role in polyphenol-carbohydrate interactions (Jakobek, 2015). Therefore, one of the ways to benefit from the health effects of polyphenols is to encapsulate them. Several polyphenols have been subjected to nanoencapsulation as follows.

4.1.1.1 Resveratrol

Resveratrol (Res) is a polyphenol (*trans*-3,5,4trihydroxystilbene) that can be isolated from grapes and by-products of the wine industry. Res demonstrated many health promoting effects including antioxidant, antiinflammatory, anticancer, neuroprotective, cardioprotective, and hepatoprotective effects (Baur and Sinclair, 2006). The Res exists in *cis* and *trans* structural isomers, but only *trans*-Res demonstrates health benefits. The potential use of Res as a food supplement is faced by its poor solubility in water. In addition, Res is a very reactive molecule, very susceptible to reaction with dissolved oxygen, producing different degradation products, as well as very easily degraded by sunlight. The solubility and stability of Res can be improved by complexation with SCN, β-lactoglobulin, Lf, and whole buttermilk. High encapsulation and retention efficiencies of resveratrol have been achieved using several biopolymer-based delivery systems. The encapsulation of resveratrol in both CS and pectin microspheres improved the chemical stability of resveratrol during storage (Peng et al., 2010; Das et al., 2010). The high encapsulation efficiency (EE), good retention capacity, and improved stability of resveratrol within biopolymer microspheres suggest that these may be good candidates for the fabrication of delivery systems suitable for use in the food industry. Res has been successfully encapsulated in NPs formed by antisolvent precipitation of hydrophobic proteins (zein or gliadin) and coated with hydrophilic copolymers (pectin or sodium caseinate). Resveratrol was best protected against UV-light when encapsulated in sodium caseinate-coated zein particles (Joye et al., 2015a; Davidov-Pardo et al., 2015a). This approach has been further developed by encapsulating Res in biopolymer NPs consisted of a zein core surrounded by a Res bound to caseinate or caseinate–dextran shell. Both NPs protected Res against UV light and enhanced its bioaccessibility, but NPs coated by caseinate–dextran were more stable to aggregation under simulated gastrointestinal conditions than those coated by caseinate (Davidov-Pardo et al., 2015b).

4.1.1.2 Tea Polyphenols

Green tea contains a large and heterogeneous group of flavonoids of which catechins account for 80%. Epigallocatechin-3-gallate (EGCG) is the most prevalent and bioactive tea catechin. EGCG inhibits the growth of cancer cells, such as human cervical cancer cells or colon cancer cells. Also, EGCG is a potent antioxidant which can act up to 10 times as effectively as L-ascorbate or β-carotene but can be easily oxidized in aqueous environments, especially at neutral and basic pHs. Several biopolymers have been used in nanoencapsulation of EGCG (Table 19.3). All these systems increased variably the solubility, stability, and bioavailability of the EGCG.

Table 19.3 Nanoencapsulation of Tea Polyphenols (EGCG/Catechins)

Polymer Used	Method	Results	References
CS-TPP	Ionic gellation	Increased stability, enhanced anticancer and antiatherogenic activities	Hu et al. (2008)
Heat denatured β-Lg	Self-assembly	<50 nm NPs, increased oxidative stability, suppress astringency and bitterness of EGCG, slow release in SGI	Shpigelman et al. (2012)
Heat denatured β-Lg	Self-assembly	Variable NP sizes, highest protection of EGCG antioxidant activity was obtained with β-Lg heated at 85°C and the molar ratio of 1:2 (β-Lg:EGCG)	Li et al. (2012)
CS/CMP	Electrostatic interaction	Less toxicity than that CS-TPP	Hu et al. (2012b)
β-Cyclodextrin	Molecular inclusion	67–470 nm NPs, formation of catechin-β-cyclodextrin supra molecule, catechin in amorphous form	Krishnaswamy et al. (2012)
BSA/CS	BSA-EGCG NP by dissolvation/ CS coat	~300 nm NPs, Improved stability at 60°C, prevent aggregation of NP at pH 4.5–5.5, increase permeability on Caco-2 monolayer	Li et al. (2014a)

(*Continued*)

Table 19.3 Nanoencapsulation of Tea Polyphenols (EGCG/Catechins) (*cont.*)

Polymer Used	Method	Results	References
Native/denatured β-Lg	Self-assembly	Denatured β-Lg, higher binding, increased stability, unchanged bioefficacy	Lestringant et al. (2014)
CS/polyaspartic acid	Self-assembly	102 nm NPs, limited release in SGF and SIF, improve rabbit atheroschlerosis	Hong et al. (2014)
Ovalbumin–dextran conjugate	Self-assembly/ heating 80°C Cross-linking	Spherical NPs of sizes 285 and 339 nm	Li and Gu (2014)
Casein-dextran conjugate	Self-assembly	Limited release of EGCG in SGF and SIF, increase permeability on Caco-2 monolayer	Xue et al. (2014)
Lf	Interact with EGCG at pH 3.5	~300 nm NPs, protection of EGCG from degradation, slow and sustained release in GI fluids	Yang et al. (2014)
CPP-CS/genipin	Electrostatic interaction/ cross-linking	Formation of NP, which had high ζ-potential, small size and soluble, entrapped catchins-Zn improved stability, slow release in GI	Hu et al. (2014)
CS-TPP	Ionic gelation	High antioxidant activity, sustained release of catchins-Zn comples	Zhang and Zhao (2015)
Gallic acid grafted CS, gallic acid grafted CS-CPP	Ionic gelation	Spherical NP ~300 nm and zeta potential of <+30 mV, high solubility under neutral and alkaline environments, strong antioxidant activity and cytotoxicity against Caco-2 colon cancer cells, controlled release of EGCG	Hu et al. (2015b)

CS, chitosan; *TPP*, tripolyphospate, *CPP*, casein phosphopeptide; *CMP*, casein macropeptide; *rCM*, reassembled casein micelles; *BSA*, bovine serum albumin; *EGCG*, epigallocatechin-3-gallate; *BSA*, blood serum albumin; *NP*, nanoparticle; *CS-TPP*, chitosan-tripolyphosphate; *CPP-CS*, casein phosphopeptide-chitosan; *CS/CMP*, casein macropeptide/chitosan.

4.1.1.3 Curcumin

Curcumin (diferuloylmethane) is a natural polyphenol isolated from rhizome of turmeric (*Curcuma longa*). Curcumin possesses antioxidant, antiinflammatory, anticancer or antitumor, and antiallergic activities (Chin et al., 2014). However, the inclusion of curcumin in functional foods is limited by its extremely low water solubility and poor bioavailability. In order to improve its solubility and bioavailability, curcumin has been encapsulated in NPs prepared from natural polymers and polymer conjugates (Table 19.4).

Table 19.4 Nanoencapsulation of Curcumin in Natural Polymers Nanoparticles

Polymer	Method Used	Particle (nm)	Results	References
Zein	Antisolvent precipitation	100–150	Good stability in GI conditions, retain 60% of curcumin adhesive activity for 150 min	Patel et al. (2010)
β-Casein	Micillization	—	Increased solubility, bioavailability, and stability	Esmaili et al. (2011)
Zein	Electrohydro dynamic atomization	175–900	Retain shape and curcumin content for 3 month storage	Gomez-Estaca et al. (2012)
Soy protein isolate	Solvent desolvation and cross-linking with glutraldehyde	220.1–286.7	Biphasic release of curcumin (slow followed by rapid)	Teng et al. (2012)
Na caseinate	Spray drying	168.7 ± 10.2	Improved solubility and biological activity	Pan et al. (2013)
Starch	Precipitation in water/oil microemulsion	87	Sustained release of curcumin	Chin et al. (2014)
Zein/pectin (core/shell)	Electrostatic interaction	Spherical ~250	Curcumin in amorphous form	Hu et al. (2015a)
Soy protein isolate	Self-assembly	74–90	Increased solubility and stability of curcumin, unchanged protein digestability	Chen et al. (2015)

GI, gastrointestine.

4.1.1.4 Quercetin

Quercetin, 3,3′,4′,5′-7-pentahydroxy flavone, is a hydrophobic bioactive compound that has many potential beneficial effects on human health. Similar to other hydrophobic bioactive compounds, quercetin has low bioavailability due to its poor solubility in aqueous solutions and stability during food processing, storage and in the intestinal tract. A complex was formed between quercetin and CS through hydrophobic interaction and precipitated by ion gelation using TPP as NPs of average particle size of 76.58 nm (Zhang et al., 2008). The entrapped quercetin exhibited increased dispersibility in water and bioavailability. Aqueous solutions of quercetin and zein were added simultaneously to Na caseinate solution to entrap quercetin in the formed colloidal particles (Patel et al., 2012). The shape of the formed particles changed from needle-like to spherical particles with the increase of the zein concentration. The particle sizes ranged from 130 to 161 nm and carried negative surface charge (–30 to –41 mV). The encapsulated quercetin retained high stability against chemical and photodegradation. CS was modified with different percentages of Linoleic acid (LA), mixed with β-Lg and precipitated by ion gelation with TPP (Ha et al., 2013). The CS-LA/β-Lg preparations were used to encapsulated quercetin in the form of spherical shaped NPs with sizes of 170–350 nm. The EE of quercetin increased with the increase in the LA charge and low temperature of encapsulation.

4.1.1.5 Tangeretin

5,6,7,8,40-Pentamethoxyflavone (PMF) is a flavonoid found in citrus fruits. The PMFs are considered as functional ingredients in foods and pharmaceuticals because of their potential beneficial activities, such as anticarcinogenic activities and antiinflammatory activities. However, the low solubility of PMFs makes them difficult to be incorporated into liquid food products. Tangeretin-loaded protein NPs were produced by mixing ethanolic solution of zein and tangeretin with an aqueous β-Lg solution and then converted into powder by freeze-drying (Chen et al., 2014b). Particles of the obtained powder had well-defined spherical shapes, with diameters ~200 nm. This powder formed a colloidal suspension when dispersed in water that is relatively stable to particle aggregation and sedimentation.

4.1.1.6 Genistein

Spherical dextran NPs (100–450 nm) were enzymatically prepared by dextransucrase under optimal conditions of pH 5.2–6 and sucrose concentration >0.5 M (Semyonov et al., 2014). The isoflavone genistein was loaded in the dextran NPs using acidification and dimethylsulfoxide (DMSO). The DMSO gave better genistein load than the acidification method. Freeze drying increased the yield of loaded genistein.

4.1.2 Carotenoids

Carotenoids are natural pigments widely found in plants and algae. They share the isoprenoid carbon skeleton, which is responsible for their high hydrophobicity and low solubility in both aqueous solutions and oils. In addition to the nutritional value offered by some of them, such as β-carotene (BC) as a precursor for vitamin A, several carotenoids exhibit variable health effects (Yi et al., 2015). Increasing the uptake of BC can decrease the risk of age-related macular degeneration, cancer, and cardiovascular diseases. In addition to its nutritional and health effect BC can be used as natural coloring agent in processed foods. Lycopene (LY) is the main carotenoid present in tomato and tomato products. LY has received great interest because of its potential health effects in the prevention of chronic diseases such as atherosclerosis, skin cancer, and prostate cancer (Xue et al., 2013). However, LY can be susceptible to oxidation during storage, especially when stored in the presence of oxygen. Lutein (LU) is an antioxidant carotenoid pigment, which protects the retinal epithelial cells from reactive oxygen species, the skin from UV-induced damage, and may reduce the risk of coronary heart disease (Hu et al., 2012a). Fucoxanthin (FUCO), a nonprovitamin A, marine xanthophyll carotenoid, exerts beneficial effects including hypolipidemic, antiobesity, antidiabetic and anticarcinogenic effects (Ravi and Baskaran, 2015). The low solubility, stability, and bioavailability of carotenoids limit their use in fortification of foods. Nanoencapsulation has been used to overcome these limitations (Table 19.5).

4.1.3 Other Phytochemicals

Indole-3-carbinol (I3C) is a hydrolysis product from the sulfur-containing phytochemicals (glucosinolates) of cruciferous vegetables. The I3C has potential health effects to influence carcinogenesis. However, it suffers from low stability and conversion to the oilgomer 3,30-diindolylmethane (DIM). The efficacy of I3C and DIM were greatly improved when they were encapsulated in zein and zein/carboxymethyl chitosan (CMCS) NPs (Luo et al., 2013). Both NPs provided controlled release of I3C and DIM in PBS medium and similar protection for the two bioactives against UV light. However, zein/CMCS NPs exhibited better protection of I3C against degradation and decreased its oligomerization to DIM under thermal conditions.

4.2.2 Nanoncapsulation of Lipids

4.2.2.1 Fat-Soluble Vitamins

Vitamin D is an essential fat-soluble vitamin for human health. It takes part in calcium and phosphate metabolism, in the formation of osteoblasts, in fetal development, and so forth. The major forms of the vitamin class responsible for human health benefits are ergocalciferol (D_2) and cholecalciferol (D_3). The vitamin D_3 is the form that is synthesized in the skin from 7-dehydrocholesterol

Table 19.5 Nanoencapsulation of Some Carotenoids in Natural Biopolymer Nanoparticles

Polymer Used	Preparation Method	Results	References
β-Carotene			
Casein-grafted-dextran	Self-assembly micelle	Improved solubility and release in GI	Pan et al. (2007)
Na caseinate, WPI, SPI	Homogenization-evaporation	WPI was better in sustained release of β-C	Yi et al. (2015)
β-Lg, β-Lg-dextran	Homogenization-evaporation	Uniform β-Lg-dextran NPs (60–70 nm) were more resistant to aggregation with good permeability and release	Yi et al. (2014)
Reassembled casein micelles	Self-assembly	Improved stability under thermal and high-pressure homogenization	Säiz-Abajo et al. (2013)
Lycopene			
Zein	Spray drying	Improved stability and retarded release in stomach	Xue et al. (2013)
Lutein			
Low-molecular-weight CS	Ion gelation using TPP	Improved stability and bioavailability	Arunkumar et al. (2013)
Zein	Solution enhanced dispersion by supercritical fluids	High loading and entrapment efficiency, particle size (198–355 nm), controlled release of lutein	Hu et al. (2012a)
Fucoxanthin			
CS/glycolipid	Ion gelation using TPP	Improved bioavailability	Ravi and Baskaran (2015)

WPI, whey protein isolate; *SPI*, soy protein isolate; *GI*, gastrointestinal; *TPP*, tripolyphosphate.

when the skin is exposed to sunlight, whereas D_2 is only synthesized by plants and fungi. Vitamin D has long been known to play an important role in bone development by promoting calcium absorption in the gut and in bone mineralization. Also, there is a relation between vitamin D deficiency and the increased risk of chronic diseases, such as diabetes mellitus, cancer, autoimmune disorders, and osteoporosis (Calvo et al., 2004). VD_3 was encapsulated in reassembled casein micelles (rCM), which offered protection of the entrapped vitamin against light (Semo et al., 2007). The method was further improved by subjecting the rCM loaded with VD_3 to high-pressure homogenization (Haham et al., 2012). This modified method was reported to improve the heat stability and bioavailability of entrapped VD_3. The conjugate of oleoyl alginate ester (OAE) was found to form self-assembled NPs at low concentrations in aqueous medium (Li et al., 2011). VD_3 was loaded in the formed NPs where the loading capacity (LC) increased with the increase of VD_3 concentration but the loading efficiency decreased. The entrapped VD_3 released from the NPs at sustained rate in GI fluids. VD_3 was encapsulated into zein NPs coated with carboxymethyl chitosan followed by cross-linking with Ca (Luo et al., 2012). The NPs (86–200 nm) had a spherical structure that provided better photostability against UV light and controlled release of VD_3 in both PBS medium and simulated gastrointestinal tract fluids. Glycol-CS was conjugated to VD_2(ergocalciferol hemisuccinate), which formed self-assembled NPs with sizes of 279 nm in aqueous solution and 50–90 nm after drying and VD_2 content of 8.4% (w/w). In vitro studies indicated the dependence of vitamin D_2 release on the solution acidity at an almost constant rate (Quiňonesa et al., 2012). VD3 was successfully incorporated into carboxymethyl CS–soy protein (CMCS/SPI) complex prepared by ionic gelation with Ca^{++} to form NPs (162–243 nm) (Teng et al., 2013a). CMCS/SPI showed better particle forming capability and increased loading efficiency than CMCS. Entrapped VD3 in CMCS/SPI NPs exhibited slower release in gastro stomach fluid (GSF) and faster release in gastrointestinal fluid (GIF) than that encapsulated in CMCS. VD was nanoentrapped in casein-maltodextrin (CN–MD) conjugate to form small (~30 nm) and stable NPs at the IP of casein (Markman and Livney, 2012). Conjugates conferred better protection against oxidation and degradation at low pH for VD. VD_2 was loaded in native casein micelles subjected to hydrostatic pressure at different temperatures (Menéndez-Aguirre et al., 2014). Loading of VD_2 reached 10.4 ± 0.2 μg/mg protein by treatment at 600 MPa and 50°C. The average diameter of VD_2 loaded micelles was 272 ± 10 nm. VD_3 was encapsulated in core-shell micelles formed from *N,N*-dimethyl-hexadecyl carboxymethyl chitosan (DCMCS). The micelles had smooth surfaces and average size of ~140 nm (Li et al., 2014b). The solubility of entrapped VD_3 was improved with higher EE (53.2%). VD_3 was initially released from the core–shell micelles rapidly and then followed by a sustained release. Vitamin D_3 entrapped in casein micelles was successfully used in the fortification of milk and cheese (Al khalidi, 2012). He found that over 90% of encapsulated VD_3 added to milk was retained and uniformly distributed in both cheddar and mozzarella cheeses, with ~8% of VD_3 loss in whey. Also, he demonstrated that the pizza packing process did not alter the bioavailability of VD_3 in mozzarella cheese (Al-khalidi, 2012).

Vitamin E (VE) is a mixture of tocopherols, the most potent of which is α-tocopherol. α-Tocopherol (TOC) is a labile compound to heat, light, and oxygen. Encapsulation was found to provide an effective way for its protection. Vitamin E was encapsulated in gliadin NP (~900 nm) with an EE of 77%, which showed dual release characteristics (Duclairoir et al., 2002). Zein has been used as nanocarrier for TOC (Luo et al., 2011) TOC-loaded zein NPs of sizes ranging from 300 to 1000 nm have been prepared by hydrophobic interaction between TOC and zein. Smaller and homogeneous TOC loaded NPs have been generated using zein/CS complex as wall material. These NPs exhibited better control over the release of the entrapped vitamin compared to the TOC/zein NPs. The formed particles were spherical with smooth surfaces and the size and zeta potential of the complex varied from 200 to 800 nm and +22.8 to +40.9 mV, respectively. Release of TOC showed burst effect followed by slow release. The zein/CS complex provided better protection of TOC against gastrointestinal conditions than the uncoated particles.

4.2.2.2 Fatty Acids

Omega-3 polyunsaturated fatty acids are important nutraceutical lipids, providing protection against cardiovascular and other diseases. Ocosahexaenoic acid (DHA), one of these acids, can bind to casein at the ratio of 3–4 DHA molecules per protein molecule with the formation of NPs with average diameter of 288.9 ± 9.6 nm (Zimet et al., 2011), and in the form of rCM when calcium and phosphate were added (at 4°C). DHA loaded in r-CM and casein NPs showed more resistance against oxidation and better keeping stability throughout storage at 4°C in comparison to the free DHA. Also, DHA was entrapped in β-Lg-pectin NPs prepared by electrostatic interaction of low methoxyl pectin at pH 5.2 (Zimet and Livney, 2009). The average size of the formed NPs was ~100 nm, which offered protection to the entrapped DHA against oxidation. Also, nanosized complexes were prepared from high amylase corn starch and flax seed oil as a source for omega-3 fatty acids (Gökmen et al., 2011). The prepared complexes were converted to powder by spray drying. Encapsulation significantly decreased lipid oxidation as measured by the formation of hexanal and nonanal in breads during baking. The DHA was encapsulated in zein ultrathin capsules produced by electrospraying (Torres-Giner et al., 2010). The capsules were spherical and had an average size of 490 ± 200 nm. The encapsulated ω-3 fatty acid showed a 2.5-fold reduction in the degradation rate constant at different conditions of relative humidity and temperature. Also, it had extended degradation induction time. A complex between nacaseinate and GA was prepared by electrostatic interaction and was used to nanoencapsulate fish oil as a source of eicosapentaenoic acid (EPA) and DHA (Ilyasoglu and El, 2014). The formed NPs (~233 nm) showed an ecapsulation efficiency of 78.88 ± 2.98%. The nanoencapsulated EPA/DHA were used to fortify fruit juice, but their stabilities under processing and storage conditions were not studied.

Conjugated linoleic acids (CLA) is a group of linoleic isomers conferred to have several health effects. Complexes between V-amylose and CLA were prepared by two methods using water/dimethylsulphoxide (DMSO) and KOH/HCl, respectively (Lalush et al., 2005). The first method yielded spherical NPs (150 nm) with superior protection to CLA against oxidation. The second method gave particles (43–160 nm) of elongated structure. All complexes showed high retention for CLA in GSF. Also, CLA was loaded in soy lipophilic protein nanoparticles (LPP), by ultrasonication (Gao et al., 2014). LPP (170 ± 0.63 nm) exhibited high LC (26.3 ± 0.40%), oxidation protection and a sustained releasing profile in vitro for CLA. Coating the surface of the CLA-loaded LPP with SCN improved its colloidal stability.

4.2.3 Essential Oils (EOs)

Essential oils (EOs) are natural volatile aromatic oils obtained from several parts of plants. They are complex blends of a variety of volatile molecules such as terpenoids, phenol-derived aromatic components characterized by their antimicrobial activities. Nanoencapsulation of EOs aims to decrease their volatility and to increase their stabilities, solubility in aqueous media and efficacy.

- Garlic oil (GO), was encapsulated in β-cyclodextrin in order to improve its stability and decrease its losses in foods due to its volatility. The formation of GO/β-CD (1:1) inclusion complex was demonstrated by different analytical techniques (Wang et al., 2011). The apparent stability and solubility of GO in the formed complex were greatly improved. Also, the prepared complex allowed for controlled release of the entrapped GO.
- Turmeric oil (TO) is an essential oil commonly used in food, cosmetic, and pharmaceutical applications due to its antimicrobial, antiinflammatory, antioxidant properties. TO was encapsulated in alginate-chitosan (Al-CS) nanocapsules by dropwise addition of O/W emulsion of TO in Al solution containing Tween 80 in CS–$CaCl_2$ solution (Lertsutthiwong and Rojsitthisak, 2011). Several factors affect the size of the formed

capsules including the concentrations of Tween 80 and CS and sequence of steps followed in the capsule formation.

- Thymol and carvacrol EOs, were encapsulated in zein NPs using the liquid-liquid dispersion method (Wu et al., 2012). After lyophilizing, only samples prepared under neutral and basic conditions formed NPs, which reduced the *Escherichia coli* counts by 0.8–1.8 log CFU/mL. Also, the antioxidant activity of the entrapped EOs was reduced in the range of 24.8–66.8% depending on the formulation.
- Thymol was loaded in WPI-maltodextrin conjugate nanocapsules by emulsion evaporation technique (Shah et al., 2012). An EE of 51.4% was obtained and the capsules formed a clear dispersion in water with diameter of <90 nm after heating at 80°C/15 min.
- Nanogels based on CS and cashew gum (CG) were prepared and loaded with *Lippia sidoides* EO(Abreu et al., 2012). Optimum nanogels (335–558 nm) that showed high loading (11.8%) and EE (70%) were prepared from ratios of matrix:oil 10:2, gum:CS 1:1, and 5% gum concentration. It exhibited slow and sustained release of the entrapped EO.
- Core/shell zein/SCN NPs have been prepared by direct pouring of SCN solution into zein solutions (Li et al., 2013). The NPs were spherical in shape, possessing strong LC for thymol. These NPs were also effective in delaying the growth of *Staphylococcus aureus*. Thymol loading resulted in a slight increase in particle size without affecting the redispersibility of the NPs. The NPs sustained the release of thymol in two-step biphasic process, that is, rapid release at first followed by slow release step.
- Eugenol EO was encapsulated in CS NPs prepared by ion gelation using TPP (Woranucha and Yoksan, 2013). The concentration of eugenol affected the LC and EE and NPs properties. Particles with LC of 12% and EE of 20% exhibited a spherical shape with an average size <100 nm. Encapsulation improved markedly the thermal stability of the EO in model system.
- Oregano essential oil (OEO) was encapsulated in CS NPs by ionic gelation with TPP (Hosseini et al., 2013). The obtained NPs were spherical in shape and exhibited a narrow size distribution ranging from 40 to 80 nm. The EE and LC of OEO-loaded CS NPs were about 21–47 and 3–8%, respectively.
- Self-assembled chitosan-benzoic acid (CS-BA) NPs were used to encapsulate thymol. The encapsulated EO showed higher inhibitory effect on *Aspergillus flavus* compared to the free thymol (Khalili et al., 2015).

- Cashew gum (CG) is a heteropolysaccharide extracted from the exudate of *Anacardium occidentale*. The structure of CG resembles that of GA (Herculano et al., 2015). *Eucalyptus staigeriana* EO was nanoencapsulated using CG as wall material (Herculano et al., 2015). The sizes of the formed NPs ranged widely from 27.70 to 432.67 nm and had negatively charged surfaces and encapsulation efficiency ranging from 24.89% to 26.80%. The NPs showed greater activity against gram-positive than gram-negative bacteria.
- *Zataria multiflora* EO was encapsulated by an ionic gelation technique into CS NPs that had sizes of 125–175 nm (Mohammadi et al., 2015). The EO-loaded NPs showed controlled and sustained release of the entrapped EO over 40 days.
- NPs smaller than 200 nm were produced from zein solution in propylene glycol and GA using a stir plate (Chen and Zhong, 2015). Both electrostatic and hydrophobic interactions contributed to the adsorption of GA on zein NPs. Encapsulation of peppermint oil in zein-GA NPs did not significantly change particle dimension and dispersion stability. Entrapped peppermint oil gradually released of from freeze dried samples at pH 2.0–8.0.

4.2.4 Water-Soluble Vitamins

Development of delivery systems for hydrophilic bioactive ingredients is lagging behind those developed for the delivery of lipophilic bioactives. The need for designing different delivery systems for lipophilic and hydrophilic bioactive can be understood from the differences in the chemical and physical properties of the two groups. The challenges associated with developing delivery systems for hydrophilic food bioactives have been recently presented and discussed (McClements, 2015). Therefore, limited studies have described the nanoencapsulation of water-soluble vitamins.

Vitamin C was encapsulated in CS NPs prepared by ionic gelation using TPP (Alishahi et al., 2011). Low-molecular-weight CS generated NPs with better size, morphology, and delivery rate. The shelf life of the vitamin C entrapped NPs increased in comparison to its free form. The release of vitamin C from the CS NPs was pH-dependent. Also, vitamin C was loaded *N*-acyl CS NPs prepared by self-aggregation method (Choa et al., 2012). The particle sizes ranged from 444 to 487 nm for NPs with various acyl chain lengths. The NPs particle sizes were reduced to 216–288 nm with vitamin C loading. Controlled release of entrapped Vitamin C occurred at pH 1.3 and 7.4 and at reduced rates with increasing the length of acyl side chain.

Folic acid (FA) and folates are essential dietary components. It has been suggested that FA can be effective in decreasing the risk for cardiovascular diseases, colon cancer, neurological dementia and Alzheimer's disease. The key role of FA is in women's nutrition before conception, during pregnancy, and lactation. Studies showed that, in spite of the hydrophilic nature of FA, it may self-assemble into unique fine structures even at low concentrations such as 0.1% (w/w) through hydrogen bonds and stacking interactions. β-lg/FA nano-complexes (mean size <10 nm) were formed at protein:vitamin molar ratio 1:10 (Pérez et al., 2014). Formation of the complexes improved β-lg dissolution near its isoelectric point but did not alter its digestibility. These complexes can be used as a vehicle for the delivery of FA in clear beverages. Casein NPs loaded with FA were prepared by coacervation method (Penalva et al., 2015). The formed NPs had an average size of 150 nm and a FA content of around 25 mg per mg NP. The FA was only released from the NPs under simulated intestinal conditions.

Vitamin B_2 was loaded in alginate-chitosan (Al-CS) NPs prepared by ionic gelation (Azevedo et al., 2014). The average sizes for Al-CS NPs without and with B_2 were 119.5 ± 49.9 nm and 104.0 ± 67.2 nm respectively. The NPs showed EE and LC values of 55.9 ± 5.6 and 2.2 ± 0.6%, respectively and the release of entrapped B_2 was affected by the polymer relaxation. The sizes of vitamin B_2-loaded NPs were more stable during storage than that of NPs without vitamin B_2.

5 Bioavailability and Toxicity

Bioavailability is defined as the fraction of unchanged bioactive component that is absorbed and eventually reaches the systemic circulation. Most studies have been directed to the nanocapsulation of lipophilic bioactive ingredients known by their poor solubility in aqueous medium and in turn with low bioavailability. Due to diversity of the wall materials used in the fabrication of the nanoparticles and the entrapped bioactive ingredients, the following general trends can be highlighted:

1. Nanoencapsulation in natural polymer may improve the bioavailability of entrapped food bioactive as a result of the increase in their solubility in aqueous medium, protection against gastrointestinal condition, and increase in the resident time in the gastrointestinal tract.
2. The NPs based on natural polymers would undergo rapid digestion within the stomach, small intestine, and colon. Therefore, it is unlikely for NPs to be directly absorbed and accumulated in the human body. Even in the case that some NPs remain

intact, their ability to penetrate the negatively charged mucous layer depends on their size, shape, and surface charge. Mucous layer inhibits particle diffusion if the particles have a diameter larger than 400 nm (Cone, 2009) or if they make a chemical complex with mucous. The reported NPs sizes in some studies exceeded this limit.

3. The natural proteins and polysaccharides used for encapsulation poses no safety threat since they have been approved as GRAS. However, chemically modified natural polymers might possess some safety challenges, especially in their nano sizes.
4. Residual organic solvents used in fabrication of some NPs may be found in the final products. This may pose a safety threat depending on the toxicity of these solvents. Therefore, preparation of NPs delivery systems for food applications should be limited to solvent-free techniques.

6 Conclusions and Future Trends

The past decade has witnessed a marriage between the growing demand for functional foods and the emerging food nanotechnology. Considerable interest has been directed to the development of nano-scale delivery systems that can satisfy the needs for efficient protection of food bioactives against environmental stresses without affecting the desirable quality of the host food. The present chapter highlighted the progress achieved in the development of suitable nanodelivery systems based on proteins and polysaccharides for food uses. Fundamental studies have enabled researchers to develop methods for nanoencapsulation of various food bioactive ingredients under controlled conditions. However, there is slow progress in several areas in this field:

1. Apart from the general <100 nm set for the size of NPs there is no accepted limits for the size and shape of food nanoparticles that can offer the optimum advantages from their use in encapsulation of the bioactive food ingredients.
2. Although the safety of proteins and polysaccharides used in the manufacture of nanoparticles, there is concern about the safety of NPs produced from the chemically modified natural polymers and those produced by some of the fabrication methods particularly those using some organic solvents.
3. Efforts have been directed mainly to nanoencapsulation of lipophilic food bioactives and very little has been done with respect to hydrophilic bioactives. Development of special nano-delivery systems for hydrophilic nutraceuticals is a need.

4. Scaling up the production of nanoencapsulated food bioactives. Although, many of the developed nanoencapsulation techniques have the potential for industrial production, no study has been cited on pilot or semiindustrial production of any of these developments.
5. Very few studies have described the behavior of the entrapped bioactives in foods during processing and storage. Redesigning the process flows and optimization of processing conditions will be a future challenge, aiming to maximize the bioavailability of bioactive ingredients used in the fortification of functional foods. This area needs much attention to exploit the potential uses of nanoencapsulated food bioactives in the development of novel functional foods.

Abbreviations

AL Alginate
BSA Blood serum albumin
CMCS Carboxymethyl chitosan
CS Chitosan
BC β-Carotene
CMP Casein macropeptide
CPP Casein phosphopeptide
CLA Conjugated linoleic acid
β CD β-Cyclodextrin
DMSO Dimethylsulphoxids
EPA Eicosapentaenoic acid
EE Encapsulation efficiency
EGCG Epigallocatechin-3-gallate
EO Essential oils
FA Folic acid
FUCO Fucoxanthin
GIF Gastrointestinal fluid
GSF Gastrostomach fluid
GRAS Generally recognized as safe
GA Gum Arabic
α-La α-Lactalbumin
β-Lg β-Lactoglobulin
Lf Lactoferrin
LC Loading capacity
LU Lutein
LY Lycopene
MD Maltodextrin
NPs Nanoparticles
DHA Ocosahexanoic acid
OVA Ovalbumin
PUFA Polyunsaturated fatty acids
PBS Phosphate buffer saline
r-CM Reassembled casein micelles
Res Resveratrol
SGI Simulated gastrointestinal fluid

SGF Simulated gastric fluid
SPI Soy protein isolate
TPP Tripolyphosphate
WPC Whey protein concentrate
WPI Whey protein isolate
VD Vitamin D

References

Abreu, F.O.M.S., Oliveiraa, E.F., Paulaa, H.C.B., de Paula, R.C.M., 2012. Chitosan/cashew gum nanogels for essential oil encapsulation. Carbohyd. Polym. 89, 1277–1282.

Acosta, E., 2009. Bioavailability of nanoparticles in nutrient and nutraceutical delivery. Curr. Opin. Coll. Interf. Sci. 14, 3–15.

Alishahi, A., Mirvaghefi, A., Tehrani, M.R., Farahmand, H., Shojaosadati, S.A., Dorkoosh, F.A., Elsabee, M.Z., 2011. Shelf life and delivery enhancement of vitamin C using chitosan nanoparticles. Food Chem. 126, 935–940.

Al-khalidi, B., 2012. Casein proteins as a vehicle to deliver vitamin D_3: Fortification of dairy products with vitamin D_3 and Bioavailability of vitamin D_3 from fortified mozzarella cheese baked with pizza. MSc thesis, Univ. of Torento, Canada, 91.

Arunkumar, R., Prashanth, K.V.H., Baskaran, V., 2013. Promising interaction between nanoencapsulated lutein with low molecular weight chitosan: characterization and bioavailability of lutein in vitro and in vivo. Food Chem. 141, 327–337.

Azevedo, M.A., Bourbon, A.I., Vicente, A.A., Cerqueira, M.A., 2014. Alginate/chitosan nanoparticles for encapsulation and controlled release of vitamin B_2. Int. J. Biol. Macromol. 71, 141–146.

Baur, J.A., Sinclair, D.A., 2006. Therapeutic potential of resveratrol: the in vivo evidence. Nat. Rev. Drug Discov. 5, 493–506.

Calvo, M.S., Whiting, S.J., Barton, C.N., 2004. Vitamin D fortification in the United States and Canada: current status and data needs. Am. J. Clin. Nutr. 80(6 Suppl):1710S–1716S.

Chen, H., Zhong, Q., 2015. A novel method of preparing stable zein nanoparticle dispersions for encapsulation of peppermint oil. Food Hydro. 43, 593–602.

Chen, L., Remondetto, G.E., Subirade, M., 2006. Food protein-based materials as nutraceutical delivery systems. Trends Food Sci. Technol. 17, 272–283.

Chen, N., Lin, L., Sun, W., Zhao, M., 2014a. Stable and pH-sensitive protein nanogels made by self-assembly of heat denatured soy protein. J. Agric. Food Chem. 62, 9553–9561.

Chen, J., Zheng, J., McClements, D.J., Xiao, H., 2014b. Tangeretin-loaded protein nanoparticles fabricated from zein/β-lactoglobulin: preparation, characterization, and functional performance. Food Chem. 158, 466–472.

Chen, F.-P., Li, B.-S., Tang, C.-H., 2015. Nanocomplexation between curcumin and soy protein isolate: influence on curcumin stability/bioaccessibility and in vitro protein digestibility. J. Agric. Food Chem. 63, 3559–3569.

Chin, S.F., Mohd, Y., Siti, N.A., Pang, A.C., 2014. Preparation and characterization of starch nanoparticles for controlled release of curcumin. Int. J. Polym. Sci. 2014, 1–8.

Choa, Y., Kimb, J.T., Park, H.J., 2012. Size-controlled self-aggregated *N*-acyl chitosan nanoparticles as a vitamin C carrier. Carbohyd. Polym. 88, 1087–1092.

Cone, R.A., 2009. Barrier properties of mucus. Adv. Drug Deliv. Rev. 61, 75–85.

Das, S., Ng, K.Y., Ho, P.C., 2010. Formulation and optimization of zinc-pectinate beads for the controlled delivery of resveratrol. AAPS Pharm. Sci. Tech. 11, 729–742.

Davidov-Pardo, G., Joye, I.J., McClements, D.J., 2015a. Encapsulation of resveratrol in biopolymer particles produced using liquid antisolvent precipitation. Part I: preparation and characterization. Food Hydro. 45, 309–316.

Davidov-Pardo, G., Pérez-Ciordia, S., Marín-Arroyo, M.R., McClements, D.J., 2015b. Improving resveratrol bioaccessibility using biopolymer nanoparticles and complexes: impact of protein-carbohydrate Maillard conjugation. J. Agric. Food Chem. 63, 3915–3923.

Dong, F., Padua, G.W., Wang, Y., 2013. Controlled formation of hydrophobic surfaces by self-assembly of an amphiphilic natural protein from aqueous solutions. Soft Matter 9, 5933.

Duclairoir, C., Orecchioni, A.M., Depraetere, P., Nakache, E., 2002. Alpha-tocoperol encapsulation and in vitro release from wheat gliadin nanoparticles. J. Microencapsul. 19, 53–60.

Esmaili, M., Ghaffari, S.M., Moosavi-Movahedi, Z., Atri, M.S., Sharifizadeh, A., Farhadi, M., et al., 2011. Beta casein-micelle as a nanovehicle for solubility enhancement of curcumin: food industry application. LWT—Food Sci. Technol. 44, 2166–2172.

FDA, 2005. GRAS Notice 000443 Shrimp-derived chitosan—Food and Drug.

Gao, Z.-M., Zhu, L.-P., Yang, X.-Q., He, X.-T., Wang, J.-M., Guo, J., Qi, J.-R., Wang, L.-J., Yin, S.-W., 2014. Soy lipophilic protein nanoparticles as a novel delivery vehicle for conjugated linoleic acid. Food Funct. 5, 1286–1293.

Gökmen, V., Mogol, B.A., Lumaga, R.B., Fogliano, V., Kaplun, Z., Shimoni, E., 2011. Development of functional bread containing nanoencapsulated omega-3 fatty acids. J. Food Eng. 105, 585–591.

Gomez-Estaca, J., Balaguer, M.P., Gavara, R., Hernandez-Munoz, P., 2012. Formation of zein nanoparticles by electrohydrodynamic atomization: effect of the main processing variables and suitability for encapsulating the food coloring and active ingredient curcumin. Food Hydro. 28, 82–91.

Gravel and-Bikker, J.F., de Kruif, C.G., 2006. Unique milk protein based nanotubes: food and nanotechnology meet. Trends Food Sci. Technol. 17, 196–203.

Ha, H.-K., Kim, J.W., Lee, M.-R., Lee, W.-J., 2013. Formation and characterization of quercetin-loaded chitosan oligosaccharide/β-lactoglobulin nanoparticle. Food Res. Int. 52, 82–90.

Haham, M., Ish-Shalom, S., Nodelman, M., Duek, I., Segal, E., Kustanovich, M., Livney, Y.D., 2012. Stability and bioavailability of vitamin D nanoencapsulated in casein micelles. Food Funct. 3, 737–744.

Herculano, E.D., de Paula, H.C.B., de Figueiredo, E.A.T., Dias, F.G.B., Pereira, V.A., 2015. Physicochemical and antimicrobial properties of nanoencapsulated *Eucalyptus staigeriana* essential oil. LWT—Food Sci. Technol. 61, 484–491.

Hong, Z., Xu, Y., Yin, J.-F., Jin, J., Jiang, Y., Du, Q., 2014. Improving the effectiveness of (−)-epigallocatechin gallate (EGCG) against rabbit atherosclerosis by EGCG-loaded nanoparticles prepared from chitosan and polyaspartic acid. J. Agric. Food Chem. 62, 12603–12609.

Hosseini, S.F., Zandi, M., Rezaei, M., Farahmandghavi, F., 2013. Two-step method for encapsulation of oregano essential oil in chitosan nanoparticles: preparation, characterization and in vitro release study. Carbohyd. Polym. 95, 50–56.

Hu, B., Pan, C., Sun, Y., Hou, Z., Ye, Y., Hu, B., et al., 2008. Optimization of fabrication parameters to produce chitosan-tripolyphosphate nanoparticles for delivery of tea catechins. J. Agric. Food Chem. 56, 7451–7458.

Hu, D., Lin, C., Liu, L., Li, S., Zhao, Y., 2012a. Preparation, characterization, and in vitro release investigation of lutein/zein nanoparticles via solution enhanced dispersion by supercritical fluids. J. Food Eng. 109, 545–552.
Hu, B., Ting, Y., Yang, X., Tang, W., Zeng, X., Huang, Q., 2012b. Nano chemoprevention by encapsulation of (−)-epigallocatechin-3-gallate with bioactive peptides/chitosan nanoparticles for enhancement of its bioavailability. Chem. Commun. 48, 2421–2423.
Hu, B., Xie, M., Zhang, C., Zeng, X., 2014. Genipin-structured peptide–polysaccharide nanoparticles with significantly improved resistance to harsh gastrointestinal environments and their potential for oral delivery of polyphenols. J. Agric. Food Chem. 62, 12443–12452.
Hu, K., Huang, X., Gao, Y., Huang, X., Xiao, H., McClements, D.J., 2015a. Core–shell biopolymer nanoparticle delivery systems: synthesis and characterization of curcumin fortified zein–pectin nanoparticles. Food Chem. 182, 275–281.
Hu, B., Wang, Y., Xie, M., Hu, G., Ma, F., Zeng, X., 2015b. Polymer nanoparticles composed with gallic acid grafted chitosan and bioactive peptides combined antioxidant, anticancer activities and improved delivery property for labile polyphenols. J. Func. Foods 15, 593–603.
Ilyasoglu, H., El, S.N., 2014. Nanoencapsulation of EPA/DHA with sodium caseinate–gum arabic complex and its usage in the enrichment of fruit juice. LWT— Food Sci. Technol. 56, 461–468.
Jakobek, L., 2015. Interactions of polyphenols with carbohydrates, lipids, and proteins. Food Chem. 175, 556–567.
Jones, O.G., McClements, D.J., 2011. Recent progress in biopolymer nanoparticle and microparticle formation by heat-treating electrostatic protein-polysaccharide complexes. Adv. Coll. Interf. Sci 167, 49–62.
Joye, I.J., McClements, D.J., 2014. Biopolymer-based nanoparticles and microparticles: fabrication, characterization, and application. Curr. Opin. Coll. Interf. Sci 19, 417–427.
Joye, I.J., Davidov-Pardo, G., McClements, D.J., 2015a. Encapsulation of resveratrol in biopolymer particles produced using liquid antisolvent precipitation. Part 2: Stability and functionality. Food Hydro. 49, 127–134.
Joye, I.J., Nelis, V.A., McClements, D.J., 2015b. Gliadin-based nanoparticles: Fabrication and stability of food-grade colloidal delivery systems. Food Hydro. 44, 86–93.
Kato, A., 2002. Industrial applications of Maillard-type protein-polysaccharide conjugates. Food Sci. Technol. Res. 8, 193–199.
Katzbauer, B., 1998. Properties and applications of xanthan gum. Polym. Degrad. Stab. 59, 81–84.
Kean, T., Thanou, M., 2010. Biodegradation, biodistribution, and toxicity of chitosan. Adv. Drug Deliv. Rev. 62, 3–11.
Khalili, S.T., Mohsenifar, A., Beyki, M., Zhaveh, S., Rahmani-Cherati, T., Abdollahi, A., Bayat, M., Tabatabaei, M., 2015. Encapsulation of Thyme essential oils in chitosan–benzoic acid nanogel with enhanced antimicrobial activity against *Aspergillus flavus*. LWT—Food Sci. Technol. 60, 502–508.
Krishnaswamy, K., Orsat, V., Thangavel, K., 2012. Synthesis and characterization of nanoencapsulated catechin by molecular inclusion with beta-cyclodextrin. J. Food Eng. 111, 255–264.
Lalush, I., Bar, H., Zakaria, I., Eichler, S., Shimoni, E., 2005. Utilization of amylose-lipid complexes as molecular nanocapsules for conjugated linoleic acid. Biomacromolecules 6, 121–130.
Lertsutthiwong, P., Rojsitthisak, P., 2011. Chitosan-alginate nanocapsules for encapsulation of turmeric oil. Pharmazie 66, 911–915.

Lestringant, P., Guri, A., Gülseren, I., Relkin, P., Corredig, M., 2014. Effect of processing on physicochemical characteristics and bioefficacy of β-lactoglobulin – epigallocatechin-3-gallate complexes. J. Agric. Food Chem. 62, 8357–8364.

Li, Z., Gu, L., 2014. Fabrication of self-assembled (−)-epigallocatechin gallate (EGCG)ovalbumin – dextran conjugate nanoparticles and their transportacross monolayers of human intestinal epithelial caco-2 cells. J. Agric. Food Chem. 62, 1301–1309.

Li, Q., Liu, C.-G., Huang, Z.-H., Fang-Fang Xue, F.-F., 2011. Preparation and characterization of nanoparticles based on hydrophobic alginate derivative as carriers for sustained release of vitamin D_3. J. Agric. Food Chem. 59, 1962–1967.

Li, B., Du, W., Jin, J., Du, Q., 2012. Preservation of (−)-epigallocatechin-3-gallate antioxidant properties loaded in heat-treated β-lactoglobulin nanoparticles. J. Agric. Food Chem. 60, 3477–3484.

Li, K.-K., Yin, S.-W., Yin, Y.-C., Tang, C.-H., Yang, X.-Q., Wen, S.-H., 2013. Preparation of water soluble antimicrobial zein nanoparticles by a modified antisolvent approach and their characterization. J. Food Eng. 119, 343–352.

Li, Z., Ha, J., Zou, T., Gu, L., 2014a. Fabrication of coated bovine serum albumin (BSA)–epigallocatechin gallate (EGCG) nanoparticles and their transport across monolayers of human intestinal epithelial Caco-2 cells. Food Funct. 5, 1278–1285.

Li, W., Peng, H., Ning, F., Yao, L., Luo, M., Qiang Zhao, Q., Zhu, X., Xiong, H., 2014b. Amphiphilic chitosan derivative-based core–shell micelles: synthesis, characterization and properties for sustained release of Vitamin D_3. Food Chem. 152, 307–315.

Liu, R.H., 2003. Health benefits of fruit and vegetables are from additive and synergistic combinations of phytochemicals. Am. J. Clin. Nutr. 78suppl, 517S–520S.

Liu, F., Tang, C.-H., 2013. Soy protein nanoparticle aggregates as pickering stabilizers for oil in-water emulsions. J. Agric. Food Chem. 61, 8888–8898.

Luo, Y., Wang, Q., 2013. Recent advances of chitosan and its derivatives in novel food applications. J. Food Proc. Beverage 1, 13–25.

Luo, Y., Zhang, B., Whent, M., Yu, L., Wang, Q., 2011. Preparation and characterization of zein/chitosan complex for encapsulation of α-tocopherol, and its in vitro controlled release study. Coll. Surf. B: Biointerf. 85, 145–152.

Luo, Y., Teng, Z., Wang, Q., 2012. Development of zein nanoparticles coated with carboxymethyl chitosan for encapsulation and controlled release of vitamin D3. J. Agric. Food Chem. 60, 836–843.

Luo, Y., Wang, T.T.Y., Teng, Z., Chen, P., Sun, J., Wang, Q., 2013. Encapsulation of indole-3-carbinol and 3,3 O-diindolylmethane in zein/carboxymethyl chitosan nanoparticles with controlled release property and improved stability. Food Chem. 139, 224–230.

Markman, G., Livney, Y.D., 2012. Maillard-conjugate based core–shell co-assemblies for nanoencapsulation of hydrophobic nutraceuticals in clear beverages. Food Funct. 3, 262–270.

McClements, D.J., 2015. Encapsulation, protection, and release of hydrophilic active components: potential and limitations of colloidal delivery systems. Adv. Coll. Interf. Sci. 219, 27–53.

McClements, D.J., Decker, E.A., Park, Y., Weiss, J., 2009. Structural design principles for delivery of bioactive components in nutraceuticals and functional foods. Crit. Rev. Food Sci. Nutr. 49, 577–606.

McSweeney, P.L.H., Fox, P.F. (Eds.), 2013. Advanced Dairy Chemistry: Volume 1A: Proteins: Basic Aspects, fourth ed. Springer Science + Business Media, New York.

Menéndez-Aguirre, O., Kessler, A., Stuetz, W., Grune, T., Weiss, J., Hinrichs, J., 2014. Increased loading of vitamin D_2 in reassembled casein micelles with temperature-modulated high-pressure treatment. Food Res. Int. 64, 74–80.

Mession, J.-L., Chihi, M.L., Sok, N., Saurel, R., 2015. Effect of globular pea proteins fractionation on their heat-induced aggregation and acid cold-set gelation. Food Hydro. 46, 233–243.

Mohammadi, A., Hashemi, M., Hosseini, S.M., 2015. Nanoencapsulation of *Zataria multiflora* essential oil preparation and characterization with enhanced antifungal activity for controlling *Botrytis cinerea*, the causal agent of gray mould disease. Innov. Food Sci. Emerg. Technol. 28, 73–80.

Nyemb, K., Guerin-Dubiard, C., Dupont, D., Jardin, J., Rutherfurd, S.M., Nau, F., 2014. The extent of ovalbumin in vitro digestion and the nature of generated peptides are modulated by the morphology of protein aggregates. Food Chem. 157, 429–438.

Pan, X., Yao, P., Jiang, M., 2007. Simultaneous nanoparticle formation and encapsulation driven by hydrophobic interaction of casein-*graft*-dextran and β-carotene. J. Coll. Interf. Sci. 315, 456–463.

Pan, K., Zhong, Q., Baek, S.J., 2013. Enhanced dispersibility and bioactivity of curcumin by encapsulation in casein nanocapsules. J. Agric Food Chem. 61, 6036–6043.

Patel, A., Hu, Y.C., Tiwari, J.K., Velikov, K.P., 2010. Synthesis and characterization of zein–curcumin colloidal particles. Soft Matter 6, 6192–6199.

Patel, A.R., Heussen, P.C.M., Hazekamp, J., Drost, E., Velikov, K.P., 2012. Quercetin-loaded biopolymeric colloidal particles prepared by simultaneous precipitation of quercetin with hydrophobic protein in aqueous medium. Food Chem. 133, 423–429.

Penalva, R., Esparza, I., Agüeros, M., Gonzalez-Navarro, C.J., Gonzalez-Ferrero, C., Irache, J.M., 2015. Casein nanoparticles as carriers for the oral delivery of folic acid. Food Hydro. 44, 399–406.

Peng, H., Xiong, H., Li, J., Xie, M., Liu, Y., Bai, C., et al., 2010. Vanillin cross-linked chitosan microspheres for controlled release of resveratrol. Food Chem. 121, 23–28.

Pérez, O.E., David-Birman, T., Kesselman, E., Levi-Tal, S., Lesmes, U., 2014. Milk protein-vitamin interactions: Formation of beta-lactoglobulin/folic acid nano-complexes and their impact on in vitro gastro-duodenal proteolysis. Food Hydro. 38, 40–47.

Quiňonesa, J.P., Gothelf, K.V., Kjemsc, J., Caballerod, Á.M.H., Schmidt, C., Covas, C.P., 2012. Self-assembled nanoparticles of glycol chitosan–Ergocalciferol succinate conjugate for controlled release. Carbohydr. Polym. 88, 1373–1377.

Rabea, E.I., Badawy, M.E.T., Stevens, C.V., Smagghe, G., Steurbaut, W., 2003. Chitosan as antimicrobial agent: applications and mode of action. Biomacromolecules 4, 1457–1465.

Ravi, H., Baskaran, V., 2015. Biodegradable chitosan-glycolipid hybrid nanogels: A novel approach to encapsulate fucoxanthin for improved stability and bioavailability. Food Hydro. 43, 717–725.

Relkin, P., Shukat, R., Moulin, G., 2014. Encapsulation of labile compounds in heat- and high-pressure treated protein and lipid nanoparticles. Food Res. Int. 63, 9–15.

Rishi, S., Munir, C., 2001. Zein: the industrial protein from corn. Ind. Crops Prod. 13, 171–192.

Säiz-Abajo, M.-J., Gonzälez-Ferrero, C., Moreno-Ruiz, A., Romo-Hualde, A., Gonzälez-Navarro, C.J., 2013. Thermal protection of β-carotene in reassembled casein micelles during different processing technologies applied in food industry. Food Chem. 138, 1581–1587.

Semenova, M.G., Moiseenko, D.V., Grigorovich, N.V., Anokhina, M.S., et al., 2014. Protein-polysaccharide interactions and digestion of the complex particles. In Boland, M., Golding, M., Singh, H. (Eds.), Food Structures, Digestion, and Health, pp. 169–192. Elsevier, Amsterdam, Netherlands.

Semo, E., Kesselman, E., Danino, D., Livney, Y.D., 2007. Casein micelle as a natural nanocapsular vehicle for nutraceuticals. Food Hydro. 21, 936–942.

Semyonov, D., Ramon, O., Shoham, Y., Shimoni, E., 2014. Enzymatically synthesized dextran nanoparticles and their use as carriers for nutraceuticals. Food Funct. 5, 2463–2474.

Shah, B., Ikeda, S., Davidson, P.M., Zhong, Q., 2012. Nanodispersing thymol in whey protein isolate–maltodextrin conjugate capsules produced using the emulsion–evaporation technique. J. Food Eng. 111, 79–86.

Shpigelman, A., Cohen, Y., Livney, Y.D., 2012. Thermally induced β-lactoglobulin-EGCG nanovehicles: Loading, stability, sensory, and digestive-release study. Food Hydro. 29, 57–67.

Sponton, O.E., Perez, A.A., Carrara, C.R., Liliana, G., Santiago, L.G., 2015. Impact of environment conditions on physicochemical characteristics of ovalbumin heat-induced nanoparticles and on their ability to bind PUFAs. Food Hydro. 48, 165–173.

Sundar Raj, A.A., Rubila, S., Jayabalan, R., Ranganathan, T.V., 2012. A review on pectin: chemistry due to general properties of pectin and its pharmaceutical uses. Open Access Scientific Reports.

Teng, Z., Luo, Y., Wang, Q., 2012. Nanoparticles synthesized from soy protein: preparation: characterization application for nutraceutical encapsulation. J. Agric. Food Chem. 60, 2712–2720.

Teng, Z., Luo, Y., Wang Q, 2013a. Carboxymethyl chitosan-soy protein complex nanoparticles for the encapsulation and controlled release of vitamin D_3. Food Chem. 141, 524–532.

Teng, Z., Luo, Y., Wang, T., Zhang, B., Wang, Q., 2013b. Development and application of nanoparticles synthesized with folic acid conjugated soy protein. J. Agric. Food Chem. 61, 2556–2564.

Torres-Giner, S., Martinez-Abad, A., Ocio, M.J., Lagaron, J.M., 2010. Stabilization of a nutraceutical omega-3 fatty acid by encapsulation in ultrathin electrosprayed zein prolamine. J. Food Sci. 75, N69–N80.

Wandrey, C., Bartkowiak, A., Harding, S.E., 2010. Materials for encapsulation. In: Zuidam, V.A., Nedović (Eds.), Encapsulation Technologies for Active Food Ingredients and Food Processing, NJ, pp. 31–100.

Wang, J., Cao, Y., Sun, B., Wang, C., 2011. Physicochemical and release characterization of garlic oil–β-cyclodextrininclusion complexes. Food Chem. 127, 1680–1685.

Weiss, J., Takhistov, P., McClements, D.J., 2006. Functional materials in food technology. J. Food Sci. 71, R107–R116.

Woranucha, S., Yoksan, R., 2013. Eugenol-loaded chitosan nanoparticles: I. Thermal stability improvement of eugenol through encapsulation. Carbohydr. Polym. 96, 578–585.

Wu, Y., Luo, Y., Wang, Q., 2012. Antioxidant and antimicrobial properties of essential oils encapsulated in zein nanoparticles prepared by liquid–liquid dispersion method. LWT—Food Sci. Technol. 48, 283–290.

Xia, W., Liu, P., Zhang, J., Chen, J., 2011. Biological activities of chitosan and chitooligosaccharides. Food Hydro. 25, 170–179.

Xiao, J., Li, Y., Li, J., Gonzalez, A.P., Xia, Q., Huang, Q., 2015. Structure, morphology, and assembly behavior of kafirin. J. Agric. Food Chem. 63, 216–224.
Xue, F., Li, C., Liu, Y., Zhu, X., Pan, S., Wang, L., 2013. Encapsulation of tomato oleoresin with zein prepared from corn gluten meal. J. Food Eng. 119, 439–445.
Xue, J., Tan, C., Zhang, X., Feng, B., Xia, S., 2014. Fabrication of epigallocatechin-3-gallate nanocarrier based on glycosylated casein: stability and interaction mechanism. J. Agric. Food Chem. 62, 4677–4684.
Yang, W., Xu, C., Liu, F., Yuan, F., Gao, Y., 2014. Native and thermally modified protein – polyphenol coassemblies: lactoferrin-based nanoparticles and submicrometer particles as protective vehicles for (−)-epigallocatechin-3-gallate. J. Agric. Food Chem. 62, 10816–10827.
Yi, J., Lam, T.I., Yokoyama, W., Cheng, L.W., Zhong, F., 2014. Controlled release of β-carotene in β-lactoglobulin – dextran-conjugated nanoparticles' in vitro digestion and transport with caco-2 monolayers. J. Agric. Food Chem. 62, 8900–8907.
Yi, J., Lam, T.I., Yokoyama, W., Cheng, L.W., Zhong, F., 2015. Beta-carotene encapsulated in food protein nanoparticles reduces peroxyl radical oxidation in Caco-2 cells. Food Hydro. 53, 31–40.
Zhang, H., Zhao, Y., 2015. Preparation, characterization and evaluation of tea polyphenol-Zn complex loaded β-chitosan nanoparticles. Food Hydro. 48, 260–273.
Zhang, Y.Y., Yang, Y., Tang, K., Hu, X., Zou, G.L., 2008. Physicochemical characterization and antioxidant activity of quercetin-loaded chitosan nanoparticles. J. Appl. Polym. Sci. 107, 891–897.
Zhong, Q., Jin, M., 2009. Zein nanoparticles produced by liquid–liquid dispersion. Food Hydro. 23, 2380–2387.
Zimet, P., Livney, Y.D., 2009. Beta-lactoglobulin and its nanocomplexes with pectin as vehiclesfor ώ-3 polyunsaturated fatty acids. Food Hydro. 23, 1120–1126.
Zimet, P., Rosenberg, D., Livney, Y.D., 2011. Reassembled casein micelles and casein nanoparticles as nano-vehicles for ώ-3 polyunsaturated fatty acids. Food Hydro. 25, 1270–1276.

20

PROCESS TECHNOLOGY OF NANOEMULSIONS IN FOOD PROCESSING

Siddhartha Singha*, Bhaswati Bhattacharya*, Santanu Basu**

**National Institute of Food Technology Entrepreneurship and Management, Kundli, Haryana, India; **Panjab University, Dr. S. S. Bhatnagar University Institute of Chemical Engineering and Technology, Chandigarh, India*

1 Introduction

Many food products, both native and processed, are either emulsions themselves (McClements, 2004) or their processing involves an emulsification step. It makes it obvious that emulsion technology is one of the evergreen areas of research and development in food processing. In sync with this, the latest addition into the array of food emulsions is nanoemulsion. Nanoemulsion is dispersion of oil/water phase into a water/oil where the dispersed phase remains as a droplet of a size ranging between 100 and 600 nm (Bouchemal et al., 2004). This is often confused with macroemulsions, which also consist of dispersed phase of nanometer size. However, it is the *stability* of the two emulsion types that segregates them and not the size of the two categories. The microemulsion uses a combination of oil, water, surfactant, and/or cosurfactants in such a way that they spontaneously form the dispersed phase of nanometer size without any input of mechanical energy and remain as emulsion forever at a particular temperature. On the other side, nanoemulsion in general uses less surfactant and/or stabilizer and does not form emulsion spontaneously. Also, nano emulsion separates into individual phases at some point in time, even when stored at a particular temperature. Though both types of emulsions apparently look similar (transparent or translucent) their behavior is different. Ostensibly, the microemulsions may appear attractive but they are susceptible

Encapsulations. http://dx.doi.org/10.1016/B978-0-12-804307-3.00020-X

to minute temperature or compositional changes. Moreover, in a food system there is a restriction of usage of type and quantity of surfactant due to safety reason and stability of microemulsions strongly depends on surfactants. Therefore, in food for producing emulsions with dispersed phase of nanometer size, nanoemulsion is a better choice. Nanoemulsion is a man-made product and there is no evidence of naturally occurring nanoemulsions (Mason et al., 2006). Such emulsions can increase the aqueous solubility, thermal stability, and oral bioavailability of many functional compounds present in the food by acting as a delivery vehicle (Huang et al., 2010; McClements and Li, 2010). In addition to that, nanoemulsions can be used for fine tuning various quality attributes of a food product like texture, taste, mouth feel, color, processability, and stability (Silva et al., 2012). Ever-expanding consumer demand for safer, nutritious, and attractive food products has driven the food scientists, technologists, and processors to develop new products or modify the existing products. Hence, research activities on nanoemulsion are increasing exponentially.

In the manufacturing of nanoemulsions from two distinct oil and water phases, two primary aspects are creation of the dispersed phase as tiny droplets and keeping then intact for a desired length of time. To create the nanoscale dispersed phases either disintegration or condensation method can be used (McClements, 2004). In other words, either mechanical equipments (high shear rotor–stators, high-pressure homogenizers, ultrasonicators) can be used to tear apart the dispersed phase to create nanoemulsion (disintegration) or membranes or proper combination of surfactants or stabilizers can be used to create the nanoemulsion (condensation). The composition of the surfactant, stabilizer, and other components decide the ease of formation of the emulsion, its size, and the stability at a proper condition in both the cases.

The mechanical disruption or so-called high-energy processes are more flexible and more adoptable in the industries, hence, they have been discussed in more detail in this chapter. The discussion should help in selecting a method of homogenization, optimizing, and scaling up the process for a given nanoemulsion. Modeling of such processes has been emphasized for making them more efficient. The so-called low energy (condensation) methods are also covered to facilitate design of new nanoemulsion systems for food (Lovelyn and Attama, 2011).

Quality of a nanoemulsion is not depicted only by a desired minimum size and very narrow distribution of size but also its stability. In food processing, requirements can go beyond size and additional functionality is required. For example, masking of taste can be by done by one emulsion better than another, irrespective of their size. Hence, a section in the chapter is devoted to characterizing nanoemulsions with respect to their functionality. Especially, the traditional techniques used for characterizing macroemulsions are not suitable for nanoemulsions. Many of the manufacturing and testing techniques used in food nanoemulsion are inspired by pharma and cosmetic industries where nanoemulsion has been explored much deeper. This text has aimed to put them together to facilitate both formulations of novel food emulsions as well as their commercialization.

2 Thermodynamics of Nanosized and Nanoemulsion Production Technologies

The way a dispersed phase distributes in a continuous phase, that is, the droplet size, size distribution, and morphology depends on the thermodynamics of the mixture of two (or more) immiscible phase, surfactant(s), and other components. Both oil-in-water and water-in-oil types of emulsions can be explained in terms of Gibbs free energy. The change in free energy for the transformation from separated phases to emulsion state (Fig. 20.1) is given by the difference in chemical potential (μ) of the two states (Eq. 20.1) (Schramm, 2006).

$$\Delta G = \mu_{\text{emulsion}} - \mu_{\text{dispersed}} \tag{20.1}$$

When ΔG is positive the formation of emulsion is not feasible thermodynamically whereas, when $\Delta G = 0$, oil and water phase remains in equilibrium. For nanoemulsions the emulsification does not happen spontaneously due to the positive ΔG of the process. At a negative ΔG, the separate oil and water phases convert to single-phase emulsion spontaneously. This reversible conversion reaches equilibrium when ΔG becomes zero. For microemulsions

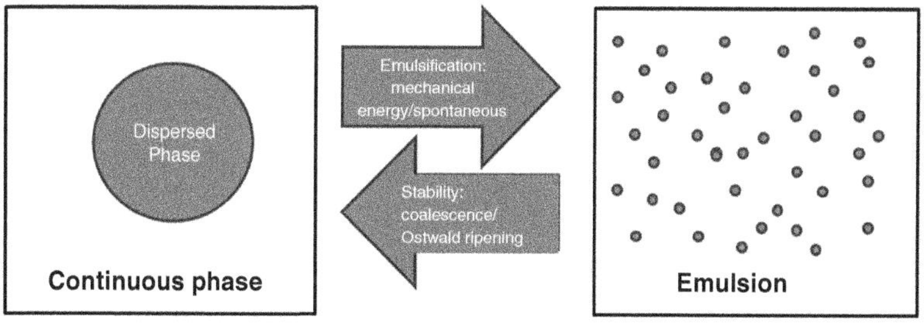

Figure 20.1. Schematics of emulsification by mechanical energy or spontaneous method.

the ΔG is negative so, as soon as the proper ratio of oil, water, surfactant (and cosurfactant) is mixed an emulsion forms spontaneously (McClements, 2012). Since the dispersal phase gets distributed as smaller droplets in the continuous phase (Fig. 20.1), the entropy change for emulsification increases following Eq. 20.2:

$$\Delta S_{\text{emulsification}} = -Nk_B\left[\ln\Phi + \left(\frac{1}{\Phi} - 1\right)\ln(1-\Phi)\right] \tag{20.2}$$

where N is the number of droplets of dispersed phase, k_B is the Boltzmann constant, and Φ is the volume ratio of dispersed phase to continuous phase. The Gibbs free energy for emulsification is the sum of the free energy ($\Delta A_{\text{interface}}$) change due to creation of new interfacial area and the energy spent due to entropy change (Eq. 20.3).

$$\Delta G = \Delta A_{\text{interface}} - T\Delta S_{\text{emulsification}} \tag{20.3}$$

If $T\Delta S_{\text{emulsification}}$ is much higher than the ($\Delta A_{\text{interface}}$) or the free energy change for creation of oil-water interfaces, then only the ΔG for emulsification would be negative. Again $\Delta A_{\text{interface}}$ depends on the interfacial tension of the oil water used in the emulsification process. Therefore, surfactants are necessary to reduce the interfacial tension and in turn $\Delta A_{\text{interface}}$. Unlike microemulsions, nanoemulsions do not form spontaneously and their formation requires intensive energy input to rupture the dispersed phase into smaller units. High-energy emulsification devices such as high-pressure homogenizer or microfluidizer are used to supply such intense mechanical energy. Moreover, when a droplet forms during emulsion the pressure difference (p) between inside and outside of the droplet is given by Laplace equation (Eq. 20.4).

$$p = \sigma\left[\frac{1}{R_1} + \frac{1}{(R_1 + d)}\right] \tag{20.4}$$

where R_1 and d are the inner radius of curvature and thickness of the droplet and σ is the interfacial tension. In presence of surfactant(s) σ reduces to a great extent, the pressure difference as well as stress on the droplet film also becomes less. Hence, less shear is required with increased concentration of an appropriate surfactant (Tadros, 2009). During a mechanical energy driven emulsions including nanoemulsions, the disruptive forces (given by high shear or high pressure homogenizers) rupture the dispersed phase into smaller droplets whereas interfacial force restricts reduction of droplet size (Fig. 20.1). Weber number is ratio of these two forces; at equilibrium its value is close to one.

In laminar shear condition this ratio is defined as We = $U\mu_c R/2\sigma$ where U is the velocity gradient, μ_c is the viscosity of continuous phase, and R is radius of the droplet. On the other hand when turbulent disruptive forces act the We is $G'\rho_c^{1/3}\varepsilon^{2/3}R^{5/3}/12.699\sigma$ Where G' is a characteristic constant of the system, ρ_c density of the continuous phase, and ε energy density added to the system externally. Above a "critical Weber number" the droplets break down until We reaches the critical value. Therefore, during processing of nanoemulsion contribution of laminar and turbulent forces vary according to the type of homogenizer. Also, the above thermodynamic discussion is valid for assessing stability of an emulsion (McClements, 2004).

3 High-Energy Production Technologies of Nanoemulsions

In high-energy nanoemulsion production methods high shear is necessary to create the dispersion of remarkably smaller size (typically 10–100 nm). Often, such high shear is accompanied by high temperature, which may complicate the quality control of the final nanoemulsion, especially in food products where such emulsion systems contain multicomponents prone to unwanted secondary chemical reactions. Therefore, choice of processing for nanoemulsions depends on composition of the emulsion phases as well as the objective of the emulsion (Sanguansri and Augustin, 2006). High shear required for such emulsions can be generated by rotor–stator, microfluidizer, ultrasonicator, high-pressure homogenizer, or a combination of these. For a particular combination of emulsion forming ingredients these technologies can result different kind of emulsion or nanoemulsion. In recent times, each of these technologies has developed at an extraordinary pace to meet the complex demand of including nanoemulsions in food processing and other processing industries (Ma and Boye, 2013). This section would facilitate selection of an appropriate process for a specific nanoemulsion system as well as it would serve as a guideline to fine-tune the efficiency of the emulsification process.

3.1 Rotor–Stator Technology

For high-shear emulsification both in macro size or submicron size range, rotor–stator technology is a well-accepted technology. It uses a fast spinning inner element (rotor) and a stationary outer sheath (stator) to draw the liquid into the space between the rotor

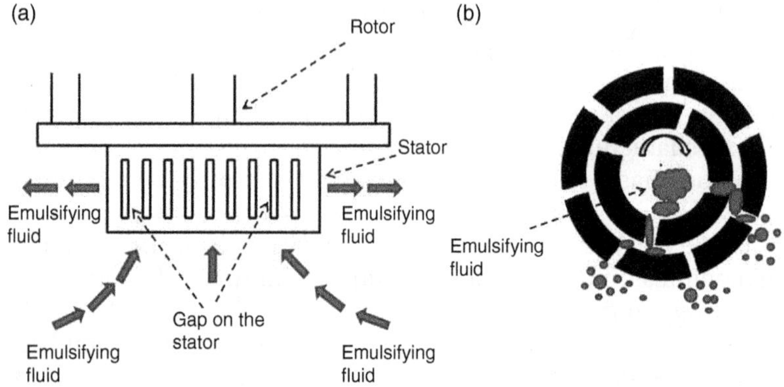

Figure 20.2. Schematics of emulsification by a rotor–stator device. (a) Front view of rotor–stator and (b) cross-sectional view of rotor–stator.

and the stator and to apply a high shear on the liquid before it comes out of the rotor–stator (Wengeler, 2007).

3.1.1 Process Definition

In this type of emulsifying device the rotating element or blade rotates at a high speed between 2,000 and 15,000 rpm to draw the emulsion forming liquid axially and push it through the opening of the stator (Fig. 20.2). Due to the high velocity added and the small gap between the rotor and stator, high intensity of shear results in rupture of the dispersed phase into smaller droplets. The process can be carried out in both batch or in-tank and continuous or in-line modes.

3.1.2 Equipment Designs

Original equipment manufacturers like Ross and Sons, Ika, Silverson, and many others are continuously expanding their portfolio of rotor–stator designs for better choice of design for a given emulsification system. These designs are different in terms of the type of the rotors as well as in terms of the stators. The available rotor heads may have a typical four-blade configuration or unique designs like Delta™ type or with intermeshing teeth. The stators can vary in type and size of opening, which controls the size of the droplets. In most designs the gap between the rotor–stator designs are adjustable. Considering these three design variables a wide variety of rotor stator systems are available for various emulsification tasks. However, for nanoemulsion, an ultra-high shear regime is required. To reduce the droplet size and improve the particle size distribution, one common strategy is using recirculation systems. Though in recirculation systems the particle size reduction takes place only in the first few passes, hence, alternative strategies are being explored. Three such patented designs are (http://www.highshearmixers.com/):

- The emulsion forming liquid flows at the center of the stator consisting of two or more concentric rows of intermeshing teeth and moves radially through channels in the rotor/stator teeth. High tip speeds of the rotor (typically up to 57 m/s) and very narrow clearance (minimum 0.7 mm) between the adjacent planes of the rotor and stator creates intense shear on the fluid in each pass.
- A design with a series of rotor–stators with the same gap is used for a large volume of emulsion but often is less effective in creating nanoemulsions.
- For imparting ultra-high shear, parallel semicylindrical grooves are made in the rotor and stator through which the emulsion fluid is forced at high velocity by pumping vanes. As the fluid flows, multiple streams generate within the grooves and rapidly collide with each other before leaving the emulsification chamber.

The rotor stator assemblies can be used in both batch and continuous mode. In batch system the liquid typically enters from the top and for smaller batches the rotor–stator itself reduces the dispersed phase size as well as does the mixing. However, to prepare large batches of nanoemulsions, additional agitators assist in mixing to obtain monodispersity of the emulsion. In continuous mode, a rotor–stator is used to impart emulsification of liquids flowing in a pipeline or in the recirculation loop (Fig. 20.3).

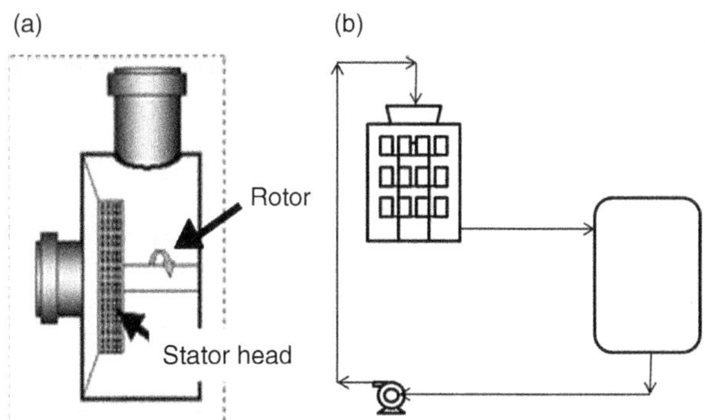

Figure 20.3. Continuous emulsification using a rotor–stator in a pipe line or in tank with a recirculation loop. (a) In-line rotor–stator for continuous emulsion and (b) continuous emulsification with recycle stream.

3.1.3 Process Modeling

Basic understanding of ultra-high shear emulsification processes is still inadequate, which reflects a trial-and-error method-based selection process, scale-up, and optimization of rotor–stator systems for nanoemulsification process. In addition most of the models developed in the industrial labs are not available in the open space. Therefore, modeling of the rotor–stator-based emulsification process is pertinent in the present context.

In rotor–stator when the fluid flows through the openings of the stator the speed is proportional to the rotor tip speed (Utomo et al., 2009). Computational fluid dynamics studies have shown that the maximum size reduction of the dispersed phase occurs at the point of discharge of the stator and not in the space between the rotor and stator (Barailler et al., 2006). Mixing in such a system is linked with the turbulence of the discharged fluid. Pacek et al. (2007) showed by numerical solution of the flow field in a Silverson rotor–stator that the radial jet velocity emerging from the stator slots is proportional to the speed of the rotor (N). Also, the simulation suggested ~70% of the energy applied by the rotor to the liquid dissipates around the mixing head and in turbulent flow the dissipated energy varies proportionally with N^3. However, assessing the extent of turbulence in such devices is tricky because if Reynolds number is used as the indicator of turbulence then at least three different values of the dimensionless number are possible. These values could be different by at least one order of magnitude.

$Re = \rho ND^2/\mu$ where the N and D are the speed and diameter of the rotor

$Re = \rho ND\delta_g/\mu$ where δ_g is the gap between the rotor and the stator

$Re = \rho NDb/\mu$ where b is the dimension of the openings of the stator

Often it is suggested that the Re of such system is not an appropriate predictor for performance of nanoemulsion formation; instead specific power is used for prediction of the emulsification performance. Assuming the centrifugal pump-like action, the power requirement of a rotor stator is calculated by $P = P_0\rho N^3D^5$ where the typical values of the power number (P_0) are between 1 and 3 (Barailler et al., 2006). The energy requirement to produce a particular emulsion of given composition and size distribution depends on the specific power or energy density of the emulsifying device. In other words, the mean diameter (d_{32}) depends on energy density (ε) and the residence time ($t_{residence}$) of the fluid in the zone of emulsification by the following equations 20.5 and 20.6,

$$d_{32} = \varepsilon^{-a_1} t_{\text{residence}}^{-a_2} \tag{20.5}$$

$$\varepsilon = \frac{\rho (ND)^3}{4b} \tag{20.6}$$

a_1 and a_2 are the characteristic parameters of the rotor–stator system. Therefore, the entire design of the rotor stator device controls the emulsion quality. The narrower openings of the stator causes even distribution of energy dissipation (ε) and in turn uniform droplet size (Brocart et al., 2002).

The above mathematical models are based on the knowledge of emulsifications in traditional rotor–stators typically used for macroemulsions. However, there is a need to carry out correlation studies to understand the breakage kinetics in the latest designs of rotor stator under various levels of the process parameters of emulsification.

3.1.4 Process Optimization

The majority of the studies devoted to optimization of emulsification conditions using a rotor–stator device are of two types. One type involved two-process parameters speed of the rotor and time of emulsification keeping the ingredient composition as well as the device design constant. Another type of studies searched for an optimal composition of the ingredients for a given emulsion where the other process parameters were kept constant. Statistical designs of experiment are convenient and reliable methods of optimization in such cases. In an optimization study a nanoemulsion of narrow size distribution with 135 nm mean diameter was prepared at an optimal condition of 36,000 rpm and 5 min processing time (Scholz and Keck, 2015). Asmawati et al. have optimized the composition of a cinnamaldehyde nanoemulsion using ultra turrax T25 operated at 12,000 rpm for 5 min (Mustapha et al., 2014). In spite of a number of reports on optimization of the nanoemulsion-making process using rotor–stator, a systematic study involving composition, process variables, and design variables is not available.

3.1.5 State-of-the-Art Rotor–Stator Technology for Food Nanoemulsion

Since nanoemulsion is an effective tool for delivery of various nutrients, flavors, colors, and other additives there are an increasing number of reports on encapsulation of many target compounds using rotor–stator technology. In general the rotor stator system is used for coarse emulsion (>μm) preparation, which undergoes a second stage of homogenizion with another homogenization/emulsification process to obtain a nanoemulsion. To improve the particle size of protein containing nanoemulsion of beta carotene a combination technique has been reported by Trentin et al. (2011). In the study a coarse emulsion was made by rotor–stator followed by membrane-assisted emulsification. Imai et al. (2008) have studied oxidative degradation kinetics of methyl linoleate in different size of the dispersed phase produced by rotor–stator. However, the emergence of new-generation of rotor–stator (see Section 3.1.1), makes it possible to achieve submicron-sized emulsion in rotor–stator homogenizers itself, provided the emulsification process is optimized for a given type of emulsion. Scholz and Keck (2015) were able to prepare an oil-in-water nanoemulsion with a mean droplet size of 135 nm and narrow size distribution using ART MICCRA D27 rotor stator at an optimal condition, that is, homogenization at 36,000 rpm for 5 min. In a separate study, palm oil ester was emulsified in presence of multiple surfactants and stabilizer using a Polytron rotor–stator to obtain a nanoemulsion of mean droplet size 126 nm (Han et al., 2014). In the emulsification process the oil and water phase were taken at 70°C and then homogenized at 6000 rpm and at 40°C for 5 min.

3.1.6 Scale Up

Turbulent forces are responsible for breaking of the dispersed phase into a nanoemulsion in a rotor stator. Hence, finding a reliable scale-up criterion to ensure same hydrodynamic flow in both small and large scale is difficult. Scale-up attempts based on geometrical similarities, same rotor tip speed, or gap widths were not very successful in such systems. Often, the circulation rate in a rotor–stator homogenizer at two different scales have been predicted considering total area of the slots of the stator, radial velocity at the stator slot, and the centrifugal force generated by the rotor (Maa and Hsu, 1996). Kamiya et al. (2010) have proposed a homogenization index as a scale-up criterion based on turbulent energy dissipation rate inside the homogenization region of the rotor stator and a circulation number. Their theory could predict the droplet diameter for a scale-up from 1.5 to 9 L. In small-scale operation, rotor–stator homogenizers are used exclusively, but in a larger batch size (>1 gal) an additional agitator is required to shorten the homogenization cycle and ensure homogeneity. This makes the process more flexible, albeit a pilot label study is needed to scale up a particular batch nanoemulsification process using a rotor–stator.

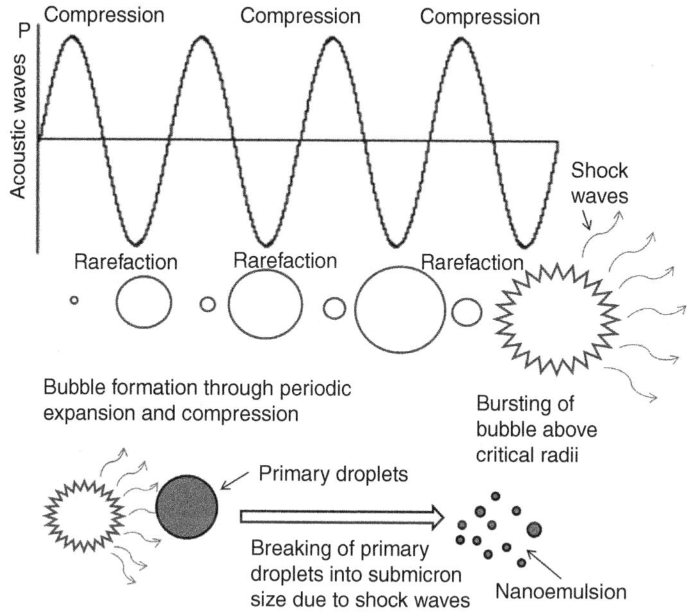

Figure 20.4. Cavitation due to ultrasound and size reduction of emulsion droplets.

3.2 Ultrasonic

Application of sound waves of higher frequency (beyond 20 kHz) in liquids creates pressure waves that agitate and mix the liquid as well as cause cavitation (Sivakumar et al., 2014) (Fig. 20.4) (Awad et al., 2012). As a result ultrasonics can be common processing in various food processing operations like microstructure modification of fat products (sonocrystallization), defoaming, food protein modification, inactivate/accelerate enzymatic reactions, microbial inactivation, freezing, thawing, extraction including emulsification. Particularly, the method is useful in supplying the high-energy intensity needed to create nanoemulsions.

3.2.1 Process Definition

High-energy ultrasound of frequencies between 20 and 500 kHz with typical intensities more than 1 W·cm^{-2} is used to disperse a liquid phase in another immiscible liquid phase in the presence of suitable emulsifying and so-called stabilizers. Such dispersion produce reasonably stable emulsions due to high shear in the liquid medium. The underlying process can be divided into two stages; the first step involves a creation of primary macro-sized droplets due to the combined effect of acoustic field and Rayleigh–Taylor instability (Kentish et al., 2008). Subsequently, shock waves generated due to cavitation breaks the primary droplets to submicron size (Fig. 20.4).

Usually, the ultrasound nanoemulsification process involves a premix process where the two emulsion-forming phases are initially mixed to produce a coarse emulsion to avoid high ultrasonic energy requirements. In the next stage ultrasound waves are applied to the premix to produce a nanoemulsion with small and uniform droplets of the dispersed phase. The second homogenization step can be performed in both batch and continuous mode based on the scale of operation.

3.2.2 *Equipment Designs*

The major component of an ultrasonic homogenizer (UH) consist of an ultrasound source which can be supported by a temperature control system, recirculation system etc. A typical batch UH is given in Fig. 20.5a. The ultrasound source usually consists of an ultrasonic transducer, signal generator, and a signal amplifier. The UH not only produces ultrasound of an appropriate particular intensity and frequency but provides a proper contact pattern with the emulsifying liquids. For larger operations ultrasonic homogenizers can be run in continuous mode (Fig. 20.5b) Based on the contact pattern five different designs of ultrasonic homogenizers

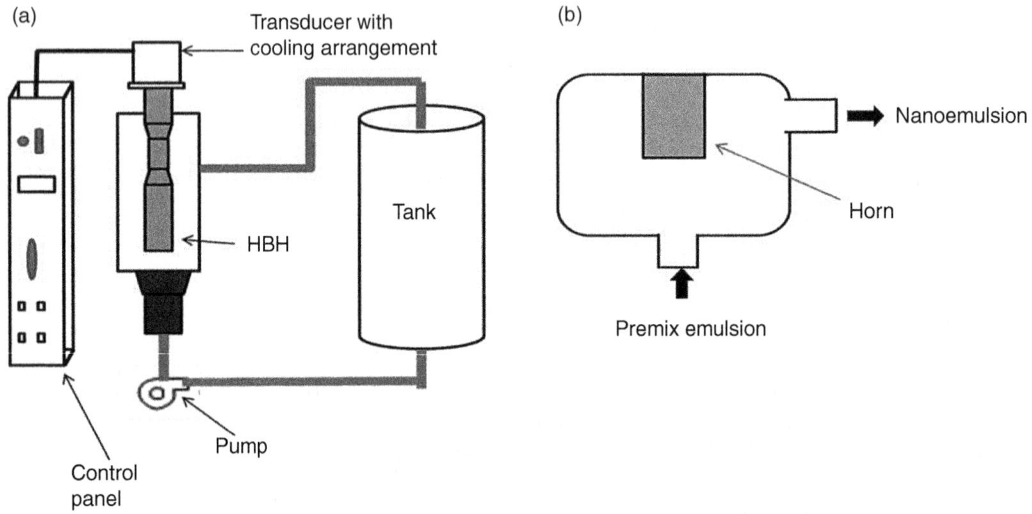

Figure 20.5. Different modes of operations of ultrasonic homogenizers. (a) Batch ultrasonic homogenizer with barbell horn type and (b) ultrasonic homogenizer in continuous mode.

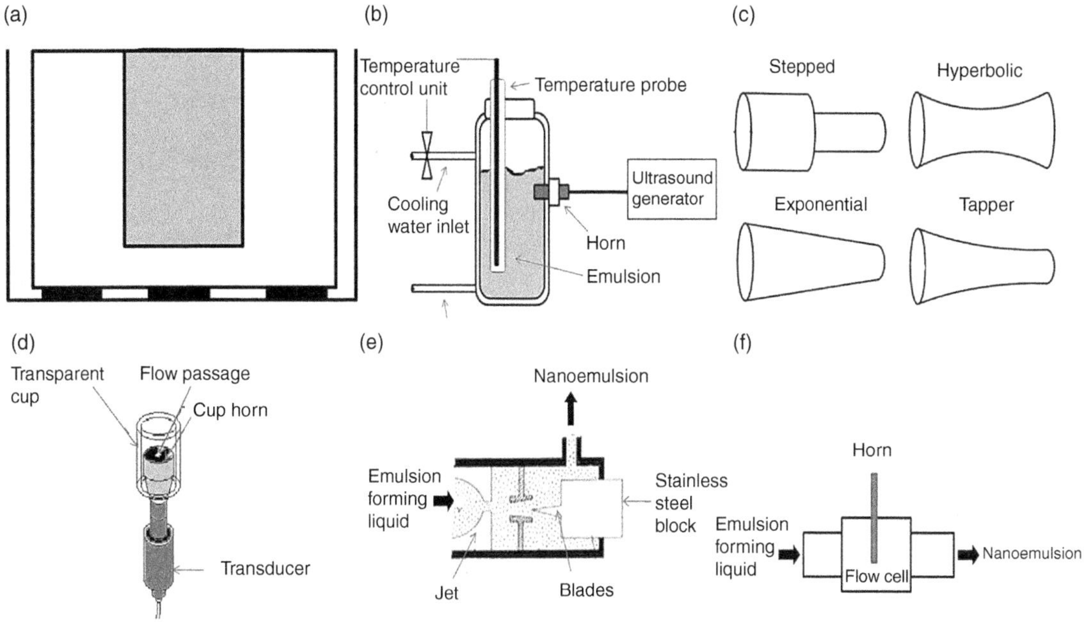

Figure 20.6. Different designs of ultrasonic homogenizer in batch and continuous mode. (a) Bath type; (b) horn type; (c) different shapes of horns; (d) cup horn type; (e) liquid whistle type; and (f) flow cell type.

are possible (Fig. 20.6a,b,d,e, and f). In horn type ultrasonic homogenizers the shape of the horn Fig. 20.6c decides the energy and the amplitude of the ultrasound generated. Table 20.1 gives brief description of ultrasonic homogenizers and their comparative features) (Luo et al., 2014).

Power requirement of ultrasonic homogenizers vary depending on scale of operation. In practice, the laboratory scale small volume homogenizers for new emulsification process feasibility studies come with a power capacity less than half a kW. For process optimization studies to evaluate optimal amplitude, operational pressure, flow rate, and so forth at pilot plant level the ultrasound equipments come with a range of power supply typically from 0.5 to 2 kW. The present ultrasonic power supplies available for industry-level nanoemulsion preparation vary from 2 to 16 kW. For some large-scale industrial applications multiple units can be employed in parallel (Hielscher, 2007).

3.2.3 Process Modeling

Information about the effect of composition and various process parameters viz., process volume, residence time in the acoustic field, ultrasonic power, and so forth on ultrasonic emulsification is important for designing equipment for ultrasonic

Table 20.1 Comparison of Different Types of Ultrasonic Homogenizers

Type of Homogenizer	Configuration	Frequency and Power Intensity	Possibility of Scaling Up	Advantages
Ultrasonic bath	Transducer at bottom or side position of bath; reaction vessel fixed at some positions in bath	15 kHz >1 MHz, normally 1–2 W/cm^2	Medium	1. Low cost 2. Commercialized
Probe reactor	Direct delivery of ultrasonic energy to liquid reactant through immerssible horn	Up to 100 W/cm^2	Medium	1. High power output 2. Concentrated energy delivery 3. Commercialized
Cup-horn reactor	Transducer fitted with a cup, the cup being the reactor	20–504 kHz, 19–270 W	Low	1. Concentrated energy delivery 2. Avoid contamination of reactant from horn tip erosion
Liquid whistle	Ultrasound generated by mechanical oscillation	5–30 kHz 1.5–2.5 W/cm^2	High	1. Low cost 2. Suitable for continuous flow reactions
Flow cell reactor	Intensified acoustic intensity by reflection and reverberation	17–45 kHz, Up to 3 W/cm^2	High	Concentrated and intensified energy delivery, uniform ultrasonic field

Source: Based on data from Luo et al., 2013.

emulsification, optimizing the process conditions, and controlling a process. The ultrasonic cavitation process and the emulsion formation involves multiple processes like the bubble formation, movement and collapse in an acoustic field as well as dispersion, reflection (by the droplet surface), and dampening of the ultrasonic waves; therefore, in a practical scenario modeling of ultrasonic emulsification is very difficult. The model should consider emergence of the bubbles due to pressure difference generated by ultrasound, coalescence of the neighboring bubbles, survival time of the bubbles before their rupture, and the way acoustic field emitted from the horn interacts with the bubble (Lauterborn et al., 2007; Leighton, 1995). The ultrasonic energy applied to a medium creates bulk motion. This motion or kinetic energy of the liquid particles converts into heat due to the viscosity of the medium (Tjøtta, 1999). There are models that correlate the two forms of energy dissipation because in the formation of bubbles of the dispersed phase the amount of these individual forms of energy plays an important role. In one of the earliest attempts based on the Kolmogorov eddy theory the maximum size of the droplet (d_{max}) due to ε ultrasonic energy dissipation per unit volume of the liquid can be calculated by the following model:

$$\varepsilon = \left[\frac{\rho_c^{-1/5} \gamma^{-3/5}}{d_{max}} \right]^{2.5}$$

Chen (2013) have proposed a model to correlate the thermal release per unit volume (ε) and the strength of ultrasonic wave (I_u) by Eq. 20.7:

$$\varepsilon = 2\alpha I_u / c \quad (20.7)$$

where α and c are acoustic absorption coefficient of the liquid and velocity of the ultrasound in the liquid, respectively.

For a batch or a continuous system acoustic intensity (I_u) data can be converted to energy density (ε), using Eq. 20.8 where S_A, V, and t are representing surface area of the tip of the horn, processing volume and time (batch time or residence time for the continuous system) respectively (O'Sullivan et al., 2015).

$$\varepsilon = \frac{I_u S_A t}{V} \quad (20.8)$$

In a comparative study, the empirical model to relate emulsion droplet size data (d_{32}) was correlated to energy density using inverse power laws for a small scale ultrasonic homogenizer setup. The models for batch and continuous system are $d_{32} = \frac{13.45}{\varepsilon^{0.85}}$ and $d_{32} = \frac{14.8}{\varepsilon^{0.77}}$ respectively.

Such simple models are often a good starting point for design of ultrasonic homogenizers and more true in a single bubble system. However, in a real system where a population of bubble exist a deeper understanding of the cavitations phenomena is required for constructing of models for designing of novel equipments or optimization of the emulsification process using ultrasound.

3.2.4 Process Optimization

In each batch or continuous configurations, optimization studies of process parameters (ie, energy density, residence time, etc.) have been done extensively for macro emulsions (Hielscher, 2007) unlike nanoemulsions where only few reports are available. For producing a nanoemulsion of curcumin, a coarse emulsions of the same composition was exposed to an optimal condition for ultrasonic. The conditions were 40% of applied power (power density: 1.36 W/mL) for 7 min, stabilized by 1.5% (w/v) Purity Gum Ultra, 0.05 volume fraction of oil (medium chain triglycerides), and 6 mg/mL of the bioactive compound (Abbas et al., 2015). Hosseini et al. (2015) have optimized the conditions for preparing an oil-in-water (O/W) nanoemulsion for polyunsaturated fatty acids (PUFA) using response surface methodology. At the optimal conditions, that is, 10:100 (V/V) linseed oil and and ratio of Tween 80 and Span 60 with the HLB of 11.7 an emulsion was obtained with a mean droplet size as small as 72±5.14 nm and viscosity 3.84±0.26 mPa·s.

3.2.5 State-of-the-Art Ultrasonic Technology for Food Nanoemulsion

Most development of emulsification process using low frequency ultrasound was for nonfood applications like paints, synthesis of polymeric nanoparticles etc. but lately the technology has been used in food processing (Freitas et al., 2006; Jafari et al., 2008). Lower production cost, ease of cleaning, and possibility of aseptic design has attracted various studies on new ultrasonic homogenizer designs and optimization of processing conditions. Another aspect of ultrasonic nanoemulsion that is attracting research focus is fabricating functional nanoemulsions like emulsions with antimicrobial activities. Peshkovsky et al. (2013) found ultrasonic amplitude played a significant role in ultrasonic nanoemulsification. To scale up the emulsification process by 10 times the authors used Barbell Horn Ultrasonic Technology. In another study, environmental pressure along with temperature influenced the nanoemulsification. They suggested total energy density followed an exponential relationship with the pressure in the sonication vessel up to 400 kPa and this relation can be exploited for formulation of better emulsification (Leong et al., 2009). Time of emulsification in an ultrasonic reactor depends on the design (geometry) of the equipment. To improve bioavailability or digestibility, specifically skin permeability, prolonged released characteristics, antimicrobial property etc nanoemulsifications were formulated using ultrasonics. Using statistical optimization technique, surfactant concentration (nonionic surfactant Tween 80) and emulsification time were optimized to achieve a stable nanoemulsion of basil oil in water with droplet diameter 29.3 nm and antibacterial activity against *Escherichia coli* (Ghosh et al., 2013). Nanoemulsion formulation with smaller molecular weight surfactants have been studied to great extent but the same with polymeric substances like protein is presently being essential because of their better acceptance in food system. O'Sullivan (2015) have studied extensively the effect of ultrasound on the protein structure individually and on emulsifying properties of protein.

3.2.6 Scale Up

All local phenomenon, that is, amplitude of the acoustic waves, intensity of cavitation, residence time of an emulsifying fluid at the cavitation zone, and so on occurring at various locations such as near the sonotrode, near periphery, and so forth is important for the scale up of ultrasonic homogenizers. Only then the quality of the final emulsion product produced in both smaller- and higher-scale ultrasonic homogenizers remain the same. However, the ultrasonic processors cannot be scaled up in terms of size alone. A large horn can generate high total power because of large surface area but provides low power densities due to low amplitudes. Therefore, existing high-power industrial-scale ultrasonic processors with horns cannot generate ultrasonic amplitudes above ~20 μm whereas much larger amplitude (~80 μm) is required to produce nanoemulsions of lesser mean droplet size and narrow size distribution. Peshkovsky and Bystryak (2014) demonstrated a scale-up study using a Half-wave Barbell Horn (BH) having a 50 mm broad tip and double radiating surfaces. The horn is capable of generating longer amplitudes up to 100 μm. Since processing capacity of a horn is proportional to its radiating area, when a Barbell Horn replaces the conventional one the processing capacity increases $2(D_{BH}/D_{CH})^2$, where D_{CH} and D_{BH} are the respectively, diameters of the tips of conventional and Barbell horns (Peshkovsky et al., 2013). This enables very high (ie, 50–60 times) scale-up emulsification processes, which facilitates single-step scale up of laboratory emulsification processes into an industrial size. The future design efforts for ultrasonic homogenizers at a larger scale should solve problems of low penetration depth of cavitation into the liquid to guarantee uniform emulsification (Mason et al., 2006). Also for effective scale-up optimization of process should be done with an auxiliary agitation system to make the nanoemulsion uniform.

3.3 High-Pressure Homogenization

High-pressure homogenization is a mechanical process, where a liquid is forced through a narrow gap at high pressure of typically more than 100 MPa value. A typical range of 150–200 MPa is used for high-pressure homogenization and 350–400 MPa for ultra-high-pressure homogenization. Here the pressures refer to the hydrostatic pressure applied by the pump to a liquid system before the liquid is restricted by an adjustable valve or a fixed geometry opening (Paquin, 1999).

3.3.1 Process Definition

Using a high-pressure pump when two immiscible liquids and appropriate emulsifier and/or stabilizer are passed through a narrow opening (often called as the homogenizing valve, nozzle, or microchannel) a combination of physical effects like high shear, turbulence, compression, acceleration, drop in pressure, and impact causes the breakdown of one of the phases and causes formation of an uniform dispersed phase (Fig. 20.7a).

3.3.2 Equipment Designs

Designwise the HPHs have many variants; however, the basic requirement for any HPH is a high-pressure positive displacement pump and a restriction assembly. Their designs may vary in terms of the type of the pump, type of the restriction assembly, and in terms of the number of restrictions (Köhlera and Schuchmann, 2011). A typical design of HPH with two-stage restriction is given below (Fig. 20.7b).

Using plunger-type pumps, high pressure can be imparted in the fluid in HPH. Suction valve of the pump withdraws the liquid mix to be emulsified to the pump and in the forward stroke of the plunger pushes the liquid through the discharge valve of the pump and homogenizing valve. The plunger can be electrically or pneumatically actuated. Such pumps can be single- or double-acting based on the discharge volume per stroke of the pump from a given volume cylinder volume of the pump. In order to reduce fluctuation in the supply of the fluid to the homogenization valve and reduce vibration, multiple plungers are used. Three to five

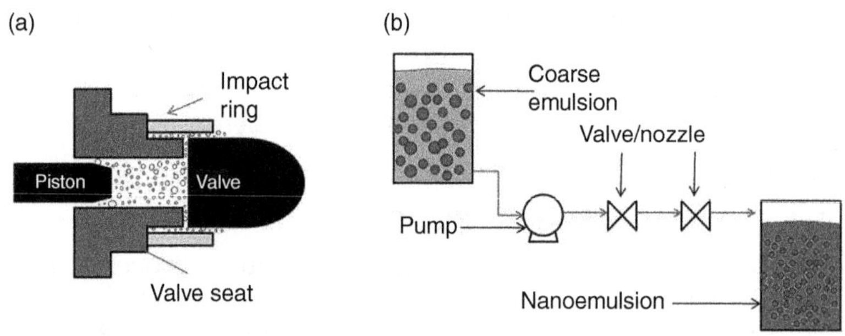

Figure 20.7. Schematics of homogenization by a high-pressure valve and schematics of a two stage homogenization system. (a) Valve-type high-pressure homogenizer; and (b) two-stage high-pressure homogenizer.

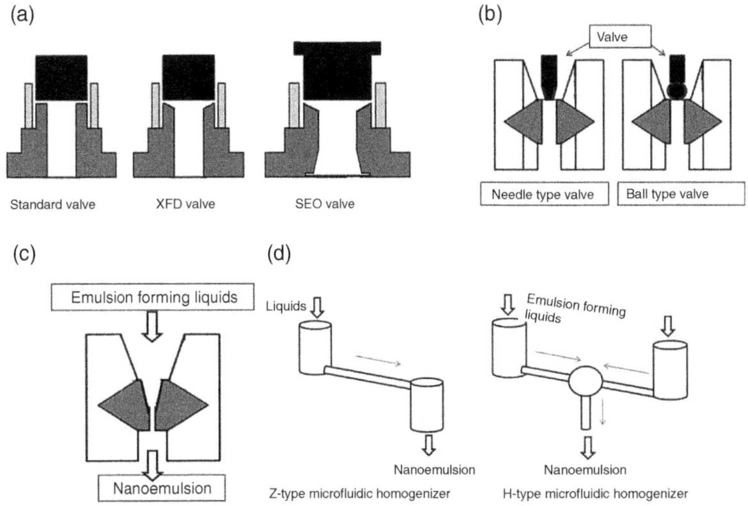

Figure 20.8. Different types of high pressure homogenizer with various adjustable valves, with nozzle, and with microchannel. (a) Valve-type HPH with different valve seat; (b) valve-type HPH with different valve types; (c) nozzle-type HPH; and (d) microfluidic-type HPH with different microchannel orientation.

plungers working in a concerted manner are common in modern HPHs. To achieve ultra-high pressures (~400 MPa) in the liquid, intensifier technology is used in the latest HPHs.

The other important part of HPHs is the restriction assembly. Restriction assembly can be of three types. They are HPH with adjustable valve (Fig. 20.8a,b), with nozzle (Fig. 20.8c), and with microchannel (Fig. 20.8d). Depending on the type of the restriction assembly, relative contribution of shear, turbulence, impact, and cavitations on emulsification changes categorically in the homogenizer. Due to the high shear produced at the restriction all these restriction assemblies require high durability and often are made up of zirconium or tungsten carbide. In a typical HPH with valve, the force applied over the valve blocks the fluid stream and builds up the homogenization pressure. There is a continuous effort by the HPH manufacturers to modify valve geometry in order to reduce the mean size of the dispersed phase and reduce the dispersity of the emulsions. Fig. 20.8a shows three different configurations of valves used for high-pressure homogenization. Industrial scale homogenizers with valve setup uses two stages of homogenization. First valve operating at a higher pressure produces crude emulsion and second valve at a lower pressure (~typically 10–20% of the first valve) offers correct degree of cavitation and turbulence to produce uniform nanoemulsion (http://web.utk.edu/). Nozzles are another type of restriction unit that can serve the purpose of high shear homogenization. The orifice diameter of the nozzle head can be selected according to the required emulsion duty (Stang et al., 2001). They are usually made up of hard and tough material like ruby, sapphire, or diamond to withstand the high pressure and remain functional for longer time. Homogenization pressure at the nozzle is controlled by the pump and/or the bends between the pump and the nozzle. Nozzle is generally a better choice over valves because adjustable valves may produce greater batch to batch variation. A further improvement in HPH design is use of microfluidic channels for homogenization. This design is often referred as Microfluidizer; however, the name is a brand name of the equipment. Essentially the technique pushes the liquid through one or more microchannels. Fig. 20.8d illustrates two geometries, Z-type and Y-type of microchannel. The Y type carries two liquid stream and the streams mixes into each other in the form of jets. Y-type of microchannels can be used for nanoemulsion due to high shear generated in the device. The branches of Y-type microchannel can be a single pair or multiple pairs to bring the liquid entering for homogenization. The design allows the shear rates in the channel to go up to 10^7 s^{-1}. The exiting fluid from the microchannels comes with a very high speed (~500 m/s) and therefore shows more uniform nanoemulsion as compared to the other two HPHs.

3.3.3 Process Modeling

Understanding of the emulsification process in a high-pressure homogenizer is still inadequate. Producing nanoemulsion with better size control in food system is often difficult due to complex rheology of the food materials as well as restricted choice of food-grade emulsifiers/stabilizers. Therefore, to choose an appropriate design of high-pressure homogenization equipment or to optimize a particular emulsification process in a given HPH, understanding the fate of the liquid in the homogenizer and its relation to the quality of nanoemulsion is essential. Modeling of the emulsification process has been attempted to address the problem. Since milk homogenization is one of the most used homogenization techniques, there are many empirical models developed for predicting emulsion efficiency in various HPH design. Particularly, modeling has been done of the mean fat globule size ($\bar{d}$) with the homogenization design parameters and other process parameters in a HPH with valve is, $\bar{d} = \frac{(g/v_0)(\sigma/\rho v)}{Re^{1/3}}$ where σ, ρ, and v are surface tension, density, and kinemetic viscosity of the liquid respectively, g is gap of the valve, v_0 and Re are the velocity and Reynolds number at the valve opening respectively,

(Kessler, 2002). This model predicts only the size of the dispersed phase but not any other quality parameter of the emulsion. Also, such a model is not universal for all kinds of valves or HPHs. Depending on the valve orientation, like valves with a flat seat and with an inclined seat, the relation would be different (Saravacos and Kostaropoulos, 2002). In a valve-type HPH, Augusto et al. (2012) have suggested two different types of empirical relations (Eqs. 20.9 and 20.10) of viscosity (μ_{dis}) after dispersion and the homogenization pressure (P_H). A similar approach can be taken for nanoemulsions, too.

$$\mu_{dis} = 3.03\left[\frac{2.55 + e^{-0.0059 P_H}}{3.55}\right] \tag{20.9}$$

$$\mu_{dis} = \frac{3.02}{1 + (0.0099 P_H)^{0.856}} \tag{20.10}$$

OEM of HPH carries out optimal sizing and design of the valve by CFD modeling. Modeling of emulsification in nozzle type and microchannel type homogenizers are less explored due to lack of understanding compared to the traditional valve type homogenizer. An example of modeling the emulsification in a nozzle type HPH is given by Eqs. 20.11 and 20.12 (Marie et al., 2002). Assuming the nozzle as a pipe with a sudden opening and turbulent flow at the nozzle, the Kolmogorov equation can be used. Hence, the largest diameter of the droplet can be modelled as $d_{max} = C\varepsilon^{-0.4}\sigma^{0.6}\rho^{-0.2}$ where C, ε, σ, and ρ are respectively a characteristic constant, power density (W m^{-3}), interfacial tension (mNm^{-1}), and continuous phase density. The power density applied by the homogenizer to the liquid near the nozzle is the function of pressure differential across the nozzle ($P_{upstream} - P_{atm}$) and the frictional loss (f) across the nozzle (Eq. 20.11).

$$\varepsilon = \frac{P_{upstream} - P_{atm} - f}{\Delta t} \tag{20.11}$$

where Δt is time of expansion

The frictional loss can be calculated by the following equation:

$f = \dfrac{kV_{min}\rho}{2g}$ $k = (1 - r)^2 + r^2/9$ is a constant dependent on nozzle and liquid recieving chamber diameter ratio (r) of the homogenizer and V_{min} minimum velocity of fluid. The minimum velocity of the fluid is given by Eq. 20.12 (Marie et al., 2002).

$$V_{min} = \frac{0.098 D^2_{nozzle} P^{0.5}_{upstream}}{\pi D^2_{pipe}} \tag{20.12}$$

Amornsin (1999) had attempted the modeling of flow rate of the emulsifying liquid and pressure of the intensifier in a microfluidizer. However, the model does not predict the product quality.

$V_{min} = \frac{2C_v}{3a\rho_c^{0.5}}\left[(at_2 + P_d)^{1.5} - (at_1 + P_d)^{1.5}\right]$ where C_v, coefficient of the throttling valve, P_d is the minimum pressure of the intensifier, a is the linear rate of the pressure variation of the intensifier.

In another study, Qian and McClements (2011) have reported a semiempirical model of emulsion formed in a microfluidizer to predict minimum size of the droplets (d_{min}) $d_{min} = \frac{6G\varphi}{C_{surfactant}}$ where G, φ, and $C_{surfactant}$ are the surfactant concentration present per unit area of the droplets created during emulsification (kg m^{-2}), volume ratio of dispersed to total volume, and surfactant concentration in the emulsion (kg m^{-3}). The mean droplet diameter ($\bar{d}$) and dispersed phase viscosity (μ_{dis}) is related in an empirical relation $\bar{d} = A\mu_{dis}^{b}$ where A and b are constants (Schultz et al., 2004). For high-pressure homogenizers, (b) varies between 0.2 and 0.9, depending on the mechanism of the dispersed phase rupture involved.

Considering the need to predict the final emulsion quality in terms of process parameters of a microfluidizer, CFD-simulations are required. Simulating such turbulent process are difficult due to the complexity of the models. The experimental data to build up empirical models of emulsification in microfluidizer are also inadequate and less reliable.

3.3.4 Process Optimization

In order to get better size distribution and stability, the optimization of emulsification process in a HPH covers two classes of variables; first one is finding the appropriate combination of operational variables like homogenization pressure, number of passes, premixing, and temperature and the second one is finding an appropriate recipe of the emulsion. The two said classes of variables are not independent and often influence the quality of emulsion in a complex fashion. Hence, statistical design of experiment is an efficient approach to decide the optimal settings of the process variables. Yuan et al. (2008) had attempted optimization of preparation of a nanoemulsion of medium-chain triglyceride and water where the oil phase contained β-carotene using a two-stage valve-type HPH. To achieve minimum size distribution and maximum stability of the emulsion, Response Surface Methodology was used in the study to determine appropriate levels of β-carotene, emulsifier, concentrations, the homogenization pressure, and temperature. Adopting superimpositions method of the contours obtained for each pair of the said independent variables an optimal value for each variable were selected. Some of the drawbacks of using RSM in this kind of optimization study are large number of variables that influence emulsion quality, multiple output variables depicting quality, and nonlinearity of the relation between the emulsion quality and the independent variables. However, optimization of such a complex process is sequential and better understanding of the process would lead to precise control over the quality of the emulsion.

3.3.5 State-of-the-Art High-Pressure Homogenization Technology for Food Nanoemulsions

The effect of composition on a nanoemulsion produced by an HPH is one of the major areas of studies in high-pressure homogenization. There are only a small number of reports on elucidating basic mechanism of the emulsification process in HPHs. Donsì et al. (2011) have studied the effect of surfactants as well as disruption chamber geometry on nanoemulsion formation. The study suggested the effect of surfactant controlled the kinetics of the emulsion condition, but an appropriate design of the homogenization chamber ensured proper mixing and narrow size distribution of the dispersed phase. Particularly, influence of small surfactant in combination with surface active protein remained the subject of many studies in HPH (Perrier-Cornet et al., 2005; Mao et al., 2010). Most trials on nanoemulsion done by HPH involves typically two stages: the primary emulsification through a mechanically agitated mixer or in a rotor–stator and secondary emulsification through an HPH. Unfortunately, there are not many studies on the effect of premixing conditions on the final nanoemulsion. Sometimes an inline mixer can be used for producing premix of 2–5 μm size distribution for HPH, which improves final nanoemulsion quality.

3.3.6 Scale Up

In general, scalability of HPH is a challenging issue. Particularly, industrial scale preparation of nanoemulsion in HPH has crippled commercialization of many promising nanoemulsion formulations. In almost all cases, a traditional scale-up strategy is followed where a pilot study is carried out to scale up a high-pressure homogenization to produce nanoemulsion. The flow filed around a narrow gap with liquid flowing at a very high velocity in HPHs rules out experimental estimation of the velocity profile.

That is why there is no established scale-up criteria of the emulsion process in HPHs. Innings and Trägårdh (2007) have proposed the two criteria be kept constant for scaling up HPHs, Reynolds number (Re), and the ratio of turbulent gap height (h) and Kolmogorov length (ξ) scale ($N_{TK} = h/\xi$). The gap (or jet) Re is defined as hU/v, where h and U are the length of the narrow gap and the velocity of the fluid at the narrow gap of the HPH.

4 Low-Energy Production Technologies of Nanosized Emulsifiation

Low-energy methods of emulsification (or condensation) method make use of phase transitions occurring when emulsion forming substances (essentially oil, water, surfactant, stabilizer) come together or droplet formation through a membrane. Unlike high-energy methods of emulsification, in such methods external energy is not used for creation of new interfaces during emulsification. Rather it converts the energy stored in the emulsion forming mixture to create the large surface area in the form of the droplets of few nanometers (Lovelyn and Attama, 2011). Therefore, they are more energy-efficient than its high-energy counterparts, where the typically 0.1% of the total energy goes for homogenization. Not only that but growing research in the past two decades have shown these methods could be a better alternative due to greater emulsion yields, ease of scale-up, and less harsher methods for emulsification (or encapsulation) of weak active molecules (Anton and Vandamme, 2009).

4.1 Membrane Emulsification

Membrane emulsification methods involve forcing the dispersed phase or a preemulsion into a continuous phase through a microporous membrane of suitable pore size (Fig. 20.9). It consists of two steps: first the droplets grow while forced through the pores and second the drops detach and go away from the pore tip (Peng and Williams, 1998). Since, in this method droplet formation does not depend on turbulent disruption the dispersed phase size can be tightly controlled. Considering these capabilities, membrane emulsification is a good choice for food nanoemulsions. The method can be used for both types of nanoemulsions viz., single emulsions (O/W or W/O) and double emulsions (O/W/O or W/O/W). Double emulsions are difficult to make due to the complex structure and inherent thermodynamic instability.

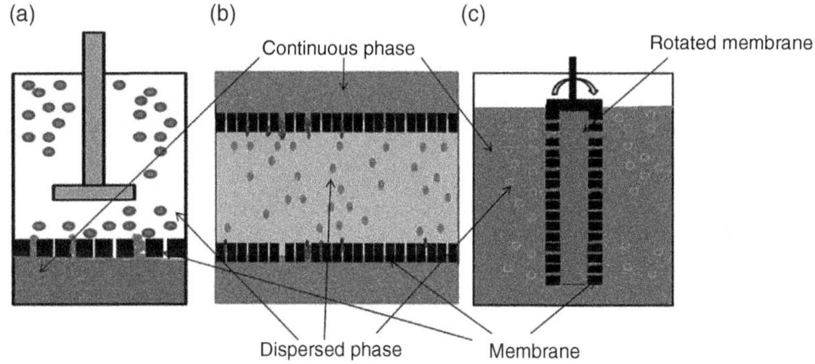

Figure 20.9. Three major types of membrane emulsification systems. (a) Agitated cell; (b) cross-flow system; and (c) rotated membrane.

The droplet formation depends on the design of the membrane assembly, membrane characteristics (pore size and shape, porosity, wall contact angle, etc.), liquid characteristics (density, interfacial tension, and viscosity), and process characteristics (temperature, pressure, etc.). Designwise there are three major types of membrane emulsification systems: (1) stirred-tank system where the liquid to be dispersed is pumped through a microporous membrane to a tank holding the continuous phase and an impeller agitates the liquid inside the tank to produce the shear for the droplet detachment from the membrane as well as the mixing in the tank (Fig. 20.9a), (2) cross-flow system involves a tubuler membrane assembly where the continuous phase flows as retentate axially in the tube and the dispersed phase/preemulsion flows as permeate into the tube to form a nanoemulsion (Fig. 20.9b), and (3) a modified cross-flow system where the membrane tube contains the dispersed phase and rotates to generate enough shear for detachment of the droplets on the continuous phase side (Fig. 20.9c). The design directly influences the quality and yield of the emulsion formed, that is, size of the dispersed phase droplets and its distribution, flux of the dispersed, and the volume percentage of the dispersed phase.

Membrane properties also control quality and yield of emulsion. The mean droplet diameter is proportional to the mean pore size of the membrane. Decrease in porosity reduces the flux of the dispersed phase whereas increase in porosity increases the probability of merging of droplets coming out of adjacent pores of the membrane. So an optimal porosity is needed to obtain near mono-disperse droplets of low mean size and considerable yield. The critical pressure (P_{crit}) required for formation of droplets for a given interfacial tension (Υ), contact angle of the dispersed phase on the membrane surface (θ), and mean pore diameter of the membrane (d_p) can be calculated by following relation (Eq. 20.13).

$$P_{\text{crit}} = \frac{4\gamma\cos\theta}{d_{\text{p}}} \quad (20.13)$$

The dispersed phase flux (J_{d}) also depends on the membrane properties. Assuming the flow through porous medium, for given pressures difference (ΔP) across the membrane the flux can be calculated using Eq. 20.14.

$$J_{\text{d}} = \frac{K\Delta P}{L_{\text{L}}} \quad (20.14)$$

where K and L are permeability and thickness of the membrane respectively, μ is the dispersed phase viscosity. Assuming a simpler structure of the membrane the pores can be mimicked as a series of (n) numbers of uniform cylinders of radius r, the permeability can be calculated by Hagen–Poiseuille equation, $K_{\text{d}} = nr^2/8\,\mu$.

Near the membrane pores the local hydrodynamics forces determine the size of the droplets and their distortion. Configuration of the membrane assembly primarily governs the local flow pattern. In membrane emulsification the continuous phase fluid flow near membrane follows laminar pattern, therefore, the shear stress (τ) acting on the droplets of the dispersed phase at the tip of the pore is given by $\tau = \mu_{\text{c}}\, U$, where U is the velocity gradient adjacent to the membrane in the continuous phase and μ_{c} is the viscosity of the continuous phase (Walstra, 1993; Adler-Nissen et al., 2004). As the droplet size decreases the shear stress at the interface of the membrane and the continuous phase liquid also increases. Also type and quantity of surfactant used in the emulsion process plays a great role in formation of the droplets and in maintaining the dispersed phase intact as an emulsion.

For scaling up and optimization of such membrane emulsification processes, a reliable model of the process is necessary. Especially, the droplet formation and detachment is a very complex process and without proper understanding of the mechanism of the process it is almost impossible to develop a model and to identify a scale-up criterion. Hao et al. (2008) have developed a model based on the various torque acting on the droplets in a cross-flow continuous emulsification system. The model identifies three major factors viz., low cross-flow velocity of the continuous phase, low transmembrane pressure, and high viscosity of the dispersed phase favors narrow size distribution of the dispersed phase.

Membrane emulsification technique is being used by food nanoemulsions, as well, even if at a slower pace. Various membranes for this purpose studied are: ceramic aluminium oxide (a-Al2O3) membranes (Schröder and Schubert, 1999), a-alumina and zirconia-coated membranes (Joscelyne and Trägårdh, 1999), and polytetrafluoroethylene (PTFE) membranes (Suzuki et al., 1998; Kanichi et al., 2002; Yamazaki et al., 2002), microporous polypropylene (Sotoyama et al., 1999) or polyamide (Giorno et al., 2003, 2005), hollow fibers membrane, and homemade silica-based monolithic (Hosoya et al., 2005), or macroporous silica glass (Fuchigami et al., 2000) membranes, and so forth. A nanoemulsion to encapsulate vitamin E was prepared from medium chain triglycerides, and a mixture of two small molecular weight surfactants, for example, Tween 80 and Brij 35. The mean size of the droplets of nanoemulsion was 78±3 nm with a span factor of 0.25±0.01 (Laouini et al., 2012). In another study various oil phase, that is, medium-chain triglycerides, soybean oil, and trimyristin, and different emulsifiers sodium dodecyl sulfate, poloxamer 188, polyglyceryl-10-laurate, and sucrose laurate were used to study membrane emulsion at two different scales. The study found a monodisperse droplet of 100 to 200 nm size (Joseph and Bunjes 2012). Trentin et al. 2011 have reported a encapsulation technique for beta-carotene where a membrane emulsification technique with microporous polymeric membranes was used. Bovine serum albumin or whey protein concentrate along with Tween 20 were used as an emulsifier–coemulsifier system to get a monodisperse emulsion.

Two major drawbacks of this method is low throughput compared to other emulsification methods and exorbitant high cost of the membrane at a large scale. In addition to that fouling reduces efficiency of these processes to a great extent. Hence, efforts are focused on novel membrane development, novel design of membrane assembly (eg, adding vibratory motion in the continuous phase), process optimization, and so forth. Such efforts would remove the challenges of industrial application of nanoemulsion preparation using the membrane emulsification technique.

4.2 Spontaneous Emulsification

The method implies addition of an oil phase (oil phase, surfactant, one or more oil-soluble active components) into an aqueous phase (Anton et al., 2008; Vandamme and Anton, 2010) in appropriate ratio. The surfactant used in the process should be water soluble with a high hydrophilic-to-lipophilic balance (HLB) number (typically >8).

The surfactant molecules condense in the water phase after the oil phase is poured into it to give rise to nanosized oil droplets (Fig. 20.10). This method of preparation sometimes is confused with microemulsion. However, the chief difference between the formation of a nanoemulsion (kinetically stable) and a microemulsion (thermodynamically stable) is the order of mixing of the

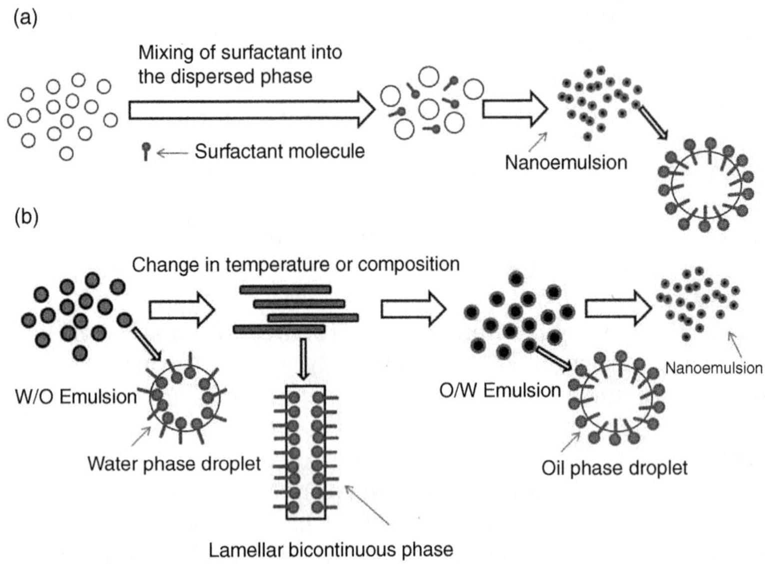

Figure 20.10. Schematics of spontaneous emulsification and phase inversion method of emulsification. (a) Spontaneous emulsification; and (b) phase inversion.

surfactant and the other two-phase forming phases. Nanoemulsions can only form if the surfactant(s) are added into the oil phase first and then the entire oil phase is added to the water phase. On the contrary, if the surfactant is first added to the aqueous phase an emulsion with macro sized droplets results. For microemulsions this order is immaterial as long as the ratio of surfactant and oil phase is appropriate (often surfactant is 20% of the total emulsion forming phase); sooner or later the mixture will form a thermodynamically stable emulsion (Anton & Vandamme, 2011). Apart from the order of mixing of the emulsion forming components, the rate of emulsification, volume of emulsion, and droplet size (mean diameter as well as the size distribution) are important characteristics for spontaneous emulsification process of nanoemulsion preparation. In the formation of such emulsions, spreading pressure, interfacial tension, viscosity of the continuous phase as well as the interface, phase transition region, structure, and concentration of the surfactants and cosurfactants, salinity, and temperature are to be examined (López-Montilla et al., 2002).

4.3 Phase Inversion Methods

Phase inversion, or transformation of the dispersed phase into continuous phase and vice-versa, can be used effectively to create nanoemulsions. Change in temperature, oil, and aqueous phase ratio, ionic strength (ie, salt concentration) or flow pattern (eg, extensional flow) can bring catastrophic change in the arrangement of the surfactants molecules at the oil-water interface (Fig. 20.10b). Based on the agent to bring such sudden inversion of phase, these methods of nanoemulsion preparation can be categorized as three different classes.

4.3.1 Phase-Inversion Temperature Methods

The Phase Inversion Temperature (PIT) method utilizes change in hydration or solubility of nonionic surfactants with temperature. Such surfactants get more lipophilic with increase in temperature. In this method, a mixture of oil, water and surfactant are prepared at their PIT (aka hydrophilic–lipophilic balance temperature), which is then exposed to sudden heating or cooling to obtain W/O or O/W emulsions, respectively. For example, when an O/W macroemulsions is heated the emulsion gets destabilized and an intermediate liquid crystalline or biocontinuous microemulsion phase occurs around PIT. In this condition the interfacial tension minimizes and the interface attains a zero curvature. A stable W/O emulsion forms with further increase of temperature due to the changes in the physicochemical properties of the surfactant. The arrangement of the surfactant molecules are dependent on a packing parameter, $p = \frac{A_{\text{lipophilic}}}{A_{\text{hydrophilic}}}$ where, $A_{\text{lipohillic}}$ and $A_{\text{hydrophillic}}$ are the cross-sectional areas of the lipophilic- and hydrophilic-group respectively. The surfactant molecules arrange themselves spontaneously and compactly in water to form a monolayer of optimum curvature due to hydrophobic forces (Israelachvili and Wennerstroem, 1992). The optimum curvature of a surfactant monolayer depends on the packing parameter; surfactants with $p < 1$, $p < 1$, and $p \approx 1$ curvature of the monolayer is convex, concave, or zero, respectively. Curvature of the monolayer at the O/W interface decides formation of O/W emulsions ($p < 1$), W/O emulsions ($p > 1$), or liquid crystalline/bicontinuous systems ($p = 1$).

4.3.2 Phase Inversion Composition Methods

The phase-inversion composition (PIC) method involves changing concentration of one of the components of the emulsion-forming components to attain the required curvature of the surfactant layer unlike the PIT method, which does the same by altering the temperature (Anton et al., 2009). For instance, in the case of an O/W emulsion prepared with an ionic surfactant, if salt concentration is increased at a particular concentration it can change the packing parameter of the surfactant from lesser than one to higher than one. Therefore, the O/W emulsion inverts into W/O emulsion due to shielding of the electrical charge of the surfactant groups (Maestro et al., 2008). A reverse phenomena, that is, inversion of a W/O emulsion with high salt concentration into O/W emulsion can be obtained by simply diluting the system. Similar result can be obtained by changing the pH of the emulsion-forming phases due to the alteration in dissociation (and solubility in oil or water phase) of the fatty acids (Solè et al., 2006; Maestro et al., 2008).

4.3.3 Emulsion Inversion Point (EIP) Methods

Another method of nanoemulsion preparation using phase inversion is by changing the water volume fraction, often called the emulsion inversion point (EIP) method. By successively adding water into oil, initially water droplets are formed in a continuous oil phase. Especially, a W/O emulsion stabilized by short-chain surfactants undergoes a spontaneous change in curvature of the surfactant assembly with the increasing percentage of water (Fernandez et al. 2004).

4.4 Secondary Emulsification Methods

To form a nanoemulsion of particular quality a coarse nanoemulsion can be modified by various secondary emulsification methods (Figs. 20.10 and 20.11).

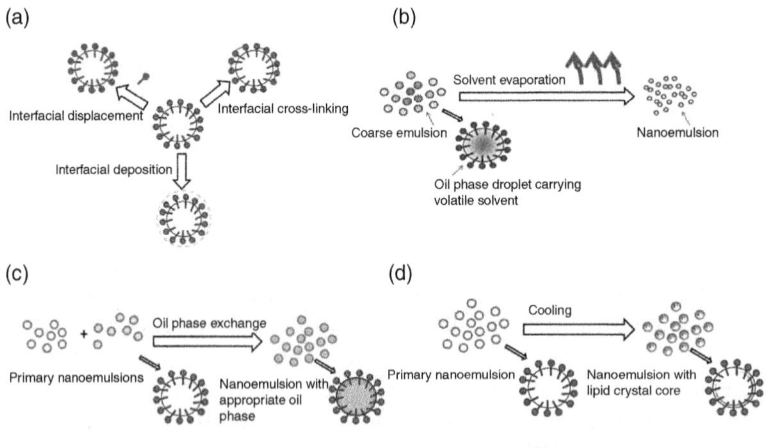

Figure 20.11. Mechanisms of secondary emulsification methods. (a) Interfacial engineering; (b) solvent displacement; (c) lipid phase exchange; and (d) lipid phase crystallization.

4.4.1 Interfacial Engineering

By the following method the surfactant layer of an existing nanoemulsion can be modified directly to achieve improved functionality or stability (Guzey and McClements, 2006; McClements, 2010).

Interfacial displacement: A preformed nanoemulsion when mixed with a solution of another surfactant, the new surfactant molecules can replace the existing surfactants at oil-water interface partially or completely (Fig. 20.11a), depending on the relative surface activities and concentration of the two surfactants (Rao and McClements, 2010).

Interfacial deposition: A multilayer droplets of the dispersed phase can be created by depositing a different type of molecule on the existing surfactant layer of droplets of an existing emulsion (Fig. 20.11a). The other type of molecule can be a polyelectrolytes or an inorganic ion that combines with the exposed ionic group of the surfactant molecules (Guzey and McClements, 2006; McClements, 2010).

Interfacial cross-linking: In a preformed nanoemulsion the surfactant molecules can be cross-linked by chemical or enzymatic method, for example, protein (Sandra et al., 2008) and polysaccharide (Littoz and McClements, 2008) emulsifiers can be cross-linked to improve the stability of the final nanoemulsion (Fig. 20.11a).

4.4.2 Solvent Displacement

A lipid material containing crystalline active materials is difficult to emulsify. Therefore, a semipolar water-miscible solvent (eg, methanol, ethanol, or acetone) can be used to dissolve the lipid phase (Ribeiro et al., 2008). If the lipid phase is added into an aqueous surfactant solution with agitation, the solvent diffuses into the water phase resulting in an emulsion of the lipid phase (Fig. 20.11b). The water miscible organic solvent can be completely removed by distillation or evaporation.

4.4.3 Lipid Phase Exchange

A nanoemulsion can be mixed with another emulsion or microemulsion to allow transfer of any mutually soluble component from/to the droplets of nanoemulsion to/from the second emulsion or microemulsion droplets (Fig. 20.11c). This molecular diffusion results in a new nanoemulsion with improved property or encapsulation of active components (Weiss et al., 2008).

4.4.4 Lipid Crystallization

In nanosized oil droplets fat crystallization occurs at a much lower temperature than it occurs in bulk condition. This fact is used to prepare a novel category of nanoemulsion by a method called lipid crystallization method. In the process, the emulsification is carried out at much higher temperature using any of the low/high energy method and then the nanoemulsion is cooled in a controlled way so that the lipid gets crystallized in the dispersed phase droplets (Fig. 20.11d).

5 Functional Characterization Processes of Nanoemulsions

In order to apply nanoemulsions effectively in food processing or food science proper characterization of the emulsion is essential irrespective of their method of production. Structure and function are two important characteristics of a nano-emulsion. There are many good reviews on structural characterization of nanoemulsions. Dynamic light scattering, zeta potential, differential scanning calorimetry, Fourier transform infrared, x-ray diffraction, nuclear magnetic resonance, small-angle x-ray scattering, optical microscopy with image analysis, atomic force microscopy, electron microscopy, and so forth are techniques that indicate size (and distribution), and some of them indicate morphology of the nanoemulsion formed. However, detailed methods of characterization are beyond the scope of this text but certain aspects of functional characterization and the rheology of nanoemulsions would be discussed. In addition, stability of the emulsion has also been given due attention.

5.1 Rheology of Nanoemulsions or Nanosized Emulsions: Aspects of Processing of Food Nanoemulsions or Nanosized Emulsions

The flow properties of a nanoemulsion are an essential physical attributes that controls their processing, physiological efficacy, toxicity, or some other functionalities. This need is driving the research focus toward measuring, modeling, and modifying these properties continuously. The basic rheology-determining parameters of an emulsion are continuous phase rheology and the nature of the droplets (size distribution, deformability, internal viscosity, concentration, and nature of particle–particle interaction). These are often appropriately understood using Krieger–Dougherty microstructure/viscosity equation. The qualitative relations between various formulation and processing variables and change of dispersed phase volume, size distribution, and particle–particle interaction as well as particle deformability is well understood.

However, quantitative understanding of rheological changes of nanoemulsion with the emulsion structure in various process and storage condition is inadequate. Effects of dispersity and particle deformability need to be understood separately (Barnes, 2004).

Rheology of a nanoemulsions of rice bran oil prepared by low-energy method and sonication method were studied using classical power law model (Sanabria, 2012). For both the emulsions consistency indexes (K) were low, which indicated the emulsion had low viscosity (or consistency). In addition, they had pseudoplastic shear thinning behavior because their flow behavior index (n) values were below one. Apparent viscosities of the low energy based emulsions were lesser than the emulsion prepared by sonication due to the smaller droplet size of the former (199 nm). Also the nanoemulsions with a droplet size range between 80 and 90 nm behaved as fluid with zero yield stress. This appropriate microstructure and low viscosities of the nanoemulsion makes them suitable for better mixing in liquid food products like beverages (McClements, 2011). Viscoelastic properties of the nanoemulsions were also estimated. Elastic or storage modulus (G') and the viscous or loss modulus (G'') are measures of solid-like and liquid-like properties of an emulsion. The oscillation test suggested the nanoemulsions had a weak structure and therefore showed higher loss modulus (G'') in comparison to the storage modulus (G'). Or in other words, the emulsions were more liquid-like.

5.2 Biological Fate of the Lipophilic

Ideas about the fate of nanoemulsions, that is, *digestion, absorption, metabolism, distribution, and excretion* of the nanoemulsion is important to know efficacy of the nanoemulsions as well as its toxicity aspects. The emulsion may contain one or more nonpolar components in the oil phase including triacylglycerols, diacylglycerols, monoacylglycerols, flavored oils, essential oils, mineral oils, fat substitutes, weighting agents, oil-soluble vitamins, and nutraceuticals (such as carotenoids, phytosterols, and coenzyme Q). These lipophilic compounds can face three different processing conditions in the human gastrointestinal (GI) tract. Compounds like triacylglycerols and diacylglycerols can be digested in GI tract, mineral oils and fat substitutes can pass through the GI tract without getting absorbed, or compounds like carotenoids and phytosterols can get absorbed by the GI tract in undigested form. The shell around nanoemulsion droplets typically contains multiple components (eg, surfactants, phospholipids, proteins, polysaccharides, and minerals). These emulsion-forming individual components may get digested at different rates and break down to different levels in the GI tract. For example, phospholipids and proteins may be digested in the stomach and small intestine itself by phospholipases and proteases but many fibres can be digested only in the large intestine.

5.3 Stability

Nanoemulsions do not possess greater physical stability than submicron-sized emulsion droplets. There are two main coarsening mechanisms for emulsions: coalescence and Ostwald ripening. The driving force for Ostwald ripening is the solubility of the dispersed phase in the continuous phase. When droplets are in the nanosize region the solubility of the dispersed phase in the vicinity of the droplet is significantly greater as a result of the Kelvin effect, leading to growth of larger droplets at the expense of the smaller ones via molecular diffusion. In fact, slowing Ostwald ripening is generally critical in being able to form a nanoemulsion with droplets less than 100 nm. This is often accomplished by combining two oils in the dispersed phase, including one with very low solubility in the continuous phase. Coalescence depends less on the curvature of the droplets and more on the curvature in the vicinity of a nucleation hole. This is controlled by the curvature of the surfactant monolayer stabilizing the droplets.

Typically stability of an emulsion is estimated by recording changes in droplet size, viscosity, pH, conductivity, and refractive index of an emulsion at various temperatures. The kinetics indicates the initiation of phase separation. To accelerate the study, the changes of parameters are estimated at a higher temperature (Alam et al., 2015). Using kinetic data for thermal degradation of an active substance thermodynamic stability can be measured. The entropy and enthalpy change of the degradation reaction calculated by the transition state provides important information about the relative stability of different nanoemulsons and their mechanism also. Pascual-Pineda et al. (2015) reported about stability of paprika oleoresin nanoemulsions prepared by ultrasonic homogenization. The mean diameters of the oil phase droplets containing paprika varied from 48 to 250 nm with varying composition. The carotenoids were more stable in smaller size droplets and at a lesser HLB value. The enthalpy–entropy compensation suggested the instability of carotenoid red fraction was enthalpy driven, whereas the stability of the yellow fraction of the carotenoid was entropy driven.

Ng et al. (2013) have proposed a novel method to determine stability of nanoemulsion in accelerated condition. The emulsion under study was exposed to centrifugal force of range between

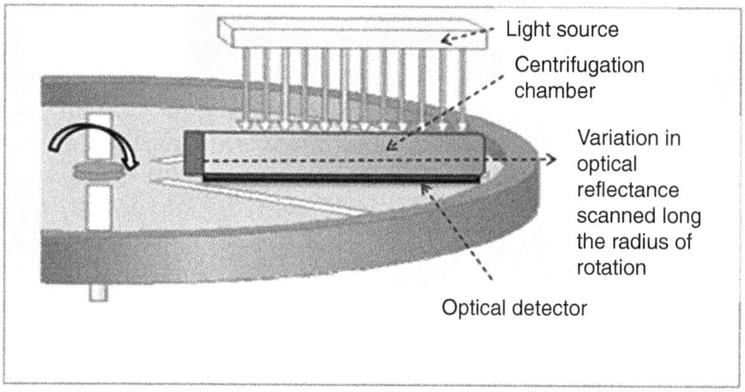

Figure 20.12. A modified centrifugation method to estimate solubility of nanoemulsions.

11 and 1140g in a vertical centrifuge. The centrifuge has an additional feature, that the transmittance of light through the liquid can be measured at various locations of the centrifuge tube (Fig. 20.12).

Often a single method like DLS is not enough to visualize the change in particle size and combination of method is better to predict stability (Grapentin et al., 2015).

6 Application of Nanoemulsion Science in Food Processing

A series of research publication on the application of nanoemulsion in food processing supports the promise of the technology as delivery vehicle for various oil soluble substances and to improve functionality of food products.

Encapsulation of lipophilic compounds by nanoemulsion can improve its ease of handling and utilization, possibility to incorporation in a food product, solubility and efficient absorption of lipophilic compounds to increase its bioavailability, controlled or environment triggered release of active components, stability against chemical degradation. However, degradation reactions at the oil-water interface may sometime accelerate due to large surface are per unit volume as well as due to photocatalysis in the transparent nanoemulsion.

Often to obtain a highly viscous or gel-like consistency of emulsion, nanoemulsion is a better option because in nanoemulsion similar consistency can be achieved at much lower concentration of fat. There are many studies on encapsulation of antimicrobial agents in nanoemulsions. Nanoemulsions can be employed to deliver one or more antimicrobial agents in the oil phase, which supplies the compounds on the cell membrane of microorganism.

Nanoemulsions have been used to reduce fat content of ice cream (Silva et al., 2012). In another interesting application a water-in-oil emulsions of droplet size of range 10–500 nm was used for uniform thawing of frozen foods in the microwave (Möller et al., 2009). Komaiko and McClements (2015) have reported a process to incorporate nanoemulsions containing bioactive components into a gelatin dessert (hydrocolloid) by use of the spontaneous emulsification process.

7 Conclusions

Food manufacturing is facing tighter regulations, fierce competition, high energy costs, consumer demands for products with broader functionality. However, at present there are only isolated examples of processes to manufacture nanoemulsions on an industrial scale. In spite of great potential of nanoemulsions, such as applicability in food products without altering the final appearance, greater stability, and improved assimilation in physiology, there are four major challenges associated with the commercialization of nanoemulsions.

1. There is a need to develop food-grade surfactants, stabilizers, cosolvents, oil phase systems like flavor oils, triglycerides, proteins, polysaccharides, and so forth, for formulating edible nanoemulsions.
2. Scalable emulsification technologies including hybrid technologies or upgradation of existing emulsification technologies.
3. Optimization of emulsification processes to reduce cost of production. Also the process of optimization should consider the final use of the nanoemulsion.
4. Characterizations of food nanoemulsions to quantify their functionality as well as toxicity in conjugation with food matrix.

References

Abbas, S., Karangwa, E., Bashari, M., Hayat, K., Hong, X., Sharif, H.R., Zhang, X., 2015. Fabrication of polymeric nanocapsules from curcumin-loaded nanoemulsion templates by self-assembly. Ultrason. Sonochem. 23, 81–92.

Adler-Nissen, J., Mason, S.L., Jacobsen, C., 2004. Apparatus for emulsion production in small scale and under controlled shear conditions. Food Bioprod. Process. 82 (4), 311–319.

Alam, M.S., Ali, M.S., Alam, M.I., Anwer, T., Safhi, M.M.A., 2015. Stability testing of beclomethasone dipropionate nanoemulsion. Trop. J. Pharm. Res. 14 (1), 15–20.

Amornsin, A., 1999. Effect of high-pressure throttling on ascorbic acid, pectin esterase activity, and limonin content in citrus juice. Masters Dissertation, University of Georgia.
Anton, N., Vandamme, T.F., 2009. The universality of low-energy nanoemulsification. Int. J. Pharma. 377 (1), 142–147.
Anton, N., Vandamme, T.F., 2011. Nanoemulsions and micro-emulsions: clarifications of the critical differences. Pharma. Res. 28 (5), 978–985.
Anton, N., Benoit, J.P., Saulnier, P., 2008. Design and production of nanoparticles formulated from nano-emulsion templates—a review. J. Control. Release 128 (3), 185–199.
Augusto, P.E., Ibarz, A., Cristianini, M., 2012. Effect of high-pressure homogenization (HPH) on the rheological properties of a fruit juice serum model. J. Food Eng. 111 (2), 474–477.
Awad, T.S., Moharram, H.A., Shaltout, O.E., Asker, D., Youssef, M.M., 2012. Applications of ultrasound in analysis, processing, and quality control of food: a review. Food Res. Int. 48 (2), 410–427.
Barailler, F., Heniche, M., Tanguy, P.A., 2006. CFD analysis of a rotor–stator mixer with viscous fluids. Chem. Eng. Sci. 61 (9), 2888–2894.
Barnes, H.A., 2004. The rheology of emulsions. Interf. Sci. Technol. 4, 721–759.
Bouchemal, K., Briançon, S., Perrier, E., Fessi, H., 2004. Nanoemulsion formulation using spontaneous emulsification: solvent, oil, and surfactant optimization. Int. J. Pharma. 280 (1), 241–251.
Brocart, B., Tanguy, P.A., Magnin, C., Bousquet, J., 2002. Design of in-line emulsification processes for water-in-oil emulsions. J. Disper. Sci. Technol. 23 (1–3), 45–53.
Chen, Z., 2013. Assessment of ultrasound field properties and the potential effects on cells. Masters dissertation, Royal Institute of Technology, Sweden.
Donsì, F., Sessa, M., Ferrari, G., 2011. Effect of emulsifier type and disruption chamber geometry on the fabrication of food nanoemulsions by high pressure homogenization. Ind. Eng. Chem. Res. 51 (22), 7606–7618.
Fernandez, P., André, V., Rieger, J., Kühnle, A., 2004. Nanoemulsion formation by emulsion phase inversion. Colloid. Surf. A 251 (1), 53–58.
Freitas, S., Hielscher, G., Merkle, H.P., Gander, B., 2006. Continuous contact- and contamination-free ultrasonic emulsification—a useful tool for pharmaceutical development and production. Ultrason. Sonochem. 13 (1), 76–85.
Fuchigami, T., Toki, M., Nakanishi, K., 2000. Membrane emulsification using sol-gel derived macroporous silica glass. J. Sol-Gel Sci. Techn. 19 (1–3), 337–341.
Ghosh, V., Mukherjee, A., Chandrasekaran, N., 2013. Ultrasonic emulsification of food-grade nanoemulsion formulation and evaluation of its bactericidal activity. Ultrason. Sonochem. 20 (1), 338–344.
Giorno, L., Li, N., Drioli, E., 2003. Preparation of oil-in-water emulsions using polyamide 10 kDa hollow fiber membrane. J. Membrane Sci. 217 (1), 173–180.
Giorno, L., Mazzei, R., Oriolo, M., De Luca, G., Davoli, M., Drioli, E., 2005. Effects of organic solvents on ultrafiltration polyamide membranes for the preparation of oil-in-water emulsions. J. Colloid Interf. Sci. 287 (2), 612–623.
Grapentin, C., Barnert, S., Schubert, R., 2015. Monitoring the stability of perfluorocarbon nanoemulsions by cryo-TEM image analysis and dynamic light scattering. PloS One 10 (6), e0130674.
Guzey, D., McClements, D.J., 2006. Formation, stability and properties of multilayer emulsions for application in the food industry. Adv. Colloid Interf. Sci. 128, 227–248.
Han, L., Li, L., Li, B., Zhao, L., Liu, G., Liu, X., Wang, X., 2014. Effect of high-pressure microfluidization on the crystallization behavior of palm stearin—palm olein blends. Molecules 19 (4), 5348–5359.

Hao, D.X., Gong, F.L., Hu, G.H., Zhao, Y.J., Lian, G.P., Ma, G.H., Su, Z., 2008. Controlling factors on droplets uniformity in membrane emulsification: experiment and modeling analysis. Ind. Eng. Chem. Res. 47 (17), 6418–6425.

Hielscher, T., 2007. Ultrasonic production of nanosize dispersions and emulsions. arXiv preprint arXiv:0708.1831. http://www.hielscher.com.

Hosoya, K., Bendo, M., Tanaka, N., Watabe, Y., Ikegami, T., Minakuchi, H., Nakanishi, K., 2005. An application of silica-based monolithic membrane emulsification technique for easy and efficient preparation of uniformly sized polymer particles. Macromol. Mater. Eng. 290 (8), 753–758.

Hosseini, S., Tarzi, B.G., Gharachorloo, M., Ghavami, M., Bakhoda, H., 2015. Optimization on the stability of linseed oil-in-water nanoemulsions generated by ultrasonic emulsification using response surface methodology (RSM). Oriental J. Chem. 31 (2), 1223–1230. http://www.highshearmixers.com/.

Huang, Q., Yu, H., Ru, Q., 2010. Bioavailability and delivery of nutraceuticals using nanotechnology. J. Food Sci. 75 (1), R50–R57.

Imai, H., Maeda, T., Shima, M., Adachi, S., 2008. Oxidation of methyl linoleate in oil-in-water micro-and nanoemulsion systems. J. Amer. Oil Chem. Soc. 85 (9), 809–815.

Innings, F., Trägårdh, C., 2007. Analysis of the flow field in a high-pressure homogenizer. Exp. Therm. Fluid Sci. 32 (2), 345–354.

Israelachvili, J.N., Wennerstroem, H., 1992. Entropic forces between amphiphilic surfaces in liquids. J. Phys. Chem. 96 (2), 520–531.

Jafari, S.M., Assadpoor, E., He, Y., Bhandari, B., 2008. Recoalescence of emulsion droplets during high-energy emulsification. Food Hydrocolloids 22 (7), 1191–1202.

Joscelyne, S.M., Trägårdh, G., 1999. Food emulsions using membrane emulsification: conditions for producing small droplets. J. Food Eng. 39 (1), 59–64.

Joseph, S., Bunjes, H., 2012. Preparation of nanoemulsions and solid lipid nanoparticles by premix membrane emulsification. J. Pharma. Sci. 101 (7), 2479–2489.

Kamiya, T., Kaminoyama, M., Nishi, K., Misumi, R., 2010. Scale-up factor for mean drop diameter in batch rotor-stator mixers. J. Chem. Eng. Japan 43 (4), 326–332.

Kanichi, S., Yuko, O., Yoshio, H., 2002. Properties of solid fat O/W emulsions prepared by membrane emulsification method combined with preemulsification. 3ième Congrès Mondial de l'Emulsion, vols. 24–27. Lyon, France (September).

Kentish, S., Wooster, T.J., Ashokkumar, M., Balachandran, S., Mawson, R., Simons, L., 2008. The use of ultrasonics for nanoemulsion preparation. Innov. Food Sci. Emerg. 9 (2), 170–175.

Kessler, H.G., 2002. Food and bioprocess engineering. Dairy Technol. Verlag A. Kessler, München.

Köhlera, K., Schuchmann, H.P., 2011. Homogenization in the dairy process: conventional processes and novel techniques. Procedia Food Sci. 1, 1367–1373.

Komaiko, J., McClements, D.J., 2015. Food-grade nanoemulsion filled hydrogels formed by spontaneous emulsification and gelation: optical properties, rheology, and stability. Food Hydrocolloids 46, 67–75.

Laouini, A., Fessi, H., Charcosset, C., 2012. Membrane emulsification: a promising alternative for vitamin E encapsulation within nanoemulsion. J. Membrane Sci. 423, 85–96.

Lauterborn, W., Kurz, T., Geisler, R., Schanz, D., Lindau, O., 2007. Acoustic cavitation, bubble dynamics, and sonoluminescence. Ultrason. Sonochem. 14 (4), 484–491.
Leighton, T.G., 1995. Bubble population phenomena in acoustic cavitation. Ultrason. Sonochem. 2 (2), S123–S136.
Leong, T.S.H., Wooster, T.J., Kentish, S.E., Ashokkumar, M., 2009. Minimizing oil droplet size using ultrasonic emulsification. Ultrason. Sonochem. 16 (6), 721–727.
Littoz, F., McClements, D.J., 2008. Biomimetic approach to improving emulsion stability: cross-linking adsorbed beet pectin layers using laccase. Food Hydrocolloid. 22 (7), 1203–1211.
López-Montilla, J.C., Herrera-Morales, P.E., Pandey, S., Shah, D.O., 2002. Spontaneous emulsification: mechanisms, physicochemical aspects, modeling, and applications. J. Disper. Sci. Technol. 23 (1–3), 219–268.
Lovelyn, C., Attama, A.A., 2011. Current state of nanoemulsions in drug delivery. J. Biomat. Nanobiotechnol. 2 (05), 626.
Luo, J., Fang, Z., Smith, R.L., 2014. Ultrasound-enhanced conversion of biomass to biofuels. Prog. Energ.Combust. 41, 56–93.
Ma, Z., Boye, J.I., 2013. Advances in the design and production of reduced-fat and reduced-cholesterol salad dressing and mayonnaise: a review. Food Bioproc. Technol. 6 (3), 648–670.
Maa, Y.F., Hsu, C., 1996. Liquid-liquid emulsification by rotor/stator homogenization. J. Control Rel. 38 (2), 219–228.
Maestro, A., Solè, I., González, C., Solans, C., Gutiérrez, J.M., 2008. Influence of the phase behavior on the properties of ionic nanoemulsions prepared by the phase inversion composition method. J. Colloid Interf. Sci. 327 (2), 433–439.
Mao, L., Yang, J., Xu, D., Yuan, F., Gao, Y., 2010. Effects of homogenization models and emulsifiers on the physicochemical properties of β-carotene nanoemulsions. J. Disper. Sci. Technol. 31 (7), 986–993.
Marie, P., Perrier-Cornet, J.M., Gervais, P., 2002. Influence of major parameters in emulsification mechanisms using a high-pressure jet. J. Food Eng. 53 (1), 43–51.
Mason, T.G., Wilking, J.N., Meleson, K., Chang, C.B., Graves, S.M., 2006. Nanoemulsions: formation, structure, and physical properties. J. Phys.: Cond. Matter 18 (41), R635.
McClements, D.J., 2004. Food Emulsions: Principles, Practices, and Techniques. CRC Press.
McClements, D.J., 2010. Emulsion design to improve the delivery of functional lipophilic components. Ann. Rev. Food Sci. Technol. 1, 241–269.
McClements, D.J., 2011. Edible nanoemulsions: fabrication, properties, and functional performance. Soft Matter 7 (6), 2297–2316.
McClements, D.J., 2012. Nanoemulsions versus microemulsions: terminology, differences, and similarities. Soft Matter 8 (6), 1719–1729.
McClements, D.J., Li, Y., 2010. Structured emulsion-based delivery systems: controlling the digestion and release of lipophilic food components. Adv. Colloid Interf. Sci. 159 (2), 213–228.
Möller, M., Eberle, U., Hermann, A., Moch, K., Stratmann, B., 2009. Nanotechnology in the food sector. Zürich: TA–SWISS 47, https://www.ta-swiss.ch. pp. 27–58.
Mustapha, W.A. W., Yusop, S.M., Maskat, M.Y., Shamsuddin, A.F., 2014. Characteristics of cinnamaldehyde nanoemulsion prepared using APV-high pressure homogenizer and ultra turrax. In: Proceedings of the 2014 UKM FST Postgraduate Colloquium, Vol. 1614. Universiti Kebangsaan Malaysia, Faculty of Science and Technology. AIP Publishing, pp. 244–250.

Ng, S.H., Woi, P.M., Basri, M., Ismail, Z., 2013. Characterization of structural stability of palm oil esters-based nanocosmeceuticals loaded with tocotrienol. J. Nanobiotechnol. 11, 1–7.

O'Sullivan, J., Murray, B., Flynn, C., Norton, I., 2015. Comparison of batch and continuous ultrasonic emulsification processes. J. Food Eng. 167, 114–121.

O'Sullivan, J.J., 2015. Applications of ultrasound for the functional modification of proteins and submicron emulsion fabrication. Doctoral dissertation, University of Birmingham.

Pacek, A.W., Baker, M., Utomo, A.T., 2007. Characterization of flow pattern in a rotor-stator high-shear mixer. In: Proceedings of European Congress of Chemical Engineering (ECCE-6). Copenhagen, pp. 16–20.

Paquin, P., 1999. Technological properties of high-pressure homogenizers: the effect of fat globules, milk proteins, and polysaccharides. Int. Dairy J. 9 (3), 329–335.

Pascual-Pineda, L.A., Flores-Andrade, E., Jiménez-Fernández, M., Beristain, C.I., 2015. Kinetic and thermodynamic stability of paprika nanoemulsions. Int. J. Food Sci. Tech. 50 (5), 1174–1181.

Peng, S.J., Williams, R.A., 1998. Controlled production of emulsions using a cross-flow membrane: Part I: Droplet formation from a single pore. Chem. Eng. Res. Des. 76 (8), 894–901.

Perrier-Cornet, J.M., Marie, P., Gervais, P., 2005. Comparison of emulsification efficiency of protein-stabilized oil-in-water emulsions using jet, high-pressure, and colloid mill homogenization. J. Food Eng. 66 (2), 211–217.

Peshkovsky, A.S., Bystryak, S., 2014. Continuous-flow production of a pharmaceutical nanoemulsion by high-amplitude ultrasound: process scale-up. Chem Eng. Process. 82, 132–136.

Peshkovsky, A.S., Peshkovsky, S.L., Bystryak, S., 2013. Scalable high-power ultrasonic technology for the production of translucent nanoemulsions. Chem. Eng. Process. 69, 77–82.

Qian, C., McClements, D.J., 2011. Formation of nanoemulsions stabilized by model food-grade emulsifiers using high-pressure homogenization: factors affecting particle size. Food Hydrocolloids 25 (5), 1000–1008.

Rao, J., McClements, D.J., 2010. Stabilization of phase inversion temperature nanoemulsions by surfactant displacement. J. Agric. Food Chem. 58 (11), 7059–7066.

Ribeiro, H.S., Chu, B.S., Ichikawa, S., Nakajima, M., 2008. Preparation of nanodispersions containing β-carotene by solvent displacement method. Food Hydrocolloids 22 (1), 12–17.

Sanabria, L.A. A., 2012. Development of a frozen yogurt fortified with a nanoemulsion containing purple rice bran oil. Doctoral dissertation, Louisiana State University.

Sandra, S., Decker, E.A., McClements, D.J., 2008. Effect of interfacial protein cross-linking on the in vitro digestibility of emulsified corn oil by pancreatic lipase. J. Agric. Food Chem. 56 (16), 7488–7494.

Sanguansri, P., Augustin, M.A., 2006. Nanoscale materials development: a food industry perspective. Trends Food Sci. Technol. 17 (10), 547–556.

Saravacos, G.D., Kostaropoulos, A.E., 2002. Mechanical processing equipment. In: Handbook of Food Processing Equipment (pp. 134–207). Springer, US.

Scholz, P., Keck, C.M., 2015. Nanoemulsions produced by rotor–stator high-speed stirring. Int. J. Pharma. 482 (1), 110–117.

Schramm, L.L., 2006. Emulsions, foams, and suspensions: fundamentals and applications. John Wiley & Sons Weinheim.

Schröder, V., Schubert, H., 1999. Production of emulsions using microporous, ceramic membranes. Colloid. Surf. A 152 (1), 103–109.

Schultz, S., Wagner, G., Urban, K., Ulrich, J., 2004. High-pressure homogenization as a process for emulsion formation. Chem. Eng. Technol. 27 (4), 361–368.
Silva, H.D., Cerqueira, M.Â., Vicente, A.A., 2012. Nanoemulsions for food applications: development and characterization. Food Bioprocess. Technol. 5 (3), 854–867.
Sivakumar, M., Tang, S.Y., Tan, K.W., 2014. Cavitation technology: a greener processing technique for the generation of pharmaceutical nanoemulsions. Ultrason. Sonochem. 21 (6), 2069–2083.
Solè, I., Maestro, A., Pey, C.M., González, C., Solans, C., Gutiérrez, J.M., 2006. Nanoemulsions preparation by low energy methods in an ionic surfactant system. Colloid. Surf. A 288 (1), 138–143.
Sotoyama, K., Asano, Y., Ihara, K., Takahashi, K., 1999. Water/oil emulsions prepared by the membrane emulsification method and their stability. J. Food Sci. 64 (2), 211–215.
Stang, M., Schuchmann, H., Schubert, H., 2001. Emulsification in high-pressure homogenizers. Eng. Life Sci. 1 (4), 151–157.
Suzuki, K., Fujiki, I., Hagura, Y., 1998. Preparation of corn oil/water and water/corn oil emulsions using PTFE membranes. Food Sci. Tech. Int. Tokyo 4 (2), 164–167.
Tadros, T.F., 2009. Emulsion science and technology: a general introduction. Wiley-VCH Verlag GmbH & Co. KGaA, Weinheim. pp. 1–56.
Tjøtta, S., 1999. Theoretical investigation of heat and streaming generated by high-intensity ultrasound. Acta Acusti. 85 (6), 780–787.
Trentin, A., De Lamo, S., Güell, C., López, F., Ferrando, M., 2011. Protein-stabilized emulsions containing beta-carotene produced by premix membrane emulsification. J. Food Sci. 106 (4), 267–274.
Utomo, A., Baker, M., Pacek, A.W., 2009. The effect of stator geometry on the flow pattern and energy dissipation rate in a rotor–stator mixer. Chem. Eng. Res. Des. 87 (4), 533–542.
Vandamme, T.F., Anton, N., 2010. Low-energy nanoemulsification to design veterinary controlled drug delivery devices. Int. J. Nanomed. 5, 867–873.
Walstra, P., 1993. Principles of emulsion formation. Chem. Eng. Sci. 48 (2), 333–349.
Weiss, J., Decker, E.A., McClements, D.J., Kristbergsson, K., Helgason, T., Awad, T., 2008. Solid lipid nanoparticles as delivery systems for bioactive food components. Food Biophys. 3 (2), 146–154.
Wengeler, R., 2007. Hydrodynamic stress induced dispersion of nanoscale agglomerates by a high-pressure process. Cuvillier Verlag, Göttingen.
Yamazaki, N., Yuyama, H., Nagai, M., Ma, G.H., Omi, S., 2002. A comparison of membrane emulsification obtained using SPG (Shirasu Porous Glass) and PTFE (poly [tetrafluoroethylene]) membranes. J. Disper. Sci. Technol. 23 (1–3), 279–292.
Yuan, Y., Gao, Y., Zhao, J., Mao, L., 2008. Characterization and stability evaluation of β-carotene nanoemulsions prepared by high-pressure homogenization under various emulsifying conditions. Food Res. Int. 41 (1), 61–68.

SUBJECT INDEX